CDX Learning Systems™

MASTER
AUTOMOTIVE
TECHNICIAN SERIES

Light Vehicle Diesel Engines

Gus Wright

JONES & BARTLETT
LEARNING

World Headquarters
Jones & Bartlett Learning
25 Mall Road
Burlington, MA 01803
978-443-5000
info@jblearning.com
www.jblearning.com

Jones & Bartlett Learning books and products are available through most bookstores and online booksellers. To contact Jones & Bartlett Learning directly, call 800-832-0034, fax 978-443-8000, or visit our website, www.jblearning.com.

Substantial discounts on bulk quantities of Jones & Bartlett Learning publications are available to corporations, professional associations, and other qualified organizations. For details and specific discount information, contact the special sales department at Jones & Bartlett Learning via the above contact information or send an email to specialsales@jblearning.com.

978-1-284-14509-0

Production Credits
General Manager: Kimberly Brophy
VP, Product Development: Christine Emerton
Content Services Manager: Kevin Murphy
Product Manager: Jesse Mitchell
Senior Vendor Manager: Sara Kelly
Marketing Manager: Amanda Banner
VP, Manufacturing and Inventory Control: Therese Connell
Composition and Project Management: Integra Software Services Pvt. Ltd.
Cover Design: Scott Moden
Rights & Media Specialist: Robert Boder
Media Development Editor: Shannon Sheehan
Cover Image (Title Page): © Joseph Coville at CCA Advertising
Printing and Binding: LSC Communications
Cover Printing: LSC Communications

Library of Congress Cataloging-in-Publication Data unavailable at time of printing.

6048

Printed in the United States of America
26 25 24 23 22 10 9 8 7 6 5 4 3 2

BRIEF CONTENTS

CONTENTS

NOTE TO STUDENTS

This book was created to help you on your path to a career in the transportation industry. Employability basics covered early in the text will help you get and keep a job in the field. Essential technical skills are built in, cover to cover, and are the core building blocks of an advanced technician's skill set. This book also introduces "strategy-based diagnostics," a method used to solve technical problems correctly on the first attempt. The text covers every task the industry standard recommends for technicians. This book will help you on your path to a successful career.

As you navigate this text book, ask yourself, "What does a technician need to know and do at work?"

This book is set up to answer that question. Each chapter starts by listing the technician's tasks that are covered within. These are the objectives for that chapter. Each chapter ends by reviewing those things a technician needs to know. The content of each chapter is written to explain each objective. As you study, continue to ask the question above. Gauge your progress by imagining yourself as the technician. Do you have the knowledge, and can you perform the tasks required at the beginning of each chapter? Combining your knowledge with hands-on experience is essential to becoming a Master Technician.

During your training, remember that the best thing you can do as a technician is learn to learn. This will serve you well because vehicles keep advancing, and good technicians never stop learning.

Stay curious. Ask questions. Practice your skills, and always remember that one of the best resources you have for learning is right there in your classroom: your instructor.

Best wishes and enjoy!
The CDX Automotive Team

CHAPTER 1

Strategy-Based Diagnostics

NATEF Tasks

- **N01001** Review vehicle service history.
- **N01002** Demonstrate use of the three C's (concern, cause, and correction).
- **N01003** Identify information needed and the service requested on a repair order.
- **N01004** Research vehicle service information including fluid type, vehicle service history, service precautions, and technical service bulletins.

Knowledge Objectives

After reading this chapter, you will be able to:

- **K01001** Describe the purpose and use of vehicle service history.
- **K01002** Demonstrate an understanding of the active listening process.
- **K01003** Demonstrate an understanding of the strategy-based diagnosis process.
- **K01004** Describe step one of the strategy-based diagnosis.
- **K01005** Describe step two of the strategy-based diagnosis.
- **K01006** Describe step three of the strategy-based diagnosis.
- **K01007** Describe step four of the strategy-based diagnosis.
- **K01008** Describe step five of the strategy-based diagnosis.
- **K01009** Explain how the three Cs are applied in repairing and servicing vehicles.
- **K01010** Describe the information and its use within a repair order.

Skills Objectives

After reading this chapter, you will be able to:

- **S01001** Use service history in the repair and service of vehicles.
- **S01002** Complete a repair order.

You Are the Automotive Technician

A regular customer brings his 2014 Toyota Sienna into your shop, complaining of a "clicking" noise when he turns the steering wheel. You ask the customer further questions and learn that the clicking happens whenever he turns the wheel, especially when accelerating. He tells you he has just returned from vacation with his family and has probably put 300 miles (482 kms) on the car during their trip.

1. What additional questions should you ask the customer about his concern, the clicking noise he hears when turning?
2. How would you verify this customer's concern?
3. What sources would you use to begin gathering information to address this customer's concern?
4. Based on what you know this far about the customer's concern, what systems might be possibly related to this customer's concern?

▶ Introduction

The overall vehicle service involves three major components. Those pieces are gathering information from the customer, the strategy-based diagnostic process, and documenting the repair. The flow of the overall service can be seen below.

1. Initial information gathering is often completed by a service advisor (consultant) and should contain details about the customer concern and pertinent history.
2. Verifying the customer concern begins the strategy-based diagnostic process. Technicians will complete this step to ensure that a problem exists and that their repair eliminated it.
3. Researching the possible cause will provide a list of possible faults. The technician will expand this list as testing continues.
4. Testing will focus on the list of possibilities. Technicians will start with broad, simple tests that look at an entire system or group of components. Testing will progressively become more narrowly focused as it pinpoints an exact cause.
5. Repairs will be made using suggested tools and recommended procedures. This is done to ensure a reliable repair and that manufacturer requirements are met.
6. Repairs must always be verified. This confirms that the technician has completed the diagnosis accurately and completely.

The repair must be documented. The technician has been doing this all along. When the customer concern is recorded, the tests are recorded, and the final repair procedure recorded, the repair has been documented.

▶ Vehicle Service History

K01001 Describe the purpose and use of vehicle service history.

N01001 Review vehicle service history.

S01001 Use service history in the repair and service of vehicles.

Service history is a complete list of all the servicing and repairs that have been performed on a vehicle (**FIGURE 1-1**). The scheduled service history can be recorded in a service booklet or owner's manual that is kept in the glove compartment. The service history can provide valuable information to technicians conducting repairs. It also can provide potential new owners of used vehicles an indication of how well the vehicle was maintained. A vehicle with a regular service history is a good indication that all of the vehicle's systems have been well maintained, and the vehicle will often be worth more during resale. Most manufacturers store all service history performed in their dealerships (based on the VIN) on a corporate server that is accessible from any of their dealerships. They also use this vehicle service history when it comes to evaluating warranty claims. A vehicle that does not have a complete service history may not be eligible for warranty claims. Independent shops generally keep records of the repairs they perform. However, if a vehicle is repaired at multiple shops, repair history is much more difficult to track and, again, may result in a denial of warranty claims.

Vehicle service history can be very valuable to the technician. This history is typically retrieved from service records kept by the shop, dealer network, **original equipment manufacturer (OEM)**, or **aftermarket** service center. This information often contains a list of services performed on a vehicle and the date and mileage at which they were completed. Not all service history contains the same information. Some histories may only contain repair information, while others include every customer concern and maintenance task performed. This information can be very helpful when diagnosing a concern. Service history may help technicians diagnose a vehicle and can also be used to prevent costly duplicated repairs.

Service history can also be used to guide repairs. Records of vehicle service history may indicate that the customer has recently been in for service and now has returned with a new concern. This all-too-common situation is usually found to be caused by error during the previous service. When working on a vehicle that has returned after a recent repair, the previous technician's work (whoever that may be) should be inspected meticulously.

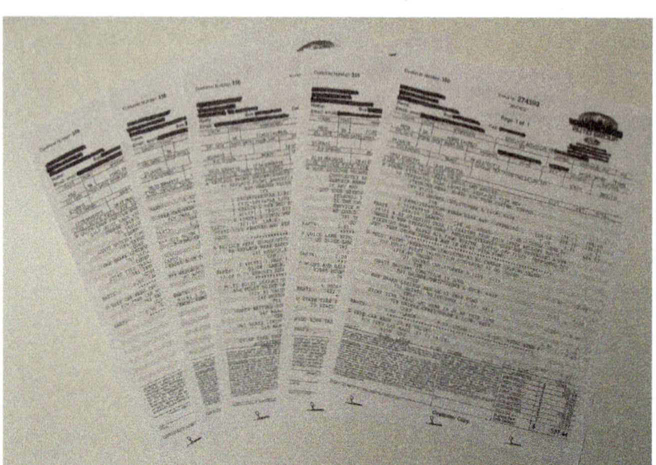

FIGURE 1-1 Print outs of completed repair order as saved in the online repair order system.

The service history may also show that the customer is returning for the same issue due to a component failure. The history might indicate when the component was installed, help the customer get their vehicle repaired, and help the shop to get paid under the component warranty. A vehicle that returns more than once for the same repair could be an indicator that an undiagnosed problem is causing these failures. The service history allows technicians to determine if the vehicle has been well maintained. This can be extremely useful when a technician suspects that lack of maintenance may be the cause of the problem.

The vehicle's service history helps technicians determine what maintenance needs to be performed, and therefore helps customers save money over time by preventing future costly repairs. Routine maintenance is essential on today's modern automobile and prevents premature failures due to contamination and component wear.

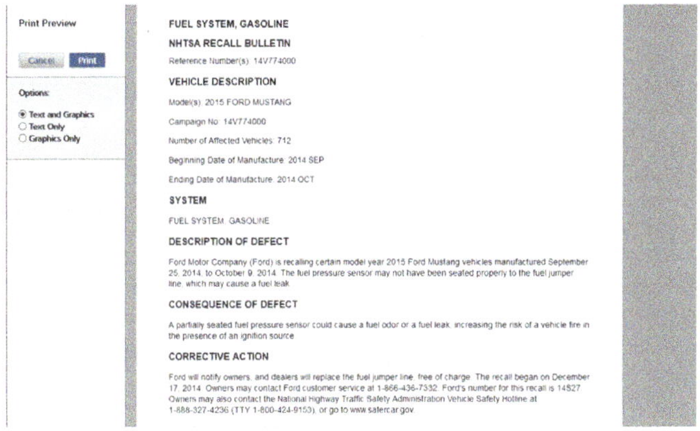
Print Preview
Cancel Print
Options:
Text and Graphics
Text Only
Graphics Only

FUEL SYSTEM, GASOLINE
NHTSA RECALL BULLETIN
Reference Number(s): 14V774000
VEHICLE DESCRIPTION
Model(s): 2015 FORD MUSTANG
Campaign No: 14V774000
Number of Affected Vehicles: 712
Beginning Date of Manufacture: 2014 SEP
Ending Date of Manufacture: 2014 OCT
SYSTEM
FUEL SYSTEM, GASOLINE
DESCRIPTION OF DEFECT
Ford Motor Company (Ford) is recalling certain model year 2015 Ford Mustang vehicles manufactured September 25, 2014, to October 9, 2014. The fuel pressure sensor may not have been seated properly to the fuel jumper line, which may cause a fuel leak.
CONSEQUENCE OF DEFECT
A partially seated fuel pressure sensor could cause a fuel odor or a fuel leak, increasing the risk of a vehicle fire in the presence of an ignition source.
CORRECTIVE ACTION
Ford will notify owners, and dealers will replace the fuel jumper line, free of charge. The recall began on December 17, 2014. Owners may contact Ford customer service at 1-866-436-7332. Ford's number for this recall is 14S27. Owners may also contact the National Highway Traffic Safety Administration Vehicle Safety Hotline at 1-888-327-4236 (TTY 1-800-424-9153), or go to www.safercar.gov.

FIGURE 1-2 Recall notice example.

Today's vehicles also require regular software updates. There are many advanced computer systems on modern vehicles. From time to time, updates will be available to fix a bug or glitch in the computer programming. These updates are often designed to eliminate a customer concern, improve owner satisfaction, or increase vehicle life. This is very similar to an update for your PC or mobile device. Service history will indicate to the technician that the vehicle may need an update. The technician will inspect the vehicle's computer system and perform any needed updates as necessary.

Service history can also be used to keep customers safe. Occasionally, manufacturers may need to recall a vehicle for service due to a safety concern that has been identified for a vehicle (**FIGURE 1-2**). This means that the manufacturer has found that the potential exists for a dangerous situation to occur, and the vehicle must be serviced to eliminate it. Depending on the nature of the problem, recalls can be mandatory and required by law, or manufacturers may voluntarily choose to conduct a recall to ensure the safe operation of the vehicle or minimize damage to their business or product image. The service history would be used to verify that the vehicle is subject to the recall and has or has not had the recall service completed. The technician would perform the service, update the service history, and return the vehicle to the customer.

▶ TECHNICIAN TIP

Technicians and service advisors should check the vehicle service history against the manufacturer's service maintenance schedule to determine if the vehicle is due for scheduled maintenance. The maintenance schedule is a guide that indicates what service is due when; it can be found in the manufacturer's service information and often in the owner's manual. Keeping the vehicle well maintained can avoid a failure that strands the customer on the roadside.

Applied Science

AS-11 Information Processing: The technician can use computer databases to input and retrieve customer information for billing, warranty work, and other record-keeping purposes.

Dealership service departments have access to databases run by manufacturers for the purposes of accessing warranty information, tracking vehicle servicing and warranty repair history, and logging warranty repair jobs for payment by the manufacturer. When a customer presents their vehicle for a warranty repair, the customer service department staff begin by consulting the database to confirm that the vehicle is within its warranty period and that the warranty has not been invalidated for any reason. Once it is confirmed that the vehicle is still under a valid warranty, the repair order will be passed to the workshop for diagnosis and repair. Any parts required for the warranty repair must be labeled by the technician and stored for possible recall by the manufacturer.

For example, a young man comes in complaining that his vehicle is "running rough." The customer service staff confirms that the vehicle is nine months old and only has 14,500 miles (approx. 23,000 km), so it is within the manufacturer's 3-year/100,000 mile (160,000 km) warranty period. They check the manufacturer's database to confirm that the vehicle's warranty has not been invalidated before handing the repair order onto the workshop. Then a technician diagnoses the fault as a defective ignition coil and fills out a warranty parts form.

Once the repair has been completed and the parts labeled, the warranty parts form and any repair order paperwork is passed back to administrative staff for processing. Processing will include billing the manufacturer for the correct, pre-approved amount of time, logging the repair on the database for payment, and ensuring that all documentation is correct for auditing purposes.

Warranty Parts Form

Customer concern: Vehicle running rough	Vehicle Information
Cause: #6 ignition coil open circuit on primary winding	VIN: IG1123456789101 11
Correction: Replaced #6 ignition coil	RO Number: 123456
Parts description: #6 ignition coil	Date of repair: 10/04/2016

SKILL DRILL 1-1 Reviewing Service History

1. Locate the service history for the vehicle. This may be in shop records or in the service history booklet within the vehicle glove compartment. Some shops may keep the vehicle's service history on a computer.
2. Familiarize yourself with the service history of the vehicle.
   a. On what date was the vehicle first serviced?
   b. On what date was the vehicle last serviced?
   c. What was the most major service performed?
   d. Was the vehicle ever serviced for the same problem more than once?
3. Compare the vehicle service history to the manufacturer's scheduled maintenance requirements, and list any discrepancies.
   a. Have all the services been performed?
   b. Have all the items been checked?
   c. Are there any outstanding items?

To review the vehicle service history, follow the steps in **SKILL DRILL 1-1**.

Active Listening Skills

K01002 Demonstrate an understanding of the active listening process.

Depending on the size of a shop, the first point of contact for the customer is the **service advisor** or consultant. This person answers the phone, books customer work into the shop, fills out repair orders, prices repairs, invoices, keeps track of work being performed, and builds customer relations with the goal of providing a high level of customer support. The service advisor also serves as a liaison between the customer and the technician who is working on the vehicle. A service advisor or consultant may advance to become a service manager. In smaller shops, a technician may perform these duties.

When the customer brings his or her vehicle in for service, the service advisor or technician should ask for more information than just the customer concern. It is important to let the customer speak while you use active listening skills to gather as many pertinent details as possible. Active listening means paying close attention to not only the customer's words, but also to their tone of voice and body language. Maintain eye contact with the customer throughout your conversation and nod to show you understand and are paying attention. Do not interrupt. Wait for the customer to finish speaking before responding, then ask open-ended questions to verify that you have heard the complaint clearly and understand the problem. An open-ended question is one that cannot be answered with a yes or no, but instead requires the customer to provide you with more information about the problem (**FIGURE 1-3**). If the shop is noisy, try to find a quieter location in which to speak with the customer. Excellent communication helps ensure that all relevant information is collected. It also makes a good first impression with customers; they are likely to feel that they were listened to and cared for.

Politely use open-ended questions to ask about any symptoms the customer may have noticed, such as:

- Under what circumstances does the concern occur or not occur?
- What unusual noises do you hear (e.g., squeaks, rattles, clunks, and other noises)?
- What odd smells or fluid leaks have you noticed?
- What recent work, service, or accessories have been added to the vehicle?
- What other recent changes or experiences have you had with the vehicle?
- What other systems seem to be operating improperly?

▶ TECHNICIAN TIP

A vehicle's service history is valuable for several reasons:

- It can provide helpful information to the technician when performing repairs.
- It allows potential new owners of the vehicle to know how well the vehicle and its systems were maintained.
- Manufacturers use the history to evaluate warranty claims.

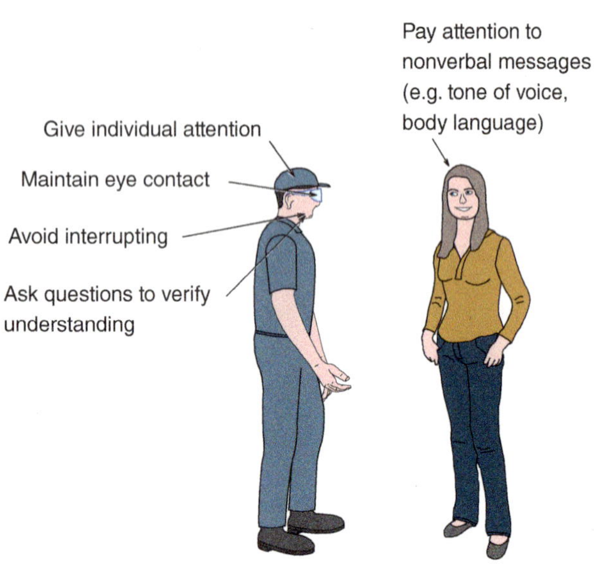

FIGURE 1-3 The active listening process.

Although problems may seem unrelated initially, when multiple systems fail at the same time, the issues are frequently related. Open-ended questions can provide valuable information to the technician who is performing the diagnosis.

▶ Strategy-Based Diagnostic Process

K01003 Describe each step in strategy-based diagnosis.

Diagnostic problems can be very challenging to identify and correct in a timely and efficient manner. Technicians will find that having a plan in place ahead of time will vastly simplify the process of logically and systematically (strategically) solving problems. The plan should be simple to remember and consistent in its approach; yet it must work for the entire range of diagnostic problems that technicians will encounter. In this way, technicians will have one single plan to approach any diagnostic situation they may encounter, and will be confident in their ability to resolve it. This problem solving plan is called the **Strategy-Based Diagnostic Process**.

The strategy-based diagnostic process is focused on fixing problems correctly the first time. It is a scientific process of elimination, which is much the same process as a medical doctor uses for their diagnosis. It begins with identifying the customer's concern and ends with confirming that the problem has been resolved. The purpose of the problem-solving process is twofold: to provide a consistent road map for technicians as they address customer concerns that require diagnosis, and to ensure that customer concerns are resolved with certainty.

This process simplifies the problem-solving portion of the repair, making the job easier for the technician; it prevents technicians from having to work on the same job more than once; and it all but eliminates customer comebacks. While repeat customers are good for business, a customer coming back with the same problem is not. The customer is likely to be upset and the technician is likely to be working for free. In order to avoid this scenario, it is imperative to address customer concerns correctly the first time.

Proper diagnosis is important to consumers and to the federal government. Federal and state law protects consumers against the purchase of vehicles with significant persistent defects. Technicians are held to a standard of reasonable repair times and limited visits for the same concern. Although the law varies from state to state, this means technicians must not return a vehicle to a customer without addressing the customer's original concern. Also, technicians cannot make the vehicle unavailable to the customer for a long period while the vehicle is being repaired. The purpose of the state and federal laws is to protect consumers buying new vehicles.

Failure to comply with the state and federal law can be very expensive for the dealership and manufacturer. Although most state laws hold the manufacturer directly responsible, dealerships are also hurt by a loss in sales revenue, a loss in repair revenue, and irreparable damage to their customer and sometimes manufacturer relationships. Many state laws hold the manufacturer responsible for full purchase price, incurred loan fees, installed accessories, and registration and similar government charges. This can be a heavy cost on top of the value of the vehicle itself.

▶ TECHNICIAN TIP

Technicians need to do their best to find the issue and resolve it; otherwise, the vehicle may be required to be bought back from the customer, costing the dealership and manufacturer significant money.

Need for the Strategy-Based Diagnostic Process

Finding the source of every customer concern can prove to be a challenge. Novice technicians frequently struggle with diagnostics situations. Even some veteran technicians have difficulty tackling diagnosis on some new technologies. However, if the strategy for solving a problem is generally the same every time, this greatly simplifies the process. Hopefully, by applying a strategy-based diagnostic process, technicians will resolve challenging customer concerns 100% of the time in an efficient manner.

Customer comebacks occur when the customer picks up the vehicle after service, only to bring it back shortly thereafter with the same concern. This situation is understandably upsetting to the customer. Typically, the end result is wasted labor time and a loss in shop productivity. The customer is left with one of the following impressions:

- The work was not performed;
- The shop is incompetent;
- Or, worse yet, the shop was trying to scam the customer.

▶ TECHNICIAN TIP

The diagnostic process makes the technician's job easier by providing a step-by-step strategy to solving the problem. It also answers the question: "Now what do I do?" As even the toughest job becomes easier, technicians will find their rate of diagnostic success increasing.

Customer comebacks are usually caused by one of two avoidable reasons:

1. The customer concern is misinterpreted or misunderstood. This results in the technician "fixing" a problem that does not exist or missing a problem altogether.
2. The technician failed to verify that the original concern was resolved. Technicians are often hurried; some will forget to ensure that the repair they had performed actually fixed the original customer concern.

Use of the strategy-based diagnostic process enables the technician and shop to make more money and satisfy more customers. This is a win-win situation for all involved. Using the strategy-based diagnostic process requires starting at the beginning and following it through to the end every time (**FIGURE 1-4**). This systematic approach will ensure the best results for each diagnostic situation.

Step 1: Verify the Customer's Concern

K01004 Describe step one of the strategy-based diagnosis.

The first step in the diagnostic process is to verify the customer's concern. This step is completed for two main purposes:

- To verify that the vehicle is not operating as designed
- To guarantee that the customer's concern is addressed

SAFETY TIP

There should be limits to recreating the customer concern. Technicians need to be careful when riding as a passenger with the customer or driving an unfamiliar customer vehicle. Technicians have died during test drives due to customers' driving or their own driving of unfamiliar vehicles. The purpose of the test drive is to verify the concern or its repair. It is not an opportunity for a thrill ride. Customers and their vehicles should be treated with respect. Additionally, customers should respect the technician. If a customer asks a technician to verify a concern in an unsafe situation, such as a high rate of speed, the technician should decline. This is for both safety and liability reasons.

Failure to complete this step may result in wasted time, wasted money, and, worst of all, an unhappy customer. The customer is probably not an experienced automotive technician. For this reason, the customer does not always accurately verbalize the problem that may be occurring. Therefore, it is very important that you have a complete understanding of the customer's concern before beginning the diagnosis. This will enable you to know with certainty that you have actually resolved the original concern after repairing the vehicle and before returning it to the customer. During this step, you may perform several of the following tasks, depending on the customer concern.

First, ask the customer to demonstrate the concern, if possible. This may necessitate a test drive (**FIGURE 1-5**). The customer should be encouraged to drive the vehicle while you ride along as a passenger and gather symptoms and details about the concern. Seeing the customer recreate the concern in real time will often provide some much needed context to the problem. Having the customer demonstrate the concern is ideal in most situations, though not always possible. In the event that the customer is not present, you must do your best to recreate the concern on your own based on the information obtained from the customer. With or without the customer present, be sure to document in writing any details about the scenario in which the concern arises.

Next, make sure that the customer concern doesn't fall outside the range of normal operation of the component or system. The manufacturer's service information provides

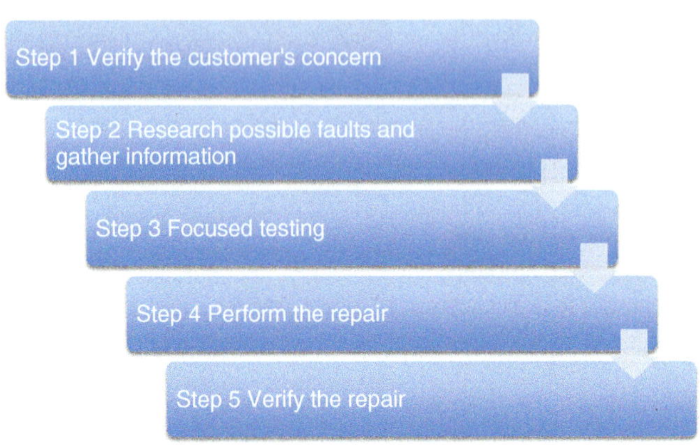

FIGURE 1-4 The strategy-based diagnostic process.

FIGURE 1-5 Ask the customer to describe the concern.

system descriptions and expected operations; technicians can use these details, provided in the owner's manual or in the vehicle service manual, to become familiar with the system and then explain its operation to the customer. Especially on new cars with many amenities, customers may not be familiar with the controls and subsequent operation. This can cause a customer to bring a vehicle in for service unnecessarily, due to unfamiliarity with the system controls. Many shops use online service (shop) manuals where you can quickly access any information related to the customer's concern (**FIGURE 1-6**). Checking to make sure that the concern is really a fault, and not a normal operation, will avoid unnecessary diagnosis time. This is also an opportunity to provide excellent customer service by demonstrating the features and their controls to the customer.

Conducting a quick visual inspection to look for obvious faults can be very helpful (**FIGURE 1-7**). However, it does not replace the need for testing and is absolutely not intended as a shortcut to the diagnostic process. With that said, the visual inspection can provide valuable information that may speed up the testing processes. The visual inspection provides an opportunity for a quick safety check by the technician and may help to avoid some potentially dangerous situations during service.

While visual inspections can be very valuable, technicians must be careful not to jump to conclusions based on what they see. For example, a customer comes in with an illuminated and flashing overdrive light on the control panel. The technician has seen this problem before and it was caused by a bad solenoid pack in the transmission. If the technician decides that this problem is also caused by a bad solenoid pack, this determination is one that was reached solely on conjecture; no actual test was performed. Although the flashing light might indicate a fault with the solenoid pack, steps in the diagnostic process should never be skipped. This guess can lead to a very costly mistake when it is discovered that the new solenoid pack does not, in fact, fix the problem. In reality, the wiring harness to the transmission is frayed and shorting out. Had the technician performed a test, the cause of the customer concern could have been confirmed or denied before a solenoid pack was put in unnecessarily. While the visual inspection is very valuable, tests must always be performed to compare suspected faults against the expectations and specifications defined in the service information.

When recreating the customer concern, the technician should operate the system in question in all practically available modes. System operation should be checked to see if there are other symptoms that may have gone unnoticed by the customer. These other symptoms can be very valuable when determining which tests to perform; they could save the technician significant time during the diagnosis. When recreating the customer concern, it is important to check the entire system for symptoms and related faults.

Recreating **intermittent faults** can be a challenge. Intermittent symptoms often stem from a component or system that is failing or one where the nature of the fault is not yet clear. In these situations, the aforementioned check of system operation can prove to be highly valuable, as it may uncover previously unnoticed but consistent symptoms. Attempting to repair

FIGURE 1-6 A technician researching service information.

FIGURE 1-7 Performing a visual inspection.

an intermittent fault without consistent symptoms, data, or diagnostic trouble codes (DTCs) is a gamble, because a technician cannot be certain that the actual problem is isolated. This means that there would be no way to confirm with certainty that a repair was effective. The fault could appear again as soon as the vehicle is returned to the customer. To avoid such a situation, look for symptoms, data, or DTCs that are repeatable or consistent. Intermittent diagnosis may require the use of an oscilloscope (a specialized tool for looking at electrical waveforms), or a "wiggle" test (as the name implies, a test instrument is monitored as the electrical or vacuum harness is manipulated by hand). This can verify the customer concern and remove some of the challenge from the diagnosis of an intermittent fault.

Lastly, but notably, save DTCs and freeze frame data. **Freeze frame data** refers to snapshots that are automatically stored in a vehicle's power train control module (PCM) when a fault occurs; this is only available on vehicles model year 1996 and newer. Intermittent faults may be found by reviewing data stored just before, during, and after the fault occurred, similar to an instant replay. When working with computer controlled systems, it is very important to save the recorded data. It may become necessary to erase this information from the computer, though that should generally be avoided. This information is absolutely critical when the technician is trying to answer the questions, "When did this happen?" and "What was going on at the time?"

▶ TECHNICIAN TIP

All too often, the customer does not have symptoms to share and their only concern is that the malfunction indicator is illuminated. In this situation, the data stored in the computer is invaluable. Record it and do not clear it out unless directed to do so in the manufacturer's service procedure. Even then, you should capture the information before clearing the memory.

What will step one look like? When information is gathered and recorded for step one, it should contain the customer concern, any symptoms, and any retrieved DTCs. View the following example from a vehicle that has no reverse. The technician verified the customer concern and recorded:

1. Vehicle will not move when shifted into reverse
2. Vehicle operates normally in all forward gears in OD, D, L2, and L1
3. Current code P0868

Notice that the technician in this example verified and recorded the customer concern. The technician also tried other functions in the system. Specifically, the technician drove the vehicle and tested the other gears in each of the gear ranges and then recorded the results. The DTC data was also retrieved from the control module and recorded. Although it was short and concise, the information will be very useful in the next step.

Step 2: Researching Possible Faults and Gathering Information

K01005 Describe step two of the strategy-based diagnosis.

N01004 Research vehicle service information including fluid type, vehicle service history, service precautions, and technical service bulletins.

The second step in the diagnostic process is to research possible faults that may be related to the customer's concern. The goal of this step is to create a list of possible faults. The list is created based on the information gathered in step one. The list will later be narrowed down by the tests performed in step three until the cause of the concern has been confirmed.

Before testing can begin, a technician must know what possible faults need to be tested. Researching possible faults should begin broadly. Especially when diagnosing electrical and electronic systems, this step should begin at the system level and work down to individual components. For example, if a vehicle engine cranks, but will not start, a technician would list these familiar possible faults: Air, Fuel, Ignition, Compression, and Security. These possible faults are not single components, but rather they are systems. This is where a diagnosis should begin. Starting a diagnosis by listing the dozens of components for each system will make the job unreasonably time intensive. However, once a test determines that there is a fault within a specific system, the list should be expanded to encompass that particular system's subsystems and components. This systematic elimination starts broadly and narrows, allowing technicians to work more efficiently.

In the second step of the diagnostic process, the technician creates a list to help focus their tests. The list may aid in a simple process of elimination by testing one possibility after the next. The list can also start broadly and narrow as testing continues. When starting a list, it may look similar to the following:

1. Air
2. Fuel

3. Ignition
4. Compression
5. Security

This list is broad and starts at the system level. As you'll soon see in the next step, the technician would eliminate possible faults with a test that is focused on analyzing the whole system. When a system is located with a fault, in the ignition system for example, the list would become more specific:

1. Spark Plug
2. Coil on Plug
3. CKP
4. CMP
5. Sensor Triggers
6. Harness
7. Control Module

The technician would again focus his or her testing on the list, seeking to eliminate possible faults until one is confirmed, repaired, and verified.

Several great sources of information are available for researching possible faults, although the best source of information is usually the manufacturer's service information system. These systems are typically found online; however, some manufactures still publish paper service manuals. The manufacturer's service information contains definitions for diagnostic trouble codes, system description and operation, electrical wiring diagrams, diagnostic steps, repair procedures, and much more. Fault diagnosis should almost always begin with the factory service information.

Other resources for identifying faults can be used in conjunction with the factory service information. As previously discussed, the vehicle service history can provide valuable insight into the past maintenance or lack thereof. It can also provide information about recent or repeated repairs. **Technical service bulletins (TSBs)** are service notifications and procedures sent out by the manufacturers to dealer groups alerting technicians about common issues with a particular vehicle or group of vehicles (**FIGURE 1-8**). Some aftermarket sources also exist for the pattern failures addressed by TSBs (**FIGURE 1-9**). Additionally, both original equipment manufacturers (OEM) and aftermarket technician support services offer hotlines, or call-in support, that specifically provide technical support to professional technicians. Some of these hotlines offer subscriptions to searchable web-based components. These resources do not guarantee a repair; that is still the responsibility of the technician. However, all of the sources mentioned here can be a huge help as technicians research possible faults.

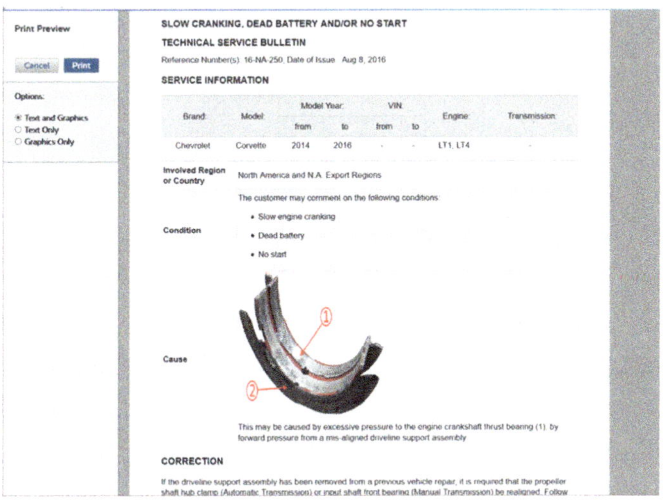

FIGURE 1-8 Technical service bulletin.

FIGURE 1-9 Aftermarket source.

While these resources are essential, the list of possible faults is just that: a list of possible faults. A technician must always be aware that steps in the diagnostic process cannot be skipped.

Step 3: Focused Testing

K01006 Describe step three of the strategy-based diagnosis.

Step three of the diagnostic process involves focused testing. In this step technicians use their testing skills to eliminate possible faults from the list they created in step two. Steps two and three work together; testing will start at a system level and work down to subsystems, then finally to individual components. The idea of focused testing should be to eliminate as many potential faults as possible with each test.

Focused testing is intended to eliminate possible causes with certainty. Each time a test is performed, the following three pieces of information must be recorded:

- a test description
- an expectation
- a result

These can be recorded on the repair order, electronic service record, or on an extra sheet of paper. Test records must be kept handy because they will become part of the documented record for this repair.

The three pieces of test information are recorded carefully for several reasons. Having an expectation before a test is performed makes each test objective and effective. The expectation is what the result is compared against, in order to determine if the vehicle passed or failed.

Many manufacturers, both original equipment and aftermarket, require that documented test results be submitted with each warranty claim. If the technician fails to document his or her work, the manufacturer will not pay the claim. The result is that the shop is out money for the parts and service, and the technician will not be paid for their work. Be sure to document the work properly (**FIGURE 1-10**).

> **▶ TECHNICIAN TIP**
>
> A repair should never be performed unless the possible fault has been verified through testing. Do not let a possible fault become a possible mistake. In some cases, the list of possible faults can be found in the service information, but many times the technician will need to produce the list based on the concern, the information gathered, and the results of the research.

> **▶ TECHNICIAN TIP**
>
> The test description must provide enough information that someone could repeat the test with the same result. This is very important!

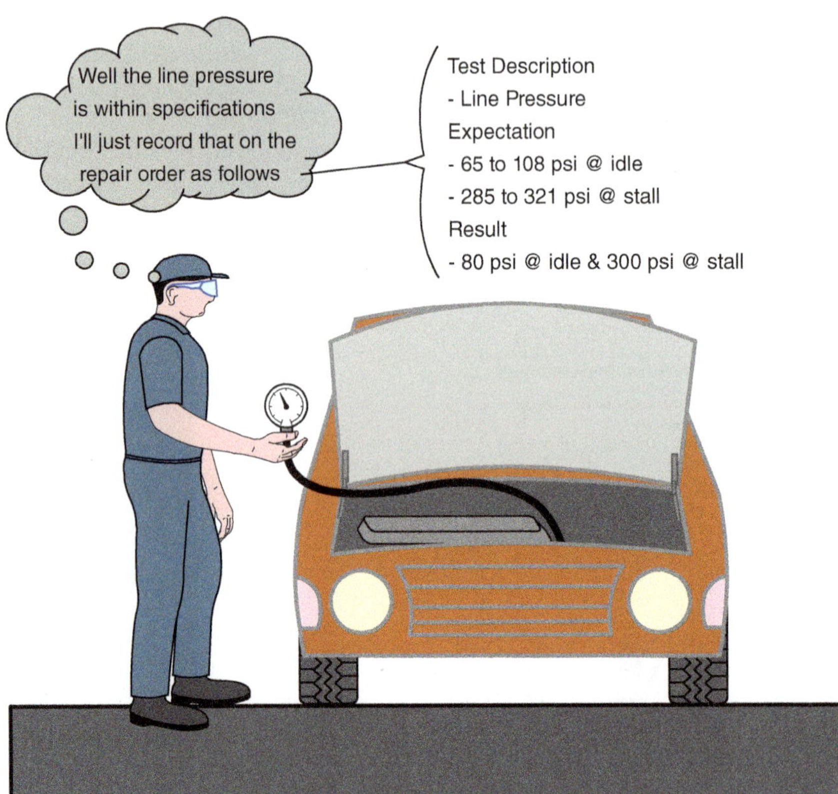

FIGURE 1-10 The test record should include the test description, expectation or specification, and the result of measurement.

1. The test description is not long, or even a complete sentence; it is simply a brief description. It allows the reader to know what test was performed and on what component or system. The test description should be accurate enough that the reader could repeat the test with the same result.
2. The expectation should describe the expected result as if the system is operating normally. The expectation could come directly from the system specifications listed in the manufacturer's service information or from system description and operation.
3. The result is the third part that must be recorded for each test. This information should accurately reflect what happened when the test was performed.

In summary, the testing is focused on isolating a fault or faults from the list of potential faults, and the results are compared to the expectation.

Testing should begin broadly and simply. Consider the following example: A light bulb circuit is suspected of having a fault. If the light bulb is easily accessible, the first test might be to check the voltage drop (i.e., voltage used to push current through the bulb). If the result of the voltage drop measurement is as expected (i.e., within specification), then the problem is in the bulb or socket. In this test, the technician is able to check the integrity of the entire electrical circuit with one test. If the result of the measurement is outside of the expectation (i.e., out of specification), the technician would know that the bulb is not the source of the problem. Further testing would isolate the problem to the ground or power side of the circuit.

The technician in the example performed a simple test with an easy expectation. The test allowed the technician to quickly determine the state of operation for the entire system/circuit and move on. If a fault had been found, then the technician would have isolated the cause of the customer's concern to that particular system/circuit and would need to perform further testing to isolate the cause to a particular component. To do that, the technician would use the service information to determine what components comprise the system and adjust the list from step two to take into account the new information. Then testing would continue.

The next test might measure voltage supply at the bulb (i.e., available voltage). In this way, the technician would be testing the power supply, the conductors, and the switch (assuming a power-side switched circuit). The technician would have an expectation for the circuit voltage and compare his or her result to this expected voltage. As we saw earlier, the technician is testing more than one component with a single test, thereby operating in an efficient manner.

This strategy—starting with broad, simple tests and moving to more complicated, pinpoint tests—makes efficient use of the technician's time while still effectively testing the possible faults.

1. A technician is investigating a customer concern of "no heat from the dash." The technician's investigation might begin with a simple list.
    a. Engine cooling system
    b. HVAC duct and controls
2. The technician would then eliminate one or the other and expand the list. The technician might verify coolant level and temperature at the inlet and outlet of the heater core. The HVAC components controlling and delivering warm air could then be used to expand the list for the next round of testing.
    a. Doors and ducts
    b. Cables
    c. Servos
    d. HVAC control head
    e. Blower motor
    f. Harness
    g. In-cabin filter or debris
3. Notice that the technician has moved from broad system tests to individual components or component groups. The technician's test continues to become more specific as the possibilities are narrowed down.

FIGURE 1-11 Select tests that have simple expectations and are easy to perform.

Technicians commonly encounter vehicles with more than one customer concern. When these concerns both originate from the same or companion systems, technicians are inclined to search for one cause to both problems. Unfortunately, trying to diagnose two faults at once can quickly become problematic and confusing. Instead, select the easier customer concern and follow it through to the end. If both problems were caused by the same faults, then both were fixed. If they were caused by two separate faults, the technician is no worse off for having fixed one concern.

When selecting tests to perform, remember that they should be simple and easy (**FIGURE 1-11**). Except when following service procedures, you should select tests that have simple expectations, are easy to perform, and provide you with the maximum amount of information. This means simple tests that inspect an entire system or circuit are ideal ways to begin testing. Simple tests have expectations and results that are quickly understood and interpreted. They are short and involve basic tools and access to areas that are comfortable to reach.

> **▶ TECHNICIAN TIP**
>
> When selecting tests, it is not a bad idea to choose those tests that might look at components of both systems (e.g., voltage drop on a shared electrical ground), but DO NOT attempt to test for both faults at once. While multiple faults within a companion or the same system often turn out to be related, they should be isolated and tested separately.

When selecting tests, prioritize your testing. First choose tests that can be performed quickly and simply, even if they do not test an entire circuit. If a preferred test is in a difficult place to access, move to another test and come back to it, if needed. The answer may be found in the meantime and the time-consuming test can be avoided. Simple and easy tests are ideal, but they must be measurable or objective.

Yes, a visual inspection is a simple and valuable test, but a technician must determine what the issue is in an objective manner, with help from the service information. A guess based only on appearance is insufficient. If the service information says, "cracks in the serpentine belt indicate that it needs replaced," the belt can be visually and objectively (yes or no) tested. The belt will either have the indicated wear or it will not. If the service information states, "Chain deflection cannot exceed 0.75," then the deflection can be measured and compared to the specification. As testing continues, it may become necessary to use advanced tests, sophisticated equipment, additional time, or tests in areas difficult to access. Keeping initial testing simple and easy will produce the quickest, most reliable, and effective results.

> **▶ TECHNICIAN TIP**
>
> When performing tests for an inspection under warranty, it is absolutely necessary to follow the manufacturers' guidelines.

When testing, use the recommended procedures and equipment. Manufacturers frequently recommend a particular procedure when testing one of their systems.

Failure to follow the specified service procedure can result in the warranty claim being denied by the manufacturer. In that case, both the shop and the technician lose money. Manufacturers may recommend a certain procedure because of the way their system is designed or monitored. Technicians must also be very careful to perform tests safely (**FIGURE 1-12**).

Beyond the mechanical dangers posed by automobiles, many of today's vehicles have dangerously high fluid pressures and deadly high voltage. It is of the utmost importance for the safety of the technician, and those working in the area, that safety procedures are always followed.

Proper test equipment and procedures are intended to test a particular component or system without causing any damage. Improper equipment or test procedures can create a second fault in the system being tested; making the technician's job even more difficult. For example, front probing an electrical terminal with the lead of a DMM can cause the terminal to spread or deform. This can create an intermittent high resistance or open within the circuit that was not there prior to the technician's test. Using the recommended equipment and procedures will help to ensure warranty claims are approved, people are safe, and testing goes smoothly.

When performing repairs, look beyond the obvious for the root cause. This simple suggestion can avoid customer comebacks. Novice technicians frequently have problems with misdiagnosing fuse-related issues. For example, a technician diagnoses a blown fuse as the cause of the customer concern. While replacing the fuse may have fixed the immediate fault, the technician did not look beyond the obvious. What causes a fuse to blow? Low

FIGURE 1-12 Always perform all tests safely.

resistance and increased amperage cause a fuse to blow. However, the technician did not test for one of these faults and the vehicle is likely to return with the same customer concern and the same blown fuse.

In another example of incomplete reasoning, a technician diagnoses a leaky transmission cooler line. The line is chaffed and leaking. This cooler line runs along the frame rail; the inner and outer tie rods are immediately below. The technician diagnoses the vehicle while it is on a lift and the suspension is unloaded (increasing the distance between the hose and steering linkage). The technician should have looked for the root cause of the chaffing, but instead the vehicle and customer come back some time later for the same concern. The technician notices several broken clips that held the flexible line into place on the frame rail. In both cases, the technician will work for free to repair the same vehicle, because time was not taken to ask the question: "Did something else cause this failure?" Testing must be focused beyond the obvious to identify the root cause of the problem and consequently avoid customer comebacks.

In summary, focused testing has several key elements. It picks up the possible faults identified in step two and begins testing each one broadly, narrowing down to more specific tests. Focused testing requires accurately documenting the tests performed, including a test description, expectation, and result, each and every time a test is performed. It should also be performed in a safe and proper manner, following manufacturers' guidelines and safety protocols. Focused testing is a safe, accurate, and repeatable method for isolating possible faults.

Step 4: Performing the Repair

The fourth step of the diagnostic process is to perform the repair. Although performing the repair is often the most straightforward step in the process, technicians must still avoid making several common mistakes. The following tips will help you to perform an effective and reliable repair.

K01007 Describe step four of the strategy-based diagnosis.

Use Proper Service Procedures

Manufacturers will often indicate what procedures are appropriate for their vehicles and components. Many design features and component materials require certain procedures be used and others avoided. Following the manufacturer's service information can prevent premature failure of the repair (**FIGURE 1-13**). For example, repair methods that are safe around the home may be unacceptable in the automotive industry. The use of twist-on wire connectors can create an unreliable and potentially dangerous electrical situation when used in a vehicle. Additionally, warranties, both original equipment and aftermarket, rely on the technicians' adherence to the manufacturer's service information. If technicians fail to do so, the warranty claim can go unpaid and the shop will lose money. Therefore, it is important for reliable repairs and warranty reimbursements that technicians follow the service information when performing repairs.

Use the Correct Tool for the Job

Failure to use the correct tool can lead to a customer comeback and injury to the technician. Proper tool selection is essential. If you are ever in doubt, refer to the manufacturer's service information. Improper tool use or selection can damage the component being installed or other components around it. For example, a technician may choose to install a pump busing with a hammer instead of using the recommended press and bushing driver. This incorrect tool selection can easily lead to misalignment, or damage to the bushing, pump, or torque converter. Using the wrong tool (or the right tool in the wrong manner) can also damage the tool and potentially injure those in the area. For example, if a technician is using a hardened chrome socket on an impact wrench, the socket may shatter, sending shrapnel flying. Using the correct tool for the job will produce better work and ensure the safety of the technician.

MAINTENANCE/SPECIFICATIONS

CHANGING YOUR WIPERS

The wiper arms can be manually moved when the ignition is disabled. This allows for ease of blade replacement and cleaning under the blades.

1. Disable the ignition before removing the blade.
2. Pull the arm away from the glass.
3. Left leading edge retaining block to release the blade. Swing the blade, away from with the arm, to remove it.
4. Swing the new blade toward the arm and snap it into place. Replace the retaining block at the leading edge of the wiper arm. Lower the wiper arm back to the windshield. The wiper arms will automatically return to their normal position the next time the ignition is enabled.

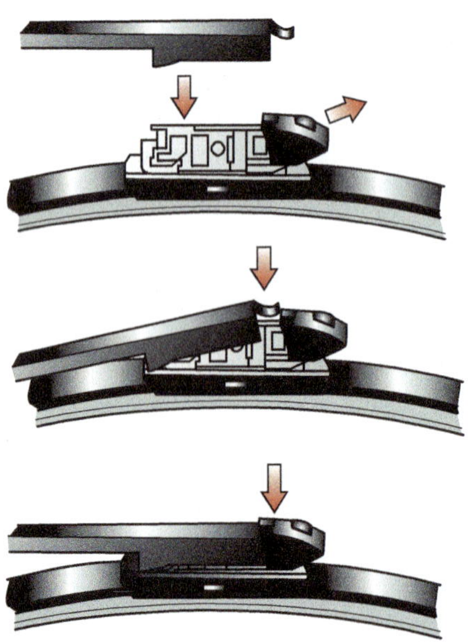

Refresh wiper blades at least twice a year for premium performance.

Poor preforming wipers quality can be improved by cleaning the blades and the windshield. See *Windows and wiper blades* in the *Cleaning* chapter.

To extend the life of wiper blades, scrape off the ice on the windshield BEFORE turning on the wipers. The ice has many sharp edges and will damage and shred the cleaning edge of your wiper blade.

FIGURE 1-13 A typical shop manual page has a task description broken into steps and diagrams or pictures to aid the technician.

Take Time to Perform the Repair Properly

Because technicians are frequently paid by the job, or a flat rate, rather than paid hourly, it is possible for technicians to feel a rush to complete their current job. Rushing increases the likelihood of a mistake. If a mistake occurs, the customer will come back with the vehicle and the technician will work for free to repair the mistake. For example, if a technician replaces a water pump and fills the coolant without bleeding the system, a potentially damaging situation can occur. The trapped gas can affect the flow of coolant and create a hot spot in the cylinder head. This can lead to warning lights, poor performance, and possible engine damage. Take a little extra time to ensure that the work is performed correctly, with the right tools and the proper service procedures. Taking time to perform the repair will ensure fewer "comebacks" and more satisfied customers.

Make Sure the Customer Approves of the Repair

This may seem trivial, but it is extremely important. Most states' laws protect consumers by preventing unauthorized services from being charged or performed. This means that technicians cannot just repair a vehicle and charge the customer for the cost incurred. If the customer is paying, shops must receive a customer's approval prior to performing repairs.

▶ TECHNICIAN TIP

It is very important to quote accurately and wait for approval before performing repairs on a customer's vehicle.

Check for Updates Prior to the Repair

It is also good practice to check for updated parts and software/firmware before performing a repair. It is possible that manufacturers have become aware of a problem with a particular component or software version and issued a software update or produced an updated component. When performing repairs it is a good idea to check for these sorts of updates (often found in TSBs), because it may prevent a customer comeback. Software updates are often downloaded from the manufacturer's website. For hard parts, the best resource is frequently the respective dealership's parts department.

Technical service bulletins also provide information related to unexpected problems, updated parts, or changes to a repair procedure on a particular vehicle system, part, or component. The typical TSB contains step-by-step procedures and diagrams on how to identify if there is a fault and perform an effective repair. Shops typically keep TSBs in a central location, or you may look them up online. Compare the information contained in the TSB with that of the shop manual. Note the differences and, if necessary, copy the TSB to take with you to perform the repair.

Pay Attention to Details

Performing the repair is straightforward but requires attention to detail. There are several things to keep in mind. Proper service procedures can be located in the manufacturer's service information. The correct tool for the job will lessen injury and ensure reliability. Use the necessary time to make sure that the repair was completed correctly. Document your work. These tips can greatly improve the likelihood of a successful repair, but the process does not stop with the repair.

Step 5: Verify the Repair

K01008 Describe step five of the strategy-based diagnosis.

The most important step of the strategy-based diagnostic process is verifying the repair. The reason that this is the most important step is straightforward. The vehicle would never have been in the shop if the customer did not have a concern. If the technician fails to address the original concern, the customer may view the trip as ineffective, a waste of their time and money. Even when a valid repair that makes the vehicle safer and more reliable was performed, the customer will still be unsatisfied if his or her original concern was not addressed. For example, a customer brings the vehicle into the shop for a sticky glove box latch. The technician identifies and repairs a dangerous brake line leak, but fails to fix the glove box. Some customers may view this trip to the shop as unsuccessful because it failed to fix their original issue. When verifying the repair, technicians must always double check their work. This is a valuable confirmation that the repair performed did fix the identified problem.

There are several ways to verify a repair, but generally, the simplest method is the best method. For example, a customer is concerned that the wipers stop moving when the switch is moved into the high position. In step one, the technician will verify that the customer concern and fault exist by turning the wiper switch to all positions. Then the technician uses the wiring diagram (step 2) to diagnose a fault (step 3) within the wiring harness and repairs it (step 4). The technician could then verify the repair by performing the last diagnostic test (from step 3) again. In most cases, the repair would be confirmed if the results had changed and were within expectation/specification.

But what if there was a second problem affecting the wipers such as worn brushes in the motor, or the wiper linkage fell off of the pivot on one side? The customer would still have issues with the wipers and would likely to be unhappy with the repair. So while performing the last diagnostic test (step 3) is a valid verification method, it is not foolproof. An easier method exists: simply return to the process used in step one to verify the customer concern. If the repair has eliminated the problem, the technician should now be able to turn the wiper switch to all positions (step 1) and confirm normal operation. Be certain to perform the same inspections used to verify the customer concern in step one after the repair is performed. This may include checking the entire system operation, not just a single function. This method of verifying that the customer concern is resolved is usually best in most scenarios because it is simple and it is exactly what the customer will do to check your work.

However, sometimes verifying the repair requires a more complicated means of verification. A common concern that falls into this scenario is as follows: The customer brings their vehicle in with a concern that the MIL (malfunction indicator lamp) or "check engine" light is illuminated. In this scenario, *NEVER* verify the repair by simply checking to see that the light is off. While this is what the customer will do to check your work, the failure of the MIL to light can often be misleading and result in a comeback for the exact same problem. This can occur because the MIL is illuminated when tests run by the computers in the vehicle fail. The computers are constantly running tests, but some tests require very specific conditions before they can be run and, hence, fail. Due to the requirements for the conditions to be right, simply checking to see if the light is illuminated is an inadequate method of verification.

For more complicated computer-controlled systems, the best method of verification is checking the test results stored on the vehicle's computer. This option will require an electronic scan tool that communicates with the vehicle's computer, along with a high level of diagnostic experience and service information to verify that the concern has been fully resolved. If the communication option is not available, the second best method of verification is repeating the last diagnostic test performed (in step 3) and confirming that the result has changed to now match the expectation/specification. Complicated computer-controlled systems require that the technician do more than verify the customer concern is eliminated. The technician will have to repeat a diagnostic test (step 3) or view test results stored on the vehicle's computer (this is the preferred method) in order to verify that the repair was effective.

Step five of the strategy-based diagnostic process is the most important. A vehicle should never be returned to a customer without this step completed.

▶ TECHNICIAN TIP

The job is not complete until you have verified that the repair resolved the customer's concern.

▶ Documenting the Repair

The first two components, gathering information from the customer and the strategy-based diagnostic process, have already been described; this section discusses documentation. The repair is documented for several reasons: accurate vehicle history, returns or comebacks, and OEM or aftermarket warranties. Keeping accurate service records will help technicians to know what services and repairs have been performed on a vehicle when it needs any future services. This can be invaluable during the diagnostic process and can also help service advisors and technicians identify what maintenance or recall work still needs to be completed. Documenting the repair also helps technicians in the event that a vehicle returns, now or in the future, with the same customer concern or fault. This can help to identify defective parts or common problems.

Warranty work is another reason that all repairs must be documented. Whether the repair is submitted to an original equipment manufacturer or to an aftermarket warranty company for reimbursement, the repair must be well documented. Warranty clerks will review the repair order to ensure that proper testing and repair procedures have been followed. Technicians must document their work to ensure that the shop, and in turn the technician, get paid for the work performed.

Finally, documenting the work provides the shop with a record that the work was initiated and completed. This is important in case the vehicle is later involved in an accident or other mishap and the shop is involved in a lawsuit. It is important to have the customer sign or initial, depending on shop policy, the repair order, to verify that the customer accepted the repair.

The Three Cs of Documentation

When documenting a repair, technicians need to remember the **3 Cs**: **concern**, **cause**, and **correction** (**FIGURE 1-14**).

K01009 Explain how the three Cs are applied in repairing and servicing vehicles.

N01002 Demonstrate use of the three C's (concern, cause, and correction).

Concern

The main focus of the 3 Cs is the customer *concern*, which is also the focus of step one of the diagnostic process. Often the concern is documented on the repair order prior to the technician receiving the vehicle. If this is the case, the technician who works on the vehicle should take time to fully understand the concern, read the repair order, and possibly talk further with the customer to understand the nature of the problem. Think through the problem and develop a strategy to attack it. Other symptoms and diagnostic troubles codes are some examples of other information that should be included in the "concern."

Cause

The second C in the 3 Cs is *cause,* which details the cause of the customer concern. This correlates to the documentation done in step three of the diagnostic process. The technician should document any tests that they perform with enough detail that they can be repeated, as well as specifications/expectations, and results. This goes for all tests, even the simple ones.

Correction

The technician should then document the last C, the correction. This must include the procedure used as well as a brief description of the correction. This information comes from the fourth step of the diagnostic process. When documenting the repair order, technicians should include the customer concern and symptoms (DTCs are symptoms); brief descriptions of tests; expectations; and results, along with the procedure and repair that were performed. The technician should also include all parts that were replaced as well, and noted if they were new or used, OEM, or aftermarket.

FIGURE 1-14 The 3Cs of documenting the repair.

Other Parts of Documentation

Additional service recommendations should also be documented on the repair order. While working on the vehicle, technicians should also be mindful of other work that may need to be performed. Technicians are obligated to make the customer aware of safety concerns that require attention. Customers may be unaware of a potential hazard or lack of maintenance. Bringing this to the attention of the customer right away can help the technician, as well as the customer. For example, if the technician is already working on the vehicle, they would not have to remove the vehicle from the service bay, bring in a new vehicle, and start all over. Repairing multiple issues in one trip to the service bay makes good use of the technician's time. It also improves customer relations by bringing the customer's attention to problems and thereby preventing possible failures.

For example, a technician may be changing the fluid in a transmission and notice that the brake friction pads are extremely low. Bringing this to the attention of the customer can result in additional work for the technician and save the customer from a potentially more costly repair. For this reason, technicians should also note safety issues and maintenance items on the repair order.

Repair Order

K01010 Describe the information and its use within a repair order.

N01003 Identify information needed and the service requested on a repair order.

S01002 Complete a repair order.

A **repair order** is a key document used to communicate with both your customers and coworkers. Thoroughly document the information provided by the customer on the repair order; every bit of it may be helpful during the diagnosis (**FIGURE 1-15**). If you are not typing this information, make sure your handwriting is clear and easy for others to read. Unfortunately, if documentation of a complaint is not done well, the technician could be led on a much longer diagnostic path, wasting everyone's time. It can also be a time-consuming process for the diagnosing technician to make contact with the customer in order to get more information that was missed the first time. From time to time, it may be inevitable that the customer will need to be contacted for further inquiry after the initial visit. However, carefully gathering information from the customer on their initial visit will save time, prevent inconveniencing the customer, and aid in the diagnostic process.

To complete a repair order using the 3 Cs, follow the steps in **SKILL DRILL 1-2**.

▶ TECHNICIAN TIP

A repair order is a legal document that can be used as evidence in the event of a lawsuit. Always make sure the information you enter on a repair order is complete and accurate. The information required on a repair order includes: date; customer's name, address, and phone number; vehicle's year, make, model, color, odometer reading, and VIN; and description of the customer's concern. Store repair orders in a safe place, such as in a fireproof filing cabinet or electronically on a secure computer network. Finally, to prevent future complications, it is a good idea to have the customer sign or initial the repair order, indicating that they understand and agree to the needed repair. Having the customer's signature will help prevent the shop from being held liable in an accident involving the vehicle later.

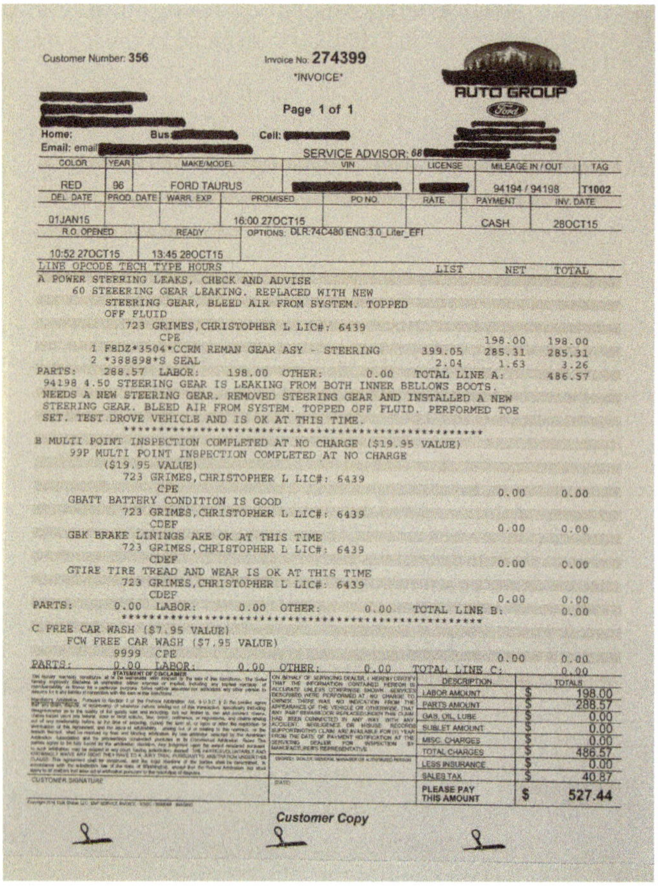
Customer Number: 356
Invoice No: 274399
INVOICE
[redacted] AUTO GROUP
[redacted]
Page 1 of 1
Home: Bus: [redacted] Cell: [redacted]
Email: email [redacted]
SERVICE ADVISOR: 68 [redacted]

COLOR	YEAR	MAKE/MODEL	VIN	LICENSE	MILEAGE IN / OUT	TAG
RED	96	FORD TAURUS	[redacted]	[redacted]	94194 / 94198	T1002

DEL. DATE	PROD. DATE	WARR. EXP	PROMISED	PO NO	RATE	PAYMENT	INV. DATE
01JAN15			16:00 27OCT15			CASH	28OCT15

R.O. OPENED	READY	OPTIONS: DLR:74C480 ENG:3.0_Liter_EFI
10:52 27OCT15	13:45 28OCT15	

```
LINE OPCODE TECH TYPE HOURS                                   LIST      NET     TOTAL
A POWER STEERING LEAKS, CHECK AND ADVISE
    60 STEEERING GEAR LEAKING. REPLACED WITH NEW
       STEERING GEAR, BLEED AIR FROM SYSTEM. TOPPED
       OFF FLUID
         723 GRIMES,CHRISTOPHER L LIC#: 6439
             CPE                                                  198.00    198.00
     1 F8DZ*3504*CCRM REMAN GEAR ASY - STEERING         399.05    285.31    285.31
     2 *388898*S SEAL                                     2.04      1.63      3.26
PARTS:    288.57  LABOR:    198.00  OTHER:      0.00  TOTAL LINE A:         486.57
 94198 4.50 STEERING GEAR IS LEAKING FROM BOTH INNER BELLOWS BOOTS.
 NEEDS A NEW STEERING GEAR. REMOVED STEERING GEAR AND INSTALLED A NEW
 STEERING GEAR. BLEED AIR FROM SYSTEM. TOPPED OFF FLUID. PERFORMED TOE
 SET. TEST DROVE VEHICLE AND IS OK AT THIS TIME.
          ***************************************************
B MULTI POINT INSPECTION COMPLETED AT NO CHARGE ($19.95 VALUE)
    99P MULTI POINT INSPECTION COMPLETED AT NO CHARGE
       ($19.95 VALUE)
         723 GRIMES,CHRISTOPHER L LIC#: 6439
             CPE                                                    0.00      0.00
    GBATT BATTERY CONDITION IS GOOD
         723 GRIMES,CHRISTOPHER L LIC#: 6439
             CDEF                                                   0.00      0.00
    GBK BRAKE LININGS ARE OK AT THIS TIME
         723 GRIMES,CHRISTOPHER L LIC#: 6439
             CDEF                                                   0.00      0.00
    GTIRE TIRE TREAD AND WEAR IS OK AT THIS TIME
         723 GRIMES,CHRISTOPHER L LIC#: 6439
             CDEF                                                   0.00      0.00
PARTS:      0.00  LABOR:      0.00  OTHER:      0.00  TOTAL LINE B:           0.00
          ***************************************************
C FREE CAR WASH ($7.95 VALUE)
    FCW FREE CAR WASH ($7.95 VALUE)
         9999  CPE                                                  0.00      0.00
PARTS:      0.00  LABOR:      0.00  OTHER:      0.00  TOTAL LINE C:           0.00
```

STATEMENT OF DISCLAIMER

CUSTOMER SIGNATURE

DESCRIPTION		TOTALS
LABOR AMOUNT	$	198.00
PARTS AMOUNT	$	288.57
GAS, OIL, LUBE	$	0.00
SUBLET AMOUNT	$	0.00
MISC. CHARGES	$	0.00
TOTAL CHARGES	$	486.57
LESS INSURANCE	$	0.00
SALES TAX	$	40.87
PLEASE PAY THIS AMOUNT	$	527.44

Customer Copy

FIGURE 1-15 A repair order.

Applied Communications

AC-23: Repair Orders: The technician writes a repair order containing customer vehicle information, customer complaints, parts and materials used (including prices), services performed, labor hours, and suggested repairs/maintenance.

A repair order is a legal contract between the service provider and the customer. It contains details of the services to be provided by you and the authorization from the customer. To make sure everyone understands clearly what is involved, a repair order should contain information about the following aspects of the repair.

- Your company or service providers: The service provider section contains the company name, address, and contact details; the name of a service advisor who is overseeing the job; and the amount of time the service technician will have to service the vehicle.
- The customer: The customer section contains the customer's name, address, and contact phone numbers.
- The customer's vehicle: The vehicle section includes details about the vehicle to be serviced. Check the vehicle's license plate before starting work. The license plate numbers are usually unique within a country. You should also record information about the vehicle's make, model, and color. This information will make it easier for you to locate the vehicle on the parking lot. You may also need to know the manufacture date of the vehicle to be able to order the right parts. The odometer reading and the date will help keep track of how much distance the vehicle travels and the time period between each visit to the shop. The VIN is designed to be unique worldwide and contains specific information about the vehicle. Many shops do a "walk-around" with the customer to note any previous damage to the vehicle and to look for any obvious faults such as worn tires, rusted-out exhaust pipes, or torn wiper blades.
- The service operations: This section contains the details of the service operations and parts.
- The first part is the service operation details. For example, the vehicle is in for a 150,000 mile (240,000-km) service, which can be done in 3 hours and results in approximately $300 of labor costs. The information about the chargeable labor time to complete a specific task can be found in a labor guide manual. In some workplaces, this information is built into the computer system and will be automatically displayed.
- The second part of this section is the details of parts used in the service, including the descriptions, quantities, codes, and prices. The codes for each service and part are normally abbreviations that are used for easy reference in the shop. Some shops may have their own reference code system.
- As you do the vehicle inspection, you may discover other things that need replaced or repaired. These additional services can be recorded in another section. It is essential that you check with the customer and obtain their approval before carrying out any additional services.
- The parts requirements: This section lists the parts required to perform the repair.

Some repair orders also contain accounting information so they can be used as invoices.

SKILL DRILL 1-2 Completing a Repair Order

1. Greet the customer.
2. Locate a repair order used in your shop and obtain or verify the customer's name, address, and phone number.
3. Obtain details about the vehicle, including the year, make, model, color, odometer reading, and VIN.
4. Ask the customer to tell you more about the *concern* by using open-ended questions, such as "When does problem occur?" "At what speed(s)?" "How do you experience the problem?" "How long has this been occurring?" "How many passengers do you typically carry?" Type or clearly write the customer's responses on the repair order.
5. Ask the customer about other changes with the vehicle, such as recent work, or recent travel. Type or clearly write the customer's responses on the repair order.
6. Remembering the lessons learned regarding the proper diagnostic process, begin to verify the customer's *concern* by first performing a visual inspection.
7. If you see nothing unusual during your visual inspection, continue to verify the customer's concern by conducting a road test of the vehicle. The customer may ride along, if possible, to help identify the issue as it occurs, or you may conduct the test by yourself. Following the test drive, after verifying the customer's concern, record it on the repair order.
8. The second step of the diagnostic process is to research the possible faults, and gather information. Access the vehicle service history to determine if the vehicle has experienced a similar problem in the past, requires a routine service maintenance, or has been serviced recently. Document this information, if applicable, on the repair order.
9. Conduct research by accessing various sources of information related to the vehicle, such as the vehicle service manual or the owner's manual. Check to see if a TSB related to the issue exists. As part of the process, rule out the possibility that the customer's concern is a normal operation of the vehicle.
10. Now that you have your broad list of possible faults related to the concern, begin step three, focused testing. Choose one of the possible broad faults you identified in step two. Now refer to the service manual to locate information that matches the concern. Service manuals usually contain diagnostic charts to aid in the focused testing process.
11. Conduct a test and record its description, your expectation, and the result on the repair order or another piece of paper. Continue to check each possible fault until you identify the *cause* of the concern.
12. Once you have identified the fault, you're ready for step four, performing the repair. You would inform the customer of your finding and obtain his or her approval to make the repair. Pending customer approval, you would then follow proper safety procedures and use the manufacturer's guidelines to correct the problem, being sure to use the correct tools and taking the time to complete the job properly.
13. Once you've made the repair, you are ready for step five, verifying the repair. The simplest way to verify that you have addressed and corrected the customer's concern is to repeat the test drive. Take the vehicle for a test drive and repeat the tests you initially performed. Is the issue gone? If so, you have verified the repair and can return the vehicle to the customer.
14. Document the correction on the repair order. If the issue is not resolved, you must return to your list of possible faults and continue testing after first alerting the customer that additional work and time will be necessary.

▶ Wrap-Up

Ready for Review

- Service history is typically retrieved from service records kept by the shop, dealer network, original equipment manufacturer (OEM), or aftermarket service center and contains a list of services performed on a vehicle and the date and mileage at which they were completed.
- The service history allows technicians to determine if the vehicle has been well maintained. This can be extremely useful when a technician suspects that lack of maintenance may be the cause of the problem.
- Failure to comply with the state and federal law can be very expensive for the dealership and manufacturer.
- Today's vehicles also require regular software updates made available to fix a bug or glitch in the computer programming. These updates are often designed to eliminate a customer concern, improve owner satisfaction, or increase vehicle life.
- The strategy-based diagnostic process is focused on fixing problems correctly the first time. It begins with identifying the customer's concern and ends with confirming that the problem has been resolved.
- The problem-solving process provides a consistent road map for technicians as they address customer concerns that require diagnosis and to make sure that customer concerns are resolved with certainty.
- Strategy-based diagnosis simplifies the problem-solving portion of the repair, making the job easier for the technician; it prevents technicians from having to work on the same job more than once; and it all but eliminates customer comebacks.
- Customer comebacks are usually caused by the customer concern being misinterpreted or misunderstood or failing to verify that the original concern was resolved.
- The strategy-based diagnostic process begins by gathering preliminary information from the customer and by reviewing the vehicle's service history.
- The first step in the diagnostic process is to verify the customer concern. This step is completed for two main purposes: verify that there is an actual problem present, and guarantee that the customer's concern is addressed.
- Visual inspections can be very valuable, but technicians need to be careful not to jump to conclusions.
- DTC's (Diagnostic Trouble Codes) and freeze frame data should always be saved and recorded on the repair order. Freeze-frame data provide a snapshot of the entire engine data when the DTC occurs, which allows for duplication of the condition so that the DTC can be replicated.
- The second step in the diagnostic process is to research possible faults. The goal of this step is to create a list of possible faults. The list will be created based on the information gathered in step 1 and narrowed down by the tests performed in step 3 until the cause of the concern has been confirmed.
- The best source of information is usually the manufacturer's service information system.
- Technical service bulletins (TSBs) are service notifications and procedures sent out by the manufacturers to dealer groups, alerting technicians to common issues with a particular vehicle or group of vehicles. Some aftermarket sources also exist for the pattern failures addressed by TSBs
- A technician must always be aware that steps in the diagnostic process cannot be skipped. A repair should never be performed unless the possible fault has been verified through testing.
- Step 3 of the diagnostic process involves focused testing, where technicians use their testing skills to eliminate possible faults from the list they created in step two. Steps 2 and 3 work together, because testing starts at a system level and works down to subsystems, then finally to individual components.
- When selecting tests prioritize your testing. First choose tests that can be performed quickly and simply, even if they do not test an entire circuit. If a preferred test is in a very difficult place to access, move to another test and come back to it, if needed.
- Following manufacturers' guidelines and safety protocols keeps technicians safe. Focused testing is a safe, accurate, and repeatable method for isolating possible faults. Once the fault has been isolated, it is time to perform the repair.
- The fourth step of the diagnostic process is to perform the repair. Performing the repair is often the most straightforward step in the process.
- Use proper service procedures when performing a repair. Manufacturers often indicate what procedures are appropriate for their vehicles and components.
- Use the correct tool for the job when performing a repair. Failure to use the correct tool for the job can lead to a customer comeback and injury to the technician.
- Take time to perform the repair properly. Technicians are frequently paid by the job, or flat rate, rather than paid hourly, it is possible for technicians to feel a rush to complete their current job. Rushing increases the likelihood of a mistake and the next time you may pay for it.
- The most important step of the strategy-based diagnostic process is verifying the repair. The reason that this is the most important step is straightforward. The vehicle would never have been in the shop if the customer did not have a concern.
- Verifying the original concern is the best method of double-checking your work and meeting your customers' expectations. The job is not complete until you have verified that the repair resolved the customer's concern.

- Documentation is key to effective and efficient repairs. Keeping all the information available to the service advisor, technician, and the customer allows for a more open dialogue which can limit the confusion of the repair process.
- The repair is documented for several reasons: accurate vehicle history, returns or comebacks, and warranties. Keeping accurate service records will help technicians to know what services and repairs have been performed on a vehicle when it needs any future services.
- When documenting a repair, technicians need to remember the three Cs: concern, cause, and correction.
- When documenting the repair order, technicians should include the customer concern and symptoms (Diagnostic Trouble Codes are symptoms) and a brief description of tests, expectations, and results, along with the procedure and repair that were processed.

Key Terms

Strategy-Based Diagnostic Process A systematic process used to diagnose faults in a vehicle.

service advisor The person at a repair facility that is in charge of communicating with the customer.

service history A complete listing of all the servicing and repairs that have been performed on that vehicle.

repair order The document that is given to the repair technician that details the customer concern and any needed information.

freeze frame data Refers to snapshots that are automatically stored in a vehicle's power train control module (PCM) when a fault occurs (only available on model year 1996 and newer).

technical service bulletin (TSB) Service notifications and procedures sent out by the manufacturers to dealer groups alerting technicians about common issues with a particular vehicle or group of vehicles.

original equipment manufacturer (OEM) The company that manufactured the vehicle.

aftermarket A company other than the original manufacturer that produces equipment or provides services.

intermittent faults A fault or customer concern that you can not detect all of the time and only occurs sometimes.

3 Cs A term used to describe the repair documentation process of 1st documenting the customer concern, 2nd documenting the cause of the problem, and 3rd documenting the correction.

concern Part of the 3Cs, documenting the original concern that the customer came into the shop with. This documentation will go on the repair order, invoice, and service history.

cause Part of the 3Cs, documenting the cause of the problem. This documentation will go on the repair order, invoice, and service history.

correction Part of the 3Cs, documenting the repair that solved the vehicle fault. This documentation will go on the repair order, invoice, and service history.

Review Questions

1. When a vehicle comes in for repair, detailed information regarding the vehicle should be recorded in the:
   a. service booklet.
   b. repair order.
   c. vehicle information label.
   d. shop manual.
2. The service history of the vehicle gives information on whether:
   a. the vehicle was serviced for the same problem more than once.
   b. an odometer rollback has occurred.
   c. the vehicle meets federal standards.
   d. the vehicle has Vehicle Safety Certification.
3. Which of the following steps is the last step in a strategy-based diagnostic process?
   a. Verifying the customer's concern
   b. Researching possible faults
   c. Performing the repair
   d. Verifying the repair
4. When possible, which of the following is the best way to understand the customer's concern?
   a. Asking the customer to guess the cause of the problem.
   b. Asking the customer to suggest a solution to the problem.
   c. Encouraging the customer to demonstrate the problem.
   d. Encouraging the customer to help you fix the problem.
5. The best way to address intermittent faults is to:
   a. look for symptoms, data, or DTCs that are repeatable or consistent.
   b. reverse the steps in the diagnostic process.
   c. ask the customer to bring back the vehicle when the fault occurs.
   d. take it up only when it is covered by warranty.
6. When the technician encounters a vehicle with more than one customer concern, and both originate from companion systems, the technician:
   a. should attempt to test for both faults at once.
   b. need not attempt to fix the second fault.
   c. should never choose those tests that might look at components of both systems.
   d. should isolate the faults and test them separately.
7. Choose the correct statement.
   a. When performing tests for an inspection under warranty, follow your intuition rather than the manufacturers' guidelines.
   b. Researching possible faults should begin with a specific cause in mind.
   c. For hard parts, the best resource is frequently the respective dealership's parts department.
   d. DTCs and freeze frame data need not be captured before clearing the memory.
8. All of the following will happen if the technician fails to document test results *except*:
   a. The manufacturer will not pay the claim.
   b. The shop is out money for the parts and service.

c. The technician will be unable to diagnose the fault.
d. The technician will not be paid for his or her work.

9. All of the following statements with respect to the 3 Cs are true *except*:
   a. Customer concern is documented on the repair order prior to the technician receiving the vehicle.
   b. The second C in the 3 Cs refers to the cause of the customer's concern.
   c. Technicians should note safety issues and maintenance items on the repair order.
   d. Additional service recommendations should never be documented on the repair order.
10. Which of the following is not one of the 3 Cs of vehicle repair?
   a. Cause
   b. Cost
   c. Concern
   d. Correction

ASE Technician A/Technician B Style Questions

1. Tech A says that when diagnosing a transmission problem, it is important to first verify the customer concern by taking the vehicle on a road test if possible. Tech B says that you should check for TSBs during the diagnostic process. Who is correct?
   a. Tech A only
   b. Tech B only
   c. Both A and B
   d. Neither A nor B
2. Tech A says that additional service recommendations should be documented on the repair order. Tech B says technicians are obligated to make the customer aware of safety concerns that require attention. Who is correct?
   a. Tech A only
   b. Tech B only
   c. Both A and B
   d. Neither A nor B
3. Tech A says the strategy-based diagnostic process is a scientific process of elimination. Tech B says the strategy-based diagnostic process begins with scanning the vehicle for DTCs. Who is correct?
   a. Tech A only
   b. Tech B only
   c. Both A and B
   d. Neither A nor B
4. Tech A says that manufacturers will often indicate what procedures are appropriate for their vehicles and components. Tech B says that the manufacturer's service information can avoid premature failure of the repair. Who is correct?
   a. Tech A only
   b. Tech B only
   c. Both A and B
   d. Neither A nor B
5. Tech A says that technicians are frequently paid by the job, or flat rate, rather than paid hourly. Tech B says rushing the repair is best for the customer, so they get their vehicle back quickly. Who is correct?
   a. Tech A only
   b. Tech B only
   c. Both A and B
   d. Neither A nor B
6. Tech A says that it is only necessary for dealerships to check for updated parts and software/firmware before performing a repair. Tech B says that it is possible that manufacturers have become aware of a problem with a particular component or software version and have issued a software update. Who is correct?
   a. Tech A only
   b. Tech B only
   c. Both A and B
   d. Neither A nor B
7. Tech A says that the customer concern is the focus of step 1 of the diagnostic process. Often the concern is documented on the repair order prior to the technician receiving the vehicle. Tech B says the technician who works on the vehicle should take time to fully understand the concern, read the repair order, and possibly talk further with the customer to understand the nature of the problem. Who is correct?
   a. Tech A only
   b. Tech B only
   c. Both A and B
   d. Neither A nor B
8. Tech A says that a repair order is only used in the shop and will be discarded when the vehicle is complete. Tech B says that the repair order is a legal document and could be used in a court. Who is correct?
   a. Tech A only
   b. Tech B only
   c. Both A and B
   d. Neither A nor B
9. Two technicians are discussing a transmission problem. Tech A says that it is important to test drive the vehicle because there may actually be no issue with the vehicle and the customer complaint is actually a normal operational characteristic of the transmission. Tech B says you should always check TSBs before performing any service of the transmission because the manufacturer may have updated a component. Who is correct?
   a. Tech A only
   b. Tech B only
   c. Both A and B
   d. Neither A nor B
10. Tech A says that experience will allow you to skip many of the steps of the diagnostic process because you will be familiar with the transmission. Tech B says that skipping steps of the diagnostic process can cause issues to be missed, or misdiagnosis of the problem. Who is correct?
   a. Tech A only
   b. Tech B only
   c. Both A and B
   d. Neither A nor B

CHAPTER 2

Introduction to Diesel Engines

Learning Objectives

After reading this chapter, you will be able to:

- 2-01 Outline the historical development of diesel engine technology.
- 2-02 Identify and explain the advantages of diesel engine technology compared to other powertrain technology.
- 2-03 Identify and describe the important elements of compression ignition systems.
- 2-04 Identify and compare major features of direct- and indirect-injection combustion chambers.
- 2-05 Identify and describe the stages of direct-injection combustion.
- 2-06 Explain the advantages and disadvantages of indirect-injection combustion chambers.

You Are the Technician

The company where you work as a lead vehicle service technician maintains and repairs a fleet of over 200 service-utility vehicles. Vehicles perform an assortment of vocational tasks. These range from regularly dispatching work crews to perform power line repairs—carrying tools and repair supplies—to using large service vans that transport one or two electricians and supplies needed to perform repairs on streetlights or electric meter replacements. The equipment division is preparing to make a large phased-in replacement of all vehicles and is asking for your input on the decision because you are a technician, not just because you know about service center operations but because of your familiarity with vehicle technology in general. The company is asking you to prepare a list that outlines the advantages and disadvantages of replacement vehicles operating with either diesel engines or gasoline engines and electric drive-only propulsion systems. As you prepare your summary of pros and cons for each technology, consider the following:

1. Rank which vehicles you expect will likely operate the longest during prolonged periods of power outages—such as after a storm or hurricane disaster recovery—and which will allow work crews to remain in the field longest during snowstorms and cold snaps.
2. Which vehicle will likely deliver the best fuel efficiency and longest service life for the company operating in severe service conditions, such as when carrying heavy loads and towing trailers of supplies, under varying speed and distance conditions?
3. Rank which vehicles you expect will require more regular servicing, such as oil changes; replacing spark-plugs, glow-plugs, fuel injectors, and filters; and other engine servicing or powertrain-related downtime.

▶ Introduction

Diesel engines are the workhorses of the heavy-duty commercial transportation industry. Their proven economy, rugged durability, and reliability have made them the engine of choice for use in over 95% of buses, heavy trucks, and off-road machinery. In contrast to the North American heavy-duty transportation sector, diesel use had relatively little representation in North America applications such as cars, light trucks, and SUVs (sport utility vehicles) until recent years.

In this market, diesel has made significant inroads, and penetration into the market continues to grow in spite of the negative fallout from the 2015 Volkswagen emission scandal. Data from 2016, a year when Volkswagen (VW) diesel sales were suspended, surprised some industry observers who expected a sharp decline in diesel sales. In that year, diesel passenger vehicle registrations in the United States alone increased by 280,000 over the previous year, even with 25% fewer models to choose from compared to 2015. Most of this increase came from take-up rates of diesels in a fast growing market for light-duty pickups and SUVs.

Adding to the diesel sales momentum since then are new models with expanded diesel powertrain choices from automakers other than VW. GM will offer 12 models of vehicles with diesel engines in its 2018 line-up. One of these is a new diesel engine in the Chevy Cruze delivering 52 mpg (U.S. highway) (4.52 L/100 km), the highest fuel efficiency for any non-hybrid or electric vehicle (**TABLE 2-1**). The same 1.6 L whisper diesel in the Equinox enables this popular crossover utility vehicle to outperform the Toyota RAV4 and Nissan Rogue hybrids in highway fuel efficiency (39 mpg US vs. 35 and 30 mpg US, respectively, or 6.03 L/100 km vs. 6.72 L/100 km and 7.84 L/100 km, respectively) (**FIGURE 2-1**). Ford will offer a new 3.0 L diesel in the F-150 pickup to complement 3.2 L five-cylinder Powerstroke diesel available in the Ford Transit and the 6.7 L Powerstroke used by close to two-thirds of its heavier pickups. Not to be outdone, GM is making an in-line 6-cylinder 3.0 L diesel available for its 1500 series pick-up trucks to join its 2.8 L V6 Duramax diesel used in the Colorado. New diesel models from Jaguar for its new F-type crossover utility vehicle (CUV), Fiat Chrysler Automobiles (FCA) Jeeps, the Kia Sorento, along with the Mazda CX-5 SKYACTIV-D SUV diesel may not have the media attention electric and hybrid electric vehicles are receiving, but diesels growing popularity, with production which currently outstrips electric and hybrid vehicle sales reflects the well-reasoned, practical choices of both consumers and manufacturers for powertrain technology (**FIGURES 2-2** and **2-3**). Table 2-1 shows North American models using the new 1.6 L whisper diesel—named for its superlative quiet operation.

TABLE 2-1 Comparing GM's 16 L Whisper Diesel's Advantages

2018 Equinox Automatic Six-Speed AWD							
Diesel Four-Cylinder Turbo 1.6 L				**Gasoline Four-Cylinder Turbo 1.5 L**			
City	**HWY***	**Combined**	**Torque**	**City**	**HWY**	**Combined**	**Torque**
28 8.4 L/100 km	38 6.2 L/100 km	32 7.4 L/100 km	236 ft-lb/ 320 N·m @ 2,000 rpm	24 mpg 9.8 L/100 km	30 mpg 7.8 L/100 km	26 mpg 9.0 L/100 km	203 ft-lb/ 275 N·m @ 2,000 rpm
Advantage—Improvement:							
+17%	+21%	+19%	+13%				
2018 Cruze Manual Six Speed							
Diesel Four-Cylinder Turbocharged 1.6 L				**Gasoline Four-Cylinder Turbocharged 1.4 L**			
City	**HWY**	**Combined**	**Torque**	**City**	**HWY**	**Combined**	**Torque**
30 7.8 L/100 km	52 4.5 L/100 km	37 6.4 L/100 km	236 ft-lb/ N·m – 320 @ 2,000rpm	27 mpg 8.7 L/100 km	38 mpg 6.2 L/100 km	31 mpg 7.6 L/100 km	177 ft-lb/240 N·m @ 2,000 rpm
Advantage—Improvement:							
+10%	+27%	+17%	+25%				

*HWY = highway

(Sourced from Environmental Protection Agency and GM Authority)

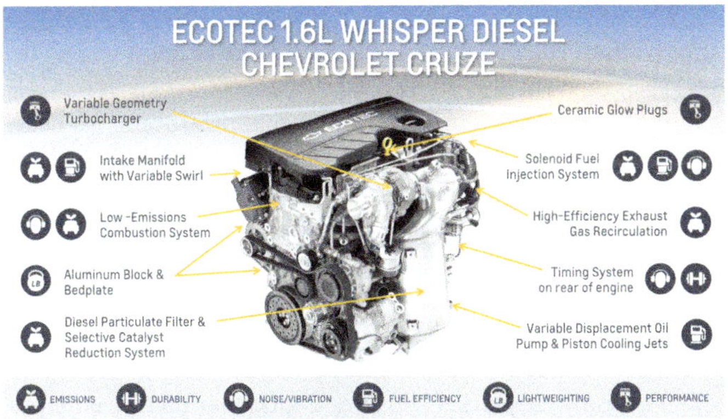

FIGURE 2-1 Technical features of the 1.6 L GM whisper diesel.

FIGURE 2-2 The availability of a 2017 emission compliant drop in replacement diesel engine for the Wrangler Jeep underlines the enthusiasm for diesel engines.

For the consumer, the primary appeal of the diesel in light-duty applications has traditionally been its superior fuel economy. Compared to similarly sized gasoline-fueled engines in light-duty automobiles, diesels are between 25% and 60% more fuel-efficient than similar displacement gasoline-fueled engines. With increasing vehicle weight, such as in pick-ups, delivery vehicles, SUV's and particularly when towing, the advantage over gasoline engine fuel economy is even greater. The economy of the diesel, combined with its long service life, low maintenance cost, and high torque output at low engine speeds, has persuaded the consumer market of its viability for use in automobiles, SUVs, and pickup trucks where power and efficiency matter. In North America's fastest growing vehicle market, this is where the diesel makes the most sense – heavier vehicles combined with better performance and fuel efficiency effectively balance out the additional premium paid to purchase a diesel. (See 2-4 Powertrain Quadrant).

FIGURE 2-3 The 5.0 L V8 from Cummins is used by Nissan and Toyota.

Benefits from the automotive use of diesel technology appear jeopardized according to some industry because of the fallout from Volkswagens 2015 diesel emission scandal. The concealment of emission defeating devices and operating strategies by one manufacturer has expanded to a general indictment of diesel technology. Cities in many western European countries where nearly half of all new cars run with diesels now have plans to ban diesels. But calls for the end of diesel engines based on arguments about their emission are not new and not without some validity. Diesel does have dirty legacy built on images of older, heavy-duty vehicles that often-leaked oil and preceded long dark trails of soot from the exhaust of poorly maintained engines or engines with tampered emission controls. Noisy engines used in crowded urban transit buses, garbage packers or dump trucks didn't boost the reputation of diesels either. Severe service operation of these type of vehicle having accumulated enormous operating distances frequently while stopping and starting on city streets added to the public unpleasant sense of diesels as toxic, dangerous and unrefined.

The emissions argument against diesel can point to more than flawed perceptions about worn-out engines in poorly maintained and noisy vehicles. Emissions legislation for diesel engines has lagged behind legislation for gasoline engines. Although gasoline fueled passenger vehicles were equipped with emissions controls beginning in the late 1950s, the first emissions standards for diesels were not established until the mid-1970s. One reason for lagging emission standards is that the number of diesel engines in use is small compared to the number of gasoline engines in use. In North America, diesel engine registrations make up approximately 4–6% of the total number of vehicle operating on the roads. Emission standards for diesel engines began to become more of a priority after emissions from gasoline-fueled vehicles dropped. Not only did the drop in emissions from gasoline engines make the diesel's more noticeable,

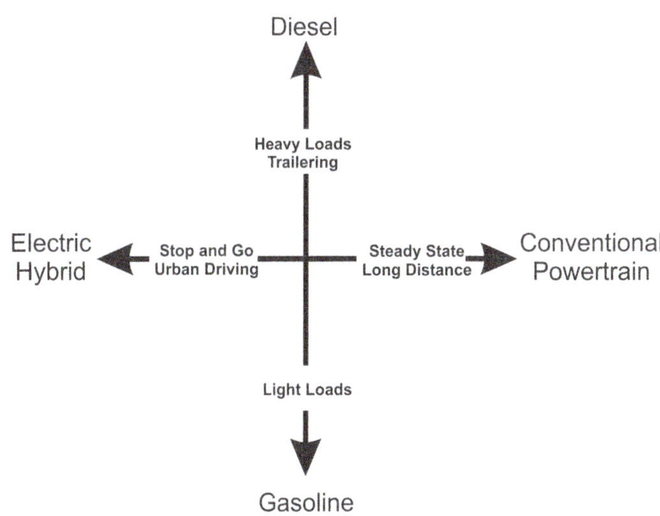

FIGURE 2-4 Powertrain Quadrant: Diesels make the most sense in terms of economy versus cost of ownership when used to tow or haul heavy vehicle loads over long distances.

the actual proportion of emissions produced by diesels rose. Data estimates based on EPA models during the 1980s and 1990s showed that diesel contributed more than 50% of the noxious air emissions from transportation sources—a disproportionate amount when you consider that diesels are only ~5% of the vehicles on the road.

These numbers do not paint the whole picture, however. Actual measurements evaluating the accuracy of air models later showed that diesel exhaust made up only 24% of overall air emissions, far lower than earlier estimates predicted by EPA models. This number is surprisingly much smaller than would be expected considering longer distances are traveled by diesel powered vehicles; 40% of total accumulated trip distances are made by diesel powered commercial vehicles such as buses, heavy trucks, and delivery vehicles. (See **FIGURE 2-5**) Also, diesel-powered vehicles carry much heavier loads than gasoline-fueled vehicles. Combining these factors with fuel consumption data by vehicles used for transportation that account for 23% of all transportation fuel consumed, a cleaner picture emerges of diesel engines. Diesels travel much farther, carrying heavier loads, while using less fuel. A diesel-powered vehicle will produce lower emissions than gasoline-powered vehicles for every ton of weight towed or carried per mile. Today's diesels are in fact as clean as, if not cleaner than, gasoline engines.

Part of the proposition for a cleaner reputation that is deserved by diesels in North America is the equivalency of emission standards for gasoline and diesel vehicles. Gas and diesel automotive engines have been required to meet the same emission standards as for more than a decade. Under intense new scrutiny, diesels do meet and exceed the worlds toughest emission standards set by the US-EPA and California regulators. In Europe, new emission test cycles introduced in 2017 measure emissions under real world driving conditions. The WLTP cycle (Worldwide Harmonized Light Vehicle Test Procedure) also called the RDE (Real Driving Emissions Test) is intended to deliver more realistic emission and fuel consumption test results than the previous NEDC procedure (New European Driving Cycle). These new procedures are expected to eliminate any attempt to circumvent emission standards by replacing the traditional dynamometer test with an on-road test procedure.

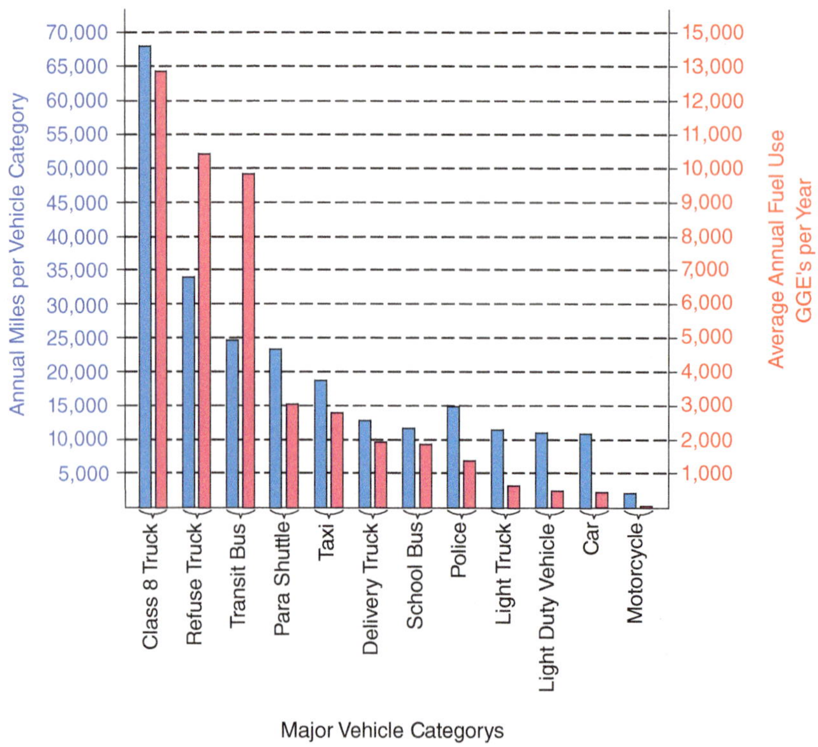

FIGURE 2-5 Fuel Consumption: Diesel powered vehicles in these categories consume 23% of all transportation fuels but produce disproportionately fewer emissions for weight carried and distance travelled.

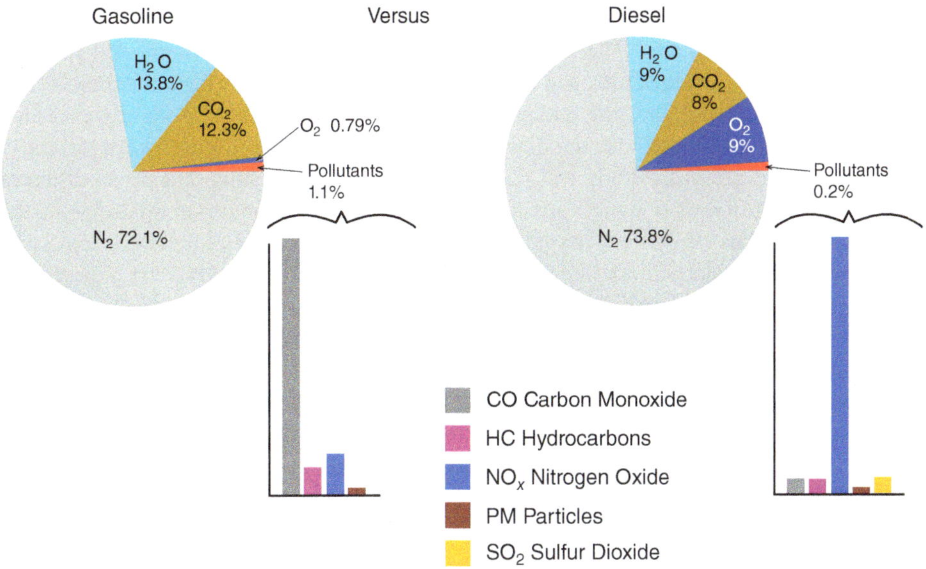

FIGURE 2-6 Comparison of exhaust gas composition of diesel- and gasoline-fueled engines.

Adapted from NGK Spark Plug Europe. Technologies in Detail. Basic Exhaust principles. Accessed at, https://www.ngk.de/en/technology-in-detail/lambda-sensors/basic-exhaust-principles/exhaust-and-harmful-emissions/

Another diesel feature that sets it apart with cleaner emissions is the physics of higher combustion pressures and temperatures in air rich combustion chambers that ensure diesels have naturally lower carbon monoxide, hydrocarbons, and cold- start emissions than gasoline engines (**FIGURE 2-6**). Without significant evaporative emissions, hydrocarbon production from evaporating diesel fuel is not even considered for regulations. However, one of the most important comparisons, and a fact that has escaped media attention during recent emission controversies is the production of particulate emissions from diesel and gasoline engines. Particulate matter emissions, which are composed primarily of microscopic carbon is arguably one of the most dangerous emissions since it is easily respirable and requires an immune system response to clear it from the body. Inflammatory respiratory illnesses such as asthma and emphysema are aggravated and in some cases triggered by particulate matter (PM) originating from vehicle exhaust. Adding to the menace are deadly carcinogens formed in engine combustion chambers that can attach to the surface of carbon particles and be delivered deep into lung tissue. To analyze the threat of PM emitted from the latest engines, a 2017 six-nation study, published in several leading scientific journals including Nature, pointed out that real world measurements of particulate emissions from gasoline cars was on average 10 times greater at 22°C and 62 times more at −7°C compared to diesel cars. (See **FIGURE 2-7**) Other similar studies testing gasoline cars found similar results with

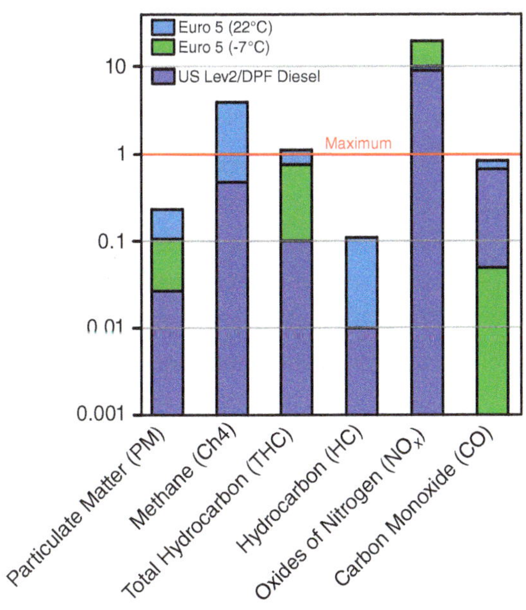

FIGURE 2-7 A rigorous 2017 scientific study done by six nations comparing the ratio of diesel to gasoline emissions. With the exception of NO_x, gasoline engines produced more emissions than diesels in real world tests.

gasoline direct injected engines emitting 10 to 100 times more fine soot particles than the number from diesel engines. The study also found the ratio between diesel and gasoline engine emissions of oxides of nitrogen (NO_x) were found to be higher from the diesels than the gasoline engines, but with the exception of only one test vehicle that slightly exceeded the emission standard threshold, the diesels were within legislated limits. It's important to note this rigorous scientific study was completed by emission testing labs in six different countries using different test methods and researchers to eliminate potential bias in the results. Results of the study have triggered a new examination of gasoline engine emission requirements. Particulate filters like those required for diesels since 2007 are expected to become mandatory for all gasoline engines.

In spite of VW's 2015 admission to cheating emission certification tests, the actions of a single manufacturer do not diminish the fundamental advantages of diesel: its exceptional fuel economy, exciting performance characteristics, rugged durability, superior towing abilities, and a more favorable environmental footprint (**FIGURES 2-8** and **2-9**). Manufacturers who are leading the shift away from the legal problems at VW, in order to refocus on the benefits of diesel with new model introductions, recognize how necessary diesels are

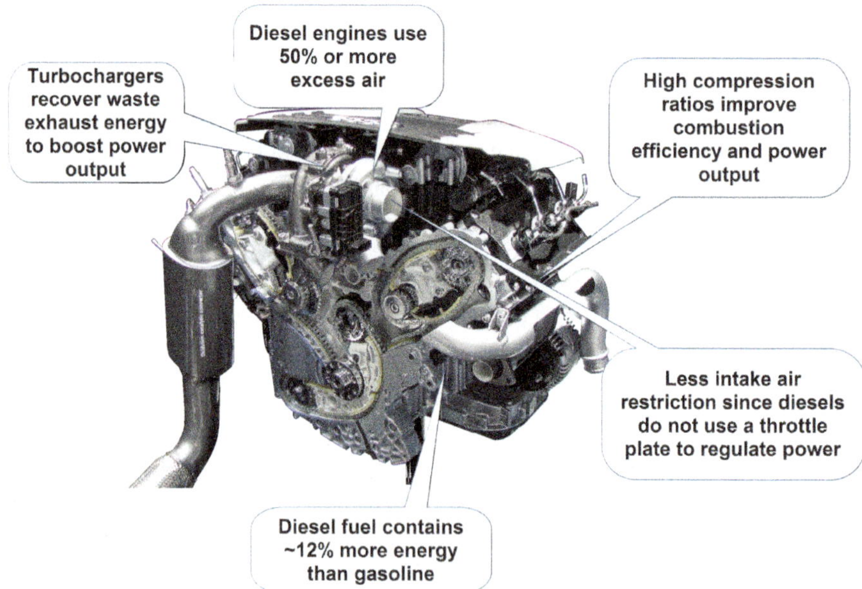

FIGURE 2-8 Features of a diesel contributing to better fuel economy.

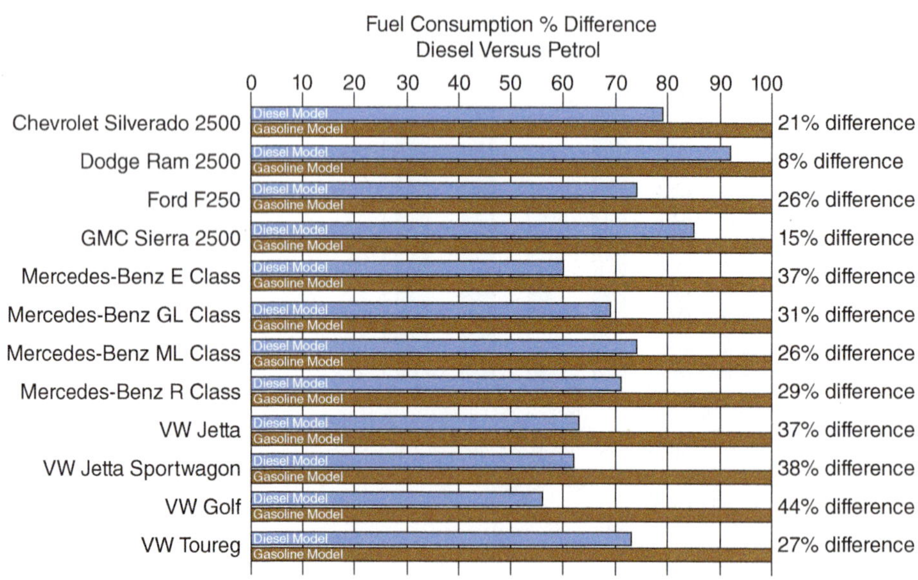

FIGURE 2-9 Comparing fuel economy advantage of diesel and equivalent gasoline engines on a model basis.

to meet legislated emission standards for limiting greenhouse gases (GHGs): emission gases believed to contribute to global warming (**FIGURE 2-10**). To meet these emission targets through reduced fuel consumption, manufacturers must choose between using more diesel, gasoline electric hybrid drive, diesel electric hybrid drives or battery electric vehicles.

Diesel's comparatively lower fuel consumption translates into an automakers ability to reach legislated standards for corporate fuel economy which indirectly regulates carbon dioxide emissions, a GHG. It was, in fact, emissions of nitrogen oxides, not carbon dioxide, and the superior fuel economy and performance of VW diesels that instigated a backlash against diesels. Commonplace tampering with emission controls in Europe by service garages contributed to a widespread problem with engine emissions in European cities too. But with a looming 52.5 US mpg (4.48 L/100 km) corporate average fuel economy mandated for the year 2025, North American manufacturers are realizing not only how essential diesel technology is to reach legislated targets but also the benefits of diesel engines for consumers.

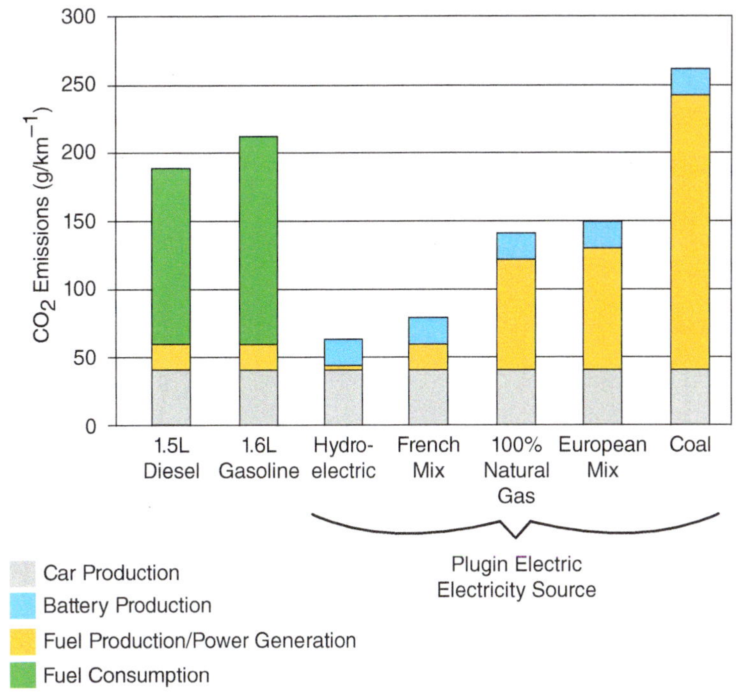

FIGURE 2-10 Electric vehicle greenhouse gas emissions are only as clean as the energy source generating electricity for charging.

Diesels are not a perfect fit for every automotive application due to their higher acquisition cost. Small lightweight cars operating in stop and go urban driving conditions are not an ideal match for diesels. But alternatives such as gasoline electric or battery electric are even more costlier than diesels in heavier vehicles. The weight of batteries and the diminishing torque output of electric motors at higher speeds make variations of that powertrain choices impractical for high speed torque needed by vocational applications (construction and delivery vehicles), pick-ups, SUV's and trailer towing. When long distance travel or commutes are factored into criteria for selecting a powertrain option, diesel makes the most sense with a fast return on investment due to fuel savings. The latest electrically heated exhaust aftertreatment systems do create potential for stop-start and hybridization of diesel engines for urban driving cycles.

A 2017 Vincentric study in Canada documented the advantages of diesel ownership: It compared the ownership costs of four cars and 56 utility vehicles and vans sold with either a diesel-fueled or a gasoline-fueled engine. Of those, 39 vehicles were less expensive to own when they were equipped with diesels. The largest savings were found in the Range Rover diesel, which according to the study's calculations, reduced costs after five years by $22,000 when compared to a gasoline-fueled model.

Vincentric, a leading automotive business consulting firm, examined major variables that factor into the cost of vehicle ownership, such as fuel purchases, vehicle depreciation, various fees and taxes, financing costs, insurance, maintenance, repairs, and other relevant expenses. Each diesel-powered vehicle was then compared to its gasoline-powered equivalent, and in every model comparison, lower fuel consumption for the diesels saved owners thousands of dollars. Interestingly, fuel savings for all the pickups in the study varied between $4,000 and almost $5,000. However, a premium price is paid to equip a vehicle with diesel engines; the technology to produce diesel is more expensive, particularly due to costly exhaust aftertreatment systems. In spite of the higher the acquisition cost every Ford truck was less expensive to own and maintain after five years as a diesel, by a minimum of $8,215 and a maximum of $8,856. The Canadian study has parallels with the ownership experience in other countries, but it cannot be duplicated in the United States, due to higher taxes and profit margins added to the retail cost of diesel fuel, compared to gasoline, which are imposed by the U.S. government, refineries, and distributors (**FIGURE 2-11**).

Encouraging automakers to offer more diesel options is the recent success of diesels in light-duty pickups. The 3.0 L six-cylinder EcoDiesel used in the Ram 1500 and the

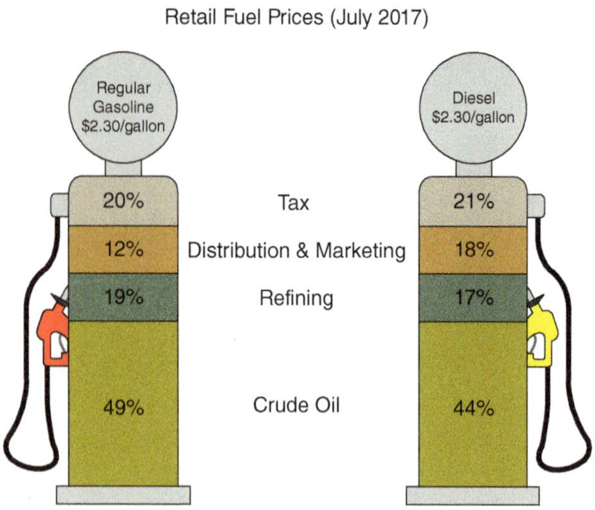

FIGURE 2-11 In spite of lower production and refining costs, taxes and higher profits are charged at retail pumps for diesel fuel throughout the United States.

Courtesy of U.S. Energy Information Administration

4-cylinder 2.8 L Duramax in the Chevy Canyon, which has the best fuel economy of pickup trucks in its specific market (22 mpg city and 30 mpg US highway or 10.7 L/100 km city and 7.84 L/100 km highway), each accounts for 15% of sales for these model in the United States (**FIGURE 2-12**). In the heavier pick-up truck market, such as the 2500 and 3500, the take-up rate is much higher when longer travel distances carrying or towing more weight is a distinct diesel advantage. (**TABLE 2-2**). Once again, the success of these vehicles is not at the expense of increased emissions, as some diesel detractors would believe. Technological advances in recent years have transformed diesels from noisy, smelly, crude machinery into a sophisticated, powerful, clean, and environmentally friendly powertrain alternative. Tier 3 emission requirements, which diesel engines and gasoline engines must meet, phase in at the beginning in 2017, replacing a decade of U.S. Federal EPA (Environmental Protection Agency) Tier 2 standards, which already mandated almost undetectable emission levels. Tier 3 lowers emissions to one-fifth or less than previous levels and requires even longer engine durability while monitoring more rigorously for potential emission-increasing conditions.

Development of Automotive Diesels

2-01 Outline the historical development of diesel engine technology.

Diesel engines did not begin to appear in cars and trucks until the early 1930s, after the development of compact fuel-injection equipment by innovators like Robert Bosch. Only when the oil crisis hit North America in the early 1970s did the thriftier diesel become regarded as a potential replacement for gasoline engines. As fuel prices declined, so did the use of diesels, which historically tend to track the cost of fuel prices.

Compression Ignition Combustion

Compression igntion engines remains the most efficient combustion system developed, converting the greatest amount of mechanical energy from burning fuel (**FIGURE 2-13**). Chemical energy stored in diesel fuel is converted to mechanical energy through the process of combustion. Essentially, **combustion** is a chemical reaction between fuel and oxygen molecules contained in the air. When Rudolph Diesel (1858–1913) patented his engine in 1892, he envisioned a machine that would operate with dramatically improved efficiency over both the steam engine and the spark ignition Otto cycle engine of his day. Until the development of the diesel, those engines extracted only 1% to 8% energy available from burning fuel into mechanical energy. This meant that very little of the energy produced by the fuel was converted into power. Diesel believed he had a better idea for a more efficient engine, leading to his development of the compression ignition combustion system. In Diesel's **compression engine**, air alone (not a spark) would be the ignition source (**FIGURE 2-14**). He had observed that air, like any gas, would increase in temperature when compressed. Diesel correctly

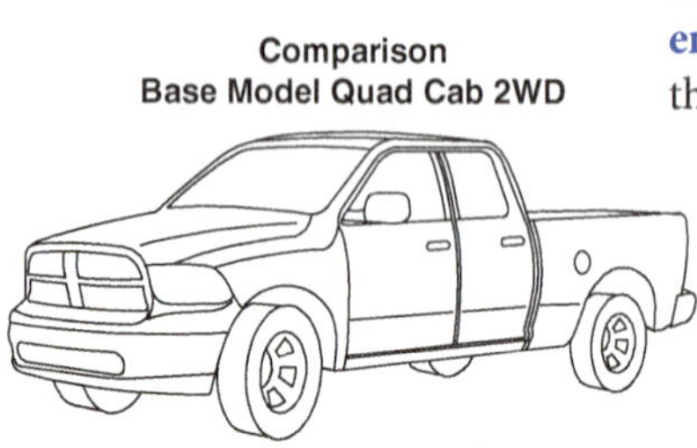

Engine Capacity		3.6 Liter V6	3.0 Liter V6
Fuel		Gasoline	Diesel
Horsepower		305	240
Torque (ft-lb)		269	420
Fuel Economy (m/g)	City	18	21
	Highway	25	29
Towing Capacity (lb)		7,750	9,200

FIGURE 2-12 Comparing gas and diesel versions of the Dodge Ram.

TABLE 2-2 Diesel Pickup Trucks

Model Year	Light Duty	Medium Duty
2008	10%	63%
2009	22%	59%
2010	30%	59%
2011	30%	62%
2012	30%	62%
2013	23%	62%
2014	19%	63%

Source: Polk (2015).

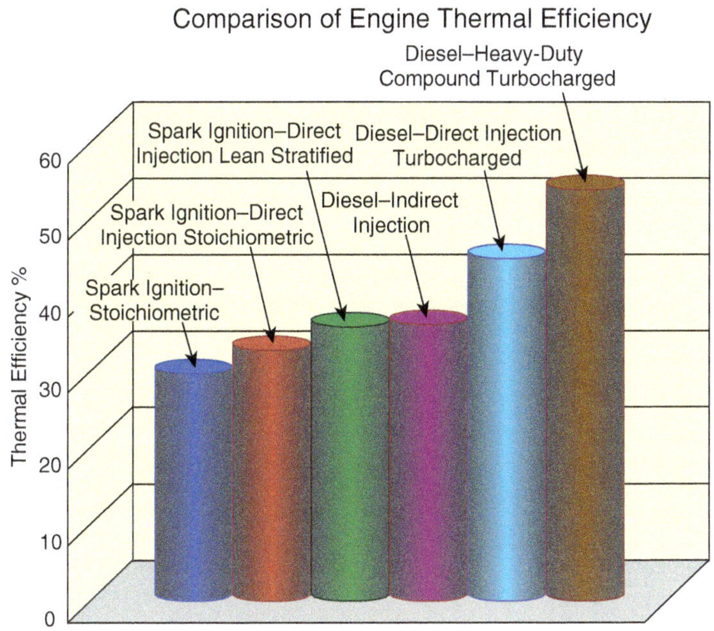

FIGURE 2-13 Various engine technologies have different levels of engine efficiency.

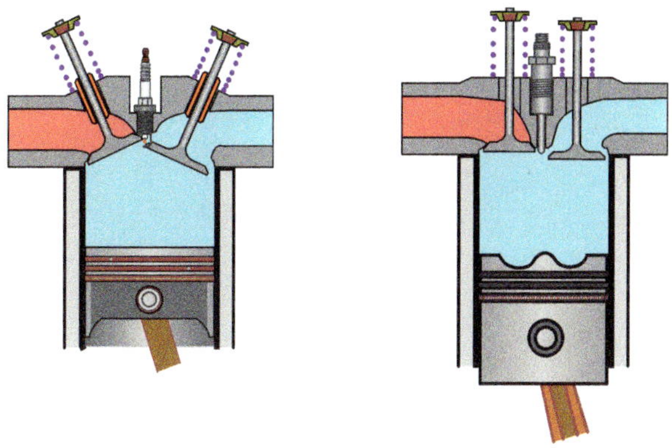

FIGURE 2-14 Engines are classified into one of two ignition systems: spark ignition (SI) or compression ignition (CI).

believed that if fuel was injected into a cylinder when air is compressed to a temperature hot enough to ignite the fuel, combustion would take place (FIGURE 2-15). The injection event he planned would take place near the end of the compression stroke near, **top dead center (TDC)** the highest point at which the piston travels in a cylinder (FIGURE 2-16). The burning fuel would cause cylinder pressures to intensify and drive the piston downward after. Diesel believed this unique idea for a combustion mechanism would deliver a considerable increase in engine efficiency and power (FIGURE 2-17).

Inspiration for the engine concept may have occurred when Rudolf Diesel observed a demonstration of an ancient device used to start fires. The device, called a fire piston, consists of a hollow, narrow tube sealed at one end and a plunger matched to the diameter of the tube. The plunger forms an air-tight seal in the tube, and a ramrod handle is attached to the plunger, so that the user can move the plunger up and down the tube. To start fires, the user places a small bit of wood or dried grass in the bottom of the tube and then repeatedly rams the plunger into the tube. Compressing the air with the plunger increases the pressure and temperature inside the tube, which ignites the kindling.

In 1893, Diesel built the first working model of his patent (FIGURE 2-18). For fuel, the engine used discarded coal dust found in abundant quantities in the Ruhr Valley of Germany. The dust was blasted into the combustion chamber by using pressurized air. Injecting coal dust proved to be too complex and cumbersome, leading to its discontinuation in favor of liquid fuels. In spite of the problems with coal dust injection, the engine operated with 26% thermal efficiency. Amazingly, this was more than three times the efficiency of steam engines of that day.

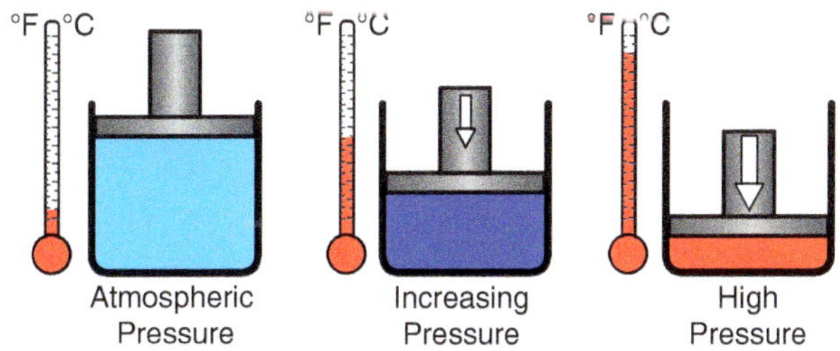

FIGURE 2-15 Compressing a gas raises its temperature. The heat of compression can ignite fuel.

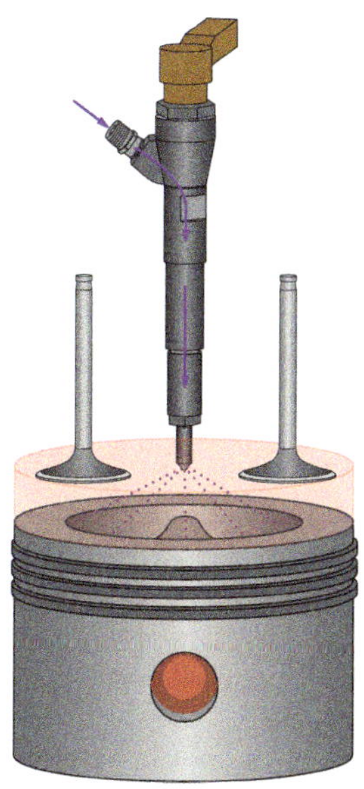

FIGURE 2-16 Diesel engines inject liquid fuel under high pressure into the combustion chamber near the end of the compression stroke. Air, which is heated during compression stroke, becomes the ignition source.

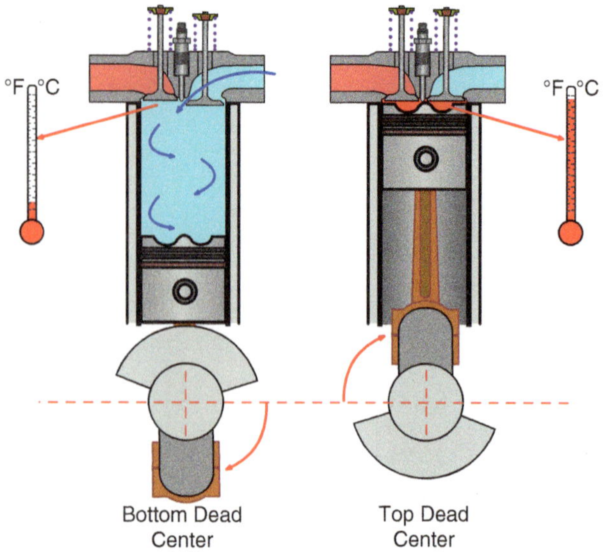

FIGURE 2-17 Compressing a gas will cause its temperature to increase. The temperature increase can be predicted mathematically.

FIGURE 2-18 Rudolph Diesel's first working engine on display in the Deutsche Museum. It was cumbersome to operate, stood over ten feet tall (a single-story high), and was practical only in stationary applications such as ships and factories.

While most diesel engines use a four-stroke cycle operation, also called the Otto cycle, two-stroke cycle engines have been produced (**FIGURE 2-19**). Current development of experimental engines, such as those by Michigan-based EcoMotors and California-based Achates Power, produce opposed-piston two strokes, delivering 30–50% better fuel efficiency than the already efficient four-stroke cycle diesels. For example, the EM100 from EcoMotors has a two-cylinder, four-piston engine with a cylinder bore of close to 4" (10 cm). But the engine weighs only an astonishing 296 lb (134 kg). Measuring 23" x 41" x 19" (58.4 cm x 104 cm x 48.3 cm), it is rated for 325 hp (242 kilowatts) at 3,500 rpm while torque is 664 ft-lb (900.3 N·m) at 2,100 rpm.

Early Automotive Diesels

Originally, diesels were slow and heavy, suitable for ships and factory applications. By the 1920s, Robert Bosch developed practical compact-injection pumps and governors that allowed smaller diesel engines to power motor vehicles. In 1929, Clessie Cummins, founder of Cummins Engine Company, built North America's first diesel-powered car. Coincidently, the Depression arrived, and with it the economic incentive for fuel-efficient transportation, which could be supplied by the diesel engine. Cummins had already popularized the reputation of the diesel as a fuel-efficient engine. In 1925, Cummins drove a seven-passenger Packard limousine retrofitted with his U-model diesel engine. His highly publicized coast-to-coast tour of the continental United States in 1931 using only $11.22 worth of fuel drew even more attention to the diesel advantage. Later, in 1936, the diesel engine first found its way into a mass-production passenger car, the Mercedes-Benz 260 D.

> **▶ TECHNICIAN TIP**
>
> In 1897, Diesel operated his first practical engine and demonstrated it at the 1898 Exhibition Fair in Paris, at the invitation of the French government. His slow-rotating engine operated at an astonishing 26% efficiency. This engine is also particularly noteworthy because it ran on peanut oil, the first biodiesel. The use of vegetable-based oil was due to the French government's interest in operating engines using renewable fuels commonly found in their colonies. The term "biodiesel" refers to fuel made from biological sources such as animal fats or vegetable oils. The use of straight vegetable oil (SVO) or biodiesel alternatives instead of petroleum-derived fuel can dramatically reduce the amount of noxious exhaust emissions.

Automotive Diesels of the 1970s

Until the oil price shocks of the 1970s, there were not many diesel-powered cars produced in North America. Diesel engine production in light-duty vehicles was limited to Mercedes-Benz, Peugeot, and Opel. The superior fuel efficiency of the diesel and the low price of diesel fuel, which then cost approximately 20% less than gasoline, helped many people to ignore the shortcomings of the engine. More widespread use of diesels was hindered by the following negative characteristics:

- poor starting in cold weather
- noisy engine operation
- visible emissions
- strong exhaust odor
- poor acceleration
- higher initial purchase costs.

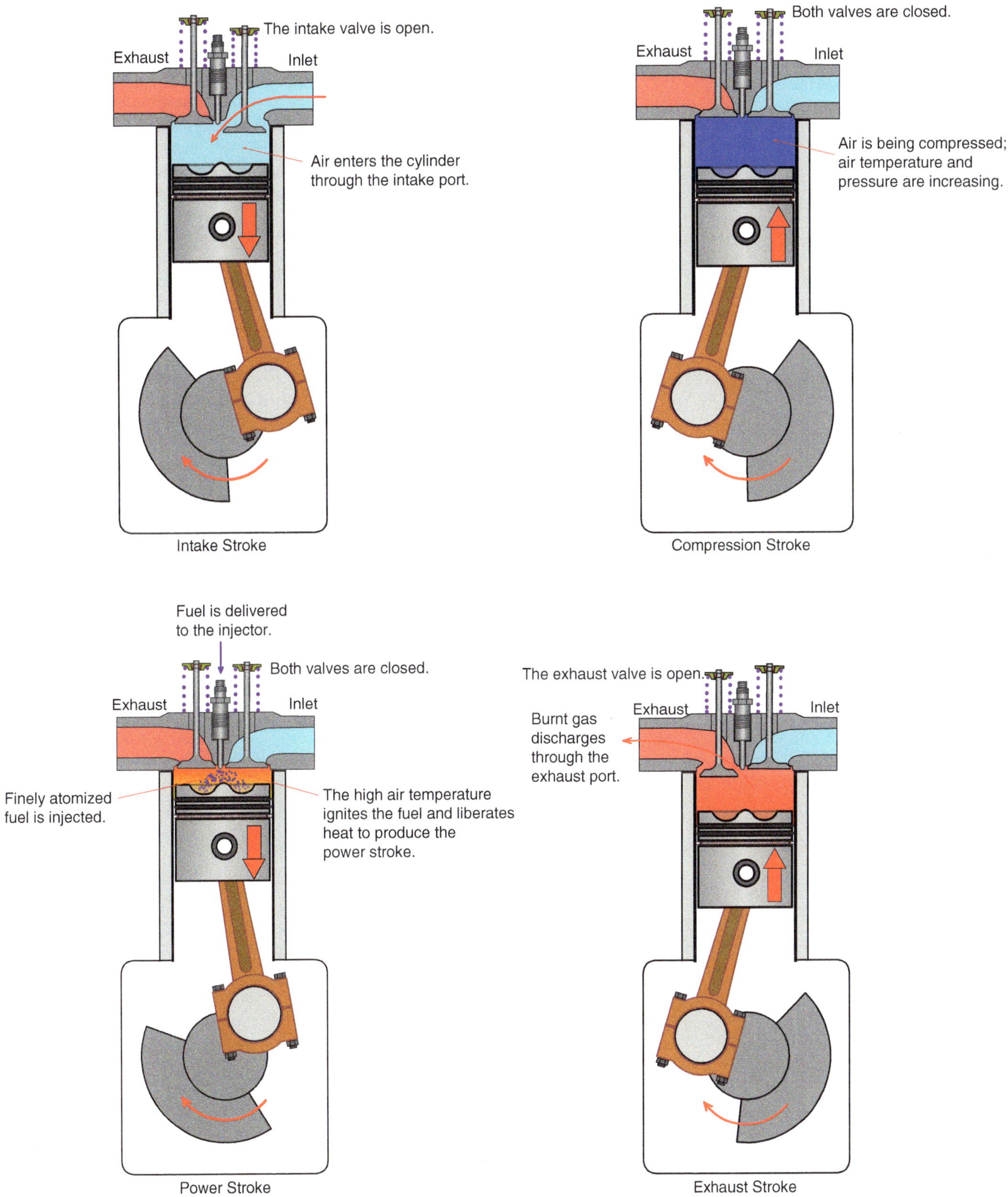

FIGURE 2-19 Four movements of the piston make up the four-stroke combustion cycle. One stroke is the movement of the piston from either TDC or bottom dead center (BDC).

2-02 Identify and explain the advantages of diesel engine technology compared to other powertrain technology.

The Diesel Advantage

Technological innovations have transformed the diesel engine, and today's engines are unlike early engines or those seen in buses and heavy trucks rumbling over the highways. In terms of emissions, diesels can meet and exceed the standards of gasoline engines. Not only are diesel emissions as clean as (if not cleaner than) gasoline engine (see emission standards mandated for 2017), but they also emit a minimum of 20–40% fewer GHG emissions. The noise, vibration, sluggishness, and smells of older diesel engines are virtually gone. Instead, microprocessor controlled injection systems and exceptional low RPM torque provide an excitement factor to the operation of diesel powered vehicles (**TABLE 2-3**). The following is a brief summary of advantages for diesel vehicle over gasoline and electric vehicles.

1. *Greater fuel efficiency*. The diesel uses substantially less fuel to produce the same horsepower from a similarly displaced gasoline-fueled engine (**FIGURE 2-20**). Typically, diesel engines will achieve between 25% and 60% better fuel economy in comparison to similar gasoline engines. Common efficiency improvement is between 25% and 40% over gasoline engines in light- and medium-duty vehicles. The high compression ratio, lean burn or air-rich combustion, turbocharging, and sophisticated direct-injection fuel system of the modern diesel engine are what primarily account for the superior fuel efficiency. Of these factors, the most important is the use of a higher compression ratio that is nearly double that of spark ignition engines. With an average compression ratio of 16:1, the diesel converts much more fuel energy into power (**FIGURE 2-21**).

TABLE 2-3 Contrasting 30 Years of Diesel Development. The newest GM 1.6 L whisper diesel engine is just over one-quarter the size and has half the cylinders of the 5.7 L GM diesel.

Model	Engine	Weight	Power	Fuel Economy	0–60 mph
1978 GM Bonneville	5.7 L diesel	4,000 lb (1,814 kg)	105 hp (78.3 kW) @ 3,600 rpm Torque 225 ft-lb (305 N·m) @ 1,800 rpm	28 mpg US combined (8.4 L/100 km)	21 seconds
2018 Chevy Cruze	1.6 L whisper diesel	3,100 lb (1,406 kg)	137 hp (102 kW) @ 3,750 rpm Torque 240 ft-lb (325 N·m) @ 2,000 rpm	37 mpg US combined (6.4 L/100 km)	8 seconds

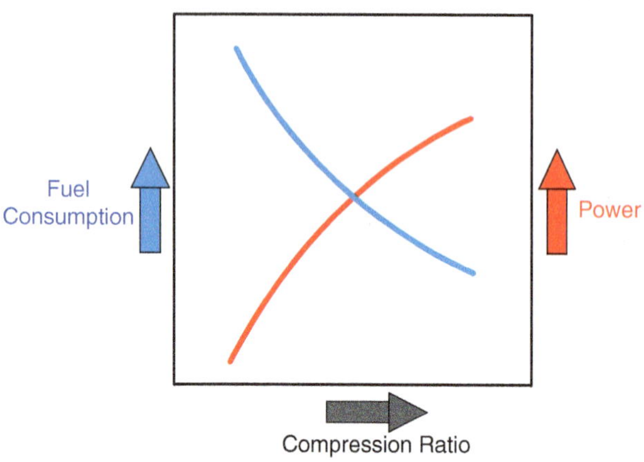

FIGURE 2-20 Higher compression ratios increase power output and decrease fuel consumption. The high compression ratio of diesel is the main reason for the extraordinary power and fuel efficiency of diesel engines.

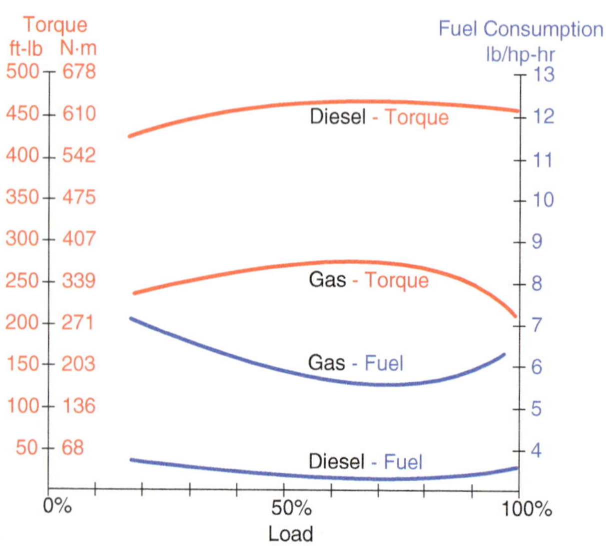

FIGURE 2-21 Compared to a 5 L (302 cubic inch) naturally aspirated gasoline engine, a 5 L (302 cubic inch) naturally aspirated diesel engine produces much more power while using far less fuel.

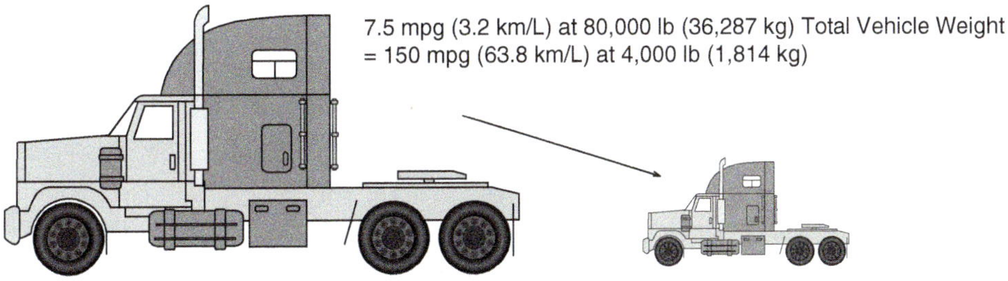

FIGURE 2-22 Diesel engine fuel efficiency, compared to gasoline engines, increases even more as vehicle weight increases.

The driving range of a vehicle is also related to fuel efficiency. Diesel will go farther on a tank of fuel and not add range anxiety associated with electric vehicles to the driver's other concerns. Unlike electric vehicles, there is no need to make frequent stops on long trips to recharge batteries that can take as long as a day to recharge when using a household current. In snowstorms and other traffic-disrupting emergencies, electric vehicles would simply leave their occupants quickly stranded on a roadway without the ability to even keep them warm.

2. *Higher torque output*. The diesel engine produces significantly higher torque output compared to a similarly displaced gasoline-fueled engines (**FIGURE 2-23**). A **crankshaft** is a major engine component: it converts the reciprocating movement of pistons into rotational movement. **Torque** is the twisting force applied through the crankshaft to the flywheel. Torque is a function of cylinder pressure and crankshaft leverage or throw. Note the illustration and the comparison in **FIGURE 2-24**, which defines **crankshaft throw** as the distance between the centers of the crankshaft's main bearing journal and its connecting rod bearing journals. The crankshaft throw operates as a lever, providing greater mechanical advantage. Longer crankshaft throws are used to produce higher compression ratios and provide a longer lever to the crankshaft to convert cylinder pressure into torque. Given that torque is the most important force moving a vehicle and its load, the diesel engine has tremendous advantage in SUVs and pickups pulling heavy loads. The torque output profile produced from diesels is also flatter, which means it has more torque available over a wider range of usable engine speeds (**FIGURE 2-25**). High torque output at low speeds gives the engine stump-pulling ability—to pull the heaviest loads at speeds just above idle. Without having to increase torque through gear

▶ TECHNICIAN TIP

Diesel engine fuel efficiency, compared to gasoline engines, increases even more as vehicle weight increases. While fuel consumption will normally increase for diesel and gasoline engines as the vehicle becomes heavier, consumption does not increase as much in diesel engines as it does in gasoline engines (**FIGURE 2-22**). Consider as an example a highway tractor-trailer combination. A semitrailer with a gross vehicle weight of 80,000 lb (36,287 kg) can obtain on average ~7.5 mpg (~31 L/100 km) or better fuel efficiency. If the weight of the vehicle were scaled back to only 4,000 lb (1,814 kg) (one-twentieth its weight or close to the curb weight of an F-150 pick-up) and the same scale were applied to fuel economy (20 times), it would get 150 mpg (1.57 L/100 km). And that's with extra rolling resistance from additional tires and wind resistance from an even larger vehicle profile.

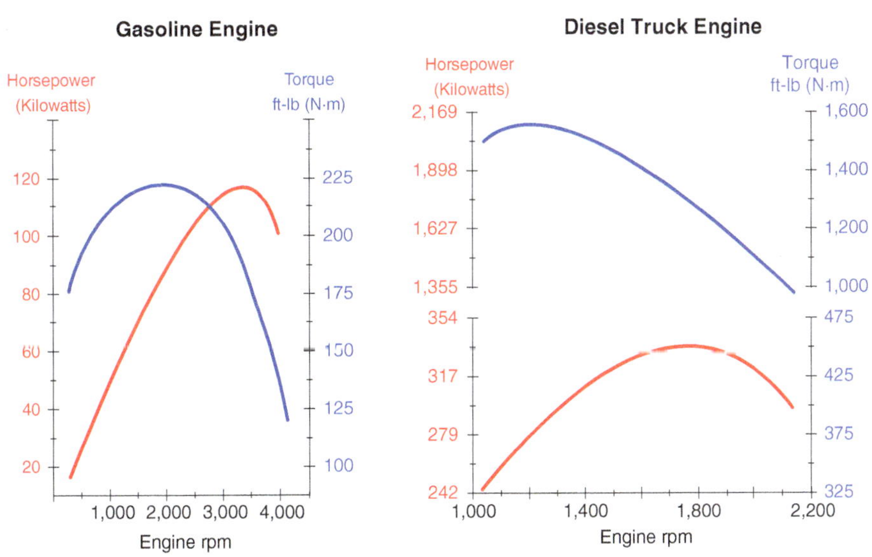

FIGURE 2-23 Compared to gasoline engines, diesel engines have a flatter torque rise curve, which means that more torque is available over a wider range of engine speeds.

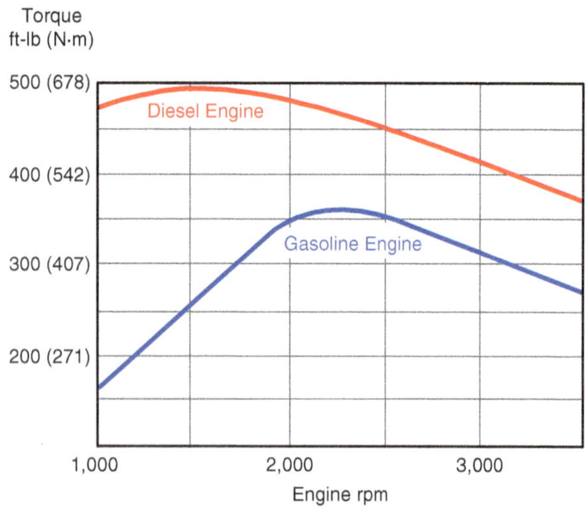

FIGURE 2-24 Diesel engines provide more usable torque than gasoline engines at low-speed operation.

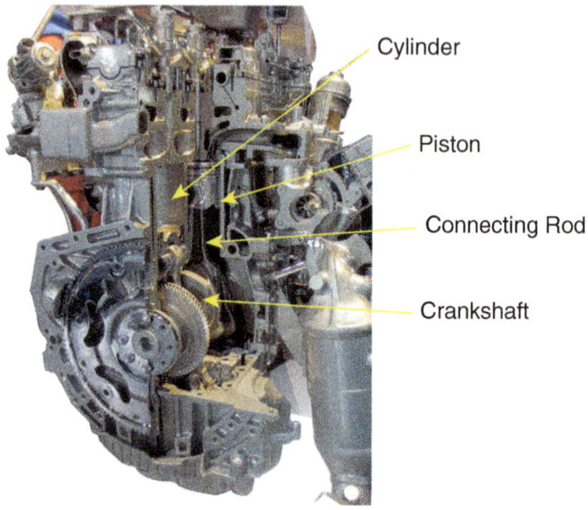

FIGURE 2-25 Diesel engines are designed to withstand higher cylinder pressures. The rugged construction features add durability to the diesel engine.

reduction combined with higher engine speeds, diesel engines deliver much better fuel economy when towing than similar gasoline powered vehicles. At highway speeds, the flat torque curve profile enables diesel vehicles to possess higher torque elasticity. Torque elasticity refers to the ability to accelerate at high speed rather than from a standing start. A tap of the throttle will quickly accelerate a diesel-powered vehicle 50–80 mph (80–130 kph) in just a few seconds. This is usually much faster than comparable gasoline engines. Torque elasticity offers a special advantage to drivers of European highways, which have much higher maximum speeds than those in North America. Over 90% of the luxury vehicles (models from Mercedes-Benz, Audi, and BMW) found on these high-speed thoroughfares are powered by diesels (**FIGURE 2-26**).

Diesel engines provide more usable torque than gasoline engines at low-speed operation. The fun-to-drive or excitement factor of diesel is provided by the additional torque, gives sling shot–like performance from a standing start.

3. *Greater durability*. Diesel engines typically operate three to five times longer than gasoline engines before wearing out. A diesel engine will travel farther and operate over a longer life cycle for several reasons. First, the engine operates at a lower rpm, which means piston speed is slower and will reciprocate fewer times inside the cylinders for every mile/kilometer that the vehicle travels. In other words, whereas an automotive gasoline engine may operate between 2,000 and 4,000 rpm at cruising speed, the diesel commonly operates at 1,500 to 2,500 rpm. It is interesting to note that at cruising speed, the diesel engine will operate more quietly than a comparable gasoline engine, because of the lower engine speed. Second, washing lubrication from the cylinder wall with fuel is another wear factor the diesel engine does not contend with to the same degree as gasoline engines do. Because there is no fuel present during the compression stroke of the diesel, lube oil on the cylinder walls can do a better job at minimizing engine friction. Third, because diesel engines must withstand higher peak combustion chamber forces than gasoline engines, more rugged construction features used to construct a diesel engine confer greater durability to it (**FIGURE 2-27**). These include stronger pistons, crankshafts, valve trains, and gear trains. **Cylinder blocks**, the largest structure of an engine, which enclose the cylinders and provide a rigid frame to support attaching the rest of the engine components, are more rigid so that they resist distortion from greater combustion forces. Crankshafts, cylinder walls, and cylinder blocks in a diesel will often be specially hardened to resist wear, which adds to diesels durability.

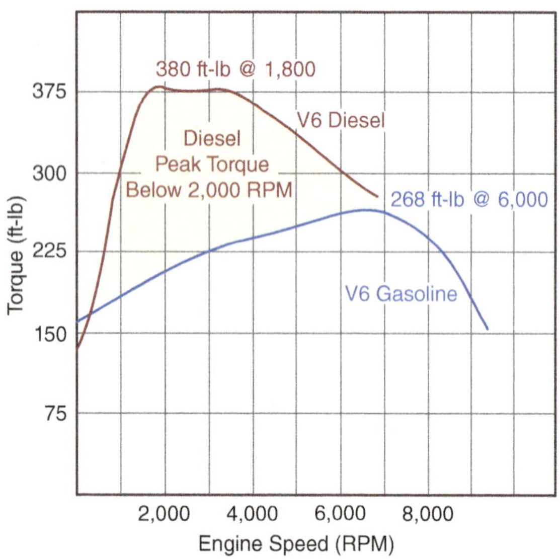

FIGURE 2-26 Comparing torque profiles of a 3.0 L Mercedes-Benz diesel and a 3.5 L Mercedes-Benz gasoline engine.

Adapted from the Martech Group LLC

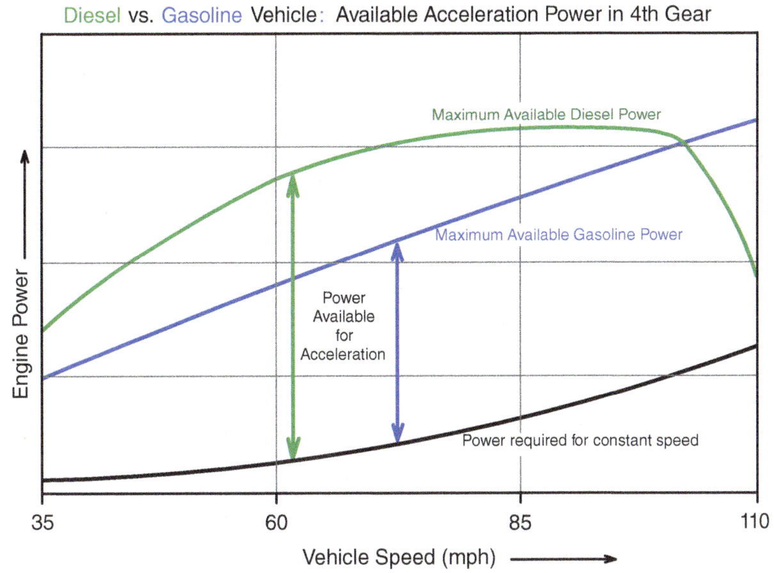

FIGURE 2-27 Torque elasticity provides diesel engines with faster acceleration at high speeds.

4. *Lower ownership costs.* The diesel also enjoys the advantage of having fewer maintenance requirements. The absence of an ignition system and spark plugs and using a rugged, durable fuel system translate into the potential for less maintenance and lower aggravation level. Diesel engine service includes only oil and filter changes for much of its life cycle. **Durability requirements** or the useful life demanded by the EPA emission legislation mean that light-duty diesel engines are expected to remain emission compliant for a minimum of 120,000 miles (193,000 km) or 11 years, as of 2007. Tier 3 emissions regulations require 150,000 miles (240,000 km) of durability or 15 years of useful life. Fuel savings over the lifetime of the vehicle also translate into lower ownership costs. Many automotive diesels begin to return the premium purchase cost after as little as two to three years. As annual distances traveled increase, the return period becomes progressively shorter (**FIGURE 2-28**).

   Although initial purchase diesel costs are higher, much of the expense is recovered at resale. Diesel-powered vehicles command substantially higher resale value than gasoline equivalents due to their durability and lower operating costs. Compared to hybrid electric vehicles, diesel vehicles generate improved cost savings. One study currently finds that a 1 mpg improvement to fuel economy typically costs \$130 with a diesel engine. With the Hybrid Electric Vehicle (HEV), a 1 mpg improvement to fuel economy costs between \$300 and \$700.

   The largest return on investment from diesel engines is moving vehicles with high fuel consumption to even incrementally lower fuel consumption. Hundreds of gallons or liters of fuel normally consumed on an annual basis can be saved by taking larger and fuel-thirsty SUVs and pickup trucks and moving them to use diesel engines. For example, a pickup delivering only 10 mpg (23.52 L/100 km) would save close to 30 gallons (132.1 L) of fuel every 1,000 miles (1,609 km) if it improved 5 mpg (7.85 L/100 km). Moving a fuel-efficient car from 35 mpg (6.72 L/100 km) to 40 mpg (5.99 L/100 km) would only save less than 5 gallons (22 L) of fuel every 1,000 miles (1,609 km). This is a major reason some energy analysts believe diesel engines for heavier vehicles should receive any available tax credits rather than smaller vehicles (**FIGURE 2-29**).

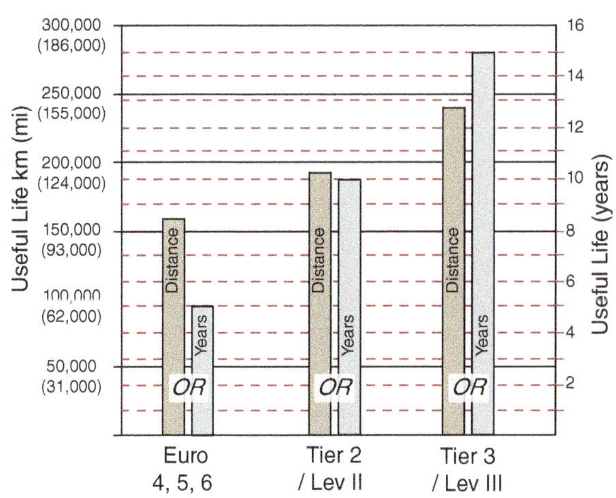

FIGURE 2-28 Tier 3 standards, which phase in beginning in 2017, extend the useful engine life further than before.

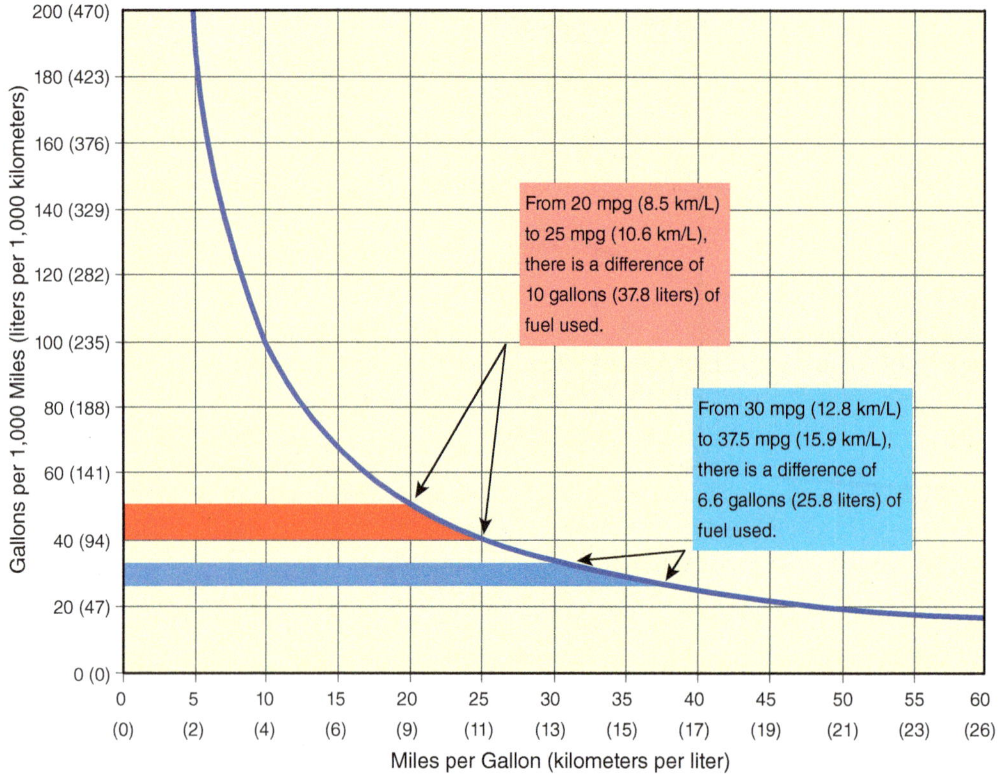

FIGURE 2-29 Even slight improvements in fuel economy in vehicles with high fuel consumption are more beneficial than making major fuel economy improvements in more fuel-efficient vehicles.

In comparison to electric vehicles, a recent study by Energy and Environment Study Institute determined the battery electric models offered for sale in the United States have an average base price of $63,000 and an average driving range 211 miles (339.6 km). Compare this against a Chevy Cruze, which has a base price of approximately U.S. $25,000 and a driving range of 712 miles (1,146 km). Comparing the cost of fuel versus electricity is not straightforward, because of a variety of assumptions that need to be included in a fair comparison. The conversion efficiencies of engines, as well as electric motors, power generation efficiencies, well-to-wheel costs, taxes, energy rates, and energy cost subsidies, should all be factored into a discussion. But considering that diesel fuel has the electrical energy equivalent of 40.7 kWh for 1 gallon (3.78 L) of diesel fuel, electricity at a nominal price of $0.15/kWh would cost $6.00 to produce the same energy as the diesel fuel. The aggravation level of finding battery charging stations, handling charging cords, disconnected charging cords and battery replacement are not easily factored into cost comparisons of electric vehicles either.

5. *Lower Emissions.* It may be a surprise given the recent diesel controversies, but diesel engines' higher compression ratios and air-rich, lean-burn combustion conditions enable them to produce comparatively less carbon monoxide (CO) and hydrocarbon (HC) emissions than spark-engine (SI), gasoline-fueled engines. The discrepancy between emissions is very large when the engines are cold, because gasoline engines produce much larger quantities of emissions until exhaust catalysts are warmed up. Evaporative fuel emissions from diesel fuel are also insignificant, because diesel fuel is much less volatile than gasoline. This means vapor recovery systems found on gasoline-fueled vehicles are unnecessary for diesels. The production of carbon dioxide (CO_2), a greenhouse gas, in a diesel-fueled engine is 20–40% lower compared to gasoline-fueled engines, due to the diesel engine's ability to extract more power from less fuel. Diesel fuel does contain more carbon than gasoline; however, the factor offsetting the carbon content difference is that less energy is required to refine diesel fuel than gasoline (**TABLE 2-4**). Diesel fuel is a by-product of gasoline production and requires less energy and less refining to produce than gasoline does.

TABLE 2-4 Fuel Carbon Content and CO_2 Emissions

Category	Gasoline	Diesel
Carbon Content	5.3 lb per gallon of fuel (635 g per liter of fuel)	6.1 lb per gallon of fuel (731 g per liter of fuel)
CO_2 Emissions	19.4 lb per gallon of fuel (2,325 g per liter of fuel)	22.2 lb per gallon of fuel (2,660 g per liter of fuel)

Lower CO_2 emissions was a primary reason for European governments to develop policies favoring diesel use. Diesel adoption rates peaked in Europe at around 55% in 2012, mainly due to the introduction of a vehicle excise duty based on CO_2 emissions. The tax policy favored diesel cars because they had lower CO_2 emissions than gasoline versions. Fewer taxes were applied to diesel fuel compared to gasoline, as well. Yet, the economic incentives that favored diesel adoption do not overshadow Europeans' high regard for diesels or European environmental sensitivities. From a cultural perspective, because of its origins in Germany, there is a European preference for diesels partly due to the sense that the diesel is more "their" engine than North Americans do. Much higher fuel costs in Europe also make the argument for paying the upfront purchase premium for diesel powertrains more appealing since the break-even return on investment through lower fuel costs happens sooner. The calculation is still valid even though Europeans drive shorter distances on an annually than North American drivers do (13,476 miles or 21,561 km per year in the United States vs. 8,125 miles or 13,000 km per year in Europe). Long distance commutes in heavier vehicles or when towing trailers still makes the best case for adopting diesel powertrains.

Debates over estimates of the actual environmental impact of electric vehicles are controversial. Electric vehicles have no emissions, but how clean a vehicle actually is depends on the energy source used to generate the electricity used to charge a vehicle. The wheel-to-well calculation adds all the energy and emission inputs required to produce the car and operate it over its lifetime. Assumptions made by some studies suggest that electric vehicles may in fact be environmentally dirtier due to battery manufacturing, sources producing electricity for charging, and inputs from mining materials to produce a vehicle. If coal-fired power plants produce the electricity, then battery electric vehicles are no better than a traditional internal combustion engine in terms of greenhouse gases. However, improvements to the emission profile of electric vehicles will only take place proportional to the use of renewable energy sources, such as solar and wind power. Calculating the negative environmental impact of a vehicle becomes more complex when the mining of metals for motors and rare metals for electrical parts and battery production is factored into the emission equation. Supporting the conclusion and assumptions made by a number of studies, a recent statement from Union of Concerned Scientists noted that after including emissions from lithium-ion battery manufacturing, electric vehicles still generate half the emissions of a traditional internal combustion engine over the course of its life.

▶ Compression Ignition Principles

2-03 Identify and describe the important elements of compression ignition systems.

In contrast to spark-ignition (SI) engines, compression-ignition (CI) engines use the heat generated through compressing air in order to ignite fuel (**FIGURES 2-30** and **2-31**). Not only does a CI engine require virtually no distinct ignition parts, but higher compression ratios have the added advantage of converting more combustion energy into usable mechanical power. **Compression ratio** refers to the comparison between the entire cylinder volume and the clearance volume. **Clearance volume** is the small remaining space left in a cylinder after all the air in a cylinder is squeezed by piston movement. The CI system was embraced by early engine developers because of the shortcomings of spark-ignition systems and gasoline properties. The most significant obstacle to making the spark-ignition engine efficient is its limited compression ratio. Air and and fuel mixed together

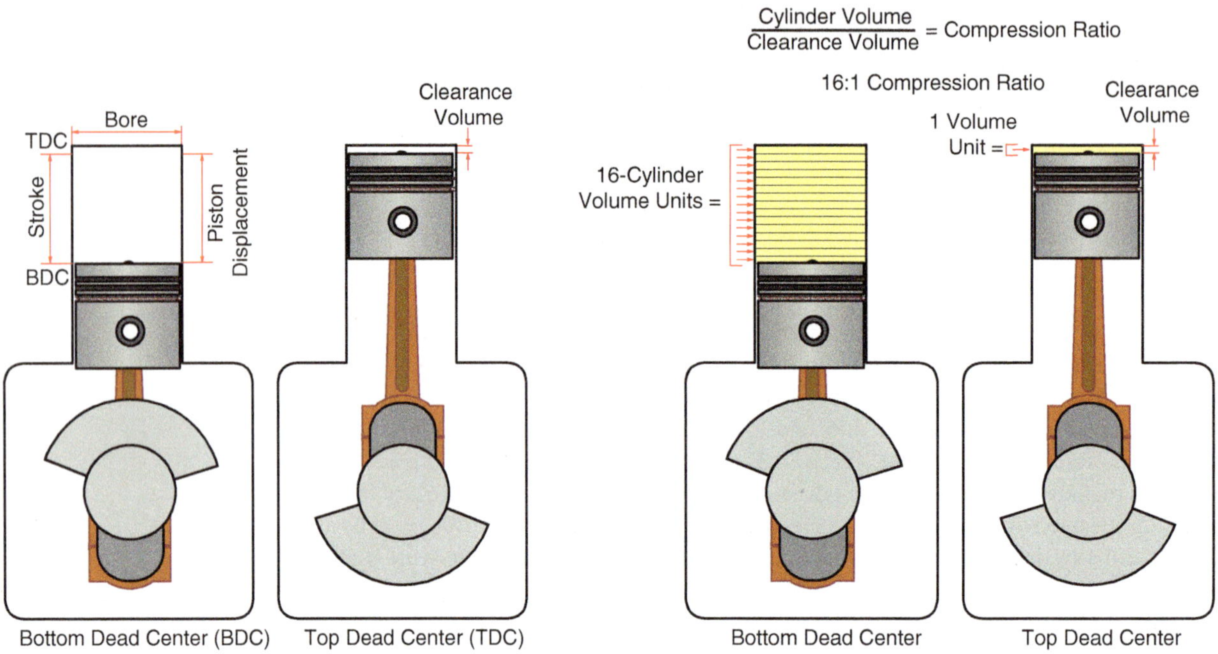

FIGURE 2-30 Compression ratio compares clearance volume and total cylinder volume.

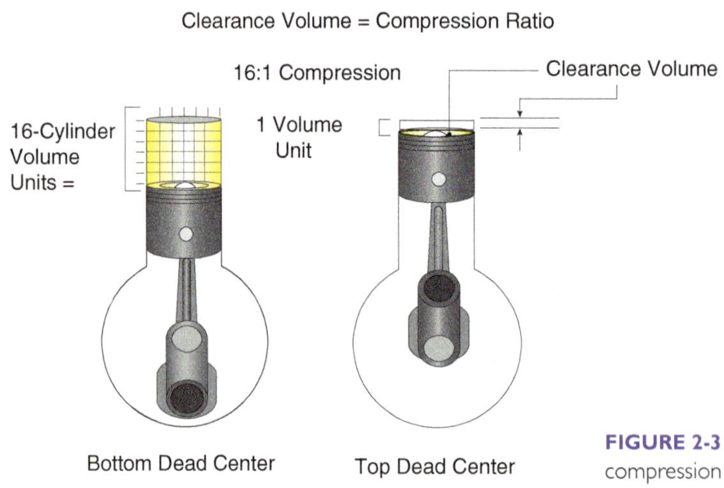

FIGURE 2-31 Calculating compression ratio.

as they are drawn into the cylinders can prematurely ignite when compressed too much (**FIGURE 2-32**). Until the development of antiknock additives for gasoline, such as tetraethyl lead in the late 1920s, the compression ratio of an air and fuel mixture was limited to 4:1. Cylinder temperatures during the compression stroke would cause detonation, which is premature ignition of air-fuel mixture before the piston reached top dead cylinder. Diesel engines do not have the same compression limits, since only air is drawn into the cylinders during intake stroke. While diesel engines do not have compression limits for combustion, compression ratios above 16:1 provide minimal power increases. In fact, any increase in the compression ratio beyond 23:1 requires an exponentially greater force to squeeze the air (**FIGURE 2-33**). This results in a decrease in engine efficiency since more energy is used compressing the air than is returned in an increase in power.

Compression Pressures and Temperatures

As mentioned, the more technically correct term for diesel engines is "CI engine." This classification is applied because diesel engines use high compression ratios to produce the heat needed to start combustion. The average compression ratio of today's diesel engine is

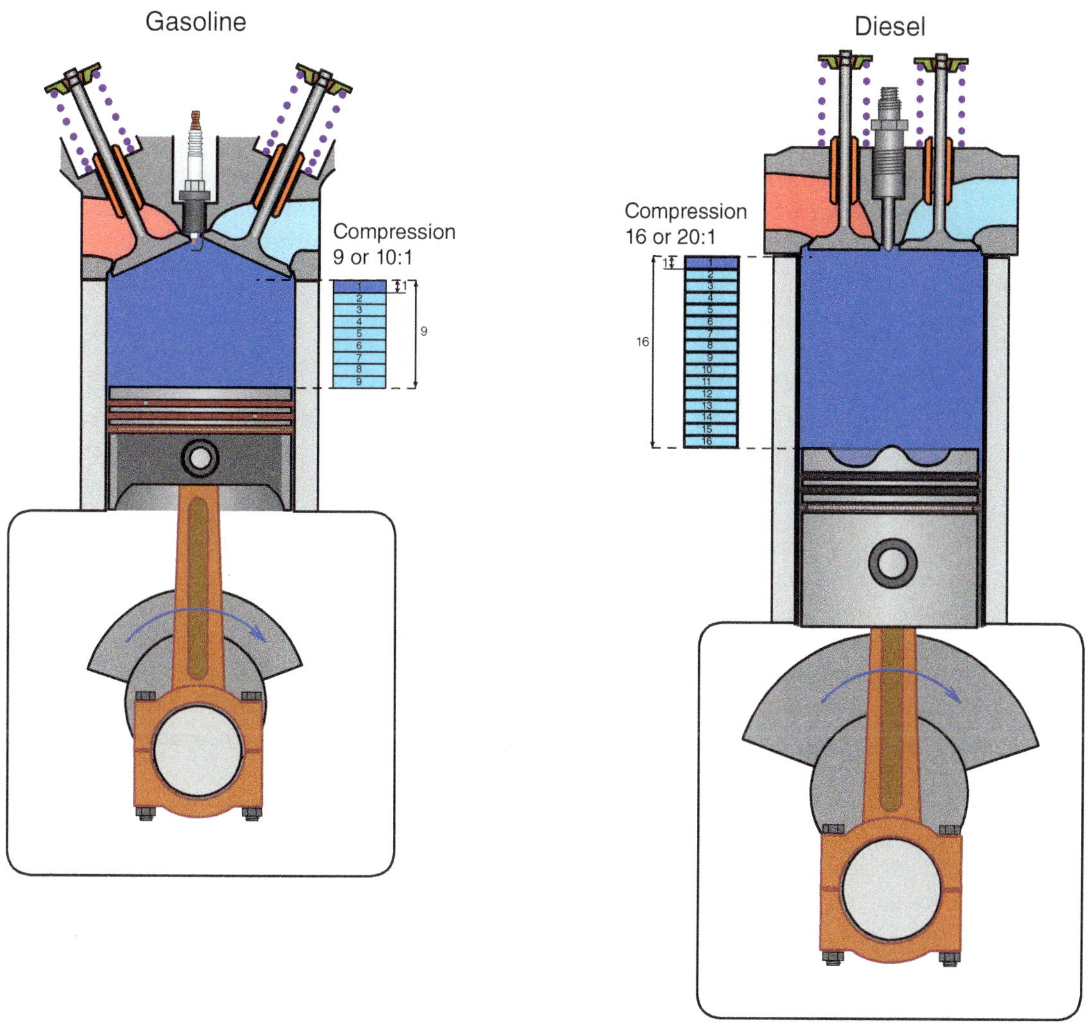

FIGURE 2-32 A comparison of spark and CI compression ratios.

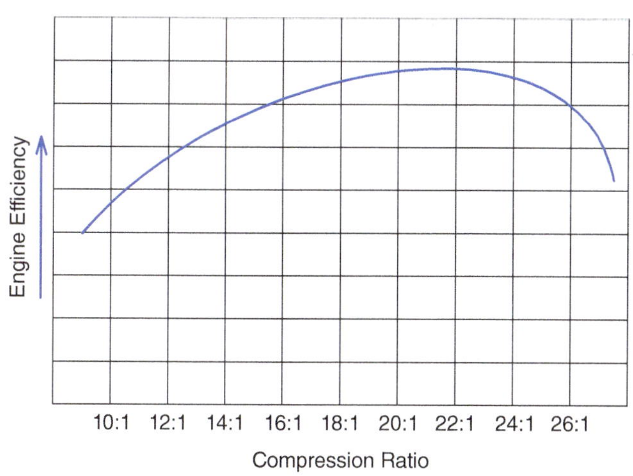

FIGURE 2-33 Engine efficiency drops after a compression ratio of 23:1 since more energy is used to compress air than is derived from increased combustion pressures.

commonly in the vicinity of 16:1. Direct-injection combustion chambers used by today's diesel engine are typically between 15.5:1 and 17:1. Indirect-injection combustion chambers (described later in this chapter) are usually between 19:1 and 23:1. Compression pressures produce an average pre-ignition temperature of 1,000°F (538°C). Fuel sprayed into the cylinders of a CI engine will break up into a fine mist, which then absorbs heat,

TECHNICIAN TIP

Smaller four-stroke diesels require a minimum cranking speed of 125 rpm to start. Without adequate cranking speed, insufficient compression and pressure, as well as heat, is available to start combustion. Normal minimum cranking speeds for diesel engines are between 200 and 250 rpm.

> **TECHNICIAN TIP**
>
> The gasoline-fueled engine is becoming more like the diesel. Recent technological developments now enable the injection of gasoline directly into the combustion chamber of spark-ignited gasoline engines. Eco-boost engines from Ford and Eco-tech engines from GM are marketing names for engines that use diesel-like combustion systems. These direct-injection engines inject gasoline near the end of the compression stroke. This fuel delivery strategy minimizes the problem of pre-ignition-limited compression inherent in turbocharged gasoline engines. Compression ratios in these engines are typically between 9:1 and 11.5:1, which is only slightly higher than comparable non-turbocharged engines. The use of turbochargers on all direct injection spark-ignition engines, which serve to boost intake air pressure to between 12 and 15 psi (between 82.74 and 103.4 kPa), increases compression pressures and, consequently, power output. Pending emission standards will likely require the use of particulate filters on these engines.

vaporizes, and mixes with air before igniting and burning. Combustion heat causes cylinder gases to expand and push down the piston. Depending on the quality and grade of the fuel, the auto-ignition temperature of diesel fuel is approximately 380°F to 500°F (200°C to 250°C). During cranking, most engines will produce a minimum of 280 to 300 psi (1,931 to 2,068 kPa) compression pressure. These compression pressures will increase when the engine is running and when engines are turbocharged. Cylinder compression pressures of 500 to 1,200 psi are easily reached when the turbocharger is providing the greatest airflow into the engine (**FIGURE 2-34**). A higher compression ratio provides the following:

- greater expansion ratios, which produces more power from less fuel
- better air-fuel mixture, for improved combustion.

Disadvantages of High Compression Ratios

The higher cylinder pressures of high compression ratio engines means head gaskets used to seal joints between the cylinder block and the cylinder head must be manufactured for a tight seal. The additional cylinder head clamping force requires more head bolts. Heavier, costlier starting systems with greater cranking torque are required because of high compression pressures encountered during cranking. More vibration and engine noise also accompanies engines with high compression ratios. During compression stroke, pistons will slow down as they approach TDC due to high cylinder pressures. The rapid rise of combustion pressure after TDC accelerates the crankshaft. Continual speeding up and slowing down of the crankshaft produces torsional vibration in the engine and driveline. Torsional vibration is unpleasant and can be transmitted to the chassis through engine mounts and the driveline. Better engine mounts and driveline features capable of dampening added vibration from diesels are needed.

Lean Burn Combustion

CI engines are also different in their air requirements for combustion. Instead of operating through the use of stoichiometric air-fuel ratios where the combustion chamber contains only the minimum amount of air for combustion, they operate fuel lean or air rich. The benefits of lean burn combustion are reduced fuel consumption and lower emissions. Overall air-fuel ratios are quite high, but concentrations vary widely throughout the combustion

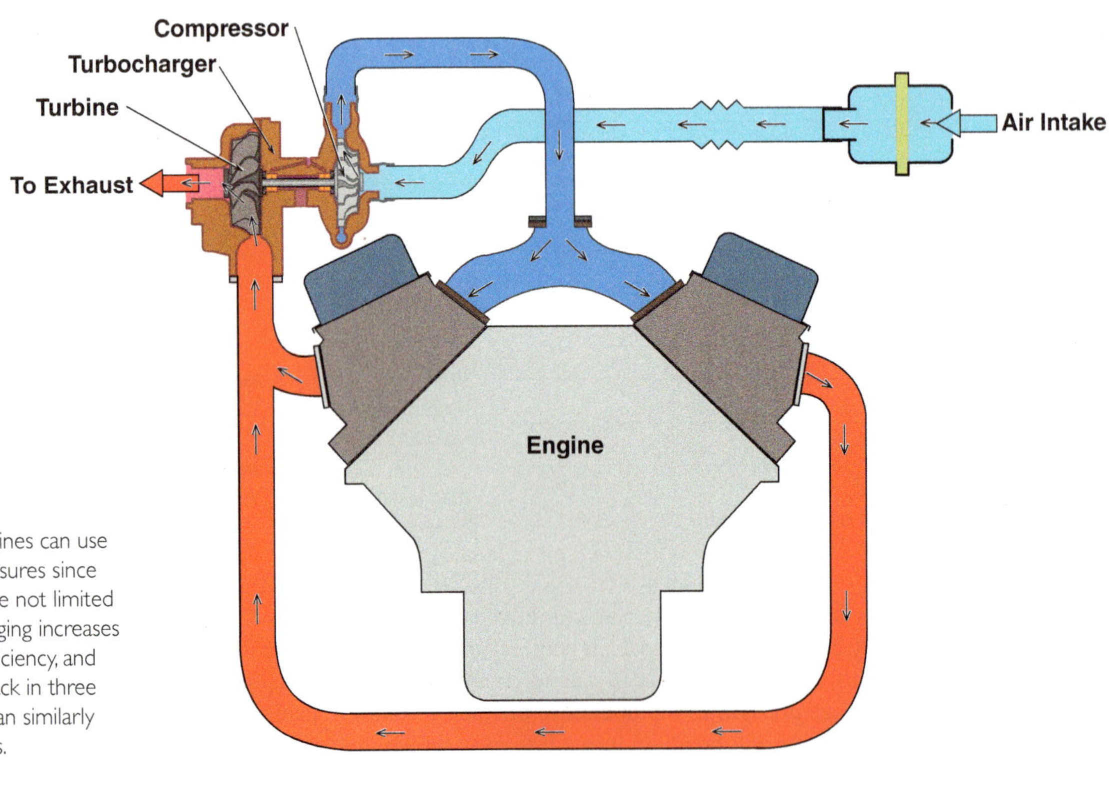

FIGURE 2-34 Diesel engines can use higher turbocharging pressures since compression pressures are not limited by pre-ignition. Turbocharging increases an engine's volumetric efficiency, and today's diesels typically pack in three or four times more air than similarly displaced gasoline engines.

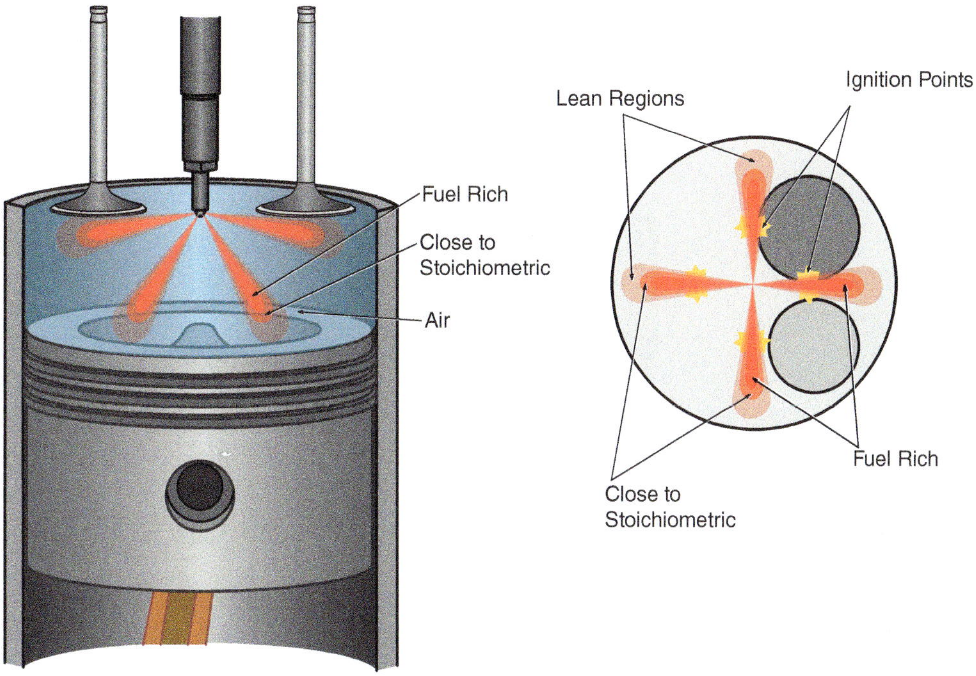

FIGURE 2-35 Diesel combustion is both lean and layered (stratified). Regions where fuel and air are mixed closest to the stoichiometric ratio of diesel fuel and where they have reached the auto-ignition temperature ignite first. Multiple ignition flame points and high pressures cause fuel to burn rapidly after combustion begins.

chamber. Spray patterns from the injector nozzle cause fuel distribution to be stratified or layered. When fuel is injected, it is not evenly distributed in the combustion chamber. Areas near the fuel spray are rich in fuel. Other regions will have little fuel or will consist only of air. Combustion begins in the regions where there is sufficient heat to ignite fuel and where the air-fuel ratio is close to the diesel fuel stoichiometric ratio (15:1 to 20:1) (**FIGURE 2-35**).

2-04 Identify and compare major features of direct- and indirect-injection combustion chambers.

▶ Diesel Engine Combustion Chambers

The combustion chamber is responsible for promoting good mixing of air and fuel to achieve efficient engine operation, low emissions, and other favorable operational characteristics. Characteristics of a good combustion chamber design include the ability to do the following:

- promote efficient combustion by enabling the formation of good air-fuel mixtures
- promote the maximum conversion of combustion energy into mechanical force
- minimize production of noxious emissions such as HCs, NO_x, CO, and particulates
- minimize noise
- provide good fuel economy
- provide smooth engine operation.

> **▶ TECHNICIAN TIP**
>
> The #2-D diesel fuel has a stoichiometric ratio of approximately 20:1. Because units for stoichiometric ratios are always measured by weight, #1-D fuel with a ratio of approximately 15:1 requires 15 lb (6.80 kg) of air to burn 1 lb (0.45 kg) of fuel. And 20 lb (9.07 kg) of air are required to burn 1 lb (0.45 kg) of #2-D fuel. However, diesel combustion chambers contain excess air and have air-rich and lean-fuel combustion. Lean air-fuel ratios improve fuel economy while lowering emissions, but this does not help ignition, because fuel molecules are spaced further apart. High compression ratios help pack fuel and air together more tightly to promote improved ignition and combustion quality. Ignition begins in the region of the combustion chamber where the mixtures are near the stoichiometric ratio and at a high enough temperature to ignite fuel.

Types of Combustion Chambers

There are two basic combustion chamber diesel engine designs but numerous variations of both. All diesel engines can be categorized as having either an indirect-injection (IDI) combustion chamber or a direct-injection (DI) combustion chamber. Each chamber design has its own unique fuel system and operational characteristics. The DI chamber is currently the most popular in all applications because of its simplicity, superior fuel economy, and good emission characteristics when combined with the electronically controlled high-pressure fuel systems (**FIGURES 2-36** and **2-37**). The IDI chamber was the most popular design in automotive applications until the early 1990s because of its high-speed capabilities, favorable emissions characteristics, and simple and inexpensive fuel-injection systems. Since the

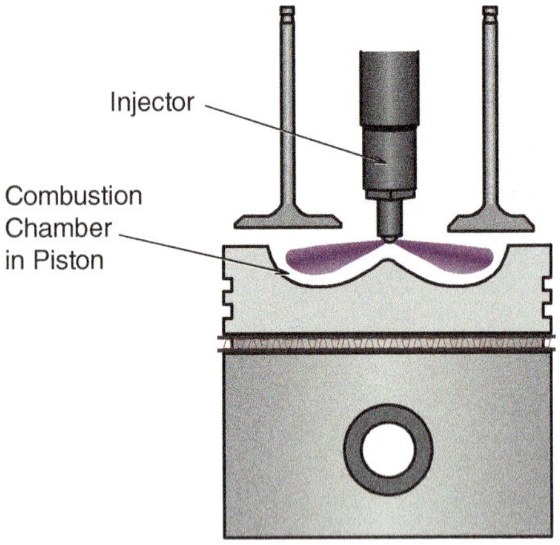

FIGURE 2-36 A Direct Injection (DI) chamber has a piston-formed combustion chamber that uses a multi-orifice injector with direct access to the combustion chamber.

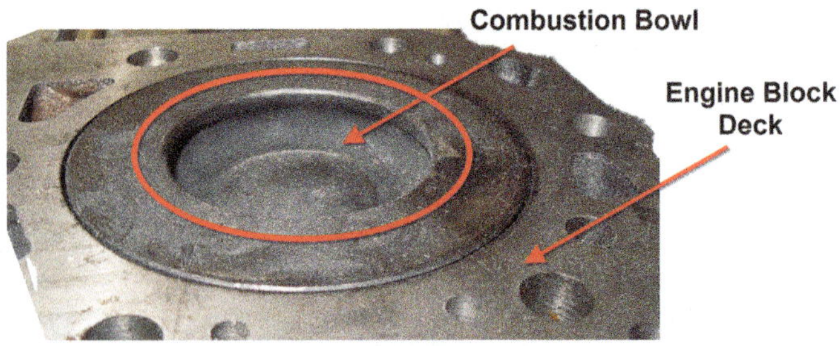

FIGURE 2-37 A combustion bowl of a DI chamber.

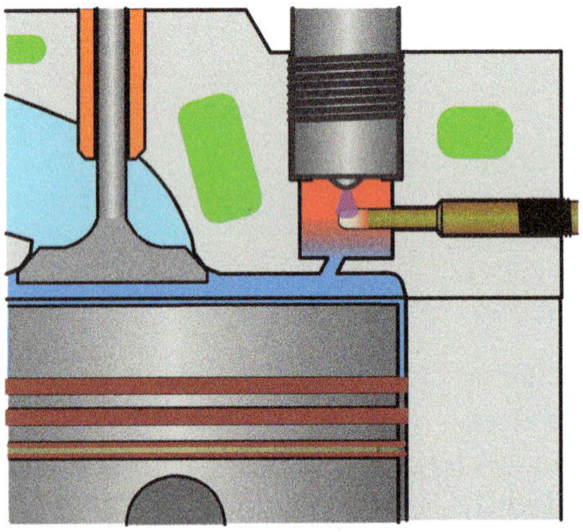

FIGURE 2-38 The IDI combustion chamber is one of the two basic configurations for diesel combustion chambers. Fuel is injected into a small pre-chamber connected to the main combustion chamber.

FIGURE 2-39 A cylinder head using IDI chambers. Note the replaceable chamber insert for the IDI chamber is removed from one cylinder.

performance characteristics of an IDI are similar to spark-ignition, gasoline-fueled engines, IDI-chambered diesels directly replaced gasoline-fueled engines without major changes to the drivetrain configuration. Furthermore, IDI fuel systems operate at lower injection pressures, so they can use less complex and relatively inexpensive fuel-injection systems (**FIGURES 2-38** and **2-39**).

Direct-Injection Features

DI chambers describe a CI combustion chamber where fuel is injected directly into a combustion chamber located between the top of the piston and the cylinder head. The combustion chamber is formed in the piston and the cylinder head is flat. The fuel injector or spray nozzle and the intake and exhaust valves are located in the combustion chamber, directly above the piston. Higher engine thermal efficiencies are achieved by using DI chambers, since there is a small ratio between combustion chamber surface areas and combustion volume. This results in 10–15% greater fuel efficiency over IDI chambers because more heat is used to produce power and less heat is rejected or absorbed by the cooling system. DI diesels use multi-orifice fuel injectors. These nozzles produce a very fine, fog like air-fuel mixture from several holes in the fuel nozzle. Fuel begins spraying from the latest injectors when nozzle opening pressures are between 3,200 and 5,500 psi (220 to 379 bar). When some injectors electronically control nozzle opening pressures, nozzle opening pressures vary between 5,500 and 20,000 psi (220 to 1,379 bar) (**FIGURE 2-40**).

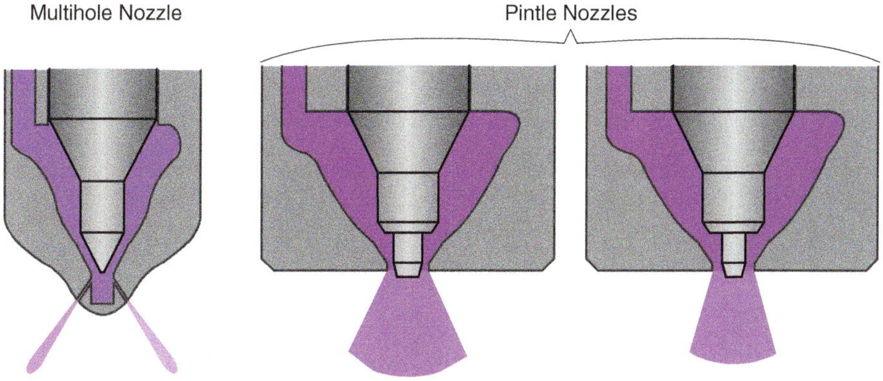

FIGURE 2-40 Types of nozzles. The left nozzle is a multi-orifice type. The center and right nozzles are pintle-type nozzles. (Courtesy of Ministry of Training Colleges and Universities–Ontario.)

DI Starting Aids

DI chambers generally require no starting aids, such as glow plugs or intake heaters. Glow plugs and intake heaters are used on some DI engines, but they are present to minimize cold-start emissions and white smoke at start-up caused by low cylinder temperatures, poor fuel atomization, and cylinder misfires during the warm-up period. Higher pre-ignition temperatures are obtained by using intake heaters, which provide faster start-up in colder ambient temperature with less soot and other emissions. Since some common rail and HEUI fuel systems do not effectively pressurize fuel at cranking speeds, a glow plug placed near the fuel nozzle will speed up ignition. This is necessary because large fuel droplets produced by the injectors will not produce enough fuel vapor to start the engine. The use of a hot glow plug will vaporize fuel and promote faster ignition.

Classification of DI Chambers

Two common DI chambers are the offset and the central types (**FIGURE 2-41**). The names refer to the location of the injector in the combustion chamber. Central DIs use four valves per cylinder, and the injector is centrally located. The centralized placement of the injector allows these DIs to distribute fuel more symmetrically for cleaner, less noxious emission characteristics. The four-valve intake and exhaust configurations also improve the engine's ability to breathe, which increases volumetric efficiency. Offset DIs use two valves per cylinder. This setup does not provide room to accommodate the centrally located injector due to larger valves diameters. Instead, the injector is leaned to one side of the cylinder head, making room for the valves. Offset injection, while less expensive to manufacture, creates uneven piston crown temperatures and thrust loads in addition to excessive noxious emissions from asymmetrical fuel distribution. In recent years, offset injection has disappeared from all clean diesels.

DI Chamber Air Motion

Turbulence is a critical factor to combustion mixture preparation in diesel engines. Without adequate turbulence, air and fuel will not mix, resulting in lost performance and efficiency and in increased emissions. The Mexican hat or semi-toroidal shape of the DI chamber is designed to mix the air and fuel. Since fuel is injected near the end of the compression stroke, very little time is available to prepare the mixture for combustion, so the turbulence created by the chamber must quickly mix air and fuel. The shape of the combustion bowl converts the twisting, toroidal airflow created during the intake stroke into many small rolling balls of air during the compression stroke (**FIGURE 2-42**). Combustion turbulence and piston bowl shape are

▶ **TECHNICIAN TIP**

The pressure spike produced during combustion causes a characteristic noise in diesel engines that is sometimes called "fuel knock." When injected, diesel fuel does not ignite and burn immediately. Time is required for fuel to atomize, absorb heat, vaporize, and mix with air before it can ignite. The delay between the point when fuel is injected and when it finally ignites is called the "ignition delay period." When fuel does ignite, the high compression pressures and heat in the combustion chamber cause the fuel to rapidly ignite and burn. Multiple ignition points and high cylinder pressures cause the fuel to suddenly burn much faster as compared to gasoline engines. Modern diesels now use a variety of strategies to minimize the length of time for ignition delay. Pilot or split-shot injection, finer fuel atomization, and improved cylinder pressure control are just a few ways that the traditional combustion noise from diesel engines is now almost eliminated.

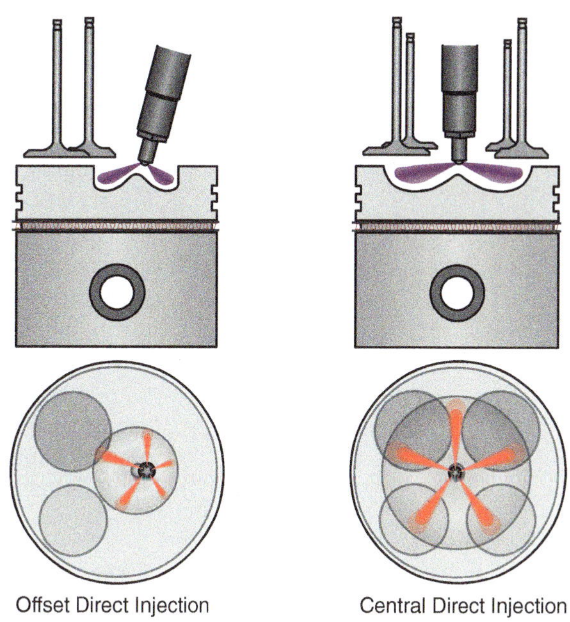

FIGURE 2-41 Two types of DI chambers are offset and central. DI chambers have piston-formed combustion bowls to which the injector has direct access.

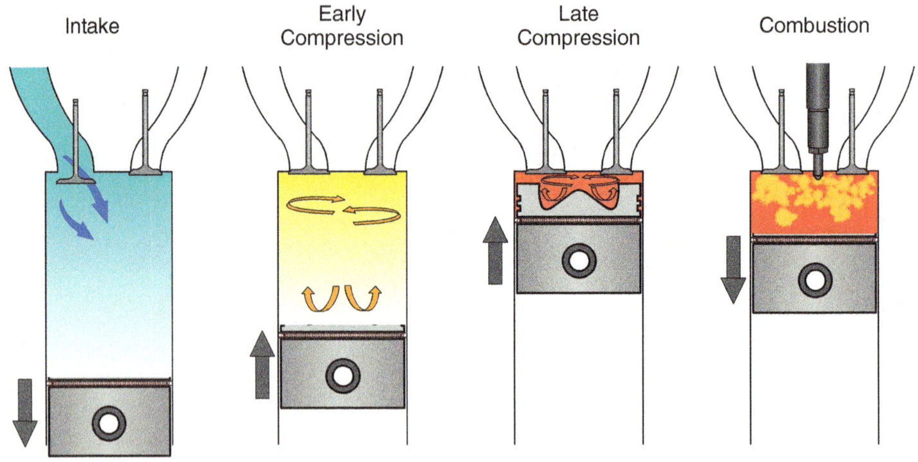

FIGURE 2-42 During intake stroke, air turns in a circular or toroidal direction. When air is compressed, the semi-toroidal shape of the piston bowl converts toroidal airflow into smaller circular balls of semi-toroidal air turbulence. Vigorous air turbulence is needed to mix air and fuel.

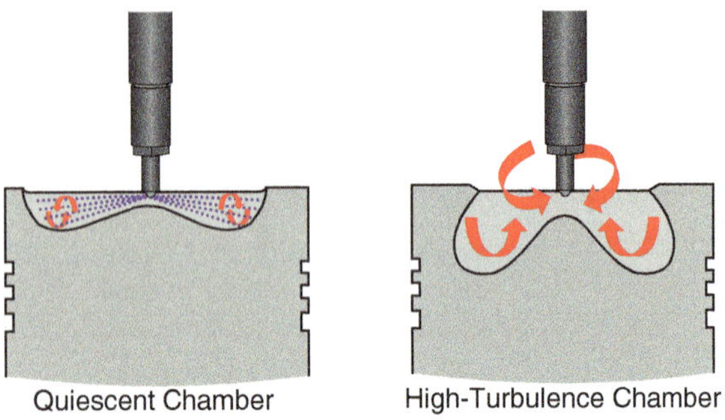

FIGURE 2-43 The combustion chamber air turbulence necessary to mix air and fuel is created by the shape of the combustion bowl. Deep bowls are more turbulent, and shallow bowls are less turbulent. Shallow, wide bowls are called quiescent bowls, from the word "quiet," since there is relatively little turbulence. Most late-model engines use quiescent bowls and very high injection pressures to compensate for reduced turbulence mixing of air and fuel.

unique to engine horsepower and fuel settings. Correctly matching injector spray patterns is required to obtain proper mixing. The geometric configuration of the piston bowl will be determined by the length of fuel spray penetration, spray angle, droplet sizing, and turbulence. For this reason, injectors and piston-formed combustion chambers must be precisely matched to ensure good performance and emission compliance. A recent trend in combustion chamber design is the use of low swirl or quiescent combustion chambers. Fuel in a high swirl chamber can readily impinge on or condense onto the cylinder walls, piston, and enter crevice volume space between the piston and cylinder wall. In those locations, fuel will not burn. Overlapping spray in highly turbulent chambers will also create denser air-fuel mixtures and increase emission levels of HC, CO, and particulate. The alternative strategy to prevent the formation of these emissions is to reduce combustion chamber turbulence, increase injection pressures, and increase the number of orifices at the injector tip. Piston bowls will be wider and shallower in the quiescent combustion chamber design (**FIGURE 2-43**).

DI Pressure Volume Cycle

Please refer to **FIGURE 2-44**, a pressure volume (PV) graph plotting pressure versus cylinder volume in a diesel engine. Ignition lag refers to the period between points A and B, and it is used to describe the delay between the start of injection and the actual point of ignition (Figure 2-40). Lag or delay occurs because fuel needs to absorb heat, vaporize, and mix with air. Ignition lag, also called ignition delay, is the factor responsible for the characteristic "knock" or combustion noise from diesel engines. This delay period and noise is more pronounced at idle. During the delay period, fuel will accumulate in the cylinder while mixing and vaporizing. When fuel does begin to burn during the uncontrolled combustion period noted between points B and C, a sharp increase in pressure produces a sound wave known as fuel knock. Ideally, the smaller the ignition delay period, the more quietly the engine will operate. A small delay is approximately 0.001 seconds. A time of 0.003 seconds or more results in a rough running engine and an exaggerated fuel knock because more fuel accumulates during the delay period. When the fuel does ignite, the larger fuel accumulations cause the pressure to rise much faster than what would occur with a shorter delay.

Factors Affecting Ignition Lag

The following are some of the factors that affect ignition lag:

- *Fuel quality*: Fuel with high cetane values ignites more quickly to produce a shorter delay time.
- *Compression ratio*: High compression ratios produce shorter ignition delay times.

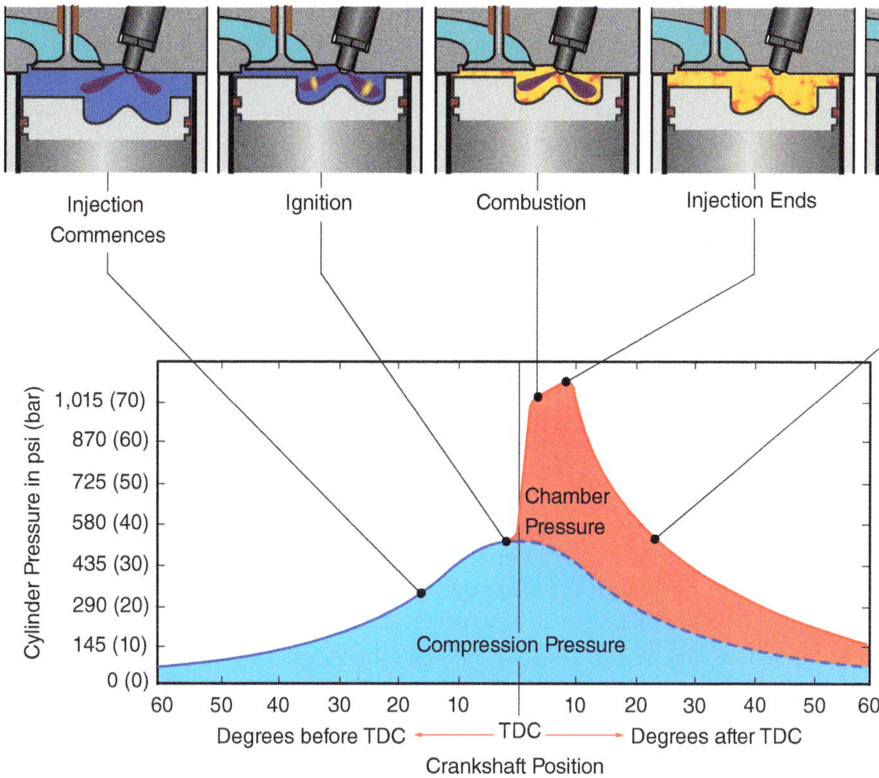

FIGURE 2-44 Pressure volume cycle in a DI engine.

- *Degree of fuel atomization*: Finer atomization will shorten delay time.
- *Pre-ignition temperature*: Colder ambient and engine temperatures will create a longer delay. This explains why diesel engines are noisier when cold (**FIGURE 2-45**).

Flame Propagation Period

The flame propagation period, also called the uncontrolled burn period, occurs between points B and C on the PV graph. This event is characterized by uncontrolled combustion in a cylinder when fuel that was injected until the end of point C is burned. Many points in the

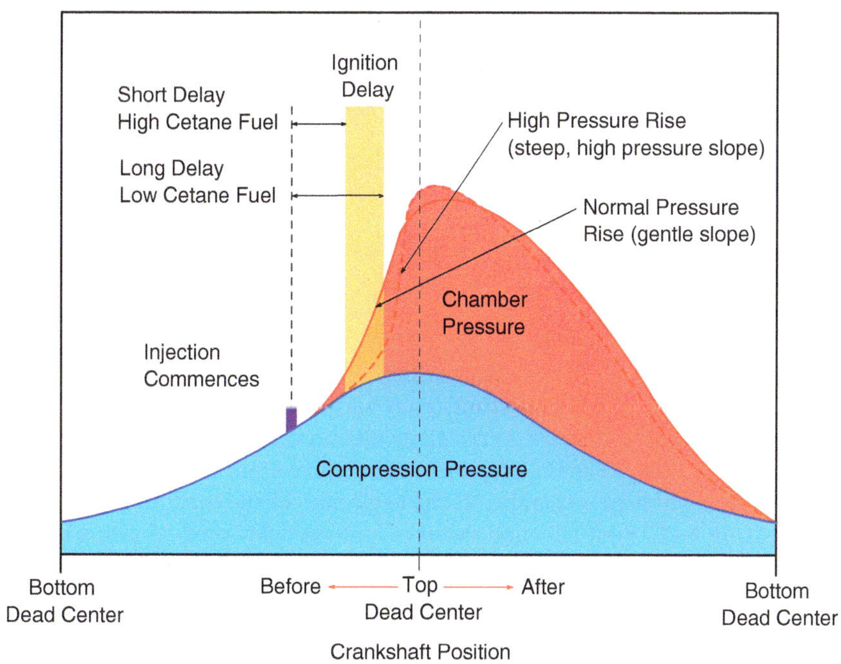

FIGURE 2-45 Ignition delay is caused by a number of factors, including ambient cold temperatures, low compression, poor fuel atomization, and low cetane fuel.

combustion chamber will reach the temperatures required for ignition of fuel and multiple flame fronts move through the air-fuel mixture. This rapid progression of multiple flame fronts causes a sharp pressure spike in the cylinder while all the accumulated fuel suddenly burns. This combustion phase is responsible for the actual knock in the engine. A longer ignition delay period creates bigger pressure spikes during this combustion event.

Controlled Burn Period

A controlled burn period refers to the phase of combustion between points C and D when delivery of the remaining injection quantity occurs. Combustion during this phase causes a gradual rise in cylinder pressure after TDC. Depending on the injection rate, fuel delivery will not stop until just after TDC. In engines with common rail fuel systems, an additional post combustion injection event may also take place to keep cylinder pressures and temperatures hot enough to reduce soot formation.

Afterburn Period

The afterburn period is the combustion phase between points D and E when any injected fuel is completely burned. This event produces pressure on the piston to keep it moving during its power stroke. The afterburn period can affect the formation of particulate emissions. Having adequate afterburn time is important to completely consume any unused or partially burned fuel. Poor mixing of air and fuel prior to this phase, or an extended ignition delay, will not allow time for complete combustion during the afterburn period, because cylinder pressure and temperatures are decreasing as the piston moves downward on the power stroke. If afterburn time is shortened more particulate will form. In engines with common rail fuel systems, an additional post combustion injection event may also take place to keep cylinder pressures and temperatures hot enough to reduce soot formation.

Rate-Shaped Injection

In older mechanical engines, fuel was injected in one large main injection event. Injection pressures were controlled by the speed of the engine or velocity of injector plunger pressurizing the fuel. Highest injection pressures are only available when plunger travel is at its highest velocity. Newer engines have benefited from research that has demonstrated that at each stage of the combustion cycle, there is an ideal quantity of fuel that should be delivered at a specific time and injection pressure. Changing injection pressure and adding more injection events delivered at a precisely timed moment during the combustion cycle will produce better fuel economy, more power, and cleaner emission characteristics for any given speed and load condition. Current engines will electronically vary injection pressures and timing of as many as nine injection events during the combustion cycle. Matching the quantity of fuel with the optimal number and time of injection events can produce superior performance with the lowest exhaust and noise emissions with the bonus of better fuel economy. Varying injection pressure along with the time and number of events changes the shape of what is called the **injection discharge curve**. When used as a fuel delivery strategy, changing the injection discharge curve during the combustion cycle is referred to as **rate-shaped injection** (**FIGURE 2-46**). Further discussion about rate shaping is provided in Chapter 8, Functions of a High-Pressure Injection System. The use of pilot or split-shot injection is the simplest type of rate shaping. Pilot injection reduces combustion chamber noise caused by ignition delay. Pilot injection involves injecting fuel 8° to 10° of crankshaft rotation before the main injection event. With pilot injection, the small quantity of fuel injected before the main injection will produce a burning flame front and a hotter combustion chamber at the beginning of the main injection event (**FIGURE 2-47**). Pilot injection shortens ignition delay time, resulting in quieter combustion noise. Shortening the ignition delay period lengthens the burn time, resulting in more complete and cleaner combustion. Timing the beginning of the main injection can also be delayed, or retarded, when pilot injection is used because fuel ignites and burns faster. By retarding the

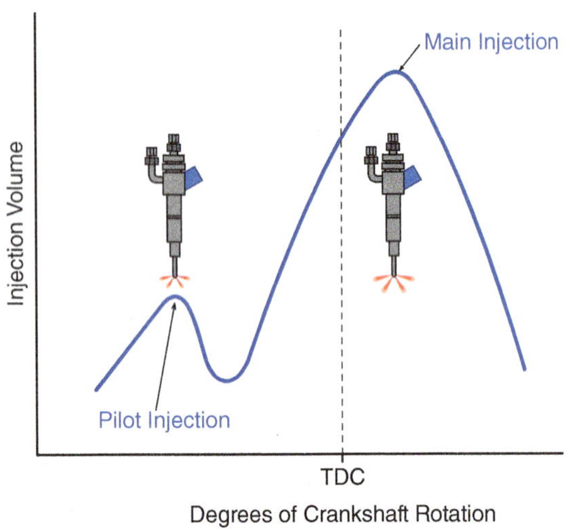

FIGURE 2-46 Rate-shaped injection refers to the shape of the fuel discharge curve during the injection event. Split-shot or pilot injection is part of the injection rate shape. The use of one or two pilot injections before the main injection reduces combustion chamber noise caused by ignition delay. The result is a dramatically quieter engine and shorter ignition delay.

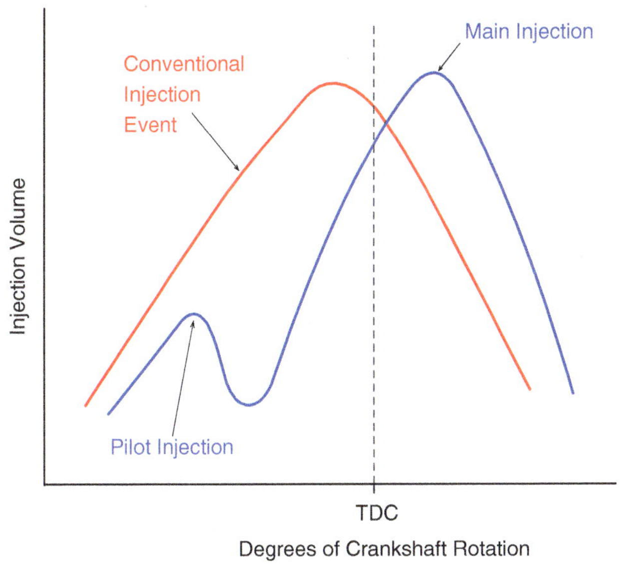

FIGURE 2-47 Pilot injection involves injecting a small quantity of fuel before the main injection. Pilot injection reduces combustion noise and permits retarded injection timing. Retarding injection timing reduces NO_x emissions.

FIGURE 2-48 The cylinder head of a Duramax 6.6 direct-injection diesel contains a glow plug. The glow plug heats fuel (not air) for faster start-ups at lower temperatures. Fuel spray vaporizes instantly after contacting the glow plug, thus shortening the ignition delay time.

beginning of the main injection event, peak cylinder pressures and temperatures are lowered near TDC, which reduces the formation of nitrogen oxide (NO_x) emissions.

Adding a glow plug to a DI chamber can further shorten ignition delay time in cold combustion chambers since fuel vapor needed to start combustion is produced more quickly. Placing the glow plug near the edge of a fuel spray hole will cause fuel to instantly vaporize, which promotes faster ignition. The glow plug will also reduce soot formation during warm-up periods because faster ignition translates into a longer burn period (**FIGURE 2-48**).

▶ Operational Characteristics of DI Combustion

Cold Idle

2-05 Identify and describe the stages of direct-injection combustion.

Injection quantities of fuel in a diesel engine at idle are very small. Combined with the cooling effect of excess air, prolonged idle has detrimental effects on the engine. A typical idle injection quantity in a diesel engine is between 4 and 10 cubic millimeters (mm^3). This compares to the size of a fuel droplet that is normally about 28 mm^3 in volume. Small quantities are adequate to create enough cylinder pressure to rotate the crankshaft but produce very little heat, some of which is transferred to engine coolant and cylinder components. Because the use of excess air in diesels dilutes the heat of combustion and lowers average cylinder temperatures, less heat energy is transmitted to the cooling system. The small combustion chamber surface area of the DI chamber further exaggerates this problem. DI chambers do not allow the engine to warm up from idle; in fact, an engine will cool down below operating temperature when idled (**FIGURE 2-49**). Passenger compartments heated with engine coolant will be cold. For this reason, some manufacturers will use exhaust backpressure devices to accelerate engine warm-up. These devices operate by restricting exhaust gases leaving the engine. Exhaust backpressure causes more heat retention by the engine around the exhaust ports, thus transferring more heat to the cooling system. Air-rich exhaust gases not scavenged or removed from the combustion chamber at the end of the exhaust stroke will form part the air charge for the next combustion event. Warm exhaust gases will increase pre-ignition and combustion temperatures, which speeds up engine warm-up. Piston rings will seal better with increased exhaust back pressure.

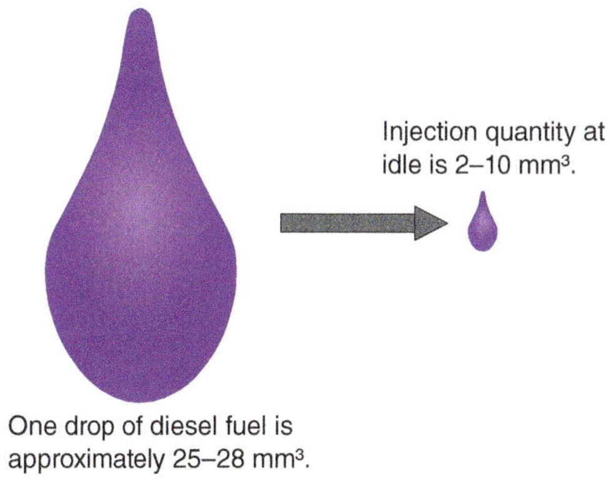

FIGURE 2-49 A comparison between the sizes of a droplet of fuel and the quantity injected at idle. Small injection quantities at idle combined with excess air in diesels do not allow the engine to properly warm up. Excess air creates low exhaust temperatures and low cylinder temperatures, which have adverse consequences on engine life and operation.

Combustion Slobber and Wet Stacking

In the absence of heat and low cylinder pressures, DI combustion qualities at idle also tend to be very poor. Excess air supplied for combustion will lower cylinder temperatures even more. The excess air factor can be as high as 1,000:1. Not all of the injected fuel will burn when temperatures and pressures are low. Fuel may condense or be quenched in certain regions of the chamber, such as between the piston crown and cylinder wall. Low piston temperatures also mean the piston has not fully expanded and will easily allow lube oil into the combustion chamber. When soot produced by partially burned fuel and liquids—such as lube oil and fuel—mix, they produce a substance known as combustion slobber. This is also called turbo slobber or wet stacking, and the name applies to the black gooey liquid that leaks from exhaust manifold joints and the exhaust pipes when engines are excessively idled.

▶ Indirect Injection

2-06 Explain the advantages and disadvantages of indirect-injection combustion chambers.

The IDI combustion chamber is the other of the two main types of diesel combustion chambers. This chamber was traditionally most common in automotive applications because of its high-speed capabilities and low-emission characteristics, relative to the DI chamber. Pickup trucks, vans, and small displacement engines used this chamber design because the engines offered gasoline-like high-speed performance with the durability and fuel economy of a diesel. The Comet Combustion Chamber and the Ricardo Comet are names for the IDI chamber that capture its operational characteristics. The IDI combustion system uses fuel injected into a smaller, spherical, highly turbulent chamber that is connected to a main chamber through a narrow passageway called the venturi or chamber throat. Fuel is mixed with air and ignited in this pre-chamber. The fireball then expands into the main chamber, where combustion is completed. In North America, the GM 6.5 L diesel was the last engine to use this design in "G" vans, until 2001. Before then, engine makers such as VW, Mercedes, Navistar, Ford, BMW, and others popularized the use of combustion chamber.

IDI Features

The following are some of the major characteristics of the IDI chambers:

- It uses a pre-combustion chamber formed in the cylinder head.
- The chamber typically holds approximately 70% of the cylinder volume.
- IDI injectors use pintle-type nozzles.
- Pintle nozzles operate with lower NOPs than multi-orifices.
- Pintle nozzle opening pressures range between 900 and 2,300 psi (62 to 159 bar).

Pintle nozzles have a single orifice at the tip to spray a single coarse stream of fuel into the chamber. The atomization of fuel critical to the DI chamber is not required in an IDI, due to the high degree of turbulence in the pre-chamber. The turbulent pre-chamber is responsible for mixing the air and fuel. It performs this job much better than the DI chamber, which accounts for its high-speed and low-emission characteristics. Vaporizing the coarse fuel droplets occurs when fuel initially contacts the incandescent glow plug during start-up and then on the hot walls of the pre-chamber during regular operation (**FIGURE 2-50**).

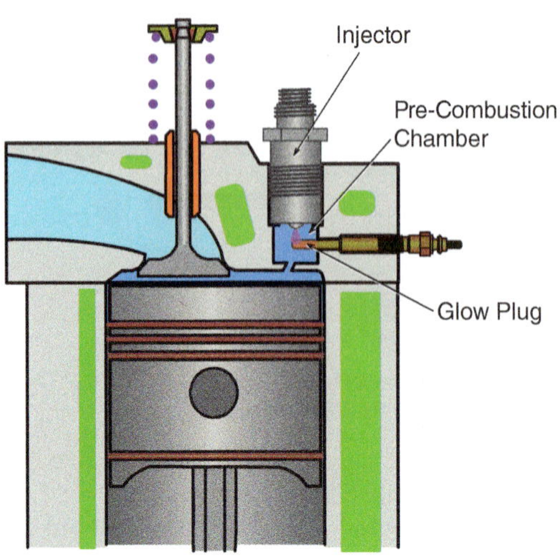

FIGURE 2-50 A glow plug vaporizes heavy coarse fuel droplets to promote faster ignition.

IDIs are capable of a faster top engine rpm limit due to a higher degree of turbulence to produce more rapid and even mixing of air and fuel. The use of a pre-chamber allows the engine to tolerate poorer grades of fuels without incurring a significantly longer ignition delay. Mixing fuel in the pre-chamber and its passage into the main chamber through a venturi enhances mixing and ensures more complete combustion. The two-combustion chamber design creates a quieter engine with less knock since it minimizes ignition delay. The reason for this once again is the turbulent pre-chamber, which promotes faster fuel mixing and fuel vaporization to ensure a rapid combustion sequence from injection to afterburn. Noise is further reduced because the expansion of the burning fuel is regulated to smoothen the pressure spike as it passes through the venturi, thus dampening the rapid rise in cylinder pressure. Higher thermal efficiency and the cleaner emissions from electronically controlled high-pressure injection systems for DI diesels have ended the use of IDI engines.

Glow Plugs and IDI Chambers

One of the most significant features of the IDI chamber is the requirement for a glow plug starting aid. Glow plugs are required to produce fuel vapor from the coarse fuel spray, which does not generate sufficient fuel vapor for ignition. When operated at an incandescent temperature, the glow plug heating element causes the fuel to vaporize on contact with the element. Fuel, not air, uses heat from the glow plug to create the necessary combustion conditions required for starting. After a few minutes of engine operation, glow plug operation is no longer required. The combustion chamber walls will provide the hot surface for fuel vaporization (**FIGURE 2-51**).

Glow Plug Control

Control of glow plugs in modern engines is delegated to a separate electronic module on both IDI and DI engines (**FIGURE 2-52**). Glow plug failure will produce excessive cold start emissions and therefore has a diagnostic monitor associated with it on vehicles compliant with OBD-II. Since conventional plugs will draw approximately 5 amps of current after they are heated, current is best switched through a relay. Glow plugs must be heated before the operator begins cranking the engine over to start. Coolant, ambient air, or oil temperature is sensed by the control circuit to determine the length of time the glow plug will be heated before the operator is prompted to crank the engine. An instrument panel–mounted "wait to start" light usually illuminates until the correct amount of time has elapsed for the plugs to reach starting temperature. Once the engine is operational, the controller will cycle the plugs on and off during an afterglow period, which lasts until the engine has warmed up. Subsequent to the afterglow period, the plugs are switched off to prevent burning out.

Instant-Start Systems

To reduce wait-to-start times, newer glow plugs have been developed that will heat to operating temperature within as little as two seconds, called instant-start glow plug systems (ISS) (**FIGURE 2-53**). The GM Duramax Ford 6.7 Powerstoke and VW TDI's use glow plugs that reach 1,832°F (1,000°C) in three seconds. This means the wait-to-start lamp will be on for

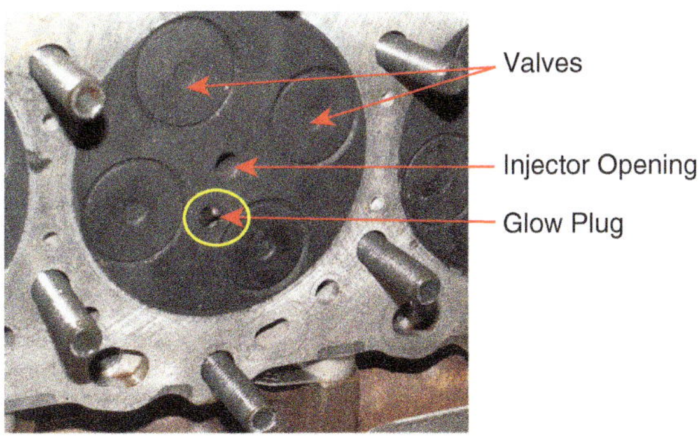

FIGURE 2-51 The glow plug in the DI 6.6 Duramax is positioned to contact only the edge of the fuel sprayed from one hole of the injector.

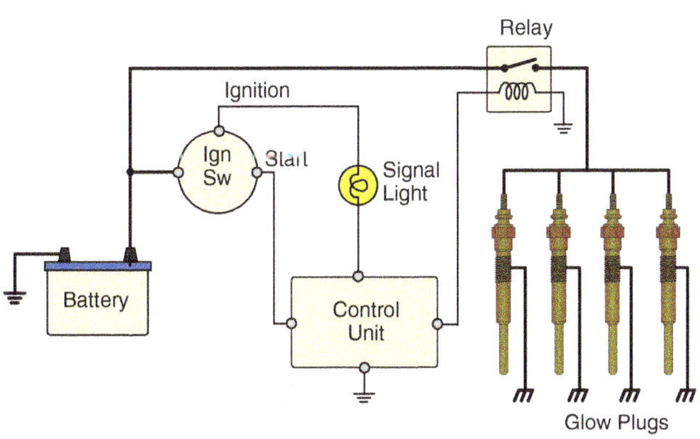

FIGURE 2-52 A schematic of a simple glow plug control circuit. (Courtesy of BERU Aktiengesellschaft.)

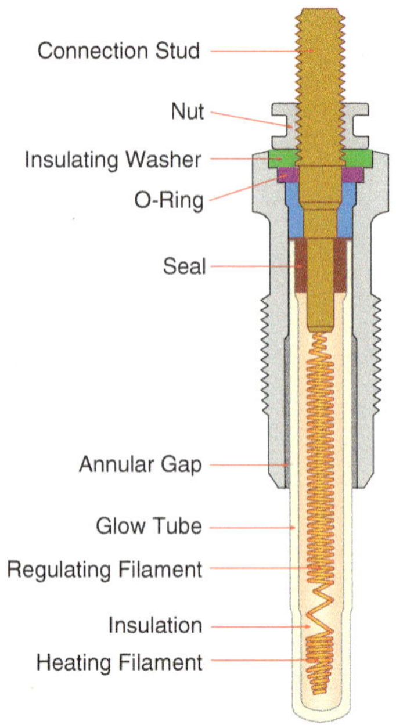

FIGURE 2-53 This glow plug has two heater elements connected in series, which helps limit current and prevents overheating or rapid burnout in the elements.

a maximum of three seconds under the coldest conditions. Initial amperage draw for eight plugs during the first 1.8 seconds will approach 110 amps. Subsequently, the current flow will taper off until only 30 amps are used for the eight plugs to remain heated to 1,832°F (1,000°C). The use of silicon nitrate ceramics and dual coil heating elements has enabled this rapid cycle time (**FIGURE 2-54**). Silicon nitrate, which has the appearance of steel, is a better heat conductor than other ceramics. A second, "intelligent" control heating element connected in series with the heating coil changes resistance, depending on temperature. The variable resistance of this second coil prevents damage to the plug from overheating due to excessive current draw. Newest innovations to glow plug mechanisms include an integrated pressure-sensing glow plug that is able to measure cylinder pressure. This input data supplied to the electronic control module (ECM) can be used to measure combustion events, which could produce excessive emissions or poor performance and thus permit modifications to engine operating strategies. Optimizing combustion events by using sensor data could lead to further refinements to combustion events, enhancing performance, durability, fuel economy, and even smoother, quieter engine operation (**FIGURES 2-55** and **2-56**).

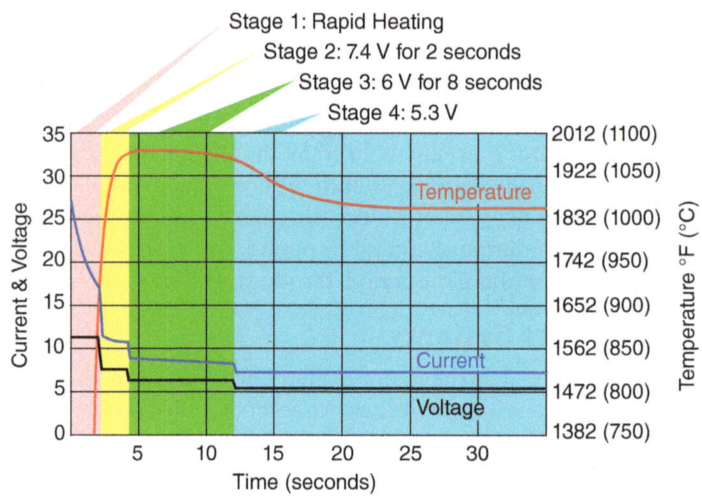

FIGURE 2-54 The time and current required to bring an ISS into operation. (Courtesy of BERU Aktiengesellschaft.)

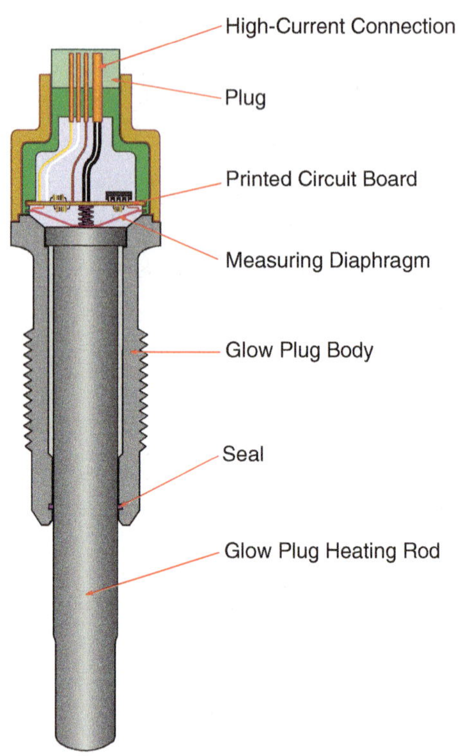

FIGURE 2-55 A cross-sectional view of the pressure-sensing glow plug used to monitor cylinder pressure. Integrated electronics are used to provide closed-loop control regarding cylinder pressure to control various functions, including injection quantity and timing. This sensor helps bring diesel emissions to almost undetectable levels. (Courtesy of BERU Aktiengesellschaft.)

FIGURE 2-56 Best light-duty applications for diesels are in SUV's and pick-up. Ford's new 3.0 L diesel for the 2019 F-150 is a version of the European Lion V-6 used by Land Rover SUV's. The 250 HP engine delivers 440 ft/lbs of torque (597 Nm) at a useable 1,750-rpm while achieving a 20/30 mpg city/hwy (11.7/7.8 L/100 km) fuel efficiency.

Wrap-Up

Ready for Review

- Diesel engines are used widely in transportation, construction, mining, emergency services, military applications, and electric power generation.
- Diesel engines are the primary engines used in heavy-duty and commercial vehicles because they produce more power while using less fuel and last longer than other types of engines.
- Diesel engine use in automobiles has varied over the years, but modern diesel engines are becoming more popular in automobiles because they achieve 25–60% greater fuel economy than gasoline-powered engines, and they have more power and provide quieter, smoother engine operation than older diesel technology. To meet GHG emission targets, manufacturers will need to use either more diesels, electric hybrids or battery electric vehicles.
- Diesel engines are the most efficient combustion system, converting burning fuel into the most mechanical energy in comparison to any other engine technology.
- Rudolf Diesel developed the concept of the CI combustion system. In his engine, heat produced from compressing air, not the heat of an electric spark, ignited fuel.
- Diesel's original concept of a highly efficient engine caught the attention of other inventors, who adapted it for use in smaller, lighter transportation applications. By the 1920s, Robert Bosch developed compact-injection pumps and governors that enabled smaller diesels to power motor vehicles.
- Diesel engines today can meet and exceed the emissions standards set for gasoline engines. Not only are the emissions of diesel engines as clean as, if not cleaner than, gasoline-fueled engines, but diesel engines also emit 20–40% fewer GHG emissions.
- A diesel engine uses substantially less fuel to produce the same horsepower (kilowatts) as a similarly displaced gasoline-fueled engine.
- The diesel engine has significantly higher torque output than a similarly displaced gasoline-fueled engine. Higher compression ratios produce the high torque output.
- Diesel engines typically operate three to five times longer than gasoline-fueled engines before wearing out. Because diesel engines must withstand higher combustion chamber forces than gasoline-powered engines, more durable construction features are used in a diesel engine. Diesel engines turn slower than SI engines and have better cylinder wall lubrication.
- Some noxious emissions are much lower from diesel engines than from gasoline-fueled engines. Diesel engines produce lower amounts of CO and fewer HCs.
- Diesel fuel contains more carbon than gasoline, but less energy is required to refine diesel fuel than gasoline.
- Emission legislation requires diesel engines to meet minimum life expectancy standards and remain emission compliant during their useful life.
- Direct-injection (DI) combustion chambers use piston-formed combustion and multi-orifice spray nozzles.
- Indirect-injection (IDI) combustion chambers are formed in the cylinder head and use pintle-type injector nozzles.

Key Terms

crankshaft The component that converts the reciprocating action of the pistons to a rotational movement and revolves inside the crankcase portion of the engine block.

crankshaft throw The distance between the centerline of the main bearing journal and crankpin journal.

clearance volume The space remaining in a cylinder when the piston is at top dead center (TDC).

combustion A chemical reaction between fuel and oxygen that releases heat.

compression ignition (CI) A combustion mechanism that ignites fuel by using the heat derived from compressing air as an ignition source.

compression ratio A comparison between total cylinder volume and clearance volume (the space left in the cylinder when the piston is at top dead center [TDC]).

cylinder block The largest structure of an engine, which encloses the cylinders and provides a rigid frame to support attaching the rest of the engine components.

cylinder displacement The volume displaced by the piston as it moves from (TDC) to bottom dead center (BDC).

Durability requirement Emission legislation that requires an engine to remain below maximum emission thresholds for its normal expected service life. Also referred to as useful life requirement.

greenhouse gases (GHGs) Gases that are believed to contribute to global warming.

Injection discharge curve The shape of a plotted graph displaying the volume of fuel leaving the injector tip compared to the engine position reported in degrees of crankshaft rotation.

Rate-Shaped Injection A fuel injection strategy that changes the profile of the injection discharge curve by varying injection pressure, the number of injection events and timing of injection events.

top dead center (TDC) The highest point at which the piston travels in a cylinder.

torque A measurement of rotational, or twisting, force transmitted from the crankshaft to the flywheel which is a function of cylinder pressure and length of crankshaft throw.

torque elasticity The ability of an engine to produce the torque needed to accelerate a vehicle from a moderate to high vehicle speed while under load.

Review Questions

1. Which diesel engine combustion chamber design will produce the fastest coolant warm-up times?
   a. Direct injection.
   b. Indirect injection.
   c. Two stroke.
   d. Hemispheric chamber.
2. What is the average compression ratio of an on-highway diesel engine?
   a. 23:1.
   b. 20:1.
   c. 16:1.
   d. 10:1.
   e. 8:1.
3. Which of the following diesel engine combustion chambers requires a glow plug to start?
   a. Direct injection.
   b. Indirect injection.
   c. Hemispheric chamber.
   d. Turbulence chamber.
4. How many valves will most likely be used by a centrally located fuel injector?
   a. Two valves per cylinder.
   b. Three valves per cylinder.
   c. Four valves per cylinder.
   d. Any number of valves.
5. At which of the following compression ratios does the efficiency of an engine begin to diminish with increasing compression?
   a. 12:1.
   b. 16:1.
   c. 23:1.
   d. 25:1.
6. Which one of the following conditions will be produced by prolonged ignition delay in a diesel engine?
   a. An engine knocking sound.
   b. Blue smoke.
   c. White smoke.
   d. Lower peak cylinder pressures.
7. Which combustion chamber design produces the best fuel economy in a diesel engine?
   a. Direct injection.
   b. Indirect injection.
   c. Swirl chamber.
   d. Wedge chamber.
8. Which combustion chamber design has traditionally operated with the least amount of noise?
   a. Direct injection.
   b. Indirect injection.
   c. Two stroke.
   d. Penta chamber.
9. Which of the following best describes the term "ignition delay"?
   a. The time required to raise the fuel pressure high enough to overcome compression pressure in a cylinder.
   b. The time lag from initial injection to actual ignition.
   c. The time required to heat up a glow plug.
   d. The time required for vaporized fuel to atomize.
10. Which type of fuel nozzle is used in a direct-injection combustion chamber?
    a. Multi-orifice nozzle.
    b. Pintle-type nozzle.
    c. Closed nozzle.
    d. Open nozzle.

ASE Technician A/Technician B–Style Questions

1. Technician A says that a diesel engine's higher compression ratio helps it to produce more power and use less fuel. Technician B says that higher compression ratios only help the diesel engine produce more torque. Who is correct?
   a. Technician A
   b. Technician B
   c. Both A and B
   d. Neither A nor B
2. Technician A says that diesel engines inject fuel into the cylinder during the intake stroke. Technician B says that diesel engines inject fuel into the cylinder near the end of the compression stroke. Who is correct?
   a. Technician A
   b. Technician B
   c. Both A and B
   d. Neither A nor B
3. Technician A says that diesel engines supply only enough air to completely burn all the fuel in the combustion chamber. Technician B says that diesel engines supply excess air to the combustion chamber to completely burn all the fuel in the combustion chamber. Who is correct?
   a. Technician A
   b. Technician B
   c. Both A and B
   d. Neither A nor B
4. Technician A says a power stroke in a modern diesel engine uses only a single injection of fuel. Technician B says late-model diesel engines use multiple injection events during a combustion cycle. Who is correct?
   a. Technician A
   b. Technician B
   c. Both A and B
   d. Neither A nor B
5. Technician A says the evaporation of diesel fuel from the tank of a diesel bus on a warm day produces more emissions than a car. Technician B says that diesel fuel does not produce a significant amount of emissions due to fuel evaporation. Who is correct?
   a. Technician A
   b. Technician B
   c. Both A and B
   d. Neither A nor B

6. Technician A says that high horsepower (kilowatts) is the most important attribute that an engine needs to pull heavy loads and accelerate quickly. Technician B says that high torque output is the most important attribute that an engine needs to quickly climb hills. Who is correct?
   a. Technician A
   b. Technician B
   c. Both A and B
   d. Neither A nor B
7. Technician A says increasing an engine's compression ratio will produce better fuel economy but less power output. Technician B says increasing an engine's compression ratio produces more power but increases fuel consumption. Who is correct?
   a. Technician A
   b. Technician B
   c. Both A and B
   d. Neither A nor B
8. Technician A says that diesel engines can produce more power because air and fuel are not mixed during the intake stroke. Technician B says that diesel engines produce more power because they use excess air to burn fuel. Who is correct?
   a. Technician A
   b. Technician B
   c. Both A and B
   d. Neither A nor B
9. Technician A says that most diesel engines used today have direct-injection combustion chambers. Technician B says that most diesel engines built today use multi-orifice spray nozzles. Who is correct?
   a. Technician A
   b. Technician B
   c. Both A and B
   d. Neither A nor B
10. Technician A says that a knocking sound in a diesel is normal and is caused by a sharp rise in cylinder pressure. Technician B says that a knocking sound in a diesel is caused by ignition delay of fuel. Who is correct?
    a. Technician A
    b. Technician B
    c. Both A and B
    d. Neither A nor B

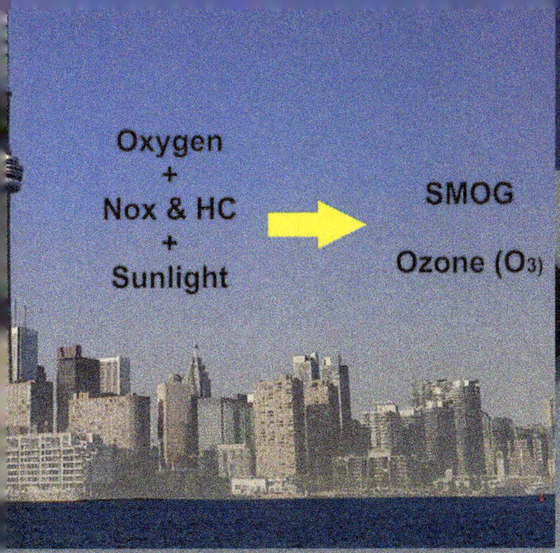

CHAPTER 3

Diesel Engine Emissions

Learning Objectives

After reading this chapter, you will be able to:

- 3-01 Identify the categories and sources of noxious emissions from diesel engines.
- 3-02 Identify and describe the gaseous emissions from diesel engines.
- 3-03 Describe the composition of PM and mechanisms for PM production from diesel engines.
- 3-04 Identify legislative standards used to regulate diesel emissions.
- 3-05 Identify and explain the purpose of onboard diagnostic systems and emission system monitoring.

You Are the Technician

As part of a new initiative, vehicle safety inspections in your shop will include a complimentary inspection of the vehicle's emission system. Since many of the vehicles serviced in your shop include diesel-powered pickups, vans, and a variety of cars, you begin to put together a list of standard emission system inspection procedures intended to guide you and other technicians who work in your shop. As you compile this list, outline answers to the following questions:

1. What is the single most reliable indicator that can be quickly checked to determine whether there is an emission system defect?
2. What are three of the most important inspection items of an OBD-II system to check as part of the inspection procedure?
3. What are two reliable indicators that a diesel particulate filter (DPF) is likely cracked or missing?

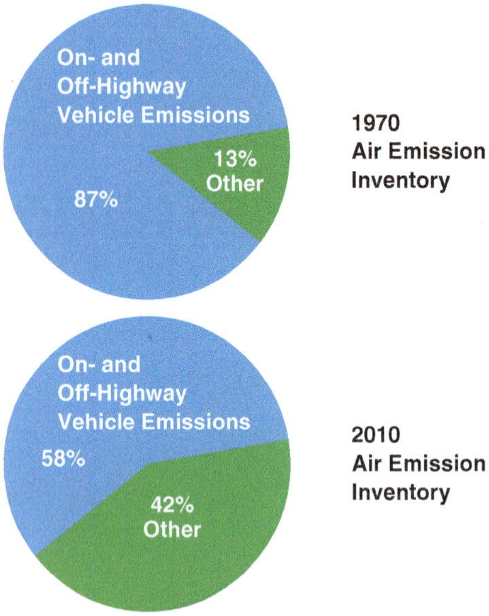

FIGURE 3-1 Technological innovations have reduced and even eliminated emissions, leading to a substantial drop in air pollution from motor vehicles.

▶ Introduction

Technological innovations for today's diesel engine are driven by emissions legislation. Legal standards reducing and eliminating the production of toxic substances from all internal combustion engines, including diesel-powered vehicles, have produced cleaner, more efficient, and more powerful engines with greater driving refinement and noise reduction than their predecessors (**FIGURE 3-1**). Although emission-related scandals have brought about an opportunity to vilify the use of diesel engines, improvements in fuel, engine design, and emissions systems technology have produced engines that achieve nearly zero emissions and maintain those almost undetectable emissions for much longer than comparable gasoline engines do. And these improvements have been in place and been achieved by many manufacturers before investigations into emission cheating were initiated. Over the past couple of decades, new technology applied to diesel engines have reduced NO_x emissions by more than 98% (**FIGURE 3-2**). Diesels in many respects are cleaner than spark-ignition, gasoline-fueled engines and are more than capable of meeting strict standards. Under intense scrutiny, new diesel models, like the Chevy Cruze, easily meet the latest Tier 3 emissions standards introduced in 2017 while achieving the highest non-hybrid highway fuel economy ratings (**FIGURE 3-3**). Tampering with emission-control systems is unfortunately a widespread problem, particularly in regions where diesel market penetration is highest and where services to defeat emission-control systems are openly advertised. Allowing emission scandals to undermine the benefits and advantages of clean diesel technology would be an unfortunate setback in meeting goals for energy conservation and cleaner air while having fun to drive, high-performance vehicles.

Because emission reductions are a crucial part of engine technology, identifying the types and categories of emissions unique to diesels, as well as the mechanisms of emission formation, enables technicians to better understand and service emission systems. Emissions standards also have implications for service practices and procedures. For example, technicians are required to maintain vehicles in compliance with emissions standards. This requires a good understanding of the categories of regulated emissions, and the mechanisms that form them. Emission legislation also prescribes many service requirements such as engine durability, service tool capabilities, onboard communication, and a vehicle's self-diagnostic function to identify potential emissions-related faults.

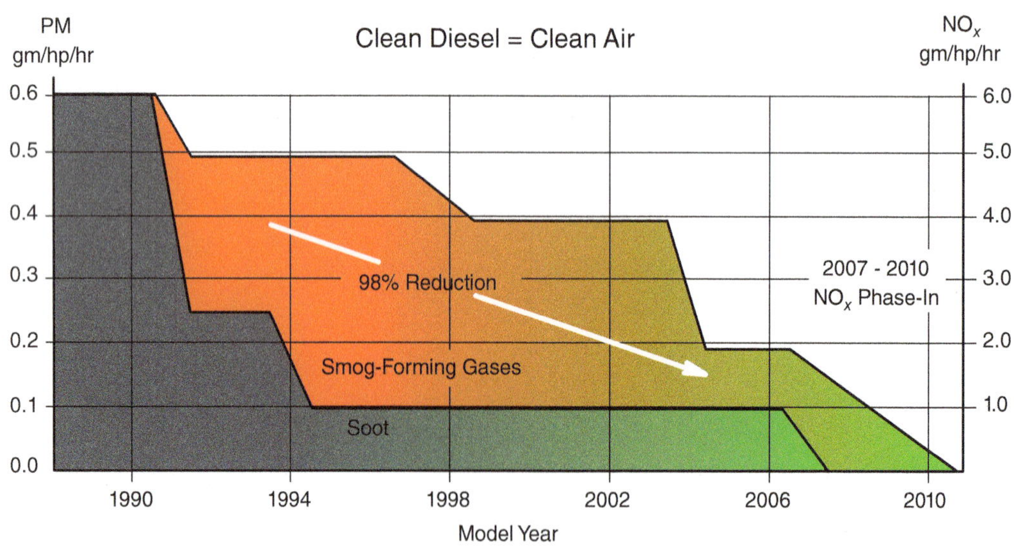

FIGURE 3-2 New technology introduced over the past couple of decades have reduced diesel engine emissions by more than 98%.

CAC - Charge Air Cooler
TC - Turbocharging
DFI - Diesel Fuel Injection
OC - Oxidation Converter
DPF - Diesel Particulate Filter
SCRC - Selective Catlyst Reduction NH3 *
RDQS - Reductant Quality Sensor
EGR - Exhaust Gas Recirculation
EGRC - Cooled Exhaust Gas Recirculation
NO_Xs - NO_X Sensor
PMS - Particulate Matter Sensor
CA OBD II California On Board Diagnostic
* SCR-U or SCRC/SCR-N or SCRC-NH3: Selective Catalytic Reduction-Urea/Ammonia

FIGURE 3-3 An emission decal for a 2017 Chevy Cruze diesel. Note the long list of emission-control technologies enabling this engine to qualify as an ultra-low-emission category engine.

▶ Classification of Emissions

3-01 Identify the categories and sources of noxious emissions from diesel engines.

Close to 2,000 types of substances are identified as engine emissions. Not all these emissions are toxic to humans or plant life. However, emissions that are hazardous to human or plant life are identified as noxious and classified into four main categories:

1. Hydrocarbons (HCs) are gases formed primarily of unburnt fuel.
2. Carbon monoxide (CO) gas is formed from partially combusted fuel.
3. Oxides of nitrogen (NO_x) are gases produced when nitrogen reacts with oxygen under high-temperature and high-pressure conditions during combustion.
4. Particulate matter (PM) refers to all forms of emissions existing as either a solid or liquid but not a gas. Black smoke, for example, is made primarily of solid carbon particles. Oil droplets from the crankcase are liquids and categorized as particulates too.

These basic categories of noxious emissions are referred to as regulated emissions, because legislation limits their production from motor vehicles and equipment (**FIGURE 3-4**).

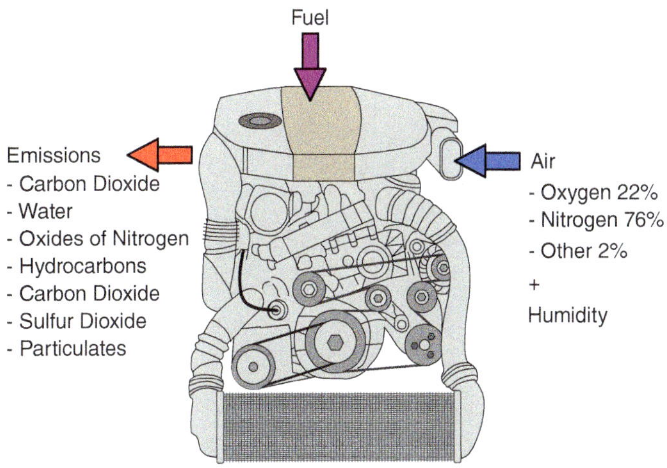

FIGURE 3-4 Emissions from diesel engines.

▶ Sources of Engine Emissions

Diesel vehicles produce emissions from two sources: the exhaust system and the crankcase (**FIGURE 3-5**). Regulatory attention until recently has been given only to exhaust emissions, since negligible emissions are produced from the evaporation of diesel fuel. Crankcase emissions from diesels started being regulated from the model year of 2007, about 40 years after gasoline engines. Regulating these emissions lagged because emissions from this area of the engine are relatively minimal in comparison to gasoline engines.

FIGURE 3-5 Sources of emissions: Diesels have no evaporative emission.

Crankcase Emissions

Crankcase emissions are composed of cylinder blowby gas and oil droplets picked up by crankcase ventilation. **Blowby** refers to the leakage of cylinder gases, past the piston rings into the crankcase area. In any engine, gas leakage past the piston rings occurs mostly during the compression stroke (**FIGURE 3-6**). This happens because piston rings cannot form a perfect gastight seal and thus permit some air to leak past. Since ring pressure against the cylinder walls is low during compression stroke, rings do not work effectively because gas pressure to help them push harder against the cylinder wall for a better seal is minimal. The opposite is true during power stroke when gas pressure behind the rings is highest. During the power stroke, little blowby takes place and almost no blowby slips past rings during exhaust and intake strokes.

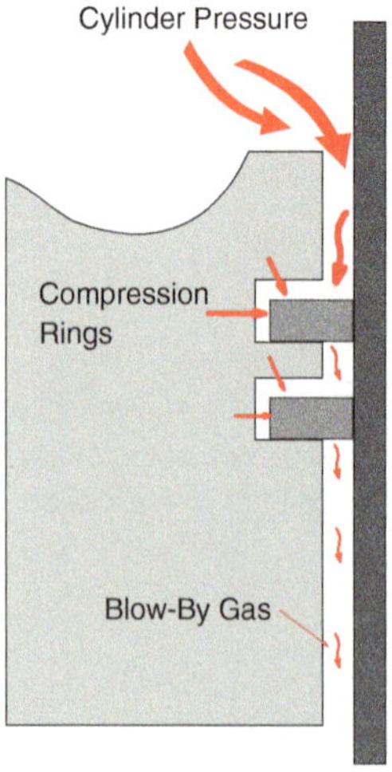

FIGURE 3-6 Blowby from diesel engines consists mostly of air leaking past the piston rings during the compression stroke. *Note*: ring pressure against the cylinder wall increases when cylinder pressure is highest.

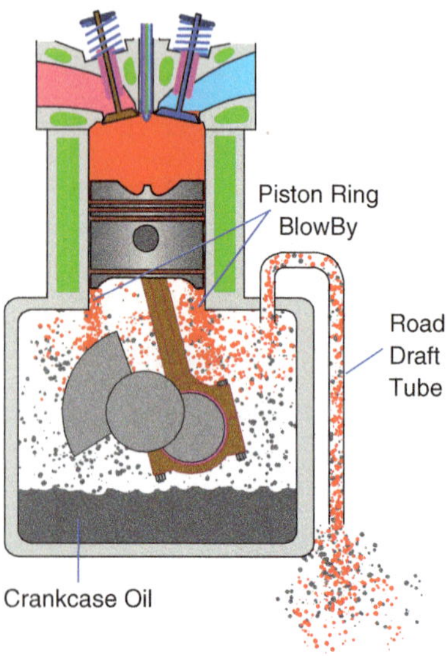

FIGURE 3-7 Blowby is classified as a particulate emission because it contains oil droplets.

Crankcase vapors are not as noxious from a diesel as they are from a spark-ignited gasoline-fueled engine, where fuel is in the cylinder during compression stroke. In contrast, only air is present in the cylinder of a diesel during compression stroke. This means most of its blowby would normally consist of air. Unless vented to the atmosphere, air pressure built up in the crankcase would cause engine seals and gaskets to leak. However, when blowby gas travels through the crankcase to the atmosphere, it picks up droplets of oil thrown from the crankshaft and piston cooling jets, forming a particulate emission (**FIGURE 3-7**). Crankcase controls for diesels involve either recirculating the crankcase vapors back into the intake air inlet or separating oil from blowby gases before it is vented to the atmosphere (**FIGURE 3-8**). Details of crankcase controls are covered in Chapter 13.

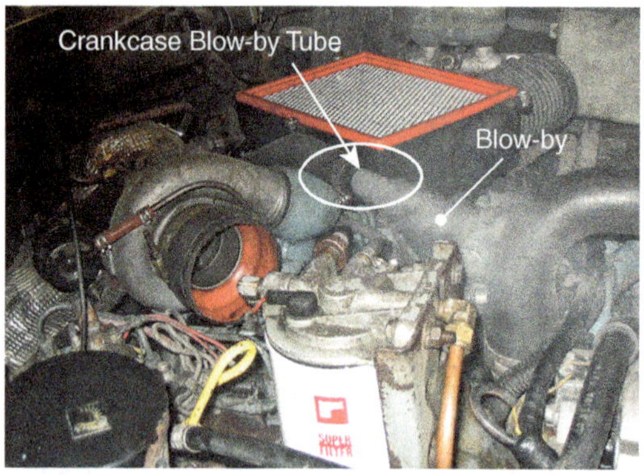

FIGURE 3-8 Droplets of vaporized oil mixed with engine blowby contribute to the particulate emission originating from the crankcase. This 7.3 L turbocharged indirect-injection (IDI) diesel engine has excessive blowby due to cylinder wear.

▶ TECHNICIAN TIP

The public perception of dirty diesel is not without foundation. Emissions legislation for diesels has lagged behind that of gasoline emissions. Emissions controls for gasoline engines were introduced in the early 1960s. However, the first diesel emissions standards for diesels were not established until 1970, when limits for peak smoke opacity of no more than 50% were first required for heavy-duty diesels. Only in 1984 were the first diesel emissions standard for NO_x introduced, followed by a PM standard in 1988. Part of the explanation for lagging emissions standards is that diesel technology was used primarily in heavy-duty buses and trucks, which make up only approximately 5% of the number of vehicle registrations (**FIGURE 3-9**). However, an estimated 50% of all emissions from transportation sources originated from diesels. The following are some explanations for the disproportionate emission contributions:

- Longer distances are traveled by heavy-duty diesels: 40% of the total accumulated trip distances are made by buses, heavy trucks, and delivery vehicles.
- Diesels carry much higher loads and travel much farther using less fuel. In 2005, 23% of fuel used for transportation was diesel fuel. Given the ratio of fuel used to distances traveled and weight carried, diesel vehicles produced fewer emissions for every ton of weight transported per mile.

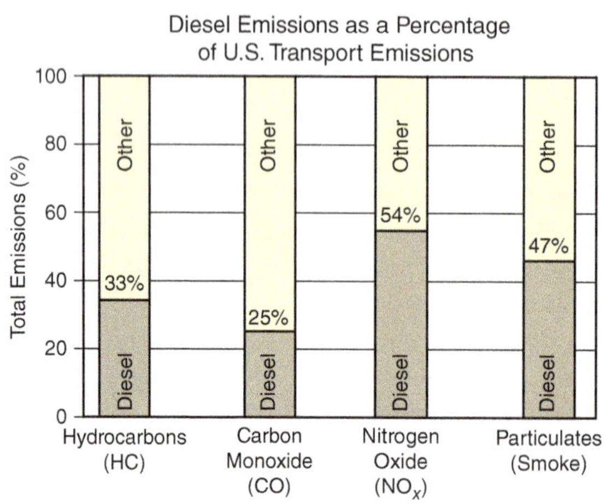

FIGURE 3-9 A 1980s estimate for the contribution of diesel emissions to total emission inventory from transportation sources.

Based on the data from U.S. Department of Energy

Gaseous Emissions Formation

In contrast to particulates that are either solid or liquid particles emitted from the crankcase or exhaust pipe, gaseous emissions are substances that can pass through the walls of catalysts and mix in the atmosphere. Some of the gaseous emissions form other noxious substances—Other gases like carbon dioxide are not noxious but is regulated becasue it is a **greenhouse gas (GHG)**. Carbon Dioxide absorbs infrared radiation from the sun, trapping and holding the heat in the atmosphere. In addition to greenhouse gases, regulated noxious gaseous emissions are broadly categorized as hydrocarbons (HCs), oxides of nitrogen (NO_x), and carbon monoxide (CO).

▶ Types of Gaseous Emissions

3-02 Identify and describe the gaseous emissions from diesel engines.

Gaseous emissions are typically identified by their chemical name. Internal combustion engines create a great many identifiable emissions, but those regulated as noxious emissions are described below.

Carbon Dioxide Emissions

Until 2009, carbon dioxide (CO_2), a by-product of normal combustion, was not a regulated emission, because it is was not considered toxic to animal or plant life. However, much attention has been paid to its role as a GHG, so much so that public interest in reducing the production of CO_2 has turned to regulating CO_2 emissions. GHGs trap heat in the atmosphere, which contributes to global warming (**FIGURE 3-10**). Beginning in January 2009, California was granted a waiver exempting it from U.S. federal emission legislation, which permitted the state to enact its own mandatory limits on carbon emissions from motor vehicles. The remainder of North America indirectly regulates CO_2 by fuel economy standards. Federal greenhouse gas legislation established by the Environmental Protection Agency (EPA) requires cars and light trucks to meet a target of 163 grams of carbon dioxide equivalent (CO_2e) per mile average across a manufacturer's fleet by the model year 2025. This standard corresponds to 54.5 mpg in gasoline-fueled vehicles (**TABLE 3-1**).

Corporate Average Fuel Economy (CAFE) standards, which are not emissions standards but fuel economy standards, were implemented in 2011. This legislation requires new passenger cars and light trucks to improve their fuel economy by 33% from the current average of 26 mpg (9.0 L/100 km) to 35 mpg (6.7 L/100 km) by 2020. Heavy-duty pickups and vans must increase fuel economy by approximately a third to 16 mpg (14.7 L/100 km) by 2027 from 12 mpg (19.6 L/100 km) in 2016. A formula calculates CO_2 emissions and fuel economy standards by using a vehicle footprint, measured in square feet. After multiplying the vehicle's wheelbase by track width to calculate the vehicle's footprint, another formula uses the data to compare the fuel economy of each vehicle sold by a manufacturer traveling the same number of miles or kilometers to score

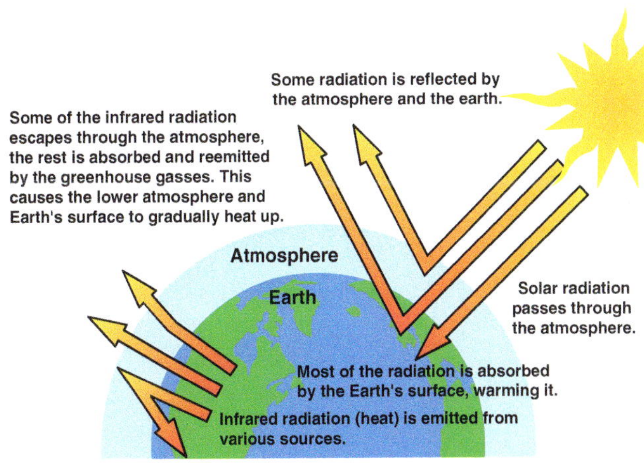

FIGURE 3-10 The cause of global warming is attributed to large amounts of CO_2 emissions produced by burning fossil fuels.

TABLE 3-1 Greenhouse Gas Emissions Standards and Café Standards

	Grams of CO_2e/mile										
Vehicle Type	2015	2016	2017	2018	2019	2020	2021	2012	2023	2024	2025
Combined Cars and Light Trucks	263	250	243	231	222	213	199	190	180	171	163
	Fuel Economy mpg (top) or L/100 km (bottom)										
Combined Cars and Light Trucks	32.6 7.2	34.3 6.9	35.1 6.7	36.1 6.5	37.1 6.3	38.3 6.1	40.3 5.8	42.3 5.6	44.3 5.3	46.5 5.1	48.7 4.8

Source: U.S. EPA

Courtesy of United States Environmental Protection Agency

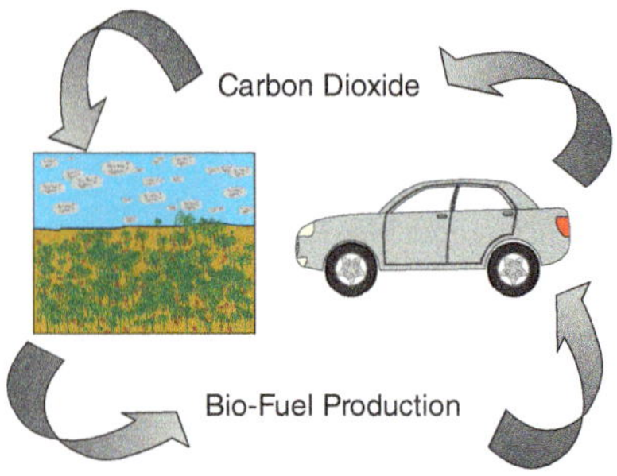

FIGURE 3-11 The use of biodiesel can reduce CO_2 emissions by more than 70%. CO_2 from the atmosphere is trapped by plants, producing oil used to make biofuel.

a corporate average fuel economy. The mandate will require a combination of various fuel economy–improving technologies, such as transmissions using more gear steps, engine friction reduction, waste heat recovery systems, low-rolling-resistance tires, axle disconnects, LED lighting, and electric- rather than engine-driven hydraulic accessories. Diesel engines are the best solution to meet fuel efficiency standards and meet reduction targets for CO_2 emissions, which from diesels are 30–40% less than an equivalently powered gasoline-fueled engine. Factoring in the use of biodiesel, a 100% renewable fuel source, the environmental footprint of the diesel engine can be even smaller (**FIGURE 3-11**). Diesel represents one of the most attractive alternatives—especially in larger-vehicle applications, where its superior fuel economy and flat fuel consumption characteristics over the entire power range further lower CO_2 emissions.

Hydrocarbon Production

If perfect combustion takes place, only harmless carbon dioxide and water are by-products (**FIGURE 3-12**). Hydrocarbon emissions are formed during imperfect combustion from unburned fuel. Fuels are classified as hydrocarbons because they are made of mostly hydrogen and carbon atoms. Hydrocarbon emissions are produced from the evaporation of fuel or from incomplete combustion. Hydrocarbons in the atmosphere will react with other engine emissions, such as nitrogen oxides, in the presence of sunlight to form photochemical smog. Exposure to hydrocarbons is also associated with human health problems, including cancer.

The following are some of the common causes of the production of HC emissions:

- inadequate combustion time
- improperly mixed air and fuel
- coarse spray droplets
- insufficient combustion heat and pressure.

Diesel engines produce comparatively little HC emissions as compared to gasoline-fueled engines. Higher compression ratios, cylinder temperatures, turbocharging, and excess air in the diesel combustion chamber ensure a greater likelihood of chemical reactions between oxygen and fuel.

While diesel combustion properties tend to allow engines to emit few to no HC emissions, fuel can hide in several areas in the cylinder, preventing it from properly burning. One major area is above the top compression ring between the piston crown and the cylinder wall. This area is called the **crevice volume** (**FIGURE 3-13**). Research has discovered close to 50% of HC emissions from diesels can originate from fuel trapped in this region, which prevents proper heating and mixing with air. To combat this, manufacturers have since moved the top compression ring closer to the piston crown and made other modifications to piston design, to minimize crevice volume clearances between the piston and the cylinder wall (**FIGURE 3-14**). For example, the use of steel rather than aluminum piston crowns reduces clearances between the piston and cylinder wall to minimize crevice volume clearances. The gap between the cylinder head and block deck where the head gasket is sandwiched is another region contributing to hydrocarbon emissions. To combat this, the fire ring of the head gasket, which seals the combustion chamber, has been also moved closer to the edge of the cylinder in modern engines.

Another potential source where HCs are formed is from fuel sprayed directly onto pistons and unheated glow plugs. This fuel does not vaporize as easily as fuel sprayed into

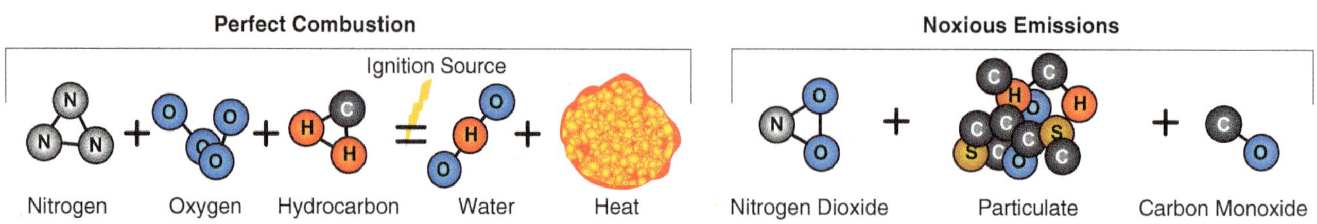

FIGURE 3-12 Perfect combustion compared to imperfect combustion. Imperfect combustion produces noxious emissions.

hot turbulent air. If incorrect injectors are used, having a spray angle mismatched with the shape of the combustion bowl, or having fuel sprayed onto an incorrectly timed movement of the piston, hydrocarbon emissions will form.

Fuel evaporation from sources such as a fuel tank can be a major source of HC emission from gasoline fuel; however, diesel fuel's low volatility means little fuel evaporation takes place from diesel fuel tanks (**FIGURE 3-15**). When diesel fuel does evaporate, it exerts very little vapor pressure, unlike more highly volatile gasoline. Low volatility and diesel fuel vapor pressure means there is no need for evaporative emissions controls on diesel engines.

Methane (CH_4) is a hydrocarbon counted in the measurement of an engine **total hydrocarbon (THC)** emissions. Methane contributes to approximately 3% of the total emissions from diesel engines, but it is not classified as a noxious emission, because it is

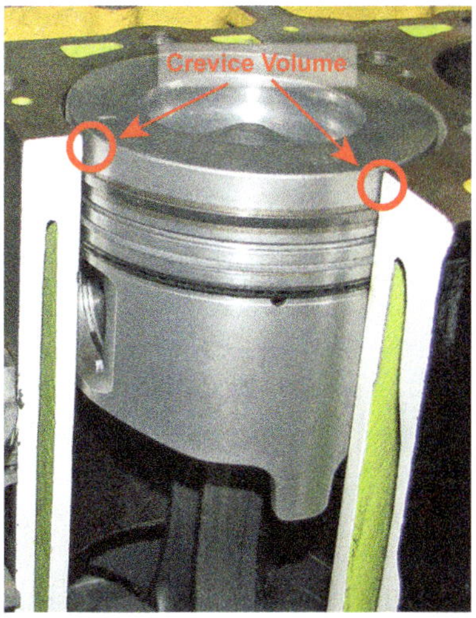

FIGURE 3-13 Crevice volume makes up the clearance between the piston crown and cylinder wall above the top piston ring.

FIGURE 3-14 Two pistons from identical engines. The piston on the left has a smaller crevice volume due to the ring position.

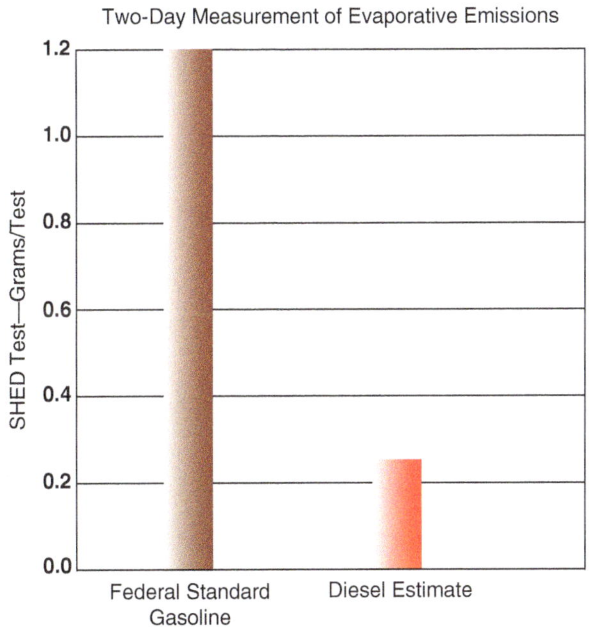

FIGURE 3-15 Comparing evaporative emissions from diesel fuel and gasoline.

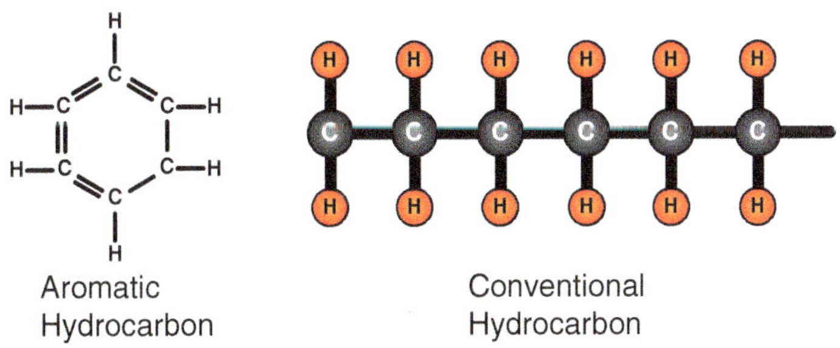

FIGURE 3-16 Aromatic fuel molecules are complex hydrocarbon molecules that are difficult to burn. In older diesels, aromatics give diesel exhaust a unique diesel smell.

▶ TECHNICIAN TIP

Difficult-to-burn fuel molecules, called aromatic molecules, cause a characteristic smell from diesel exhaust at idle and low speed (**FIGURE 3-16**). These molecules are complex hydrocarbons which sometimes are not completely combusted, especially during idle, low-speed, and load conditions, when cylinder pressure and temperatures are reduced. The aromatic content of diesel fuel is regulated by legislation to reduce hydrocarbon and CO emissions. Damaged, missing, or cold catalytic converters are common causes for the strong odor from exhaust and fuel with high aromatic content.

nonvolatile. That means it doesn't easily react with other substances to form noxious emissions and that it is considered generally harmless. However, methane is lighter than air and does have a strong **global warming potential (GWP)**.

Carbon Monoxide

Carbon monoxide (CO) is another noxious emission formed from the incomplete combustion of fuel. CO is a colorless, odorless, and tasteless gas that is poisonous in concentrations of as little as 0.3% of volume. Carbon monoxide is formed when the fuel has only partially reacted with oxygen. With adequate heat, pressure, and combustion time, CO would continue to change into CO_2. Low combustion temperatures and pressure, lack of oxygen, or insufficient burn time will typically contribute to increased CO production from an engine. Diesel engines using late or retarded injection timing as a strategy will produce more CO.

Oxidation catalytic converters are used on many late-model light and midrange diesels to clean up CO emissions. Diesel engines produce small amounts of CO in comparison to gasoline-fueled engines. Unlike gasoline engines, diesel engines do not operate near stoichiometric air-fuel ratios. Generally, diesel combustion has excess oxygen, which means a greater likelihood of contact between fuel and oxygen limits CO production. High compression ratios and excess oxygen in the diesel combustion chamber also ensure a greater likelihood of chemical reactions between oxygen and fuel.

Like HCs, very few CO emissions are produced from diesel combustion, because high compression ratios and combustion temperatures will push fuel and oxygen molecules more tightly together and cause faster reactions (**FIGURE 3-17**). Excess air factor in diesel engines also means fuel molecules are more likely to find oxygen molecules during the short combustion period (**FIGURE 3-18**).

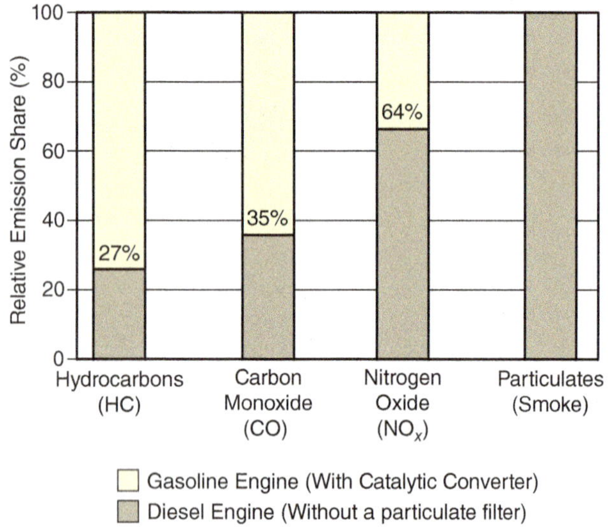

FIGURE 3-17 Comparison of hydrocarbon emissions between a late-model gasoline spark-ignited engine with a diesel engine. Excess air, higher combustion temperatures, and higher pressures ensure more complete combustion of fuel.

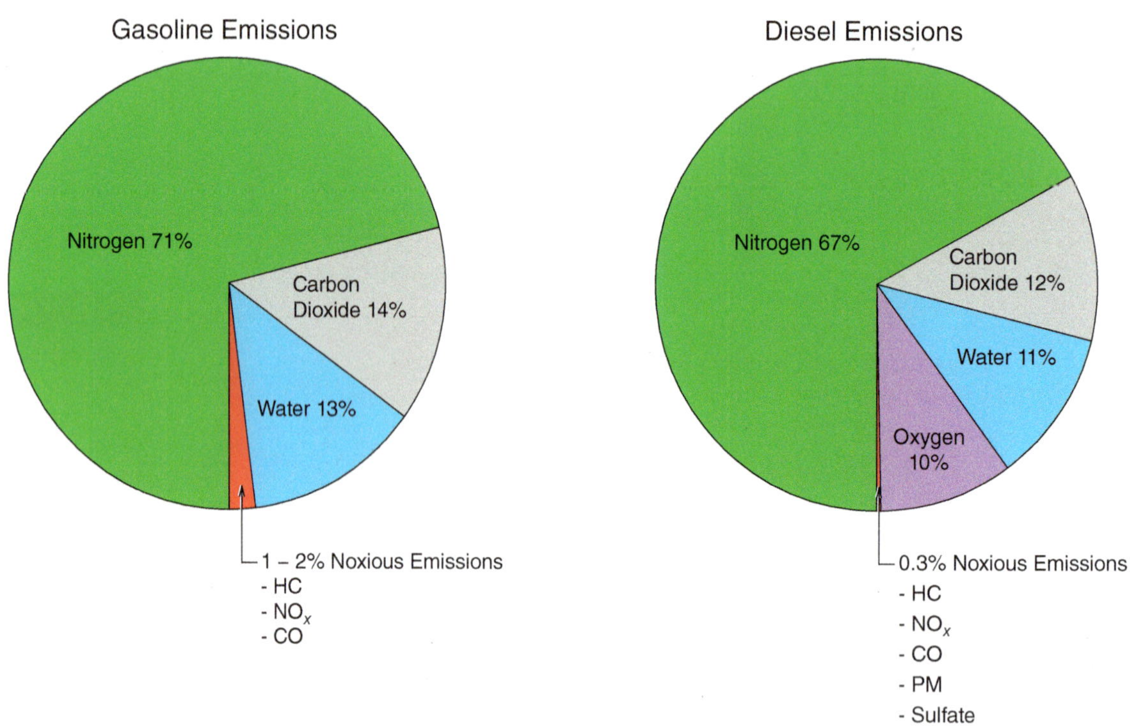

FIGURE 3-18 Even from an older engine, noxious emissions from diesel engine exhaust consists of 0.3% compared with 1–2% of the exhaust from a catalyst-equipped gasoline engine.

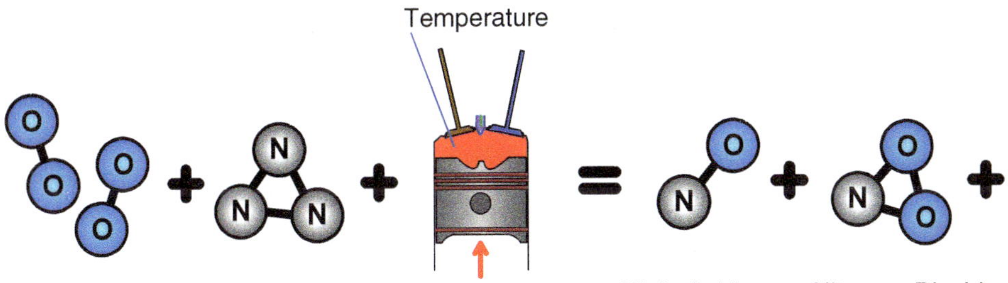

FIGURE 3-19 Nitric oxide (NO) is a clear but very unstable gas. In the presence of oxygen and water vapor, NO readily and quickly converts to NO_2, which produces the red-brown haze visible in smog.

Oxides of Nitrogen (NO_x)

NO_x is an abbreviated term for a category of emissions formed from combining nitrogen with oxygen. While nitrogen is a relatively inert gas and has many industrial applications where this property is useful (e.g., shock absorbers, welding shielding gas, air conditioning leak detection), it is an essential ingredient to the formation of smog. Without NO_x emissions from engines, photochemical smog would simply not form. On warm days in the presence of sunlight and volatile organic compounds (VOCs), which are chemically reactive molecules containing carbon, NO_x breaks down to form ground level ozone (O_3). Ozone can damage lung tissue, sting eyes, irritate the nose, and aggravate respiratory problems such as asthma and bronchitis. NO_x emissions can also form nitric acid, contributing to acid rain.

There are a number of types or species of nitrogen oxides, but two are most commonly found in engine emissions (**FIGURE 3-19**). Nitric oxide (NO) is a colorless, odorless, and tasteless gas that quickly converts to NO_2 in the presence of oxygen. NO_2 is a red-brown poisonous gas. The red-brown haze observed over roadways and cityscapes on smoggy days evidences the presence of NO_2 in the atmosphere. These gases should not be confused with nitrous oxide N_2O, which is used as an anesthetic and is otherwise known as laughing gas. Nitrous oxide (N_2O or NOS) is also used to boost engine performance when delivered to the combustion chamber through the intake manifold or fuel. N_2O is another GHG with a high GWP and is present in diesel exhaust. The presence of this gas is, however, associated more with the type of catalytic converter technology used in the aftertreatment system.

NO_x is one of the most abundant of noxious emissions from diesel engines, and cleaning up NO_x from an engine is difficult. High combustion temperatures and pressures eventually cause reactions between nitrogen and oxygen. Since almost 75% of the atmosphere contains nitrogen, its sheer volume means a great potential exists to enter combustion reactions. Unlike HCs and CO emissions, diesel engines have inherent in their design the capability to produce far more NO_x than gasoline-fueled engines. Combining high combustion pressures with combustion temperatures above 2,500°F (1,371°C) are the prerequisite conditions for NO_x formation. Increases in temperatures above this point cause NO_x production to increase exponentially. The diesel engine's higher compression ratios, cylinder pressures, temperatures, and use of turbocharging are the primary contributors to the disproportionate levels of NO_x produced from the diesel engines. In cylinder strategies to reduce the likelihood of NO_x formation run counter to increasing engine combustion efficiency because they reduce the amount of combustion force applied to the top of a piston (**FIGURE 3-20**).

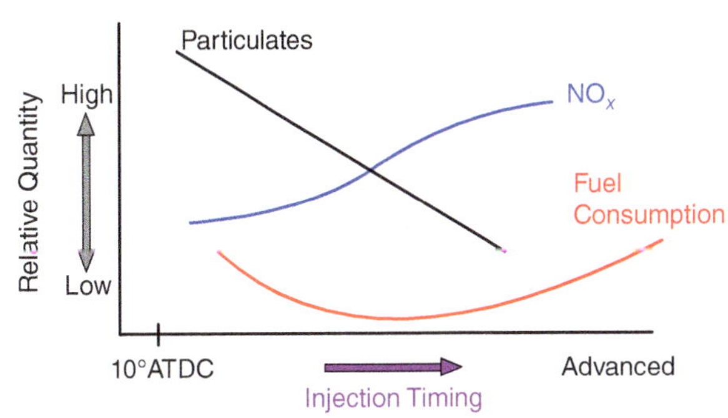

FIGURE 3-20 The NO_x particulate trade-off is a dilemma for diesels. Increasing cylinder pressure and temperature burns fuel more efficiently and with less PM emissions but produces more NO_x.

Sulfate (SO_x) Emissions

When diesel fuel is refined from petroleum extracted from the earth, it is called mineral petroleum. Varying quantities of sulfur naturally combine with mineral oil. Burning fuel containing sulfur produces sulfate (SO_x) emissions, which contribute to the formation of more particulate emissions. SO_x also produce hazardous sulfur dioxide (SO_2), emissions, and acid rain. Since the mid-1990s, sulfur content in fuel has been progressively reduced to minimize SO_x production. In 2007,

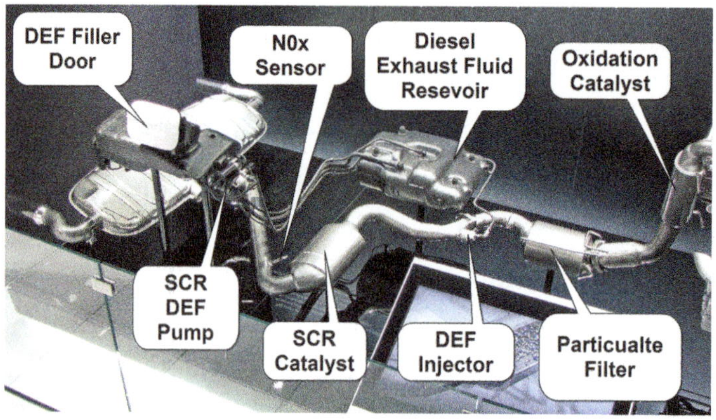

FIGURE 3-21 Aftertreatment system catalysts can be poisoned by sulfur.

sulfur content dropped from a high of 500 ppm to a low of 15 ppm, a drop from 0.5% to 0.0015% in weight. Further reductions are anticipated for fuel regulations. Sulfur removal also enables the use of sophisticated exhaust emissions in treatment systems, such as particulate filters, selective catalyst reduction systems, and NO_x absorbers (**FIGURE 3-21**). High sulfur content in fuel will poison these devices, rendering them ineffective.

▶ Particulate Matter Production

3-03 Describe the composition of PM and mechanisms for PM production from diesel engines.

Particulate, as the name suggests, refers to the combination of liquids and solid particles emitted from the exhaust pipe and crankcase of a diesel engine (**FIGURE 3-22**). Before diesel, particulate filters became mandatory, heavy amounts of particulate originating from the exhaust could sometimes be observed as soot particles emitted during engine acceleration (**FIGURE 3-23**). Black carbon is the primary constituent of this emission. Though less visible, liquids in the exhaust, such as oil droplets and sulfate compounds, contributed to particulate emissions. Oil droplets make up the bulk of particulate originating from the engine crankcase.

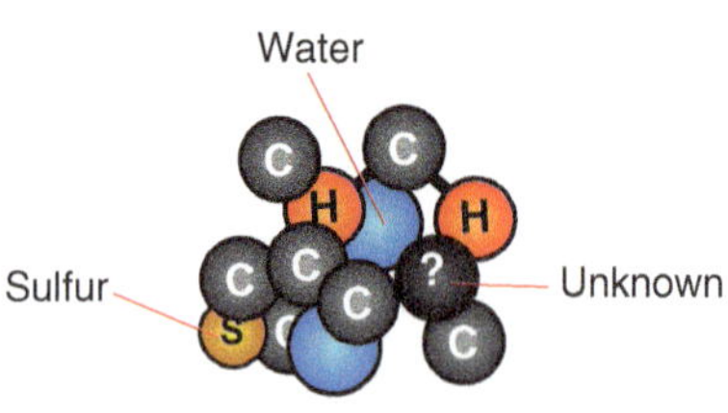

FIGURE 3-22 Composition of a typical particle of soot found in diesel exhaust. Exhaust particulate has had almost no sulfate emissions since the 2006 introduction of ultra-low sulfur.

PM is not a gaseous emission so it is regulated and classified by particle size. The four categories used are PM-50, PM-10, PM-2.5, and ultrafine PM. PM-50 is the only visible particulate. PM-10 is particulate matter less than 10 microns in diameter, PM-2.5 is particulate 2.5 microns and smaller, and so on. Although PM-2.5 is a subset of PM-10, it is categorized separately because it is considered more hazardous and requires separate accounting. Ultrafine particles, or nanoparticles, are those between 5 and 50 nanometers (nm) in diameter (**FIGURE 3-24**). While ultrafine PM accounts for the bulk of particulate in the exhaust, typically between 50% and 90% of the number of particles in diesel exhaust, it is only 1% to 20% of the total particulate mass.

FIGURE 3-23 Black sooty exhaust smoke is evidence of particulate emissions. This engine had a fuel system defect.

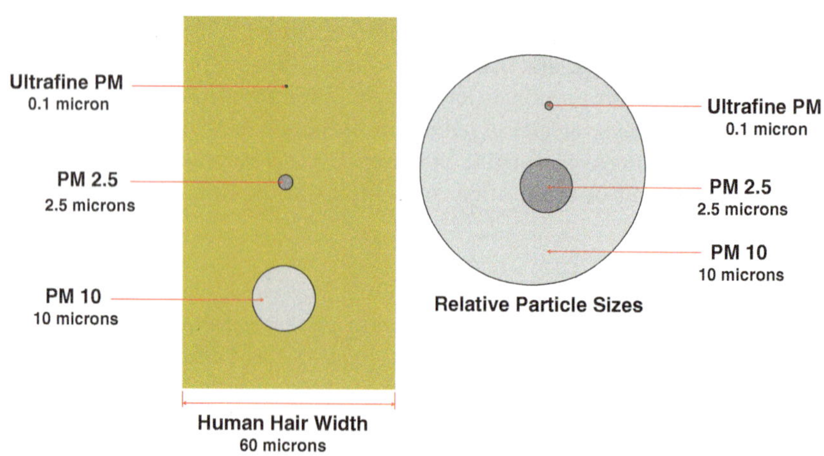

FIGURE 3-24 The size of a micron.

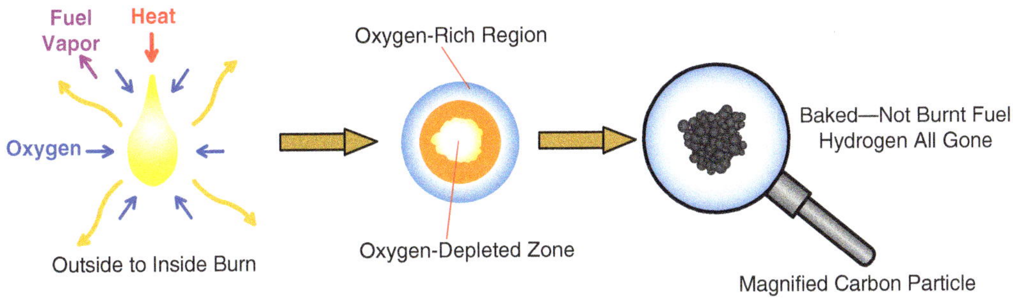

FIGURE 3-25 Black carbon particulate is formed primarily because diesel fuel burns from the outside in. The lack of oxygen surrounding the fuel droplet after air around the droplet is used up causes the droplet to burn incompletely, or simply bake rather than burn.

PM forms in the diesel combustion chamber from the tiny drops of fuel sprayed into the chamber near the end of the compression stroke. It's important to remember that these are liquids and not vapors. Vapors—not liquids—burn. Time and heat are needed to convert diesel fuel into vapor. The problem with PM formation during diesel combustion is the limited time for vaporization to take place since fuel is injected near the end of compression stroke. Late injection means fuel has limited time to vaporize and tends to burn from the outside in. Burning of fuel from the outside in, with limited time, heat, and air, will produce a small carbon nucleus, which is the primary component of particulate. In a sense, the fuel droplet is baked and not burned. Carbon rather than hydrogen tends to be left over since it reacts more slowly with oxygen due to its multiple molecular bonds (**FIGURE 3-25**).

The short amount of time to inject, vaporize, distribute, mix, and burn the fuel at the end of the diesel's compression stroke means PM production is comparatively higher from a diesel engine than in a gasoline engine. In a port-injected engine, gasoline will vaporize and mix with air quickly during the intake and compression strokes, which also permits more time for improved mixture preparation. Creating smaller droplets of diesel fuel by using finer atomization is a common strategy to reduce particulate formation. This is done by injecting fuel at higher pressure through smaller, more numerous injector nozzle spray holes (**FIGURE 3-26**). Smaller droplets catch fire faster and burn more quickly and completely, thus leaving less particulate compared to larger droplets. More complete combustion delivers better fuel economy since energy isn't wasted by sending black carbon out the exhaust pipe. Higher combustion pressures and temperatures can also be used, and they can accelerate the burn rate, which in turn causes more complete oxidation of the carbon content of the diesel fuel. Advancing injection timing that injects fuel earlier creates higher cylinder pressures near top dead center (TDC) and a longer burn time before the exhaust valve opens, ending the power stroke. The result is lower PM but higher NO_x emissions (**FIGURES 3-27** and **3-28**).

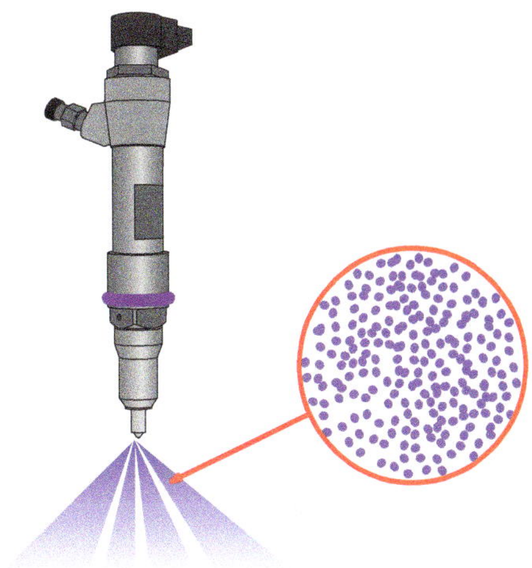

FIGURE 3-26 Producing smaller droplets by injecting at higher pressures and using smaller more numerous holes in the injector tip results in less particulate production.

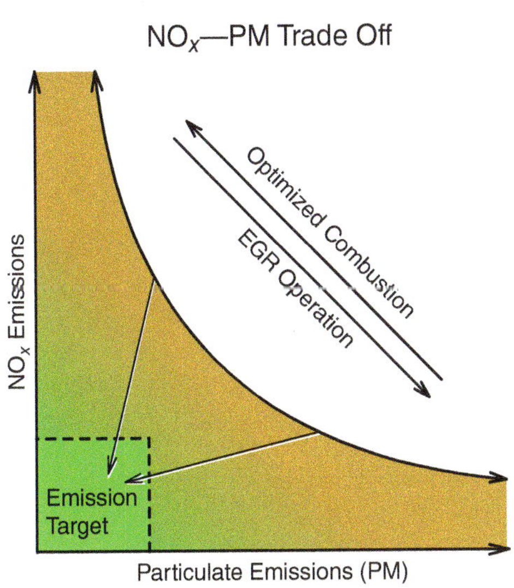

FIGURE 3-27 Higher injection pressures combined with smaller nozzle spray holes produces finer atomization and less particulate.

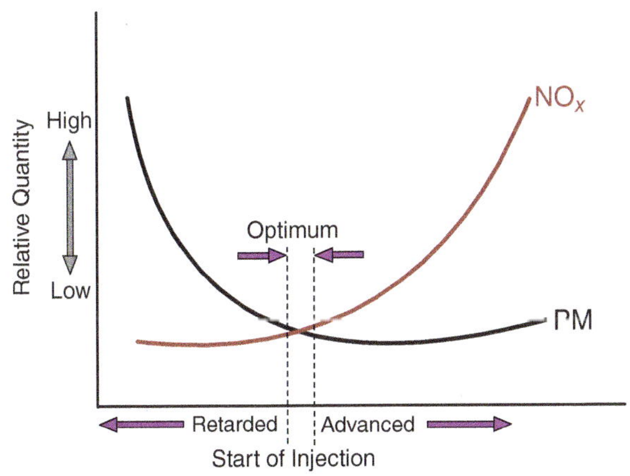

FIGURE 3-28 The beginning of injection timing is a function of finding the optimal trade-off for engine position, where both NO_x and PM production are lowest. Fuel economy is better with advanced injection timing.

Particulate and the Environment

Particulate in the atmosphere can be partly responsible for the hazy appearance of the sky on smoggy days. Airborne particulate has many sources other than diesel engines, but estimates suggest that transportation sources make up nearly a third of that total. The small size of these particles is what makes diesel particulate hazardous. The largest proportion of the exhaust, ultrafine PM and PM-2.5, is not even visible to the eye since it is microscopic in size. Particles less than 10 microns are easily respirable and can penetrate into the deepest parts of the lung. These small particles within the lungs aggravate respiratory conditions such as asthma and bronchitis because they require an immune system response to clear and are not simply coughed up. Particulate is also believed to carry other toxic, combustion-formed compounds with it into the lungs and bloodstream. Compounds found in soot, such as polycyclic aromatic hydrocarbons (PAHs), are carcinogenic; in fact, diesel soot itself is classified by many government agencies as either a probable or known cancer-causing agent. The California Air Resources Board has concluded that diesel soot is responsible for 70% of the state's risk of cancer from airborne toxics.

To help diesel engines achieve clean emissions, wall flow particulate filters have been mandatory since 2007 in North America. Particulate filters allow the gas portion of exhaust to pass through the wall of an exhaust system catalyst, removing between 85% and 100% of exhaust soot, enabling particulate emissions to decrease to 0.0016 grams/mile (0.001 g/km) or less (**FIGURE 3-29**).

A crack or broken particulate filter will allow soot to leak past the catalyst and into the exhaust. Exhaust pipe tips will have a black sooty appearance if this happens. To inspect the particulate filter, follow the steps in **SKILL DRILL 3-1**.

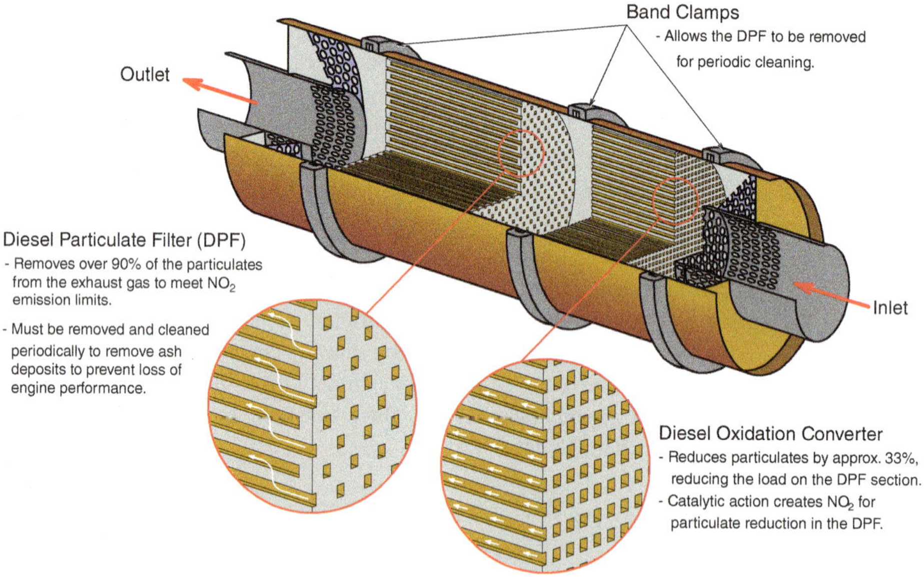

FIGURE 3-29 DPFs remove soot and other solid particles from the exhaust gas stream.

SKILL DRILL 3-1 Inspecting the Particulate Filter

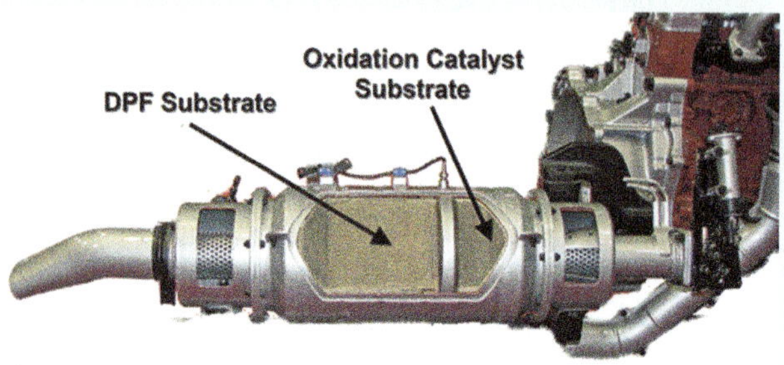

1. With the engine off, visually examine the exhaust pipe outlet for accumulations of soot. If heavy soot is present, the filter is likely cracked or missing.
2. Wipe the exhaust pipe outlet with a tissue or light-colored cloth. No soot should be detected. Rust or gray metal oxidation is acceptable.
3. If the particulate filter is suspected of being missing, check for malfunction indicator lamp (MIL) codes for missing or cracked catalyst substrate. Also check the PM monitor to see if it has successfully run and its status is ready.

PM Emissions and Engine Speed

Maximum engine speed is often part of the emissions information on diesel engine emission decals for a good reason (**FIGURE 3-30**). Also note that tachometers in diesel-powered vehicles have a lower rpm limit than gasoline-fueled engines have. Diesel engines do not rotate as fast as gasoline engines, not because they cannot, but because they run out of time to properly combust fuel at high speeds. In other words, the maximum rpm limit for diesel engine speed corresponds to the point where excessive emissions begin to be produced due to the inability of the combustion chamber to completely burn fuel. To understand this important point, consider first that fuel is injected near, or even after, the piston reaches TDC. Before fuel burns, it must penetrate highly dense air in the chamber. Next it must atomize, vaporize, and then mix with air in correct proportions before fuel can ignite and burn. All these steps take time. At high speeds, the available combustion time may be as little as 0.003 seconds. Furthermore, maximum combustion pressure should occur no later than 10°–15° after TDC. Given diesel fuel's low volatility, it's not hard to imagine why diesel engines can run out of time to efficiently complete all the steps during a single combustion event. The graph in **FIGURE 3-31** illustrates an engine operated at three different rpm but with identical loads. Note that when the engine operates at a higher rpm, darker exhaust smoke appears sooner.

FIGURE 3-30 The rpm limit for this V8 diesel is much lower than its gasoline spark-ignited equivalent.

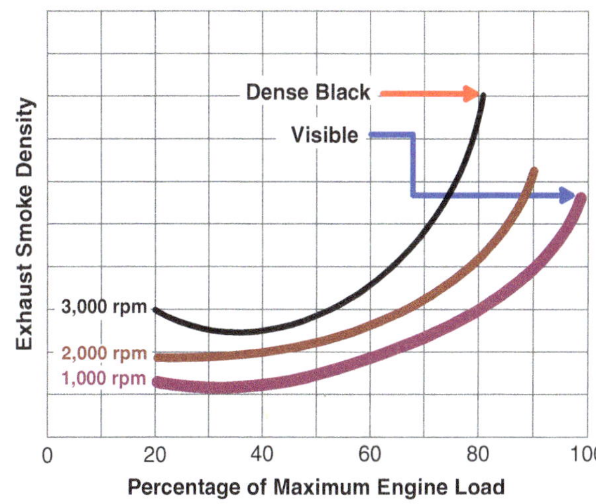

FIGURE 3-31 Excessive emissions are produced at high speed and load because diesel combustion events run out of time to complete. Limits on rpm are imposed on diesel engines to limit emissions formed when inadequate time is available to completely burn fuel.

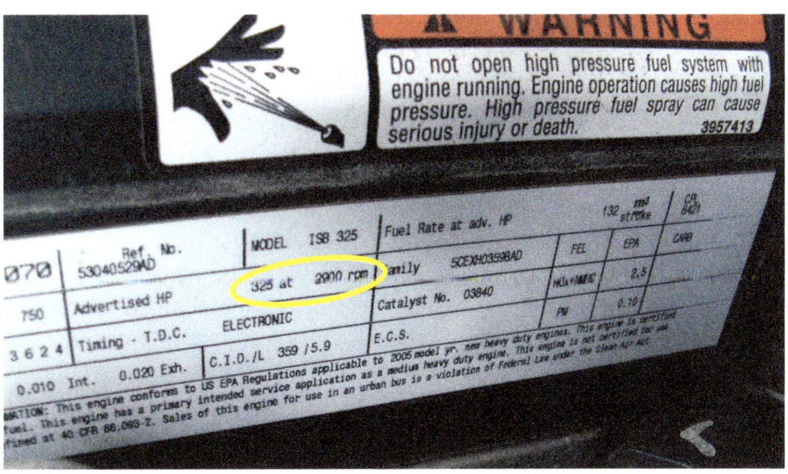

FIGURE 3-32 Maximum engine speed is often listed on engine emission data plates. Excess speeds will produce particulate and other emission increases.

▶ TECHNICIAN TIP

Adjusting engine governors beyond maximum the limits prescribed by the manufacturer and listed on the emission data plate will cause excess engine emissions (**FIGURE 3-32**). Engine speed limits are established not to prevent mechanical damage from over speed but because combustion chambers have smoke limits. Diesels will simply run out of time to completely burn fuel-efficiently if engine speed is excessive. On vehicles and machines without a DPF, excess smoke will appear at the exhaust pipe outlet.

▶ Emissions Standards

3-04 Identify legislative standards used to regulate diesel emissions.

Around the world, concerns about the negative environmental and health impacts of vehicle emissions have produced legislated standards for noxious emissions. The European Union (EU) and the United States have developed two emissions standards adopted by most other countries in the world (**FIGURE 3-33**). The greatest number of countries use emission legislation passed by the European Parliament, which was phased in between 1993 and 2014. In the EU, there are six progressive stages for reducing emissions beginning with EURO 1 until the most current EURO 6 from 2014.

North American Emissions Standards

Two sets of emission regulations are used in the United States—a federal standard and a state (California) standard. Federal standards are set by the U.S. EPA; California standards are set by the California Air Resources Board (CARB). California standards are stricter, due in part to high population density and geographical factors which contribute to poorer air quality. CARB, which developed the standards for the onboard diagnostic system, are much more technically detailed in contrast to the EPA's, which are more performance orientated. Other states have adopted California's emissions standards, but California is the only state permitted by federal law to enact stricter vehicle emissions standards than those set by the federal government. Other states can then choose California's standards, but cannot be the first to surpass those set by the U.S. federal government. The EPA adapts CARB standards, but generally, they not aligned at the same time, with the EPA lagging CARB by as much as several years. For example, the latest EPA standard, called Tier 3, beginning in 2017 for light- and medium-duty vehicles was introduced in California in 2012 as Low-Emission Vehicle III (LEV-III). Most car and truck makers will certify their vehicles to CARB standards to permit sales in all 50 U.S. states. Canada has harmonized it emissions regulations with EPA Federal standards. Mexico and Peru are the other two countries. Separate U.S. EPA standards are established for light-medium duty emissions, and heavy-duty vehicles according to vehicle weight. Below a GVW of 14,000 lb (6,350 kg), a vehicle is categorized as light or medium duty. Vehicles 14,001 lb (6,350 kg) GVW and over are required to meet heavy-duty standards.

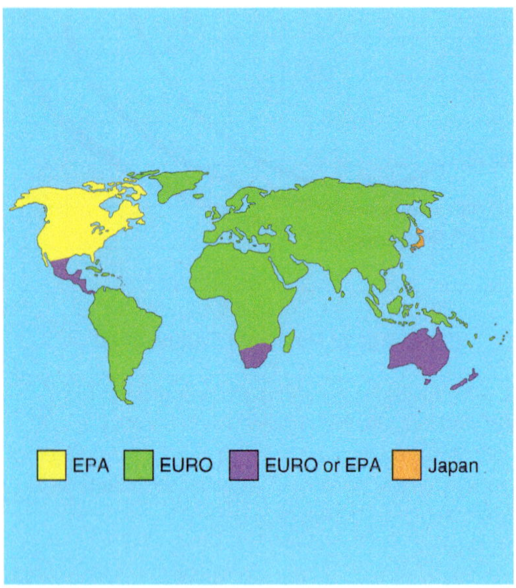

FIGURE 3-33 Either EPA or EURO standards are used by most countries around the world.

Tier 2 Standards

EPA Tier 2 emissions standards used by on-road vehicles up to 14,000 lb gross vehicle weight (GVW) were phased in between 2004 and 2010 (**TABLE 3-2**). They continued to be in use until the 2016 model year. A complex set of regulations permits phasing in increasingly tighter standards from 2004 to 2010. This accounts for many of the discrepancies between emissions equipment found on identical models with different GVW ratings. Since pickup trucks are now used like cars, all passenger vehicles are regulated to the same standards. That means vehicles fueled by gasoline, diesel, or alternative fuels; cars; minivans; light-duty trucks; and sport utility vehicles (SUVs) will all have close to the same maximum emissions limits. When fully implemented in 2010, the average NO_x emissions of the entire light-duty vehicle fleet sold by each manufacturer met the average NO_x standard of 0.07 g/mile (0.04 g/km). This was a major reduction from the Tier 1 standard for diesels of 1.0 g/mile. The T2B5 federal standard also corresponded to the CARB California Low-Emission Vehicle Standard II (CA LEV-II).

Tier II standards categorize vehicle emission levels into "bins." Initially, 11 bins were used, with bin 1 being the cleanest—a standard met only by zero-emission vehicles—and 11 the dirtiest. Manufacturers can produce vehicles that fit into any of the available bins, but they still must meet average emission targets for their entire fleets. Since Tier 2 bin 5 (T2B5) has an NO_x limit of 0.07 g/mile (0.04 g/km), which is equal to the fleet average NO_x standard, most manufacturers certified their vehicles to this

TABLE 3-2 U.S. EPA Tier 2 and CA LEV-II Standards (g/mile)

	Durability 50,000 Miles or 5 Years				Durability 120,000 Miles or 11 Years			
Emission Category	NO_x	PM	CO	HCs	NO_x	PM	CO	HCs
EPA Tier 2 Bin 5	0.05	0.01	3.4	0.015	0.07	0.01	4.2	0.018
CA—LEV	0.05	0.01	3.4	0.015	0.07	0.01	4.2	0.018
CA—ULEV	0.05	0.01	1.7	0.008	0.07	0.01	2.1	0.011

Source: U.S. EPA website
Durability requirements and emission standards

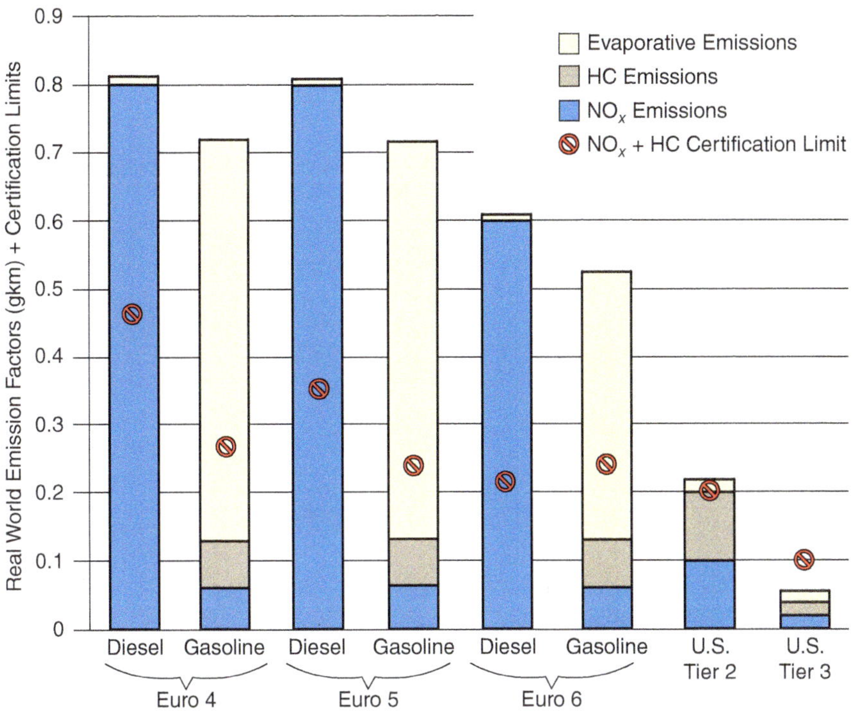

FIGURE 3-34 U.S. Tier 3 standards implemented beginning in 2017 are the toughest in the world. No difference between emissions standards for gasoline and diesel engines are permitted.

level (**FIGURE 3-34**). The T2B5 federal standards exceed EURO standards but correspond to the CARB California Low-Emission Vehicle Standard II (CA LEV-II) (**FIGURE 3-35**). An emission system durability requirement was also introduced under Tier 2 rules. A useful life period for vehicle emission systems required vehicles to maintain compliance with the emissions standards for a minimum of 120,000 miles (193,000 km) or 10 years of operation. The useful life period is also referred to as the durability requirement.

Because light- and medium-duty vehicles are certified by using a chassis dynamometer, and emissions standards are set according to body type, emissions standards are expressed in mass of emissions per mile or kilometer (**FIGURE 3-36**). Heavy-duty diesels, which can be used in anything from a road grader to highway tractor, are certified on an engine dynamometer, and emissions are measured according to grams produced per brake horsepower per hour or grams per kilowatt per hour.

One other major difference between light and heavy duty emission standards is the not to exceed emission (NTE) limit that only applies to HD diesels and not to light duty vehicles. Given the recent scrutiny over diesel engine emission testing, it is important to note LD emission limits for any light-duty vehicle are exceeded only when the average measurement of emissions collected over a three part test cycle exceeds CARB or US-EPA threshold limits. During the most commonly used emission test cycle, vehicles will simulate driving 11.04 miles (17.77 km) in mostly urban driving conditions at an average speed of 21.2 mph (34.12 km/h) for close to 31 minutes. Each of the test phases for emission collection lasts about 8 or 14 minutes. Emissions may temporarily exceed the threshold limits during the test, but the weighted average of emission mass measured during each phase of the test cannot exceed the threshold limits. This means there are potentially many permissible circumstances when a vehicle can exceed the emission limits such as when aggressively accelerating or during the time it takes for an emission monitor to operate. In contrast, HD emissions are averaged over a period as short as thirty seconds and then compared with a calculated NTE emission limits value. This NTE standard is in place to ensure emissions from HD diesels are controlled over all reasonably expected engine load and speed ranges.

Tier 3 Standards

The Tier 3 standards introduced in 2017 closely align with California LEV-III standards (**TABLE 3-3**). Maximum emission thresholds are being phased in between 2017 and 2025 (**TABLE 3-4**). Like Tier 2 standards, bins are used to group vehicles according to maximum

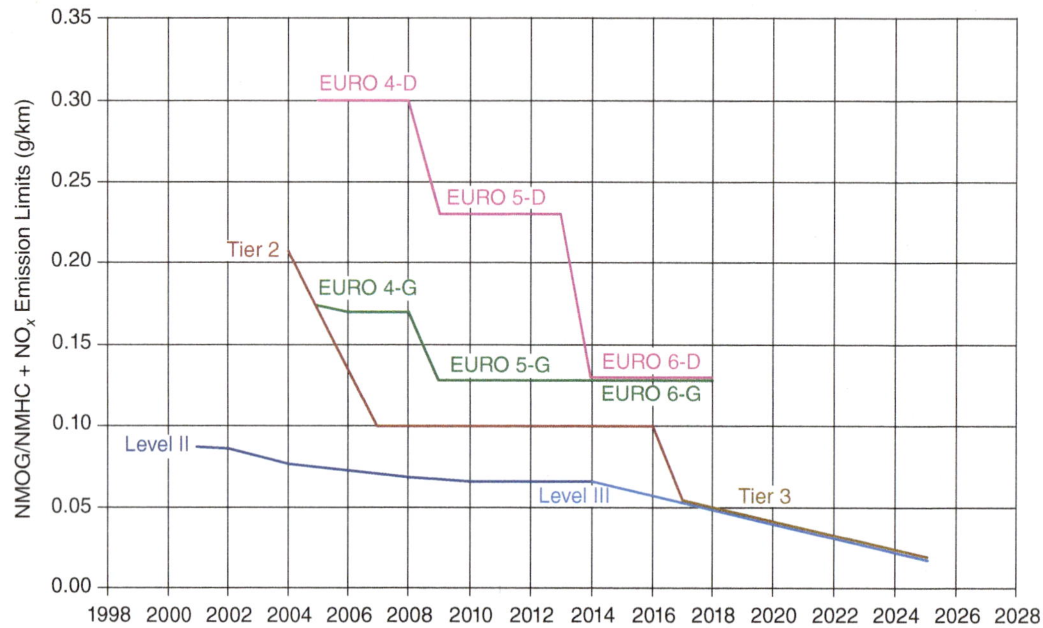

FIGURE 3-35 Tier 2 bin 5 standards have lower cut points for NO_x emissions as compared to EURO standards.

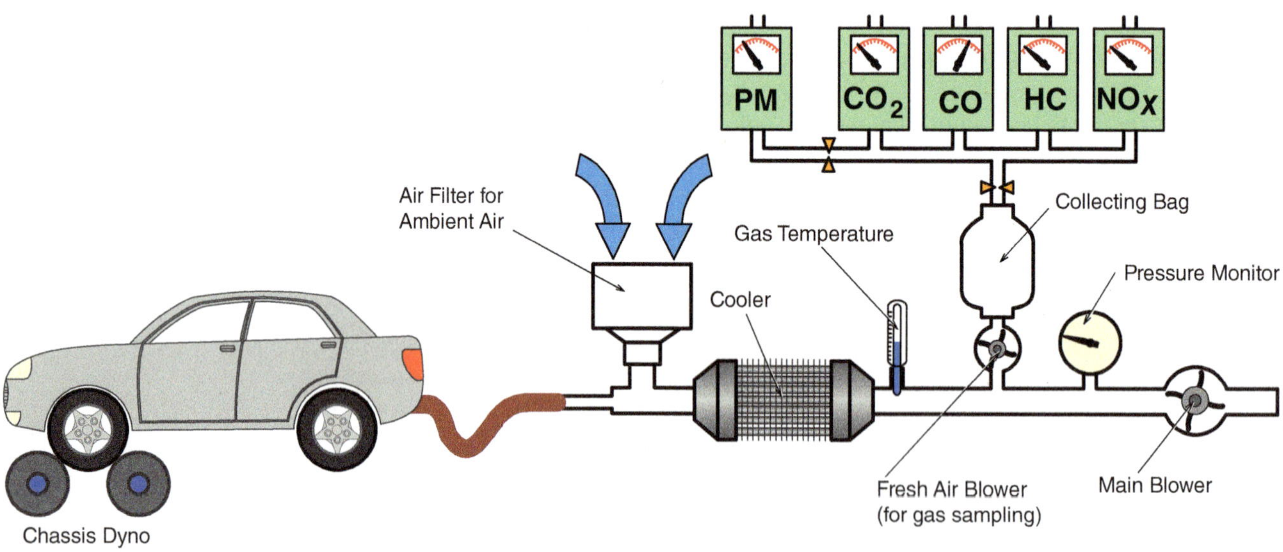

FIGURE 3-36 Vehicles up to 1,000 lb (6,350 kg) GVW are emission certified on a chassis dynamometer.

TABLE 3-3 Tier 3 Certification Bin Standards (FTP; 150,000 miles (241,400 km))

Bin	NMOG + NO_x		CO		PM*	
	mg/mile	mg/km	g/mile	mg/km	mg/mile	mg/km
Bin 160	160	0.99	4.2	0.0260	3	0.019
Bin 125	125	0.78	2.1	0.013	3	0.019
Bin 70	70	0.44	1.7	0.010	3	0.019
Bin 50	50	0.31	1.7	0.010	3	0.019
Bin 30	30	0.19	1.0	0.006	3	0.019
Bin 20	20	0.12	1.0	0.006	3	0.019
Bin 0	0	0	0		0	0

Source: U.S. EPA website

TABLE 3-4 California LEV-III/EPA Tier 3 150,000-Mile (241,400 km) Exhaust Emission Limits

Vehicle Classes: Passenger Cars (PC) Light-Duty Trucks (LDT) Medium-Duty Passenger Vehicles (MDPV)	Certification Level	$NMOG + NO_x$ (mg/mi)	CO (g/mi)	PM mg/mi
PC and LDT up to 10,000 lb (4,545 kg) GVWR	* LEV 160/Bin 160 g/mile	160 g/mile	4.2 g/mile	3 g/mile
MDPVs up to 8,500 lb (4,545 kg) GVWR for PCs, LDTs	0.99 mg/km	0.99 mg/km	0.026 mg/km	0.019 mg/km
	**ULEV 125/Bin 125	125 0.78 mg/km	2.1 0.013 mg/km	3 0.019 mg/km
	ULEV 70/Bin 70	70 0.44 mg/km	1.7 0.010 mg/km	3 0.019 mg/km
	ULEV 50/Bin 50	50 0.31 mg/km	1.7 0.010 mg/km	3 0.019 mg/km
	***SULEV 30/Bin 30	30 0.19 mg/km	1.0 0.006 mg/km	3 0.019 mg/km
	SULEV 20/Bin 20	20 0.12 mg/km	1.0 0.006 mg/km	3 0.019 mg/km
MDVs (8,501–10,000 lb (3,864–4,545 kg) GVWR)	LEV 395/Bin 395	395 0.254	6.4 0.04 mg/km	10 0.062 mg/km
	ULEV 340/Bin 340	340 0.211 mg/km	6.4 0.04 mg/km	6 0.048 mg/km
	ULEV 250/Bin 250	250 155 mg/km	6.4 0.04 mg/km	6 0.048 mg/km
	ULEV 200/Bin 200	200 0.124 mg/km	4.2 0.026 mg/km	6 0.048 mg/km
	SULEV 170/Bin 170	170 105 mg/km	4.2 0.026 mg/km	6 0.048 mg/km
	SULEV 150/Bin 150	150 0.093 mg/km	3.2 0.020 mg/km	6 0.048 mg/km
MDVs (10,001–14,000 lb (4,545–6,364 kg) GVWR)	LEV 630/Bin 630	630 0.391 mg/km	7.3 0.045 mg/km	10 0.062 mg/km
	ULEV 570/Bin 570	570 0.354 mg/km	7.3 0.045 mg/km	6 0.048 mg/km
	ULEV 400/Bin 400	400 0.248 mg/km	7.3 0.045 mg/km	6 0.048 mg/km
	ULEV 270/Bin 270	270 0.167 mg/km	4.2 0.026 mg/km	6 0.048 mg/km
	SULEV 230/Bin 230	230 0.142 mg/km	4.2 0.026 mg/km	6 0.048 mg/km
	SULEV 200/Bin 200	200 0.124 mg/km	3.7 0.023 mg/km	6 0.048 mg/km

* LEV: Low-Emission Vehicle

** ULEV: Ultra-Low-Emission Vehicle

*** SULEV: Super-Low-Emission Vehicle

Source: U.S. EPA website

emission cut points, where 0 is the cleanest. Significant changes include combining maximum NO_x and hydrocarbon emissions together to form a new category called $NMOG + NO_x$. **Non-methane organic gases (NMOG)** plus NO_x have a single combined standard rather than two separate standards as Tier 2 (**TABLE 3-5**). Tier 3 bin 19 are equivalent to Tier

TABLE 3-5 Tier 3 Comparison

Tier 3 Bins (FTP)	NMOG+ NO_x (mg/km)	PM[a] (mg/km)	CO (mg/km)	HCHO (mg/km)
Tier 2 Bin5	56 + 44 = 100	6	2,610	11
Bin 160	100	2	2,610	2
Bin 125	78	2	1,305	2
Bin 70	44	2	1,057	2
Bin 50	31	2	1,057	2
Bin 30	19	2	622	2
Bin 20	12	2	622	2
Bin 0	0	0	0	0

Source: U.S. EPA

2 bin 2. Tier 3 bin 160, the dirtiest category, is equal to Tier 2 bin 5. Durability requirements of the emission system have been extended from T2B5 120,000 miles to 150,000 miles (193,000 km to 241,400 km). During this period, the vehicle emissions must not exceed the maximum thresholds as the vehicle ages and accumulates mileage. Longer-lasting components are required, and the ability to compensate for system deterioration must be built in to meet durability requirements. A new high-altitude standard allows manufacturers some latitude in maintaining emissions below maximum limits when air density is low. This provision has been granted since catalytic converters have a more difficult time warming up and staying hot when air pressure is lower.

Off-Highway Emissions Standards

Standards for off-road equipment are different from those for on-highway equipment (**FIGURE 3-37**). Not to be confused with on-highway references to tiers, these standards are referred to as Tier 1, 2, 3, and 4 emissions standards and are harmonized worldwide with European standards. Tier 1, established in 1994 and phased in between 1996 and 2000, represents the earliest phase in increasingly stricter emission limits, with Tier 4 being the final, cleanest standard of the four phases. These standards are set according to horsepower (kilowatt) ratings, with larger horsepower (kilowatt) engines taking longer to reduce emissions. Off-road diesels have emissions standards applied depending on the power and year in which the engine is produced. Emissions standards lag on-highway engines but approach

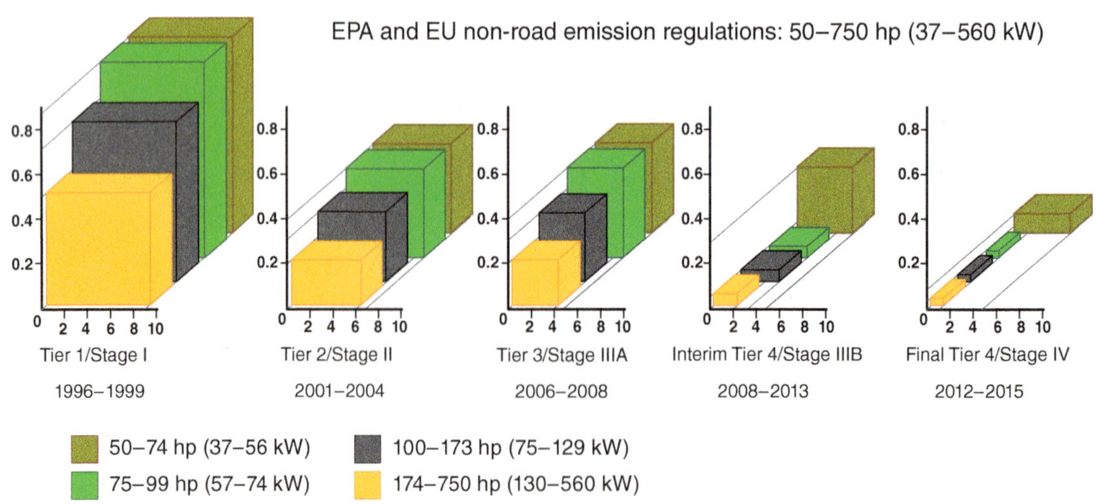

FIGURE 3-37 Off-highway emissions standards.

the same levels by the 2015–2020 phase-in period. Tier 1 standards remained in effect until the Tier 2 and/or Tier 3 standards took effect.

Emissions Decals

Vehicles sold for on-highway use in North America must have an emissions decal affixed to the vehicle or engine (**FIGURE 3-38**). Before an engine or vehicle can be sold, it must pass standardized emissions test by the EPA for the particular engine family. The decal contains emissions information indicating to what standard the vehicle or engine is certified and important information for the technician about part replacement and adjustment procedures, software calibration files, and other relevant data that could affect emissions production. For example, on a diesel engine, the decal will contain information about valve lash settings, injection timing, fuel rates, and a list of emissions system components the vehicle was built with (**FIGURE 3-39**). Since heavy-duty engines emissions are certified in grams per brake horsepower per hour (g/kWh) on an engine dynamometer and not a chassis dynamometer, emission decals are affixed to the engine. Because light-duty vehicle emissions are certified by vehicle type and measured in mass units for every mile or kilometer traveled, mandatory emission decals are placed in the vicinity of the engine compartment.

Fuel Standards

Minimizing the production of engine out emissions rather than cleaning the exhaust was the traditional emission reduction strategy used until 2007, when diesel particulate filters (DPFs) were introduced for all on-highway diesels. Until then, the problem of diesel exhaust's relatively cool temperatures and oxygen-rich output prevented the use of conventional catalytic converter technology. To meet the 90% reduction in NO_x output for 2007 T2B5 engines, several selective catalyst reduction (SCR) systems were introduced. Before then, a few light- and medium-duty diesels used low-efficiency oxidation catalysts to reduce some PM, HCs, and CO, but they were primarily intended to minimize exhaust odors caused by aromatic content in fuel. Because sulfur content in fuels could poison the most catalytic converters, ultra-low sulfur fuel was introduced in 2006 (**FIGURE 3-40**). Prior to that, in 1993, low-sulfur fuel reduced the typical 5–7% content down to less than 1%. Reducing sulfur content also helped reduce particulates and minimize both corrosion

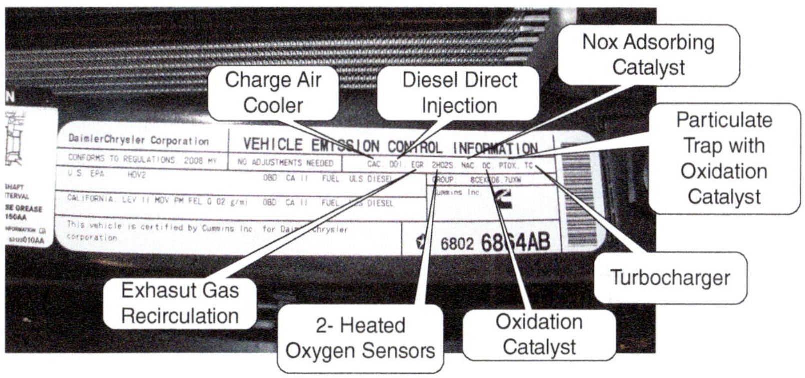

FIGURE 3-38 Light-duty vehicle emissions decals are located on the chassis. Heavy-duty engine emission decals are located on the engine.

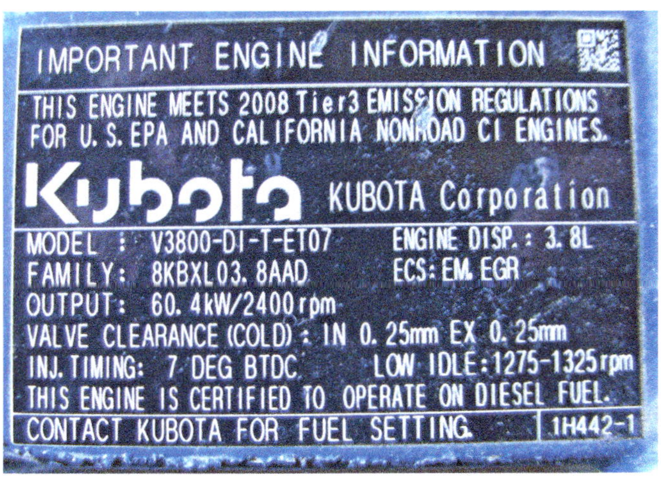

FIGURE 3-39 Note the off-road engine decal. Emission decals contain important service information.

FIGURE 3-40 Ultra-low-sulfur diesel is important technology for exhaust aftertreatment systems.

and deposit formation in fuel system components. Maximum aromatic content was limited simultaneously in 1993. European manufacturers reduced sulfur content much earlier and introduced SCR systems before using DPFs. SCR systems optimized engine power output while reducing particulate emissions close to North American standards without the fuel penalty: especially important in the European market, which is sensitive to higher fuel prices.

▶ Emission-Control and Emission-Monitoring Systems

3-05 Identify and explain the purpose of onboard diagnostic systems and emission system monitoring.

Emissions standards require that manufacturers both limit emissions and monitor the engine to detect conditions that could potentially cause emission increases 1.5 times the maximum of certification levels. This system, called On-Board Diagnostic II, is required for diesel engines, but manufacturers are relatively free to determine how they will control emissions to meet standards effectively, as long as they do meet the standards. This section will detail several common emission-monitoring and control systems.

Confirming emission system components are present, connected, and operational are important inspection routines for emission system inspection and maintenance programs. To pass an emission system inspection, no fault codes can illuminate the MIL light and all emission system monitors must all be ready. To inspect emission systems, follow the steps in **SKILL DRILL 3-2**.

Diesel Engine Emission Controls

Diesel emissions can be limited and even eliminated by using a combination of two basic strategies. The first is in cylinder control method. These systems prevent the formation of emissions during combustion conditions inside the cylinder. Temperature, pressure, gas composition, and fuel injection variables are manipulated to provide maximum combustion efficiency while minimizing emission formation. The second strategy is to clean up emissions in the exhaust system after they are formed. Exhaust aftertreatment enables more efficient engine operation at the cost of more complexity through the use of expensive exhaust system catalysts. Common emissions control systems will be discussed in detail in

SKILL DRILL 3-2 Inspecting Emission Systems

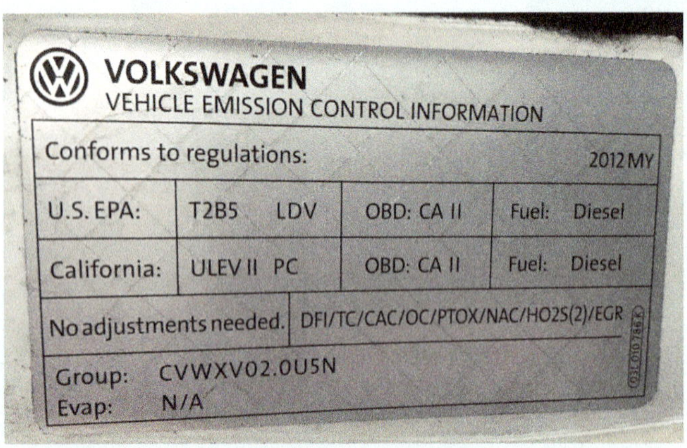

1. Locate the emission decal somewhere beneath the hood of a vehicle. Make a list of the emission components identified on the emission label.
2. If the emission decal is missing due to the replacement of a hood or front-end components after an accident, use an online service or an online maintenance manual from the original equipment manufacturer (OEM) to obtain a list of the required emission equipment.
3. Locate each component listed on the decal according to the acronyms on the label.
4. Inspect each component to determine whether all electrical, vacuum, exhaust, fluid air, and/or other connections are made to the component. Make sure proper shrouding is surrounding the charge air cooler and exhaust heat shields are in place.
5. Switch on the key and check to determine whether the MIL light is operational and illuminates.
6. If codes are present, the MIL will typically turn on, then off, and then back on again.
7. Connect an electronic service tool (scanner, Bluetooth transceiver, or laptop cable) to the diagnostic connector or OBD-II port. The device must be capable of displaying Mode 2 information about the emission system monitors.
8. With the key switched on, check the readiness codes for emission monitors. All late-model diesels should display that each monitor is "ready."
9. Navigate to Mode 7 to check for pending fault codes. No pending faults should be detected.
10. If all monitors are set to ready, if there are no MIL codes, and if all devices are present, connected, and operational, then release the vehicle.

all other chapters in this textbook. The following are some examples of engine technology required to meet current Tier 3 emissions standards:

- common rail injection with injection rate shaping
- unit injection with injection rate shaping capabilities
- low- and high-pressure cooled exhaust gas recirculation (EGR)
- variable geometry turbochargers (VGTs)
- charge air cooling (CAC)
- two-way oxidation converters (DOCs)
- DPFs
- SCR
- SCR-coated particulate filters
- lean NO_x traps (LNT) and NO_x absorbers (NAC)
- PM sensors
- exhaust gas sensors—wide range oxygen, NO_x, and ammonia (NH_3)
- sophisticated electronic control units and software
- adaptive strategies to compensate for component wear and deterioration.

Onboard Diagnostics (OBD)

To ensure compliance with legislated emissions standards, the CARB began developing regulations in 1982 that required vehicles sold in that state use a diagnostic system to detect emission-related failures. The system alerted the driver through a dash-mounted MIL that a potentially excessive emission-producing problem exists with the engine or transmission control system. As technology and emission-control systems evolved, a new emission-monitoring system called the On-Board Diagnostics II (OBD-II) standard was developed. Introduced in the mid-1990s, OBD-II has several aims, the primary being a feature capable of identifying emission system–related faults that could potentially cause noxious emissions to exceed between 1.5 and 1.75 times the Federal or Californian emissions standards. OBD-II is also a more comprehensive set of legislative standards for monitoring, maintaining, and servicing vehicle emission systems. Note also that OBD-II systems are designed with protocols for use by emission inspection and maintenance (IM) programs conducted across North America. OBD-II features provide faster, more comprehensive, and more convenient emission-testing results.

OBD-II standards are developed by the Society of Automotive Engineers (SAE) and reviewed every two years. The U.S. EPA and CARB then adopt the recommendations made by the SAE and incorporate them into the Code of the Federal Registry, CFR 40-86.

OBD-II has many specific standards governing everything from the style, type, and location of electrical system and diagnostic connections, to communication standards, emissions thresholds, fault code structure, storage, and emission-monitoring strategies (**FIGURE 3-41**). The MIL in the instrument panels is a critical feature used to alert the driver to the existence of an emission-related fault. Whenever conditions are detected that could allow emissions to exceed a maximum threshold level, the MIL is required to illuminate and warn the driver so that emission system service can be promptly performed.

FIGURE 3-41 The OBD-II connector is the connection point for electronic service tools to read fault codes and obtain other diagnostic information.

Diesel Engine Emission Monitors

Major engine systems which influence the production of emissions are monitored by the OBD-II system (**FIGURE 3-42**). Malfunctions in the following systems unique to diesel engines are evaluated regularly by the engine control system to ensure they are operating correctly.

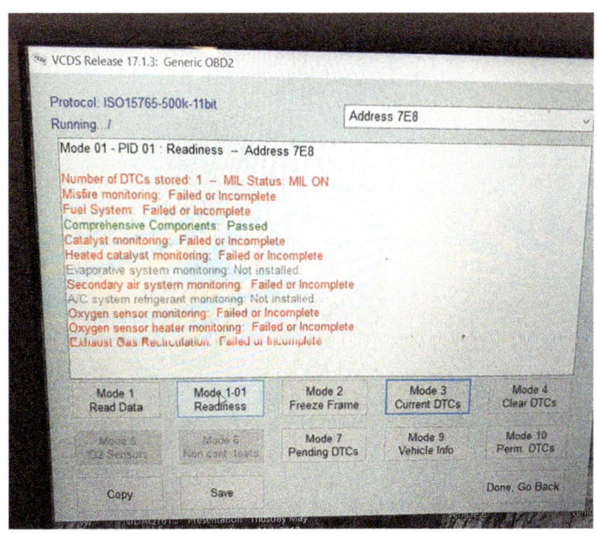

FIGURE 3-42 A list of emission systems monitors available in mode 1 on a 2013 diesel.

Major system monitors include the following:

1. Fuel system

   Typical monitor checks include the following:

   - fuel rail pressure sensor circuit
   - fuel rail pressure sensor range
   - injector code missing or invalid
   - fuel system injection pressure control
   - fuel rail pressure
   - injection timing
   - injection quantity
   - zero fuel calibration
   - feedback controls.

2. Misfire monitoring

   Typical checks by the monitor include the following:

   - detecting misfire occurring continuously in one or more cylinders.

3. Exhaust gas sensor monitor

   - air-fuel ratio sensors: tailpipe NO_x
   - O_2 sensor control module.
   - Soot sensor
   - Ammonia Slip Sensor

4. EGR system

   Typical monitor checks include the following:

   - EGR rate
   - EGR cooler efficiency/EGR cooler bypass
   - EGR system slow response
   - EGR closed-loop control limits
   - mass airflow closed-loop control limits.

5. Boost pressure control system

   Typical monitor checks include the following:

   - under- and over-boost malfunctions
   - waste gate operation
   - slow response (VGT systems only)
   - charge air under cooling

6. Glow plugs or cold start warm-up strategy

   Typical monitor checks include the following:

   - on time
   - glow plugs enabled.
   - Intake air heater

7. Diesel engine exhaust aftertreatment

   Typical monitor checks include the following:

   - DPF filter efficiency and missing substrate
   - DPF regeneration frequency
   - DPF incomplete regeneration
   - DPF feedback control
   - DPF restriction

   The monitor requires potential faults to be detected before emissions exceed 1.75 times the standards for any of the following systems:

   - oxidation catalyst
   - lean NO_x catalyst

- SCR catalyst
- NO_x trap
- PM trap.

8. Comprehensive component monitoring (CCM)

   The CCM is part of the OBD-II system. It monitors electrical circuits operating powertrain components that

   - can cause a measurable emissions increase during any reasonable driving conditions
   - are used for other OBD monitors
   - are required to monitor input components for circuit and rationality faults
   - are required to monitor output components for functional faults.

The CCM is not tied to emission thresholds. Any electrical problem related to emission systems will cause the MIL to illuminate.

Opacity Testing

Before the use of particulate filters, opacity testing was the common measurement for evaluating smoke density from diesel engines (**FIGURE 3-43**). This is based on the principle of light extinction by exhaust smoke. A beam of red light is passed through a sample of smoke measures the percentage of light blocked. The darker the smoke, the higher the percentage of opacity. A measure of 75% opacity means 75% of the light is blocked, while 25% is transmitted. The procedure used by law enforcement agencies to measure smoke emissions is based on the SAE J1667 Snap Acceleration Test. This is a test for nonmoving vehicles that can be conducted along the roadside, at a truck stop, at a vehicle repair facility, or at inspection maintenance test facilities. The Snap Acceleration Test is intended to be used on both light-medium-duty and heavy-duty vehicles. The procedure is intended to provide an indication of the state of maintenance and/or tampering of the engine and fuel system relative to the factors that affect exhaust smoke. Surprisingly, some manufactures may still include opacity testing as part of a diagnostic strategy for today's vehicles. To perform opacity testing on late-model vehicles, the particulate filter may need to be removed to observe smoke density.

While electronic test equipment with specialized software is available to precisely measure smoke opacity, opacity can be evaluated visually. Cards with various shades of gray scale are commercially available, and a chart called a Ringleman scale is available to visually approximate the percentage of light blocked by exhaust smoke from older engines not equipped with DPFs. To visually evaluate opacity, follow the steps in **SKILL DRILL 3-3** as a guideline.

FIGURE 3-43 Opacity, the measure of smoke density, is based on the percentage of light blocked by exhaust smoke.

SKILL DRILL 3-3 Measuring Opacity

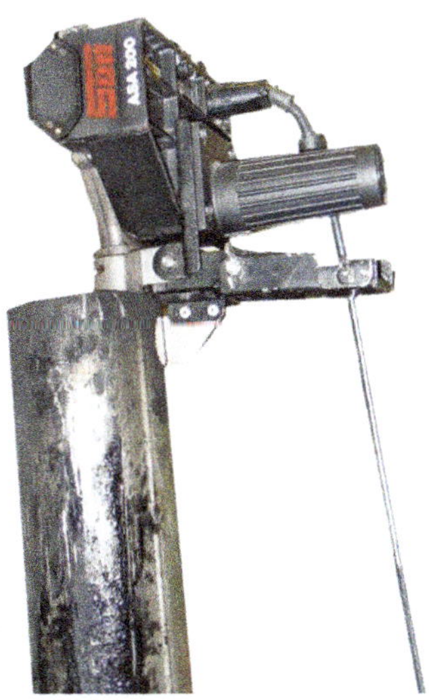

1. With the engine at operating temperature, accelerate the engine quickly to maximum engine speed.
2. Have an observer examine the exhaust plume against a light-colored background or bright sky.
3. Using a chart, match the darkness of the exhaust plume against an appropriate scale-measuring opacity and record the results.
4. When exhaust smoke is just visible, opacity corresponds to approximately 10%.
5. Visible exhaust smoke for more than 15 seconds over a 1-minute period indicates a major defect in the engine combustion system.

Wrap-Up

Ready for Review

- Emissions legislation that requires reducing vehicle emissions drives most changes in engine technology.
- Emissions from modern diesel engines are so low that only test instruments with the highest sensitivity can measure their exhaust emissions.
- Diesel engines produce lower emissions per ton of weight carried than gasoline engines.
- Noxious emissions from diesels comprise 0.2–0.3% of untreated exhaust compared to as much as 3% from spark-ignition, gasoline-fueled engines.
- Diesel engines produce more NO_x and particulate emissions than spark-ignition engines, but they produce lower levels of hydrocarbons (HCs) and carbon monoxide (CO) emissions.
- Most diesel exhaust particulate emissions are composed of black carbon or soot.
- NO_x formation is a major emission problem with diesel engines. Higher compression ratios leading to increased combustion temperatures and pressure help form more NO_x.
- Diesel fuel does not produce significant evaporative emissions, and diesel fuel systems are not regulated for evaporative emissions.
- Crankcase emissions are classified as particulates because they contain oil droplets.
- Excess air, higher combustion temperatures, and pressure in diesel engines mean lower amounts of HCs and CO emissions are produced.
- Diesel fuel properties and the short time that fuel has to mix with air and burn contribute to higher amounts of particulate formed in engine cylinders.
- Hard-to-burn fuel molecules, called aromatic molecules, produce the distinctive odor of diesel exhaust.
- Reducing sulfur content in fuel reduces particulate formation.
- Finer atomization and higher pressurization of fuel reduces particulate formation.
- Carbon dioxide (CO_2) emissions are greenhouse gases (GHGs) and thus are regulated indirectly through fuel economy standards.
- The Environmental Protection Agency (EPA) and the EURO 1–6 set the emissions standards for most of the world.
- The use of ultra-low-sulfur diesel fuel has enabled the use of exhaust aftertreatment systems such as selective catalyst reduction (SCR) systems and particulate filters.
- Emission decals indicating compliance with emissions standards are required by all on- and off-road diesel engines.
- Diesel engines must have onboard diagnostic systems that detect potential conditions that could lead to increased NO_x emissions.
- Diesel engine speed is lower than spark-ignition, gasoline-fueled engines because excess emissions are produced at high engine speeds. The engine will run out of time to burn fuel at high speeds.

Key Terms

carbon dioxide (CO_2) An inert gas that is a byproduct of normal combustion.

carbon monoxide (CO) A noxious gas that is a byproduct of abnormal combustion.

diesel particulate filter (DPF) A device installed in the exhaust system to filter out black carbon soot and other particulates from the exhaust stream.

emission monitor An organized procedure designed to test the functionality or operation of the emission systems and it's components.

greenhouse gases (GHGs) Gases that are believed to contribute to global warming.

hydrocarbons (HCs) Fuel molecules made of both hydrogen and carbon atoms.

malfunction indicator lamp (MIL) A dash mounted warning light required by OBD-II legislation to alert the driver of a potential emission related fault.

heavy-duty onboard diagnostics (HD-OBD) The equivalent of the light-duty OBD-II system for heavy duty vehicles.

opacity A measurement of exhaust smoke density expressed as a percentage of light blocked by exhaust particulate.

oxides of nitrogen (NO_x) A group of noxious emission gases produced during combustion when nitrogen reacts with oxygen. NO_x collectively refers to NO, NO_2 and NO_3.

ozone A noxious molecule made from three oxygen atoms (O3) which helps forms photochemical smog.

particulate matter (PM) A category of engine emissions formed from solids or liquids.

selective catalyst reduction (SCR) A NO_x reduction system that selects oxygen for removal from NO_x molecules rather other exhaust oxygen compounds.

smog An early name for photochemical smoke combined from the words SMOKE and FOG.

Tier 1, 2, 3, and 4 emissions standards EPA emission standards for on and off road vehicles and engines.

Tier 2 bin 5 emissions standards Emission standards for light and medium duty vehicles phased in beginning in 2004 until 2009. The standard was in effect until 2016.

Tier 3 emissions standards EPA emission standards for light and medium duty vehicles phased in beginning in 2017 and applicable until 2025.

volatile organic compounds (VOCs) Engine emissions containing carbon compounds that react easily with other substances.

Review Questions

1. Categories of noxious emissions are referred to as regulated emissions because they __________.
   a. are greenhouse gases
   b. are carcinogenic
   c. are limited by legislative standards
   d. each have an emission monitor to detect the production of a noxious gas
2. Ultra-low-sulfur fuel has been used since 2006 because __________.
   a. it produces fewer emissions
   b. high-sulfur diesel is more costly and difficult to produce
   c. it provides increased power and improves fuel economy
   d. it enables the use of diesel catalysts in the exhaust system
3. During which of the following conditions are the highest NO_x levels are produced?
   a. High combustion temperatures, low compression ratios.
   b. Low combustion temperatures, high compression ratios.
   c. High combustion temperatures, high compression ratios.
   d. Low combustion temperatures, low compression ratios.
4. In which of the following categories of emissions do crankcase emissions belong?
   a. Hydrocarbons.
   b. Oxides of nitrogen (NO_x).
   c. Particulates.
   d. Carbon monoxide (CO).
5. Ultrafine particulate is the most hazardous emission because it __________.
   a. makes up the greatest mass of all particulate categories
   b. quickly settles and falls to ground level
   c. penetrates deeply into lung tissue
   d. is a visibility hazard
6. The useful life requirement for a light-duty 2018 diesel is __________.
   a. 150,000 miles (241,400 km)
   b. 185,000 miles (292,700 km)
   c. 435,000 miles (700,000 km)
   d. 100,000 miles (161,000 km)
7. The 2018 emission-monitoring system responsible for detecting potential faults that could cause light- and medium-duty diesel emissions to rise above maximum legislated levels is known as __________.
   a. On-Board Diagnostics 1 (OBD)
   b. On-Board Diagnostics 2 (OBD-2)
   c. Comprehensive Component Monitor (CCM)
   d. Emission-Monitoring System (EMS)
8. Diesel engines do not use evaporative emissions controls because __________.
   a. fuel tanks are sealed and not vented to the atmosphere
   b. vapors from diesel fuel are not toxic
   c. diesel fuel has low volatility and exerts very low vapor pressure
   d. there are fewer diesel-powered vehicles on the road and emissions are of little consequence
9. Diesel's high compression pressure and excess combustion air increases which of the following exhaust emissions?
   a. Oxides of nitrogen (NO_x).
   b. Particulates (PM).
   c. Hydrocarbons (HCs).
   d. Carbon monoxide (CO).
10. Measurement of engine emissions from a light- or medium-duty diesel to certify its compliance with EPA and EURO standards is __________.
   a. an opacity test
   b. performed on a chassis dynamometer
   c. performed by the OBD-II system
   d. performed by following the SAE J-1667 test protocol

ASE Technician A/Technician B–Style Questions

1. Technician A says that creating clean diesel engines has resulted in diesel engines using more fuel with less power output compared to older engines. Technician B says that even the latest engine technology will not enable diesels to operate as cleanly as spark-ignited gasoline-fueled engines. Who is correct?
   a. Technician A
   b. Technician B
   c. Both A and B
   d. Neither A nor B
2. While discussing the reasons for banning diesel engines, technician A says that diesel engines make up only 5% of the vehicles on the road but produce 24% of the transportation emissions, an amount of emissions that is disproportionately higher than what spark-ignition, gasoline-fueled vehicles produce. Technician B says that diesels will produce fewer emissions for every ton of weight hauled compared to spark-ignited, gasoline-fueled engines. Who is correct?
   a. Technician A
   b. Technician B
   c. Both A and B
   d. Neither A nor B
3. While discussing greenhouse gas (GHG) emissions and diesels, Technician A says that carbon dioxide emissions are regulated through fuel economy standards. Technician B says that there are no direct carbon dioxide (CO_2) standards in North America for diesel-powered vehicles. Who is correct?
   a. Technician A
   b. Technician B
   c. Both A and B
   d. Neither A nor B
4. Technician A says that carbon monoxide (CO) emissions are higher from diesel engines than from gasoline engines. Technician B says that diesel engines produce little

or no carbon monoxide gas compared to gasoline engines. Who is correct?
a. Technician A
b. Technician B
c. Both A and B
d. Neither A nor B

5. Technician A says that black smoke from the exhaust of a diesel engine is evidence of particulate emission. Technician B says that black smoke is just soot, which will harmlessly fall to the ground. Who is correct?
a. Technician A
b. Technician B
c. Both A and B
d. Neither A nor B

6. Technician A says that crankcase vapors have oil droplets that categorize the emission as a hydrocarbon (HC). Technician B says that only gases are considered hydrocarbon emissions and oil droplets qualify crankcase emissions as a type of particulate emission. Who is correct?
a. Technician A
b. Technician B
c. Both A and B
d. Neither A nor B

7. Technician A says that diesel exhaust has a noxious portion that is 5 to 10 times greater than the noxious component of spark-ignited, gasoline-fueled engines. Technician B says that the noxious portion of diesel exhaust is many times smaller than the noxious portion of spark-ignited, gasoline-fueled engines. Who is correct?
a. Technician A
b. Technician B
c. Both A and B
d. Neither A nor B

8. Technician A says that diesel engine crankcase vapors are made up primarily of air leaking past the pistons rings and can be vented directly to the atmosphere since it has no hydrocarbon or fuel in the air. Technician B says that crankcase vapors are classified as particulate emissions and are not vented to the atmosphere on late-model engines. Who is correct?
a. Technician A
b. Technician B
c. Both A and B
d. Neither A nor B

9. Technician A says that high crevice volume in a diesel engine is responsible for producing excessive amounts of particulate emissions. Technician B says that high crevice volume produces excessive amounts of NO_x emissions. Who is correct?
a. Technician A
b. Technician B
c. Both A and B
d. Neither A nor B

10. Technician A says the malfunction indicator lamp (MIL) is used only to indicate emission-related faults. Technician says the MIL replaces other warning lamps to indicate any vehicle-related fault B. Who is correct?
a. Technician A
b. Technician B
c. Both A and B
d. Neither A nor B

CHAPTER 4

Cylinder Blocks and Crankshafts

Learning Objectives

After reading this chapter, you will be able to:

- **4-01** Identify and explain the important factors determining diesel cylinder block design.
- **4-02** Identify and describe the construction and composition of diesel engine cylinder blocks.
- **4-03** Identify and describe the construction methods used to increase cylinder block strength and rigidity.
- **4-04** Identify and describe the methods used to form cylinders in an engine block.
- **4-05** Identify and explain methods used to correct cylinder wear.
- **4-06** Identify and explain unique construction features of diesel engine crankshafts.
- **4-07** Identify and explain the function of vibration dampeners.
- **4-08** Identify flywheel types and construction features.
- **4-09** Identify and explain construction features and functions of engine bearings.

You Are the Technician

The construction fleet operation where you work is beginning to see a number of engines in ¾ - 1 ton service trucks wearing out after accumulating more than 500,00 miles (800,000 km) of service. You have been tasked to make service recommendations for a program to replace or rebuild all engines in-house rather than subcontract the work to an independent engine repair facility. The fleet is a mixed operation: a third of the trucks come from three major manufacturers. As part of the task, you are to develop a set of guidelines to determine the economic viability of either repairing or replacing each engine. Using the guidelines, technicians will perform inspection procedures and determine whether parts will be replaced or the entire engine exchanged for a rebuilt unit. As you set out to prepare the guidelines consider the following:

1. What type of engine blocks do you expect to encounter? Explain your answer.
2. What parts do you expect will have the greatest wear?
3. What would you expect are some of the greatest limiting factors determining the economic viability of rebuilding versus replacing the engines?

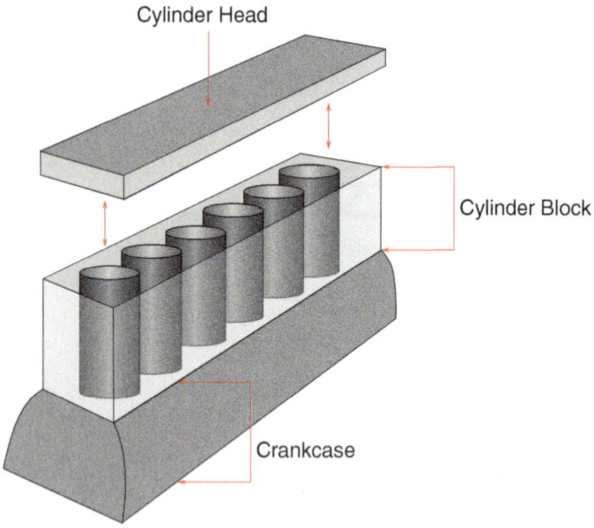

FIGURE 4-1 The cylinder block forms the main structure of the engine. It encloses the cylinder components, and other major parts are attached to it.

4-01 Identify and explain the important factors determining diesel cylinder block design.

▶ Introduction

The cylinder block, which encloses the cylinder bores, is the largest part and the main structure of the diesel engine. Since the diesel's combustion characteristics and high cylinder pressure generates substantially more power output per cubic inch of displacement, it produces extraordinary vibration that requires unique design and construction features. The blocks design and other major engine components integral to the block—such as crankshafts, balance shafts, vibration dampeners, and flywheels—are examined in this chapter as well (**FIGURE 4-1**).

▶ Cylinder Block Design

The cylinder block can be described as the largest single part and the main structure, or backbone, of the diesel engine. Its primary function is to support cylinders and liners (if equipped), as well as major engine components such as the crankshaft and camshaft. The transmission, cylinder head(s), and other engine parts are bolted or connected to the cylinder block. Diesel engine blocks have cylinders arranged in-line, V banked, or horizontally opposed (**FIGURE 4-2**). The number of cylinders used by light- and medium-duty blocks are commonly configured in four, five, six, or eight. However, many single-cylinder, two-cylinder, and three-cylinder blocks are produced for the off-road equipment market. High cylinder pressures combine with a rapid combustion phase to create extraordinary vibration and torsional twisting forces in diesel blocks. What makes diesel engine blocks so distinctive are the specialized features to reduce noise, vibration, and harshness caused by these combustion forces. Many of the unique features found in diesel engine blocks are simply there to provide exceptionally stronger and more rigid block designs. These features complement, vibration-free, lightweight, durable construction also necessary for engines to perform well in automotive and other light- and medium-duty applications.

Cylinder Block Structure

The diesel engine cylinder block consists of these three distinct sections:

1. the main block structure, which encloses the cylinder components
2. the cylinder component assemblies, which include the pistons, rings, wrist pins, and connecting rods
3. the crankcase, which is the area below the cylinders that encloses the crankshaft, main bearing bores, seals and crankshaft bearings.

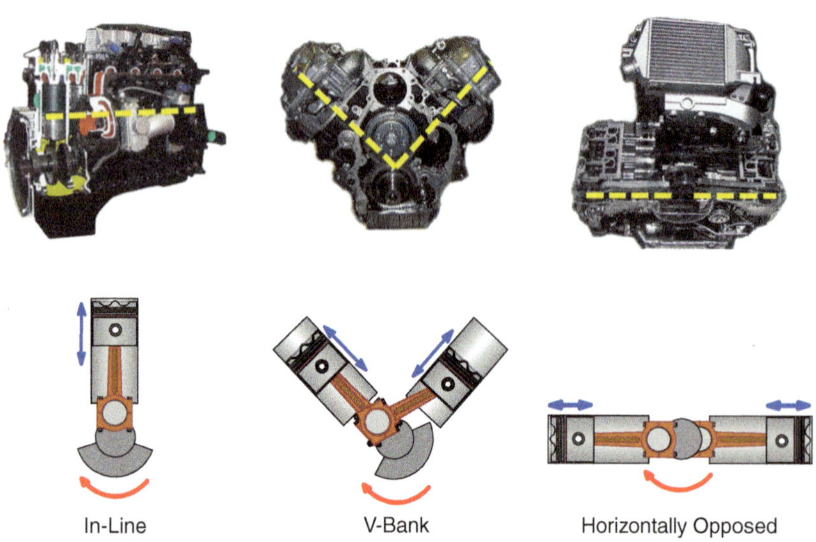

FIGURE 4-2 Engine block cylinder banks are configured in at least three basic ways: in-line, V banked, and horizontally opposed.

The geartrain may be located on either the front or the rear of the engine. Geartrain rattle is a problem in diesel engines, affecting not only noise characteristic but also engine durability. More recent engine designs place the geartrain at the rear of most engines, to reduce noise and wear caused by geartrain rattle (**FIGURE 4-3**). Torsional vibration is the primary cause of gear rattle. Torsional vibration is produced by alternating firing impulses and compression events in the cylinders causing sharp variations to the crankshafts rotational velocity. A rhythmic speed change takes place during every combustion as the crank speeds up during poer events and slows during compression events (**FIGURE 4-4**). Accelerating and decelerating crankshaft results in camshaft

FIGURE 4-3 Rear-mounted geartrains are quieter, with less gear wear and more accurate valve and unit injector operation.

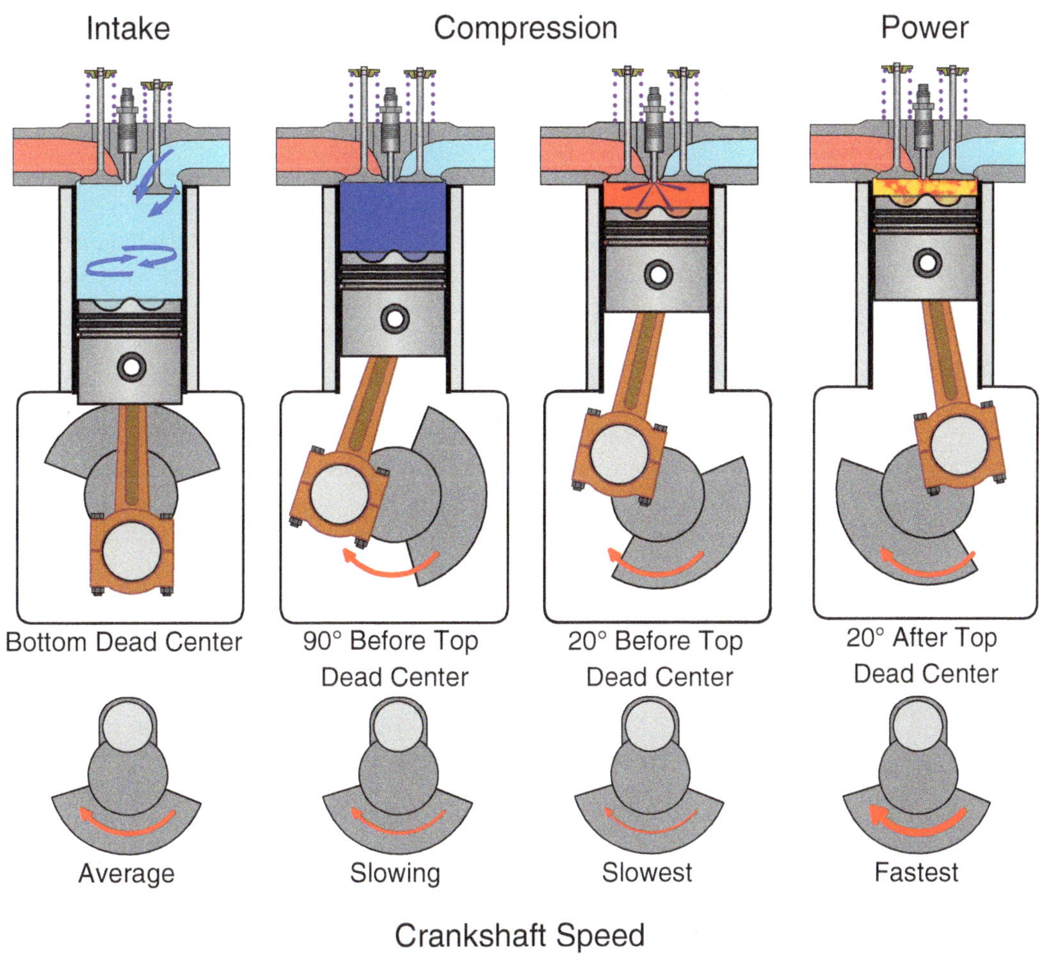

FIGURE 4-4 Torsional vibration causes a rhythmic change in engine speed. High compression ratios and combustion pressures create stronger torsional vibration in diesels.

and crankshaft speed variation of between 3% and 6% during each crank rotation. Changing cam and crankshaft speed causes the face of each contact tooth in the geartrain to hit the driven and back sides of opposing gear teeth face each time the engine crankshaft changes speed. The "slapping" back-and-forth action of teeth rapidly moving in and out of contact with one another produces noise and increases wear (**FIGURE 4-5**). Moving the

▶ TECHNICIAN TIP

When an engine operates, the wave-like movement of accessory drive belts may casue them to appear loose. They may not in fact be loose but instead are changing in length due to torsional vibration. Torsional vibration refers to the speeding up and slowing down of the crankshaft, caused by alternating compression and power strokes of the engine cylinder. High compression ratios and even higher cylinder pressures in a diesel engine produce intense torsional vibration or rhythmic twisting forces. Torsional vibration is observed by watching the back-and-forth oscillation of belt tensioners. The change in speed of the crankshaft causes drive belts to alternately stretch and shorten, thus causing tensioner movement and belt vibration.

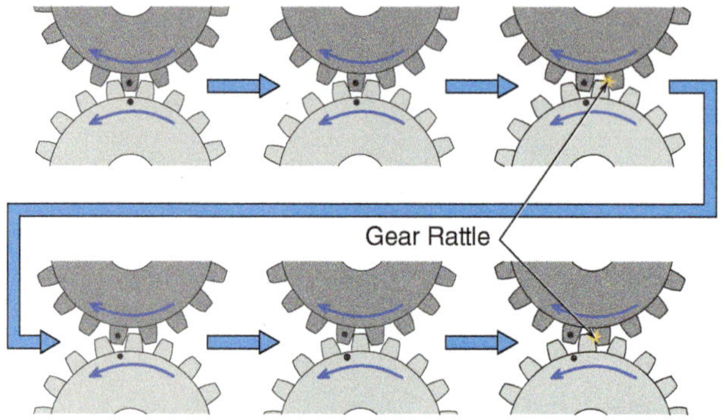

FIGURE 4-5 Gear rattle is caused by alternating, back-and-forth contact between gears turning at different speeds.

geartrain to the rear of the engine, where torsional vibrations are reduced due to the heavier flywheel mass, can dramatically reduce geartrain noise. Gear life and valve timing accuracy is improved.

4-02 Identify and describe the construction and composition of diesel engine cylinder blocks.

▶ Cylinder Block Materials

Most diesel cylinder blocks today are manufactured from cast iron but aluminum and a newer block material made from **compacted graphite iron (CGI)** are now popular in many light-duty engines. Ford's 6.7 L Powerstroke and the Lion V6 for the Ford F-150 use a CGI block with aluminum cylinder heads to create a stronger but lighter block than is possible with ordinary cast iron (**FIGURE 4-6**). Various European diesels from Mercedes and Volkswagen (VW) also use CGI blocks.

Cast Iron Blocks

In contrast to CGI, ordinary cast iron is a metal made from iron ore with a carbon content of between 3% and 5%. The carbon is scattered throughout the metal in flakes, creating metal with a crystalline-like structure. The metal is poured or cast into molds commonly made from sand or fiberglass. The main advantages of cast iron are that it absorbs vibration well and that it is easy to manufacture and machine, inherently strong, and corrosion-resistant and heat-resistant. The high carbon graphite content of iron makes it an ideal wear surface for moving parts such as pistons. To improve strength, hardness, and corrosion resistance, manufacturers may alloy the iron by adding nickel, molybdenum, and chromium. Cast iron is still not as strong as steel, so bolt threads are usually cut with a deeper coarse thread pitch to minimize the problem of stripping threads. Main bearing caps are often hardened, made from steel, or use special alloys to provide improved block strength.

FIGURE 4-6 The Ford 6.7 L Powerstroke uses a CGI cylinder block with aluminum cylinder heads.

Casting and Seasoning Cast Iron Blocks

Sand casting blocks made from sand molds require the removal of sand from inner passageways (**FIGURE 4-7**). **Core plugs**, sometimes called frost plugs, are used to cover holes made in the block to flush sand out of small passageways. Coolant heaters are often installed in the core plugs, but the core plug is not used to prevent damage to a block caused by frozen coolant. The plugs are mistakenly called frost plugs since they are pushed out by frozen coolant. Engine block damage is still often the result even after frozen coolant pushes out a core plug.

After rough casting, blocks tend to warp. To minimize distortion, diesel blocks are often stored at high temperatures for a day or two to heat-treat, or as it is occasionally described "season," them. Prolonged heating tends to relieve stress from the block and reduce any future distortion. Metallurgical changes to make the iron less brittle also take place while heat-treating the block. Subsequent to seasoning, castings are bored, tapped, and machined to finish. To

FIGURE 4-7 Core plugs are installed in a block only after casting sand, which is used to form the blocks internal passageways, is cleaned out of the block passage through pressure washing.

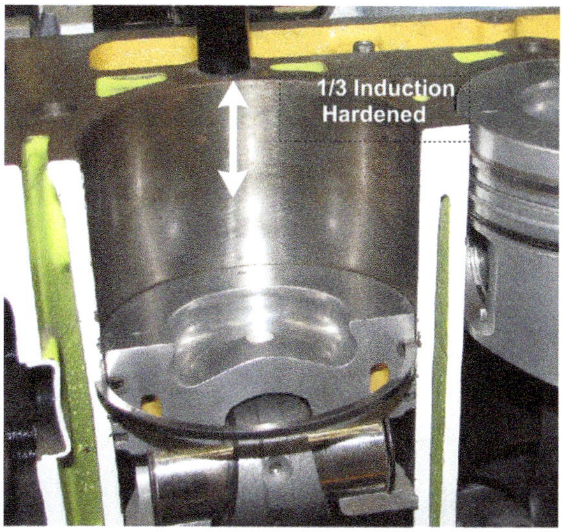

FIGURE 4-8 The arrow shows an area of discoloration where the upper third of the 6.6 L GM Duramax cylinder has been induction hardened in the ring turnaround area, where there is the least amount of cylinder lubrication.

increase the hardness and durability of cast iron cylinder blocks, they may be induction hardened. Induction hardening is a heat treatment process that involves passing alternating electric current through coils of heavy-gauge wire surrounding the material to be hardened. Through the principle of magnetic induction, intense localized heating takes place on the surface of the metal, which is then quenched with water to produce a hard, wear-resistant metal surface. GM's Duramax 6.6 L blocks induction harden the upper third of the cylinder walls. Induction hardening increases engine life by minimizing cylinder wall wear in the ring turnaround area, where there is the least amount of cylinder lubrication (**FIGURE 4-8**).

Disadvantages of Cast Iron

Cast iron is heavy, and engines made from cast iron add considerable weight to the vehicle in comparison to lighter materials such as aluminum or CGI. Heavy engines can negatively affect vehicle handling and braking characteristics. These heavier vehicles tend to dive more when braking, steer less nimbly, and require heavier front suspension systems. Cast iron does not transfer heat as readily as aluminum. Cast iron cannot be easily repaired by welding, like steel or other metals can. Most often, broken or cracked parts require replacement. However, some block porosity problems, such as those causing external oil leaks, can be repaired with special epoxies. Original equipment manufacturer (OEM) repair recommendation for external block porosity is to use epoxy in areas where mechanical stress is not significant. In high-value blocks, welding techniques to repair cracks involve preheating the block and using a welding rod with high nickel content or a powdered iron welding rod.

Aluminum

Aluminum is a block material that is being used more commonly for automotive diesels. Advances in manufacturing and hardening techniques have enabled this lighter material to be used in high-performance applications. In comparison to cast iron, aluminum in small, six-cylinder diesel engines can eliminate more than 77 lb (35 kg) of weight. But until the late 1990s, aluminum engine blocks were used only in gasoline-fueled engines because high-performance diesel engines are required to withstand higher combustion temperatures and three times more cylinder pressure than spark-ignition, gasoline-fueled engines. Aluminum also has challenges in diesel applications, in terms of rigidity, strength, and reliably bolting the cylinder head and main bearings. Although it transmits heat well,

FIGURE 4-9 This Mercedes 3.0 L Bluetech engine is a die cast aluminum block with a "cast-in-place" iron cylinder sleeve.

aluminum has an expansion coefficient of four times the rate of iron. This can cause critical dimensions such as the crankshaft bearing bores and cylinder bore clearances to enlarge. Installing head and main bearing cap bolts with enough clamping force is another problem. With aluminum, there is greater potential to strip bolt threads if fasteners are tightened too much or cylinder pressures forcing the crankshaft downwards through the oil pan. To overcome this problem, a technique similar to seasoning cast iron blocks, **precipitation hardening**, also called particle hardening, is a widely used heat treatment technique used to increase the strength of aluminum. By alloying the aluminum with a small percentage of unique, proprietary particles, combined with prolonged heat-treating and slow cooling, aluminum blocks can achieve additional strength and hardness.

While aluminum can be hardened, the greatest challenge with aluminum is that it cannot withstand the friction between pistons and the cylinder wall without rapidly wearing. However, because of aluminum's weight-saving advantage, other manufacturing techniques have been employed to overcome these difficulties. For example, in aluminum blocks, cylinder walls and bearing bores are made from using cast iron liners to provide durable, low-friction wear surfaces (FIGURE 4-9). Another technique is to apply another very hard material to cylinder wall surfaces. Spray welding, called plasma welding, applies an alloyed iron coating over the cylinder wall. Tool steel hardness coatings of iron carbon alloys can be applied by plasma spraying the cylinder walls using a rotating plasma torch and metal alloy powder. The coating which is a few tenths of a millimeter thick to give the cylinder walls great wear resistance and durability.

Nanoslide Technology

To reduce friction and increase wear resistance, Mercedes-Benz developed a spray welding technique to coat cylinder walls of the 3.0 L block with a tool-grade hardness iron alloy (FIGURE 4-10). The technique, known as Nanoslide technology, first melts wires of iron/carbon alloy in an electric arc. Pressurized gas deposits only a 0.1–0.15 mm thick lining of the melted material onto the cylinder walls (FIGURE 4-11). This ultra-fine nano-crystalline coating is then finished to a smooth mirrorlike surface by using a special laser honing process that melts the wall and permits nitrogen gas to blend into the coating. In spite of the extreme smoothness, the nitride hardening of the cylinder wall during honing opens pores in the material that are able to retain oil for lubricating the piston and cylinder wall. The result is an ultra-low-friction surface where mechanical losses due to friction are reduced by up to 50% in comparison to gray cast iron cylinder liners. In addition, extremely high wear resistance is imparted to the cylinder walls. Other advantages include lower engine weight, reduced fuel consumption, and lower emissions.

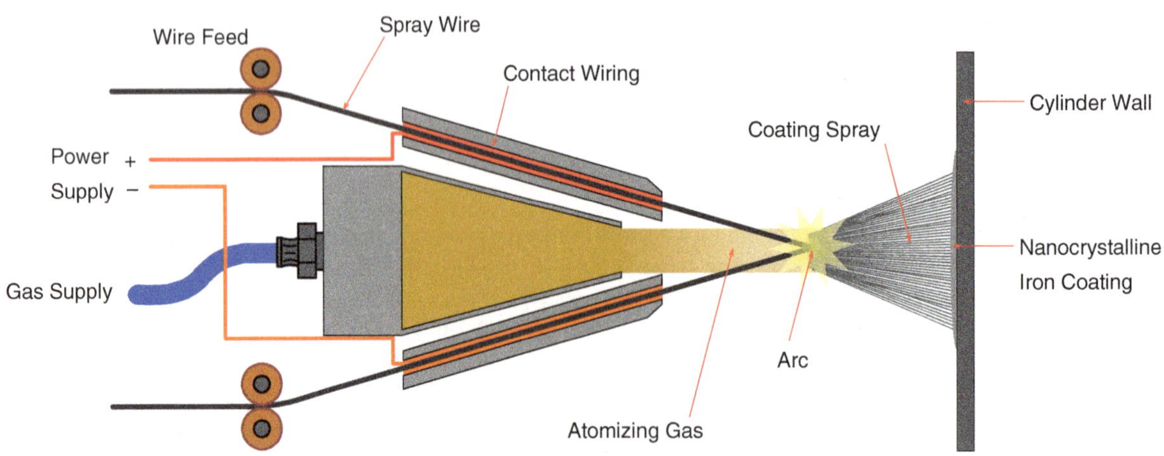

FIGURE 4-10 Nanoslide technology involves spraying a thin coating or extremely hard iron alloy onto the cylinder walls of an aluminum engine. The coating is further hardened through nitriding while laser honing of the cylinder wall.

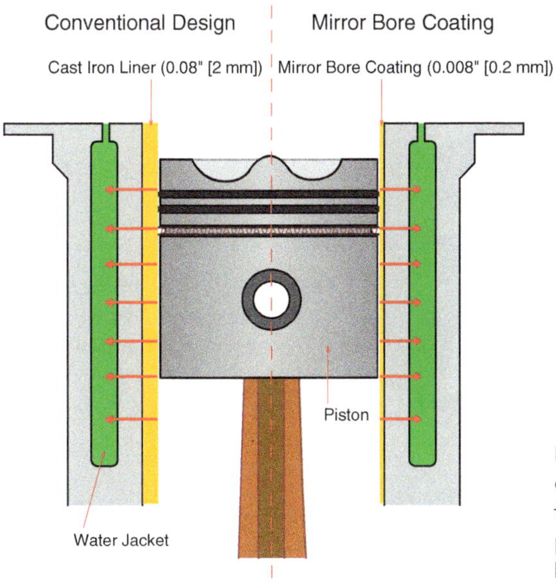

FIGURE 4-11 The cylinder walls of a Mercedes 3.0 L engine have a thin layer of hardened iron alloy to prevent wear on the aluminum cylinder block walls.

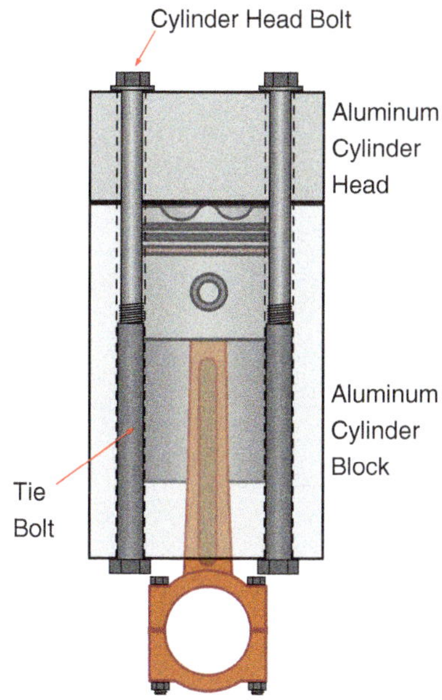

FIGURE 4-12 Aluminum cylinder blocks can use a warp anchor principle to prevent distortion of the block from cylinder head bolts. This means the cylinder head and the cylinder block are bolted together by tie bolts. Sliding steel sleeves, which are locked in the block, accept the cylinder head bolt from one side and the tie bolt from the other, preventing the block from being distorted.

Cast-in-Place Cylinder Sleeves

Lightweight aluminum engine blocks can use hardened sleeves to withstand the friction from reciprocating pistons and rings. Aluminum blocks with hardened iron sleeves are constructed with a **cast-in-place sleeve**, where the aluminum block is molded around a cylinder sleeve that is placed into the mold before molten aluminum is poured to form the cylinder block. GM's 1.6 L Chevy whisper diesel uses an aluminum block construction with a cast iron sleeve. A cast iron plate bolted to the bottom of the crankcase further stiffens the block.

Aluminum Block Design Variations

Since aluminum blocks are softer metal, one problem is the potential of stripping bolt threads. Small aluminum blocks cannot withstand the tensional loading on bolt threads without stripping threads out of the block. To overcome this problem, blocks can be made with steel thread inserts. One technique is to tie the cylinder head to the block with steel tie bolts inserted through the bottom of the block. This construction technique, called the **warp anchor method**, clamps the cylinder head to the block by using head bolts threaded into steel sleeves locked into the block (**FIGURE 4-12**). The block accepts the cylinder head bolt from one side and the tie bolt from the other. Another technique, called the **tension bolt design**, is almost identical to the warp anchor bolt construction, except that it anchors the bolt through the main bearing cap or main bearing girdle, which is a steel plate attached to the bottom of the crankcase on ladder block designs (**FIGURE 4-13**).

Open Versus Closed Block Decks

Aluminum engine blocks are commonly die cast. This involves squeezing molten aluminum between two steel halves of a stamping die to produce a finished part. The advantages include improved block dimensional accuracy, smother surfaces, the ability to use thinner block walls, and eliminating secondary machining in comparison to cast iron or aluminum blocks. In fact, die casting is a faster production method and a less expensive method than casting blocks using sand or fiberglass molds. Sleeves and steel inserts can also be cast in place. Honda, Mercedes, and GM are examples of OEMs that use die cast blocks. Honda uses a proprietary ASCT (Advanced Semisolid Casting Technology) for it's die cast aluminum 2.2L 4-cylinder i-CTDi (Intelligent Common rail Turbo Diesel Injection) engine. Aluminum used for the casting has a "mushy" consistency when it is injected into dies and pressurized before the dies are separated. Using this method, cylinders can be separated with aluminum that is as thin as 8 mm (0.315") to form a very compact engine that is 30% lighter than cast iron. These blocks are specially heat-treated and particle hardened, followed by a plasma sprayed coating of hard iron alloy applied to the cylinder walls. Cast iron main bearing caps are used in this engine that delivers a combined fuel economy of 40.5 US-mpg (5.8 L/100 km) in the Accord wagon.

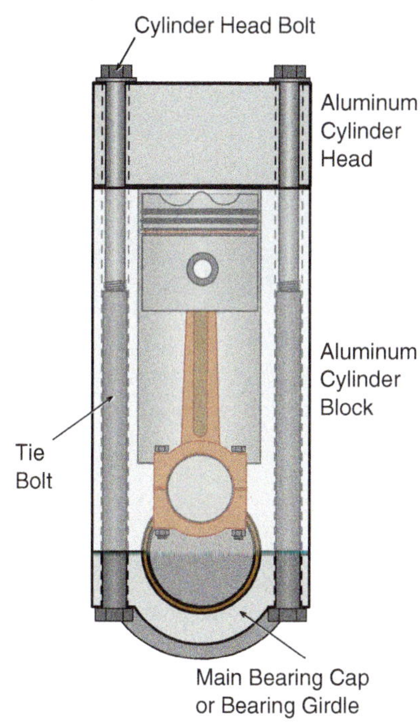

FIGURE 4-13 Tension bolt blocks anchors the cylinder head bolts through the main bearing cap or main bearing girdle on ladder block designs.

FIGURE 4-14 This die cast block uses an open-block deck. There is no support at the top of the cylinder that bridges the cylinder walls with the rest of the block.

FIGURE 4-15 Closed blocks connect the top of the cylinder wall with the block deck through the use of a metal bridge. This bridge closes the gap between the cylinder and the rest of the block.

In contrast to cast iron blocks, die cast blocks have what is known as an **open-deck design** (**FIGURE 4-14**). Open decks are easily identified since the block deck—the top surface of the block, in contact with the cylinder head—has no connecting metal between the block walls and the upper portion of the cylinder wall. Although coolant can better circulate in open deck blocks, the design has the disadvantage of an absence of reinforcement in the upper cylinder walls. Without a metal bridge between the cylinder walls and block deck the weakly supported cylinder walls are capable of moving during service. Closed decks connect the block deck with the cylinder walls to form a stronger, more rigid cylinder structure (**FIGURE 4-15**).

Compacted Graphite Iron

Using CGI, or sintered graphite, for a block material is the latest engine block construction technique. CGI engine blocks provide blocks with greater strength, rigidity, and durability at a reduced weight and brittleness compared to cast iron. Theoretically, a CGI block can be constructed with the strength of steel but with almost the weight of aluminum. The stronger, steel-like-strength material enables the use of thinner block and cylinder walls without sacrificing stability or strength. CGI starts with powdered iron and graphite alloys squeezed into molds at high pressures and then heated to bond the metal particles together. This construction technique yields blocks with at least 80% higher tensile strength, 45% greater stiffness, and approximately double the fatigue strength of conventional gray cast iron and aluminum. The weight of an engine block made from CGI alone can be reduced by 22% through a 15% reduction in the thickness of the block walls. Noise transmission through the block is dramatically reduced as well, because of the vermicular (wormlike) grain structure of the material, which has sound-absorbing properties. By reducing the block size and weight while increasing strength, CGI materials allow engine designers to improve power density, performance, fuel economy, and durability while reducing engine weight and noise. Many light-, medium-, and heavy-duty diesels are now made from CGI, including blocks from Audi, Cummins, VM Motori (Dodge 3.0 L Eco diesel), and Fords F-150 3.0 L introduced in 2018 and 6.7 L Powerstroke introduced in 2010. Ford claims that the 6.7 L Powerstroke is a made from patented SinterCast CGI material that is 40% more rigid and 100% more fatigue-resistant than gray cast iron (**FIGURE 4-16**). CGI has allowed Ford designers to reduce the engine wall thickness by 15% in comparison to the thickness that would be needed if it used gray cast iron. A weight reduction of 160 lb (70 kg) is achieved in comparison to the earlier cast iron 6.4 L engine.

FIGURE 4-16 The EcoDiesel 3.0 L from Fiat Chrysler Automobiles (FCA) used in the Dodge Ram uses a CGI engine block.

CGI engine blocks have their own unique service challenges in terms of machining. CGI is steel-like in hardness but also abrasive and "gummy" when machined. Special machine tools with extraordinary harness are needed to cut the material. Honing cylinders are much harder, and a coarse grit honing stone will produce a smoother finish than the same stone will produce in a cast iron block. These blocks cannot be repaired when damaged, so they must be replaced.

▶ Block Crankcase Construction

4-03 Identify and describe the construction methods used to increase cylinder block strength and rigidity.

Blocks must be designed to be as compact as possible to reduce weight and size while maintaining strength and rigidity. Unnecessary weight is the enemy of good vehicle dynamics, such as braking, precise handling, and quick acceleration. However, engine weight translates to strength and rigidity, two important features of an engine to reduce vibration, noise, and harshness (**FIGURE 4-17**).

Three basic types of block construction are used to best turn weight into strength. Construction techniques used by these blocks reinforce the crankcase part of the

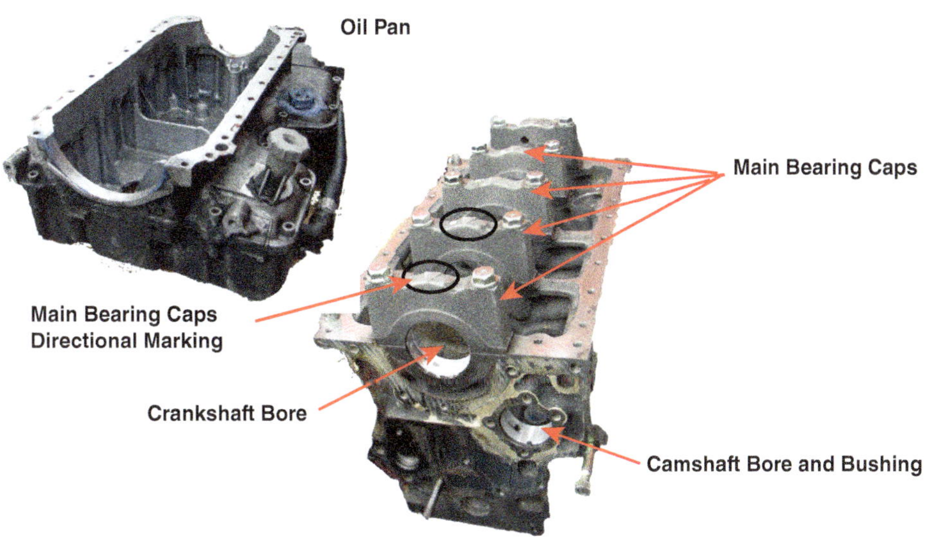

FIGURE 4-17 A low power output cast iron block design. The block walls extend only to the main bearing caps to reduce the height of the block. Only the oil pan forms the crankcase and reinforces the block and crankcase structure.

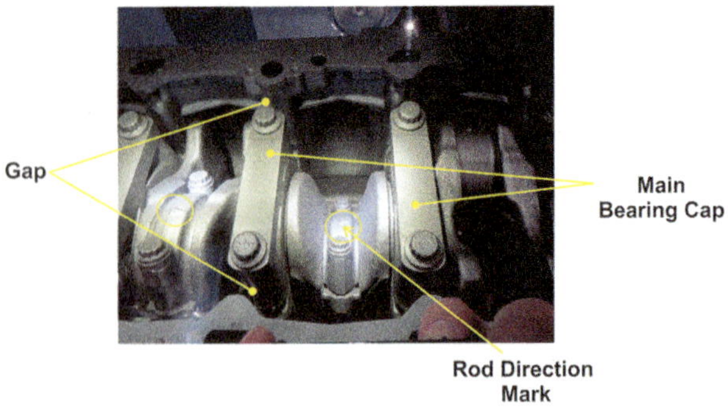

FIGURE 4-18 Deep-skirt blocks extend the block walls below the main bearing caps. Note also the direction indicators for the road bearing caps point toward the front of the engine.

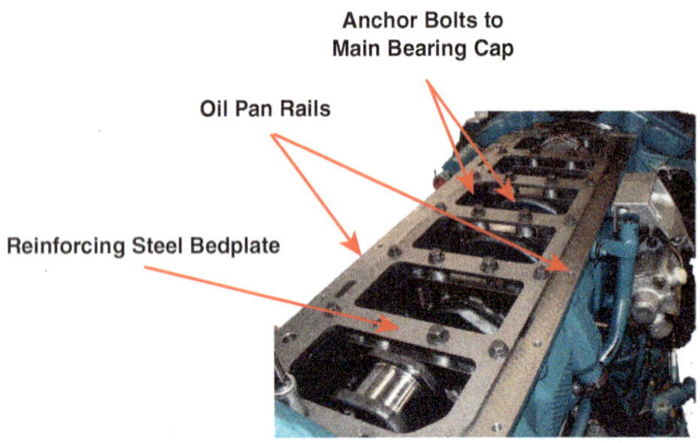

FIGURE 4-19 A steel bedplate attached to the oil pan rails and the main bearing caps of this high-torque-output, deep-skirt engine block adds extra rigidity and strength to the block.

cylinder block, which in turn will increase rigidity to resist bending and noise generating vibration as the power strokes on the crankshaft transmits energy capable of distorting bearing bores.

Deep-Skirt Blocks

A **deep-skirt block** is a design that refers to a block configuration where the bottom edge of the block or crankcase walls extends well below the crankshaft's centerline (**FIGURE 4-18**). The sides of the crankcase are separated from the main bearing caps to minimize transmitting the noise produced by combustion events out from the main bearing caps. Bearing caps are held in place with two or four vertically placed bolts. Lengthening the block walls below the main bearing caps enhances structural rigidity and operating smoothness which allows for a larger mating surface with the transmission. A reinforcing steel plate may also be attached to the oil pan rails and main bearing caps to enhance rigidity and strength of the block. The plate, whose thickness is approximately ⅜"–½" of steel, ties the oil pan rails to the main bearing caps to add further structural rigidity to the block (**FIGURE 4-19**).

Reducing vibration and increasing block rigidity can be further enhanced by adding curves to the deep-skirt crankcase walls. Also known as serpentine convolutions, the curves make the crankcase less flexible than flat surfaces would be. Adding ribs and webbing to the crankcase area is a design technique that minimizes crankcase distortions that would cause noise vibration and harshness in all types of blocks (**FIGURE 4-20**).

Crossbolted Blocks

Advancing beyond the traditional deep-skirt block technique is the crossbolted block. Also known as tie-bolted or bolster-bolted blocks, this design uses additional horizontal placed bolts to connect crankcase walls of the block to the main bearing caps (**FIGURE 4-21**). Crossbolted

FIGURE 4-20 The crankcase of this 5.0 L Cummins diesel used by Nissan pickups has a serpentine wall. The design adds rigidity and strength to its deep-skirt design.

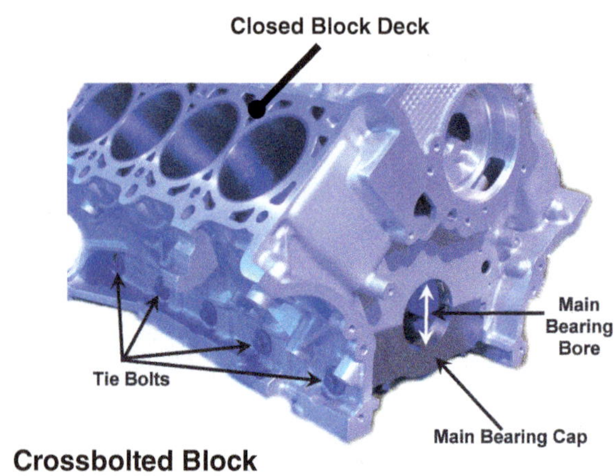

FIGURE 4-21 A crossbolted main bearing cap in an aluminum block with cast iron liners. Horizontal crossbolts or tie bolts attach the main bearing caps to the crankcase wall. A pair of vertical bolts also ties the main bearing cap to the block. Crossbolting is used to enhance engine block strength and rigidity while reducing noise and lowering engine weight.

blocks permit the use of more than the traditional two or four vertical bolts to hold the main bearing caps. Main bearing caps fit tightly between the short walls of the crankcase and are bolted both vertically and horizontally. Crossbolting adds the structural strength of crankcase walls to resist crankshaft flexing, and it stiffens the engine's structure, to reduce overall vibration. This type of block reinforcement technique permits further weight reductions because the crankcase walls can be shorter than in a deep-skirt block. Block height can also be lowered.

Ladder-Frame Blocks

A number of newer engine designs are incorporating ladder-frame construction features, which can strengthen a block while reducing weight, noise, and vibration (**FIGURE 4-22**). In ladder-frame blocks, the sides of the block extend exactly to the centerline of the crankshaft bearing bore. A separate additional section of the block structure, called a ladder frame, attaches to the crankcase and the oil pan (**FIGURE 4-23**). The main bearing caps are all integrated in this lower single-piece section of the block. Removing one bearing cap requires removing the entire ladder frame, which contains all the bearing caps. It is called a ladder-frame construction due to the ladder-like appearance of the crankcase-mounted bearing girdle (**FIGURE 4-24**). The ladder frame may also contain balance shaft housings and the oil pump. Although ladder frames add considerably to block rigidity and strength, they also can add to the cost and complexity of an engine. Another drawback, depending on construction technique, is that the gasket joint created by the device has the potential for leakage. Engine bearings cannot be serviced in frame or in chassis. Removing and replacing a bearing requires completely removing of the engine from the vehicle.

FIGURE 4-22 The 6.0 L Powerstroke and 6.4 L Powerstroke have a ladder-frame engine block. A ladder frame uses an additional block structure located between the cylinder block and oil pan, which incorporates the main bearing caps into a one piece component resembling a ladder, sometimes called a girdle.

Tunnel-Bore Blocks

Tunnel-bore blocks are one of the most smoothly operating and strongest of any cylinder block design. Integrating the main bearing bores into a single-piece block structure provides the highest strength and rigidity to the block. In a tunnel-bore block, as the name suggests, the crankshaft and main bearing enclosures are surrounded by the "tunnel" cast into the block (**FIGURE 4-25**). The crankshaft bearings are held in place by bearing support structures. After the bearing support structures are bolted around the crankshaft journals, the entire crankshaft assembly is lowered into the tunnel bore of the

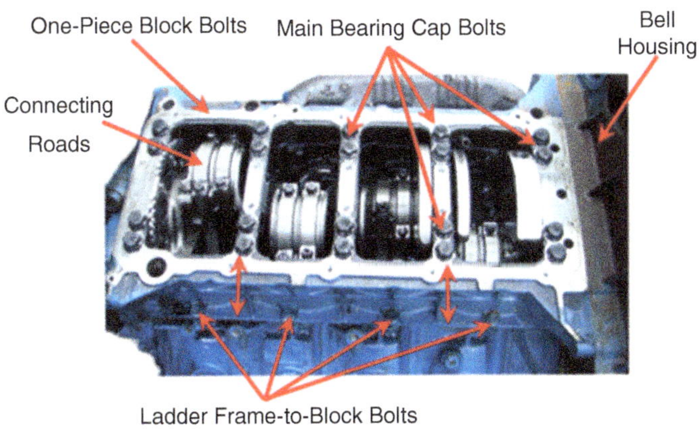

FIGURE 4-23 This 6.0 L ladder-frame block uses a two-piece crankcase. The upper section incorporates the main bearing caps into the crankcase structure.

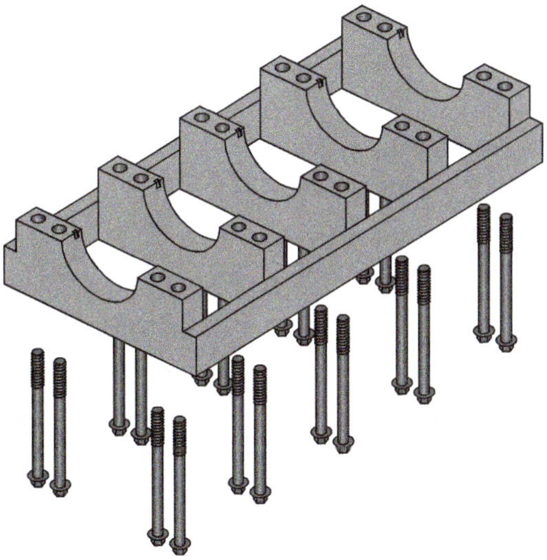

FIGURE 4-24 The ladder-frame block incorporates the main bearing caps into a single structural component, resembling the rungs of a ladder. The frame not only supports the crankshaft but also adds more structural rigidity and strength to the block, producing substantially less noise and vibration.

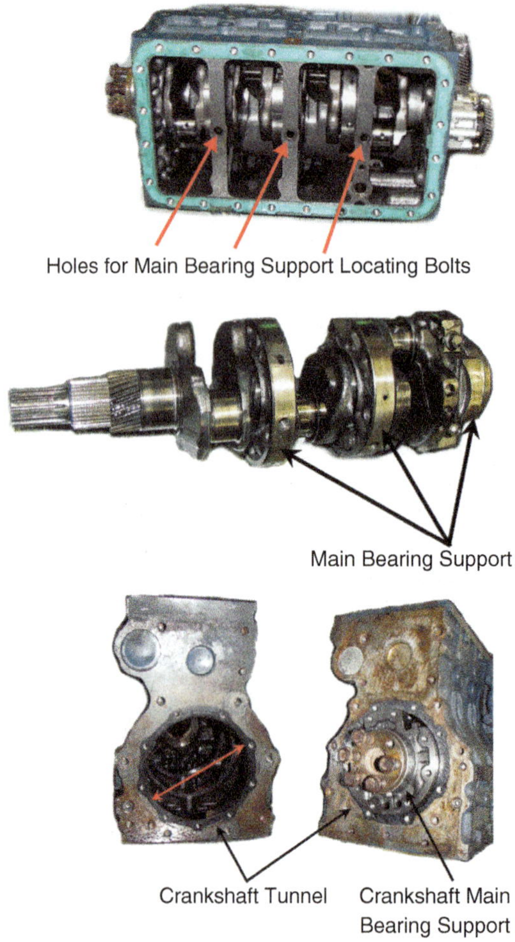

FIGURE 4-25 In this tunnel-bore engine block, the crankshaft and its main bearing supports are dropped into the tunnel during assembly. Locating bolts in the block aligns oil supply holes and prevents the main bearing supports from rotating with the crank.

block. Locating bolts in the block aligns oil passageways in the bearing supports structures with passages in the block. The locating bolts also lock the main bearing supports into position. No automobile use these blocks, but Kubota supplies these blocks for use in compact off-road equipment.

▶ Engine Block Cylinder Configurations

4-04 Identify and describe the methods used to form cylinders in an engine block.

The cylinder support section of the engine block uses several configurations, depending on the application, production cost, and expected vehicle lifecycle. Common cylinder construction types include the parent bore, wet sleeve, and dry sleeve, as well as the cast-in-place sleeve previously mentioned in the section aluminum blocks.

Parent Bore Block

A parent bore block, also commonly called a no-sleeve block, has holes cast and bored in the block for the cylinders. The pistons are inserted directly into these cylinder holes. No provision is made to add a cylinder sleeve to this type of block. The cylinder walls form a part of the block structure, giving it rigidity and strength. Parent bore blocks can be either the open-deck or the closed-deck type, with open-deck blocks fabricated by using forging dies rather than constructed through casting techniques (**FIGURE 4-26**).

Advantages

The major advantage of parent bore blocks are the lower initial cost of construction because machining and fitting cylinder sleeves is not required. Unlike sleeved blocks, the cylinder walls are structural members of the casting and add strength and rigidity to the block while

FIGURE 4-26 The 6.7 L Cummins on the right is an example of a parent bore block. A parent bore block forms the cylinders and block from a single casting. Water jackets and other internal passageways are cast into the block.

using less block mass. Thin cylinder walls with minimal material thickness between cylinders permit compact, lightweight construction. These blocks are best used in applications where chassis life and expected engine life are approximately equal. Water jackets formed inside the block permit cooling of cylinder walls all the way to the top of the block, unlike sleeved engines. This feature reduces carbon formation on pistons and other problems related to excess heat near the top of the cylinder.

Disadvantages

A major disadvantage of parent bore blocks is that while rebuilding or repairing the engine, a worn cylinder must be re-bored and honed. Re-boring requires special equipment available only at a machine shop. The engine must usually be removed from the chassis and disassembled completely to perform the machining process. The number of times the engine can be rebuilt is limited to available piston oversizes and the thickness of the cylinder walls. Parent bore blocks have reduced engine block life as compared to sleeved engines. Part of the reason is the block materials form the cylinder walls. Without costly alloys and hardening techniques, the cast iron cylinder walls will wear at a faster rate than hardened cylinder sleeves will. However, parent bore CGI blocks do not wear as fast as cast iron blocks. Since conventional cast cylinder walls are irregularly thick, heat transfer differentials cause varying cylinder dimensions and accelerated wear. Cast iron cylinder wall surfaces are not as easily hardened as sleeved engines.

> ▶ **TECHNICIAN TIP**
>
> Coolant temperatures are hotter at the rear of the engine, where coolant has to travel the farthest to remove heat. A 10–20°F temperature differential exists between the temperature of cooler coolant near the front of the block and the rear of the block, which is farthest from the radiator. Some engine rebuilders recommend a larger piston-to-cylinder wall clearance in the rearmost cylinders of some V8 diesel engines, to accommodate the greater thermal expansion of parts.

Cylinder Sleeves

Sleeves—or liners, as they are commonly called—are replaceable wear surfaces making up the cylinder wall (**FIGURE 4-27**). Sleeved engines are not used in light duty diesels due to their high cost, but the large numbers of in service sleeved engines justifies a brief discussion in a survey of diesel block construction techniques. Constructed to the exact inside diameters and lengths, the cylinder sleeves may be replaced individually if they become worn or damaged prematurely. Cylinder sleeves are typically induction hardened and manufactured using special alloys to produce durable, wear-resistant surfaces. Sleeves are commonly used

FIGURE 4-27 Replaceable cylinder sleeves enable a worn-out engine to be quickly rebuilt and restored to factory specifications.

in medium and large bore engines operating under high-mileage, extended service conditions. Today's heavy-duty diesel truck with sleeved cylinders will operate for up to 1 million miles or close to 1.6 million km before needing overhaul. Long distance travel means that an engine can wear out faster than the chassis. To avoid expensive engine or vehicle replacement, replacing the liner sleeves by using cylinder kits made up of pistons, rings, and liners restores engine to factory specifications.

Dry Sleeve Block

A dry sleeve block is designed with a bored or honed hole in the block that allows no coolant contact with the cylinder sleeve. The sleeve is inserted into the bored hole as either a slip or press fit. The liner is installed with an interference fit that will not usually have a flange and is machined flush with the block deck. A slip fit liner will use a counter bore machined into the block to support the bottom of the liner. Pressure applied by the cylinder head to a slight protrusion of liner flange will prevent liner movement.

Advantages

A replaceable cylinder sleeve can be used during block manufacturing to restore the cylinder block to its original specification after worn sleeves are removed and new ones installed. Since the sleeve is fitted into a bored hole in the block, it does not have contact with coolant that could cause sleeve deterioration. Additionally, the absence of coolant contact means there is no need to seal the sleeve to prevent coolant leakage into the crankcase. Cylinder sleeves may be replaced individually if they become worn prematurely or damaged.

Disadvantages

Since the coolant is not in direct contact with the dry sleeve, heat transfer from the combustion chamber to the coolant water is not as rapid as it would be with a wet-type sleeve. This slow heat transfer may potentially result in short engine life, cylinder damage, and limitations to power output per cylinder. Extra steps in manufacturing and machining the block are required for sleeve installation. Dry sleeved blocks tend to weigh more than parent bore blocks.

Wet Sleeve Block

A wet sleeve block is designed with a number of large holes into which the cylinder sleeves are inserted. Coolant has direct contact with the outside of the sleeve, and there is no supporting cylinder bore structure, like a dry sleeve around the wet sleeve. Wet sleeves are thick enough to withstand the higher heat loads and forces of combustion in high horsepower engines (FIGURE 4-28).

FIGURE 4-28 Wet liners make direct contact with coolant. O-rings prevent coolant from leaking out of the water jackets and into the oil pan.

Advantages

The major advantage of wet sleeve liners is the direct contact of coolant with the sleeve that enables rapid heat transfer from the cylinder to the coolant. Sleeves are easily removed and installed during engine rebuilding to restore the cylinders to original specifications.

Disadvantages

The major disadvantages with wet sleeve liners are the problems associated with preventing internal engine coolant leaks. The O-ring seal (if used) at the lower part of the liner can deteriorate and leak coolant into the lubrication oil. Additionally, liner vibration caused by combustion forces can exaggerate the effects of cavitation. Greater attention to maintaining the

FIGURE 4-29 Pin holes in this liner are caused by cavitation erosion. Cavitation erosion produces pinholes in cylinder block walls, injector tubes, water pumps, and cylinder heads.

cooling system is required in wet sleeve engines, to prevent liner cavitation. **Cavitation erosion** is a condition caused by liner vibration, producing pinhole size holes in the sleeve (**FIGURE 4-29**). The holes enable coolant to enter the cylinder or combustion gases to enter and pressurize the cooling system. A second problem is ensuring that liner protrusion above the block deck is within specification and even across all cylinders. Without correct protrusion, combustion gases will leak into the coolant; head gaskets will leak; and other structure problems can result.

Further information about various types of liners and liner service are covered in the textbook *Fundamentals of Medium/Heavy Duty Diesel Engines*.

▶ TECHNICIAN TIP

Parent bore and sleeved diesel engines have a unique problem, known as cavitation erosion. Cavitation erosion is caused by the collapse of tiny water vapor bubbles formed when coolant vaporizes on hot cylinder wall surfaces. These vapor bubbles collapse when cooled by surrounding coolant or when pressure increases around the vapor bubble. Pressure exerted by the force of bubble collapse is reported to be as high as 60,000 psi (413,700 kPa). The force of vapor bubble implosion against cylinder walls, cylinder liners, and other components can eventually perforate them. For this reason, diesel engines require the addition of supplemental coolant additives to the cooling system, which contains nitrite. The nitrite chemical additive forms a sacrificial barrier to protect the coolant side of the cylinder wall or liner against pitting due to cavitation erosion, which can quickly destroy the cylinder walls and liners.

▶ Cylinder Block Service

4-05 Identify and explain methods used to correct cylinder wear.

Cylinder block service primarily deals with wear caused by high engine hours, accumulated distances, and contaminants. Diesel engines can outlast spark-ignition counterparts for several reasons. The primary reason is slower engine speeds. Maximum engine speed is lower and torque output can reach its peak at engine speeds just above idle. This eliminates the need for additional gearing to boost torque while keeping engine rpm low. A second reason is that gasoline-fueled engines tend to wash friction reducing lubrication oil from cylinder walls. Since fuel does not enter the cylinders of a diesel during intake stroke, washing and diluting cylinder wall lubricant does not take place, which enables greater engine longevity. More durable cylinder and block materials can withstand higher combustion forces, which helps to further extend engine life. Small bore diesels can last

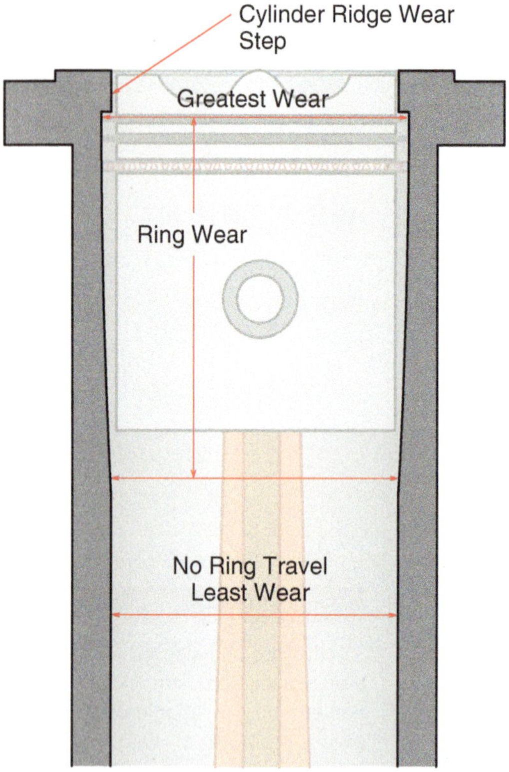

FIGURE 4-30 Wear patterns in a cylinder. A carbon ridge is formed from carbon deposits on the untraveled portion of the cylinder wall surface above the top ring. The ridge must be removed prior to piston removal to prevent damage to the piston and rings as they pass through the ridge.

three to four times longer than spark-ignition engines, but eventually, cylinders wear and require service (**FIGURE 4-30**). Dirt ingestion through leaking air intakes or coolant contamination of engine oil can take place, which also dramatically accelerates engine wear. For this reason, cylinder block service consists primarily of repairing and restoring the cylinder wall finish.

Cylinder Wall Finish

Cylinder walls have a crosshatch finish made by honing stones, which serves two purposes. The first is to retain oil. Lubricating oil is left in the fine lines of the crosshatch to help lubricate rings and pistons. The fine lines will intersect one another at an angle of between 45° and 60° and are deposited with lubricating oil by the pistons oil control ring (**FIGURE 4-31**). A second purpose for a crosshatch finish is to provide a surface to enhance sealing piston rings and the cylinder wall. During the initial engine break-in period, the finer lines are more easily worn away than a solid wall of iron. Because a cylinder is not perfectly symmetrical, the crosshatch will allow initial wear of the ring face material into the cylinder wall to compensate for irregular ring – wall contact, which eventually produces a tighter seal between the two parts.

Highly polished cylinder walls having no crosshatch will not allow residual oil to cling to the cylinder walls. Without the critical oil film on the cylinder walls, a gas-tight piston ring seal cannot be achieved. Consequently, poor sealing between the piston ring and cylinder wall causes compression and power loss due to blowby and piston and ring scuffing. Poor ring sealing also leads to high oil consumption because oil is not scraped from the cylinder wall and can flood the combustion chamber (**FIGURE 4-32**).

Cylinder crosshatch is installed when an engine is manufactured and during proper cylinder reconditioning processes during an engine rebuild. Honing stones are used to produce the crosshatch finish of the cylinder walls after they are bored to a nominal size.

To reduce the time and distance required for engine break-in, manufacturers use a process called plateau honing. This involves honing the cylinders and then removing the peaks of the cylinder crosshatch with specialized brushes generally made of nylon-like material. Plateau honing also reduces engine oil consumption during break-in.

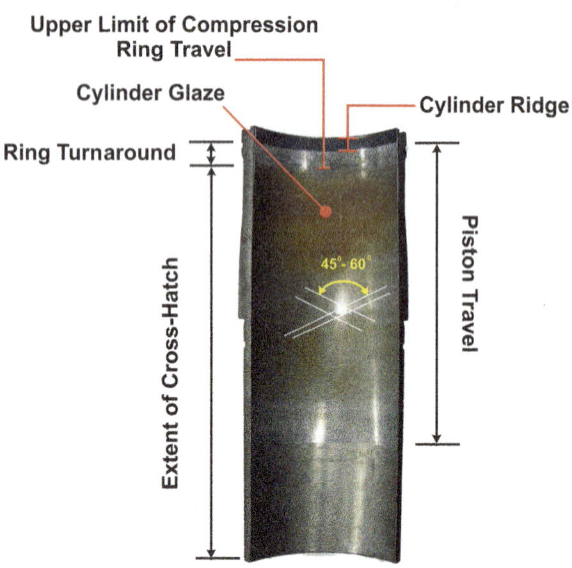

FIGURE 4-31 Crosshatch cylinder finish lines intersect each other and retain lubricating oil.

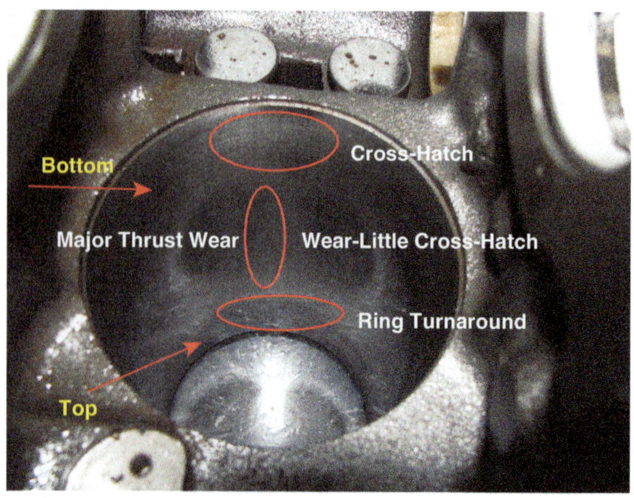

FIGURE 4-32 A missing crosshatch is evidence of excessive cylinder wear. Having a crosshatch on the cylinder wall is important to provide a surface for piston rings to wear into, or seat, during the break-in period.

Cylinder Wall Wear

Diesel engine cylinder walls last longer than their gasoline counterparts, for several reasons. Slower engine speeds reduce the amount of piston travel. Diesel fuel can also act like a lubricant to minimize wear from friction—unlike gasoline, which is a solvent for lubricating oil. Special alloys and processes that harden cylinder walls further increase the durability of diesel blocks. However, cylinder block walls progressively wear with increasing engine hours and distance traveled. The greatest amount of wear occurs near the uppermost part of the compression ring travel and diminishes downwards over the ring swept area of the cylinder. This gives a worn cylinder a slight bell-shaped wear pattern, with the widest dimension observed at the top of the cylinder. The reasons for the greatest amount of wear near the top of the cylinder wall causing the cylinder wall to taper include the following:

- Higher cylinder pressures near the top part of piston travel increase ring pressure against the cylinder walls. (Gas pressure behind the rings forces the piston rings outward from their grooves.)
- Higher cylinder temperatures make the cylinder wall softer and more prone to wear.
- The lack of upper cylinder wall lubricant is due to the fact that lubrication oil can be carried into this area only on the face of the compression ring. The oil control ring is lower on piston and below the compression ring.

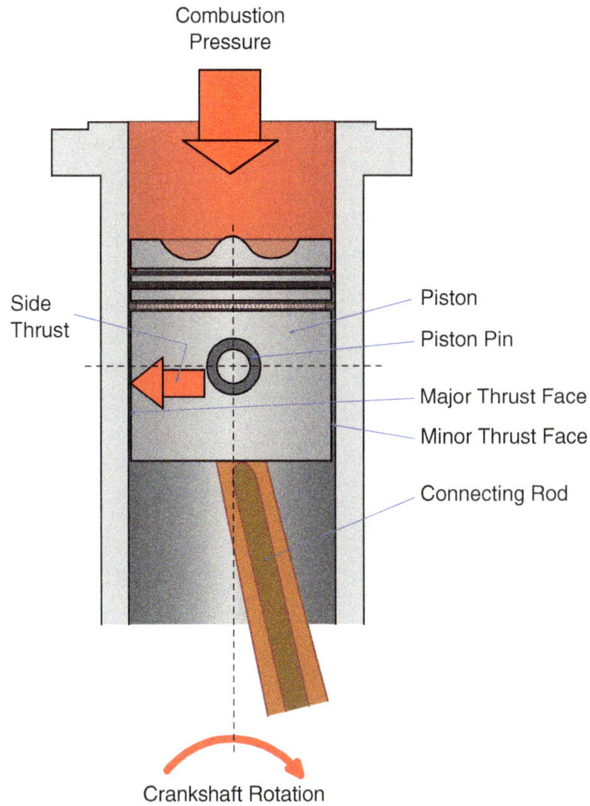

FIGURE 4-33 The angle formed between the connecting rod and the piston contributes to side thrust wear of the cylinder wall.

Long stoke engines will have more wear mid-stoke as well, due to greater forces of major thrust. The side thrust is produced by the angle of the connecting rod with the piston, which tends to push the piston sideways into the cylinder wall when cylinder pressures are high (**FIGURE 4-33**). The angle formed between the connecting rod and the piston forces the piston against the cylinder wall during compression and power stroke. During power stroke, the force is greatest, producing a major thrust force and a minor thrust force during compression stroke. The longer stroke distance of the diesel makes for higher thrust angle forces.

Dusted Cylinders

Diesel engines are air-breathing machines that consume tremendous quantities of air due to their lean burn combustion principles. Any leak in the air intake system can allow dust and dirt into the cylinders, where it can stick to the oil film on the cylinder walls. Lubricating oil provides a sticky surface to allow the dirt to adhere to the cylinder wall, which forms an abrasive lapping compound for the cylinders and pistons rings. When engine wear is rapidly accelerated due to air intake leaks, the condition is referred to as "dusting out." Coolant may accelerate wear too, but it more rapidly destroys engine bearings before contributing to significant cylinder wall wear.

Measuring Cylinder Wall Wear

Several methods can be used to measure the amount of cylinder taper inside a cylinder:

1. One is to measure the amount of piston ring end gap in a cylinder. By measuring the end gap at the top of the cylinder, in the ring turnaround area, the middle, and the bottom, cylinder taper can be calculated. End gap is measured after "squaring" the ring inside the cylinder with a piston. A 0.001" (0.025 mm) increase in cylinder wall diameter will produce approximately a 0.003" (0.076 mm) increase in ring end gap. A typical maximum taper wear of a cylinder is 0.001" (0.025 mm) per inch (2.54 cm) of cylinder diameter up to 0.005" (0.127 mm) maximum wear in larger cylinders.
2. A quick and very effective method of evaluating excessive cylinder wear is to determine the percentage of missing cylinder wall crosshatch in comparison to the remaining

crosshatch. The presence of crosshatch on the cylinder wall is important to provide a surface to seat piston rings and to retain oil for piston and ring lubrication. When crosshatch is absent, approximately 0.005" of wear has taken place. This means that highly polished walls without crosshatch indicate excessively worn cylinders, which are not capable of properly sealing the piston inside the cylinder bore. If 20% or more crosshatch is missing, the engine has 0% service life. Similarly, 10% missing crosshatch in the ring turnaround area of the cylinder wall indicates 50% or less remaining engine life. To visually evaluate cylinder wear, follow the steps in **SKILL DRILL 4-1**.

3. A dial bore gauge can be used to measure cylinder taper but not cylinder diameter. These gauges, which are dial indicators mounted on a bar that extends into the cylinder, are used when honing or performing other machining processes to ensure a symmetrical consistent dimension is produced (**FIGURE 4-34**).
4. Using a micrometer and inside telescoping gauges is another method. An inside telescoping gauge is used to gauge the cylinder diameter. An outside micrometer is then used to measure the width of the inside telescoping gauge.

SKILL DRILL 4-1 Visually Evaluating Cylinder Wear

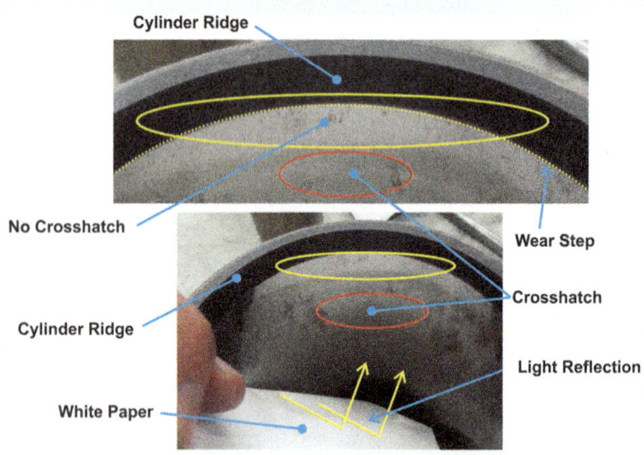

1. One effective method to visually evaluate cylinder wear is to examine the ring turnaround area.
2. Using a piece of paper to reflect light directly into the crosshatch helps make the cylinder wear more visible.
3. If 20% or more of the crosshatch is missing, 0% service life remains. If 10% of the crosshatch is missing, 50% service life remains. An engine with a 5" (13 cm) stroke and 0.25" (6 mm) of polish has 75% service life remaining.

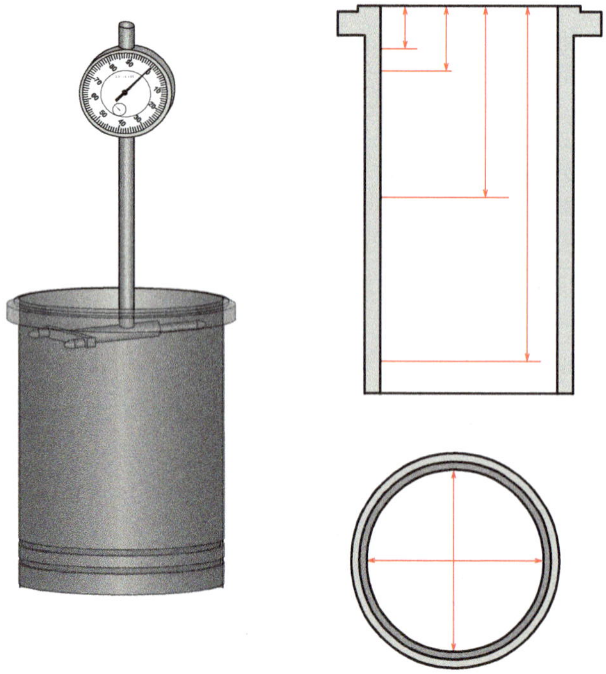

FIGURE 4-34 A dial bore gauge is used to measure cylinder wear due to taper.

▶ Crankshafts: Construction and Function

4-06 Identify and explain the purpose of unique features used to construct diesel engine crankshafts.

The main purpose of the crankshaft in a reciprocating engine is to convert the linear motion of the connecting rod assemblies to rotational movement. While having a rugged function, the sophistication of the crankshaft is reflected in manufacturing costs, which makes the component one of the most expensive parts of an engine after the fuel system and turbocharger. The mechanical stress caused by rotational speeds, combustion pressure, thrust loads from the geartrain, and other mechanical vibrations require the crankshaft to be especially strong and precisely manufactured. Two connecting surfaces are the main and connecting rod bearing journals. Main bearing journals are located along the crankshaft centerline. Main bearing caps support the crankshaft as it rotates. Connecting rod journals are offset from the crankshaft centerline main journals. The distance of rod journal offset from the main bearing journals is known as the throw. The length of throw determines the stroke length of the engine. The arrangement of the rod journals, or throws, allows an even amount of spacing between power impulses. For example, a six-cylinder engine will have 120° of spacing between the rod journals (720° of rotation/six cylinders) (**FIGURE 4-35**). The throws are further arranged to evenly distribute the firing order over the length of the shaft to balance out vibration and stresses. For example, in in-line, six-cylinder engines, the throws of cylinder are paired 1-6, 2-5, 3-4. This distributes power impulses evenly throughout the crankshaft.

When the engine is operating, tremendous inertial forces are generated by the reciprocating mass of the connecting rod and piston assembly. To balance these forces, large counterweights are formed on the crankshaft 180° opposite the rod journal (**FIGURE 4-36**).

Fillet radii are another critical feature often unique to a diesel crankshaft. The fillet radius refers to the circular-shaped radius formed in the 90° intersection between the crankshafts rod and main bearing journals and the cheeks of the crankshaft. Torsional twisting of the crankshaft causes this area to be one of the most stressed areas of the crankshaft. Leaving only a sharp 90° angle between the two surfaces could induce a crack through a stress riser. Imparting a smooth circular radius to this area strengthens the crankshaft and minimizes the possibility of a fracture.

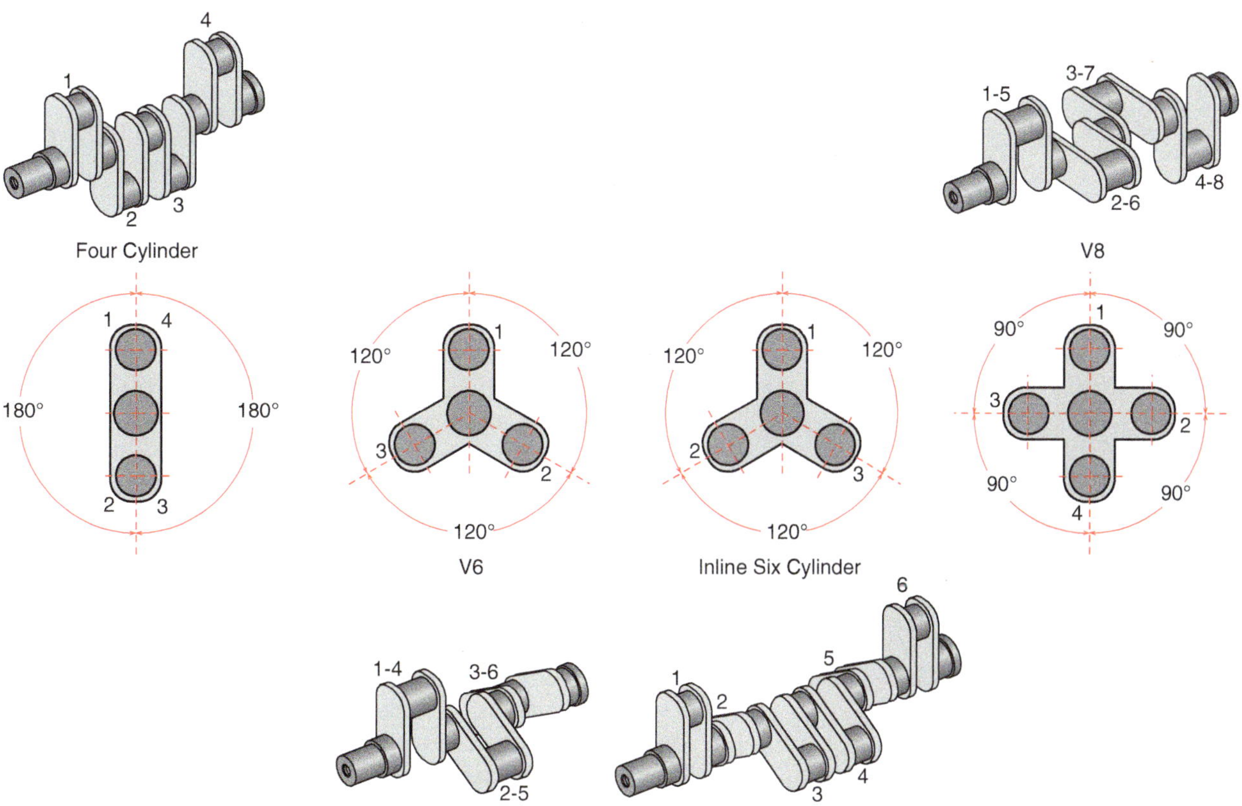

FIGURE 4-35 Arrangement of crankpins spaces firing impulses evenly. In-line, six-cylinder engines space crank throws at 120°.

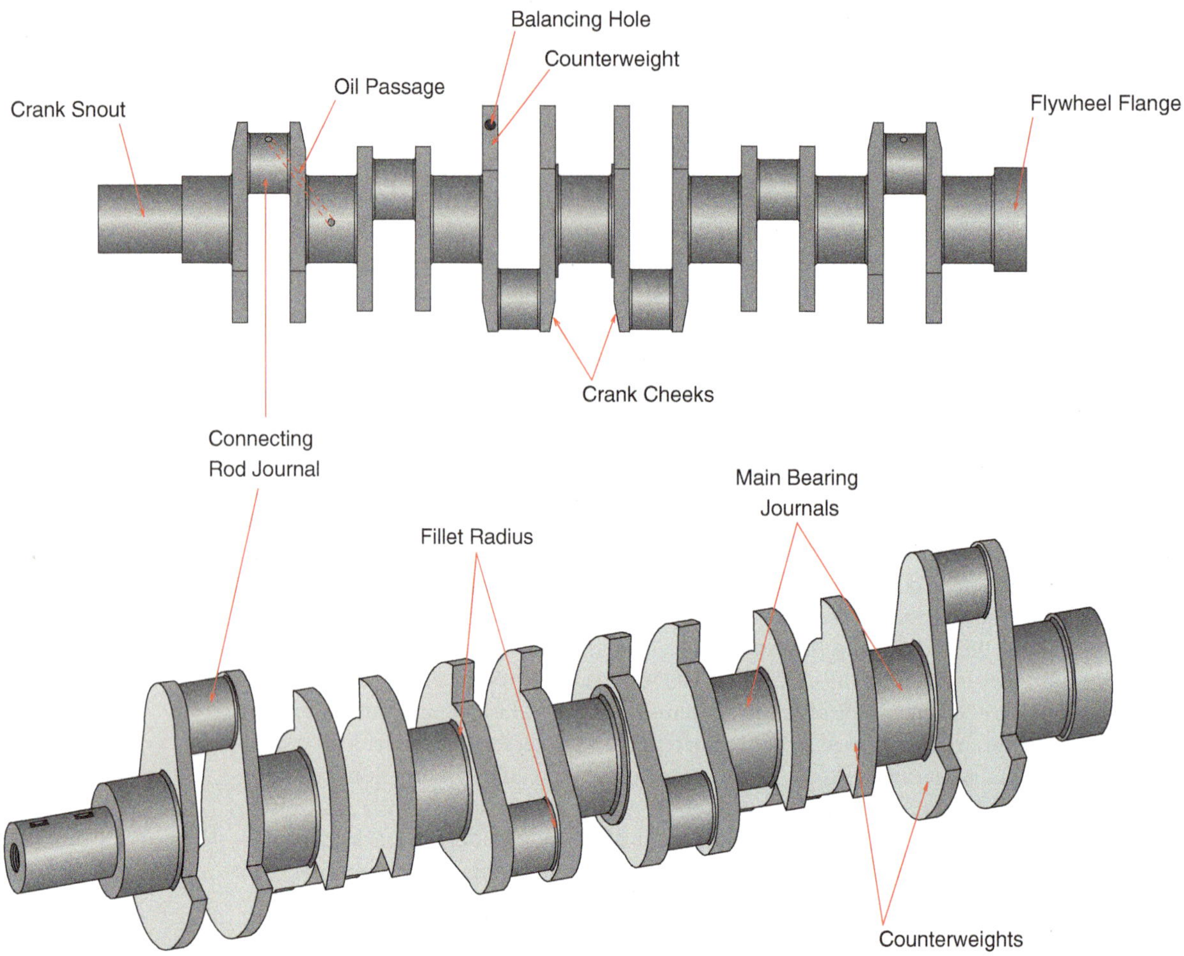

FIGURE 4-36 Crankshaft terminology. Crankshaft counterweights offset the mass of the connecting rod and piston assemblies.

Crankshaft Journal Hardening

Crankshafts in diesel engines are commonly forged from high carbon steel alloys because forging produces a denser and stronger component. Following the machining of crankpins, counterweights, and other surfaces, journal surfaces are hardened to improve the durability of a diesel crankshaft. The two most common processes used to harden bearing surfaces of the crankshaft are induction hardening and nitriding.

FIGURE 4-37 Induction hardening causes discoloration of metal surfaces hardened through this process. This crankshaft has discoloration around hardened journals, which is not evidence of damage from overheating but is caused by heating during induction hardening.

Induction Hardening

To induction harden a crankshaft, the journal is wrapped in a coil of wire that has high amperage, alternating current passed through it. The journal surface will have current induced into it by mutual induction, much like the secondary winding of an ignition coil. This current flows over the journal surface, heating the metal surface and changing its molecular structure (**FIGURE 4-37**). Water quenching the heated surfaces produces a very hard surface finish. Typically, induction hardening is effective to no more than 0.030" (0.762 mm) deep. It may go deeper, but it becomes softer with increasing depth. This is one reason that explains why machining diesel crankshafts is not a good work practice: The hardening material is lost and most rebuilders are not able to restore it during rebuilding and repair. Spray welding the crankshaft with high alloy steel is a more effective technique for crankshaft restoration. This process is performed by the OEM by using advanced techniques and facilities.

Nitriding

Another crankshaft hardening technique is nitriding. Chevy diesels, which include the Duramax and the older 6.2 L and 6.5 L, use this technique. The process involves heating the shaft and introducing cyanide salt or liquid (CN_3) into the heating chamber with the crankshaft. Slowly cooling or quenching the crank with these substances in its heated condition causes the nitrogen and carbon to molecularly bond with metal surfaces to create an extremely hard journal. This hardening finish does not extend more than 0.012" (0.305 mm) deep but is one of the hardest journal surface finishing techniques.

Crankshaft Lubrication

To lubricate the main and rod bearings, engine oil needs to be directed in sufficient quantities to lubricate, cool, and clean the crankshaft journals and bearings. Engine oil is delivered to the oil galleries in the block and then distributed to the main bearings by using holes present in the upper main bearing shells. A cross-drilled hole in the crank main bearing journal uniformly distributes oil around the crankshaft journal (**FIGURE4-38**). From there, the crankshaft has holes drilled from main bearing cross-drilling to the rod journal. Cross-drilling rod and main bearing journals means each main bearing is connected to an adjacent rod bearing through an oil passageway. The exception to this would be the rear main, which in some cases has a smaller oil hole into the block only for lubrication of one main journal (**FIGURE 4-39**). The requirement to supply only one main bearing journal relieves lubrication oil pressure exerted against the rear main seal, which minimizes potential leaks and oil weepage from the rear main seal.

> ▶ **TECHNICIAN TIP**
>
> Grinding diesel crankshafts to undersize dimensions to repair damaged journal surfaces is common in gasoline-fueled engines. Diesel cranks, however, are almost always hardened, and grinding to repair journal damage will remove the surface hardening. Furthermore, a fillet radius ground into each crankshaft journal is difficult for most machine shops to restore. For these reasons and because most manufacturers do not recommend the practice, undersize crankshaft bearings are not commonly available for diesel crankshafts.

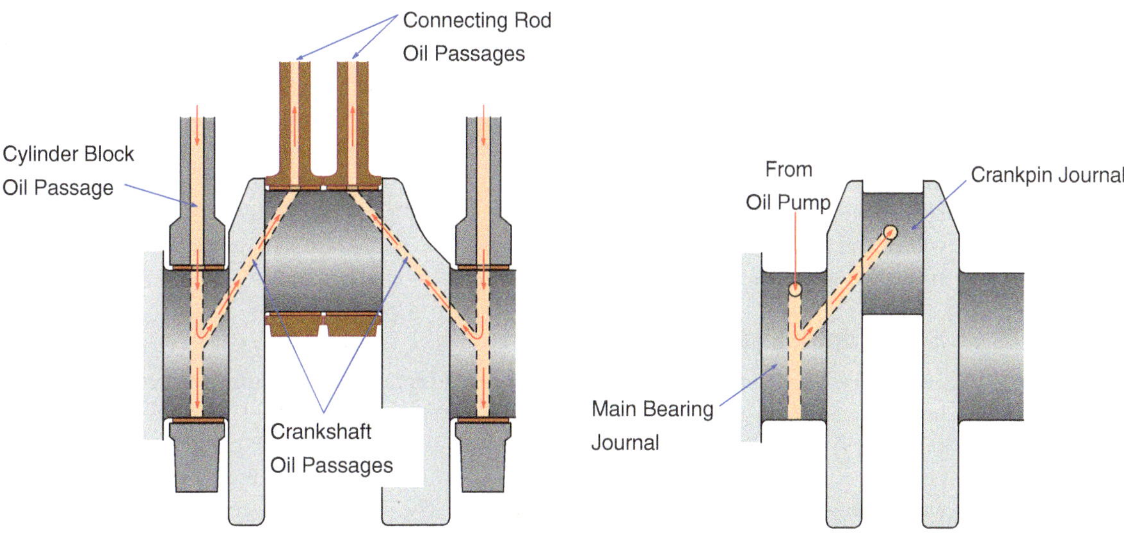

FIGURE 4-38 Drilled passageways through the crankshaft distributes pressurized oil to the rod bearings and piston pins in some engines. Passageways drilled between the oil rail and the main bearing supply oil to the crankshaft.

FIGURE 4-39 The location of a rear main seal, which prevents oil from leaking out of the crankcase.

Crankshafts and Engine Balance

As noted in previous paragraphs outlining techniques to construct strong durable block designs, it's important to understand that combustion forces operate to push apart the pistons and crankshaft from the block. Understanding this concept helps explain why in-line engines are preferred in vocational and heavy-duty diesel applications since the crankshaft is supported by more main bearings than in V-type configurations (seven main bearings in in-line, six-cylinder vs. four in a V-banked cylinder engine). This means the in-line block is better able to handle higher power output with greater durability than V-block configurations. The advantage of a V-block is that it permits lower hood profiles due to a more compact engine design. The dilemma OEM's have selecting light duty diesels such as Ford's 3.0 L V-6 and GM's 3.0 L inline 6-cylinder is choosing between an engine that can be packaged better for use in a vareity of chassis' or choosing a block configuration with potentially greater durability and power output. A V-6 can be installed longitudinally or transversely in rear or front wheel drive vehicles. An inline engine can generally fit into only a rear wheel drive pick-up or large delivery van. Cylinder banks leaned 60°, 72°, or 90° result in an engine with less overall height (**FIGURE 4-40**). The spacing of cylinder banks also helps reduce imbalance forces in engines caused by the spacing of firing impulses. For example, firing impulses of a V8 engine occurring every 90° of crank rotation are evenly spaced 90° apart in cylinder banks with 90° spacing. In V6 configurations, a 60° bank spacing separates firing impulses taking place every 120° by an even 120° of crankshaft rotation. Mercedes-Benz uses a 72° bank spacing, plus the use of a counter-rotating balance shaft in between the cylinder banks for its Bluetech 3.0 L engine. The crankpins are offset by 48° to eliminate vibration problems from uneven spacing between combustion events, making the engine evenly spaced when firing with 72° bank spacing.

Primary Balance

Primary balance is achieved when the crankshaft counterweights offset the weight of the piston and connecting rod assembly. Most engines use a counterweight on the crankshaft for each cylinder to counteract the inertia of the piston and connecting rod assemblies to minimize engine vibration. In in-line, six-cylinder engines, crankshaft counterweights can be removed from cylinders 2 and 5 since the inertia forces generated by other piston pairs will cancel imbalance forces. Externally balanced engines will use weights on flywheels or ring gears to help achieve good primary balance.

Secondary Balance

Secondary balance is achieved when the inertia forces of one piston counterbalances the inertia of another (**FIGURE 4-41**). Since all the pistons are in one plane in in-line engines, secondary balance is easily achieved. Inherent in the in-line, six-cylinder design is the advantage pairing pistons, which move simultaneously in the same direction and plane against those moving in the opposite direction. The dynamic forces of piston weight can be canceled out by the forces of pistons moving in the opposite direction. Horizontally opposed engines also have the best secondary balance. In contrast, V-engines have a natural secondary imbalance due to the pistons operating in different planes. Piston movement may cause engine vibration because pistons do not move in the same plane simultaneously. The direction of movement may be 60°, 72°, or 90° apart. An in-line, six-cylinder engine is

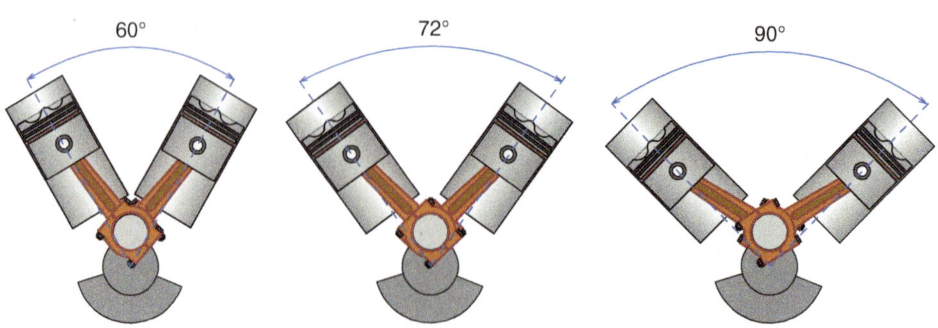

FIGURE 4-40 Common cylinder bank spacing arrangements.

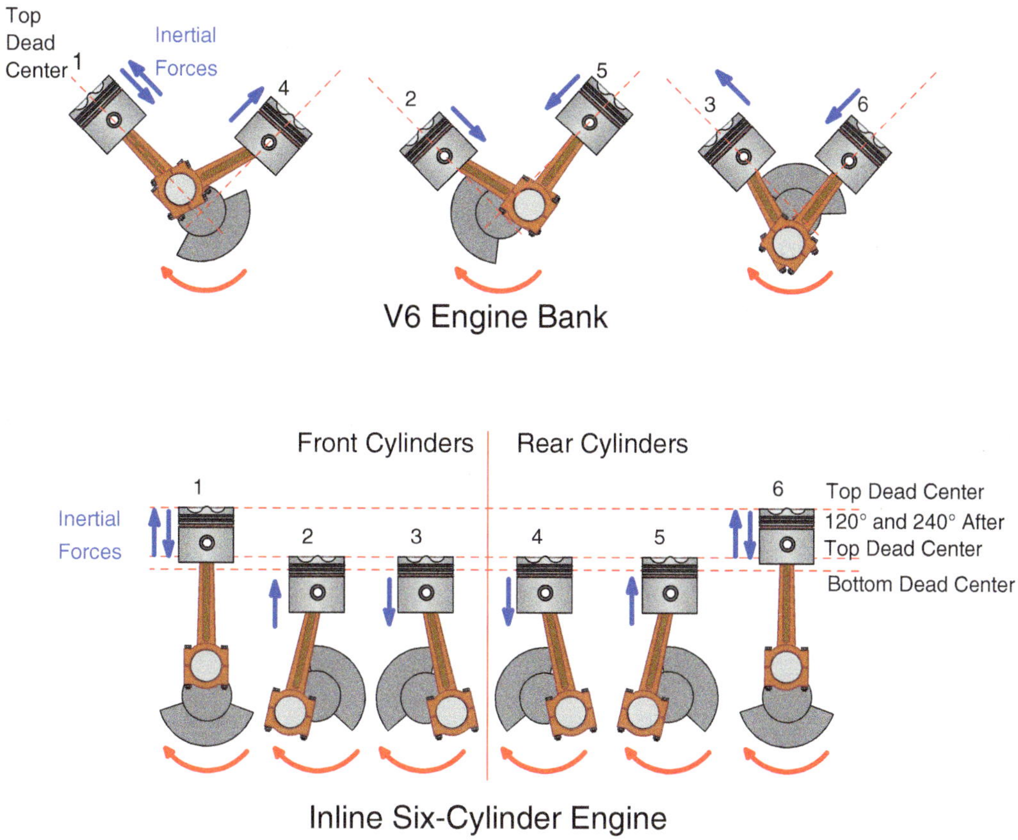

FIGURE 4-41 Secondary balance requires the inertia forces of piston pairs to cancel out the inertia of pistons traveling in the opposite direction.

FIGURE 4-42 Firing order of a Duramax V8.

also among the most balanced configurations for the internal combustion engine, possessing excellent primary and secondary balance characteristics.

Firing Orders

Firing orders and cylinder numbering are important to technicians when locating cylinders with fault codes associated with their operation. Cylinder numbering begins from the front of the engine with the most forward cylinder being the number 1 cylinder and ends at the rear cylinder with the last number in the firing order. The leading bank of a V-type engine identifies cylinder 1 (**FIGURE 4-42**). Since the flywheel is the reference point for engine rotation, the front of the engine is opposite the flywheel. Firing orders are also arranged to produce even spacing between power impulses and to distribute loads evenly along the crankshaft. This arrangement minimizes damaging torsional or twisting forces along the crankshaft. An even number of degrees of crankshaft rotation between cylinder power impulses minimizes engine vibration.

▶ Vibration Dampeners

4-07 Identify and explain the function of vibration dampeners.

Two potentially destructive forces applied to a crankshaft are harmonic and torsional vibration. As previously mentioned, torsional vibration is observed as the deceleration and acceleration of the crankshaft during rotation and is caused by the alternating combustion and compression pressure in each cylinder. At specific engine speeds, accessory drive belts can be spontaneously thrown and driveline vibration can occur due to excessive, uncontrolled torsional vibration. Alternators, air conditioning compressors, and other accessories will loosen and fall off due to excessive torsional twisting of the crankshaft transmitted to components through the gear train and drive belts. Cylinder pressure applied against the piston and connecting rod assembly is also intense enough to deflect and twist the crankshaft while it turns. The twisting event eventually results in a rebound of the crankshaft as the journal snaps back

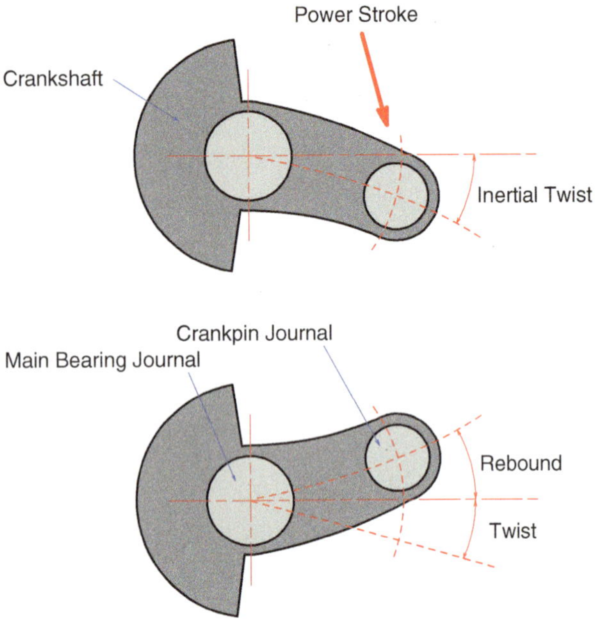

FIGURE 4-43 Harmonic twisting is different from torsional vibration. A crankshaft will actually deflect under power impulses, producing harmonic wave. Torsional vibration involves the rhythmic speeding up and slowing down of the entire crankshaft.

in the opposite direction of rotation. The effect of this harmonic stress is telegraphed through the geartrain and driveline, and if not properly controlled, harmonic and torsional vibration causes rapid main bearing and geartrain wear and possibly causes the crankshaft and camshaft to break as both torsional and harmonic to forces are transmitted through the geartrain. Harmonic vibration or waves are like the plucking and subsequent vibration of a guitar string. The frequency of these harmonic vibrations allows pressure waves to be transmitted back and forth along the length of the crank, much like an instrument string or a long wire of a clothesline. Nodal points—the place where the amplitude of the wave form is reversed—are formed by harmonic waves along the crank and are points of potential failures due to harmonic vibration (**FIGURE 4-43**).

To control the destructive effects of both torsional and harmonic vibrations, diesel engines are equipped with a vibration dampening devices, which reduce both torsional and harmonic vibrations. The mass of the flywheel can substantially suppress torsional vibration at its end of the crankshaft. At the flywheel, changes in crankshaft velocity are minimized by the inertia of the flywheel that resists speeding up and slowing down with each change in cylinder pressure. The large diameter of the flywheel dampens harmonic waves, which tend to bend the crankshaft away from the crankshaft centerline. To reduce torsional and harmonic vibrating forces at the opposite end of crank, a vibration damper is located at the front of the crank. Like the dampening of a soft glove on a ringing metal bell, the front vibration dampener quickly absorbs harmonic vibrations.

Balance shafts are another vibrating dampening mechanism used to cancel out rhythmic forces produced by long or uneven delays between firing impulses from an engine. In smaller three- or four-cylinder engines, the torsional vibrations are more pronounced since the crankshaft is allowed more time to decelerate between firing impulses happening every 180° or longer in comparison to 120° in six-cylinder engines or 90° in V8 engines. Balance shafts can also help counterbalance the fore and aft engine movement in V-6 engines casued by the inertia forces of pistons changing direction in an engine (**FIGURE 4-44**).

The following are four types of dampeners used on diesel engines:

1. elastomer dampeners
2. viscous or fluid dampeners
3. combination dampeners
4. rotating pendulum vibration absorbers.

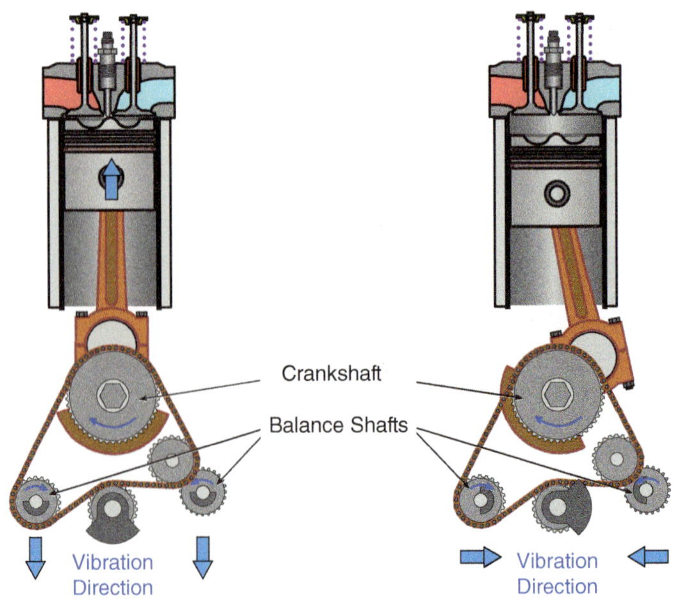

FIGURE 4-44 Balance shafts help correct for secondary imbalances caused by changes in piston direction.

Elastomeric Dampeners

Elastomer dampeners refer to vibration dampeners using rubber as the primary material to absorb harmonic vibrations and reduce torsional stress on the crankshaft (**FIGURE 4-45**). These are made of three parts. The inner hub is attached to the crankshaft, which connects to a heavier, outer iron inertia ring through a rubber or elastomeric ring. The dampening action occurs as the outer ring oscillates with speed changes relative to the inner ring movement. For example, when the crankshaft slows, the inner ring is sped up by the inertia of the outer ring. This means that when the inner ring slows down, the outer ring transmits inertial force through the rubber ring to speed up the inner ring (**FIGURE 4-46**). The opposite action occurs as the inner ring accelerates and resistance to the sudden speed change is transmitted through the rubber ring. The back and forth movement of the two rings out of phase with one another results in an elastomeric dampener naturally tuned to vibrate out of phase with the lowest torsional frequency of the engine.

Alignment and verification of the proper operation of the two inertia rings is confirmed by the use of a mark scribed in the face of the outer and inner ring. If the marks are not aligned, the two rings have debonded from the elastomer ring, so the dampener is defective (**FIGURE 4-47**).

Viscous Dampeners

Silicone fluid is used in viscous dampeners to tune out harmonic and torsional vibrations. Higher horsepower (kilowatt) engines, those over 300 hp (223.7 kW), will possess higher vibrational energy, which can lead to overheating and shortening of an elastomer dampener's life. In a viscous damper, the inertia ring rotates inside of a sealed case connected to the crankshaft. A thin film of highly viscous silicone fluid fills the small gap between the inner inertia ring and the outer case (Figure 4-47). During changes in crankshaft speed, the silicone fluid drags the inertia ring, permitting it to rotate at approximately the same rpm as the crankshaft. If no torsional vibration is present, the inertia ring and outer casing will rotate at identical speeds. However, changes to crankshaft speed due to torsional vibrations cause the inertia ring and outer case to shear the silicone fluid. This converts the relative motion between the inertia ring and the outer case into heat energy and is effective at dampening torsional and harmonic vibrations caused by defection over the face of the inertia ring. The oscillating (back-and-forth) movement between the ring and the case allows the viscous dampener to dissipate more energy than the same size elastomer-type damper, permitting more effective dampening of torsional vibrations.

> **TECHNICIAN TIP**
>
> Dampeners should never be removed with a puller applied to the outer inertia ring. Most dampeners are designed to be removed with hand force only. Pulling on the outer ring or the case of a viscous dampener will prevent the dampener from functioning properly.

> **TECHNICIAN TIP**
>
> Timing marks located on the outside of elastomer vibration dampeners may be incorrect if the outer hub has slipped relative to the inner hub. Oil on the dampener and age are common factors causing slippage, so a technician must carefully inspect the dampener for evidence of slippage.

FIGURE 4-45 This elastomeric vibration damper from a 3.2 L Ford diesel has two rubber elastomeric bands to balance torsional vibration from 5 cylinders.

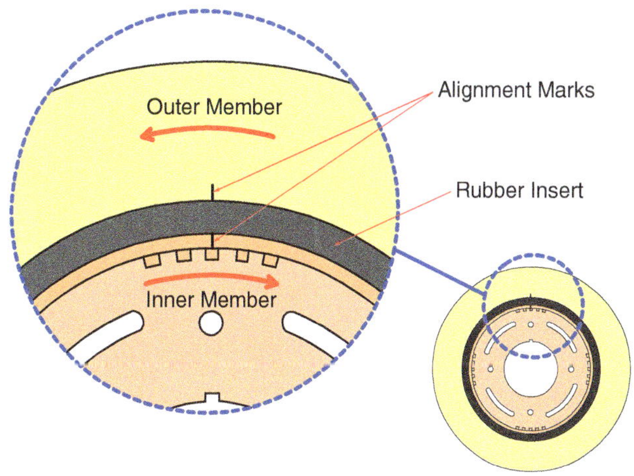

FIGURE 4-46 How the elastomeric-type vibration dampener operates. The outer inertia ring and inner hub move in opposite directions during a torsional impulse. The movements help minimize changes to crankshaft speed by accelerating the crank when it slows and by slowing it down when it speeds up.

FIGURE 4-47 Elastomeric- and viscous-type dampeners. Lines on the outer and inner hub of an elastomer dampener should always line up. If they don't, the dampener is defective.

The shearing effect of silicone fluid actually produces heat that can be used to quickly check the viscous dampener operation. After running an engine for 10–20 minutes, the case should feel quite warm to touch if the viscous dampener is operating correctly.

Combination Dampeners

Combination dampeners combine the dampening advantages of both viscous and elastomeric dampeners. Elastomeric-type dampeners provide good low-speed harmonic and torsional dampening in comparison to viscous-style dampeners. However, at higher engine speeds, the reverse is true, where viscous dampeners provide better vibration dampening.

Rotating Pendulum Vibration Absorber

Unlike regular balancers, pendulum dampeners provide torsional control by producing forces that directly cancel the forces producing torsional vibration. Steel rollers, called centrifugal pendulums, fit loosely into a specific number of holes in either a harmonic balancer or an overhead cam drive gear. The rollers store and release energy back into the crankshaft rather than convert the mechanical energy into heat energy as dampers do. An engine specific mathematical algorithm is used to calculate roller size, hole size, and gear size regarding how the rollers will move forward during compression strokes when the crank slows and rolls backward during the power stroke. The back-and-forth movement of the rollers minimizes engine speed changes, keeping torsional vibration low. Manufacturers claim that the pendulum dampener provides an engine that runs more smoothly, increases valve train stability while offering more accurate injection timing. Pendulum dampeners also have the advantage of being lighter than conventional dampeners and can be used on either the crankshaft or camshaft of overhead cam engines.

Balance Shafts

Balance shafts are shafts with eccentric weights attached to rotating shafts that are used to dampen engine vibration caused by the inertia forces of pistons changing direction at top dead center (TDC) and bottom dead center (BDC). Because the inertia force is generated twice in every stroke, the vibration occurs at twice the engine speed. This explains why balance shafts are driven at two times engine speed and when two balance shafts are used, they rotate in opposite directions of one another at twice the engine speed, countering secondary imbalance forces. Balance shafts help eliminate vibrations that drivers and passengers might otherwise feel transmitted through the engine mounts to the steering wheel, seats, floor pan, or instrument panel. Three- and four-cylinder engines commonly use balance shafts. But V-6 engines can benefit from a fore and aft or front to back vibration caused by pistons changing over at TDC at slightly different times and in different planes.

▶ Flywheels

Flywheels are like large plates bolted to the rear of the engine, and they serve several purposes. Their primary purpose is to provide some inertia between power impulses to keep the engine rotating. Storing energy in the heavy flywheel produced during each power impulse helps smooth out engine speed. A toothed ring gear fastened to the circumference of the flywheel provides a drive point for the starting motor pinion gear. To crank the engine, the starter will engage the engine through the flywheel ring gear. Turning tools are often temporarily engaged with the flywheel to rotate the engine when performing service procedures.

4-08 Identify flywheel types and construction features.

A flywheel also provides a mounting surface for a clutch in the case of a manual transmission and a torque converter for automatic transmissions. Flywheel shape and construction can vary, depending on the type of clutch used and whether torsional dampening capabilities are built into the flywheel.

Dual Mass Flywheels

Torsional vibration dampening is important to drivetrain durability and reduction of noise, vibration, and harshness. The GM 6.2 L; the Ford 6.0 L, 6.9 L, 7.3 L, and 7.3 L Powerstroke; and some engines manufactured by Volkswagen, BMW, and others incorporate specialized torsional dampening springs in their construction. Known as dual mass flywheels (DMFs), when connected to manual transmissions, they enable the use of lighter transmissions and drivelines by minimizing torque spikes produced by torsional vibration (**FIGURE 4-48**). Transmissions shift more easily and transmission gear rattle is virtually eliminated with DMFs.

DMFs are made from two separate sections, which can be disassembled and inspected during clutch servicing. The clutch surface and ring gear are each separate halves of the flywheel. DMFs can also be recognized by a series of springs located around the flywheel circumference on the engine side of the flywheel (**FIGURE 4-49**). One set of bolts near the center hub connects the flywheel to the crankshaft, and another set holds both halves of the flywheel together.

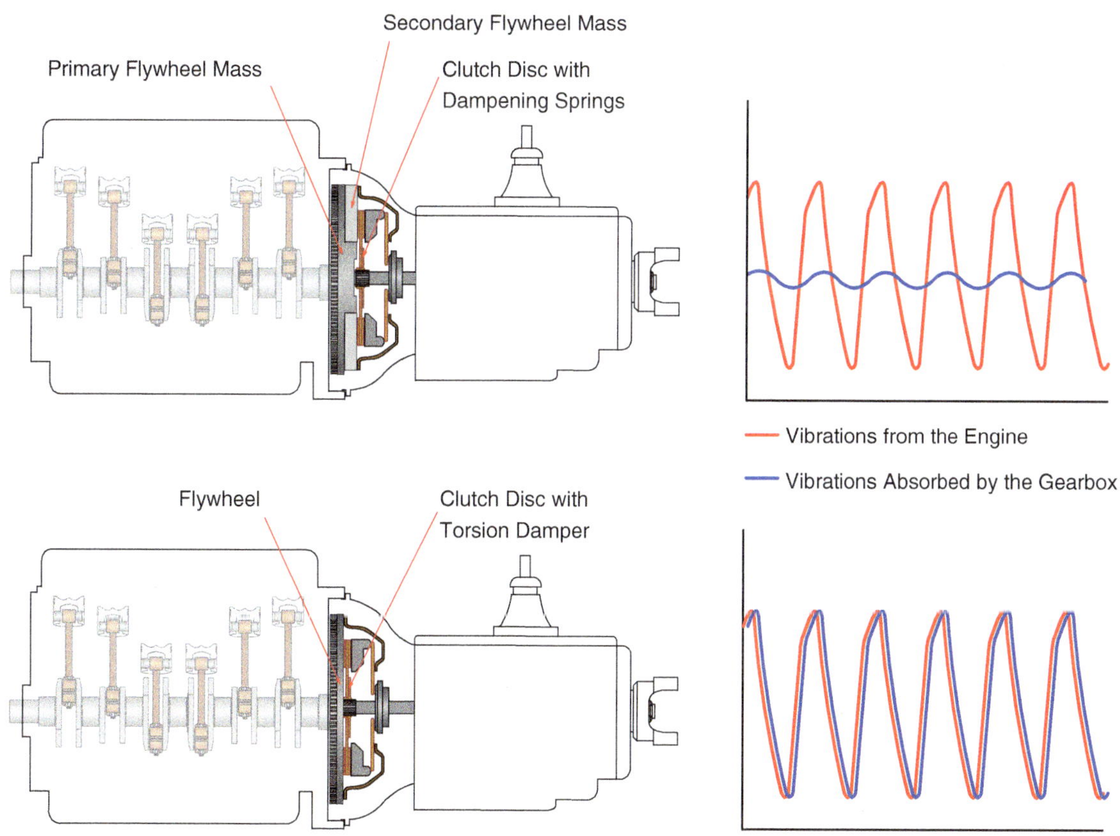

FIGURE 4-48 DMFs minimize the transfer of torsional vibration to transmissions and drivelines.

FIGURE 4-49 (Rear Secondary Plate) A DMF is constructed of two major parts. This rear section contains dampening springs used to absorb torsional vibration. Energy from the crankshaft is transmitted through the springs and into the friction material on the face of this section.

FIGURE 4-50 (Front Primary) A flat steel clutch plate (removed) bolts to this section front section of the flywheel. Torque is not transmitted directly from the crankshaft to the clutch plate. Instead, torsional dampening springs transmit torque from the outer disc to the inner disc and then to the clutch plate.

Power is transmitted to the flywheel clutch mounting surface through a series of springs located around the circumference of the rear plate, which attaches to the crankshaft. These springs act like the dampening springs of a clutch disc to reduce torsional vibration. On the front clutch disc is friction material that transmits drive torque to the front clutch disc from the rear disc through torsional dampening springs. This friction ring will slip under high torque input, which further prevents transmission damage from excessively high torque inputs caused by loading the vehicle beyond its gross vehicle weight (GVW) capacity (**FIGURE 4-50**). The friction ring is made of a material similar to brake lining and can wear out if excessive torque loads are continuously applied. If the customer complains of any unexplained engine noises, vibration, or clutch slippage, then the possibility of a defective DMF flywheel should be investigated. Friction material between the flywheel plates could be burned, and springs could be broken, missing, or worn out. To inspect a DMF, follow the steps in **SKILL DRILL 4-2**.

SKILL DRILL 4-2 Inspecting a Dual Mass Flywheel

1. Operate the engine and listen for unusual noises originating from the flywheel. Pay particular attention to flywheel spring rattling and clunking sounds during engine start-up or shutoff. Also listen for excessive gear rattle
2. After removing the clutch from the flywheel, unbolt the DMF clutch plate cover and check the drive plate for any damage. Pay attention to straps and the friction material. Recommend replacement if the friction material is burned or glazed or if any springs are missing.
3. Inspect the spring retainer for damage and recommend replacement if large gaps exist between the springs and nylon retainers.
4. Reinstall the clutch cover and measure the flywheel free play. This can be done by turning the flywheel clutch cover all the way to a stop point in either a clockwise or counterclockwise rotation. Mark the flywheel cover and line up the mark with one of the starter's ring gear teeth. Rotate the clutch cover again, and mark the limit of flywheel rotation by lining up the mark on the cover with another ring gear tooth.

Continued

5. Count the number of teeth the clutch cover moved and compare with specifications. There are specifications for each flywheel, but generally, more than 1" of free play travel indicates that replacement is required.
6. Check for secondary plate rock. This test checks the axial movement. The secondary plate is the plate behind the clutch cover plate and no axial movement should be observed on the secondary plate. Push on one edge of the plate. Check to see whether the opposite side also moves away when one side is pushed. If it does, the flywheel is defective.

▶ Engine Bearings

4-09 Identify and explain the relationship of construction features to the functions of engine bearings.

Engine bearings are replaceable wear surfaces used on the rotating journals of the crankshaft and camshaft. Bearings used on the crankshaft journals are referred to as main bearings. The upper shell of a main bearing always has a hole to receive lubrication and a circumferential groove to channel oil from the oil supply hole to the rest of the bearing. Rod bearings are used on the crankpin journals or connecting rod journal surfaces (**FIGURE 4-51**). The engine's connecting rods and main bearing caps could be designed to operate without bearings, but those parts would require much more expensive and intensive labor to replace. The materials that bearings are made from are also chosen to extend the life of the engine by reducing friction between the moving surfaces. Experimentation with various types of bearing materials and construction techniques have helped diesel engines evolve to the point where more than 1.5 million miles (2,414,000 km) can accumulate on heavy-duty diesels before the bearings need replacement. Hardened bearing shells, called sputter bearings, are the latest bearing technology used to extend bearing life even further (**FIGURE 4-52**). Sputter bearings use a construction technique that hardens the bearings top overlay surface to increase their durability. A much harder layer of aluminum and tin is applied using Physical Vapor Deposition (PVD) technique. The hardened sputter overlay, which may have a blackened or red oxide appearance, is 3 times harder than traditional lead-tin alloys. Loading pressures are the highest of any bearing having up to 100-120 MPa (14500-17400 psi) load bearing capacity. Operating clearances on these bearings are also much smaller—typically just over 0.001" (0.025 mm) of oil film compared to as much as 0.0035" (0.089 mm) in conventional bearings. Cast copper based bearings or high strength aluminum based bearings are commonly plated by sputter overlays.

Load carrying capacity of sputter bearings is highest of all bearing materials, being in the region of 100-120 MPa (14500-17400 psi). While bearings have increased engine longevity, technicians still need to be aware of different types of bearings and their construction, to be ready to diagnose problems with bearings and replace them when necessary.

▶ TECHNICIAN TIP

The latest engine bearing called sputter bearings use hardened bearing surfaces and operate with much smaller oil clearances. It is imperative that low viscosity oils such as 5w-30 or 10w-30 FE grade oils only are used with these bearings. Using thicker, higher viscosity oils will potentially cause oil starvation leading to catastrophic engine damage.

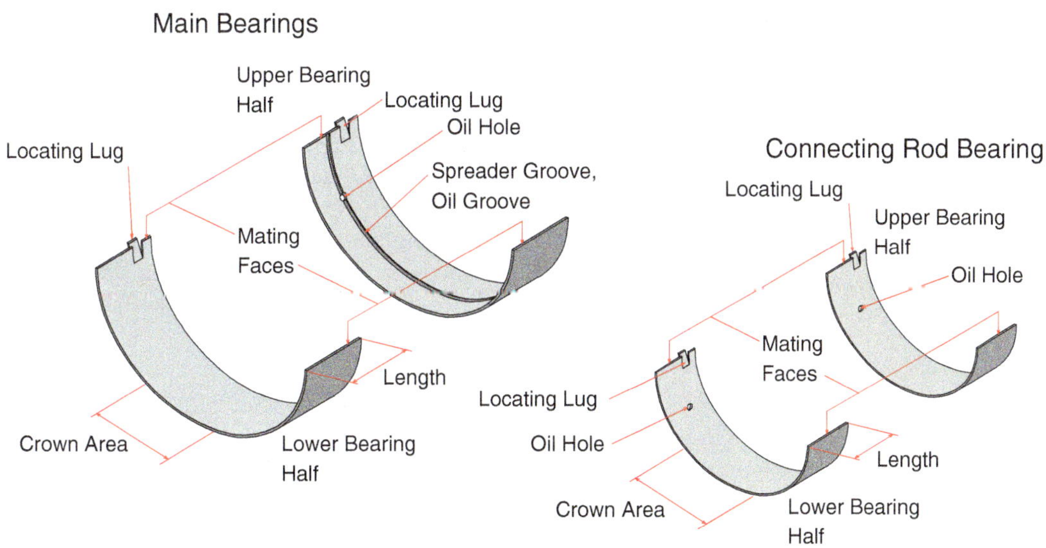

FIGURE 4-51 Rods and main bearings are constructed differently. An oil hole with an annular groove to distribute oil is typical for the upper main bearing shell. An oil hole may be present in the rod bearing if pressurized oil is used to lubricate the piston pin.

FIGURE 4-52 Sputter bearings use a hardened bearing shell. The darker bearing shell is hardened.

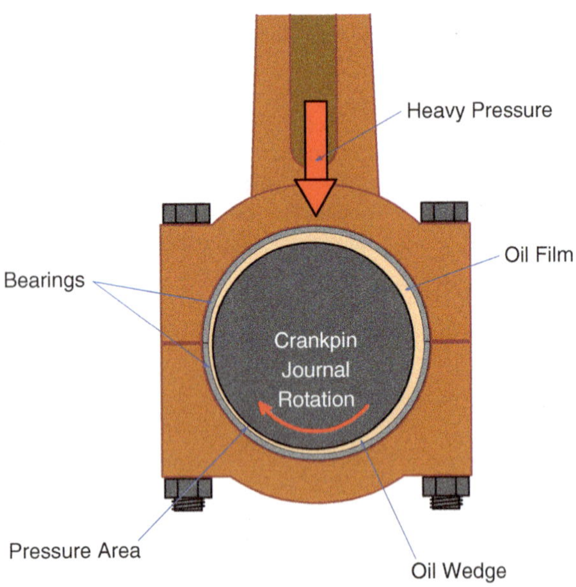

FIGURE 4-53 Formation of the hydrodynamic wedge.

Plain Insert Bearing

Engine bearings are known by a number of names: Precision inserts, plain bearings, and tri-metal bearings are a few common ones. Bearings used around the crankshaft are split-type insert bearings because the bearing is made in two pieces. Manufacturing the bearing into halves allows the bearing to be easily serviced and allows the use of a one-piece crankshaft. Camshaft bushings are examples of plain bearings that are one piece. Plain bearings also are lighter and more compact for their load bearing capacity than roller types are.

Bearings perform a number of important functions inside the engine:

- reducing friction
- supporting moving parts under load
- serving as replaceable wear surfaces.

Reducing friction is one of the most obvious functions that bearings perform, along with the lubrication system. Bearing surfaces should minimize friction and heat generation. It should be noted that plain bearings with a pressurized oil film have less high-speed friction than an antifriction type (roller, ball bearing, etc.).

To achieve the engine life that bearings have today, plain bearings are the best to use. A Cummins 6.7 L has a B-50 life of 350,000 miles (~550,000 kms). That means that at that distance there is still 50% bearing life remaining. Fords 6.7 L V8 B10 bearing life gives it's engines a 90% probability of operating an to 500,000 miles (~800,00 kms) without requiring the removal of the cylinder heads or oil pan for service.

To achieve the long durability the characteristics of the lubrication film between the rotating shaft and bearing are critical. The term "hydrodynamic wedge" is the explanation given to understand why bearings can last so long but wear so little. Essentially, the oil, bearing, and rotating shaft are separated from one another by hydraulic pressure. Engine bearings depend on a film of oil to keep the shaft and bearing surfaces separated (FIGURE 4-53). Bearings fail when the oil film breaks down or when the bearing is overloaded. The oil film pressure is generated by shaft rotation. At rest, the shaft and bearing are in contact. On start-up, the shaft contacts the bearing briefly. As it runs, the shaft pulls oil from the clearance space into the wedge-shaped area between the shaft and bearing. Oil is pulled in because of the attraction oil has to the metal shaft. The oil film pressure is generated by shaft rotation to wedge or lift the shaft off its bearing shell and support it during engine operation. During normal operating conditions, a continuous supply of clean oil will keep the shaft and bearing surfaces separated. Bearings fail when the oil film breaks down or when the bearing is overloaded.

At rest, the shaft and bearing are in contact. On start-up the shaft may briefly contact the bearing. While running, the shaft pulls oil from the clearance space into the wedge shape area between the shaft and bearing. Oil is pulled in because of the electrostatic attraction oil has to the metal shaft. The oil wedge lifts the shaft off its bearing and supports it during engine operation. During normal operating conditions, a continuous supply of clean oil between the shaft and bearing surfaces will ensure that the hydrodynamic wedge separates the two surfaces.

Bearing Construction

In early automotive history, bearings were made of a lead-tin alloy called "Babbitt." This alloy was effective because of its low frictional characteristics: the ability to distort under severe load and embed dirt into its shell so that the crankshaft would not be damaged. However, one characteristic these bearings did not have was fatigue strength: the ability to withstand high loading for long periods of time.

The following is a summary of those important bearing characteristics:

- *Compatibility*: a bearing's ability to allow friction without excessive wear or friction. Dissimilar metals have better compatibility. Copper to steel and bronze or lead to steel have good compatibility or lower wear when used against one another. A bronze crankshaft with steel bearings could be combined to minimize wear, except the crank isn't as strong as steel.
- *Fatigue strength*: a bearing's ability to carry a load. A bearing will crack or be wiped away if overloaded by shaft pressure.
- *Conformability*: a bearing's ability to conform to irregularities on of a journal surface (which works against fatigue strength).
- *Embeddability*: a bearing's ability to absorb particle contamination. Dirt particles will scratch a shaft and ruin it (**FIGURE 4-54**).

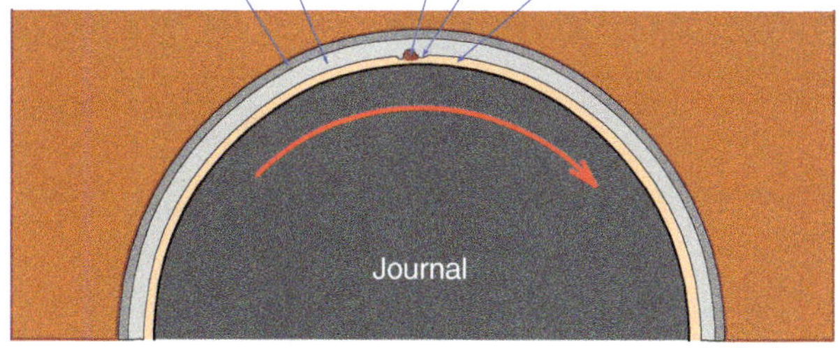

FIGURE 4-54 Embeddability refers to the ability of dirt particles to embed themselves into the soft lead layer of Babbitt to prevent scoring the crank journal.

A variety of construction techniques are used to achieve different balances among the characteristics that will adapt to each engine's unique operating conditions. Tri-metal bearings with copper alloys for the intermediate layer are used in diesel engines. A steel back with a copper alloy intermediate layer gives the strength, fatigue resistance, and conformability characteristics required for a diesel engine bearing. The lead-tin Babbitt layer provides good wear and compatibility features. A nickel barrier plating prevents galvanic reactions between the Babbitt and the copper alloy surface beneath it that would lead to corrosion. Flash lead-tin plating protects and provides a finish for the engine break-in period. This finish can be easily removed when it's new, by touching the bearing or wiping it with a shop rag (**FIGURE 4-55**).

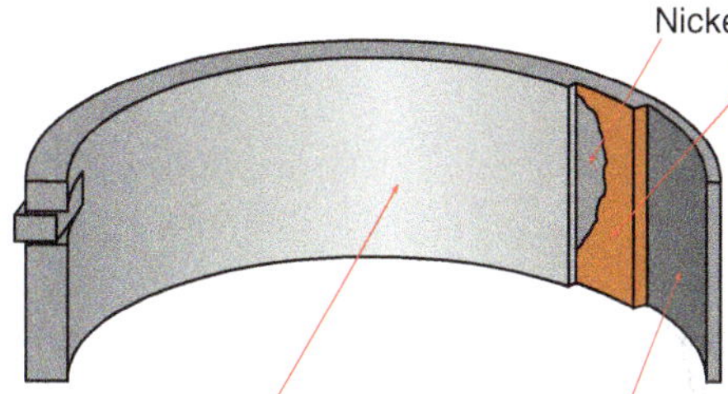

FIGURE 4-55 A tri-metal plain insert bearing is made up of three layers of metal with a nickel layer between each.

Bearing Lubrication

Every bearing receives lubrication through holes drilled into the crankshaft. Main bearings receive oil first from the oil pump and main oil gallery. The oil enters the main bearing journal and drilled passages inside the crankshaft to flow to the rod bearing journals and bearings. Some main bearing journals are cross-drilled so that the oil holes align twice for every crankshaft revolution. Oil would otherwise be cut off, so the main bearings have a groove down the center of the top bearing shell that allows oil to flow around the main bearing journal. The amount of oil clearance is important to maintaining the hydraulic pressure of the oil wedge separating the rotating shaft and bearing. Too little clearance, and inadequate oil will flow into the bearing for lubrication and cooling. Too much clearance, and the oil wedge will collapse, resulting in shaft to bearing contact. Excessive bearing clearance is a primary reason to have low oil pressure. If the oil pump is required to pump extra volume to fill the clearances while even greater quantities of oil are thrown of a shaft, loss of engine oil pressure will result. (recall that pumps produce volume, not pressure; lubrication systems require restrictions to produce pressure.) Higher shaft speeds also produce greater oil wedge pressure.

Plastigauging

To evaluate the oil clearances of engine bearings, the use of Plastigauge is recommended. Plastigauge consists of two essential parts:

- an oil-soluble plastic material
- packaging with a printed thickness gauge.

Measuring bearing clearances with Plastigauge is very simple when the engine removed an is turned upside down while on an engine stand. After the main bearing cap has been removed and the oil wiped from the crankshaft and bearing shell, a piece of Plastigauge is laid across the crankshaft journal. The bearing cap is reinstalled with a torque recommended by the engine manufacturer. When the bearing cap is tightened,

FIGURE 4-56 Plastigauging a main bearing journal determines oil clearances and is typically done only when a replacement crank or service crank is used. The Plastigauge is squeezed by the bearing and cap. Afterward, the width of the Plastigauge is compared against a gauge on the package.

the pressure causes the Plastigauge to be flattened. The less clearance there is, the greater the flattening and the wider it will be. When the bearing cap is removed, the width is measured by direct comparison with the graduated scale on the Plastigauge envelope. The numbers on the graduated scale indicate bearing clearance in thousandths of an inch or millimeter on the reverse side of the paper gauge. Plastigauge is available in four different sizes and differentiated by four colors—red, green, yellow, and blue—each covering a particular bearing clearance range (**FIGURE 4-56**).

Bearing Location

Bearings must not move within the bearing bore and must have good contact within the bore for good heat transfer. Several bearing construction features are common to ensure these functions.

Bearing spread, refers to the parting face of a bearing being made wider than the diameter of the cap or web, assists in bearing retention and ensuring good contact with the cap or web of the bearing bore for heat transfer. Bearings require being "snapped" or pushed into place (**FIGURE 4-57**).

Bearing crush is the distance that a bearing is higher than the cap or web. In fact, the diameter of the bearing shells is larger than the bearing bore. This feature assists in locking the bearing in place and heat transfer from the bearing shell into the bearing bore. The dark spots observed on the back of a bearing indicate where greatest heat transfer takes place. For this reason, bearings should always be installed dry on the backside with no grease or threadlockers applied to the back of the shell. Using this material would impede the transfer

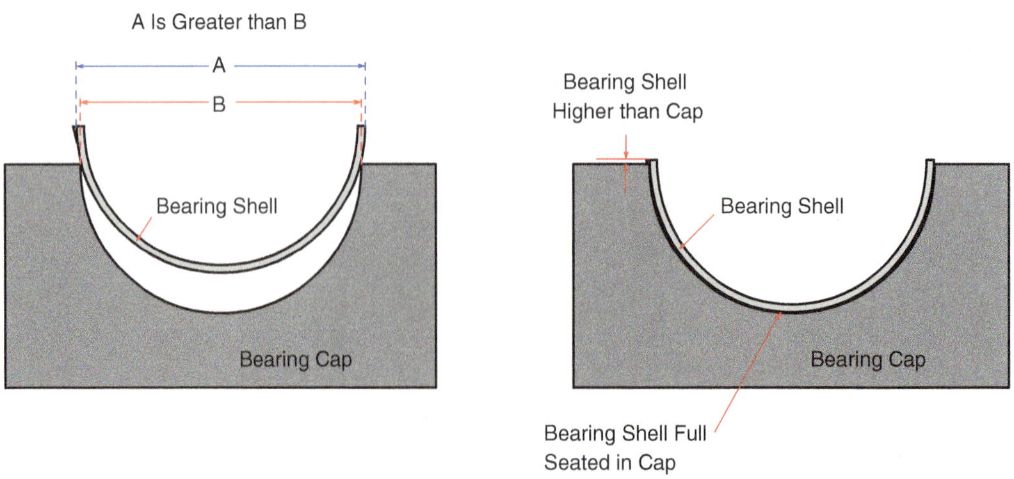

FIGURE 4-57 The concepts of bearing crush and spread. Bearings must seat tightly into the bearing bore to prevent movement and to transfer heat.

of heat from the bearing. Locating lugs are used on bearings to prevent movement and to properly center the bearing in the bore. To ensure that bearing caps and lugs are not incorrectly matched, assemble the caps together, locating tab to tab. Make sure identification numbers match for main bearing caps location in block and on connecting rods to caps.

Undersize and Oversize Bearings

An undersize bearing is thicker than a standard bearing. It does not describe the bearing but the shaft it fits. If a crankshaft is worn or is ground to an undersize, an undersize bearing is used to fit the journal. If a spun main bearing damages a block, the bearing bore can be re-bored to an oversize and an oversized bearing can be installed. Cam bearings can typically outlast two sets of mains since there is less loading; however, excessively worn and bleeding cam bearings with excessive oil throw-off can starve mains.

Thrust Bearings

Thrust bearings control end play in a crankshaft (**FIGURE 4-58**). The forces of disengaging the clutch, helical gearing on the geartrain, and thrust from automatic torque converters can easily push/pull a crankshaft to the point where severe damage to the connecting rods and journals occur. Thrust bearings are used to control end play (**FIGURE 4-59**). Washers or bearings with flanges ride against a specially machined journal on the cheeks of only one main bearing journal. To measure crankshaft end play, follow the steps in **SKILL DRILL 4-3**.

Bearing Failures

Bearing failures are indicated by the following conditions:

- a drop in lubricating oil pressure
- excessive oil consumption (excessive oil thrown onto the cylinder walls)
- noises that can be minimized during a cylinder cutout procedure (**FIGURE 4-60**).

FIGURE 4-58 A thrust bearing is used on a main bearing journal to control crankshaft endplay.

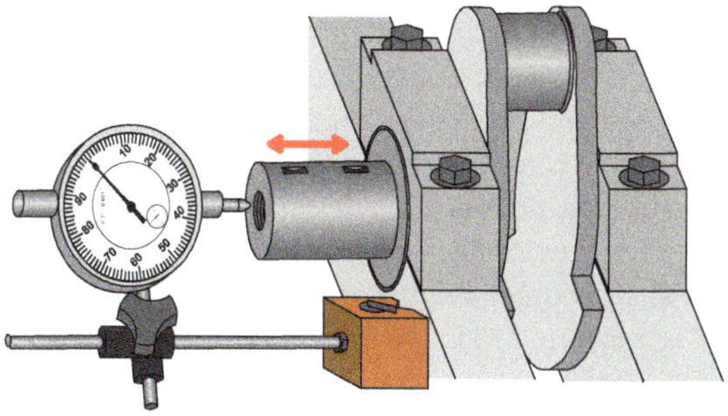

FIGURE 4-59 Measuring crankshaft end play with a dial indicator. The crank is pushed all the way back, the dial indicator zeroed, and then the dial is pushed forward to measure travel distance.

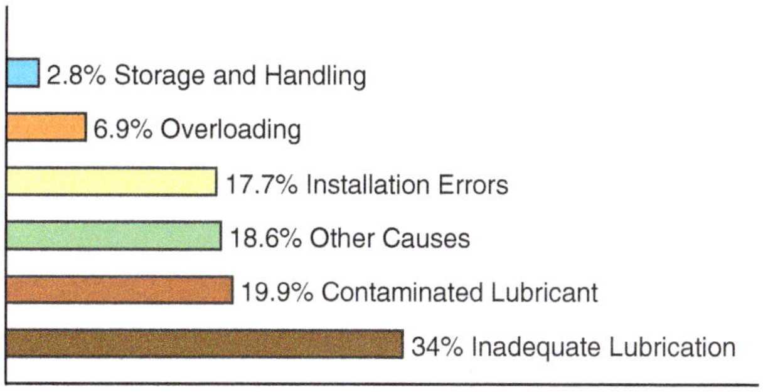

FIGURE 4-60 Causes of bearing failures.

SKILL DRILL 4-3 Measuring Crankshaft End Play

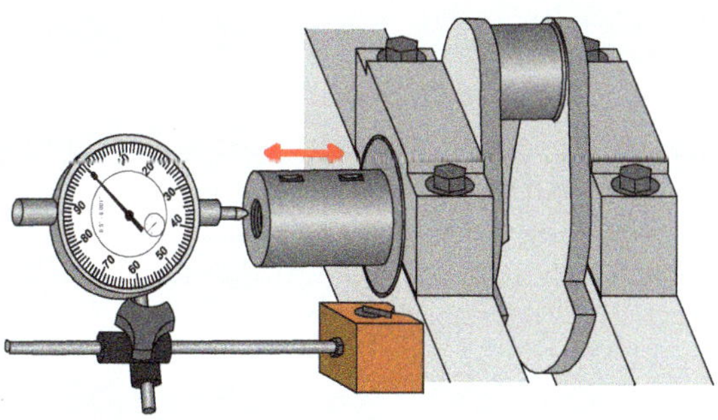

1. Place a dial indicator magnetic base on the engine block. Place the needle indicator on the vibration dampener.
2. If the vehicle is equipped with a manual transmission, push on the clutch. With the clutch depressed, zero the dial indicator. Alternatively, a pry bar between the flywheel and flywheel bell housing inspection hole can be used to pull the flywheel rearward.
3. Push the flywheel forward using a pry bar between the flywheel and the inspection hole in the flywheel housing.
4. Observe and record the measurement on the dial indicator. Specifications are generally in the range of 0.004–0.019" (0.102–0.483 mm). The specifications may seem large, but bearing clearances must be wide enough to allow oil to flow on both sides of two thrust washers or bearing flanges.

Wrap-Up

Ready for Review

- The cylinder block is the largest single part and main structure of the diesel engine, and its primary function is to support cylinders and liners (if equipped) and major engine components, including the crankshaft and camshaft.
- Diesel engine blocks have cylinders arranged in-line, V-banked, or horizontally opposed and can be configured with different numbers of cylinder bores.
- Cylinder blocks can be constructed from cast iron, aluminum, or compacted graphite iron (CGI); each of these materials has advantages and disadvantages.
- Cylinder block walls progressively wear with increasing engine hours and distance traveled; a dial bore gauge is used to measure cylinder wear due to taper.
- The main purpose of the crankshaft in a reciprocating engine is to convert the linear motion of the connecting rod assemblies to rotational movement.
- Primary balance, which is achieved when the crankshaft counterweights offset the weight of the piston and connecting rod assembly, and secondary balance, which is achieved when the movement of one piston counterbalances the movement of another, result in engines that run more smoothly.
- Counter balance shafts are used more commonly in three- and four-cylinder engines to compensate for engine vibration produced by changes in the direction of pistons.
- The mechanical stresses placed on a crankshaft require it to be especially strong, so crankshafts are hardened to improve durability; the two most common processes are induction hardening and nitriding.
- Engine oil is used to lubricate, cool, and clean the crankshaft journals and bearings.
- To control the destructive effects of both torsional and harmonic vibrations, diesel engines are equipped with vibration dampeners.
- Blocks must be designed to be as compact as possible, to reduce weight and size while maintaining strength and rigidity; common block designs include deep-skirt blocks, crossbolted blocks, ladder-frame blocks, and tunnel-bore blocks.
- There are several different configurations for engine block cylinders; which one is best depends on the vehicle's application, production cost, and expected vehicle lifecycle.
- Replaceable cylinder sleeves enable a worn-out engine to be quickly rebuilt and restored to factory specifications.
- Bearings perform a number of important functions inside the engine, including reducing friction, supporting moving parts under load, and serving as replaceable wear surfaces.
- The characteristics important for engine bearings are compatibility, fatigue strength, conformability, and embeddability; a variety of construction techniques are used to achieve different balances among the characteristics that will adapt to each engine's unique operating conditions.
- The amount of oil clearance between the rotating shaft and the engine bearing is important to maintaining the hydraulic pressure of the oil wedge that cools and lubricates these parts.
- Bearings must not move within the bearing bore and must have good contact within the bore for proper heat transfer.
- Thrust bearings control end play in a crankshaft.
- Dual mass flywheels (DMFs) enable manufacturers to use lighter transmissions and avoid damage caused by torsional vibration.

Key Terms

cast-in-place sleeve A block construction technique that casts aluminum block material around an iron cylinder sleeve.

cavitation erosion Pinholes produced in cylinder block walls, heads, and liner sleeves as a result of the collapse of tiny water vapor bubbles formed when coolant vaporizes on hot cylinder wall surfaces.

compacted graphite iron (CGI) A material produced from powdered iron alloys, which is squeezed into molds at high pressures and then heated to bond the metal particles together; also known as sintered graphite.

crossbolted block A variation of the deep-skirt block that uses additional horizontally placed bolts to connect the crankcase walls of the block to the main bearing caps; also known as a tie-bolted or bolster-bolted block.

deep-skirt block A block configuration with a crankcase wall that extends well below the crankshaft's centerline.

dry sleeve block A block designed with a bored or honed hole in the block that allows no coolant contact with the cylinder sleeve.

dual mass flywheel (DMF) A two-piece flywheel design that incorporates specialized torsional dampening springs.

fillet radius A round shaped machined intersection ground into the surface between the journal and the crankshaft cheek that strengthens the crankshaft and minimizes the possibility of a fracture.

harmonic vibration A vibration that sends pressure waves moving back and forth along the crankshaft.

induction hardening A heat treatment process that involves passing alternating electric current through coils of heavy-gauge wire surrounding the material to be hardened; through magnetic induction, heat is produced in the metal, which is then quenched with water to produce a hard, wear-resistant metal surface.

ladder-frame block A block design that uses a ladder shaped component incorporating main bearing caps into a section that attaches to the crankcase and the oil pan.

Nanoslide technology A spray welding (or plasma welding) technique developed by Mercedes-Benz to coat cylinder walls with a tool-grade hardness iron alloy.

nitriding A crankshaft hardening technique that involves heating the shaft and introducing cyanide salt or liquid into the heating chamber with the crankshaft; slowly cooling or quenching the crank with these substances in its heated condition causes the nitrogen and carbon to molecularly bond with metal surfaces to create an extremely hard journal.

parent bore block A block design that has holes cast and bored in the block for the cylinders with the pistons inserted directly into these holes; also known as a no-sleeve block.

pendulum vibration absorber A dampener that provides torsional vibration control by producing forces that directly oppose and cancel the forces producing torsional vibration.

primary balance Balance achieved when the crankshaft counterweights offset the weight of the piston and connecting rod assembly.

scissor gear Two separate spring-loaded gears incorporated into a single unit to reduce gear rattle caused by torsional vibration.

secondary balance Balance achieved when the movement of one piston counterbalances the movement of another, resulting in engines that run more smoothly.

sputter bearing The latest technology in bearing overlay material, which deposits a metal overlay surface onto a bearing backing that is three times harder than conventional overlay. This complex process involves spray welding in a vacuum and is used to make a bearing that has the greatest ability to carry a load over any other bearing.

torsional vibration The speeding up and slowing down of the crankshaft caused by alternating compression and power strokes in the engine cylinder.

tunnel-bore block A block that has the main bearing bores and block formed into one solid structure for maximum block rigidity and strength.

warp anchor A block design that has the cylinder head and the cylinder block bolted together by using tie bolts; sliding steel sleeves, which are locked in the block, accept the cylinder head bolt from one side and the tie bolt from the other.

wet sleeve block A block designed with a number of large holes into which the cylinder sleeves are inserted; coolant has direct contact with the outside of the sleeve, and there is no cylinder bore structure.

Review Questions

1. What type of tapped threads is most commonly used in cylinder block castings?
   a. Fine.
   b. Coarse.
   c. Rolled.
   d. Metric and imperial.
2. Which of the following is the manufacturer's purpose for core plugs?
   a. To clean sand from blocks after casting.
   b. To install block heaters.
   c. To prevent freezing damage to cylinder blocks.
   d. To add inspection holes for internal block castings.
3. Which of the following is the primary reason for locating geartrains on the rear of the engine?
   a. To minimize engine noise.
   b. To ease of service.
   c. To make it easier to mount accessories such as power steering pumps and air compressors.
   d. To minimize engine torsional vibrations.
4. The presence of cavitation on a replaceable cylinder liner is most likely caused by ___________.
   a. liner vibration
   b. low cylinder pressures
   c. excessive cooling system pressure
   d. improperly conditioned coolant
5. Which of the following is the purpose of bolster bolts or tie bolts?
   a. To hold the main bearing caps down tightly.
   b. To reduce torsional vibrations.
   c. To increase engine block rigidity and strength.
   d. To secure the crankcase bedplate to the engine block.
6. Which of the following features will a ladder-frame engine block have?
   a. A block skirt extending past the main bearing caps.
   b. A steel plate bolted to the oil pan rail and main bearing caps.
   c. Main bearing caps integrated into a one-piece casting.
   d. A lightweight engine block.
7. Torsional vibration resulting in "jumped" accessory drive belts, excessive geartrain wear, and broken camshafts is commonly the result of which of the following defective components?
   a. Dual mass flywheel (DMF).
   b. Vibration dampener.
   c. Imbalanced crankshaft.
   d. Balance shaft.
8. The lightest, strongest, and quietest engine blocks are ___________.
   a. made from aluminum
   b. parent bore blocks
   c. dry sleeve blocks
   d. made from compacted graphite iron (CGI)
9. V8 engines have cylinder banks arranged at which of the following angles to allow for even crankshaft rotation between firing impulses?
   a. 60°.
   b. 90°.
   c. 37.5°.
   d. 45°.
10. Crankshaft counterweights are used to minimize engine vibration caused by which of the following?
    a. The inertia of the piston/connecting rod assembly.
    b. Uneven firing impulses.
    c. Torsional vibration.
    d. Angular velocity changes of the crankshaft.

ASE Technician A/Technician B–Style Questions

1. After discovering coolant is leaking into the engine oil, Technician A says the leak will rapidly accelerate cylinder wall wear. Technician B says the coolant leak is likely casue more damage to engine bearings first. Who is correct?
   a. Technician A
   b. Technician B
   c. Both A and B
   d. Neither A nor B
2. Technician A says that polished cylinder walls will cause low power and more blowby. Technician B says cylinder walls with more than 90% remaining crosshatch have less than 50% remaining service life. Who is correct?
   a. Technician A
   b. Technician B
   c. Both A and B
   d. Neither A nor B
3. Technician A says that Plastigauge that becomes wider indicates a smaller oil clearance between a crankshaft and main bearing. Technician B says the Plastigauge will become narrower when bearing oil clearances are smaller. Who is correct?
   a. Technician A
   b. Technician B
   c. Both A and B
   d. Neither A nor B
4. While installing new rod bearing inserts, the shells protruded very slightly above the margin of the bearing cap. Technician A says this is normal. Technician B says that wrong bearings produce this result. Who is correct?
   a. Technician A
   b. Technician B
   c. Both A and B
   d. Neither A nor B
5. Technician A says that rear geartrains are used to shorten the length of the engine block. Technician B says that rear geartrains reduce engine vibration. Who is correct?
   a. Technician A
   b. Technician B
   c. Both A and B
   d. Neither A nor B
6. Technician A says that cast-in-place iron liners are commonly used in aluminum cylinder blocks. Technician B says that only closed-deck blocks will use cast iron liners. Who is correct?
   a. Technician A
   b. Technician B
   c. Both A and B
   d. Neither A nor B
7. While examining a thick, heavy, flat steel plate bolted to the oil pan rails and main bearing caps, Technician A suggests it is used to stiffen the crankcase to reduce torsional twisting. Technician B says the plate is used to prevent oil from splashing up onto the cylinder walls and causing high oil consumption. Who is correct?
   a. Technician A
   b. Technician B
   c. Both A and B
   d. Neither A nor B
8. Technician A says that crankshaft end play is controlled by the fillet radius of the crankshaft journals. Technician B says that the main bearings limit crankshaft end play. Who is correct?
   a. Technician A
   b. Technician B
   c. Both A and B
   d. Neither A nor B
9. Technician A says that a defective vibration dampener will cause the accessory drive belts of an engine to jump off the drive pulleys. Technician B says a defective vibration dampener will cause the crankshaft or even the camshaft to break. Who is correct?
   a. Technician A
   b. Technician B
   c. Both A and B
   d. Neither A nor B
10. Technician A says that nitriding bearing journal surfaces produces the deepest amount of crankshaft hardening. Technician B says that induction hardening produces the hardest journal finish. Who is correct?
    a. Technician A
    b. Technician B
    c. Both A and B
    d. Neither A nor B

CHAPTER 5

Cylinder Components

Learning Objectives

After reading this chapter, you will be able to:

- 5-01 Identify and describe the operating factors that affect how pistons are constructed.
- 5-02 Identify and explain the functions of major piston construction features.
- 5-03 Identify and explain the functions of piston ring construction features.
- 5-04 Identify and explain the purpose of piston ring material and shape according to ring function.
- 5-05 Identify and describe diagnostic techniques used for evaluating the condition of a piston ring–to–cylinder wall seal.
- 5-06 Identify and explain the function of piston pins.
- 5-07 Identify and explain the function of connecting rod construction features and mechanical stresses imposed on connecting rods.

You Are the Technician

A 3500 series (class 2) truck arrives at your shop, and the driver complains that the engine lacks power and is running rough. During an inspection, you observe a knocking sound in the engine. There are many significant oil leaks from the gaskets and seals on the engine. In addition, a considerable amount of oil-laden blowby is being emitted from the crankcase tube connected to the turbocharger intake pipe, and oil has fouled the turbocharger compressor wheel. The engine oil level is low as well. When the throttle is snapped to accelerate the engine, visible black and gray smoke are observed at the tailpipe. Although the odometer reading is only 279,617 miles (450,000 km), the vehicle is close to 14 years old and is used primarily to pick up and deliver scrap metal in and around the city on a daily basis. Before you can prepare an estimate for the cost of repairs, you need to validate whether the base engine system is in satisfactory condition to justify any repairs or adjustments to the fuel system and other engine systems.

1. What techniques would you use to estimate the remaining service life of the engine, and why would you recommend them?
2. The accumulated distance on the odometer is relatively low for a diesel engine to be worn out and in need of an overhaul. What factors could contribute to premature wear of this truck's engine? What observations could you make that would support your answer?
3. If the engine is overhauled or replaced and the vehicle is returned to service, what operating recommendations would you make to the driver to extend the life of the engine?

▶ Introduction

Cylinder components are the engine components first used to convert combustion forces into mechanical power, which makes them a logical starting point to examine diesel engine construction. These components include pistons, piston rings, piston pins (also called wrist pins), piston pin bushings, and connecting rods (also called con-rods) (**FIGURE 5-1**). These components convert and transfer the forces of combustion to the crankshaft. They are the most highly stressed parts of the diesel engine. The flame temperature of diesel fuel is over 3,900°F (2,149°C), and cylinder pressures can momentarily reach 3,000 psi (207 bar) in turbocharged diesels compared to less than 1,000 psi (69 bar) in naturally aspirated spark-ignition engines (**FIGURE 5-2**). Because of these stresses, the design, construction, and operation of cylinder components are critically important to engine performance and durability. In addition, cylinder components, together with other engine systems, have features needed to reach emissions targets.

Emission legislation imposes new design quality on cylinder components through durability requirements. Legislated standards for the engine durability require noxious emissions to remain below maximum thresholds through the expected useful life of an engine. To keep emissions low throughout an engine's anticipated lifecycle, cylinder components are now constructed to last longer than ever. Tier 3 emission standards, which are similar to Euro 6 and California LEV-III, require light- and medium-duty diesels to have a useful life of 150,000 miles (241,400 km) or 15 years (**TABLE 5-1**). A shorter useful life of 120,000 miles (193,000 km) or 10 to 11 years is permitted only if vehicles certify to lower initial limits for NO_x and hydrocarbon emissions. Cylinder components in contemporary diesels have undergone numerous refinements to help meet higher standards for power density, performance, life expectancy, and low emission output. Major engine repairs involve replacing or determining the serviceability (remaining service life) of these parts, so technicians need to familiarize themselves with the construction, operating principles, and failure modes of these parts. Engine failures often involve damage to these components, so understanding failure modes is necessary in order to correct the root causes of damage to these components.

FIGURE 5-1 Cylinder components include the piston, piston rings, piston pin, and connecting rod.

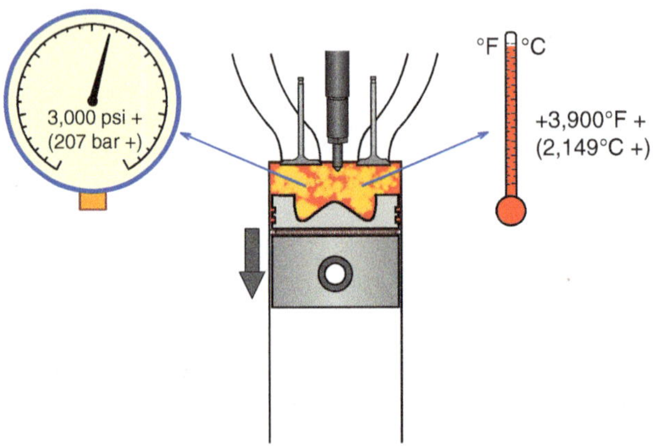

FIGURE 5-2 High combustion pressures and temperatures require pistons to be capable of withstanding high thermal loads and cylinder pressures.

TABLE 5-1 Increasing Emission Durability Standards Requires More Durable Engine Components

Emission Standard	Distance	Time	Maximum Limit
Tier 3 (2017–2025)	150,000 miles (241,400 km)	15 years	NMOG + NO_x
Tier 2 bin 5 (2007–up to 2019)	120,000 miles (193,000 km)	11 years	FTP* Standards
EPA** (LEV-III) (2015–2025)	120,000 miles (193,000 km)	11 years	NMOG + NO_x × 0.85

* FTP: Federal Test Protocol (A standardized emission testing procedure)

**EPA: Environmental Protection Agency

Source: U.S. EPA

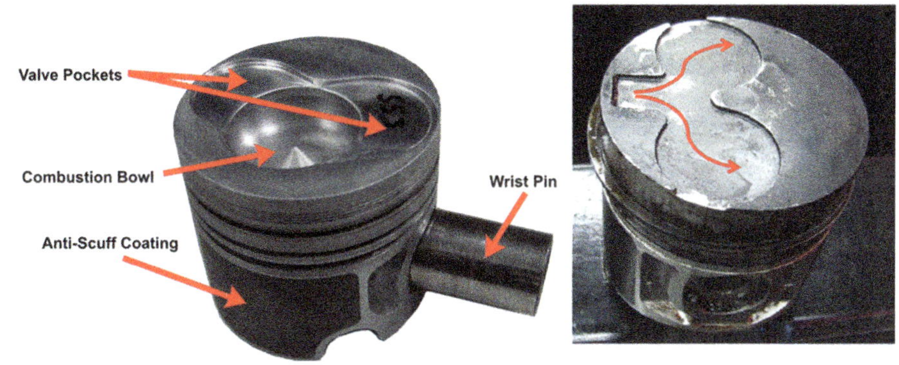

FIGURE 5-3 The piston crowns of DI and IDI pistons.

▶ Piston Classification and Design

5-01 Identify and describe the operating factors that affect how pistons are constructed.

Since pistons are relatively uniform in their function, their classification often follows the construction details of the component. For example, the shape, materials, or construction techniques help differentiate the different types of pistons. In direct-injection (DI) diesel engines, the combustion chamber is formed in the piston. Indirect-injection (IDI) engines shape the piston crown to direct the flow of burning combustion gases from the pre-chamber to the main chamber (**FIGURE 5-3**).

Tremendous heat, pressure, and inertia forces are sustained by pistons as they capture combustion energy and transmit it to the other reciprocating and rotating parts of the engine (**FIGURE 5-4**). In order to operate reliably for thousands of hours under extreme conditions critical construction features are built into piston design. Piston design is a major factor in determining a variety of engine characteristics such as power output, emission characteristics, maximum speed, oil consumption, and durability.

Before examining any specific features of each type of piston, it's helpful to identify the parts common to all pistons. Pistons have three main sections (**FIGURE 5-5**):

1. the crown, or top of the piston, which is subjected to cylinder pressure and heat
2. the ring belt, which is where piston ring grooves are located
3. the skirt, which is the surface that slides directly against the cylinder wall to stabilize piston movement.

In the uppermost part of the skirt is the piston pin. Additional piston material reinforces the area around the pin; this material is called the pin boss (**FIGURE 5-6**). Steel struts are often cast into the pin boss for improved strength.

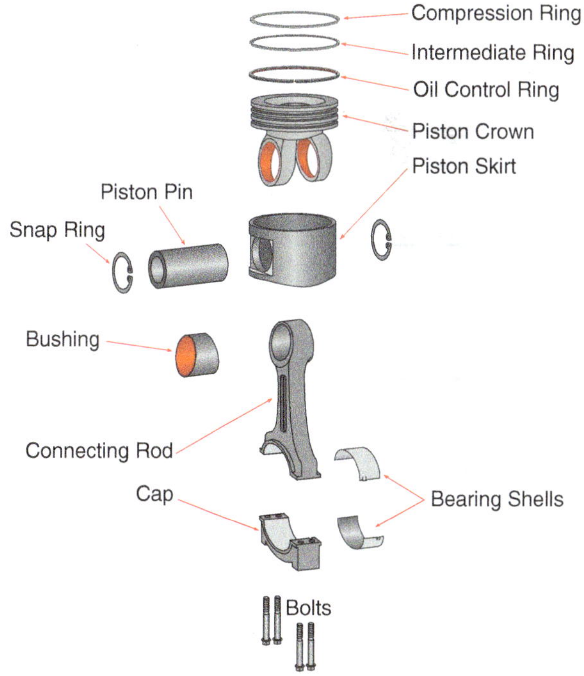

FIGURE 5-4 Identification of various piston and connected cylinder components.

FIGURE 5-5 The main parts of a piston.

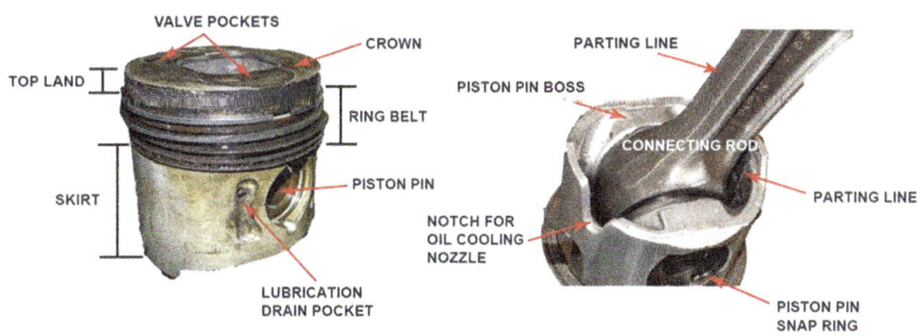

FIGURE 5-6 Piston nomenclature. The connecting rod and piston are forged.

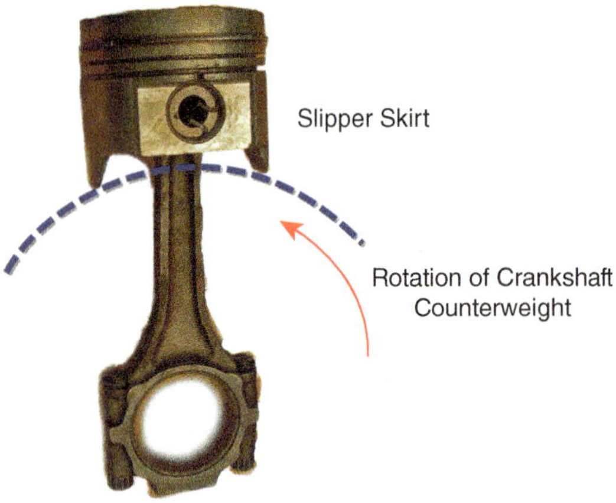

FIGURE 5-7 This piston uses a slipper-skirt-piston design.

Types of Pistons

Four basic piston designs are used in today's diesel engines:

1. trunk
2. slipper skirt
3. articulated or low crevice volume (LCV)
4. Monosteel.

Slipper Skirt Pistons

In the design of a slipper skirt piston, a portion of the skirt is removed on both non-thrust sides of the piston. This provides clearance for the turning radius of the crankshaft counterweights when the piston approaches bottom dead center (BDC) (FIGURE 5-7). Removing skirt material enables slipper skirt pistons to use a shorter connecting rod, the smaller combined length of the rod and piston assembly enables vertically shorter, lighter, more compact engine blocks to be manufactured around it. This is important when the profile of the engine must be kept small to fit in a smaller vehicle. More compact engine size translates to lighter front end designs for better vehicle dynamics. When engine mass and inertia are lighter, the vehicle will depend less on heavier front end components needed to support engine weight, the front end will dive less during braking, and become more responsive turning when engines are lighter. Slipper skirt pistons are essential in smaller V-block diesel engines, where compact engine design is vital.

FIGURE 5-8 A trunk-type piston having a full piston skirt.

Trunk-Type Pistons

Trunk-type pistons describe a piston construction that has a full skirt (FIGURE 5-8). The full-skirt design is stronger, resisting side thrust loads better, so that cylinders can support higher cylinder pressures. Trunk-type pistons are used in more rugged applications, such as in-line-sixes with high horsepower (kilowatt) output. **Side thrust** is the force of a piston against the side of a cylinder wall produced by the angle of the connecting rod with the piston, which pushes the piston sideways when cylinder pressures are high. Cylinder pressure during power stroke produces **major side thrust**, while compression pressure produces **minor side thrust** (FIGURE 5-9).

Articulating Pistons

Another innovation in piston technology is the use of **articulating pistons** (FIGURE 5-10). These pistons are used in many medium- and heavy-duty diesel engines, but they are not yet used in light-duty diesels. Articulated pistons are constructed by using two pieces: a

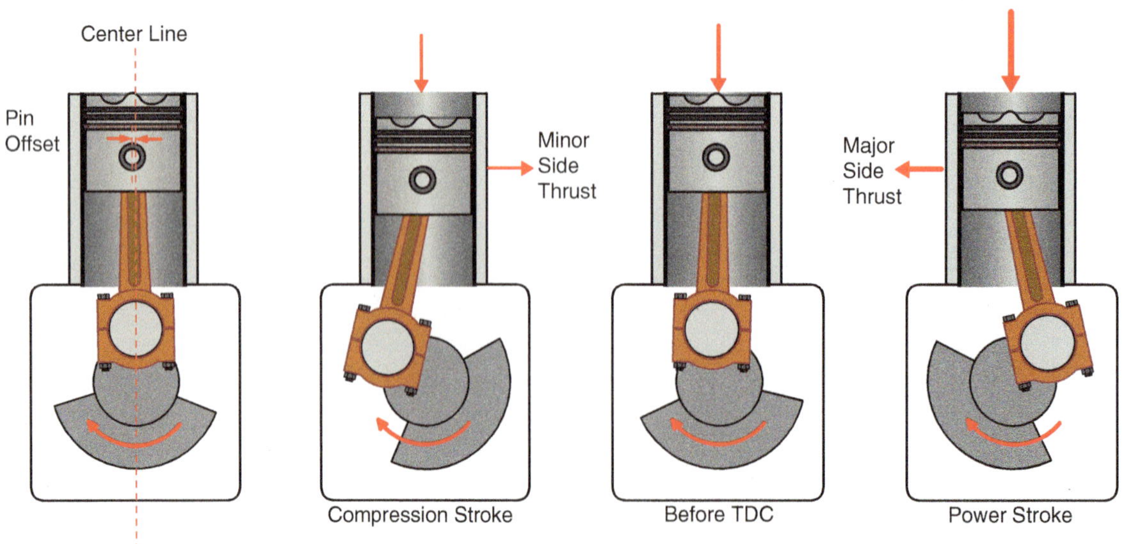

FIGURE 5-9 Major and minor thrust forces.

FIGURE 5-10 An articulating piston is made of two pieces: a steel crown connected to an aluminum skirt by using a piston pin.

FIGURE 5-11 Crevice volume refers to the area between the piston and cylinder wall above the top piston compression ring.

separate aluminum skirt connected to an alloy steel crown by using a piston pin. The use of steel as a construction material helps these pistons resist damage from high temperatures while supporting greater cylinder pressures. Steel piston crowns also reduce the volume of the area above the top ring and between the piston and cylinder wall (**FIGURE 5-11**). This region between the piston crown and cylinder wall above the top compression ring is called the **crevice volume**. The strength of a steel crown enables the top ring to be placed closer to the crown without causing a fracture between the ring groove and crown. Pistons with this feature are referred to as LCV or Low Crevice Volume pistons. Manufacturers' studies have found that this change alone improves fuel economy by as much as 5%. Eliminating the dead air of the crevice volume area lowers hydrocarbon (HC) exhaust emissions by as much as 50%. Trunk pistons are also made with reduced crevice volume, but aluminum pistons cannot reduce the crevice volume to the same extent as pistons with steel crowns can.

Monosteel Pistons

Monosteel pistons—also known as Ecotherm or Monotherm pistons—are the newest innovation in diesel piston design (**FIGURE 5-12**) now used by even small-bore light-duty diesels. The most noticeable feature of these one-piece alloy steel design is its use of less material between the piston crown and piston pin (**FIGURE 5-13**). Shortening the piston achieves nearly a 50% reduction in the distance between the centerline of the pin bore and the top of the piston—a dimension referred to as the **compression height**. Thinner skirt walls, a large cooling oil passageway, and the lower compression height help make these pistons comparable in weight to aluminum pistons used in similar applications. The smaller compression height has the potential to enable the use of shorter connecting rods and reduced block height to lower engine size and weight (**FIGURE 5-14**).

FIGURE 5-12 A Monosteel piston.

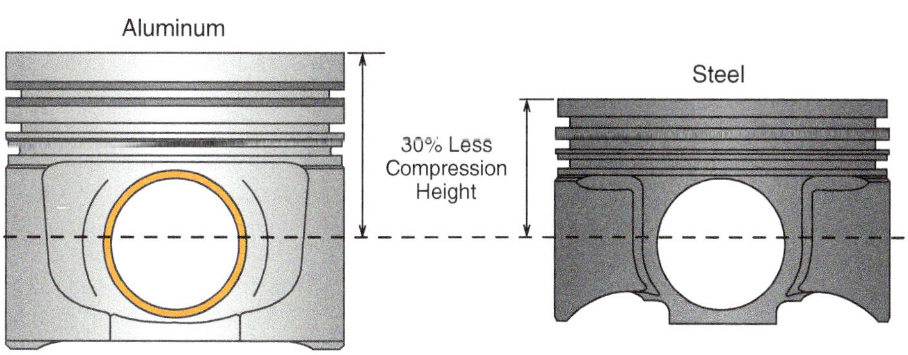

FIGURE 5-13 The use of steel pistons can reduce a piston's compression height.

FIGURE 5-14 Comparing aluminum- and steel-type pistons.

In comparison to aluminum, which expands at a rate of four to seven times faster than steel, Monosteel or articulated pistons can operate with tighter operating clearances. Operating clearances are required to allow the piston to expand when heated and allow oil to lubricate moving parts. With reduced operating clearances, both piston slap noise at cold start-up and low-load conditions are minimized. Using steel as a piston construction material essentially reduces blowby and oil consumption because of the tighter operating clearances.

Using steel pistons in a diesel engine has decreased fuel consumption by as much as 2% in test conditions. Much of the improvement in efficiency is achieved through a reduction in friction and tighter operating clearances. Low expansion ratios and a smaller piston skirt area due to the use of stronger steel provides 1% higher power output and thus 1% lower fuel consumption. This improvement also decreases turbocharger speed by 1,000 rpm in some engines, leading to improved turbocharger durability. More important are the benefits of increased durability and the ability to withstand higher temperatures and pressures than aluminum pistons. Current maximum cylinder pressures for these pistons is 3,770 psi (260 bar). This advantage alone increases power density from an engine, which means more power can be produced from each cubic inch (or cubic cm) of engine displacement. Higher piston temperatures reduce carbon formation and improve engine thermal efficiency.

The following are some of the other advantages of using steel pistons:

- Alloyed steel construction means that tear apart forces during catastrophic failure are lower, leading to less engine destruction.
- No piston pin bushing is required by the steel construction. Instead, the wrist pin bore is coated in phosphate to reduce friction. A phosphate coating reduces the possibility of corrosion in exhaust gas recirculation (EGR) engines having more acidic oil.

▶ Piston Construction

5-02 Identify and explain the functions of major piston construction features.

A piston may be made through a casting process or a forging process. In a casting process, molten metal is injected into a mold from the bottom up. This minimizes the likelihood of trapped air pockets forming in the casting. Steel reinforcing struts are placed in the pin boss area, and hardened nickel alloy inserts for ring grooves are also fixed in aluminum castings during this process. When cooled, the piston features are machined to dimensions and to accept rings and wrist pins.

Pistons are also made by using a forging process to produce a much stronger but heavier type of piston. When forged, a metal billet is heated to a plastic-like state. The part is placed into a mold with two halves, where it is squeezed by high pressure into a finished shape. Final machining is done afterward, including trimming away the flashing or excess metal squeezed out between the halves of the mold, called parting lines (**FIGURE 5-15**). This process of forging compacts the metal structure more tightly, forming a denser, heavier piston and realigns the grain structure of the metal. Forged pistons are able to withstand

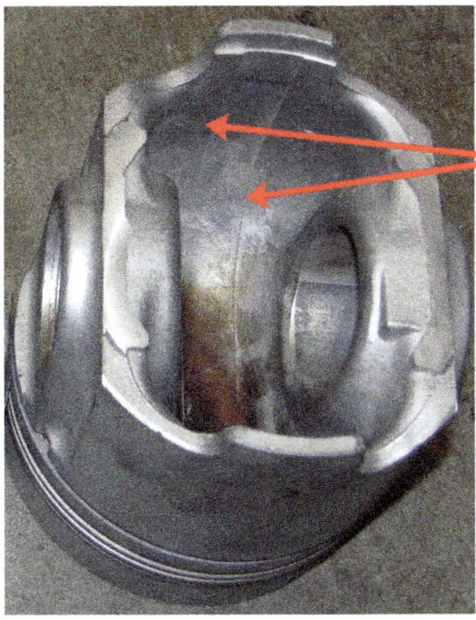

FIGURE 5-15 Parting lines on a forged piston.

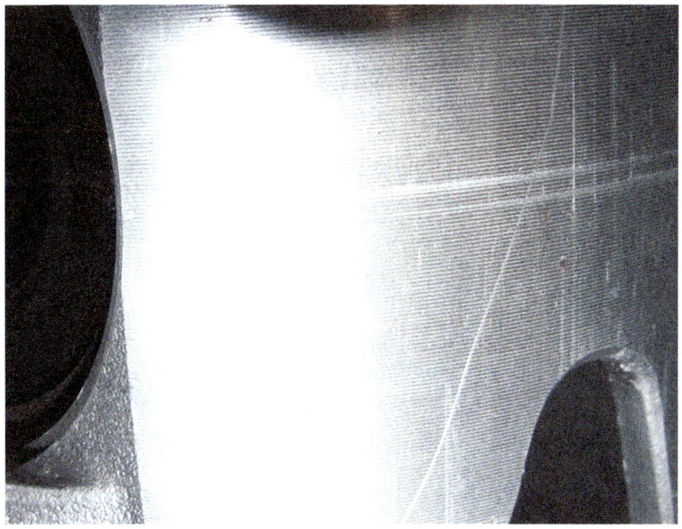

FIGURE 5-16 The finely machined lines on this piston skirt are used to retain oil for lubrication.

higher cylinder pressures and temperatures than cast pistons can. The denser structure will conduct heat away faster than a cast piston will, reducing its operating temperature by 20% compared to a cast piston. However, a disadvantage to forged pistons is that they have a greater expansion coefficient than cast pistons, thus requiring larger running clearances in cold engines.

Pistons may have the crown and skirt manufactured in two separate processes. The skirt may be forged and the crown cast, but both are bonded together by welding or by simply casting the crown onto the forged skirt.

Subsequent to shaping, skirts made of aluminum pistons are anodized to resist wear. **Anodizing** is an electrochemical process in which aluminum is combined with oxygen to create a tough aluminum oxide finish. A micro-finish etching or machining of fine lines is commonly added to the piston skirt to retain oil (**FIGURE 5-16**). Oil trapped on the skirt's micro-finish is used to help lubricate the piston.

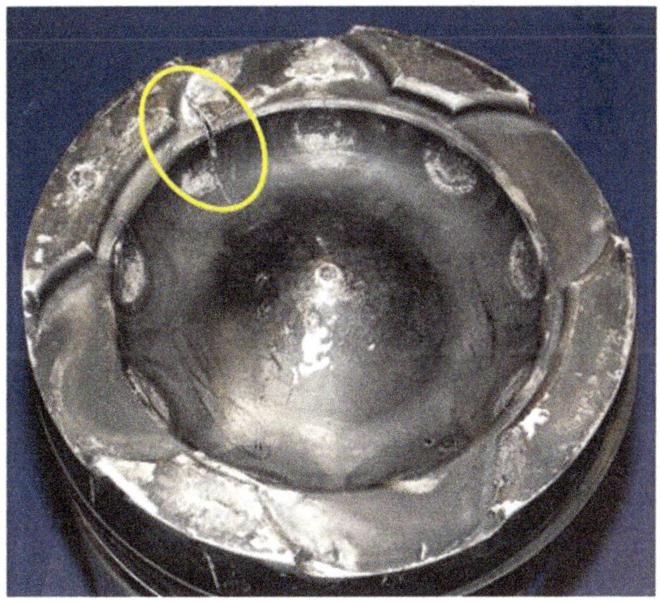

FIGURE 5-17 A cracked piston crown, where the crack was caused by inadequate cooling during the valve overlap period. This engine's air filter was plugged.

Piston Cooling

High cylinder temperatures and pressures found in diesel engines exert high heat loads on pistons. To prevent piston failure, several mechanisms are used to transfer heat away from the pistons. First, piston rings can transfer some heat to the cylinder walls through the piston rings and skirt. As much as one-third of the heat can be removed this way. Heat is also removed during the valve overlap period. During the valve overlap period, when the intake and exhaust valves are both open at the end of exhaust stroke and the beginning of intake stroke, fresh, cool air mass flowing between the intake and out the exhaust valve can cool the piston crown (**FIGURE 5-17**). Most of the cooling for diesel engine pistons, however, is accomplished through spraying the pistons with oil under the crown surface (**FIGURE 5-18**). Passageways formed in the pistons will also help transfer heat to the oil (**FIGURE 5-19**). Heat absorbed by the oil is transferred to the cooling system through the oil cooler. This piston cooling feature helps explain why during engine assembly and service, the aim of the cooling nozzles to the piston is critical to ensure adequate heat transfer and prevent piston failure.

> **▶ TECHNICIAN TIP**
>
> Whenever cylinder components are removed, always remove the cooling nozzle first. Failure to first remove these nozzles may result in damage or misalignment during service, resulting in an improperly cooled piston. Incorrectly aimed nozzles and plugged or restricted nozzles will result in an overheated piston.

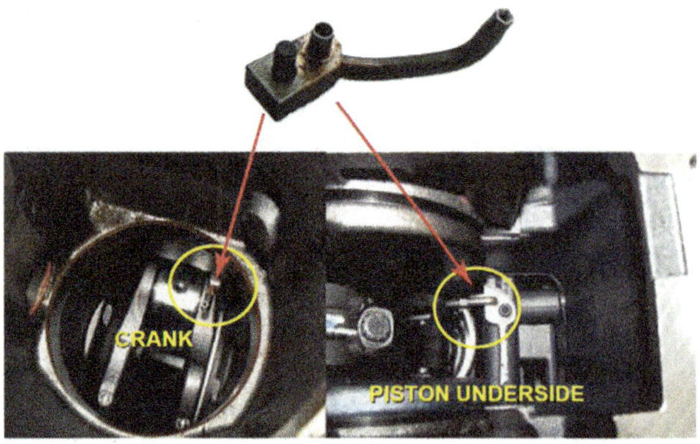

FIGURE 5-18 Oil cooler nozzles spray lubricating oil to the underside of the piston crown, removing excess heat.

FIGURE 5-19 Passageways inside the piston allow oil to circulate and remove heat from the piston.

FIGURE 5-20 This aluminum piston crown melted due to a failure in a common rail injector. After the crown overheated, cylinder pressure pushed a hole through the piston.

Aluminum Piston Materials

Due to the high speed of a reciprocating engine and the tremendous inertial forces of the piston at the end of each stroke, aluminum alloys are the material of choice for most light-duty diesel pistons. Lower inertial forces translate into less stress on other engine components, such as connecting rods and crankshafts, which extends engine life. Using aluminum alloyed pistons minimizes engine vibration and harshness. In addition to being one-third lighter than steel or iron, lightweight aluminum conducts heat about four times faster than iron. This means more heat will be transferred from the piston to the cylinder wall and engine oil. Aluminum is relatively easy to machine as well. However, when aluminum is used as a piston material, it comes with a number of disadvantages.

1. *Aluminum has a low melting point.* Since aluminum has a melting temperature of about 1,250°F (675°C) and the flame temperature of diesel fuel is 3,900°F (2,149°C), aluminum can potentially melt during extreme engine operating conditions. To overcome this, aluminum is alloyed with about 10% of other materials, such as silicon, copper, zinc, and chromium. Silicon, which makes up quartz or beach sand, gives the metal higher temperature resistance and minimizes thermal expansion. A hypereutectic piston has 16–20% silicon content, whereas other pistons have 8–11% (the maximum saturation point, or eutectic point, of aluminum is 12% silicon) (**FIGURES 5-20**, **5-21**, and **5-22**).

FIGURE 5-21 This melted aluminum piston has a grainy texture along the edges of the piston crown, demonstrating the high silicon content. The overheating may have been caused by over-fueling, advanced injection timing, or inadequate piston cooling.

FIGURE 5-22 The underside of a diesel piston crown will show evidence of baked oil from high crown temperatures. The right pistons discoloration is at the limit of normal.

FIGURE 5-23 Location of a Ni-Resist insert in a Ford five-cylinder 3.2 L diesel.

2. *Aluminum is soft and has poor wear characteristics.* This problem can be partially overcome by alloying aluminum. One of the areas of the piston that experience the greatest wear is the top ring groove. The high combustion temperatures, combined with ring movement in the top ring groove, can cause rapid wear in this area. To prevent this, a band of special stainless steel alloy is cast into the top ring land or even both ring lands. The band is machined with a groove to accept the piston ring. This band is commonly known as a Ni-Resist insert (**FIGURE 5-23**). A Ni-Resist insert further reinforces the top ring piston land, which will prevent the top of the weaker aluminum of the ring land from breaking away. Installing this land comes with the disadvantage of adding complexity and cost to the manufacturing process.
3. *Aluminum has a high expansion coefficient.* Aluminum pistons have a high expansion coefficient, which means they will expand four to seven times more than ferrous metal when heated. Aluminum alloy pistons will require large operating clearances between the piston and cylinder wall to prevent the piston from seizing after the piston warms up to operating temperature. If this expansion is not controlled, the piston will swell and score the cylinder and piston or even seize in the cylinder. Using excessive clearances is not an alternative, since this will result in piston slap or noise when the engine is cold. This noise is the result of the piston running loosely within the cylinder. Collapsed skirts, poor sealing, and oil control are some other undesirable results of running excessive clearances.

Cam Ground Pistons

A number of strategies are used to minimize the expansion problems of pistons. One method is to alloy the piston with silicon. Aluminum pistons are alloyed with approximately 20% silicone to add heat resistance and minimize expansion. At this high percentage of alloying, aluminum pistons are occasionally referred to as hypereutectic pistons. However, the method that is used almost universally to control expansion is to manufacture the piston in an elliptical or oval shape. Elliptically shaped pistons permit expansion when

> ▶ **TECHNICIAN TIP**
>
> Piston slap refers to a noise pistons make that can be heard in an engine during warm-up and light-load operation, when the piston has not fully expanded in the cylinder bore. The noise should disappear when the piston is warmed up.

> ▶ **TECHNICIAN TIP**
>
> Aluminum alloy pistons are common in light-duty diesels and operate with larger clearances between the cylinder wall when the engine is cold. It is important to warm up the pistons before heavily loading an engine. This can be accomplished through operating the engine at high idle for a couple of minutes. It is important to engage light-load operation for the first few minutes to prevent piston damage from collapsed piston skirts.

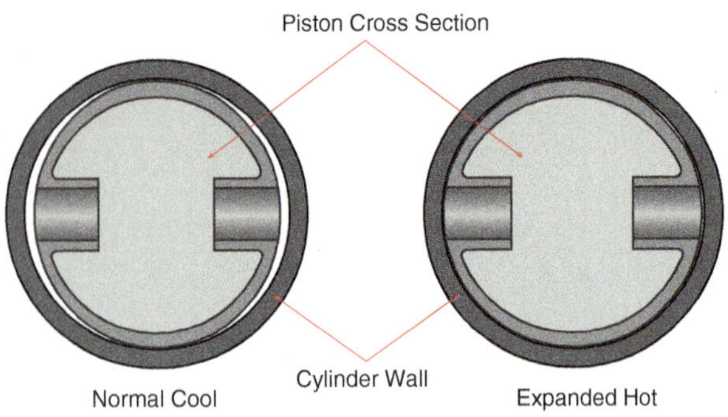

FIGURE 5-24 Pistons are machined not to be perfectly round but instead to be elliptically shaped, so that they can expand while reducing the looseness of the piston in the cylinder.

heated yet allow for minimal operating clearances to reduce noise and slap. When pistons are manufactured into an elliptical or oval shape, the expansion can be controlled so that the piston can expand to a symmetrically round shape. This piston configuration is also commonly known as a cam ground piston (**FIGURE 5-24**).

Most of this expansion takes place in the dimensions where the most metal and heat are. This means the crown will expand the greatest amount when heated. Because the wrist pin bore is reinforced, it will expand to a large extent as well. Areas that expand the least are the bottom of the skirt and thinner skirt areas around located at 90° from the wrist pin. This area corresponds to the major and minor thrust surfaces of the piston. Knowing this helps explain why measuring the piston diameter should be done at 90° opposite to the piston pin bore since it is the widest cross section of the piston.

Aluminum pistons also possess a barrel or taper shape. The crown and ring belt parts of the piston are significantly smaller than the bottom circumference. A slight bulge in the middle section is built in to minimize noise and slap and to enhance sealing capabilities. Since the cam ground piston is irregularly shaped, rings and not the piston are responsible for sealing the clearance between the cylinder walls when the engine is operated cold. This explains why piston rings on aluminum pistons will generally be thicker and require greater ring tension to ensure a gas-tight seal.

Aluminum Pistons and Emissions

Aluminum trunk-type pistons were satisfactory for many years and continue to be used today in many engines. However, other disadvantages to aluminum pistons have arisen as manufacturers have strived to reduce engine emissions and increase engine longevity and horsepower (kilowatts) all at the same time. Because aluminum pistons cannot withstand the high cylinder pressures and temperatures produced by high engine power output, new materials have been sought to replace it. Ceramic-coated pistons and ceramic alloy crowns are two such replacement materials. These pistons can not only withstand greater thermal loading but help reduce heat rejection loss from the combustion chamber. Reducing how much heat transfers to the coolant or oil means more fuel energy is converted into mechanical energy (**FIGURE 5-25**). Of course, new trunk-type pistons made entirely of steel are being introduced for a number of reasons.

> ▶ **TECHNICIAN TIP**
>
> A diesel engine will often run rough, accelerate slowly, and lack power if it has idled for extended periods of time. This condition is caused by cylinder glazing. Even as little as 10 or so minutes of idle can produce enough glazing to cause a significant change in engine performance. Tow trucks, ambulances, and emergency vehicles, which spend a lot of time idling, are especially prone to cylinder glazing. To prevent this, excessive engine idle should be avoided. When it does happen, operating the engine under load at highway speed will clear the glaze from the cylinder walls and restore normal engine performance.

Engine Damage at Idle

It is important to know that idling a diesel engines for prolonged periods of time shortens engine life and damages cylinder components. Idling damages cylinder components in several ways.

1. *Slobber production*. Combustion and exhaust temperatures at idle in diesels are relatively "cold." This happens because excess air combines with small quantities of fuel injected at idle to produce low cylinder temperatures and pressures. Pistons do not expand properly when cold, resulting in poor cylinder sealing. When pistons are insufficiently sealed,

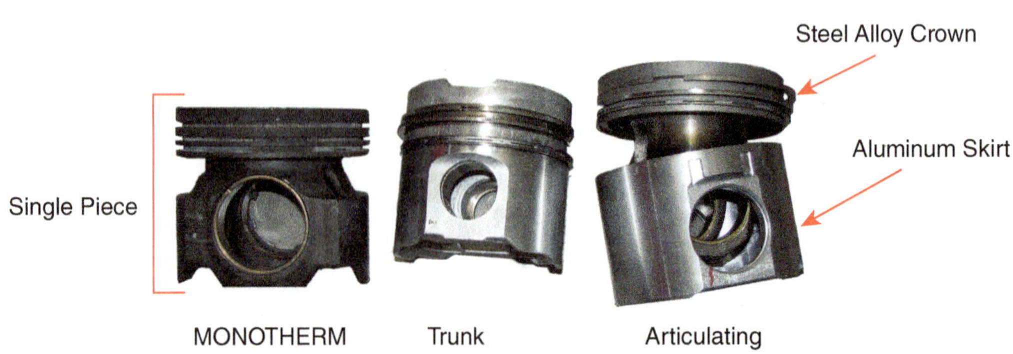

FIGURE 5-25 Comparing an aluminum piston with Monosteel and articulated. Note the distance between the top ring and the crown. Aluminum pistons have larger clearance volume.

compression pressure slips past the rings but allows more lubricating oil to enter the combustion chamber. Unburned oil, fuel, and soot are produced from low-temperature, low-pressure combustion, which when mixed forms engine "slobber." This slobber will accumulate in the exhaust system and stick around the exhaust valves, around the piston rings, and in the ring belt area. If this slobber gets around the piston rings and hardens, the rings will stick in their grooves and become almost ineffective. Slobber is one of the most common causes of premature EGR valve failure because it will contaminate or "gum up" EGR valves, causing them to stick.

DI engines are especially prone to slobbering because of their high thermal efficiency and low heat-transfer rate to engine coolant. Slobber is a substance that must be purged from the cylinder during combustion or prevented from forming; otherwise, slobber residue can accumulate and cause engine damage. During loaded engine operation, heat hardens the deposits into very abrasive carbonized crystals that eventually score the cylinder walls (**FIGURE 5-26**).

2. *Excessive soot loading in engine oil due to idling and prolonged low engine load operation.* Soot loading accumulating in engine oil is the result of poor combustion quality when the engine is at idle. Soot clings to the cylinder wall and is scrapped down into the crankcase to contaminate lubrication oil. A large amount of soot loading in the oil shortens the maintenance intervals between oil changes and requires that oil filters need to be changed more frequently. Soot is also an abrasive compound that will cause premature engine wear if allowed to reach more than 5% contaminant loading.
3. *Cylinder glazing.* Oil fills the crosshatch finish of the cylinder walls during idle and is baked onto the cylinder wall. Piston and ring travel smoothen the baked oil into a mirrorlike finish on the cylinder wall, which is then unable to retain oil. This process, called glazing, prevents the piston rings from effectively sealing leading to a loss of compression through blowby. After an engine is rebuilt, it is especially critical to prevent the cylinder walls from becoming glazed, by placing the engine under load soon after the engine starts. The deep cylinder wall crosshatch and poorly seated new rings will allow an engine to quickly glaze to the point where it is very hard to start and has low power. Shortly after starting, applying a heavy load and running the engine for at least 45 minutes to an hour helps seat the rings into the cylinder wall. This process, called break-in, uses high cylinder pressure to push the rings into the cylinder wall, where friction will smooth both ring face and cylinder wall surfaces to ensure that the rings better conform to the cylinder wall. Rapid break-in after rebuilding will ensure reliable and powerful new engine performance (**FIGURES 5-27**, **5-28**, and **5-29**).

FIGURE 5-26 This piston has operated for prolonged periods at idle, leading to excessive carbonizing in the ring belt area.

▶ TECHNICIAN TIP

Engines that operate under severe conditions, such as prolonged idle conditions in cold weather, require shorter service intervals between oil changes. Under these conditions, engine lubricating oil will contaminate faster, with soot produced by poor quality combustion. Because soot contaminates the lubricating oil of diesel engines, that oil will blacken faster than gasoline engine lubricating oil.

FIGURE 5-27 Crosshatch is observed by the fine intersecting lines on the cylinder walls. Crosshatching provides a cylinder wall finish, which retains engine oil needed for piston lubrication and to promote a gas-tight piston ring seal.

FIGURE 5-28 A ball-type hone can be used to apply new crosshatch finish to a cylinder.

▶ TECHNICIAN TIP

To avoid engine damage, diesel engines should never be operated at idle speed with no load for prolonged periods. If an engine must be idled, it is best to operate it at high idle. Manufacturers are using engine calibrations that increase engine speed and change injection timing if oil and coolant temperatures are low. Similarly, many late-model diesels have limited engine speed when cold. This ensures cold, soot thickened oil is flowing to critical engine parts and that the pistons are adequately warmed before fully loading the engine. Even automatic transmissions are now adding supplemental loads on the engine to speed up warm-up. Adding cetane booster to fuel in cold weather regions can help reduce the slobber production that contaminates turbocharger actuators and EGR valves.

FIGURE 5-29 Honing stones used to apply a crosshatch finish to a cylinder wall.

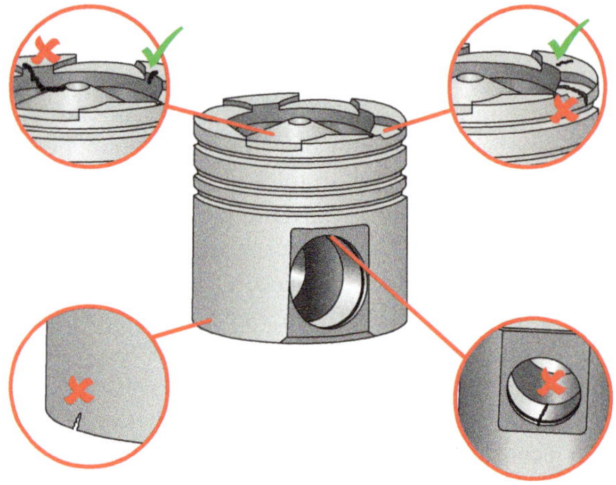

FIGURE 5-30 Inspection points on pistons.

▶ TECHNICIAN TIP

The small clearance volume in diesel engines makes the engine vulnerable to hydrostatic locks. This occurs when liquids such as oil, coolant, or fuel enter the cylinder during servicing or through leaking head gaskets or water ingestion through the intake or porous cylinder walls. A bent connecting rod or locked engine will be the result of hydrostatic locks. In severe situations, cranking the engine can result in broken cylinder walls and bent valves. If a hydrostatic lock is suspected, barring the engine over in the opposite direction of rotation will usually allow the engine to rotate 360°. During normal rotation, the engine will stop during the compression stroke of the cylinder that contains the liquid. Liquids can be vacuumed through injector holes in the cylinder head. Slightly holding an exhaust valve open while barring the engine over will also clear a lock.

Piston Cleaning and Inspection

Pistons have a micro-finish on the skirt that allows them to retain oil to improve wear characteristics. Some pistons have anodized skirts. Anodizing is a hardening process electrically imparted to the aluminum. Anodizing combines oxygen with the aluminum to form aluminum oxide, a very tough, corrosion-resistant material. Therefore, it is important to remember that this finish should not be damaged during servicing. The temptation to clean pistons with abrasive cleaners, wire wheels, and other things should be resisted. Soaking aluminum pistons in a mild aluminum-compatible cleaner followed by steam cleaning the pistons is usually the best acceptable method for cleaning. Hard carbon deposits on the crown may be scraped off by using a plastic scraper.

Visually inspect pistons for evidence of cracking, overheating, and wear (**FIGURE 5-30**). A check for the presence of vertical or horizontal cracks in the piston bowl or wrist pin bore is important when performing failure analysis (**FIGURE 5-31**). Horizontal cracks usually indicate a bonding problem between the piston skirt and crown. Vertical cracks often indicate high cylinder temperatures.

Worn pistons or pistons with a collapsed skirt, which is a deformed skirt caused by lugging a cold engine result in engine noise, high oil consumption, excessive blowby, and power loss. To inspect whether a piston is reusable or damaged or to inspect the fit of a new piston in a cylinder, follow the steps in **SKILL DRILL 5-1**. Clearances between the cylinder wall and piston can be checked by using a long feeler blade between the piston and cylinder wall at 90° to the pin bore. A worn piston or cylinder wall or a collapsed skirt is indicated when clearances exceed 0.004" (0.1016 mm).

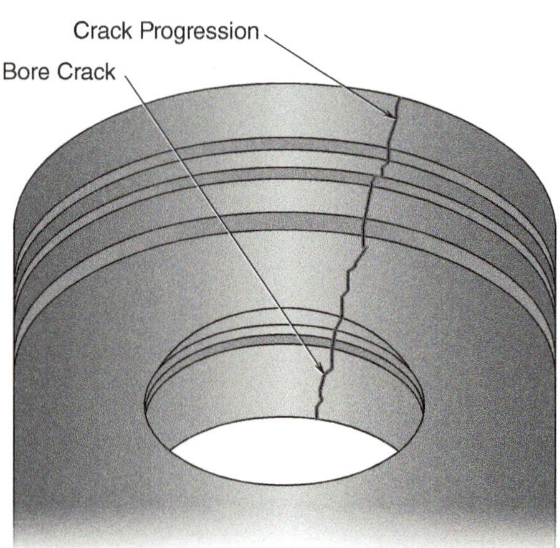

FIGURE 5-31 Piston cracks between the piston pin bore and the crown indicate abnormally sharp spikes in cylinder pressure caused by conditions such as the improper use of ether as a starting aid or hydrostatic locks.

SKILL DRILL 5-1 Measuring Piston to Wall Clearance

1. Before installing piston rings, insert the inverted piston into a clean and properly honed or prepped cylinder.
2. Place a feeler blade between the piston and the cylinder wall at 90° to the piston pin bore. Using the largest feeler blade that will slide with a reasonable amount of drag, measure the clearance between the skirt and cylinder wall.
3. Compare the thickness of the feeler blade with the specifications the original equipment manufacturer (OEM). Generally, the clearance is ideally 0.0035" (0.0889 mm) between the piston and cylinder wall. This clearance allows for just an adequate film of lubricating oil.

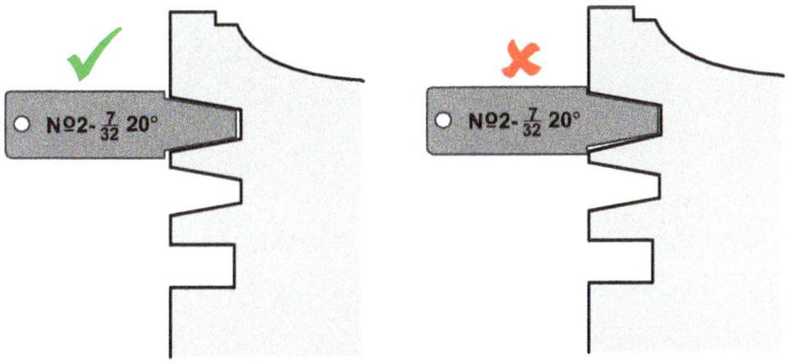

FIGURE 5-32 Measuring ring groove wear with a go/no-go gauge.

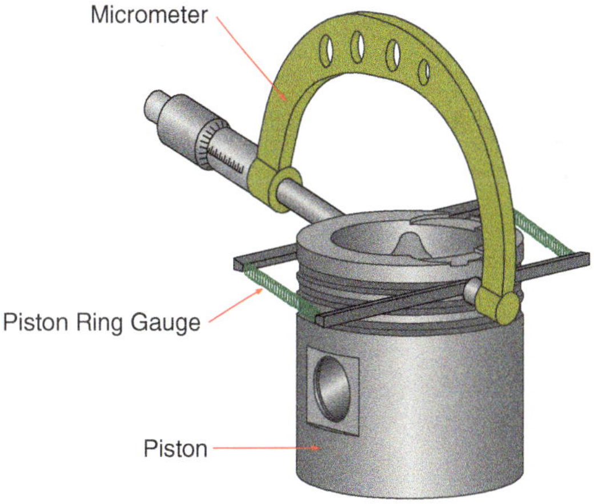

FIGURE 5-33 Gauges with the same profile as a ring are used to inspect piston ring grooves for wear.

Wear in the pistons ring groove dimensions are typically checked with specialized OEM gauge pins or go/no-go gauges (**FIGURE 5-32** and **FIGURE 5-33**). Worn ring grooves will cause excessive oil consumption.

▶ Piston Ring Construction and Functions

5-03 Identify and explain the functions of piston ring construction features.

The piston rings are some of most critical components to ensuring engine durability and efficient operation. Rings wear faster than other internal engine parts, thus limiting engine life. Worn rings cause a loss of compression, resulting in low power and excessive blowby.

Piston rings perform the following functions:

1. They form a gas-tight seal between the piston and the cylinder wall. The gas-tight seal is formed in combination with an oil film on the cylinder walls, which enhances the sealing between the ring and cylinder wall. Ensuring that the cylinder has a gas-tight seal is necessary to provide adequate compression pressure and to minimize power-robbing blowby, which also leads to oil contamination and excessive crankcase emissions (**FIGURE 5-34**).
2. Assist in cooling the piston by transferring heat to the cylinder wall.

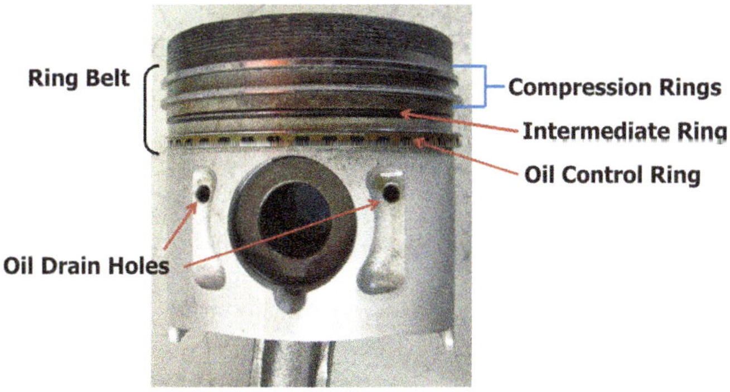

FIGURE 5-34 Identifying piston rings.

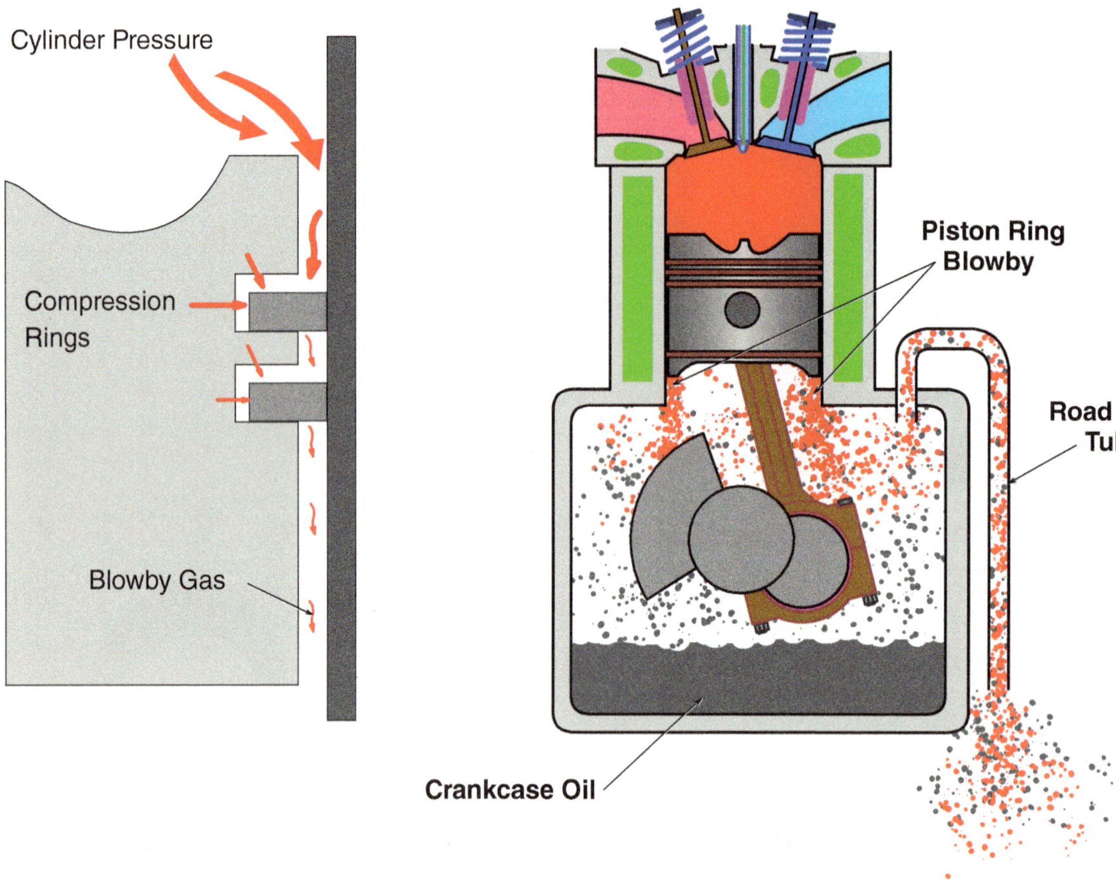

FIGURE 5-35 Piston rings provide a gas-tight seal in the cylinder to minimize loss of compression and blowby.

3. They apply a film of lubrication oil on the cylinder wall of the engine. The film must provide adequate lubrication properties, yet not be excessive, because excessive lubrication will cause high oil consumption and excessive emissions.
4. They form a compatible, replaceable wear surface with the cylinder wall. Compatibility refers to the ring's ability to operate with as little friction as possible.

The rings must perform these jobs under extreme temperatures and for the entire life of the engine. In fact, one of the primary determinants of engine life is the ring's ability to do the above jobs efficiently. Ensuring long ring life is a matter of ring design, correct service practices, and proper engine operation (**FIGURE 5-35**).

▶ Classifying Piston Rings

5-04 Identify and explain the purpose of piston ring material and shape according to ring function.

There are three types of piston rings found in most engines:

1. compression
2. oil control
3. intermediate (a combination of compression and oil control).

Exhaust Gas Recirculation Engines and Ring Longevity

Refinements to ring design, cylinder bore finishing techniques, and engine oils have helped produce engines that can operate for thousands of hours and for more than a million miles or kilometers of use before rings need replacement. Ring design and the performance of engine oil are even more critical with the use of abrasive EGR-equipped engines. Higher cylinder temperatures and pressures combined with reintroducing soot and corrosive exhaust gases found in EGR engines can affect ring longevity. For this reason, ring technology, which was relatively stable, has changed to compensate for these new wear factors. New materials and manufacturing techniques have produced some of the most sophisticated and durable rings ever made (**FIGURE 5-36**).

FIGURE 5-36 Using EGR reintroduces exhaust gases into the cylinders, which contain soot. Soot is an abrasive compound that can accelerate engine wear, including cylinder wall and ring wear.

Compression Rings

Compression rings are located nearest the piston crown (**FIGURE 5-37**). Their primary function is to prevent gases from leaking past the piston during compression and power stroke. When gas leaks past these rings, referred to as blowby, it robs power since it reduces compression pressures and may even allow combustion gases to leak into the crankcase.

Gas Sealing

Compression rings seal in a number of ways. One way is through the static force of ring tension against the cylinder wall. Since rings are manufactured, having a larger relaxed diameter than the diameter of the cylinder they are to be installed in, the ring's spring tension seals by expanding against the cylinder wall. In older rings, static ring force may easily be over 20 lb (7.465 kg). However, in an effort to minimize friction losses in an engine, high ring tension is not a desirable feature in a ring. Thinner rings can in fact seal better using a knife-edge like sealing. This means more pressure is directed against a small surface area. Today, ring tension may be as low as 4–8 lb (1.5–3 kg) of ring force. Low-tension rings can significantly improve fuel economy by decreasing internal friction and cylinder wear.

Another way compression rings seal is through the use of gas pressure during compression and power stroke to push the ring out harder against the cylinder wall (**FIGURE 5-38**). To accomplish this, the thickness of the ring is made to be slightly smaller than the height of the ring groove. With this arrangement, gas pressure is able to get behind the ring and push the ring out against the cylinder wall harder than ring tension can. Gas pressure also pushes down on the ring against the bottom of the groove for better sealing. This means the higher the cylinder pressure, the greater the ring's ability to seal. During power stroke, the compression and intermediate ring seals more effectively. This explains why most blowby occurs during compression stroke.

Piston speed and size are often factors in determining how many compression rings are used. Slower speed engines require more rings to minimize the loss of compression pressure. Compression rings, like all piston rings, are split for installation and to accommodate thermal expansion. Gases can escape through the ring end gap, so two compression rings are sometimes used on pistons.

Piston Ring End Gap

A minimum ring end gap is critical to prevent losing compression and excessive engine blowby. Too small of a ring gap will cause the ends to butt and prevent expansion. Excessive gap will obviously cause a loss of compression and excessive blowby. A general specification for ring end gap is 0.003–0.004" per inch (0.0762–0.1016 mm per 2.54 cm) of cylinder bore. One of the primary reasons for ring breakage in worn engines is flexing of the ring as it moves from a narrow cylinder diameter to a larger diameter and back again, repeatedly.

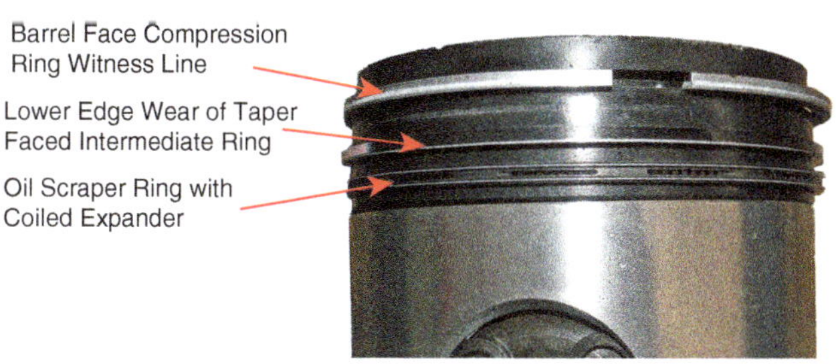

FIGURE 5-37 Identification of common ring configuration.

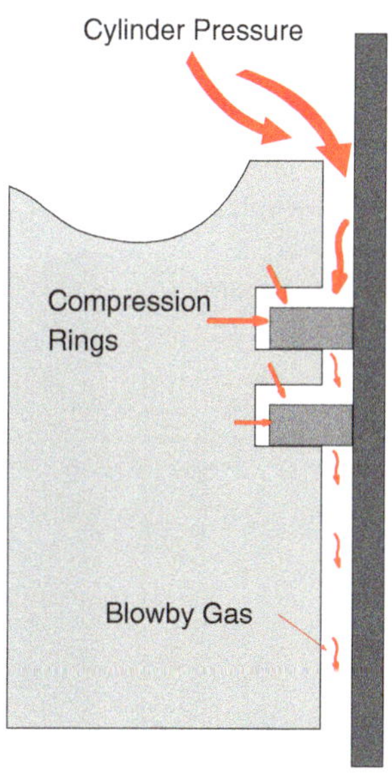

FIGURE 5-38 Gas pressure above the compression and intermediate ring forces the ring out against the cylinder wall. Making the ring groove slightly larger than the ring and notching the backside of the ring enhances this effect.

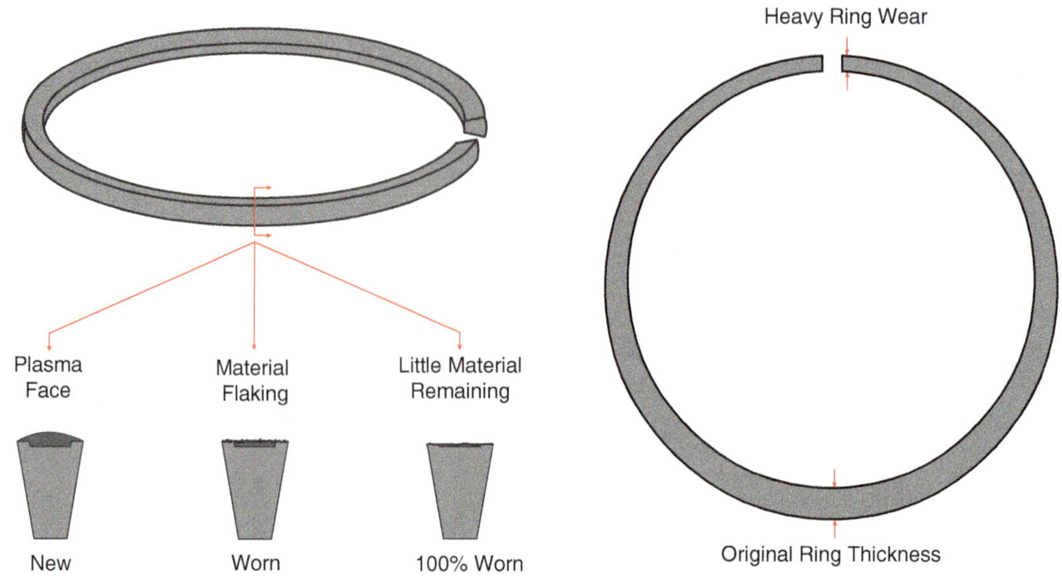

FIGURE 5-39 This ring is severely worn from dirt entering the engine. Wear like this is referred to as engine "dust out."

For every 0.001" of cylinder taper, the ring end gap will expand slightly more than 0.003". This flexing will eventually fatigue the ring and cause it to break. Maximum cylinder taper is limited in most engines to 0.001" per inch of bore. No more than 0.005" of taper is generally allowed in a cylinder. A dial bore gauge is generally used to measure this taper. An alternative method is to measure the ring end gap with the ring at several locations in the cylinder: near the top, middle, and bottom.

Compression Ring Construction and Function

Cast iron rings were used for ring material in early automotive days. Iron is an ideal ring material because of its ability to break in easily and conform to cylinder irregularities and because of its high heat resistance. Today, cast iron rings may be only occasionally used when rebuilding gasoline engines and never used for diesel engines. Ductile iron or steel rings are the most common ring materials used today, with special coatings applied to the ring face to improve hardness or for oil retention.

Turbocharged diesel engines with high temperatures and cylinder pressures require more durable ring materials. Malleable iron was often used as the ring base material since it will bend and not fracture like cast iron rings. Chrome plating was used on the ring face to improve wear and heat resistance. This material started to be used in World War II to prevent engines in desert tanks from **dusting out** due to ingesting desert sand. During dust out, sand enters the cylinder and causes rapid abrasive wear (**FIGURE 5-39**). Chrome has one disadvantage, however: its smooth face will not retain oil for lubrications. The consequence is higher cylinder wall wear in the ring turnaround area since upper cylinder wall lubricant is nonexistent above the intermediate ring (**FIGURE 5-40**). During the 1970s, ring manufacturers found that a coating of molybdenum on the ring improves wear characteristics. Wear resistance is improved because "moly"-faced rings have a porous finish. This porosity keeps oil on the ring face so that oil can be carried into the ring turn-around area, thus reducing upper cylinder wall wear (**FIGURE 5-41**). A combination of chrome and molybdenum was used for decades for compression rings.

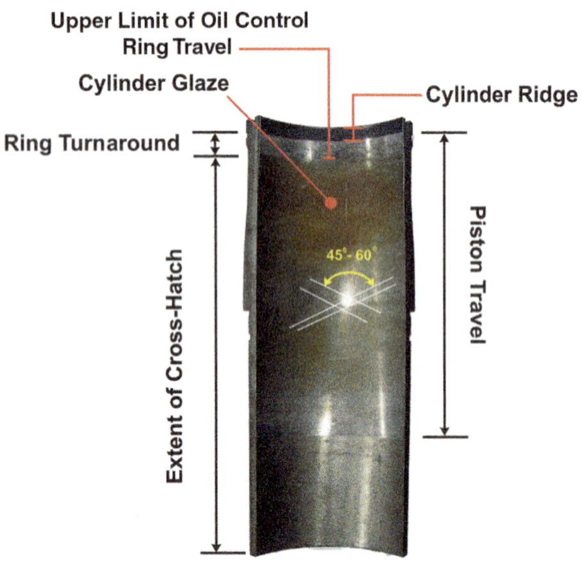

FIGURE 5-40 Most cylinder wear takes place at the top and tapers toward the bottom. A lack of lubrication oil above the oil control and intermediate ring is the primary reason for most wear taking place in the ring turnaround area.

Steel Compression Rings

A much better material for compression rings than cast iron and ductile iron is steel. This material can be more than three times as strong as ductile iron. Since steel is also the best material for withstanding the pressure

TABLE 5-2 Comparing Piston Ring Alloys

Material Hardness	Tensile Strength
Gray cast iron	20,000 psi (140 mPa) minimum
Ductile iron	40,000 psi (275 mPa) to over 90,000 psi (620 mPa)
Steel	200—2,100 mPa 29,008—304,579 psi

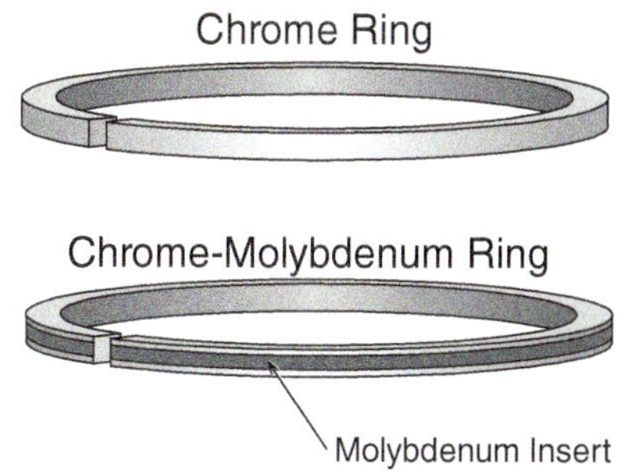

FIGURE 5-41 This illustration of a molybdenum-faced (or "moly"-faced) ring has a porous finish which is used to retain oil aiding upper cylinder wall lubrication. Note that chrome rings are shiny.

and temperature loads found inside high-compression, turbocharged diesels, this material has been used for 30 years or more in many heavy-duty diesel applications (**TABLE 5-2**).

Generally, steel compression rings have the following advantages:

- better breakage resistance
- improved heat resistance
- better mechanical stress resistance
- reduced ring side wear
- reduced groove side wear
- longer service life
- reduced cost and complexity when being manufactured (made from coiled wire).

With the advent of EGR, ring face materials need to be even harder. Combinations of ceramic metal alloys are now being used. Rings are made harder by nitriding them. Nitriding is a hardening process that impregnates the surface of the steel rings with nitrogen. The hardening extends into the ring to a depth of about 0.001". This hardening technique achieves almost 50% greater wear-resistance than steel rings and four times that of gray cast iron rings. This improves the ring wear resistance to side wear and face wear to the point where ring wear is negligible. Cylinder walls will wear out before these rings do.

One other exotic diesel engine ring material uses a combination of plasma moly and ceramics such as chromium carbide. Ceramics are extremely hard and wear resistant, but do not conduct heat well. Moly-Cermet is one such material. It is composed of 80% molybdenum and 20% chromium carbide ceramic, which makes for durable ring face material in a turbocharged diesel engine.

▶ TECHNICIAN TIP

The durability requirement for engine life established by emission regulations means that rings must last longer. Rings can be made harder for improved wear resistance but the disadvantage is prolonged break-in periods. Both the ring and cylinder wall must conform to each other's shape during the initial engine run-in period for optimal cylinder sealing. At one time, break-in would takes place in hours with softer ring materials. Some heavy-duty engines now take as long as 12,000 miles (19,300 km) before they are considered to have been "broken in." Blowby, low power, and the excessive production of engine slobber are complaints arising from longer break-in periods.

Compression Ring Shapes

Keystone-shaped rings are the ring shape of choice for most engine manufacturers (**FIGURE 5-42**). Rectangular-shaped rings were once the standard ring shape, but they have a problem with sticking when combustion and baked oil and fuel residues carbonize in the ring groove. Keystone-shaped ring action helps keep the ring land clean and helps prevent ring sticking.

A barrel face is often ground on the face of compression rings to ensure the ring is properly sealed and to minimize friction. The barrel face forms line contact with the cylinder wall to form tight and relatively friction-free contact. Prior to packaging compression rings, the manufacturer laps rings in a cylinder identical to the one that the rings will run in. A witness line left by the lapping process shows the contact with the cylinder. This same line will grow wider as the ring wears into its installed cylinder. The width of the witness line can be used as a visual indication to determine the amount of ring wear.

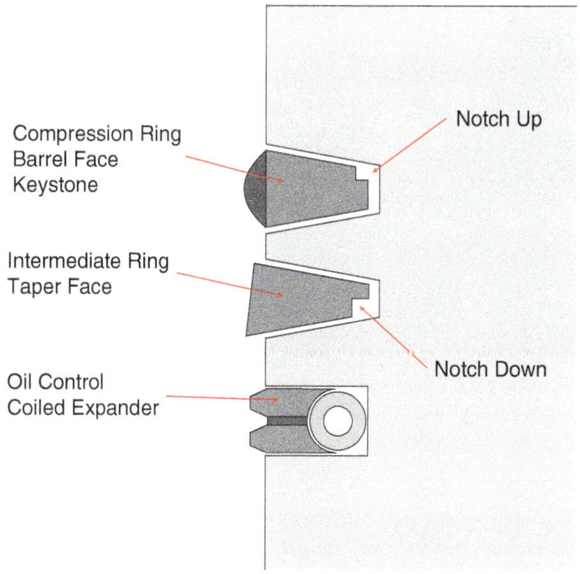

FIGURE 5-42 A keystone-shaped compression ring.

Intermediate Rings

Intermediate rings might also be called compression rings. However, they do more than compress and seal. While intermediate rings seal compression gases that escape past the top rings, they also assist in oil control. A small, metered quantity of oil is required on the cylinder walls for the piston and rings, but any excess quantity will lead to high oil consumption and emissions. The ring shape of choice for the intermediate ring is a taper-faced

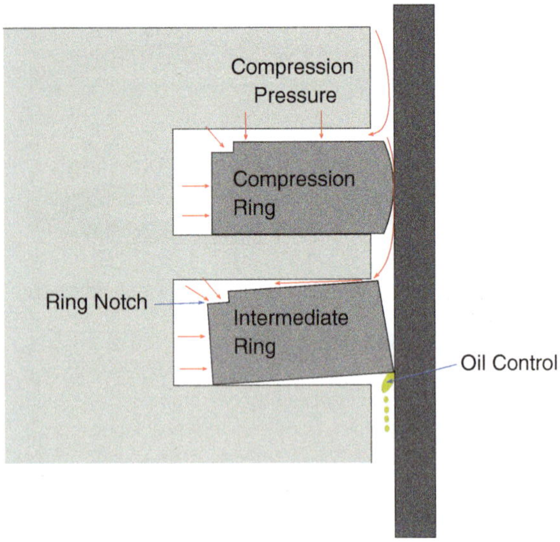

FIGURE 5-43 Depending on the twist of the intermediate ring, the ring's oil control or compression sealing function can be enhanced.

ring. This shape allows the ring to glide over the oil film on the upstroke, enhancing the sealing effect of oil between the ring and cylinder wall. On the downstroke, the ring will scrape oil from the cylinder back into the crankcase.

A variation of this design lets the ring twist in the groove. Using a ring with a groove machined in the bottom or top inside edge of the ring permits gas pressure to cause the ring to twist during either compression or power strokes. Either compression sealing or oil control can be enhanced, depending on the direction of twist (**FIGURE 5-43**).

Oil Control Rings

The last ring type found on a ring pack is the oil control ring. This ring is usually cast iron with scraper rails that remove excess oil from the cylinder wall while leaving a metered quantity for lubricating the cylinder wall. This oil film thickness is typically no greater than four-millionth of an inch (0.000004" or 0.0001016 mm) thick.

The oil control ring is usually a low-tension-type ring and has no gas pressure acting on it in a significant way. For this reason, a metal expander is placed behind the ring to maintain constant uniform pressure against the cylinder walls by the scraper rails. The expander ring may be corrugated or slotted or a wire coil.

Like the intermediate and compression rings, the oil control ring is vulnerable to premature wear from dirt ingestion into the engine. Scraper rails will show premature wear if abnormal amounts of dirt contaminate the oil through the intake air (**FIGURE 5-44**).

▶ TECHNICIAN TIP

It is important to note that compression rings and intermediate rings have directional markings to guide the technician during ring installation. Markings are usually located on the top of the ring and orientate the ring to the top of the piston. For the intermediate ring, the direction of orientation is critical since an incorrectly orientated intermediate ring can lead to high oil consumption, excessive blowby, and low compression.

Ring Installation

Care must be exercised when installing rings onto a piston and into the cylinder since rings are fragile. A common cause of ring breakage is overstretching the ring during installation. Ring expanders are tools that stretch a ring just far enough for it to be removed but not too far to cause it to bend or break. To prevent loss of compression and damage to the cylinder wall, ring end gaps must be staggered 120° apart in a three-ring pack configuration. End gaps should not be aligned near the wrist pins (**FIGURE 5-46**). If rings are reused such as after performing warranty repairs on engines having accumulated low travel distances, extra caution should be exercised when removing pistons from cylinders to prevent ring damage. The wear step or cylinder ridge found above the ring travel area in the cylinder should be first removed to prevent damage to the rings as well as the piston when both the piston and rings are pushed up over the ridge during removal. Failure to remove the cylinder ridge by cutting or grinding it away will score the piston and even break the rings.

Use ring compressor to install a piston into the cylinder and prevent damage to the rings (**FIGURE 5-45**). A number of designs are available to ease ring installation.

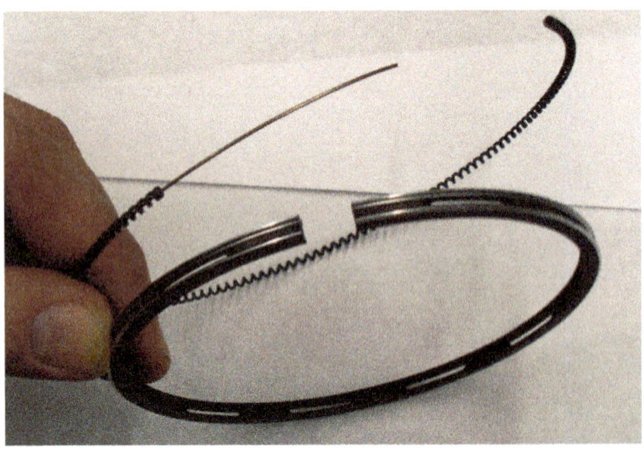

FIGURE 5-44 Oil control rings consist of scraper rails and an expander ring to maintain rail contact with the cylinder wall.

▶ Cylinder Component Service

5-05 Identify and describe diagnostic techniques used for evaluating the condition of a piston ring–to–cylinder wall seal.

While diesel engines provide two to four times longer service than comparable spark-ignition engines, the engines are often installed in a chassis or equipment with even longer life. Off-road equipment and on-highway vehicles operating long hours or distances every day will accumulate high engine mileage before the rest of the equipment will wear out. Engine repair or overhaul on high-value machines is justified in these circumstances. Occasionally, engine failures may need to be corrected at a component level too. More often than not, labor costs, warranty policy, the cost of downtime, and available skill level for repairing the engine are guiding factors for the decision as to whether to repair or replace. However, technicians should be familiar with some of the common service and inspection procedures for cylinder service if they find themselves called upon to repair rather than replace an engine.

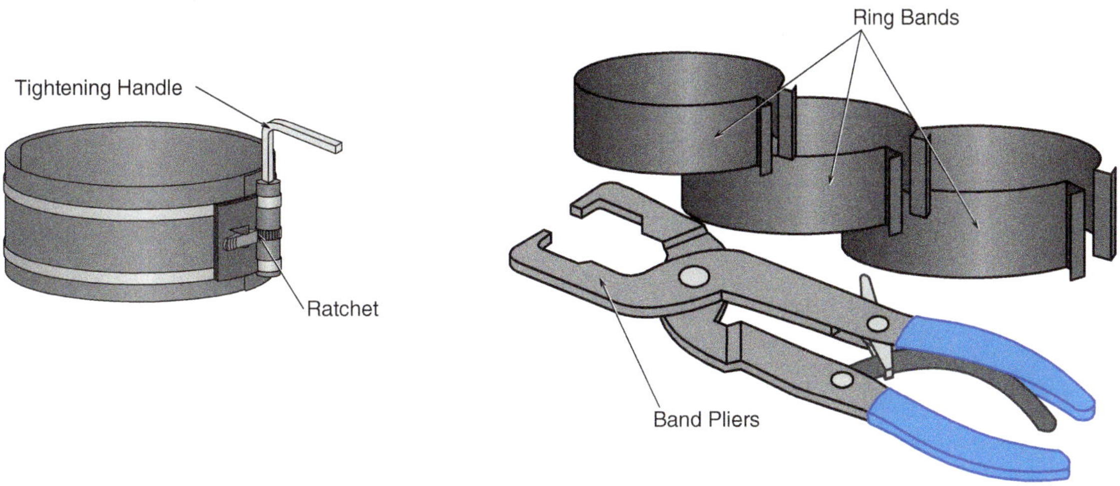

FIGURE 5-45 Various types of piston ring compressors are used to squeeze rings on the piston while inserting the piston into the cylinder.

Evaluating Cylinder Sealing

The effectiveness of seal between the rings and cylinder wall is an important measure of engine life. A poor seal means the engine is worn out or has internal damaged. The question a technician needs to answer to recommend whether a repair should be made is, how well are the rings sealing against the cylinder wall seal? This means that the method to quantify or measure sealing effectiveness is an diagnostic technique. Three methods are available to the technician to evaluate the condition of piston ring–to–cylinder wall seal.

1. *Compression testing.* Compression testing is rarely used on diesel engines, due to the lack of clearance volume that the cylinders have and due the prospect of the cylinders being hydraulically locked with oil. Wide variations of compression results are possible due to the design of piston rings and low clearance volume in a diesel. Furthermore, it's cumbersome to remove an injector to install a compression gauge into a cylinder. However, glow-plugs in smaller, light-duty diesel engines can be removed to test compression pressure. Other methods to evaluate cylinder sealing are more effective though.

   Maximum compression pressure is measured with a pressure gauge installed in the injector or glow-plug hole. When the engine is cranked over, maximum pressure values are observed for each cylinder and compared against manufacturer's specifications as well as for cylinder-to-cylinder variations. Compression pressures should not vary by more than 10% between cylinders. Compression testing is more commonly used in spark-ignition engines (**FIGURE 5-47**).
2. *Cylinder leak-down testing.* A more accurate and meaningful measurement of the cylinders sealing capabilities can be obtained from a cylinder leak-down test. This test can be performed separately or in conjunction with compression testing and is more practical for diesel engines. Unlike compression testing, air is fed into a cylinder at pressures between 80 and 120 psi (between 551.6 and 827.4 kPa). The engine should be warm to minimize clearances between the cylinder walls and the piston. With the valves closed, the piston is positioned at top dead center (TDC) and the engine locked in this position. The cylinder leakage tester is connected to the cylinder and

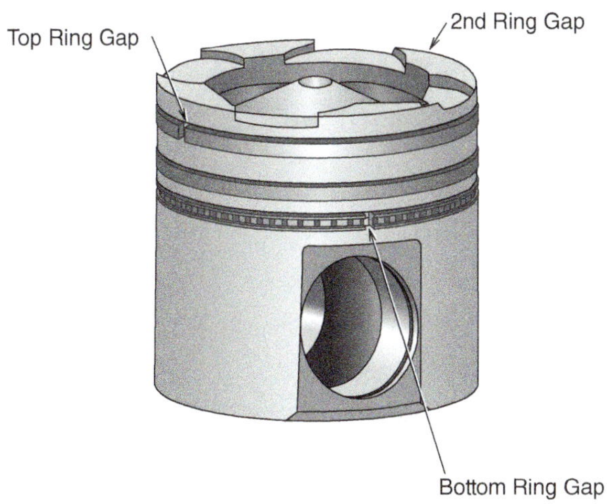

FIGURE 5-46 Piston ring end gaps must be staggered 120° apart to prevent gases from easily leaking past the rings.

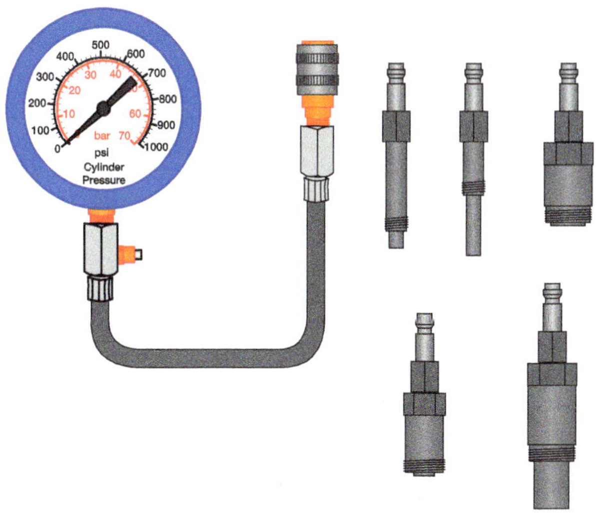

FIGURE 5-47 Diesel compression testing gauge with a variety of glow-plug and injector adapters.

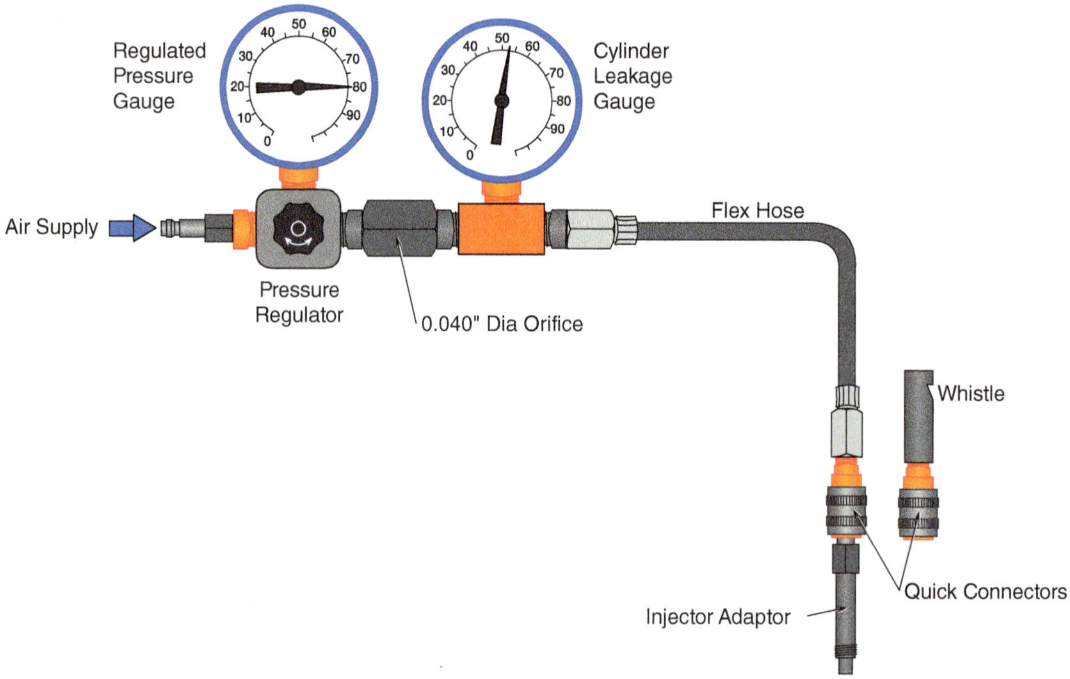

FIGURE 5-48 A leak-down tester supplies regulated air pressure to a cylinder and measures the percentage of regulated air pressure maintained in the cylinder. The percentage of leak-down is an indicator of cylinder and ring condition.

pressurized with regulated shop air. A second gauge on the leakage tester will record the percentage of air leaking into the cylinder (**FIGURE 5-48**).

The rate of leakage is compared against manufacturers' specifications to determine whether the amount of leakage is acceptable. Engine problems can be detected by listening for large volumes of air leaking out of the turbocharger, intake or exhaust system, or even the oil filler tube.

An engine in good condition should generally show only 5–10% leakage in the pressure of air supplied to the cylinder after a minute. Some engines in acceptable condition may indicate as high as 20% leakage. But more than 30% leakage indicates problems.

A cylinder that has poor compression but minimal ring leakage may indicate a valve-train problem such as a worn cam lobe, broken valve spring, inadequate valve lift due to a defective cam follower, or even a bent push rod. If all the cylinders have low compression but show minimal leakage, the most likely cause is incorrect valve timing. A slipped notch on a timing belt or timing chain or an incorrectly installed cam gear may be the problem in this instance. A damaged valve will leak air from either the intake or the exhaust pipe.

If compression is good and leakage is minimal but a cylinder is misfiring or shows up weak in a power balance test, it may indicate that the diesel engine has a defective injector or injection pump.

3. *Crankcase blowby*. One of the best recommended practices for evaluating the condition of piston rings in a diesel engine is to test crankcase pressure. Since blowby gas pressure in the crankcase is a function of the effectiveness of the ring-to-cylinder wall seal, the quantity of blowby and crankcase pressure is a good measure of ring-to-cylinder wall condition. Crankcase blowby pressure is usually very low—just a few inches (centimeters) of water column pressure when the engine is under full load (rings are seated). When the rings and cylinder walls are worn, blowby volume increases, which drives up crankcase pressure.

Blowby testing is a fast, simple, and unobtrusive technique to evaluate the condition of piston ring-to-cylinder wall sealing. Pressure is measured at high idle, no-load operation. When piston rings seal well, only a little blowby gas enters into the crankcase. When cylinder pressures are higher, such as during high-idle-rpm operation, the rings seal much better

SKILL DRILL 5-2 Performing a Crankcase Blowby Test

1. Before starting the crankcase pressure test, it is important to run the engine until it reaches normal operating temperature. A cold engine will produce an abnormally high pressure reading.
2. Install the crankcase orifice restrictor tool in a road draft tube or a crankcase pressure test adapter in the oil fill hole. Use the manufacturer's prescribed tool and restriction diameters.
3. Ensure that the oil dipstick tube seal and any other gaskets and seals on the engine are functioning properly.
4. Connect the Magnehelic gauge that has plastic tubing to the restriction tool. There are two ports on the gauge: pressure and vacuum. Connect the restriction tool attached to the crankcase outlet to the pressure port of the Magnehelic gauge.
5. Make sure the gauge tubing is clean and dry so that the gauge does not become damaged. It is good practice to blow clean, dry compressed air through the tubing before each use.
6. Zero the Magnehelic gauge.
7. Chock the wheels of the vehicle and apply the brakes. Place the transmission in neutral or park.
8. Start the vehicle's engine and run it at wide-open throttle under no load. Maintain this engine speed for at least 30 seconds and take a stabilized reading. Make sure the hole in the top of the crankcase orifice restrictor tool is never blocked when the engine is running.
9. Record the results and compare them against specifications. Any readings of more than 4" (102 mm) of water during the crankcase pressure test indicates a worn engine. It is likely that base engine mechanical concerns exist. Make a service recommendation to overhaul the engine if engine hours and distance traveled are high.

than they would at low-engine rpm. In a good engine, crankcase pressure is more often higher at idle than at high idle. However, glazed cylinders are indicated when crankcase pressure is significantly higher at idle and drops as engine speed increases. To perform a crankcase blowby test, follow the steps in **SKILL DRILL 5-2**.

Pressure in the crankcase is measured in inches of water column or kilopascals by using a water manometer or Magnahelic gauge (**FIGURE 5-49**). Measurement units for crankcase pressure are very sensitive measurements of pressure. The units of measurement and the tool used to make the measurements are also often not familiar to many technicians. To develop a sense of how sensitive they are, and to avoid the mistake of using other pressure gauges or units of measurement, see **TABLE 5-3**. A calibrated restriction fitting is placed over the crankcase breather tube. If equipped, the crankcase depression regulator will be blocked during the test. After the engine coolant and oil are warmed to operating temperature, the engine is put to full throttle, no load and crankcase pressure is measured. Usually only a few inches of pressure are measured in an engine in good condition. Small-bore diesels will have a maximum limit of 10" of water column pressure (2.49 kPa). Larger engines may be allowed as much as 15–18" of pressure (3.73–4.48 kPa) (**FIGURE 5-50**).

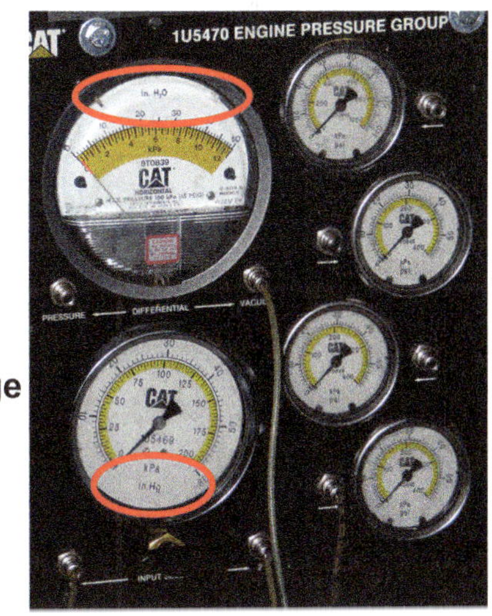

FIGURE 5-49 A Magnahelic gauge measures pressure in inches of water column (or kilopascals). These gauges are much more sensitive to pressure changes than measuring using with mercury.

Measuring Cylinder Wear

In addition to unobtrusive measurements to evaluate the effectiveness of the ring-to-cylinder wall seal, direct measurement of the cylinder can indicate the amount of engine wear to provide a service recommendations.

TABLE 5-3 Pressure Equivalents

1 bar =	1.02 atmospheres	29.53" of mercury	401" of water	14.5 psi	1,000 kPa

FIGURE 5-50 Measuring crankcase blowby pressure involves the use of a restrictor orifice and a Magnahelic gauge measuring inches of water column pressure or kilopascals. Blowby is measured on a warm engine operated at high-idle rpm.

Tapered and Out-of-Round Cylinders

In slightly tapered and out-of-round cylinders, oil consumption is controlled by the pistons and rings. However, with increased cylinder taper and out-of-round conditions, good oil control in the cylinders is difficult to maintain due to a combination of several factors. First, increased piston clearances allow pistons to rock in the worn cylinders. This condition allows an abnormally large volume of oil to enter on one side of the piston. If the rings are also tilted as the piston rocks, oil can enter the combustion chamber and burn when the piston reverses direction and moves downward (**FIGURE 5-51**). Second, at high piston speeds, the rings may not have adequate time to conform perfectly to all the worn parts in the cylinders on every stroke. Whenever this occurs, the engine burns larger quantities of oil that have remained on the cylinder walls.

Distorted Cylinders

Cylinders that are distorted so that they are out of shape from other causes, such as unequal heat distribution or unevenly tightening the cylinder head bolts, which have distorted the cylinders. Cylinder distortion limits the piston ring's ability to scrape the cylinder walls completely. Another type of distortion is cylinder taper, which is caused by cylinder walls undergoing uneven wear. Taper, which is widest near the top of the cylinder where the compression ring turns around, becomes progressively smaller near the bottom of the cylinder. Taper wear is caused by several factors. The most significant is the absence of upper cylinder wall lubrication, where temperatures and compression ring pressure are highest (**FIGURE 5-52**). In the ring turnaround area, above the oil control ring's uppermost limit of travel, only lubricant on the compression and intermediate ring face reduces cylinder wall friction. Gas pressure behind the rings creates the highest unit loading of the ring face. Below the ring turnaround area, the presence of more lubricant on the cylinder wall reduces friction and wear.

Cylinder taper and out-of-round conditions can be evaluated by measuring cylinder dimensions with a dial bore gauge or ring end gap method. By installing a ring into a cylinder in three different places—one in the bottom, middle, and ring turnaround area—and then measuring the end gap, an approximate measurement of cylinder wear can be determined. The ring should be squared or perpendicular in the cylinder by using a piston

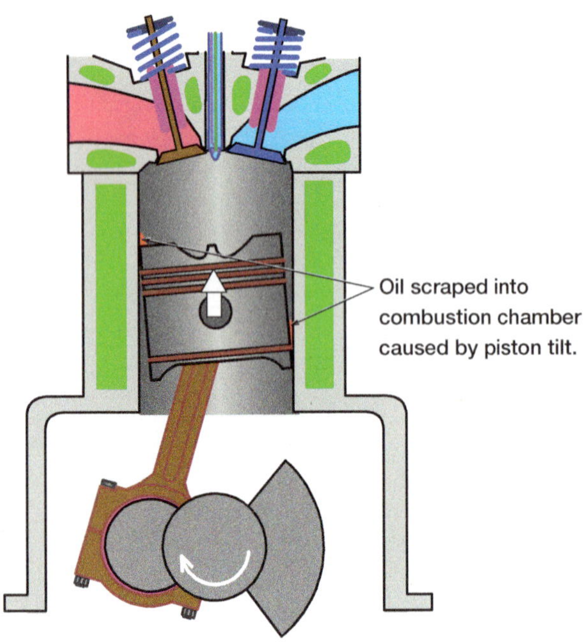

FIGURE 5-51 Wear points on a worn piston. Worn pistons tilt and scrape oil up into the combustion chamber.

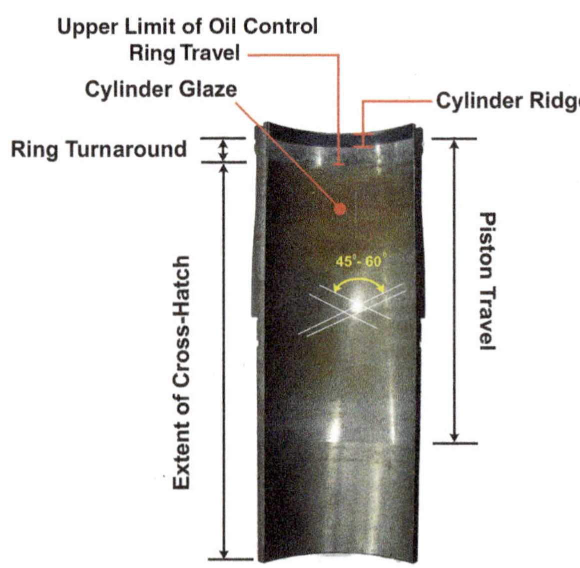

FIGURE 5-52 The area above the intermediate ring but below the compression ring experiences the greatest wear due to lack of upper cylinder wall lubrication. This is the ring turnaround area.

SKILL DRILL 5-3 Measuring Piston Ring End Gap and Ring Groove Wear

1. With the piston removed, clean the ring groove by using a soft nylon bristle brush and a non-caustic cleaner. Alternatively, using a broken piece of ring, clean carbon from the ring groove by working a section of broken ring through the groove. Be sure to wear protective gloves.
2. Using a new ring, place the outside circumference of the ring in the ring groove.
3. Using flat feeler blades, measure the side clearance of the ring by sliding the largest possible feeler blade between the top of the ring and the groove.
4. Compare specifications against the service manual. Generally, small-bore engines allow no more than 0.002" or 0.003" (0.0508 mm or 0.0762 mm) and larger pistons will be allowed as much as 0.01" (0.254 mm).
5. To measure ring end gap, place the ring inside the cylinder.
6. Square the ring in the bore with the top of a piston so that the ring is perpendicular with the cylinder wall.
7. Select the largest thickness of feeler blade that will fit between the ends of the piston ring.
8. Compare the end gap measurement against specifications. Generally, 0.004" (0.1016 mm) of end gap is required for every inch (2.54 cm) of cylinder bore diameter.

before measuring the end gap with a feeler blade. Since the ratio of cylinder diameter to radius is 3.14, the differences between the maximum and minimum end gap should be divided by a factor of 3.14. In other words, a difference of maximum and minimum end gap of 0.008" ÷ 3.14 = 0.0025" taper.

Piston rings can wear and their condition can be evaluated by examining the ring face. A barrel-faced ring will have a wide line worn into it in comparison to the narrow contact line on a new ring. Taper-faced rings will similarly have a wider, "shinier" line in comparison to a new ring. Rings ends will be thinner and will taper rather than have a square appearance. However, ring grooves can wear due to the continuous flexing of rings from a combination of reciprocating action and gas pressure. Several types of gauges are available to use to inspect the groove shape and wear, but one alternative method is to measure the clearance between a new ring and the ring groove. To measure piston ring end gap and inspect for excessive compression ring groove wear, follow the steps in **SKILL DRILL 5-3**.

▶ Piston Pins

5-06 Identify and explain the function of piston pins.

Piston pins are the connecting joint between the connecting rod and the piston. Piston pins may be either hollow or solid and are fabricated from high-alloy steels. The area of the piston pin bore of a piston is referred to as the pin boss. This area is reinforced since the thrust forces of combustion are transmitted through the piston to the pin through that area.

Semi-floating Pins

There are a variety of piston pins designs. One design is the semi-floating wrist pin. This design requires the piston pin to be pressed into the connecting rod and piston bore. In this arrangement, only the piston rotates on the pin while the pin is fixed by an interference fit with the rod. Assembling and disassembling the pin requires several tons of force and using a special fixture to prevent damage to the piston. The advantage of this pin design

FIGURE 5-53 Hollow and solid wrist pins. Lower cylinder pressures mean that hollow pins can be used.

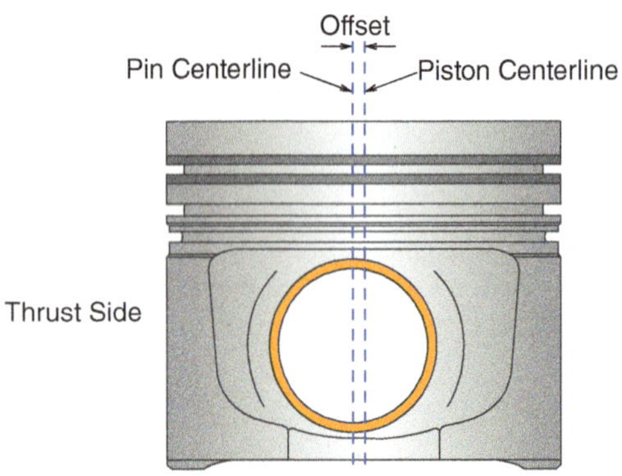

FIGURE 5-54 The piston pin axis offsets from the cylinder centerline to eliminate noises caused the piston rocking at TDC. Doing this allows the piston to change sides before it reaches TDC, minimizing noise.

is that it does not have the problem of clips loosening and falling out. This configuration is not commonly used on diesel engines, since wear is localized only on the piston.

Full Floating Piston Pins

The piston pin design common to most diesel engines is a full floating piston pin. Both the piston and the connecting rod rotate on the pin. Although this requires a bushing in the small end of the rod, more surfaces are rotating with less load with an oil film to better distribute wear. This type of pin is held in place by spring clips or, in the instances of some newer engines, Teflon caps that prevent the pin from coming out of the bore and scratching the cylinder. Some wrist pins may have an interference fit when at room temperature but the pin bore will increase in diameter with temperature in a running engine to allow the pin to move freely on both piston and con-rod. This type of pin needs to be heated while removing and replacing it (FIGURE 5-53).

Wrist Pin Offset

Although the piston moves linearly in the cylinder, the angle formed between the connecting rod and the crank throw pushes the piston sideways as it converts the reciprocating piston motion to crankshaft rotational motion. This push is known as side thrust and is divided into a major and minor thrust. The amount of thrust is a function of the angle formed between the connecting rod and the piston and cylinder pressure. Major thrust is produced during power stroke and minor thrust during compression stroke. Long stroke engines, like diesels, have greater thrust because of the angles formed between the long crank throw and longer connecting rods.

The problem of the side thrust combined with higher cylinder pressures can cause three problems in diesels engines:

1. Higher side thrust forces can lead to excessive cylinder wall wear.
2. Cylinder wall and liner cavitations may appear on the coolant side of the cylinder wall. This is a problem unique to diesel engines, where pitting on the coolant side of the cylinder wall can penetrate the cylinder. This phenomenon is explained in more detail in Chapter 7, on cooling systems.
3. Noise can be heard as the piston crosses over from pressing against one side of the cylinder to the other near TDC. Clearances between the sides of the piston instantly switch from one side to the other side as the piston moves through TDC (FIGURE 5-54).

To minimize the degree of side thrust and minimize noise caused by piston slap, engine manufacturers often offset the centerline of the wrist pin with the centerline of the piston. Because the centerlines of both the piston and wrist pin are not the same, the piston must have a specific orientation in the engine. This means there is a side of the piston that must be oriented to the left or right side of the engine. The purpose of wrist pin offset explains why pistons have markings to ensure that they are installed in the correct direction in the engine. Markings are usually present on the piston to indicate its orientation toward the front, rear, or side of the engine (FIGURE 5-55).

Another strategy to reduce the thrust angles in the engine is to offset the centerline of the crankshaft with the engine centerline. By moving the centerline of the crankshaft and wrist pin slightly to one side, the thrust angle is reduced. What this means for the technician is that connecting rods and pistons have a specific front/back left right orientation in an engine. For example, pistons must be orientated typically with a reference to the camshaft or front of the engine. Similarly, connecting rods often have directional indicators or stamping marks facing the front of the engine.

FIGURE 5-55 Note the directional arrow on this piston crown. Thrust forces are minimized by offsetting the wristpin's centerline. For this reason, pistons have a specific orientation in the engine.

▶ Connecting Rods

5-07 Identify and explain the function of connecting rod construction features and mechanical stresses imposed on connecting rods.

Connecting rods link the piston to the crankshaft enabling the conversion of the pistons reciprocating motion into rotational movement of the crankshaft. Connecting rods are the most stressed component of the engine since they experience centrifugal tensioning and compression loading every time the piston changes direction (**FIGURE 5-56**).

Compression and Tensional Loading

Tensional loading occurs at the end of the end of upstrokes at TDC, where the weight of the piston assembly would want to continue in a straight line through the cylinder head. Without the cushioning effect of pushing exhaust gases out or compressing the gases in the cylinder, the tensional forces would be even greater.

Compression loading occurs during power stroke, where the combustion forces push the connecting rod down with forces, in excess of 15 tons of force (206.8 mPa). For this reason, connecting rods are built to typically withstand forces greater than 16 times what is normally encountered. These forces could be from abnormal combustion conditions or engine over-speeding. Engines using heavier steel or articulated pistons have larger-sized con-rods to handle the increased tensional forces (**FIGURE 5-57**).

Connecting Rod Materials

Since the con-rods are highly stressed, they are commonly forged from high-alloy steel. Rods are then shot peened to relieve stress risers and add additional fatigue strength. Bolts used on the caps are made from proprietary alloys of grade-eight metals. Rolled threads are used on these bolts to increase their strength (**FIGURE 5-58**).

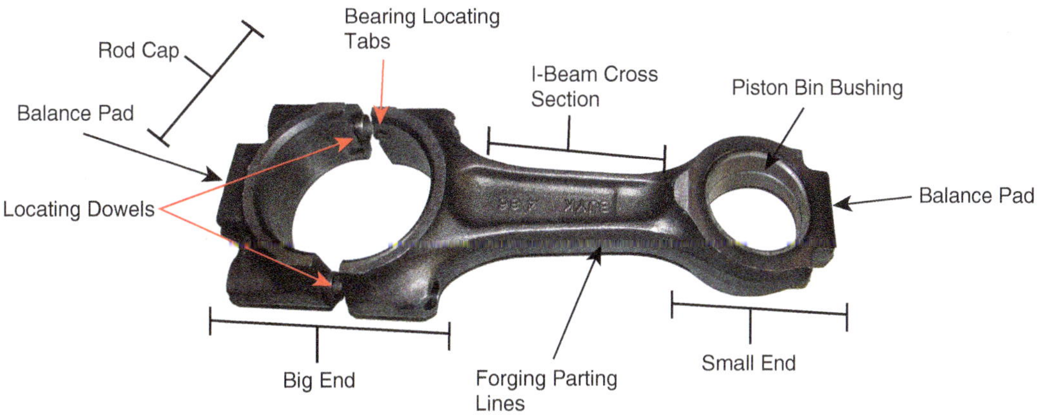

FIGURE 5-56 Connecting rod features and nomenclature.

FIGURE 5-57 Connecting rods use bolts with stronger rolled threads and propriety grades of hardness. Rolled threads have a larger root diameter than the equivalent conventional threaded bolts.

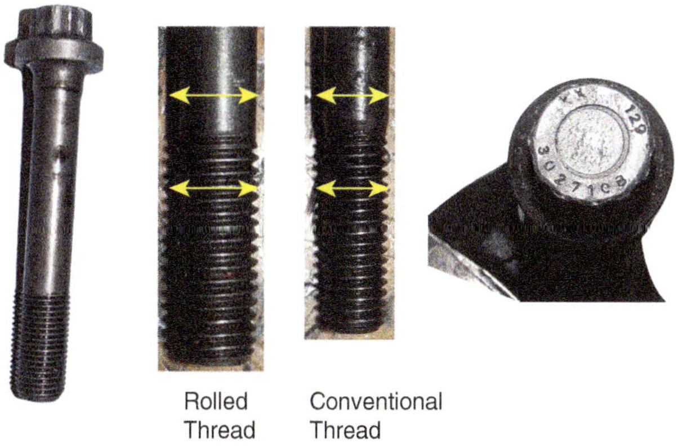

FIGURE 5-58 The root diameter of a rolled thread is larger than a conventional thread.

FIGURE 5-59 The teepee style of connecting rod is used most commonly in diesel engines since it possesses better wear and loading characteristics.

Connecting Rod Features

Connecting rods will use an I-beam style cross section, which has good weight-to-strength characteristics and resists bending. The smaller end of the rod will have a bushing pressed into it if it is used with a full floating piston. When the small end is shaped like a triangle, it is called a teepee-style rod (**FIGURE 5-59**).

The weight of a set of connecting rods is critical to engine balance. If connecting rods vary in their mass, an engine vibration will result. To adjust the weight of connecting rods, balance pads are located at each end for grinding, to balance rod weight between rods and around its center point. Select fit rods are often used together in sets in an engine. This means that rods are weighed after they are manufactured and that those closest in weight are used together in sets. These can be identified by the paint marks that uniquely identify the weight variation of the rod. For example, a group of connecting rods may have white, yellow, red, or other paint markings on them designating whether they belong to a particular weight classification. Such rods sets should be kept together when assembling an engine. If a service rod is required, it will often have a paint marking that is in the middle of the weight distributions.

Fracture-Split Connecting Rods

Fracture-split connecting rods are the most commonly manufactured type of connecting rods today. These rods have the big end of the rod parting surfaces that are split by fracturing rather than machining the mating surfaces between the rod and bearing cap. Using a fracture-split rod cap offers several advantages:

- more precise alignment of the rod and cap than a machined rod and cap
- improved control of oil clearances
- reduced production cost since machining steps are eliminated
- greater retention strength between the cap and rod.

Fracture-split rods allow the rod and cap to be separated along a predetermined fracture line. Instead of cutting the rod cap away from the rod, then machining and re-boring the rod and cap, a hydraulically operated wedge pushes the two pieces away from one another. A line scribed where the split should take place provides a weakened point to control the split. Since the big end of the connecting rod is fracture-split, the rough-edged of the big end and cap have greater retention strength than a smoother conventional flat-machined surface (**FIGURE 5-60**).

Fracture-split rods can be made to be about 17% lighter, reducing engine vibration while maintaining the strength of traditionally manufactured rods. Fracture-split rods are also produced more economically and by using less energy during production.

FIGURE 5-60 Fracture-split rods have a coarse grainy appearance on the bearing cap parting halves.

Connecting Rods and Other Unique Features

One common feature is that rod ends can have an offset or split of the rod cap at 45°–60° rather than at a conventional 90° (**FIGURE 5-62**). This feature is to accommodate larger rod journal sizes while allowing the rod to pass through the cylinder during assembly. If the rod were not split at an offset angle, the big-end size would not pass through the cylinder, because of the direction of the orientation of the rod bolts. The mass of the large end can be somewhat minimized by using this feature. Diesel engines will often have a drilled passageway through the rod to the wrist pin bushing to supply oil to the wrist pin.

In V6 and V8 engines, connecting rods share a common crankpin journal, which means rods must be correctly orientated. A chamfered edge in the rod cap is installed toward the crankshaft fillet radius—the edge between the journal and crank cheek. The flat sides of each rod match one another, allowing a surface for each rod to glide over one another.

Connecting Rod Servicing

Connecting rods should be checked for twists, bends, wear, nicks, and cracks, as well as how clear the oil passageways drilled through the rod are (**FIGURE 5-64**).

Twists and bends are usually due to hydrostatic locking. A hydraulic lock occurs when liquids entering the combustion chamber eliminate the little clearance volume remaining in a diesel engine. When the piston compresses the liquid, it will stop moving, and the crankshaft continues to move, causing the con-rod to bend and twist (**FIGURE 5-65**). When this happens, the engine is said to be locked up.

Sources of liquids entering the combustion chamber include head gasket leakage; block porosity (i.e., cavitation erosion); leaking injectors or injector tubes; or water entering the

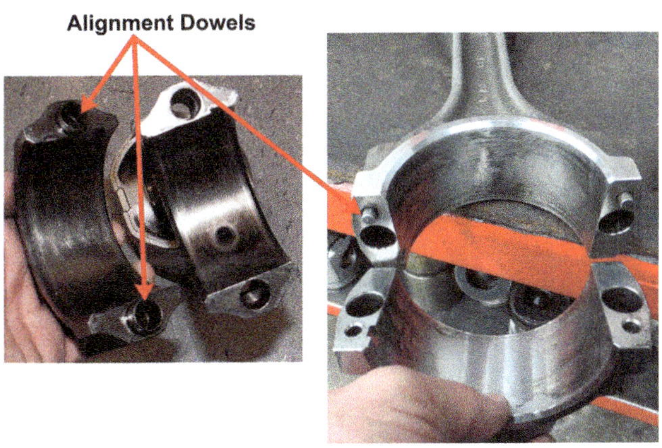

FIGURE 5-62 An offset split rod permits a larger crankpin diameter and a stronger rod cap retention. Loads are carried by the bolts and fillets machined into the rod.

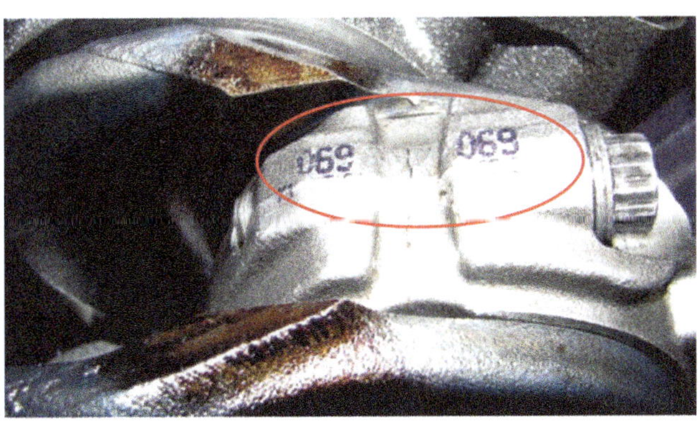

FIGURE 5-63 When assembling any rod and cap, markings must match one another and align together.

▶ TECHNICIAN TIP

If the con-rod cap of a fracture-split rod is fitted incorrectly and tightened, then the connecting rod must be replaced. This is necessary because the unique profile of the mating surfaces will have been damaged when the cap tightened causing distortion of the two mating surfaces. The cap will no longer locate correctly, even if it is returned to its correct position. To minimize the possibility of this error, connecting rod bolts or cap screws are located off-center, to ensure that the cap is fitted to the rod in the correct orientation (**FIGURE 5-61**).

FIGURE 5-61 Locating dowels rod cap and rod must match one another and line up together to properly assemble a cap and rod.

▶ TECHNICIAN TIP

Connecting rods caps are line bored (drilled with a large drill bit) when manufactured and cannot be reversed during installation; otherwise, the bore will not be concentric (**FIGURE 5-63**). Technicians must carefully observe match marks used to put the cap and rod together properly during assembly. Other markings are used to indicate a front-to-back orientation in an engine since an offset angle may be present in the con-rod. On V-block engines, two connecting rods share a single crankshaft journal. One side of the con-rod is chamfered to prevent damage to the crankshaft fillet radius. The other side of the same rod will be smoothly machined to allow the connecting rods to rotate against one another.

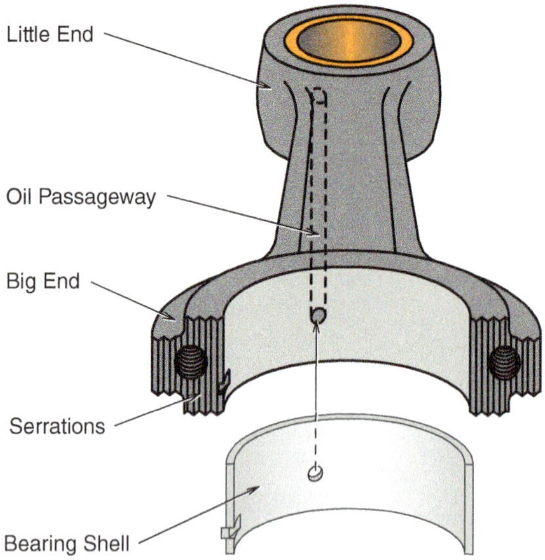

FIGURE 5-64 A piston pin lubrication hole drilled through some rods needs to be inspected and cleaned.

FIGURE 5-65 This connecting rod was bent due to a hydrostatic lock in the cylinder.

> **▶ TECHNICIAN TIP**
>
> Whenever any major engine work is performed, always turn the engine over manually first before using the starter. This can identify incorrectly installed, misadjusted components or the presence of hydraulic locks, which could potentially damage the engine if the starter is used.

exhaust system through an open exhaust stack or water, snow, oil, or fuel entering through the intake system. For this reason, when servicing an engine with a procedure that could cause a hydrostatic lock, such as during an injector replacement, it is important to bar the engine over manually and determine whether the engine is locked up.

Inspecting Connecting Rods

To evaluate whether a connecting rod is bent or not, a technician may first locate a misfiring cylinder due to low compression. Sometimes a rod may be bent far enough that the skirt hits the crankshaft counterweight, creating a tremendous bottom-end noise. If the cylinder head is removed, a simple check of piston protrusion above or below the block deck may be enough. All pistons should have relatively the same amount of protrusion if the rods are the same length (**FIGURE 5-66**).

Finally, a rod can be removed and measured in a machine that will evaluate the length and degree of parallelism between the big and small end.

Crack Detection

Connecting rods may be checked for cracks by using dye penetrant or magnetic particle detection. Stresses encountered during operation can lead to stress risers opening and a catastrophic engine failure. Machine shops can also check connecting rod bores for roundness and wear. Worn bushings in the small end can be replaced.

FIGURE 5-66 Measuring piston protrusion above the block deck is one method to check for a bent connecting rod. A bent rod will have less protrusion than other cylinders.

▶ Wrap-Up

Ready for Review

- Cylinder components consist of the pistons, piston rings, piston pins, and connecting rods. Because these components convert combustion energy into mechanical power, they are the most highly stressed parts of the diesel engine.
- Cylinder components need to have a useful life that meets or exceeds EURO 6 and EPA emission standards for durability. The EPA requires 150,000 miles (241,400 km) or 15 years for Tier 3 engines to remain emission compliant.
- Basic piston designs used in diesel engines include slipper skirt, trunk, articulating (medium-duty), and Monosteel (light-, medium-, and heavy-duty) pistons.
- Slipper skirt pistons are used in lighter, more compact diesels, whereas Monosteel pistons are used in the a few light-duty but mostly medium- and heavy-duty engines. Articulating pistons have a two-piece design and use an aluminum skirt attached to a steel crown with a piston pin. Trunk pistons are longer than they are wide.
- Low crevice volume (LCV) pistons reduce emissions by minimizing the space between the piston crown and cylinder wall above the compression ring. Articulating and Monosteel pistons are classified as LCV pistons because the top ring is closer to the crown and because the clearance between the cylinder wall and crown is reduced.
- Although aluminum pistons are traditionally lighter than steel pistons, which minimizes inertia forces at the end of each stroke, aluminum pistons have several disadvantages.
- Aluminum pistons need high crevice volume and larger operating clearances, and this has given rise to pistons with steel alloy crowns, which can withstand higher cylinder temperatures and pressures, to displace aluminum pistons in many applications.
- Pistons are cooled by several mechanisms, including by transferring heat through the piston rings during valve overlap, when the intake air mass crosses over the piston, and by oil cooler nozzles, which spray engine oil aimed at the underside of the piston.
- Piston rings help form a gas-tight seal in a cylinder, distribute oil over the cylinder wall, transfer heat from the piston, and provide a low-friction, replaceable wear surface.
- Three types of piston rings are used in an engine: a compression ring, an intermediate ring, and an oil control ring. Each ring has its own unique shape and materials based on its primary function.
- Compression rings help form a gas-tight seal in the cylinder. Intermediate rings form a compression seal and control oil distribution. Oil control rings distribute a thin coat of oil onto the cylinder walls.
- The introduction of low-tension, low-friction piston rings is the most recent change in ring technology to reduce engine wear and friction.
- Steel piston rings are more commonly used. Longer break-in periods are required due to harder ring face materials.
- Piston pins are offset from the piston's centerline to reduce noise when the piston crosses over TDC.
- Piston rings, pistons, and connecting rods all have directional markings that indicate which way they should be installed in an engine.
- Rolled threads, which are stronger than machine-cut threads, are used on connecting rod cap bolts.
- Prolonged engine idle produces engine slobber, which is a combination of lubricating oil, unburned fuel, and soot. Slobber can cause piston rings to stick and leave damaging deposits on pistons.
- Prolonged engine idle time can also increase cylinder glazing and soot loading in engine oil. Glazing takes place when the fine lines on the cylinder wall, called crosshatch, are filled with oil that is then baked and smoothed over by the piston rings. When glazed, cylinder walls cannot properly retain oil, which leads to increased cylinder blowby and power loss.
- Crankcase pressure testing is the most effective way to unobtrusively evaluate how much an engine's cylinders are worn.
- Most wear takes place in the cylinders in the ring turnaround area, which is the place in the cylinder above the highest point of travel in the oil control rings and below the compression ring. Cylinder wear is tapered, with the greatest wear at the top of the cylinders.
- After rebuilding an engine, the engine must be run in under a heavy load to seat piston rings and minimize the likelihood of cylinder glazing.

Key Terms

anodizing A process used to harden aluminum by electrochemically reacting oxygen with aluminum.

articulating piston A two-piece piston design that uses a separate aluminum skirt connected to an alloy steel crown through a piston pin.

break-in period The operation of an engine after it is initially assembled or rebuilt when piston ring, cylinder wall, bushing, and bearing surfaces have high initial wear as the moving surfaces conform to each other.

cam ground piston An elliptically shaped piston that expands to a round, symmetrical shape after it is warmed up.

compatibility A property that allows two metals to slide against one another with minimal friction or wear.

compression testing A measure of the maximum pressure of engine cylinders when cranking.

crankcase pressure test A measurement of the amount of cylinder blowby. This indicates the sealing ability of the piston rings and cylinder walls.

crevice volume The area between the piston crown and cylinder wall above the top compression ring.

crosshatch A cylinder wall finish of fine intersecting lines that are used to retain oil.

cylinder glazing A condition that occurs when lubricating oil is first baked and then smoothed into the crosshatch of the cylinder wall. Cylinder glazing prevents the cylinder wall from retaining oil and reduces ring's ability to seal.

cylinder leakage The percentage of gas leakage past the rings. High leakage rates are usually due to worn cylinders and rings.

dial bore gauge A precision measuring tool used to measure taper wear in a cylinder.

durability requirements Legislated standards for engine durability that require noxious engine emissions to remain below set thresholds through the expected useful life of an engine.

dusting out A condition where dirt is drawn into an engine. This causes premature abrasive wear on the cylinder walls and rings.

forged pistons Pistons made from aluminum alloy billets that are stamped into shape by forging dies.

hydrostatic lock A condition that occurs when fluids in the engine's cylinder prevent the engine from rotating.

hypereutectic piston A piston that has high silicon content (16–20% silicon).

Magnehelic gauge A gauge that measures both pressure and vacuum by using inches or a water column or kilopascals as a measuring unit.

major side thrust A piston side thrust caused by cylinder pressure and the angle of the connecting rod during power.

micro-finish etching A piston skirt finish where many fine lines are machined into the skirt to retain oil.

minor side thrust Piston side thrust caused by compression pressure and the angle of the connecting rod.

Monotherm piston A one-piece piston design made entirely of alloyed steel, which has a compact height and a large reduction of material between the skirt and crown.

Ni-Resist insert A stainless steel–nickel alloy insert placed in aluminum pistons in the compression ring groove. The insert minimizes groove wear caused by ring movement.

nitriding The process of hardening a metal's surface by heating the metal and quenching it with cyanide salts.

piston slap A noise in the engine produced by large operating clearances between a piston and cylinder wall. Piston slap is most often heard when an engine is cold.

shot peened A technique that uses small steel balls to blast metal surfaces in order to close up any small cracks or pores, which have the potential to become larger.

slipper skirt piston A piston design that has a portion of the skirt removed on both non-thrust sides of the piston to provide clearance for the crankshaft counterweights.

Review Questions

1. A piston's compression ring seals best during a(n) __________.
   a. compression stroke
   b. power stroke
   c. intake stroke
   d. exhaust stroke
2. Which ring will cause excessive oil consumption if it is installed upside-down?
   a. The compression ring.
   b. The intermediate ring.
   c. The oil control scraper ring.
   d. Both the intermediate and compression rings.
3. If an engine has a noise when it is cold but the noise disappears when the engine is warmed up, what is a likely cause of the noise?
   a. A piston installed on the connecting rod is in the wrong direction.
   b. An incorrectly installed piston skirt on an articulating piston.
   c. Piston slap.
   d. A loose connecting rod bushing.
4. Which of the following is the practice recommended to evaluate how well a diesel engine's piston rings seal?
   a. A cylinder leak-down test.
   b. A compression test.
   c. A wet-dry compression test.
   d. A measurement of the crankcase pressure.
5. Which shape is most often used for compression rings in medium-duty diesel engines?
   a. Square.
   b. Rectangular.
   c. Bevel faced.
   d. Keystone.
6. Where is a nickel insert used in an aluminum alloy piston used by diesel engines?
   a. Pin boss.
   b. Intermediate ring groove.
   c. Skirt.
   d. Compression ring groove.
7. Which ring face material is used on compression rings to aid upper cylinder wall lubrication?
   a. Steel.
   b. Chrome.
   c. Molybdenum.
   d. Cast iron.
8. If an engine has 30% loss of crosshatch near the top of each cylinder, approximately what percentage of remaining service life does the engine have?
   a. 30%.
   b. 50%.
   c. 20%.
   d. 0%.

9. What kind of useful information is supplied by measuring the distance a piston protrudes above the cylinder block deck?
   a. The amount of liner protrusion.
   b. The length of the stroke.
   c. Whether the connecting rod is bent.
   d. The required thickness required for a head gasket.
10. Under which of the following conditions is the most engine slobber produced?
   a. Idle conditions.
   b. Light-load engine operation.
   c. Full-load engine operation.
   d. High-speed, light-load engine operation.

ASE Technician A/Technician B–Style Questions

1. Technician A says that when a diesel engine idles excessively, it will cause its cylinder walls to glaze and piston rings to stick. Technician B says idling an engine regularly for long periods of time causes piston rings to break. Who is correct?
   a. Technician A
   b. Technician B
   c. Both A and B
   d. Neither A nor B
2. Technician A says that oil control rings are designed to scrape as much oil as possible from the cylinder walls. Technician B says oil control rings leave a metered amount of oil on the cylinder walls. Who is correct?
   a. Technician A
   b. Technician B
   c. Both A and B
   d. Neither A nor B
3. Technician A says compression rings in a diesel engine are rectangular to extend engine life. Technician B says that the top compression rings have a barrel face to provide effective sealing. Who is correct?
   a. Technician A
   b. Technician B
   c. Both A and B
   d. Neither A nor B
4. Technician A says that piston ring markings identify the type of ring and its location in a piston ring groove. Technician B says ring markings are used to ensure the rings are installed facing downward. Who is correct?
   a. Technician A
   b. Technician B
   c. Both A and B
   d. Neither A nor B
5. Technician A says that after rebuilding an engine, the engine should be operated at low load and low speed to prevent any catastrophic damage. Technician B says that an engine should be operated under a heavy load as soon as possible after ring replacement. Who is correct?
   a. Technician A
   b. Technician B
   c. Both A and B
   d. Neither A nor B
6. After discovering broken rings in an with more than a five thousand engine hours, Technician A says that the rings were most likely broken due to improper installation. After discovering the same broken rings in an engine, Technician B says the rings were broken due to excessive cylinder taper. Who is correct?
   a. Technician A
   b. Technician B
   c. Both A and B
   d. Neither A nor B
7. Technician A says that most high-horsepower (high-kilowatt) diesel engines use trunk-type pistons. Technician B says that most high-horsepower (high-kilowatt) diesel engines today use steel pistons. Who is correct?
   a. Technician A
   b. Technician B
   c. Both A and B
   d. Neither A nor B
8. Technician A says a scuffed piston skirt and burned piston crown in one cylinder of a recently overhauled engine is likely due to a damaged or improperly aligned oil cooler nozzle. Technician B says the engine was overheated. Who is correct?
   a. Technician A
   b. Technician B
   c. Both A and B
   d. Neither A nor B
9. While two technicians are discussing the cause of a cracked piston crown with a vertical crack through the inner edge of the combustion bowl, Technician A suggests that the piston has a manufacturing defect. Technician B suggests that using too much starting fluid cracked the piston. Who is correct?
   a. Technician A
   b. Technician B
   c. Both A and B
   d. Neither A nor B
10. Technician A says that a compression test of an engine's cylinders is the best method to use to diagnose a worn-out engine. Technician B says that measuring crankcase pressure is the best method to identify a worn-out engine. Who is correct?
   a. Technician A
   b. Technician B
   c. Both A and B
   d. Neither A nor B

CHAPTER 6

Cylinder Heads and Valve Train Mechanisms

Learning Objectives

After reading this chapter, you will be able to:

- 6-01 Identify and classify cylinder heads according to their construction, composition, and valve arrangement.
- 6-02 Identify and explain the function of cylinder head components and valve trains.
- 6-03 Identify and describe the construction and function of valve train actuation mechanisms.
- 6-04 Explain the function of valve timing in a diesel engine.
- 6-05 Outline the procedures for servicing cylinder heads and components.
- 6-06 Identify and describe the function and construction of cylinder head bolts and gaskets.

You Are the Technician

Several school buses in the fleet operation where you work are encountering premature head gasket failures. Some of the head gasket failures are producing external coolant leaks while others are internal and pressurizing the cooling system. After some investigation, you discover several of the engines were previously repaired to correct head gasket leaks before you arrived at the repair facility. You realize that more engines will likely need a third gasket replacement. As you consider how you will precede with replacement of the cylinder head gaskets, consider the following:

1. Outline some diagnostic procedures you could follow to identify the engines that need immediate attention and head gasket replacement.
2. Outline the inspection procedures you should undertake to determine whether the cylinder heads and valve train gears are in serviceable condition or require replacement.
3. Outline the procedures that should be followed when replacing the head gaskets.

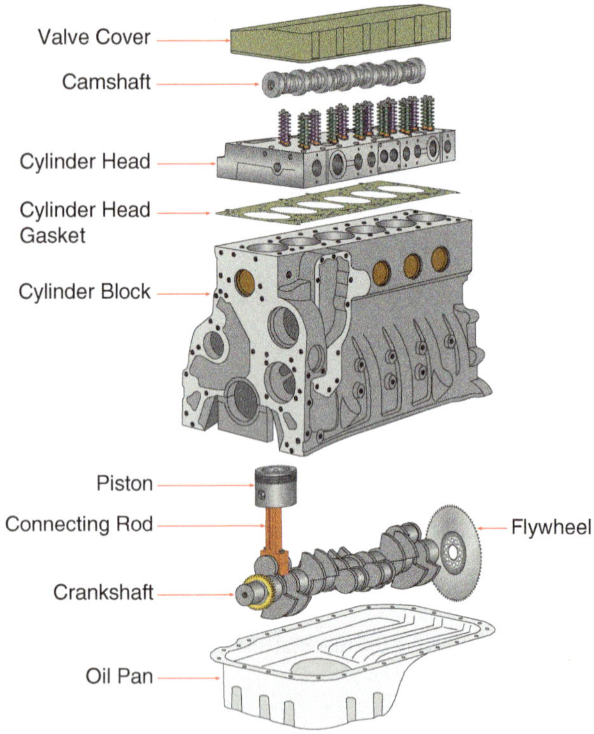

FIGURE 6-1 Major engine components. The cylinder head is a major engine casting that is bolted to the top of the cylinder block. The cylinder head gasket seals the block to the head.

▶ Introduction

The cylinder head is a metal casting that forms a major engine part and attaches to the top of the cylinder block (**FIGURE 6-1**). Cylinder heads are needed to seal the combustion chamber so that the engine can develop and contain compression and combustion pressure. Because compression and combustion pressures are much higher inside a diesel engine, cylinder heads are ruggedly designed and use high clamping forces to secure the head to the block (**FIGURE 6-2**). Many of the components of the valve train and fuel-injection mechanism are incorporated into the cylinder heads, which adds complexity to their design and service. As a major structural element of the engine, the cylinder head also supports other components critical to the operation of the engine.

Components that attach to the cylinder head include the following:

- valve train components
- fuel injectors
- camshaft(s) in overhead cam engines
- camshaft bearing cradles
- valve guides and seats for valves
- intake and exhaust manifolds.

The cylinder head also contains drilled bores and cast passages for injectors, nozzles, injector tubes, coolant, fuel, and oil passageways; provides intake air and exhaust passageways; and incorporates pre-combustion chambers (**FIGURE 6-3**). Various sensors, such as the cam position, coolant temperature, and accelerometer pilot control (APC), the diesel equivalent of a knock sensor, may be located in the head as well.

Cylinder heads are classified by the type of material used, by the way exhaust and intake ports are arranged, by the number of cylinder heads used on an engine, and by the valve train arrangement.

▶ Types of Cylinder Heads and Valve Trains

6-01 Identify and classify cylinder heads according to their construction, composition, and valve arrangement.

Because the cylinder head has to withstand high combustion temperatures and pressures without distortion, it is essential that it be constructed with high levels of strength and temperature resistance. Cooling the head to keep NO_x emissions low and the addition of temperature-sensitive components, such as injectors and electronic sensors, have also become particularly important in the latest engines. Air and exhaust gas flowing through the cylinder head is another design concern. The shape and position of the intake and exhaust ports, the position of the valves, and the location of the camshaft(s) determine not only the overall geometry of the cylinder head but also performance characteristics. In addition to these considerations, manufacturers must be able to construct cylinder heads that require the least amount of maintenance, are economical to produce, and promote outstanding performance, accompanied by low emissions and fuel consumption. A wide variety of cylinder head design, construction techniques, and service procedures are currently in use to meet the constantly improving benchmarks of engine performance.

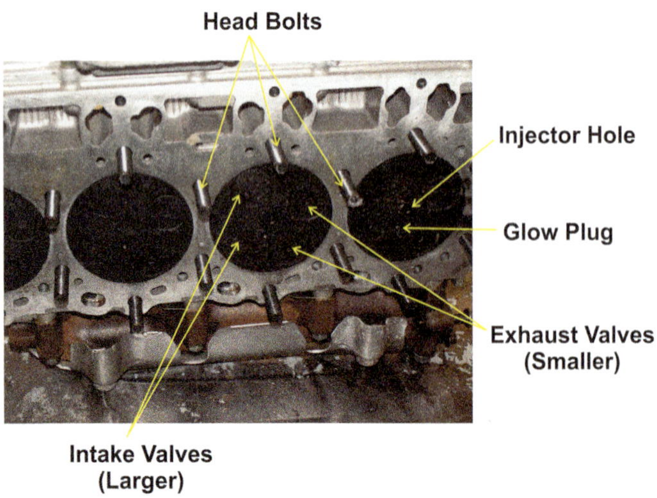

FIGURE 6-2 This cylinder head from a 6.6 L GM Duramax engine is made from aluminum. Note that there are six head bolts per cylinder to supply the clamping force needed to seal the combustion chambers.

Classification by Materials

Potentially corrosive coolants, the flow of viscous oils, high-horsepower (high-kilowatt) engine applications (in which cylinder temperatures and pressures are high), and the demand for lightweight engine designs are just a few of the many factors involved

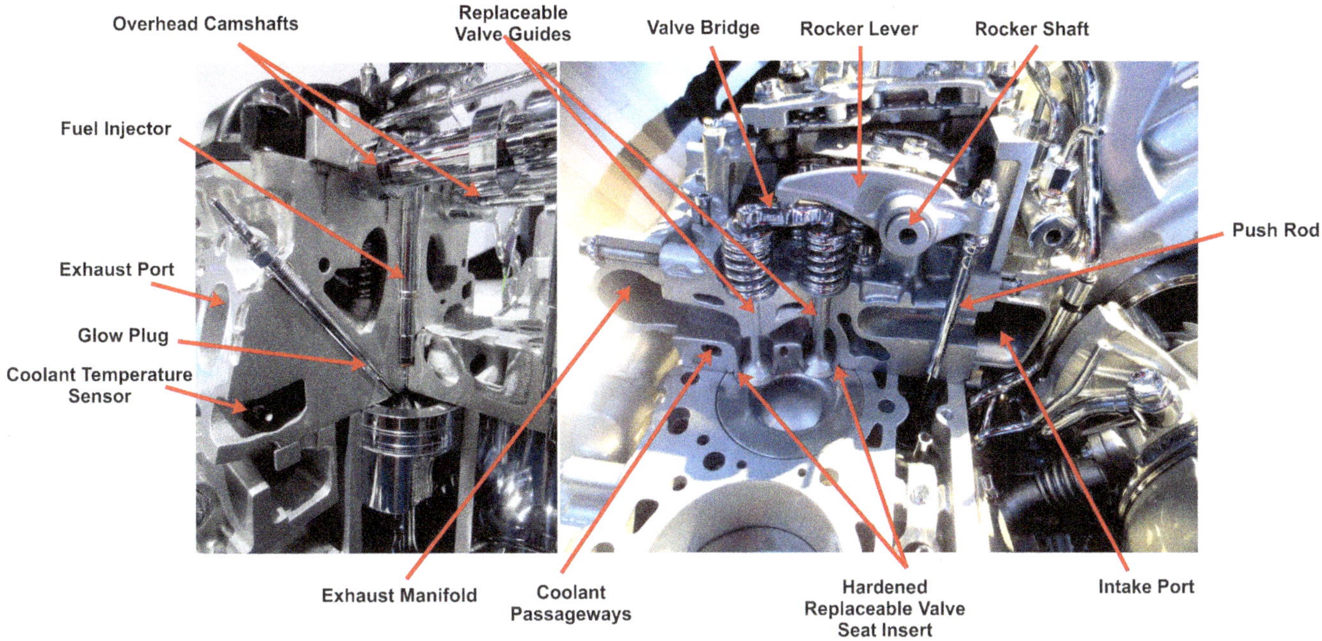

FIGURE 6-3 Many engine components are supported by the cylinder head. Passageways for coolant, fuel, air, and exhaust are incorporated into the head casting.

in selecting cylinder head materials. While traditional materials such as cast iron have provided good strength and wear characteristics for valve train mechanisms, new materials and construction techniques are improving engine performance. Lighter-weight designs now have greater durability and lower maintenance requirements while improving fuel economy.

Aluminum Versus Cast Iron

Cylinder heads are commonly made of cast iron alloy because of its superior strength, ease of manufacture, machinability, and capacity to withstand high temperatures. Aluminum is another material that is commonly used as a material of choice. At one time, aluminum was not considered, for several reasons:

- It had a tendency to warp and crack especially when overheated.
- It had difficulty tolerating abrasive dirty oil, which can quickly wear away softer aluminum friction surfaces.
- It required extra manufacturing steps to install cast iron valve guides, valve seats, and bushings for camshaft (if used).
- It caused greater electrochemical corrosion in engines with cast iron blocks, making them unable to tolerate old contaminated coolant in the same way as cast iron.
- It was more costly to cast than iron was.

Aluminum's higher expansion rate at operating temperature compared to iron's is another problem. The expansion coefficient of aluminum, which is four to seven times greater than that of cast iron, has specialized requirements when selecting or replacing head gasket materials. The higher expansion rate of aluminum requires different methods and greater precision in tightening procedures for cylinder head fasteners. Valve lash can be similarly affected by expansion. The expanding and contacting dimensions of aluminum will alter the clearance between rocker levers in contact with valve stems, particularly if the camshaft is located in the engine block. Cylinder head bolts must also stretch more than bolts used in cast iron heads and generally require non-reusable torque-yield-type bolts.

Aluminum's primary advantage, however, is its lightweight quality. Unnecessary engine weight is the enemy of good braking, handling, and quick acceleration. Because aluminum is 50% lighter than cast iron, aluminum cylinder heads can easily save more than 60–100 lb (27–45 kg) in some larger diesel engines. Aluminum also lends itself well to having large intake and exhaust ports when manufacturing cylinder head castings, to produce more powerful engines.

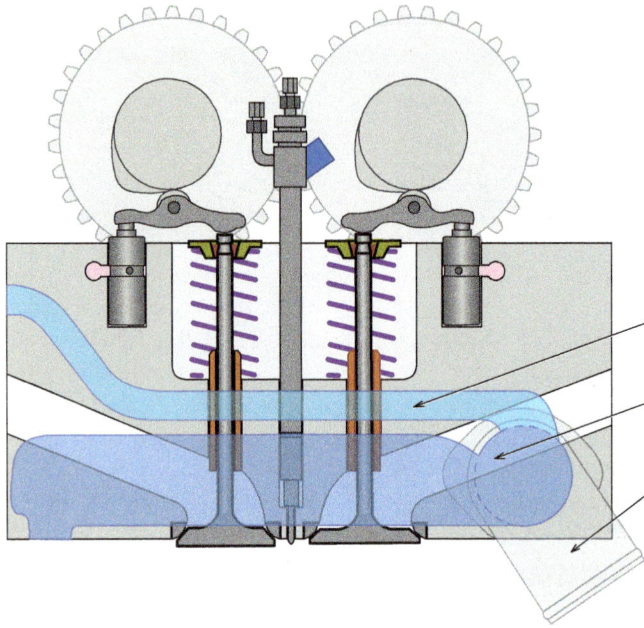

FIGURE 6-4 The most recent EA288 version of the Volkswagen (VW) 2.0 L TDI engine uses an upper and lower cylinder head coolant jacket. The higher volume and lower jacket in contact with the combustion chambers transfers heat more efficiently.

Another favorable attribute of aluminum is its higher heat conductivity, which improves heat transfer from the combustion chamber to coolant. This is especially important for emissions. Valve temperatures can be as much as 200°F (93°C) cooler in aluminum cylinder heads. This not only lowers the variations in valve lash but also enhances valve seat durability and increases reliability of the engine as a whole. Aluminum carries away excess heat more rapidly, and localized hotspots are eliminated with aluminum cylinder heads; thus, higher combustion pressures and temperatures can be sustained. Similarly, because high combustion chamber temperatures cause increases in NO_x production, this emission can be reduced with the use of aluminum heads since it carries heat away more quickly from combustion chamber surfaces (**FIGURE 6-4**).

Aluminum is now a material of choice primarily because improved hardening techniques make it very durable and heat-resistant. A hardening process performed shortly after the head is cast, called T6 treatment, relieves casting stresses and creates a more uniform metallurgical structure. T6 treatment involves heating the casting to 1,000°F (538°C) for about six hours and then quenching the casting in water for a few seconds. The cooled casting is then seasoned in a 320°F (160°C) oven for approximately five hours and then allowed to cool to ambient temperature. The result is an aluminum casting with a much greater scale of hardness. Another name for this process is precipitation or particle hardening. Particles of specialized alloys, such as copper, are mixed into the molten aluminum during the casting process. The rapid quenching enables the copper particles to interfere with the arrangement of aluminum atoms in a crystalline lattice, which in turn makes aluminum less brittle but harder at the same time.

FIGURE 6-5 Dodge's 3.0 L EcoDiesel uses CGI material for its cylinder heads.

Composite Graphite Iron

A material forming many new diesel cylinder block designs, **composite graphite iron (CGI)** is also used for cylinder heads. CGI, also called sintered graphite, is a material produced from powdered iron alloys squeezed into molds at high pressures and then heated to bond the metal particles together. When made from CGI, diesel cylinder heads have approximately the same weight as aluminum but also possess the higher strength, rigidity, and sound-dampening characteristics of CGI. The 3.0 L EcoDiesel used in Dodge Ram and Jeep products is a popular application of CGI for cylinder head material. CGI cylinder heads are reported to increase strength and rigidity by 75%, with a 200% increase in thermal fatigue resistance (**FIGURE 6-5**).

Classification by Intake and Exhaust Port Arrangement

Because intake and exhaust port arrangements influence gas flow and turbulence or swirl into the cylinders, how these ports are configured affects engine performance and emissions. The **cross flow cylinder head** is the most popular (**FIGURE 6-6**). In this configuration, the intake and exhaust manifolds are located on opposite sides of an in-line engine to improve engine-breathing characteristics. Air enters one side of the engine and exhaust exits the other side. Also, because the exhaust ports and manifolds are opposite one another, intake air heats up only a little. This design provides cooler air for efficient combustion, improved fuel economy, and reduced emissions. The drawback to this configuration is that exhaust and intake manifolds on opposite sides of the cylinder head create a wider temperature gradient across the head, with the exhaust side having much higher temperatures than the intake side. This situation creates greater potential for head gasket leaks and head cracking.

The **parallel flow head**, sometimes referred to as the uniflow design, places the intake and exhaust manifolds on the same side of the engine (**FIGURE 6-7**). In this design, joining together two adjacent short but large ports provides a more compact engine design with adequate airflow to the cylinders. Cylinder heads incur less thermal stress because the cooler intake ports are placed alongside the exhaust ports.

Reverse-flow cylinder heads used on V-block engines place the exhaust ports on the inboard side of the engine. This shortens the connection between the exhaust ports and turbocharger, resulting in a quicker turbocharger response. Without outboard exhaust manifolds, reverse-flow cylinder heads occupy a smaller footprint beneath the hood of a vehicle, where space is limited (**FIGURE 6-8**). This configuration is used by Ford 6.7 L Powerstroke engines. It was originally used by GM's ill-fated 4.5 L Duramax V8, designed with Isuzu to use as an option for its sport utility vehicle (SUV) lineup.

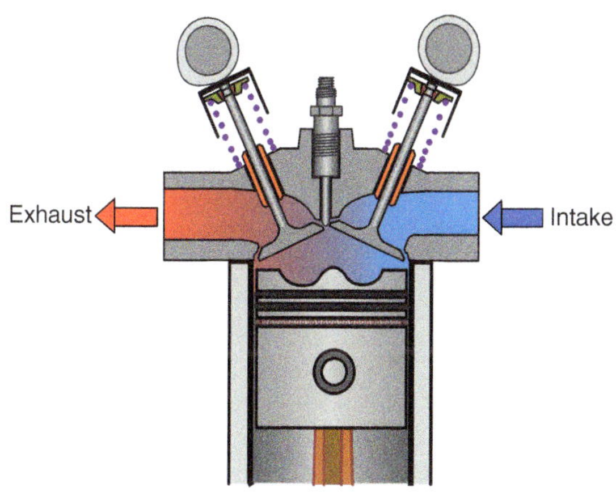

FIGURE 6-6 A cross flow cylinder head arranges the intake and exhaust ports on opposite sides of the cylinder head.

Variable Swirl Intake Ports

To promote better combustion at low speed and low engine loads, more turbulence is needed to better mix intake air and fuel. When the air and fuel are better mixed, less particulate emissions are likely to form. Variable intake charge and motion systems, as they are called, have a flap deactivation system to change the turbulence or swirl characteristics of intake air (**FIGURE 6-9**). This feature provides a variable amount of swirl to optimally match the intake air turbulence created by intake air entering the combustion chamber with engine load and fuel-injection rates. At low speeds, more turbulence is needed than at high speeds, where maximum intake flow and

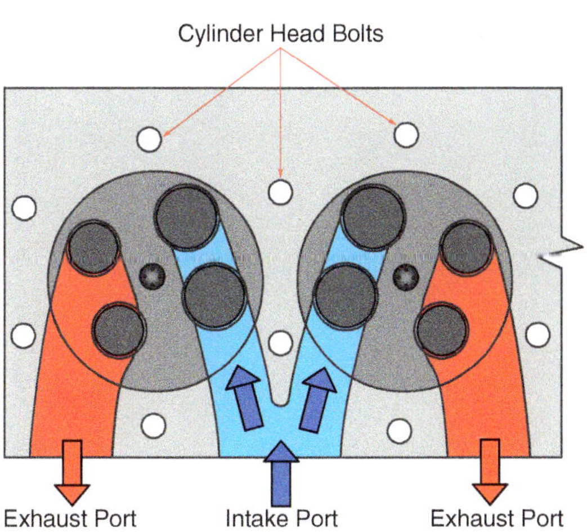

FIGURE 6-7 A parallel port cylinder head arranges the intake and exhaust ports on the same side of the engine.

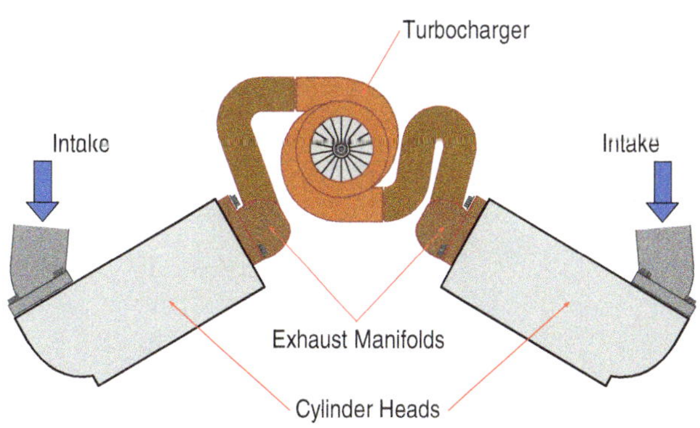

FIGURE 6-8 The 6.7 L Ford Powerstroke uses reverse-flow cylinder heads.

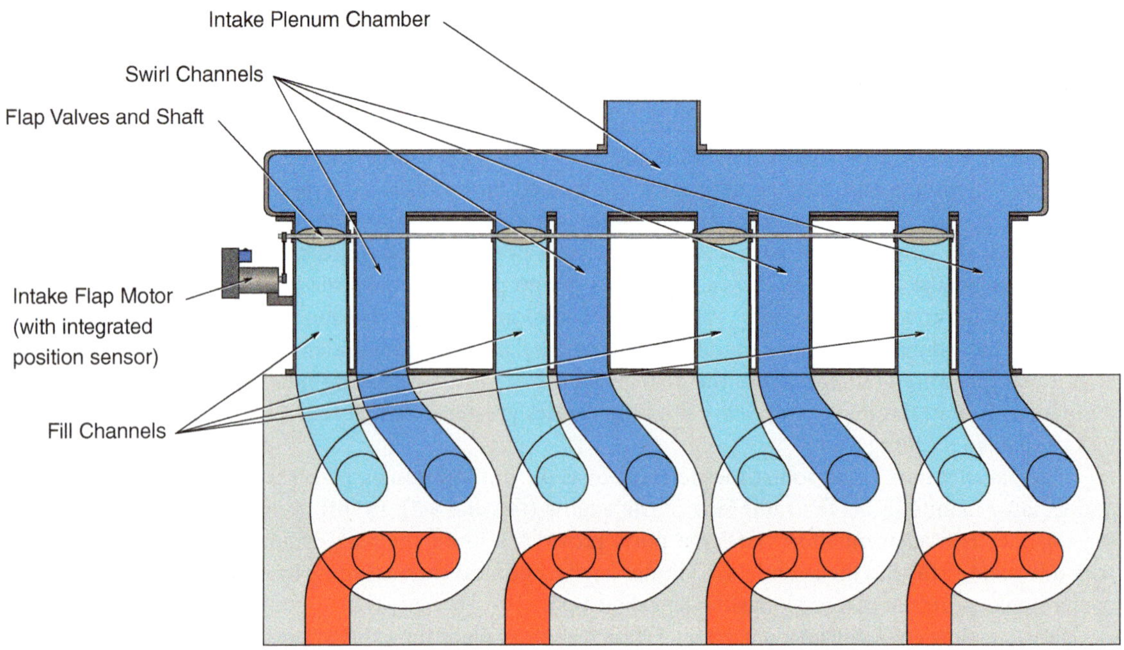

FIGURE 6-9 A throttle plate in one of the intake runners will vary between open and closed to change the degree of air turbulence entering the cylinder.

air velocity into the cylinders creates the best turbulence and combustion quality. The shape of the intake port is also shaped to impart a twisting or spinning motion to the air as it enters the cylinder, to increase turbulence (**FIGURE 6-10**). Valves can be placed in what is sometimes called a "star" configuration, rather than simply placing the inlet and exhaust valves in line with one another (**FIGURE 6-11**). The tangential valve arrangement, whose valves have a diagonal or star-shaped placement, and the uneven length of the intake ports (or intake runners) operate to spin the air entering the cylinders. When the engine is operating at low speeds, the flap or throttle plate in the intake runner will be almost entirely closed, but it progressively opens to admit more airflow into the cylinder as the engine speeds up.

Variable intake swirl systems using an electronically controlled throttle flap are more common in light-duty automotive diesel engines. A more recent trend is to use a cross flow

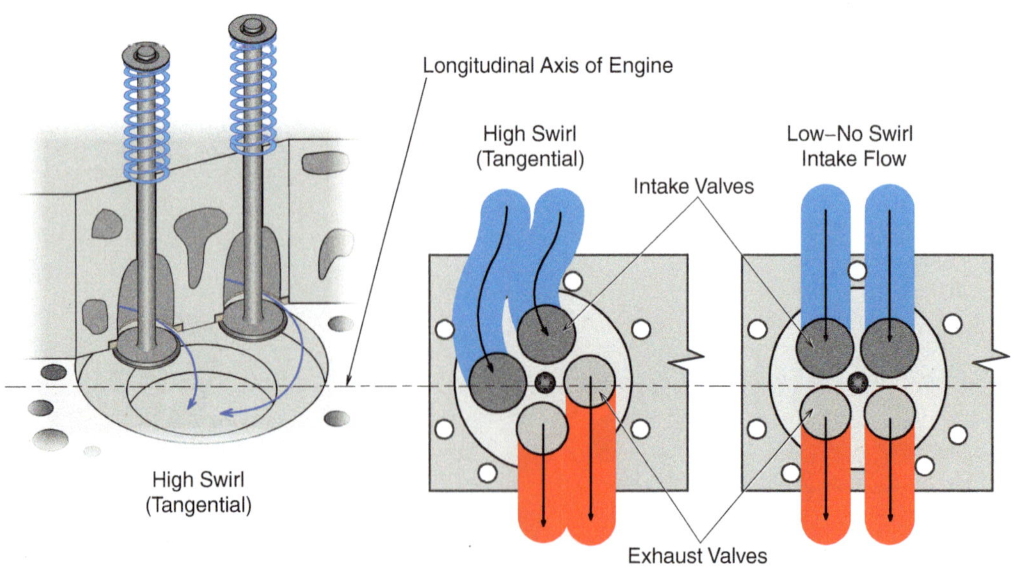

FIGURE 6-10 A tangential or star-shaped arrangement of the intake valves turns the valve placement 45°. The placement helps impart more turbulence to intake air entering the cylinders because it enters at an angle.

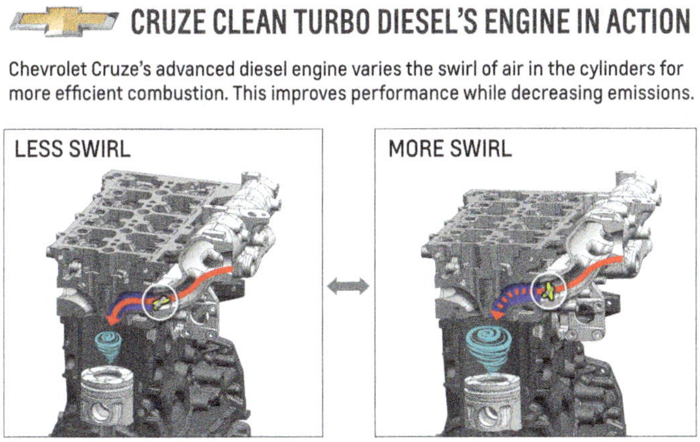

FIGURE 6-11 The shape of the intake port flap deactivation system on Chevy Cruze diesel engines increases combustion turbulence for better combustion quality at low speed and under low load.

valve port arrangement in the cylinder head and use two camshafts: one on each side of the head. This system increases power density from smaller engines as engines that have become progressively downsized in displacement for efficiency reasons.

Classification by Number

The number of cylinder heads used by a diesel engine is influenced by a variety of factors. The strength and rigidity of the engine is enhanced by a single-piece casting, but this type of cylinder head does not have the flexibility to adapt to multiple engine platforms that use modular engine designs. Gas flow, serviceability, and the location of the camshaft are other considerations that need to be factored in when selecting the type of cylinder head casting.

Single-Piece Cylinder Heads

Single-piece cylinder head castings cover the engine block completely by using one head. Simplistic, one-piece castings are relatively easy to manufacture, and their solid rigidity adds structural strength to a block.

Single-piece castings do, however, have the greatest amount of creep when thermally cycled, which means they expand and contract significantly across the block deck by sliding over the head gasket. This can translate into more head gasket failures at the rear of the engine, where coolant temperatures are hottest. Cylinder head gaskets often use special features in these areas to resist leakage in areas where there are high amounts of "head creep."

Multiple-Piece Cylinder Heads

Multiple-piece cylinder heads use a separate cylinder head for a group of two or three cylinders in a multiple cylinder engine. Multiple-piece cylinder heads are ideal for modular engine designs, where part castings shared between different engines can be used to construct various engine configurations (**FIGURE 6-12**). For example, cylinder head for a single cylinder can be used to build single-, two-, four-, six-, or eight-cylinder engines. Multi-piece small bore diesel engines are common in marine applications. Servicing is simplified if only one or a few of an engine's cylinders are being serviced. However, multiple cylinder heads tend to move around more, wearing the top of liners or block decks.

FIGURE 6-12 This Cummins B series engine 3.9 L/4BT is an example of a modular engine used on off-road applications and is popular in commercial delivery step vans. The engine was available in 3.3 L, 5.9 L, and 6.7 L models.

Classification by Valve Train Arrangement

Cylinder heads are further classified by the arrangement of the valve train mechanism. For example, the camshaft arrangements in diesel engines are commonly one of two types: overhead camshaft or I-head. **Overhead camshaft engines** have the camshaft

FIGURE 6-13 This Mazda Skyactive D diesel engine uses an overhead camshaft. The camshaft is located in the cylinder head.

FIGURE 6-14 Older I-head or pushrod engines place the camshaft in the engine block. Pushrods are used to actuate the valves by using rocker levers.

located in the cylinder head. Elimination of push rods enables the highest injection pressures by using camshaft actuated unit injectors, because push rods, which can bend at high pressure, are eliminated. Valve timing is also more accurate, and the valve train mechanisms have less inertia, which is especially valuable at high-speed automotive operation (**FIGURE 6-13**). **I-head engines** have only the valves, rocker levers, and bridges located in the cylinder heads, and the camshaft is located in the engine block (**FIGURE 6-14**). The camshaft actuates the valves through a pushrod and rocker lever.

Overhead Cam Advantages

Overhead cam engines are now a popular design, with a number of advantages:

1. *Simplicity*. Pushrods and several other valve train components are eliminated and fewer parts are required. This also means there are fewer wearing surfaces, due to eliminating several parts, resulting in fewer adjustments required to maintain valve clearances.
2. *Easier inspection and service*. All the components are easier to inspect and remove. For example, camshaft inspection can be performed visually, and a camshaft can be replaced without disassembling the engine front cover, removing the cylinder head, and performing other steps.
3. *More accurate valve train action*. Shortening the length of the valve train pathway that is transmitting force minimizes its thermal expansion and reduces its inertia. This permits higher engine speeds and closer valve train clearances, resulting in more accurate valve timing. A stiffer and more rugged valve train can be used without the penalty of added mass and length associated with a pushrod.
4. *Optimized intake and exhaust passageways for easier breathing*. Because there is no obstruction from push tubes and castings to interfere with intake and exhaust port design, the size, location, and shape of intake and exhaust ports are optimized for improved engine breathing characteristics.
5. *Higher injection pressures*. A shorter, stiffer valve train means that unit injectors can be actuated to higher injection pressures (30,000 psi [206,843 kPa]) with more accurate injection timing. When pushrods are used to transfer movement from the camshaft to injectors, they tend to bend and add considerable inertia to injector actuation. For this reason, overhead cam diesel engines that do not use pushrods will have minimal injection lag, which is an important consideration for emissions.

▶ Components in Cylinder Heads and Valve Trains

Valve trains refer to the mechanisms that control the gas flow in and out of cylinders through the opening and closing of the intake and exhaust valves. With only one exception, the valve trains include a crankshaft-driven camshaft operating at half the engine speed. This next section examines the principal components of the valve train mechanisms.

6-02 Identify and explain the function of cylinder head components and valve trains.

Valves

Poppet valves are the most common type of valve used in diesel engines. Poppet valves are characterized by a large head and a long stem. Intake valve heads are usually larger than exhaust valves, because pressure forcing the air charge into the cylinder is much lower than forces pushing exhaust gases out. Exhaust gases can be pushed out much more quickly through the smaller valves than intake air driven into cylinders under only turbocharger boost pressure. On diesel engines, the valves are often designed in two pieces that are spun welded together (**FIGURE 6-15**). The valve stem is typically chrome-plated steel to reduce its requirement for lubrication and minimize wear and friction. The head is made of specialized alloys to withstand high temperatures and mechanical forces that slam the head into the valve seat. Stellite or Inconel are alloys commonly used to make the valve head. These are materials made from a combination of chromium, nickel, tungsten, and cobalt to give the metal great strength and wear resistance at high temperatures.

Two and four valves per cylinder are common, but a number of engines may use three and five valves. An extra intake valve, usually the third valve (such as the 7.2 L used in the Caterpillar 3126b), is used to improve intake flow to lower emissions. Four smaller valves, rather than two large intake and exhaust valves, are used in central (direct-injection) DI combustion chambers to accommodate an injector nozzle. A centrally located injector provides better uniform fuel distribution for lower emissions (**FIGURE 6-16**). In two-valve engines, the injector is offset to one side of the cylinder and leaned over to make room for the exhaust and intake valves.

FIGURE 6-15 A broken weld of a two-piece valve is evidenced by a valve head embedded in the top of a piston. Overheating a valve can cause the weld to break.

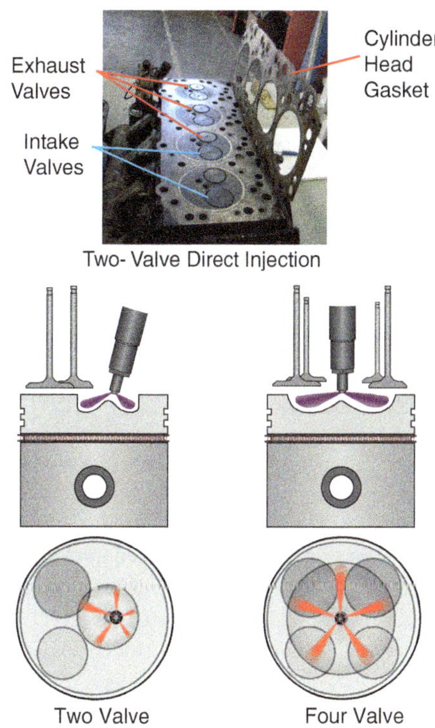

FIGURE 6-16 Two-valve engines use a larger intake valve and a smaller exhaust valve. Fuel distribution is not symmetrical and creates more noxious emissions compared to a four-valve engine.

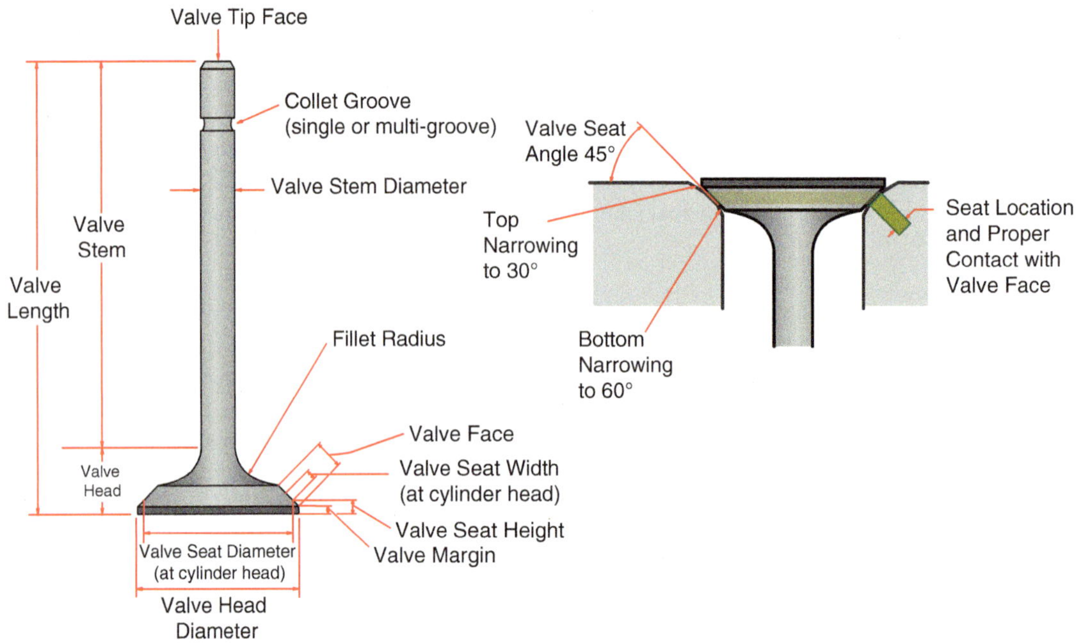

FIGURE 6-17 Terminology associated with poppet valves. Note that the valve seats are typically cut with three angles, to improve gas flow across the seat.

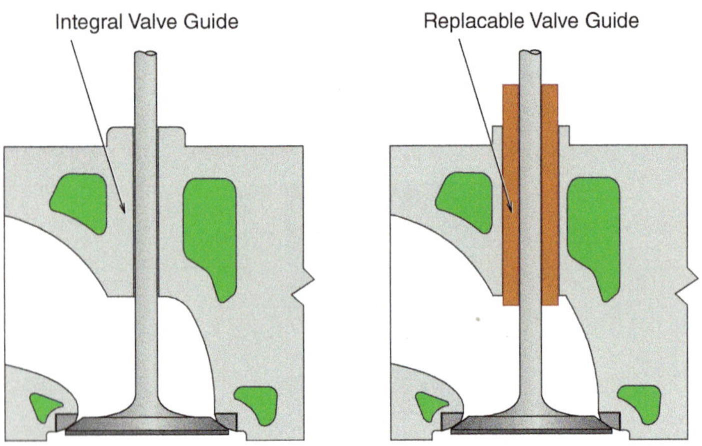

FIGURE 6-18 Valve guides can be either cast into the cylinder head or replaceable and made from an iron alloy insert pressed into the cylinder head.

Valve Angles

Valve faces are machined at an angle to wedge themselves into the seat to improve sealing. The degree of angle determines the valve port's flow characteristics (**FIGURE 6-17**). Valve face angles are measured from the horizontal angle. A 45° angle is common and is better at self-centering the valve at closure while wedging the valve more tightly to the valve seat. Some manufacturers use alternate angles. Flatter angles of 30° or 40° can be used on diesel exhaust valve faces for better heat transfer. Some exhaust and intake valve face angles are between 20° and 40° for better flow over the edge of the seat. This flatter angle also creates less wear, resulting in fewer overhead adjustments.

Valve Guides

Valves are located in guides that are either integral, meaning that they are cast into the cylinder head or removable, which means they can be replaced (**FIGURE 6-18**). Premium engines have replaceable valve guides that can be easily serviced during overhaul. The condition of a valve guide is critical to oil consumption control. A limited amount of lubrication oil must pass between the guide and valve. However, worn guides can allow oil to be drawn into the combustion chamber or exhaust stream because of the Venturi effect. This happens as a low-pressure area is created at the base of the valve guide by high-velocity gases that draw oil into the exhaust or intake stream, causing high oil consumption that may be accompanied by blue smoke (**FIGURE 6-19**). Older diesel engines, particularly turbocharged engines, did not use seals on the intake guides, because the intake manifold was pressurized and had little or no vacuum. However, positive umbrella seals are now used on all valve guides and stems to minimize the entrance of oil into the guide (**FIGURE 6-20**). The high temperatures of the guide can cause oil to coke and accelerate abrasive wear (**FIGURE 6-21**). Oil entering the exhaust stream can contribute to engine emissions. The newest guides are made from sintered iron, which is porous and can retain oil in the guide better while using smaller stems to guide clearances.

▶ TECHNICIAN TIP

Worn guides and missing valve stem seals lead to high oil consumption and blue smoke during deceleration. Oil in the guides can carbonize, accelerating stem wear. Also, any oil consumption in an engine leads to particulate emissions forming. Long exhaust systems, such as those found on school buses, create a longer and more intense period of low pressure at the valve guide in the exhaust port, especially during deceleration. More oil can be dragged down the stem during this time, which explains the higher oil consumption often encountered in school buses.

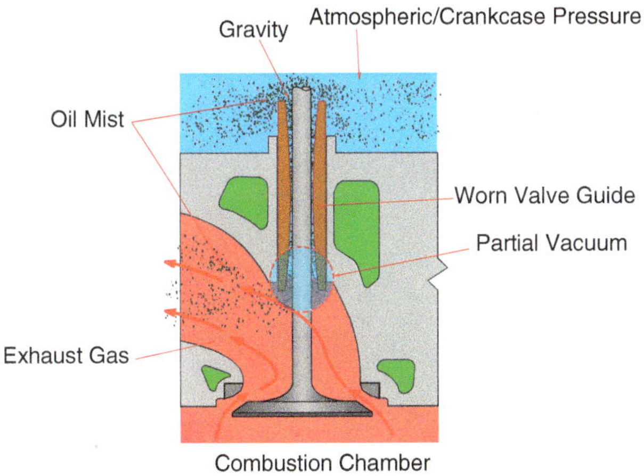

FIGURE 6-19 Valve stem seals are used to minimize the entry of oil into the valve guide. Worn guides can allow oil to enter the exhaust or intake stream.

FIGURE 6-20 A positive valve stem seal prevents oil from entering the valve guide.

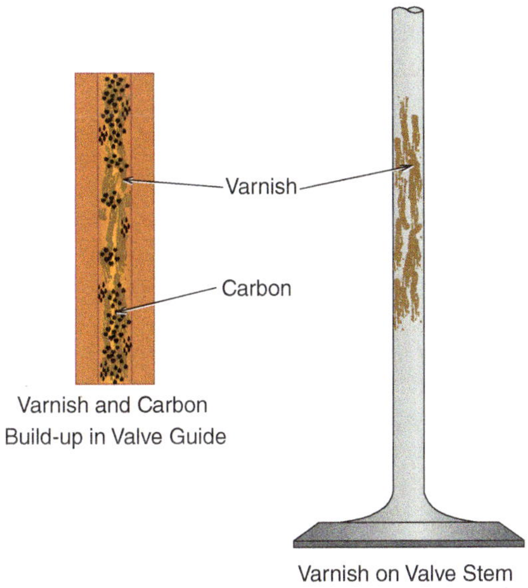

FIGURE 6-21 Coking oil in a valve guide can cause premature valve stem and guide wear.

▶ TECHNICIAN TIP

Combustion quality in diesels at idle is often poor due to low combustion pressures and temperatures. Engine slobber, a residue consisting of unburnt fuel, soot, and engine oil, can coat the back face of valves during prolonged idle. When left to cool, slobber will solidify on the back face of an open valve, preventing the valve from closing during engine start-up (**FIGURE 6-22**).

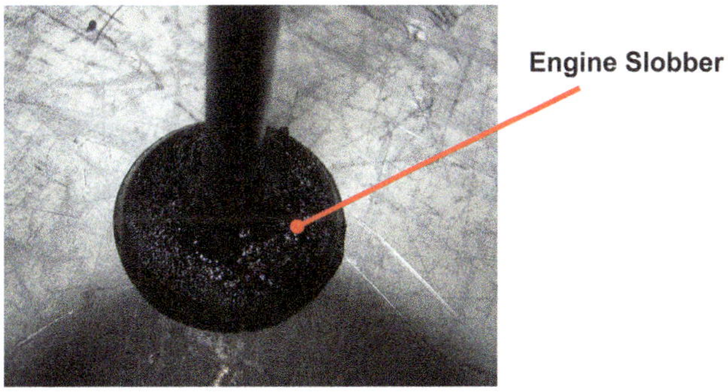

FIGURE 6-22 Slobber has accumulated on the back side of this exhaust valve.

Valve Seat Inserts

Many premium diesel engines use replaceable valve seat inserts. Typically, these are manufactured from Stellite-like material. Replaceable inserts eliminate the need for complex machining processes to install a new valve seat into a damaged or rebuilt head. Their wear-resistant hardness and heat resistance add considerable durability and prolong an engine's operating life. Valve seats have a minimum of three angles to ensure good gas flow and proper heat transfer from the valve face. A narrow 45° seat promotes clean valve seats and good sealing. The three angle cuts on the seat provide improved gas flow and narrow contact between the valve face and seat to promote best heat transfer (**FIGURE 6-23**).

Valve face-to-seat contact width must be adjusted to within typically 0.0625–0.125" (1.5875–3.175 mm) for optimal heat transfer. Too wide a seat contact diminishes pressure per square unit of area and will

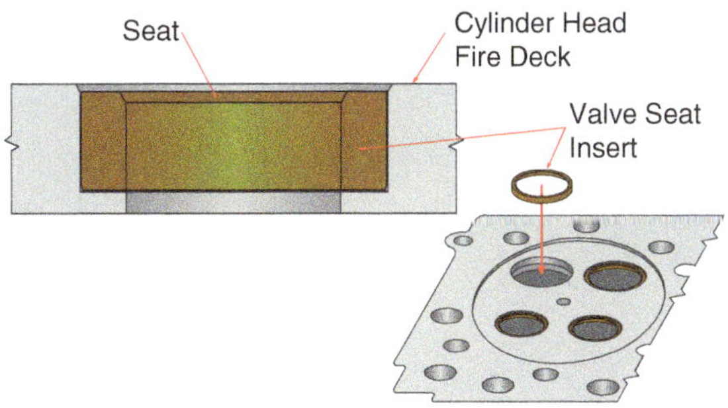

FIGURE 6-23 Valve seat inserts are made from extremely durable materials to prolong engine life. Valve seats and guides are replaceable when cylinder head components get remanufactured.

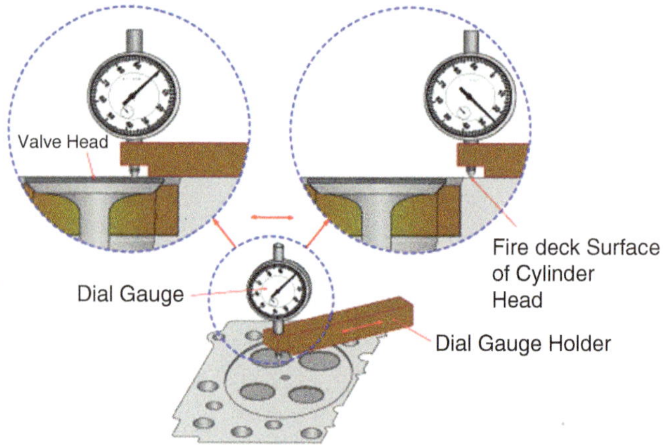

FIGURE 6-24 Measuring valve protrusion or intrusion is critical after replacing valves and seats. Too much protrusion will allow contact between the piston and valve; too little will prevent proper gas flow.

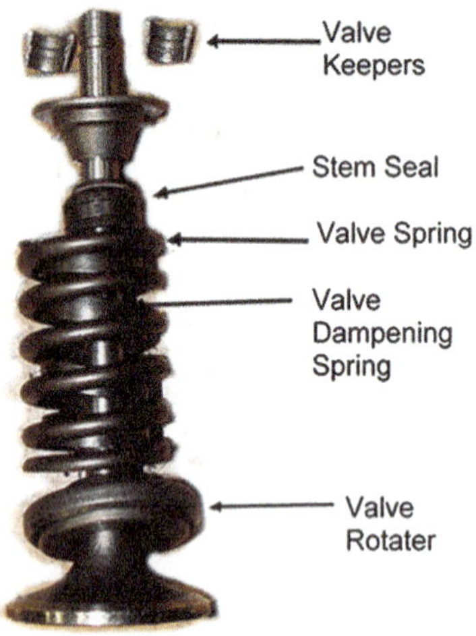

FIGURE 6-25 Parts of a valve mechanism. This valve from a high-speed diesel engine uses dual springs to prevent valve float.

result in less heat transfer from the face to the seat. This may seem counterintuitive, but high-pressure contact transfers heat better than low-pressure contact.

When servicing valve seats and faces, the technician must ensure that the correct amount of valve protrusion or intrusion exists (**FIGURE 6-24**). A valve sunk too low into the head will prevent proper gas flow and require a new valve seat. A valve sitting too high in the seat could potentially contact the piston, causing catastrophic engine damage.

Valve Springs

Valve springs hold the valve closed on its seat and must maintain tension on the valve train when the valve is open (**FIGURE 6-25**). If valve train inertia is great enough, weak valve springs will allow the valve to stay open after the cam rotates onto its inner base circle. This condition where valves stay open and don't return to their seat is known as **valve float**. However, if spring tension is excessive, it will slam the valve against the seat, prematurely wearing the valve, stretching the stem, and wearing cam lobes and other parts due to high valve opening pressures. Variable rate valve springs with coils wound with different spacing help reduce valve spring tension at the beginning and end of valve lift. Dual valve springs with valve springs that have different tensions can further minimize the incidence of valve float in light-duty diesel engines, which rotate at higher speeds than heavy-duty diesel engines.

Valve springs should be inspected for free length, cracks, nicks, pitting, straightness, and spring tension. A valve spring testing apparatus, which operates like a weigh scale, is used to check spring tension. The spring is compressed to a specified height and the spring scale reports the valve spring's tension. For example, if a spring is compressed to 1" (2.5 cm), a corresponding manufacturer's specifications would indicate that an acceptable valve spring pressure of 150 lb (68 kg) or greater is required. If spring tension falls below this, the valve spring must be discarded (**FIGURE 6-26**).

FIGURE 6-26 A valve spring tester measuring a spring's free length and compressed tension.

Valve Cooling

Because exhaust valve temperatures can exceed 1,200°F (649°C), which is the temperature at which valve damage can occur while under heavy prolonged load, excess heat must be removed to prevent burning, warping, and leakage. Good valve face-to-seat contact helps this and is provided by a three-angle valve grind and is important for heat transfer. Also, rotating the valve slightly each time it opens can eliminate localized hot spots that can lead to valve burning. Valve rotators installed on either the bottom or top of the valve produce this rotation (**FIGURE 6-27**). Rotators will rotate the valve slightly each time the valve lifts and closes. A ratchet-like mechanism inside the rotator moves the valve incrementally with each lift and close cycle.

Valve-Actuating Mechanisms

6-03 Identify and describe the construction and function of valve train actuation mechanisms.

Transferring camshaft movement to the valves is performed by the valve train. In I-head engines, the camshaft will initiate valve action by transferring force through the cam follower, then to the pushrod, rocker lever, and finally the valve. Multivalve setups will use a valve bridge to open two valves with one rocker lever (**FIGURE 6-28**). One interesting exception is the Ford 6.7 L Powerstroke engines, which use a dedicated hydraulic valve lash adjuster for each intake and exhaust valve for the four valves per cylinder configuration. Four separate hydraulic adjusters are used for the configuration of four valves per cylinder. A patented actuation mechanism uses two rollers to lift four hydraulic adjusters. One roller in contact with the camshafts intake lobe lifts two intake valve hydraulic lash adjusters. Because of this configuration, the length of two intake valve pushrods can be adjusted independently of one another. Another roller following the camshafts exhaust lobe to lifts two hydraulic lash adjusters for a pair exhaust valves. This enables independent adjustment of the length of the exhaust valve pushrods. In total, two cam followers lift four pushrods, one for each intake and exhaust valve, that move four rocker levers (**FIGURE 6-29**). Most valve-actuating mechanisms will use a roller-type cam follower or lifters and roller rocker levers to reduce friction. Not only does this reduction in friction increase the durability of the components, but it also can add 1.5–3% higher power output. Reducing valve actuation friction is especially important in diesel engines because camshaft high loading occurs

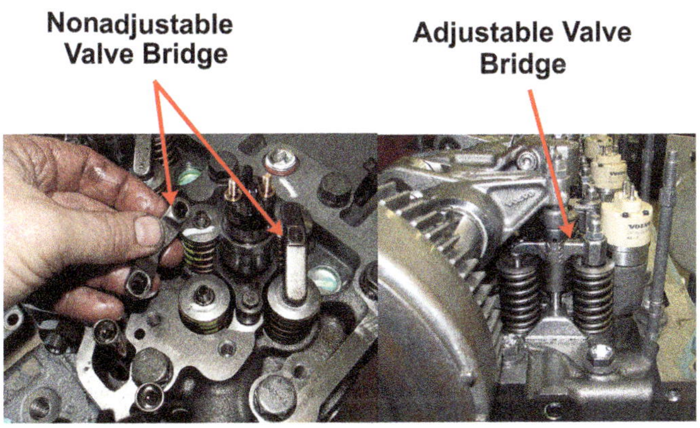

FIGURE 6-28 Valve bridges actuate two valves with a single rocker lever. Older bridges were adjustable. Newer engines use nonadjustable bridges.

FIGURE 6-27 The top of a dual valve spring. The inner spring rubs against the outer spring to minimize valve float, and the rotator turns the valve each time it opens and closes to help cool the valve.

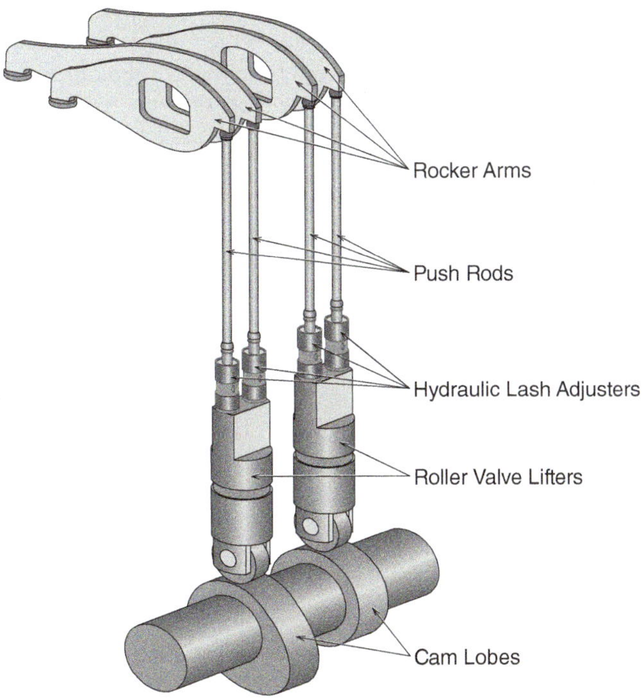

FIGURE 6-29 Each 6.7 L Powerstroke roller lifter contains two hydraulic lash adjusters, and each valve is individually actuated through its own pushrod and rocker arm.

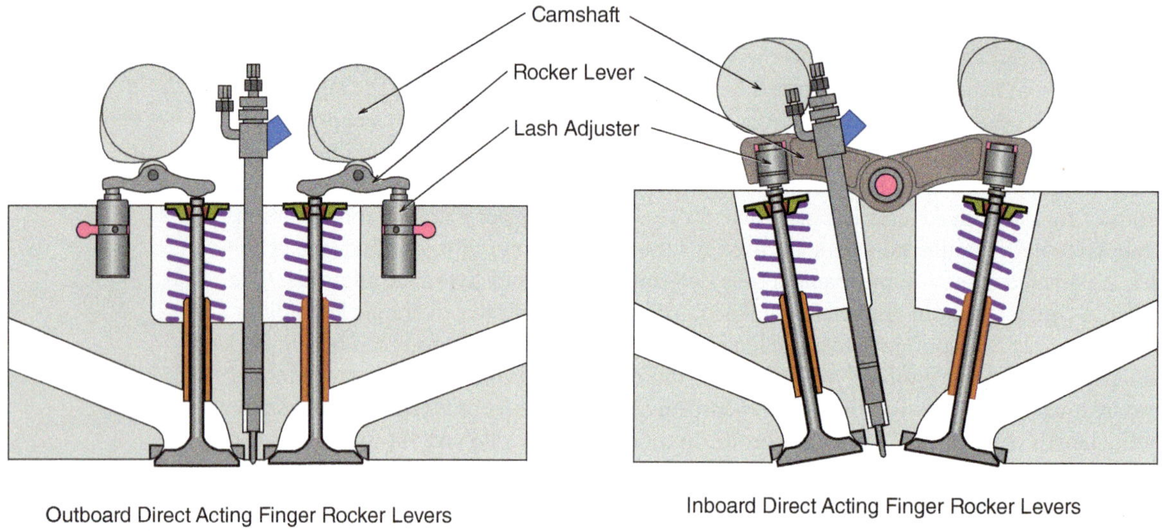

FIGURE 6-30 Placement of inboard and outboard lash adjusters for direct-acting, finger-type cam followers.

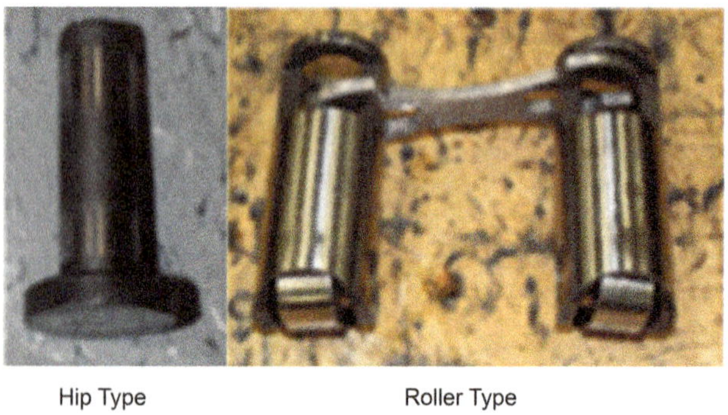

FIGURE 6-31 A comparison of two hip and roller mechanical lifters. The fork used on the roller lifter prevents the lifter from spinning in its bore.

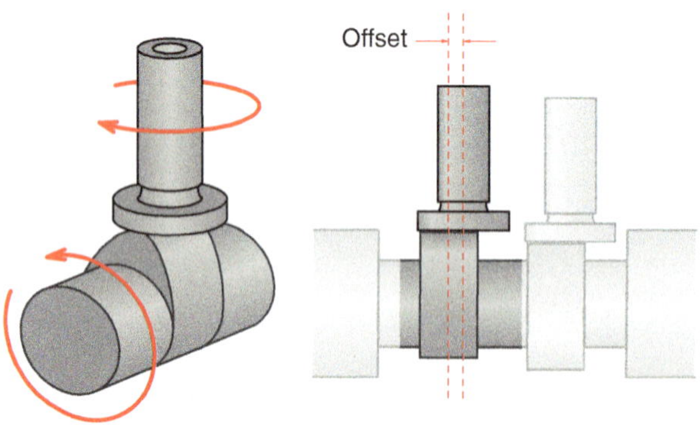

FIGURE 6-32 Hip valve lifters follow the cam profile and have a centerline slightly offset from the cam lobe centerline. When the cam turns, the lifter rotates on its axis, reducing friction between the lobe and lifter.

when the exhaust valve opens against cylinder pressure. Cylinder pressures remaining at the end of the power stroke are still high and exert tremendous forces along the exhaust valve train when the valve opens to begin exhaust stroke. The combination of pressure and friction in the exhaust valve train can cause premature camshaft exhaust lobe and other valve train wear and failures. Pushrods can be eliminated through the use of direct-acting, finger-type cam followers that incorporate a hydraulic lash-adjusting mechanism (**FIGURE 6-30**).

Roller-type followers, large diameter, and wider cam lobes are commonly used in combination with short exhaust rocker arm ratios to increase the durability of exhaust valve train components. **Rocker arm ratios** compare the distance between the rocker shaft or fulcrum point of the rocker arm and the distance to each end of the rocker arm. A short distance between the push rod contact point of the rocker arm and the rocker shaft in comparison to a longer distance between the rocker shaft and valve contact point supplies a high rocker arm ratio. Roller-type lifters and followers reduce friction between the camshaft and the follower. When hip or flat lifters are used, the cam may have a special angle ground laterally into the cam lobe to rotate the lifter as the cam wipes along the bottom of the lifter (**FIGURE 6-31**). This rotation of the lifter, rather than using a simple wiping action, helps minimize premature lifter and camshaft wear. Offsetting the lifter slightly from the lobe centerline will cause lifter rotation (**FIGURE 6-32**).

Hydraulic Lifters

Hydraulic valve lifters, or valve lash adjusters, are an automatically adjusting mechanisms for maintaining zero valve clearance between components in the valve train (**FIGURE 6-33**). The clearances between rocker levers and valve stems, push rods, and camshafts will vary with the engine's temperature. Thermal expansion and contraction will cause changes to the length of all the parts forming the valve train mechanism. This means the valve train can be noisy when cold, due to clatter from excess valve to rocker lever clearances. When hot, a mechanism that is adjusted too tightly can leave a valve partially open, interfere with proper valve timing or have too much tension on the mechanism to allow oil between moving parts in the valve train. A conventional solid lift will have a larger clearance when cold, which becomes smaller when hot. But the major problem

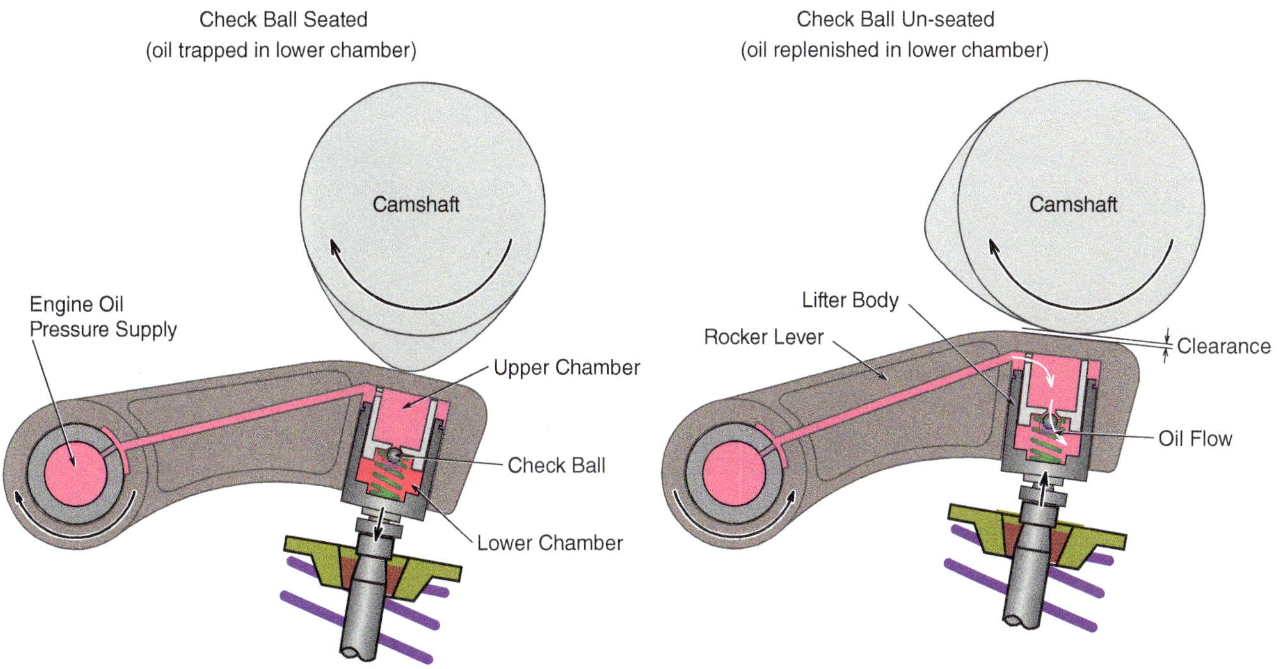

FIGURE 6-33 Typical components of a hydraulic lash adjuster built into a rocker lever. Note that the lifter check ball regulates oil in and out of the lifter body to eliminate rocker lever to camshaft clearance.

with a solid-type lifter is that it requires periodic adjustment and can be noisy when cold. More commonly in diesels, valve face to seat recession eliminates clearance and the valves become too tight which interferes with proper valve timing. To overcome this problem, the hydraulic lifter is designed to continuously correct the clearances in the valve train by expanding to eliminate clearances and contracting to adjust for length increases. Located between the camshaft and each intake and exhaust valve, the basic structure of the valve lifter body consists of a hollow steel tube with an internal piston. A slight clearance exists between the piston and lifter body to allow oil below the piston and lifter body to slowly leak out between the parts when the lifter must contract or shorten.

Oil under constant pressure is supplied to the lifter through an oil gallery and passes into the lifter through a small hole in the lifter body. Oil will separate the lifter body from the internal piston by passing through a check valve located between the piston and lifter body. When the engine valve is closed, oil fills the cavity between the lifter body and piston, causing the lifter body to expand and take up any clearances in the valve train. When the camshaft rotates, noncompressible oil is trapped between the two parts by the one-way check valve, and the lifter behaves like a solid body since oil is compressed between the two parts. Oil pressure between the piston and lifter body is low so that only enough pressure is available to separate the inner piston and body, allowing only excess clearances in the valve to be removed. When the valve train lengthens or expands due to heat, the additional pressure applied to the lifter when it returns to the cam's inner base circle squeezes oil from between the piston and lifter body to shorten the lifter, thus enabling the clearances in the valve train to remain constant whether the engine is warm or cool.

Pushrods

Pushrods transmit the cam actuation force from the followers to the rocker levers (**FIGURE 6-34**). Longer pushrods are usually hollow to reduce weight and inertia at high speeds. When servicing, the pushrods should be checked for wear on the ends and bending caused by maladjustment or high-speed valve float. Some pushrods will use a hardened end at the rocker lever to prevent wear caused by friction at this point. If the weld at the ball socket end fractures, oil can enter and accumulate inside the pushrod. The easiest way to check for oil entering the pushrod is to listen to the sound made when it is dropped. Shorter, solid pushrods capable of transmitting larger forces with less flexing

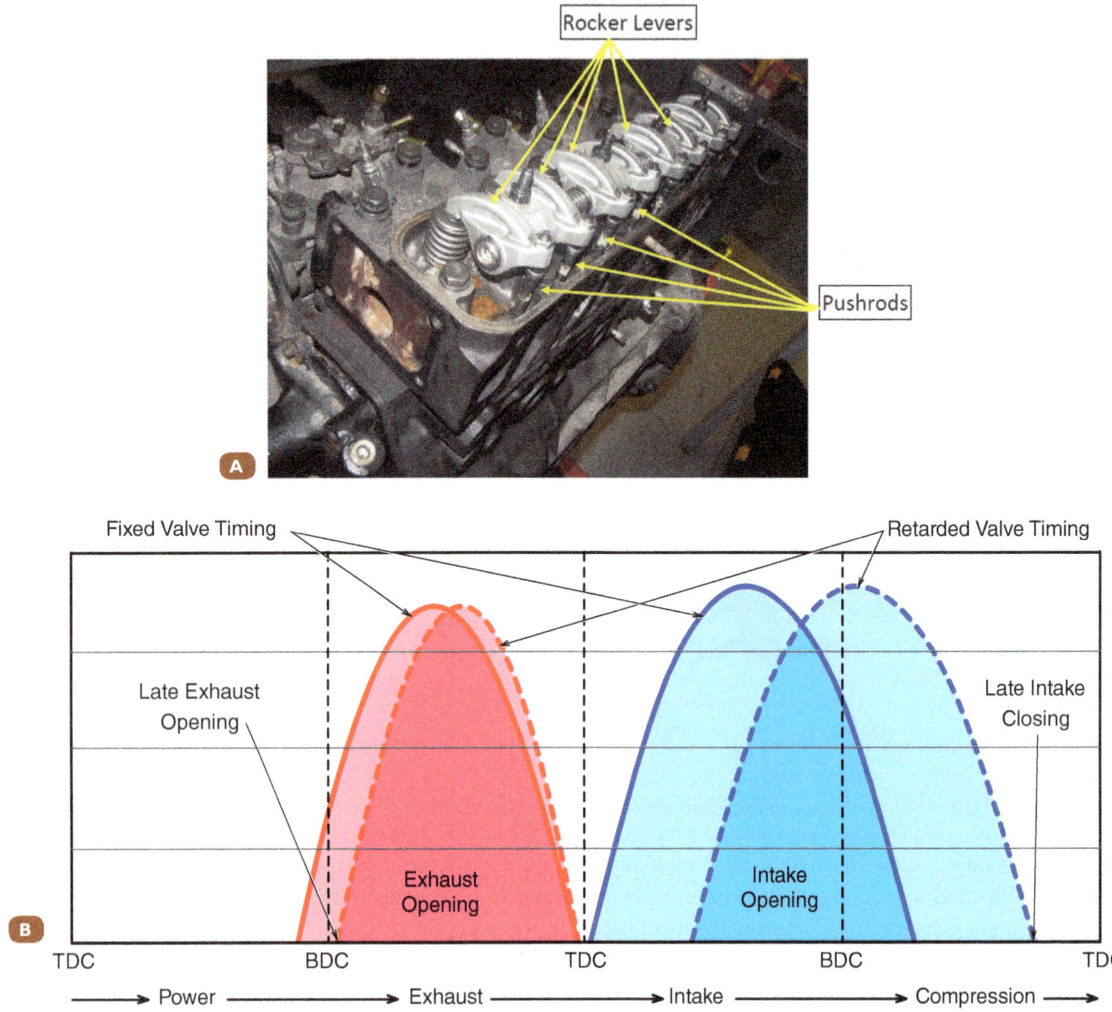

FIGURE 6-34 **A.** A two-valve industrial engine using pushrods and lightweight aluminum rocker levers. **B.** Retarding valve timing produces late intake and exhaust valve closing for more efficient engine operation. The compression function is transferred to the turbocharger during late intake vavle closing.

are also commonly used with high-mounted camshafts. High-mounted camshafts place the camshaft close to the block deck so that shorter lighter pushrods can be used. Higher injection pressures and reduced injection lag when unit injectors are used are made possible by using solid, short pushrods.

FIGURE 6-35 A different rocker arm ratio is often used on the intake and exhaust valve. Rocker arms are heat-treated to resist wear. Discoloration of rocker levers from induction hardening hardening is not evidence of damage.

Rocker Arms

Rocker arms convert the linear change of the cam profile into the reciprocating motion used to open and close the valves. The arms may be constructed of aluminum, stamped steel, cast, or forged steel. A roller is incorporated into the levers of overhead cam engines to reduce friction. A much wider roller is used on the exhaust valve rocker lever, because cylinder pressures are higher when the exhaust valve opens which casues loading of the roller to be much higher than on the intake valve. The roller for an injector lobe is the widest; injection pressures can reach as high as 33,000 psi (227,527 kPa) in unit-injected engines. Hardened wear pads may be welded to the arm, or the arm may be induction hardened for increased durability. If rocker arms are of different lengths on the engine, the exhaust rocker lever is often longer than the intake lever (**FIGURE 6-35**). This is due to the cam profile of the exhaust cam lobe. A less aggressive profile with a smaller lift is used on the camshaft's exhaust lobe to minimize wear. The rocker arm ratio means that a greater mechanical advantage is used on the lever to create the necessary valve lift. This translates into less friction on the exhaust cam lobe.

Crossheads, Yokes, and Valve Bridges

To actuate more than one valve with a single rocker lever, most four-valve engines use a crosshead to exert the camshaft force against the valve stem. Crossheads are also called yokes and valve bridges. There are two varieties of crossheads: guideless crossheads, which are nonadjustable, and adjustable crossheads, which use a guide pin to locate the device in the cylinder head.

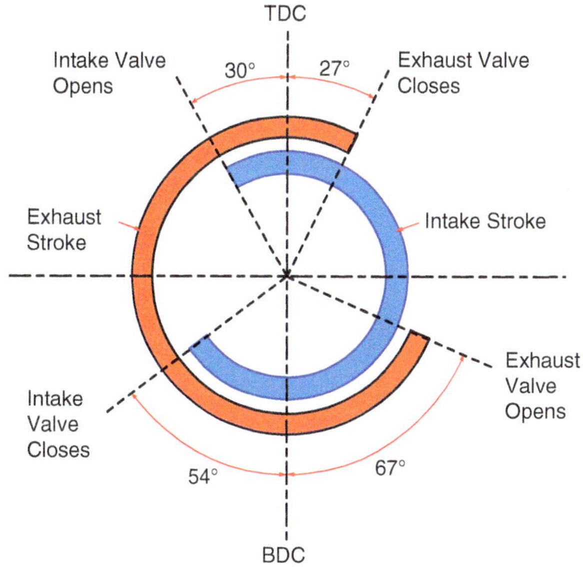

FIGURE 6-36 Valve overlap is the number of degrees of crankshaft rotation when both the intake and exhaust valves are open. Large valve overlap is characteristic of turbocharged diesel engines, due to their long stroke design and ability to admit a larger volume of air intake. Large overlap produces superior breathing characteristics at higher rpm and under heavy load conditions.

▶ Valve Timing in a Diesel Engine

6-04 Explain the function of valve timing in a diesel engine.

Unique valve timing characteristics are another important feature that gives diesel engines their superior power output. The addition of a turbocharger and the absence of fuel in the intake air during intake stroke provide the diesel with the capability to increase power output by using a prolonged valve overlap period. Valve overlap is a valve timing term that refers to the number of degrees of crank rotation when the intake and exhaust valves are open together (**FIGURE 6-36**). Overlap occurs at the end of the exhaust stroke and the beginning of the intake stroke when the piston is at top dead center. A longer overlap period allows more air to be packed into the cylinder so that more fuel can be burned, thus increasing engine torque without increasing displacement. Because fuel does not enter the cylinders with air during intake stroke, no fuel escapes out the exhaust valve during overlap, enabling diesel engines to use large amounts of valve overlap and not waste fuel or generate hydrocarbon emissions. If fuel were mixed into the intake air, long overlap would result in wasted fuel and high emissions since the air-fuel mixture would leave the cylinder through the open exhaust valve.

A large valve overlap helps create a power profile that delivers higher torque at high speed and under heavy load. Large overlap helps an engine to breathe better by taking advantage of high intake air velocity, boost pressure and gas inertia. Even though little piston movement takes place near top dead center (TDC), air can still rush into the cylinder and fill the cylinder with more air if the turbocharger is positively pressurizing the intake manifold. Increased overlap also produces soot loading in the intake manifold and more intake noise since exhaust pulses can travel into the intake manifold at low speed during overlap. Fewer degrees of valve overlap delivers higher torque at lower engine speeds, but the power will drop off as the engine speed and load increase. A turbocharged diesel engine benefits from a larger valve overlap because there is more time to charge and scavenge a cylinder.

Variable Valve Timing

Diesel engines have very little clearance between the piston and valves when the piston is close to TDC and the valve is open. This means that the valves can have little or no lift during overlap which requires valve pockets to be built into pistons. It also means that the use of a variable valve timing systems (VVT), common on spark-ignition engines, is limited. VVT systems, which advance or retard the rotation of camshafts relative to crankshaft position, can only minimally advance intake valve opening (IO) and minimally retard exhaust valve closing (EC) before valve-to-piston contact will occur.

Another reason for the limited use of VVA in diesel engines is the volumetric efficiency of an engine is already optimized by the valve timing (opening and closing in relation to crank position). Since fuel delivery primarily regulates engine speed rather than air flow, improvements to engine performance provided by VVT in spark-ignition engines have not been used.

One major exception is VW's 2nd generation 2.0L TDI which completely replaced the earlier TDI by 2015 model year. Like spark ignition VVT mechanisms, the VW diesel VVT uses a hydraulically operated camshaft position adjuster to move both camshafts simultaneously up

▶ TECHNICIAN TIP

Soot loading in an intake manifold is common on diesel engines even without an exhaust gas recirculation (EGR) system. During low-speed, low-load operation, some exhaust gases will back up into the intake manifold when both intake and exhaust valves are open. Soot can accumulate in the intake, which is a relatively harmless condition unless buildup is excessive. EGR and closed crankcase ventilation systems can also contribute to soot loading and sludge formation in the intake manifold which can restrict airflow in severe cases.

to 50-degrees relative to the crankshaft. Because VW's engine uses four valves with the intake and exhaust valve arranged together on each side of the cylinder head, each of the 2 camshafts actuates both an intake and exhaust valve. Only one hydraulic position adjuster is used since only one camshaft is driven by the crankshaft. Camshaft rotation is synchronized between the 2 camshafts by a drive gear. The overlap period cannot change since the camshaft phasing, which refers to the movement of the cams relative to one another is not possible since they are synchronously driven together.

Electronic control of valve timing is optimized to produce maximum compression pressures for good starting and low emissions during the warm-up period. However, to improve fuel economy and reduce emissions, cylinder compression pressures are controlled by closing the intake valve later than it normally would during the upward piston movement during compression stroke. By advancing camshaft rotation relative to the crankshaft, the intake and exhaust valves will close later than they would in an engine with fixed valve timing. Pushing intake air out the cylinder during compression stroke reduces energy consumption by shortening the time when intake air is compressing. Adequate air mass and temperatures are still available to properly burn fuel, but unnecessary compression of air mass and the energy it uses is avoided. To achieve an adequate amount of air mass and preignition temperatures with a shortened compression stroke, more of the compression function is transferred to the turbocharger. This means turbocharger boost pressure will increase as the camshaft valve timing is advanced. Exhaust energy driving the turbocharger rather than mechanical energy from the crankshaft supplies compression pressure. Additional cooling of intake air supplied by the turbocharger through the charge air cooler helps lower preignition temperatures further to avoid NO_x production.

Further efficiency improvements are achieved by opening the exhaust valve later during the downward piston movement of power stroke which lengthens the power stroke. Residual combustion pressures can act on the piston for a greater number of degrees of crankshaft rotation. See additional Figure 6-34B.

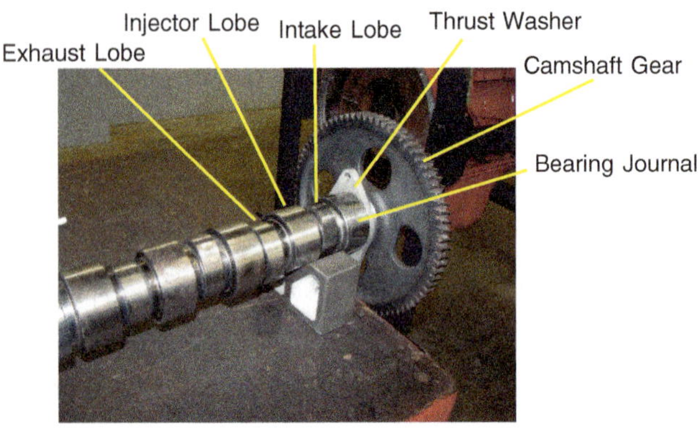

FIGURE 6-37 A typical diesel camshaft. Note the three lobes per cylinder: One is for the injector, and the other two are for the valves.

Camshafts

Camshafts are critical components to engine performance because they control the gas flow in and out of cylinders or engine-breathing characteristics. They are commonly made from nodular cast iron. Camshaft lobes may be occasionally induction hardened or even nitrided to withstand high unit loading. Cam profiles are ground to create quieting ramps that prevent high valve velocities when seating valves. Valve timing, valve lift, duration, and lobe separation are controlled by the cam profile. If a camshaft is shared between different power ratings on the same engine model, offset keyways on the cam gear may change valve timing (**FIGURE 6-37** and **FIGURE 6-38**).

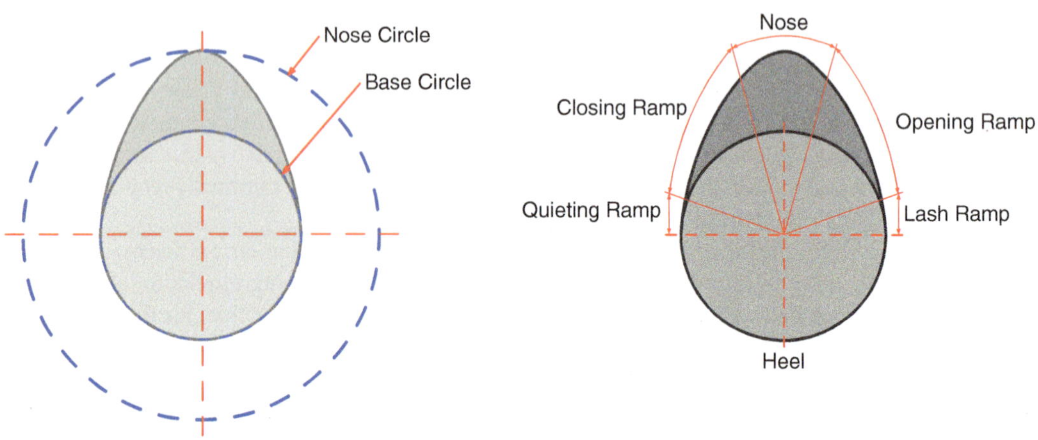

FIGURE 6-38 Terminology of a camshaft lobe.

FIGURE 6-39 This Ford 3.2 L engine uses dual overhead camshafts.

Overhead camshaft engines will often use two camshafts to optimally locate valves and ports on either side of the cylinder head (**FIGURE 6-39**). Two camshafts eliminate the need for long rocker levers and rocker shafts. Less reciprocating mass is present in the valve train, which in turn reduces the potential for valve float. In dual overhead cam engines, several valve and camshaft arrangements are used. One of the two camshafts may actuate an intake and exhaust valve in each engine's cylinders. The pair of camshaft will operate at the same speed and perform the same function, but on the opposite sides of the cylinder head. Alternatively, one of the camshaft may actuate all the intake valves only while the other operates all the exhaust lobes. The use of hollow camshafts with cam lobes, which are pressed onto the camshaft, provides weight reduction while minimizing engine noise vibration and harshness produced by heavier solid camshafts. If fuel injectors are camshaft driven, a third lobe will operate the injector pumping plunger. A High degree of fuel pressurization requires a wider camshaft lobe to distribute the injection actuation force.

Dual overhead camshafts (DOHCs) are often located in their own cam housing. A separate housing can help create a less complex modular cylinder head design while still enabling accurate synchronization of both camshafts with each other (**FIGURE 6-40**). A drive gear will connect the camshafts together and allow only one camshaft to be driven by the crank. Needle bearings or the cam bushings can be used for camshaft bearings support and are integrated into the cam housing. In L-head engines, cam bushings support the camshaft bearing journals in the engine block.

All camshafts require correct timing to the engine to ensure that the valves open at precisely the correct time in relation to piston position. Timing marks are stamped into the geartrain and are aligned during assembly to establish correct camshaft-to-engine timing (**FIGURE 6-41**). An engine block–mounted camshaft will typically use a single camshaft gear that rotates at half the engine speed to link to the engine's crank gear.

Geartrain Mechanisms

To synchronize the movement of the crankshaft with the camshaft and the valves, a geartrain or valve timing mechanism is used. The geartrain may also transfer power from the crankshaft to operate fuel-injection equipment lubrication, power steering and coolant pumps, accessory drive mechanisms, and even power takeoff pumps (**FIGURE 6-42**). The high driving forces and potentially destructive torsional vibration caused by the speeding up and slowing down of the crankshaft means that gear drive mechanisms are best used to

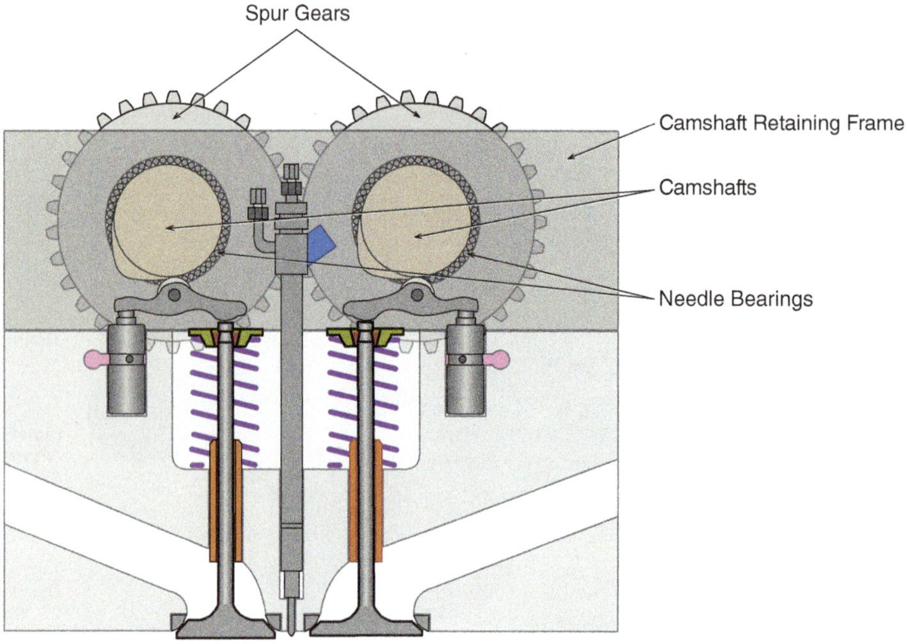

FIGURE 6-40 DOHCs for this VW EA-288 TDI are in a separate housing and are timed together using spur-type drive gears.

provide the longest possible service. However, chain drives and even belts are commonly used on automotive and light-duty diesels.

The use of a timing belt or chain eliminates the need to transfer crankshaft torque to the camshafts through multiple noisy gears. Because the distance between the crankshafts and camshafts is much farther on overhead camshaft engines, belts and chains can more simply bridge that distance. This is particularly important in engines that use DOHCs on V-bank engines, where there are multiple camshafts and where the distance between the crankshaft and camshafts can be even farther apart (**FIGURE 6-43**). When chain drive mechanisms are used, only two sprockets—a driven and a drive sprocket—plus a

> **▶ TECHNICIAN TIP**
>
> Exhaust cam lobes and exhaust valve train mechanisms have the greatest loading pressure, which produces comparatively greater wear in these components. The loading pressure is caused by the high force required to open the exhaust valves at the end of the power stroke. The wide profile of exhaust cam lobes and rollers helps distribute the mechanical forces over a larger surface area. Injector lobes require even wider than exhaust lobes, to pressurize fuel and prevent camshaft wear.

FIGURE 6-41 Timing gears are aligned to synchronize cam and crank rotation by using timing marks.

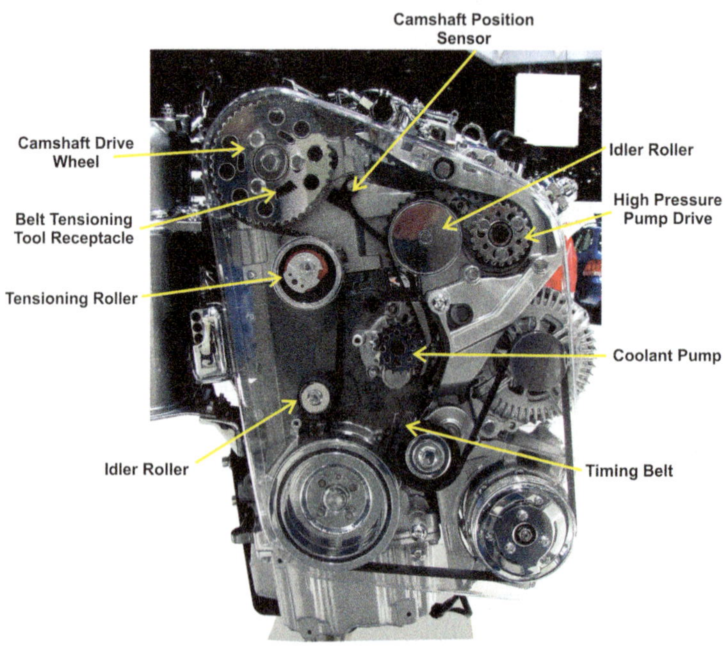

FIGURE 6-42 This VW TDI has the coolant pump and high-pressure injection pump driven by a rubber timing belt.

roller-type chain are required to synchronize cam and crank rotation. This is unlike a gear drive mechanism requiring multiple gears to transmit rotation. Chains are narrower than belts or gears, a characteristic that provides a more compact engine footprint. However, chain and gear drive systems require lubrication and must be enclosed in a case to lubricate and to prevent oil leakage. To maintain tension and prevent error in synchronization between the crank and camshaft, both belts and chain require a tensioning mechanism. Belt and chain whipping on the slack side of the drive mechanism is minimized when the non-driving side of the chain or belt is tensioned. Oil pressure will apply tensioning force to the chain through a guide bar. Some shock loading can be absorbed by the chain tensioner, and chain wear is automatically compensated for because the tensioner force is constant whenever the engine is running. Since belts can operate in spaces outside the engine and do not require lubrication, a spring-type tensioning mechanism that automatically adjusts belt tension to compensate for belt wear is used.

FIGURE 6-43 The I-4 2.8 L Duramax diesel uses a rubber timing belt to drive two overhead camshafts.

Although chains and gears may be more durable, rubber composite belts are quieter during operation and are less expensive to produce when compared to a gear or chain system (**FIGURE 6-44**). Timing belts do not require lubrication, but they do have a much shorter service interval, requiring replacement at approximately 100,000 miles (160,000 km).

Timing belts are typically covered by metal or plastic timing belt covers that require removal for inspection or replacement. Engine manufacturers recommend replacing belts at specific intervals and may also recommend replacing other parts, such as the water pump, idler wheels, and belt tensioners, when the timing belt is replaced. Timing belt failures, which are more likely to happen with a rubber belt system, can cause catastrophic damage because engine valves will be struck by the pistons if the crank and cams are not synchronized.

> **▶ TECHNICIAN TIP**
>
> Timing belts must be replaced at regular intervals to prevent belt breakage or slippage, which would result in catastrophic valve-to-piston contact. Exposure to heat, oil, and antifreeze can accelerate belt deterioration. When replacing the belt, a variety of methods are used to establish precise camshaft-to-crank synchronization. Timing marks and the use of timing pins to lock gears in place are a couple of methods used to establish correct belt timing. It is critical to follow the manufacturer's instructions when replacing belts, to prevent engine damage at start-up. Before starting an engine, always bar an engine over by hand to verify that no contact will take place due to an installation error.

FIGURE 6-44 Four separate timing chains are used at the rear of this VW-Audi 3.0L TDI engine. three alignment pins lock the 2 camshafts and crankshaft during installation.
1. Bank 1 (Right) camshaft drive (Camshaft to right intermediate sprocket)
2. Bank 2 (Left) camshaft drive (Camshaft to left intermediate sprocket)
3. Central Drive Chain (Crank sprocket to intermediate gears for camshaft drives)
4. Second Drive Chain (Oil pump to balance shaft driven by crank sprocket).

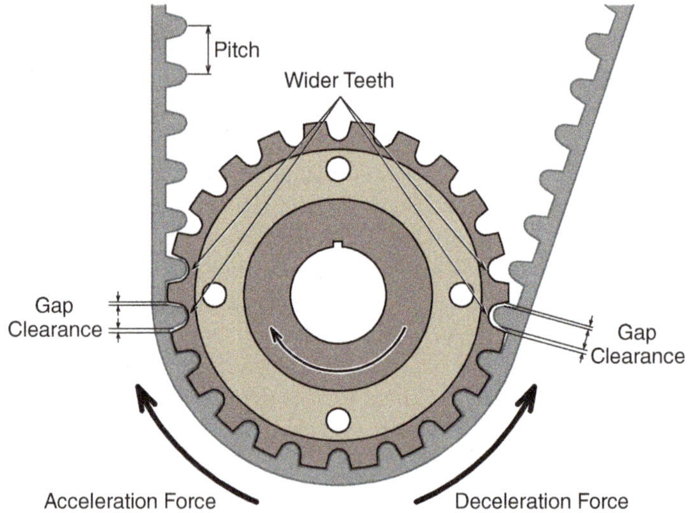

FIGURE 6-45 Two wider teeth on opposite sides of the crank gear reduce timing belt wear caused by the torsional vibration.

Timing belts are typically made from molded polyurethane, neoprene rubber with high-tensile strength fiberglass, or Kevlar cords running the length of the belt. Cogs, which are like gear teeth, engage the crankshaft and camshaft wheels or gears. The distance between the centers of two adjacent belt cogs on the timing belt is referred to as the pitch.

To lengthen belt life, Volkswagen (VW) uses a unique belt designed to reduce the load on the toothed belt during the injection cycle. When an injection pump is driven by the belt, the toothed wheel driven by the crankshaft uses two pairs of teeth, 180° apart, that have a larger gap between adjacent teeth compared with the other teeth (**FIGURE 6-45**). During the injection cycle, the injection pump's high driving forces exert a heavy load on the toothed belt and stretch it. Combustion events, in turn, stretch the belt even further. Stretching the belt temporarily increases the belt tooth pitch (distance between teeth). To compensate for the change in tooth pitch, the two wider teeth are used to minimize excess friction between the belt cogs and the gear teeth. Belt wear is reduced because the same cogs on the timing belt engage the same drive teeth every time there is a combined combustion/injection event.

In addition to the accumulated distance, which determines when the belt should be changed, other factors can shorten the interval. These include unlogged engine idle times, such as idling in traffic and extreme temperature operation. Belt replacement can be a complex task that requires a number of specialized tools and procedures, as well as adjustments to the injection pump timing by using original equipment manufacturer (OEM) software.

Geartrain Noise

Noise from gear rattle caused by torsional vibration can be dampened by a **scissor gear** (**FIGURE 6-46**). These are two-piece gears with a pair of gears that are spring loaded to keep meshed gears in a constant state of tension. Spring loading the two drive gears eliminates

FIGURE 6-46 A scissor gear helps reduce geartrain rattle noise. Gears operate with zero lash because the two gears' teeth are forced apart by spring tension.

FIGURE 6-47 A worn valve face has a dished appearance.

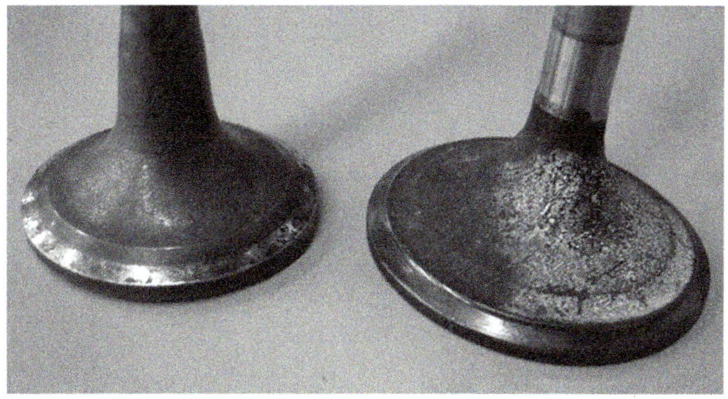

FIGURE 6-48 The intake and exhaust valve faces have worn away with accumulated distance. Most valves recede into the valve face and seat, often causing valve adjustments to tighten up.

gear lash, which minimizes geartrain noise that is heard as geartrain rattle. Rear geartrains are now the most common geartrain available on contemporary diesel engines, because torsional vibration is much lower at the flywheel end of the engine. Geartrain noise and gear wear is kept to a minimum with a rear geartrain. More accurate valve timing and injection action is possible because camshaft speed is more consistent when torsional vibration is reduced.

▶ Maintaining Cylinder Heads and Valve Trains

6-05 Outline the procedures for servicing cylinder heads and components.

Cylinder heads often need maintenance when coolant or combustion leaks occur. In addition, valve seats and faces can become worn or damaged, which means that a cylinder head will need to be rebuilt or replaced (**FIGURE 6-47**). The valve faces and seats may suffer continuous pounding whenever valves are out of adjustment; this prevents the quieting ramps on the camshaft from working effectively, which slows down valve velocity as they seat. Valve faces become recessed when worn out. Excessive oil consumption and extended idle time can contribute to valve seat and face pitting, which eventually result in valve leakage (**FIGURE 6-48**). Very long periods of extended idle can lead to too much slobber forming on the back of the valve. If the slobber solidifies during a cold shutdown period while the valve is open, the valve and piston will make contact and the valve will bend. To perform a valve adjustment, follow the steps in **SKILL DRILL 6-1**.

SKILL DRILL 6-1 Performing a Valve Adjustment

Cylinder	1		2		3		4		5		6	
Valve	E	I	E	I	E	I	E	I	E	I	E	I
Cylinder 1 TDC	X	X	O	X	X	O	O	X	X	O	O	O
Cylinder 6 TDC	O	O	X	O	O	X	X	O	O	X	X	X

1. Clean the upper engine with a pressure washer.
2. Locate OEM service literature for valve adjustment procedures.
3. Unless the manufacturer indicates that their valve adjustment specifications are for warmed engines, the lash should be adjusted on a cold engine. If the OEM indicates that a warm engine is required, bring the engine operating temperature to 140°F (60°C).
4. Remove the valve cover or rocker housing cover(s). Position the engine at TDC unless otherwise indicated. Use the companion cylinder method to find TDC for the end of compression stroke for cylinder 1. This is done by positioning the engine at TDC. Cylinder 6, the companion of cylinder 1, should have two tight rocker levers because the valves should both be open as the engine moves through overlap position. If not, rotate the engine 360° and recheck. Cylinder 1 should have both valves closed because it is finishing compression stroke and beginning exhaust stroke. At TDC position, half of the valves can be adjusted. Use the chart in this box to find the adjustable valves.

(Continued)

5. With the engine at cylinder 1 TDC (end of compression stroke), check the valve clearance by inserting the correct thickness gauge between the rocker lever and the valve stem or bridge. The blade should slide between the two machined surfaces easily, but a light amount of drag should be felt when pulling and removing the blade. If the setting is correct, mark the rocker with a yellow crayon marker. Follow this sequence for all valves that can be adjusted at cylinder 1 TDC.
6. Rotate the engine 360° in the correct direction of rotation until cylinder 6 is at TDC (end of compression stroke). Cylinder 1 should be on overlap, and both rocker levers should be tight because the valves are both open. Both rocker levers should be loose on cylinder 6.
7. Repeat the procedure for checking the adjustment for the remaining intake and exhaust valves.
8. If the lash is too loose or tight, it needs to be adjusted to the correct clearance. To adjust valve lash, loosen and back off the locknut on the rocker adjusting screw. Initially back off the adjusting screw for the rocker lever as well.
9. Screw down the adjusting screw until the rocker lever clearance just disappears. Next, back off the adjusting screw enough to insert the gauge of the correct thickness.
10. Turn the adjusting screw down again until a moderate amount of resistance is felt after the screw makes contact with the valve stem or bridge. This step actually overtightens the screw and reduces the amount of valve clearance.
11. Check the thickness gauge drag. If the drag is acceptable, snug up the locknut while holding the screwdriver to prevent the adjusting screw from rotating and to maintain correct lash clearance. Tightening the locknut will pull on the threads of the adjusting screw and slightly increase clearances. After tightening the locknut, the correct valve lash clearance should be established.
12. Torque the locknut to the correct OEM specifications.
13. Reinstall rocker housing cover(s) by following the manufacturer's procedure. Most current on-highway diesel engines use rocker housing cover gaskets that are reusable provided that they are installed when clean and are torqued correctly.

▶ Cylinder Head Bolts and Gaskets

6-06 Identify and describe the function and construction of cylinder head bolts and gaskets.

Because of their higher cylinder pressures, consistent bolt clamping force is a critical maintenance issue for diesel engines. The size, number, and grade of bolts are increased in four-stroke diesels. Typically, five to seven bolts are used per cylinder, but as few as four per cylinder may be used to distribute the clamping force through the block.

Cylinder head bolts use specialized rolled threads like those used to retain main and rod bearing caps. Unlike cut threads, which have material removed during fabrication, rolled threads are instead formed by using a tool that pushes into and displaces the metal while the bolt is turned. This technique does not diminish the root diameter of a bolt. Because no material is removed and the grain of the bolt is uninterrupted by cutting, rolled threads are stronger and less prone to damage (**FIGURE 6-49**). Usually the bolt, thread, and grade are proprietary, which means the bolt diameter, thread pitch, and strength are unique to the manufacturer.

Care must be taken to inspect bolts for damage before reuse. Nicks, corrosion, or thread damage can lead to stress risers and breakage (**FIGURE 6-50**). Bolts should also be inspected for necking out or stretching using a go/no-go gauge. If the bolt is longer than the gauge permits, then the bolt has been stretched and should be discarded (**FIGURE 6-51**).

Bolt threads should be cleaned and lightly oiled before installation, to ensure maximum and consistent clamping force. Oiling the bolts ensures that more of the twisting force used to tighten a bolt turns the bolt and is not instead lost to friction on the bolt threads. Large variations between different bolts' clamping forces can lead to premature head gasket failure and head warping. Research has revealed that tightening the head bolts with only a torque wrench to torque specification results in over 30% variation in clamping force. Because of this, several service techniques have evolved to ensure consistent gasket clamping forces.

1. **Torque to yield (TTY) bolts** are cylinder head bolts that have been tightened past their yield strength or elastic point. When tightened to specification, they stretch to a point just over a precisely calibrated yield strength but not over the tensile strength where the bolt will break. When stretched to its yield point, the bolt becomes more elastic, allowing it to accommodate changes in the dimensions of the cylinder head caused by temperature fluctuations between cold and high-operating temperatures. The bolts grade of metal and root diameter determines its clamping force. All engines with aluminum cylinder heads use TTY bolts, as do some cast iron engines. These bolts must be discarded after a single use.

FIGURE 6-49 Comparison of rolled threads (left) and cut threads (right). Rolled bolts have a wider root diameter than conventional bolts.

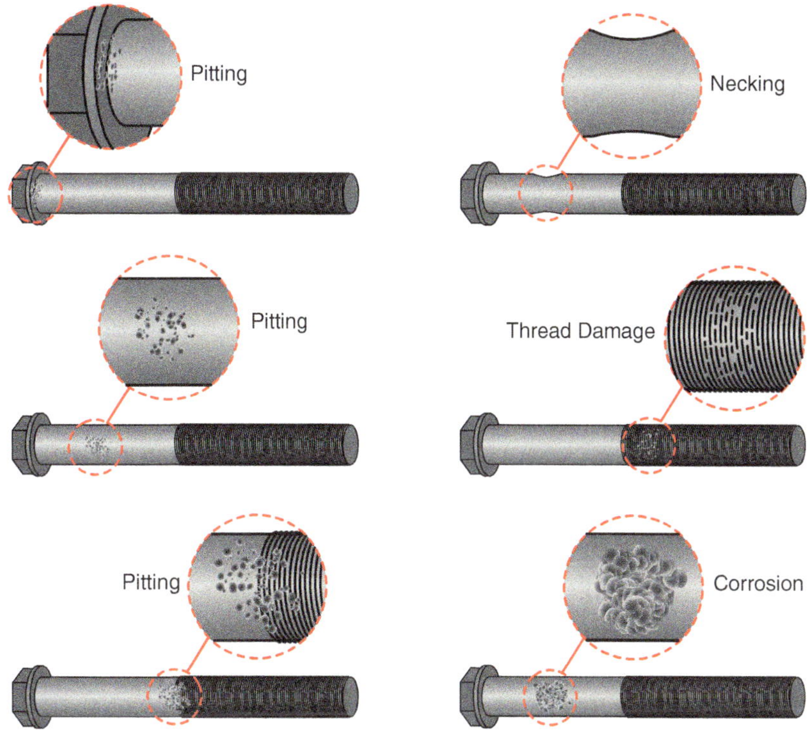

FIGURE 6-50 Don't reuse head bolts if any of the conditions shown here are discovered during inspection.

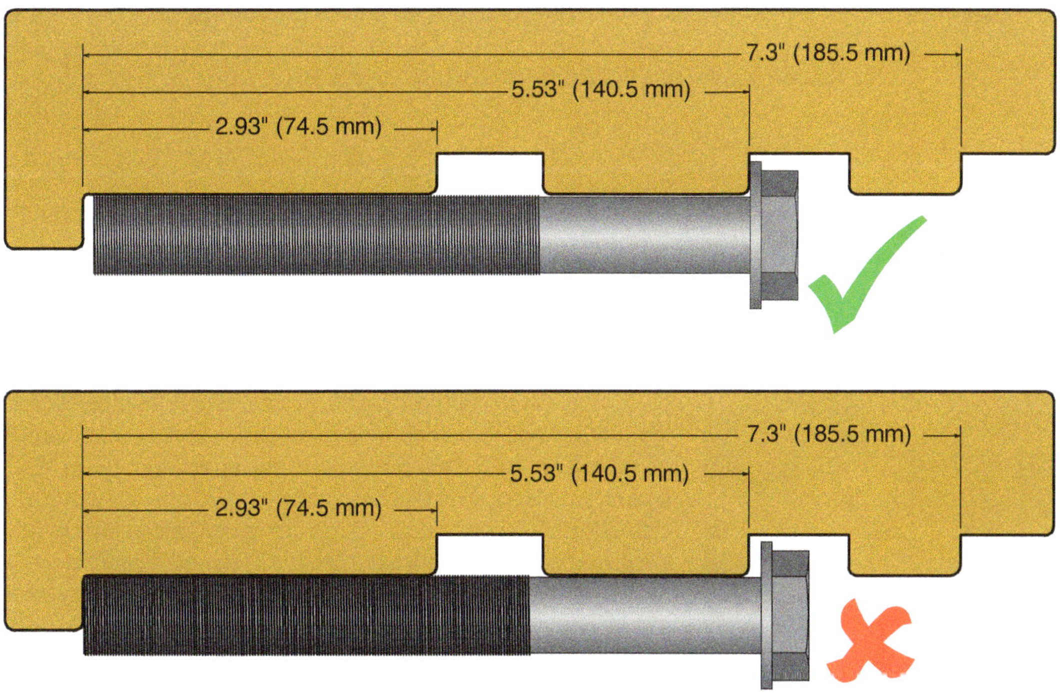

FIGURE 6-51 A go/no-go gauge used to check cylinder head bolts for stretching.

2. The torque-plus-angle or **torque-turn method** requires initial bolt tightening with a torque wrench and then turning the bolt an additional number of degrees past this point (**FIGURE 6-52**). This turn after torquing may be measured in degrees of rotation, number of turns, or a specified number of flats on the bolt. Clamping forces with this method are the most consistent and widely preferred method. When tightening down a

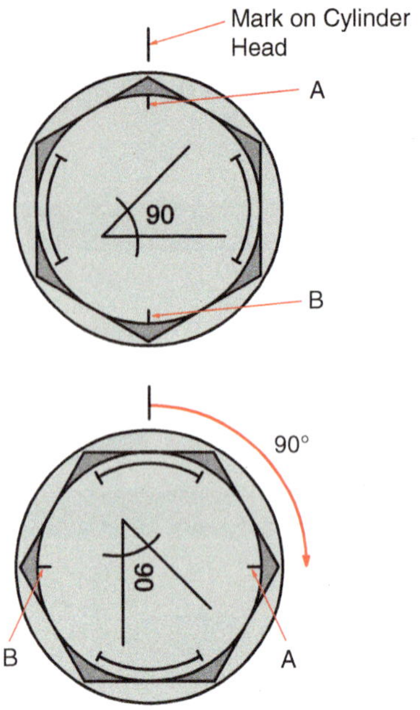

FIGURE 6-52 The torque-turn method requires that a bolt receive an initial preload by using a torque wrench. Then the bolt is turned several additional degrees, flats, or turns.

bolt by using the torque-plus-angle method, it is helpful to mark a socket with whiteout to gauge the amount of turning. Special markings are embossed into the cap of torque-plus-angle bolts, designating the specialized tensile strength of the bolt. While the bolt usually is higher than a grade 8 tensile strength, the threads are annealed to soften them so that they strip less easily.

Studs Versus Bolts

Some engine applications or high-performance modifications use studs rather than bolts to clamp cylinder heads. There are several advantages to using studs. First, studs are often made to higher tensile strengths than stock bolts, which provides higher clamping forces. Also, a stud installed in the cylinder head is far less likely to break when removed.

Studs also can provide an even more consistent clamping force than bolts. With threads at each end, studs are initially finger tightened into the block. After the cylinder head and gasket are installed, torque is applied to a nut. Unlike a bolt, which is reacting to two separate twisting forces on the bolt head and threads in the block, the stud will stretch only on the vertical axis when the nut is tightened. Rotational force on the stud is avoided, and clamping force is directly translated from the torque applied to the nut. This provides a higher and more consistent clamping force on the cylinder head.

Cylinder Head Gaskets

The **cylinder head gasket** is arguably the most important seal in the diesel engine. The head gasket must maintain the seal around the combustion chamber at peak operating temperatures and pressures, as well as seal air, coolants, and engine oil at their respective peak temperatures and pressures. The design and materials used must be thermally and chemically resistant to the products of combustion and to the coolants and oils used in the engine.

The following are some of the most widely used materials:

- steel
- stainless steel
- fiber-based composition materials
- graphite
- numerous chemical formulations that contain silicone, fluoropolymers, nitriles, neoprenes, polymeric resins, polytetrafluorothylene (PTFE or Teflon), and others.

FIGURE 6-53 An MLS gasket with an elastomeric sealing material between layers of laminated steel. The elastomer seals liquids such as coolant and oil.

Traditional copper and brass gasket materials were replaced by metal and asbestos gaskets during the 1950s. Composite metal fiber or graphite composite gaskets superseded these gaskets in the 1980s. Since the early 1990s, those gasket systems have largely been replaced by the development of the **multilayer steel (MLS) gasket**. An MLS gasket is the preferred method of sealing between cylinder head and engine block. Today, an estimated 80% of new engines are designed with MLS gaskets as standard equipment. In the MLS system, multiple thin layers of cold-rolled, spring-grade stainless steel are coated with elastomeric (rubber) material. The resilient elastomer is necessary in the gasket structure to micro-seal metal surface imperfections while making the gasket chemically resistant to aggressive combustion gases, oils, and coolants (FIGURE 6-53).

Using an MLS gasket has many advantages. It has

- uniform loading distribution on sealing beads
- constant operation thickness
- high crush strength
- low relaxation coefficient (it doesn't shrink)

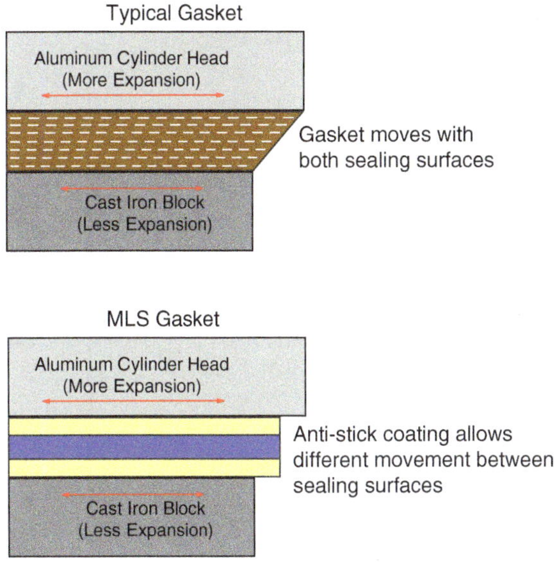

FIGURE 6-54 This MLS gasket has elastomeric sealing material between layers of laminated steel. The elastomer seals liquids such as coolant and oil.

- a reduced clamp load requirement resulting in less block and cylinder bore distortion
- high strength for increased durability
- no need to be retorqued after engine run-in.

Cylinder bores are commonly sealed by a ring of stainless steel positioned over a wire around the perimeter of the combustion chamber. This is known as the fire ring because it is directly exposed to combustion. This ring is critical to sealing the combustion chamber in high-pressure conditions. Graphite or rubber facings are applied to a head gasket for sealing liquids. Graphite is especially useful on aluminum-headed engines because it allows the head to slide or creep easily across the gasket. This is especially important when an aluminum head is used over a cast iron block because they have different expansion coefficients (**FIGURE 6-54**). Surface finishing the block deck and cylinder head is critical when installing gaskets. If the surface is too smooth, the gasket will be unable to properly grip the cylinder head and block deck surfaces, and the gasket will tend to slip, resulting in combustion leaks. If the surface is too rough, the gasket may be unable to fill in the voids between the metal peaks and valleys. These voids can form leak paths for fluids and combustion gases.

Head Gasket Replacement

To prevent cylinder heads from warping, bolts and threads from being damaged, and gaskets from leaking, cylinder heads must be consistently and adequately clamped. A traditional technique for clamping cylinder heads to the block requires using a torque wrench and a spiral pattern sequence for torqueing the head bolts down in the correct sequence. Starting at the center of the cylinder head and working out, this technique prevents the head gasket from being pinched, by ensuring that the gasket is evenly spread out along the block deck surface. Any bolt-tightening sequence should take the cylinder head down in a minimum of three progressive steps of torque, and taking as many as five or six steps is not uncommon (**FIGURE 6-55**). Reversing this pattern when removing a cylinder head will prevent the head from warping. It is especially important to follow this practice in the case of aluminum cylinder heads.

When tightening head bolts, the threads must be cleaned and lubricated with oil. The underside of the head bolt and any washers should be lubricated as well. This ensures consistent clamping force on the head bolt because more torque will be used up attempting to overcome resistance from dry threads in poor condition (**FIGURE 6-56**).

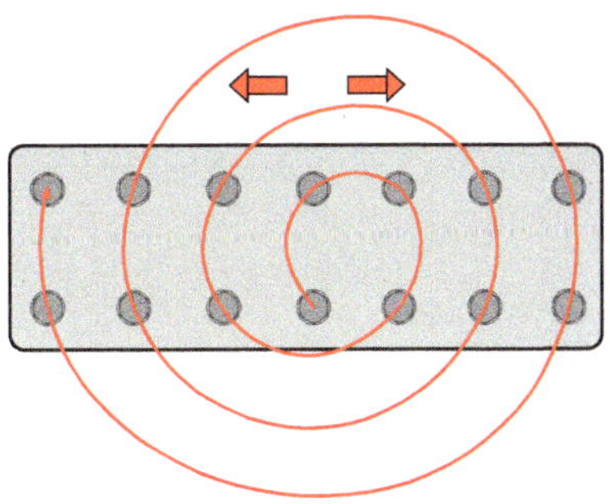

FIGURE 6-55 A spiral pattern of three to five progressive steps of torque is used to tighten head bolts.

> **TECHNICIAN TIP**
>
> Cracked cylinder heads and leaking head gaskets can cause a loss of coolant. Coolant can find a leak path through a head gasket and leak externally or into the engine's crankcase. An overpressurized cooling system will push coolant out of the surge tank overflow, leading to further overheating. To differentiate between a leaking head gasket that is causing coolant loss and combustion gases leaking into the cooling system, two inspection procedures can be performed. One is to check for excessive bubbles or pressure in the cooling system, described in the section on combustion leakage. Another check is to remove the oil pan while the cooling system is pressurized. A leaking head gasket coolant leak path can be observed to leak coolant into the crankcase around the camshaft side of the engine or drain holes from the block deck.

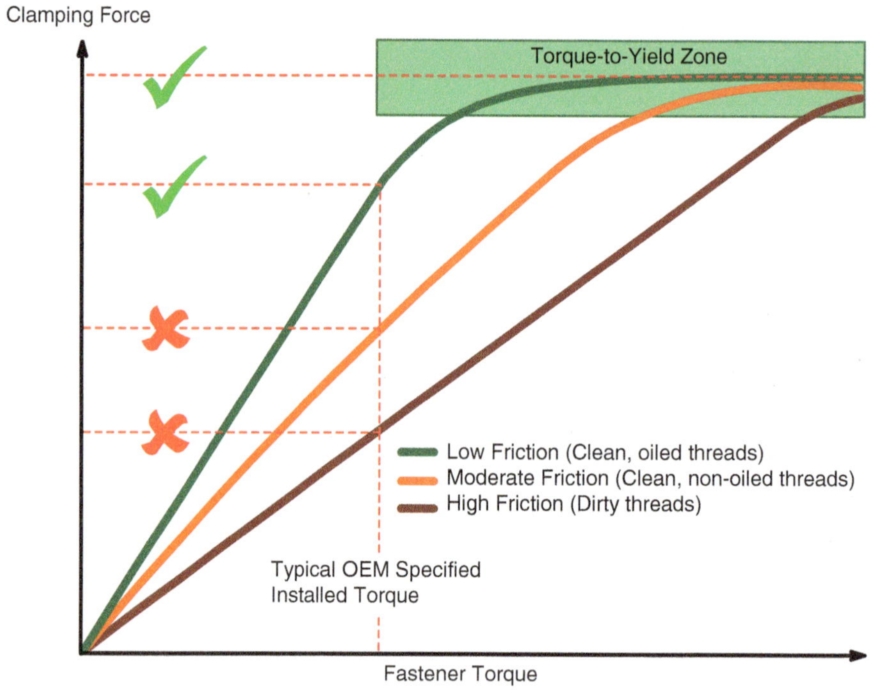

FIGURE 6-56 Bolts with clean, lubricated threads achieve higher clamping force.

FIGURE 6-57 Check a cylinder head for warpage by using a straight edge and a thickness gauge.

After removing a cylinder head, it should be checked for warping. Generally, a cylinder head should not be allowed to warp more than 0.006" (0.015 mm) (**FIGURE 6-57**). To measure cylinder head warpage, follow the steps in **SKILL DRILL 6-2**.

Manufacturers' specifications and techniques will vary. Diesel engine block decks and cylinder heads cannot be milled to straitened warped surfaces. (**FIGURE 6-58**).

Combustion Leakage

An effective method for checking for head gasket and combustion leakage into coolant from a cracked cylinder head is to check for a continuous release of gas bubbles from the cooling system. After warming up the engine, a hose is connected to the coolant overflow pipe at the pressure cap. The other end of the hose is then submerged in a bottle of water, and the amount of gases continuously released from the surge tank or radiator overflow is observed (**FIGURE 6-59**). In a normal engine, there will be a small number of gas bubbles released in a steady stream as some coolant continuously vaporizes. However, a large and steady stream of bubbles indicates that combustion gases are leaking into the cooling system, which requires further diagnosis. Cavitated liners can also leak combustion gases into the cooling system, but they tend to produce hydrostatic locks after the engine is shut down when hot. It's important to remember that chemical detectors used on gasoline engines identify combustion leakage by checking for hydrocarbons in coolant. Because diesel engines do not have fuel present in the cylinders during compression stroke, hydrocarbons are not present in coolant, so these chemical detection methods will not work.

A head gasket leaking coolant into the engine can quickly cause catastrophic engine damage because coolant will easily displace oil and allow oil film shearing or separation on engine bearings. To check for an internal head gasket leak, a scheduled oil sampling can detect coolant in the oil. However, to pinpoint the source of a leak, such as a leaking liner O-ring or head gasket, the engine oil is drained and the oil pan removed. After pressurizing the cooling system with a radiator pressure tester, the internal parts of an engine can be inspected from the crankcase to pinpoint where coolant is leaking.

> **TECHNICIAN TIP**
>
> When removing a cylinder head, the best technique to prevent cylinder head from warping is to remove the head bolts, starting at each end of the cylinder head. Alternately removing a bolt from each end while working toward the center in a spiral-like pattern minimizes the likelihood of the cylinder head lifting, twisting, and warping. This sequence is the reverse of the sequence used to torque a cylinder head.

SKILL DRILL 6-2 Measuring Cylinder Head Warpage

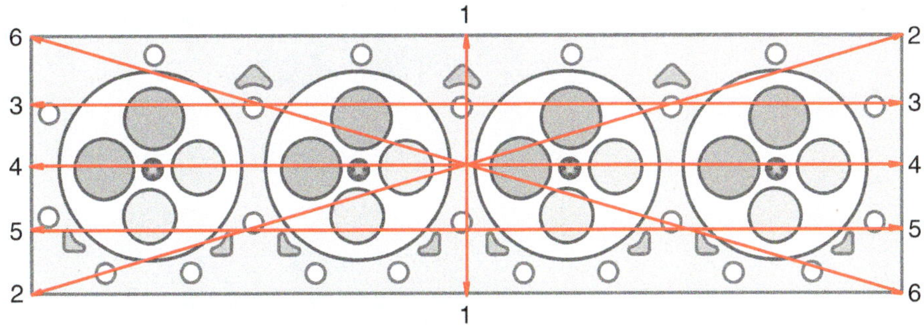

1. Thoroughly clean the cylinder head and the block deck surfaces with solvent. Remove any foreign material, gasket, or silicone by using a cleaning pad. Choose a cleaning pad that removes soft material but not metal; this will protect the casting surfaces from damage that more aggressive abrasives and tools can cause.
2. Use a purpose-made straight edge. Lay the straight edge along the flat cylinder head surface. Try to slide a feeler gauge or thickness gauge between the straight edge and the casting. Any gaps between the straight edge and the surface indicate head deck warpage. A general specification allows no more than 0.006" (0.152 mm) of warpage in 12" (30 cm) of casting surface when using MLS gaskets. This means a 0.006" (0.152 mm) thickness gauge should generally be used to check for warpage, but check the manufacturer's specifications to be sure. Much smaller clearances are only permissible on the smaller heads used in automobile engines.
3. Check for warpage in the longitudinal and diagonal directions as well as across the head deck. Follow the chart in this box when measuring.
4. If the feeler blade slipped beneath the straight edge and there is little resistance when pulling on the blade, it indicates a depression in or warpage of the head surface. If the clearance is outside specifications, make a service recommendation for the cylinder head to be resurfaced by a machine shop. Note, though, that due to the small clearance space between valves and pistons on diesel engines, very little, if any, material can be machined from the cylinder head to correct warpage. Always check specifications for head thickness if a head has been machined.
5. Check the cylinder block deck by using the same procedure as checking the cylinder head.

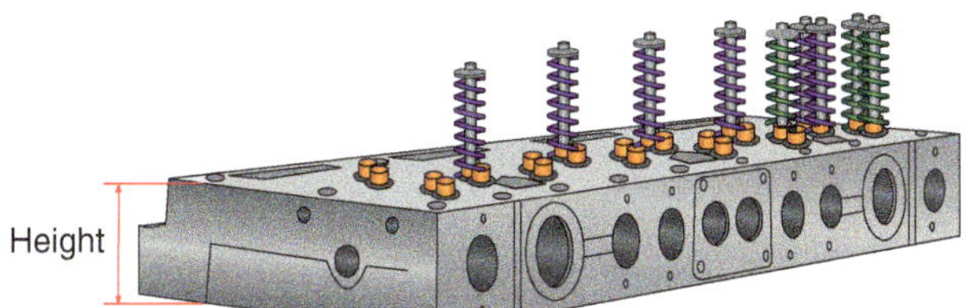

FIGURE 6-58 Because there are very small clearances between the valves and pistons in a diesel engine, very little material can be milled from a cylinder head to correct warpage.

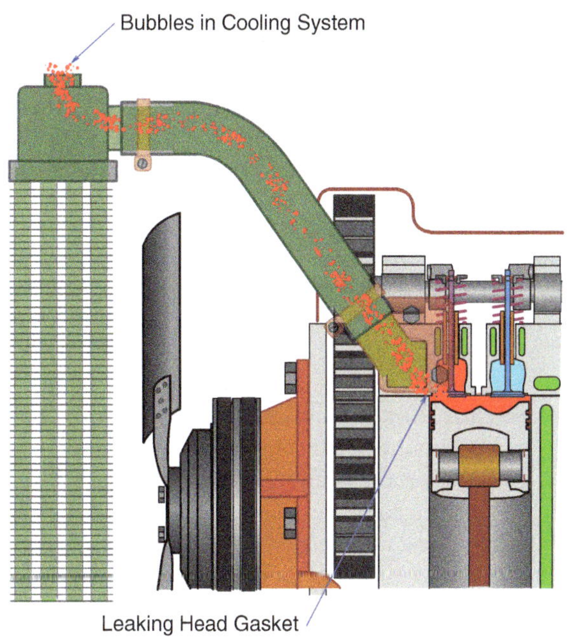

FIGURE 6-59 A defective head gasket will allow pressurized combustion gases to pass into the cooling system.

Inspecting for Cracks in the Cylinder Head

Cylinder heads should also be inspected for cracking. Several techniques are available to the technician:

- dye penetrant
- magnetic flux
- pressure testing.

If a cylinder head is suspected of being cracked or leaking, often the overpressurization of the cooling system presents itself only after an engine has run very hard for some time and thermal expansion has opened a crack. If this is the case, the cylinder head can be removed and pressure tested in a hot tank with special fixtures that supply pressurized air into the intake and exhaust ports, where leak paths are formed. In the tank, the head is sealed by the fixtures to prevent leakage through the valve ports, and then it is carefully inspected for any leaking air bubbles. Cracks are easily detected by observing where the bubbles originate from in the head.

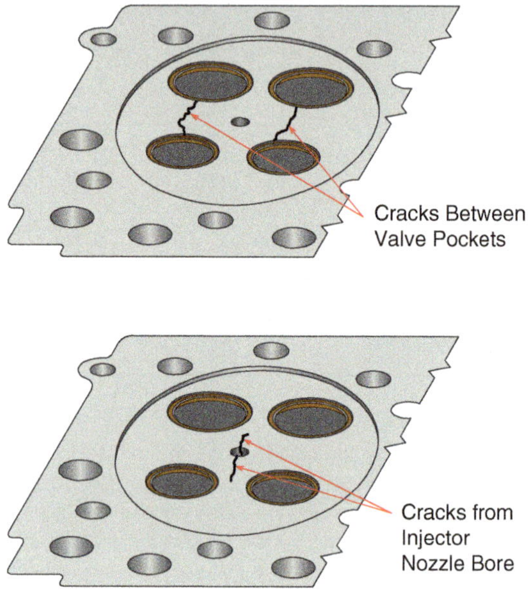

FIGURE 6-60 Cracks often develop between valve seats and around an injector hole when the engine has overheated.

▶ TECHNICIAN TIP

A loss of engine coolant or excessive cooling system pressure can often be attributed to a leaking cylinder head gasket or a cracked cylinder head. Air bubbles originating from the engine block will appear in the radiator, and pressure will build rapidly in the cooling system, which can be measured with a radiator pressure tester. Technicians can check for hydrocarbons or carbon monoxide in the gas bubbles on gasoline engines, but not diesels, because only air is present in the cylinders during compression stroke. Test kits that measure carbon dioxide gas in coolant are used to determine whether gas bubbles are caused by combustion leakage in diesel engines.

Cylinder head cracks often develop between the valve seats after an engine has overheated (**FIGURE 6-60**). When the head is removed, the valves are also removed and the seat areas visually inspected. Dye penetrant testing, which is a three-step process that allows dye to penetrate cracks to make them more visible, is a convenient way to identify casting cracks. A magnetic flux technique is another method that uses a powerful magnet and iron filings to identify cracks. In this method, a cylinder head is magnetized by either using alternating current and a specialized piece of equipment or using a strong magnet placed on either side of a suspected crack. Iron filings are then blown onto the area by using a small squeeze bulb filled with filings. Because cracks will develop a north-and-south polarity, the filings will line up end to end along a crack, so that the crack can be easily visible.

▶ Wrap-Up

Ready for Review

- The cylinder head is a metal casting that attaches to the top of the cylinder block and seals the combustion chamber so that the engine can develop and contain compression and combustion pressure.
- Cylinder heads are commonly made of cast iron alloys because of their superior strength, ease of manufacture, machinability, and capacity to withstand high temperatures; aluminum and composite graphite iron (CGI) are other materials commonly used for cylinder heads.
- Because intake and exhaust port arrangements influence gas flow and turbulence, or swirl, into the cylinders, the configuration of these ports is used to classify cylinder heads.
- Cylinder heads can also be classified by their number of pieces; single-piece cylinder head castings and multiple-piece cylinder heads have advantages and disadvantages.
- Cylinder heads are further classified by the arrangement of the valve train mechanism. The camshaft arrangements in diesel engines are commonly one of two types: overhead camshaft or I-head.
- Intake and exhaust valves allow air into and out of an engine; diesel engines usually have two or four valves per cylinder, but a number of engines may use three or five valves.
- Pushrods transmit the cam actuation force from the followers to the rocker levers.
- Rocker arms convert the linear change of the cam profile to the reciprocating motion used to open and close the valves.
- To actuate more than one valve with a single rocker lever, most four-valve engines use a crosshead, also called a yoke or bridge, to exert the camshaft force against the valve stem.
- Camshafts are components that are critical to engine performance because they control the movement of air and exhaust in and out of the cylinders or engine-breathing characteristics.
- Because of the higher cylinder pressures, consistent bolt clamping force is a critical maintenance issue for diesel engines.
- In some applications or high-performance modifications, studs are used instead of bolts.
- The cylinder head gasket is arguably the most important seal in the diesel engine. To prevent cylinder heads from

warping, bolts and threads from being damaged, and gasket from leaking, cylinder heads must be consistently and adequately clamped.

- To coordinate the movement of the crankshaft with the camshaft and the valves, a geartrain or valve timing mechanism is used. The geartrain may also transfer power from the crankshaft to operate fuel-injection equipment, lubrication, power steering and coolant pumps, accessory drive mechanisms, and even power takeoff pumps.

Key Terms

composite graphite iron (CGI) A material produced from powdered iron alloys squeezed into molds at high pressures and then heated to bond the metal particles together; also known as sintered graphite.

cross flow cylinder head A head design with the intake and exhaust manifolds located on opposite sides of an in-line engine to improve engine-breathing characteristics.

cylinder head gasket The component that maintains the seal around the combustion chamber at peak operating temperatures and pressures and keeps air, coolants, and engine oil at their respective peak temperatures and pressures.

I-head engine An engine that has only the valves, rocker levers, and bridges located in the cylinder heads above the piston; the camshaft is located in the engine block.

multilayer steel (MLS) gasket A method of sealing the cylinder head to the engine block by using multiple thin layers of cold-rolled, spring-grade stainless steel coated with elastomeric (rubber) material.

overhead camshaft engine An engine that has the camshaft located in the cylinder head.

parallel flow head A head design that features intake and exhaust manifolds on the same side of the engine and short, large ports that are joined together to provide a more compact engine design with adequate airflow to the cylinders; also known as the uniflow design.

pushrod engine An engine with the camshaft in the engine block and pushrods that are used to actuate the valves by using rocker levers.

reverse-flow cylinder heads A cylinder head with no exhaust manifolds on the outside of the cylinder head; instead, short exhaust runs are fed directly to the turbocharger located in the V between the cylinder banks.

Rocker Arm Ratio A comparison between the length of the rocker lever and the distance between the rocker shaft fulcrum point and contact point with either a camshaft or pushrod.

scissor gear Two separate gears that are incorporated into a single unit to reduce gear rattle caused by torsional vibration.

timing gear A gear that synchronizes the timed events of rotational engine components like camshafts and fuel-injection pumps.

torque-turn A method that requires a bolt to receive an initial preload by using a torque wrench; afterward, the bolt is turned several additional degrees, flats, or turns.

torque to yield (TTY) Cylinder head bolts that are tightened past their yield strength or elastic point; when tightened to specification, they stretch to a point just over precisely calibrated yield strength and not to the tensile where the bolt will break.

torsional vibration The speeding up and slowing down of the crankshaft caused by engine cylinder alternating compression and power strokes.

Valve float The condition when valves do not seat and remain partially open.

valve overlap The angle in crankshaft degrees during which both the intake and exhaust valves are open; overlap occurs at the end of the exhaust stroke and the beginning of the intake stroke.

Review Questions

1. Which of the following reasons best explains why diesel engine cylinder heads use more head bolts per cylinder?
   a. The engines are larger.
   b. Diesel fuel burns at a hotter temperature.
   c. Cylinder pressures are higher.
   d. The stroke length is longer.
2. The purpose of a variable intake swirl mechanism is to __________.
   a. meter airflow into the cylinder
   b. increase the exhaust temperature
   c. reduce particulate emissions
   d. reduce intake air noise
3. Which of the following cylinder head valve train arrangements can provide the highest injection pressures?
   a. Overhead camshaft.
   b. L-head.
   c. Overhead valves.
   d. Roller type.
4. Valve overlap occurs during __________.
   a. the beginning of intake stroke and end of compression stroke
   b. the beginning of intake stroke and end of exhaust stroke
   c. the beginning of compression stroke and end of intake stroke
   d. the beginning of intake stroke and end of power stroke
5. A valve failure with a break between the head and stem was likely caused by ________.
   a. a defective weld
   b. excessive valve guide wear
   c. contact between the valve and piston
   d. an overheated valve
6. Excessively worn valve train components are more commonly found on which valve actuation mechanism?
   a. Intake valves.
   b. Exhaust valves.
   c. Engine brakes.
   d. Rocker levers.
7. The maximum allowable warpage for a cylinder head is approximately __________.
   a. 0.004–0.006" (0.102–0.152 mm)
   b. 0.04–0.06" (1.02–1.52 mm)
   c. 0.001–0.005" (0.025–0.127 mm)
   d. 0.1–0.2" (2.5–5.1 mm)

8. Which valve guide in a turbocharged diesel engine will more likely not have valve stem seals?
   a. Intake.
   b. Exhaust.
   c. Integral.
   d. Replaceable.
9. The most common valve angle for poppet valves is ______.
   a. 45°
   b. 30°
   c. 20°
   d. 60°
10. Which of the following is the most common method for achieving consistent head gasket clamping force on today's diesel engines?
   a. Torquing bolts to a fixed value.
   b. Using torque-yield head bolts.
   c. Using the torque-turn method.
   d. Using MLS head gaskets.

ASE Technician A/Technician B–Style Questions

1. Technician A says that the cam gear rotates two times faster than the crank gear. Technician B says the cam gear rotates at half the engine speed. Who is correct?
   a. Technician A
   b. Technician B
   c. Both A and B
   d. Neither A nor B
2. While examining a valve spring from an engine that is being overhauled, technician A says that valve springs should be checked for free length and cracks. Technician B says the spring should be checked for compressed spring pressure, nicks, and straightness. Who is correct?
   a. Technician A
   b. Technician B
   c. Both A and B
   d. Neither A nor B
3. While installing a cylinder head, technician A says the threads in cylinder head bolts must be lubricated before installation. Technician B says that only the underside of the bolt head requires lubrication and that the threads only need be clean. Who is correct?
   a. Technician A
   b. Technician B
   c. Both A and B
   d. Neither A nor B
4. While discussing variable valve timing in diesel engines, technician A says that only dual overhead camshafts (DOHCs) use variable valve timing. Technician B says variable valve timing is not commonly used in diesel engines. Who is correct?
   a. Technician A
   b. Technician B
   c. Both A and B
   d. Neither A nor B
5. When adjusting the valve clearances on the engine, technician A says the valves are closed when the valve lifter or follower is on the inner base circle of the camshaft lobe. Technician B says that valves are always adjusted when closed while the lifter or follower is on the camshaft lobe outer base circle. Who is correct?
   a. Technician A
   b. Technician B
   c. Both A and B
   d. Neither A nor B
6. Technician A says that before reusing any head bolt, it should be inspected for nicks and pitting. Technician B says the service manual should be consulted before reusing because some bolts can only be used once. Who is correct?
   a. Technician A
   b. Technician B
   c. Both A and B
   d. Neither A nor B
7. Technician A says that rear geartrains are used on engines because they operate with less noise. Technician B says that rear geartrains are used because they have less wear than a front geartrain mechanism. Who is correct?
   a. Technician A
   b. Technician B
   c. Both A and B
   d. Neither A nor B
8. Technician A says that overhead cam engines with unit injectors can provide the greatest actuation force for the highest injection pressures. Technician B says that pushrod engines that use a low-mounted camshaft can provide the same force as an overhead camshaft engine to its unit injectors. Who is correct?
   a. Technician A
   b. Technician B
   c. Both A and B
   d. Neither A nor B
9. Technician A says that fiber gaskets are the best and most commonly used gaskets in modern diesel engines. Technician B says that MLS gaskets are now the most commonly used head gaskets. Who is correct?
   a. Technician A
   b. Technician B
   c. Both A and B
   d. Neither A nor B
10. Technician A says that whether combustion gases are leaking into the engine can be detected by measuring whether fuel hydrocarbons are present in the coolant. Technician B says that excessive gas pressure in the cooling system can indicate a head gasket leak. Who is correct?
   a. Technician A
   b. Technician B
   c. Both A and B
   d. Neither A nor B

CHAPTER 7

Diesel Engine Cooling and Lubrication Systems

Learning Objectives

After reading this chapter, you will be able to:

- 7-01 Identify and describe the functions of the diesel engine cooling system.
- 7-02 Identify and describe the construction and operation of cooling system components.
- 7-03 Identify and describe the functions of a lubrication system.
- 7-04 Identify and explain the classification systems for engine oil according to oil properties and characteristics.
- 7-05 Identify and explain factors that limit oil service intervals.
- 7-06 Identify and explain the construction and operation of pressurized oil circuits.

You Are the Technician

A number of delivery trucks you service at your repair center ship time-sensitive loads, which means there is almost no tolerance for breakdowns on the road. To ensure these vehicles do not experience any kind of mechanical problems when they're operating, you have been tasked with developing a preventive maintenance schedule and identifying inspection and service criteria to present to customers as part of a maintenance agreement. Part of preventive maintenance involves cooling and lubrication system service. As you prepare the maintenance schedule, consider the following:

1. Identify potential problems with the cooling and lubrication system that could lead to a breakdown or road service call.
2. Outline maintenance practices and inspection points of the cooling and lubrication system that will require attention during preventive maintenance services. Explain your service recommendations.
3. Outline some of the more important factors of cooling and lubrication system maintenance that promote extending engine life and reliability.

▶ Introduction

Diesel combustion characteristics, the type of fuel burned, turbocharging, and high cylinder pressures create additional tasks for the cooling and lubrication systems. These engine support systems have unique construction features and operating requirements to enable the engine to perform all it's functions well and deliver long and reliable engine operation.

▶ Functions of a Cooling System

7-01 Identify and describe the functions of the diesel engine cooling system.

The following points outline the functions of a cooling system.

1. It removes excess heat. The flame temperature of diesel fuel is approximately 3,900°F (2,150°C). Continuous engine operation at temperatures like this without some means to remove heat would quickly result in damaged and destroyed engine components. Operating clearances between moving engine parts would disappear and cause engine seizure as components thermally expand. Components such as pistons and valves would soften and permanently distort under the high temperatures. Lubrication oil would lose viscosity and fail to maintain separation between moving parts (**FIGURE 7-1**). Since diesel thermal efficiency is almost double that of gasoline engines, less heat from the combustion chamber is transferred to the cooling system. However, even though diesel engines transmit less heat to engine coolant, diesel engines have additional heat loads that are absorbed by the cooling system (**FIGURE 7-2**). These additional heat loads include:
   - *Cooled exhaust gas recirculation.* Up to 30% of intake air mass in current diesel engines will include recirculated exhaust gases, which can reach 1,200°F (650°C). These temperatures are reduced to 400°F (204°C) or less before being mixed with fresh intake air.
   - *Lubrication oil.* Since every turbocharged engine cools the underside of the piston with engine oil, hotter engine oil loses its viscosity, thus thinning too much to support loads and properly lubricate parts. Diesel engines will use oil coolers to reduce oil temperatures. Oil temperature sensors are often used to derate engine power or warn the operator if oil exceeds 250°F (121°C) (**FIGURE 7-3**).
   - *Air intake system.* Intake air compressed by the turbocharger easily heats to over 400°F (204°C). Engines using series turbocharging will drive intake temperatures even higher than this. Engine coolant is often used to help reduce charge air temperatures. For example, the Powerstroke diesel of the 2010 model year and afterward use a liquid charge air coolers. A separate cooling circuit with its own water pump and radiator is used to lower the temperature of pressurized intake air.

FIGURE 7-1 This crankshaft journal was damaged because of a lack of lubrication. The oil feed hole is plugged with bearing material, which melted during engine operation.

2. The cooling system regulates temperature. If temperatures climb even as little as 12°F (5°C) above normal range (195°F–207°F or 91°C–97°C), cylinder wall wear begins to

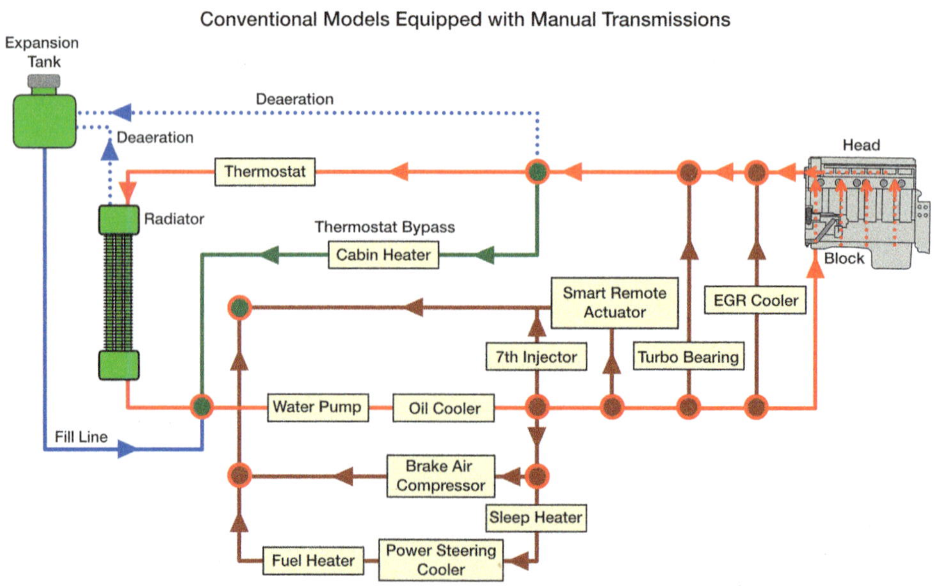

FIGURE 7-2 Diesel cooling systems absorb additional heat loads from components. High-capacity exhaust gas recirculation (EGR) coolers, air compressors, oil coolers, and turbochargers are just a few additional heat loads.

FIGURE 7-3 Oil cooler nozzles spray the underside of all engine piston crowns with oil to remove heat. The heat absorbed by the oil is removed by the cooling system through the oil cooler.

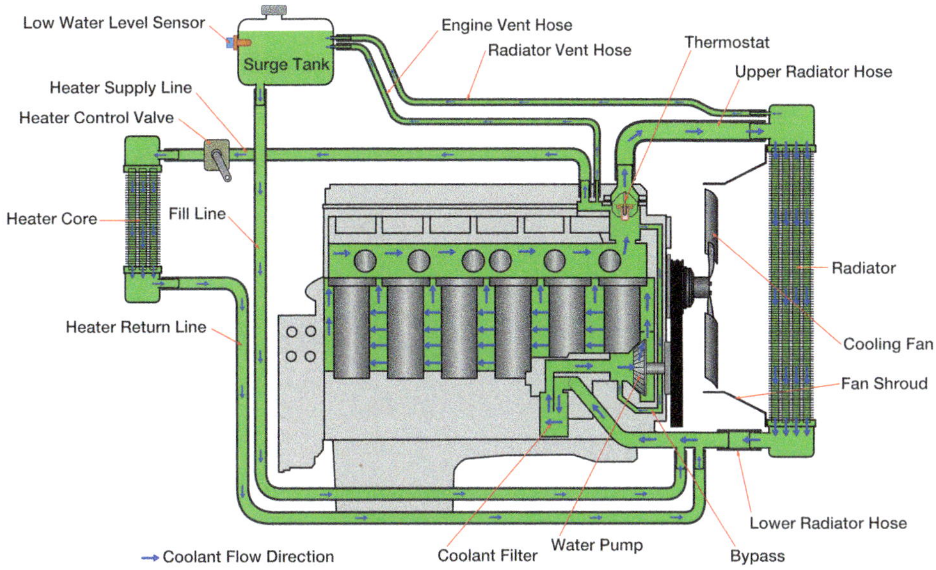

FIGURE 7-4 Cooling system components and coolant circulation patterns through a typical I6 heavy-duty diesel engine.

dramatically increase. Conversely, cold cylinders will prevent engine parts from operating with optimal clearances, causing excessive oil consumption, carbon deposits on pistons, poor combustion quality, and oil sludging due to water condensation. In current diesel engines, the coolant must reach a minimum temperature necessary to enable onboard diagnostic (OBD) monitors or any emission control strategy to operate or execute within a reasonable time. In fact, an OBD cooling system monitor is dedicated to alerting the driver if the warm-up of the cooling system becomes irregular.

3. The cooling system provides for coolant expansion. As engine coolant warms, it expands. Cooling systems provide overflow reservoirs so that coolant can expand (**FIGURE 7-4**).
4. The cooling system pressurizes engine coolant. Cooling systems are pressurized for two important reasons. The first is to maintain a greater-than-atmospheric pressure at the water pump inlet. Without pressure to the water pump inlet, pumps operate less efficiently and can cause cavitation. Pressurizing coolant also raises its boiling point, thus preventing coolant from being lost due to boil-over (**FIGURE 7-5**).

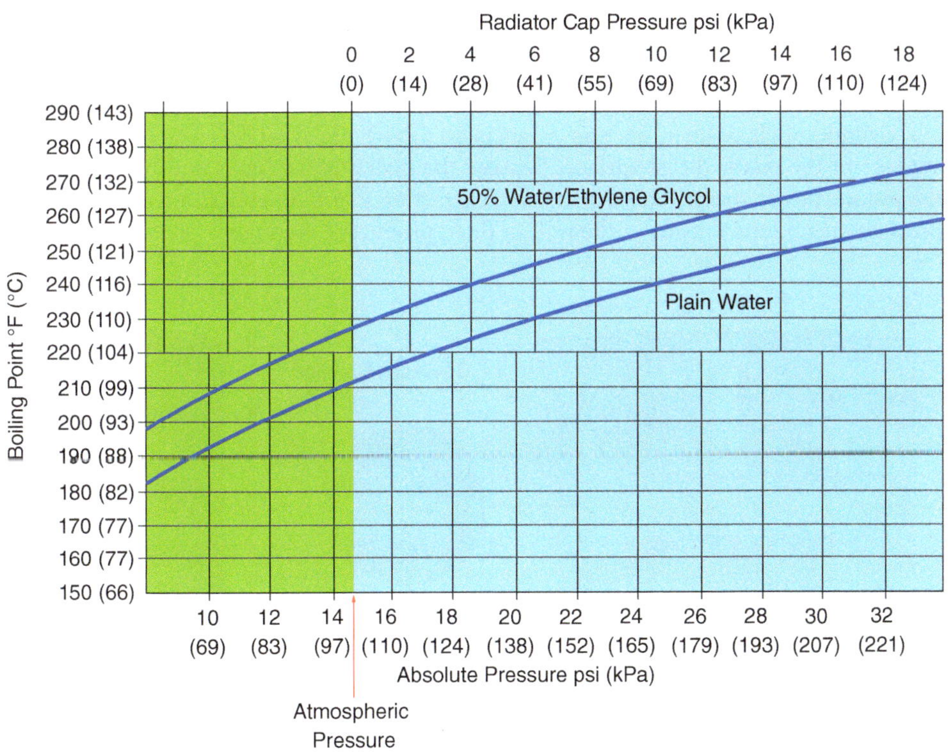

FIGURE 7-5 Increasing the pressure of the cooling system increases the boiling point of the coolant. Adding antifreeze to the coolant further increases the boiling point of the coolant.

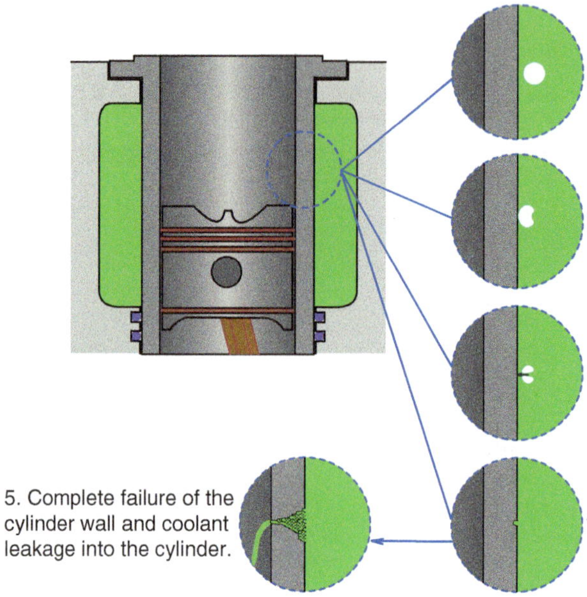

FIGURE 7-6 Hot coolant vaporizes, forming a bubble on the side of the cylinder wall. The force of a collapsing vapor bubble is great enough to remove material from the cylinder wall.

5. The cooling system deaerates coolant. Cooling systems must deaerate coolant to prevent pumps and coolant passageways in the cylinder head from becoming air bound. Without deaeration, vapor from hot coolant collects into large steam pockets, preventing heat transfer.
6. The cooling system minimizes corrosion. When dissimilar metals come in contact with engine coolant, the contact causes electrochemical reactions. Some metals will be severely and quickly corroded through these chemical reactions. Diesel antifreeze contains a number of additives that minimize the electrochemical reactions that cause corrosion.
7. The cooling system provides freeze and boil protection. Using antifreeze, also referred to as coolant, provides freeze and boil protection.
8. The cooling system minimizes cavitation erosion. A condition unique to diesel engines is cylinder wall and liner cavitation. Cavitation erosion is caused by the collapse of tiny water vapor bubbles formed when coolant rapidly heats next to hot surfaces and/or when the coolant is subjected to a rapid drop in pressure. Liquid will vaporize in both of these situations, producing gas bubbles. When these vapor bubbles collapse, the pressures are as high as 60,000 psi (413,700 kPa). The force of implosion against cylinder walls, cylinder liners, water pump inlets, injector tubes, and other components easily blasts holes into the components (**FIGURE 7-6**, **FIGURE 7-7**, **FIGURE 7-8**).

Cylinder walls or liners have an exaggerated pattern of cavitation erosion on the coolant side of the major thrust surfaces because the liner surface flexes and relaxes during the piston's power stroke. This expansion–contraction vibration creates a temporary low-pressure area on the outside of the wall or liner, producing even more cavitation erosion.

> **▶ TECHNICIAN TIP**
>
> Core plugs or frost plugs are used to seal passageways in cylinder blocks and head castings, which function to flush sand from casting after manufacturing. Although these plugs may be displaced by frozen coolant, they are not designed to prevent engine damage.

> **▶ TECHNICIAN TIP**
>
> Coolant in the cylinder of a diesel engine will prevent the engine from rotating because the cylinder has very little clearance volume and cannot compress coolant. This condition is called hydrostatic lock. Coolant may enter through a leaking cylinder head gasket, intake air systems, or a pinhole-sized opening in the cylinder caused by cavitation erosion. An engine turned in the correct direction of rotation will lock up if liquids are in the cylinder. To confirm whether the engine is hydrostatically locked, the engine when rotated in the opposite direction of correct rotation will push liquids out the exhaust valve and the engine will freely rotate in that direction only.

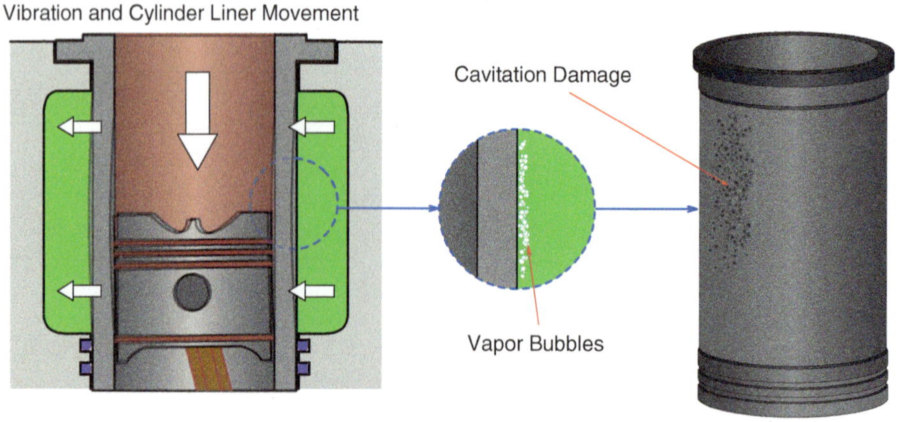

FIGURE 7-7 Cavitation erosion is unique to diesel engines because of higher cylinder pressures and longer stroke lengths. Movement of the cylinder wall due to high combustion pressure and the thrust angle of the connecting rod creates a momentary void between the coolant and the cylinder wall.

Note, parent bore blocks are not immune to cavitation erosion. Chemically treating the cooling system is a preferred method for minimizing cavitation erosion. Both new and old formulations of antifreeze for diesel engines use nitrite as an additive to minimize cavitation erosion. Nitrite added to the coolant forms a thin, sacrificially protective film to prevent cavitation erosion. Vapor implosion will instead damage only the nitrite layer, which is replenished from the nitrite dissolved in the coolant. Depending on the formulation of antifreeze, coolant will use varying amounts of nitrite, which requires periodic testing during service inspections to maintain the correct level of nitrite.

FIGURE 7-8 An injector tube damaged by cavitation erosion.

▶ Cooling System Components and Construction

7-02 Identify and describe the construction and operation of cooling system components.

Durability of automotive-type diesel engines depends on the efficiency of liquid cooling systems. Water mixed with antifreeze plus additives to protect the cooling system make up the cooling medium used to absorb engine heat. A heat exchanger or radiator transfers heat from coolant to the atmosphere. The diesel cooling system includes several unique features, to be discussed in this section (**FIGURE 7-9**).

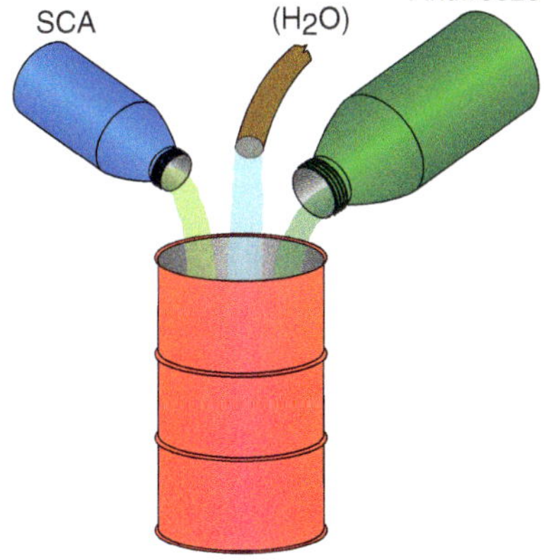

FIGURE 7-9 Diesel engine coolant is made up of water, antifreeze, and supplemental coolant additives (SCA)—also called diesel coolant additive (DCA); SCA and DCA contain nitrite.

Coolant

Diesel engine coolant, while similar in some respects to gasoline-fueled engine coolant, is not the same. Diesel coolant consists of water and antifreeze, and contains a specialized set of corrosion inhibitors referred to as supplemental coolant additives (SCA) or diesel coolant additive (DCA).

Water

Water is the most efficient fluid used to transfer heat. However, water will cause metal to corrode when in contact with the cooling system, so it needs an additional corrosion inhibitor.

Antifreeze should be added to a minimum of 40% concentration because of the corrosion inhibitors it contains. Too high a concentration of antifreeze, however, will reduce the ability of coolant to absorb and release heat. Results comparing the specific heat values of antifreeze water mixtures show it absorbs and releases only 50% to 70% as much heat as pure water. Poor engine cooling, overheating, and inadequate passenger warm air heating results when antifreeze is overconcentrated. For this reason no more than a 60% concentration of antifreeze is permissible, and more than a 65% concentration can also cause the coolant to lose its freeze protection. In cold climates, a 50/50 mixture of ethylene glycol–type antifreeze is recommended (ethylene glycol can be abbreviated as EG). Propylene glycol–based antifreeze requires a higher concentration to achieve the same freeze protection as EG antifreeze (propylene glycol can be abbreviated as PG). For boil protection, antifreeze concentrations in engines operating above 195°F should not be less than 50% (**TABLE 7-1**).

TABLE 7-1 Freezing Points of EG and PG Antifreeze

Concentration of Antifreeze by Percentage of Volume	Freeze Point of Coolant			
	EG		PG	
0% (water only)	32°F	0°C	32°F	0°C
20%	16°F	−0°C	19°F	−7°C
30%	4°F	−16°C	10°F	−12°C
40%	−12°F	−24°C	−6°F	−21°C
50%	−34°F	−37°C	−27°F	−33°C
60%	−62°F	−52°C	−56°F	−49°C
80%	−57°F	−49°C	−71°F	−57°C
100%	−5°F	−22°C	−76°F	−60°C

> ▶ TECHNICIAN TIP
>
> EG, which is the base used for the majority of antifreezes, is toxic. Less than 4 oz (113 g) will kill a 150 lb (68 kg) person. Because of its toxicity, EG is an environmental hazard; therefore, old antifreeze must be properly disposed of by a licensed recycler. Service garages are required by law to separate and store old antifreeze in specialized containers while waiting for recycling.

> ▶ TECHNICIAN TIP
>
> Important criteria for selecting antifreeze includes whether the formulation is compatible with gaskets, coolant hoses, and other cooling system materials. Using the wrong type of antifreeze for a cooling system can slowly or quickly lead to deteriorated hoses and engine gaskets leaking. And cooling system materials can become corroded when the wrong type of antifreeze is used.

> ▶ TECHNICIAN TIP
>
> Diesel engine and gasoline engine Type I "green" antifreezes are different and should never be interchanged. Gasoline engines use Type I antifreeze, containing higher silicate levels. Only low-silicate antifreeze should be used in diesel engines, since the addition of nitrite to diesel coolant will cause silicate dropout. Dropout or the appearance of green "goo" can lead to plugged radiators, water pump seal leaks, and an engine overheating.

Ethylene Glycol Antifreeze

Coolant has evolved considerably since the early days of automotive technology. At one time, coolant included mixtures of alcohol and water, as well as water mixed with honey and soluble oils. In the 1940s, EG was developed to provide a combined freeze and boil-over protection. When initially manufactured, EG is a clear, syrup-like fluid. However, green dye was initially added for identification since it is extremely poisonous. The majority of antifreezes currently manufactured use EG as a base. Today, EG-based antifreeze is colored with other dyes than green to designate specific formulations of corrosion inhibitors blended with the EG.

Propylene Glycol

PG is made available for diesel engine cooling systems to minimize the problems associated with the toxicity of EG, an environmentally friendly antifreeze base. PG is an ingredient in many cosmetics and foods (e.g., hand cream, ice cream, milkshakes). It has slightly different physical properties than EG does. Since the density and concentration required for freeze protection is different from EG, more PG is mixed with water to obtain adequate freeze protection (**FIGURE 7-10**, **TABLE 7-2**).

Antifreeze Formulations

The mixing of EG or PG antifreeze and water alone is not effective to prevent cooling system components from corroding or eroding. When exposed to air, EG will become acidic; PG, on the other hand, is corrosive to certain metals. Antifreeze formulation develops as an evolutionary process driven by the need to prevent corrosion, ensure material compatibilities, and minimize maintenance. While the bases of all modern antifreezes are PG or EG, the additive packages vary considerably, which can lead to confusion and severe maintenance problems if not properly understood and the correct formulation used. At least six different types of coolant formulations are used today. The additive package added to the antifreeze base differentiates between the various types of EG- and PG-based coolants. Coolants are usually dyed according to The Maintenance Council (TMC) recommendation to help distinguish between the various formulations. Note, however, that some coolant manufacturers do not follow TMC recommendations.

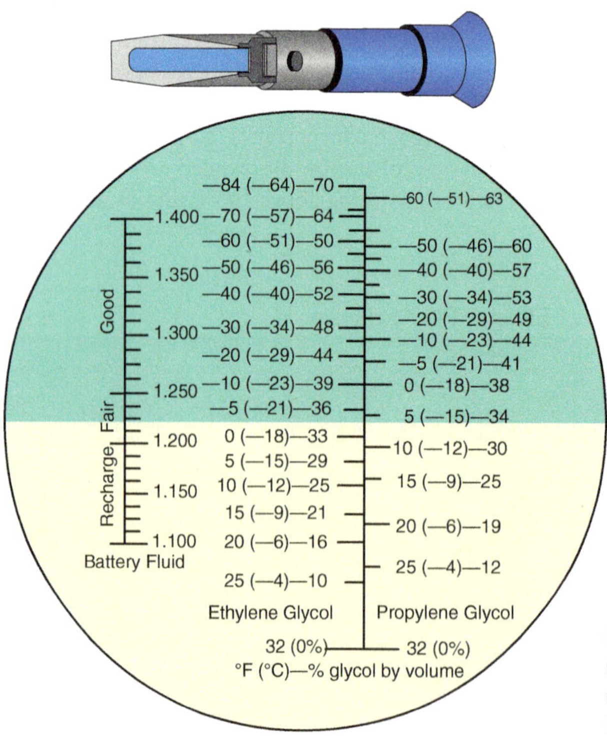

FIGURE 7-10 A refractometer screen has two different scales for measuring the freeze protection of PG and EG coolants.

TABLE 7-2 Antifreeze Types

Antifreeze Type	Performance Specification	Suggested Color
Type I conventional low silicate using inorganic acid technology (IAT)	Technology & Maintenance Council (TMC) RP-302a	Green
Type II fully formulated EG extended life using organic acid technology (OAT)	Technology & Maintenance Council (TMC) RP-3298	Purple or pink
Type III fully formulated PG extended life using OAT	Technology & Maintenance Council (TMC) RP-330	Blue
Type IV OAT	Technology & Maintenance Council (TMC) RP-338	Red
Hybrid organic acid technology (HOAT)	Combination of IAT and OAT with nitrites added	Ford—yellow; Daimler-Chrysler—orange
Nitrated organic acid technology (NOAT)	Similar to HOAT	No color designated

Type I: Conventional Low-Silicate Antifreeze

Antifreeze contains corrosion inhibitors to protect cooling system materials such as cast iron, aluminum, copper, brass, solder, steel, and many nonmetallics such as nylon and silicone. Type I conventional antifreeze uses older additive technology and contains corrosion inhibitors such as borate, phosphate, and sodium silicate. These types of inhibitors are known as inorganic additive technology (IAT) and are identified by a green dye. Sodium silicate is used primarily to protect aluminum and is found in a much higher concentration when used in gasoline engines that have aluminum cylinder heads. A problem with sodium silicate in diesel engines, however, is the incompatibility of silicate with nitrite, which is added to protect against cylinder wall cavitation erosion. High concentrations of both these additives together will cause a condition known as silicate dropout. Dropout has a characteristic appearance of green "goo" or gel, which plugs radiator and heater core tubes, causing the engine to overheat and even catastrophic failure (**FIGURE 7-11**). For this reason, Type I antifreeze is blended differently for diesel engines than it is for gasoline engines. Foam suppressants are added to maintain coolant contact with engine parts since steam and other vapor bubbles do not transfer heat well.

Type I additives deplete quickly and need to be monitored. Additives, such as phosphorous, drop out with silicate and coat cooling system components with a layer of slime-like deposits. It is important to drain and replace Type I antifreeze that contains this corrosion inhibitor package every two years (**TABLE 7-3**). To maintain the correct level of additives, the concentration of nitrite in Type I coolant needs to be tested, typically using test strips, to ensure the correct levels of nitrite and other inhibitors are present. Between 3% and 5% of DCA or SCA are added after the initial mixing of antifreeze and water to bring the additive package for diesel engines to the correct protection level (**FIGURE 7-12**). Monitoring the additive package is necessary every time the engine is serviced, to maintain the correct levels of the various corrosion inhibitors.

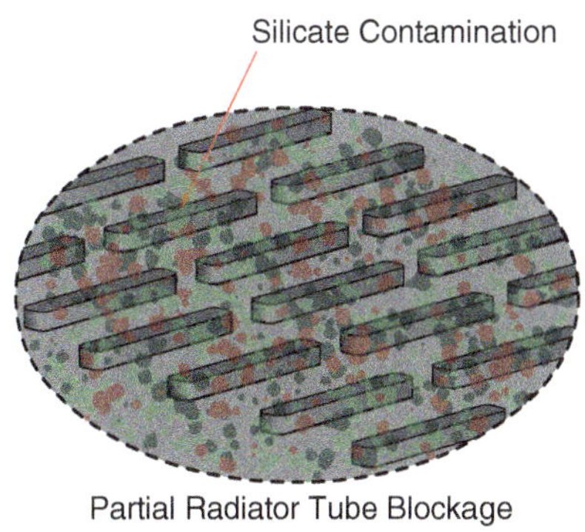

FIGURE 7-11 Silicate dropout forms a green gooey substance in the cooling system that blocks coolant tubes in radiators and heater cores.

TABLE 7-3 Type I Antifreeze Inhibitor Additives

Type I Antifreeze Inhibitor	Protects
Phosphate	Iron pH control (keeps coolant alkaline between 8.5 and 10.5 pH)
Borate	Iron pH control (keeps coolant alkaline)
Sodium Silicate	Aluminum
Nitrite	Cast iron Steel Aluminum Solder
Mercaptobenzothiazole (MBT) and Tolytriazole (TT)	Copper Brass

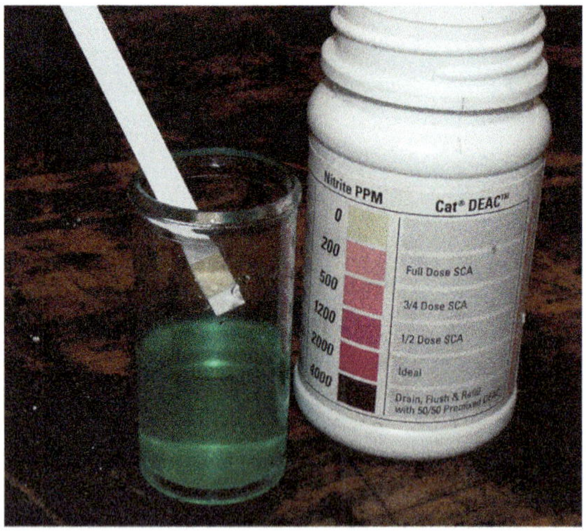

FIGURE 7-12 This test strip checks the nitrite levels in either Type I or extended-life coolant. A different scale on the side of the bottle is used for each type of antifreeze.

Extended-Life Antifreeze

Given the high maintenance requirements of Type I antifreeze and its short two-year service interval, manufacturers have sought to develop new coolant technology to minimize cooling system maintenance and extend the service interval, called extended-life coolant (ELC). To achieve this, a new category of antifreezes has been developed with corrosion inhibitors made from organic acids. Organic acids are different from inorganic acids in that they contain carbon. Referred to as organic acid technology (OAT), these inhibitors are generally low or free of nitrite, nitrate, phosphate, silicate, and borate (**TABLE 7-4**). OAT antifreeze has the advantage of being more chemically stable, which allows for exceptionally long service intervals. The formulation provides maximum protection of the six basic metal alloys found in most cooling systems. Since the coolant generally contains no phosphates or silicates, deposits in the cooling system are almost eliminated. The low level of abrasive solids dissolved in antifreeze results in improved water pump seal life. Antifreezes using this coolant chemistry are categorized also as long life, OAT, hybrid organic technology, and nitrited organic acid technology. Variations in the inhibitor package differentiate between these types of antifreezes.

To maintain these fully formulated antifreezes at every preventive maintenance inspection or at least twice per year, the color and freeze point and nitrite level of the coolant is checked.

> ▶ **TECHNICIAN TIP**
>
> Sensors and other electrical devices exposed to engine coolant have the potential to short out through the coolant. Electrical current passing though the coolant will quickly corrode brass and aluminum oil coolers, injector tubes, and other metals in the cooling system. To minimize this problem, engine blocks use a number of ground straps not only to lower resistances for engine-mounted electrical devices but also to minimize electrical conduction through engine coolant. Electrical conduction through coolant quickly eliminates nitrite additives in coolant, leading to further engine damage.

Type III: Low Toxicity Antifreeze

Type III antifreeze is used mostly in off-road applications due to its low toxicity. In the event of a coolant leak in off-road situations, PG's low toxicity will not contaminate soil to the extent that EG would (**FIGURE 7-13**).

Type IV: OAT Antifreeze. The OAT-type antifreeze uses non-carboxylate acids, such as benzoate, from benzoic acid, to form the additive package. This red antifreeze is commonly used in today's light- and heavy-duty diesel and gasoline engine.

Hybrid Organic Acid Technology

HOAT is a combination of IAT and OAT with nitrites added. This makes HOAT suitable for use in both light- and heavy-duty systems. Versions of this antifreeze are dyed between orange and a straw-yellow color. This antifreeze is often marketed as universal antifreeze.

Nitrated Organic Acid Technology

NOAT antifreeze is similar to HOAT. NOAT is an OAT with nitrates added. This chemical combination makes it suitable for use in both light- and medium-duty diesel cooling systems and is often marketed as a universal antifreeze. NOAT and HOAT are very similar in performance characteristics. Mixing IAT with OAT, HOAT, or NOAT antifreeze will not damage the engine's cooling system, but it will nullify the extended-life characteristics of these formulations. Depending on the type of formulation, acceptable contamination

TABLE 7-4 Type II Antifreeze Additives and Their Purposes

Type II Antifreeze Inhibitor	Protects
Potassium Soap of Dibasic Carboxylic Acid	Iron Solder Aluminum
Potassium Soap of Monobasic Carboxylic Acid	Aluminum Iron
Nitrite	Cast iron Steel
Molybdate	Iron
Tolytriazole	Copper

can range from 10% to 25%. Because NOAT can easily be confused with the variety and complexity of antifreeze formulations available for use, the technician and diesel engine owner must be fully aware of what the vehicle manufacturers' requirements are for antifreeze and follow those recommendations carefully. Since ELC additives can deplete with time, bottled SCA additives, called extender, which are specifically formulated for different types of OAT antifreeze, can be added to lengthen the coolant change interval by six to eight years.

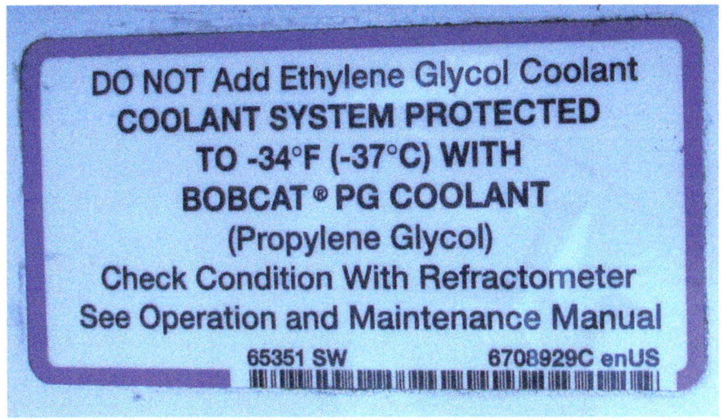

FIGURE 7-13 PG is a nontoxic alternative to EG-based antifreeze.

Radiators

Radiators are heat-exchanging devices that release heat absorbed from the engine to the air. Coolant can flow through the radiator in two ways. In down flow radiators, the coolant will enter through a hose connected to the top tank that is carrying hot coolant from the water pump outlet. The coolant will then flow from the top tank, through cooling tubes, and into the bottom tank. Alternatively, in cross flow radiators, coolant will flow from a top-connected hose, through a side-mounted tank, into cooling tubes, and across the radiator to another side-mounted tank before reaching a bottom-mounted hose (**FIGURE 7-14**). Cross flow radiators are preferred in vehicles with low-profile aerodynamic hoods because they can be made wider with a lower height profile.

Tubes inside the radiator core provide the surface area to exchange heat. Heat dissipates from the coolant through the tube wall and then through the fins. Air passing through the fins carries away heat, thereby allowing tubes and fins to absorb more heat from the coolant (**FIGURE 7-15**).

Early radiator construction used copper brass radiator components. Copper brass construction seemed the obvious choice for the radiators because of its superior heat conductivity, how easy it is to form, and how easy it is to repair. However, it was learned that increasing

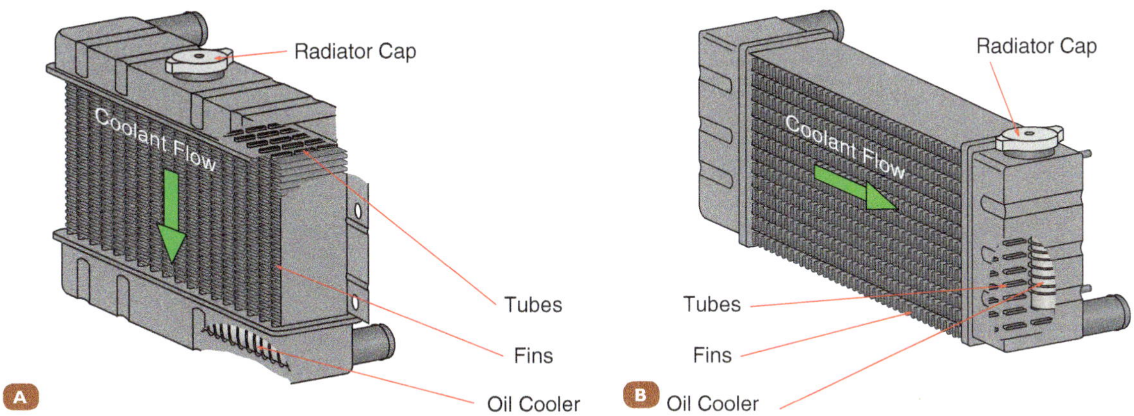

FIGURE 7-14 **A.** A down flow radiator. **B.** A cross flow radiator.

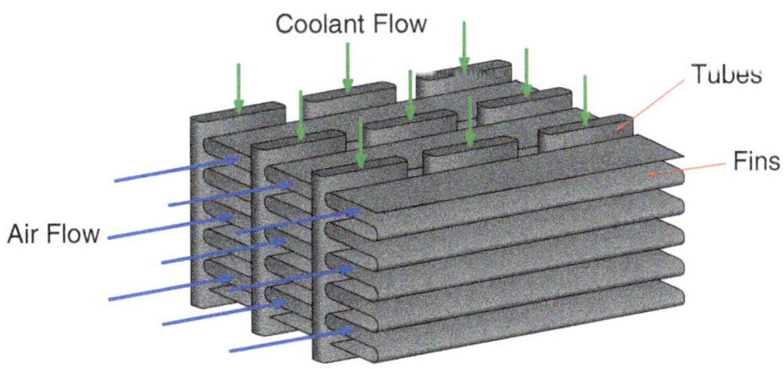

FIGURE 7-15 A tube-and-fin construction of a radiator.

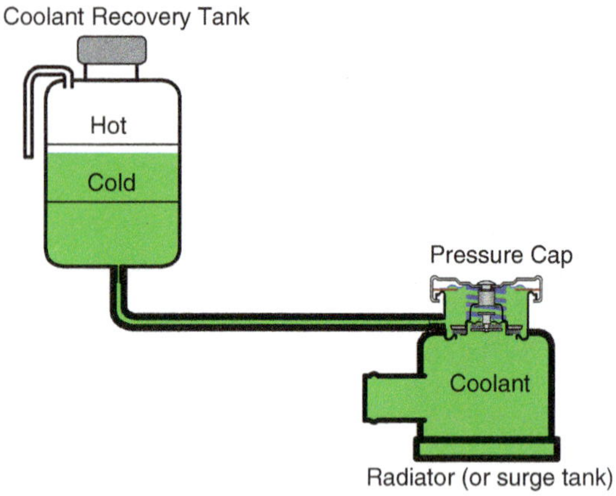

FIGURE 7-16 When coolant is heated, it expands. The overflow reservoir or surge tank collects coolant leaving the cooling system through the overflow pipe below the neck of the radiator cap. After the engine cools, coolant is drawn back into the cooling system.

the diameter of tubes in a radiator made radiators more efficient. This required thicker tube walls to provide mechanical strength but radiator construction techniques using copper and brass prevented its use in improved radiator designs.

Copper is a good heat conductor but the solder required to bond the tubes to the fins creates an insulation point that prevents some heat transfer. Another problem is all the mechanical stresses are borne by the solder in the joints between the radiator tank and the tube core. Solder can be attacked if the coolant pH level becomes too alkaline through the excessive use of DCA/SCA. Today, the preferred method to construct a radiator core is to use aluminum tubes, which transfer heat well. Aluminum tubes are welded or crimped rather than soldered to the aluminum or plastic tanks. This type of joint provides a more efficient conductor for transferring.

Surge Tanks

Additional volume is necessary for the cooling system coolant so that the coolant can expand. Surge or overflow tanks provide the space for coolant and vapor to move into when the engine coolant is hot. When the engine cools, the stored overflow volume returns to the radiator. Sometimes, large radiator top tanks will perform the same function as a separate reservoir. Overflow tanks use a line connected to the radiator just below the pressure cap, allowing coolant to move back and forth during the cooling system's thermal cycles (**FIGURE 7-16**).

Surge and overflow tanks provide a coolant reserve for the gradual loss of coolant. Most of the leakage originates from radiator hose clamps. Constant torque spring clamps are the best clamps to use to minimize coolant loss, because they change dimensionally with the expansion and contraction of hoses and connections.

SAFETY TIP

Never open the radiator cap of a hot engine: the coolant will be depressurized and boil over. The sudden conversion of hot water into steam pressure in the engine will cause water to explosively erupt from the radiator with projectile force. Anyone near the radiator can be scalded.

Radiator Caps

Whereas early radiator caps simply prevented coolant from spilling out of the cooling system, today's radiator caps have several additional functions. First, the calibrated pressure valve inside the radiator cap increases the pressure of the cooling system to raise the boiling point of coolant. This feature prevents coolant loss from boil-over (**FIGURE 7-17**).

Second, a vacuum valve incorporated into the cap allows air or coolant to reenter the cooling system when the pressure drops after the engine cools (**FIGURE 7-18**). Faulty

FIGURE 7-17 An example of the pressure rating of a radiator cap at which the pressure-relief valve opens is stamped on the cap: 15 psi (105 kPa).

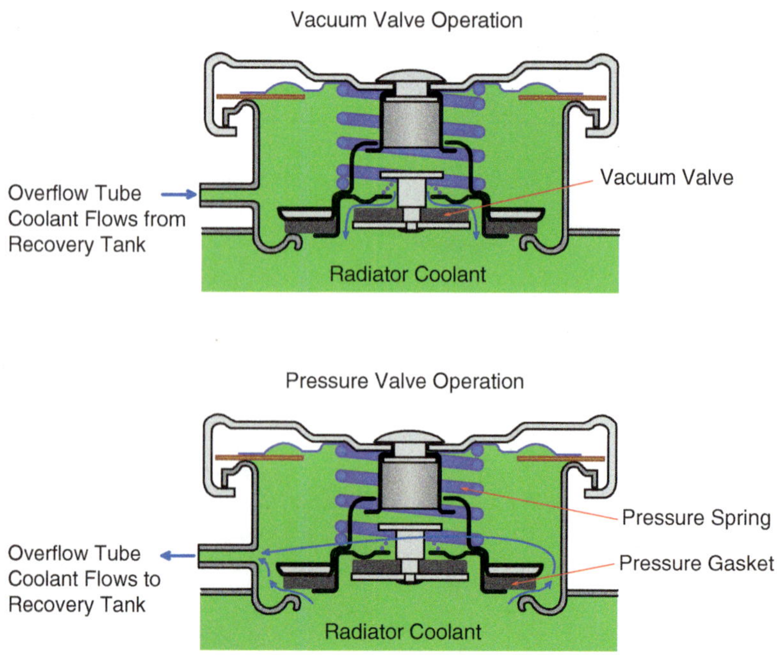

FIGURE 7-18 Operation of the pressure-relief function and vacuum valve operation of a radiator cap.

vacuum valves are often identified when the upper radiator hose collapses after the engine cools or when coolant is lost. A coiled spring inside the lower radiator hose, or a solid metal tube, is required to prevent it from collapsing during high engine speeds, when accelerating due to negative pressure at the water pump inlet.

The radiator cap is also important for water pump efficiency. When the cooling system is pressurized, the pump operates more efficiently than the 85% efficiency without pressure cap. Finally, pressurizing the cooling system minimizes the degree to which the entire cooling system is cavitated, because vapor bubbles caused by low pressure in the cooling system are less likely to form.

Water Pumps

To ensure that the coolant is positively circulated, an engine-driven water pump moves coolant through the cooling system. Pumps can be belt- or gear-driven. Although a gear-driven pump does not rely on a belt, which minimizes service requirements and is more reliable than a belt for long distances, if the weep hole fails to drain coolant leaking past a defective water pump shaft seal, a gear driven pump is more likely to allow coolant to enter the engine. All water pumps are centrifugal turbine pumps, which means they become more efficient the faster they turn. The inlet to the water pump is the lower radiator hose or some point near the lower radiator hose. Water drawn into the pump is discharged into the engine. Under fast acceleration, internal engine block pressures between the pump and thermostat can reach as high as 40 psi (276 kPa). To reduce energy losses from driving water pumps, manufacturers are introducing electrically controlled variable speed water pumps to help reduce fuel consumption to meet greenhouse gas (GHG) targets (**FIGURE 7-19**). A clutch that controls the speed of the water pump is signaled by the electronic control module (ECM) by using a pulse-width–modulated (PWM) electrical signal. A speed sensor on the pump provides data for closed-loop feedback control of the water pump speed.

The impeller, which transfers the pump's energy to the coolant, is located in the coolant and may be made of plastic or metal (**FIGURE 7-20**). Impellers can wear if a significant amount of abrasive circulates through the cooling system. Impellers can also become separated from the shaft and cause poor coolant flow. This problem is usually identified when the engine overheats but the lower radiator hose, which is normally hot to the touch, is instead cold or lukewarm because radiator coolant is not being pulled by the pump into the engine.

A special seal separates coolant from the bearings supporting the water pump shaft (**FIGURE 7-21**). If coolant reaches the bearings, the water pump will suffer a major failure. To prevent this from happening, manufacturers add a weep hole between the seal and the bearing (**FIGURE 7-22**). If the seal is worn and damaged, the weep hole allows coolant to leak from the seal cavity inside the water pump to prevent bearing damage. Some coolant leakage from a weep hole is normal, and some water pumps hold any small leakage of coolant in a small reservoir, where it evaporates before leaking out.

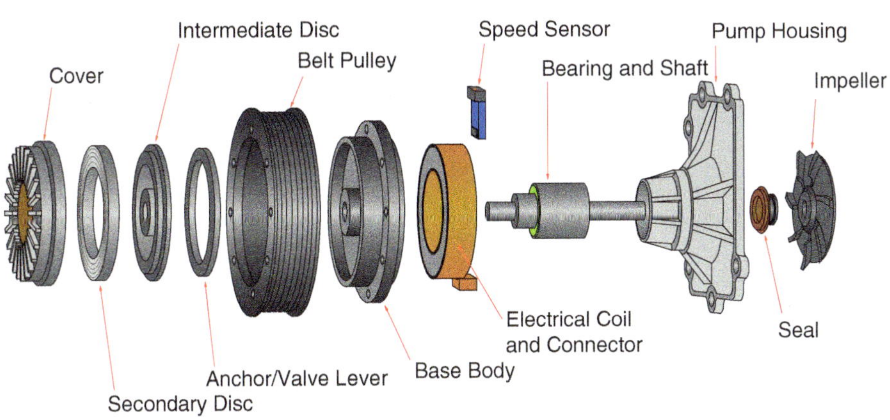

FIGURE 7-19 An electronically controlled, variable speed water pump reduces engine drag when coolant flow is minimal.

FIGURE 7-20 A centrifugal metal water pump impeller.

▶ TECHNICIAN TIP

A coolant leak from the weep hole of a water pump indicates that the water pump seal is leaking. The weep hole prevents coolant reaching the water pump shaft's support bearing. If coolant were to reach the bearing, a catastrophic bearing failure could occur. On gear-driven pumps, the weep hole prevents coolant from entering the engine and mixing with lubrication oil. When inspecting an engine for leaks, or when pressure testing a cooling system, keep in mind that a small amount of coolant leakage from a pump is acceptable, because the seal cannot form a perfect liquid-tight barrier. To pressure test the cooling system and check for leaks, follow the steps in **SKILL DRILL 7-1**.

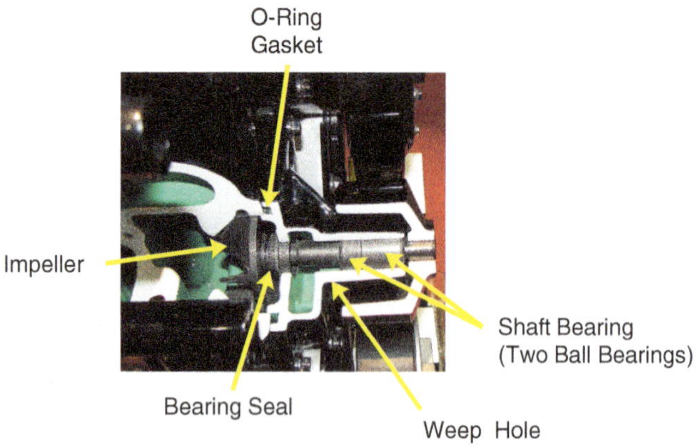

FIGURE 7-21 The water pump seal prevents coolant from contacting the pump shaft bearings.

FIGURE 7-22 The water's pump weep hole on a 6.5 L GM diesel. Some weeping from the hole is permissible but should not drip when the cooling system is pressurized.

> **▶ TECHNICIAN TIP**
>
> Silicate, phosphates, and borates used as corrosion inhibitors are dissolved solids. In time, they crystallize and fall out, or precipitate out, of the glycol and water solution. These crystals can enter the area between the water pump shaft and seal, acting as an abrasive and causing the seal to wear prematurely. The result is a coolant leak from the weep hole of the water pump.

Adding cold water to hot engine operating with low coolant will thermally shock the engine and damage ceramic or sintered metal water pump seals. Cold coolant added to a hot engine can also crack in all of the engine castings. If an engine cannot be left to cool down before adding water, it should be added only when the engine is running.

Thermostats

Thermostats help accelerate engine warm-up and regulate the temperature of the engine coolant by modulating or controlling the volume of coolant circulated to the radiator. Accelerating warm-up is particularly important for direct-injection (DI) combustion chamber engines since much less heat is transferred to the cooling system than in spark-ignition (SI) engines. A prolonged warm-up period is required at idle speed, and the engine will not reach operating temperature unless some auxiliary device is used to help transfer more heat to the engine coolant. Exhaust back pressure regulators are one such device. Exhaust back pressure regulators restrict exhaust flow out of the cylinders at idle, causing combustion temperatures to increase and higher exhaust temperatures to linger at the exhaust ports of the engine. The coolant thermostat speeds up warm-up when the engine is below its normal operating temperature, by blocking coolant flow from the engine to the radiator. At a predetermined point—that is, 180°F (82°C)—the thermostat will begin to open, and it will continue to open further as the

SKILL DRILL 7-1 Pressure Testing the Cooling System and Checking for Leaks

1. With the engine off, completely fill the engine coolant level with premixed coolant.
2. With the engine off, connect the radiator pressure tester by replacing the pressure cap with the pressure testing adapter. Larger caps, which are found on heavy-duty cooling systems, or unique original equipment manufacturer (OEM) pressure cap designs may not fit with a standard pressure cap tester. An adapter may be required to seal the tool to the neck of the surge tank or radiator. Pressurize the system to 15–17 psi (103–117 kPa).
3. Observe the pressure readings for 10 minutes. During this time, visually inspect the cooling system hoses, all hose connections, auxiliary coolant heaters, cab heaters, EGR coolers, EGR valves, turbocharger coolant lines, the water pump weep hole, all

Continued

gasket sealing surfaces, and any other components that have coolant flow.

4. Check the engine oil level and verify that the oil has a normal color. Oil that looks milky likely has coolant leaking into it.
5. Inspect the seals on the pressure cap and pressure test the cap. The seals should be intact and in good condition without any cracks. The same pressure tester used for the engine's cooling system is used to test the cap. An adapter is placed between the cap and test tool and the tester is pumped once again. The cap's release pressure should be indicated on the cap, and it should hold pressure until the tester applies the amount of pressure specified on the cap.
6. Refer to the OEM service information to check whether the pressure cap is the correct pressure rating for the vehicle. If the cap will not hold or cannot reach specified pressure, or if it does not vent at the specified pressure, then replace it with a new cap, one that has the correct pressure rating.
7. Record and report the results.
8. If there is a significant drop in pressure and the complaint investigated is coolant loss, the engine may have an internal coolant leak. The engine should be first checked to determine whether the cooling system is excessively pressurized during operation by using a separate procedure. The EGR cooler should also be inspected by using a separate procedure.
9. To check for internal leaks, drain the engine oil and remove the oil pan.
10. While the cooling system continues to be pressurized, observe whether coolant is internally leaking or if coolant is originating from a head gasket leak. In push tube engines, head gasket leaks will drip coolant from the cam side of the engine.

coolant temperature continues to increase: 5–10°F (2–8°C) later, the thermostat will be fully open (**FIGURE 7-23**).

A variety of thermostat designs are commonly used in diesel engines. Choking-type thermostats are most commonly used in light- and medium-duty engines. These thermostats have a valve that opens and closes to regulate coolant flow through the engine; two thermostats are used in larger engines with higher coolant flow. Choking-type thermostats operate by using a special wax pellet combined with a powdered metal tightly squeezed into a copper cup that is equipped with a piston inside a rubber boot (**FIGURE 7-24**). Heat causes the wax pellet to expand, which pushes the piston upward, opening the valve (**FIGURE 7-25**). In engines not equipped with vent lines, a small jiggle or venting valve is located in a chocking thermostat to allow air and steam to escape past the thermostat and into the radiator (**FIGURE 7-26**).

Faulty thermostats will cause the engine to overcool by staying partially open. An engine that will not warm up to operating temperatures, is very slow to warm up, or warms only under heavy load likely has a thermostat that is stuck partially open. Failsafe mechanisms

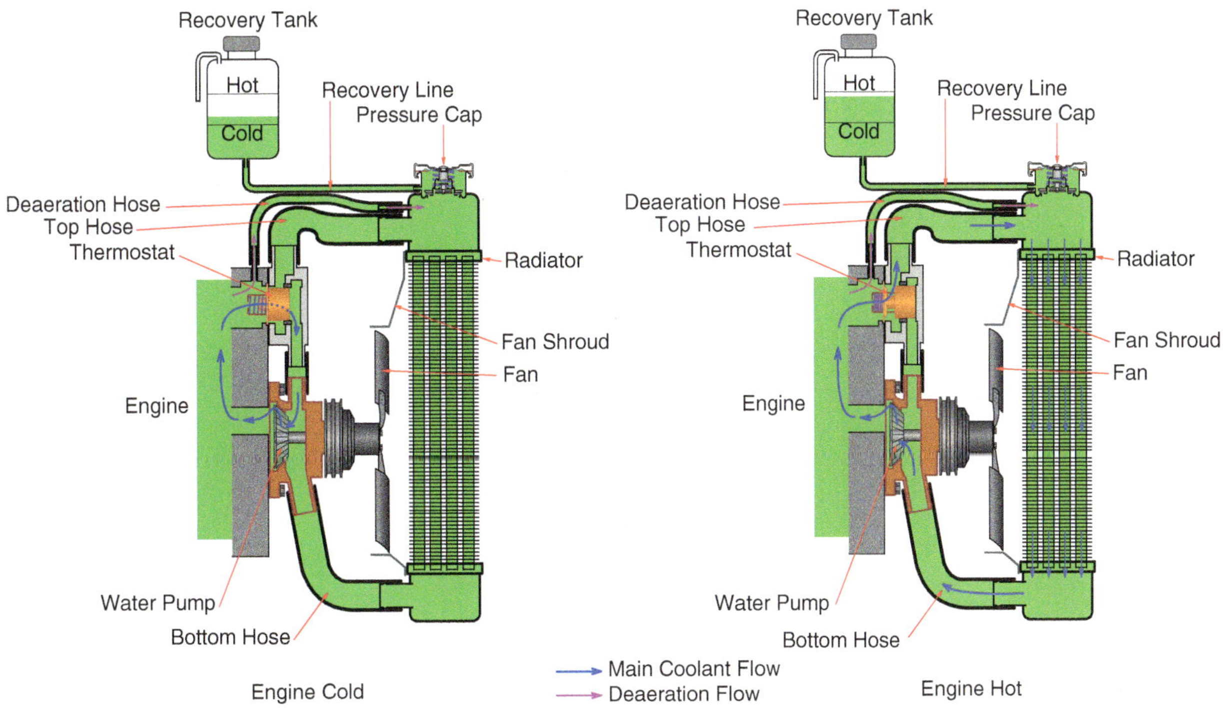

FIGURE 7-23 Coolant flow through a cooling system when using a full blocking thermostat.

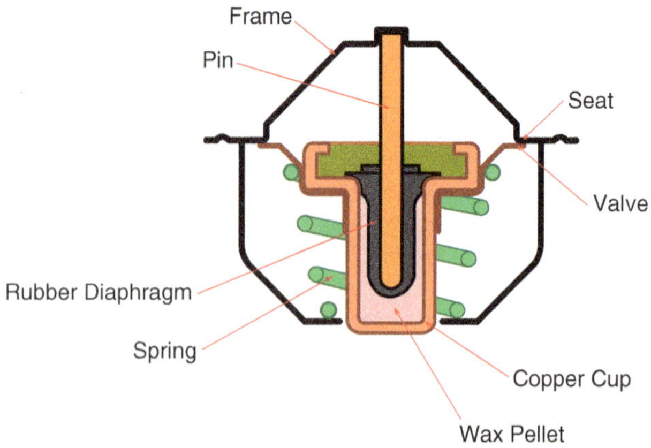

FIGURE 7-24 Cross section of a choking thermostat.

built into the wax pellet of contemporary thermostats prevent the thermostat from sticking closed, causing catastrophic engine damage. When overheated, the temperatures that the thermostat is exposed to will lock in the completely open position, but it still must be replaced because it cannot close again.

Fans and Fan Drives

Engine fans are designed to increase the airflow across the radiator, which improves the efficiency of heat removal from coolant. The number of blades, the fan speed, and the fan pitch will determine the volume of air moved by the fan. For most vehicles, engine fans pull air through the radiator in the same direction as airflow when traveling in a forward direction. When the vehicle is traveling, its movement usually creates enough airflow, so the fan is not required. At low-speed operation and at idle, air movement is often inadequate, and a fan is required for the cooling system and to move air across the air conditioning condenser and any other front-mounted heat exchangers. To reduce the parasitic loss of power from the engine, a fan drive mechanism is used to decouple or reduce the fan speed when airflow is not required.

Viscous Fan Clutches

One of several common methods of controlling fan operation is through a temperature-controlled modulating fan clutch. These fans are coupled to the drive through a highly viscous silicone fluid. The degree to which the clutch slips can be controlled by several mechanisms, including a temperature-sensitive bimetal control valve regulating the volume of silicone fluid in the clutch coupling. When the clutch is cool, little silicone fluid enters the clutch and the fan freewheels. As the temperature of air in the fan clutch area increases, more fluid enters the clutch mechanism, which creates more drag or silicone fluid shear within the coupling mechanism. Because the fan speed does not correspond to engine speed and is considered to be modulated or has a heat-dependent speed adjusted according to the heat load, it absorbs from air passing through the radiator (**FIGURE 7-27**). These clutches are quieter because fan speed is kept proportional to heat load. Because their operational speed is proportional to engine temperature, they conserve energy and extract significantly less energy from the engine compared to an engine with a directly coupled fan.

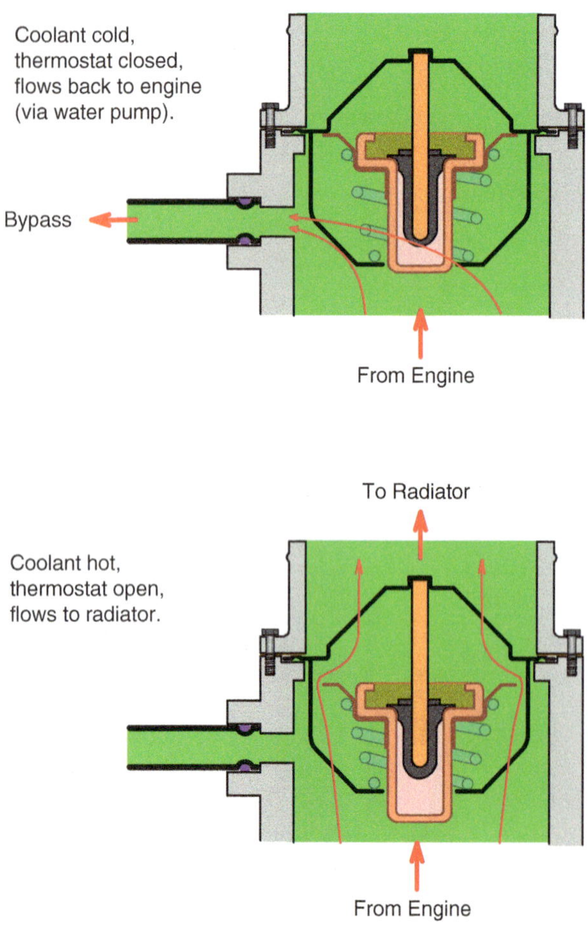

FIGURE 7-25 Operation of a choking engine thermostat to regulate engine temperature.

FIGURE 7-26 A jiggle valve in a choking thermostat enables trapped steam to vent through a closed thermostat.

FIGURE 7-27 This silicone cooling fan hub freewheels when the hub is cool. As more heat is absorbed by the hub, internal drag causes the fan speed to increase closer to the driven speed of the hub. Note the bimetal spring in the center of the hub, which is connected to a valve that controls fluid movement.

FIGURE 7-28 An electrically controlled variable speed engine fan reduces parasitic power loss and provides optimal ECM control of fan speed. The bimetal spring is replaced with an electrically controlled vavle.

Variable Speed Engine Fan

To reduce parasitic loss caused by operating a cooling fan, many late-model engines use a variable speed cooling fan. The fan is controlled by the ECM by using a PWM signal to regulate the fan speed. An electrically operated control valve is used instead of the bimetallic heat-sensing spring in conventional viscous cooling fans. Slower fan speed also translates to less engine noise (**FIGURE 7-28** and **FIGURE 7-29**).

FIGURE 7-29 A variable speed engine fan connects the inner and outer hub by shearing silicone fluid between discs.

▶ Construction and Functions of a Lubrication System

The lubrication systems of diesel engines are designed to perform the following tasks:

- reduce friction between moving parts (**FIGURE 7-30**)
- cool internal engine parts such as pistons
- remove dirt, abrasives, and contaminants from inside the engine
- assist in sealing the combustion chamber by forming a film between the piston rings and the cylinder wall
- absorb shock loads between bearings and gears to protect engine parts while reducing engine noise
- store an adequate supply of oil
- minimize corrosion on internal engine components (**FIGURE 7-31**).

7-03 Identify and describe the functions of a lubrication system.

Emission legislation and durability requirements impose new demands on the lubrication system too. Oils now need to be compatible with the latest exhaust aftertreatment systems and biofuels while meeting customers' expectations for longer oil drain intervals

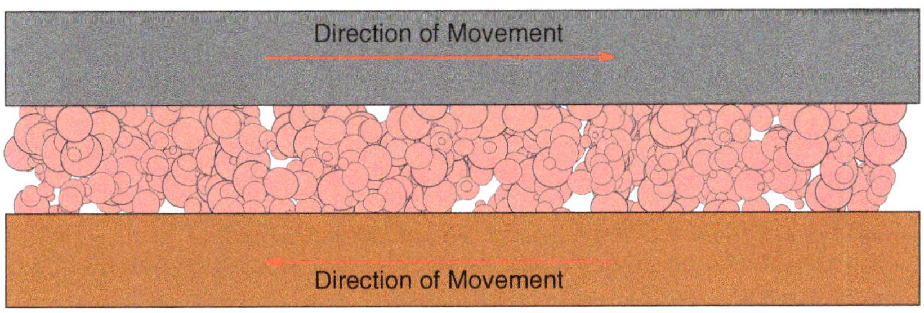

FIGURE 7-30 Oil lubrication behaves like ball bearings, enabling metal surfaces to slide over one another without metal-to-metal contact.

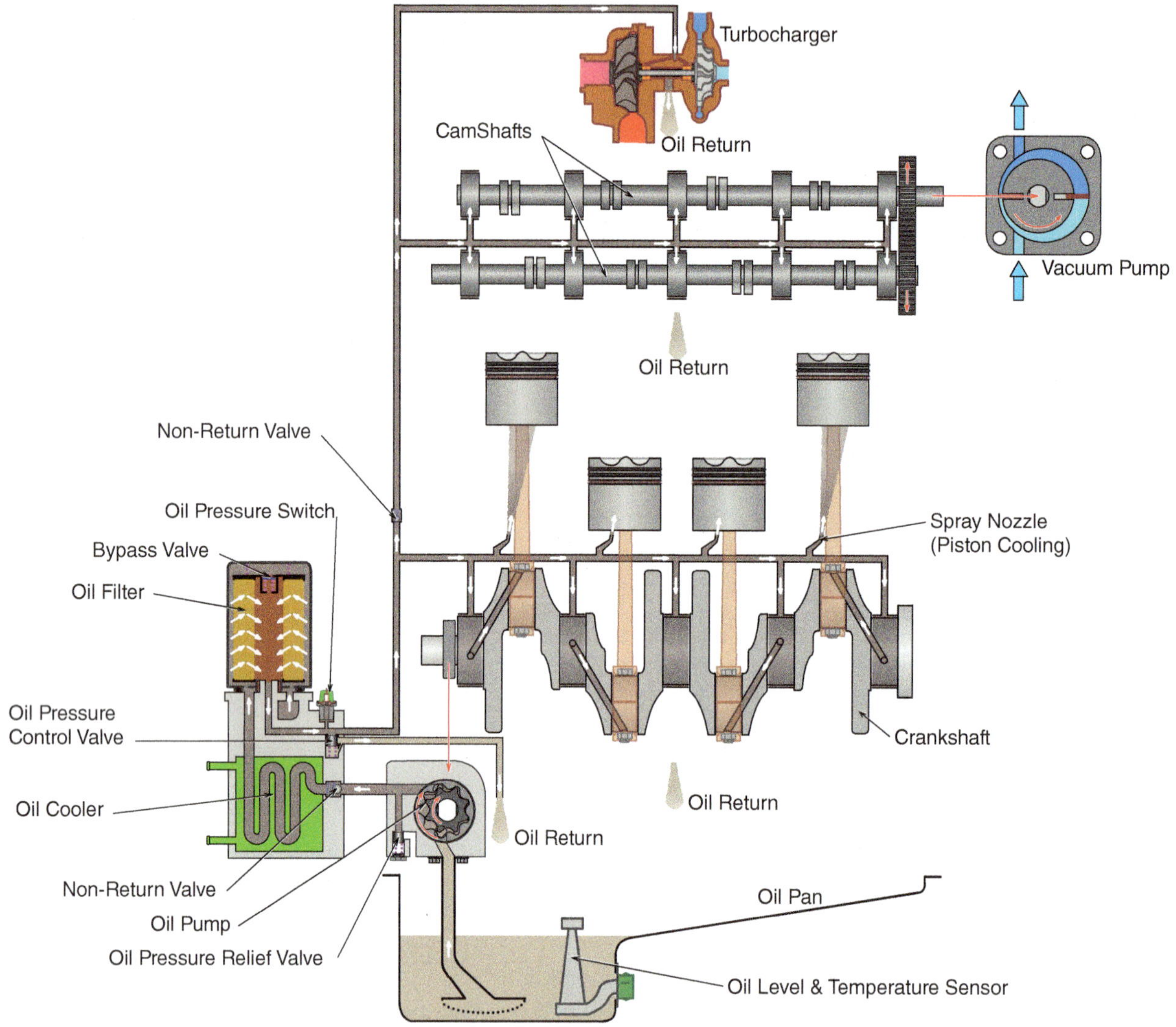

FIGURE 7-31 Oil circuits and the placement of lubrication system components in an in-line four-cylinder turbocharged diesel engine.

to reduce operating costs. The following major lubrication system components function to meet these demands.

- engine oil
- oil pump
- oil pan
- oil cooler
- oil filter(s)
- oil-pressure regulating and pressure-relief valves
- oil level dipstick.

▶ Engine Oil Classification

7-04 Identify and explain the classification systems for engine oil according to oil properties and characteristics.

Performance standards for oil are established by the American Petroleum Institute (API) and the Society of Automotive Engineers (SAE). The SAE establishes standards for measuring oil viscosity, called the SAE viscosity rating, and the API sets the remaining standards. To be approved for the operating conditions and in use protecting unique technologies of various manufacturers, engine oils classified by the API will carry the API designation, indicating their suitability for an engine application. For example, engines that use high-speed turbochargers and that recirculate exhaust gases into the combustion chamber will place different demands on oil than oil used to lubricate a naturally aspirated engine with little or no emission controls. Two distinct classifications of oils are recognized by the API. S series oils are designed to meet performance standards for spark igntion (SI) engines SI; C series oils are designated for use in compression-ignition engines. The C and S series oils are distinctly formulated due to the combustion by-products of gasoline and diesel fuel combustion. The API donut-shaped service symbol indicates which standards an oil meets. Engine oil carrying the API service symbol is certified only after specialized testing to determine whether it meets minimum API standards (**FIGURE 7-32**).

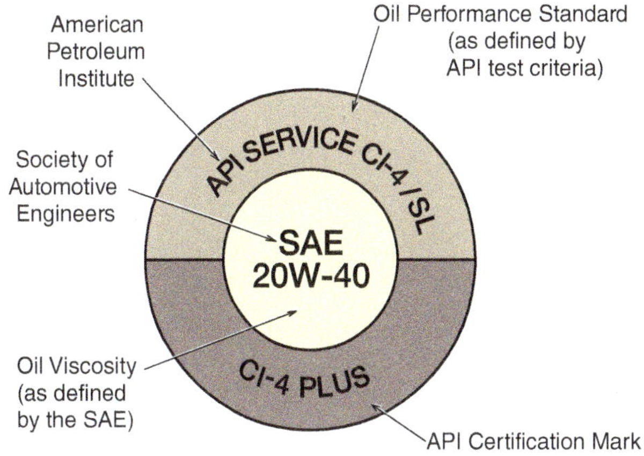

FIGURE 7-32 1. The top half designates the oil's performance standard set by the API test criteria. 2. The center identifies the oil's viscosity, which is a standard defined by the SAE. 3. The bottom half is the API certification mark. Separate API designations are used for gasoline engines and diesel engines. Gasoline engines may carry a starburst symbol, whereas diesels do not. 4. The shaded FA-4 designation helps distinguish the oil from CK-4 oils.

API Certification Mark for Diesels

API certification marks for diesel engine oils meeting the API CK-4 and the new FA-4 standards are the latest standards for diesel oil introduced in 2016. CK-4 is designed for compatibility with biofuels and the latest exhaust aftertreatment systems. It is backward compatible with CJ-4 and CI-4 oil. CJ-4 was the previous oil standard, which was developed for engines to meet the 2010 model year EPA (Environmental Protection Agency) Tier 5 bin 2 emission standards. These engines typically used EGR diesel particulate filters (DPF) and ultra-low-sulfur diesel (ULSD) fuel. Since combustion by-products enter the oil from cylinder blowby, engines that use ULSD and EGR need oils capable of holding more soot produced by the EGR containing less acidity derived from burning fuel with almost no sulfur. By-products of oil burned in the cylinders will end up in the particulate filter, where oil additives leave ash residues that must be physically cleaned out of the filter. The 2010 and later oils produce minimal ash residues, which will block exhaust gas flow through a DPF and cannot be removed by regeneration. Substances capable of deactivating catalyst material are removed from this category of oil.

FA-4 oils are a new category for use in high-speed four-stroke cycle diesel engines designed to provide better engine protection and help meet the 2017 model year on-highway GHG and Tier 4 non-road exhaust emission standards. The "F" in FA-4 designates the oil as a fuel-efficient oil, which is expected to reduce fuel consumption by up to 1.6%. A future oil would be designated FB-4. The new oil represents a large divergence in oil classifications to date and is not necessarily backward compatible with any CJ-4 or CI-4 oils. Because these new oils have much lower viscosity (are much thinner) to help reduce parasitic power losses caused by pumping and shearing thicker oil, they could potentially damage an engine that needs heavier, more viscous oil film. The term "high-temperature, high-shear (HTHS) viscosity" refers to FA-4 oils' energy consumption–reducing properties,

> ▶ **TECHNICIAN TIP**
>
> Diesel engine oil turns darker more quickly than oils used in gasoline engines. Soot produced during diesel combustion will cling to the oil film on the cylinder walls and is scraped into the oil pan. Soot, which is composed primarily of black carbon, turns engine oil black. Newer engines with finer fuel atomization and engines operating consistently at high load and speed conditions will take much longer to turn dark since less soot is produced.

> ▶ **TECHNICIAN TIP**
>
> Combustion by-products such as soot and other residues stick to oil on the surface of the cylinder wall, where they are scraped into the oil pan. Using EGR can introduce more soot into the oil and requires oil to be better able to dissolve and hold soot in the lubricating oil until it can be removed by the filter. Extra detergents in engine oil cause soot to stay in suspension better than oils with little or no detergent. Note too that excessive idling and stop and start operations also contribute to higher levels of soot loading in engine oil. Soot-thickened oil contributes to accelerated engine wear since soot is an abrasive. Soot also combines with water to form sludge, and thickened oil increases engine cranking resistance when cold (**FIGURE 7-33**).

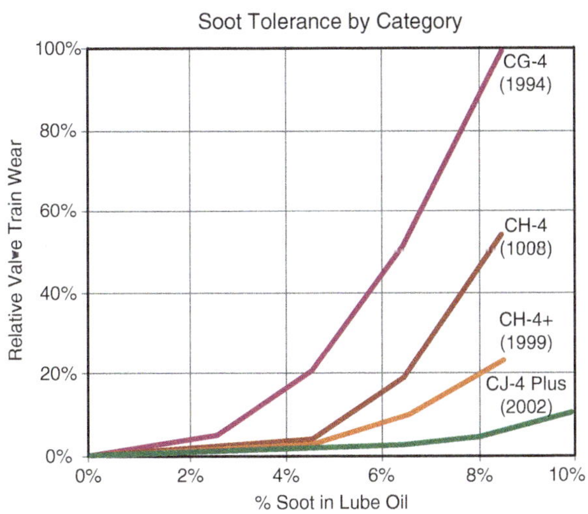

FIGURE 7-33 Diesel engine oils are formulated to tolerate the increased quantity of soot from diesel combustion. Retarded injection timing and using EGR contribute to higher soot levels in diesel engine lube oil. Adding extra detergents in oils formulated for diesels helps keep the soot suspended in lubricating oil, where it can be removed by the oil filter. Without the detergents, soot would settle and cover internal engine parts and form heavy sludge capable of blocking oil passageways for the lubrication system.

which help engines meet lower GHG targets for CO_2 emissions. More precisely, CK-4 oils are referred to as high HTHS engine oils, and FA-4 oils are being referred to as low HTHS oils. These oils will operate only in engines with relatively new hardened bearing shell technology, running with extraordinarily smaller oil clearances, which maintain higher oil film strength.

API FA-4 oils are also designed to operate in much hotter engine environments, which means that they need to be better able to prevent oil oxidation caused by high temperature, which leads to thickening. The oil can also minimize viscosity loss due to molecular shear—cutting up chains of oil molecules by bearings and shafts. Additional antifoaming agents prevent FA-4 oil aeration. And like previous oils, when consumed by the engine, the oils do not contain additives that can poison exhaust catalysts or block particulate filters.

Oil Viscosity

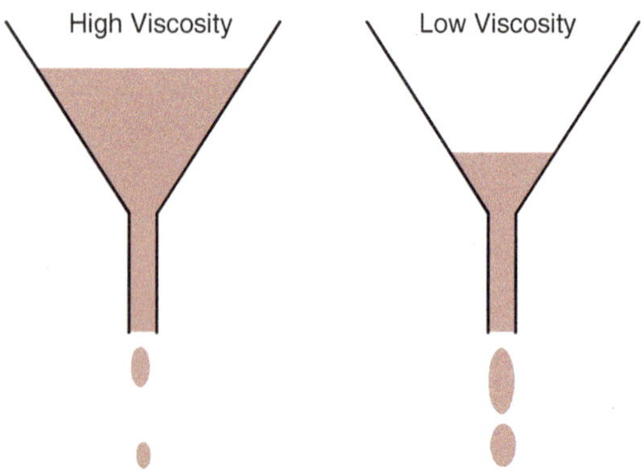

FIGURE 7-34 Viscosity is a liquid's resistance to flow. Low-viscosity oil flows more easily than high-viscosity oil.

The term "viscosity" refers to a liquid's resistance to flow (**FIGURE 7-34**). A numerical designation developed by the SAE is used to measure oil viscosity. The numbers indicate whether a particular oil flows easily or slowly. For example, engine oils that have a viscosity with low numbers such as 0, 5, or 10 flow more easily than viscosities of 20, 30, or 50. In other words, the higher the number, the thicker or more viscous the oil. Oil viscosity decreases when hot. Likewise, oil thickens and the viscosity increases as its temperature decreases (**FIGURE 7-35**). Oil viscosity measured at 0°F (−18°C) is given a W for a suffix, which is short form for winter viscosity rating. Oil viscosity measured at 212°F (100°C) is referred to as a hot or summer viscosity and no letter accompanies the viscosity grade. Oils with a single number are called a straight grade or single grade (**FIGURE 7-36**). Oil viscosity is indicated by two numbers (e.g., 15W and 40 in 15W-40), which means viscosity is measured under both hot and cold conditions. Referred to as multigrade or multiweight viscosity oils, these oils can be a blend of several different oils with different viscosities or blended oil with viscosity-changing additives. The numbers provide viscosity information

FIGURE 7-35 Temperature changes an oil's viscosity. Multigrade oils have a lower viscosity index than single-weight oils. That means multigrade oil viscosity does not change as dramatically with temperature.

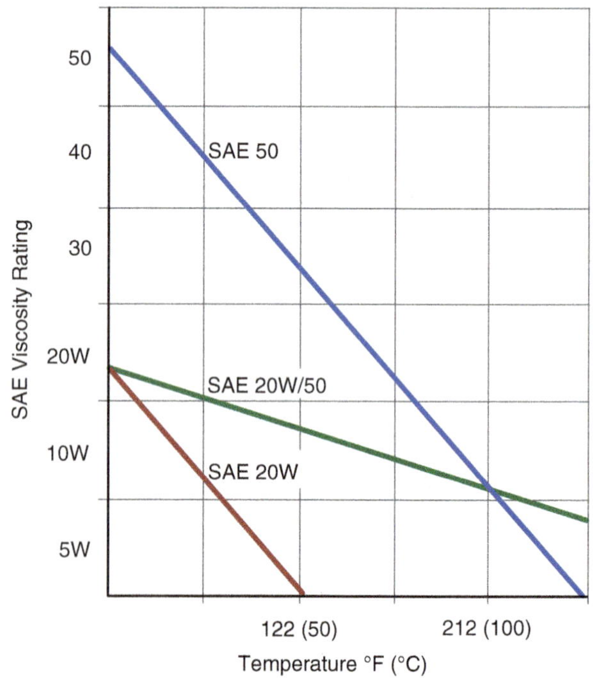

FIGURE 7-36 Temperature changes an oil's viscosity. Additives and blends of oils with different viscosities can alter the rate at which oils change viscosity with temperature.

about the oil's performance under a wide range of operating conditions. The low-temperature viscosity (the first number: e.g., 15W in 15W-40 oil) helps predict how well an engine might crank in cold temperatures when resistance to cranking is high due to thick oil. Generally, the lower number means the engine will start more easily in cold weather and flow more quickly to critical parts. The high-temperature viscosity (the second number: e.g., 40 in a 15W-40 oil) provides information about the oil's thickness or body, for good lubrication and load-bearing characteristics at normal engine operating temperatures. The main advantage of multigrade oils is improved cold-starting characteristics with less engine drag.

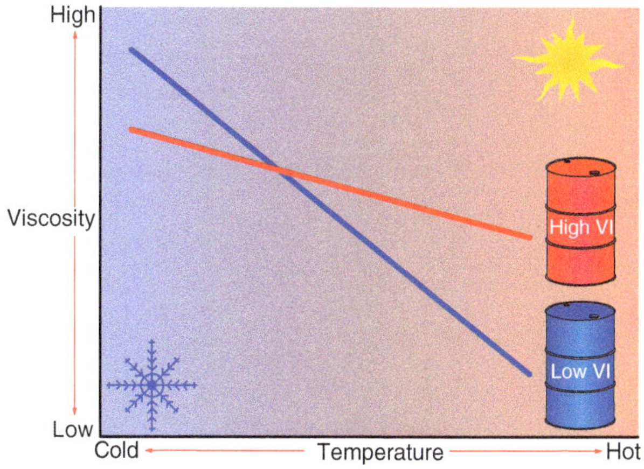

FIGURE 7-37 The VI of oil refers to how much its viscosity changes with temperature. A high VI changes little, whereas a low VI changes considerably. Ideal oils have a high VI.

Viscosity Index Change and the Viscosity Index Improver

Looking at the viscosity numbers of multigrade oil—e.g., 15W-40. It appears that the oil is thicker when hot and thinner when cold. However, this grade of oil is still thicker when cold than when hot, because a cold 15W winter viscosity scale measures viscosity differently than the summer scale. This means 40W is still thicker than a hot 40 weight and a 15W is more viscous than 15 hot. Another feature is built into multiweight oils to control viscosity change, called the viscosity index (VI) improver. The VI of an oil refers to how much the oil's viscosity changes with temperature. The viscosity of an oil with a low VI does not change much with the temperature, in contrast to an oil with a high VI (**FIGURE 7-37**). The VI improver is a special hydrocarbon polymer or, more simply, a molecular chain that changes shape with temperature (**FIGURE 7-38**). When cold, VI improver polymer chains will coil into a ball-like shape. This causes the oil to be less viscous than oil without the VI improver molecule. When heated, the polymer chain stretches into a long rod shape. This action tends to thicken the oil, counteracting the natural tendency of the oil to thin. When added to oil, the VI improver will counteract the tendency of oil to thin when hot and thicken when cold. Because oil in diesel engines gets very hot taking on the job of cooling piston crowns, this viscosity adaptation feature is especially important.

Eventually, VI additives wear out as the polymer chains are broken through the mechanical action of shearing the oil molecules. Moving oil through pumps and between bearings at high pressure and scraping oil from cylinder walls literally chop or shear molecular chains of VI improver, causing oil to lose VI improver properties. Shearing VI improvers explains why oil levels remain initially stable for a few thousand miles but suddenly drops with extended use. When VI improvers are sheared, they can no longer thicken oil when hot, and then the oil thins out. Thinner, less viscous oil will pass around piston rings and valve guides more easily and burn (**FIGURE 7-39**).

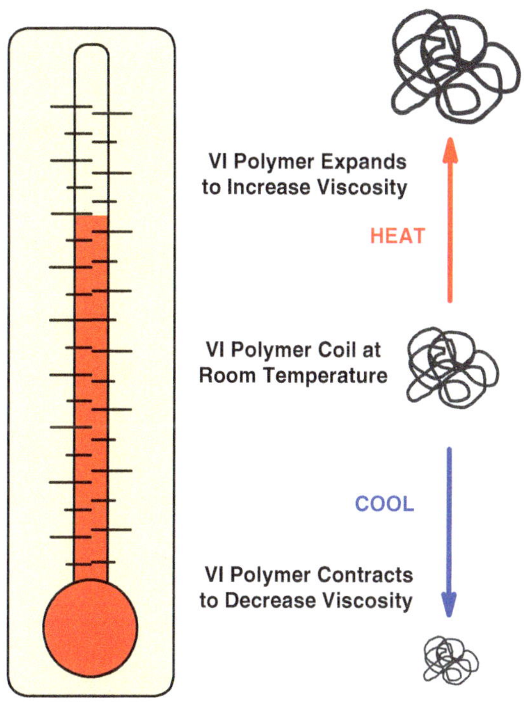

FIGURE 7-38 The volume of a VI improver molecule expands with temperature to thicken engine oil as its temperature increases.

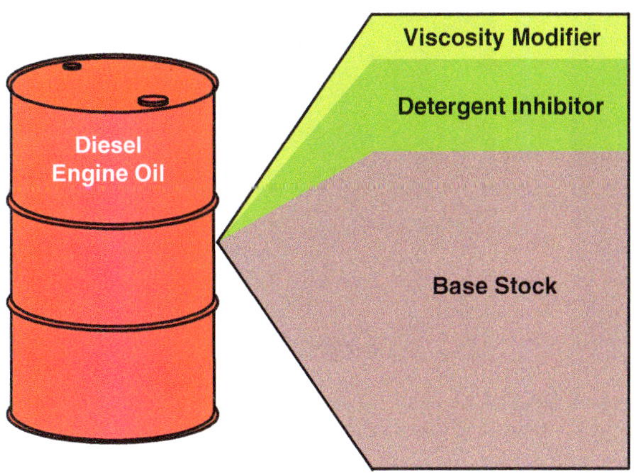

FIGURE 7-39 Relative values of viscosity improvers and detergents in diesel engine oil.

Common Oil Additives

Additives are blended with diesel engine oil to help it meet API standards. The following are some of the common additives:

- Pour point depressants help oil flow in cold conditions.
- Antiwear additives protect against metal-to-metal contact under high-pressure conditions, when oil viscosity is low.
- Detergents and dispersants are used to keep internal engine components clean and prevent the oil from soot thickening and sludging.
- Oxidation inhibitors are added to maintain oil viscosity stability between changes. (Oil reacted with oxygen in the air will thicken, particularly at higher temperatures, and form deposits.)
- Corrosion and rust inhibitors needed to protect engine parts against the effects of condensation and acid.
- Antifoaming additives—minimize oil foaming and the accompanying loss of system pressure due to oil becoming more elastic when pressurized.
- Emulsifiers are used to prevent water from combining with engine oil.

Reserve alkalinity additives are used to protect the engine from acidic oil caused by the products of combustion. When sulfur in diesel fuel combines with water vapor, they will produce acidic oil over an extended period of time. The total base number (TBN) is the measurement of a lubricant's reserve alkalinity, which aids in the control of acids forming during the combustion process. The higher an oil's TBN, the more effective it is at preventing the formation of wear-causing contaminants. The total acid number (TAN) refers to oils level of acidity, which can increase proportional to fuel consumption. TAN increases as TBN decreases.

Detergents and Diesel Engine Oil

Diesel engine oil turns blacker more quickly than oil in gasoline engines since diesel combustion produces more soot. Soot sticks to the oil film in the cylinders and is scraped off the cylinder walls and into the crankcase by the rings. Diesel engines require more detergent and dispersant additives to prevent the soot from clumping and thickening the oil. Filters are able to remove soot from the oil when it stays in suspension. Using oil without detergents or inadequate level of detergents will allow oil to thicken and possibly block oil passageways to bearings (**FIGURE 7-40** and **FIGURE 7-41**). In contemporary oil classification, the +4 designation to CI, CJ, or CK oils indicate the oil has high amounts of detergent.

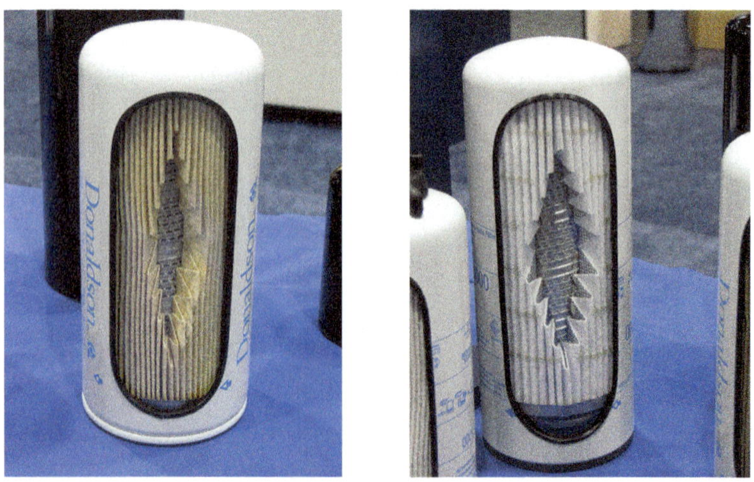

FIGURE 7-40 Detergents in oil enable soot and other solids to stay in suspension where the filter can remove them. Two types of filter media: cellulose—yellow paper; microglass—white media.

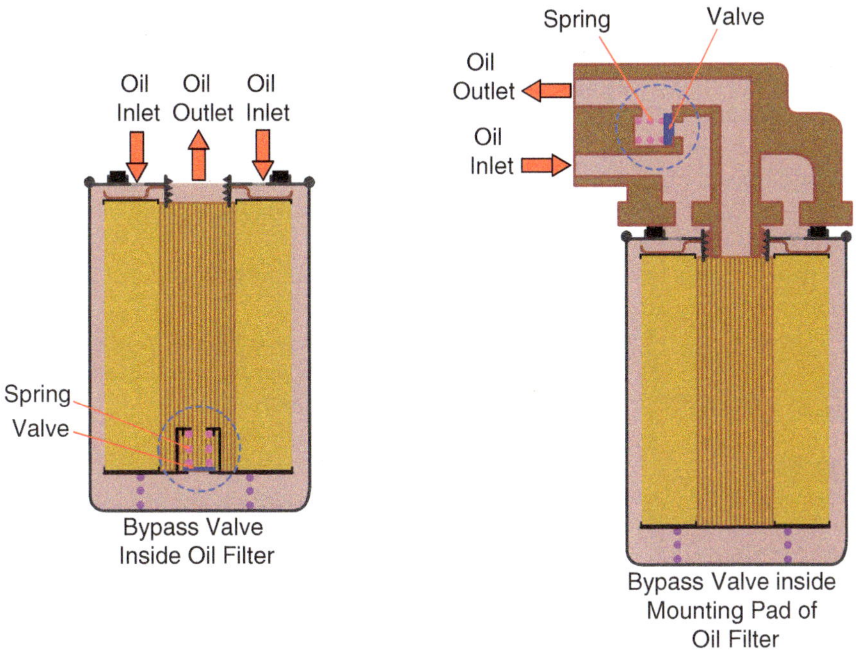

FIGURE 7-41 A filter bypass mechanism will allow oil to flow around filter media that are plugged with soot and solids.

▶ Factors That Limit Engine Oil Life

7-05 Identify and explain factors that limit oil service intervals.

It's in the interest of owners and the environment to extend oil change intervals. Owners are motivated to sample engine oil to determine whether any oil properties are depleted. Oil needs not be changed if oil properties are intact and extending oil change intervals lowers operating costs. Important clues about impending engine failures are detected before a catastrophic failure occurs. From an environmental viewpoint, extending oil change intervals reduces environmental waste. The EPA reports that of the 1 billion gallons of engine oil consumed every year, about 185 million gallons of it is disposed of improperly after is has been used. Likely the oil is dumped into the ground, tossed into landfills, or poured into sewers. The EPA now encourages vehicle owners to follow manufacturers' oil change recommendations and not the oil change industry's. Research has demonstrated oil changes can be extended by using higher quality oil and using oil analysis.

Synthetic Oils

Synthetic-base lubricants have been used in aviation and special-purpose engines since World War II. Tanks used in World War II needed synthetics to allow the oil to properly lubricate the engines in extreme cold. High temperatures encountered in jet engines would burn petroleum-based oils, necessitating the use of synthetics. Synthetic lubricants did not begin to make inroads into automotive applications, however, until the late 1970s.

Synthetic oils generally refer to oils whose base stock is synthetically derived or manufactured. Most brands of synthetic oils are extracted from ultra-refined petroleum oil, and the most popular base stock for synthetic oil is made from a petroleum-derived oil molecule called polyalphaolefin (PAO). PAO molecules are smaller and more consistent in size, and no impurities are found in this oil, because it is derived through a chemical process, making it very chemically stable and flexible compared to conventional oil. Approximately 70% of the base stock of synthetic oils is made from PAO. Chemical additives are then blended with the synthetic oil base stock, similar to the process of making conventional oils. However, synthetic oils usually have more additives than conventional oils. No API or SAE standards exist yet for defining what synthetic oil is, but they do need to meet minimum API specifications.

Synthetic oils are of tremendous benefit to high-speed turbocharged diesels for at least nine reasons:

1. Turbochargers and rocker shafts receive lubrication faster than petroleum-based lube oils do. Since synthetics flow easily at cold temperatures, engine components, particularly the turbocharger, do not starve for oil during cold start-up conditions.

2. Synthetic oils are resistant to coking. Coking occurs when oil is burned and forms carbon deposits. In turbochargers during hot shutdown, exhaust heat in the turbine housing moves into the center bearing housing, which can cause the turbine shaft bearing to heat and coke oil. Shaft bearings are damaged and the turbine shaft scored because of the abrasive carbon residue left from coked oil.
3. Improved shear strength of more flexible molecules occurs. Engines that use unit injectors, for example, exert greater stress to oil through shearing. Synthetic oil viscosity remains stable over a longer period of time.
4. Synthetic oils have a lower VI index. They flow better at low temperatures and resist thickening at higher temperature.
5. Synthetic oils have a lower friction coefficient. Most high-quality synthetics reduce friction, improving fuel economy and lowering the engine oil temperature.
6. Synthetic oils decrease oil consumption: When synthetic oil is used, 42% less oil consumption occurs. This is due to less evaporation and due do its resistance to deposit formation.
7. Synthetic oils have a greater resistance to chemical breakdown, oxidation, coking, and deposit formation.
8. Synthetic oils have a lower coefficient of friction (they are slipperier). Users of synthetic oil have claimed a 2–5% increase in their fuel economy.
9. Synthetic oils have longer intervals between oil changes due to their higher resistance to oxidation.

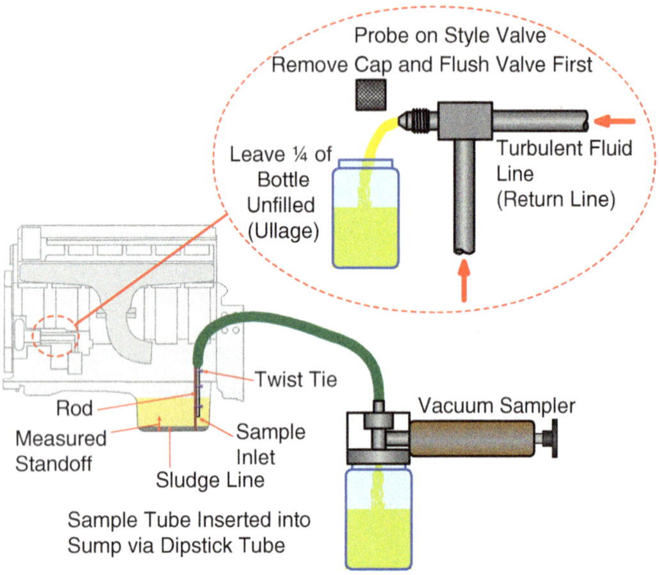

FIGURE 7-42 A clean representative sample of oil is removed for analysis. During SOS, oil can be removed from the oil pan or engine oil gallery.

Scheduled Oil Sampling

In the absence of any sensor capable of comprehensively measuring oil quality, scheduled oil sampling (SOS) is recommended. A sample of engine oil is obtained from the engine and sent out for analysis (**FIGURE 7-42**). SOS can provide an extremely detailed analysis of the condition of the engine and quality of the engine oil (**TABLE 7-5**).

TABLE 7-5 Measures of Oil Quality for Analysis

Indication in Oil	Possible Cause
Viscosity increase	Soot loading Oxidation Water contamination
Viscosity decrease	Fuel dilution VI depletion from shear Overheated oil
Depleted additive	High operating temperature Extended oil change intervals Addition of aftermarket additives Incorrect oil
High solids content	Dirt contamination Excessive metal wear
Water contamination	Extended idle time Coolant leak
Antifreeze contamination	Head gasket leakage Cylinder liner O-ring leak
Fuel contamination	Injector damaged Cylinder misfire Injector O-ring damage
Metal contamination	Excessive wear or damage indicated by type of metal contaminant (i.e., lead from bearings, chrome from rings, tin from pistons, iron from gears or block material, silicon from coolant additive, etc.)

Oil Quality Sensors and Level Sensors

Oil quality sensors are now available to evaluate the condition of lubrication oil but are not mature in terms of technological sophistication to perform condition-based oil changes. Such sensors, however, can help to alert the operator to a situation in which oil quality falls outside specified parameters, which could be either due to the oil exceeding its useful life or a symptom of an engine problem. One of several sensor technologies can accurately measure soot content, which absorbs oil additives and contributes to engine wear. The sensor operates like a variable capacitor, using oil as the dielectric element. By measuring alternating current conductivity across the sensor at frequencies between 2 and 5 MHz, the sensor evaluates the dielectric strength of oil by measuring the time needed to charge the sensor's capacitor plates. This variable will change in proportion to the soot loading in oil.

Oil level sensors are used to alert the operator to low oil levels. One type can use ultrasonic sound waves to monitor the oil level in the oil pan during the brief initial key-on-engine-off period. The sensor monitors the correct engine oil level to avoid an overfill or a low oil level while driving or at the key-on. Typically, the sensor is mounted on the bottom of the oil pan and has a piezoelectric transducer that sends ultrasonic sound pulses into the oil. Wave reflection at the oil surface and then back to the sensor measures the time the echo is received from the oil level surface to calculate the distance and hence oil level (**FIGURE 7-43** and **FIGURE 7-44**). To locate an oil leak, follow the steps in **SKILL DRILL 7-2**.

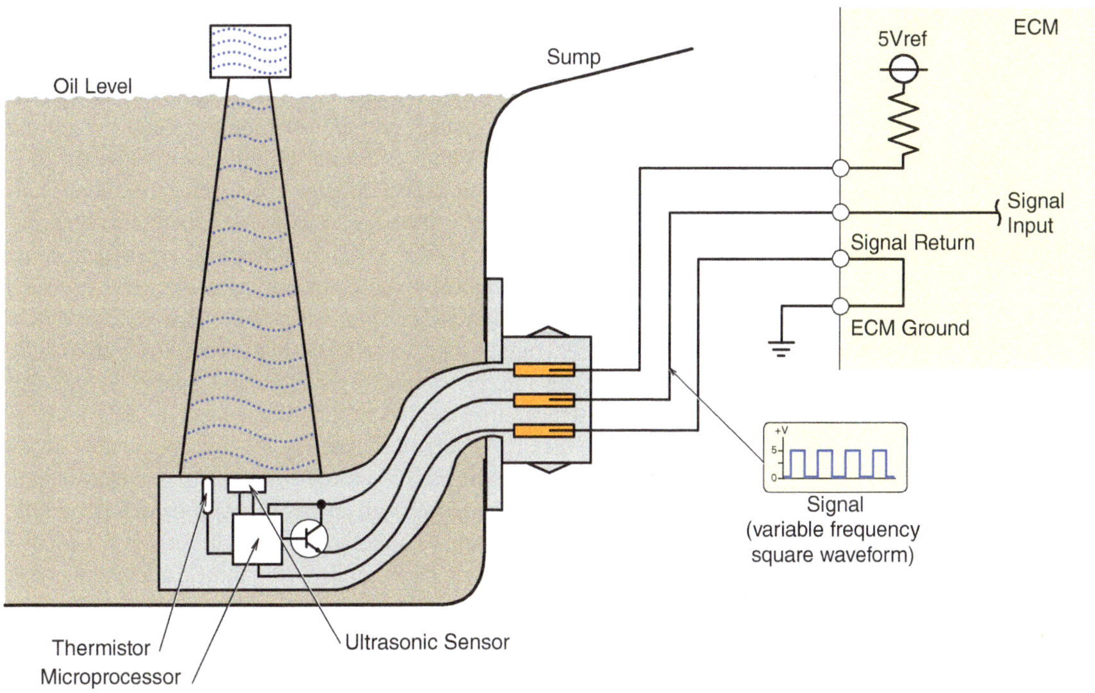

FIGURE 7-43 An oil level sensor uses ultrasonic sound waves that reflect back from the surface of the oil level in the oil pan.

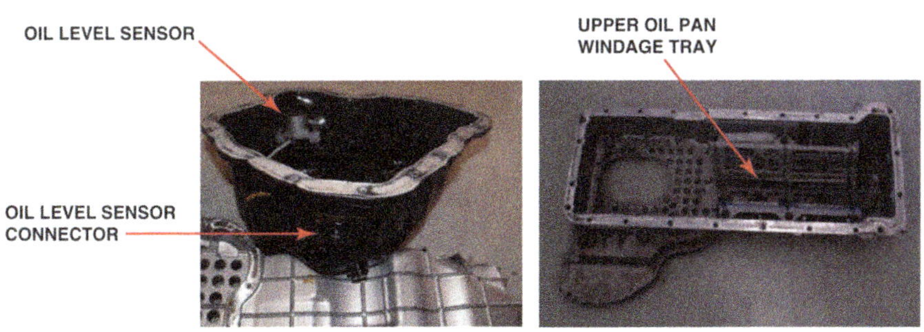

FIGURE 7-44 Location of an oil level sensor and connector on a 6.6 L Duramax oil pan.

SKILL DRILL 7-2 Locating an Oil Leak

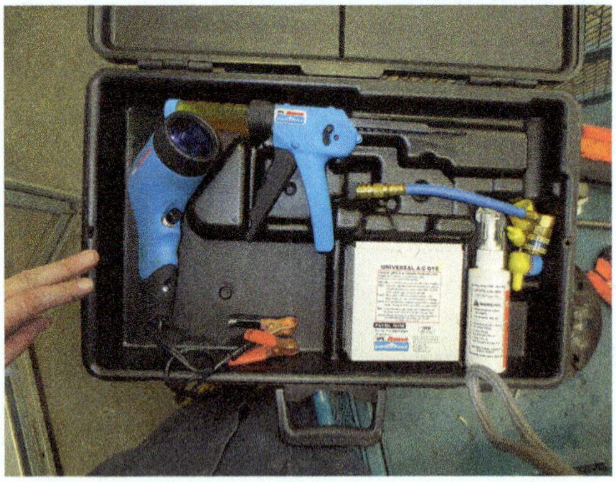

1. Obtain the correct type of fluorescent tracing dye. There are dyes that are compatible only with oil, fuel, coolant, or refrigerant. Choose the correct type.
2. Bring the engine to operating temperature, and add the prescribed amount of dye to the engine oil as recommended by the dye manufacturer.
3. Road test the vehicle with the engine under load. When checking for a leak, it is important to get the engine oil to operating temperature. High oil pressure and hot oil will reveal leaks faster than cold, low-pressure oil.
4. After running the engine for 20–40 minutes, shut the engine off and examine the engine by using an ultraviolet (UV) light. Some kits use alligator clips connected to the vehicle's battery to power compact 12-volt lights. The newest dyes reflect a bright yellow-green at a leak path when using the UV light. The use of glasses with yellow lenses further enhances the appearance of the dye, making even small leaks easier to detect in tight spots.
5. If it is difficult to verify the actual spot that's leaking, clean the area with an aerosol solvent and recheck the area while the engine is running.
6. Record all observations, and make a repair recommendation based on the findings.

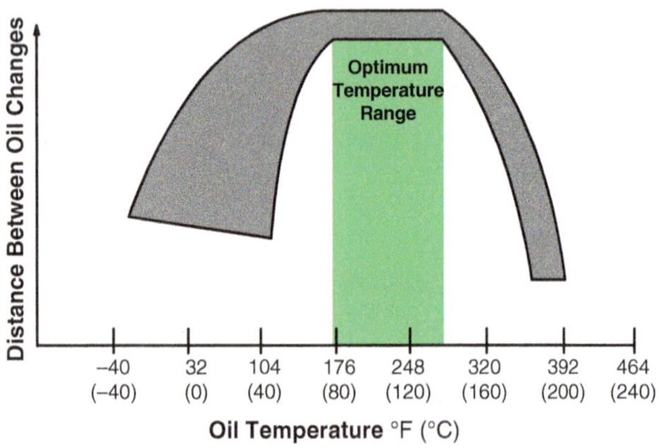

FIGURE 7-45 Engine oil life is influenced by its operating temperature.

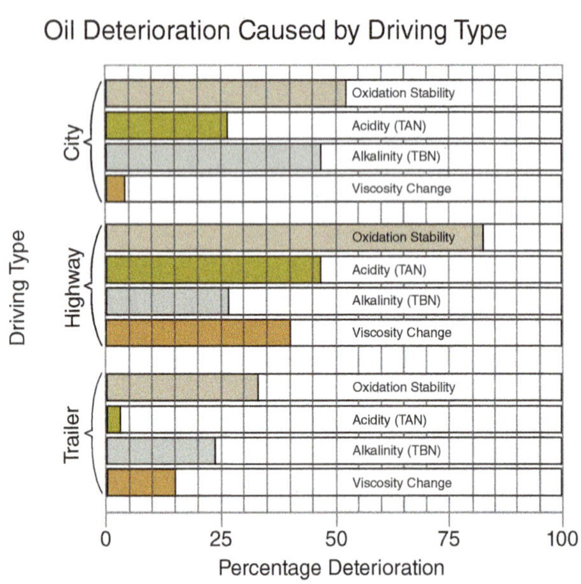

FIGURE 7-46 Oil chemistry changes depend on the type of driving cycle the vehicle encounters.

Oil Life Algorithm Systems

Software-based calculations to determine when to change oil use a formula or algorithm with input variables from the engine operating conditions. This means a computer continuously monitors engine operating conditions to determine when to change oil. Since oil degradation is associated with combustion events, logging combustion conditions can accurately determine when oil life has reached its maximum value. For example, when a cylinder fires, a small amount of oil exposed to combustion is destroyed. Combustion gases containing acids, unburned fuel, water, and other by-products leak past the piston rings and react with engine oil. Diesel engines tend to generate much more soot and acidic combustion blowby into the crankcase than gasoline engines do. It is also known that oil temperature exerts a strong influence on the deterioration of oil quality. As the oil temperature increases, the rate of oxidation accelerates, which can lead to the oil thickening. Turbochargers subject engine oils to high temperatures and make an engine more prone to form engine deposits. Colder oil temperatures increase the concentration of oil contaminants, such as fuel and water. Depending on the vehicle and engine type, each engine receives its own unique algorithm to determine engine oil life (**FIGURE 7-45** and **FIGURE 7-46**). The following are some of the other primary factors used to determine the limits to oil life, divided into subsections.

Percentage of Solids

The percentage of solids refers to the total amount of particles that are suspended in the oil. Combustion soot, dirt, and oxidized oil are typically the major solids found in oil. The percentage of solids becomes a limiting factor to oil life when the particles interfere with the lubricating abilities of the oil. Engine manufacturers suggest that an oil change to remove dissolved solids should occur before the solids rise above 5% by weight; however, improvements in wear reduction are best achieved if the solids are kept below 2%.

Detergents are added to the oil to prevent the solids from clumping and forming damaging deposits inside the engine. Like hand cleaner, these detergents surround the particles and suspend them in the oil until it is drained or the filter removes the particles. It is normal for high-detergent oil to appear dirty after a short time. Just like dishwashing soap cleans dishes and keeps dirt suspended in the water, so too do oil detergent additives hold solids in suspension. More solids are removed through filtering oil with detergents. Chemical dispersants, called emulsifiers, are also used to keep these particles in suspension throughout the oil. Synthetic oils will require change intervals just like conventional oils if the percentage of solids is the limiting factor to engine oil life.

Sulfur and Combustion Acids

Sulfur acids are present in engine oil whenever there is sulfur in the diesel fuel and corrosive acids produced by using recirculated exhaust gas are in the cylinders. Since sulfur combines with water produced during combustion, highly corrosive acids are formed in the oil. To counteract this condition, oil manufacturers add alkali, or base substances, to the oil to neutralize the effect of sulfuric acids. The use of more alkaline oil properties by diesels distinguishes oils used in SI engines from compression-ignition engines. The TBN is the measure of the engine oil's ability to neutralize acid.

Oxidized Oil

Oil is oxidized when it is exposed to air and the oxygen joins the oil molecules. Conventional oils generally thicken and form deposits if they are highly oxidized. This process occurs more rapidly at high temperatures. The problem with oxidation is that oil loses some of its lubricating properties as it oxidizes, thus causing accelerated engine wear. Synthetic oils are at an advantage since they do not oxidize as fast as conventional oils do. This means that if oil oxidation due to high engine operating temperatures is the limiting factor on lube oil life, then using synthetic oil could extend the oil change interval.

Biodiesel and Engine Oil

For years biodiesel has been considered as a lubricity additive to keep moving parts operating smoothly inside fuel-injection systems. Dilution of engine oil with biodiesel is another matter. Biodiesel actually promotes dilution of engine oil with fuel to a much greater extent than petroleum diesel fuel does. Dilution of oil with biodiesel primarily occurs when a post-combustion injection of fuel into the cylinders is used to regenerate the particulate trap or NO_x adsorber. All the fuel is expected to vaporize in the cylinder and not combust until it reaches the exhaust catalysts. Heavier, less volatile fractions of biofuel, however, do not vaporize during post-combustion injection, and liquid biofuel clings to the cylinder walls. Since biodiesel has a higher distillation temperature and boiling point, it will not evaporate from engine oil and it tends to dilute the engine oil disproportionately to its blend ratio in the fuel. As biofuel passes by the pistons and accumulates in the crankcase, it changes oil properties. Once there, biofuel polymerizes (joins together to form long, complex-shaped molecules), producing sludge-thickened oil. Engine damage can result from oil passageways that get blocked by sludge. Deposits also form in the ring belt area, leading to ring stickage and increased crankcase blowby, which further aggravates oil becoming contaminated by increased levels of soot. Raw or refined vegetable oils produce the greatest harm, which may not become evident until a significant amount of damage has occurred over an extended period. Petroleum-based fuel does not produce these same effects at all, since it readily evaporates when the oil is heated. Biodiesel blends higher than B20 (20% bio source) cause a larger amount of unburned fuel to slip by the piston rings and condense in the lubricating oil. Engine oil change intervals may need to be shortened significantly if blends of biodiesel higher than B20 are used (**FIGURE 7-47**). Biodiesel also interferes with the action of zinc dialkyl dithiophosphate (ZDP), an antiwear additive. The chemical properties of biodiesel cause the fuel to compete with ZDP for surface area on metals, thus displacing ZDP on metal. And detergents do not work as well when oil has been diluted with biodiesel.

Other Limiting Factors

Engine problems such as coolant leaks into the oil from sources such as a leaking head require an oil change immediately after a repair and before the engine is run. A defective air filter that permits dirt to enter the engine, fuel leaking into the oil, wrong fuel (e.g., gasoline

> **▶ TECHNICIAN TIP**
>
> Whenever gasoline has been mistakenly used in a diesel engine, it is critical to drain and flush the fuel system, change fuel filters, and change the engine oil. Misfiring cylinders and gasoline will wash into the oil and dilute it. Similarly, always change the engine oil after any major repair to an engine, because the oil can be contaminated with dirt, coolant, or fuel. Never use an air gun to blow contaminants away from inside an engine. The contaminants are usually blown into the oil. Use a vacuum cleaner to remove contaminants from an open engine. Always cover an engine that is under repair when between work shifts, to prevent parts from being contaminated and to prevent the lube oil from eventually being contaminated with dirt. Leave parts in bags or boxes until just prior to installation.

> **▶ TECHNICIAN TIP**
>
> Engine oil is supposed to remain chemically neutral to prevent internal engine corrosion. Combustion gases dissolving in engine oil can make oil corrosive. Sulfur content is one element that can contribute to this change. To prevent these, alkaline additives are added to diesel engine oil. The TBN is used to measure the engine oil's acid neutralizing ability. The TAN is another measure of a lubricant's pH factor.

FIGURE 7-47 This oil filler cap from a 6.7 L Powerstroke diesel indicates that the engine oil must be biodiesel compatible and identifies the grade of oil required for the engine.

contaminating diesel fuel), water in the oil, or a mechanical failure that produces internal debris are all circumstances that necessitate an oil change. Many engines have experienced catastrophic oil-related failures subsequent to seemingly minor repairs when the oil has not been changed.

▶ Oil Pumps and Pressure Circuits

Oil pumps are used to pressurize the lubrication system of a diesel engine. They are positive displacement pumps, usually constructed with a closed gear or gerotor design. A gerotor pump is constructed of an inner rotor and an outer rotor. The inner rotor turns inside the outer rotor ring, which also turns. The inner rotor has one fewer lobe than the outer ring and is positioned off-center from the outer ring. As the lobes slide up and over one another, oil is drawn through a fixed passageway connected to the oil pan pickup tube. Oil is squeezed out into the main oil gallery through the outlet passageway behind the rotors (FIGURE 7-48). The external gear pump is another common configuration for an engine oil pump. This pump has two opposing gears: A drive shaft turns one gear, which in turn rotates the opposite gear. Oil is trapped and carried around the inside of the pump housing between the gear teeth. A seal is formed between the gear teeth and the oil pump housing (FIGURE 7-49).

The lube oil is supplied under pressure to moving parts through drilled passageways in the engine called oil galleries. Pressurized oil is supplied to most moving parts, for lubrication. However, cylinder walls are lubricated by oil thrown off of engine bearings and from oil cooler nozzles. Oil is drawn into the pump through a pickup tube and screen located in the oil pan sump (FIGURE 7-50).

7-06 Identify and explain the construction and operation of pressurized oil circuits.

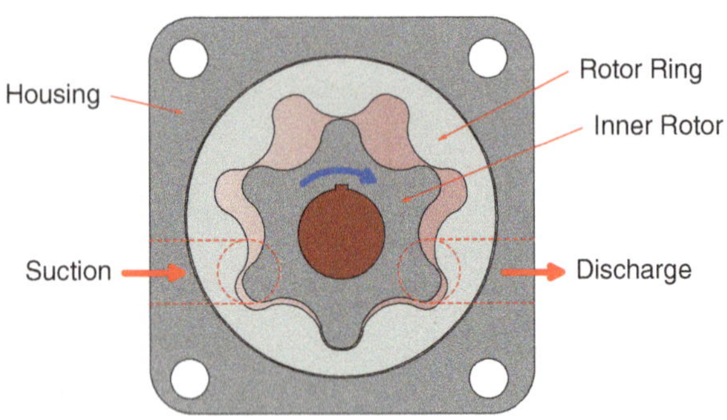

FIGURE 7-48 A gerotor oil pump.

Oil-Pressure Regulating and Pressure-Relief Valves

Because the oil pump can produce more oil flow than the engine can use, an oil-pressure regulating valve is used to control oil pressure. Oil pressure that is too high can blow gaskets on the spin-on oil filter and burst or make the oil filter swell into a pumpkin shape. Excessively high oil pressure can also erode engine bearings. Keeping the oil pressure regulated to between 20 and 40 psi (138 and 276 kPa) when the engine oil is warm can reduce

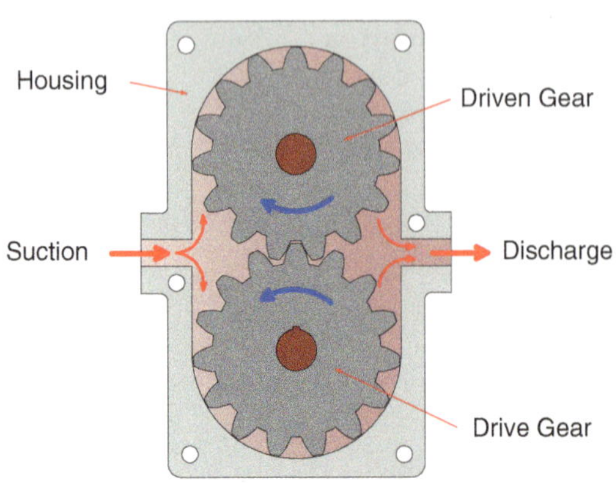

FIGURE 7-49 An external gear oil pump.

FIGURE 7-50 The oil pickup tube and screen is located in the oil pan sump.

parasitic power loss that is caused by pumping more oil than is necessary to regulate oil at a higher pressure (**FIGURE 7-51**).

Energy-Saving Oil Pumps

The oil pump is a common target for achieving energy reductions in today's diesels. Many new engine designs use flow rate–controlled oil pumps or two-stage pumps that operate at either a high-pressure or low-pressure setting, depending on operating requirements. Data from the oil-pressure sensor and ECM calculations will electrically switch the oil pump between the two pressure stages, depending on engine load, speed, and oil temperature. Power consumption by the oil pump is reduced when driving in low-speed urban or off-road conditions when high oil pump pressure is not required. For example, late-model Volkswagen (VW) and Ford 3.2 L Powerstroke engines use a flow rate–controlled vane pump that regulates pump delivery volume by rotating an eccentric adjustable ring that surrounds the pump vanes. The space between the pump vanes is increased or decreased to change pump output volume. A hydraulic control valve rotates the ring against spring pressure to vary the space swept by the pump vanes (**FIGURE 7-52**).

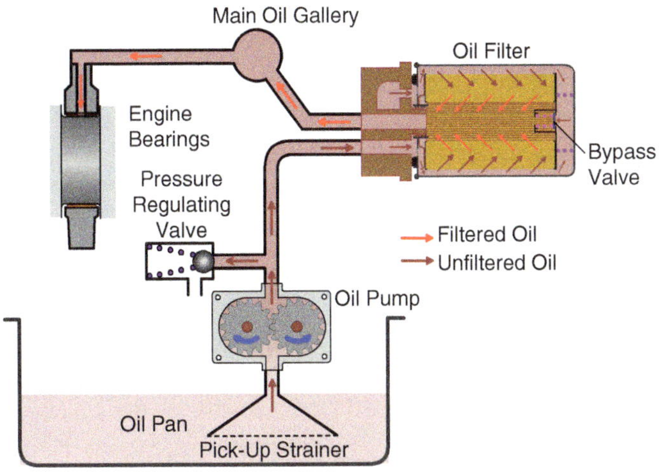

FIGURE 7-51 Operation of the oil-pressure regulating valve. When oil pressure exceeds the calibrated spring tension of the regulator, the check ball will uncover a passageway, permitting excess oil to return to the sump.

Oil Coolers

Oil coolers are heat exchangers used to remove excess heat from engine oil. Diesel engines use oil coolers since pistons and turbocharger center housings are cooled by lubricating oil. Without a cooler, engine oil can easily become overheated while absorbing heat from internal parts and turbochargers. Engine oil should never exceed 250°F (139°C). Many electronic engines use oil temperature sensors to shut down or derate engine power if oil temperatures exceed 250°F to 260°F (139°C to 143°C). Normal maximum engine oil temperature is maintained at 30°F to 40°F (16°C to 22°C) above the temperature of the coolant. This means an engine operating between 190°F and 205°F (between 91°C and 96°C) could have a normal oil temperature of as high as 220°F to 245°F (122°C to 136°C) under load. Plate-type and tube-type oil coolers are the most commonly used design. Generally, oil coolers have a thermostatically controlled inlet for either coolant or oil that will not allow oil cooling until the engine oil or coolant has reached operating

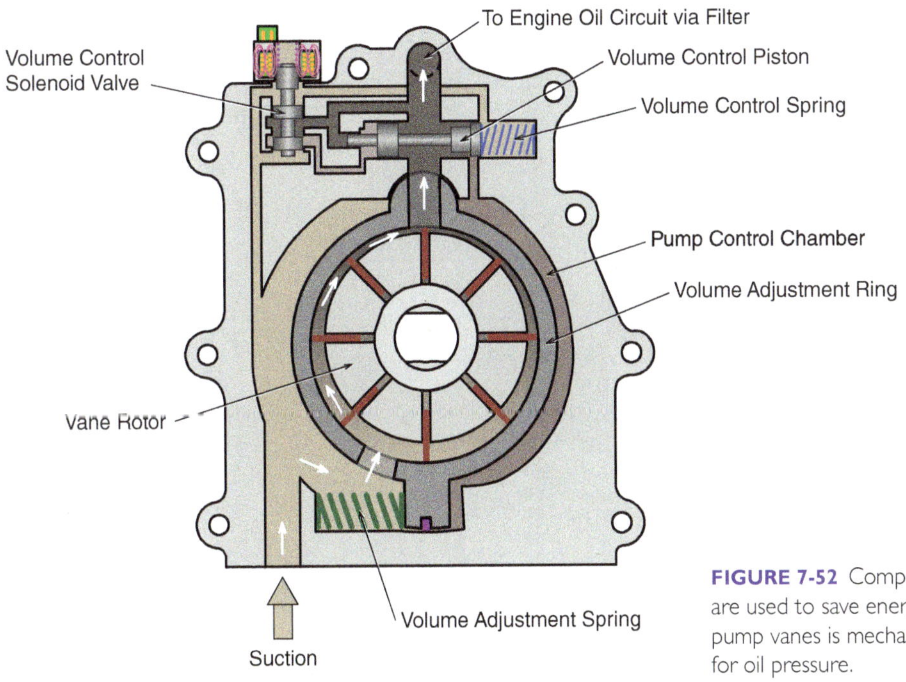

FIGURE 7-52 Components of a variable displacement oil pump that are used to save energy. The area of the delivery space above the pump vanes is mechanically varied in accordance with engine demand for oil pressure.

FIGURE 7-53 This plate-type oil cooler is located near the outlet of the oil filter.

temperature. This hastens the oil getting warmed up, which improves engine lubrication (FIGURE 7-53).

The cooler may also be equipped with a pressure-relief valve that allows oil to bypass the cooler if it becomes plugged. The bypass valve will open if a substantial pressure drop occurs across the cooler, such as when it becomes plugged with sludge or oil contaminants. Generally, since most engines are designed to allow oil to pass through the cooler first before being filtered, dirt, gasket pieces, silicone, and other materials can restrict how well oil flows through the cooler (FIGURE 7-54).

Cartridge Versus Spin-On Filters

Spin-on oil filters, introduced during the 1950s, made changing oil easier and less messy. Since the late 1980s, however, European and now North American engine manufacturers are reverting to the cartridge-style filter for several reasons, one being underhood space. Because space is at a premium in the engine compartment of today's vehicle, the oil filter can be very difficult to locate and change. New-style cartridge filter housings are usually conveniently located on the top or side of the engine compartment, making them accessible from under the hood without needing to raise the vehicle. Cartridge-type filter housings are typically designed with a screw-on-type cap and a single sealing gasket. The housings are ventilated when opened, and then the oil flows out of the filter back to the sump through a separate drain system. The messy spills associated with spin-on filters are minimized this way. Cartridge-type filters are less expensive than spin-on filters. It is reported that almost 80% of the cost of a spin-on filter is in the steel canister and valves. With a cartridge filter, only the pleated media and a gasket are replaced during filter service. The second, but most important, reason for the change from spin-on to cartridge filters is the problems of spin-on filter disposal. Disposing of the filter element with even the smallest quantity of oil is not only environmentally unfriendly but prohibited by law in many jurisdictions in North America (FIGURE 7-55). To replace a spin-on oil filter, follow the steps in SKILL DRILL 7-3.

▶ TECHNICIAN TIP

Oil filters are considered hazardous waste and must be disposed of accordingly. Both national and state environmental laws regulate how used filters are to be disposed. Usually the hot crushing method is required for oil filters to separate oil from the filter before disposal. Fuel filters must be crushed and drained before disposal.

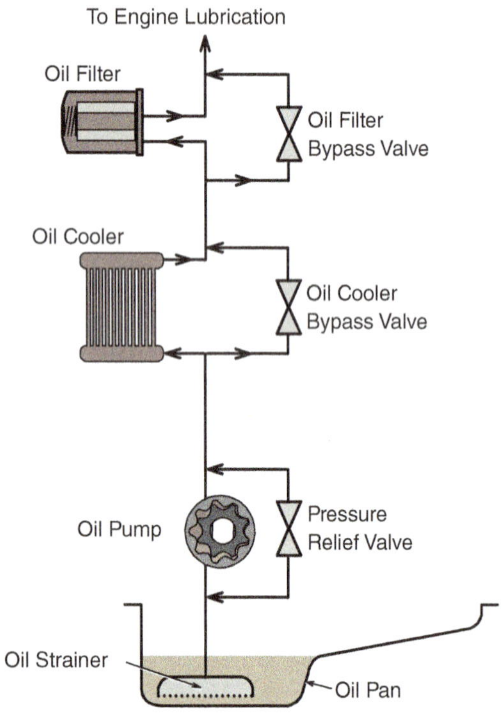

FIGURE 7-54 Bypass valves in the lubrication system prevent either over-pressurization of oil or low oil pressure caused by restricted coolers or filters.

FIGURE 7-55 Cartridge filters are quickly replacing spin-on filters, due to environmental regulations, space and accessibility requirements.

SKILL DRILL 7-3 Replacing a Spin-On Oil Filter

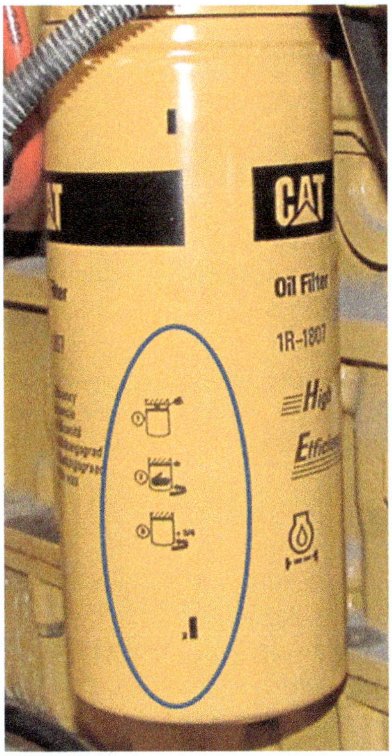

1. With the engine off, remove the filter by using a filter wrench. A large belt-type filter wrench assisted by a power bar can be used to remove a tight filter. Occasionally, chain-type belt wrenches are used to remove tight filters. To prevent hot oil from scalding hands when the filter is removed, the filter may have a hole punched into the bottom to allow oil to drain before removing it.
2. Visually inspect the filter base for any damage or warpage. Ensure that the old filter gasket is removed.
3. Fill the new filter with oil. Put oil in the unfiltered outer holes of the cartridge (the center filter hole returns filtered oil to the engine).
4. Lubricate the filter gasket by using oil or a smear of grease. This will help prevent the gasket from sliding on the filter base and prevent the gasket from buckling when the filter is tightened. It is also necessary to lubricate the gasket so that it's easy to remove during the next service.
5. Spin the new filter on hand tight. Note the arrows or markings at the top of the filter canister, which indicate how far the new filter should be tightened after snugging the filter to the base by using hand force only. Most filters are tightened a quarter or half turn beyond hand tight. Tightening the filter further could potentially distort the filter base, which is often made of aluminum. The filter can also be damaged and internal leak paths formed if it is overtightened. If the filter is too loose, it will leak.
6. If the engine is being started for the first time after an overhaul, the engine should be prelubed with pressurized oil through an oil gallery plug. As an additional precaution to prevent damage to bearings and other moving parts from oil starvation, disable the fuel system by disconnecting an ignition fuse or crank and cam sensor or by using a remote starter button. Crank, but do not start, the engine until oil pressure appears on the direct reading gauge in the instrument cluster. Electronic gauges can be read if only the cam and crank sensors are disconnected.
7. After the engine is started and oil pressure is stabilized, check for leaks. Reset any maintenance monitor to indicate when the next scheduled filter and oil service is required.

▶ Wrap-Up

Ready for Review

- Diesel engine heat loads include exhaust gas recirculation (EGR), lubrication oil, and the air intake system.
- Diesel coolant consists of water, antifreeze (which contains a special corrosion inhibitor package), and supplemental coolant additives (SCA) that minimize cavitation erosion. Water is the most efficient fluid to transfer heat; however, water will cause metal parts in contact with the cooling system to corrode and therefore needs corrosion inhibitors added to it.
- An important job of the cooling system is to raise engine coolant temperature quickly to keep engine component wear to a minimum.
- As engine coolant warms, it expands, and cooling systems must provide room for this expansion.
- Cooling systems are pressurized to maintain greater-than-atmospheric pressure at the water pump inlet and increase the boiling point of water.
- Cooling systems must deaerate coolant to prevent pumps and coolant passageways in the cylinder head from becoming air bound.
- Antifreeze contains a number of additives that minimize electrochemical reactions, and these additives prolong the durability of engine and cooling system components.
- Freeze and boil protection are both supplied to the engine by antifreeze; pressurizing the cooling system provides additional boiling protection.
- Cylinder wall and liner cavitation is a condition unique to diesel engines; chemically treating the cooling system is a preferred method for minimizing cavitation erosion.
- Most antifreezes are made with a base of either ethylene glycol (EG) or propylene glycol (PG).
- Radiators are heat-exchanging devices that release heat absorbed from the engine to the air.

- While early radiator caps simply prevented coolant from spilling out of the cooling system, today's radiator caps have additional functions, such as increasing the pressure of the cooling system to raise the boiling point of coolant and to allow air or coolant to reenter the cooling system when the pressure drops after the engine cools.
- Surge or overflow tanks provide space for coolant and vapor to move when the engine coolant is hot.
- Vent lines, located at the highest points on the engine and radiator, bleed any steam and air out of the engine and into the surge tank.
- To ensure that coolant is positively circulated, an engine-driven water pump is used to move coolant through the cooling system.
- The lubrication system reduces friction between moving parts in an engine and is essential for cooling, cleaning, sealing, and absorbing shock loads inside the engine.
- Lubrication systems in today's diesels have new demands because of emission legislation and durability requirements.
- Oil works by forming a separating film between parts, which minimizes direct contact between moving parts, reducing friction and heat buildup caused by friction.
- Engine oils are classified by the American Petroleum Institute (API) and carry an API designation that indicates their suitability for an engine application.
- Viscosity is a liquid's resistance to flow; a numerical designation developed by the Society of Automotive Engineering (SAE) is used to measure oil viscosity.
- Additives improve the original properties of the base stock oil; enhancing the performance of motor oil base stocks is necessary to adjust the performance of the oil to suit its intended application.
- The major lubrication system components function to meet increasing demands for compatibility and owners' expectations: the oil pump, oil pan, oil cooler, oil filter(s), and oil-pressure regulating and pressure-relief valves.
- The oil pump can produce more oil flow than the engine can use; an oil-pressure regulating valve is used to control oil pressure.
- Oil coolers are heat exchangers used to remove excess heat from engine oil.
- The main factors that can limit oil life are the percentage of solids suspended in the oil, sulfur and combustion acids, and oxidized oil.

Key Terms

aeration A condition in which excessive amounts of air or steam bubbles are dissolved in coolant, diminishing the coolant's effectiveness.

additives Chemicals that improve the original properties of the base stock oil.

American Petroleum Institute (API) An organization that has developed specifications to define engine oil performance standards.

base stock The raw mineral processed from crude oil.

cartridge filter A filter element that consists of only filter media unenclosed by a metal container.

cavitation Erosion in cylinder block walls, heads, and liner sleeves as a result of the collapse of tiny water vapor bubbles formed when coolant vaporizes on hot cylinder wall surfaces.

coolant extender An additive package that is used only with ELC, which is added at the midpoint of the coolant's life.

diesel coolant additive (DCA) An additive used to treat cooling systems to reduce the effects of cavitation erosion.

ethylene glycol (EG) The base chemical from which the majority of antifreezes are made.

extended-life coolant (ELC) Several types of long-life coolant formulations that contain an anticorrosion additive package that does not deplete; also known as long-life coolant (LLC).

heat exchanger A system that transfers heat from coolant to the atmosphere.

heat rejection The transfer of heat into the cooling system from the combustion chamber.

hybrid organic acid technology (HOAT) A combination of IAT and OAT with nitrites added, making it suitable for use in both light- and heavy-duty systems.

inorganic additive technology (IAT) A diesel engine cooling system conditioner containing non-carbon-based corrosion inhibitors, such as phosphates, borates, and silicate.

multigrade oil A blend of a several different oils with different viscosities; also known as multiweight oil.

nitrated organic acid technology (NOAT) An ELC that uses OAT with nitrates added.

oil quality sensor An electrical device that measures the amount of soot loading in engine oil.

organic acid technology (OAT) A category of ELC that contains carbon-based corrosion inhibitors.

polyalphaolefin (PAO) An artificial base stock (synthetic) used in place of mineral oil. PAO molecules are smaller and more consistent in size, and no impurities are found in this oil, because it is derived through a chemical process.

propylene glycol (PG) An antifreeze base that is nontoxic and environmentally friendly.

scheduled oil sampling (SOS) An extremely detailed analysis of the condition of the engine and quality of the engine oil.

Society of Automotive Engineers (SAE) viscosity ratings Oil performance criteria that indicates oil's flow characteristics.

spin-on filter Filter media enclosed in a metal can that is threaded onto a filter header.

supplemental coolant additive (SCA) An additive used to treat cooling systems to reduce the effects of cavitation erosion and optimize cooling system performance.

synthetic oil Oil made from base stock that is synthetically derived or manufactured.

thermal efficiency The ability of an engine to convert the energy content of fuel into mechanical force.

total acid number (TAN) The acidity of an oil. Certain contaminants cause engine oil to increase in acidity, which is measured by using the TAN index.

total base number (TBN) The measurement of a lubricant's reserve alkalinity, which aids in controlling acids forming during the combustion process.

viscosity A measure of oil's resistance to flow.

viscosity index (VI) A measurement of the total amount of change in an oil's viscosity due to temperature.

viscosity index (VI) improver An additive that prevents oil from thickening when cold and thinning when hot.

Review Questions

1. Which series of engine oils is used in a diesel engine?
   a. S series.
   b. ASTM.
   c. C series.
   d. SAE.
2. What is the maximum percentage of soot loading that means engine oil needs to be changed?
   a. 2%.
   b. 5%.
   c. 10%.
   d. 20%.
3. Which of the following is the normal operating oil temperature of a diesel engine?
   a. 20°F to 30°F (5°C to 14°C) lower than coolant temperature.
   b. Close to cooling system operating temperature.
   c. 30°F to 40°F (−1°C to 4°C) above cooling system temperature.
   d. Approximately 250°F (121°C).
4. The 15W in 15W-40 engine oil stands for which of the following?
   a. Multiviscosity.
   b. Viscosity at 0°F (−18°C).
   c. Viscosity at 212°F (100°C).
   d. The thickness of the oil when cold.
5. Which of the following will likely cause excessively high oil pressures?
   a. A plugged oil filter.
   b. A defective oil filter bypass valve.
   c. A defective oil-pressure regulating valve.
   d. A defective pressure-relief valve.
6. Which of the following ingredients is added to diesel engine coolant to prevent cavitation?
   a. Nitrite.
   b. Silicate.
   c. Ethylene glycol (EG).
   d. Borate.
7. Which of the following is the most recent engine oil made for a diesel engine?
   a. SG-4.
   b. CJ-4.
   c. SJ.
   d. FA-4.
8. Diesel engine oil has which job of the following extra jobs to perform, which gasoline engines are not necessarily required to perform?
   a. High-speed lubrication.
   b. Cooling piston crowns.
   c. Absorbing shock loading that engine parts can incur.
   d. Sealing piston rings.
9. The presence of cavitation on the coolant side of the cylinder wall is most likely caused by __________.
   a. lubrication breakdown
   b. excessive cylinder pressures
   c. excessive cooling system pressure
   d. improperly conditioned coolant
10. Which antifreeze will require the highest concentration to maintain 0°F (−18°C) freeze protection?
    a. Ethylene glycol (EG).
    b. Propylene glycol (PG).

ASE Technician A/Technician B–Style Questions

1. Technician A says that blue antifreeze is long-life antifreeze. Technician B says that blue antifreeze is propylene glycol–based antifreeze. Who is correct?
   a. Technician A
   b. Technician B
   c. Both A and B
   d. Neither A nor B
2. Technician A says that a drop in the coolant level in the surge tank or coolant overflow reservoir when the engine cools indicates an internal engine leak. Technician B says that the drop in the level is normal. Who is correct?
   a. Technician A
   b. Technician B
   c. Both A and B
   d. Neither A nor B
3. Technician A says that the cooling system pressurizes to enable coolant to remove more heat from the engine. Technician B says pressurization is needed to keep the coolant from boiling. Who is correct?
   a. Technician A
   b. Technician B
   c. Both A and B
   d. Neither A nor B
4. Technician A says that antifreeze is added to coolant to protect the coolant from freezing and to lower its boiling point. Technician B says antifreeze lowers the freeze point and raises the boiling point of coolant. Who is correct?
   a. Technician A
   b. Technician B
   c. Both A and B
   d. Neither A nor B
5. Technician A says the problem with new long-life antifreeze is it can produce silicate dropout, which plugs cooling system passages. Technician B says that the problem

with some types of long-life antifreeze is that it is not compatible with all cooling system materials. Who is correct?
   a. Technician A
   b. Technician B
   c. Both A and B
   d. Neither A nor B
6. Technician A says that low-sulfur oil needs to be used today to prevent damage to exhaust aftertreatment systems. Technician B says that today's oils should have low ash content. Who is correct?
   a. Technician A
   b. Technician B
   c. Both A and B
   d. Neither A nor B
7. Technician A says that viscosity index (VI) improvers thicken oil when it becomes hot and thins it when it becomes cold. Technician B says that VI improvers thin oil when it's hot and thicken it when it's cold. Who is correct?
   a. Technician A
   b. Technician B
   c. Both A and B
   d. Neither A nor B
8. Technician A says that diesel engine oil turns black faster than oil in an SI engine because the oil composition and quality are different between the two oils. Technician B says that there is no difference between diesel engine and SI engine oil. Who is correct?
   a. Technician A
   b. Technician B
   c. Both A and B
   d. Neither A nor B
9. Technician A says the oil-pressure regulator automatically reduces oil pressure at low-speed and low-load conditions. Technician B says that two-stage pressure regulation is performed electronically to reduce fuel consumption. Who is correct?
   a. Technician A
   b. Technician B
   c. Both A and B
   d. Neither A nor B
10. Technician A says that spin-on oil filters in metal cans are being eliminated because they are less convenient and messy. Technician B says that cartridge-type filters are becoming more commonly used to allow them to be more easily disposed in a more environmentally friendly way. Who is correct?
   a. Technician A
   b. Technician B
   c. Both A and B
   d. Neither A nor B

CHAPTER 8

Diesel Fuel and Low-Pressure Fuel Systems

Learning Objectives

After reading this chapter, you will be able to:

- 8-01 Outline the development of diesel fuel properties and characteristics.
- 8-02 Describe the production methods used to produce diesel.
- 8-03 Identify and describe the purpose, functions, important properties and characteristics of diesel fuel quality.
- 8-04 Identify and describe the properties and characteristics of biodiesel fuel and compare them to conventional petroleum diesel fuel.
- 8-05 Identify and describe the classification method used to grade petroleum diesel and biodiesel fuels.
- 8-06 Identify and explain the purpose and applications of various common diesel fuel additives.
- 8-07 Identify and describe the components and construction of the low-pressure fuel supply system.
- 8-08 Outline the purpose and methods used to diagnose faults in the low-pressure fuel circuits.

You Are the Technician

As a diesel technician, you encounter a variety of fuel–quality-related problems, and many customers and drivers seek you out for advice about purchasing fuel and fuel additives. Because your repair shop is located in a colder region of the country that has dramatic seasonal temperature changes, you can anticipate a number of fuel-related problems that result from colder temperatures. The following are some of the cold weather and other fuel-quality-related problems you have encountered:

- drivers entering the region with fuel purchased in warmer regions
- performance complaints such as low power and hard starting from using fuel purchased locally in the winter
- drivers unnecessarily using fuel additives and incorrectly using those additives
- stale fuel
- the use of biodiesel purchased in regions where biodiesel is inexpensive and widely available
- water and ashphaltine contamination
- plugged fuel filters and lines when the temperatures are very low.

1. For each of the above situations, outline the advice you would provide to customers, explaining either the potential fuel-related operating problems that a driver could expect to encounter or the cause for each of the service conditions.
2. Outline the procedures you would follow to evaluate the fuel quality used by an engine that you suspect has a fuel-related problem or complaint. Explain your reasons for using each of those procedures.
3. Make a list of service recommendations that you would make to identify and correct problems with low-pressure, fuel-system-related performance.

▶ Introduction

The proper supply and quality of fuel is vital to the operation of diesel engines. Even small problems with fuel quality will cause the supply to have outsized effects on engine operation and performance. It is important, then, for skilled technicians to understand the characteristics and properties of diesel fuel and how they influence engine operation. Not only do diesel engine performance and durability depend on good-quality fuel, but the supply of fuel to the high-pressure injection system is just as critical. Precision high-pressure injection systems are highly sensitive to dirt, water, and other contaminants. This inescapable feature of diesels places enormous requirements on a low-pressure fuel system to transfer clean fuel, with adequate volume and pressure, from the fuel tank to the high-pressure injection stage of the fuel system. Fuel tanks, pumps, lines, and filters are the most visible components of the low-pressure fuel system examined in this chapter. A thorough understanding of the construction and operation of these components is important since diesels have a disproportionate number of maintenance and repair procedures associated with the system compared to other fuels. Technicians should also be prepared to make sound service recommendations regarding the use of fuel additives and the proper selection of fuels that are more widely available to operators of diesel-powered vehicles and equipment.

▶ Diesel Fuel Fundamentals

8-01 Outline the development of diesel fuel properties and characteristics.

Diesel fuel properties and characteristics are matters a technician can do little to change. However, diesel fuel properties exert an enormous influence on engine performance. A knowledgeable and competent diesel technician should have the expertise to make recommendations about the use of different grades of diesel fuel and fuel additives. It's important to have a good understanding of diesel fuel properties and characteristics in order to properly select the appropriate fuel and fuel additives for the wide variety of applications for diesel engines.

Development of Diesel Fuel

While the term "diesel fuel" bears the name of the engine's inventor, it is a generic term that refers to a range of fuels burned in compression-ignition (CI) engines. Each fuel has its own uniform designation as well as its own formulation. Some diesel fuel products are made from petroleum, and others are processed from biological sources and even waste products. The following section surveys the development of diesel fuel since the invention of the diesel engine.

Early Diesel Fuels

Coal dust was the original diesel fuel. Rudolf Diesel's prototype engine, based on his 1892 patent for a CI engine, used this resource found in plentiful quantities around the mines and factories of Germany's industrialized Ruhr Valley. The fuel did not work well, and Diesel searched for an alternative liquid fuel. In 1900, the demonstration of an early prototype engine took place at the Paris World Exhibition in France. This engine was fueled by the first biodiesel fuel, made of peanut oil. Encouraged by the French government to support its colonies around the world, the intention was to develop a renewable nonpetroleum-based fuel that would permit small industry, farmers, and other producers to effectively compete with the monopolized oil industry of the time. In fact, until the 1920s, many diesel and other CI engines were powered by biologically derived fuels.

▶ TECHNICIAN TIP

Biodiesel refers to diesel fuel derived from biological sources, such as vegetable oils or animal fats. In contrast, petroleum diesel is refined from crude oil. Crude oil is considered a mineral oil because it is extracted from the earth. Biodiesel is fuel made exclusively from biological sources, such as plant oil or animal fat, or a combination of petroleum-based fuel and biological fuel sources. The number following the letter "B" indicates the percentage of biologically sourced fuel combined with petroleum fuel. Thus, biodiesel made from 20% biological fuel sources and 80% petroleum fuel sources would be designated B20.

Diesel Fuel Standard Set by American Society for Testing Materials

For a number of reasons, the more favored alternative to biologically based diesel engine fuels are derived from petroleum. **Petroleum distillates**—that is, fuel recovered from distilling crude oil—could readily burn in the diesel engine and were widely available because there was plenty of this byproduct left over after refining gasoline used to fuel automobiles and other vehicles. The same fuels commonly used for lanterns and heating homes (kerosene and other light hydrocarbons) could also be used to operate diesel engines. Until the 1930s,

▶ TECHNICIAN TIP

In the United States, diesel fuel properties are regulated by ASTM International and the Environmental Protection Agency (EPA). While most characteristics of diesel fuel are defined by ASTM, the EPA also imposes additional environmental standards to minimize the formation of toxic emissions.

these petroleum products varied widely in composition and quality, so their performance was unpredictable. Manufacturers began to develop diesel fuel specifications after observing numerous engine failures associated with fuel composition. Today, **ASTM International** (formerly known as the American Society for Testing Materials) establishes diesel fuel standards. While most countries possess their own separate and distinct performance standards for diesel fuel, the ASTM-defined fuel standards are widely recognized around the world. Petroleum diesel fuel is defined in ASTM standard D975, and biodiesel specifications are set in ASTM standard D6751. ASTM diesel fuel grades are denoted as 1-D (for Diesel #1) and 2-D (for Diesel #2); 1-D and 2-D are both commonly used by on- and off-road diesel engines. In North America, 2-D is commercially retailed at pumps and 1-D and 2-D are both available at pumps outside North America (**FIGURE 8-1**). ASTM grade 4-D is used primarily as home heating oil, but it is also used by some large-displacement, stationary, low-speed diesel engines in applications predominantly involving constant speed and load, such as water pumps and power generators. Because 4-D is not refined in the same manner as 1-D and 2-D, it has impurities and other characteristics that will quickly damage on-highway engines, so it should never be used in them. There is no ASTM 3-D designation.

FIGURE 8-1 The two common ASTM diesel fuel grades for on-highway use are 1-D and 2-D. High-speed automotive diesels, which are constantly changing engine speed, operating in urban conditions are more suited to use #1-D.

▶ Production of Diesel Fuel

8-02 Describe the production methods used to produce diesel.

Fuels such as diesel or gasoline are derived from crude oil. Crude oil is a mixture of various types of hydrocarbon molecules, which are composed of hydrogen and carbon (**FIGURE 8-2**). Some of the molecules are small, having higher hydrogen-to-carbon ratios and are therefore lighter. Others are heavier, having higher carbon-to-hydrogen ratios. High-carbon molecules are typically larger and heavier molecules. By boiling the oil and collecting oil vapors through a process known as **distillation**, oil molecules are separated into **fractions**, or cuts, based on the boiling point of each fraction. Smaller, lighter hydrocarbon molecules have a lower boiling point, whereas larger, heavier molecules have a higher boiling point. The boiling and vapor separation process produces petroleum products ranging from propane and butane (called light ends) to gasoline and **middle distillates**, which include diesel fuel (**FIGURE 8-3**).

Diesel and gasoline fuels are different in that diesel fuels contain heavier hydrocarbon molecules, which have a higher boiling point than gasoline and a higher carbon-to-hydrogen ratio. Kerosene, home heating, and jet fuel account for another proportion of

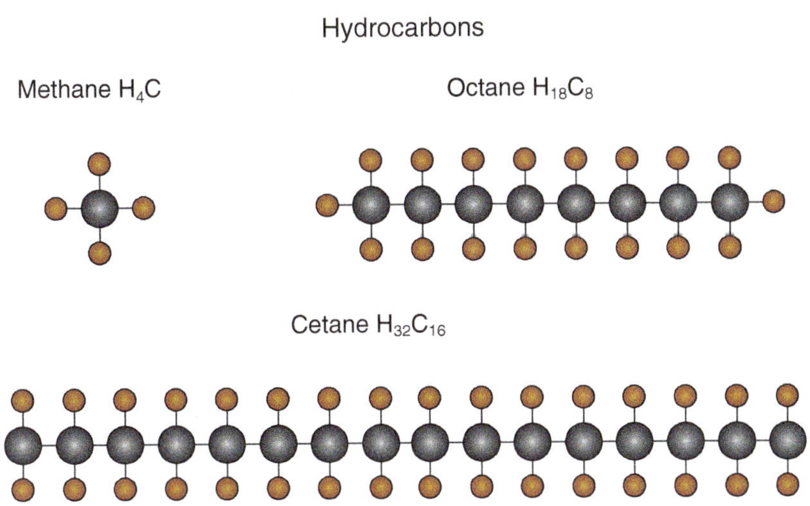

FIGURE 8-2 Fuels properties often depend on different carbon-to-hydrogen ratios. The number of atoms in a fuel molecule and the type of chemical bonds it has determine a fuel's properties.

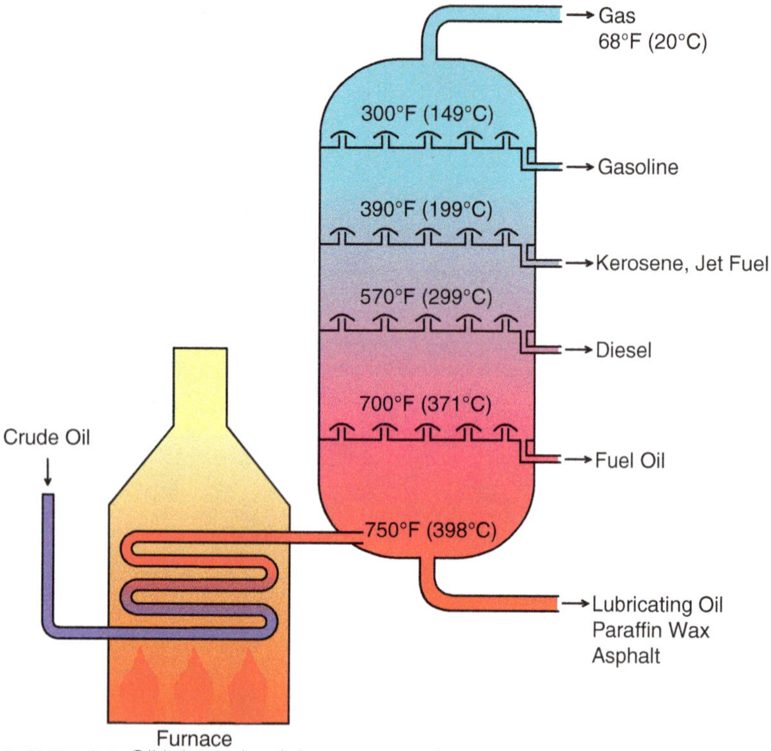

FIGURE 8-3 Oil is heated and the vapors condensed to distill various types, or fractions, of fuel. Diesel fuel is a heavier fraction than gasoline and a similar fraction to home heating and jet fuel.

middle-distillate production, and they have specifications and physical properties similar to diesel fuel. However, while similar in properties, they must be refined for their specific applications. Fuel specifications developed to operate a camp stove, fly an aircraft, and power an on-highway engine are different and are not interchangeable.

Diesel fuel is also produced by liquefying hydrocarbon gases, such as methane, and modifying the chemical structure of the hydrocarbon molecules. Such **gas-to-liquid (GTL) diesel fuel** is manufactured by using a chemical process invented in Germany during the 1920s called the **Fischer–Tropsch reaction**. Coal, carbon monoxide (CO), and gases produced by decomposing wood and other plant material can all be used to manufacture this kind of **synthetic diesel fuel** (that is, diesel fuel not derived from petroleum). Testing such fuels manufactured from this process has successfully demonstrated the viability of this fuel. However, the high operation and maintenance costs associated with its use, plus volatile oil prices, have prevented greater commercial production of this fuel.

Historically, diesel fuel has been cheaper to buy than gasoline because it is a byproduct of gasoline production. With new standards required to produce diesel fuel with lower emissions, its production costs have increased slightly in recent years in North America. In Europe, diesel fuel is typically still cheaper than gasoline because it has favorable taxation policies to encourage diesel's use as a more economical fuel source. In the US, production costs for diesel fuel are less than gasoline, but higher taxes and profits made on diesel fuel sales push its price higher than gasoline in most regions.

Diesel Fuel Color

Quality commercial diesel fuel is normally clear to light amber in color, depending on the refining process and the source of the oil (**FIGURE 8-4**). Certain diesel fuels are dyed to differentiate their sources and applications. Aviation fuel, for example, is dyed blue. Diesel fuels intended for home heating oil and agricultural machinery are dyed red because they are taxed at a lower rate and are not permitted in vehicles for on-highway use. The dyes have no effect on fuel performance or emissions.

Diesel fuel may darken after prolonged storage, due to oxidation. This darkening is often accompanied by performance problems due to a change in the ignition quality of the

FIGURE 8-4 Fresh, clean petroleum-based diesel fuel is clear when commercially retailed in North America. Dyed fuels are intended for specific uses and taxation categories.

fuel. If the darkening is accompanied by the formation of sediment, the fuel can foul and plug fuel filters (**FIGURE 8-5**).

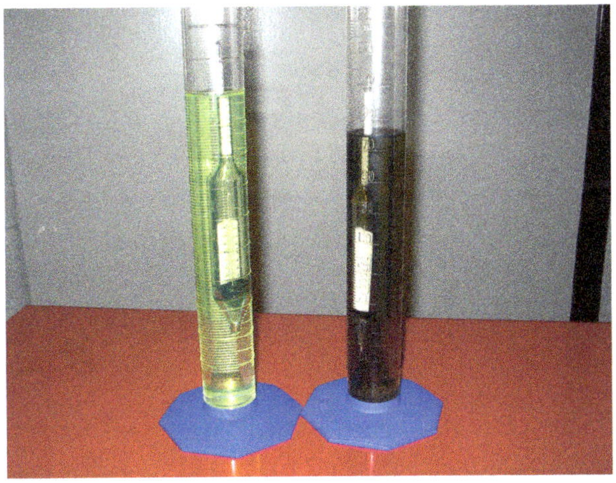

FIGURE 8-5 Older fuel darkens and becomes cloudy as it absorbs water. It will also develop a varnish- or turpentine-like odor.

Sulfur and Diesel Fuel

Sulfur naturally occurs in much of the crude oil that is pumped from underground deposits. (Biodiesel has no sulfur content.) Sulfur not removed during the refining process is present when the fuel burns. The sulfur content of diesel fuel affects particulate matter (PM) emissions because some of the sulfur in the fuel is converted to sulfate particles in the exhaust. How much sulfur is converted to PM varies from one engine to another, but reducing sulfur decreases PM in almost all engines. For this reason, the EPA began to limit the sulfur content of on-road diesel fuel in 1993 to 0.05% mass or 500 ppm maximum, which mandated the production of **low-sulfur diesel (LSD) fuel**. Such regulations are not necessary for biodiesel, because it has no sulfur content.

Reducing the sulfur content in fuel can increase engine longevity. Sulfur can deposit on rings, which causes premature wear and sticking. Sulfur can also combine with water and other substances to damage fuel injectors and contaminate engine oil, resulting in the production of corrosive compounds that damage bearings and other metal parts. Lower sulfur content in the exhaust stream also translates into reduced corrosion of the exhaust system.

> ▶ **TECHNICIAN TIP**
>
> When evaluating diesel fuel quality, in addition to color and the American Petroleum Institute (API) rating, the fuel's odor is another indicator of its quality. Fresh fuel has a strong, pungent, characteristic odor. Old or stale fuel lacks a strong odor when exposed to air and smells like varnish or turpentine.

Ultra-Low-Sulfur Diesel

Ultra-low-sulfur diesel (ULSD) fuel has its sulfur content reduced from approximately 500 ppm to 15 ppm. This cleaner diesel fuel, designated S15, which is appended to the fuel grade (e.g., 1-D S15), became available beginning in 2006, before new diesel vehicles were introduced to meet the EPA 2007 emissions standards (**TABLE 8-1**). Emissions reductions from diesel vehicles and equipment require the use of exhaust aftertreatment devices, such as particulate filters (DPFs), NO_x adsorbers (NACs), liquid selective catalyst reduction (SCR), and oxidation catalysts. ULSD is necessary to avoid damaging or "poisoning" the catalytic materials in these devices (**FIGURE 8-6**). In contrast to North American standards, the European Union has used a 10 ppm ULSD standard since 2009.

Aromatic Content

Aromatic content refers to a particular type of hydrocarbon molecule that contains sulfur and nitrogen. Aromatic molecules are large, complex, and difficult to burn. The characteristic odor of diesel exhaust is attributed to the aromatic content of diesel fuel. Because aromatic fuel molecules are hard to break apart and burn, their presence in fuel contributes to increases of emissions (**FIGURE 8-7**). Current environmental legislation limits aromatic content to 15% by volume, with some jurisdictions

TABLE 8-1 Benefits of Using Sulfur-Reduced Fuel

Diesel Fuel Type	Sulfur Limit	U.S. EPA Year of Mandate
On highway	500 ppm	1993
	30 ppm	2004
	15 ppm	mid-2006
	2,500+ ppm	pre-1993
Off road	500 ppm	mid-2007
	15 ppm	mid-2010
	500 ppm	mid-2007
Railway and marine	15 ppm	mid-2012

FIGURE 8-6 ULSD fuel is mandatory for contemporary exhaust aftertreatment systems. This green color fuel cap indicates that the required fuel is ULSD.

opting for 10%. California has limited aromatic content to 10% since 1993, with a 20% exemption for small refineries. ASTM standards limit aromatic content to a maximum of 35%.

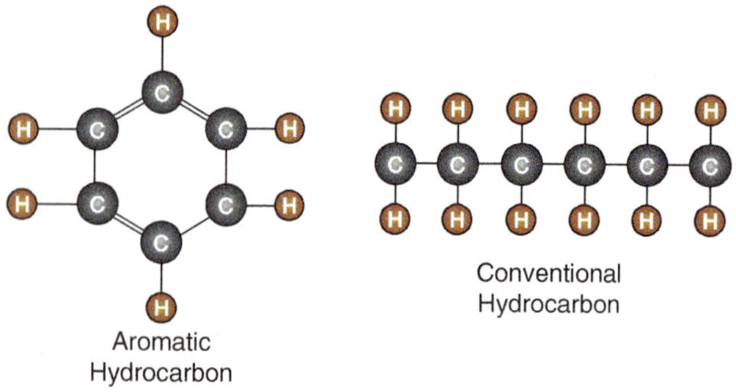

FIGURE 8-7 Aromatic fuel molecules are complex bonded compounds that are difficult to burn. Unburned aromatic molecules give diesel exhaust its distinct odor.

8-03 Identify and describe the purpose, functions, and important properties and characteristics of diesel fuel quality.

▶ TECHNICIAN TIP

Hydro-treating fuel, which is a chemical process used by refineries to remove sulfur, simultaneously reduces the aromatic content of fuel. Fuels with lower aromatic content may cause fuel system leaks if fuel system components have operated and absorbed fuels with high aromatic content. High aromatic content fuels have molecules that may be absorbed by some fuel system seals and O-rings. The use of low aromatic fuels may cause the some seals and O-rings to lose these molecules and shrink, resulting in fuel system leaks. B100 biodiesel has no aromatic content.

▶ Performance Requirements for Diesel Fuel

Diesel engines place a number of performance requirements on the fuels:

1. *Starting ease*: Diesel fuel needs to ignite at the low temperatures and pressures found in the combustion chamber during cranking. Immediately after the engine starts, a relatively cool combustion chamber can extinguish combustion flames, which causes small fuel droplets to be left unburned and forms white smoke. Diesel fuel should have properties that promote complete combustion at low pressures and temperatures.
2. *Lubricating properties*: The lubricating quality of fuel must aid in reducing engine and fuel system wear. For example, high-pressure injection systems have many moving parts that operate with very small clearances, which are lubricated only by diesel fuel. Fuel should minimize wear by possessing good lubricating properties. (See the Lubricity section in this chapter.)
3. *Efficient production of power*: Engine design factors are the most critical for extracting energy from the combustion process. Diesel fuel requires sufficient heat content to produce power ratings published by the manufacturer. Waxes and other heavy hydrocarbon compounds contained in diesel fuel produce the greatest heat release from diesel fuel.
4. *Operation during low-temperature conditions*: Diesel fuels contain wax-like molecules that solidify at low temperatures. Gelled fuel will foul or plug filters and fuel lines and prevent fuel from reaching the engine. To minimize this problem, refiners of diesel fuel blend fuels and additives to suit local climate conditions where engines are operated.
5. *Low noise production*: Prolonged ignition delay time leads to increased combustion noise from the diesel engine. Diesel fuel properties can affect noise by increasing or decreasing ignition delay time.
6. *Long filter life*: Diesel fuel should be free of particles that cause abrasive wear in the fuel system. Other particles not dissolved in fuel include gums, resins, and asphaltenes that can quickly foul and plug filters.
7. *Good fuel economy*: Fuel economy and power are proportional to the heat content of diesel fuel. Increasing the density of fuel will increase its heat content and improve fuel economy.
8. *Low emissions*: Fuel chemistry can have an impact on engine emissions. Aromatic fuel and the sulfur content in fuel, as well as other chemical properties, can change an engine's emission output. Diesel fuel contains more paraffin and other wax-like compounds than gasoline does. These waxes will congeal when the fuel is cool and affect engine performance. For this reason, additives are blended into fuel by refineries according to geographical and seasonal requirements. This means fuel purchased in a cold region may not perform well if the vehicle travels to a warmer, distant geographical location. (See the Pour Point and Cloud Point section in this chapter.)
9. *The flash point*: The **flash point** of a fuel is defined as the temperature to which the fuel must be heated to produce sufficient vapor that will ignite when exposed to a spark or open flame (**FIGURE 8-8**). If the flash point of a fuel is too low, the fuel is a fire hazard that is prone to accidental ignition and even explosion. For most diesel fuels, the flash point is 100°F–125°F (38°C–52°C),

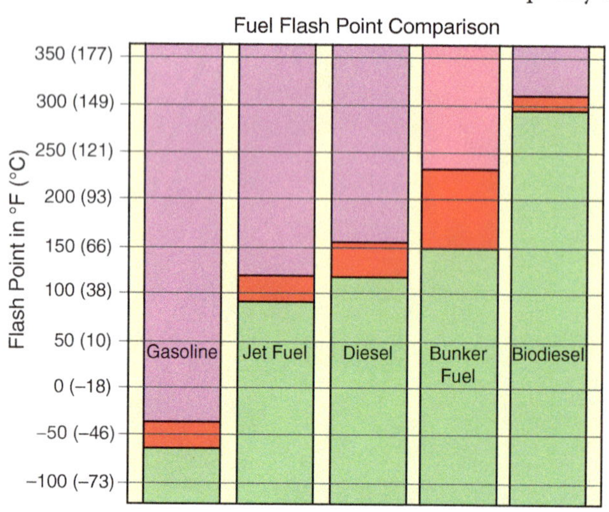

FIGURE 8-8 The flash point of diesel fuel is much higher than gasoline, making it safer to handle. This also means fuel vapors are not a significant evaporative-type noxious emission.

depending on the grade of diesel fuel. This contrasts with gasoline, which has a flash point of at least -45°F (-43°C) or lower. If both diesel and gasoline are at room temperature, a lighted match could likely be extinguished in the diesel fuel. The same action with gasoline would likely produce an immediate fire. (Do not attempt to try this!) Because the flash point of diesel is so high, it is a safer fuel to handle and store (unless it is contaminated with gasoline or alcohol). Diesel fuel's low volatility and high flash point are why diesel fuel pump islands often appear dirty and greasy. Diesel fuel is slow to evaporate, whereas gasoline quickly turns into a vapor (**FIGURE 8-9**).

10. *Auto-ignition temperature*: The **auto-ignition temperature** of diesel fuel is different from the flash point. This is the temperature at which a fuel will ignite when heated. The lowest permissible auto-ignition temperature of diesel fuel sold in North America is 292°F (144°C). Typically, diesel fuel ignites at temperatures of 39F2–482°F (200°C–250°C). Ignition temperatures are even lower when under the compression pressure inside a combustion chamber. It should be noted again that although gasoline is more volatile and possesses a lower flash point than diesel fuel, it has a higher auto-ignition temperature (**FIGURE 8-10**). Thus, if gasoline is mixed with diesel fuel, it produces a longer ignition delay time, causing excessive combustion noise, excessive exhaust smoke, and other problems.

FIGURE 8-9 Diesel fuel's low volatility makes it difficult to evaporate. Spilled, unevaporated diesel fuel makes diesel fuel pump islands appear dirty and greasy.

SAFETY TIP

Not only is mixing alcohols or gasoline with diesel fuel a harmful practice for the engine and fuel system, but it also lowers the flash point of the fuel. This means the fuel will ignite more easily if exposed to an open flame and will become more hazardous to store and handle. Alcohols are also reactive and can corrode metal inside the fuel system and compromise the lubricating characteristics of diesel fuel.

Volatility

Volatility refers to the quantity of fuel evaporating at a particular temperature. Diesel fuel boils at 302°F–662°F (150°C–350°C). Volatility is expressed as a percentage of the fuel evaporating between those boiling temperatures. For example, to express the volatility standard of 2-D fuel, 90% would be evaporated at 550°F–650°F (288°C–343°C).

Evaporative emission systems are not required on diesel-fueled vehicles, because its volatility is low. Even in warm weather, the vapor pressure that diesel fuel exerts inside a storage tank is insignificant when compared to gasoline (**FIGURE 8-11**).

▶ TECHNICIAN TIP

Diesel fuel filling stations may often have a lingering smell of diesel fuel because of the higher flash point and distillation temperature. A diesel fuel spill takes days to vaporize and dissipate, unlike gasoline, which takes minutes.

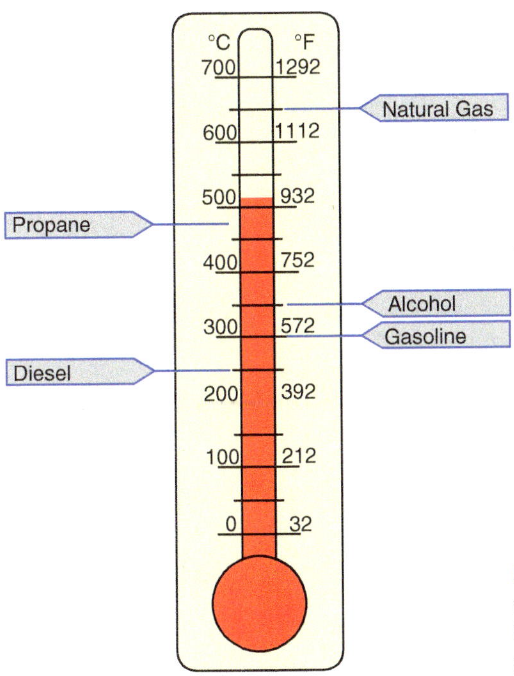

FIGURE 8-10 The auto-ignition temperature of diesel fuel is lower than that of gasoline and gaseous fuels. This means diesel fuel ignites sooner inside a pressurized combustion chamber.

FIGURE 8-11 Unlike gasoline tanks, diesel fuel tanks can be vented directly to the atmosphere because diesel fuel is not volatile.

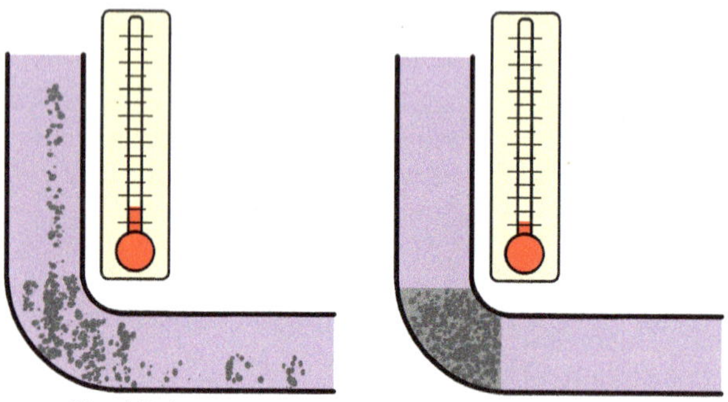

FIGURE 8-12 Cold temperatures will cause the wax content of diesel fuel to form crystals at cloud point and to congeal at pour point.

Pour Point and Cloud Point

Pour point and **cloud point** refer to the temperatures at which wax molecules in fuel crystallize. More specifically, cloud point, as the name suggests, is the temperature at which wax begins to give fuel a hazy or milky appearance. The pour point occurs when the fuel becomes even colder. Pour point is the temperature at which diesel fuel thickens—the so-called fuel-gelling phenomenon—and no longer pours. When fuel reaches pour point, it cannot pass through lines and filters (**FIGURE 8-12**).

While some engines fail to run at the cloud point, due to filter plugging, all engines will fail to operate at the pour-point temperature. Usually, in North America, the cloud point and pour point are at least on average 20°F (-6°C) apart. The actual difference depends on a variety of factors, including the origins of the fuel and the pour-point temperature depressants blended into the fuel by the production refinery. Without additives and conditioners, the cloud point for 2-D fuels are closer. Less pour-point depressant, which raises the temperature of cloud, and pout points are needed in the summer, whereas more is blended into fuel made for colder winter seasons. How much additive a refinery blends into the fuel is adjusted on a geographical and seasonal basis, adjustments that take the base stock fuel properties into account. In a cold climate, winterized fuel is a blend of 1-D and 2-D or jet aviation fuel (Jet A), which is fuel with low wax content designed to operate at high altitudes in very cold temperatures. Diesel 1-D has a lower pour point, which allows it to move through lines and filters more easily, but it also has a lower viscosity, which means it can more easily leak past operating clearances between high-pressure injection system components. Low-power problems can occur when using winter grade fuel because it has less energy content. The cloud and pour points of different grades of diesel and biodiesel vary (**TABLE 8-2**).

Low-End Design Temperature

To operate diesels year round and in cold temperatures, the properties of seasonal diesel fuel are adjusted throughout the year according to historical temperature data of the region where the fuel is sold. For example, in Ontario, Canada, a "Muskoka blend" is prepared for Southern Ontario and an "Arctic blend" is used in Northern Ontario. The refinery that produces the fuel will blend additives and even other grades of fuel together to produce fuels that best match seasonal variations in temperature according to the geographical location for where the fuel is to be retailed.

Lubricity

Lubricity refers to a fuel's lubricating quality. Many fuel system components rely on the lubricating properties of the fuel to prevent "galling" (i.e., the scarring of metal surfaces) and prevent moving parts from seizing. It is important for fuel to have good lubricating properties, to enhance the fuel system's durability. The lubricating properties of diesel fuel are provided by compounds such as fats and waxes that are already naturally present in diesel fuel.

TABLE 8-2 Typical Cold-Flow Properties of Three Grades of Diesel and Biodiesel

Diesel Fuel	Cloud Point	Pour Point	Cold Filter Plugging Point
2-D	5°F (−15°C)	−31°F (−35°C)	−4°F (−20°C)
B100 (Canola Formulated)	26°F (−3°C)	25°F (−4°C)	24°F (−4°C)
B100 (Soybean Formulated)	38°F (3°C)	25°F (−4°C)	28°F (−2°C)

A scuffing load ball-on-cylinder lubricity evaluator (SLBOCLE) test evaluates a fuel's ability to lubricate, which is expressed in grams of load. The higher the SLBOCLE value, the better the lubricating ability. Another test that determines lubricity is the high frequency reciprocating rig (HFRR). This test measures the depth and width of a scar left on a metal surface after repeated friction with another metal instrument. Depending on the lubricating properties of a fuel sample, a larger or smaller scar is produced.

Kinematic Viscosity

Kinematic viscosity refers to a liquid's resistance to flow. High viscosity means the liquid is thick and slow flowing; low viscosity means the opposite. Fuel viscosity is measured in centistokes. The higher the centistoke measure of a liquid, the thicker and more viscous it is. For example, water has a centistoke of value of 1 and 20W oil is 4.3 centistokes. The property of fuel viscosity affects injector lubrication and fuel atomization (**FIGURE 8-13**). Fuels with low viscosity may not provide adequate lubrication for the close clearances of precision-fit fuel system components, because parts are separated with only thin lubricating film barriers. If fuel's viscosity is too low, it can easily leak and slip around the high-pressure system's plunger and barrels. The result is that less fuel is injected, which reduces power output and increases wear of the fuel system's components. Fuels that are too thick or viscous will also cause other performance-related problems. Because fuel atomization is affected by fuel viscosity, fuel with high viscosity forms larger droplets during atomization. Larger droplets take longer to burn and mix poorly with air, which leads to poor combustion quality and increased exhaust emissions.

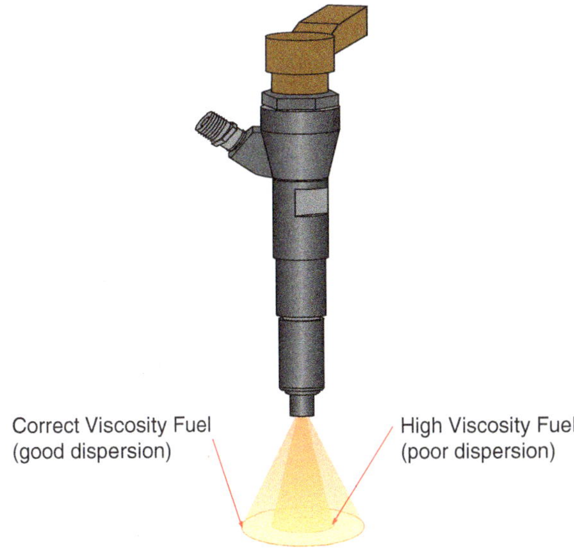

FIGURE 8-13 Fuel viscosity can affect the spray pattern and fuel droplet size from an injector. Highly viscous fuel will not atomize well and will negatively affect combustion quality.

> **TECHNICIAN TIP**
>
> Removing sulfur from diesel fuel does not directly cause fuel to lose lubricating qualities. The process used in North America to remove sulfur from diesel fuel also removes molecules from the fuel that give it its natural lubricating qualities. Low-sulfur (LSD S-500) and ultra-low-sulfur (ULSD S-15) diesel fuel are associated with greater fuel system component wear. Manufacturers have hardened fuel system components with a variety of ceramic alloy metals to better withstand the effects of the poorer lubricating properties of fuel with lower sulfur content.

Mixing Gasoline and Diesel Fuel

Gasoline and diesel fuels should never be mixed. One percent or less gasoline will lower the flash point of a gasoline-diesel fuel blend below the minimum specification for diesel fuel. This does not dramatically affect the engine performance, but the fuel is more hazardous to handle (**TABLE 8-3**).

Without a proper understanding of each fuel property, an incorrect recommendation could be made, such as to mix gasoline or alcohol and diesel fuel together in the winter to minimize fuel gelling and improve combustion. There are several problems with this. First, although gasoline and alcohols do have an anti-gelling effect on diesel fuel, they reduce diesel fuel lubricity. Reduced lubricity can increase fuel system wear and shorten fuel system life. The second problem is that the cetane index or ignition qualities of t gasoline and diesel fuel are inversely related to one another: gasoline burns more slowly than diesel, which contributes to a prolonged ignition delay in the combustion period. Similarly, gasolines require higher temperatures to ignite in a diesel combustion chamber. Gasoline contamination, even in small amounts, will dramatically reduce diesel fuel's ability to quickly ignite, causing a longer ignition delay to take place and pronounced engine knock. This is why gasoline contamination of diesel fuel produces a lot of smoke, causes engines to run roughly, and increases engine noise. Diesel fuel that is mixed with gasoline or alcohol simply does not have enough time to completely burn, because ignition begins later in the combustion cycle. So, while some may think mixing diesel fuel with gasoline, which has higher volatility,

TABLE 8-3 Typical Ignition Quality of Gasoline and Diesel Fuel

Property	Gasoline	#2D
Flammability limits (volume in air)	1.4–7.6%	0.6–5.5%
Auto-ignition temperature	536°F (280°C)	494°F 256°C
Flash point	−45°F (−43°C)	100°F–125°F (38°C–52°C)

> ▶ TECHNICIAN TIP
>
> If diesel fuel is inadvertently contaminated with gasoline, it should be drained from vehicle storage tanks, lines, and filters. After replacing the tainted fuel with good-quality diesel fuel, the engine lube oil should be changed if the engine has been operated with contaminated fuel. This step is needed because engine misfires and poor combustion quality will dilute engine oil with unburned fuel that has washed down the cylinder walls and into the crankcase oil.

will improve a diesel's ability to start, the opposite is true. Gasoline needs more heat and time to burn in the combustion chamber, which makes the diesel engine harder to start and run very roughly.

▶ Biodiesel Fundamentals

8-04 Identify and describe the properties and characteristics of biodiesel fuel and compare them to conventional petroleum diesel fuel.

Diesel engines burn not only those fuels made from petroleum oils but also fuels made from biological sources, such as animal fat or plant matter. Some diesel engines will run using pure vegetable oil for fuel. With legislation requiring the use of more renewable fuels, the use of biodiesel will become more widespread. Prior to 2011, almost all manufacturers endorsed the use of B5 blends of fuel, but B20 biodiesel was commonly available. Many countries have mandated the use of biodiesel fuel. In the United States, for example, the **Renewable Fuel Standard (RFS)**, part of the **Energy Independence and Security Act (EISA)** of 2007, encourages the use of biofuel by requiring manufacturers to certify engines to meet emission standards using B20 biodiesel fuel. The EISA also requires that 30% of all fuels originate from renewable sources by 2022.

Pure biodiesel can be used as fuel (such as B100) or blended with any amount of petroleum diesel to add the beneficial characteristics of biodiesel to petroleum diesel (such as B5 and B20).

Biodiesel Advantages

Biodiesel offers a number of benefits, both economic and environmental, for energy security, job creation, agriculture, rural development, energy supply, and public health. Although there are drawbacks to biodiesel, studies indicate that biodiesel can achieve as much as a 78% reduction in the amount of carbon dioxide (CO_2), a greenhouse gas, produced by automobiles burning fuel over the lifecycle of production and use. Furthermore, while production of biodiesel requires more energy input than petroleum-derived diesel fuel, biodiesel yields 3.24 units of energy for every 1 unit of biodiesel used during production. The following are some of the additional benefits of biodiesel:

- It is a renewable resource.
- It is biodegradable.
- It is not toxic to the environment if spilled.
- It contains little or no sulfur.
- It requires no engine modifications to use it.
- It produces lower hydrocarbon (HC), CO, and PM emissions than petroleum-based diesel.
- It has enhanced lubricating properties as compared to LSD and ULSD petroleum diesel.
- It has no aromatic content.
- It has a naturally high **cetane number (CN)**, a measure of the ignition quality of fuel.

Biodiesel Performance

Operationally, biodiesel performance is similar to conventional diesel in terms of power, torque, and fuel consumption without having to modify the fuel system. Biodiesel can also be blended into petroleum-based diesel, and for the most part, no engine modifications are required to use blends of biodiesel and petroleum diesel. The shape and composition of biodiesel fuel molecules provide other advantages that make it appealing as an alternative to conventional diesel fuel:

- Biodiesel has no sulfur content, which means fewer emissions and increased engine longevity.
- Biodiesel has no aromatic content, which improves emissions and reduces the risk of carcinogens.
- Biodiesel has about 11% oxygen content (petroleum-based diesel contains no oxygen) for improved combustion characteristics.
- Biodiesel has a naturally higher CN, commonly 50–55.

- Biodiesel has better lubricity. Conventional diesel fuel has a minimum standard of 109 ounces (3090 grams) SLBOCLE, whereas biodiesel has 176 ounces (4990 grams). Even mixing biodiesel with conventional fuel in a concentration as low as 1% can improve lubricity by 65%.
- Biodiesel has a higher flash point, which means the fuel is safer to handle.

Because biodiesel is made from vegetable oil or animal fat, it is comparable to table salt in toxicity and biodegrades as fast as sugar. This means B-100 it is not hazardous to the environment if it is spilled or leaks from an underground storage tank.

Biodiesel has favorable emission characteristics (**FIGURE 8-14**). Biodiesel fuel has been rigidly evaluated for its production of emissions and potential health effects, as required of all alternative fuels by the U.S. EPA under the Clean Air Act. Using B100 fuel, for example, will result in the following emission benefits:

- a 67% reduction in unburned HCs
- a 48% reduction in CO
- a 47% reduction in PM emissions.

Biodiesel Disadvantages

Although biodiesel has many advantages, it also has certain disadvantages:

- *Cost*: Biodiesel is currently more expensive than petroleum diesel fuel due to the oil content of its sources, such as the soybean, which is only 20% oil.
- *Energy input*: Biodiesel requires a high input of energy to produce compared to conventional diesel fuel. For example, soybeans require energy to sow, fertilize, and harvest.
- *Fuel system incompatibility*: Biodiesel is an excellent solvent and leaves fuel systems cleaner. However, biodiesel fuel can damage some types of rubber found in fuel hoses, gaskets, and wire insulation used in some vehicles, particularly in vehicles built before 1994. Currently, fuel system part manufacturers are switching to components suitable for use with low-sulfur and biodiesel fuel.

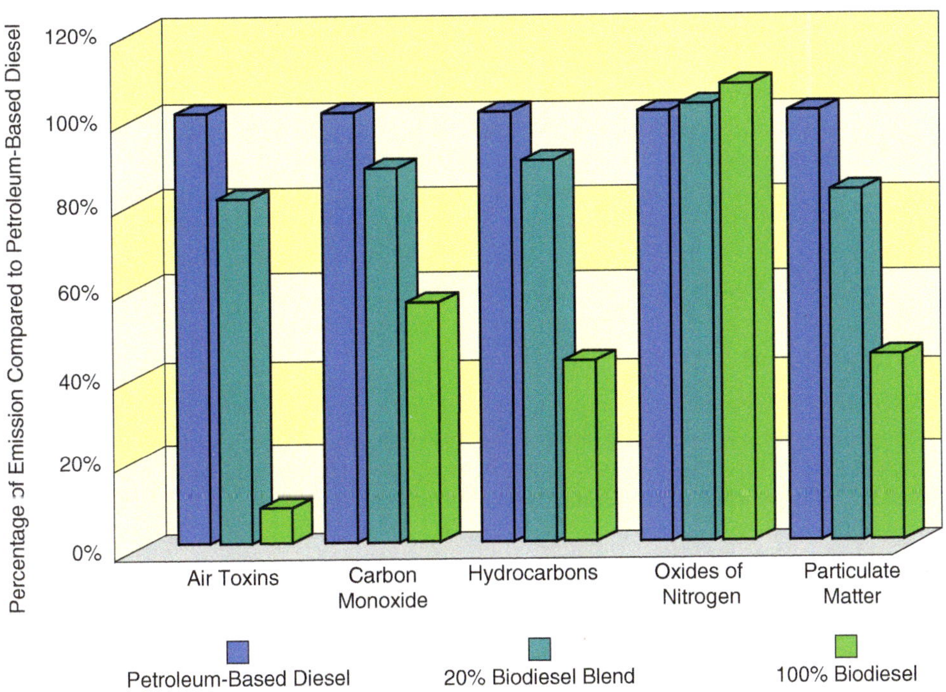

FIGURE 8-14 Biodiesel generally produces fewer noxious emissions compared to conventional diesel fuel. Note the increase in NO_x emissions with biodiesel.

> **▶ TECHNICIAN TIP**
>
> Before switching to biodiesel, technicians should check with a vehicle manufacturer to ensure materials used in the fuel system are compatible with biodiesel use. When biodiesel fuels are used in vehicles built before 2010, seals, hoses, gaskets, and even wire coatings require regular monitoring. Currently, most vehicle manufacturers will repair fuel-related failures under warranty only if B5 fuel (conventional diesel plus a 5% mixture of biodiesel) is used.

- *Filter plugging upon switching*: Plugging typically occurs when switching from conventional diesel fuel to biodiesel. Due to the solvent action of biodiesel, the dirt removed can quickly foul filters, which should be changed for several hours after engine operation.
- *Distribution infrastructure*: Biodiesel is not distributed as widely as traditional petroleum diesel. Production, blending, and distribution are performed on a smaller scale than conventional petroleum diesel. However, distribution infrastructure is improving, and many fleets use a blend of biodiesel for their operations.
- *API standards*: Despite attempting to meet ASTM D6751 standards, numerous problems with biodiesel stem from its manufacturing processes. For example, if too much methanol is used, it will cause the aluminum and zinc components in the fuel system to corrode.
- *Aromatic content*: Because there is no aromatic content in biodiesel fuel, it can cause rubber and elastomer components to crack and deteriorate.
- *Heat content*: Biodiesel has less heat content than petroleum diesel fuel. Fuel economy can be reduced by as much as 10% when using pure biodiesel.
- *Pour point*: Biodiesel has a higher pour point (5°F to 10°F [−15°C to −12°C]) than conventional diesel fuel. This means it can thicken and gel more readily in cold weather. To remedy this condition, biodiesel can be blended with conventional 2-D or 1-D diesel fuel. Biodiesel can also be treated with pour-point depressants. (Fuel system heaters and heated storage are also recommended.)
- *Shelf life*: The shelf life of biodiesel is not as long as conventional diesel. What makes biodiesel biodegradable also makes that biodiesel fuel—in its pure form—unstable. Over time, its acid content increases, which can damage metal components. Fuel aging is also accelerated in the presence of heat, oxygen, water, and metal ions. Storing biodiesel for longer than six months is not recommended. If biodiesel is stored for longer than six months, it should be reevaluated to ensure that it meets ASTM D6751.
- *Polymerization (gelling) of engine oil*: The most serious problem with the use of biodiesel is its effect on engine oil. On engines built in 2007 and later that are equipped with DPFs, which are actively regenerated by an additional injection event during the exhaust stroke, some fuel will cling to the cylinder walls. Rings will sweep the fuel into the oil, where it will accumulate because biodiesel's higher flash point will not allow it to boil out of oil, but petroleum-based fuel does. The fuel will cause chemical changes to take place in the oil that will link up chains of hydrocarbons and turn the oil into gel. Lacking proper lubrication, the engine will catastrophically fail. The biodiesel also attacks an important oil additive: zinc dialkyl dithiophosphate (ZDP). Even before the engine is damaged, the oil loses some of its lubricating properties. The solution to this problem is to change how DPF dosing takes place. Rather than supplementally heating the DPF by using a late post-combustion in-cylinder injection event, the latest engines perform hydrocarbon dosing of the DPF by using a fuel injector located in the exhaust system, just after the turbocharger.

Biodiesel and NO_x Emissions

One notable exception to the emission benefits of biodiesel is that it can produce higher levels of NO_x from some engine families. Because engines built prior to 2010 were not certified to operate by using biodiesel, some engines may in fact produce more NO_x emissions by using biodiesel than conventional petroleum-based diesel.

Production of Biodiesel

Biodiesel is manufactured primarily from seeds or grains. Such vegetable oil sources include peanuts, soybeans, rapeseeds (i.e., canola oil), cottonseeds, safflower seeds, linseeds, corn, and sunflower seeds, among others. Animal fats, such as beef tallow, recycled cooking oil, and the like, are also used in some formulations. Using **straight vegetable oil (SVO)**, or unrefined vegetable oil, poses two problems as an acceptable alternate biodiesel fuel source. First, it can lead to coking in the injectors. Coking is

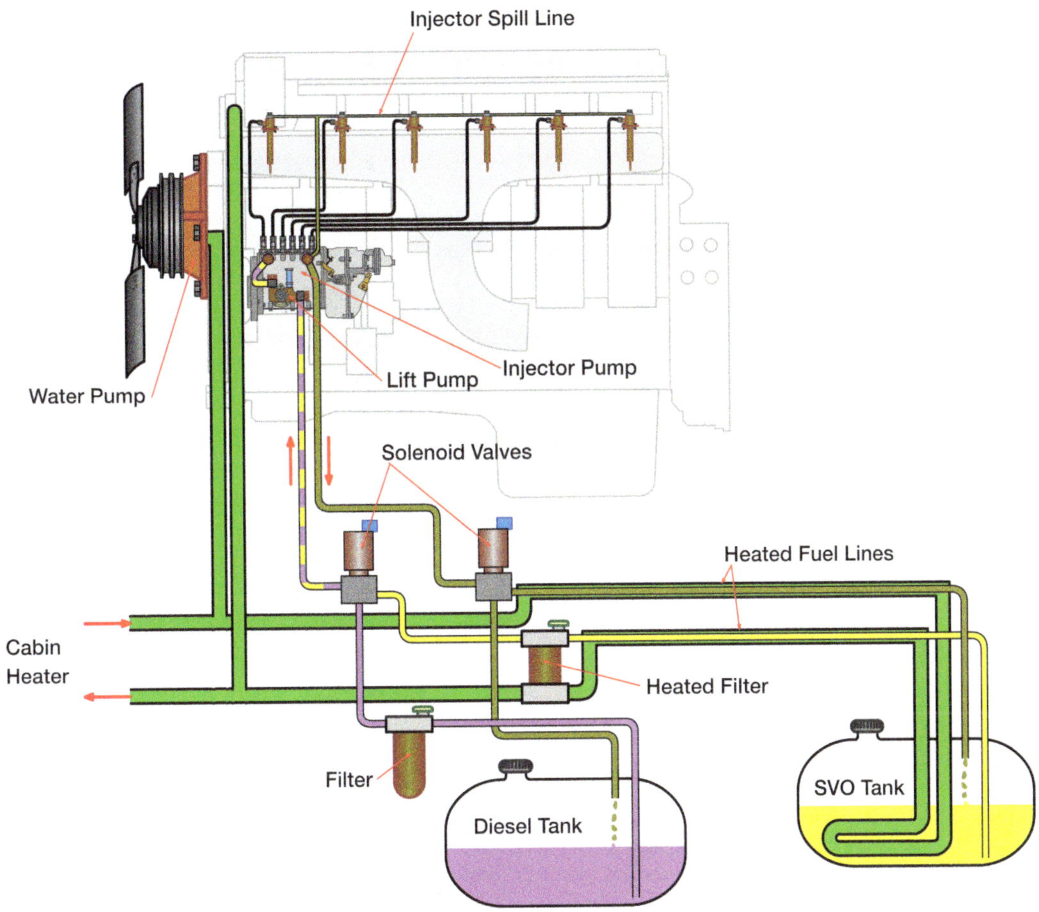

FIGURE 8-15 A typical low-pressure fuel system configuration for using SVO as a fuel source. The oil requires heating to reduce its viscosity to near that of diesel fuel.

the buildup of carbon in the injector nozzle that in turn leads to reduced fuel flow and irregular injector spray pattern. Second, at approximately eight times the viscosity of conventional diesel fuel, SVO is too viscous to move effectively through lines, filters, and other fuel system components. Heating the fuel to thin it out is one solution to this problem (**FIGURE 8-15**). Another is to remove glycerin from the vegetable oil molecule and replace it with an alcohol molecule, which is what happens during the process of transesterification.

Biodiesel Processing

Typically, for biodiesel derived from soybeans and like sources, the vegetable oil is extracted from grains after crushing them and separating solids from oils by using solvents such as hexane. Gummy substances and impurities are removed by mixing the oil with 2–3% water and heating it to 122°F (50°C) while mixing the material. Heavier gums, water, and impurities will settle out of the mixture, and the process of removing the fatty acids for fuel production can take place. In this step, the vegetable oil is reacted with an alcohol such as **methanol** or **ethanol** by using a catalyst.

Transesterification

Several techniques are used to convert animal fats and vegetable oils into biodiesel fuel. Even algae are considered a viable source of oil to produce biofuels. Microproducers, from small companies to individuals, also have processes and formulations to make biodiesel fuel for agricultural use and even private home use.

In the most common method to produce biodiesel, methanol alcohol is used to react with vegetable oils and replace glycerin with alcohol. A catalyst, such as sodium or

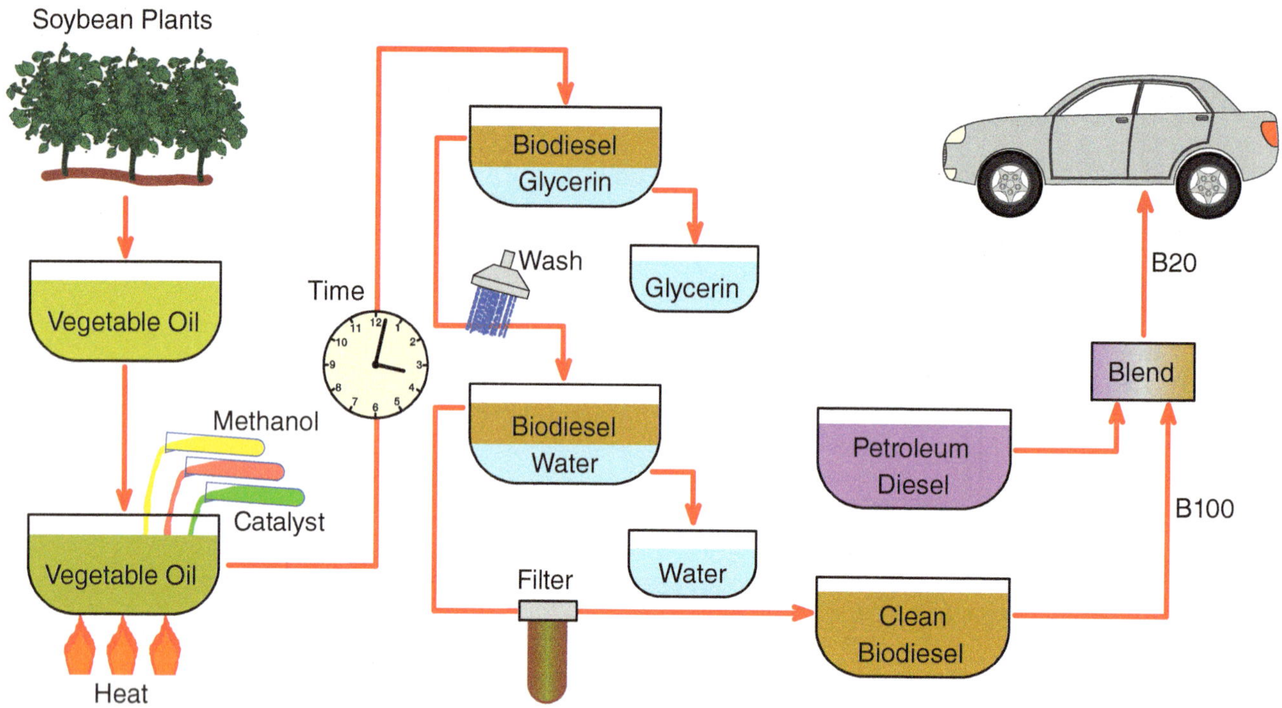

FIGURE 8-16 The basic steps used to convert vegetable oils into usable biodiesel.

potassium hydroxide, speeds the process. The glycerin separates usable fuel from the oil and sediment after some cleaning processes. Replacing the glycerin with alcohol molecules is a process called **transesterification**, that is, the substitution of methanol for fatty acids to change the oil into fuel. **Fatty acid methyl ester fuels (FAMEs)** is another name given to biodiesel produced through transesterification. FAME processes that use potassium hydroxide or sodium hydroxide as a catalyst, however, can cause corrosion in the fuel system, injector plugging, and deposits to form if any catalyst is left in the fuel (**FIGURE 8-16**).

Types of Diesel Fuel

Diesel fuel is manufactured in a variety of formulations to serve different kinds of on- or off-highway vehicles, agricultural equipment, railroad locomotives, and marine vessels (such as river barge tugs, pleasure craft, and ocean-going vessels), among others. Special diesel fuels are also used for stationary engines used to power irrigation equipment and emergency generators, such as those that hospitals rely on during a power outage. This section surveys the different types of diesel fuel, including biodiesel.

▶ ASTM Diesel Fuel Properties

8-05 Identify and describe the classification method used to grade petroleum diesel and biodiesel fuels.

To a large extent, diesel engine development has depended on the development of fuels that contribute to reliable operation and satisfactory performance. The ASTM is responsible for diesel fuel specifications and test methods for petroleum-based diesel fuel and biodiesel fuel (see the Biodiesel Fundamentals section in this chapter). These specifications are defined and explained in ASTM standard D975 and they can differ in characteristics such as sulfur content, flash point, and auto-ignition (**TABLE 8-4**).

Grades of Diesel Fuel

Diesel fuel is classified according to its viscosity and density. As discussed earlier in this chapter, commercially available diesel fuel is numbered as 1-D, 2-D, and 4-D, where 1-D is the less dense and viscous and 4-D is the heaviest and most viscous.

TABLE 8-4 Specifications for Diesel Fuel Grades and Gasoline

Diesel Fuel Grade	1-D Low Sulfur	2-D Low Sulfur	B100 Biodiesel	2-D ULSD	Gasoline	4-D
Auto-Ignition Temperature	410°F–210°C	494°F–256°C	608°F–320°C	604°F–662°F (318°C–350°C)	475°F–536°F (246°C–280°C)	N/A
Cetane Number (CN) (minimum)	40	40	48–65	40	N/A	30
Flash Point (minimum)	100°F (38°C)	126°F (52°C)[b] 109°F (43°C)[c]	340°F–390°F (171°C–199°C)	266°F (130°C)	−49°F (−45°C)	131°F (55°C)
Heat Value	135,000 Btu 142.4 MJ	140,000 Btu 147.7 MJ	117,000 Btu 123.4 MJ	134,000 Btu 141.4 MJ	124,000 Btu 130.8 MJ	N/A
Aromatic Content by Volume (maximum)	35%[d]	35%[d]	N/A	27–32%	N/A	N/A
Sulfur Content (by weight)	0.05%	0.05%	NA	0.0015%	N/A	2.00%
Water Content (maximum volume)	0.05%	0.05%	0.05%	0.05%	N/A	0.50%
Distillation Temperature[e]	550°F (288°C)	730°F (388°C)	680°F (360°C)	730°F (388°C)	N/A	N/A
Viscosity at 104°F (40°C) (maximum)[f]	1.3–2.4 cSt	1.9–4.1 cSt	4.7–6.0 cSt	1.9–4.1 cSt	N/A	13.1–29.8 cSt
Lubricity (maximum microns)[g]	0.018" (460 microns)	0.018" (460 microns)	0.012" (314 microns)	0.018" (450 microns)	N/A	N/A
Density at 59°F (15°C)	6.76 lb/gal. (810 g/L)	7.05 lb/gal. (845 g/L)	7.328 lb/gal. (878 g/L)	7.206 lb/gal. (863 g/L)	6.14 lb/gal. (736 g/L)	N/A
API rating	35–40	26–37	28–40	32–37	56–59	N/A
Reid Vapor Pressure at 100°F (38°C)	N/A	<0.2 psi (<1.4 kPa)	<0.04 psi (<0.28 kPa)	N/A	8–15 psi (55–103 kPa)	N/A

a. See cetane number section.

b. May through October.

c. November through April.

d. 90% volume recovered.

e. 10% California Air Resource Board standard in California.

f. 140°F (60°C); cSt stands for centistoke, the standard measurement for viscosity.

g. The HFRR test is at 140°F Lower numbers have better lubricating properties.

1-D is usually a special purchase by bulk users, such as transit fleets and municipalities with their own fuel distribution systems. In Europe, it is a common grade of diesel fuel available at the fuel pump for use in high-speed automotive diesel engines. 1-D is preferred whenever the lowest emission output and superior cold weather performance is required.

2-D is the most commonly retailed grade of diesel fuel. The higher heat content of 2-D provides better fuel economy. More 2-D is produced from a barrel than 1-D, which translates to lower prices at the pump.

4-D is heating fuel for oil burner furnaces, but it can be tolerated by indirect-injection (IDI) combustion chambers. 4-D has many other properties that making it undesirable for use as an engine fuel. Because it is not taxed like 1-D and 2-D, it is not permitted to be used for on-highway transportation and is not for sale at retail fuel pumps.

Biodiesel is defined as a fat-based fuel made from vegetable and animal sources. Pure biodiesel is designated as B100, with the "B" designating it as biologically derived and the 100 as the percentage of its biologically based source. A 20% mixture of biodiesel with conventional diesel would be identified as B20. Biodiesel is most commonly blended with conventional diesel fuel; it is uncommon to see B100 retailed. (See the Biodiesel Fundamentals section in this chapter.)

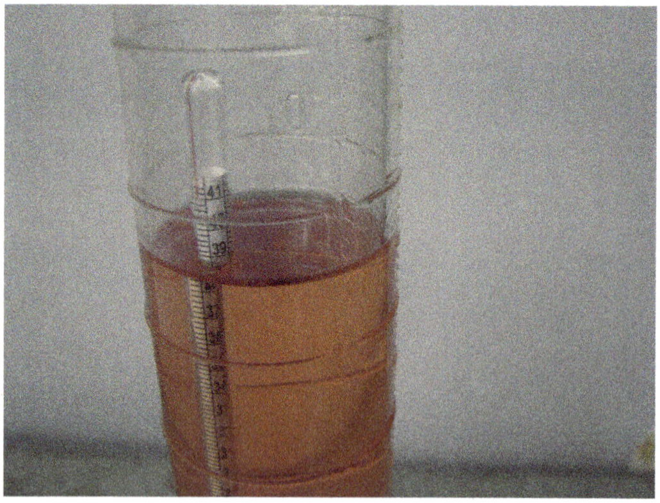

FIGURE 8-17 A fuel hydrometer measures the API number that corresponds to the specific gravity of fuel. When the API number is known, the approximate grade of diesel fuel can be determined.

Determining a Fuel's ASTM Grade

Determining the ASTM grade and API number of a fuel is an important starting point when diagnosing low-power complaints. Seasonally blended fuels with low density will not have the same energy content or characteristics as fuels prepared for warmer seasons and regions. Low-density fuels also typically have low fuel viscosity, which can allow fuel to more easily leak past clearances between plungers and barrels in high-pressure injection systems, which results in reduced fuel delivery. To determine the grade and density of fuel, a temperature-compensated hydrometer is used, which measures a fuel's API number (**FIGURE 8-17**, **TABLE 8-5**). To determine a fuel's ASTM grade, follow the steps in **SKILL DRILL 8-1**.

Cetane Number

The CN is one of the most significant properties of diesel fuel. The CN is a measure of the ignition quality of fuel and refers to the fuel's ignitability under compression. CN can be compared to the **octane number** used to measure gasoline's ignition characteristics (**FIGURE 8-18**). Fuel with a high CN ignites sooner and burns faster than a fuel with a low CN. When pre-ignition temperatures inside a combustion chamber are reduced by variables such as lower compression pressure, ambient temperature, and coolant temperature, an engine requires an increasingly higher CN fuel to start easily and burn quickly.

The CN of a particular diesel fuel is calculated by comparing the sample fuel against a reference fuel with a CN of 100. This means if the fuel has a CN of 45, then the fuel will ignite 45% as well as 100% pure cetane. The ignition quality of the fuel is analyzed by using a single piston engine operated on the sample fuel.

TABLE 8-5 The API Number and Heating Value of Each Grade of Fuel

ASTM Grade	API Number	Density	Btu Heating Value
1-D	38–45	6.675–6.950 lb/gal. (800–833 g/L)	132,900–137,000 per gallon (35,108–36,192 per liter)
2-D	30–38	6.960–7.296 lb/gal. (834–874 g/L)	137,000–141,800 per gallon (36,192–37,460 per liter)
4-D	20–28	7.396–7.787 lb/gal. (886–933 g/L)	143,100–148,100 per gallon (37,803–39,124 per liter)

SKILL DRILL 8-1 Determining a Fuel's ASTM Grade

ASTM Grade	API Number	Density	Btu Heating Value
1-D	38–45	6.675–6.950 lb/gal. (800–833 g/L)	132,900–137,000 per gallon (35,108–36,192 per liter)
2-D	30–38	6.960–7.296 lb/gal. (834–874 g/L)	137,000–141,800 per gallon (36,192–37,460 per liter)
4-D	20–28	7.396–7.787 lb/gal. (886–933 g/L)	143,100–148,100 per gallon (37,803–39,124 per liter)

1. Obtain a fuel sample from the vehicle or equipment fuel tank and place the sample in a graduated cylinder.
2. Using a purpose-made fuel hydrometer, measure the fuel's temperature and API number.
3. Using a chart provided by the hydrometer's manufacturer, look up the temperature-corrected API number for the fuel. Because the density of a fuel is affected by temperature, the corrected API value provides the API number that the fuel would have at room temperature.
4. Using the ASTM standard, compare the API number with the ranges for 1-D, 2-D, and 4-D ASTM grades of fuel.
5. Report the information on a worksheet or work order.

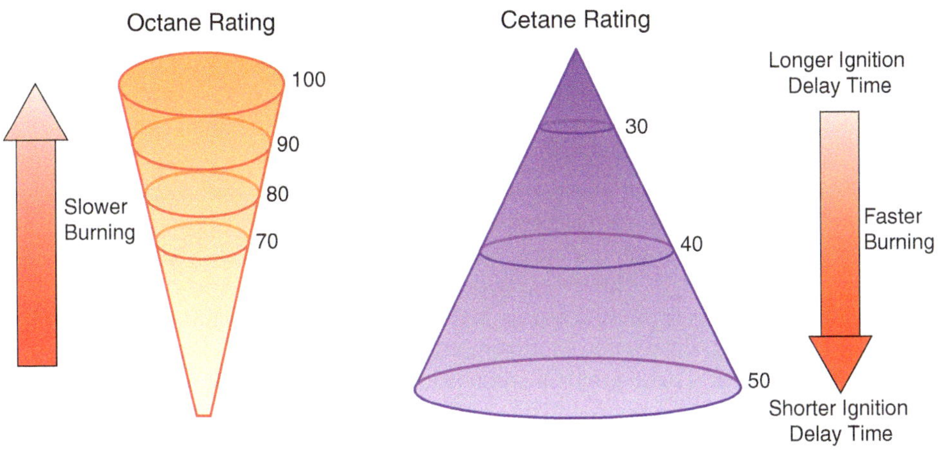

FIGURE 8-18 CN is used to measure the ignition characteristics of diesel fuel. The higher the CN, the faster the diesel fuel will ignite and burn.

The CN for a fuel can be determined only by using specialized test equipment that measures ignition quality. (A correlation between the weight or API number of the fuel measured with a hydrometer and cetane is not an accurate measure.) CN ratings apply only to diesel and middle-distillate fuels, not gasoline.

The octane rating for gasoline is a measure of ignition quality, but the numbering system moves in the opposite direction of CN. Slower igniting and burning gasoline used in high-compression-ratio gasoline engines needing high octane fuels to prevent pre-igntion, whereas slower igniting and burning diesel fuels have lower CNs.

The 2011 ASTM minimum requirement for 1-D and 2-D diesel fuel is CN 40. For biodiesel, the ASTM standard is a minimum CN of 47. When engines operate below 32°F (0°C), a higher CN is recommended. Engines that operate at slower speeds can use lower CN fuel. Large, low-speed engines, such as those used in locomotives and ships, can operate on fuels with a CN as low as 20, whereas most late-model, high-speed passenger car diesel engines operate best using CN 55 fuel. The higher CN fuel results in shorter ignition delay, providing improved combustion quality, lower combustion noise, easier cold starting, faster warm-up, less white smoke, and, in many engines, a reduction of emissions. The type of fuel molecules in high-CN fuel break apart more easily too, but they tend to release less energy. This means that a naturally high-cetane fuel, such as 1-D, produces less power and increases fuel consumption relative to 2-D. In fact, as of 2011, 2-D fuels sold at retail pumps in the United States have an average CN of approximately 47–52, depending on seasonal variations. In Europe, the minimum CN used for passenger automobiles is 55.

High aromatic content lowers the natural CN of fuel. Removing aromatic content increases the CN of diesel fuel. B100 biodiesel, for example, has no aromatic content.

Previously, many engine manufacturers did not test engines that have any blend of biodiesel. Failures caused by or attributed to the use of the fuel above a B5 blend were not covered by warranty by the original equipment manufacturers (OEMs) of parts. Because of engine and emission legislation, compatibility is legislated for B20 blends (beginning in 2011), so warranties are now honored when using biodiesel blends that meet ASTM standards.

▶ TECHNICIAN TIP

Fuels with a higher CN have lighter, less complex molecules that are easier to burn than molecules found in low-cetane fuels. Complex and heavier fuel molecules, on the other hand, have more molecular bonds that release greater amounts of heat when burned. Complex, difficult-to-burn fuel molecules are collectively referred to as aromatic fuel molecules. Fuels with a *higher* natural CN generally have *less* aromatic content, will produce *less* power, and have slightly *poorer* fuel economy in comparison to low-cetane fuels. Fuels with high aromatic content—that is, fuels with larger, heavier, and more difficult-to-burn fuel molecules—will have a lower CN but have greater heat content.

▶ Diesel Fuel Additives

8-06 Identify and explain the purpose and applications of various common diesel fuel additives.

A wide variety of diesel fuel additives are used to generally improve the quality and performance of diesel fuel when fuel quality is compromised. Normally, when clean, good-quality fuel is used, there is no need to add anything to fuel. Nevertheless, many aftermarket additives are sold to meet many different applications. Most of these products are added to the

FIGURE 8-19 This quart (0.95 L) of conditioner treats 250 gallons (950 L) of fuel.

▶ **TECHNICIAN TIP**

Many fuel producers and OEMs of engines do not recommend the use of any fuel additives to their fuels.

diesel fuel to improve engine performance and increase mileage. The formulations tend to be highly concentrated, and for this reason, a little goes a long way (**FIGURE 8-19**).

Additive Maker Claims

In the United States, all manufacturers of fuel additives must register their product with the EPA, which in turn evaluates the products for adverse human health effects or negative environmental impact.

However, registration with the EPA does not substantiate marketing claims made by additive manufacturers. Many of the claims of additive manufacturers cannot be proven. For example, a group of combustion catalysts, which claim to increase fuel economy and lower emissions, were evaluated by an independent research facility, an evaluation that revealed no significant differences in fuel economy or exhaust soot levels. Such results are not surprising, because diesel engine combustion efficiency is typically greater than 98% even when using fuel without additives. If a technician believes that an additive is necessary, they should thoroughly research the product before purchasing and using it.

Types of Additives

The large number of available additives can be simply grouped according to function (**TABLE 8-6**). The following are four of the major categories:

1. engine performance enhancers
2. fuel stability
3. fuel handling
4. contaminant control.

Engine Performance Enhancers

Additives in this category are commonly cetane boosters, detergents, lubricity additives, and smoke suppressants.

Cetane boosters, also called ignition accelerators, are supposed to reduce combustion noise and smoke, improve starting in cold weather, and increase acceleration response.

TABLE 8-6 Fuel Additive Types and Benefits

Additive	Benefit
Cetane number (CN) boosters	Improve ignition quality by increasing the CN for improved starting and reduced white smoke
Lubricity improvers	Improve lubricating properties of fuel
Detergents, dispersants, and antioxidants	Prolong storage life; minimize oxidation; reduce gums, resins, and sediments; and improve injector spray patterns
Stabilizers	Inhibit oxidation to extend storage life
Metal deactivators	Deactivate metal ions such as copper compounds in fuel, to promote longer storage life
Biocides	Minimize bacterial and fungal growth, which helps prevent fuel filter plugging
Pour-point depressants	Lower temperature operation Improve cold-flow properties Reduce pour and cloud point
Defoamers	Reduce foaming when filling tanks and transferring fuels through pipelines
Smoke suppressants	Promote more complete combustion and reduce exhaust smoke
Rust inhibitors	Reduce formation of rust in fuel storage

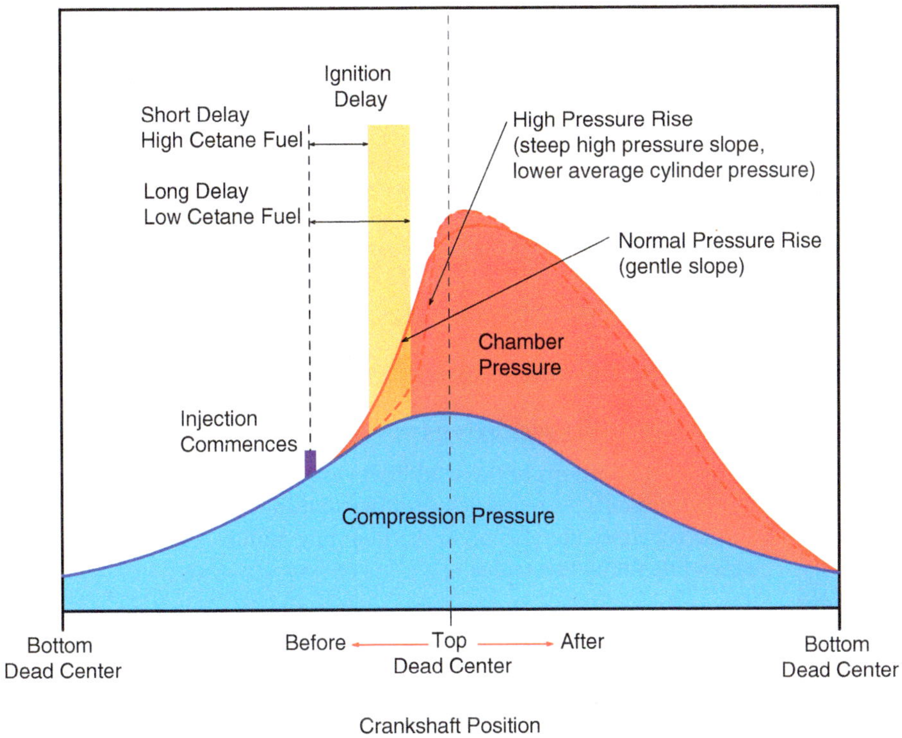

FIGURE 8-20 Fuel with a low CN will lengthen the ignition delay phase and cause an engine to be harder to start, run rougher, and have noisier combustion.

Generally, fuels that contain cetane booster will ignite and burn faster. By minimizing the ignition delay, this additive improves cold starting and acceleration and reduces smoke emissions. However, user reports indicate that an overconcentration of cetane boosters can also produce the opposite effects—excessive exhaust smoke emissions, low power, and poor fuel economy.

The majority of these CN booster compounds are nitrate based, which can increase the CN of the fuels by as much as seven points when used in small concentrations (0.2%). Cetane boosters may be added by fuel refiners or end users (**FIGURE 8-20**).

> **▶ TECHNICIAN TIP**
>
> A lot of a good thing is not always better. An excessive concentration of fuel additives in fuel can cause adverse and unintended changes to fuel chemistry. For example, an overconcentration of some cetane boosters may in fact cause excessive black smoke. Some additives will even damage fuel system materials such as elastomer seals and nylon components.

Detergents

Fuel and crankcase lubricants can leave gummy carbon deposits in the fuel system that can decrease engine performance. This is especially true of nozzle valve deposits. Detergent additives operate as solvents to dissolve these resins, allowing them to be flushed through the fuel system. Optimal concentrations of detergents are typically 50–300 ppm.

Detergent additives are also used to treat asphaltene, a contaminant that is increasingly common as refineries use new methods to increase yield from a barrel of oil (**FIGURE 8-21**). Asphaltene formation increases when fuel is stored for a long period of time and when the fuel is heated after circulating through an engine. Asphaltene can cause abrasive wear and fuel filter plugging (**FIGURE 8-22**).

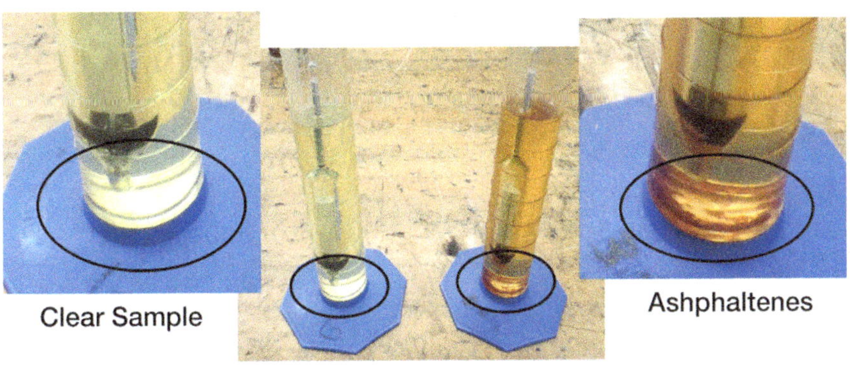

FIGURE 8-21 Ashphaltenes are small, hard particles of asphalt that develop in diesel fuel when it is exposed to high temperatures for an extended time period. Stagnant fuel will also develop ashphaltenes.

FIGURE 8-22 Asphaltene detergent conditioner dissolves small, hard crystals of tar-like asphalt compounds in diesel fuel.

Lubricity Additives

When fuel is severely hydro-treated to remove sulfur, other compounds that provide lubricating properties are also removed. The service life of injection pumps, injectors, fuel transfer pumps, and other fuel system components can be prolonged through the use of lubricity additives.

Smoke Suppressants

A number of products have been used to minimize black smoke. In the 1960s, a barium-based substance was used before the Clean Air Act became effective. This compound was later banned because of concerns about the human health impact of barium. Other suppressants, such as cerium or platinum, have been proposed to add to fuel as combustion catalysts.

Fuel-Handling Additives

End-user applications of these additives are primarily for de-icing and pour-point depressants. At refineries, additives to minimize fuel foaming and reduce drag in pipelines are also used. Antifoaming additives are used to minimize foam while filling fuel tanks.

De-icing Additives

Diesel fuel is both hygroscopic and hydrophilic, which means it attracts water and absorbs water (**FIGURE 8-23**). Because water dissolved in diesel fuel can freeze, fuel lines and filters can plug with ice. To prevent this, small quantities of glycol can be added to fuel. This substance dissolves in water rather than fuel, resulting in the water in diesel having a lower icing temperature.

▶ TECHNICIAN TIP

Water is a fuel system's biggest enemy, because it reduces fuel's lubricating ability and causes corrosion. Moisture will readily condense on cooler surfaces of diesel fuel, which subsequently absorbs water. Keeping fuel tanks full will minimize the condensation of warm, moist air inside a fuel tank. The best practice to prevent fuel from absorbing water content is to fill a fuel tank in the evening or late afternoon rather than in the morning. This prevents large quantities of warm, moist air from condensing in the fuel tank.

Pour-Point Depressants

Paraffin is one of the major types of HCs that make up diesel fuel. At low temperatures, these wax molecules can clump together, clouding and eventually gelling diesel fuel. The typical temperature range at which treated fuels thicken and no longer flow is 20°F (−6°C) below the cloud point. To solve the problem of gelling, fuel can be blended with kerosene or 1-D. Additives are manufactured that can change the size and shape of wax crystal formation.

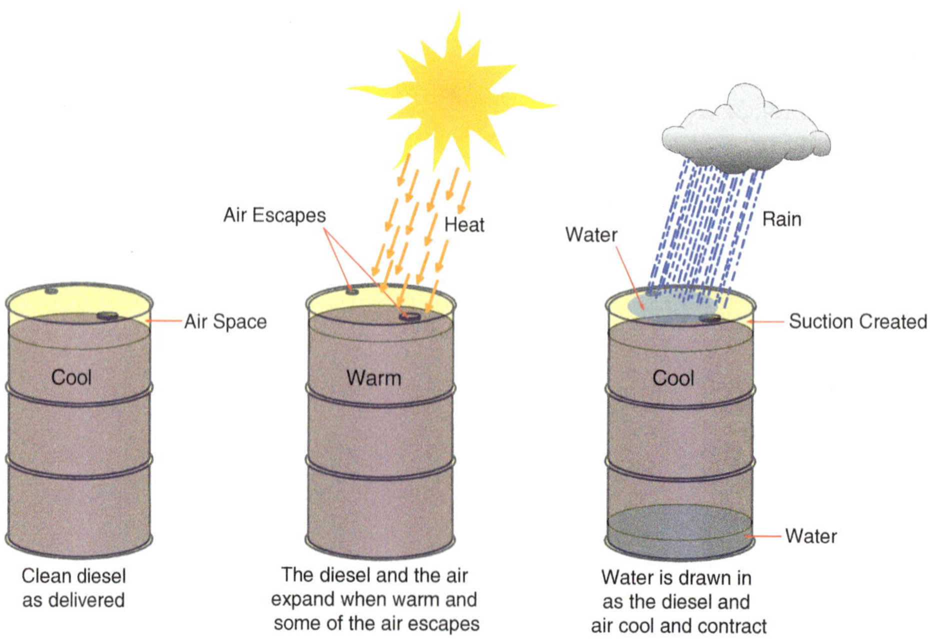

FIGURE 8-23 Diesel fuel will attract and dissolve water. Constant temperature changes during humid weather cause water to build up in fuel tanks.

Fuel Stabilizers

After refining diesel fuel, chemical reactions between air and compounds in the fuel can take place. Heat will, in turn, speed up some of these reactions, which can lead to darkening color, particles forming in the fuel, and fuel quality generally deteriorating.

Fuel refineries use fuel stabilizers to reduce oxidation and other chemical reactions between compounds in the fuel. Dispersants can also be added to keep particles dissolved in fuel rather than clustered together, eventually fouling and plugging filters.

Corrosion Inhibitors

Because many fuel tanks and lines are made of steel, the interior and both metal fittings and filter housings can rust or corrode. Eventually, corrosion can penetrate steel walls, creating fuel leaks. Rust particles can also plug fuel filters and cause abrasive wear to fuel system components. Corrosion inhibitors are compounds that form a barrier on metal surfaces that prevents chemical reactions between corrosive substances in the fuel.

Copper is another metal that diesel fuel should not corrode. Due to the sulfur content of diesel fuel, it can darken materials made of copper and bronze, forming a black, gooey coating on these parts. ASTM standard D975 has specifications for diesel fuel compatibility with copper.

Contaminant Control

Contaminant control additives are remedies to problems associated with prolonged fuel storage. Fuel can absorb water if it remains stagnant for long periods in humid conditions. As fuel absorbs more water and remains in suspension, it is referred to as bound fuel. Fuel tanks can rust and corrode as fuel absorbs water, where it remains either free or bound.

Biocides

Additives that control biological contaminants are also marketed. The most important, and useful, are **biocides**, which are poisons that kill bacteria and microorganisms that thrive in fuel that contains free water. Free water is water that is visible in fuel, which is unlike water that is bound or dissolved in fuel and is invisible. Typically, fuel that has been stored for prolonged periods and in warm humid climates is most vulnerable to this kind of contamination. When organisms are present inside a fuel tank or system, they can foul fuel filters and fuel lines.

Detecting Water in Diesel Fuel

A small amount of bound and free water in diesel fuel is normal. Some fuel specifications allow for up to 2.5% water contamination and some OEMs allow up to 4%. However, large amounts of water in fuel tanks, lines, injectors, and filters will freeze more readily than fuel without water. Microorganisms can also grow in water in diesel fuel. Excessive injector wear, filter plugging, power loss, and corrosion on or in engine fuel system parts can also take place when there is too much water in diesel fuel. Injector tips can be blown off if excessive water is present in the fuel. Checking for water in diesel fuel is an important part of monitoring fuel quality at the point of supply or distribution, as well as in the fuel tanks of equipment. Water-detecting pastes are commercially available, which when dipped into fuel that contains water, will turn color from a light brown to bright red. Food coloring added to a fuel sample will pass through fuel and color only water. To get an idea of the amount of water in fuel, some artificial coloring is sometimes added to a fuel sample (**FIGURE 8-24**). Fuel deliveries should be checked before fuel is supplied to the engines used in critical vehicles or equipment, such as backup power generators. To detect water in diesel fuel, follow the steps in **SKILL DRILL 8-2**.

FIGURE 8-24 Some blue food coloring was added to a water-contaminated fuel sample. Food coloring will not color the fuel—only the water.

Maintenance and Diesel Fuel

Diesel fuel, whether it is petroleum derived or biodiesel, has specialized maintenance needs that the technician should take into consideration for the kinds of vehicles and engines that they maintain. Fuel suspected of causing problems must be properly analyzed, not only for servicing the fuel system but also when informing fuel suppliers of a problem. This can be especially crucial when fuel is purchased in bulk. The operation of an entire company—both in costs and efficiency—can be jeopardized by bad fuel and the problems that it can cause.

SKILL DRILL 8-2 Detecting Water in Diesel Fuel

1. Take a fuel sample and examine it against a lighted background. Look for haziness, which indicates high amounts of bound water in fuel.
2. Using a good-quality water paste, dip the paste into a fuel sample to check for water. For vehicle fuel tanks and underground reservoirs, the paste can be applied to the end of a long dipping rod or stick. The color of the paste will change instantly when it contacts water.
3. Place a 16.9 ounce (500 mL) sample of fuel in a clear glass beaker or jar. Slowly add 1.7–3.4 ounces (50–100 mL) of methanol to the jar. Only methyl alcohol (methanol) should be used, because it does not mix with diesel fuel. (Ethanol can be mixed with diesel fuel and will dissolve water in the fuel.) The methanol will float on the diesel fuel.
4. Allow the two liquids to settle and separate.
5. Mark the container wall with a line at the point of liquid separation.
6. Vigorously shake and mix the alcohol and fuel together for one minute. Let the sample stand and settle for approximately five minutes.
7. Note the new separation line between the alcohol and fuel, and mark it with another line on the container wall.
8. Because methanol will absorb water, its volume will expand if it dissolves any water from the fuel. If water was present in the fuel sample, the line of liquid separation after mixing will be below the first line marked on the container wall. The difference between the two lines of separation is proportional to the fuel's water content. The water is now irreversibly bound with the alcohol.
9. Report the findings. If necessary, make a service recommendation to drain free water from fuel tanks.

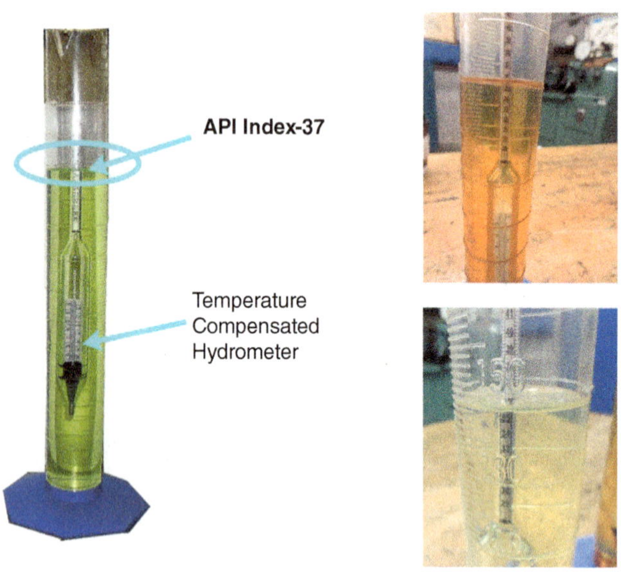

FIGURE 8-25 A temperature-compensated hydrometer measures fuel density in API units. The higher the API number, the less dense the fuel is.

Specific Gravity: API Index

For diesel service technicians, the specific gravity or density of a fuel is a good indication of the grade of fuel—but not its CN. Fuel density is an important property of fuel because it affects engine power and fuel economy. Less dense fuel, which has lower specific gravity, weighs less than denser fuel and has less energy content.

Fuel density is measured by using a **fuel hydrometer** (**FIGURE 8-25**). A fuel hydrometer will report the density in API units—that is, the **API number**. API stands for the **American Petroleum Institute**, a regulatory body that develops standards for measuring fuel density. The API index corresponds to the grade of fuel and its heat content. Generally, there is a 3–5% decrease in the thermal energy content, measured by using the unit BTU. (or MJ), of fuel for every 10-point increase in API specific gravity. Determining the API index is important when investigating symptoms of low power or high fuel consumption. Lower specific gravity can indicate whether the fuel is blended with kerosene or 1-D. Blended fuels will have a high API number, meaning the fuel is lighter or less dense. The API number will change with seasons and geography. For example, during the summer months in Ontario, Canada, retailed 2-D diesel will weigh between 6.951 lb/gal. (833 g/L) (API 38) and 7.076 lb/gal. (848 g/L) (API 35). In winter in this region, the same diesel may weigh as little as approximately 6.87 lb/gal. (823 g/L) (API 40).

Evaluating Fuel Quality

Diesel fuel can become stale and contaminated with water, oil, microorganisms, dirt, sediment, ashphaltenes, and particles of corroded metal. If fuel has become stale and contaminated, its smell and color will change. Filters will quickly plug with contaminated fuel and overwhelm the fuel–water separator. More significantly, diesel fuel's ignition quality will change when it becomes stale and will thus adversely affect engine operation. Although a

SKILL DRILL 8-3 Evaluating Fuel Quality

1. Obtain a fuel sample and allow the fuel to settle in a beaker for a few minutes.
2. Place a piece of white paper behind the beaker to reflect through the fuel, and visually examine the fuel.
3. Record the color and clarity of the fuel. Observe any sedimentation or separation of fuel from other liquids or particles in the beaker. Fuel becomes darker as it ages. Fuel contaminated with bound water that has also gone stale turns brown. Microorganisms may even be visible in fuel.
4. Note the odor of the fuel. Fuel, as it ages, will gradually lose its intense smell and begin to change to having a smell similar to turpentine, varnish, plasticine, or paint. The presence of microorganisms will give the fuel a foul, rotten smell.
5. Record any observations and make a service recommendation. A fuel system with significant contamination should be drained and flushed, and the filters should be replaced. A cetane booster may be recommended for fuel that has become slightly stale but not contaminated. Recommend a fuel stabilizer if fuel is to remain stored for several months, but do not recommend that biodiesel be stored for a long time.

fuel sample can be taken and sent to a lab for analysis, technicians can make some simple observations to quickly evaluate the quality of fuel. To evaluate fuel quality, follow the steps in **SKILL DRILL 8-3**.

Biodiesel and Engine Oil

After the introduction of low-sulfur fuels, biodiesel has been considered for use as a lubricity additive to keep moving parts operating smoothly inside fuel-injection systems. The dilution of engine oil with biodiesel is another matter. Biodiesel actually promotes fuel diluting in engine oil to a much greater extent than petroleum-based fuel. Oil dilution with biodiesel occurs when a post-combustion injection of fuel into the cylinders is used to reduce combustion soot or to regenerate a particulate trap or NO_x adsorber. Heavier, less volatile fractions of fuel, however, do not vaporize during post-combustion injection, so liquid fuel clings to the cylinder walls. Because biodiesel has a higher boiling point, it tends to dilute the engine oil disproportionately to its blend ratio in the fuel.

Fuel passes by the pistons and accumulates in the crankcase. Once there, fuel polymerizes (joins together to form long, complex-shaped molecules) with oil and produces sludge. Engine damage can result from oil passageways getting blocked by sludge. Deposits also form in the ring-belt area, leading to ring stickage and increased crankcase blowby, which further aggravates oil being contaminated by increased levels of soot.

Raw or refined vegetable oils produce the greatest harm. However, the harm may not become evident until a significant amount of damage has occurred over an extended period. Petroleum-based fuel does not produce this effect at all, because it readily evaporates when the oil is heated. Biodiesel blends higher than B20 will enable larger amounts of unburned fuel to slip by the piston rings and condense in the lubricating oil. Oil change intervals for the engine oil may need to be shortened significantly if engine uses blends of biodiesel higher than B20.

Biodiesel also interferes with the action of zinc-based antiwear additives and detergents used in engine oil. Engine oil diluted with biodiesel can displace such additives on metals and chemically destroy the additive. This means that even before the engine oil thickens and turns to gel, lubricating oil diluted with biodiesel will begin to lose its ability to properly lubricate parts under extreme pressure.

8-07 Identify and describe the components and construction of the low-pressure fuel supply system.

▶ Introduction to Low-Pressure Fuel Systems

Diesel engines need a low-pressure fuel supply to transfer fuel, with adequate volume and pressure, from the fuel tank to the high-pressure injection stage of the fuel system. Fuel tanks, pumps lines, and filters are the most visible components of the low-pressure fuel system. But fuel supply and storage are not the only important functions. Properties of diesel fuel and the sensitivity of the injection system to dirt, water, and contaminants means fuel must be conditioned to ensure that the high-pressure injection system has long-term durability (**FIGURE 8-26**). Removing vapor bubbles and supplying fuel at the correct temperature range is as vital as cleaning contaminants and abrasives from fuel.

Low-pressure fuel systems also include features that inform the operator of fuel tank levels and the need to service fuel filters and filters that separate water from fuel (see the Fuel–Water Separators section in this chapter). Operating and servicing the low-pressure supply system properly is critical to quick starting, maximum engine power output, engine durability, and getting reliable, trouble-free operation from diesel-powered vehicles (**FIGURE 8-27**).

Requirements of Low-Pressure Fuel Systems

Much of the work an engine technician performs is often directly or indirectly related to the low-pressure fuel system. Diagnosing low-power complaints, identifying problems and failures with components in the high-pressure injection system, and day-to-day work performing preventative maintenance service all require a comprehensive understanding of low-pressure fuel system components and operating principles. This next section outlines the construction and operation of the low-pressure fuel system and operating principles the system, which is essential to developing good service practices and skills.

The low-pressure fuel system has many functions:

- storing and supply fuel free of contaminants and vapor
- supplying an adequate volume of fuel at the correct pressure to the high-pressure injection system

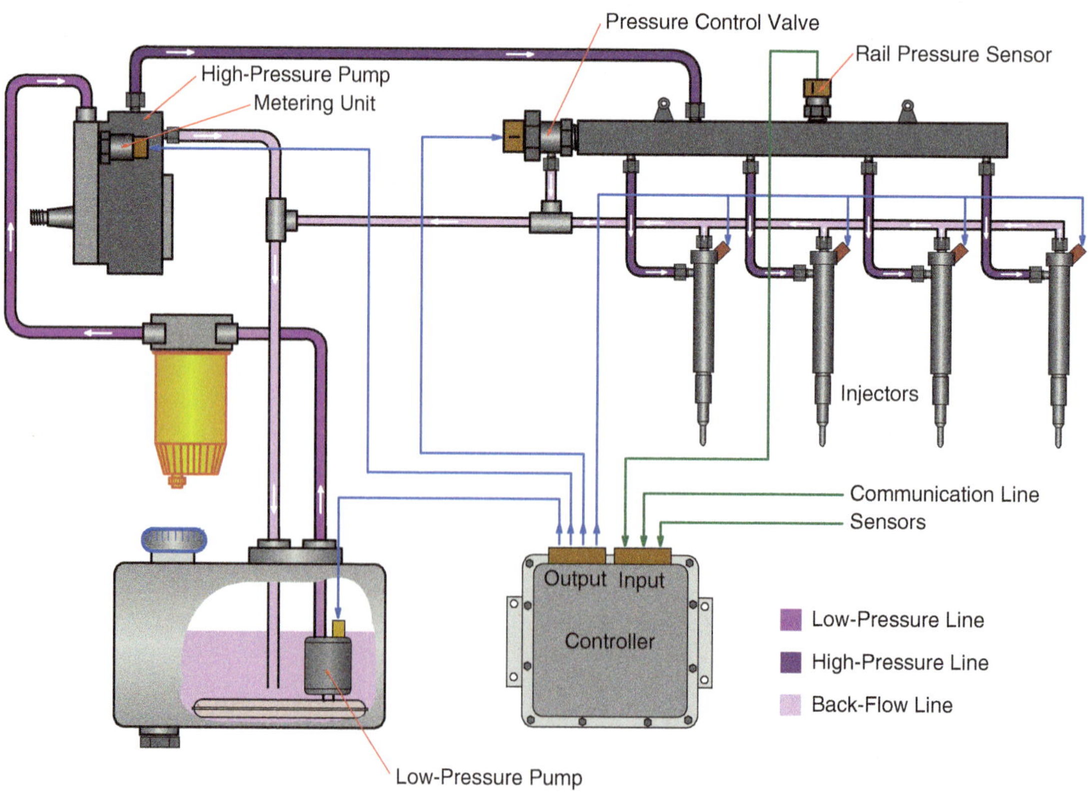

FIGURE 8-26 Typical layout for a low- and high-pressure fuel system on a four-cylinder common rail engine.

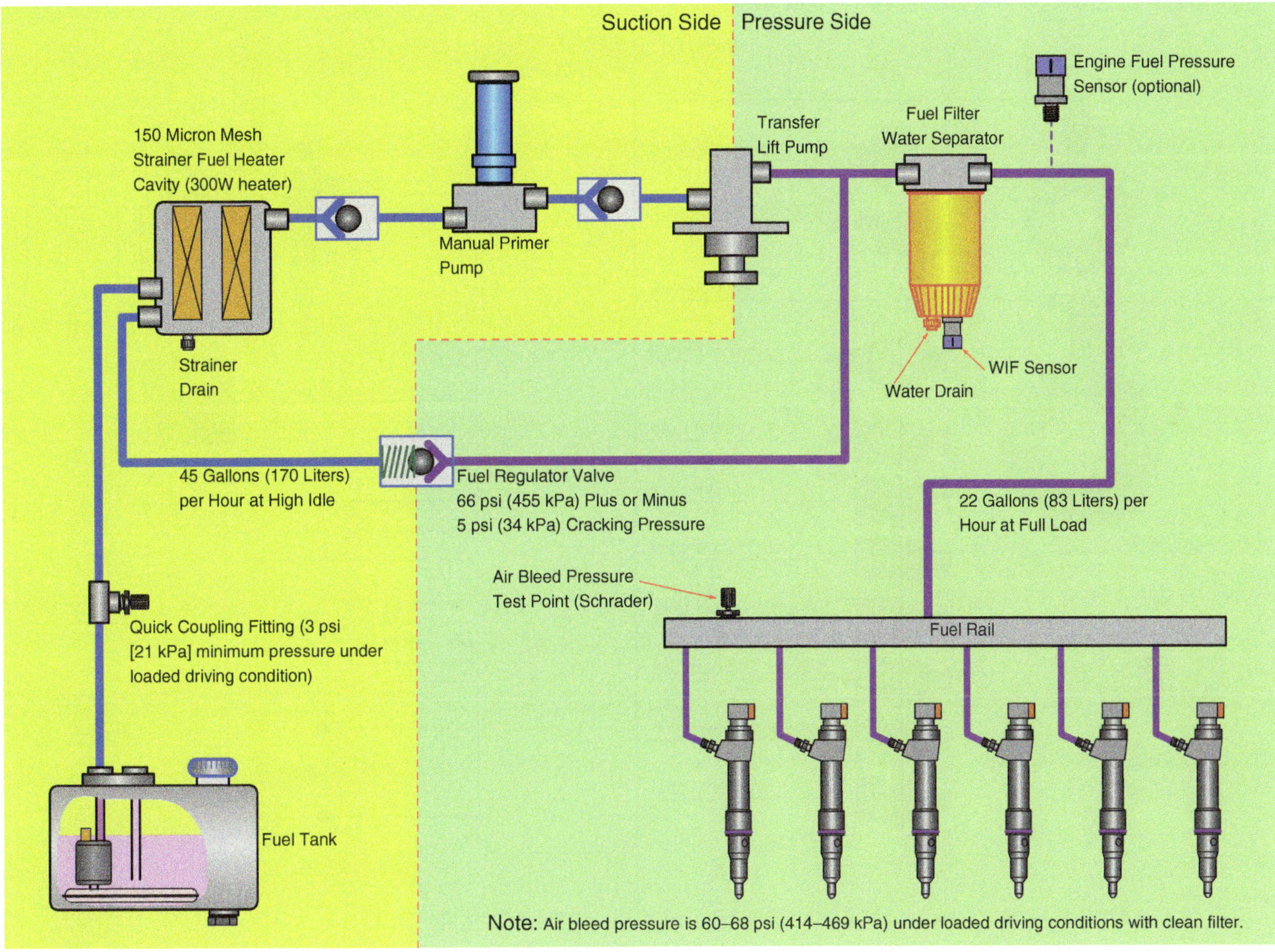

FIGURE 8-27 A number of sensors and additional devices are used by the low-pressure system to condition fuel for delivery to the high-pressure injection system.

- removing water contamination
- preventing wax gelling
- regulating fuel temperature.

Fuel Temperature

Because fuel is used to cool and lubricate high-pressure system components, high fuel temperatures can change fuel viscosity and density to the point where fuel loses its ability to lubricate and cool components. High fuel temperatures also affect engine power output as fuel density decreases. Power loss happens because fewer fuel molecules are present in every cubic millimeter in the now lower-density fuel (**FIGURE 8-28**). At low temperatures, the **paraffin** content—that is, the wax that exists in the fuel—can cause the fuel to become too viscous to properly flow through filters and lines. Restricted and plugged supply system passages also lead to power loss and engine stalling. This condition, called **gelling**, can be prevented by warming the diesel fuel.

Components of Low-Pressure Fuel Systems

The following are some of the typical components in the fuel transfer system:

- fuel tank(s)
- low-pressure fuel lines
- fuel level sending units and tank pickup
- fuel filters

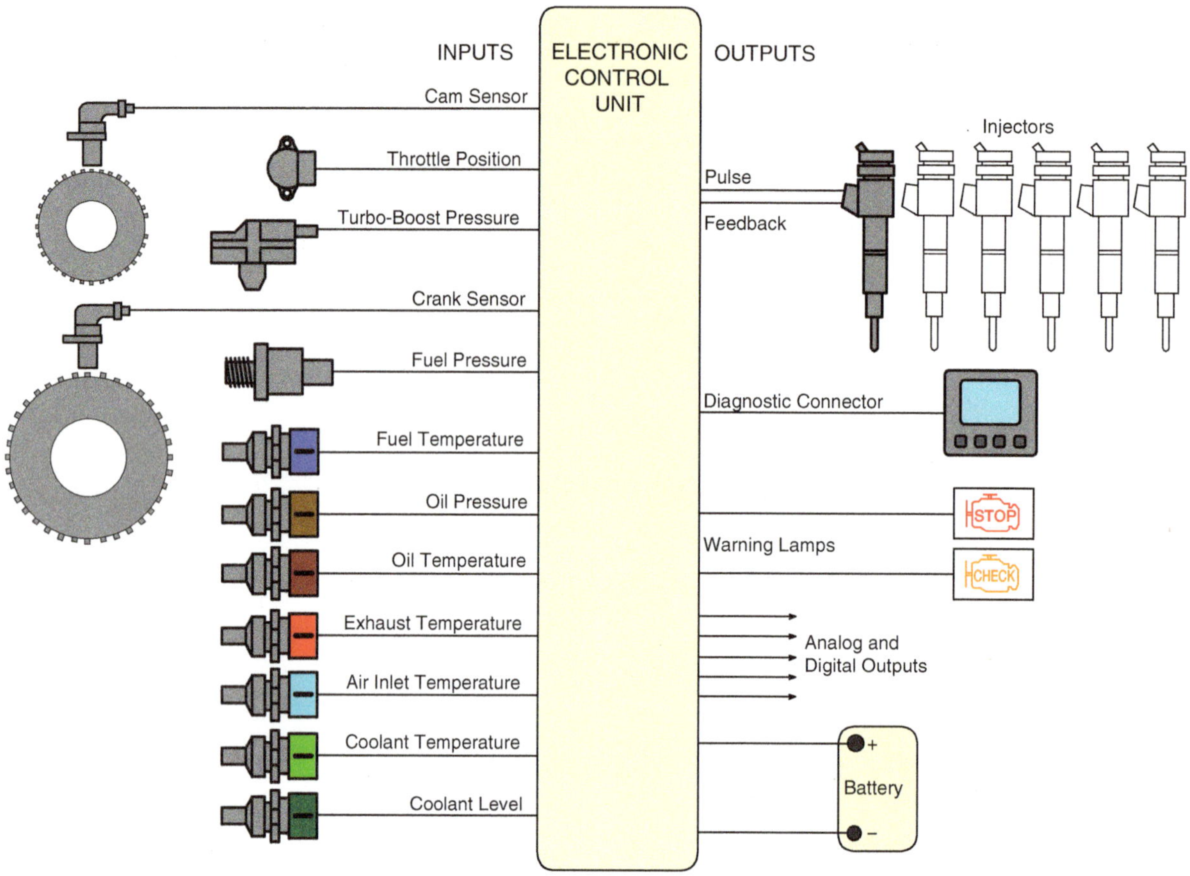

FIGURE 8-28 The fuel temperature sensor is an important input. Increasing fuel temperatures will result in power loss unless the electronic control module (ECM) can compensate for the change in fuel's density.

- fuel heaters
- water filtration
- fuel coolers
- fuel transfer pumps.

Fuel Tanks

The most obvious function of the fuel tank is to store an adequate supply of fuel for the engine to operate. A secondary function is to allow fuel that has returned from the engine to cool after absorbing heat from the engine and fuel injectors. Cooling hot fuel is important for a number of reasons. First, fuel loses its viscosity and density as it becomes hotter. This translates into a loss of film strength necessary to lubricate injection system components. More significantly, the change in density will correspond to smaller injection quantities, causing power loss. Specifically, a 2% power loss occurs for every 10°F of fuel temperature above 115°F (46°C). Most electronically controlled injection systems have the capacity to monitor fuel temperature and correct the quantity of fuel injected based on fuel density change accompanying temperature change (**FIGURE 8-29**).

Tank Materials

Fuel tanks should be made of a noncorrosive material that remains free of leaks and is chemically compatible with diesel fuel. Aluminum, plastics, and steel alloy are popular materials. Fiberglass is a strong, long-lasting, resilient material that is immune to corrosion and is easily formed into custom designs (**FIGURE 8-30**). However, the potential exists for fiberglass to disintegrate and plug fuel filters. For this reason, fiberglass is not a material of choice for tanks in mobile applications. Copper is not suitable for tanks or lines, because

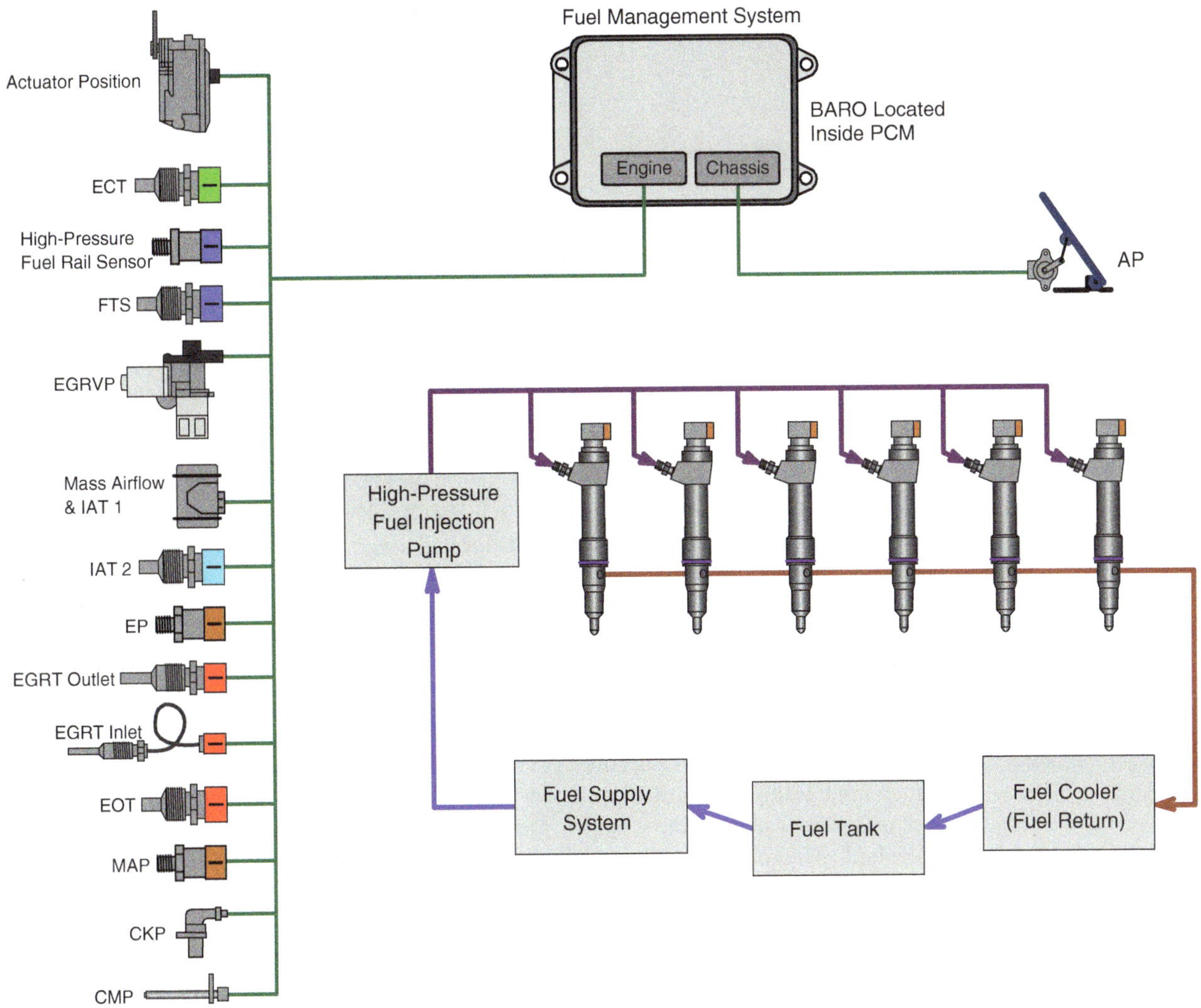

FIGURE 8-29 Fuel coolers are needed on some injection systems to prevent power loss and excessively low fuel viscosity.

chemical reactions between diesel fuel and copper cause corrosion that produces a gum-like resin that clogs injectors. Galvanized steel and zinc are also not suitable materials, because diesel fuel reacts with the zinc to form a powdery abrasive residue that clogs and damages the fuel system.

The fuel tank should be located far enough from the engine to prevent ignition and any fire hazard in the event of an accident. The Federal Motor Vehicle Safety Standards (FMVSS) and Canadian Motor Vehicle Safety Standards (CMVSS) are safety regulations that include standards for the protection and shielding of fuel tanks, ground clearance, and the like on truck, bus, and light-duty vehicle chassis.

Fuel Tank Ventilation

A completely sealed fuel tank may prevent leaks, but tanks require venting to the atmosphere. The reasons for this are simple. First, returning fuel may contain fuel vapor in the form of bubbles, which are released inside the fuel tank. Second, when the level of fuel drops in the tank as the engine uses it, a negative pressure is created, which can slow, if not stop, fuel delivery to the engine.

FIGURE 8-30 The fuel tank of this late-model pickup is made from plastic. Note the tank filler tube vent line and rollover valve at the tank vent outlet.

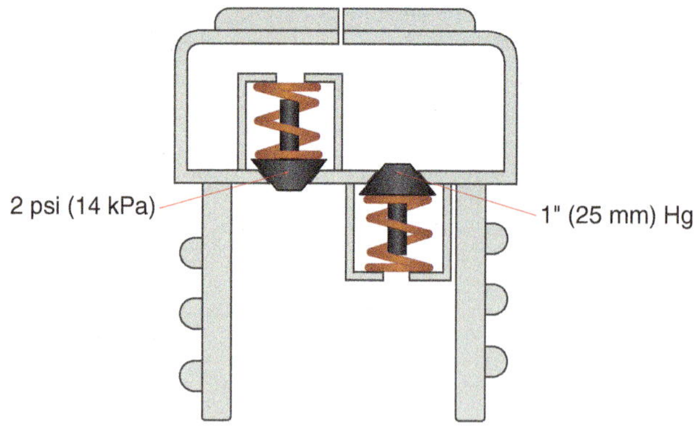

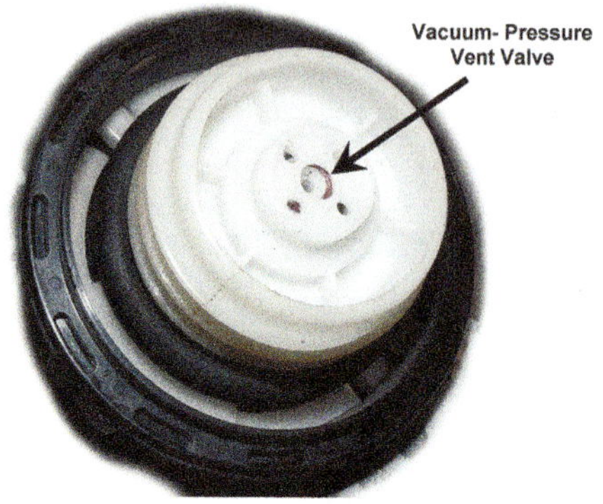

FIGURE 8-31 The fuel tank cap contains a vacuum valve to admit air and a pressure relief valve.

SAFETY TIP

Gasoline and diesel fuels have very different flash points and volatility characteristics. Diesel fuel must warm up to almost 100°F (38°C) to produce enough vapor to burn when exposed to an open ignition source such as a match or spark. Gasoline will easily vaporize and burn at −45°F (−43°C) or colder. Gasoline will also completely evaporate at much lower temperatures than diesel fuel and exert higher vapor pressure than diesel fuel. Because fuel vapors are a source of HC emissions, gasoline fuel systems require evaporative emission systems to prevent fuel vapor from escaping into the atmosphere; diesel fuel systems do not.

Tank venting is necessary to allow air into the tank and some fuel vapors to escape. A vent or pressure-compensating valve is usually located at the highest point in the tank, often in the tank fuel cap. Venting can also be incorporated into the sending unit pump module. While the vent valve allows bidirectional movement of air and tank vapors from the tank, it also must prevent dirt from entering the tank and fuel from leaking out. In the event of an accident and a vehicle rollover—or simply when fuel sloshes around while the vehicle is in motion—the valve must prevent fuel leakage from the fuel vent line.

Fuel tank caps on light-duty vehicles often use pressure-vent and vacuum-vent valves (**FIGURE 8-31**). The fuel tank **rollover valve**, which connects to the tank vent line, prevents fuel from leaking from a tank if the vehicle is overturned. The device is a pressure-sensitive valve attached to a rubber hose. The valve may be clipped to the frame or near the tank. The rollover valve vents the tank to the atmosphere as long as no fuel is in contact with the rollover valve. Contact with fuel closes the valve to prevent fuel from leaking from the tank.

Fuel tanks are designed to have an air pocket near the top to allow fuel to expand, which naturally takes place as fuel warms. The volume is 5–10% of the total tank volume. On medium-duty trucks and pickups, the filler neck may have an inner and outer tube that is designed to prevent the tank from being completely filled and to allow air to channel between the two tubes and to escape from the tank as it is being filled with fuel. A separate fuel-tube tank vent line is most often used on these tanks.

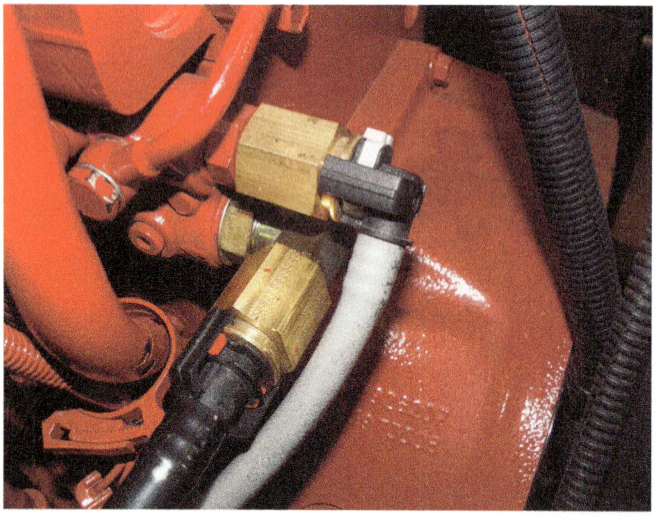

FIGURE 8-32 These snap-to-connect (STC) plastic supply and return lines are typical for diesel engines. The lines are separated after the clip is removed.

Fuel Lines

The low-pressure fuel system is connected by low-pressure fuel transfer lines. These can be constructed of braided steel, corrosion-resistant metal tubes, and diesel fuel–compatible rubber material. Such flexible tubing is designed to pass through the different sections of chassis to transfer fuel from the fuel tank to the engine. Compared to existing rubber and steel fuel lines, new grades of nylon plastic fuel lines (e.g., Synflex) are durable, lightweight, lower in cost, and corrosion-, vibration-, and impact-resistant. Nylon plastic lines are chemically compatible with diesel fuel and mold well with specialized connectors for easy assembly and service (**FIGURES 8-32** and **8-33**).

Fuel lines used on low-pressure systems must be routed to where they cannot be mechanically damaged. As well, if fuel leaks do occur, dripped or evaporated fuel must not be allowed to accumulate or ignite. For safety reasons, fuel lines are not allowed to be routed through a vehicle's passenger or driver compartment. When a vehicle or engine twists and bends, the fuel line must be flexible enough to repeatedly allow this without deteriorating.

Fuel Tank Pickup and Sending Unit

Fuel tank pickups are usually integrated with a fuel level **sending unit** (an electrical device, usually a variable-resistance rheostat, that supplies a voltage signal to an analog-type dash gauge proportional to the fuel level) in medium-duty applications. In most applications, an electrically operated fuel pump is also incorporated with the pickup and sending unit to form a module. Modules may include some or all of the following components, which may or may not be separately serviceable:

- fuel filter
- fuel pressure regulator
- electric fuel pump
- fuel pump reservoir
- a separate in-tank fuel pickup filter (strainer)
- pressure relief/rollover valve
- fuel gauge sending unit (i.e., fuel level sensor)
- an auxiliary non-pressurized fuel supply fitting.

FIGURE 8-33 These clear plastic lines help identify whether there is any air in the fuel system.

Tank Pickup

Unlike gasoline fuel tanks, diesel fuel pickup units almost always touch the bottom of fuel tanks, which range widely in size from 10 to 300 gallons (from 37 to 1,136 L). Many use a spring-loaded design to ensure the pickup contacts the tank bottom. The purpose of the design prevents the accumulation of contaminants, such as water, sediment, and microorganisms, in the fuel tank. Contaminants are generally picked up and removed by the fuel filter instead. One method to do this involves the use of nibs at the bottom of the pickup. These nibs draw water fuel and dirt into the tube by means of the Venturi effect. The venturi is created by rubber nibs that contain small restrictions that draws the fuel and water as well (**FIGURES 8-34** and **8-35**). Water is later removed from the fuel by the fuel–water separator. Other pickup modules will use one larger venturi that pulls water and fuel together by passing fuel from an in-tank electric pump over a single venturi connected to a tank pickup tube.

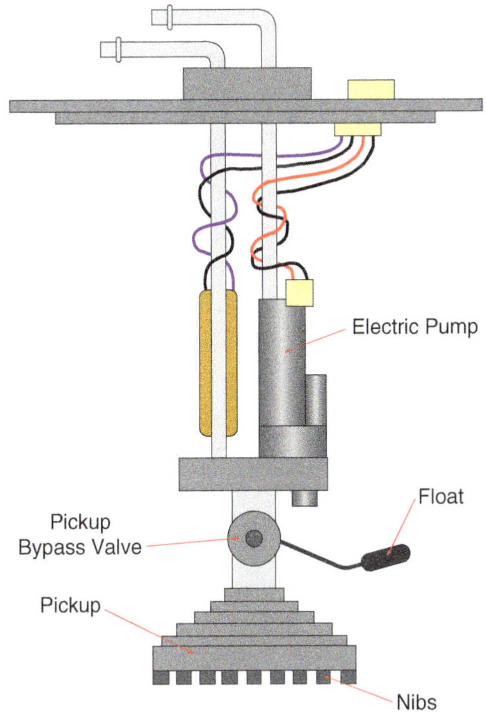

FIGURE 8-34 This tank pickup is designed to draw water in with fuel.

Fuel Reservoir Pickup

To provide adequate fuel tank sizing and meet packaging requirements, Volkswagen (VW) and Mercedes-Benz use fuel tanks with two with two separate reservoirs that straddle the driveline. In the Benz-style tanks are two separate fuel pumps, both located in only one of the separated reservoirs (**FIGURE 8-36**). Also used are two tank floats in each reservoir. Both electric in-tank pumps supply a

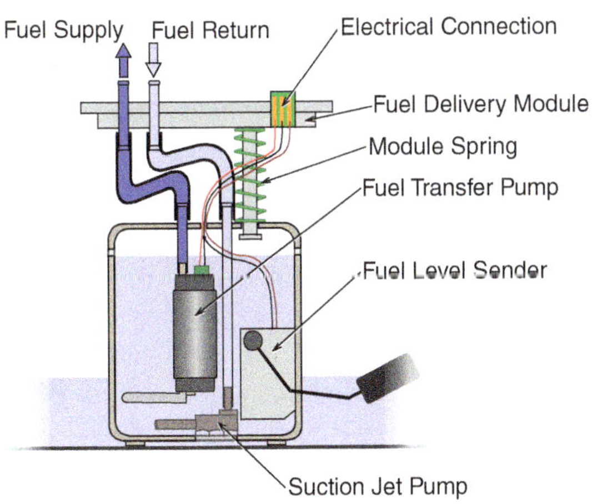

FIGURE 8-35 Note that the fuel module is spring loaded to hold the pick-up next to the bottom of the tank.

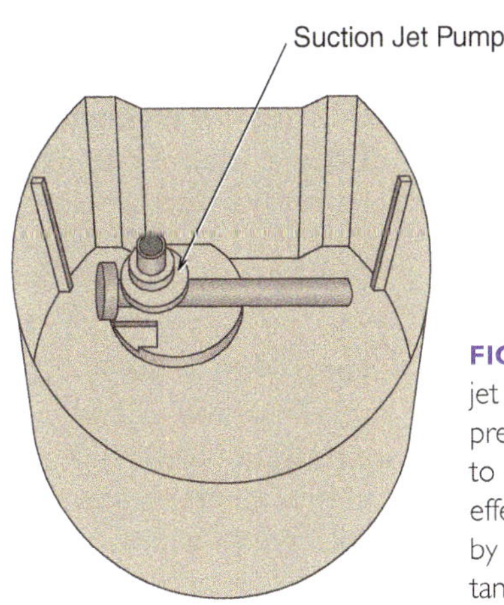

FIGURE 8-36 This suction jet fuel pick-up tube uses fuel pressure supplied by a lift pump to create a low-pressure venturi effect. Negative pressure created by the venturi pulls fuel from the tank bottom.

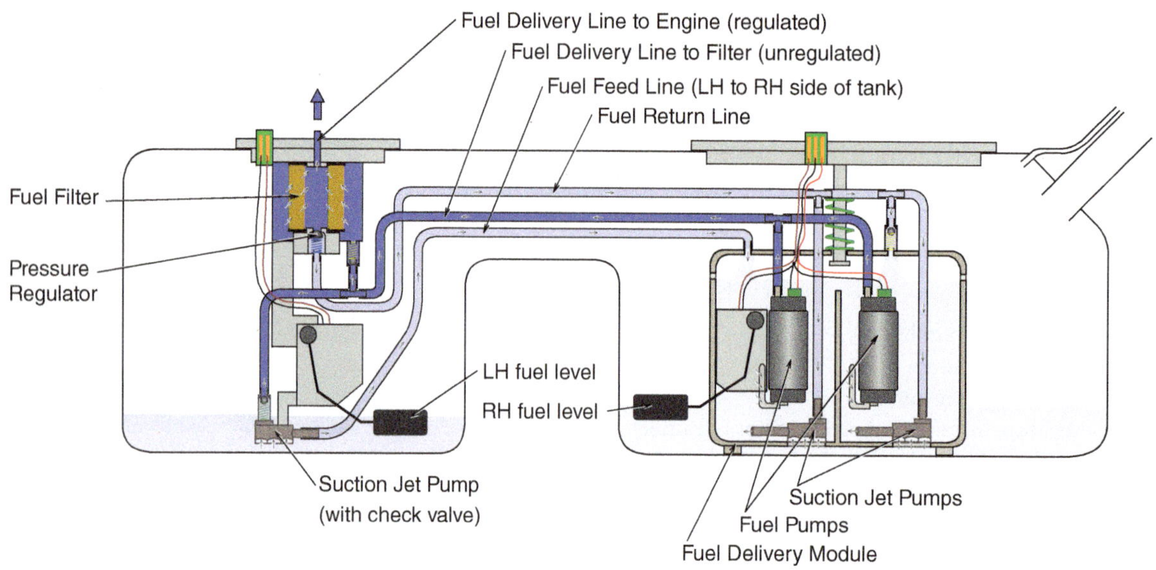

FIGURE 8-37 A venturi- or jet-style pickup tube in one reservoir has pressurized fuel pass through an open orifice to pick up fuel from the reservoir without a fuel pump.

single tank-out fuel line. To move fuel from the reservoir without a pump, the outlet of both fuel pumps connects to a venturi-style pickup to pump fuel from the adjacent tank (**FIGURE 8-37**). Fuel is passed over the top of an open orifice-like pickup and creates a low pressure at the base of the pickup tube. This idea is similar to blowing compressed air over a tube to pull liquids up a tube. However, the fuel supplied by the electric pumps and the fuel pulled up from the orifice style pick-up both mix and return to the reservoir, where the pumps are suspended in fuel.

With two tank floats, a virtual calculation is performed by an ECM, reporting fuel level to the dash gauge. Problems with tank float calibration can result in a gauge displaying no fuel when in fact the tank is three-quarters full. A common procedure to recalibrate the tank is to reduce the fuel tank level to below half tank. A fuse to the control module is briefly removed and then reinstalled to recalibrate float level readings.

Fuel-Return Circuits

Diesel fuel systems almost universally allow fuel to circulate through the injectors and other fuel system components to remove vapor and to lubricate and cool fuel system components. Pressurizing fuel also increases its temperature. It is normal for a **fuel-return circuit**, which is the fuel circuit that collects fuel from the injectors, fuel pump, and relief valves and which returns the fuel to the tank, to have a slight amount of vapor present, which is mostly light fuel fractions that have absorbed heat and changed into a vapor. The latest fuel systems minimize the amount of fuel returned to the tank to improve fuel economy and meet greenhouse gas (GHG) emission standards. No return flow means that less energy is used to pump fuel that the engine doesn't use. Reducing this parasitic power loss improves engine efficiency, enables pump downsizing, and eliminates the need for the additional expense of a return-flow fuel cooler.

Pressurized Fuel-Return Circuits

Third-generation Bosch Piezoelectric fuel injectors used by many automotive diesel engines, such as the 2011 Duramax (and later Duramax diesel engines), Ford 6.7 L Powerstroke, VW, and others, require a constant minimum pressure on the return side fuel circuit from the injectors. Typically, a minimum pressure of 0.4 to 1.1 megapascals (mPa) or 58 to 160 psi must be maintained to enable the injectors to operate. The return fuel pressure is necessary to operate a hydraulic coupling, amplifying the slight movement of the piezo electric stack and a control valve inside the injectors (**FIGURE 8-38**). Without the proper

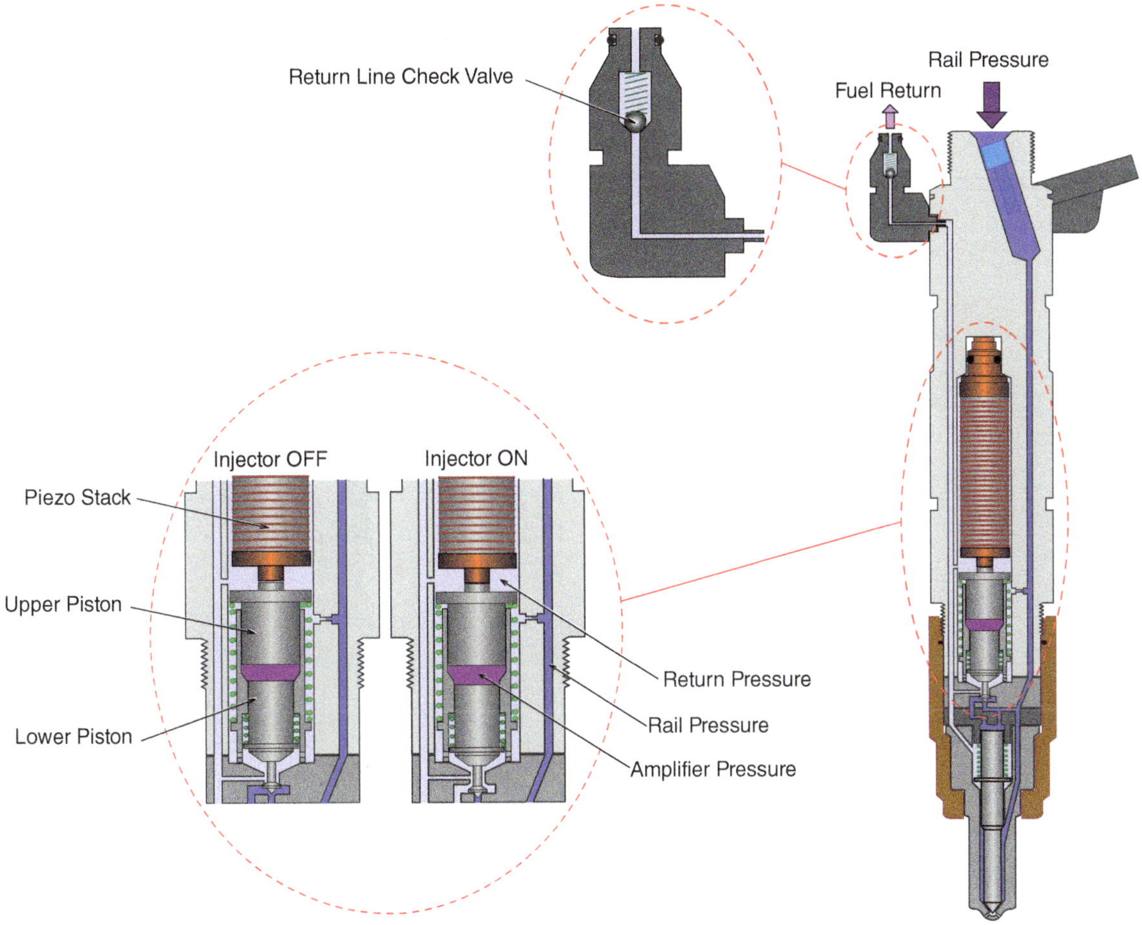

FIGURE 8-38 The hydraulically coupled pistons inside the injector amplify the force of the piezo stack. The different diameters of the upper and lower piston increase the operating stroke. A minimum fuel pressure of close to 58 psi (4 bar) in the fuel-return line is necessary to start the engine.

operation of the hydraulic coupling, the injector nozzle cannot open and spray fuel into the cylinders. Each manufacturer uses a pressure regulation device with its own unique name (**FIGURE 8-39**). General Motors (GM) uses a constant pressure regulator connected to the fuel rail return lines (**FIGURE 8-40**). If the valve fails, the engine cannot start even though fuel rail pressure is adequate. To prevent injector damage when the engine runs out of fuel, or after the fuel system is serviced and the filter is empty, the low-pressure side of the fuel transfer pump backfills the injectors with fuel through the return lines. The low-pressure fuel line is opened to backfill only if pressure falls below 43 psi (3 bar) in the injector return lines or if fuel filter restriction is excessive.

Sending Units

Each fuel tank is equipped with an electrical sending unit—typically a variable-resistance rheostat—attached to a float that transmits the fuel level to the instrument panel in the operator's cab (**FIGURE 8-41**). Current from the instrument panel or fuel gauge module will connect to the sending unit. The resistance to the ground to complete this circuit is determined by the position of the float. One end of the float arm operates an electrical wiper moving across resistive wires connected to the chassis ground. Current will pass through the wiper and then to ground through the resistive wires. On most sending units, high resistance to ground corresponds to a full tank, while low resistance signals an empty tank. Disconnecting the wire to the sending unit should cause the fuel gauge to register full because circuit resistance is high. Grounding the wire will cause the fuel gauge to indicate empty because resistance is low.

▶ TECHNICIAN TIP

Third-generation common rail injectors must have a minimum of 58 psi (4 bar) pressure supplied to the fuel-return side of the injector. This condition is necessary to fill a hydraulic coupler inside the injector that amplifies the linear movement of the piezo crystal stack, which in turn opens and closes the injector delivery control valve. Without fuel pressure, the injector cannot operate and may even be damaged. Ford uses a fuel pressure sensor that prevents the engine from starting without adequate low-pressure fuel and will derate the engine power if it falls below acceptable pressure. Unlike other fuel systems, piezo injectors with hydraulic couplers require inspection and testing during no-start conditions. A variety of components are involved in ensuring this pressure is maintained, and careful step-by-step procedures unique to each OEM are essential for an efficient diagnosis.

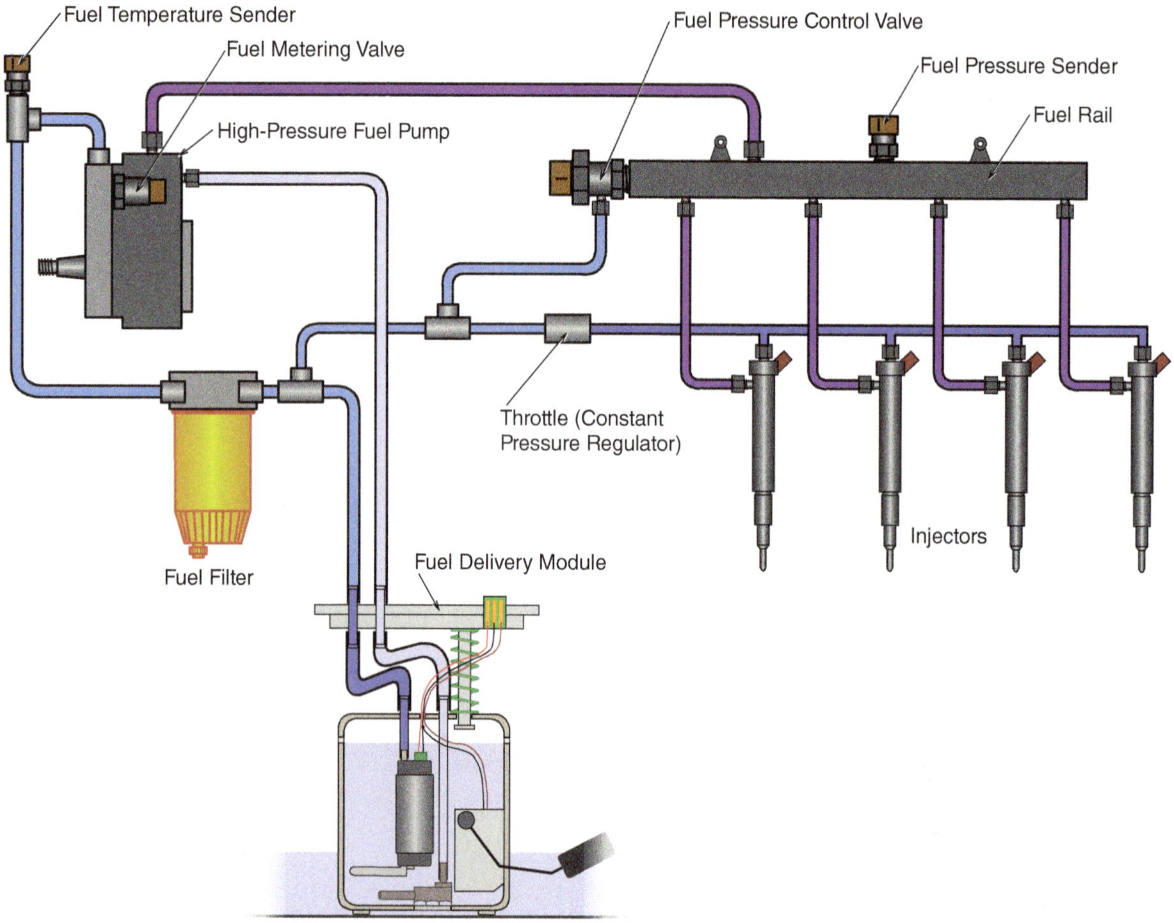

FIGURE 8-39 A constant pressure regulator maintains pressure inside the fuel-return lines to the injectors in third-generation Bosch common rail injectors.

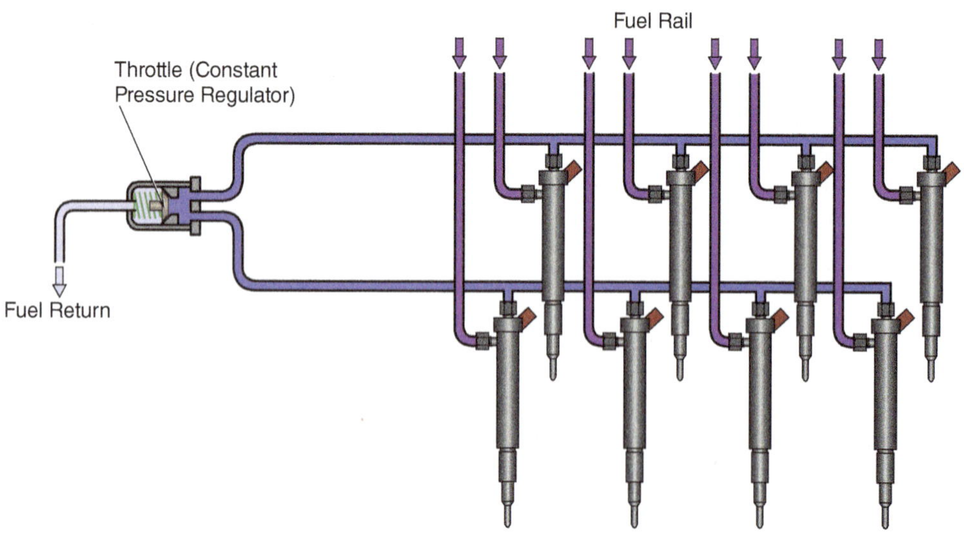

FIGURE 8-40 This VW low-pressure fuel-return circuit uses a throttle as a device to maintain fuel pressure inside the return line to the injectors.

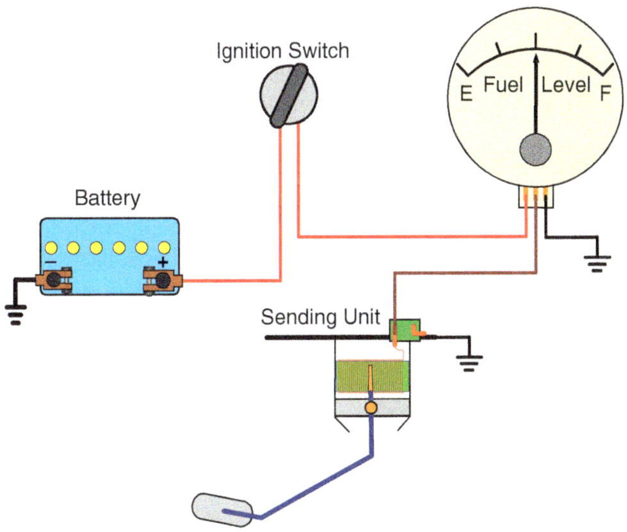

FIGURE 8-41 A rheostat sending unit is connected to the arm of the fuel tank float to vary current resistance to the dash gauge.

For vehicles that comply with On-Board Diagnostic II (OBD-II) legislation, fuel level data is more critical because input from the sending unit is used to differentiate engine misfires caused by low fuel level from other kinds of misfires. For example, if the fuel level is too low on some OBD-II-compliant vehicles, a misfire code will not be sent, because it may have been caused by the engine running out of fuel. Because most instrument clusters use stepper motor-type gauges and receive signals to operate from the controlled area network data bus (CAN-bus), an input is required to an ECM, electrical system, or body controller about fuel level (**FIGURE 8-42**). Using this input, an audible alert with a warning light informs the operator that the fuel level is low. (Newer circuits use stepper-type motors for a dash gauge but still use a variable-resistance sending unit.) Onboard telematics can also use the data to help plot a route to the nearest fuel station.

Fuel Filters

Fuel filters are one of the most common intersection points between the technician and the low-pressure fuel system. Although the concept of using a fuel filter is straightforward, it is important to understand how vital they are to the fuel system, what contaminants they remove, and how the contaminants are removed by fuel filtration components.

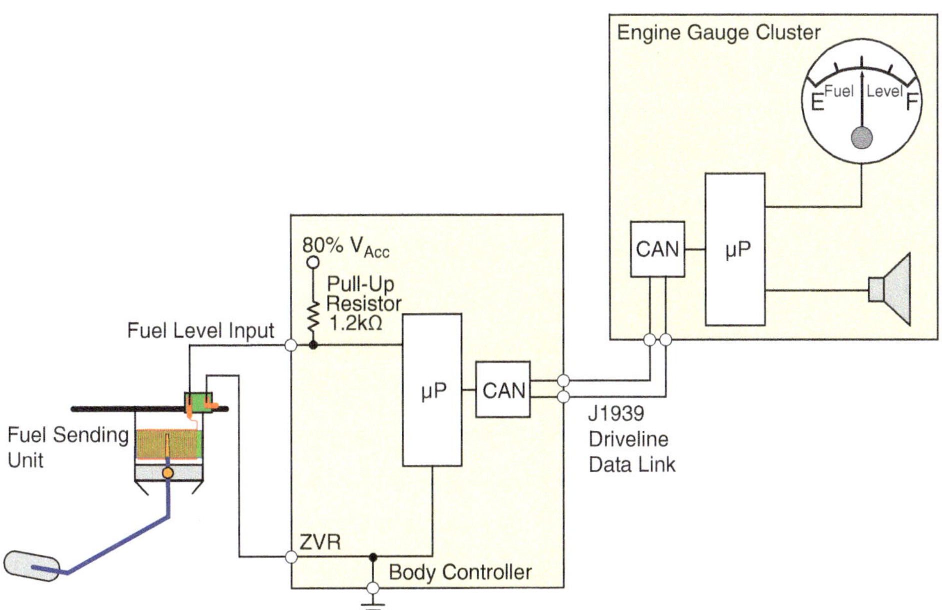

FIGURE 8-42 A fuel tank sending unit with a CANbus control instrument cluster. The OBD system requires fuel level data to prevent it from falsely detecting cylinder misfires.

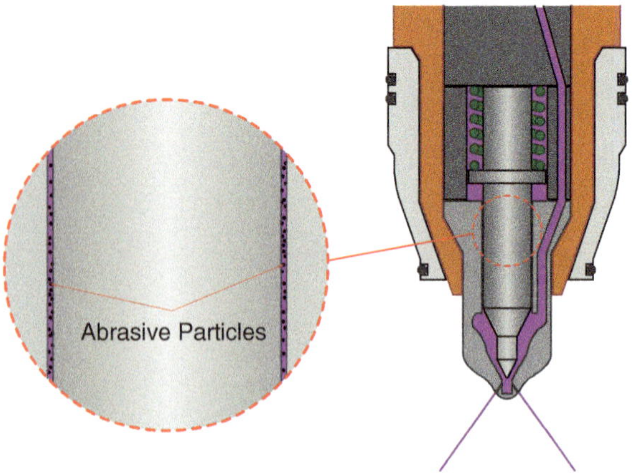

FIGURE 8-43 Abrasive particles that are 5–7 microns cause the greatest damage to a fuel system.

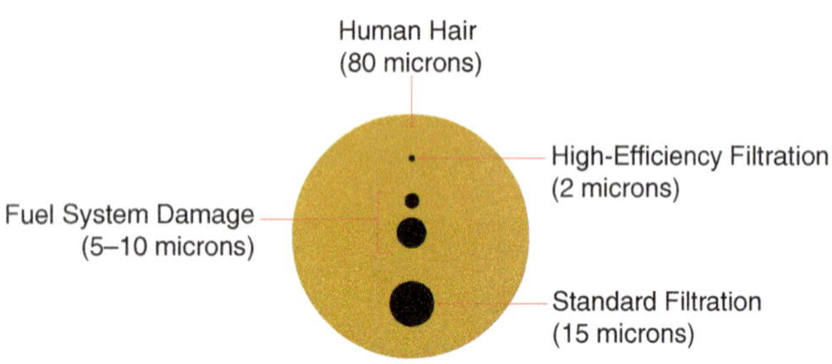

FIGURE 8-44 Micron-sized particles in relation to the width of a human hair.

Filter Requirements

The reliability and durability of diesel fuel-injection systems depends largely on the quality of fuel, which can often be contaminated by fine abrasive wear (**FIGURE 8-43**). Left free in the high-pressure injection system, dirt contaminants cause abrasive wear of injection components operating at tolerances of only a few thousandths of a millimeter. Damage-causing dirt contamination is not readily apparent when only visually inspecting the fuel. Manufacturers' analyses of component failures have discovered that most fuel system wear takes place with particle contamination of 5–7 microns in diameter, a **micron (μ)** being 0.001 millimeters (mm). The smallest particle visible to the human eye is 40 μ. A grain of salt is about 70 μ in diameter; 5 μ is 0.0002" or 0.005 mm in diameter (**FIGURE 8-44**).

Wear of fuel system components results in poor pressurization and atomization of fuel, eventually causing the following problems:

- high fuel consumption
- hard starting
- rough idle
- reduced engine power
- increased smoke and other exhaust emissions.

Contamination Sources

Contamination usually occurs after fuel is pumped into a vehicle's fuel tank and during the course of operations. Another overlooked area is on off-road use around construction sites, where airborne contaminants can be drawn into a fuel tank through its vent tube if the vent tube lacks an air filter. Likewise, a fuel tank cap vent can ingest dust if it is not properly filtered or sealed.

As mentioned already, asphaltenes, which are carbon crystals, are common contaminants that affect the low-pressure fuel supply system. These are found in the bottom of fuel tanks and filters, and they are becoming a more common form of abrasive as fuel refinery techniques change to improve yields from each barrel of oil. Asphaltene particles are soft, deformable, and approximately 1.5–2 μ in diameter. Concentrations can increase through fuel oxidation and from high heat. Although large asphaltenes are considered harmless to the injection system components, they constitute the most abundant of all substances plugging fuel filters. Smaller asphaltenes can produce abrasive damage in the fuel system. This abrasive also forms naturally when fuel is heated and stored over time.

Water is another huge problem contaminant: It not only damages fuel systems, but when high amounts of water are in fuel, it also encourages the growth of various organisms in fuel. A warm climate, where moisture condenses in fuel, contributes to the formation of biological contaminants. Biodiesel also lends itself well to the formation of biological contaminants, because biodiesel can support the growth of various organisms. Water is arguably the greatest problem to fuel systems, not only because it is the most common type of contaminant but also because it is the most destructive. Water exists either in emulsion, meaning it is dissolved and mixed with fuel, or freely as droplets. Water is often introduced into the fuel supply as warm moisture-laden air that condenses on the cold metal walls of fuel tanks or condenses on the surface of cooler diesel fuel. Diesel fuel has **hydrophilic** and **hydroscopic** properties, which means that it attracts and dissolves water. The effects of water in diesel fuel are serious. Because water cannot pass easily through nozzle orifices, water will accumulate, vaporize, and cause a fuel injector tip to blow off. Water will also seize up injector plungers and nozzle valves (**FIGURE 8-45**).

- Water causes the film strength of fuel to separate in close tolerance assemblies such as plungers and barrels.
- The loss of fuel lubricity caused by water contamination can cause seizure and scoring on moving metal surfaces.
- Water can combine with sulfur in the fuel to form strong corrosive acid.

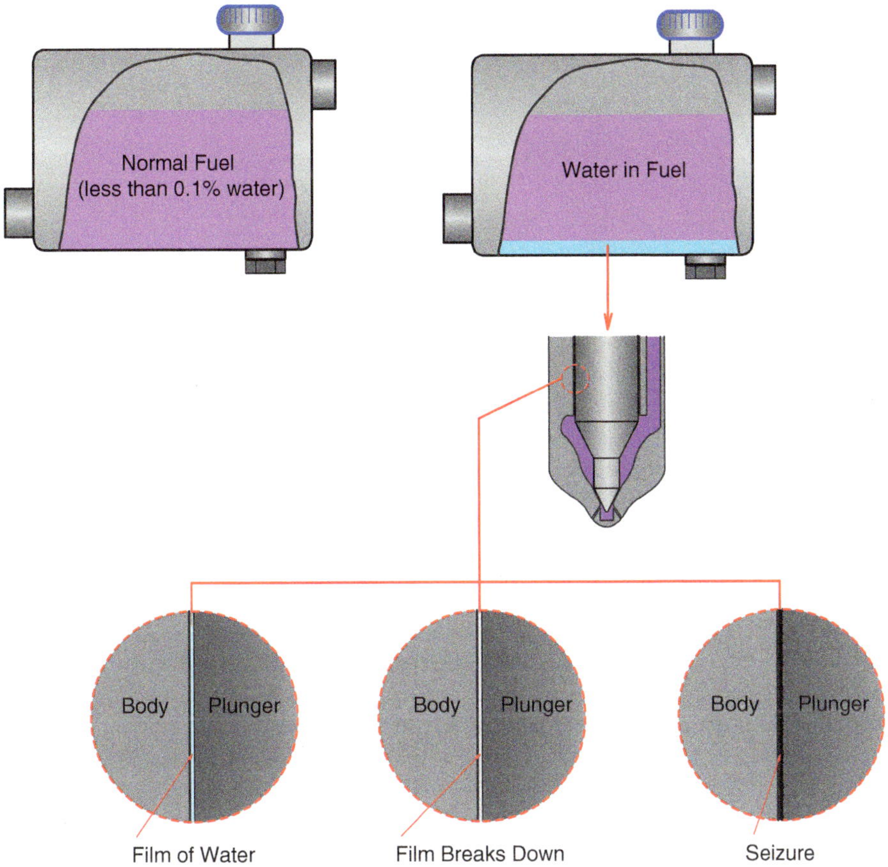

FIGURE 8-45 Water displaces the fuel film on injection system components. The loss of the lubricating film causes parts to wear and seize.

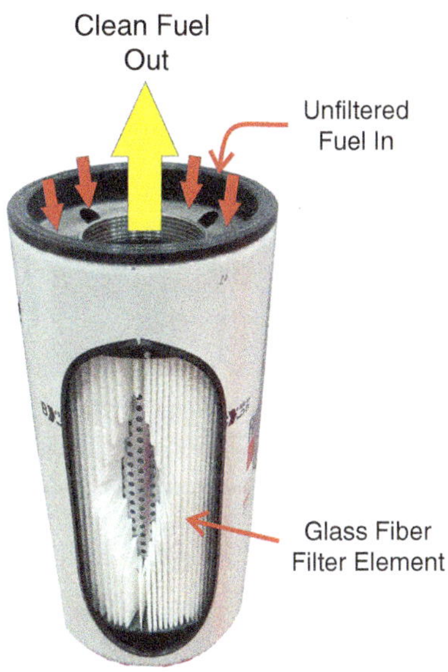

FIGURE 8-46 Unfiltered fuel enters the filter on the outer side of the media, and clean fuel leaves through the center opening.

- Water causes iron components to rust, which in turn produces abrasive iron oxide particles. Significant quantities of these particles contribute to premature wear on moving fuel system components with close tolerances.
- Water in fuel contributes to the growth of microorganisms that can multiply and eventually plug fuel filters.

Fuel–Water Separators

Fuel and water are separated by specialized filter media and coalescing filters. In a regular canister filter, unfiltered fuel enters the filter through holes located near the outer edge of the filter, which passes through the media, and then clean fuel leaves through an opening in the filter's center. Dirt and water settle in the housing at the bottom if they are not trapped in the filter media (**FIGURE 8-46**). Coalescence can be achieved by spinning fuel in a fuel bowl, which separates emulsified and unbound or free water from fuel using centrifugal force (**FIGURE 8-47**). Filter media can also remove emulsified water by coalescence, a process where smaller water droplets merge to form heavier droplets. Because water is heavier than diesel fuel, it collects on the bottom of a standard fuel filter or **fuel–water separator**–a filtration device that removes water from fuel, which extends the life of a filter (**FIGURE 8-48**). To alert the operator via the instrument display of there being too much water in the fuel filter housing or fuel–water separator filter, a **water-in-fuel**

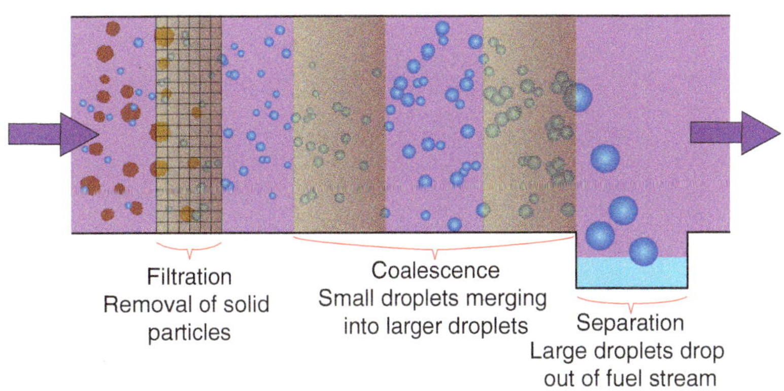

FIGURE 8-47 Coalescence is a process in which the molecules of emulsified water join into larger droplets.

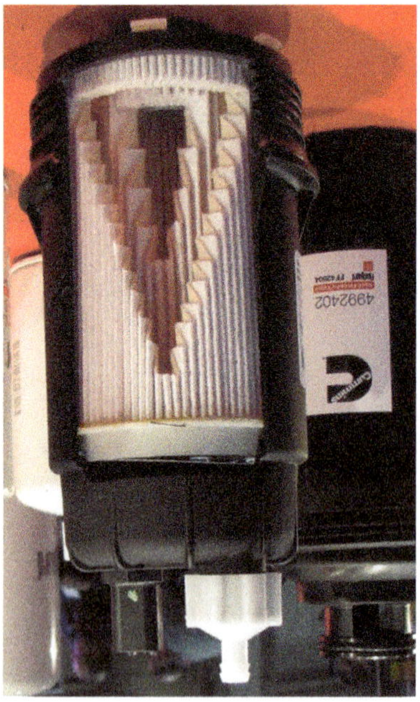

FIGURE 8-48 Specialized fuel filter media in this fuel–water separator help convert emulsified water into larger, free droplets. Water can be removed from the sediment bowl though the drain valve.

FIGURE 8-49 The WIF warning light alerts the operator to drain the fuel–water separator to prevent fuel system damage and performance problems.

(WIF) sensor is used (FIGURE 8-49). WIF sensors use an electrical conductivity probe to sense a specific level of water in fuel and cause a dash-mounted light to illuminate. This light alerts the operator of the need to drain water from the fuel–water separator. Because the electrical conductivity of water is greater than fuel, the small current passing between the fuel–water sensor probes needs to be amplified by an electronic module. The **water-in-fuel (WIF) module** is the actual device that illuminates a warning light that informs the operator to drain or service the fuel–water separator. It may be a standalone module of a circuit incorporated inside the ECU.

One type of WIF sensor used by GM's 6.6 L Duramax diesel engines uses a float that can be lifted only by water, which is denser than fuel. When high enough, the float will electrically activate a warning light on the instrument display. Other sensors measure the electrical conductivity in the bottom of the filter (FIGURE 8-50). Because water is more electrically conductive than fuel, the electrical signal can differentiate water from fuel (FIGURES 8-51 and 8-52).

▶ TECHNICIAN TIP

Alcohol should never be added to diesel fuel systems to absorb water or dissolve wax. Alcohol dramatically lowers the CN of diesel fuel, which makes it more difficult to ignite and burn and more dangerous to handle. Alcohols are volatile, and they ignite at low temperatures. Diesel fuel can be easily ignited if it contains alcohol. The most hazardous issue caused by alcohol in fuel is that it can dissolve water in fuel and carry it through the fuel system. Ethanol, which dissolves easily in diesel fuel, will irreversibly mix with water, which then cannot be removed from the diesel fuel. Water must be separated and not mixed. Alcohol has very poor lubricity and leads to galling seizure and to the fuel system incurring premature wear.

FIGURE 8-50 **A.** A WIF sensor that uses a float with an electric switch. **B.** A WIF sensor that uses an electrical conductivity probe in the bottom of the filter.

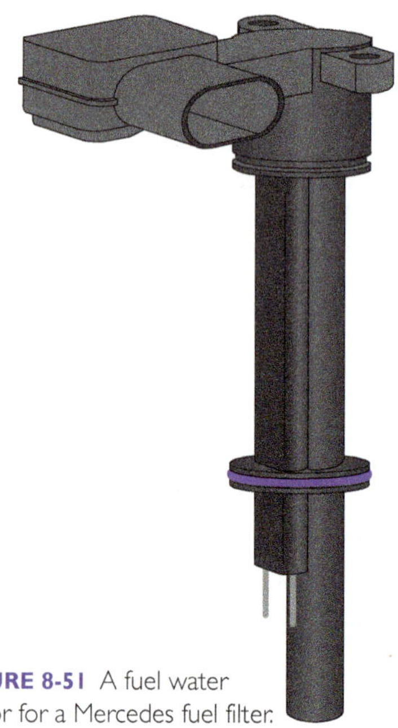

FIGURE 8-51 A fuel water sensor for a Mercedes fuel filter.

Wax

Wax or paraffin is desirable as a source of energy in fuel. However, in cold weather, wax that is normally dissolved in the fuel as a liquid begins to congeal, forming crystals. Unless properly treated or blended, 2-D fuel can begin to cloud at 32°F (0°C). When these wax crystals form, it is known as the cloud point. Fuel temperatures below the cloud point will result in wax precipitation and filter plugging.

To prevent wax formation, the cloud point of fuel must be at least 10°F (6°C) below the lowest anticipated outside temperature. Fuel suppliers normally blend diesel fuel based on local cold weather conditions. However, particular attention should be given to diesel fuel purchased outside an area where the equipment is to be operated. For example, fuel purchased in the Southern United States during the winter is not likely to be suitable for operating conditions in the American Midwest or Canada. Various types of **fuel heaters** are used to heat the fuel. These are typically thermostatically controlled devices that prevent the fuel from waxing and gelling. Electric fuel heaters are often located in fuel filter housings (**FIGURES 8-53** and **8-54**). Larger vehicles with high fuel flows may use coolant-type fuel heaters. These thermostatically controlled heaters use hot engine coolant to warm the fuel through a heat exchanger. Many diesels do not require heaters unless temperatures are consistently very low since the engine returns significant fuel flow to the tank from injectors and pumps.

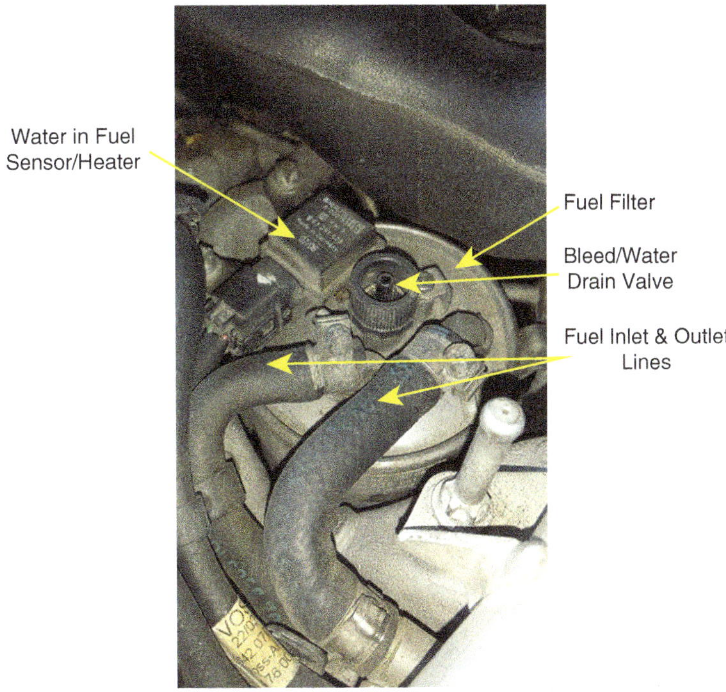

FIGURE 8-52 WIF combined with a fuel heater and pressure sensor.

Fuel Heaters

To prepare fuel for cold weather use and thus avoid operating difficulties, the low-pressure supply system is often equipped with a variety of devices to condition the fuel by removing water and dissolving wax. Engine power loss during winter operation conditions can be a problem for operators of diesel-powered vehicles. When properly diagnosed, the power loss is often associated with gelled fuel caused by fuel wax that forms a restrictive coating on the filter element, resulting in low fuel pressure. The same process occurs when starting an engine in cold weather. Fuel filters can become covered with a restrictive layer of wax.

This layer may permit enough fuel to flow and allow the engine to idle, but at full load, the engine cannot maintain operating speed (rpm) and may stall due to inadequate fuel pressure. Filter media impregnated with contaminants will result in this condition, and any

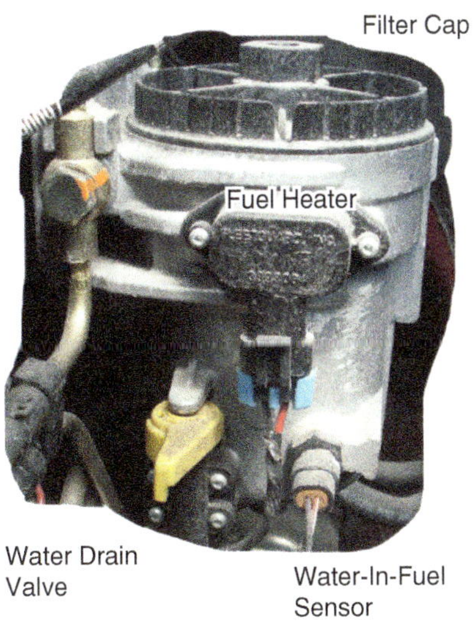

FIGURE 8-53 Location of the electric lead for the fuel heater of a Dodge Ram fuel-conditioning module.

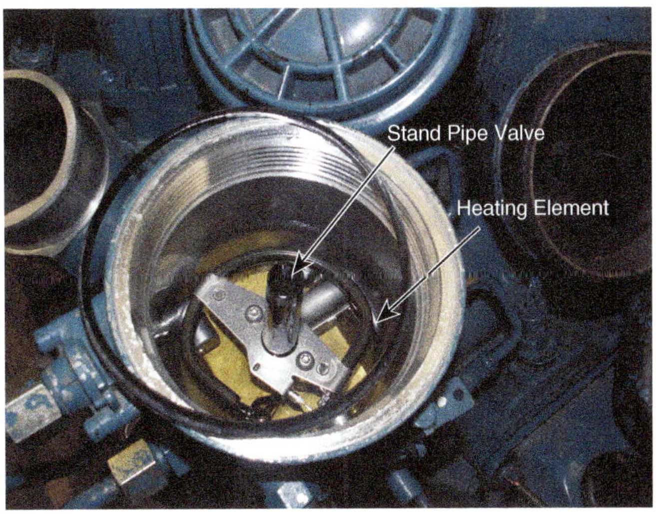

FIGURE 8-54 Location of an electric heating element for a 6.0 L Ford Powerstroke.

clogged filter must be changed as the remedy. The root cause can also often be the temperature of the diesel fuel and its effects.

Two methods are available to prevent fuel gelling: One is to chemically treat the fuel with pour-point depressants; another is to install a fuel heater, a thermostatically controlled device that prevents fuel from gelling and overheating. A fuel filter simply warms the fuel and keeps the wax in a dissolved state. Fuel heaters can use electric current or warm coolant to heat fuel. Electric in-line heaters can also be placed in line just before fuel enters the filter housing. Other electric heaters use heating elements located inside the fuel–water separator to warm fuel and to prevent icing. Coolant-type heaters provide greater temperature increases, but only after the engine coolant has warmed up.

When the water in fuel turns to ice due to freezing cold weather, like gelling, it too can cause filters to clog. In general, removing water is critical to extending fuel system service life. Most designs for water separators operate by coalescing or clumping smaller water droplets into larger ones that eventually fall out of the less dense fuel into a sump. Gravity or centrifugal force facilitates this process of coalescence in many separator designs.

Fuel–Water Separators and Heater Modules

Fuel–water separators, WIF sensors, fuel heaters, and filters are often integrated into a single compact unit. These units reflect the importance of space conservation as a design criterion. In fact, some modules also incorporate fuel pumps and pressure regulators. Heating and fuel cooling functions are commonly incorporated into the modules. Warm return fuel is blended with cooler tank fuel to ensure fuel at the correct temperature reaches the injectors. Ford uses two types of conditioning modules: one that's mounted horizontally along the vehicle frame called the horizontal fuel-conditioning module (HFCM) and a vertical or upright module simply called the fuel conditioning module (FCM) (**FIGURE 8-55**). Both

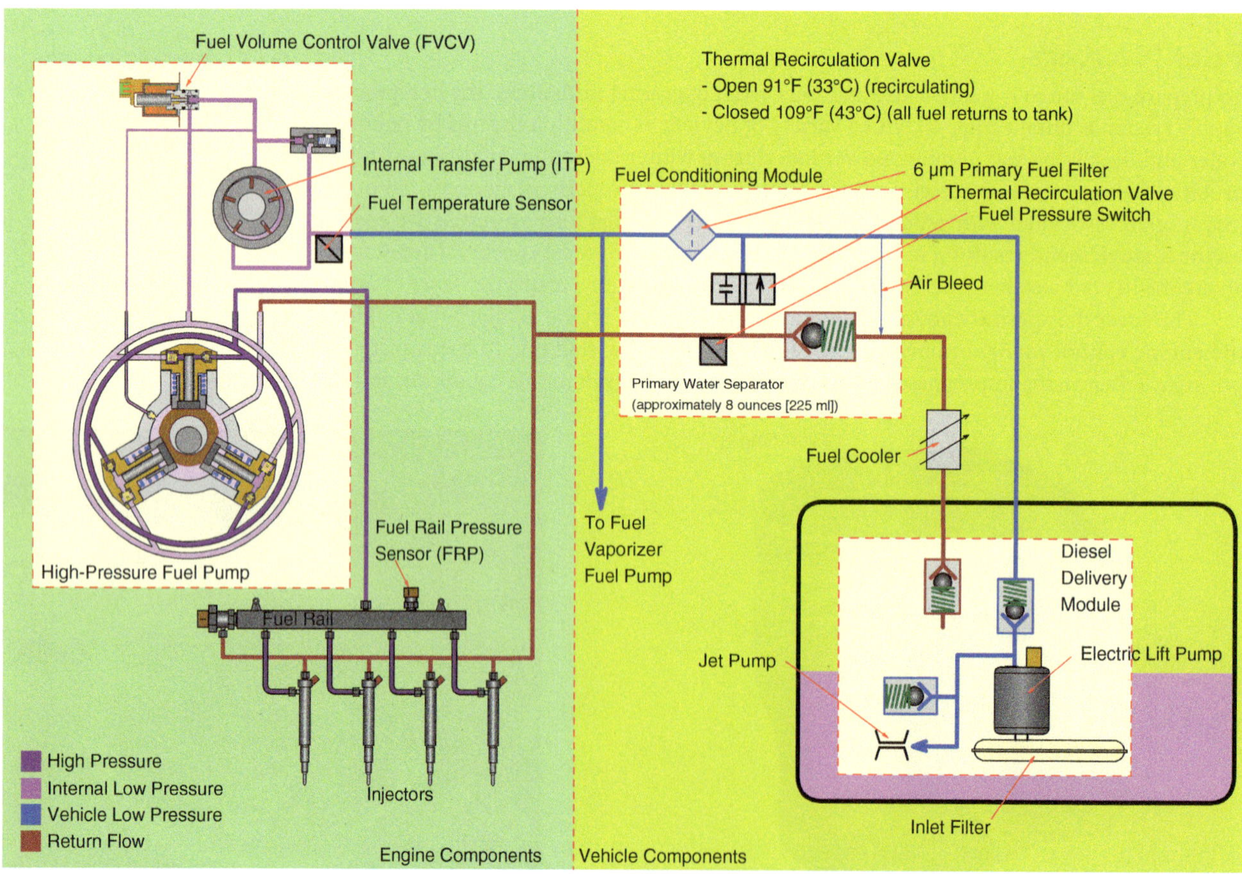

FIGURE 8-55 The FCM used by Ford contains a filter, a fuel temperature regulator, a low-pressure switch, an air bleed, a WIF sensor, and a water drain valve.

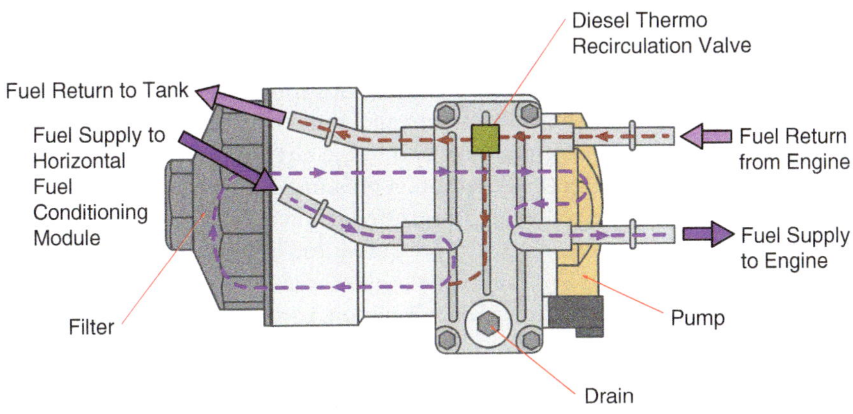

FIGURE 8-56 The HFCM regulates fuel temperature, filters fuel, and separates water. It also contains an electric fuel pump.

modules have multiple functions, but the HFCM also contains an electric fuel pump that regulates pressure to 24 psi (1.7 bar). They both contain a fuel filter, a water separator, and a thermal recirculation valve to regulate fuel temperature. Temperature regulation is performed by blending warmer engine fuel return with cooler fuel from the tank. Depending on whether fuel needs to be warmed or cooled, the modules can blend fuel to regulate fuel temperature supplied to the high-pressure system (**FIGURE 8-56**).

The FCM for the latest injection systems also contains an injector backpressure valve to regulate pressure in the return circuit for the hydraulic coupler inside the piezo injectors (**FIGURE 8-57**). It also has a low-pressure switch and a thermal recirculation valve—a

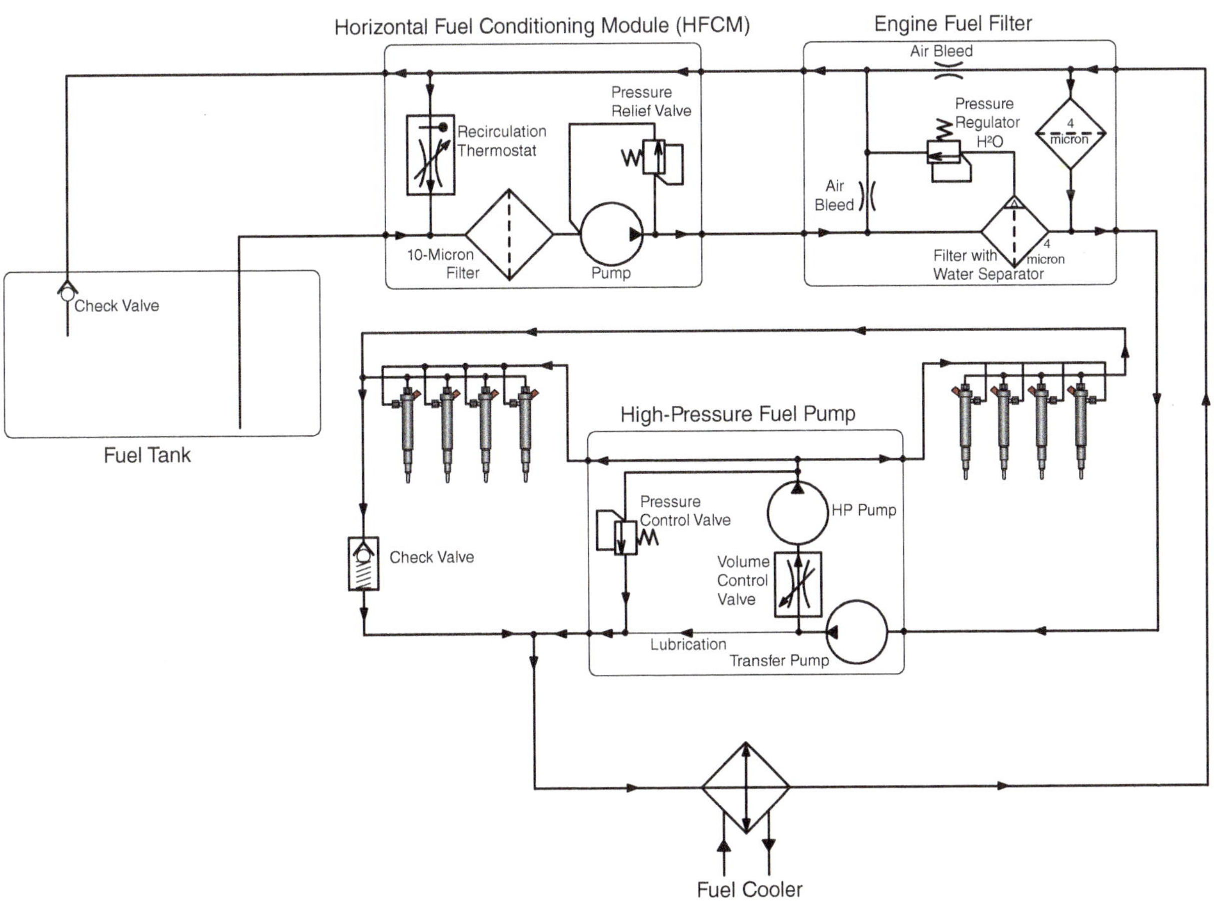

FIGURE 8-57 A Ford FCM contains a fuel pressure switch to protect the injectors from damage caused by low fuel pressure. All fuel returns directly to the tank above 109°F (43°C).

valve that blends return circuits and supply from the fuel tank. The fuel filter cleans dirt and separates water from the fuel. A small internal bleed orifice in the fuel filter housing on the dirty side of the filter allows entrained air to return to the fuel tank through the return line. The thermal recirculation valve is a temperature-sensing bimetal valve that allows warm fuel from the engine to enter directly into the dirty inlet side of the filter to help melt wax. Once filtered, fuel passes over a fuel pressure switch used to verify that adequate fuel pressure is supplied to the high-pressure fuel pump and the fuel pump for the aftertreatment systems fuel dosing valve. If pressure is too low, a warning light alerts the driver. Low fuel pressure also sets fault codes, which can be caused by a defective in-tank electric pump, restriction in the fuel supply circuit, or a restricted fuel filter.

Fuel Coolers

High-pressure common rail systems pressurize fuel, which results in a large temperature increase for the fuel. Return fuel temperatures can exceed 140°F (60°C), which is the lower end of the flash point for diesel fuel. It is important to keep fuel temperatures lower than this because heating causes fuel to lose its viscosity and density, resulting in low power and poor atomization. The greatest danger of overheating fuel is that it can not only change fuel properties but cause the formation of ashphaltenes. To prevent fuel from overheating, fuel coolers are more often used in contemporary fuel systems. Some FCMs regulate fuel temperature by mixing cooler fuel from the tank with warmer return fuel. This strategy helps warm up cooler fuel pulled from the tank. But to lower the temperature of fuel returning from the engine, fuel coolers are used. These coolers use engine coolant to reduce fuel's temperature and heat cold fuel. One example is Ford's 6.4L Powerstroke use of the horizontal fuel-conditioning module (HFCM), which works together with a secondary filter module located in the engine compartment. Fuel temperature is kept in a consistently even range by using engine coolant stored in a separate coolant reservoir in the engine compartment (**FIGURE 8-58**).

FIGURE 8-58 A separate coolant reservoir is used as a fuel cooler for the common rail 6.4 L Powerstroke.

Filter Media

A variety of materials are used to fabricate filters. Materials include paper (cellulose), microglass, cotton, felt, wood, and the like. The material should be made thick enough to trap and absorb contaminants in the filter fibers as fuel passes through filter pores. Pleating the filter media increases the surface area and adds to a filter's ability to retain contaminants. Indeed, ideal filter media remove the smallest particles with great efficiency while having a tremendous capacity to store these particles to prevent filter restrictions. Paper or cellulose media are chemically treated with resin or silicone to repel water. The coating is used primarily to prevent damage to filter media and does not necessarily prevent water from passing through the media. It should be remembered that alcohol-based fuel additives will degrade the performance of fuel–water separators. Microglass or glass-fiber, a synthetic media, is considered the best filter media, trapping the smallest particles and absorbing with the greatest capacity. Paper cellulose filters, however, are usually the least expensive (**FIGURE 8-59**).

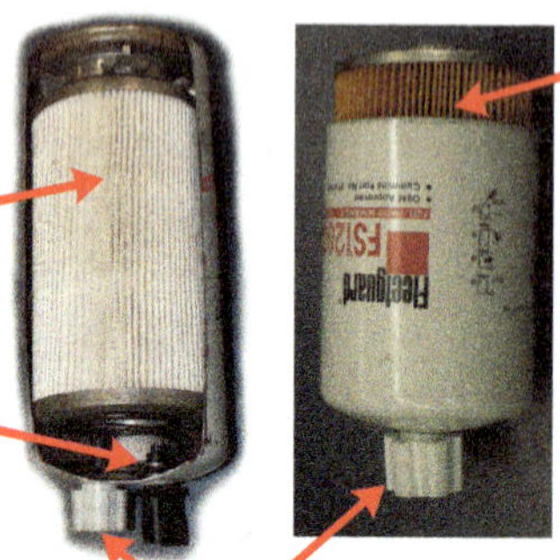

FIGURE 8-59 Comparing microglass filter media with cellulose filter paper.

Efficiency

Filters are classified not only by filtration media but by efficiency too. The quality of a filter is based on the nominal micron size of the particles it traps and its efficiency (the percentage of particles the filter can trap for a given size). Each filter has an **absolute micron rating**, which is an efficiency rating expressed as a percentage of the smallest size of particles that can be trapped by the filter. For example, a filter element may have a nominal size 5 μ filter element but it operates at only 50% efficiency. That means the filter will trap only half the particles 5 μ and larger. A nominal size 5 μ filter operating at 100% efficiency will trap all the particles 5 μ and larger (**FIGURE 8-60**).

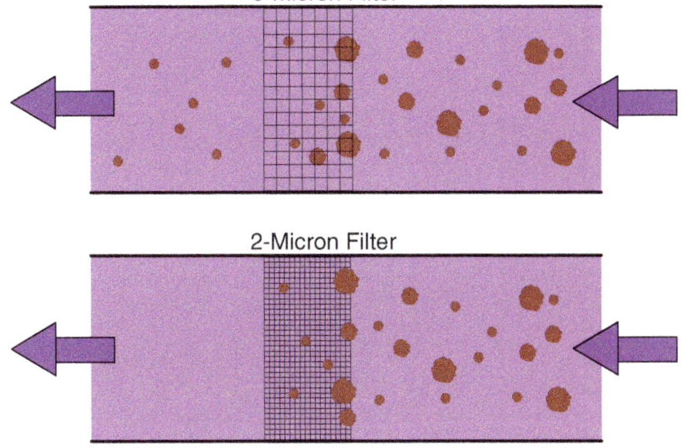

FIGURE 8-60 The absolute micron rating of a filter refers to the percentage of the smallest particles that a filter traps.

Primary and Secondary Filters

Filters are also classified by location in the fuel stream (**FIGURE 8-61**). The first filter in the fuel stream after the fuel tank is the **primary filter**. Primary filters are usually placed before any transfer pump in the system, to remove any large particles that could prematurely wear the pump and cause check-valve damage. The job of a primary filter is to remove the bulk of free water and contaminants from the fuel. Using a primary filter extends the life of the secondary filter, which needs to be effective for finer contaminants, by preventing premature clogging. Typical primary filter efficiency is 95–98% at 10–100 μ.

The **secondary filter** provides additional filtration for particles that are even smaller. The secondary filter is located between the primary filter and the high-pressure fuel-injection system. Typical secondary filter efficiency is 95–98% at 2–10 μ. Newer high-pressure fuel-injection systems need even the smallest measurable hard particles—smaller than 5 μ—to be removed. Most damage in a fuel system is done by particles from 5 to 10 μ. A properly sized two-filter system will last longer and operate more efficiently than a similarly rated single-filter system.

Cartridge Filters

While the use of spin-on fuel filters has been popular for decades, many jurisdictions have banned their disposal from landfills for environmental reasons. Today, spin-on metal-cased filters, at a minimum, must be drained of all free-flowing fuel and oil before they are disposed. Currently, in the United States, spin-on fuel and oil filters are exempt from hazardous waste regulations if the oil filter has met one of the following conditions:

- It has been punctured through the dome end or anti-drain-back valve and hot-drained.
- It has been hot-drained and crushed.
- It has been hot-drained and dismantled.
- It has been hot-drained by using an equivalent method to remove used oil.

> ▶ **TECHNICIAN TIP**
>
> Whenever a fuel filter is replaced, the fuel system most often requires priming to remove any air that may have entered the fuel system. Fuel filters should generally be installed empty (without fuel) and the fuel system primed with either an electric or hand primer pump (depending on the system), to prevent introducing unfiltered fuel into the system, which can cause damage. Some systems use a hand primer pump to prime; others an electric pump. A bleeder screw is often located in the fuel system to allow entrapped air to escape from the filter when priming. Alternatively, service software enables the electric low-pressure pump to be energized with a service routine to bleed filters with the key on engine off (**FIGURE 8-62**).

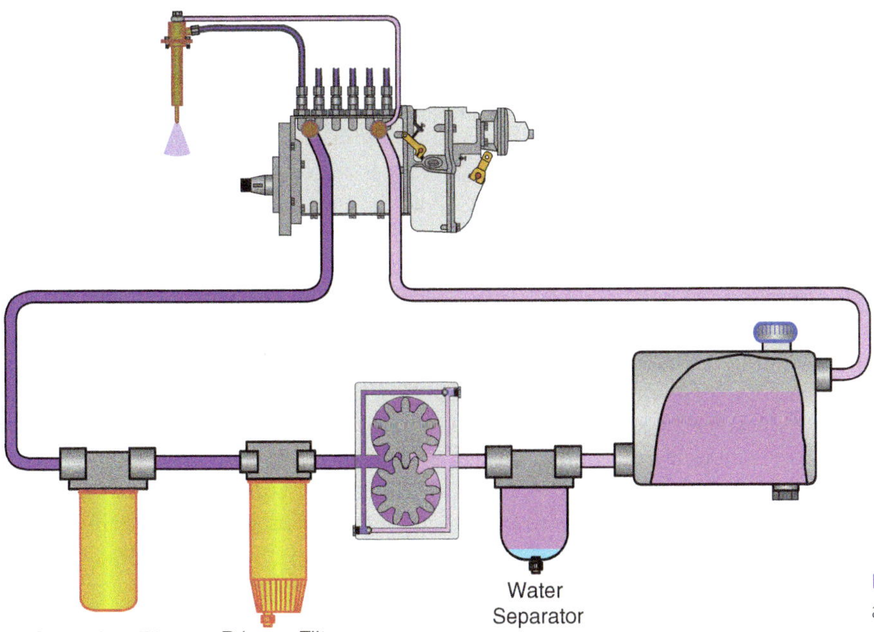

FIGURE 8-61 The location of primary and secondary filters. Secondary filters capture smaller, finer particles that escape past the coarser primary filter.

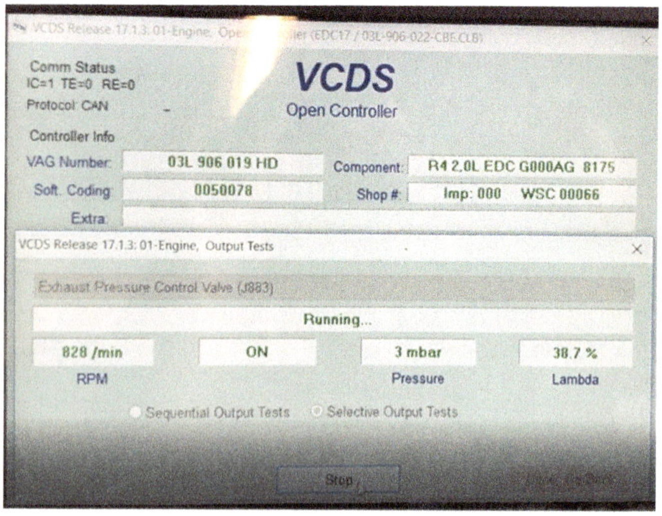

FIGURE 8-62 Priming a fuel filter using a hand operated vacuum pump. Applying a vacuum to the filters outlet fitting pulls fuel through the filter.

FIGURE 8-63 Cartridge-type filters are a more environmentally friendly choice of filter.

FIGURE 8-64 The standpipe of many filter housings uses a check valve that must be opened by the physical presence of a fuel filter to enable fuel flow. Without a filter properly installed, the filter housing will not supply fuel.

Hot draining refers to draining engine oil at near operating temperatures but no less than 60°F (16°C). Because most spin-on fuel filters are indistinguishable from oil filters and are disposed together, cartridge-type filters have become far more popular with manufacturers (**FIGURE 8-63**). In addition to issues associated with their disposal, cartridge filters have other advantages. For example, most cartridge-type fuel filter housings are designed to prevent the fuel system from operating without a fuel filter (**FIGURE 8-64**). Also, because no metal components are usually involved in manufacturing cartridges, the likelihood of metal entering the fuel system and causing damage is diminished.

Fuel Transfer Pumps

Fuel transfer pumps are used to move fuel from the fuel tanks to the high-pressure fuel system. The fuel must be supplied with adequate volume and at the correct pressure for every operating condition. A variety of pump types exist to match the unique requirements of each high-pressure fuel system.

Gear Pumps

A **gear pump** is a positive displacement pump (a pump that delivers a fixed quantity of fluid for every revolution) with two counter-rotating gears that mesh as they rotate (**FIGURE 8-65**). Fuel is drawn around the circumference of the pump gears and then trapped between the pump gears and walls when the pump rotates. Fuel is then delivered to the pump outlet as the gears mesh again. Gear pumps produce volume. However, if restrictions to space or volume at the pump outlet diminish, the pump will build pressure proportional to the force driving the pump. The seal contact point between the gear teeth and pump housing act as a one-way check valve to prevent fuel from draining back into the tank. This means a one-way flow of fuel is integral to this pump design. Check valves at the pump inlet allow fuel to bypass the pump when a hand primer pump is installed (**FIGURE 8-66**).

A positive displacement pump depends on engine speed to produce pump pressure. The most suitable applications for gear pumps are injection systems that depend on increasing pump pressure as the engine speed increases. Unit injection systems typically use gear pumps. Hundreds of pounds of pressure are capable of being produced by this pump, so pressure regulation is required. A restriction orifice or pressure-regulating valve on the fuel-return side of the system or the use of a pressure regulator generally accomplishes this. If this valve fails, it will cause the engine to lose its fuel prime, which means fuel will drain from the injectors. Air will enter the injectors, and the engine will be hard to start.

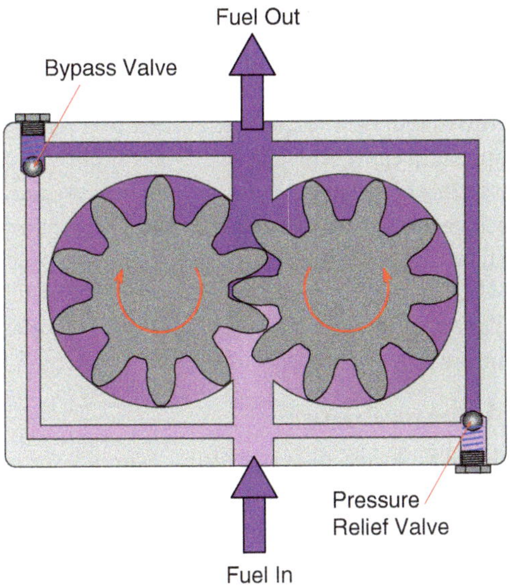

FIGURE 8-65 A gear pump is a positive displacement–type pump. Internal bypass valves help regulate fuel pressure.

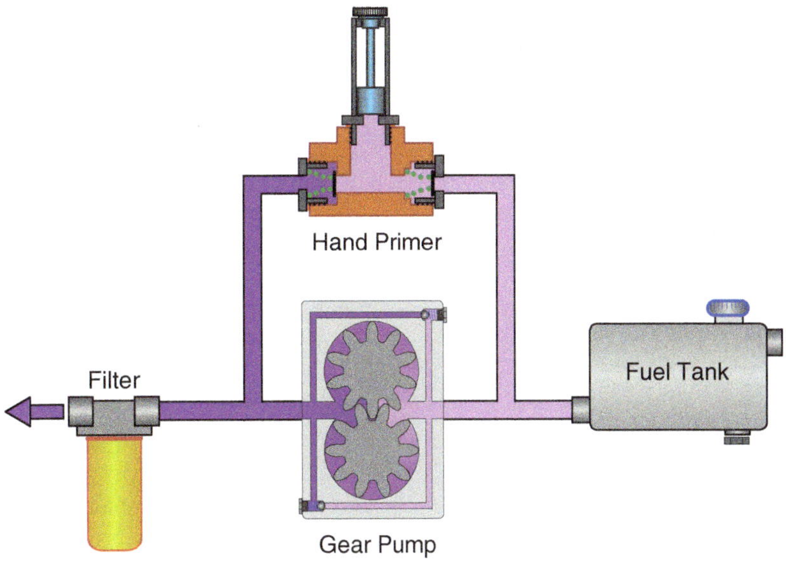

FIGURE 8-66 A hand primer pump may be installed to either bypass the gear pump, such as pictured here, or bypass valves in the gear pump, to allow a hand primer pump to pull or push fuel around the gears.

Advantages of the gear pump include both its high volume and pressure output and its simple design and durability. Gear pumps are usually maintenance-free, but they need priming if no fluid is in contact with the gears.

Mechanical Diaphragm Pump

Another common type of fuel transfer pump is the mechanically operated **diaphragm pump**. This pump is used to supply fuel at low pressure, usually not more than 7 psi (48 kPa) with a variable output volume (**FIGURE 8-67**). The engine camshaft activates this

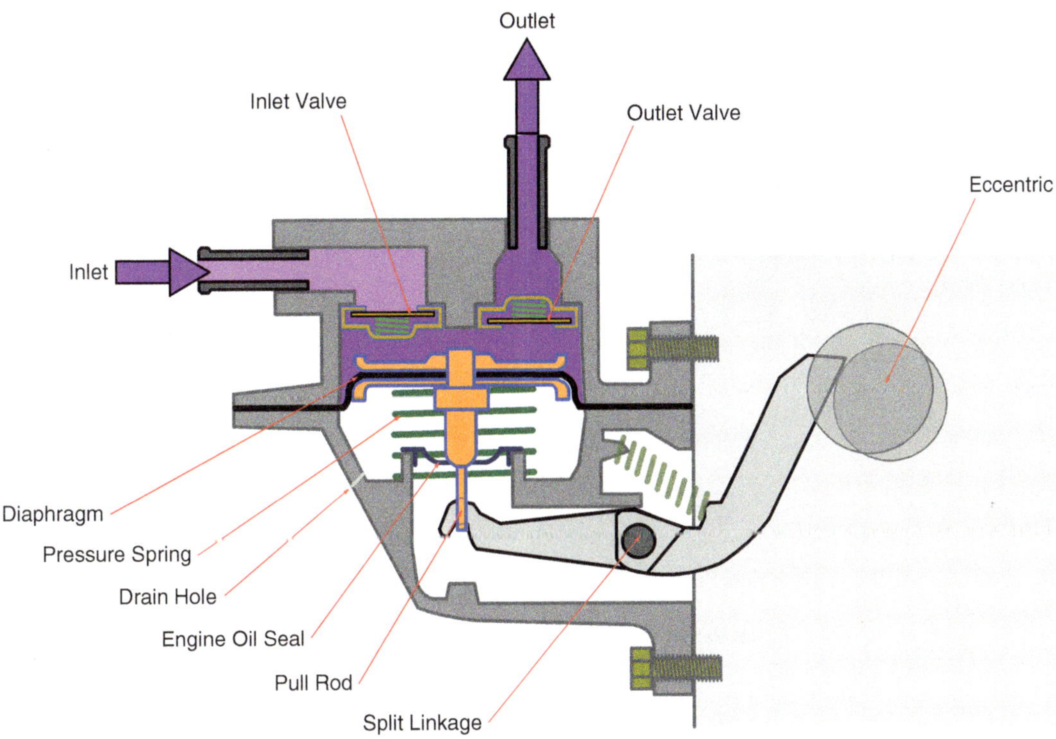

FIGURE 8-67 A diaphragm pump with a priming lever that's used during filter changes to bleed the air from the fuel system.

pump through a pushrod or lever operating against an eccentric cam lobe. As the eccentric moves the rod or lever back and forth, its linkage pulls down a diaphragm located between the upper and lower pump halves. While the reciprocating lever or rod movement stretches the diaphragm, the spring pressure beneath the diaphragm forces it to return to a relaxed position. This reciprocating motion of the diaphragm causes the chamber volume above the diaphragm to alternately increase and decrease, thus creating both a high- and low-pressure area above the diaphragm. Two one-way check valves, one located at the inlet of the pump and the other at the outlet of the pump, operate to control the direction of fuel flow through the pump. When the diaphragm is pulled down, the inlet check valve opens and the outlet closes. Relaxing the diaphragm causes the inlet valve to close and the outlet valve to open and deliver fuel at pressure no greater than the spring force.

Variable Displacement in a Diaphragm Pump

The output pressure from the pump is determined by spring pressure. When output volume from the pump exceeds fuel system requirements, the pump uses a unique arrangement of its internal linkage that allows the diaphragm to remain in a stretched or extended position while allowing the reciprocating action of the linkage. (An elongated slot on the diaphragm pull rod allows the pump lever to move back and forth.) No additional volume of fuel is delivered in this position. This means that the pump pressure and output is self-regulating, and it will deliver fuel only at the volumes required and at no more pressure than the force supplied by the spring.

Many fuel transfer pumps have an integral hand-prime button or, in the case of a diaphragm type, a lever. In the case of an external lever, it is connected to internal pump linkage that mechanically pulls down the diaphragm to draw fuel into the pump. Spring pressure pushes back the diaphragm, and the lever can be pulled down once more to charge the fuel system. After the fuel system has been serviced, the technician can use this external lever to fill fuel filters and manually prime the injection pump or purge trapped air from the system.

Piston-Type Pumps

Piston-type pumps are similar in operation to the diaphragm pump and use a similar system of check valves and spring operation (**FIGURE 8-68**). However, instead of a diaphragm

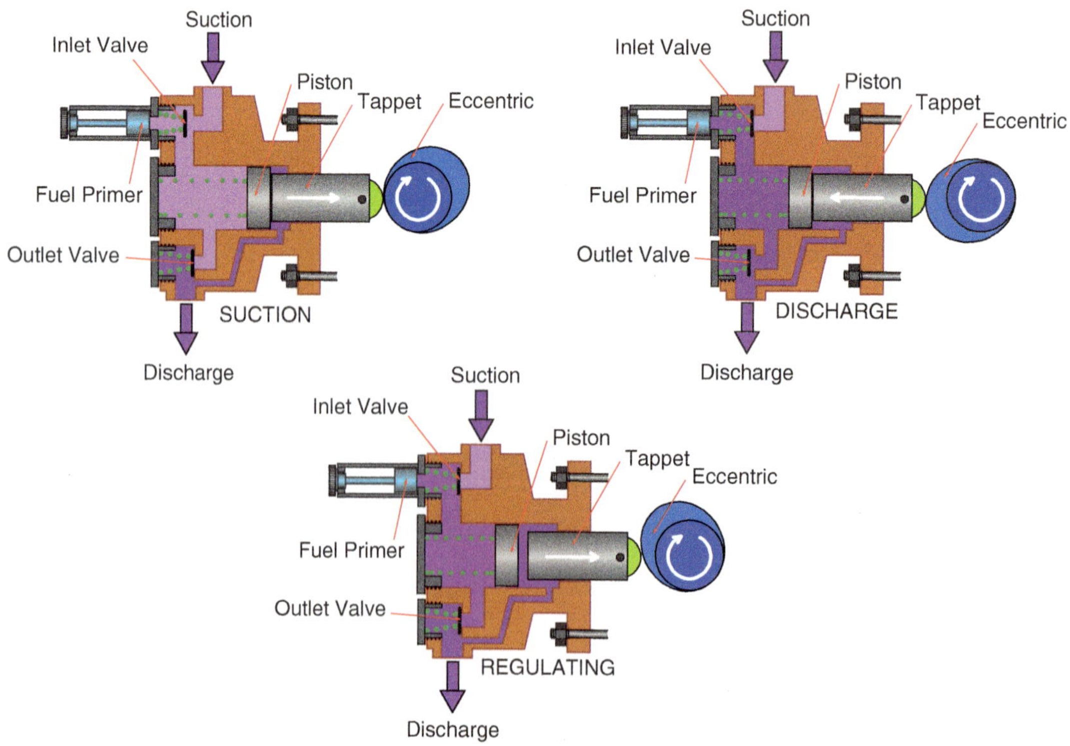

FIGURE 8-68 A piston-type transfer pump is a variable displacement pump.

to develop pressure differentials, a piston is used, which allows these pumps to operate at a higher pressure and with greater durability than diaphragm pumps.

Common applications of this pump style include fuel transfer pumps—for which power is supplied by a cam lobe—and hand primer pumps. Combination primer pumps and transfer pumps are used on mechanical in-line pumps.

Output pressure will be determined by spring pressure beneath the piston. Output volume will depend on engine speed and spring pressure. The piston, as in the diaphragm pump, will rest in a position where output pressure is balanced against spring pressure. This means that output volume is variable, because as outlet pressure increases, there will be less piston movement in these kinds of pumps.

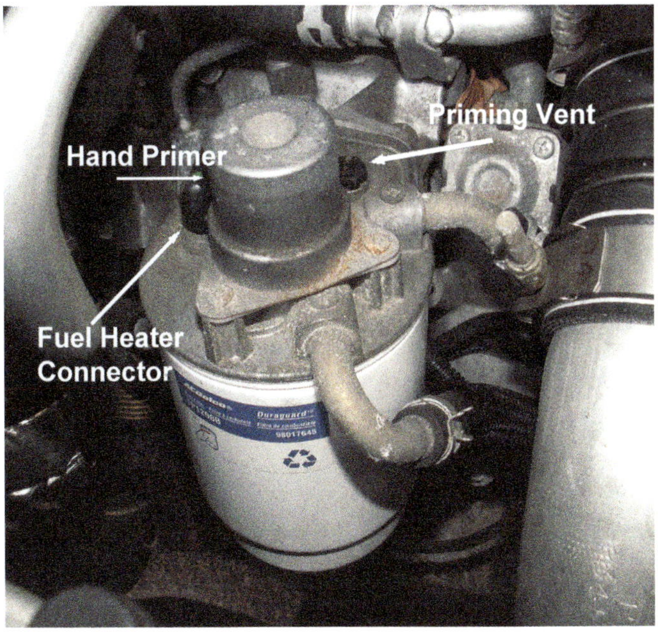

FIGURE 8-69 A hand primer button located on the filter housing of a 6.6 L Duramax diesel. Note the location of the priming air vent bleed.

Hand Primers and Priming Buttons

When the engine is not running and the fuel system needs to be purged of air or a fuel filter needs to be filled, then a hand primer or priming button, depending on the design, provides the piston action to develop pressure differentials needed to move the fuel (**FIGURE 8-69**). Hand primers are also required for priming newly installed (i.e., dry, unfilled) fuel filters. This procedure minimizes the possibility of dirt contaminating the high-pressure fuel when using potentially dirty fuel from a can or other apparatus to fill fuel filters.

Electric Pumps

Most light-duty diesel used in automobiles and light trucks use electrically operated transfer pumps located in the fuel tank or on the frame near the tank (**FIGURE 8-70**). Putting the pump inside the tank has the advantage of pressurizing the entire delivery system. This eliminates the likelihood of positive pressure eliminates suction-side leaks, which cause air to enter the fuel. Fuel filters are also able to operate with longer life because pressurized fuel delivery will make them less sensitive to pressure loss from restrictions. Because diesel fuel is less volatile than gasoline, pressurization is not necessary to prevent vapor locks in diesel fuel.

Electrically operated fuel pumps have the other advantage of reducing the inconvenience of priming filters manually during filter changes. Cycling the ignition on and off or bypassing the fuel pump relay allows filters to be installed dry with far less service time. Furthermore, when switching from one fuel tank to another, an electric pump is simply turned off while another is activated. In the event of an accident or if there is a fault with the vehicle,

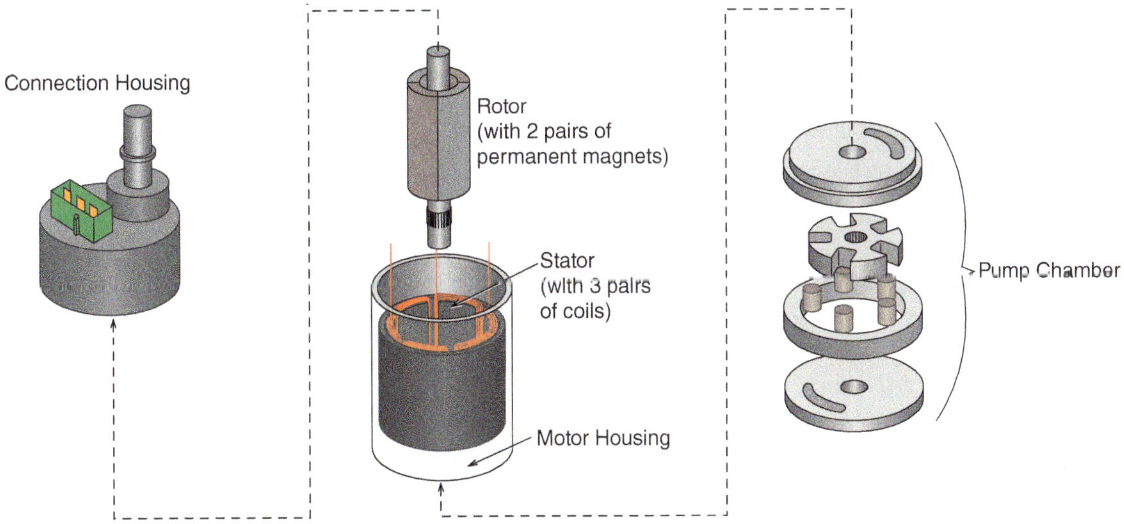

FIGURE 8-70 Components in an electric transfer pump.

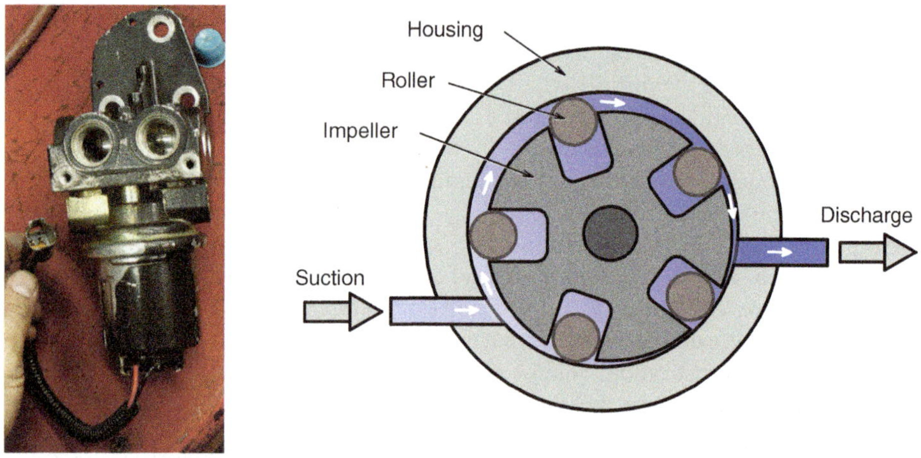

FIGURE 8-71 This frame-mounted pump uses a pulse-width–modulated pressure control.

> ▶ **TECHNICIAN TIP**
>
> If a diesel fuel system is inadvertently contaminated with gasoline, the engine should not be operated. Fuel should be drained from the tank and all fuel filters replaced. Electric tanks can be drained by bypassing the fuel pump relay to operate the in-tank or frame-mounted pump. When completed, the system and filters should be refilled with clean, fresh diesel fuel. If an engine has operated for some time with contaminated fuel, the lube oil and filters should also be changed. Note that sensitive valves such as the return side pressure control are easily damaged by gasoline.

an electric pump can easily be controlled by the vehicle ECM and shut off. The speed of the pump and output volume or pressure can also be regulated by changing the length of time a pulse-width–modulated (PWM) signal supplies electrical current to the pump.

Roller Vane Pumps

Most in-tank electric pumps are the positive displacement, roller vane–type using permanent magnet–type electric motors. A permanent magnet electric motor rotates rollers or vanes that are spring loaded and moveable in the slots on a rotor shaft. These vanes operate in an offset position inside a circular cavity. As the rotor shaft rotates, the pressure from the fuel forces the vanes to trap fuel in a decreasing volume of the pump cavity. The result is fuel pressurization and fuel movement from the inlet to the outlet of the pump. Two check valves are usually located in the pump, one on the pump inlet and one on the pump outlet to ensure one-way flow of fuel and to prevent fuel from draining back, which can lead to hard and prolonged starting. A centrifugal impeller may also be used in parallel with these pumps to improve fuel delivery (**FIGURE 8-71**).

Vane-Type Pumps

Some engines use mechanical **vane-type pumps** located inside the injection pump (distributor-type pumps) or electric vane external transfer pumps. When controlled by the ECM, the external electric vane transfer pump uses a pulse-width modulation signal to provide two duty cycles: full speed when the engine is running and half speed during cranking to prevent the injection pump from overpressurizing, which can keep the injection pump from operating. Excessive internal pump pressure during cranking can prevent the injection pressure pumping plungers in some distributor-type pumps from being forced outward, which limits the full delivery quantity of pressurized fuel during cranking.

During cranking, the pump pressure is limited to 7 psi (48 kPa); it then increases to 10 psi (69 kPa) after the engine starts. If the engine shuts down without the key being shut off, a safety feature prevents the pump from operating, to prevent fuel leaks if a broken fuel line had been the cause of the shutdown.

▶ Maintenance of Low-Pressure Fuel Systems

8-08 Outline the purpose and methods used to diagnose faults in the low-pressure fuel circuits.

Evaluating how well the low-pressure supply system is operating is necessary when investigating low-power problems, hard or prolonged starting, rough engine operation, and high-pressure injection component failure. Fuel pressure and volume can be checked by using a variety of mechanical or electronic pressure gauges to measure transfer pump output—as well as to check specific locations in the low-pressure fuel system. Volume from mechanical pumps is measured by cranking the engine with the ignition circuit disabled and then measuring the quantity of fuel delivered—usually a 30-second cranking interval—into an appropriate and graduated receptacle (**FIGURE 8-72**).

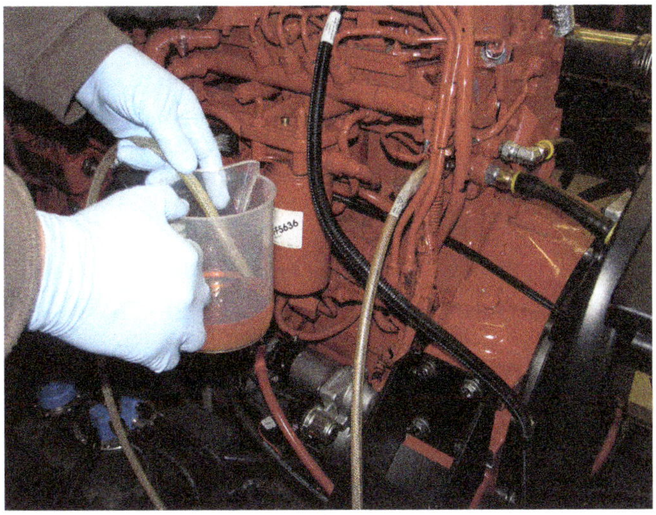

FIGURE 8-72 Performing a volume test by using a graduated cylinder for volume measurement.

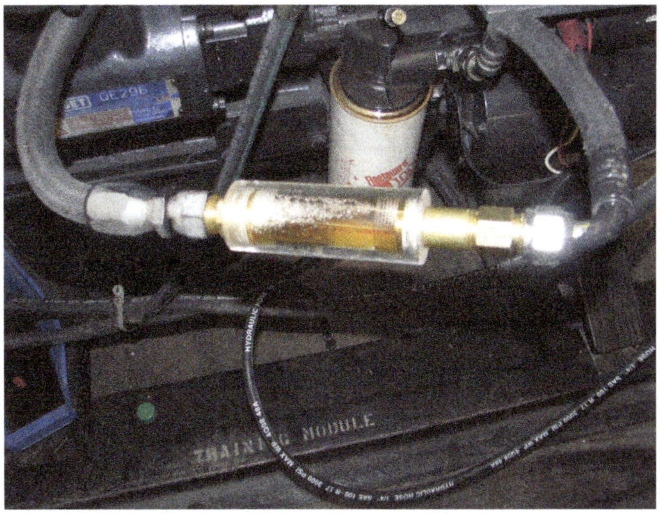

FIGURE 8-73 An in-line site glass can be used to inspect for air in the suction side of the fuel system.

Checks of the low-pressure suction system include vacuum testing and leak testing by using a site glass to visually inspect for air bubbles in the fuel (FIGURE 8-73). Since vapor can pressurize, the bubbles make the fuel elastic or take on sponge-like characteristics. This means the high-pressure fuel system cannot properly pressurize fuel or meter correct fuel volumes, which causes the engine to run rough and have low-power. A clear plastic fuel line installed after a filter will indicates whether bubbles are present in the fuel, which are caused by air leaks in the suction side of the fuel system. It is normal to have bubbles in the return fuel line. Fuel heated in the injectors or cylinder head will vaporize and return to the tank.

Filter restrictions can be evaluated by using a pressure gauge for primary filters on the suction side of the transfer system or output pressure from filters located on the outlet side of transfer pumps (FIGURE 8-74). Measurements should be compared with the manufacturer's specifications.

> ▶ TECHNICIAN TIP
>
> Air can easily enter the suction side of a fuel system through cracked pickup tubes in fuel tanks, leaking filter base gaskets, and loose fuel line connections. Not only will air in fuel cause low-power complaints, but also the engine will run rough, stall, and be hard to start.

Fuel Filter Service

Fuel filters naturally build resistance to the flow of fuel as they remove unwanted contaminants from the fuel system. Larger pores fill with particles first, leaving smaller pores open for filtration. Thus, filter efficiency improves with use. Fuel filters, unlike engine oil filters,

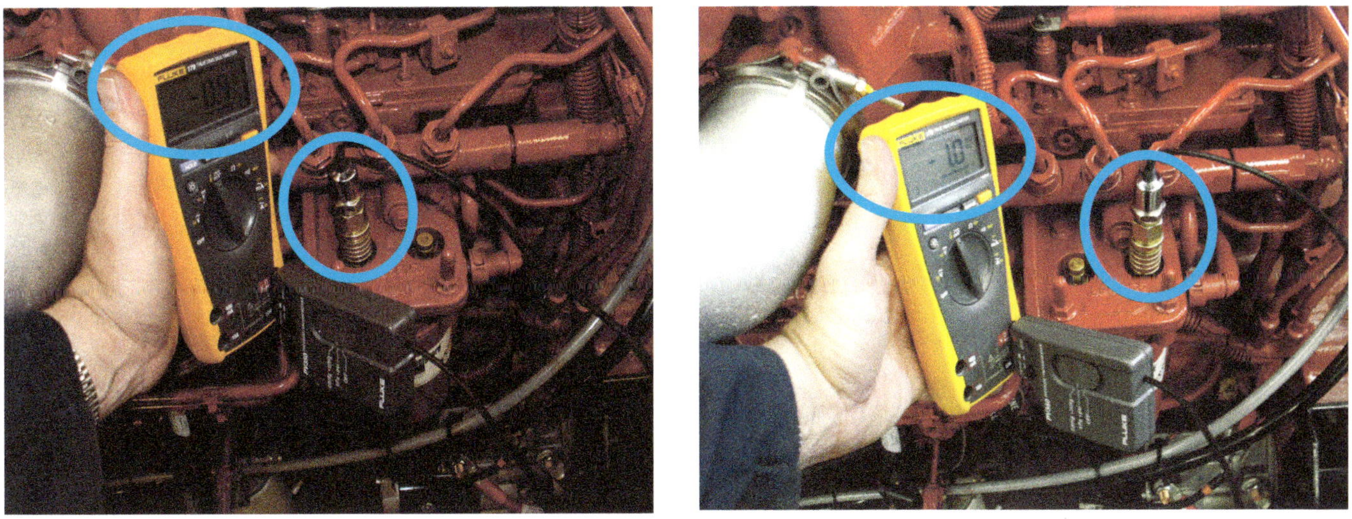

FIGURE 8-74 Measuring fuel filter restriction with an electronic pressure gauge.

have no bypass when they become restricted. Consequently, fuel flow through the filter decreases until the point is reached where the fuel supply system does not provide enough fuel at the correct pressure to properly operate the engine. This results in reduced power output from the engine, which is usually noticed first when the engine is placed under load, such as during hard acceleration or when climbing a hill.

Fuel filters should be replaced at their recommended service interval. Alternatively, to avoid overservicing filters, measure filter restriction to obtain the best indication of when a filter requires replacement. Restriction is the pressure differential created between the clean side and dirty side of the filter created when the filter begins to load with contaminants. Some vehicles include a restriction gauge or warning lamp for excessive filter restriction built into the dash. For the technician, a vacuum gauge connected with a T-fitting or service port between the transfer pump and primary filter provides an accurate indication of filter contamination. Manufacturer's specifications should be compared with actual data. Generally, restrictions should not be more than 12" (305 mm) of mercury (Hg) when the engine is operated at high idle. Most systems will run at a maximum of 4–8" (102–203 mm) Hg. Similarly, restrictions in the fuel tank lines and the pickup can be measured by connecting a vacuum gauge to a T-fitting on the suction side of the primary filter. Restrictions at this point should be generally smaller than 1" (25 mm) Hg. Kinked fuel lines or a restricted fuel pickup will increase negative pressure in the line to the point where fuel flow slows or exceeds a pump's ability to pull fuel from the tank. Only output pressure from the secondary filter is measured to determine whether it is restricted (**FIGURE 8-75**).

Replacing a Fuel Filter and Priming a Fuel System

When a cartridge fuel filter is replaced, on some engines the filter housing will contain air after the cartridge is installed. Any air in the fuel system will prevent the fuel from properly pressurizing, which will cause the engine to quit, if it starts at all. It is particularly important to bleed air from the fuel system because many systems today use very

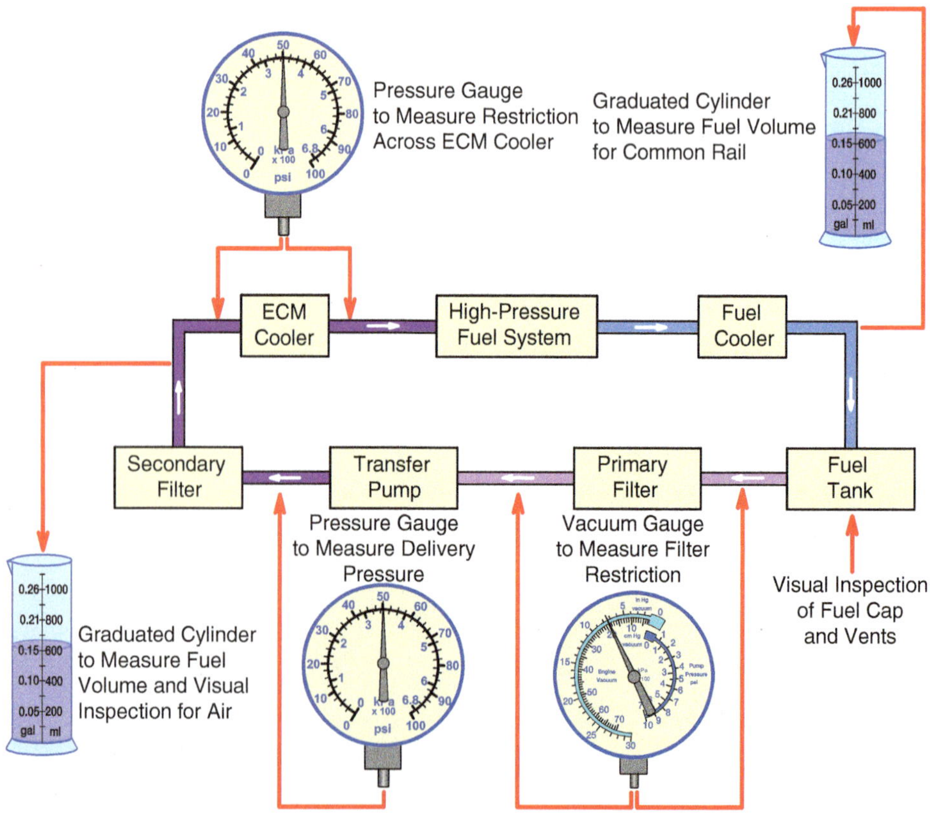

FIGURE 8-75 Common test points and tools used to check the low-pressure fuel system.

little fuel return to the tank, which can carry air and vapors back to the tank. Any air trapped in fuel lines requires a significant amount of service time and line disconnection to bleed once air has been allowed to enter high-pressure lines. These low-return fuel systems are designed to prevent parasitic power loss from pumping more fuel than is needed and overheating the fuel in the tank by continuously returning fuel with heat picked up from the engine.

Performing a Primary Fuel Filter Suction-Side Pressure Test

The use of both primary and secondary fuel filters to provide long service intervals between filter changes can also provide the cleanest fuel to the high-pressure system. The primary filter incorporates a fuel–water separator, and occasionally an additional fuel–water separator is used before the primary filter. A primary fuel filter has several important jobs to perform. First, it must protect the transfer pump from damage caused by fuel contaminants. Second, it removes larger contaminant particles and both free and bound water from fuel, which reduces the loading in the finer secondary filter. It is important for the primary filter to provide low restriction to fuel flow because it is located on the suction side of the fuel transfer pump. This means only atmospheric pressure pushes fuel through the primary filter and that contaminant buildup will reduce fuel flow more drastically than a secondary filter restriction because the secondary filter is located on the pressure side of the transfer pump. In fact, a pressure drop of as little as 0.5 psi (3 kPa) will reduce fuel output volume by approximately half. Primary fuel filters typically have a **nominal micron rating**, which is the minimum particle size a fuel filter is expected to remove, of 10–30 μ, but 40 μ and 100 μ filters are also used in the fuel–water separator. Filter restriction is measured to determine whether the filter needs servicing and to pinpoint a cause for low-power complaints. To perform a primary fuel filter suction-side pressure test, follow the steps in **SKILL DRILL 8-4**.

Measuring Primary Filter Restriction

If the primary filter suction test indicates an excessively high negative pressure, the primary filter is restricted or a restriction exists between the filter and the fuel tank. A more comprehensive test is required to compare the outlet pressure to the inlet pressure of the primary filter. To measure primary filter restriction, follow the steps in **SKILL DRILL 8-5**.

SKILL DRILL 8-4 Performing a Primary Fuel Filter Suction-Side Pressure Test

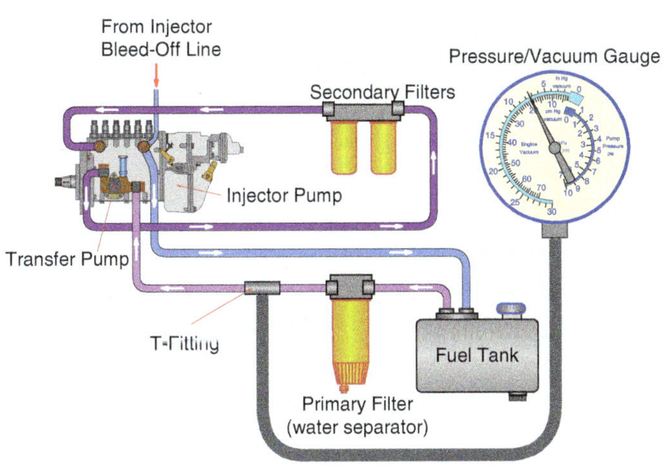

1. Obtain an accurate, high-quality pressure-vacuum gauge capable of measuring pressure in inches (millimeters) of Hg or a Magnehelic gauge that measures in inches (millimeters) of water (H_2O). An electronic pressure transducer or a mechanical vacuum-pressure gauge is best.
2. Connect the gauge to the outlet side of the primary filter by using a quick-connect fitting, which may need to be installed in the system if one was not equipped by the OEM. A T-fitting may also be used in the fuel line between the primary filter and transfer pump, but there is usually a test port and plug in the filter system.
3. Start the engine and run it at high idle (full throttle, no load). Observe and record the pressure reading. A slight negative pressure is normal, but a high negative pressure means the filter is restricted or that there is a suction-side restriction before the filter. Generally, if the negative pressure is less than 8–10" (203–254 mm) Hg or the filter is operating satisfactorily and still has service life remaining. If the restriction is 12–18" (305–457 mm) Hg, recommend that the filter be replaced. Always refer to the manufacturer's specifications.

SKILL DRILL 8-5 Measuring Primary Filter Restriction

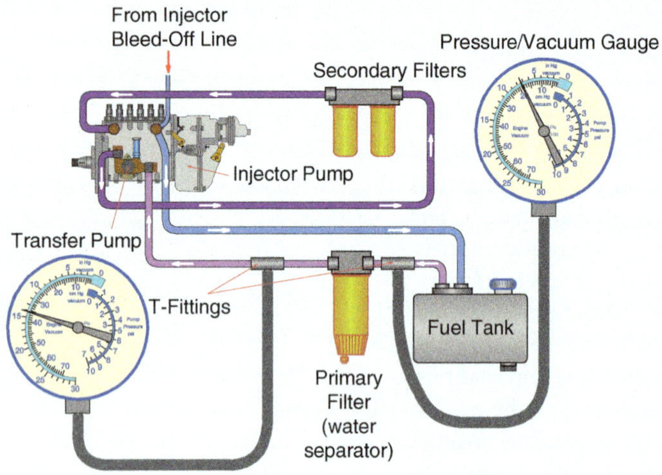

1. Install quick-connect fittings and attach pressure-vacuum gauges to both the inlet and outlet sides of the primary filter. An electronic pressure transducer can be used. Alternately, use the same gauge and take measurements on both sides of the filter during two separate sequences.
2. Start the engine and operate it at the rated speed.
3. Observe and record the pressure differences between the two gauges. Most clean filters will have less than a 0.1 psi (0.7 kPa) difference between the two readings.
4. If the filter restriction is approaching 5–8" (127–203 mm) Hg or -2.5–-4 psi (17–28 kPa), recommend filter replacement.

> **▶ TECHNICIAN TIP**
>
> If the pressure drop across the primary filter is acceptable but the suction-side pressure is too low, a suction-side restriction test should be performed. Pinched lines, undersized lines, or a pickup tube with more than one fuel line attached will produce suction-side pressures that are too low to provide adequate fuel flow. To perform a primary fuel filter suction-side restriction test, follow the steps in **SKILL DRILL 8-6**.

Measuring Secondary Filter Output Pressure

Secondary fuel filters are connected between the pressurized transfer pump outlet and the high-pressure fuel system, which means they operate with higher pressures than primary filters. Secondary fuel filters are designed to remove finer particles of dirt and other contaminants than primary filters and typically have a nominal micron rating of 2–10 μ. Because secondary filters operate at a higher pressure and the primary filters generally remove the bulk of contaminants, some fleets choose to replace secondary filter after every other service interval in which the primary filter is replaced. Changing a filter sooner than necessary does not help keep fuel cleaner because filters become more efficient with use. This happens because larger pores and openings in the filter media plug up first, and then progressively smaller pores are blocked by dirt and contaminants as the filter remains in service. The best method to determine when a secondary filter requires replacement is to measure the output pressure. To measure secondary filter output pressure, follow the steps in **SKILL DRILL 8-7**.

SKILL DRILL 8-6 Performing a Primary Fuel Filter Suction-Side Restriction Test

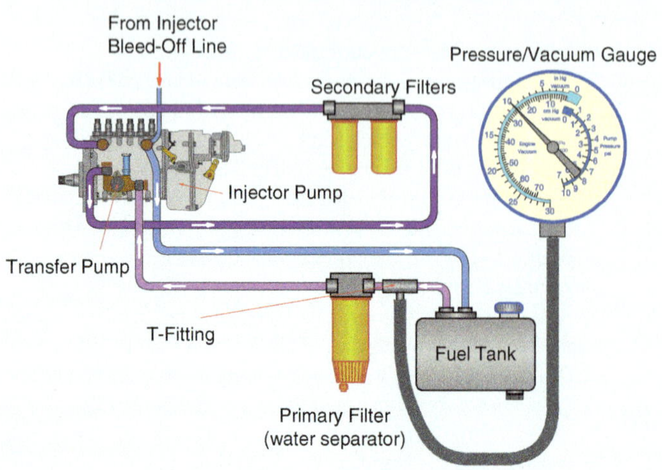

1. Install a quick-connect fitting and attach a pressure-vacuum gauge to the inlet side of the primary filter. An electronic pressure transducer can be used.
2. Start the engine and operate it at the rated speed.
3. Observe and record the pressure reading. A pressure lower than 10" (254 mm) Hg is too low, so in that case, the suction line to the primary filter should be inspected for problems. Recommend that the fuel tank pickup tube be inspected for restrictions.

SKILL DRILL 8-7 Measuring Secondary Filter Output Pressure

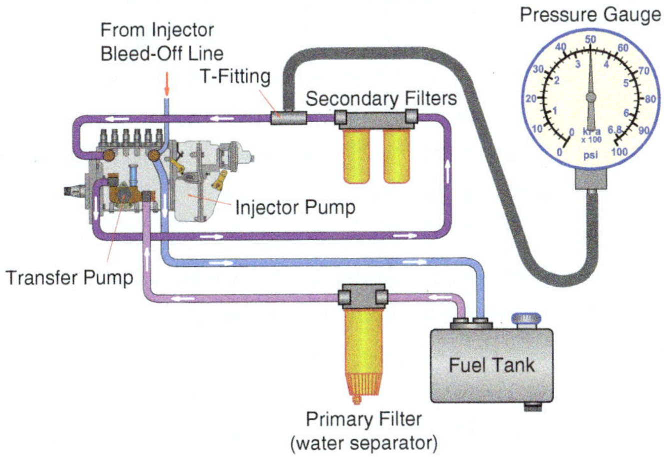

1. Obtain an accurate, high-quality pressure gauge capable of measuring pressure in psi (kPa). An electronic pressure transducer of a mechanical pressure gauge are best.
2. Connect the gauge to the outlet side of the secondary filter by using a quick-connect fitting, which may need to be installed in the system if one was not equipped by the OEM. There is always a test port or plug to measure the secondary fuel pressure.
3. Start the engine and run it at high idle (full throttle, no load). Observe and record the pressure reading. Most secondary filters operate at pressures of 50–80 psi (345–552 kPa). Compare the observed pressures with the manufacturer's specifications.
4. Make a service recommendation to replace the secondary filter if the output pressures are approaching or below the manufacturer's specifications.

Air can lead to the injector tips of newer high-pressure nozzles being damaged. When fuel that would normally cushion the seating of the nozzle valve inside the injector is absent, the nozzle valve impacts the nozzle seat with tremendous force, damaging the nozzle tip (**FIGURE 8-76**).

Diaphragm Pump Problems

Pump volume and pressure decreases when the check valves deteriorate or have trapped dirt that is preventing them from closing, or when there are broken springs. More significantly, fuel can drain back from the fuel system, which leads to prolonged cranking and hard-starting conditions. Some pumps with this condition will allow the engine to briefly start on fuel contained in the high-pressure system and then quit. More cranking is necessary to supply fuel and bleed air from the high-pressure system before the engine can restart.

Drain-back valves are often installed in filter modules and cylinder head fuel line fittings to prevent hard or prolonged starting caused by fuel draining back to the tank. Other service problems with these pumps include weak or broken diaphragm springs causing low pump output pressure and volume. A quieting spring on the rocker lever can break, causing a noise when the diagram is not retracted.

Check valves may also leak and allow fuel to drain back to the tank, causing a loss of fuel priming. Fuel system additives are often the culprit for valve deterioration as well as dirt

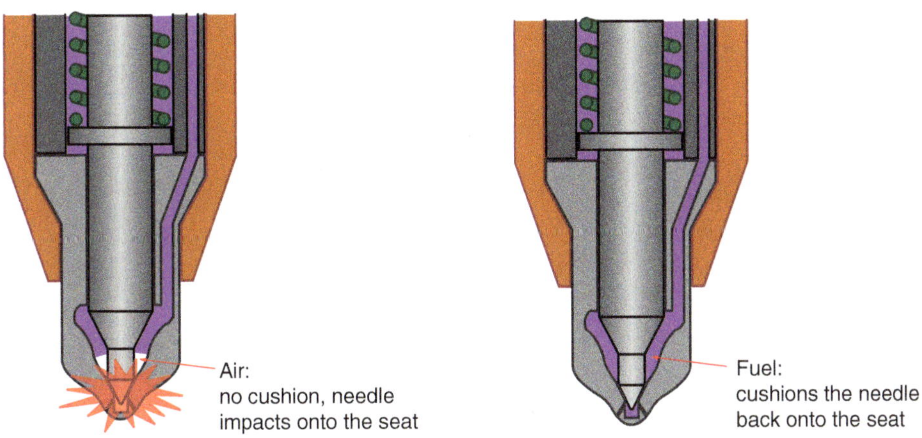

FIGURE 8-76 Fuel cushions the nozzle valve on the nozzle seat at the end of injection. Air in fuel causes the nozzle valve to damage the nozzle tip.

FIGURE 8-77 Weep holes to check for fuel or oil leaks from a diaphragm pump.

beneath the valve seats. A diaphragm has been perforated when fuel is detected leaking out the weep hole, which is designed to prevent fuel from running into the crankcase (**FIGURE 8-77**). Oil leakage from the weep hole is normal. Heavy leaks may indicate excessive crankcase blowby.

Malfunctioning Fuel Tank Vent

A malfunctioning fuel tank vent can cause negative pressure inside the tank that is apparent when removing the filler cap. Air rushes into the tank, and in some cases, the tank can be heard to expand when the cap is released. This condition can lead to low power and stalling after a few minutes or hours of vehicle operation. Releasing the tank cap restores normal power to the engine.

Air Leaks on Suction Lines

The lines connecting the tank pickup to the transfer pump must have airtight seals because any leak on the negative-pressure or suction side of the fuel system will draw air into the system. Because air expands and compresses, air in fuel will affect fuel flow to the transfer pump, which in turn causes hard-starting, low-power, and unstable engine operating conditions.

Oil Contamination in Fuel

A number of sources can cause oil to contaminate fuel. Gear-type fuel transfer pumps with damaged seals can draw engine lube oil into the pump and recirculate it with fuel. Return fuel will in turn contaminate the fuel tank with lube oil. Hydraulically actuated electronic unit injection (HEUI) injectors and their O-rings can also fail, causing oil to contaminate fuel. If the fuel is suspected of being contaminated by oil, the condition can be verified by dropping a sample of fuel on a paper towel. Oil and fuel will diffuse at different rates into the paper, where lube oil will leave a darker blotch in the center, and the fuel, which is lighter, will leave a clear stain on the outside of the stain. Scheduled oil sampling (SOS) can also be used to detect the presence of fuel in oil.

> ▶ TECHNICIAN TIP
>
> Common rail fuel systems are very difficult to bleed air and fill with fuel. This is particularly difficult at the factory. For this reason, some systems contain single-use valves or ports, called first fill valves. Ford's factory fuel filters for 6.7 L Powerstroke diesel engines have a third port on the secondary filter that is not found on replacement service filters. GM's 6.6 L Duramax diesel engines also uses a capped port in the engine compartment for the same reason.

Priming Fuel Systems

Whenever fuel filters are replaced, or if an engine has run out of fuel, the system usually needs to have air purged from the filter and lines in order to start. On mechanical transfer systems, this involves locating and opening a bleeder valve on the pressure side of the fuel filter. The technician should operate the hand pump until fuel flows continuously from the bleeder screw with no air and then close the bleeder screw. Electric systems often involve the same procedure, but the key can be cycled on and off or fuel pump relays jumped until the electric pump purges air from an opened bleeder screw.

Hard-Starting Problems

Prolonged cranking or no-starts related to the fuel transfer system typically involve the following:

- a fuel leak back to the tank
- air in the fuel system
- low or no fuel pressure
- inadequate fuel delivery volume
- defective constant pressure valves for return fuel.

Loose lines, cracked or blocked pickup tubes, and loose fuel filters can be common points that allow air into the fuel system. If air is present, fuel cannot be properly pressurized. Priming pumps become "spongy," and engines run irregularly and rough at idle when air is in the low-pressure system.

Fuel can bleed back to the tank because of defective transfer pump or priming pump check valves. An important point to inspect on unit-injected and common rail engines is the check valve for the fuel return from the cylinder head. Valves that are stuck open allow fuel to drain from the fuel rail in the cylinder head, which enables air to enter the injectors.

System restrictions can lead to performance problems under load when the engine requires maximum fuel delivery. For example, a kinked or undersized fuel line will allow fuel pressure to drop drastically when high volumes of fuel are needed. Fuel lines tied to accessories, such as refrigeration units or supplemental power systems, can drop fuel pressure for the engine supply.

Work Practices to Ensure Fuel System Cleanliness

Diesel fuel systems are extremely sensitive to dirt contamination. Even the smallest particle of dirt if it gets stuck under a control valve will cause major problems with an engine. It is absolutely vital for technicians to practice clean work practices when working on diesel engines. The following are some of the important practices for maintaining a clean engine:

- Clean dirt, dust, and other debris from the engine compartment by using a pressure washer before removing filters and other fuel system components.
- Tightly cap or plug all openings, fittings, and fuel lines immediately during repairs. This should be done even if the system is opened for just a few minutes.
- Use recommended solvents and cleaning fluids when cleaning pats. Use the recommended procedures for drying components. This almost never involves the use of compressed air blown over parts.
- Never place fuel system components directly on the ground or on dirty work benches.
- Keep new replacement parts in their original packaging until needed.
- Do not reuse seals, gaskets, or sealing washers. Always replace them.
- Seal components in clean plastic bags whenever they are removed from the engine.
- Technicians must use nitrile gloves and change them regularly, not only to protect themselves but to prevent transferring skin oils to parts.

▶ Wrap-Up

Ready for Review

- Fuels such as diesel or gasoline are derived from crude oil, which is a mixture of various types of hydrocarbon molecules. Through the processes of distillation and condensation, oil molecules are separated into fractions or cuts based on the boiling point of each fraction.
- The performance requirements of fuels include starting ease, low wear characteristics, providing sufficient power, low-temperature operability, low combustion noise, long filter life, good fuel economy, and low emissions.
- The grade of diesel fuel can be identified by a technician by measuring fuel's density with a hydrometer, which determines its API number.
- The cetane number (CN) is one of the most significant properties of diesel fuel; this rating is a measure of the ignition quality of fuel and how quickly diesel fuel ignites and burns.
- Biodiesel is fuel made from biological sources, such as plant oil or animal fat or a combination of petroleum-based fuel and biological fuel sources.
- The use of biodiesel offers a number of benefits, both economic and environmental, including to energy security, job creation, agriculture, rural development, energy supply, and public health.
- Operationally, biodiesel performs very similarly to conventional diesel in terms of power, torque, and fuel consumption without needing to modify the fuel system.
- Biodiesel has several disadvantages, including cost, lubricating oil gelling, poor cold weather performance, lower power output, and reduced fuel economy.
- Because of their abundance, vegetable oils are the preferred biological source for biodiesel fuel; oil derived from seeds or grains are the most common sources for fuel production.
- A wide variety of diesel fuel additives are used to generally improve the quality and performance of diesel fuel when fuel quality is compromised; when clean, good-quality fuel is used, there is no need to introduce additives to fuel.
- Many of the claims of additive manufacturers cannot be properly substantiated.
- The large number of available additives are grouped according to function; the four major categories include engine performance enhancers, fuel stability, fuel handling, and contaminant control.
- Determining a fuel's API index is important when investigating a low-power or high fuel consumption problem.

- Diesels need a low-pressure fuel supply to transfer fuel from the fuel tank with adequate volume and pressure to the high-pressure injection stage of the fuel system.
- Operating and servicing the low-pressure fuel supply system properly is critical to quick starting, engine durability, and getting reliable, trouble-free operation from diesel-powered vehicles.
- The functions of the low-pressure fuel system include storing and supplying fuel free of contaminants and vapor at the correct pressure to the high-pressure injection system, removing water contamination, preventing fuel wax from gelling, and regulating fuel temperature.
- While the most obvious function of the fuel tank is to store an adequate supply of fuel for the engine's application, a secondary function is to allow fuel returned from the engine to cool after absorbing heat from the engine and fuel injectors.
- Low-pressure fuel transfer lines connect the fuel circuits of the low-pressure transfer system.
- Fuel tank pickups can be integrated with a fuel level sending unit and fuel transfer pump; these modules may include several components that may or may not be separately serviceable.
- Fuel filters remove dirt contamination from fuel that could abrasively wear injection components.
- The most common contaminants found in today's fuels are asphaltene, air, water, microorganisms, and wax.
- To prepare fuel for cold weather use and avoid operating difficulties, the low-pressure supply system is often equipped with a variety of devices to condition the fuel by removing water and dissolving wax.
- Fuel transfer pumps are used to move fuel from the fuel tanks to the high-pressure fuel system; the fuel must be supplied with adequate volume at the correct pressure for every operating condition.
- A variety of pump types exist to match the unique requirements of each high-pressure fuel system.
- The operation of the low-pressure supply system will need to be evaluated when investigating complaints regarding low-power, hard, or prolonged starting, rough engine operation, and failures in high-pressure injection components.
- Whenever fuel filters are replaced, or if an engine has run out of fuel, the system usually needs to have air purged from the filter and lines in order to start.

Key Terms

absolute micron rating A rating of a fuel filter that refers to the largest sized particle that the fuel filter media will allow to pass.

American Petroleum Institute (API) A regulatory body that develops standards for a measuring fuel density.

API number A measurement of a fuel's density.

aromatic content The portion of fuel composed of a particular type of hydrocarbon (HC) molecule that is difficult ignite and burn.

ASTM International An organization that establishes today's diesel fuel standards; it was formerly known as the American Society for Testing and Materials (ASTM).

auto-ignition temperature The temperature at which a fuel will ignite when heated.

biocides A fuel treatment that kills microorganisms.

bleeder screw A component located in the fuel system that allows trapped air to escape from the filter when priming.

cetane boosters An additive that is marketed to reduce combustion noise and smoke, improve startability in cold weather, and increase acceleration response; also known as an ignition accelerator.

cetane number (CN) A measure of the ignition quality of fuel; also known as cetane rating or cetane value.

cloud point The temperature at which the wax in diesel fuel begins to congeal.

diaphragm pump A common type of fuel transfer pump used to supply fuel at low pressure.

distillation The process of boiling petroleum oil to separate oil molecules into fractions or cuts based on the boiling point of each fraction.

Energy Independence and Security Act (EISA) Legislation that requires the increased use of renewable fuels.

ethanol Alcohol-based fuel made from starches and sugars.

fatty acid methyl ester (FAME) fuels Biodiesel produced through transesterification.

Fischer–Tropsch reaction A process used to create synthetic liquid diesel fuel from vaporized hydrocarbons (HCs).

flash point The temperature to which a fuel must be heated to produce a vapor that will ignite when exposed to a spark or open flame.

fractions The different petroleum products that make up a barrel of oil, which are separated by the distillation process.

fuel heater A device that warms fuel to keep wax dissolved in the fuel.

fuel hydrometer A tool used to measure fuel density and identify its API number.

fuel-return circuit The fuel circuit that collects fuel from the injectors, the fuel pump, and relief valves and returns the fuel to the tank.

fuel–water separator A device that removes water from fuel by coalescing small water droplets into large ones that eventually fall out of the less dense fuel into a sump.

gas-to-liquid (GTL) diesel fuel Fuel derived from using the Fischer–Tropsch reaction.

gear pump A positive displacement fuel supply pump that sweeps fuel around the gear housing between pump gear teeth and can pressurize the output side to very high pressure.

gelling When the paraffin or wax content of the fuel causes fuel to become too viscous to properly flow through filters and lines due to low temperatures.

hydrophilic Water attracting properties.

hydroscopic The ability to mix and absorb water.

kinematic viscosity A measure of fuels viscosity as a ratio of its density.

low-sulfur diesel (LSD) fuel Diesel fuel with a maximum sulfur content of 0.05% mass or 500 ppm.

lubricity A fuel's lubricating quality.

methanol An alcohol made from wood or cellulose.

micron (μ) A unit of measure equal to 0.001 mm.

middle distillates A fraction or cut produced in the distillation process, such as jet fuel or diesel.

nominal micron rating Refers to the smallest sized particles trapped by a filter.

octane number A measurement of the ignition characteristics of gasoline.

paraffin A type of wax dissolved in diesel fuel.

petroleum distillates Fuel produced from distilling crude oil.

pour point The temperature at which fuel will no longer flows through lines and filters because of wax.

primary filter The first filter in the fuel stream from the fuel tank.

priming A process that removes any air that may have entered the fuel system and prevents the introduction of unfiltered fuel into the system.

Renewable Fuel Standard (RFS) Part of EISA passed in the United States in 2007 that indirectly mandates the use of B20 fuel, starting in 2011.

rollover valve A valve located on the fuel tank that prevents fuel from draining from the fuel tank vent line in the event of a rollover accident.

secondary filter The second filter in the fuel stream from the fuel tank.

sending unit An electrical device, usually a variable-resistance rheostat, that supplies a voltage signal to an analog-type dash gauge proportional to the fuel level.

straight vegetable oil (SVO) Unrefined vegetable oil used as fuel made from plants such as soybeans, and canola oil.

synthetic diesel fuel Fuel made through the Fischer–Tropsch reaction.

transesterification A process that replaces glycerin in vegetable oil or animal fat with alcohol molecules.

ultra-low-sulfur diesel (ULSD) Fuel that has a maximum sulfur content of 15 ppm.

vane-type pump A pump that uses rotating vanes or rollers to pressurize fuel.

viscosity A measure of a fluid's flow characteristics, or thickness.

volatility The quantity of fuel evaporating at a particular temperature

water-in-fuel (WIF) module A device that illuminates a warning light that instructs the operator to drain or service the fuel–water separator.

water-in-fuel (WIF) sensor A detector that uses an electrical conductivity probe to sense a specific level of water in fuel and cause a dash-mounted light to illuminate; this light or warning alerts the vehicle operator of the need to drain water from the fuel–water separator.

Review Questions

1. What effect does gasoline contamination have on diesel fuel?
   a. It lowers the auto-ignition temperature of the fuel.
   b. It lowers the flash point of the fuel.
   c. It lowers the cloud point of the fuel.
   d. It shortens the ignition delay time of the fuel.
2. Diesel fuel with a cetane number (CN) of 50 or greater has which of the following characteristics?
   a. It is 1-D and burns quickly.
   b. It is 1-D and burns slowly.
   c. It is 2-D and burns quickly.
   d. It is 2-D and burns slowly.
3. The temperature at which wax crystals begin to form in diesel fuel is referred to as its ________.
   a. cloud point
   b. pour point
   c. CN
   d. viscosity index
4. Which of the following properties does B5 diesel fuel have?
   a. It is red and has 5% sulfur content.
   b. It is 5% animal fat.
   c. It has 5% petroleum content.
   d. It has 5% biologically based content.
5. Which color is fresh, good-quality 2-D diesel fuel when it is commercially retailed?
   a. Red.
   b. Clear.
   c. Blue.
   d. Dark green.
6. Which of the following methods is best suited to treating diesel fuel that contains free water while minimizing any potential damage to the fuel system?
   a. Add alcohol to keep the water dissolved and to prevent freezing.
   b. Heat the fuel with a fuel heater to prevent icing.
   c. Separate the water from the fuel through coalescence.
   d. Treat the fuel with kerosene.
7. Which of the following is the best factor to use when determining whether a fuel filter should be replaced?
   a. Time in service.
   b. Distance traveled.
   c. Operating conditions.
   d. Restriction measurement.
8. What effect does increasing fuel temperature have on diesel engine operation?
   a. It increases power output.
   b. It reduces power output.
   c. It seems to increase fuel consumption because the fuel occupies more volume.
   d. It causes fuel to become more viscous, resulting in poor fuel atomization.

9. Which of the following is the result of water contaminating fuel?
   a. Lower fuel density.
   b. Fuel system components rusting or corroding.
   c. Cleaner injector nozzles.
   d. Cooler injector operation.
10. A primary fuel filter will typically have ________.
   a. a nominal micron rating of less than 5 microns
   b. an efficiency rating of 50% with a 10-micron rating
   c. a micron rating of 2–10 microns
   d. a micron rating of up to 100 microns

ASE Technician A/Technician B–Style Questions

1. Technician A says that the primary filter helps extend the life of the secondary filter. Technician B says that the primary filter is located between the engine's high-pressure injection system and the secondary filter. Who is correct?
   a. Technician A
   b. Technician B
   c. Both A and B
   d. Neither A nor B
2. Technician A says that placing a fuel heater in a filter housing is the best way to prevent wax and ice crystals from restricting fuel flow. Technician B says that ethanol can be used to dissolve wax and prevent water from crystalizing in fuel filters. Who is correct?
   a. Technician A
   b. Technician B
   c. Both A and B
   d. Neither A nor B
3. Technician A says that a float sensor can be used in a filter housing to measure the level of water, to determine when the filter should be drained. Technician B says that a probe in the bottom of a fuel filter, which measures the difference between the conductivity of water and fuel, is used to switch on the water-in-fuel warning lamp. Who is correct?
   a. Technician A
   b. Technician B
   c. Both A and B
   d. Neither A nor B
4. Technician A says that primary filters typically use chemically treated paper filter media with a micron rating of more than 10 microns. Technician B says that synthetic filter media is used to separate bound, or coalesced, water from fuel inside a primary filter. Who is correct?
   a. Technician A
   b. Technician B
   c. Both A and B
   d. Neither A nor B
5. Technician A says that coolant is used to heat fuel in fuel–water separators. Technician B says that electrical heating elements keep water and wax from plugging the filter in a fuel–water separator. Who is correct?
   a. Technician A
   b. Technician B
   c. Both A and B
   d. Neither A nor B
6. Technician A says straight vegetable oil (SVO) can be burned in a diesel engine without causing any long-term problems for engine durability. Technician B says SVO will damage injectors. Who is correct?
   a. Technician A
   b. Technician B
   c. Both A and B
   d. Neither A nor B
7. Technician A says that adding alcohol to diesel fuel is a good practice because it keeps fuel from waxing when it is cold and it improves engine starting. Technician B says wax in fuel should be removed at the refinery because it causes owners to make complaints about low power in hot weather. Who is correct?
   a. Technician A
   b. Technician B
   c. Both A and B
   d. Neither A nor B
8. Technician A says that adding a little gasoline to diesel fuel will help a diesel engine start faster in cold weather, reduce combustion noise, and improve power output. Technician B says that adding gasoline to diesel fuel will make starting worse, increase combustion noise, and accelerate fuel system wear. Who is correct?
   a. Technician A
   b. Technician B
   c. Both A and B
   d. Neither A nor B
9. Technician A says that 2-D fuel provides the most power and best fuel economy for an on-highway diesel engine. Technician B says that 1-D is the best fuel for producing power and good fuel economy. Who is correct?
   a. Technician A
   b. Technician B
   c. Both A and B
   d. Neither A nor B
10. Technician A says that diesel fuel's low volatility is why so many diesel fuel pump islands appear to have a lot of spilled fuel. Technician B says that fuel spills at diesel pump islands are more common than those at gasoline pumps because truck drivers are careless when refueling. Who is correct?
   a. Technician A
   b. Technician B
   c. Both A and B
   d. Neither A nor B

CHAPTER 9

Functions of High-Pressure Fuel Systems

Learning Objectives

After reading this chapter, you will be able to:

- 9-01 Identify and explain the purposes and functions of a high-pressure injection system.
- 9-02 Identify and describe the methods used to meter fuel in diesel engines.
- 9-03 Identify and explain the purposes and functions of a diesel engine governor control.
- 9-04 Describe and explain the purpose of changing injection timing.
- 9-05 Define and describe the function of injection rate control.
- 9-06 Define and explain the function of fuel atomization.
- 9-07 Define and describe the functions of fuel distribution.

You Are the Technician

A rough-running engine with a cylinder misfire and a noise is brought to your shop for diagnosis and repair. You begin by first performing a number of visual inspections and then performing various diagnostic tests: checking fuel and fluid levels and inspecting the engine for oil, fuel, and coolant for leaks; the electrical harnesses for damage; and the exhaust system for leaks or damage. A cylinder cutout test performed at idle and at 1,000 revolutions per minute (rpm) identifies a misfire in cylinder 3. A knocking sound in the engine also disappears when cylinder 3 is cut out. When you follow up the cylinder cutout test with a relative compression test or cylinder balance test, cylinders 1 and 6 are found to be weak. While checking other operating parameters, the injector pulse width is a little longer in the weak cylinders, and the pilot-injection event is taking place 8°–10° before the main injection event. As you decide which steps you'll take next, consider the following:

1. How could a defective injector be differentiated from a mechanical condition such as a damaged piston or noise caused by a loose wrist pin?
2. List some probable causes for a noise that would disappear from an engine when a cylinder is cut out for testing.
3. List and explain five possible reasons for the low-power cylinder contributions from cylinders 1 and 6.

▶ Introduction

Diesel engine fuel systems have two main divisions. One is the low-pressure transfer system that stores and supplies fuel for the injection system. The second is the high-pressure injection system that delivers fuel to the combustion chambers. A variety of types and designs of high-pressure injection systems exist and are radically different from one another in terms of their construction, complexity, and operating principles (**FIGURE 9-1**). Studying the differences between the various types of high-pressure injection systems can be overwhelming. A simpler approach to understanding the construction and operation of various high-pressure injection systems is to examine its functions that are common to all high pressure injection systems, rather than looking at each type individually. Analyzing high-pressure injection systems in terms of why (its purpose) and how (it's function) described in this chapter is not only helpful for identifying and understanding injection system operations but also necessary for developing skills related to diagnosing, servicing, and maintaining diesel fuel–injection systems. This chapter provides foundational information for understanding the operation and construction of specific fuel-injection systems presented in greater detail elsewhere in this textbook.

▶ Fundamentals of High-Pressure Fuel Systems

9-01 Identify and explain the purposes and functions of a high-pressure injection system.

The high-pressure fuel-injection system is a diesel engine's most distinctive feature—and performs one of the most essential functions of compression-ignition (CI) engines-supplying fuel to the combustion chamber (**FIGURE 9-2**). Efficient diesel engine operation

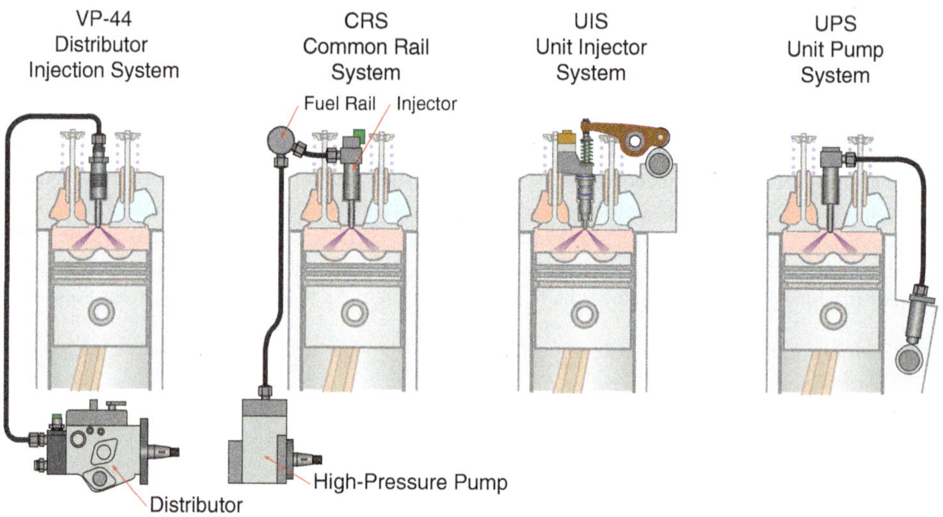

FIGURE 9-1 Four of the most common diesel fuel–injection systems: distributor, common rail, unit injection, and unit pump.

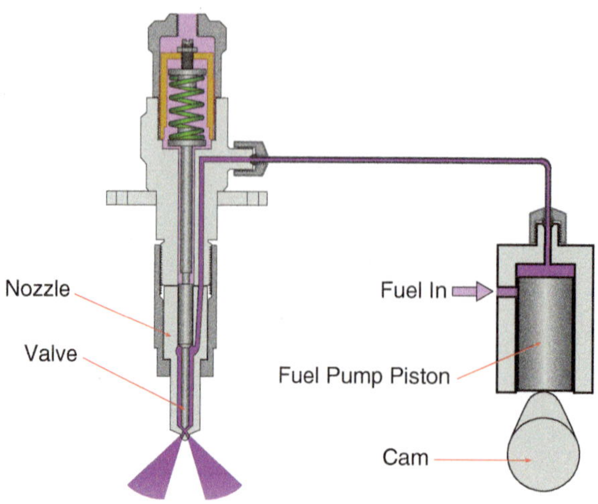

FIGURE 9-2 The high-pressure fuel system injects, distributes, and atomizes the fuel while overcoming the air resistance of cylinder pressure.

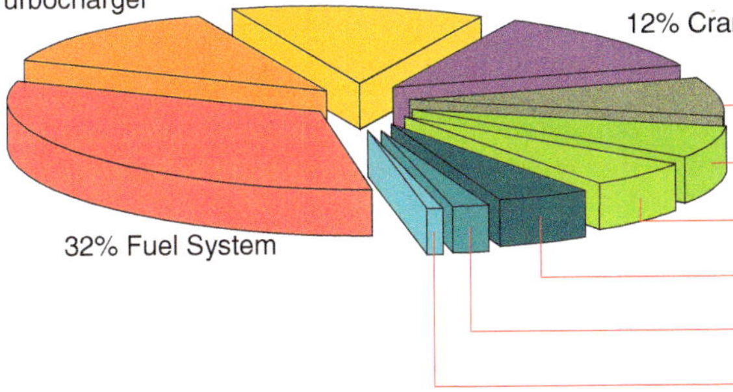

FIGURE 9-3 Compared to other systems, the high-pressure fuel system is the most expensive system to manufacture in a diesel engine.

relies on a precise quantity of fuel being injected at high pressure and at the correct time to the combustion chamber. In a sense, the fuel-injection system is the heart of diesel engine operation. The complexity and sophistication of the fuel-injection system is reflected in that it is the costliest engine system and requires the greatest technical refinement (**FIGURE 9-3**).

Purpose of a High-Pressure Fuel Systems

Fuel needs to be highly pressurized for several reasons:

1. Fuel must have sufficient pressure to penetrate the dense combustion chamber air mass. Air at high pressure in the chamber makes it even more difficult to distribute fuel through tightly packed air molecules.
2. At high rotational speeds, only a short amount of time is available for the fuel to be injected. Fuel must be atomized, distributed, and mixed with air; then it must absorb heat, convert to vapor, and begin burning near the end of the compression stroke, at top dead center (TDC). Because fuel is injected only near the end of the compression stroke, these events must be performed quickly to allow adequate time for combustion. High-pressure injection enables the combustion event to take place in less time than it takes a camera to flash. This is especially important as engine speed increases. The time available for injection shortens as the engine speed increases, so increasing injection pressures proportionally with engine speed is necessary to quickly deliver fuel into the combustion chamber.
3. High pressure is necessary to break up both small and large quantities of liquid fuel into a fine mist. This process, called **atomization**, is enhanced with high injection pressures (**FIGURE 9-4**). Higher pressurization produces increasingly smaller droplets of fuel that readily absorb heat and burn. Improving atomization results in increasingly efficient combustion, as well as better fuel economy, greater power output, and lower emissions (see the Fuel Atomization section in this chapter).

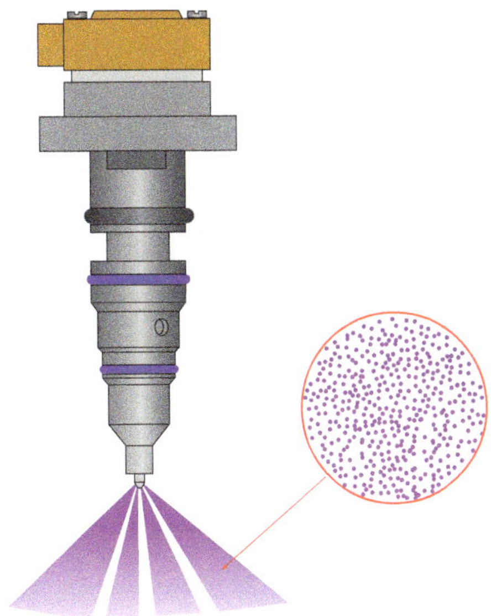

FIGURE 9-4 When fuel is better atomized into smaller, finer droplets, it produces a cleaner and faster burn.

In today's engines, injection pressures range from a low of around 3,200 psi (220.6 bar) during idle conditions to over 37,000 psi (2,551 bar) at high speed under full engine loads. (Some light-duty diesels have even lower pressures at idle speed, to reduce combustion noise.) Higher peak injection pressures are planned for the future, with developmental engines now achieving as much as 60,000 psi (4,137 bar). Decades of innovation and refinement by dozens of manufacturers around the world have given rise to a wide range of high pressure system designs. However, all these systems have specific purposes that when broken down into common recognizable tasks, can simplify understanding how a particular fuel system performs, as well provide insight for strategies to maintain, diagnose, and repair each different system.

Functions of High-Pressure Fuel Systems

All high-pressure injection systems perform the following tasks:

- metering the correct quantity of fuel for injection
- timing the injection of fuel

- controlling the rate of fuel delivery
- atomizing the fuel
- distributing the fuel throughout the combustion chamber.

When any of these functions is deficient, engine performance–related and emission-related problems are present, and further mechanical problems will likely develop.

9-02 Identify and describe the function of fuel-metering methods in a diesel engine.

▶ Fuel Metering

Metering is the process of measuring and delivering the correct quantity of fuel to each cylinder required for various speed and load conditions from the engine (**FIGURE 9-5**). Diesel engines only meter fuel and, unlike gasoline fueled engines, do not use a throttle plate in the air intake to regulate engine speed and power. Throttle plates, however, are used for other purposes, mostly for thermal control of exhaust aftertreatment systems on engines that have been built since 2007. Restricting excessive flow of intake air with a throttle plate increases a diesel engine's naturally cool exhaust system temperature so that heat-dependent chemical reactions can take place in the exhaust catalysts.

Unthrottled air intake into a diesel allows maximum airflow into the cylinders. To vary vehicle speed and power output, pressure in the cylinders must increase or decrease. To accomplish this, high-pressure systems adjust the quantity of fuel metered and delivered into the cylinders for combustion. Increasing combustion pressure inside the cylinders increases engine torque output (**FIGURE 9-6**). So, if the operator demands more power—for example, when the road grade becomes steeper or additional weight and passengers are added

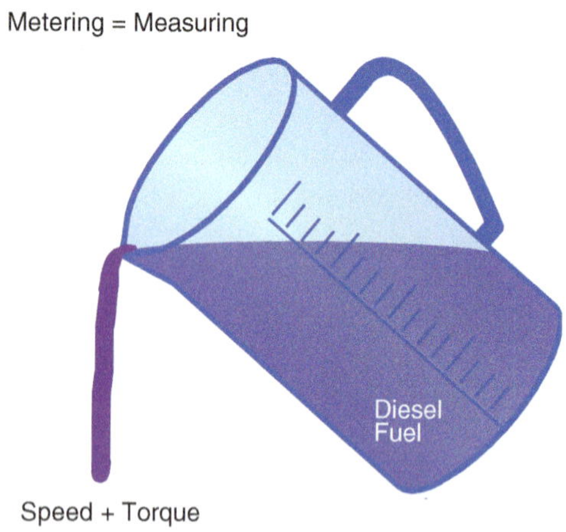

FIGURE 9-5 Metering is the process of measuring the correct quantity of fuel for each combustion cycle. Speed and torque output are controlled by metering fuel.

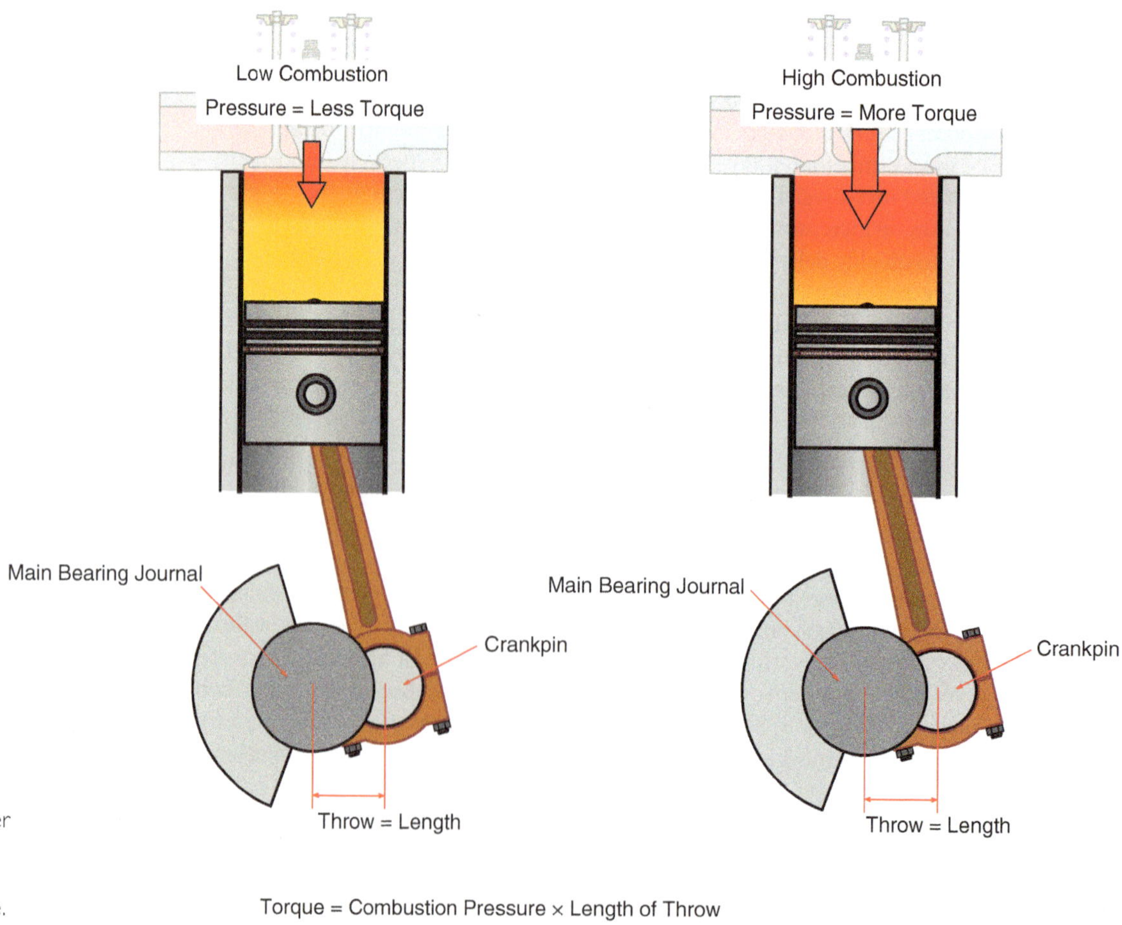

FIGURE 9-6 Injecting more fuel produces higher combustion pressure. Higher combustion pressure increases torque.

to the vehicle—more fuel is injected to increase cylinder pressure. Note again, adding fuel to maintain engine speed only increases cylinder pressure which is a direct function of torque output. Speed will not increase or decrease unless resistance to the crankshaft rotation changes.

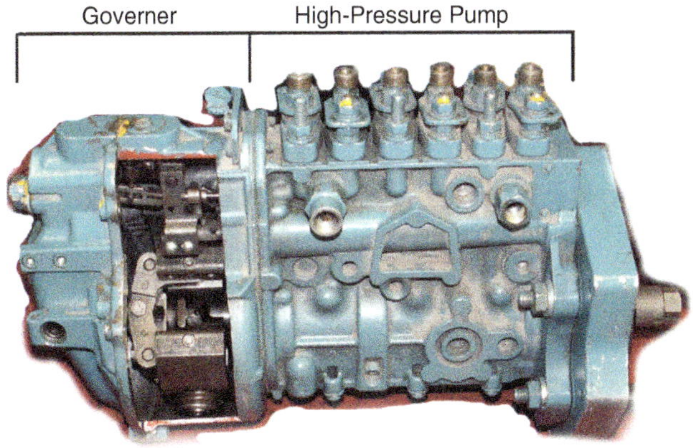

FIGURE 9-7 Whether it is an electronic or hydromechanical injection system, the governor functions to control the operation of the high-pressure injection system.

Metering and Governors

All diesel engines use a **governor** to regulate the quantity of fuel injected into the cylinders (**FIGURE 9-7**). Because it regulates the quantity of fuel delivered into an engine's combustion chambers, a governor is ultimately responsible for controlling an engine's speed and torque output. Whether it is mechanical or electronic, the governor receives, at a minimum, inputs from the vehicle's throttle pedal, engine speed, and intake manifold pressure or boost pressure data. Based on input data, the governor uses linkage or software to calculate or move fuel control linkage to change the amount of fuel injected. A diesel governor's equivalent in a gasoline or Otto cycle engine is the throttle plate. In those engines, the operator senses vehicle data, such as speed and load, and opens and closes the throttle plate by using a foot-operated accelerator pedal. The operator—or the cruise control system when engaged—determines engine power output. In a diesel engine, the operator's accelerator pedal or throttle lever is only one of several inputs to the engine governor, and the operator doesn't directly control the amount of fuel injected. Again, the amount of fuel injected is a function of several other inputs (**FIGURE 9-8**). See Chapter 10 for a further discussion of the governor and its functions.

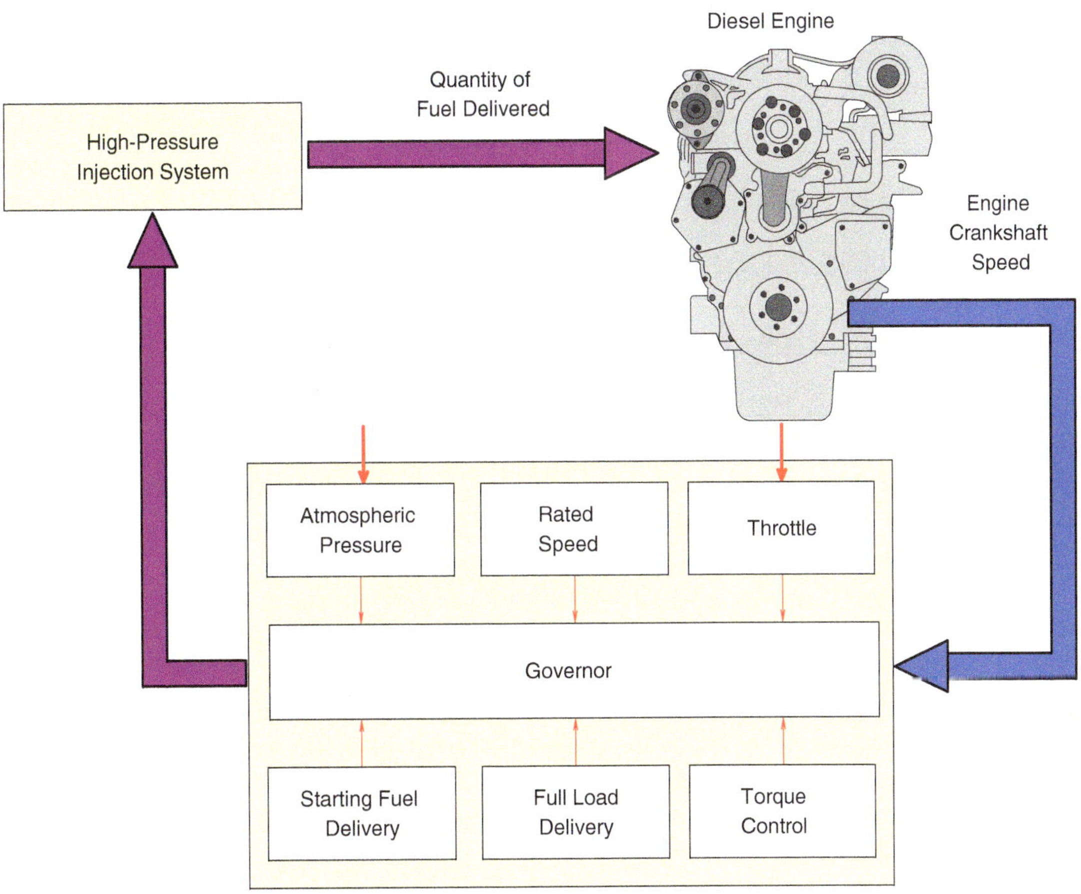

FIGURE 9-8 Governors have several inputs, but only one, the throttle, is controlled by the operator.

▶ Electronic Governing and Metering Controls

9-03 Identify and explain the purposes and functions of a diesel engine governor control.

An electronic governor which is a function or tool used by the high pressure injection system uses software to control the injection of a precisely metered amount of fuel. The governor is integral to engine software that adjusts injection timing and injection rate, which is the quantity of fuel metered into the combustion chambers over time (**FIGURE 9-9**). For example, inseparable from the engine control system is the operation of the governor subsystem. Engine sensor data is used by both systems. Like the engine control system, an electronic governor receives data from the fuel temperature sensor, as one of several sensor inputs used to calculate the final injection quantity. Fuel temperature data is needed to correct or compensate for changes to fuel density accompanying temperature changes. But fuel temperature data may also be used by the engine control system for other purposes, such as rationality checks by the OBD-II system or controlling the flow of fuel through fuel coolers and high-pressure pumps.

To meter the correct quantity of fuel for injection, engine speed and stroke for cylinder 1 need to be identified by the cam and crank sensor. These inputs are used to make cylinder-to-cylinder adjustments to the fuel metered into the combustion chamber. The crank-position sensor will also identify piston position to accurately time the injection event for the remaining cylinders. Input data from the air temperature sensor assists in calculating changes to injection timing and protecting the engine from excessive intake temperatures, which can thermally overload cylinder components. Other sensor inputs, such as coolant level, coolant temperature, oil temperature, and oil pressure, can influence metering and governor operation by signaling the need to derate the engine or even shut it down to protect the engine from catastrophic failure. Of course, all these sensors are involved in much more complex operational strategies, but these are the minimum contributions the sensors make to governor operation.

The immediate benefit introduced by electronic refinement of engine and governor controls is lower engine emissions, improved fuel economy, increased engine reliability, and

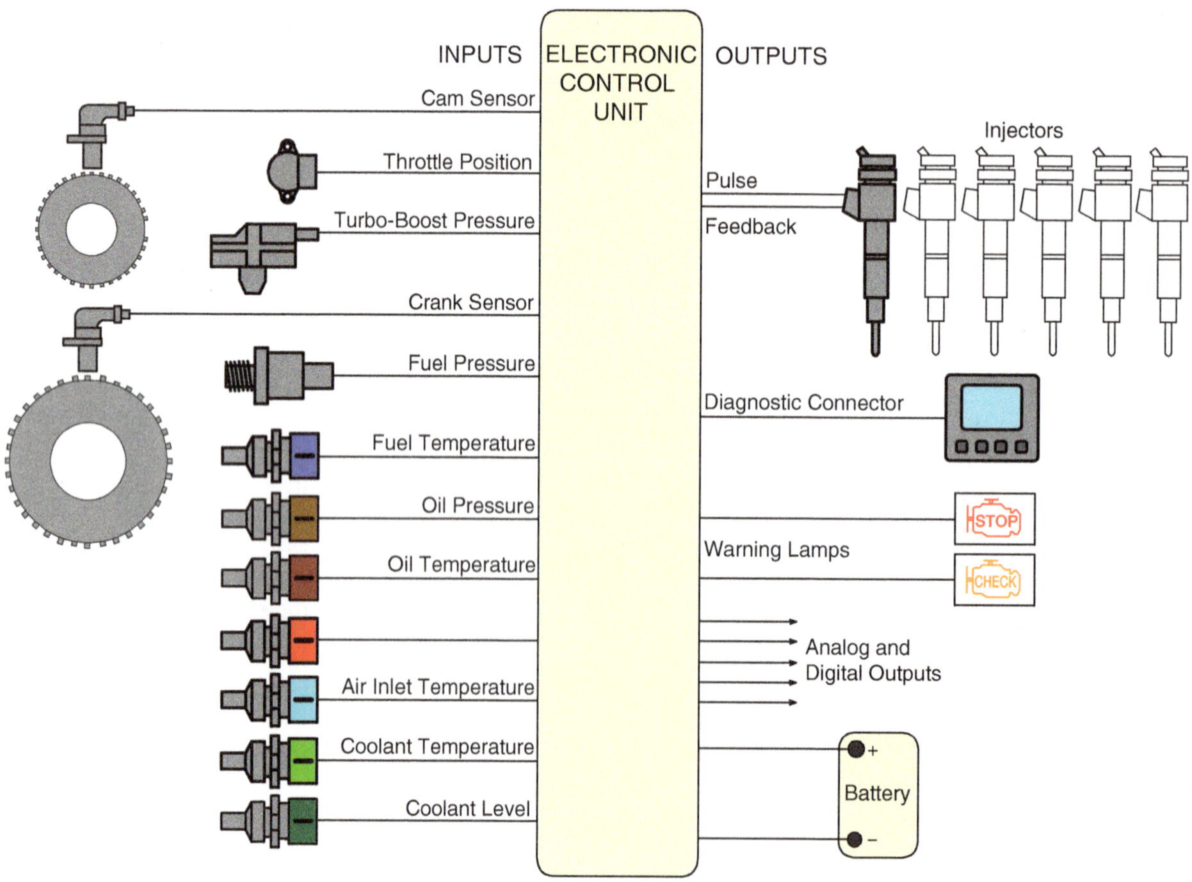

FIGURE 9-9 An electronic governor has many more inputs than a mechanical governor.

enhanced performance. Continual advances in designing engine control systems means these smarter engines deliver ever-increasing amounts of power from smaller displacements, with quieter operation and longer service intervals.

An electronic fuel governor relies on an **electronic control module (ECM)** to process electrical signals for the data it receives (inputs) and to control fuel injection (outputs) (**FIGURE 9-10**). The ECM contains a microprocessor, various microcontrollers, memory, and software to execute instructions to operate the engine. Output drivers, which are transistors inside the ECM, are also used by the ECM to control the flow or switching of electric current that small microprocessors cannot handle. Information is collected from engine and vehicle sensors that are processed by the ECM to produce an output signal. Basic operational data regarding engine speed, load, driver commands, ambient conditions, and a lot of other data are used to optimize engine operation to produce the lowest emissions while delivering the highest power output and the best fuel economy. For example, the ECM sends electrical signals through output drivers to electromechanical devices, such as the fuel injectors, using a pulse-width–modulated (PWM) signal. The length of the electrical pulse corresponds to the energization time of the injectors, which ultimately changes the quantity of fuel delivered by the injectors. Injection timing is controlled by the same electrical signal produced by the ECM that energizes the injectors. Optimal metering of injection quantities and fuel timing are calculated by using mathematical **algorithms** (mathematical formulas used to solve a problem) developed by a manufacturer. To help understand what an algorithm does, consider the following example of the data necessary to calculate vehicle speed. The ECM needs vehicle speed data to produce the correct torque for a given speed and load condition. Speed is calculated by measuring distance traveled divided by elapsed time. Tire sizes and rear-axle ratios are stored in ECM memory to help solve the mathematical problem of what the vehicle speed is. To finalize the calculation for vehicle speed, the ECM's internal clock also keeps track of the time, while a pulse-generating sensor measure driveline rotational speed to the rear axle. An algorithm stored in the ECM to calculate speed (speed = distance/time) is used along with data tables, also called lookup tables or calibration files, to supply supplementary data needed to solve the mathematical equation for vehicle speed (**FIGURE 9-11**).

Other algorithms used by the ECM are configured to calculate when to fire an injector, determine how long to energize an injector, correct a fault condition, or calculate when to

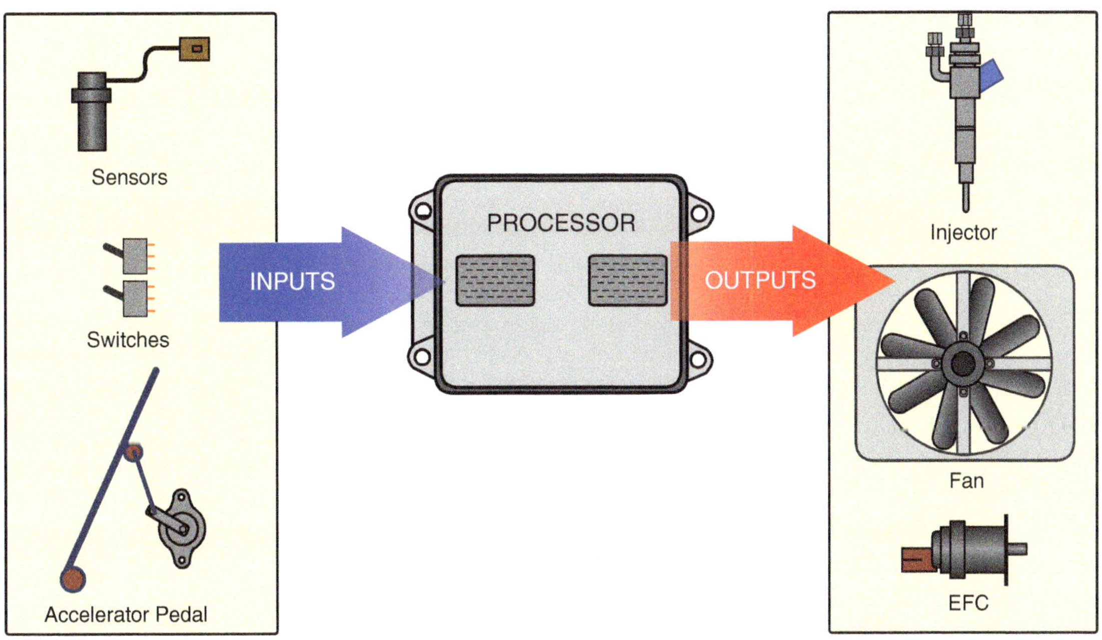

FIGURE 9-10 The engine ECM collects sensor input, processes the information by using mathematical algorithms, and then produces output signals to control injectors.

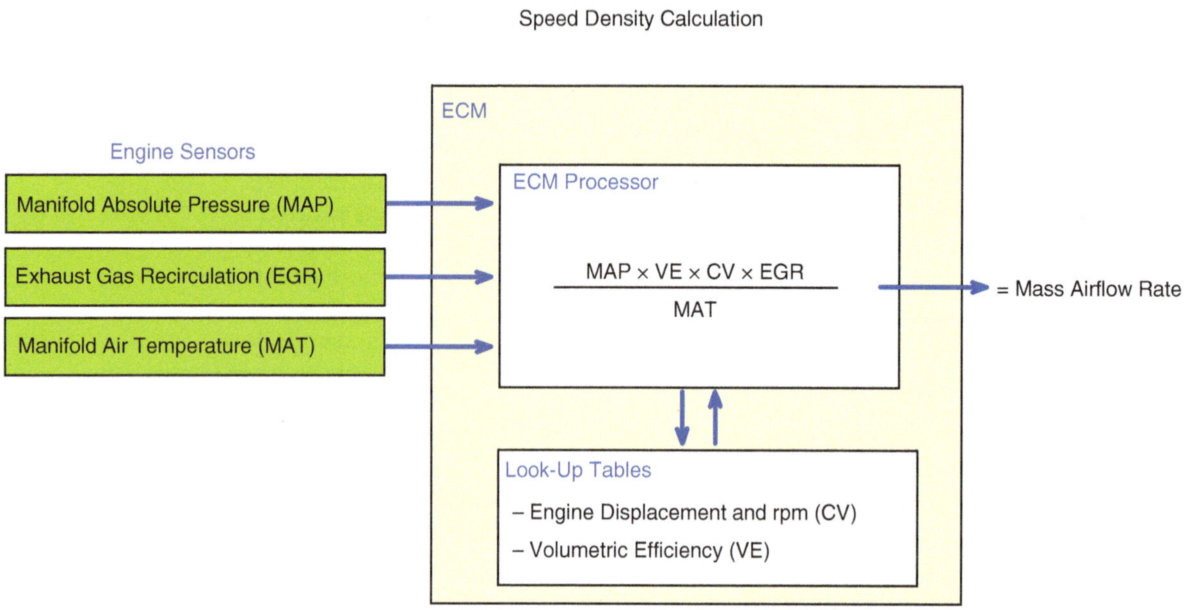

FIGURE 9-11 A speed density algorithm is an example of a mathematical formula used to calculate air mass entering the cylinders during a combustion cycle.

activate a malfunction indicator light are very complex. In the end, however, algorithms are simply large mathematical equations used by a complex calculator—the ECM—to control engine operation.

▶ TECHNICIAN TIP

A mathematical algorithm is simply a mathematical equation used to solve a problem. A simple example of an algorithm would be to calculate vehicle average speed. The mathematical formula of speed = distance/time is used to determine how fast a vehicle is traveling.

Metering and Torque Control

People often ask why a diesel engine can produce more torque at lower engine speeds than spark-ignition engines can. Part of the explanation is that unlike spark-ignition engines, diesel engines do not use a throttle plate, and maximum airflow into the engine is always possible at any speed. Turbochargers increase the maximum mass of air available for combustion too. In contrast, maximum torque from a four-stroke spark-ignition engine is only possible at wide-open throttle. That is because the maximum fuel rate needed to produce the highest torque output is limited by the mass of air available to support complete combustion. An additional point to understand about the throttle and engine efficiency is that without a throttle plate, less power is wasted in trying to pull air past a partially closed throttle plate. The energy required to pull air past a throttle plate or move gases in and out of the cylinder is known as pumping loss.

▶ TECHNICIAN TIP

All diesel engines use a governor, whether it is mechanical or electronic. The most basic function of a governor is to regulate the amount of fuel injected into each cylinder. In a gasoline or Otto cycle engine, the throttle plate serves this purpose. In such engines, the operator opens and closes the throttle plate by using a pedal in the cab. In a diesel engine, the driver's accelerator pedal (or throttle lever) is only one of several inputs to the engine governor and does not directly control the amount of fuel injected. Engine speed and load are two other inputs needed by all diesel governors to properly meter fuel into the cylinders.

▶ TECHNICIAN TIP

High compression ratios, turbocharging, and longer crankshaft throws are just a few of the major reasons why a diesel engine produces higher cylinder pressures and greater torque output when burning the same amount of fuel as a spark-ignition gasoline engine. Another major reason for higher torque at low engine speed is the absence of a throttle plate. Without restriction to airflow, maximum air volume can enter the cylinders and provide more oxygen to properly combust greater amounts of fuel.

Metering and Compensation

All diesel engine governors function to maintain a minimum engine speed and limit maximum engine speed. For example, changing engine loads at idle, such as engaging the engine-driven power steering pump, increasing the electrical load on an alternator, or switching on the air conditioning results in the need for the governor to add more fuel to prevent the engine from stalling and to maintain a consistent idle speed. A governor is also needed by a diesel engine to limit maximum engine speed because compression-ignition engines will run out of time to completely burn fuel at a much lower engine speed than spark-ignition gasoline engines. Contrary to the misconception that diesel engine speeds need to be limited because inertial forces from heavier pistons and crankshafts will tear the engine apart, diesel engine speed is in fact limited to prevent excessive emissions.

Another unique feature common to all diesel governors wrapped up in the term **compensation**. Compensation is observed when one or more cylinders are cut out yet the engine maintains a set idle speed (**FIGURE 9-12**). An increase or decrease in engine speed is sensed by the governor, and fueling is changed accordingly. So, if fuel delivery is limited or prevented by a defective injector or if a cylinder has low power contribution because of a mechanical condition, the governor responds to the drop in engine speed or power contribution by increasing fueling to the remaining cylinders to maintain engine speed. In other words, the governor compensates for uneven or inconsistent power contributions from engine cylinders. The engine's rpm may drop momentarily, depending on governor sensitivity, but it will quickly resume the set or preprogrammed idle rpm.

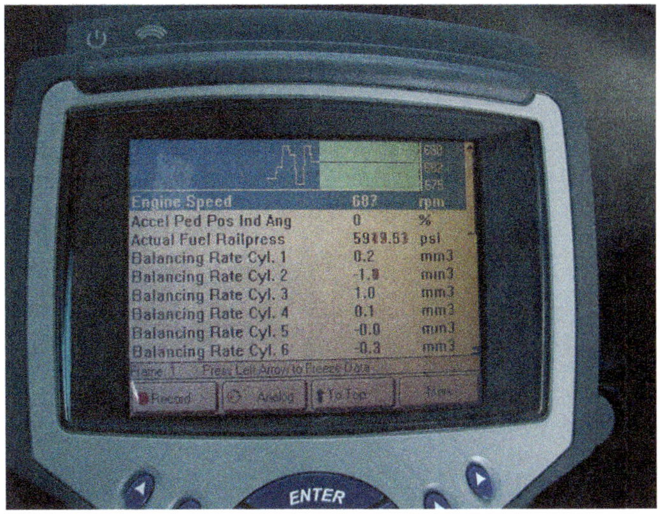

FIGURE 9-12 Results of a cylinder cutout test at idle. Each cylinder had to work harder to compensate for the cutout cylinder to maintain a set idle speed of 750 rpm.

> ▶ **TECHNICIAN TIP**
>
> If one or more cylinders are not properly contributing to engine power, the governor compensates for a lack of engine speed by increasing fuel delivery to the remaining cylinders. However, excessive cylinder heat can be localized in only a few cylinders if they receive a greater quantity of fuel than others. The excessive thermal loads will then cause heat damage to pistons and valves. This principle explains why it is important not to operate an engine for long when it has one or more misfiring cylinders.

Hunting

Irregular or inconsistent fuel delivery causes the governor to rapidly readjust the quantity of fuel metered and correct or compensate for changes in the engine speed. An engine operating condition, called **hunting** or **loping**, may be observed as a rhythmic change in engine rpm at idle speed. As the governor, whether electronic or mechanical, attempts to compensate for changes in engine rpm to maintain idle speed, it responds to a cylinder speeding up the engine by cutting back on fuel delivered to the next cylinders in the firing order. As a result, the engine rpm can fall below the governed idle rpm. This in turn causes the governor to increase fuel delivery to the next cylinders, or the one after, and so on. Before long, the engine rhythmically speeds up and slows down. The problem is made worse if only one cylinder is receiving more or less fuel and the engine has a highly sensitive and responsive governor that quickly corrects engine speed changes caused by the uneven power contributions from each cylinder. Once again, alternately reducing and increasing fuel delivery produces the rhythmic changes to engine speed referred to as hunting.

Metering and Idle Adaptation

The smoothest engine idle is a goal of every diesel engine manufacturer, the end-users, and the technicians who repair diesel engines. Variations in the amount of power contributed from cylinders, caused by even the slightest changes to fueling, results in a rough-running engine and in uneven engine operation, vibration, and emissions. Uneven power contributions are caused by a wide variety of factors:

- differences to the quantity of fuel injected from cylinder to cylinder and injection shot to injection shot inside the same cylinder
- variations in compression and combustion pressures caused by a large variety of mechanical conditions, such as engine wear, valve action, and exhaust gas recirculation (EGR) delivery
- manufacturing variations in injectors that result in different flow rates between the same injectors in an engine
- fuel system component deterioration.

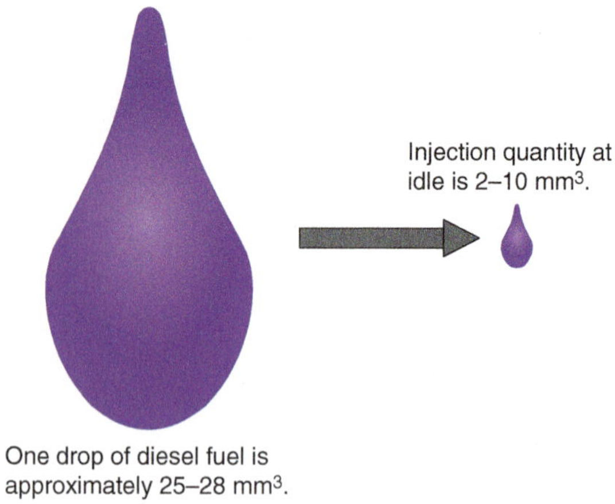

FIGURE 9-13 At idle, diesel engines use very little fuel compared to how much air they use.

Even very small changes to the injected quantities of fuel at idle produce rough-running engines and unstable engine speed. Consider that a droplet of fuel has a volume of 25–30 mm^3. A typical idle quantity of injection for most engines is 3–8 mm^3 of fuel (**FIGURE 9-13**). A variation of even 0.5 mm^3 corresponds to a 10% variation in delivery quantity and substantial changes in engine operation. In the newest engines, variations to the injection quantities at higher rates of fuel delivery are kept well below ± 0.1 mm^3. Compare this metering accuracy to the volume of a pin head, which is close to 1 mm^3.

To minimize problems associated with uneven cylinder power contribution and to ensure that all cylinders have the same consistent combustion pressure, most electronically controlled engines that have been built since the early part of the first decade of 2000s use a feature called idle adaptation (**FIGURE 9-14**). Idle adaptation involves making adjustments to injection quantities based on changes to crankshaft speed produced during every power impulse. The crank-position sensor can detect slight velocity changes between each cylinder firing that enables the ECM to subtract or add fuel to each cylinder, based on the measured variations to crankshaft acceleration during power events. Data is used by the ECM to calculate and apply a correction factor to the quantity of fuel injected into a cylinder from one combustion cycle to the next (**FIGURE 9-15** and **FIGURE 9-16**). If the correction to fuel delivery quantities becomes excessive, fault codes associated with cylinder contribution are generated. Fault codes are generated for loss of compression or uneven cylinder contribution, and power output depends on whether the measurement of the deceleration rate of the crankshaft takes place during compression events or the acceleration rate during power stroke. Low power contributions from a cylinder do not accelerate a crankshaft as fast as a cylinder with good combustion quality does. Cylinders with bent connecting rods or worn-out rings do not slow the crankshaft as fast during compression stroke either. When and how much the crankshaft speed changes are key to differentiating between cylinders with burned valves, bent connecting rods, worn-out rings, or faulty injectors.

Faster, more powerful ECMs introduced by most manufacturers in 2010 have allowed the same strategy of balancing cylinder power contributions at idle to be used throughout the entire engine operating range. The term "adaptive cylinder balance" refers to the fueling strategy used to adjust power contributions from each cylinder to achieve identical pressure from all cylinders. "Idle adaptation" and "adaptive cylinder balance" are terms used interchangeably to describe the correction made to the energization time of each injector to produce identical cylinder pressures from each cylinder either throughout the engines entire operating speed or only at idle.

While the crank-position sensor detects unbalanced cylinder contributions, the cam position sensor is used as additional data to identify which cylinder is firing and which stroke each of the cylinders is moving through. Most engines that have been built since 2002 generally have the capability to perform adaptive cylinder balance at idle. Since 2010, stricter emission legislation has required ECMs to possess faster processing capabilities to enable adaptive cylinder balance to take place throughout the engine's entire operating range. Reduction in all emissions takes place when cylinder pressures are consistent with one another.

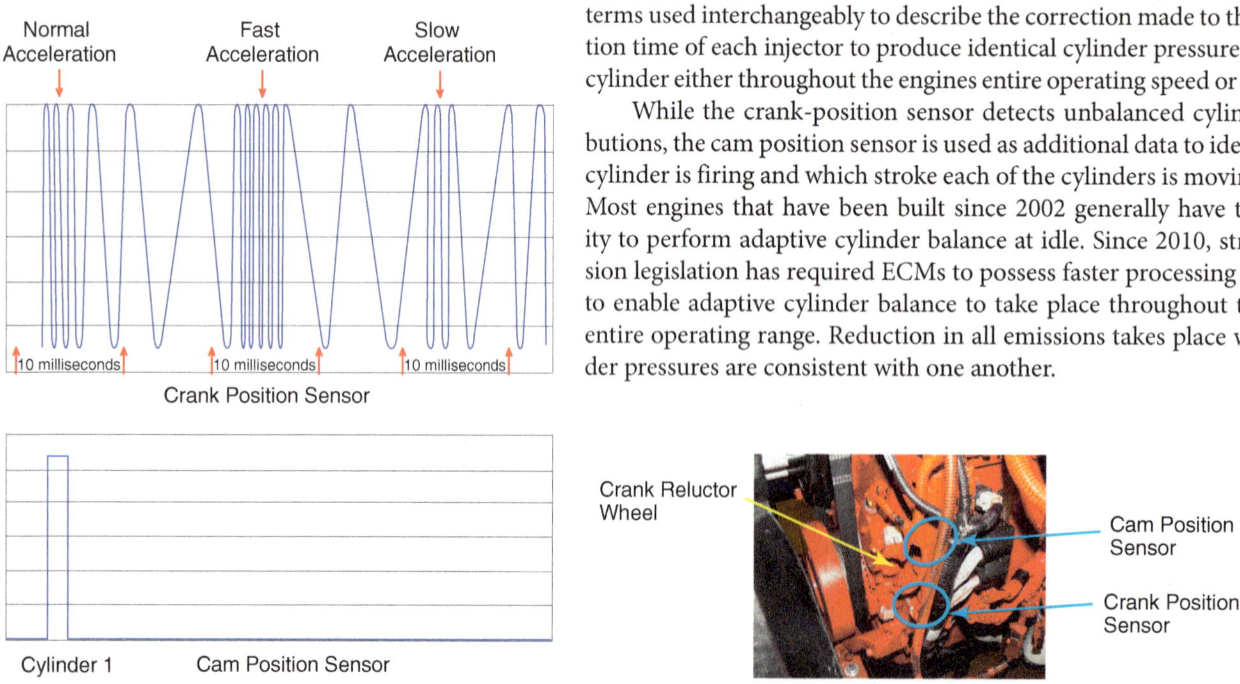

FIGURE 9-14 In idle adaptation strategy, consistent cylinder pressure is achieved from signals generated by the cam and crank sensors. Fuel is added or subtracted depending on how quickly the crankshaft rotates through combustion and compression events.

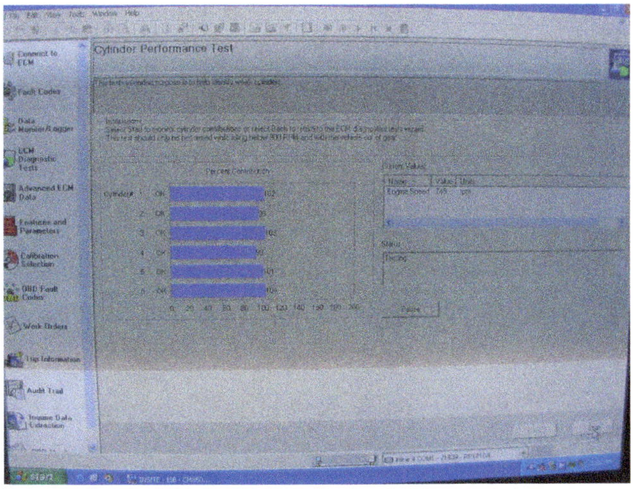

FIGURE 9-15 Adaptive cylinder balancing allows the ECM to calculate and deliver a different injection quantity for each combustion event to maintain consistent cylinder pressures. This screenshot averages all cylinders to 100%.

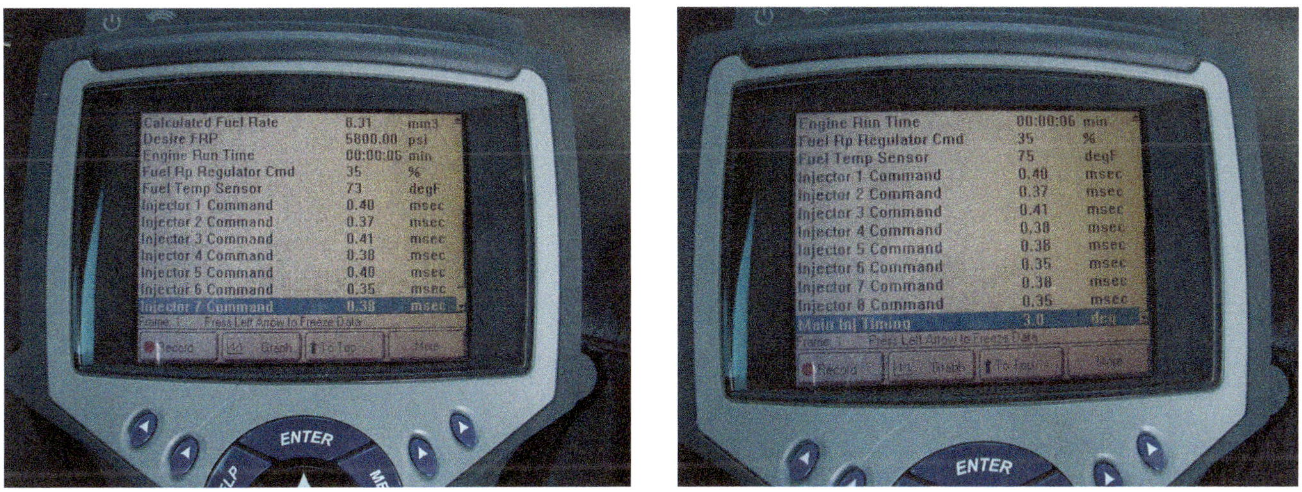

FIGURE 9-16 Injector energization time is changed to adjust delivery quantity, which enables adaptive cylinder balance.

Cylinder firing order and stroke are calculated after data from the crank-position sensor for TDC cylinder 1 is synchronized with the cam sensor. By measuring the time elapsed between each tooth passing by either the cam or crank sensor, crankshaft velocity changes caused by each compression and power event can be measured. Because the camshaft turns at one half engine speed, using a single raised pin on a cam gear, a missing tooth, or uniquely shaped window in a camshaft sensor ring that corresponds to TDC cylinder 1 enables the camshaft sensor input to the ECM to precisely identify the position and stroke of each cylinder (**FIGURE 9-17** and **FIGURE 9-18**). With only a TDC signal from the crank gear, cylinder 1 or cylinder 6 in a six-cylinder engine could be either ending compression and beginning power stroke or ending exhaust stroke and beginning intake stroke. The camshaft TDC indicator will always correspond to the stroke of only one particular cylinder ending compression and beginning power stroke. Synchronizing camshaft and crankshaft position data provides a more precise data set for piston and engine position than either sensor can provide alone. Once camshaft and crankshaft data have been synchronized during the initial cranking-starting period, the camshaft data can be ignored. A failure of either sensor will enable the engine to start and run since both sensors can provide engine position information. However, the engine may have a longer cranking period with a defective camshaft position sensor since the ECM must "guess" or attempt to start the engine by using the TDC indicator to alternate the firing order to begin on either cylinder 1 or cylinder 6 in a six-cylinder engine. Gear lash or timing chain slack

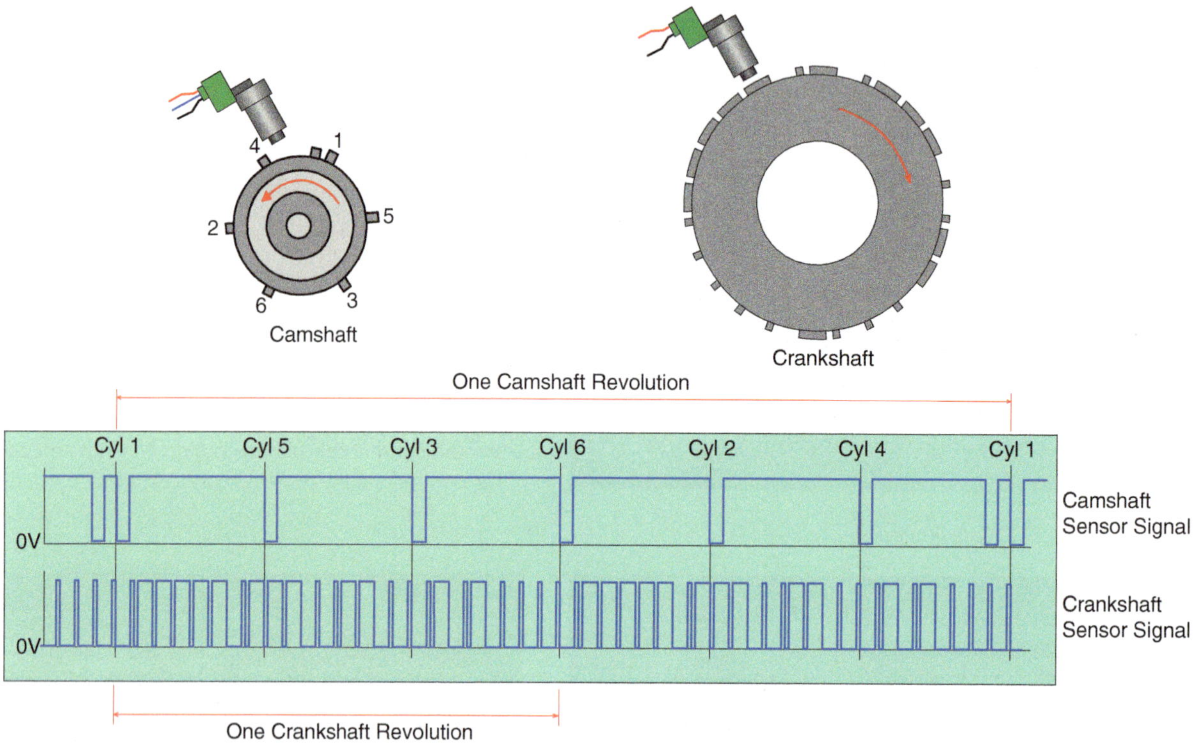

FIGURE 9-17 The cam and crankshaft position sensors are critical inputs for calculating injection timing. The arrangement of teeth identifies the engine position and stroke in each cylinder.

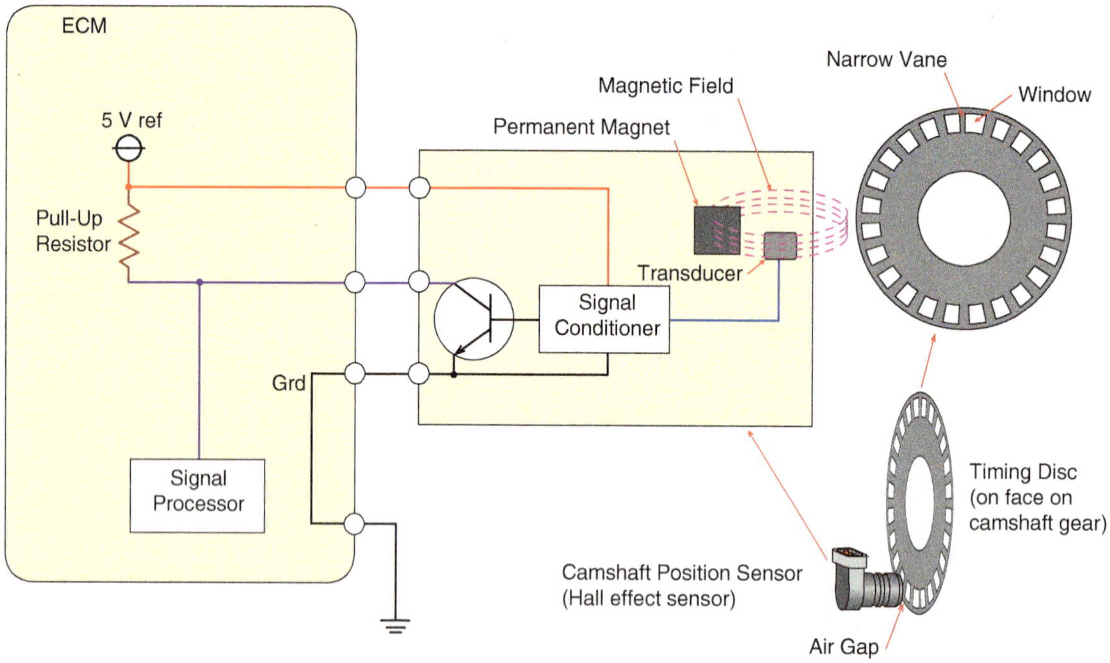

FIGURE 9-18 A different width of an opening on the face of a cam gear or cam sensor wheel is used to identify stroke position in an engine.

will introduce some error into the ECM's determination of the precise crankshaft position if a crank sensor has failed. Some operating strategies, such as adaptive cylinder balance, may not be available. After only the camshaft signal is detected and engine position is only approximately identified, the engine may operate roughly or have limited maximum engine rpm.

Metering and Noxious Emissions

A more important problem of uneven cylinder pressures is the emissions produced from inconsistent combustion pressure between each cylinder. Because formations of various noxious emissions, such as hydrocarbons (HCs), oxides of nitrogen (NO_x), particulate matter (PM), and carbon monoxide (CO), are dependent on cylinder temperature and pressure, uneven cylinder pressures can exaggerate emission-related control strategies. For example, introducing EGR gas into a cylinder that has lower-than-average combustion pressures increases HC, CO, and PM emissions. In a cylinder with above-average pressure, not enough EGR will be delivered and NO_x production will be above an acceptable range. To ensure good combustion quality and to keep emissions low, it is crucial to maintain consistent and even cylinder pressures in today's low-emission engines.

Minimizing shot to shot and cylinder to cylinder combustion temperatures and pressures will enable emission reduction strategies applied the same way to all cylinders, such as the use of EGR or retarded or advanced injection timing, will produce consistent results.

Adaptive Cylinder Balance

Since 2010, almost all engines have been able to adjust fueling not only at idle conditions but throughout the entire engine's operating range. Faster ECMs, operating with 32-bit microprocessors, provide the speed needed to make rapid calculations at high engine rpm. This means cylinder pressures are consistent all the way to maximum rpm and emissions, which form differently, depending on cylinder pressures and temperatures, and can be better controlled with emission reduction strategies applied the same way to all cylinders.

Metering and OBD

Fuel-injection pressure, cylinder misfiring, injector firing voltages, cylinder power imbalances, injection timing, injector deterioration, and the use of incorrect calibration codes are just a few of the high-pressure injection system variables monitored by the On-Board Diagnostics II (OBD-II) diagnostic system manager to detect conditions that could potentially lead to excessive emissions. Monitoring, too, must occur during the entire speed and load range of an engine. Failure or deterioration of any fuel system component can permit emissions to exceed maximum allowable levels.

Metering and Calibration Codes

During manufacturing, machining, and other production processes, many manufacturing tolerances—such as wire resistance in electrical actuators and even the concentricity of the nozzle valve seat—result in deviations to fuel delivery quantities. Even when injectors are manufactured on the same assembly line, no two are exactly identical in terms of the quantity of fuel delivered for any particular energization time or fuel pressure supplied to the injector. To correct this problem, injectors are flow tested on a fuel-injector test bench after production at an original equipment manufacturer (OEM) facility, where the characteristics of each injector are measured. Delivery quantities at different energization times and pressures are measured and graphed. A calibration code is generated based on the injector performance data. After the information is entered into an engine or fuel system control module during assembly or service, the ECM can correct fuel delivery based on the calibration code. The ECM uses the code make a global or wide range adjustment to the energization time to compensate for flow variations from an average value (**FIGURE 9-19**).

For the latest injectors, a calibration code is generated for each injector adding a correction factor for energization time dependent on injection pressure. The calibration code is stamped or etched on the injector (**FIGURE 9-20** and **FIGURE 9-21**). When this code is uploaded to the engine or other fuel system–related ECM, the control unit can apply a correction factor to the calculated fuel quantity

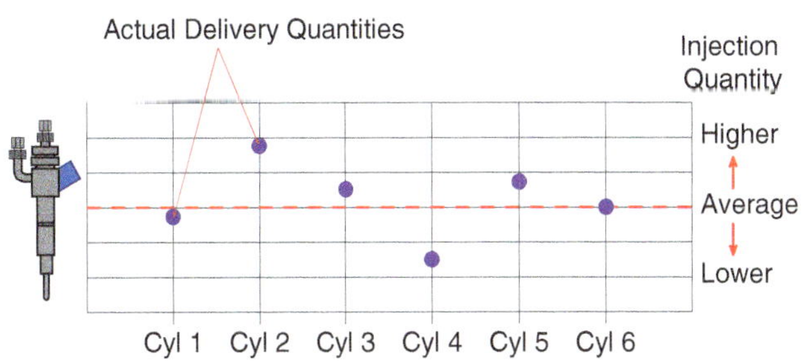

FIGURE 9-19 A calibration code is derived by comparing an injector's actual fuel delivery volume with a nominal, or average, value.

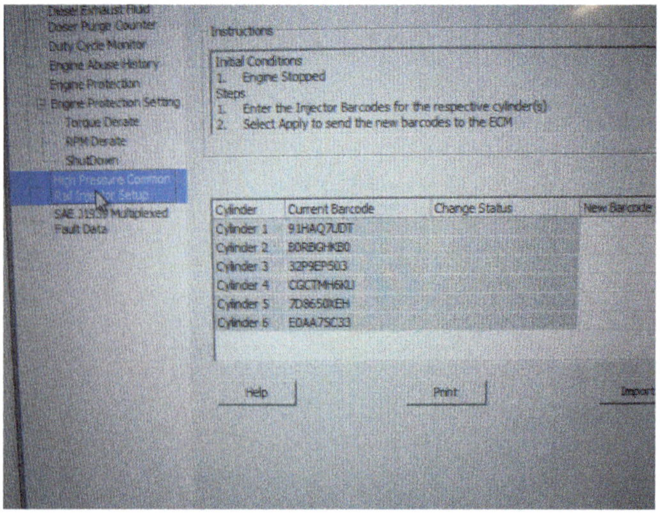

FIGURE 9-20 The alphanumeric injector codes are recorded from the injector and uploaded into the engine or fuel system ECM.

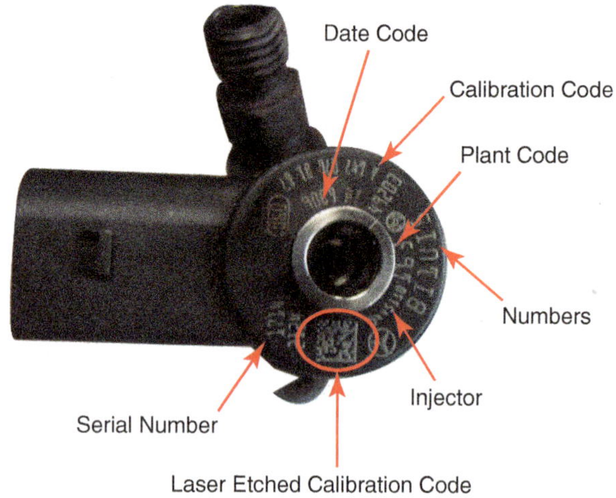

FIGURE 9-21 Calibration codes for a rail injector are laser etched during engine assembly.

> **▶ TECHNICIAN TIP**
>
> Injection quantity adjustment codes are assigned to each injector after flow testing during production. Codes are printed on the injector and are recorded by the technician when an injector is installed. The alphanumeric code is stored in the ECM by using a an electronic service tool or OEM software Failure to store the correct value for each cylinder can result in a rough-running condition and a fault code that illuminates the malfunction indicator lamp (MIL) light. The injector calibration code is also used as a reference point for fueling correction factors caused by deterioration and wear. Using the calibration code as a starting point, incremental adjustments to injection timing and metering are made throughout the injector service life. This explains why the calibration codes need to be updated whenever an injector or ECM is replaced.

for each cylinder (**FIGURE 9-22**). More sophisticated corrections are applied to the injector energization time so that variations to cylinder pressures can be further minimized by using a pressure corrected calibration code. The appropriate calibration code must be entered into the ECM each time a new injector is installed or an injector is moved. Failure to do so will result in rough engine operation and fault codes associated with cylinder balance and injector calibration (**FIGURE 9-23**). Calibration codes are given different names by manufacturers. Other common names for this feature include **injection quantity adjustment (IQA)**, the E-Trim, the adaptation value, or simply the injector calibration code. Older mechanical unit injectors (that is, non-electronic injectors) on some engines can also be flow tested as well. The best practice for a hydromechanical injection system requires injection replacement in matched sets, with injectors that have similar flow rates paired together.

Some injectors with calibration codes can be refurbished only by the OEM that produced them. Equipment used to assign calibration codes and benchmark flow rate values are proprietary, meaning they are known only to the OEM. Replacement parts such as nozzles, springs, and other internal parts change the original flow rate and consequently

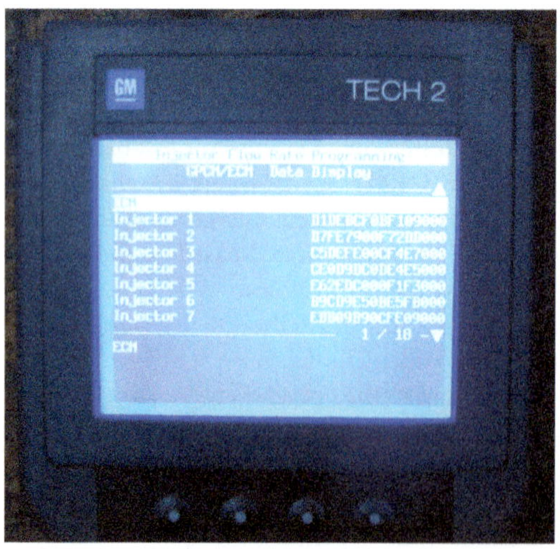

FIGURE 9-22 Injector calibration codes for each cylinder are programmed into control modules with codes obtained from each injector.

FIGURE 9-23 The calibration code on an injector must be entered into the ECM whenever injectors are replaced or relocated to another cylinder.

the injector calibration code. Failure to accurately generate and assign a correct calibration code to an injector can cause the emissions to increase beyond legislated standards. Injection pumps are also carefully calibrated to ensure that each cylinder delivers a precise fuel quantity under different speed and load conditions. Comparator benches are used at manufacturing and service facilities to measure the delivery volumes from injection pumps. Such benches, as their name suggests, compare delivery volumes from each delivery port of the pump after a fixed number of strokes or revolutions are made. Adjustments are made to ensure that the delivery quantities match specifications and that variations between cylinders are corrected.

Changing an Injector Calibration Code

Whenever an injector is replaced or relocated to another cylinder, its calibration code must be changed. To change an injector calibration code, follow the steps in **SKILL DRILL 9-1**. The procedure shown is a general description. Follow the manufacturer's instructions for products from other OEMs.

Troubleshooting Cylinder Misfires

To differentiate between a misfire caused by either a defective injector or damaged or worn cylinder components, the injector in a properly functioning cylinder can be swapped with the injector in the suspect cylinder. If the misfire code moves to another cylinder that comes after the suspect injector is moved, this movement indicates a defective injector and not a problem with cylinder components.

To identify a misfiring cylinder that causes engine noise and vibration, a cylinder cutout or balance test is useful. On hydromechanical engines that use pump-line-nozzle fuel systems, an injector line can be loosened, which prevents fuel from being injected into that cylinder. Scan tools or computers running software capable of bidirectional communication for running service routines can also electronically cut out fuel delivery to individual cylinders. A measurement can then be made of the extent to which more or less fuel is injected by other cylinders to compensate for the lost contribution from a particular cylinder. The acceleration and deceleration speeds of the crankshaft are also monitored to identify cylinders with low combustion pressure.

SKILL DRILL 9-1 Changing an Injector Calibration Code

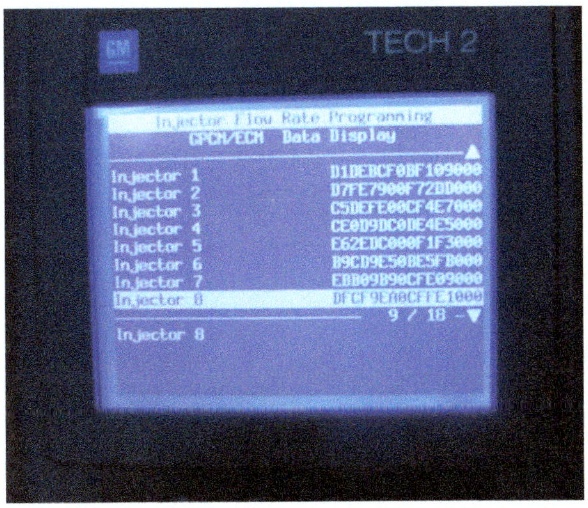

1. Place the vehicle's transmission in neutral and apply the parking brakes.
2. Connect OEM software or a scanner to the vehicle datalink connector and switch on the ignition key.
3. Navigate to "Retrieve injector calibration" in the menu bar and bring up the Injector Calibration window.
4. Remove the valve cover, and record all injector calibration codes if this has not already been done.
5. Compare the actual injector calibration codes against the calibration codes in the Injector Calibration window.
6. If there are any discrepancies between the actual injector calibration codes and codes in the Injector Calibration window, correct them. If any injectors have been replaced, update the calibration code in the Injector Calibration window by applying the code from each new injector to the corresponding cylinder where the injector is installed. Enter the engine security password in the OEM software.
7. Select the function to upload any calibration code changes to the ECM.
8. Run the engine until it reaches operating temperature and confirm that engine operation is smooth.
9. Road test the vehicle after codes have been cleared to run the fuel system monitor. Before returning the vehicle to the customer, make sure the fuel system monitor status is flagged as "ready".

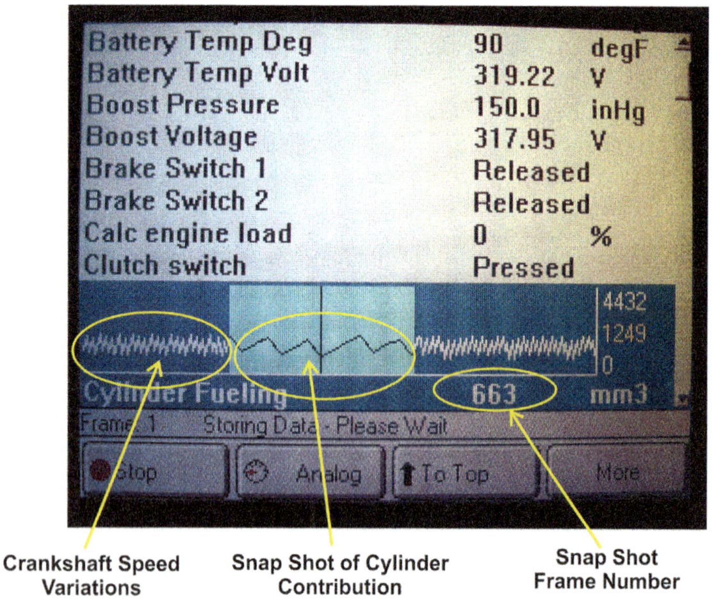

FIGURE 9-24 A cylinder cutout test from a Detroit S50 engine. Energization time is reported in degrees of crankshaft rotation.

A cylinder cutout test detects misfiring or a cylinder with a low power contribution (**FIGURE 9-24**). In Figure 9-24, the cylinder balance or relative compression test results are observed. The long sequence of accelerations and deceleration speeds of the crankshaft are seen. A closer look at a single frame provides a clearer view of crankshaft speed changes. In this particular engine, note the slope and firing event after the cursor vertical line. The longer interval or delay between the time when maximum crankshaft speed is recorded compared to the other cylinders was due to a worn cylinder. A defective EGR valve caused premature wear in several cylinders due to excessive soot. The wear which slowed the build-up of compression pressure is evidenced by a longer upward slope during the power stroke. Low cylinder compression pressure did not accelerate the cylinder as rapidly as the previous two power events. The next cylinder to fire did not slow down the engine as much relative to the other cylinders. The power stroke that followed the compression event did not accelerate the crankshaft as quickly as other cylinders.

Performing a Cylinder Balance Test

To identify weak cylinders that are contributing to low power complaints and causing rough-running conditions, the software diagnostic routines of all late-model diesel engines include some variation of a cylinder balance test. A cylinder balance test measures the crankshaft velocity changes for each cylinder compression and power stroke to identify cylinder misfiring, unbalanced power contributions, and mechanical problems with an engine, such as a bent connecting rod or burned valve. Many fault conditions produce unique patterns of velocity changes in the crankshaft. For example, low compression due to a mechanical fault will fail to slow the crankshaft down on compression stroke in comparison to a good cylinder. The crank position when deceleration occurs will also be delayed. Weak injectors can be identified by measuring how much correction the ECM needs to make to the injector pulse width by controlling the quantity of fuel each injector delivers. Crankshaft acceleration data during power stroke is used to calculate the correction factor. On most late-model engines, the idle-speed compensation number derived during a cylinder balance diagnostic routine indicates how much correction to the injector pulse-width signal is required to produce the same crankshaft acceleration from each cylinder. To perform a cylinder balance test, follow the steps in **SKILL DRILL 9-2**

SKILL DRILL 9-2 Performing a Cylinder Balance Test

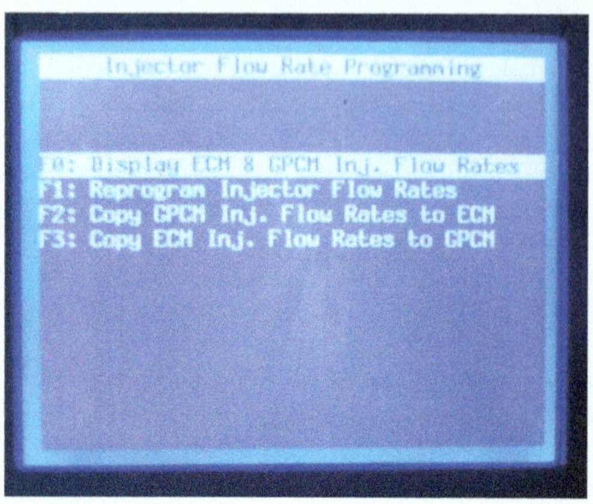

1. Start and run the engine until it reaches operating temperature. Place the transmission in neutral and apply the parking brake.
2. Connect OEM or aftermarket software to the vehicle datalink connector.
3. Navigate to the diagnostic test tab or menu function.
4. Observe and record the idle-speed balance value for each cylinder while the engine is at idle.
5. Analyze the results. The ECM will apply a positive or negative correction factor to the length of the injector pulse width. A positive value means more fuel is injected to compensate for low cylinder pressure. A negative value indicates that a shortened energization time is used to remove some fuel from the cylinder. A 100% change means that the ECM cannot add or remove any fuel and that the cylinder contribution will not be the same as other cylinders. Generally, any correction that requires more than 20% indicates a problem with the cylinder. These problems include needing to adjust a valve adjustment and defects in the injector, high-pressure fuel system, or a base engine mechanical condition.

▶ Injection Timing

For efficient combustion, clean emissions, and good engine performance, precisely timed injection events are crucial. Efficient fuel systems must have the capability to change the point when fuel is injected, based on a variety of engine operating conditions and control algorithms.

9-04 Identify and explain the purposes and functions of changing injection timing.

Variables Affecting Injection Timing

Operating and combustion conditions cause significant changes to the fuel burn rate and the time available for combustion. Temperature variables such as those for fuel, intake air, coolant, and oil affect combustion events, as do many other factors, such as boost pressure, EGR rates, engine load, and speed. To understand the factors influencing injection timing, consider that in most engines, peak cylinder pressure should occur approximately 10°–15° of crankshaft rotation after TDC (**FIGURE 9-25**). But before the piston can be pushed downwards by combustion pressure, there is an ignition delay, followed by a short, uncontrolled combustion period. Several variables influence how long each of these periods is. Take, for example, the ignition delay period. This is the time period between the injection of fuel and its actual point of ignition. There is a time lag, expressed in degrees of crank rotation, for this period. Fuel must be injected, distributed, atomized, absorbed heated, vaporized, mixed with air in the correct proportions, and allowed to reach adequate temperature before it ignites. These processes take time and are affected by factors such as air intake temperature, pre-ignition temperatures in the cylinder, fuel droplet sizing, fuel properties, combustion chamber turbulence, EGR gas in the cylinders, and other factors. Because the amount of time that the ignition delay period lasts for varies, changes to injection timing should be made to make sure cylinder pressures peak at the optimal crank position just after TDC. Without this, engine performance and emissions will be affected.

Engine Load and Injection Timing

The two most important factors influencing burn rate are cylinder pressure and temperature. Temperature and pressure affect the speed of any chemical reaction, including combustion.

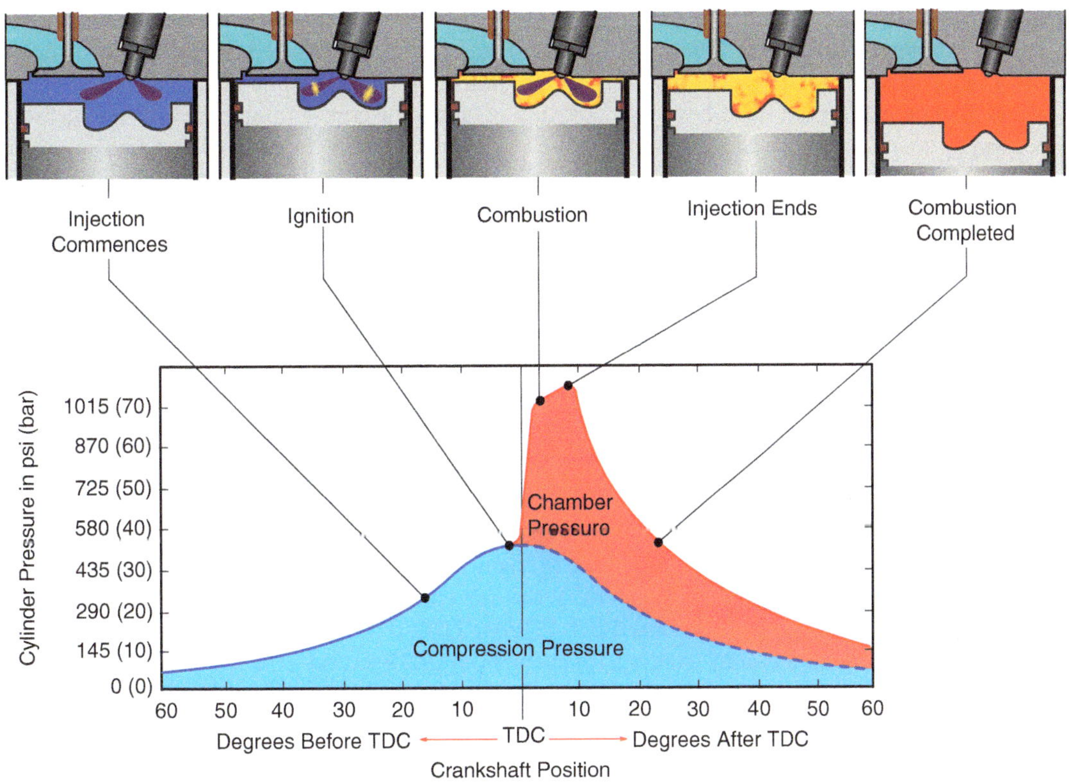

FIGURE 9-25 Injection timing has a major effect on combustion events.

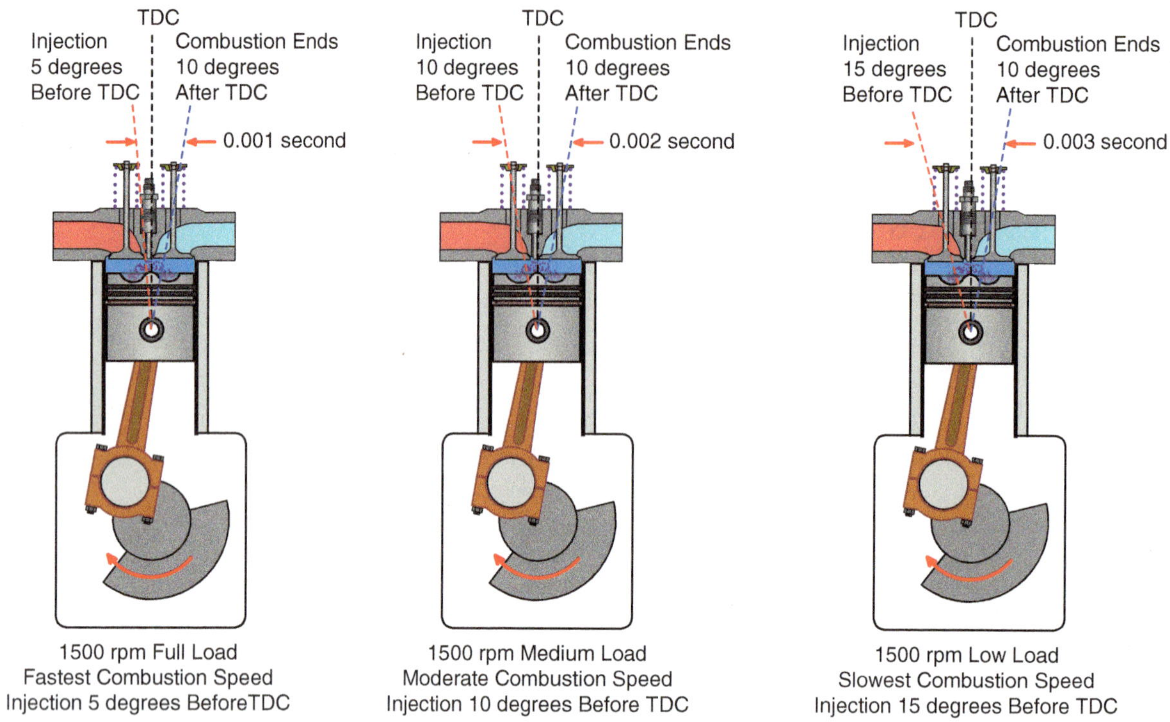

FIGURE 9-26 Increasing engine load requires adding more fuel to the cylinders to maintain engine speed. Injection is retarded because combustion time is shorter.

▶ TECHNICIAN TIP

High cylinder temperatures and pressures accelerate the chemical reactions of combustion. More fuel injected into a cylinder means higher temperature and pressure, leading to shorter combustion times. Under load, higher cylinder pressures and temperatures cause fuel to burn more completely, and only a "heat haze" or "shimmer" of hot air should be observed from an engine exhaust system. Observing any smoke from an engine under load, even without an exhaust aftertreatment, means the engine has a defect.

Higher pressures and temperatures tend to speed up reactions because molecules have more energy and are packed more closely together. These conditions increase the likelihood of combustion reactions and the speed of chemical reactions. In the combustion chamber, the addition of fuel causes temperatures and pressures to rise proportionally to the quantity of fuel burned. This means the addition of fuel shortens combustion time. Naturally, the opposite is true: When small amounts of fuel are injected, combustion times are proportionately longer. Under load, more fuel is injected, and less fuel is injected when light loads or no loads are applied. To correct for the variation in combustion time as engine load changes, the fuel system must have the capability to adjust injection timing based on fuel delivery quantities. When large quantities of fuel are injected, leaving initial timing unchanged would cause peak cylinder pressures to be reached sooner or even quite a few degrees before TDC. Similarly, under light load operation, when fuel burns slower, peak cylinder pressure occurs long after the optimal crank position. So, under load, fuel systems should retard the injection timing (**FIGURE 9-26**). Retarding means fuel is injected later. Conversely, smaller quantities of injected fuel require advancing injection timing. Advancing injection timing means fuel is injected earlier (**FIGURE 9-27**).

Engine Speed and Injection Timing

As the engine speeds up, injection and combustion events do not speed up proportionally. So, while the crankshaft may be turning through more degrees of rotation per second, combustion and injection events cannot be completed as quickly because they require a fixed amount of time. To enable efficient combustion, an injection system needs the ability to change injection timing in response to engine speed changes.

To understand this, consider that the best engine efficiency is achieved when the peak cylinder pressures are produced 10°–15° after TDC. If the time required for the combustion process stays the same—taking, for example, at 0.003 second—the injection of fuel should be timed to produce peak cylinder pressures just after TDC. At idle speed, the injection may require 15° of crank rotation to finish ignition delay and burn to produce peak pressure. However, as the engine speed increases, the number of degrees the crankshaft rotates per second is much greater. So, the crank turns farther, and the piston moves farther in one

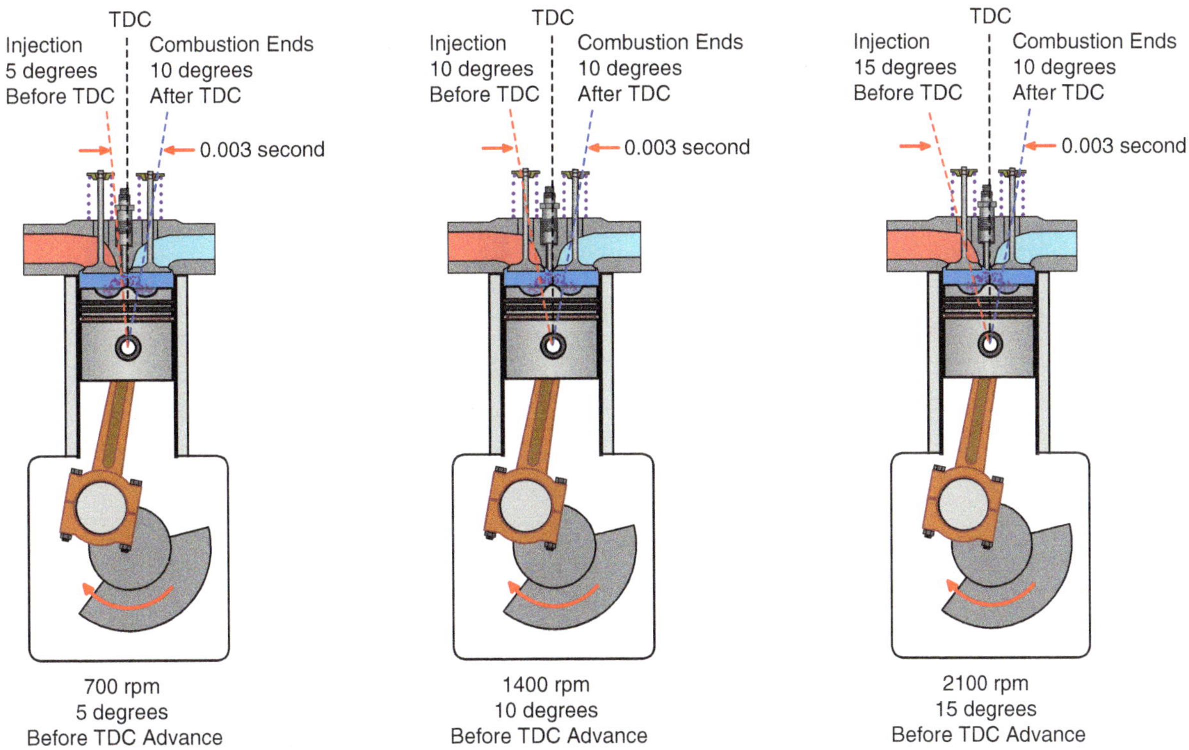

FIGURE 9-27 Increasing engine speed without a change in load requires advancing injection timing or injecting fuel earlier.

second at higher engine speed than at idle. To produce the peak cylinder pressure after TDC in this scenario, the injection needs to take place earlier, or advance. How much earlier depends on the amount of time it takes for the fuel to burn.

Consider the following example. Burn time is expressed in degrees of crankshaft rotation. The burn time would be subtracted from 10° after TDC. So, if the fuel takes 0.003 second to burn and that translates into 15° of crankshaft rotation, then injection should take place 5° before TDC. If the engine turned faster, the number of degrees of crankshaft rotation would be greater per second, and thus the timing should move closer to TDC.

Example 1:

Burn time of 0.003 second = 15° of crankshaft rotation at idle 650 rpm

10° after TDC – 15° of burn time = injection beginning at 5° before TDC

Example 2:

Burn time of 0.003 second = 30° of crankshaft rotation at 1,300 rpm

10° after TDC – 30° of burn time = injection beginning at 20° before TDC

When the engine speeds up with no change in load or injection quantity, timing is advanced. Similarly, when the engine slows down, timing is retarded.

Injection Timing and Cold Starting

Cold cylinder components and low ambient temperatures create exceptional operating conditions that require adjustments to injection timing. The primary problem of cold conditions is that they cause a longer ignition delay period. Cylinder pre-ignition temperatures are lower, so fuel requires more time to absorb sufficient heat to initiate combustion. The ignition delay causes a more pronounced knocking sound in the engine and affects combustion quality. Also, cold engine parts absorb compression heat near TDC too. When running, injection timing is advanced to compensate for the longer ignition delay period caused by

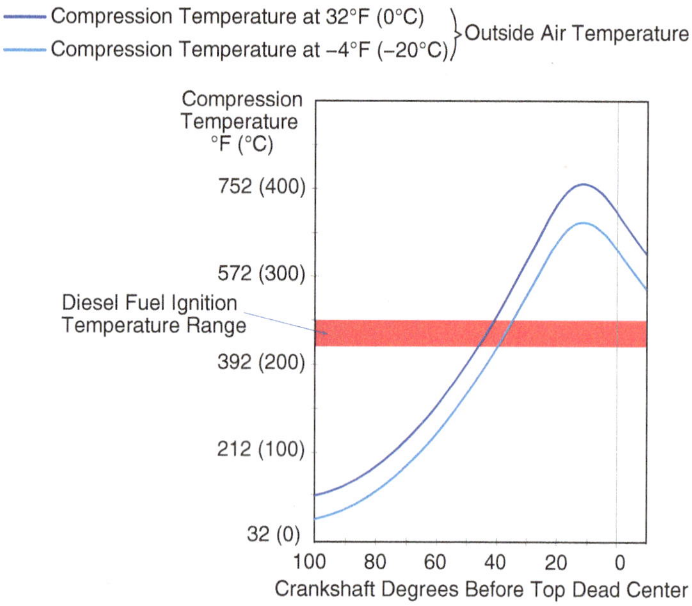

FIGURE 9-28 Cold outside temperatures produce colder pre-ignition compression temperatures.

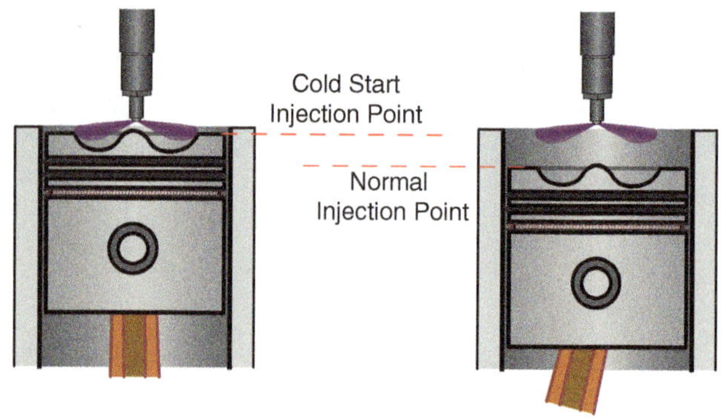

FIGURE 9-29 Retarding injection timing during cold cranking promotes a greater likelihood of combustion for improved starting.

cold temperatures (**FIGURE 9-28**). When the engine is running, the delay between injection and actual ignition can be so long that it causes black smoke emissions the result of incompletely burned fuel. This condition is due to prolonged ignition delay, which shortens the total time available for combustion. Shorter combustion time and lower cylinder pressure produce more PM, CO, and HC emissions. During cranking, igniting fuel is more difficult because insufficient fuel vapor is produced inside cold cylinders. Heat content in the cylinders is low, and fuel tends to condense onto cold cylinder component surfaces; poor vaporization results. To minimize the effects of hard starting and a prolonged cranking period, accompanied by smoke emissions at start-up, two separate injection timing strategies are often used. Injection timing is retarded only during the cranking period, and timing is advanced further than normal during cold running conditions (**FIGURE 9-29**). The benefit of retarding injection timing during cold cranking is that it provides more time for the piston to rise farther in the cylinder and generate increased pre-ignition temperatures. When fuel is finally injected, the cylinder is hotter and improved vaporization of fuel occurs. Thus, there is a greater likelihood of ignition taking place and the engine starting. Once the engine starts, cold temperatures and cylinder components still demand additional time for ignition to occur after injection. If injection timing remains retarded after start-up, white smoke changing to black and grey smoke will be emitted. White exhaust smoke is produced from fuel vapors generated during a cold misfire condition. Retarding the injection timing would minimize misfiring, but black exhaust smoke is emitted as the cylinder only slowly warms up because fuel is partially combusted. To prevent any of these conditions, fuel-injection timing maps programmed into the ECM—or the cold start mechanisms of mechanical systems—retard the beginning of injection timing during cranking but quickly advance injection timing after the engine starts.

Injection Timing and Emissions

Diesel engine emissions and performance characteristics are significantly affected by changes to injection timing (**TABLE 9-1**). Slightly advancing the start of injection may reduce smoke, increase power, and reduce fuel consumption; however, it typically results in higher NO_x **emissions** (**FIGURE 9-30**). (In fact, optimal timing for emissions negatively affects fuel economy in North American engines.) The increase in NO_x emissions is due to the higher cylinder pressures and combustion temperatures produced when peak combustion pressure is closer to TDC rather than 10°–15° after TDC. Because combustion chamber

TABLE 9-1 Influence of Injection Timing on Combustion Events

	Change to Injection Timing	
Engine Parameter	**Retarded**	**Advanced**
Cylinder pressure	Lowered	Raised
Oxides of nitrogen (NO_x)	Reduced	Increased
Particulate matter (PM) peak torque	Decreases	Increases
Particulate matter (PM) rated speed	Increases	Decreases
Particulate matter (PM) (low speed and load)	Increases	Decreases
Hydrocarbons (HC)	Increases	Decreases
Fuel consumption (BSFC)	Increases	Decreases
Intake manifold boost pressure	Increases	Decreases
Exhaust temperature	Increases	Decreases
Starting	Easier	Harder
Coolant temperature (heat rejection)	Increases	Decreases
Combustion noise	Decreases	Increases

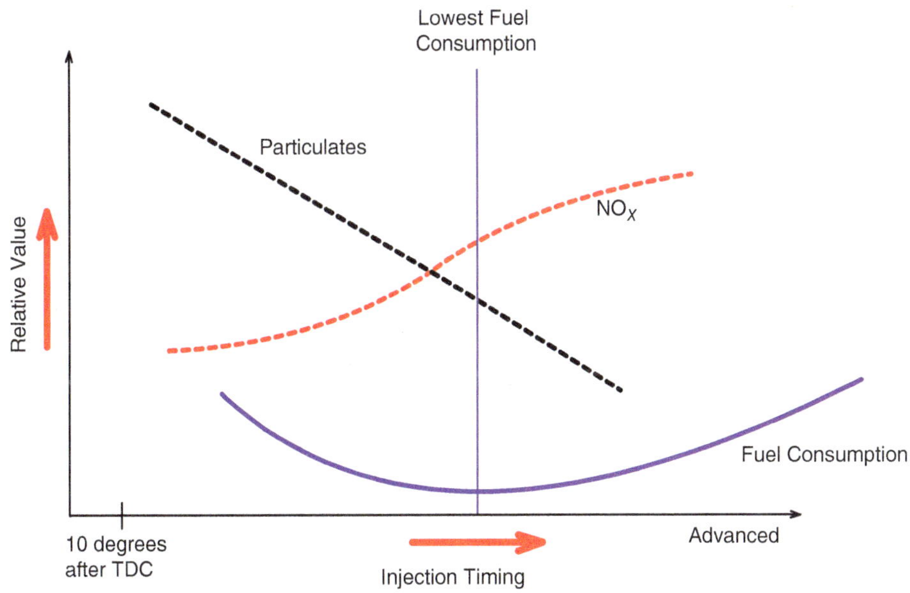

FIGURE 9-30 Advancing timing lowers particulate emissions but increases NO_x.

volume is smaller nearer TDC, increased cylinder pressure and temperature take combustion conditions above the threshold where NO_x is formed. Conversely, retarded timing increases the production of particulate and hydrocarbon emissions because combustion time is shortened and combustion is incomplete. To minimize both NO_x and particulate production, a manufacturer's goal is for finer atomization of fuel that also shortens ignition delay time: Timing can be retarded, and at the same time, the combustion process is more resistant to particulate formation when atomization is finer. On late model electronically controlled engines, timing is still drastically retarded compared to older, mechanically governed engines. Today, the beginning of a main injection event may occur after TDC or just a few degrees before TDC, compared to the 15°–20° of advance of earlier pre-emission standard engines. Retarded timing produces a gentler rise in cylinder pressure and lower peak cylinder pressures near TDC, which minimizes NO_x formation. On mechanical systems, injection timing needs to be set to within one-half degree of the manufacturer's specifications for correct emission control (**FIGURE 9-31**). Electronic engines control injection timing variations to less than 0.1° of precision.

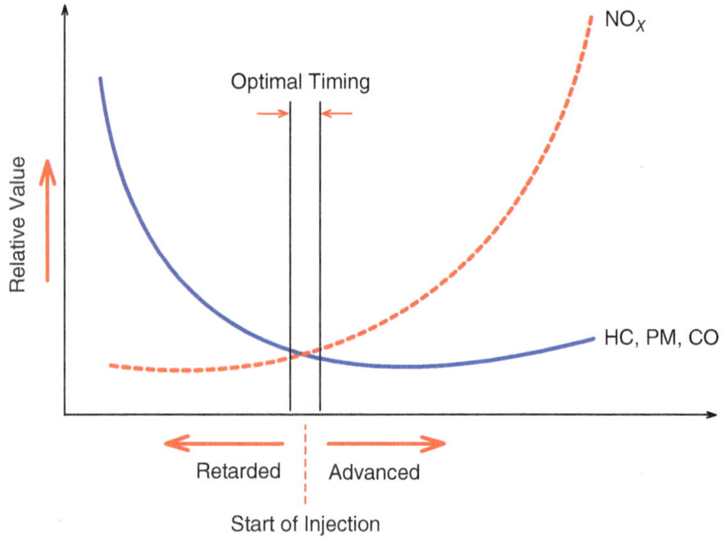

FIGURE 9-31 Injection timing on older mechanical engines has to be adjusted to within ±0.5° for lowest emissions and to less than 0.1° on electronic engines.

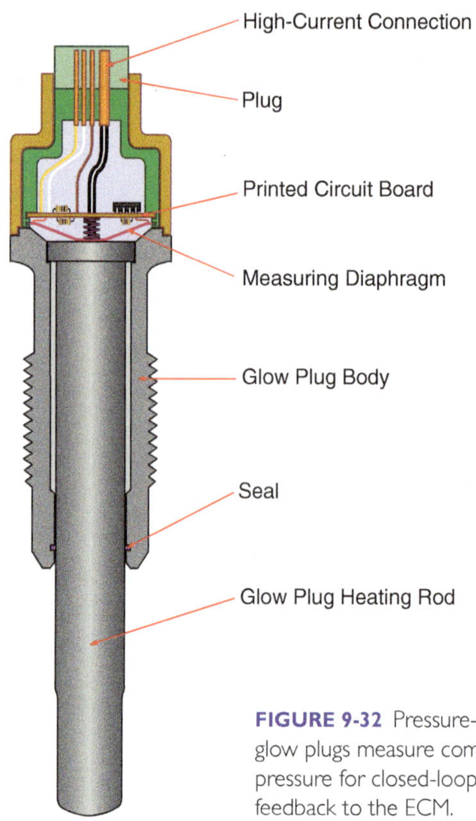

FIGURE 9-32 Pressure-sensing glow plugs measure combustion pressure for closed-loop feedback to the ECM.

Injection Timing and OBD

Injection timing has always had a significant impact on emissions; injection timing is a part of the fuel system monitor on OBD-II engines. A variety of strategies are used to determine the precise beginning of injection to ensure that it coincides with expected values. Beginning in 2009, some engines, such as the VW TDI, have used cylinder-pressure sensors to measure the rise in combustion pressure (**FIGURE 9-32**). Knowing the cylinder pressure for every degree of crank rotation can provide feedback to the ECM, enabling it to change injection timing and the injection rate in order to minimize emissions and increase engine efficiency. VW also uses this sensor to analyze variations in fuel quality and cetane rating, which can affect the ignition delay period and other combustion characteristics. Other engines analyze electrical waveforms from the injector that change when control valves open and close. Most engines use data from the crank-position sensor to detect the point when crankshaft acceleration begins due to a combustion event.

▶ Fuel-Injection Rate Control

9-05 Define and describe the function of injection rate control.

When discussing modern fuel-injection systems, the term "injection rate" has two meanings. For the purposes of this next section, injection rate will refer to the quantity of fuel injected per degree of crank angle rotation. When plotted as a graph, the injection rate is reported in milliseconds or degrees of crankshaft rotation on the horizontal (or X) axis, compared with the delivery volume of the injection event on the vertical (or Y) axis. The ability to control the fuel-injection rate, also called rate control, is most significant advancement of contemporary injection technology because it affects fuel consumption, exhaust emissions performance, and combustion noise. "Fuel rate" is an identical term that may seems indistinguishable from to rate control, but it refers to fuel consumption expressed as the quantity of fuel burned per hour. When the ECM accounts for changes in fuel density due to temperature changes, fuel rate is also reported in milligrams per stroke.

Rate-Shaped Injection

Rate-shaped injection refers to the shape of the injection discharge curve that is created through varying of injection pressure and the use of multiple injection events during a

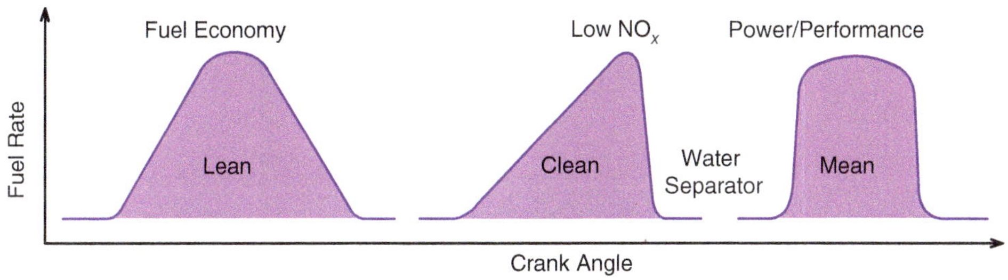

FIGURE 9-33 Three distinct shapes of injection discharge provide unique operating characteristics: *lean* promotes fuel economy, *clean* reduces NO_x, and *mean* produces optimal power.

single combustion cycle (**FIGURE 9-33**). The newest injection systems can provide several rate discharge profiles by varying injection pressures and the number and timing of injection events. Manufacturers of common rail systems can currently provide as many as nine separate injection events during one combustion cycle. These events can include several pre and post combustion events in addition to the main injection event. Post combustion injection events are used to dose the diesel particulate filter (DPF) or lean NO_x trap (LNT)—also called NO_x adsorber catalyst (NAC)—for regeneration. To keep fuel consumption and soot emissions low, the entire quantity of the main injection event should be delivered as quickly as possible. This leads to a rapid buildup in cylinder pressure and a longer, harder push on the piston after TDC. Although combustion may be efficient and large quantities of power may be derived from gas expansion, the downside is greater NO_x emissions and noise are produced. To achieve lower noise and NO_x emissions, other rate-shaped injection strategies have been developed. One basic strategy to lower NO_x and noise is to inject fuel slowly or to incrementally ramp up fuel delivery at the beginning of the injection event. After the piston travels past TDC, the fuel rate should rise sharply after TDC and then abruptly stop. This rate of discharge leads to a relatively slow increase in combustion pressure and lower noise levels. More importantly, the lower cylinder pressure and temperature produced from this ramped injection rate minimizes NO_x formation.

Mechanical Rate Shaping

Earlier, mechanically governed fuel systems relied on throttle pintle injector nozzles to achieve some injection rate shaping (**FIGURE 9-34**). These nozzles generate a spray pattern that produces a gradually rising rate discharge curve. This discharge shape is achieved by using a stepped-type nozzle valve. When the nozzle valve begins to lift, a small, narrow, low-volume spray jet is injected first into the combustion chamber. As the needle valve lifts higher because of increasing injection volume, more fuel is allowed out of the nozzle because the nozzle-valve diameter narrows. At this point, the spray pattern changes into a wider, higher-volume cone pattern. Using these nozzles substantially reduces noise and reduces emissions with its ramped injection rate. A unique multi-stepped nozzle-valve shape is still used in the latest electronically controlled injectors to improve rate shaping by mechanical means. Bosch and its licensees also use a double-spring nozzle to control the discharge rate. Each spring is calibrated to lift at a different level of fuel pressure. The first spring lifts, sprays fuel, drops injection line pressure, and seats again during the initial stage of injection. A control piston above the valve assists this process. As the injection volume increases for the remainder of the injection cycle, increased fuel volume and pressure to the nozzle is needed to lift both springs to supply a larger delivery volume during the main injection. This injector operation is illustrated in Chapter 11, on hydraulic nozzles. The first electronically controlled fuel system to use rate-shaped injections is the hydraulically actuated electronic unit injector (HEUI) built after 1996. These injectors use a groove machined into the fuel plunger that pressurizes fuel to produce a pilot-injection event through a mechanical method. When fuel pressurization begins, the plunger travels downward and then uncovers a port that spills pressurized fuel. This momentarily stops fuel delivery. As the fuel plunger continues to move downward, the spill port is covered by the plunger again and the injection event resumes. This pilot-injection feature is a type of an injection rate–shaping strategy. Variable pressurization of oil to the injectors, which supplies the force to operate the injector, also varies

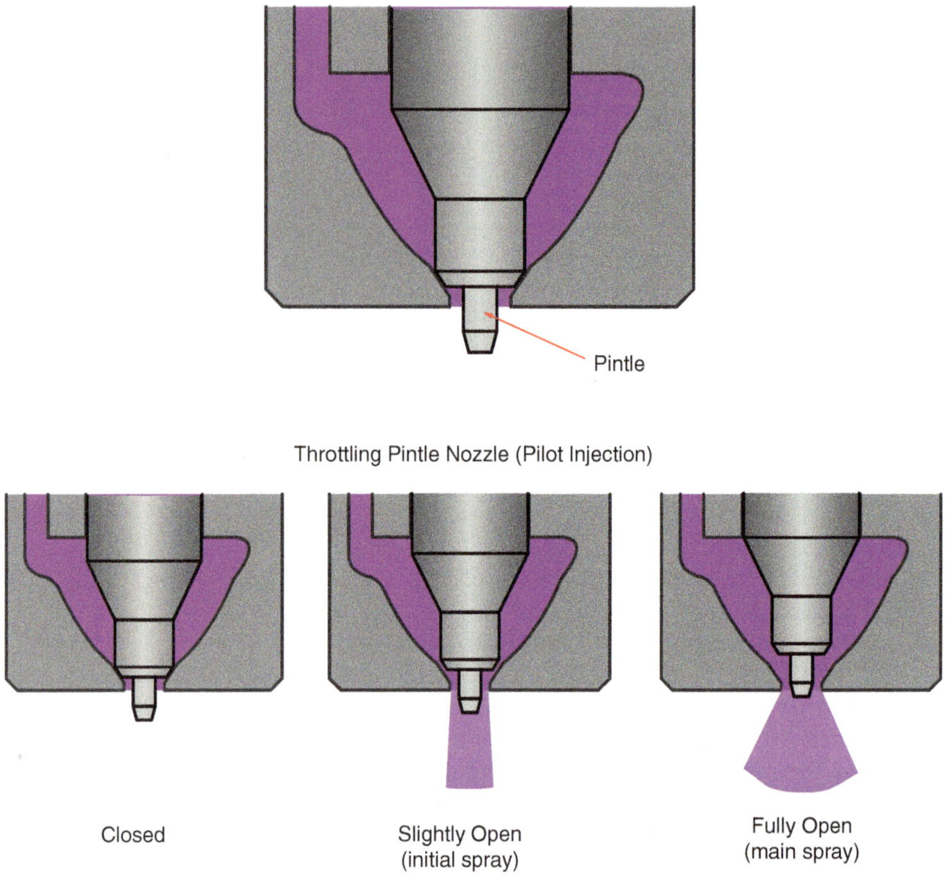

FIGURE 9-34 Throttle pintle nozzles mechanically shape the rate of injection through the profile of the needle valve.

the injection rate. Gen II HEUIs produce three injection events by using electrical signals applied to two solenoids rather than mechanically controlling the injection. This feature is illustrated in further detail in Chapter 12, on HEUI systems.

Pilot Injection

Pilot injection, also known as split-shot injection, is another feature of rate shaping, which is one of the most effective strategies for minimizing NO_x and noise emissions while optimizing engine performance. Like a pilot light in a gas appliance or furnace, this rate-shaping strategy delivers a small quantity of fuel to the combustion chamber 8°–10° before the main injection. Because the cylinder is warmed up by the ignition of a pre-injection event of fuel, fuel injected during the larger main injection event is quickly ignited. Ignition delay is minimized, and little combustion noise is produced. This approach to fueling produces exceptionally quiet engine operation, such that the traditional diesel clatter is gone and the noise level is comparable to that of gasoline-powered vehicles. Because fuel is ignited more quickly by using pilot injection, more time is available for combustion. The result is main injection timing that can be retarded to lower NO_x emissions without a penalty to power or fuel economy. The most recent injection systems use two or more pilot-injection events, rather than a single pilot-injection event, for even greater improvements to combustion quality.

Post-Combustion Injections

Research proves that an injection after the main combustion event reduces the production of soot inside the cylinder (**FIGURE 9-35**). The post–main injection event taking place shortly after the main injection adds additional heat and pressure to the power stroke that in turn consumes the carbon that forms soot. The number and pressure of injection events vary with speed and load. At lower speeds, there is more time available for a couple of

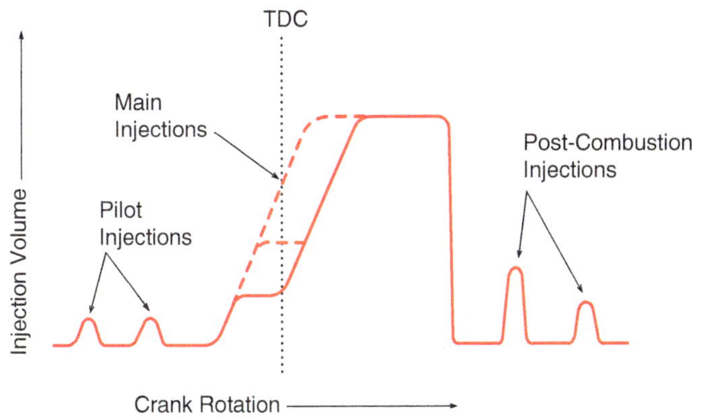

FIGURE 9-35 Two pilot injections enable the use of retarded injection timing while maintaining adequate combustion time. The first reduces soot formation by maintaining high cylinder temperatures.

pilot-injection events that are not possible to use at high speed and under heavy load, when injection timing is already retarded. High cylinder pressures and temperatures at high speed reduce the need for post–main combustion events. Diesel particulate traps require supplemental heating to burn away trapped soot. Common rail fuel systems can supply an additional injection event during the exhaust stroke. The HC in turn enables flameless combustion to supplementally heat the exhaust oxidation catalyst. Heat in this catalyst is used to combust soot trapped in the downstream particulate filter. NO_x adsorbers and lean NO_x traps require that oxygen be completely depleted from the exhaust system for a few seconds and require supplemental heating to break down stored NO_x emissions. When signaled by the engine or exhaust aftertreatment controller, additional fuel is injected during the post-combustion period, in addition to the use of other regeneration strategies.

▶ Fuel Pressurization

9-06 Identify and describe the purposes and methods used for pressurizing fuel for injection.

The main reasons for why diesel engines need highly pressurized fuel has already been explained. High injection pressure is necessary as well, for the following reasons:

- Diesel engines compress dense air mass that requires higher pressure to penetrate the dense compressed air with fuel.
- The brief time available for combustion events taking place near TDC requires rapidly executing pre-combustion events, such as distribution—that is, the mixing of fuel and air in the cylinders through atomization, mixing, and vaporization.
- Improved atomization requires higher pressurization for improved emissions and performance.

Traditional fuel systems have relied on some version of the camshaft-and-plunger system to pressurize fuel (**FIGURE 9-36**). While this configuration worked well for decades, it has a number of drawbacks. The most significant is that the cam-and-plunger injection pressure is tied to engine speed. This means that at slower engine speeds, when camshaft velocity and injection volume are low, only low levels of fuel-injection pressure is possible. If an engine needs to rapidly accelerate, large injection quantities of high-pressure injection pressure for good combustion quality and clean emissions are not available. Under other conditions—high engine speed, accompanied by small injection quantities—fuel is distributed too far and fast in the combustion chamber to provide the best combustion conditions. To meet emission standards while optimizing engine performance, injection systems for on-highway use are now electronically controlled with variable injection pressure capabilities and with maximum injection pressure reached independently of engine speed (**FIGURE 9-37**).

SAFETY TIP

Fuel pressurized to thousands of pounds per square inch (bar) can easily penetrate skin, causing bodily harm, blood poisoning, and even death. When checking for leaks from a high-pressure fuel system, use a small piece of cardboard or paper instead of a hand to detect the location of a leak.

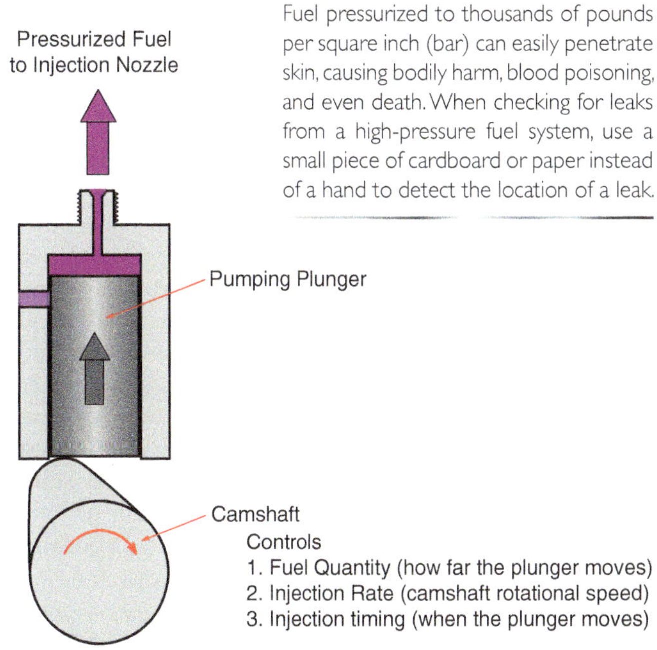

FIGURE 9-36 Traditional cam-and-plunger systems control many injection events.

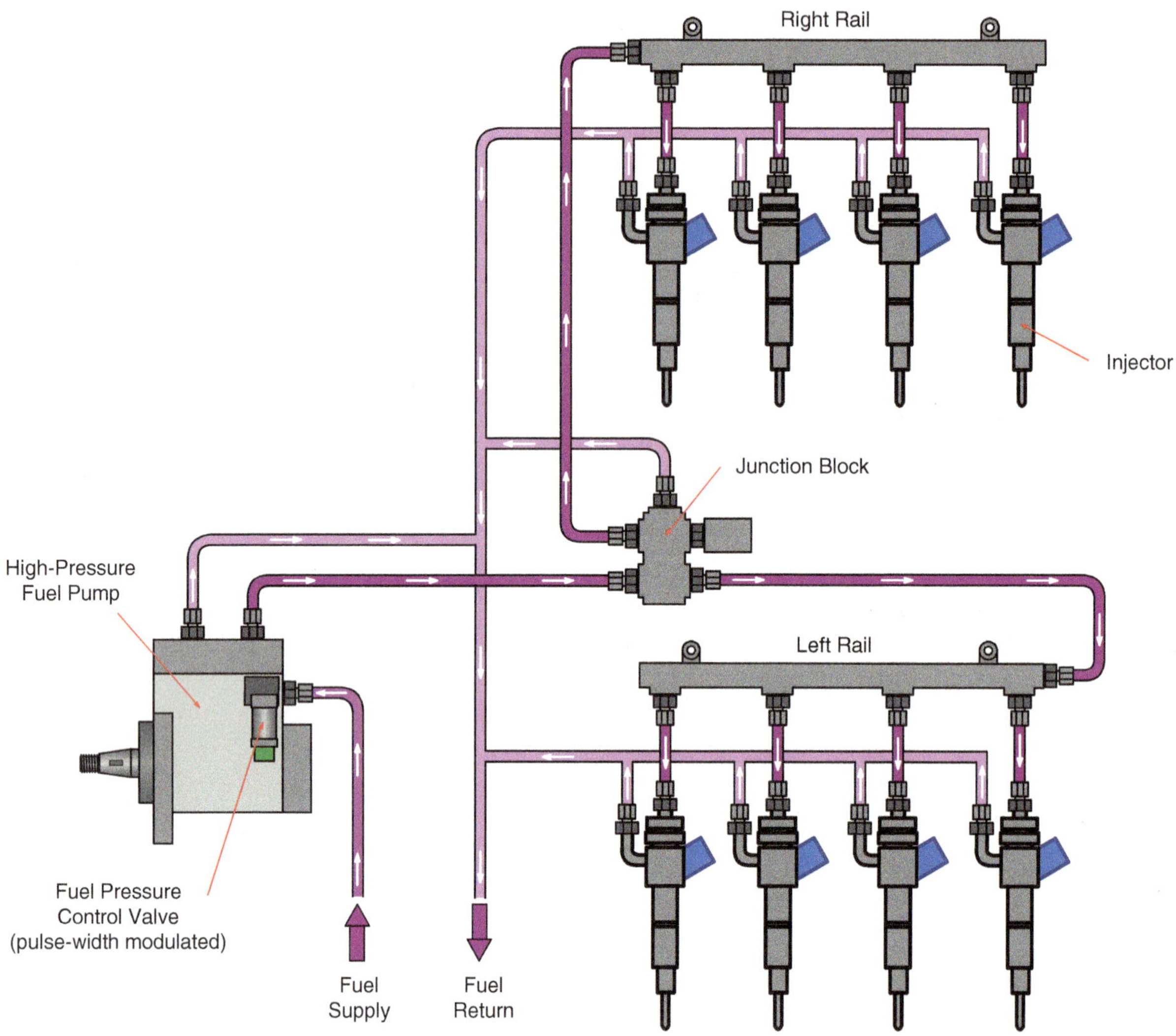

FIGURE 9-37 This electronically controlled variable injection pressure system can select exactly which injection pressure is best for any speed, load, and operating condition.

▶ Fuel Atomization

9-07 Define and explain the function of fuel atomization.

As discussed earlier in the chapter, atomization refers to the breaking up of liquid fuel into smaller, finer liquid droplets. Atomization must take place in order to vaporize and ignite fuel. When liquid fuel leaves the nozzle tip, it breaks up into smaller droplets after encountering friction with air. Breaking the fuel into smaller droplets enhances the fuel's ability to absorb heat, convert to vapor, and mix with air. Fuel droplet size affects combustion qualities, which makes proper atomization a critical function of the injection system. Just as a fire is kindled with small pieces of wood, efficient combustion requires exceedingly fine particles of fuel. Finer atomization means fuel ignites sooner, burns faster, and burns more completely. This is why worn or damaged injector nozzles can result in poor performance and emissions (**FIGURE 9-38**). The ignition delay period is also shortened, leading to lower levels of combustion noise. Better atomization promotes more complete combustion of fuel while producing lower levels of particulate emissions and improving fuel economy. Finer atomization also means the air-fuel mixture is better able to tolerate the use of recirculated exhaust gas.

Fuel Droplet Size and Particulate Matter Emissions

Fuel vapor forms around the fuel droplet as it absorbs heat from the combustion chamber. This means fuel burns from the outside to the inside. A smaller fuel droplet size converts

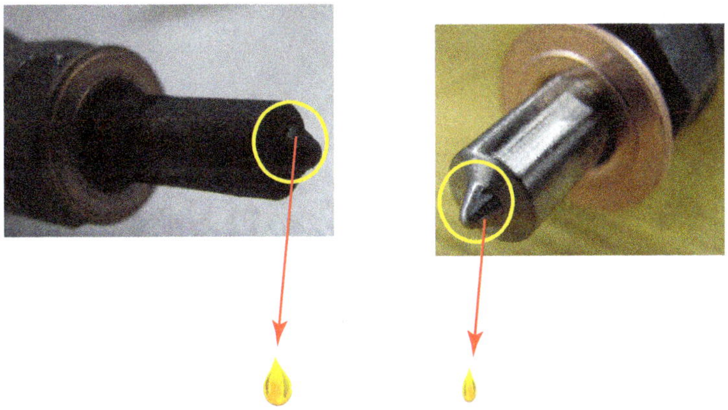

FIGURE 9-38 Worn nozzle orifices (left) produce large droplets that take longer to burn. Undamaged nozzles (right) result in finer drops that burn efficiently.

liquid to vapor faster, which in turn results in shorter ignition delay times and longer combustion time. To achieve the lowest particulate emissions, the finest atomization ensures that each droplet makes adequate contact with air for complete combustion. Using fine droplets also means that retarded injection timing can be used, to produce low NO_x emissions. The smallest droplets are produced by using smaller and more numerous injector orifices, along with high pressurization.

Particulate Formation

Ideally, only a heat wave should be observed from the exhaust pipe of a diesel engine when under load. When black or gray exhaust smoke is present, it is because of excessive particulate emissions. Although visible black and gray smoke are the most common emission-related problem associated with diesel engines—and for that reason visible during on-highway operation—they also indicate a problem with engine operation. Excessive particulate emissions are produced typically under the following conditions, with their respective causes:

1. Large fuel droplet size
    - worn nozzles and spray holes
    - low injection pressures
    - nozzle dribble.
2. Poor mixing of air and fuel
    - inadequate combustion turbulences
    - restricted airflow
    - excessive fueling of combustion chamber
    - incorrect valve timing or adjustment
    - poor distribution of fuel in chamber.
3. Inadequate pre-ignition temperatures
    - low compression pressures
    - low intake temperatures
    - excessive ignition delay.
4. Inadequate combustion time
    - incorrect injection timing
    - excessive engine speed
    - incorrect valve timing or injection timing
    - excessive ignition delay time
    - cold ambient temperatures.

▶ TECHNICIAN TIP

Not all diesel smoke means engine trouble. Some older engines lack exhaust aftertreatment devices. During acceleration or when shifting gears, these engines momentarily produce a puff of smoke and, within a few seconds, produce a clear exhaust. Anything more than this indicates some engine fault. Law enforcement agencies often use standards of no more than 15 seconds of visible emissions during one minute of operation, as the threshold for assessing emission related penalties.

▶ Distribution of Fuel

9-08 Define and describe the functions of fuel distribution.

Distribution of fuel throughout the combustion chamber is another function of high-pressure injection systems. The angle of the spray holes, number of nozzle orifices, their size, and injection pressure are some of the important variables that affect how fuel is

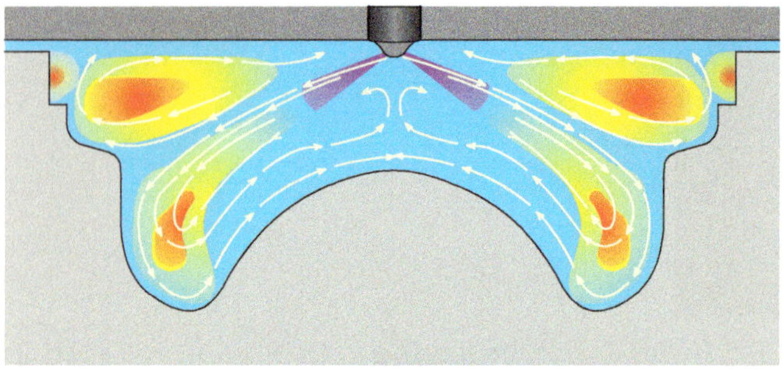

FIGURE 9-39 Fuel and air are mixed by bowl turbulence to provide low emissions while delivering the best power and fuel economy.

distributed in order to for it mix with turbulent air. Larger, heavier droplets penetrate a chamber farther and faster because they have more mass, but they do not burn as well as smaller droplets. Modern diesel engines typically feature more injector spray holes than older engines—with as many as 13 spray holes in injectors used by new engines, compared with only four or five in older engines. Even the length of a spray hole changes the velocity of fuel leaving the injector tip. Longer holes push fuel at a higher velocity than shorter holes. The angle of the spray holes is also important and must match the shape of the combustion bowl to prevent fuel from simply hitting the piston bowl. In this situation, fuel simply does not burn (see the Symmetrical Fuel Distribution section in this chapter) (FIGURE 9-39).

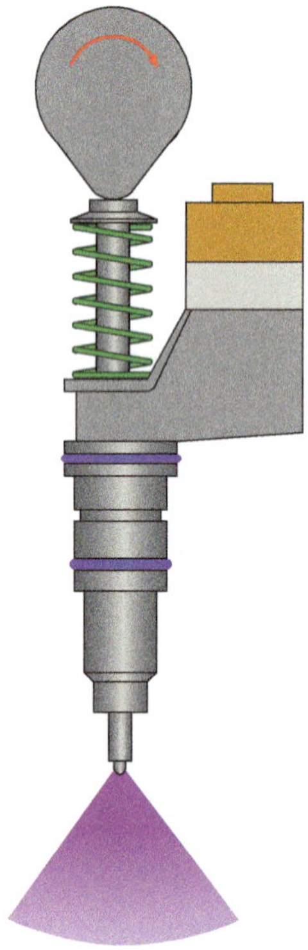

FIGURE 9-40 Lag is minimized in overhead cam-injected engines. With no pushrods to flex, injection takes place quickly after pressurization begins.

Problems with fuel distribution can have a significant impact on combustion quality. For example, in low-emission, high-speed engines, a slow start or ending of the injection period affects atomization in both the initial and final portions of the injected fuel. If fuel delivery is not stopped abruptly, the drop in injection pressures at the nozzle tip creates larger-diameter spray droplets. The effect is poor atomization and penetration of fuel into the combustion chamber, causing high emissions and poor combustion quality. Dribbling fuel, caused by low pressurization, also wastes fuel. For this reason, the design of any fuel system incorporates features to ensure a sharp stop and beginning of injection that prevents lag, the time between the beginning of pressurization and actual fuel delivery (see the Injection Lag section in this chapter) (FIGURE 9-40). To properly distribute fuel, the beginning and ending of injection pressure must be consistent from beginning to end.

Fuel systems incorporate design features to ensure the abrupt beginning and ending of injection. Using delivery valves in high-pressure lines connecting pumps and nozzles is one example (FIGURE 9-41). These valves operate as one-way checks to allow fuel to remain at relatively moderate pressure in the fuel-injection lines, typically around 800 psi (55 bar) when the engine is running. The design of the valves ensures fuel pressure drops far enough below nozzle opening pressure to prevent secondary injections caused by reflecting fuel pulsations in the injector line. These pulses occur when the needle valve in the nozzle suddenly closes, causing a pressure wave. The hammer sound heard when shutting off a water faucet in a house creates a similar condition found in the injection line when the needle valve closes. Delivery valves also prevent the nozzle from dribbling fuel. Nozzles are designed to have higher pressure to lift the needle valve than is required to close the valve. So, whenever pressure begins to drop in the fuel line, the needle valve rapidly closes until injection pressure is increased high enough to reopen the valve.

FIGURE 9-41 A delivery valve minimizes the backflow of fuel into an injection system in order to maintain line pressure between injections.

Injection Lag

Injection lag describes the time delay that occurs between the start of fuel pressurization and the moment when injection actually occurs. In electronic systems, lag refers to the time between the moment that an injector is energized until the point of crank rotation when it begins to deliver fuel for the injection event. There is a substantial time lag, or injection lag unintentionally built into some injection systems. Injection lag influences the accuracy of injection timing controls and the ability of the fuel system to achieve an abrupt beginning and ending of injection.

A number of conditions cause lag, and good fuel system design minimizes those factors. For example, while fluids are not considered compressible, time is required to transmit the pressure wave from the pumping plunger to an injector nozzle. The wave does travel fast—approximately 4,500 feet (1,372 meters) per second—but it can make slight changes to timing at high-speed operation. A more significant factor causing lag is the swelling of injection lines. In spite of lines having thick steel walls, lines do swell when pressurized fuel travels through them. Enough expansion and contraction of the lines can crack them if they are not properly secured with vibration-absorbing insulators (**FIGURE 9-42** and **FIGURE 9-43**). Fuel volume intended for delivery is lost through swelling line walls. For this reason, high-pressure fuel lines are made not only thick but also the same length and diameter (**FIGURE 9-44**). This minimizes variations to delivery quantities and injection timing. Unit injectors, which have no lines between the pressurization stage and the nozzle, have the least amount of lag caused by swelling injection lines because they have none. However, if unit injectors are actuated by using camshaft-driven push rods, the flexion of the rods during injection causes lag. One major source of injection lag is injectors that use magnetic actuators. A phenomenon, called reactive inductance or reactive resistance, causes a magnetic field to build up relatively slowly in the injector's electrical solenoid. The reactive inductance takes place because an opposing electrical current is induced in any coil of wire when an electrical current flows through it. This happens as the magnetic field developed by the initial inflow of current cuts across other conductors in the solenoid's windings and induces current flow in them. The two sources of current—initial and induced—travel in opposite directions, which produces a high initial electrical resistance in the solenoid. Not only is the magnetic field slow to build up and open the injector's control valve, but the field also takes time to collapse and close the control valve in the injector. An injector's slow response means a large conventional

FIGURE 9-42 Vibration isolators clamped to these injection lines are necessary to prevent lines from cracking.

FIGURE 9-43 Vibration isolators clamped to these common rail fuel lines are necessary to prevent lines from cracking. Continuous expansion and contraction is caused by high-pressure injection pulsations.

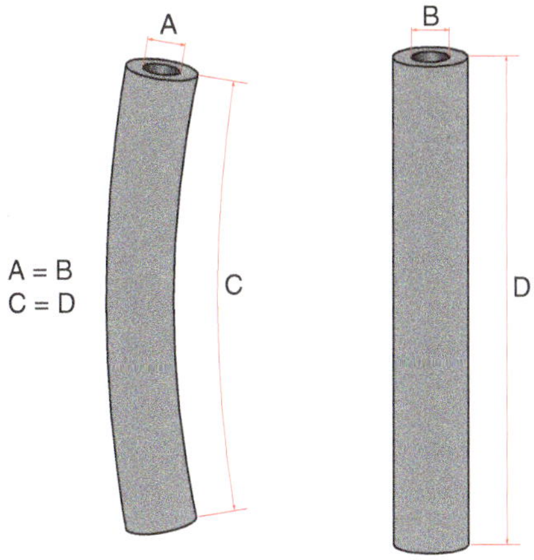

FIGURE 9-44 Fuel-injection lines must have the same diameter and length to prevent variations in injection quantities and injection timing between each cylinder.

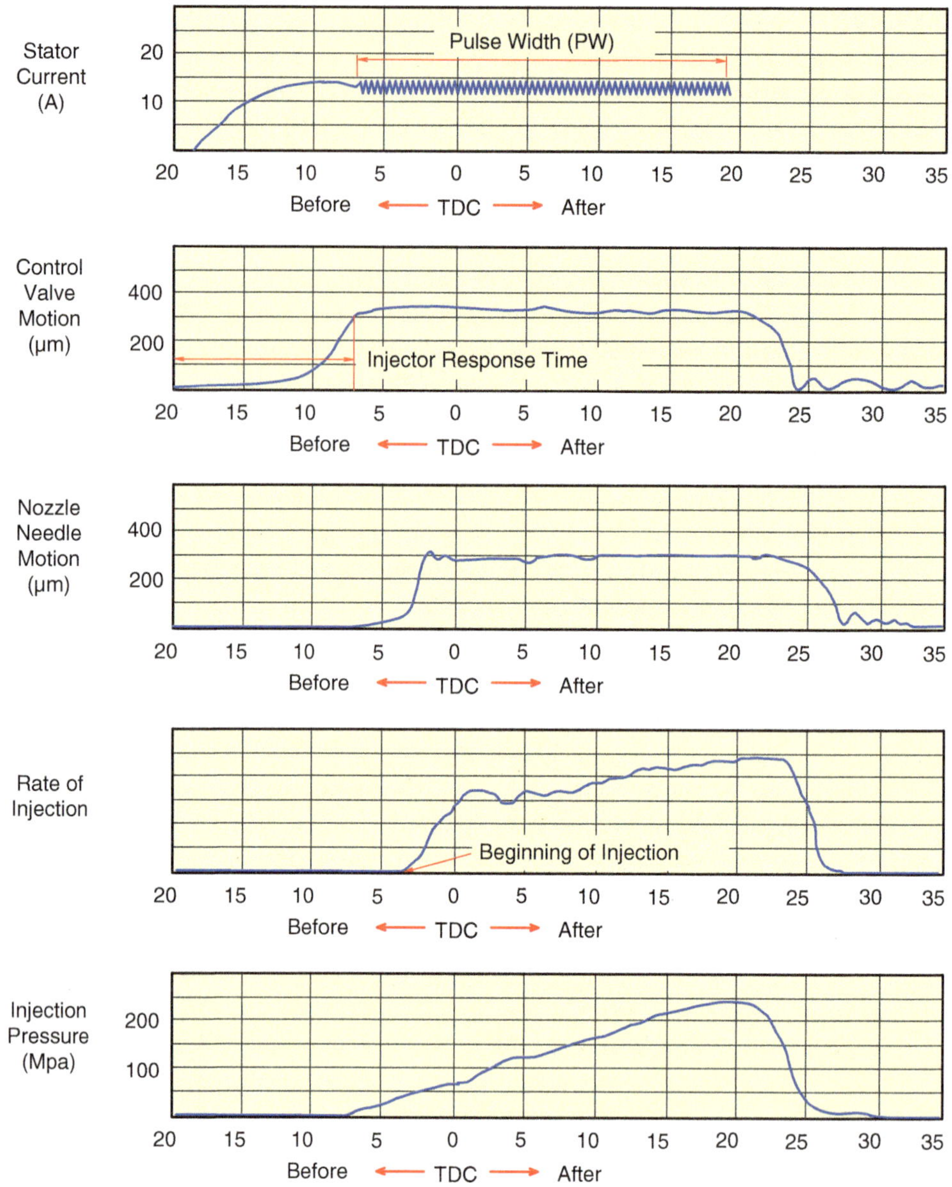

FIGURE 9-45 Injection lag in an electronic unit injector. Notice the injector is energized beginning at approximately 18° before TDC. Actual injection takes place at approximately 3°. Lag time is close to 15° of crank rotation, due to the electrical and hydraulic characteristics of the injector.

electromagnetic actuator is capable of only one injection event in a combustion cycle (**FIGURE 9-45**). Piezoelectric actuators, which have no magnetic fields, are replacing magnetic actuators in newer high-speed diesels. The following are some of the other factors that contribute to injection lag:

- The upward movement of the delivery valve in injection pumps causes fuel intended for delivery to the combustion chamber to be absorbed through displacement of the delivery valve.
- The upward movement of the needle valve in the nozzle causes the moving needle valve to displace some of the fuel intended for injection (**FIGURE 9-46**).

Symmetrical Fuel Distribution

Fuel must penetrate the combustion chamber air mass evenly in all directions to mix correctly with air. Without even distribution, the mixture will not form properly, resulting in areas of excessively rich and lean air-fuel ratios and increased emissions. To distribute fuel for optimal mixture preparation, the combustion chamber design

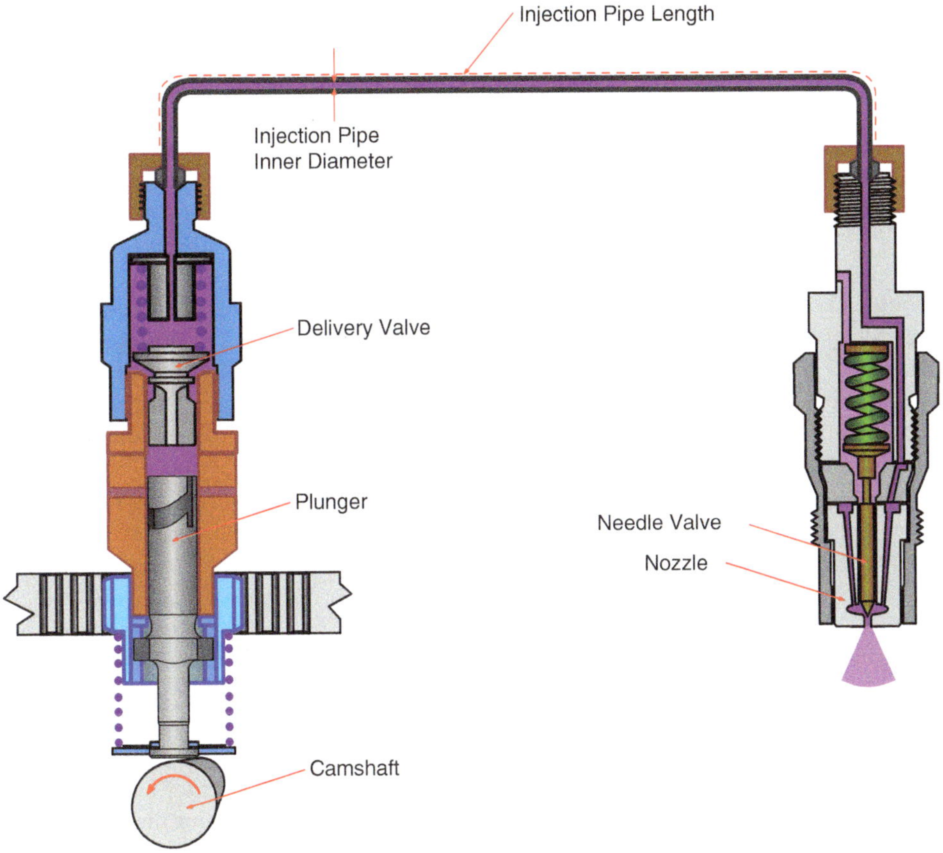

FIGURE 9-46 Swelling fuel lines, pressure wave conduction time, and lifting nozzle and delivery valves are a few of the sources of injection lag in hydromechanical injection systems.

and fuel system components must be correctly matched. Nozzle design, nozzle orifice sizing, combustion chamber turbulence, and specified fuel pressurization are some of the more significant factors that affect fuel distribution and mixture formation. Nozzles can also wear and deteriorate because of long hours of operation or distance traveled. Plugged nozzles cause uneven distribution patterns, resulting in performance and emission problems.

Wrap-Up

Ready for Review

- Diesel engine fuel systems have two main divisions: the low-pressure transfer system, which stores and supplies fuel for the injection system, and the high-pressure injection system, which delivers fuel to the combustion chambers.
- Efficient diesel engine operation relies on injecting a precise quantity of fuel, injected at high pressure, at the correct time, into the combustion chamber.
- High-pressure injection system operation is better understood by examining functions common to all injection systems.
- Ensuring that fuel is highly pressurized is necessary for several reasons: fuel must have sufficient pressure to penetrate the dense combustion chamber air mass; at high rotational speeds, only a short amount of time is available to inject fuel; and high pressure is necessary to break up the liquid fuel into a fine mist.
- All high-pressure injection systems perform the following tasks: meter the correct quantity of fuel for injection, time the injection of fuel, control the rate of fuel delivery, atomize the fuel, pressurize and distribute the fuel throughout the combustion chamber.
- Diesel engines meter only fuel, so that they can control engine speed and torque. Metering is the process of measuring and delivering the correct quantity of fuel required for each cylinder for various speed and load conditions of the engine.
- To vary vehicle speed and power output, pressure in the cylinders must increase or decrease by changing the amount of fuel injected and burned. Increasing combustion pressure inside the cylinders increases engine torque output.
- All diesel engines use governors to regulate the quantity of fuel injected into the cylinders; a diesel governor

equivalent in a gasoline or Otto cycle engine is the throttle plate controlled by a driver.

- Even very small changes to the quantities of fuel injected at idle produce rough-running engines and unstable engine speed. To minimize problems associated with uneven cylinder power contribution, most new electronically controlled engines use idle adaptation to make cylinder pressures consistent.
- Consistent combustion pressures from all cylinders are necessary to achieve low engine emissions; emission reduction strategies achieve better results when they are applied to all cylinders operating with identical pressure.
- To correct individual variations in flow rates, many newer injectors are assigned calibration codes to compensate for manufacturing tolerances to injector flow rates.
- For efficient combustion, clean emissions, and good engine performance, precisely timed injection events are crucial. Efficient fuel systems must have the capability to change the point when fuel is injected, based on a variety of engine operating conditions and control algorithms.
- The factors that influence injection timing include pre-ignition temperatures in the cylinder, fuel droplet size, combustion chamber turbulence, fuel quality, and more.
- To adjust for the variation in combustion time when an engine load or speed changes, the fuel system must be able to change injection timing. When an engine load increases, injection timing is retarded. Speeding up an engine with no change in load requires advancing injection timing.
- Injection timing is a function of both speed and load changes. Both are opposing factors with increasing loads requiring a retarding of timing and speeding-up advancing timing.
- Cold cylinder components and low ambient temperatures create exceptional operating conditions that require adjustments to injection timing.
- Diesel engine emission characteristics are significantly affected by changes to injection timing; for example, slightly advancing the start of injection may reduce smoke, increase power, and reduce fuel consumption, but it typically results in higher NO_x emissions.
- Injection timing is a part of the fuel system monitor for onboard diagnostic (OBD-II) engines; a variety of strategies are used to determine the precise beginning of injection to ensure it coincides with expected values.
- Injection rate control is the most significant advancement of contemporary injection technology because it affects fuel consumption, exhaust emissions performance, and combustion noise.
- The shape of the injection discharge curve can be optimized for fuel economy, low NO_x formation, and power. A unique injection rate shape is needed for each speed and load condition.
- Earlier, mechanically governed fuel systems used throttle pintle injector nozzles that generate a spray pattern that provided a gradually rising rate discharge curve.
- Pilot injection, also known as split-shot injection, is another method of rate shaping; it is one of the most effective strategies for minimizing NO_x and noise emissions while optimizing engine performance.
- An injection event after the main combustion event reduces the production of soot inside the cylinder by adding additional heat and pressure to the power stroke, which in turn consumes soot.
- To meet emissions standards while optimizing engine performance, electronically controlled, variable injection pressure independent of engine speed is used in all current on-highway engines.
- Atomization, or breaking fuel into smaller droplets, enhances the fuel's ability to absorb heat, convert to vapor, and mix with air.
- To achieve the lowest formation of particulate emissions, the finest atomization is required to ensure that each droplet makes adequate contact with air for faster and more complete combustion. To properly distribute fuel, the injection pressure must be consistent from beginning to end; fuel systems incorporate design features to ensure an abrupt beginning and ending of injection.
- Injection lag is the time delay that may occur between the start of fuel pressurization and the moment when injection actually occurs; this influences the ability of the fuel system to achieve an abrupt beginning and ending of injection.
- To distribute fuel for optimal mixture preparation, both the combustion chamber design and fuel system components must be correctly matched.

Key Terms

algorithm A mathematical formula used to solve a problem.

atomization the process of breaking up liquid fuel into a fine mist.

compensation A response by the fuel system governor to a change of power contribution from one or more cylinders. The governor increases or decreases the quantity of fuel injected into the other cylinders to maintain idle speed.

delivery valve Valves that operate as one-way checks and allow fuel to remain at moderately high pressure in the fuel-injection lines. The valves prevent secondary injections and help reduce injection lag.

distribution The mixing of fuel and air in the cylinders during the injection event.

electronic control module (ECM) A device containing a microprocessor that processes electronic input signals according to software-based mathematical algorithms. The ECM then produces output signals based on processed information to control engine operation.

fuel rate Fuel consumption expressed as the quantity of fuel consumed per hour.

governor A device that regulates the quantity of fuel injected into the cylinders.

hunting A rhythmic change in engine rpm at idle speed caused by uneven delivery of fuel; it is also known as loping.

injection lag The time delay that occurs between the start of fuel pressurization and the moment when injection actually occurs.

injection quantity adjustment (IQA) A fuel system feature that allows software to compensate for pressure variations in cylinders; it is also known as the E-trim, adaptation value, or injector calibration code.

injection rate The quantity of fuel injected per degree of crank angle rotation. It is reported as the position of the crankshaft when the injection begins.

metering A process of measuring the correct quantity of fuel for each cylinder required for various speed and load conditions demanded of the engine.

pressurization Compressing fuel to the degree required for an injection event.

pumping loss The energy required to pull air past a throttle plate or to move gases in and out of the cylinder.

rate-shaped injection An injection strategy that carefully regulates the amount of fuel injected into a cylinder per degree of crank angle rotation.

timing In the high-pressure injection system, the beginning of the injection event in the combustion cycle relative to crank rotation.

Review Questions

1. Which of the following statements best describes ignition delay in a diesel engine?
   a. The time required to raise fuel pressure high enough to overcome compression pressure in a cylinder.
   b. The time lag from initial injection to actual ignition.
   c. The time required to heat up a glow plug.
   d. The time required for vaporized fuel to atomize.
2. Which of the following is the most significant benefit of pilot injection?
   a. Better fuel economy.
   b. Lower combustion chamber noise.
   c. Longer ignition delay.
   d. Precise injection timing.
3. The term "retarded injection timing" refers to injection timing occurring ________.
   a. later than normal
   b. earlier than normal
   c. at top dead center (TDC)
   d. at bottom dead center (BDC)
4. How does pilot injection, also known as split-shot injection, reduce exhaust emissions?
   a. Main injection timing can be advanced.
   b. Combustion time enables a more complete burn of poorly atomized fuel.
   c. Cylinder temperatures increase near top dead center (TDC).
   d. Main injection timing can be retarded.
5. Consider a diesel engine that has an increased load applied to it while its speed remains constant. Which of the following changes to injection timing would be expected?
   a. An advance in injection timing.
   b. A delay in injection timing.
   c. No change to injection timing.
   d. Retarded injection timing.
6. In which of the following ways does the rate at which combustion occurs change when additional fuel is added to a cylinder?
   a. Combustion time is shortened.
   b. Combustion time is lengthened.
   c. Combustion time remains the same.
   d. Combustion time is advanced.
7. Which engine exhaust emission can be significantly reduced by using pilot injection (split-shot injection)?
   a. Particulate matter (PM).
   b. NO_x.
   c. Carbon monoxide (CO).
   d. Carbon dioxide (CO_2).
8. Consider an engine that is satisfactorily operating on all cylinders. If one cylinder is cut out for testing, what will happen to the pulse width (electrical signal time) of the remaining injectors?
   a. The pulse width will shorten to maintain engine speed.
   b. The pulse width will lengthen to maintain engine speed.
   c. The pulse width will shorten to compensate for the missing injector.
   d. The pulse width will lengthen to compensate for overfueling the injectors.
9. What injection strategy is best used to start, but not run, a cold engine?
   a. Retarding injection timing.
   b. Advancing injection timing.
   c. Using a post-combustion injection.
   d. Eliminating the pilot injection.
10. A slow buildup of the fuel-injection rate (ramp-shaped injection rate) for the main injection reduces which of the following emissions?
    a. Oxides of nitrogen (NO_x).
    b. Particulate matter (PM).
    c. Carbon monoxide (CO).
    d. Hydrocarbons (HCs).

ASE Technician A/Technician B–Style Questions

1. Technician A says a long injector response time from a magnetic solenoid is due mostly to the speed of the electric current through long wire windings. Technician B says a long injector response time is caused primarily by heavy, high-inertia control valves. Who is correct?
   a. Technician A
   b. Technician B
   c. Both A and B
   d. Neither A nor B

2. Technician A says a slow buildup of the injection rate, which provides a ramp-shaped injection discharge rate shape for the main injection event and reduces CO emissions. Technician B says the ramp-shaped injection rate shape reduces NO_x emissions. Who is correct?
   a. Technician A
   b. Technician B
   c. Both A and B
   d. Neither A nor B
3. While discussing the steps to perform when changing an injector in a late-model engine, Technician A says that uploading a fuel injector calibration code helps the engine to run more smoothly. Technician B says that replacing the old code with the one on the new injector enables the ECM to compensate for small variations to injection quantity unique to each injector. Who is correct?
   a. Technician A
   b. Technician B
   c. Both A and B
   d. Neither A nor B
4. While checking contributions from each cylinder for a late-model electronically controlled engine, Technician A says that the purpose of adaptive cylinder balance is to reduce combustion noise so that diesels run more quietly. Technician B says adaptive balance reduces emissions and helps an engine run more smoothly. Who is correct?
   a. Technician A
   b. Technician B
   c. Both A and B
   d. Neither A nor B
5. Technician A says that when a diesel engine has an increased load and its speed remains constant, injection timing must advance. Technician B says that under those conditions, injection timing should retard. Who is correct?
   a. Technician A
   b. Technician B
   c. Both A and B
   d. Neither A nor B
6. While discussing possible causes of black smoke from the exhaust of an older heavy-duty diesel engine, Technician A says that black smoke from an engine under acceleration indicates the engine is producing plenty of power. Technician B says that smoke could indicate that the injector nozzle spray holes are worn out. Who is correct?
   a. Technician A
   b. Technician B
   c. Both A and B
   d. Neither A nor B
7. After assembling an engine and discovering one of the vibration-insulating high-pressure fuel line clamps was missing, two technicians discussed potential consequences of the clamp remaining missing from the injector line. Technician A says that the line will crack due to vibration caused by pressure pulses. Technician B says that a single missing clamp on thick-walled steel lines will have no effect on the line. Who is correct?
   a. Technician A
   b. Technician B
   c. Both A and B
   d. Neither A nor B
8. Technician A says that multiple injection events from an injector to reduce emissions results in less torque produced from an engine. Technician B says that multiple injection events during a combustion cycle are used to reduce heat loads in a cylinder. Who is correct?
   a. Technician A
   b. Technician B
   c. Both A and B
   d. Neither A nor B
9. While discussing potential ways to increase power output from an engine, Technician A says that advancing injection timing a few degrees will increase power output. Technician B says that advancing injection timing a few degrees will reduce fuel consumption. Who is correct?
   a. Technician A
   b. Technician B
   c. Both A and B
   d. Neither A nor B
10. Technician A says that cutting out one cylinder at idle to check for a misfire condition will reduce idle speed by close to 100 rpm in a six-cylinder engine idling at 600 rpm if the cylinder is operating properly. Technician B says it is not possible to detect a misfiring cylinder at idle by noting a change in idle speed during cylinder cutout testing for any cylinder. Who is correct?
    a. Technician A
    b. Technician B
    c. Both A and B
    d. Neither A nor B

CHAPTER 10

Governors and Electronic Control Systems

Learning Objectives

After reading this chapter, you will be able to:

- **10-01** Identify and explain the purpose and functions of a diesel engine governor.
- **10-02** Explain the operating principles and control strategies of electronic governors.
- **10-03** Outline the procedures used to evaluate the operation of a governor.
- **10-04** Identity and explain the purpose and functions of electronic engine controls.
- **10-05** Identify and explain the differences between Proportional Integrative Derivative (PID) and Model Predictive Control (MPC).

You Are the Technician

An automobile with a light-duty diesel has repeatedly returned to your shop because of a low-engine-power problem and because it intermittently shuts off. Other technicians have spent many hours checking the engine for possible causes of low power, such as restrictions in the low-pressure supply system, plugged air filters, damaged turbochargers, exhaust gas recirculation (EGR) system defects, and incorrect valve and injector settings. No fault codes or other identifiable problems were found. It was suggested that the vehicle's electronic control unit (ECU) may be defective, so the ECU was removed and internally inspected. After opening the ECU, water damage and some burned components were identified, which justified replacing the ECU. To verify the complaint, and verification is needed because of the considerable cost of a new ECU, a used ECU from another, identical vehicle was installed. However, after installing the ECU, the vehicle would not start or even crank. As you begin to investigate further, consider the following:

1. What symptoms would likely indicate that the old ECU is defective?
2. What are some common procedures that should be used to properly diagnose a defective ECU?
3. Would the ECU software programming, software calibration number and verification number likely match the old ECU if the vehicles were in fact identical?
4. What vehicle security features might have prevented the engine from starting?

▶ Introduction

Diesel engines, unlike gasoline-fueled engines, do not use a throttle plate to regulate power output. Instead, engine speed and torque are controlled by the quantity of fuel injected into the cylinders by the high-pressure fuel system (**FIGURE 10-1**). Monitoring the engine speed and adjusting the fuel quantity required by the engine for a given speed and load condition are the primary jobs of the diesel engine governor. The importance of these tasks makes the governor one of the most crucial systems for providing responsive engine operation, good performance, and emission control. Governors are an electronic control subsystem integrated into the engine's electronic control module (ECM) and operate according to programmed instructions in engine calibration software (**FIGURE 10-2**). Integrating all the vehicle's electronic control systems requires far more enhanced governor functions in today's vehicles though. This integrated system approach is so sophisticated that the term "governor" has fallen into disuse. However, the governor control function is still there even though other engine controls operate invisibly by using electronic signal–processing strategies (**FIGURE 10-3**). Information provided in this chapter will enable diesel technicians and operators to properly understand the basics of governor operation in order to develop diagnostic strategies and reprogram electronic governors.

FIGURE 10-1 Governors directly control changes to engine fueling. The operator only indirectly controls fueling through one of several inputs to the governor. Changing quantities of injected fuel produces speed and torque changes.

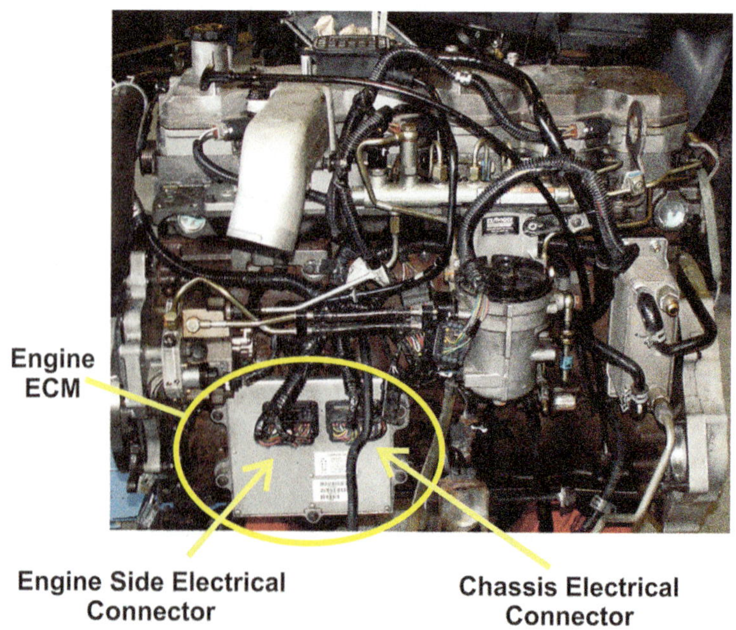

FIGURE 10-2 Governors are an engine control subsystem integrated into the engine ECM and operate according to programmed instructions in engine calibration software.

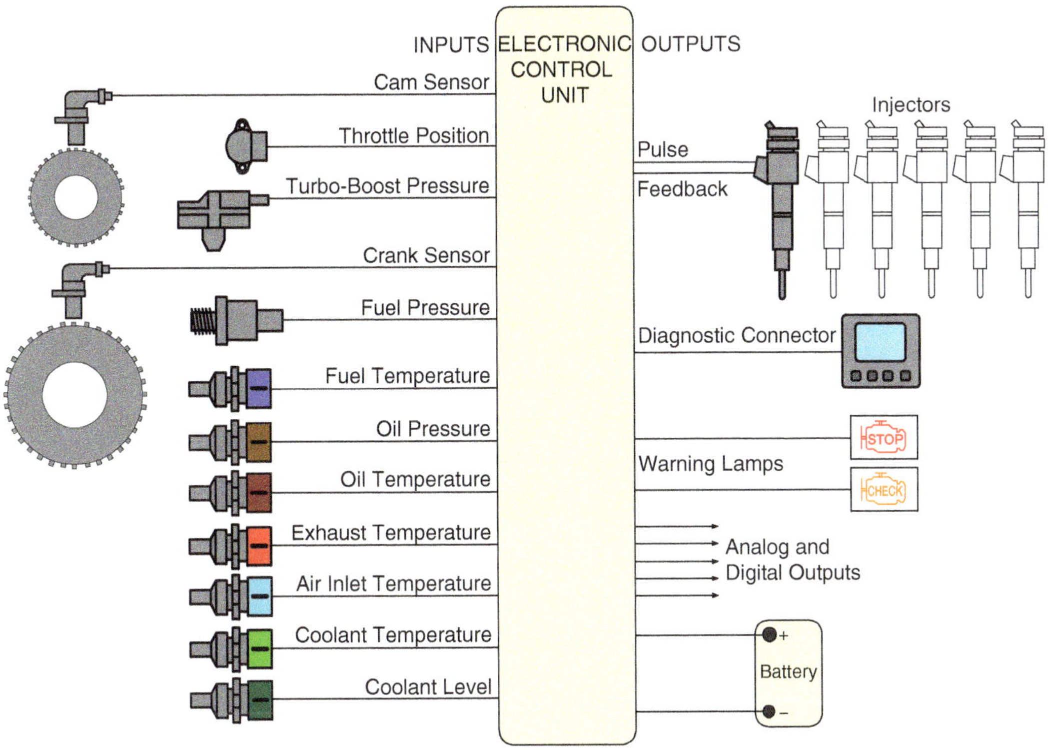

FIGURE 10-3 While the high-pressure injection system meters the fuel supplied to the combustion chamber, the governor controls the quantity of fuel injected based on a variety of inputs and software instructions.

▶ Fundamentals of Diesel Governors

10-01 Identify and explain the purpose and functions of a diesel engine governor.

The governor is one of the most crucial mechanical or electronic control subsystems needed by a diesel engine. Without the governors' ability to regulate fuel delivery for varying speed and load conditions, engine operation would at best be erratic and at worst result in rapid engine destruction. While technicians typically do not often encounter governors that need service or repair like other engine systems, its influence on engine operation cannot be overstated. The governor is essential to the functioning of the fuel system. Technicians can be expected to make adjustments to governors or change electronic control parameters, so technicians must understand how a governor functions and know all relevant governor terminology.

Governor Functions

Because governors directly control fuel system operation, it's important to understand the principles of governor operation and various terms associated with governor functions. Comprehending governor terms and principles is helpful to technicians when they analyze engine performance or emission-related problems. Problems do arise that can be corrected with an adjustment or recalibration, so understanding governor operating principles will help build effective diagnostic and service skills. The terminology and functions described in the next section apply to both electronic and mechanical governors.

Adjusting Fueling to Engine Load and Operator Demand

There are several reasons unique to diesel engine operation that make the use of governors necessary. First, on mechanical or electronically controlled high-pressure injection systems, very small changes in injection quantity produce very large changes in engine speed. And unlike gasoline engines with throttle plates, no single position of the fueling mechanism, such as a control rack inside an injection pump or the length of an electrical signal applied to an injector, is acceptable for all of the specific speed and load conditions. If only throttle linkage were connected in a way that directly controls fuel delivery, such as a control rack inside a high-pressure pump, the slightest movements of the throttle pedal would produce

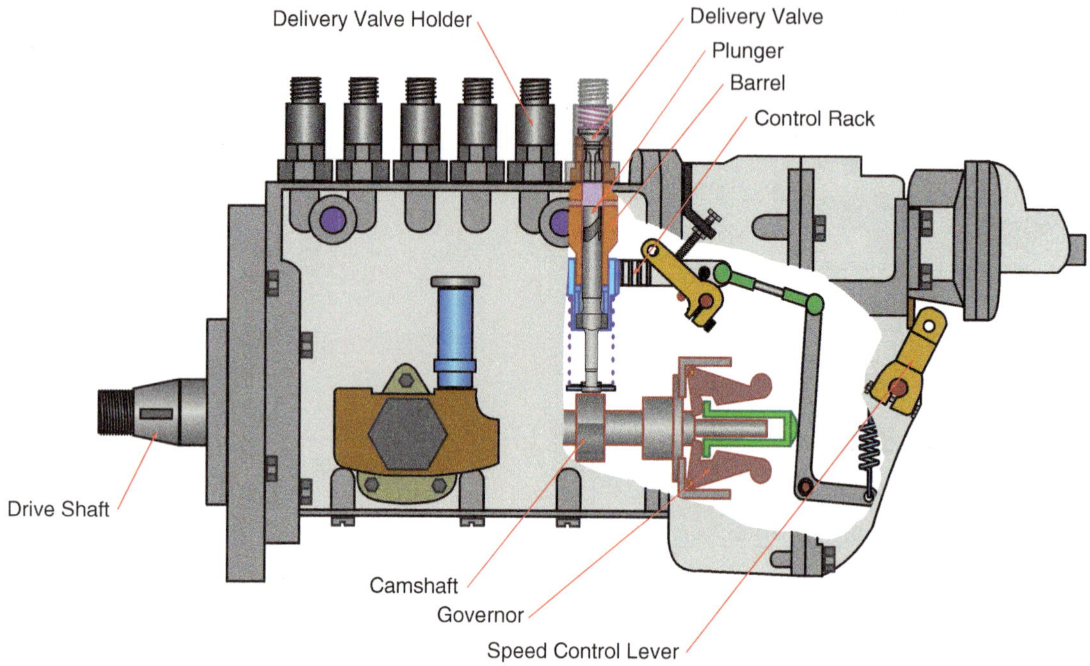

FIGURE 10-4 The control rack in a mechanical injection pump connects the governor mechanism to the high-pressure section of the pump. Moving the control rack backward or forward changes the quantity of fuel injected into the cylinders.

excessively large changes in engine speed and power output (**FIGURE 10-4**). It is impractical to have the vehicle operator continually make the type of fine adjustments required by the throttle pedal to maintain engine speed and power output as load and speed vary.

The same holds true for electronic governors. The slightest change to engine loads or speeds requires an adjustment to engine fueling; otherwise, the engine will either stall or accelerate rapidly and wildly. To better understand this, consider a cold engine that has just started. Changes to engine friction take place continuously because lubricating oil thins or because the load changes when accessories such as the alternator, cooling fan, or air conditioning compressor cycle on and off. If there were a direct mechanical linkage to the injection pump from the operator's pedal that controlled engine speed, the operator would be required to make too many small, incremental adjustments to maintain a constant, stable idle speed or any engine speed. For example, **Idle** speed is an engine's minimum operational speed. Also remember that at idle, only a very small amount of fuel is injected, typically 2–8 mm^3. A droplet of fuel is approximately 25 mm^3. Increasing the engine speed to its maximum from idle only requires less than 1 mm^3 of additional fuel for each combustion cycle in each cylinder. Controlling changes to such small injection quantities by using a direct mechanical connection to the throttle pedal is simply not possible.

Controlling Engine Acceleration

A second important reason for the use of governors in diesel engines is that diesel engines possess a naturally high acceleration rate. Because the air intake is unrestricted and because the fuel injection rate can go to maximum delivery quantity almost instantly, the rate of acceleration of a diesel engine can be as high as 2,000 rpm/sec. That means that for each second of uncontrolled fueling, the engine speed will increase 2,000 rpm. If this happens, it can easily cause the engine to go beyond critical rpm in a short time. To prevent this, governors have speed-sensing and fuel-regulating capabilities to minimize the likelihood of a runaway engine.

Limiting maximum engine speed and acceleration rate is also critical to emission control. A diesel will run out of time to complete combustion events when operating at excessive speed. As engine speed increases, the time available for diesel combustion events is shorter. More exhaust smoke is produced at high engine speeds because the engine runs out of time to properly complete combustion. (**FIGURE 10-5**).

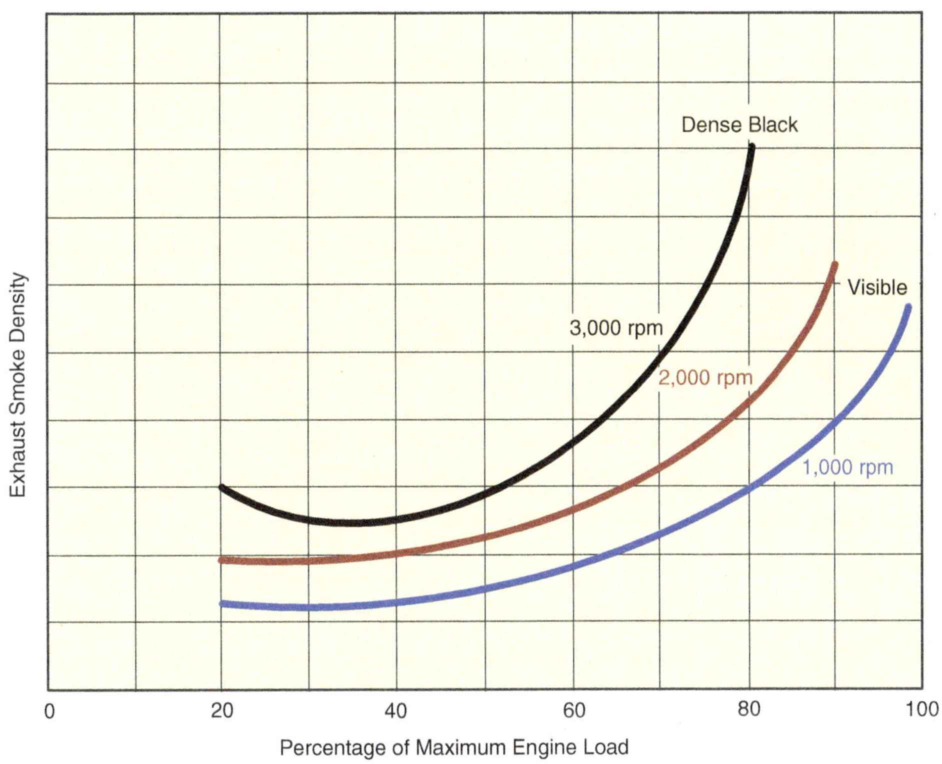

FIGURE 10-5 Limiting maximum and minimum engine speed is a function of all diesel governors. This diagram illustrates the situation in which increasing the engine speed increases exhaust smoke density even when engine load factors are low.

▶ TECHNICIAN TIP

Governors play an important role in minimizing the formation of PM emissions (black smoke). The quantity of fuel injected to control engine speed and torque is regulated by the governor. Exceeding the rated speed of an engine will produce smoke emissions from incompletely burned fuel, because the time available for combustion events becomes too short at high speeds. Similarly, air-fuel ratios are also controlled by the governor to ensure that an adequate mass of air is available to properly combust injected fuel. Air and fuel quantities are carefully calculated by the manufacturer. Tampering with governor controls to increase the quantity of fuel for a given speed or load condition will increase torque output, but it will likely produce excessive emissions.

Maximum engine speed is limited by all diesel engine governors, and the rate of acceleration is carefully regulated by a gradual increase in fueling. Emission decals on diesel engines indicate maximum engine speed or rated rpm. Exceeding the rated rpm on a diesel can produce excessive particulate matter (PM) emissions, because at higher rpm, the combustion process runs out of time to completely burn fuel (**FIGURE 10-6**). Because diesel engines must slow the acceleration rate to minimize exhaust smoke emissions, some engines may seem "lazy" during acceleration.

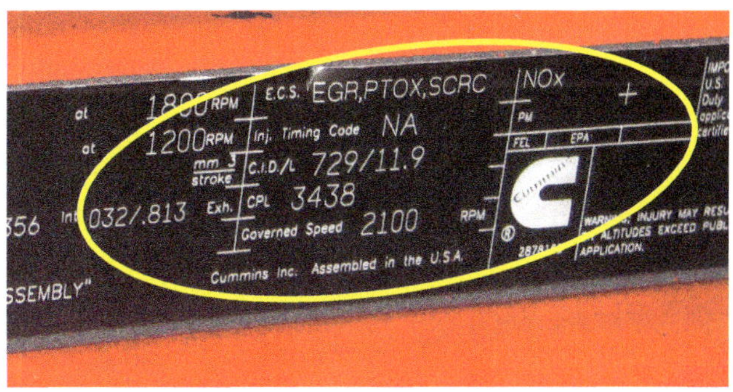

FIGURE 10-6 An emission decal on a diesel engine indicates maximum engine speed or rated rpm.

Managing Air-Fuel Ratio Controls

Governors perform a critical function in emission reduction. Incorporated into governor operation are emission-control mechanisms that regulate the engine's air-fuel ratio. Air-fuel ratio controls minimize the production of PM emissions (black smoke) produced during engine acceleration (**FIGURE 10-7** and **FIGURE 10-8**). Smoke occurs because airflow can lag behind the injection of pressurized fuel. To prevent smoke, the injection quantity of fuel cannot exceed the amount of air mass filling the combustion chamber that is needed to properly burn the fuel. A governor operates to limit the quantity of fuel injected if there is not enough air to support complete combustion. Diesel engines are a lean-burn combustion system, which means excess air is available for combustion. Air-fuel ratios will not fall below 300:1 in a direct-injection combustion chamber while under load. Air-fuel ratios of less than 100:1 will produce smoke.

At high altitudes, fuel systems may need to compensate for the loss of air density and pressure. Barometric compensation can be built into mechanical governors on engines operating at these higher altitudes. On turbocharged engines, the barometric sensor may also reduce fueling to prevent turbochargers from overspeeding, which occurs at high altitudes. When turbochargers normalize intake boost pressure (i.e., maintaining intake

FIGURE 10-7 The torque plate limits maximum air-fuel ratio control movement. Steps on the plate set maximum distance for control rack travel.

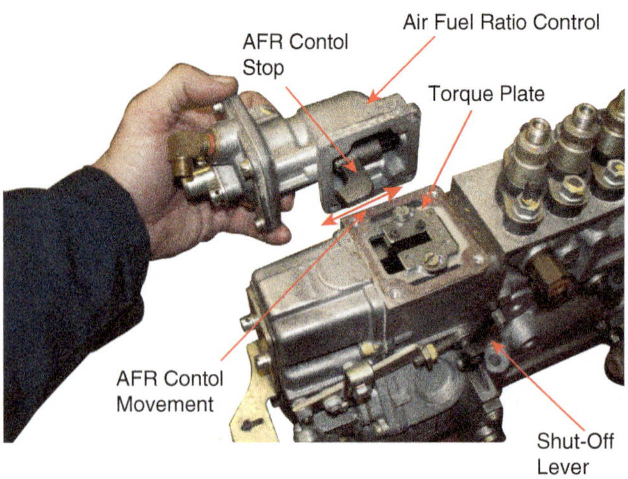

FIGURE 10-8 The air-fuel ratio control device limits maximum rack travel proportional to intake boost pressure.

manifold boost pressure)—a concept explained in Chapter 15, on fixed geometry and wastegated turbochargers—it can lead to turbocharger overspeed at high altitudes.

Other Governor Operations

Governors perform other jobs than the basic functions of measuring engine speed, sensing operator demand, and regulating the quantity of fuel injected into engine cylinders. At a minimum, *all governors* will do the following:

- limit maximum engine speed
- maintain idle speed to prevent stalling.

Depending on the type of governor, application, and sophistication of the control, governors may also do the following:

- control engine speed between idle and top engine speed limit or **rated speed**
- vary fuel-injection quantity to increase or decrease torque output when the load applied to the engine changes
- decrease fueling when the engine speed increases beyond peak torque
- have the ability to shut down the engine
- provide sufficient fuel for starting
- minimize PM emissions by regulating the delivery of fuel to match airflow or boost pressure to engine cylinders
- provide barometric compensation for fuel-injection rate
- dampen engine surge when engine loads rapidly change.

Electronic governors can provide additional features, which include but are not limited to the following:

- fast idle speed
- shutdown timers
- cruise control
- vehicle speed limitation
- **programmable parameters**, which are measurable values about engine system operation that can be changed by using service software. Examples of common programmable parameters include idle speed, maximum engine and road speed, cruise control limits, engine protection limits, features that encourage efficient transmission shifting (such as progressive shifting and **gear-down protection** for manual transmissions), anti-tampering, load-based speed control, torque limiting, cold-weather protection, and adaptive terrain fueling (smart cruise)
- remote throttle control for power takeoff devices, such as rear engine power takeoff (REPTO) or a power takeoff (PTO)
- surge or bucking dampeners, which provide smoother vehicle deceleration
- integration with all other vehicle ECUs through the controlled area network (CAN) to communicate with the antilock brake system (ABS), transmission, stability control, climate control, injection control modules, and body electrical control systems.

▶ TECHNICIAN TIP

When learning to understand engine power measurements, the difference between engine torque and horsepower can be confusing. Horsepower is a function of torque and engine speed, whereas torque is only the product of cylinder pressure applied against the crankshaft throw. Stated another way, torque is a twisting force exerted by the crankshaft of an engine.

Governor Terminology

To understand the principles of diesel engine governing, it is essential to recognize a number of basic terms associated with governor operation and control.

Torque

Ultimately, torque is the force that moves the vehicle or equipment load. It is the result of cylinder pressure applied against the crankshaft throw. Torque is a twisting force exerted by the crankshaft of an engine. Expressed mathematically in imperial units, torque = force (pounds) × length of lever (feet), expressed in pounds per foot (or foot-pounds [ft-lb], as it is more commonly called). The imperial measuring unit for torque is the ft-lb; the SI (International System) measuring unit is the newton meter (N·m) (torque = force [newtons] × length of lever [meters], expressed in N·m) (**FIGURE 10-9**). No motion is necessary to produce torque. Pushing on a one-foot-long wrench with 100 pounds of force will produce 100 ft-lb of torque, but if the bolt or nut is seized, no motion takes place. Torque is developed by the engine's crankshaft and is a function of the length of the crankshaft throw and the force or pressure inside the cylinder. A longer throw or stroke and/or greater cylinder pressure will produce increasing amounts of torque from the engine.

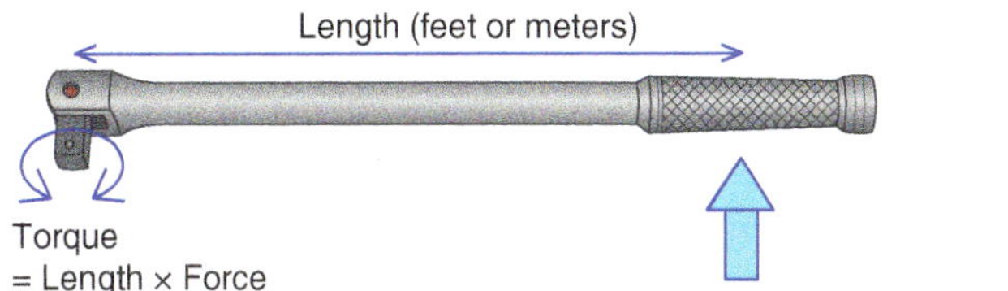

FIGURE 10-9 The twisting force of a crankshaft is like torque transmitted by a wrench. Combustion pressure and throw length create torque.

Diesel engines produce more torque compared to an equivalently displaced gasoline-fueled engine because diesel engines' higher compression ratios produce greater cylinder pressure (force) and longer crankshaft throws (lever length) due to longer stroke length.

An engine produces virtually no torque at idle or high idle. Only enough cylinder pressure is produced to overcome engine friction and inertia to keep a crankshaft turning at idle rpm. When crankshaft or flywheel movement is resisted by applying a load to the engine, the fuel system is designed to produce more torque. **Torque rise** will occur in most engines when the engine is loaded while travelling at highway speeds. When climbing a hill or encountering a steeper road grade, torque rise helps maintain road speed. Torque rise is the difference between peak torque and torque at rated speed. An increase in torque is the result of injecting greater quantities of fuel into the engine, which increases cylinder pressure.

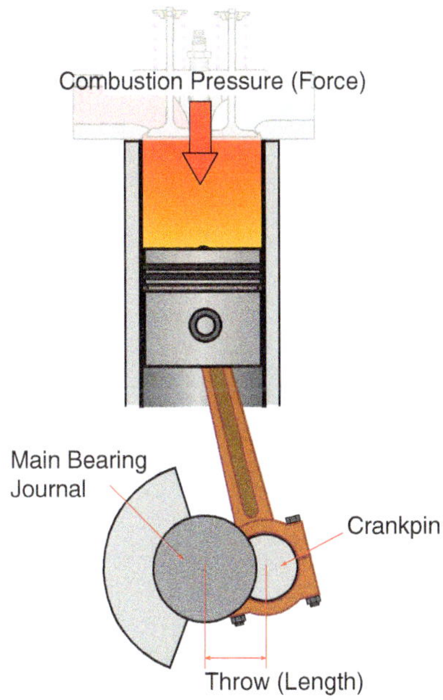

FIGURE 10-10 Engine torque is a function of cylinder pressure multiplied by the length of the throw.

Horsepower

Horsepower is a function of torque and engine speed. It is calculated by multiplying torque and engine rpm and then dividing by a mathematical constant that represents one horsepower. That means horsepower has a unit of time in its measurement. For example, two engines can produce an identical amount of torque, but if one produces it at a faster speed, the horsepower increases.

High-Idle Speed

Also known as maximum no-load speed, **high-idle** speed is the highest rpm at which the governor permits the engine to operate with no load. This occurs when the throttle is at maximum fuel position, also known as wide-open throttle (WOT), and the vehicle is in neutral. To improve engine efficiency, it is important to note that on many engines, a programmable feature limits engine speed when no vehicle speed is detected. This is a different features from engine fault de-rating when engine faults will also limit maximum engine speed to encourage a driver to have emission-related faults repaired.

Low-Idle Speed

Low-idle speed is the rpm at which the engine operates when the throttle pedal is released. Idle speed is regulated by the governor, which operates to maintain a constant stable idle speed even as engine loads change. Stable, consistent idle speed uses the governor feature referred to as compensation—which is discussed in Chapter 9, on the functions of high-pressure fuel systems. Compensation means that the loss or gain of power from one or more cylinders will cause the governor to increase or decrease fueling automatically to maintain idle speed.

Rated Speed

Rated speed is the rpm at which the most horsepower (kilowatts) is produced from a diesel engine. Rated speed coincides with the maximum horsepower that the engine will produce while under load. Rated speed is required on the emissions decal because exceeding rated speed will produce excessive emissions. The injection quantity to produce rated speed is also usually indicated on the emission decal to help technicians make governor make or check adjustments and diagnose reports of low power.

Speed Droop

Simply explained, upper speed **droop** is the change or difference in engine speed caused by an increase in load near rated speed. Upper speed droop occurs as engine fueling decreases

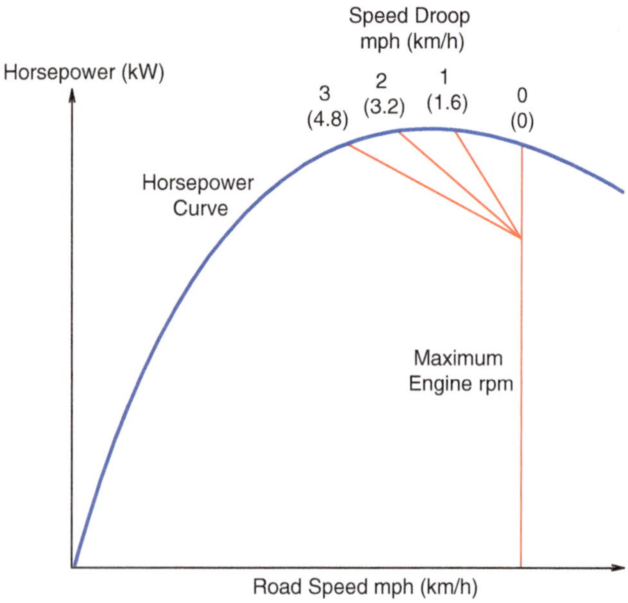

FIGURE 10-11 Upper speed droop occurs as engine fueling decreases when the engine approaches its maximum speed. As the percentage of droop increases, fuel economy improves and emissions may drop, but engine power is reduced.

when the engine approaches its rated speed. Trying to reach rated speed under heavy load can consume enormous amounts of fuel, so it's better to allow the engine speed to drop than it is to increase the quantity of fuel delivered. In mechanical governors, upper speed droop occurs because greater spring force inside the governor is required to move a control actuator, such as a control rack, to the full fuel position needed to operate an engine at rated speed. In electronically controlled engines, speed droop is a programmable feature that may or may not be accessible to a technician with service software. As the percentage of droop increases, fuel economy improves and emissions may drop; however, engine power is reduced (**FIGURE 10-11**). While it may be possible to eliminate upper droop it can lead to increased emissions and poor fuel economy.

Droop is usually expressed as the difference in engine speed measured at high-idle rpm compared to engine rpm after a load is placed on the engine. This means that if an engine has a high-idle speed of 3,200 rpm and the speed drops by 250 rpm after it is loaded, the change from 3,200 to 2,950 rpm would mean it has close to 8% speed droop.

Droop is an operating factor that has a significant impact on performance and fuel economy. Consider that during normal operation, as a vehicle climbs an uphill grade, the injection systems should provide progressively increasing fueling to the engine in order to maintain vehicle speed. Continued uphill climbing, with WOT, could eventually result in the maximum fueling rate being attained with little effect on vehicle speed. Fueling without a significant speed increase will result in wasted fuel and potentially excessive emissions. Allowing the engine speed to taper off, or droop, in this speed/load range will result in improved fuel economy. Any loss in power or performance is often disproportional to the improvement in fuel economy. Vehicle speed can be maintained, but with a penalty to fuel economy, by using a governor with little or no droop. Vehicles equipped with governors that have a high percentage of droop (i.e., more than 5%) are perceived as having poor performance ("it won't pull" or "it has low power"), but they achieve better fuel efficiency. The tradeoff from increasing the percentage of droop or allowable drop in vehicle speed is the driver's perception that the vehicle has poor power output and cannot maintain speed when loaded. (**TABLE 10-1**).

Adjusting droop is possible on a medium-duty electronically controlled engine by using original equipment manufacturer (OEM) software. However, for emission-certification reasons, these parameters are not typically made on chassis-certified light-duty engines (**FIGURE 10-12**). However, technicians should be aware of the influence of droop between diesel engines that appear to have the identical torque and horsepower (kilowatt) ratings. It is useful to understand droop when explaining some reports of low power to customers who are hauling trailers and other heavy loads. Technicians should be aware that increasing the amount of maximum upper speed droop will reduce the time the engine spends at

TABLE 10-1 Adjusting Droop Settings

Setting	Upper Droop Performance vs. Fuel Economy	Lower Droop Performance vs. Speed Control
0 mph (0 km/h)	Best Performance No Fuel Economy Benefit	Best Speed Control No Performance Benefit
1 mph (1.6 km/h)	Good Performance Moderate Fuel Economy Benefit	Good Speed Control Slight Performance Benefit
2 mph (3.2 km/h)	Decreased Performance Moderate Fuel Economy Benefit	Fair Speed Control Moderate Performance Benefit
3 mph (4.8 km/h)	Reduced Performance Best Fuel Economy Benefit	Reduced Speed Control Best Performance Benefit

maximum fueling, improving fuel economy. Decreasing the amount of upper speed droop will allow an engine to maintain its speed, but it operates closer to its maximum fueling position. Performance, torque, and road speed are increased or enhanced with less governor upper speed droop.

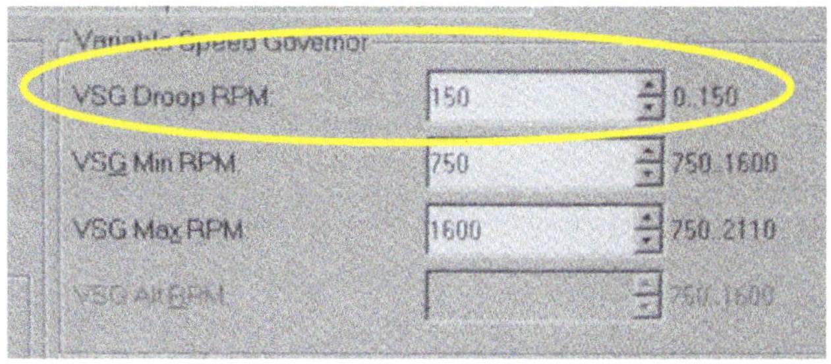

FIGURE 10-12 An OEM software menu showing speed droop and minimum idle speed can be electronically changed by a technician.

Droop Curve

Droop curve is the slope of the speed decrease near rated engine speed. Expressed as a percentage of high-idle speed, the droop curve is the difference between an engine's high-idle speed minus maximum engine speed without a load, and rated speed engine minus maximum speed with load. A 100 rpm difference in an engine with a high-idle speed of 2,000 rpm is a 5% droop. The droop curve is controlled electronically in today's engines. Governor spring tension controls droop in mechanical engines.

Governor Cutoff

Governor cutoff is the speed at which the governor cuts off fueling, meaning no additional fuel is supplied for injection. Cutoff occurs at rated or high-idle engine speed. Cutoff also happens when the throttle is backed off while the vehicle decelerates and no injection takes place.

Hunting

Hunting refers to the rhythmic change in engine idle speed, often caused by unbalanced fuel delivery in one or more engine cylinders. Engines using highly sensitive governors are prone to hunting. Both electronically and mechanically governed engines can hunt under a variety of operating conditions. Various fuel and engine system faults can cause hunting. If an electronic engine cannot adaptively balance cylinder contributions due to some sensor failure, it will often hunt.

Overrun

Overrun is the inability of a governor to keep the engine speed below the high-idle speed when it is rapidly accelerated.

Overspeed

Overspeed is any speed above high idle.

Peak Torque

Peak torque is the engine speed at which maximum cylinder pressure is produced. The highest cylinder pressures are produced during peak torque because the fuel-injection rate is greatest. For on-highway diesel engines, peak torque is commonly found at two-thirds of the rated speed (i.e., 2,000 rpm for a 3,000 rpm rated speed engine or 1,200 rpm for an 1,800 rpm rated speed engine). Peak torque is the point in the **performance curve** (a line graph that plots engine torque, fuel consumption, and horsepower [kilowatts] against engine speed) where the engine pulls best.

Torque Rise

Torque rise is the increase in engine torque as engine speed is decreased from rated speed to the engine speed where peak torque occurs. When an engine is operated at a speed above peak torque, the drop in engine speed results in increased torque or torque rise as the engine speed slows while encountering a hill or moving a heavier load. When measured, torque rise is the difference between torque at rated engine speed and the engine speed where torque output is highest (**FIGURE 10-13**). Because it is the slope of an engine's power curve that is measured, torque rise is expressed as a percentage. Cylinder pressure is highest at peak torque because the

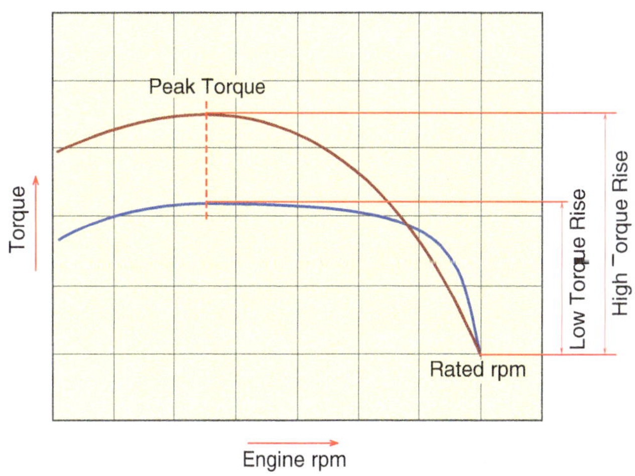

FIGURE 10-13 Torque rise is the difference between peak torque and torque at rated speed.

maximum amount of fuel is injected, which is supported by high turbocharger boost pressures. Highest turbocharger boost pressures, however, are matched with peak horsepower (kilowatts). A high percentage of torque rise means a greater amount of torque is generated as vehicle speed drops while under load, such as when climbing a hill. In other words, torque rise is a measure of an engine's lugging ability to increase pulling power under load as engine rpm drops. A distinction should be made between lugging ability and lugging. Lugging ability is an imprecise, but common, phrase used to describe load-pulling abilities. As it is used here, **lugging** is a technical term that describes an abuse condition when an engine operates under heavy load below an acceptable operating speed range. Lugging an engine can damage engine bearings because high cylinder pressures are accompanied by low oil pressure that could allow metal-to-metal contact between engine bearings and the crankshaft.

Depending on its primary application, diesels used in pickups and automobiles will generally have a flat torque profile between peak torque and its command point—the point where peak horsepower (kilowatts) and torque intersect. There are many vehicle models that are exceptions to this generalization, but a flat torque curve provides adequate and consistent power over a wide engine operating speed range. Whether the vehicle is operted mostly at highway speeds on the highway or operating in stop-and-go traffic, the flat torque profile will provide predicable amounts of acceleration and load-carrying ability (**FIGURE 10-14** and **FIGURE 10-15**). Figure 10-14 and Figure 10-15 illustrate the performance curves for two identical engines. Figure 10-14 calls out the following:

1. peak torque: expressed in ft-lb or N·m, this is where the maximum cylinder pressures occur because the quantity of injected fuel and air are greatest.
2. peak torque rpm: the speed where the greatest torque is generated. Both the rpm and peak torque are listed together on the engine data plate.
3. rated speed: the engine rpm at which maximum engine power is developed when the engine is under load. This number is also referenced on the engine data plate.
4. maximum horsepower (kilowatt) rpm: also referred to as advertised horsepower and always measured at rated speed.

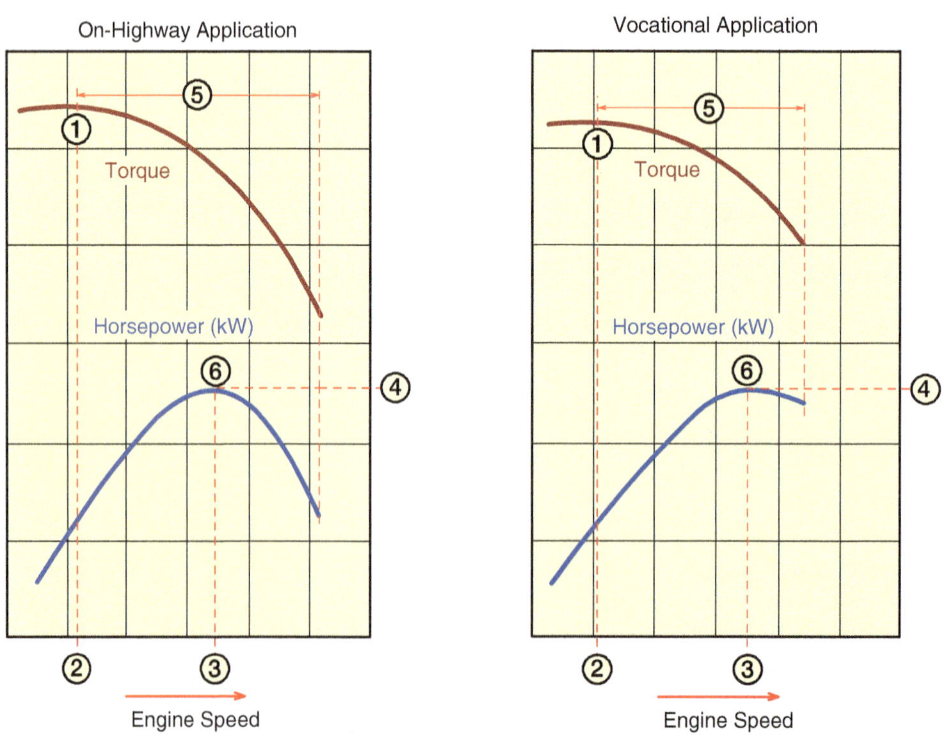

FIGURE 10-14 Performance curves for two diesel engines with identical horsepower. The left curve is best for vehicles operating mostly at highway speeds. The right is a flatter curve for best for vehicles operating in urban stop and go driving condition such as vocational vehicles. See the text for specific data points.

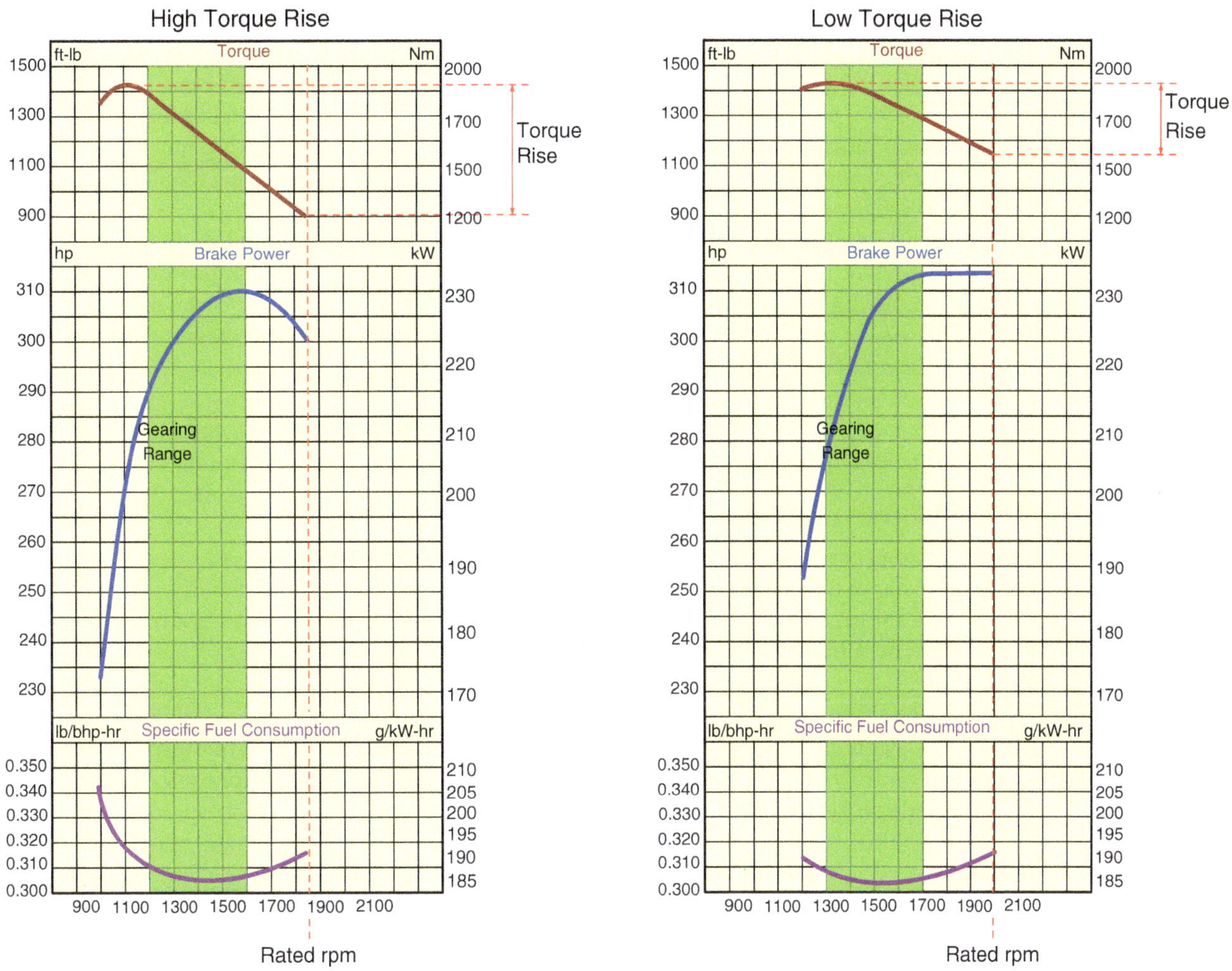

FIGURE 10-15 Performance curves and fuel consumption for two identical 300 hp (224 kW) diesel engines. The left curve is best for driving mostly on the highway at cruising speeds of 55–65 mph (90–100 kph). The right is a flatter curve is best for vocational uses, such as pickups and delivery vehicles operating in urban stop-and-go traffic conditions.

5. torque rise: the change in the engine torque from peak torque to rated speed. A small difference would be a low torque rise. A high torque rise engine develops more torque as it is loaded or lugged down from rated speed to peak torque.
6. command point: the point where maximum fuel economy and horsepower (kilowatts) coincide. Rear-axle ratios and tire sizes are calculated to coincide with the command point, which corresponds to the point where a vehicle is most often expected to be at cruising speed.

High torque rise governing allows the transmission to downshift less often when climbing a grade or when under heavy load. There is a perception that the engine "pulls better" because torque increases as the engine speed drops. This is important because downshifting, which is needed to increase torque, increases fuel consumption because it causes the engine to turn faster and slows the vehicle. High torque rise can actually help increase fuel economy. Staying in a direct drive or overdrive gear range for longer translates into lower fuel consumption. Because a vehicle's axle gear ratio and tire size are selected to enable the engine to cruise in an engine speed range for best fuel economy, which is above peak torque, a drop in speed while climbing a hill may cause engine speed to drop but torque to increase. And that increase in torque is what a driver needs in order to remain in high gear longer before downshifting.

Compared to a low torque rise engine, very high torque rise engines may have the disadvantage of producing less, or inconsistent, power output at engine speeds outside of the peak torque. City buses, delivery vehicles, work trucks, and off-road equipment engines use governors with lower and flatter torque rise profiles to make more torque available over a wider engine operating range. Accelerating in city traffic or moving some building materials in city traffic requires more torque availability at lower engine operating speeds. The tradeoff is reduced torque output at higher engine speeds to obtain a wider range of torque over all engine speeds. Low torque rise engines will also need to use more transmission gear steps to help maintain fuel economy caused by the downshifting required to increase engine torque.

To calculate torque rise, the peak torque is divided by the torque available at rated speed or horsepower. It is expressed as a percentage. For example, if the torque rise is calculated at 15%, the engine develops 15% torque increase for every 100 rpm drop from rated speed to peak torque. Alternatively, peak torque minus torque at command point divided by command torque equals torque rise expressed in units of torque: e.g., 950 - 900 ft-lb ÷ 900 × 100 = 5.55% torque.

High torque rise engines are generally considered to be those with more than a 30% increase. Automobiles and pickups will generally use less than 10% torque rise.

Governor Sensitivity

Governor sensitivity refers to the governor's ability to respond to and maintain a set rpm without fluctuation as load changes. In mechanical governors, sensitivity is often achieved at the expense of governor stability. This means engines with sensitive governors are prone to hunting conditions.

Types of Governors

Governors are differentiated and classified in a number ways. One way is by type of construction. For example, a governor may be mechanical, electronic, hydraulic, or pneumatic. Another classification system is by its function. Variable-speed, isochronous, or all-speed governors are used in off-road equipment where the engine generally operates at steady speed and needs to develop torque quickly when loaded. Automotive governors, or min/max governors, are used in cars, pick-ups and other vehciles driven on-highway (**FIGURE 10-16**). While a large number and variety of governors exist, this section will review the construction and operation of mechanical and electronic governors commonly found in light- and medium-duty automotive applications.

Construction Classification

While mechanical governors have long been discontinued in on-highway applications due to their inability to meet emission standards, their sheer numbers and continued use in off-road applications demand that a well-rounded engine technician be familiar with their construction, operation, and adjustment procedures. Electronic governors are integrated into the engine electronic controls.

Mechanical

In **mechanical governors**, the functions of sensing engine speed, load, and driver input and of processing and making changes to engine fueling are performed by a mechnical mechanisms. A mechanical governor typically uses the **centrifugal force**, or the force pulling outward on a rotating body of flyweights to sense engine speed. Driver input is sensed by the governor through a mechanical throttle linkage. Output to the injection system is also done through mechanical linkage.

Construction	Function	Vehicle
Mechanical	Minimum/Maximum Speed Control	
	Variable Speed All Speed Control	
Combined		
Pneumatic		

FIGURE 10-16 The type of governor used depends on the application.

Principles of Operation

Mechanical governors consist of two basic mechanisms:

1. a speed-sensing mechanism
2. a fuel-changing mechanism.

The speed-sensing mechanism detects engine speed changes. The fuel-changing mechanism increases or decreases the amount of fuel supplied to the engine. Together, these governor mechanisms operate to vary the quantities of fuel metered into the engine.

Fuel Control Rack

Before the specifics of the governor mechanism can be fully understood, it is important to understand that engine fueling takes place by moving an actuator such as the fuel control rack, which in turn changes the quantity of fuel injected into the cylinders (**FIGURE 10-17**). The governor is located at the rear of an in-line injection pump. The governor changes fueling because it is connected to the injection pump through linkage to control fueling. This linkage, called a rack in in-line injection pumps, extends from the injection pump housing into the governor housing (**FIGURE 10-18**). In the injection pump, the fuel control rack meshes with the injection pump's pinion gear segments. The control rack rotates the pumping plungers together in unison, which in turn changes the quantity of fuel metered for injection (**FIGURE 10-19**). Inputs from the operator, engine shutoff device, engine speed,

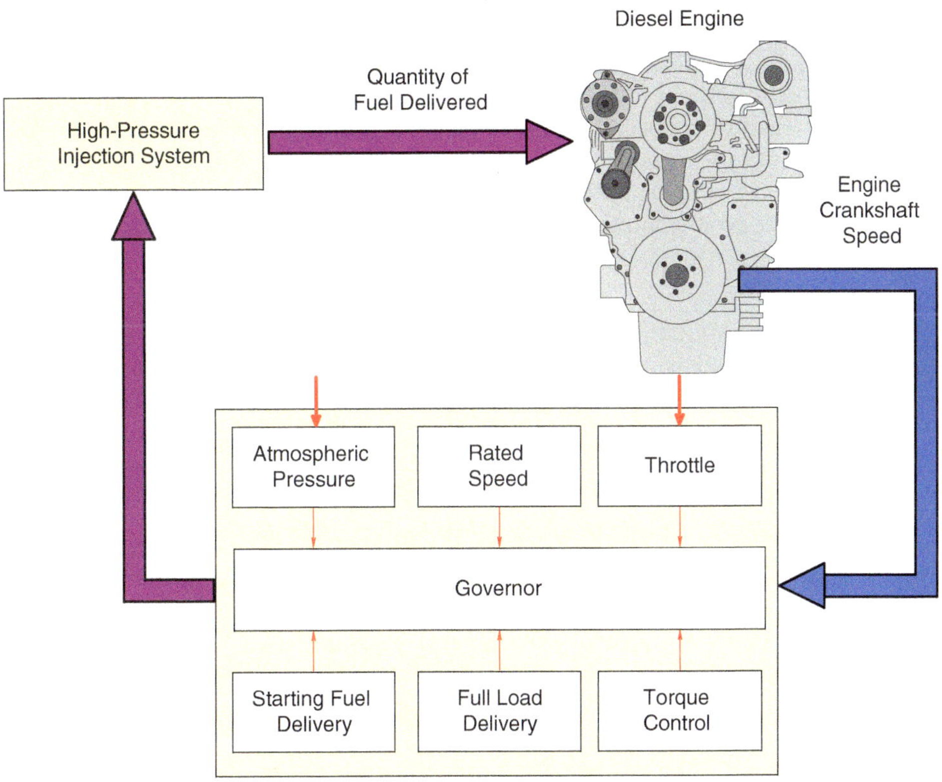

FIGURE 10-17 Mechanical governors regulate the quantity of fuel injected by the high-pressure injection pump through processing a variety of inputs and moving the control rack.

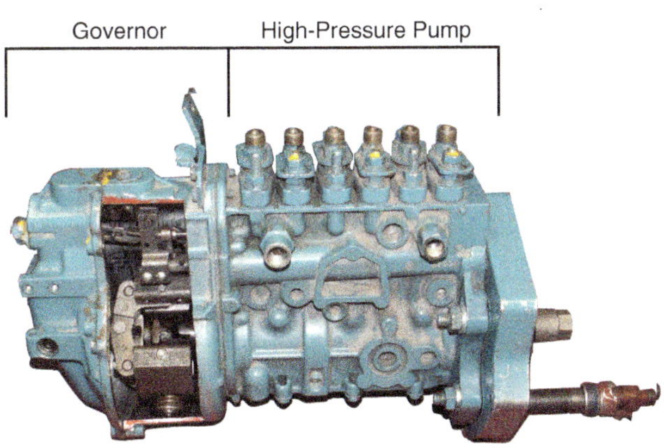

FIGURE 10-18 The governor is typically located at the rear of an in-line injection pump. Linkage connects the governor and the high-pressure metering system to vary engine fueling.

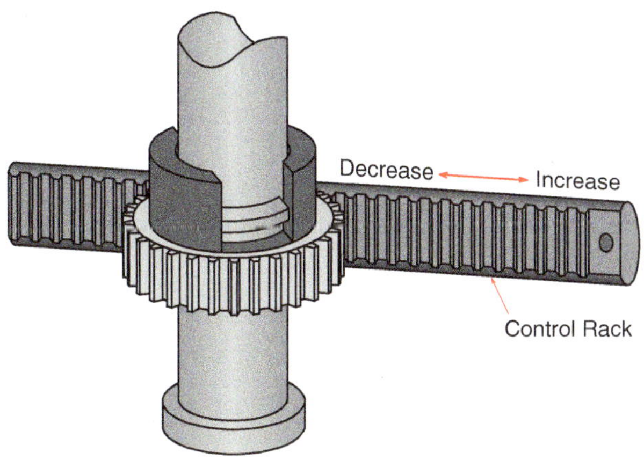

FIGURE 10-19 The control rack rotates the pumping plungers together in unison, which in turn changes the quantity of fuel metered for injection.

and intake boost pressure (if equipped with a turbocharger) are received by the governor. Because small volumes of fuel are delivered at low speed, even the slightest unintended movement in the fuel rack will cause unstable engine operation. For this reason, precision operation and adjustment of the governor is critical to good vehicle performance.

Speed-Sensing Mechanism

Mechanically operated governors speed sensing mechanism typically consist of three main parts:

1. a gear drive from the engine
2. flyweights
3. a governor spring.

The speed-sensing mechanism is responsible for measuring engine speed accurately and sensing operator throttle input. Centrifugal flyweights are the primary speed-sensing mechanism in mechanical governors. Flyweight force is proportional to engine speed. Flyweights inside a mechanical governor are rotated by the injection pump driveshaft. Centrifugal force causes the flyweight to move outward, applying force to the thrust collar. The end of the rack is connected to the thrust collar (**FIGURE 10-20**). Electronically controlled engines will use an engine speed sensor, which is either a variable reluctance or a Hall-effect sensor, generating analog or digital signals from a series of teeth on the crankshaft. A camshaft position sensor can also provide engine speed input to the electronic governor.

Governor Spring Force

Opposing the centrifugal force of the flyweight is the governor spring. The ball arms of the flyweights move outward with increasing engine speed. However, the spring force acts against the force of the rotating flyweights. A throttle lever connected to the governor control adds or removes spring tension. Therefore, the amount of spring force tension opposing the flyweight force is used to regulate the desired engine speed. Stated another way, more governor spring force increases the quantity of fuel injected.

Fuel-Changing Mechanism

There is a thrust collar mechanism connected to the end of the flyweight ball arms. The movement or force of the ball arms is summed or directed at this point. The fuel control

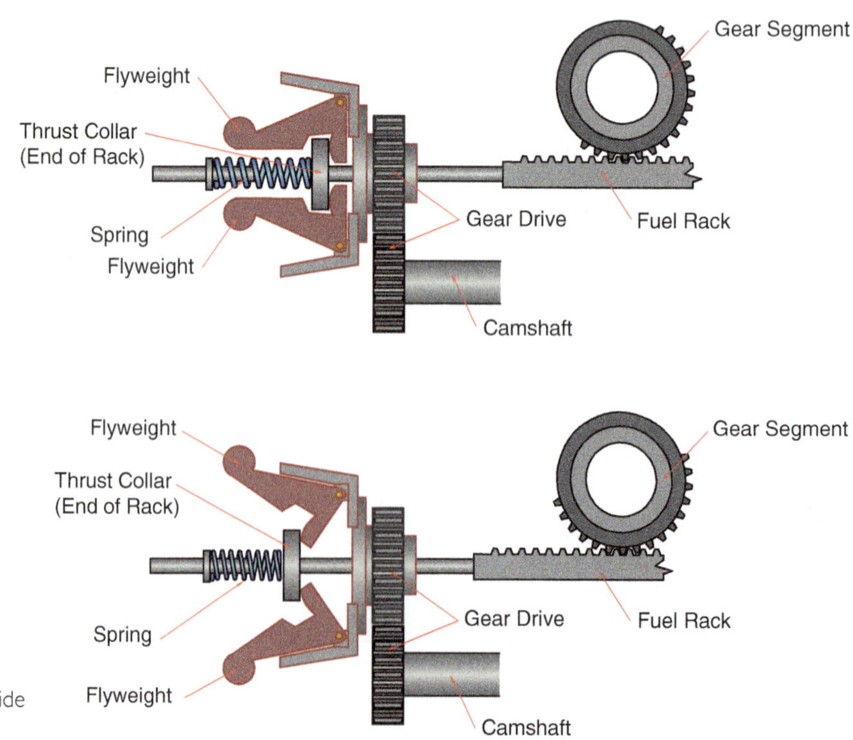

FIGURE 10-20 Flyweights inside a mechanical governor.

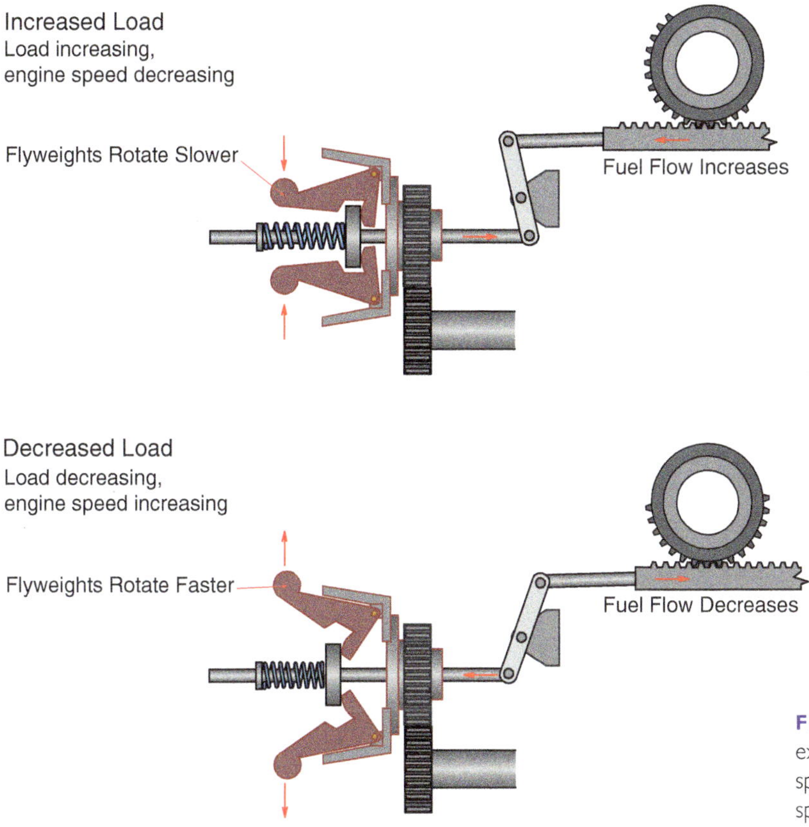

FIGURE 10-21 Changing engine speed changes the force exerted by the flyweights against spring tension. Increasing spring force increases the quantity of fuel injected. Decreasing spring force decreases the quantity of fuel injected.

rack or linkage of the high-pressure injection pump is connected to the thrust collar. Movement of the thrust collar will translate into motion in the fuel control rack.

Changing engine speed also changes the force exerted by the flyweights against spring tension. A **state of balance** exists when flyweight force equals spring force. When a state of balance exists, the quantity of fuel injected, and therefore engine speed, remains steady.

If flyweight force is greater than spring force, *fueling will decrease*. This occurs when engine speed or power output exceeds operator requirements. Because flyweight force increases, it overcomes spring force, thus moving the fuel control rack to decrease fuel to the engine.

If spring force is greater than flyweight force, *fueling will increase*. This occurs when operator demand is greater than engine speed or power output. Because spring force is greater than flyweight force, the control rack moves to increase the fuel injected into the engine (**FIGURE 10-21**).

The operator adjusts spring force tension through the throttle lever, so there is no direct mechanical connection between the throttle lever and the control rack or governor output actuator. In some governors, the throttle lever is directly connected to the governor spring to increase or decrease spring tension. In a more popular Bosch RQ governor, changing the state of balance between spring and flyweight force is done by moving a **fulcrum** (the point around which a lever rotates and that supports the lever and the load) between the spring and flyweight forces by using the throttle lever (**FIGURE 10-22**). The force exerted by the spring depends on the throttle lever's position. Moving the throttle lever from idle to high idle will add increasing tension to the governor spring. Greater flyweight force or engine speed is necessary to overcome spring force/tension. At idle, the throttle exerts little tension on the governor spring. This means it is easier for flyweight force to be greater than spring force, which causes the control rack to move to a minimum-fuel position. Moving the throttle increases the governor spring tension,

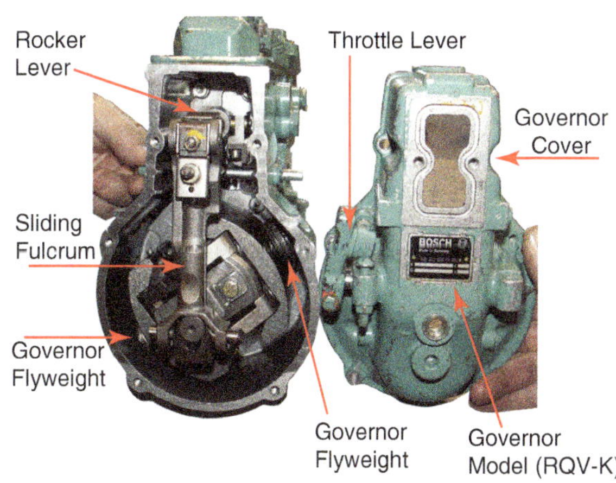

FIGURE 10-22 In this governor, a sliding fulcrum is used to change the state of balance between flyweight and spring force.

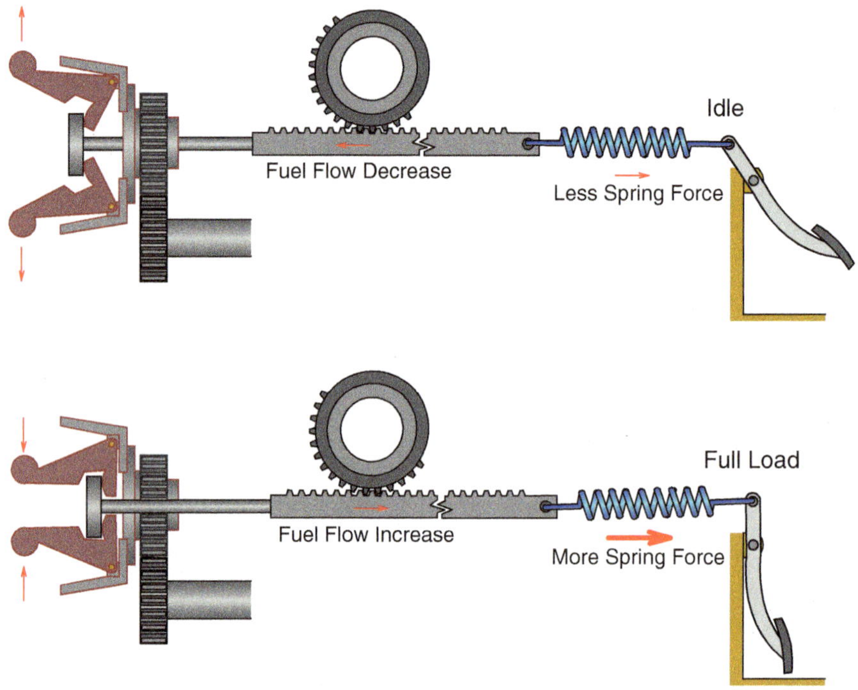

FIGURE 10-23 Flyweight vs. spring force in a simple mechanical governor.

> ▶ **TECHNICIAN TIP**
>
> Some shops sell customized performance **torque plate** products, or full-load stops, designed to increase fueling above the engine's stock power curve. These plates increase torque and power output by extending the distance of allowable maximum control rack travel. However, alteration of injection pumps beyond engine manufacturers' design limits will increase engine emissions and can lead to driveline damage. Automatic transmissions unable to handle excessive torque will slip and burn clutch plates. In standard transmissions, the excess torque will push the mainshaft and countershaft gears apart with enough force to split the transmission case.

> ▶ **TECHNICIAN TIP**
>
> Air-fuel ratio controls prevent excess smoke production by limiting the quantity of fuel injected until adequate air supply is available to support combustion. While air-fuel ratio controls are used on naturally aspirated engines, they are even more critical on turbocharged engines. Because turbocharger boost pressure will lag delivery of fuel, the excessive and prolonged production of black smoke can occur on engines with malfunctioning or incorrectly adjusted air-fuel ratio controls.

which becomes greater than flyweight force. The increased spring force causes the control rack to move to a position where more fuel is injected (**FIGURE 10-23**).

Summary of Spring Force and Flyweight Force Dynamics

Mechanical governors use flyweights that act through mechanical linkage and respond to centrifugal force. As the engine speeds up, the flyweights will move out and operate to decrease engine fueling.

Acting against flyweight force is spring force. Spring force tension is indirectly varied by the position of the throttle pedal. As the throttle lever is moved to increase fueling, the spring force is increased. Remember that flyweight force always operates to reduce fueling and that spring force operates to increase fueling.

Torque Control

Torque control refers to the governor's ability to regulate the maximum amount of fuel injected into the cylinders for a given engine speed. Generally, torque control is necessary to prevent black smoke caused by inadequate air or combustion chamber design limits. As an engine turns faster, less time is available for combustion; too much fuel will lead to visible emissions. However, the torque control device will add or subtract fuel above peak torque (**FIGURE 10-24**).

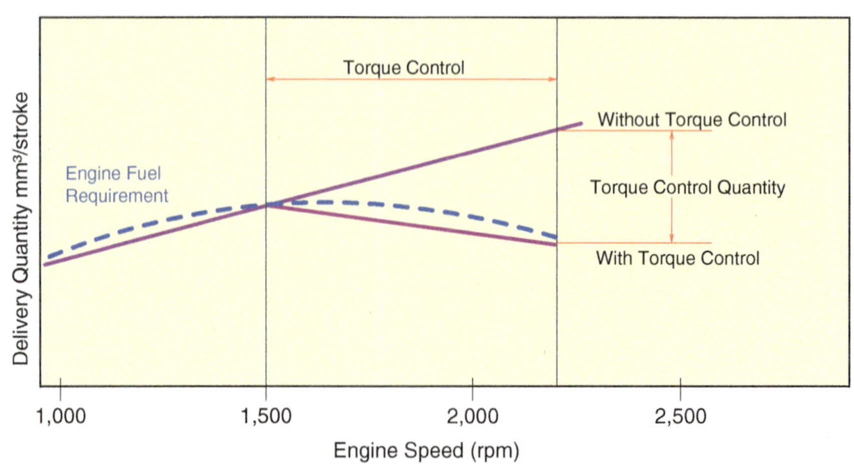

FIGURE 10-24 Torque control involves adding or subtracting fuel above peak torque.

Active Pulse Dampening

When the throttle pedal is pressed, the amount of fuel can quickly increase, causing a sudden load change on the engine. The load changes generated during acceleration produce pulsing or jerking movements by the vehicle as the driveline winds up and then releases stored energy. To reduce the jerking, a governor mechanism will usually incorporate active pulse dampening to delay the full quantity of fuel delivered for a short time and gradually increase delivery (**FIGURE 10-25**). During vehicle deceleration, abruptly cutting off fuel supply will also cause the driveline to produce vehicle speed pulsations, causing it to jerk back and forth. In this case, the abrupt decrease in the quantity of fuel injected is also minimized to dampen driveline pulsations.

These devices simply slow down control rack travel during acceleration and deceleration. By slowing down the speed of change for injection rates, enough time is also provided for the turbocharger to spool and boost, ensuring air supply does not significantly lag behind fuel delivery. These mechanisms, known as **load controllers** or pulse dampeners, improve ride quality and reduce emissions. In electronic governors, the same function is performed by algorithms in ECMs, which slowly modulate increases in fuel delivery during acceleration and deceleration events.

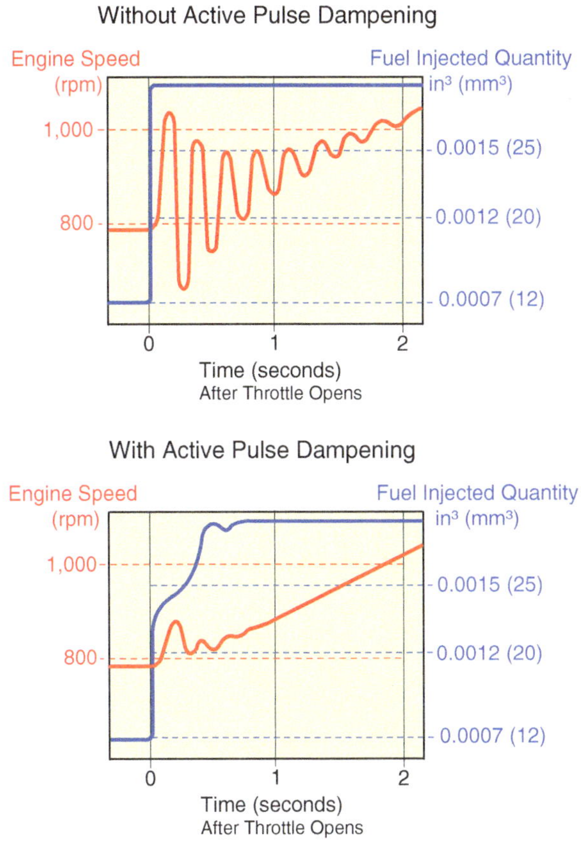

FIGURE 10-25 Active pulse dampening modulates changes in fueling delivery to prevent vehicle bucking and jerking during acceleration and deceleration.

Air-Fuel Ratio Control

In turbocharged engines, the rate of fuel injection must be matched to the pressure in the intake manifold. Intake manifolds are connected via a line to air-fuel ratio controls (**FIGURE 10-26**). Without the correct matching of adequate air mass to fuel rate, over-fueling will result in emissions of black exhaust smoke. Essentially, air-fuel ratio controls limit full-load fuel delivery until an air mass adequate for good combustion exists in the cylinders. A leaking or broken connector line will produce a low-power complaint, because without pressurized intake air, the governor will not permit full fueling.

▶ Electronic Governors

10-02 Explain the operating principles and control strategies of electronic governors.

Electronic engine governors use principles of electronic signal processing to regulate fueling. Sensor inputs supply data to an ECM, where software stored in the ECM processes the information. Like other information-processing systems, the governor collects sensor data and processes the signals by using rules contained in software. The output is the amount of fuel required for each cylinder. The ECM produces output signals used to control output devices such as injectors, fuel pressure regulators, and turbochargers (**FIGURE 10-27**). Electronic governors can be divided into two categories: partial-authority and full-authority systems.

Programmable Electronic Governors

Today, electronic governors and engine controls do more than ensure that combustion processes meet emission standards and expectations for power and fuel economy. Onboard networks communicate with other vehicle systems to improve vehicle productivity, efficiency, and safety and to enhance operator comfort. The vehicle and driving habits can be monitored through software to meet business objectives, such as to provide reports on driving habits and prevent abuse conditions.

Governors in electronically controlled engines can control fueling in more sophisticated ways that far exceed mechanical governor capabilities. In fact, the term "governing" has fallen out of use and governor adjustments has been replaced with the term adjusting "programmable parameters"; Now almost any measurable variable affecting either engine or vehicle operation can be adjusted through programmable software (**TABLE 10-2**).

FIGURE 10-26 The line connecting the intake manifold to the air-fuel ratio control device.

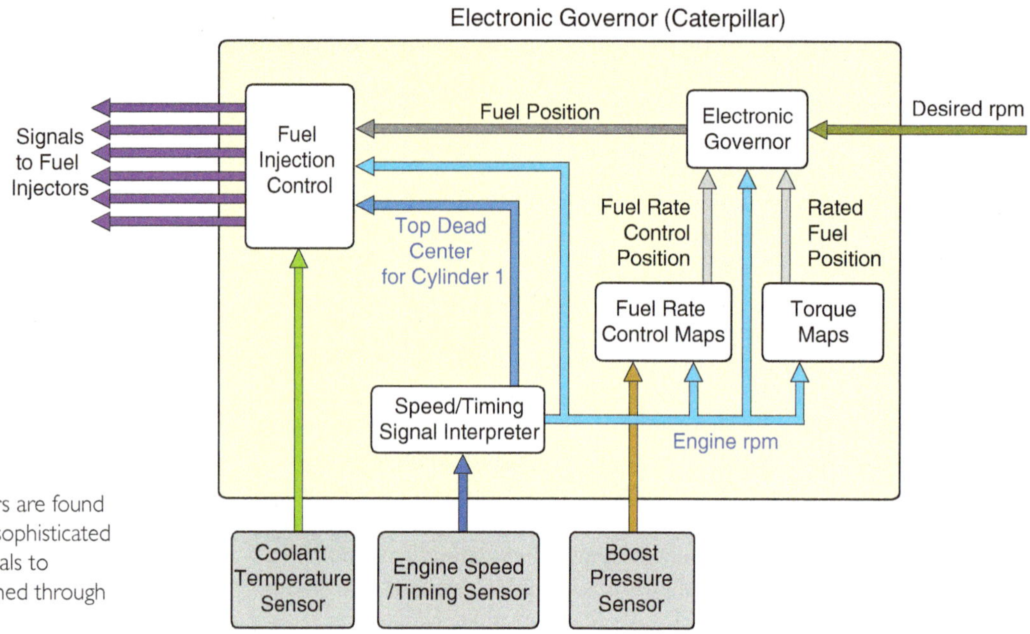

FIGURE 10-27 Electronic governors are found in the engine's ECM and use more sophisticated electronics to process electrical signals to duplicate what was once accomplished through mechanical devices.

Programmable governor features fall into two basic categories: factory specified and customer specified.

The following are some of the factory-specified parameters and features:

- engine OEM–specific features
- chassis manufacturer–specific features.

Engine OEM–specific features include variables such as horsepower (kilowatts), maximum torque output, and torque rise. These are usually not adjustable without factory authorization and specialized engineering or aftermarket software.

Chassis manufacturer–specific parameters include features such as calibrations for the tachometer, integration of chassis controllers to the onboard vehicle network such as the ABS, traction control, or specialized equipment needed to operate a boom truck (**FIGURE 10-28**). Adding an accessory, such as an exhaust brake, may also be specified by the engine OEM.

The following are some of the customer-specified parameters generally available on medium-duty engines:

- specialized OEM features chosen when the equipment is ordered
- optional programmable features unique to the equipment application and normally adjusted after the truck or bus has been delivered.

The following are the five most frequently used customer-specified parameters:

1. vehicle speed limiting
2. progressive shift (for manual transmissions only)
3. cruise control
4. engine protection
5. idle shutdown timer and PTO shutdown.

The following are a few examples of other categories of customer-specified parameters:

- maintenance monitors (**FIGURE 10-29**)
- vehicle security
- vehicle activity
- trip reports (**FIGURE 10-30**)
- engine brake or retarder controls
- dedicated PTO control features
- driver reward
- low gear torque limit.

TABLE 10-2 Typical Programmable Parameters in a Governor

Feature	Range	Setting	Setting
Maximum Engine Speed	2,120–2,120 rpm		
Maximum Rated Speed	1,800 rpm		
Rated Peak Torque	1,200 ft-lb (1627 N·m) at 1,800 rpm		
Torque at Rated Speed	1,167 ft-lb (1582 N·m) at 1,800 rpm		
Road Speed Governor			
Throttle Pedal Maximum Road Speed	30–120 mph (48–193 km/h)	65 mph (105 km/h)	1,700 rpm
Maximum Engine Speed Without VSS	1,400–3,000 rpm	900 rpm	
Maximum Engine Speed with VSS	1,400–3,000 rpm	1,800 rpm	Enabled
Lower Droop	0–3 mph (0–4.8 km/h)	1 mph(1.6 km/h)	
Upper Droop	0–3 mph (0–4.8 km/h)	3 mph (4.8 km/h)	
Transmission and Rear Axle			
Gear Down Protection Enabled			
Gear Down Protection rpm Limit	1,200–1,700 rpm	1,600 rpm	
Gear Down Protection Turn on Speed	40–65 mph (64–105 km/h)	55 mph (89 km/h)	
Lower Gear Engine Speed Limit	700–2,000 rpm	950 rpm	
Lower Gear Turn Off Speed Limit	5–15 mph (8–24 km/h)	9 mph (14 km/h)	
Torque Limit in First Gear	500–1,000 ft-lb (678–1356 N·m)	800 ft-lb (1085 N·m)	
Rear Axle Ratio	1.92–15.97	4.11	
Tire Size	300–700 rev/mile	421 rev/mile	
Engine Idle			
Idle Engine Speed	600–850 rpm	700 rpm	
Idle Shutdown			Enabled
Idle Shutdown Timer	1–100 minutes	2 minutes	Enabled
Idle Shutdown Lower Ambient Temperature Air Temperature Override	0–100°F (−18–38°C)	32°F (0°C)	Enabled
Idle Shutdown Upper Ambient Temperature Air Temperature Override	0–100°F (−18–38°C)	85°F (29°C)	
ECM Fan Control			Enabled
Minimum Fan on Time	0–1,000 seconds	120 seconds	
Cruise Control/Engine Brakes			
Maximum Cruise Control Speed	30–102 mph (48–164 km/h)	65 mph (105 km/h)	
Cruise Control Upper Droop	0–3 mph (0–4.8 km/h)	0 mph (0 km/h)	
Cruise Control Lower Droop	0–3 mph (0–4.8 km/h)	2 mph (3.2 km/h)	
Engine Brake Cruise Control Activation	Enabled/Disabled	Enabled	
Engine Brake Minimum Vehicle Speed	0–35 mph (0–56 km/h)	15 mph (24 km/h)	
Engine Brake Delay	0–10 seconds		Disabled
Engine Brake Service Brake Activation	Enabled/Disabled		Enabled
ABS Module		J-1939	Enabled
Engine Protection			
Engine Protection Shutdown Feature		Warning/Derate/Shutdown	Derate
Engine Protect Restart Inhibit		Enabled/Disabled	Enabled
Manual Override		Enabled/Disabled	Enabled

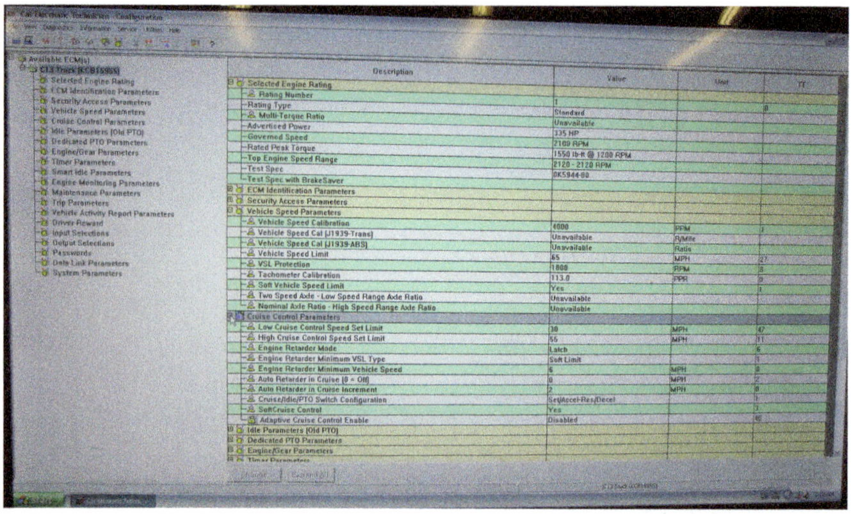

FIGURE 10-28 Programmable parameters from the chassis manufacturer reference chassis equipment such as axle two-speed shift motors, rear-axle gear ratios, and tachometer calibrations.

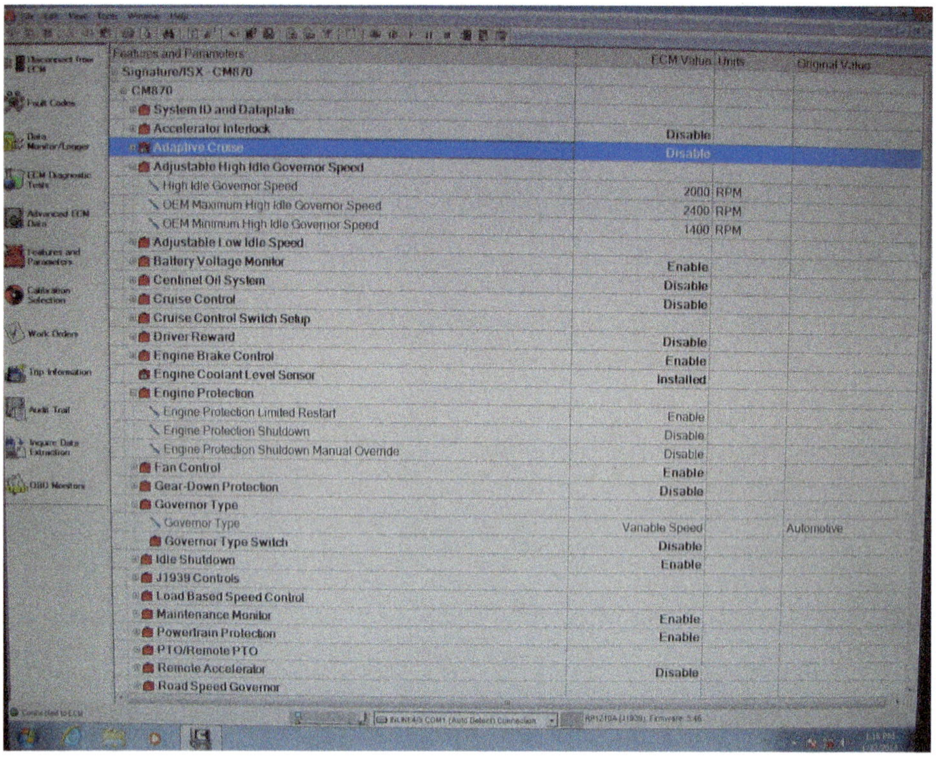

FIGURE 10-29 A screenshot of programmable parameters.

Partial-Authority Systems

Partial-authority fuel systems have adapted some electronic control to mechanical injection systems extend the fuel system and engine design usefulness, where stricter emissions regulations require cleaner combustion. Partial-authority systems range from electronically controlling just cruise control functions to electronically governing all fuel metering and injection timing controls while using a mechanical high-pressure injection system. Electronically controlling injection quantity is performed by a partial-authority governor, which uses a mechanical actuator, rather than a direct electrical signal, as an output device to move the metering control device inside a high-pressure pump. Moving the metering control device increases or decreases fueling. More sophisticated partial-authority systems incorporate features to change injection timing as well. An electronically controlled solenoid can also be used to vary injection timing in a mechanical system (**FIGURE 10-31** and

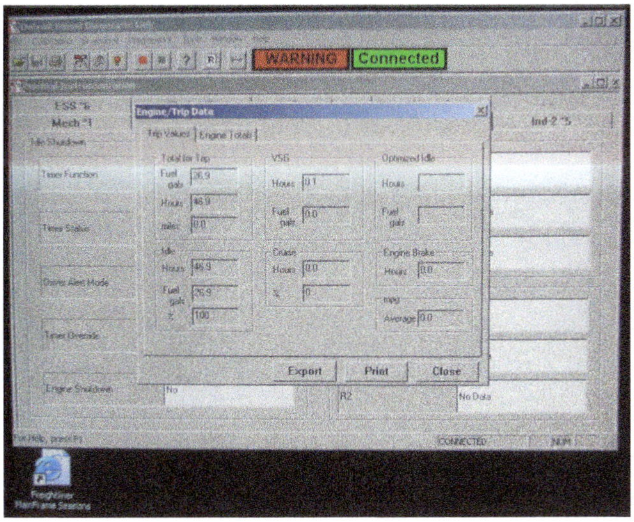

FIGURE 10-30 Modifying a programmable parameter for fuel consumed during a trip. Note the report for the amount of fuel used during non-productive idle.

FIGURE 10-31 This Bosch P7100 in-line injection pump uses an electronically controlled actuator to move the control sleeve inside the high-pressure section of the pump.

FIGURE 10-32). All partial-authority systems are drive by wire, meaning a throttle-position sensor or linkage is connected to a sensor that provides driver input to the governor.

Full-Authority Systems

Full-authority governing is more commonly referred to as full electronic engine control. All metering and timing functions use principles of electronic signal processing to regulate engine fueling (**FIGURE 10-33**). Rather than using a mechanical actuator to control output from a high-pressure injection pump, fueling is regulated using pulse-width–modulated (PWM) electrical signals. For example, an injector in a full-authority system will use a solenoid to determine the fuel quantity required and timing for an injection event. Full-authority governing systems are also drive by wire, and electronic signal–processing techniques are exclusively used to collect input data to control the

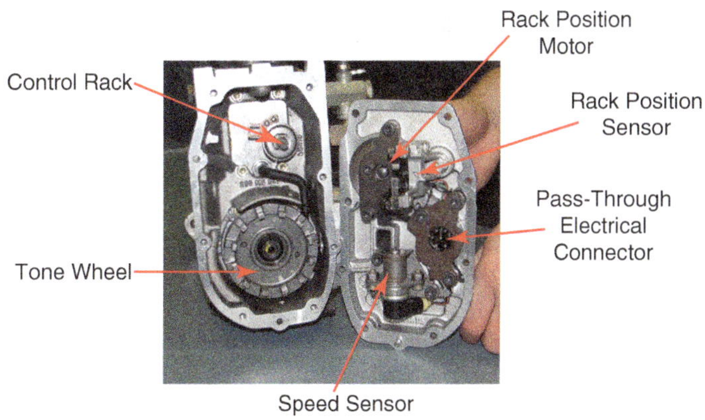

FIGURE 10-32 Components inside a partial-authority governor that are used to sense engine position and coordinate control rack travel.

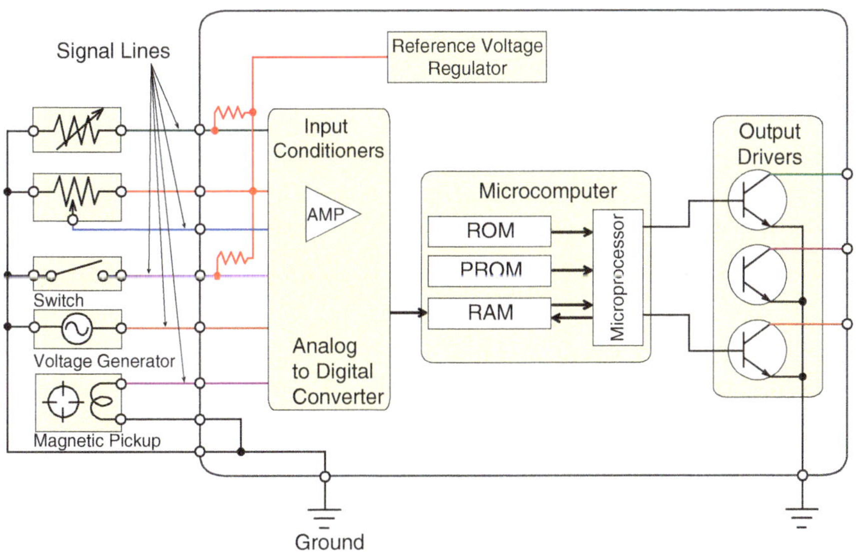

FIGURE 10-33 Full-authority governors only electronically control the fuel system to govern engine operation.

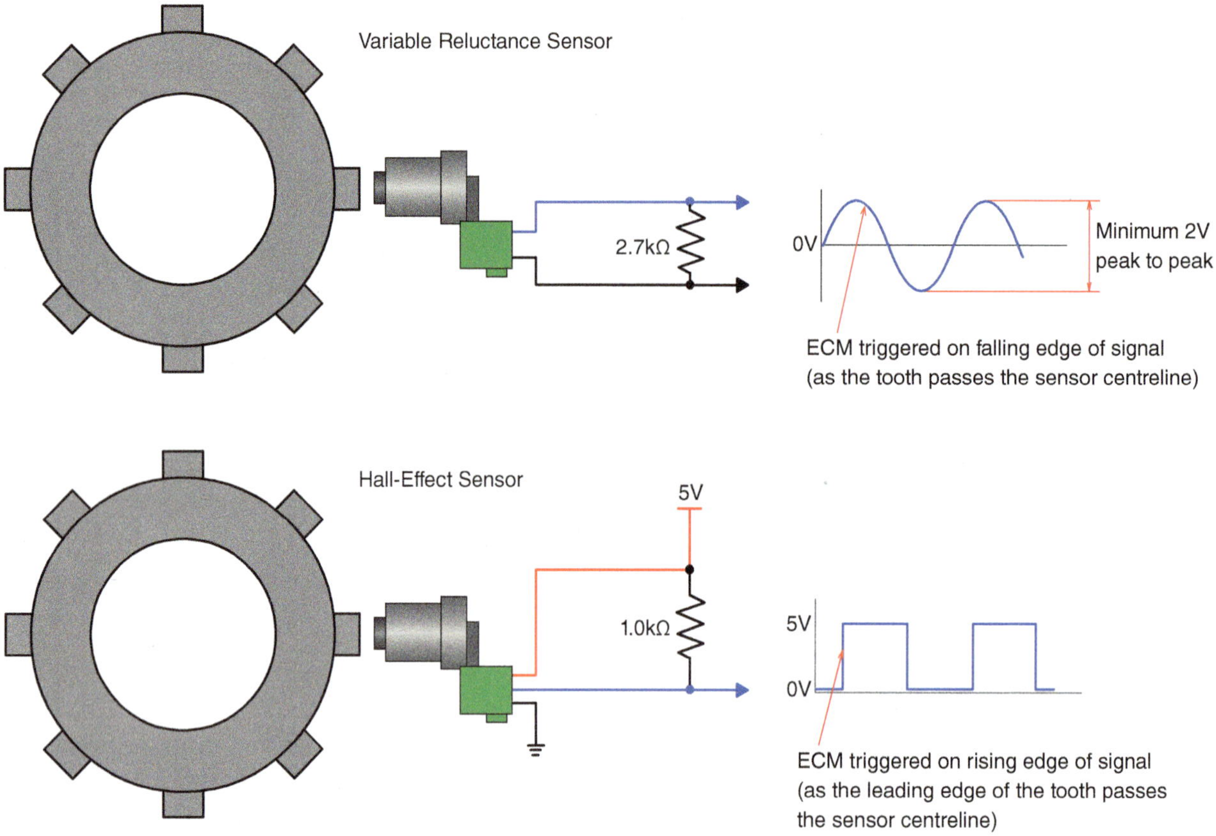

FIGURE 10-34 The crankshaft and camshaft sensors collect engine speed and position data. Digital signals are produced by Hall-effect sensors. Analog data from variable reluctance sensors.

injection event. Engine speed and position data are collected from the crankshaft and cam position sensors rather than sensed by flyweight force (**FIGURE 10-34**). All on-highway diesel engines manufactured today are full-authority systems. But some contemporary off-road engines are still manufactured with partial-authority systems.

Fueling Strategy Classification

Governors are further classified according to their response to the throttle pedal. The common fueling strategies described in the following section can be achieved either mechanically or electronically.

Automotive Governor

Another name for the **automotive governor** is limiting speed (LS) or min/max (which refers to only minimum and maximum engine speed limiting governors). The automotive governor is common in passenger and commercial vehicle applications and has an accelerator response similar to throttle plate–controlled, gasoline-fueled engines. When a load is applied to the engine, the engine speed will drop, so the driver has to push the accelerator in order to increase injection fuel quantity. Likewise, to reduce speed or compensate for a load reduction, the throttle is backed off, which decreases fueling.

Automotive governors have the advantage of enabling better fuel economy in comparison to other types of governors. The explanation for this lies in the way fueling changes occur. With automotive governors, sudden load changes are responded to by the driver rather than a mechanism that automatically and instantly changes the quantity of fuel injected to maintain engine speed and torque. Typically, this results in less aggressive engine fueling. A driver who is "easier" on the throttle should obtain better fuel economy when

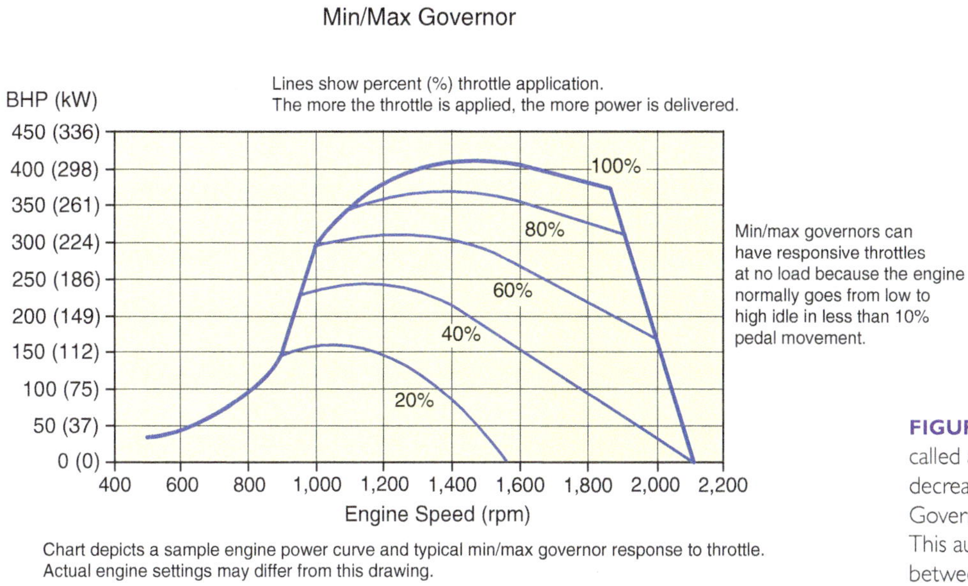

FIGURE 10-35 A min/max governor, also called an automotive governor, increases or decreases fueling only in response to driver input. Governors shape the torque curve of an engine. This automotive governor shows some droop between 1,990 and 2,100 rpm.

using this governor. Having noted this, its important to remember that the fuel economy of diesel engines is generally less sensitive to driver behavior than spark-ignition gasoline engines because the mechanical governor or ECM still has a greater degree of control over the fuel system operation than in gasoline fueled engines.

Mechanisms inside the automotive-type governor regulate only the engine idle speed and maximum engine speeds. Fueling between those points is regulated by the driver by moving the throttle pedal. In other words, the power output of an engine with an automotive governor is determined by the position of the accelerator pedal. To increase torque output, the throttle is depressed farther. To decrease engine speed and torque requires the releasing the throttle pedal or moving the throttle lever closer to a idle speed position (**FIGURE 10-35**).

Variable-Speed Governor

Sometimes referred to as an all-speed governor, the **variable-speed governor** automatically changes the amount of fuel injected into the combustion chamber to regulate engine speed, which is set according to accelerator pedal position. With this type of governor, the operator can hold engine speed steady by maintaining a constant accelerator pedal position. A given amount of accelerator pedal travel will correspond to a specific engine speed. As engine loading increases or decreases, the governor, not the operator, will change fueling to attempt to maintain that engine speed (**FIGURE 10-36**). The operation of a variable-speed governor is best observed when a vehicle is using cruise control: the engine speed is set, and the governor changes fueling to maintain a consistent road speed. Torque control is managed by the governor and not the operator's movement of the throttle pedal. By fully depressing the accelerator pedal of a variable-speed governor, the operator commands the maximum engine speed. Torque control is provided on an as-needed basis according to engine load. A variable-speed governor establishes engine idle speed, high idle, and any speed in between these points proportional to accelerator pedal position.

Off-road equipment, stationary engines, and some on-highway trucks are common applications for this type of governor. A PTO uses this type of governor because a change in load on the PTO will not produce wild swings in engine speed. This is unlike a automotive governor where changes to load on a PTO will produce wild swings in engine speed.

Combination Governors

As the name suggests, combination governors are mechanical governors that can combine the characteristics of the automotive governor and the variable-speed governor. Typically, this means that at medium-speed to high-speed operation, the accelerator pedal will control

> ▶ **TECHNICIAN TIP**
>
> When starting cold diesel engines, automotive governor mechanisms used by mechanical injection pumps often require the accelerator pedal to be partially or fully depressed to the floor once to achieve maximum fueling position for starting. Like setting the choke in old carbureted engines, the mechanical governor moves to a maximum fueling position only when the throttle is depressed. Maximum fueling is needed to ensure adequate fuel vapor is available during cold cranking/starting.

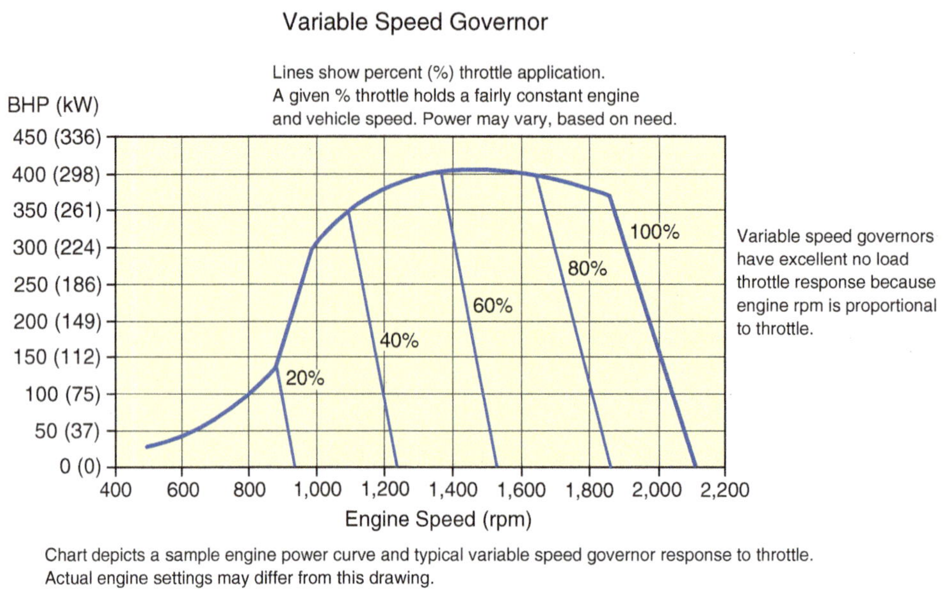

FIGURE 10-36 A variable-speed governor automatically increases or decreases fueling in response to a change in engine speed. The driver does not move the throttle in response to load changes.

fueling, because automotive governing takes over. At low-engine to mid-engine speeds the governor will behave like a variable-speed governor. This configuration makes it convenient to adapt PTO devices to the engine devices. Because PTO devices are on or off and vary the load applied to the engine according to the work they are doing, such as moving hydraulic cylinders or powering a water pump, it is desirable to use a governor with the ability to automatically adjust fueling to maintain a constant engine speed. Many medium- and heavy-duty on-highway diesel engines using mechanical governors have combination governors.

Choosing between the automotive governor and the variable-speed governor (which can be selected on some electronically controlled engines) depends on vehicle configuration and operator preference. While automotive governors generally get better fuel economy than variable-speed governors, the disadvantage is that automotive governors tend to have poor startability, meaning they cause a sudden surge or lurch when the vehicle is loaded (e.g., when pulling away from a stop with a heavy load). Lurch or surge may cause driveline damage such as snapped driveshafts or broken axles. Automotive governors do not adapt well to the use of PTOs, because engine speed can rise and fall dramatically with varying load conditions imposed by the PTO device. Operators may perceive that an engine equipped with an automotive governor is not as responsive or as powerful as an engine equipped with a variable-speed governor. The use of a PTO switch separate from the engine throttle control on electronically controlled engines allows the use of variable-speed governors and automotive governors.

To provide additional governor flexibility, some electronically controlled vehicles are equipped with a cab switch that allows the operator to switch from one type to another in response to changing vehicle functions, operating conditions, or preferences (**FIGURE 10-37**). However, in most applications, the steering wheel–mounted cruise control switches can be used to increase or decrease engine speed when the vehicle is parked. Optional fast idle switches can be configured on many vocational pickups and vans by connecting a single wire to an ECM pin.

Isochronous Governor

The **isochronous governor** is able to maintain more precise engine speed control than other types of governors. A common application for this type of governor is in diesel-powered electrical generators, or gen-sets, used for alternating current (AC) power distribution. Here, AC frequency (e.g., 60 Hz) is regulated by engine speed and has to be precisely maintained. Generally, this means an engine has to hold 1,800 rpm to produce 60 Hz AC from a three-phase AC generator (1,550 rpm produces 50 Hz AC). This governor operates with virtually

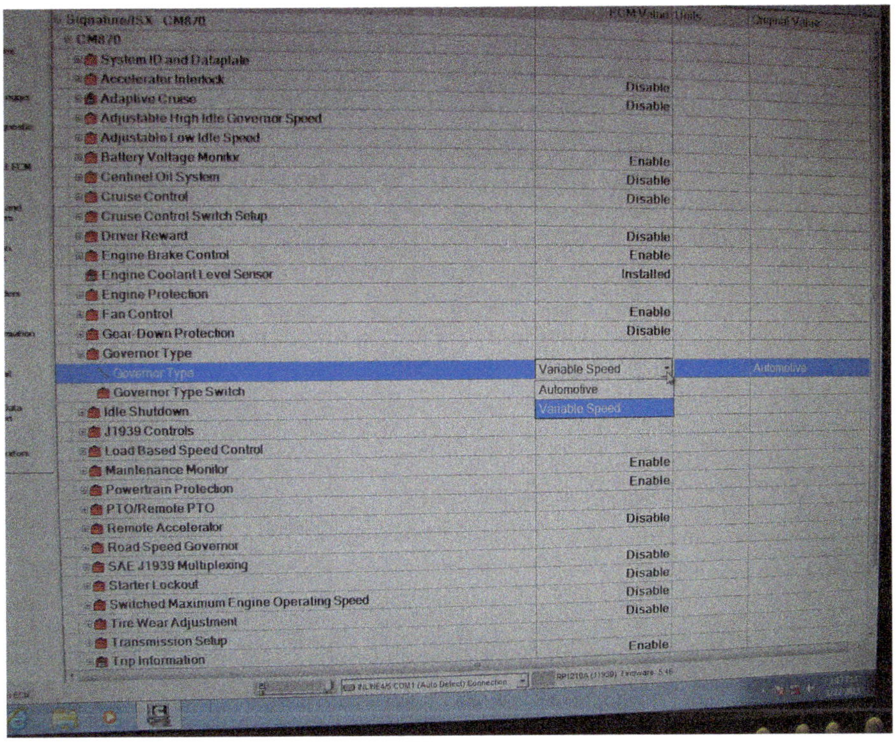

FIGURE 10-37 OEM software used to modify governor settings. Note the choice between variable-speed and automotive governor settings.

zero droop. A zero-droop governor is very sensitive to speed change and corrects speed variations almost instantaneously, maintaining engine speed regardless of engine load.

▶ Governor Maintenance

10-03 Outline the procedures used to evaluate the operation of a governor.

With a few exceptions, mechanical governors are not serviced in the field. They require specialized repair facilities equipped with sophisticated instruments and staffed by technicians with in-depth training on governors. It is useful, however, to learn what adjustments are made and how they are performed.

Mechanical Governor Adjustment and Testing

Smoke, low power, hunting, engine stalling, poor acceleration, and excessive engine speed are some of the problems that indicate a governor may require evaluation. In mechanical governors, springs can lose their tension and wear can occur between moving parts. Specialized injection repair shops will replace and adjust governor components such as flyweight assemblies, spring sets, spring retainers, bell cranks, fuel stops, and torques control devices. Wire ties sealed with lead tags cover critical governor adjustments to prevent tampering. The imprint on the lead seal identifies the manufacturer or repair shop servicing the pump (**FIGURE 10-38**). Precision components like this should be disassembled and carefully inspected before any major pump overhaul.

The most basic field test of governor wear is simply accelerating the engine gradually to its maximum no-load speed and comparing the engine speed to OEM specifications. If the engine does not reach maximum rpm, the governor spring is weakened. This happens because governor mechanisms are designed for wear and deterioration to reduce engine speed and torque, not increase it. Air-fuel ratio devices are pressure tested to check for diaphragm leaks. The devices typically have two adjustments. One adjustment is for limiting

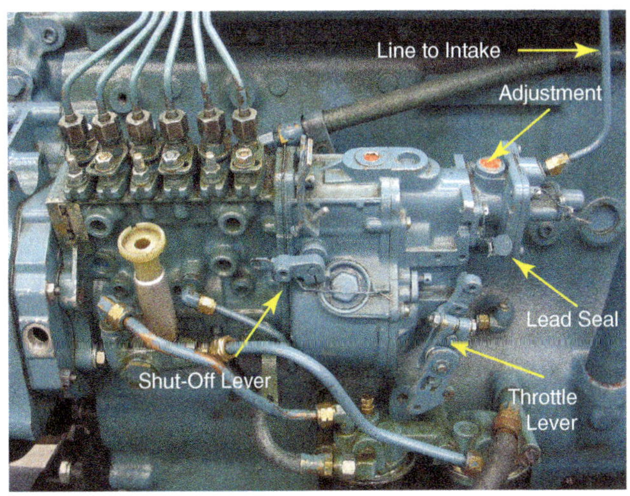

FIGURE 10-38 Wire ties sealed with lead tags cover critical governor adjustments to prevent tampering. The imprint on the lead seal identifies the manufacturer or repair shop servicing the pump.

FIGURE 10-39 The bumper screw is adjusted to prevent hunting and to change governor stability.

maximum fuel according to boost pressure. The second adjustment will control the initial starting point where maximum fuel limiting begins. The second adjustment seems unnecessary, but if fuel limiting were to take place immediately at idle speed, the engine would accelerate very slowly.

Engine idle speed is externally adjustable on most mechanical governors. An adjustment screw on the pump linkage will move the throttle lever to produce a desired idle speed. Maximum engine speed is adjusted by increasing or decreasing the tension of governor flyweights (**FIGURE 10-39**).

Most pumps also have an adjustment for torque control that is accessed externally. Moving the screw in and out will vary the maximum quantity of fuel provided for a given engine speed condition.

Testing Maximum Engine Speed for a Mechanical Governor

On mechanical governors with centrifugal flyweights, the governor springs can become weak because of age, heat, and fatigue from constant compression and release cycling. When this happens, the engine will fail to reach rated rpm and the driver will complain about low power. Before testing the governor, observe the linkage with the accelerator pedal pressed to the floor in order to ensure that the throttle is reaching full throttle. Most often, breakover linkage is used for the throttle lever. This linkage is a two-piece, spring-loaded lever; after contacting the maximum speed stop, it will allow a little more lever movement through a spring. Configuring linkage like this minimizes throttle shaft force and bushing wear on the throttle shaft when a driver pushes the pedal to the floor. A properly adjusted throttle will allow the throttle linkage to break over and should be inspected when checking for full throttle accelerator pedal travel. To test maximum engine speed for a mechanical governor, follow the steps in **SKILL DRILL 10-1**.

Adjusting Maximum Vehicle Speed

An electronically controlled engine has an adjustable vehicle speed limiting (VSL) parameter that sets the maximum speed of the vehicle. VSL cuts off fuel to the injectors when the engine exceeds the programmed parameter speed. On many engines, the VSL can be exceeded by adjusting the programmable cruise control limit to a higher value than the VSL parameter. Setting the VSL lower than the maximum speed permitted in cruise control encourages more operation in cruise control mode, which can better control engine fueling than the operator can. Using the VSL feature allows the vehicle to take advantage of a diesel's high torque output at low engine speed because limiting vehicle speed keeps

SKILL DRILL 10-1 Testing Maximum Engine Speed for a Mechanical Governor

1. After checking for full accelerator pedal travel, start the engine and allow it to run at idle rpm.
2. Check the emission decal on the engine or the engine's service literature to find its high-idle speed or rated rpm.
3. Slowly push on the throttle pedal to bring the throttle to full accelerator pedal travel.
4. Observe and record the maximum engine rpm and release the accelerator pedal.
5. Return the vehicle to idle, allow the turbocharger speed to slow, and then turn off the engine.
6. If the engine reached specified high-idle rpm, the governor spring condition is acceptable. The engine should typically reach 5–10% above rated rpm speed with no-load acceleration. If the engine did not reach target speeds, recommend governor and injection pump overhaul or replacement.

SKILL DRILL 10-2 Adjusting Maximum Vehicle Speed

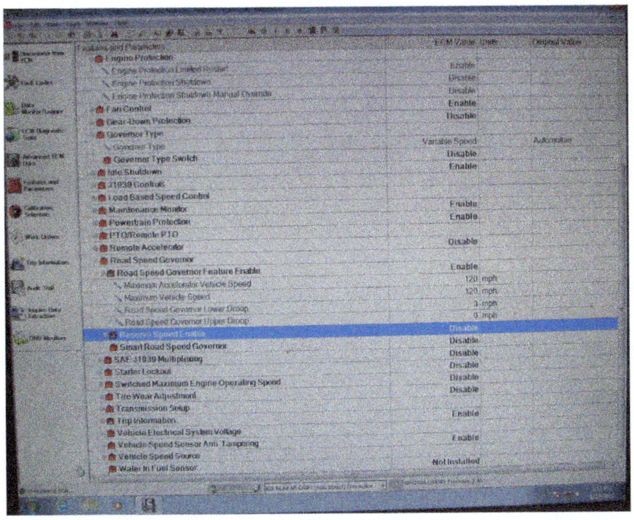

1. Connect a datalink adapter to the vehicle's datalink connector.
2. Using OEM software, locate the group of programmable parameters for vehicle speed. A separate menu is provided for making adjustments in OEM software.
3. Identify the VSL that is appropriate for the vehicle, and change the VSL limit number. A customer password may be required to lock and unlock the parameter.
4. Check and ensure that the chosen VSL is greater than or equal to peak torque rpm plus 100. If an acceptable vehicle speed is not attainable, recommend another axle ratio or transmission to increase engine rpm. If an acceptable vehicle speed is not attained, engine power and fuel economy will be adversely affected.

the engine operating in it's most efficient range. Another phrase that describes this type of operation is downspeeding–or the gear fast, run slow strategy—which optimizes the balance between fuel economy and engine performance. Because most diesel-powered vehicles are capable of very high vehicle speeds, setting VSL allows the operator to consistently drive at fuel efficient and safe vehicle speeds. To adjust maximum vehicle speed, follow the steps in **SKILL DRILL 10-2**.

Engine Protection Parameters

Engine protection is a critical governor programmable setting. Faults that could potentially cause engine damage will cause the governor to do one of three things:

1. derate power
2. shut the engine down
3. only warn the driver by using an instrument cluster warning lamp.

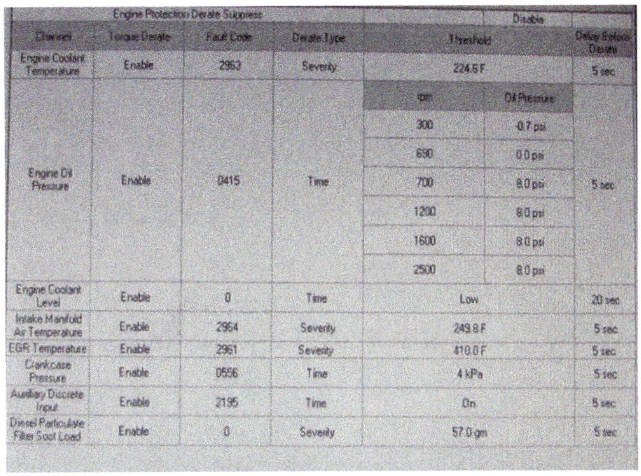

Engine Protection Derate Suppress — Disable

Channel	Torque Derate	Fault Code	Derate Type	Threshold		Delay Before Derate
Engine Coolant Temperature	Enable	2963	Severity	224.6 F		5 sec
				rpm	Oil Pressure	
Engine Oil Pressure	Enable	0415	Time	300	0.7 psi	5 sec
				690	0.0 psi	
				700	8.0 psi	
				1200	8.0 psi	
				1600	8.0 psi	
				2500	8.0 psi	
Engine Coolant Level	Enable	0	Time	Low		20 sec
Intake Manifold Air Temperature	Enable	2964	Severity	249.8 F		5 sec
EGR Temperature	Enable	2961	Severity	410.0 F		5 sec
Crankcase Pressure	Enable	0556	Time	4 kPa		5 sec
Auxiliary Discrete Input	Enable	2195	Time	On		5 sec
Diesel Particulate Filter Soot Load	Enable	0	Severity	57.0 gm		5 sec

FIGURE 10-40 The governor will limit or cut off fueling when a programmable threshold for potential engine damage has been reached.

The engine will monitor important engine protection parameters, such as fuel temperature, coolant level, oil pressure and temperature, intake air, fuel, oil, coolant, and particulate filter temperature. These values are compared to the allowable maximum limits to determine when a critical fault has been reached. The service software can be used to adjust how the governor responds to the engine protection faults (**FIGURE 10-40**).

If only a warning is specified, the vehicle operator is responsible for taking action to avoid engine damage. No shutdown will occur. If shutdown is specified in response to a fault, the engine will stop after a short time of continued operation. For example, coolant overheating might shut the engine down after 60 seconds. If it is a problem related to oil or the particulate filter, it might take only 30 seconds to shut down. If it is equipped with a stop-engine warning lamp, then the stop lamp will typically flash for 20 to 30 seconds before engine shutdown occurs, to give the operator time to pull the vehicle over to the shoulder of the road. A stop engine override (SEO) switch is available to delay engine shutdown if the operator believes safety is compromised without more time to maneuver the vehicle to a safe stopping point.

Derate warnings can be adjusted to progressively reduce engine power based on the severity of the condition. For example, if the coolant temperature reaches 225°F (107°C), the engine power may be reduced by 20%. At a higher coolant temperature like 235°F (113°C), a 60% derate will take place.

▶ Electronic Engine Control

10-04 Identity and explain the purpose and functions of electronic engine controls.

In the late 1980s, mechanical controls on diesel engines had reached their limit of efficiency, so the next logical step was to apply electronic controls already used on gasoline engines (**FIGURE 10-41**). The immediate benefits of these refinements to engine operation are lower engine emissions, improved fuel economy, increased reliability, and enhanced performance. Smarter engines continue to deliver ever-increasing power from smaller displacements, quieter operation, and longer service intervals in addition to needing less overall maintenance. The increased costs of some of these features are offset through improved engine efficiency (**TABLE 10-3**). Many of the electronic control systems have in fact lowered

TABLE 10-3 Increase in Power Output per Cubic Feet of Displacement

Engine Model	1988 – 7.3L IDI* Diesel	2015 – 6.7L Powerstroke
Horsepower	180 hp (132.4 kw)	440 hp (328.1 kw)
Torque	338 ft/lb @ 1,600 rpm 458.3 N·m	860 ft/lb @ 1,600 rpm 1,166 N·m
NO_x emissions	2.5 grams/bhp	0.07 grams/bhp
Intake air flow @ 3,330 rpm	360 cubic feet/minute 10.19 meters³/minute	732 cubic feet/minute 20.73 meters³/minute
Exhaust flow @ 3,300 rpm	1,080 cubic feet/minute 30.58 meters³/minute	1,499 cubic feet/minute 41.03 meters³/minute
Fuel system	Stanadyne mechanical distributor pump	Bosch piezoelectric injectors CP4.2 common Rail Pump Electronic engine controls

*Indirect injection

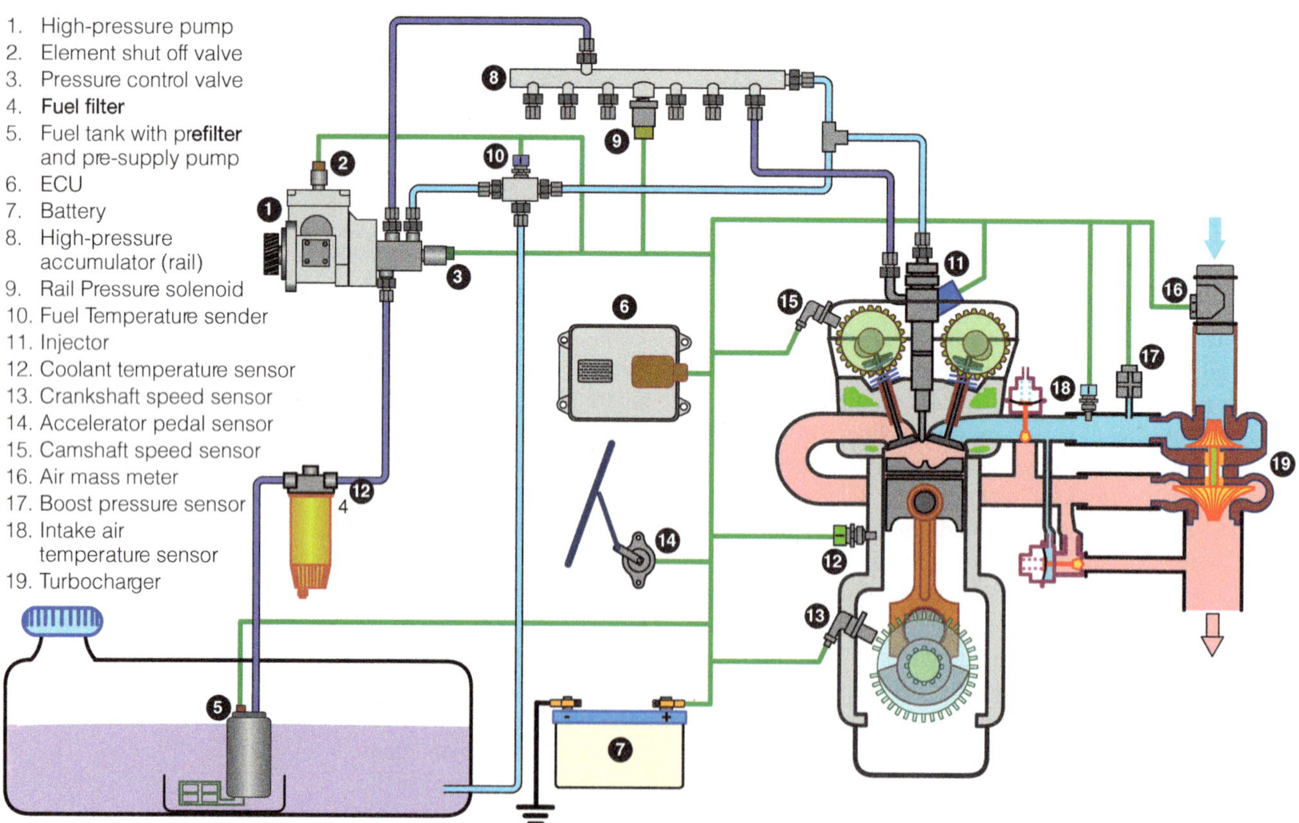

FIGURE 10-41 An overview of components used for the engine management system of a common rail diesel engine. Extensive use of electronics translates into precise control of combustion events for low emissions, superior performance, and fuel efficiency.

the cost of vehicle production while adding more features with the improved operating benefits. In comparison to mechanical controls, electronic controls enable far greater flexibility to adjust fuel-injection metering, injection rate, and timing over a large number of operating conditions. When engine operational problems that would lead to excess emissions do occur, the self-monitoring and self-diagnostic capabilities of electronic controls can identify the problem, alert the operator, and even switch to operating modes that minimize noxious emission production (FIGURE 10-42).

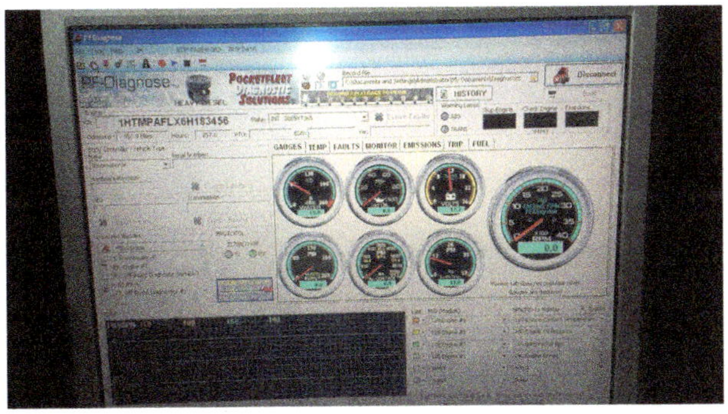

FIGURE 10-42 A screenshot from generic aftermarket diagnostic software.

Elements of Electronic Signal–Processing Systems

At first glance, operation of electronic control systems appears mysterious, because they use a variety of sensors, wires, electrical actuators, and electronic modules moving with invisible electrical signals. However, the functions of electronic engine control systems can be understood by breaking them down into three major divisions (FIGURE 10-43):

1. sensing
2. processing
3. output or actuation.

Sensing Functions

Sensing functions collect data about operational conditions or the state of a device by measuring some value, such as temperature, position, speed, pressure, flow, gas concentration, or angle. Sensors are devices designed to collect specific data in an electrical format, usually a voltage signal or digital-type signal (FIGURE 10-44).

Processing

Processing refers to the control system element that collects sensor data and determines outputs based on a set of instructions or program software. Operational algorithms, which are simply mathematical formulas used to solve problems, are included in the software that determines the steps taken when processing electrical data.

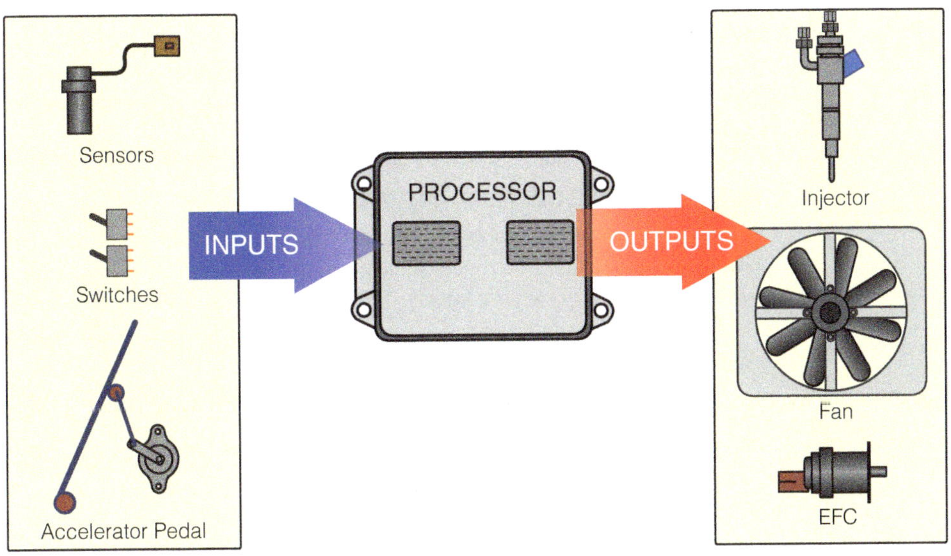

FIGURE 10-43 All engine management systems process electrical signals in three distinct stages: data collection from sensor inputs; data processing inside an ECM; and output devices, which are electrically operated.

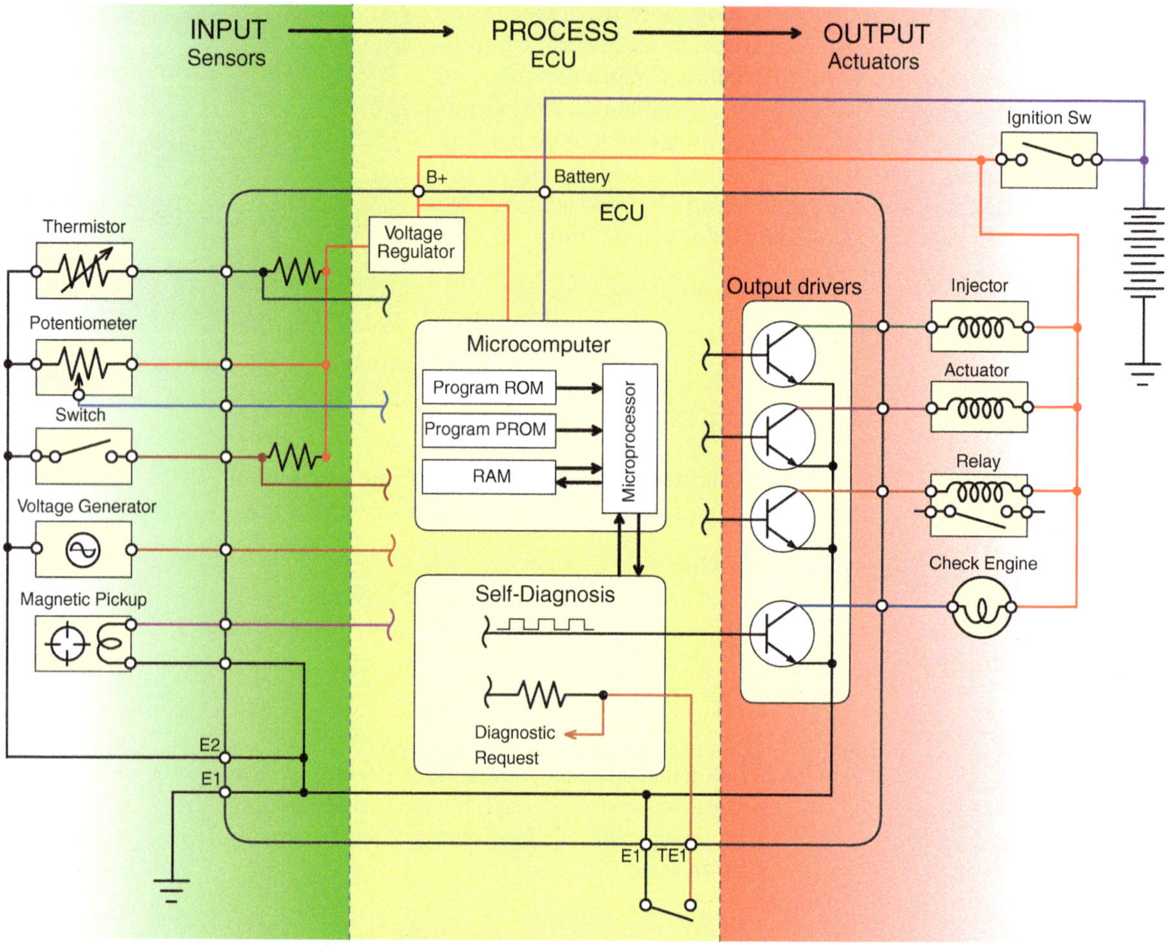

FIGURE 10-44 Three stages of signal processing: Sensors form input signals, software-controlled microprocessors are used to manipulate data, and electrically operated output devices carry out the instructions of the processor.

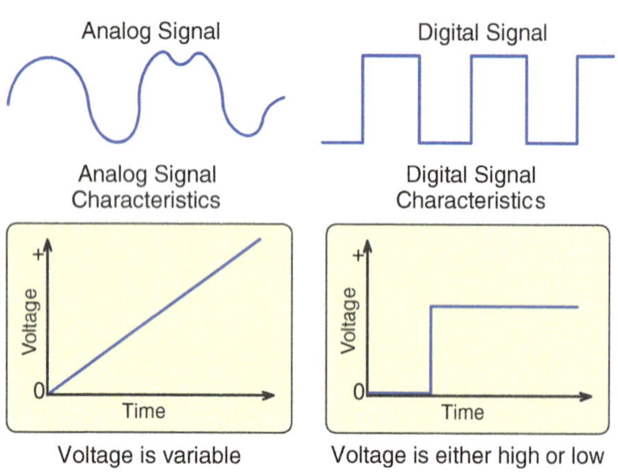

FIGURE 10-45 Electrical signal waveforms of two basic types of electrical signals: digital and analog.

Outputs

The outputs of an electronic control system are electrical signals produced by the processing section of the system. Output electrical signals are generated to operate anything from digital displays or a current to operate solenoids, injectors, electromagnetic actuators, motors, lights, or other electromechanical devices.

Types of Electrical Signals

Before looking at the elements of the electronic management system, it is first important to look at types of electrical signals used in information-processing systems (**FIGURE 10-45**). The following are three types of electrical signals commonly used as either inputs and/or outputs in electronic engine control applications:

1. analog
2. digital
3. pulse-width modulation (PWM).

Analog Signals

An analog signal is electric current that is proportional to a continuously changing variable. Analog signals then will have a changing value of voltage, amperage frequency, or amplitude. For example, temperature can change continuously and can have an infinite number of steps between each degree of change. A thermometer measuring temperature change can represent every possible temperature with the movement of liquid in a glass or a hand on a dial. Similarly, an analog electrical signal would represent the smallest change in temperature proportional to the movement of liquid or hand on the dial.

Measuring alternating electrical current is another example of an analog signal. Variable reluctance-type sensors, such as wheel speed or some engine position sensors, will produce an alternating current (AC). Changing wheel or engine speeds will continuously alter the frequency of current polarity change leaving the sensor. The intensity of the voltage will also continuously vary with speed.

A throttle-position sensor is an example where analog data can is commonly collected. The electrical signal produced by the sensor varies proportionally to pedal angle. The voltage output from the sensor will vary with the driver input or pedal angle. (**FIGURE 10-46**).

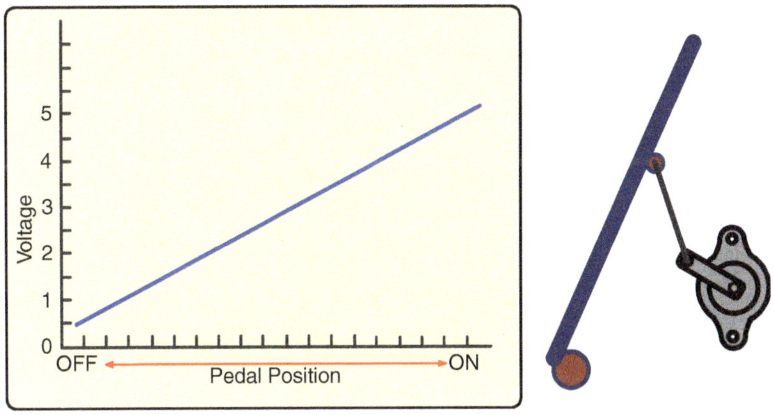

FIGURE 10-46 The signal voltage from this throttle-position sensor is a type of analog data. An infinite number of values for voltage exist between idle and WOT.

Outputs can be analog as well. The intensity of a light or a sound of an output device, such as a lamp or speaker, can be produced by varying the voltage and frequency of an electrical signal. A light is dimmed or brightened by increasing or decreasing current to a bulb. An analog signal representing sound produces loudness and pitch by varying the voltage and frequency current to a speaker.

Digital Signals

In contrast to analog signals, digital signals do not vary in voltage, frequency, or amplitude. Instead, they are electrical signals that represent data in discrete, finite values. This means the data is broken down into separate or smaller meaningful values. For example, the movement of hands on an analog clock will represent time in every possible value. However, a digital watch represents time in finite values, such as minutes and seconds. A digital multimeter (DMM) represents data in the same way as a digital watch. The numerical display for an electrical measurement is represented as a fixed number (**FIGURE 10-47**). In contrast, an analog meter would measure the same electrical value by using a sweeping needle on a scale. A more common understanding of digital signals describes them as representing data by using only two conditions or values. This can be on or off, yes or no, 1 or 0, open or closed, up or down, etc.

FIGURE 10-47 This DMM data represents resistance as a fixed precise value. Smaller changes in resistance—several places to the right of the decimal point—are not measurable.

Binary code is an example of a digital signal. Every number from 0 to infinity and the letters of the alphabet are represented by a combination of 0s and 1s. Binary code easily lends itself to use in microprocessor circuits, where processing large amounts of alphabetic or numerical data, represented in strings of 0s or 1s, is performed. Computerized powertrain control systems process information electronically by using digital signals and binary code. This means that all information, whether analog or alphabetical, is converted into 1s and 0s. Using long strings of 1s and 0s may seem cumbersome, but just as Morse code tapped out on telegraphs could send information by using only dots and dashes, the 1s and 0s of binary code can satisfactorily report all kinds of information (**FIGURE 10-48**). The difference between digital and Morse code is

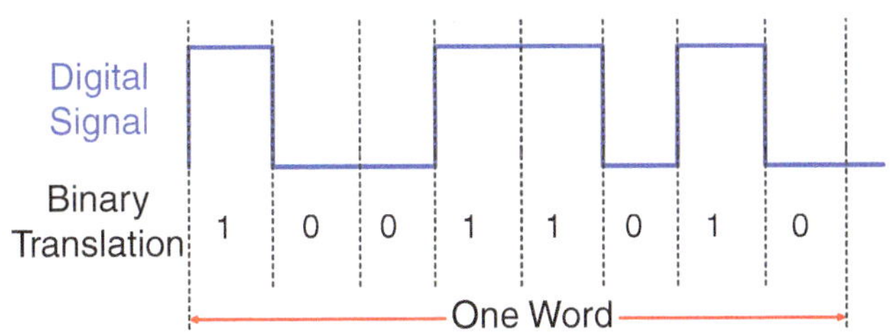

FIGURE 10-48 A bit is the smallest piece of digital information that can be either a 1 or a 0. A byte is a unit of 8 bits. Binary code represents letters and numbers in strings of 1s and 0s. Digital data can be represented as 1s and 0s.

FIGURE 10-49 This injector driver uses a 32-bit microprocessor plus a number of microcontrollers to operate the fuel injectors of a hydraulically actuated electronic unit injector (HEUI) engine. **A.** Field effect transistor FET output drivers. **B.** DC-DC voltage step-up. **C.** Microprocessor. **D.** Memory. **E.** Microcontroller CAN transceiver.

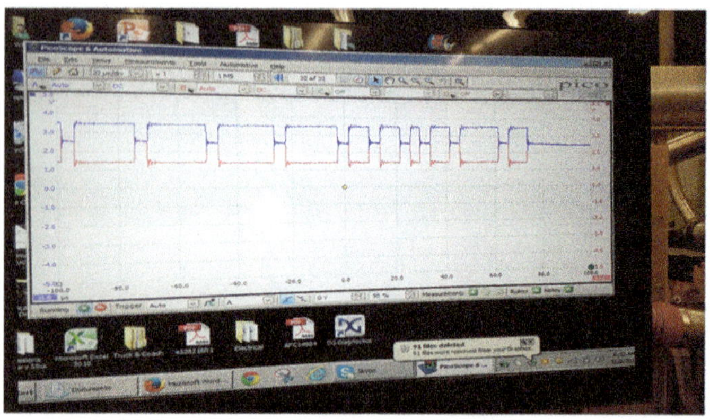

FIGURE 10-50 A datalink waveform showing serial data. Serial data transmits a series of 1s and 0s and has a digital form. The wide part of the waveform represents 1 or a string of 1s, whereas the narrow part of the wave represents 0 or a string of 0s.

in the speed and accuracy of electronic processing. Processing millions and billions of 1s and 0s per second is something digital electronics can do to compensate for the cumbersomeness of using only 1s and 0s to communicate alphanumeric data.

Bits and Bytes

A bit is a shortened term for binary digit. This is the smallest piece of digital or binary information and is represented by a single 0 or 1. As illustrated in FIGURE 19-15, a byte is a combination of 8 bits. The engine speed data is processed in the engine control module, also called ECM or ECU, and is measured in bits. The number of bits it can process during one central processing unit (CPU) clock cycle classifies powertrain ECMs and computers. Desktop or laptop computers often use 64-bit processors. A Pentium IV processor is 32-bits, whereas a late-model ECM will have 16-bit or 32-bit capability. A 3.0 MHz processor will have three million clock cycles per second, which is used as a signal to start and stop processing steps—much like how a drum beat maintains a rhythm. This means a 32-bit, 3.0 MHz processor processes 96 million bits of data per second. While the clock speeds and bit size of the processors in an engine ECM are smaller than those in an average desktop, so is the programming code. The capabilities of an engine ECM may appear to lag a personal computer, but the PC operates by using hundreds of complex software programs. An engine or powertrain control microprocessor that operates by using only one program with much simpler software code to process information and produce output signals has enormous processing capability (**FIGURE 10-49**). Today's vehicle processors have many times the digital processing capabilities of the onboard computers that were used to send Apollo rockets to the moon.

Serial Data

Understanding serial data is also important when discussing binary code and digital signals. The term "serial data" originates from the way data is transmitted. It is in series, one bit after another along a single wire or a pair of wires. When serial digital data is transmitted along a pair of wires, each wire will transmit a voltage pulse represented as a rectangular waveform. The wires will have a differential voltage, which means the voltage on the wire pair is a mirror opposite voltage when transmitting serial data (**FIGURE 10-50**). A large differential voltage pulse represents a 1, whereas a small differential voltage pulse represents a 0. Serial data is used to transmit information from one electronic module to another. Onboard data networks share information and control vehicle operation by using serial data. More important to the technician, electronic service tools will use serial data to receive and send data. The rate at which serial data is transmitted is referred to as the baud rate. Baud rate refers to the number of data bits transmitted per second.

Analog-to-Digital Conversion

Because electronic processing units can handle only binary digital data, analog signals are converted to digital signals in a process called analog-to-digital conversion (**FIGURE 10-51**). To convert analog signals to digital binary information, special circuits—known as buffers or analog-to-digital (AD) converters—are used (**FIGURE 10-52**). To convert an analog signal, the electronics do a couple of things. First, the changing analog signal is sampled or divided up into segments like a loaf of bread. In one second of time, the varying analog

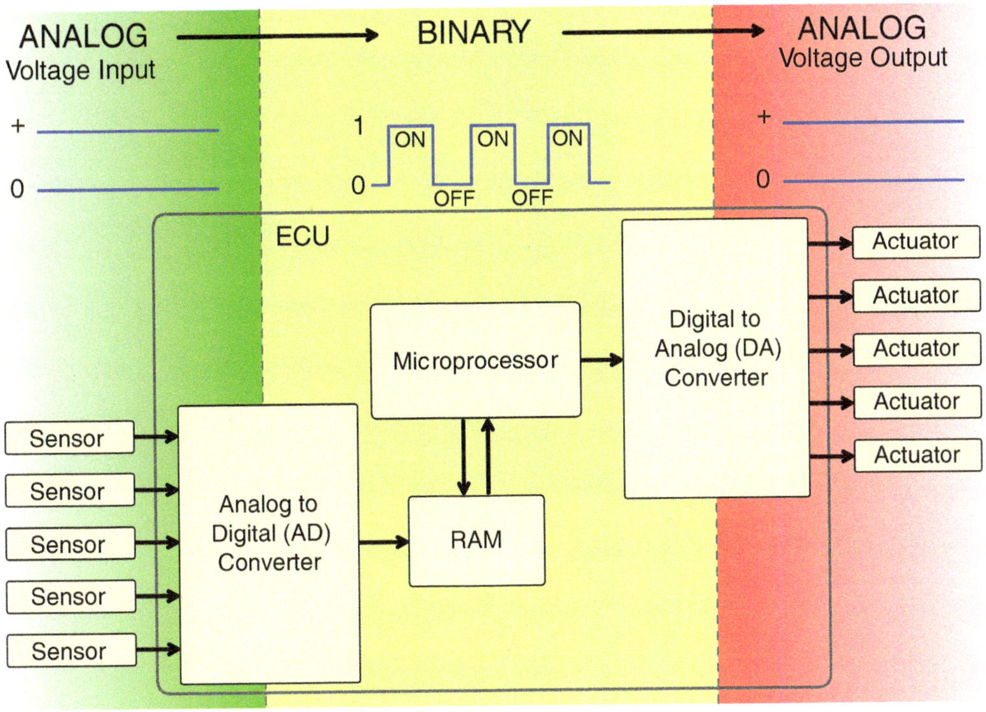

FIGURE 10-51 AD conversion occurs in both of the input circuits of the ECM. Digital-to-analog signal conversion may also occur in the output circuits of the ECM.

signal could be sampled 10, 100, or even 1,000 times (**FIGURE 10-53**). Each of these segments will represent a specific voltage value. The finer or more accurate the processor wants the data to be, the more frequent the sampling rate, resulting in better signal resolution, also called fidelity. Each of the segments will be assigned a digital value that is translated into a binary number. MP3 sound files are an example of a conversion from analog (wave file) to digital. A digital wave file could be sampled 64K times a second or 128K times a second or more. The higher fidelity—or faithfulness to the original analog signal—is achieved at the more frequent sample rate.

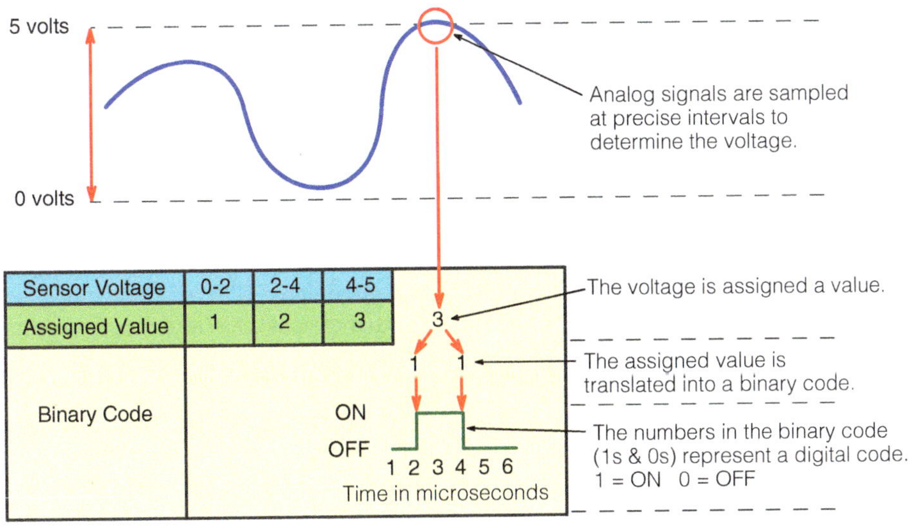

FIGURE 10-52 An AD conversion. An analog waveform is sampled and measured many times a second to generate a digital representation of the waveform. This process is identical to forming MP3 files. The more frequently the signal is sampled—128K, 256K, or more times—the higher the signal quality.

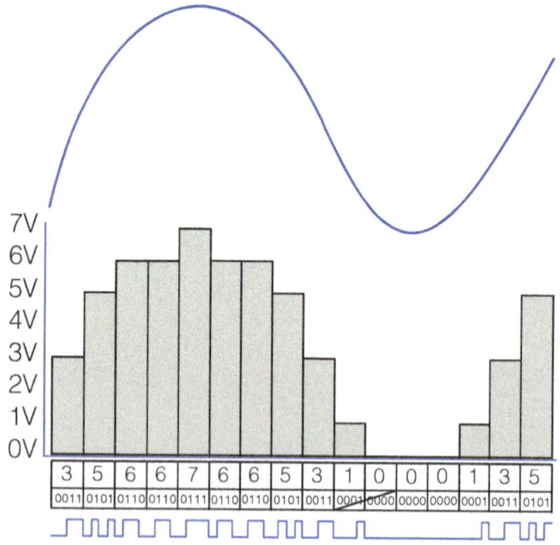

FIGURE 10-53 Analog signals from engine sensors are converted into digital signals for processing by the ECM.

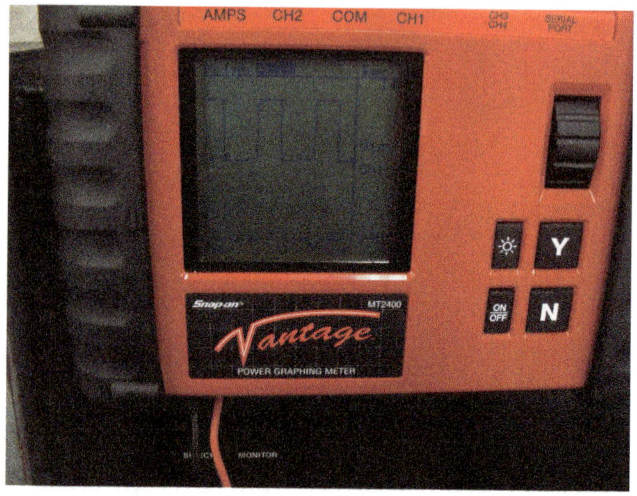

FIGURE 10-54 A PWM signal displayed on a graphing meter. Notice that the widths of the pulses are similar.

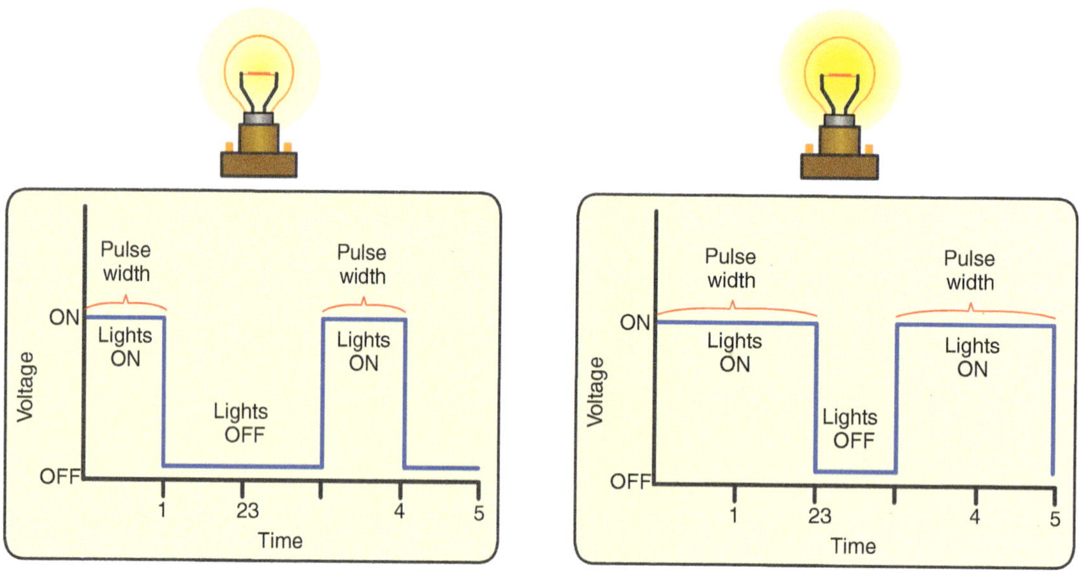

FIGURE 10-55 The longer the pulse width, the brighter the light, since more current flows through the circuit when pulse-width on time lengthens.

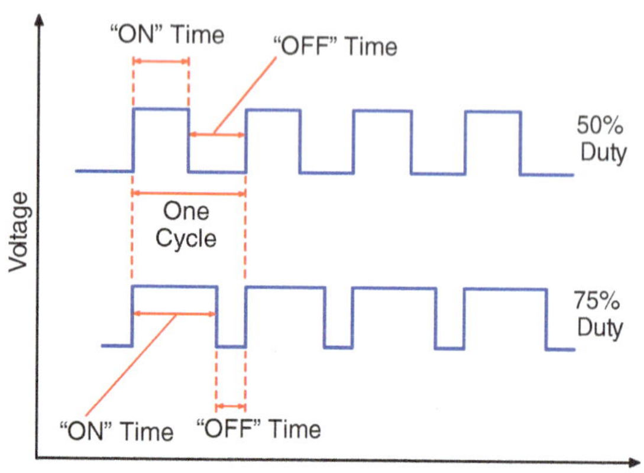

FIGURE 10-56 Duty cycle is a comparison of on time to off time in one cycle. A cycle can be 1 second, 500 msec, or any length of time, but the cycle time is fixed when measuring pulse width.

Pulse-Width Modulation

An electrical signal that shares similar characteristics with both a digital and analog signal is the pulse-width–modulated (PWM) electrical signal (**FIGURE 10-54**, **FIGURE 10-55**, and **FIGURE 10-56**). PWM refers to a signal that varies in on and off time. That means it is digital in one aspect because it represents data in two states only—either on or off, or switched high or low. However, information is also conveyed by the amount of time the signal stays on or off. Time on or off is variable, which gives it an analog characteristic. The units for measuring pulse width are always expressed in units of time. Time is the measure of how long the signal is switched high or on. To understand PWM, consider a light illuminated by a PWM signal. In one second of time, the light may be cycled on and off once. If the signal is switched on for one-quarter of a second, the pulse width would be 0.25 seconds wide (**FIGURE 10-57**). Common examples of devices using

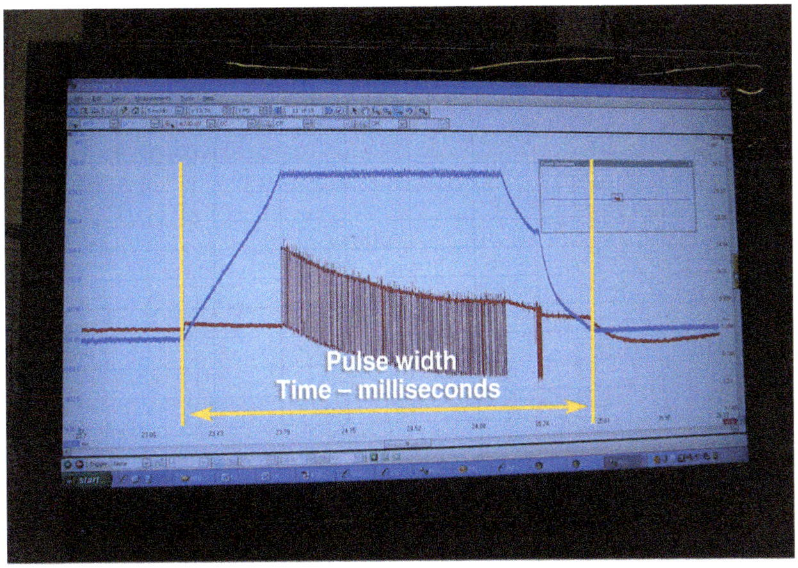

FIGURE 10-57 The pulse width of the energization time of this HEUI injector solenoid is measured in milliseconds.

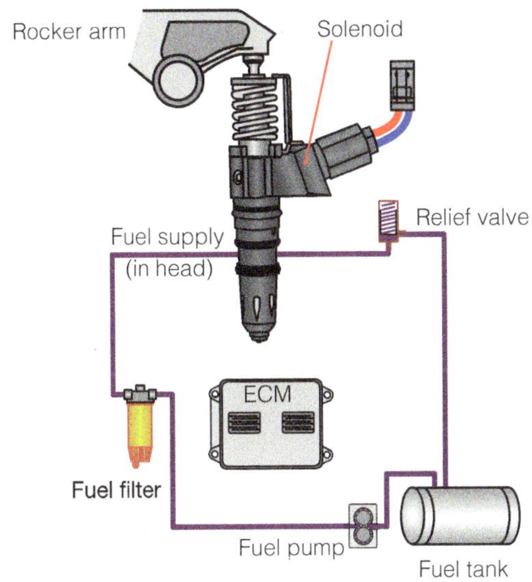

PWM signals are solenoids, injectors, and light circuits. A PWM signal's units of measurement are typically milliseconds (msec). PWM signals are commonly used as an output signal of an ECM. For example, the current supplied to a fuel injector or the pressure regulator of a HEUI or common rail pump is changed by varying the amount of on time (time during which it is energized) of the electromagnetic control valve (**FIGURE 10-58** and **FIGURE 10-59**). Output drivers of microprocessors are types of switches, usually switching transistors, which produce PWM signals to operate devices in an On or Off state (**FIGURE 10-60** and **FIGURE 10-61**). The microprocessor device can also

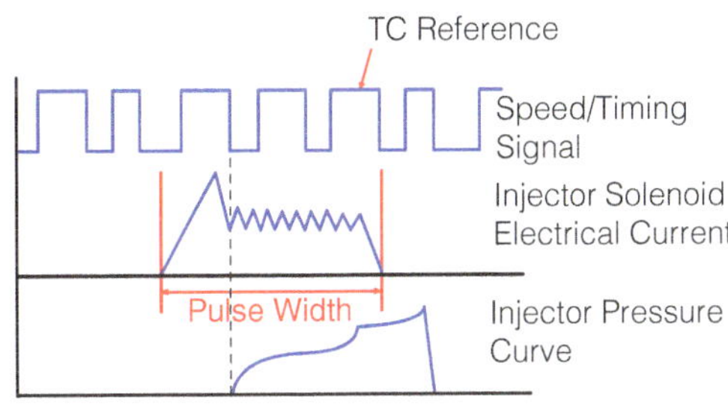

FIGURE 10-58 Varying the length of time it takes for electrical current to energize an injector solenoid will change the quantity of fuel injected into a cylinder.

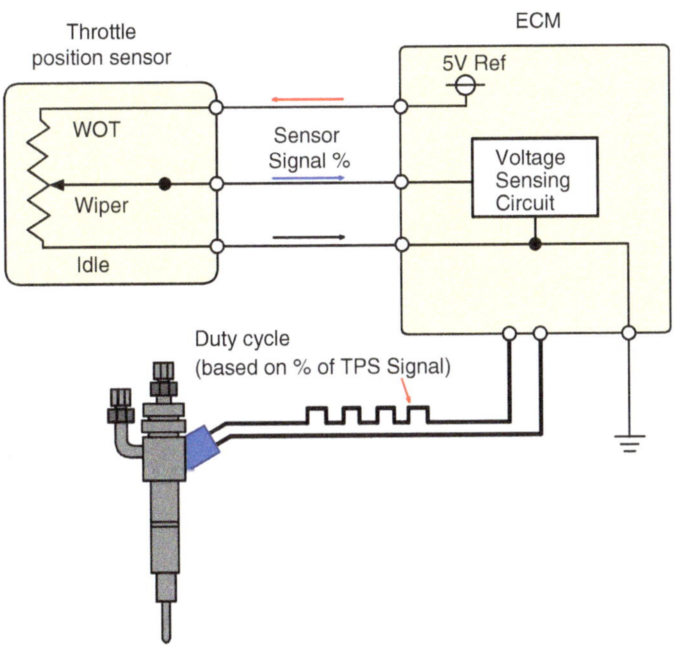

FIGURE 10-59 The ECM will use the throttle-position sensor as one of many sensors to calculate the pulse width applied to an injector solenoid.

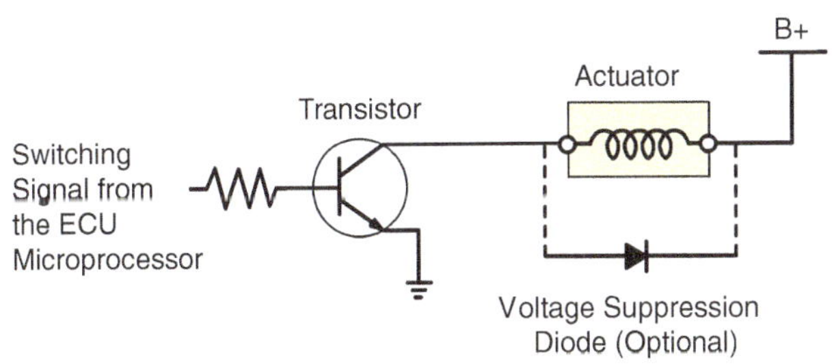

FIGURE 10-60 An output driver of an ECM is usually a switching-type transistor. When the microprocessor applies a small amount of current to the base of the driver, a larger amount of current flows through the transistor to the output device.

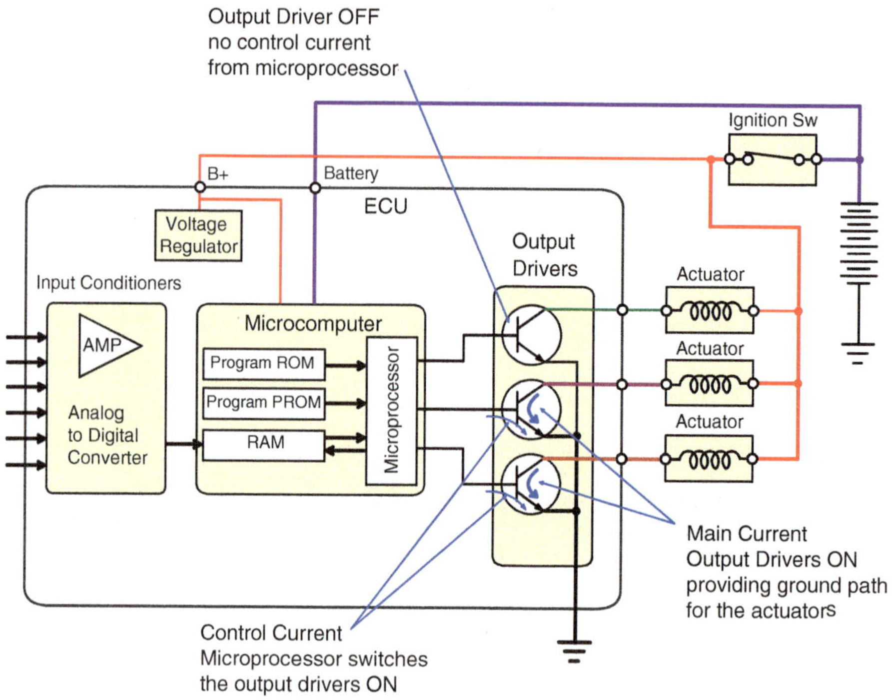

FIGURE 10-61 The microprocessor controls the operation of the output drivers, which switch current flow on and off to electrical devices.

easily vary the time during which a transistor-type output driver conducts electrical current. Some manufacturers use sensors that use PWM signals to transmit data. A PWM-type throttle-position sensor is unaffected by voltage drops encountered through long runs of wiring harnesses and multiple connectors between the sensor and ECM.

Duty Cycle

Related to the term "pulse width" is duty cycle (**FIGURE 10-62**). Duty cycle is another unit for measurement for PWM signals. While a pulse width is measured in time, duty cycle is measured as a percentage—on time versus off time. Duty cycle refers to the percentage of time during which a PWM signal is switched high or on, in comparison to off time (**FIGURE 10-63**). The duration of one on time to one off time for a PWM signal represents one cycle. Duty cycle units are expressed as a percentage of cycle time. For example, if the pulse width is 0.8 seconds and the off time is 0.2 seconds, a cycle is 1 second in length.

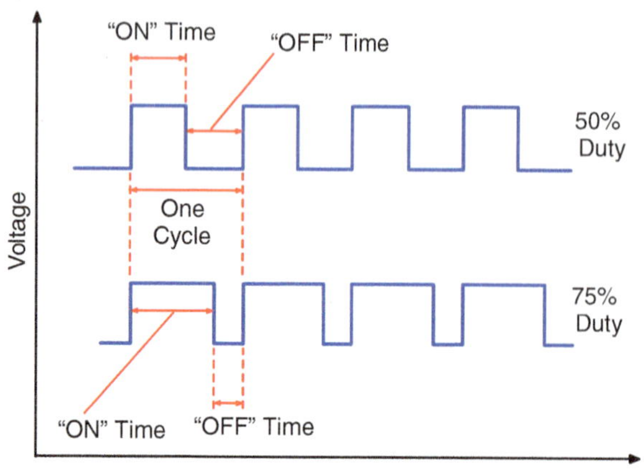

FIGURE 10-62 Duty cycle is a comparison of on time to off time in one cycle. A cycle can be 1 second, 500 msec, or any length of time, but the cycle time is fixed when measuring duty cycle.

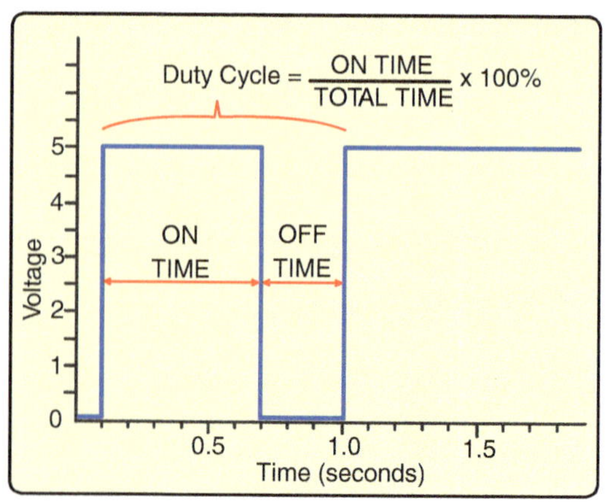

FIGURE 10-63 The measurement units for a PWM signal can also be a duty cycle. Duty cycle is expressed as a percentage of on time versus off time.

This means the duty cycle is 80%. A 100% duty cycle means the signal is on all the time, whereas a 0% duty cycle is off. Another way of expressing this relationship is signal off time versus signal on time. A signal that is applied for three-quarters of a cycle is 75% duty cycle. What is different about duty cycle in comparison with PWM is where the signal is used. Duty cycle is commonly used to measure the time a signal is applied to an output device operating at a fixed cycle frequency, whereas pulse width measures a signal applied to devices operating at a varying frequency interval. For example, an engine may speed up and slow down, so the pulse width or time an injector is energized will vary with speed. It is practical to measure actuation or energization time only, since it is difficult to always know the frequency of a cycle—rpm, in this instance. Depending on engine speed, 10 or 20 injections may take place in 1 second, making it practical to measure only pulse width. Consider, however, an electrohydraulic pressure regulator. This device will have a PWM signal applied to close a valve and increase pressure. Removing the signal will cause pressure to decrease. The time the signal is applied is broken into fixed time intervals. Therefore, a solenoid for this device may be on for 0.20 seconds out of fixed 1 second intervals. This would give it a pulse width of 0.20 seconds but a duty cycle of 20%. Pulse width could increase or decrease, changing duty cycle (**FIGURE 10-64**). To interpret system operation practically, a measurement of duty cycle is more meaningful.

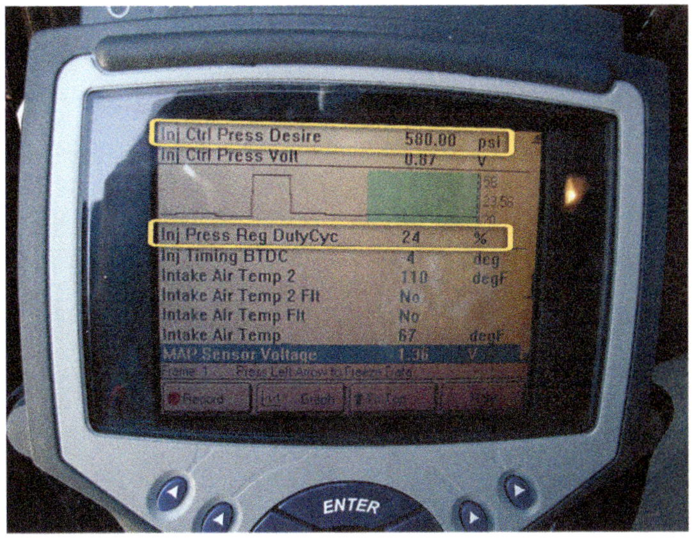

FIGURE 10-64 The duty cycle of this injection control pressure regulator for a HEUI fuel system is graphed and measured in duty cycle.

Processing Function

Processing electronic signals used in vehicle control systems is the function of ECMs. Referred to as the ECM, a microprocessor (or microcontroller) is the heart of the control unit. Made from hundreds if not hundreds of thousands of transistors contained in a semiconductor chip, the integrated circuit chips making up microprocessors and various controllers will contain a minimal amount of memory, plus input and output circuits. Microprocessors have more memory, which gives them the capability to perform advanced calculations and follow software-based instructions (**FIGURE 10-65**). In contrast, microcontrollers are less capable and carry out specific functions built into the chip design. An ECM will contain several types of memory and output drivers that control the operation of electrically operated devices, such as injectors and relays, as well as complex signal conditioning circuits for information-processing functions. An ECM will also contain a CAN transceiver, which enables it to receive and send communication signals to an onboard vehicle network.

FIGURE 10-65 Integrated circuits used on a late-model Mercedes Benz engine control module. A. Microcontroller. B. RAM and ROM. C. Flash memory. D. Microprocessor.

Several types of integrated circuit devices or "chips" are on board a typical ECM, which are essential to processing and to ECM operation:

- the clock—which starts and stops a processing cycle to keep processing functions synchronized inside an ECM
- microprocessor—which uses software instructions to execute complex programing steps and solve complex mathematical problems
- microcontrollers—which have a very limited, specific function
- AD converters—analog-to-digital converters and digital-to-analog converters
- memory—information and data storage.
- CAN transceiver—microcontrollers that send and receive serial data over vehicle networks

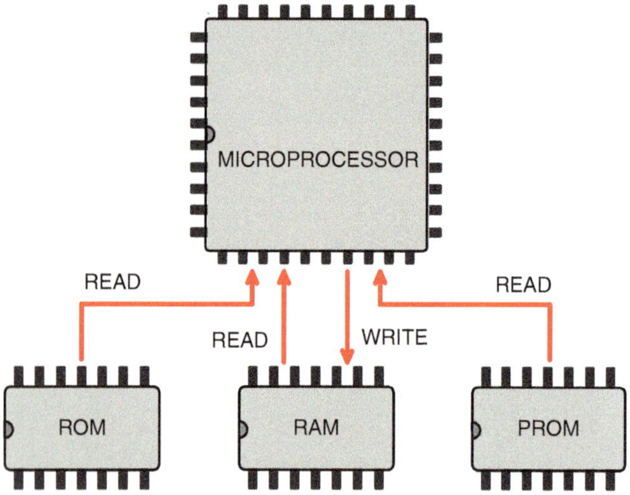

FIGURE 10-66 Several types of memory support the processing functions of the microprocessor. PROM stores specific operational instructions.

Computer Memory

Several types of memory are used in an ECM, depending on its application. Some memory is used to store data from sensors since the ECM cannot process all sensor data simultaneously (**FIGURE 10-66**). Other types of memory are required to store the instructions for operating the microprocessor. This memory would store software code or lookup tables to give the engine its unique operational characteristics. The following are some of the common categories of memory:

- read-only memory (ROM)
- random-access memory (RAM)
- programmable read-only memory (PROM)
- electrically erasable programmable read-only memory (EEPROM)
- flash memory or nonvolatile RAM (NVRAM), which is a ROM/RAM hybrid that can be written to but that does not require power to maintain its contents.

Read-Only Memory

ROM is used for permanent storage of instructions and fixed values used by the ECM that control the microprocessor. Information stored in ROM would include algorithms such as how to calculate the pulse width for the injectors or the lookup tables used by algorithms to solve mathematical problems. One example would be stored values for horsepower ratings for the vehicle. Other fixed values would identify a maximum factory engine rpm, the temperature value for an engine overheat condition, or the type of transmission. The ECM reads the instructions, but it cannot write to or change the instructions contained in ROM. ROM data is stored by the manufacturer. ROM is permanent and is not lost even if power to the computer is interrupted. This means the memory is nonvolatile (**FIGURE 10-67**).

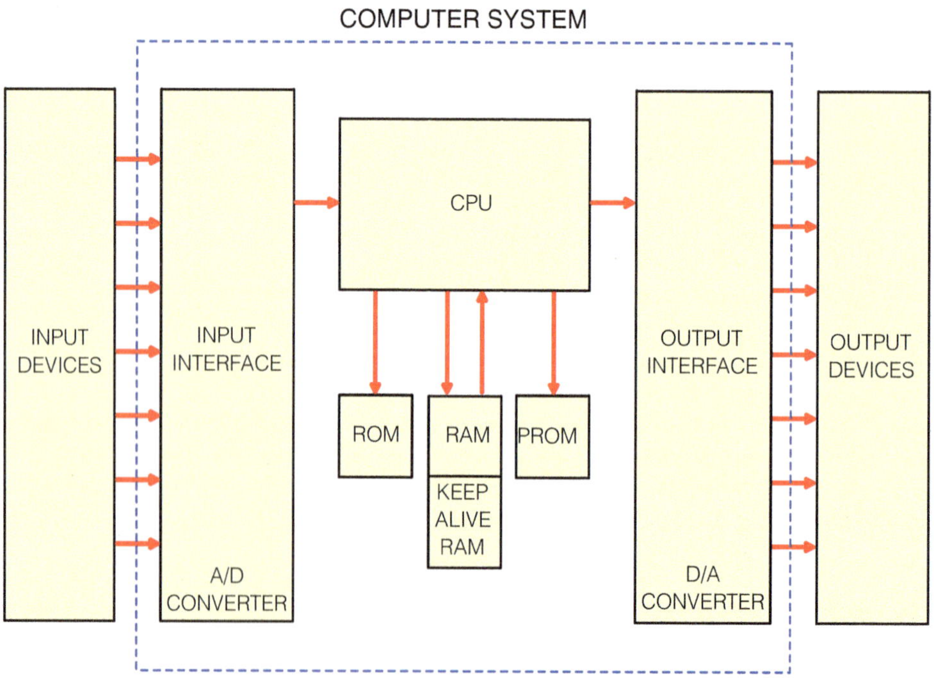

FIGURE 10-67 "Nonvolatile," in the context of computer memory, means the information is not lost when power to the ECM is disconnected or the ignition is switched off. Keep-alive memory refers to memory that is retained only due to a constant supply of current to the ECM when the ignition is switched off.

Random-Access Memory

RAM is a temporary storage place for information that needs to be quickly accessed. Input data from sensors is commonly stored in RAM waiting processing by the ECM. RAM memory is both readable and writable. Most RAM memory is designed to be lost when power is interrupted, such as turning the ignition key off. That is why RAM is often referred to as temporary memory storage. However, RAM can be stored in the ECM after the key is shut off. Nonvolatile RAM holds its information even when the power has been turned off. Volatile RAM will be erased when the power has been turned off. If volatile RAM receives its power from the ignition key, its memory is lost when the ignition is turned off. If battery power is used to keep the RAM memory intact when the key is off, it is also known as keep-alive memory (KAM).

EEPROM and Flash Memory

EEPROM was developed to allow manufacturers to change the software that operates the ECM electronically rather than physically fix it into the ECM during its design and construction. In recent years, flash memory, which is nonvolatile, is EEPROM memory. This is almost identical to the type used in a USB memory stick. It has enormous shock resistance and durability and is able to withstand intense pressure, extremes of temperature, and immersion in water, which are conditions found in vehicle ECMs. It also offers the convenience of easily reprogramming or recalibrating the ECM, also known as flashing or flash programming.

The software stored in the ECM, also known as a calibration file, is stored in EEPROM and contains a number of lookup tables specific to each output device. Calibration files control interactions between sensor inputs, injection systems, emission controls, and other output devices. Flash programming involves installing lookup tables and algorithms in the ECM. Lookup tables are used by the ECM to solve mathematical problems, called algorithms. An example of a simple algorithm is one that would determine how fast the vehicle is traveling. The mathematical formula for speed, which is distance/time, is an example of a simple algorithm. More complex algorithms involve calculating how much fuel to inject, when to inject fuel, or how long the injector should be energized. The lookup table provides specific data to help solve the problems for a specific engine. Lookup tables are a lot like tables in the back of older math textbooks that would, for example, list a decimal equivalent of an angle for a trigonometry question or an amortization rate for a mortgage. In the latest ECMs, lookup tables are no longer stored by the manufacturer but instead are generated by manufacturer-developed algorithms. These virtual lookup tables are updated whenever the vehicle driving conditions change or a fault is identified or to compensate for component wear and deterioration.

▶ Electronic Control Module Processing Strategies

10-05 Identify and explain the differences between PID and MPC.

Until recent years, ECMs for most part processed signals using a traditional strategy of looping and closed-loop control. Closed-loop control involves monitoring the operation of an output device or actuator such as fuel pressure regulator or turbocharger wastegate position with a boost pressure sensor. Sensor data supplied to the ECM will be plugged into a stored mathematical formula used by the microprocessor to solve a problem such as how much fuel should be injected or when should injection begin. The ECM will collect sensor data from the throttle, engine speed and boost pressure to plug this information into a mathematical formula calculating the answer. Next, the answer from the problem is analyzed using a series of "if-then operators." For example, if fuel-injection pressure is above or below a value established in a calibration lookup table, then do this. (Increase or decrease the pulse width of the signal to the pressure regulator). If it is equal to the lookup-table value, then do not change the signal (**FIGURE 10-68**).

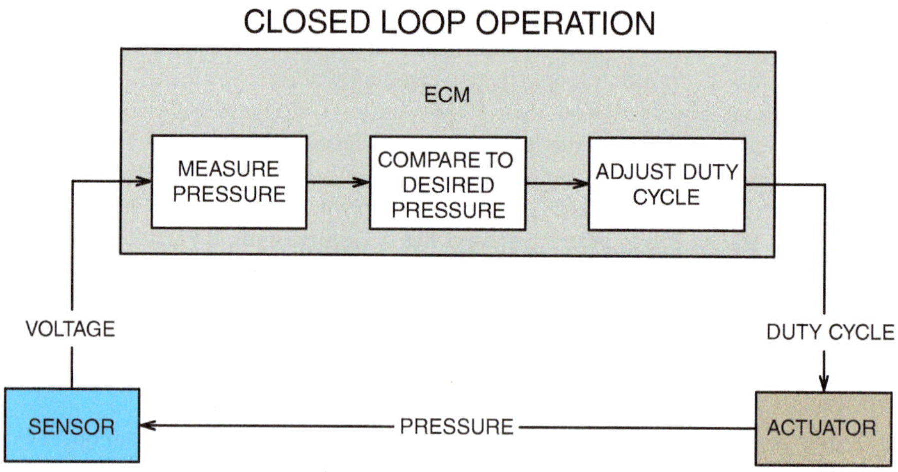

FIGURE 10-68 Closed-loop operation involves regulating the operation of an output device using a sensor to supply the ECM feedback about the output device operation.

Looping Controls

Closed-loop control, one common processing routine, involves a sensor providing direct feedback to the ECM regarding the state or position of an output actuator. Actuators are electromechanical devices controlled directly by electrical signals from the ECM. The ECM will compare sensor data with an expected or commanded value provided by a calibration lookup table and ask a series of "if" questions. If the lookup-table value matches the sensed measurement, then the ECM does not change the electrical signal supplied to the actuator. However, if the value is above or below a commanded or lookup-table value supplied by the calibration files, then the electrical signal changes until the actuator position matches lookup-table values.

Open-loop control is also used to operate devices where the relationship between input and the output state can be easily modeled by a mathematical formula. Several inputs may be used to calculate an output signal. For example, a signal to an injector will be calculated by using multiple sensor inputs and a lookup table, but the ECM would not receive direct feedback from anything like an injector position sensor to adjust the signal (**FIGURE 10-69**).

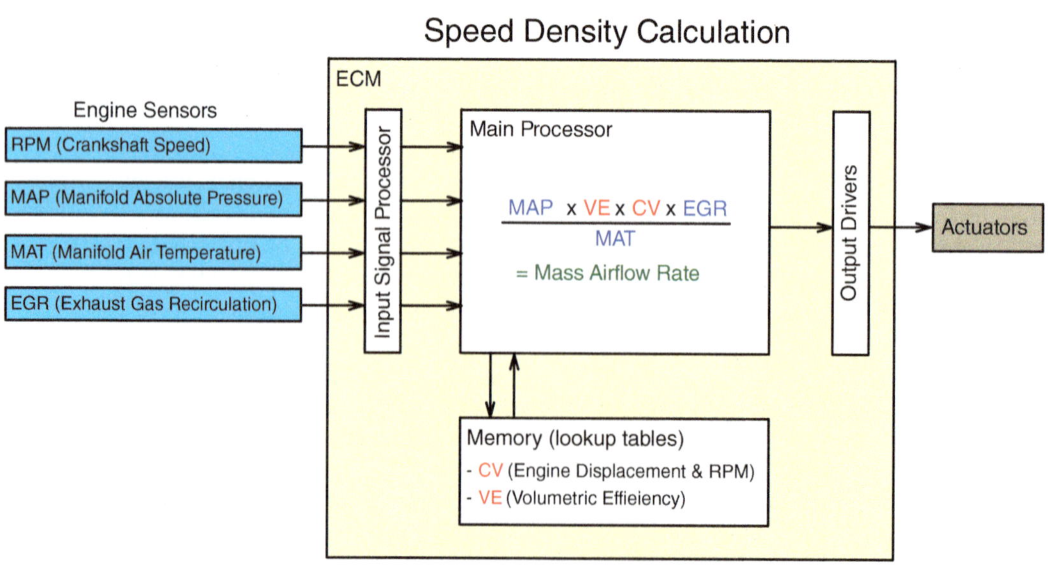

FIGURE 10-69 An open-type loop uses more sensor input and a complex mathematical equation (algorithm) to control engine operation.

Proportional-Integral-Derivative Control

The dependence of ECM engine control on a fixed set of lookup tables contained in an engine calibration file is used by engine control strategy called **proportional-integral-derivative (PID)** control. PID control will make constant corrections to engine operation based on fixed look-up tables stored as part of the ECM engine calibration file. These calibration files are developed in experimental testing of engines and vehicles operating on lab dynamometers (**FIGURE 10-70**).

Conventional PID engine control has several problems. The primary fault is with PID's dependency on fixed lookup tables. Lookup-table data is experimentally derived on a test engine operated at a specific rpm or load and under a specific test condition, such as outside air temperature, fuel rate, etc. When the lookup table is uploaded to an ECM, it is used to help solve mathematical problems such as how much fuel the ECM commands for injection or what the boost pressure and EGR rate should be for a production engine (**FIGURE 10-71**). Although the concept is simple enough, there are many problems with the arrangement. Considering just the parameter of air mass flowing into the engine, which is crucial because of its influence on emission formation, precise measurement can have a significant error. This happens due to an assortment of factors. For instance, proper airflow is often disrupted by exhaust gases lingering in cylinders due to problems such as misadjusted valves or worn camshaft lobes. Compression leaks, inconsistent EGR rates, and uneven airflow through the intake runners can cause cylinder-to-cylinder variations in the air mass filling each cylinder. A lookup table with airflow numbers may also not account for interference with other variables related to combustion, such as moisture content in the air, temperature differences, or other gas impurities. Furthermore, a fixed lookup table can't account for brief unexpected changes in engine operation caused by a wide variety of factors such as operator behavior, atypical driving conditions, sudden changes in throttle positon or anomalies in EGR valve or turbocharger actuator operation. Unaccounted for conditions in lookup-table data sets, plus errors due to unpredictable transient conditions, translate into increased emissions and fuel consumption. Good PID calibrations can account for and correct for error, but some error in adjusting output signals is almost always present in PID type control.

The fact that PID controls for engines using an increasing number of output devices on newer engines is also problematic. ECMs require more processing capabilities than did previous generations, to analyzes a larger number of inputs and output signals faster, using algorithms that are far more complex. Mandated reductions in emissions, along with goals to improve engine performance while reducing fuel consumption, require more than

MAP kPa

Engine RPM	15	21	27	34	40	46	52	58	64	70	76	82	88	94	100	106
400	671	936	1265	1341	1415	1490	1591	1627	1663	1714	1767	1792	1816	1840	1867	1892
800	807	1102	1390	1466	1541	1615	1709	1762	1794	1826	1847	1872	1896	1920	1947	1972
1200	1094	1366	1494	1806	1878	1940	2042	1961	1984	2020	2031	2048	2073	2096	2122	2146
1600	1128	1510	1638	1954	2026	2063	2140	2053	2069	2116	2131	2131	2154	2234	2241	2265
2000	1197	1555	1691	2022	2080	2132	2176	2088	2116	2148	2163	2196	2218	2306	2296	2269
2400	1249	1601	1723	2042	2108	2163	2230	2123	2160	2194	2216	2238	2257	2323	2322	2296
2800	1368	1664	1772	1878	1922	1962	2029	2058	2086	2117	2138	2159	2160	2149	2156	2146
3200	1564	1789	1903	1966	1980	2021	2097	2107	2145	2153	2174	2203	2211	2235	2247	2247
3600	1684	1850	1948	2039	2084	2120	2168	2187	2204	2238	2249	2261	2269	2275	2284	2291
4000	1791	1977	2084	2159	2201	2235	2275	2290	2305	2330	2341	2365	2368	2370	2374	2376
4400	1940	2105	2180	2277	2323	2383	2429	2439	2460	2489	2492	2529	2525	2520	2517	2513
4800	1898	2072	2170	2269	2319	2373	2408	2428	2439	2462	2465	2493	2489	2484	2481	2476
5200	1857	2040	2161	2261	2315	2362	2387	2416	2418	2435	2438	2457	2453	2447	2445	2440
5600	1816	2007	2151	2253	2311	2350	2401	2375	2396	2408	2410	2420	2417	2411	2409	2404
6000	1774	1974	2142	2245	2307	2337	2344	2354	2375	2381	2383	2384	2381	2375	2373	2368
6400	1733	1942	2132	2237	2303	2323	2323	2332	2353	2354	2356	2348	2345	2339	2336	2332
6800	1692	1909	2123	2229	2300	2309	2302	2311	2332	2327	2329	2312	2308	2303	2300	2296

FIGURE 10-70 PID control involves using control loops that use fixed lookup table to continually adjust engine actuator output.

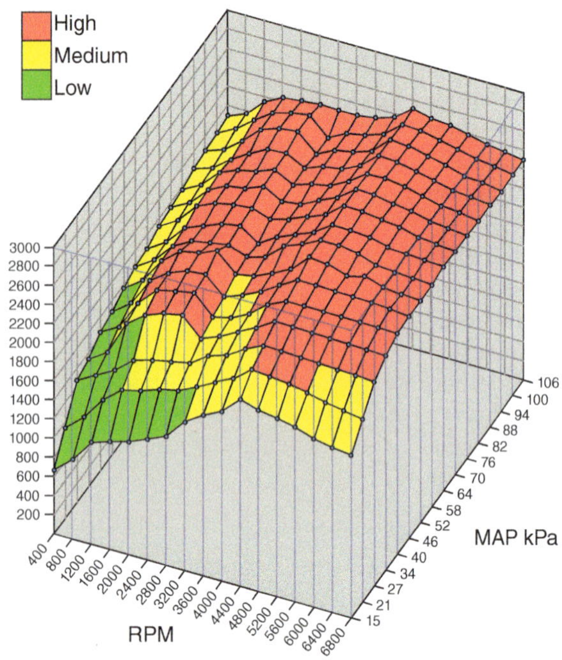

FIGURE 10-71 A lookup table is also represented by a fuel or injection timing map. The table is used to determine what actuator outputs should be under specific conditions.

refinements to only the engine components and emission-control devices. To achieve these outcomes, a substantial change in the traditional PID method for engine control is required (**FIGURE 10-72**).

Model Predictive Engine Control

MPC is the most commonly used engine calibration and control strategy in today's ECMs. Rather than make continual adjustments to engine output devices by using looping-type PID methods, model predictive control relies on just a few fast-responding sensors to generate a complete set of outputs for the engine air pathway and fuel-injection system. The lookup tables are not fixed but rather mathematically generated by using coefficients produced by algorithms supplied with data from only a few selected sensors. Since lookup tables are neither fixed nor stored in the ECM, MPC lookup tables are considered virtual lookup tables because they appear to be real, like PID tables but are continuously generated by algorithms and not stored in the ECM.

MPC replaces lookup tables with hundreds of coefficients used by mathematical equations to predict optimal values for the control system signals and operation. The model for engine operation will generally have features allowing it to adapt to changes in operating conditions, correct for component wear and unusual or atypical system responses. Fixed, predetermined lookup tables and their older generation ECMs weren't able to calculate on the fly, but virtual lookup tables can quickly adapt to uncontrolled system disturbances, component faults and correct for signal output error. The abilities of the faster microprocessors, which have a larger memory storage capacity, and the desktop computer make this possible. Tests comparing newer MPC strategies with lab generated lookup tables, used by PID control have consistently demonstrated MPC is able to achieve substantial reductions in emissions, improved fuel economy, and enhanced drivability performance. Much of the efficiency increase happens because output devices can respond more quickly to an ideal position commanded by the ECM than the slower actuator responses generated by PID control methods. Manufacturers have further benefited from the MPC strategy's shorter calibration development time and from dispensing with error-prone older engine calibration methods (**FIGURE 10-73**).

Described more simply, the MPC calibration strategy makes a decision about a particular *future* control step or sequence based on a current data sample. Like predicting the weather or traffic, the MPC strategy will use past or receding data patterns to determine the next best step of control. Since the prediction horizon for optimal control continually shifts forward, modeled predictive control is also called receding horizon

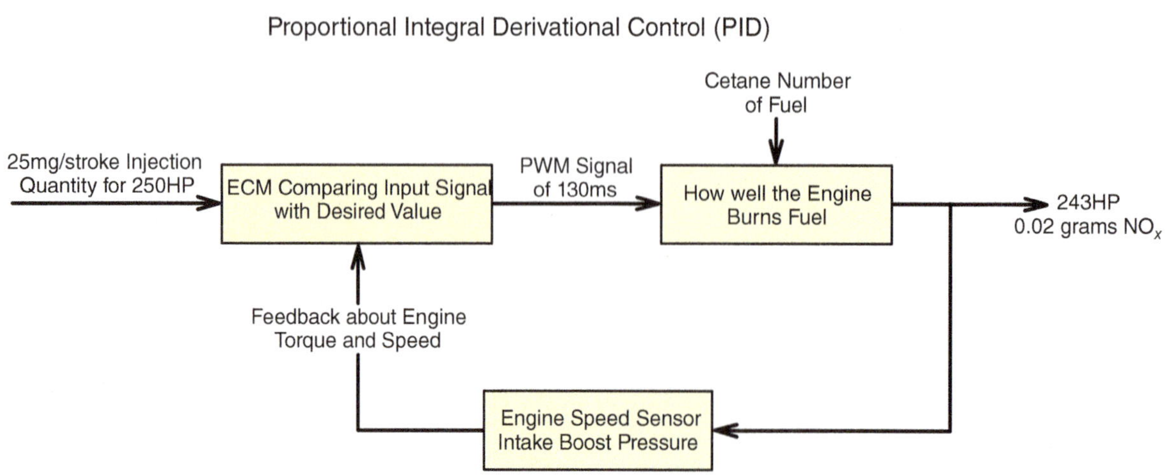

FIGURE 10-72 An example of a PID control that makes continual corrections to engine operation based on fixed lookup-table values.

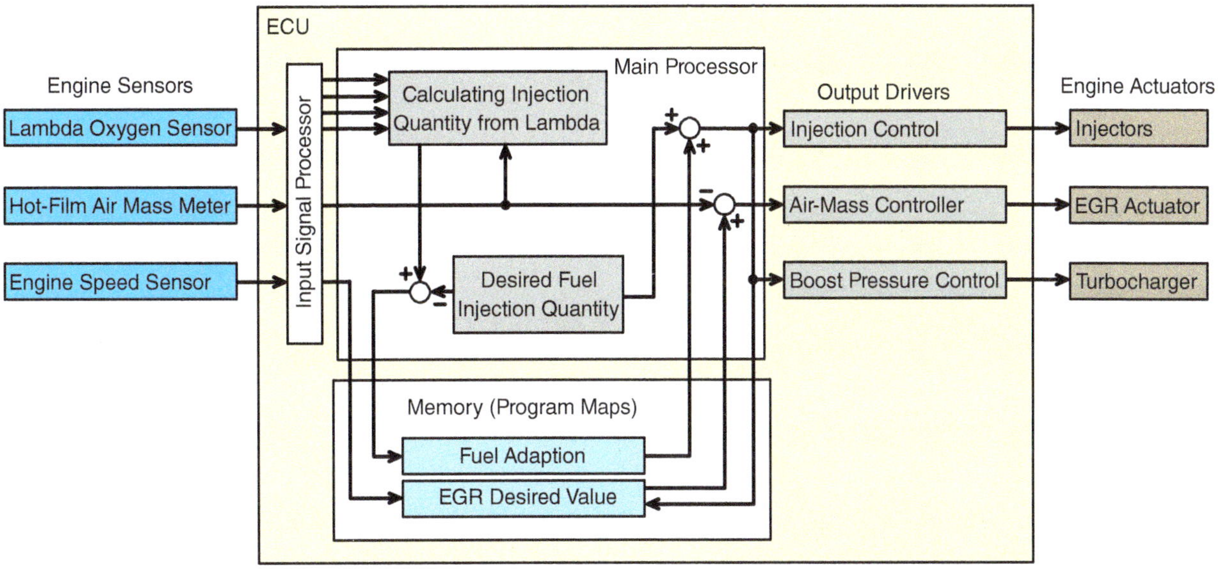

FIGURE 10-73 The use of the oxygen sensor or lambda value is a primary input of most manufactures for MPC of the air pathway and fuel system. Note that just three sensors determine the model selected for model-predicted control of engine operation—rpm, lambda (oxygen sensor), and intake air mass.

control (RHC) (**FIGURE 10-74**). Sampled data from just a few sensors is used to select a particular mathematical coefficient for the MPC algorithms to rapidly calculate all major engine control output signals and produce a virtual lookup table for each actuator. These outputs commonly include turbocharger boost pressure, EGR rates and fuel system operation. Coefficients established for optimal target-operating points are typically derived by acceptable tradeoffs between emissions, efficiency, performance, smooth operation, noise, and other relevant considerations. Once again, the model will have the capability to correct for output errors, plus adapt to system changes due to wear and unexpected outside influences or disturbances.

Explained another way, MPC is a lot like predicting which route to drive home, or how much time the drive will take by using the day and time of the week as input for the decision. Depending on whether its rush hour, a holiday, or the weekend, a driver may predict what driving conditions can be expected and then make decisions about an optimal driving route *before* beginning the journey (**FIGURE 10-75**). This technique contrasts with PID control, which makes changes and corrections to the trip route *during* the journey home. With MPC, changes are made much faster and based on a prediction, not using continuous sensor feedback but instead using a few selected sensors to generate a virtual lookup table for control decisions. A manufacturer may have 10 or more models used to predict optimal

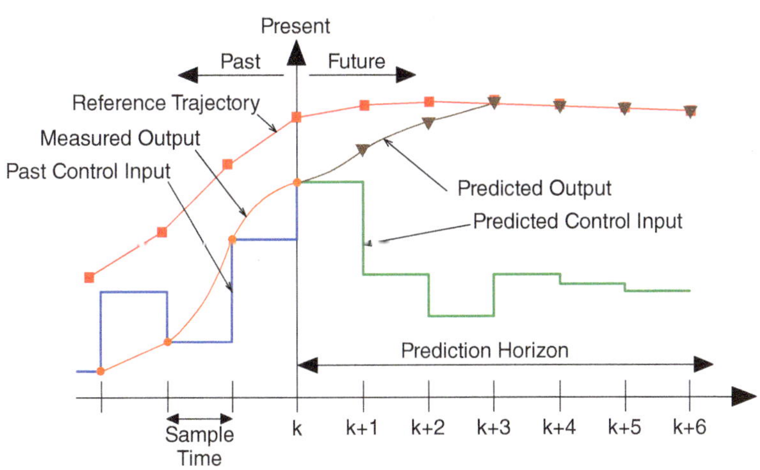

FIGURE 10-74 MPC adapts engine operation for future decisions based on past data samples.

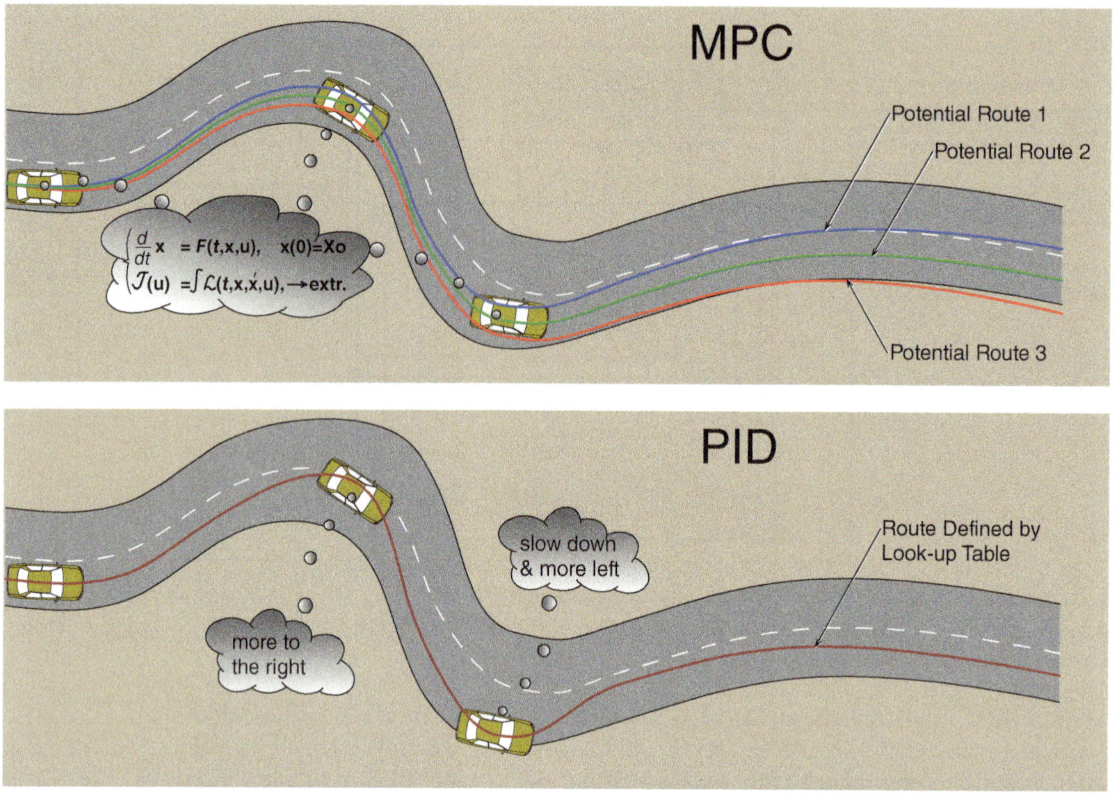

FIGURE 10-75 An analogy of MPC versus PID control. When using MPC, the driver chooses between three possible predetermined routes to adjust steering. Using PID control, the driver consults a lookup table to adjust steering wheel angle.

actuator output states. These models can include highway driving cycles, stop-and-go urban driving conditions, cold-weather operation, sport and economy driving modes, and potentially emission test cycles. In many diesel engines, the throttle-position sensor, vehicle speed sensor, and oxygen or other exhaust gas sensors are typically used for calculating engine load or torque demand, which is the best predictor for adjusting actuators in the injection system. Other MPC actuators in the engine air pathway would include the turbocharger vane position, EGR valve, throttle plate, and exhaust system control valves (such as back pressure control valves). The demand for torque best predicts the driving conditions that an ECM operating with MPC uses to adjust engine output devices (**FIGURE 10-76**).

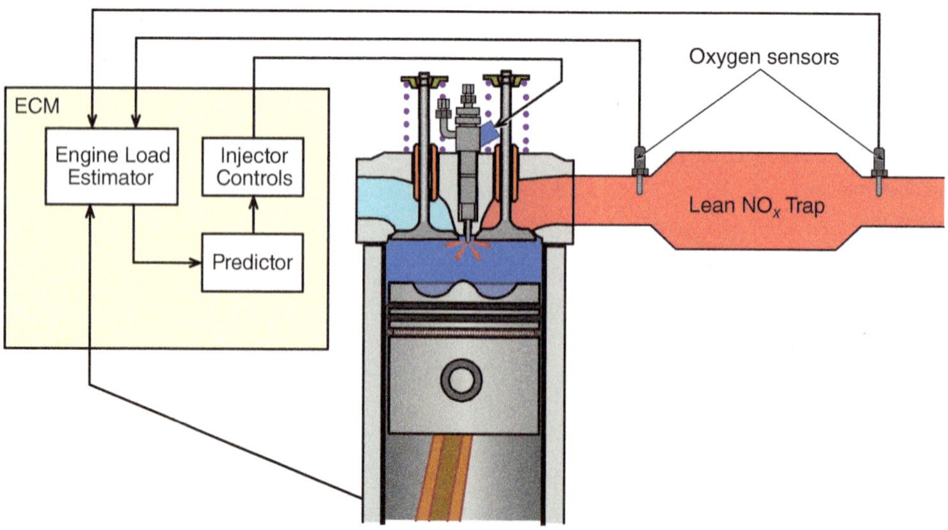

FIGURE 10-76 Oxygen sensors are the common input device to measure torque demand and are used to predict the ideal output state of engine actuators, such as the injectors, turbocharger vane, and EGR valve positions.

Wrap-Up

Ready for Review

- In diesel engines, engine speed and torque output are controlled entirely by the quantity of fuel injected into the cylinders by the high-pressure fuel system. The regulation of the high-pressure injection system is performed by the governor.
- The diesel engine governor monitors the engine speed and adjusts the fuel quantity required by the engine for given speed and load conditions.
- On mechanically or electronically controlled high-pressure injection systems, even small changes in injection quantity produce large changes in engine speed.
- In a diesel engine, because the air intake is unrestricted and fuel-injection rate can instantly go to maximum delivery quantity, the rate of acceleration can go as high as 2,000 rpm/second; this ability can easily cause the engine to go beyond critical rpm in a few seconds.
- Governors perform a critical function in emission reduction by regulating the air-fuel ratio; air-fuel ratio controls minimize the production of PM emissions (black smoke) produced during engine acceleration.
- At a minimum, all governors limit maximum engine speed and control idle speed to prevent stalling.
- Governors are differentiated and classified in a number ways, according to the governor's type of construction, operational characteristics, and fueling strategies.
- The mechanical governor uses mechanical mechanisms to sense engine speed, load, and driver input and to make changes to engine fueling.
- Mechanical governors consist of two basic mechanisms: a speed-sensing mechanism and a fuel-changing mechanism.
- Engine fueling takes place by moving an actuator, such as the fuel control rack, which in turn changes the quantity of fuel injected into the cylinders; the governor can change fueling because it is connected to the injection pump through linkage to control fueling.
- Typically, in mechanically operated governors, the main parts are a gear drive from the engine, flyweights, and a governor spring.
- In a mechanical governor, flyweight force always operates to reduce fueling, whereas spring force operates to increase fueling.
- Torque control is necessary to prevent PM emissions (black smoke) caused by inadequate air or combustion chamber design limits.
- Electronic engine governing uses principles of electronic signal processing to regulate fueling. Electronic governors can be divided into two categories: partial-authority and full-authority systems.
- Partial-authority systems are fuel systems that have adapted some electronic control to extend the design usefulness of a mechanical fuel system, particularly where stricter emissions regulations require cleaner combustion.
- In full-authority governing, all metering and timing functions use principles of electronic signal processing to regulate engine fueling.
- An automotive governor is the most common in light- and medium-duty commercial vehicle applications; it features an accelerator response similar to a throttle plate–controlled, gasoline-fueled engine.
- A variable-speed governor controls engine speed and not torque based only on accelerator pedal position.
- A combination governor combines characteristics of the automotive governor and the variable-speed governor.
- An isochronous governor is used by diesel electric generators and is able to maintain more precise engine speed control than other types of governors.
- Smoke, low power, hunting, engine stalling, poor acceleration, and excessive engine speed are some of the problems that indicate a governor may require evaluation.
- Electronic systems that use microprocessor-based and microcontroller-based control provide operational capabilities far exceeding those of any mechanical system.
- The dominance and sophistication of electronic control makes skill development related to servicing this technology one of the most important priorities for successful technicians.
- Diesel engines were the first commercial vehicle systems transformed by electronic controls. The immediate benefits of these refinements to powertrain control include lower engine emissions, improved fuel economy, increased reliability, and enhanced performance.
- Service technicians, particularly in fleet operations that use medium-duty engines, can access a wealth of diagnostic and service information much faster and with more precise detail than before.
- Built-in electronic management systems allow vehicles to check the operation of circuits and electrical devices, evaluate the rationality of data, and identify problems as they occur. The presence of faults is communicated through the malfunction indicator lamps, electronic service tools, or computer-based diagnostic software.
- Electronic control systems can be broken down into three major divisions: sensing, processing, and output or actuation.
- Sensing functions collect data about operational conditions or the state of a device by measuring some value, such as temperature, position, speed, pressure, or flow.
- Processing collects sensor data and determines outputs based on a set of instructions or program software.
- The outputs of a system are functions performed in response to electrical signals produced by the processor.
- Three types of electrical signals commonly used as either inputs or outputs in electronic engine control applications are analog, digital, and PWM.

- Serial data is used to transmit information from one electronic module to another. Onboard data networks share information and control vehicle operation by using serial data.
- Because electronic processing units can handle only binary digital data; analog signals are converted to digital signals by special circuits known as buffers or AD converters.
- The PWM electrical signal shares similar characteristics with both a digital signal and an analog signal. PWM refers to a signal that varies in on time and off time. A PWM signal is typically measured in milliseconds (msec).
- Duty cycle is another unit of measurement for PWM signals, and it refers to the percentage of time a PWM signal is on versus the time it is off.
- Duty cycle is commonly used to measure the time a signal is applied to an output device operating at a fixed frequency, whereas pulse width measures a signal applied to devices operating at a varying frequency interval.
- ECMs are microprocessors or microcontrollers that process electrical signals. Several types of integrated circuit devices that are on board a typical ECM are essential to processing and ECM operation.
- The central processing unit (CPU) clock is an oscillator inside the microprocessor that controls how fast instructions stored in memory are processed. It is like the drum beat that controls the pace of the work in the microprocessor. The clock speed is measured in hertz (Hz), megahertz (MHz), or gigahertz (GHz).
- Several types of memory are used in an ECM, depending on its application. Some memory is used to store data from sensors because the ECM cannot process all sensor data simultaneously. Other types of memory are required to store the instructions for operating the microprocessor.
- Common categories of memory include ROM, RAM, PROM, EEPROM, and flash memory (a ROM/RAM hybrid that can be written to but that does not require power to maintain its contents).
- Model predictive control (MPC) is the most commonly used engine calibration and control strategy used by today's ECMs.
- Proportional-integrative-derivative (PID) control will make constant corrections to engine operation based on fixed calibration lookup tables stored in the ECM, which are developed in experimental testing of engines and vehicles operating on lab dynamometers.

Key Terms

analog signal An electric current that is proportional to a continuously changing variable.

analog-to-digital (AD) conversion The process when an analog waveform is sampled and measured many times a second to generate a digital representation of the waveform.

automotive governor A governor that features an accelerator response similar to a throttle plate–controlled, gasoline-fueled engine; it is also known as a limiting speed (LS) governor or a min/ max governor. The operator has to change the throttle position in response to engine load and speed changes.

baud rate The rate at which serial data is transmitted.

bit The smallest piece of digital information that is either a 1 or 0.

byte A unit of 8 bits.

centrifugal force A force pulling outward on a rotating body.

closed-loop feedback An ECM control strategy for an output device based on direct sensor feedback, which is used to adjust electrical signals supplied by the ECM to the actuator.

differential voltage The voltage difference on a wire pair when one wire's voltage is the mirror opposite voltage. A wide separation between the voltage pulses represents a 1, and a narrow separation represents a 0.

digital signals Electrical signals that represent data in discrete, finite values. Digital signals are considered binary, meaning that they are either on or off, yes or no, high or low, or 0 or 1.

droop The change or difference in engine speed caused by a change in load.

duty cycle The percentage of time a PWM signal is on versus off.

electrically erasable read-only memory (EEPROM) Nonvolatile memory technology that is used to store operating instructions or programming for an ECM.

flashing Reprogramming or recalibrating the ECM with new information to store in the ECM's memory

fulcrum The point around which a lever rotates and that supports the lever and the load.

full-authority governing A system in which all metering and timing functions use principles of electronic signal processing to regulate engine fueling; this is also known as electronic engine management.

gear-down protection A parameter that limits engine speed if the engine speed is too high for a particular road speed.

high idle The maximum speed at which an engine turns without a load.

hunting A rhythmic change in engine rpm at idle speed caused by uneven fueling of cylinders.

idle An engine's minimum operational speed.

isochronous governor A governor that is able to maintain more precise engine speed control than other types of governors.

keep-alive memory (KAM) Memory that is retained by the ECM when the key is off.

load controller A mechanism that improves ride quality and reduces emissions by modulating increases and decreases in fueling to prevent vehicle bucking and jerking during acceleration and deceleration; it is also known as a pulse dampener.

load-based speed control A feature that supplies increased torque and sometimes more engine rpm when climbing hills with heavy loads, and it also reduces torque when it senses the vehicle is unloaded or only lightly loaded.

lugging An abuse condition where an engine is operated under heavy load below an acceptable operating speed range.

mechanical governor A governor that regulates the quantity of fuel injected by the high-pressure injection pump by using flyweights and spring tension.

microcontroller A special-purpose integrated circuit semiconductor processor with limited capabilities, designed to perform a set of specific tasks.

model predictive control (MPC) An ECM processing strategy that adjusts engine actuators based on virtual lookup tables for a predicted optimal output state for actuators.

nonvolatile memory Memory that is not lost when power is removed or lost.

open-loop control A control strategy where an electrical signal for an output device is calculated based on several sensor inputs and no direct feedback from a sensor.

overrun A governor's ability to keep the engine speed below the high-idle speed when it is rapidly accelerated.

partial-authority system A fuel system that has adapted some electronic control to extend the design usefulness, particularly where stricter emission regulations require cleaner combustion.

peak torque The engine speed at which cylinder pressures are highest.

performance curve A line graph that plots engine torque, fuel consumption, and horsepower (kilowatts) against engine speed.

programmable parameter A measurable value about engine system operation (e.g., maximum vehicle speed) that can be changed by using service software.

programmable read-only memory (PROM) Memory that stores programming information and cannot be easily written over.

progressive shifting A programmable parameter that limits engine speed until a minimum road speed has been reached. Progressive shifting encourages a driver to shift gears sooner to save fuel.

proportional-integral-derivative (PID) control A control that makes constant corrections to engine operation based on fixed calibration lookup tables stored in the ECM, which are developed in experimental testing of engines and vehicles operating on lab dynamometers.

pulse-width modulation (PWM) An electrical signal that varies in on time and off time.

random-access memory (RAM) A temporary storage place for information that needs to be quickly accessed.

rated speed An engine's maximum speed with a load.

read-only memory (ROM) Memory used for permanent storage of instructions and fixed values that are used by the ECM and that control the microprocessor.

state of balance The state of mechanical governor flyweights when flyweight force equals spring force.

torque control The regulation of fuel entering the cylinders to produce an appropriate amount of torque for a given engine speed.

torque plate A device that limits maximum control rack travel in a mechanical injection pump; it is also called a full-load stop.

torque rise The difference between peak torque and torque at rated speed.

variable-speed governor A governor that controls engine speed based only on accelerator pedal position; it is also known as an all-speed governor.

volatile memory A type of data storage that is lost or erased when the ignition power is switched off.

Review Questions

1. Which type of governor requires the operator to adjust engine speed when the load that is applied to an engine changes?
   a. Variable speed.
   b. Limiting speed (LS).
   c. Hydraulic.
   d. Servo.
2. Which type of governor senses changes in engine load and automatically adjusts the fuel rate to compensate for load changes?
   a. Variable speed.
   b. Automotive.
   c. Pneumatic.
   d. Overspeed.
3. Which type of governor will be commonly used to control the fueling of a diesel electric generator?
   a. Isochronus.
   b. Hydraulic.
   c. Load limiting.
   d. Overspeed.
4. During which one of the following conditions will the tension of the starting spring or governor spring of a mechanical governor usually be greater than flyweight force?
   a. During cranking.
   b. While the engine is at idle.
   c. While the engine is at high idle.
   d. While the engine is at rated rpm.
5. The difference between maximum engine speed with a load applied and a no-load speed is often known as __________.
   a. speed droop
   b. stability
   c. underrun
   d. overrun
6. Which one of the following would likely result if the line that connects mechanical air-fuel ratio control device to the turbocharger compressor outlet were broken?
   a. Heavy black smoke during acceleration.
   b. Poor fuel economy.
   c. Excessive engine speed.
   d. Low power.
7. What type of engine emission is controlled by the air-fuel ratio device?
   a. Oxides of nitrogen (NO_x).
   b. Carbon monoxide (CO).
   c. Particulate matter (PM).
   d. Hydrocarbons (HCs).

8. Which of the following torque rise and transmission configurations is best used in an automobile?
   a. A high torque rise engine with few transmission gear steps.
   b. A low torque rise engine with many transmission gear steps.
   c. A high torque rise engine with overdrive transmission gear steps.
   d. A low torque rise engine with few transmission gear steps.
9. Which of the following features is used to prevent driveline windup or jerking during vehicle deceleration?
   a. Load controllers.
   b. Air-fuel ratio control device.
   c. Isochronous governor.
   d. Variable-speed governor.
10. Controlling an output device by using feedback from a single sensor is referred to as __________.
    a. model predictive control (MPC)
    b. proportional-integral-derivative (PID) control
    c. closed-loop feedback
    d. open-loop feedback

ASE Technician A/Technician B–Style Questions

1. Technician A says that any type of governor will always attempt to maintain engine idle speed even if two or three cylinders are misfiring. Technician B says that only a variable-speed governor will compensate for a misfire loss from one or two cylinders and maintain idle speed. Who is correct?
   a. Technician A
   b. Technician B
   c. Both A and B
   d. Neither A nor B
2. Technician A says that any type of governor will work well when pulling heavy loads in off-road conditions without surging and causing driveline damage. Technician B says that an automotive governor is best suited to those operating conditions. Who is correct?
   a. Technician A
   b. Technician B
   c. Both A and B
   d. Neither A nor B
3. Technician A says that a cam position sensor can sense engine speed and position. Technician B says that a crankshaft position sensor can sense engine speed and position Who is correct?
   a. Technician A
   b. Technician B
   c. Both A and B
   d. Neither A nor B
4. While discussing possible causes of a low-power problem in an engine that uses a mechanical governor, Technician A says that weak governor springs can cause the problem. Technician B says that a leaking tube connecting the intake manifold and air-fuel ratio control device can cause the problem. Who is correct?
   a. Technician A
   b. Technician B
   c. Both A and B
   d. Neither A nor B
5. While advising a customer about the consequences of increasing maximum engine rpm with an after market software program that alters the engine calibration and increases power output. Technician A says that the engine will produce more smoke. Technician B says the problem with emissions can be corrected by adjusting the air-fuel ratio control. Who is correct?
   a. Technician A
   b. Technician B
   c. Both A and B
   d. Neither A nor B
6. Technician A says that the units on a scanner for a fuel pressure regulator energized by a PWM signal will be reported in volts and amps of electrical current. Technician B says that the units for a PWM signal may be reported as a percentage or in milliseconds Who is correct?
   a. Technician A
   b. Technician B
   c. Both A and B
   d. Neither A nor B
7. Technicians A and B are discussing which engine would be best to use to pull tree stumps from some land on a farm. The choice is between two engines in two different medium-duty pickup trucks. Technician A says the engine with a peak torque of 850 ft-lb (1,152 N·m) and 325 horsepower (242 kW) is ideal for pulling stumps. Technician B says the other truck, with an engine with a peak torque of 650 ft-lb (881 N·m) and 400 horsepower (198 kW) is better. Who is correct?
   a. Technician A
   b. Technician B
   c. Both A and B
   d. Neither A nor B
8. Technicians A and B are making programmable parameter adjustments for the governor used by an engine to drive a PTO that is operating a hydraulic pump on a landscaping truck. Technician A says that the PTO throttle switch should be configured so that the engine governor that drives the PTO is a variable-speed governor. Technician B says that the engine should be set up to use an automotive-type governor when the PTO is operating. Who is correct?
   a. Technician A
   b. Technician B
   c. Both A and B
   d. Neither A nor B

9. Technician A says that when starting a mechanically governed diesel engine, it is important to push the accelerator pedal to the floor once and then release it. Technician B says the injection pump will automatically go to part throttle when cranking, and so nothing needs to be done. Who is correct?
   a. Technician A
   b. Technician B
   c. Both A and B
   d. Neither A nor B

10. Technician A says that a diesel engine needs a governor to prevent a commercial vehicle from wasting fuel. Technician B says that a diesel engine needs a governor to regulate fuel delivery because diesels are very sensitive to slight changes in the amount of fuel injected and can overspeed. Who is correct?
    a. Technician A
    b. Technician B
    c. Both A and B
    d. Neither A nor B

CHAPTER 11

Hydraulic Nozzles

Learning Objectives

After reading this chapter, you will be able to:

- 11-01 Identify and describe the purpose and operating principles of hydraulic nozzles.
- 11-02 Identify and describe the types, styles, construction, and applications of hydraulic nozzles.
- 11-03 Identify and explain inspection and diagnostic procedures for evaluating hydraulic nozzles.

You Are the Technician

When there is little repair work in technicians' shop, they often use the downtime to rebuild engines and recondition a variety of engine components for use later. You have been asked to remove, test, and recondition the hydraulic nozzles from a number of scrap engines with pump-line-nozzle fuel systems. One of the things you do is carefully record injector part numbers and match them with the engine model and serial numbers. That will help match the injectors with the correct engine application later. After carefully cleaning the injectors with solvent and removing the copper sealing washers, you also clean the injector tips with a brass brush to avoid damaging the injector spray holes. You begin to test the nozzles in a pop tester to sort which injectors meet specifications and which need to be disassembled and rebuilt with replacement parts. Before starting your work, consider the following:

1. Outline all the tests you will need to perform to properly evaluate the hydraulic nozzles for serviceability.
2. What will be your service recommendation for nozzles that do not "chatter" when pop testing? Explain your answer.
3. Of the injectors that failed the pop tests, which parts do you believe will most likely need to be replaced? List parts in order of most likely to need replacement.

▶ Introduction

Injector nozzles are a basic element of the fuel-injection system of all diesel engines. They are used to spray fuel into the combustion chamber and are incorporated into every type of fuel-injection system. The simplest function of the nozzle is to open and close, permitting highly pressurized fuel to squeeze through small orifices in the nozzle tip, which effectively atomizes and distributes fuel into the combustion chamber. The design and operation of a nozzle is therefore critical to combustion efficiency, fuel consumption, and emission formation. Because they operate by using hydraulic pressure, nozzles are also termed hydraulic injector nozzles or simply injectors. Given the importance of injector nozzles, diesel technicians must be able to identify nozzle types, construction features, and operating principles. With this information, they can understand nozzle operation and develop diagnostic strategies and service and maintenance practices associated with these components.

▶ Fundamentals of Hydraulic Nozzles

11-01 Identify and describe the purpose and operating principles of hydraulic nozzles.

The term "injector" is often used to encompass the operation of a hydraulic nozzle. More precisely, though, a diesel engine's fuel injectors are larger components that deliver fuel to the combustion chamber and operate by using widely different high-pressure injection system technologies (**FIGURE 11-1**). Examples of these systems include various types of common rail, unit injection, unit pumps, and hydraulically actuated unit injection. Although injectors vary vastly in construction and operation, nearly all of them have something in common: a hydraulic nozzle incorporated into their design.

Functions of Nozzles

Hydraulic nozzles are composed of a nozzle valve and nozzle body. The **nozzle valve**, or needle valve, is the valve in the nozzle that seals the end of the nozzle tip when it is not injecting fuel. The two basic types of nozzles are the **multi-orifice nozzle**, which has several holes in the tip, and the **pintle nozzle**, which has a single hole (**FIGURE 11-2**). A nozzle's primary function is to atomize and distribute fuel with the help of combustion chamber turbulence. The performance, efficiency, and emissions of diesel engines are largely determined by the fuel **atomization** and distribution process. By breaking fuel into small droplets with more surface area, the combustion chamber is better able to convert fuel into a vapor state, which increases combustion efficiency, reduces emissions, and improves fuel economy. While atomizing fuel, nozzles are required to provide an abrupt beginning and end of injection to prevent fuel droplets from dribbling, rather than spraying into the combustion chamber.

Properly distributing fuel is especially critical to reducing the formation of noxious emissions. Particulate matter (PM), including soot, and oxides of nitrogen (NO_x) are produced from uneven fuel distribution in the combustion chamber. Localized regions inside the combustion chamber that are rich in fuel will typically produce higher temperatures and therefore more NO_x. Soot is also produced when the air-fuel ratio is excessive and there is not enough air making contact with fuel. Large fuel droplets that do not completely burn form soot, which is seen as black smoke emitted from the exhaust system.

FIGURE 11-1 With rare exceptions, hydraulic nozzles are used by all types of injectors. Injectors are built using a wide variety of designs.

Fuel Distribution

Fuel nozzles work with combustion chamber turbulence to distribute and mix fuel throughout the combustion chamber (**FIGURE 11-3**). Because combustion chambers have different degrees of turbulence, the injector nozzle's job of distributing fuel to prepare an air-fuel mixture that burns well varies between engine designs. For example, indirect-injection (IDI) combustion chambers have very high turbulence compared to direct-injection (DI) chambers. This means the type of nozzles used and their operation are very different depending on the

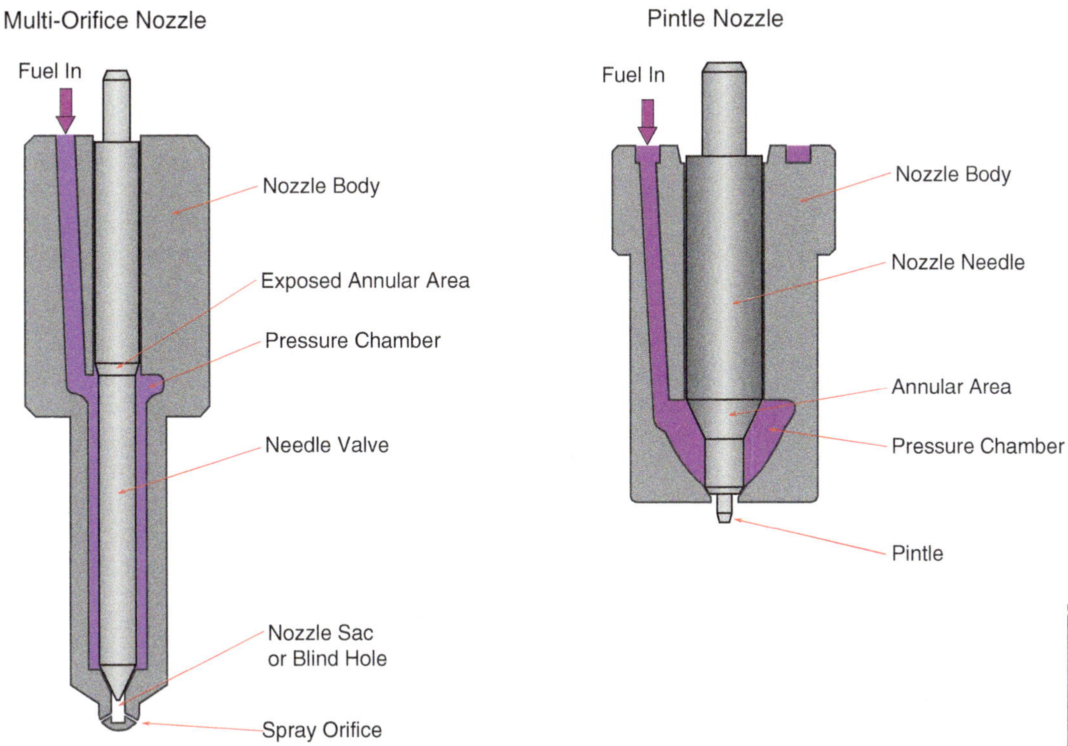

FIGURE 11-2 The two basic types of nozzles are multi-orifice nozzles and pintle nozzles.

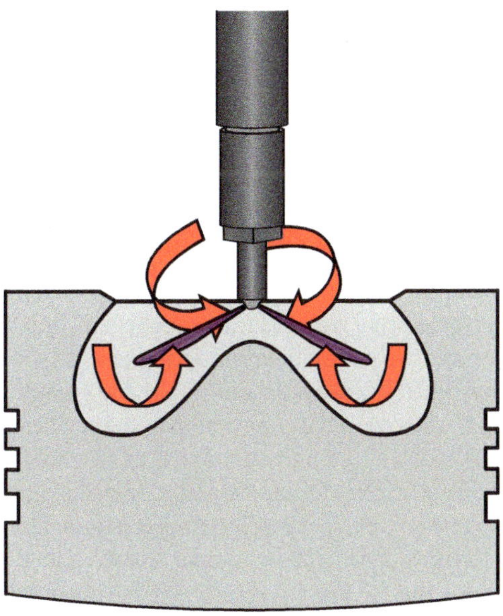

FIGURE 11-3 The injector works with turbulence in the combustion chamber to distribute and mix fuel with air.

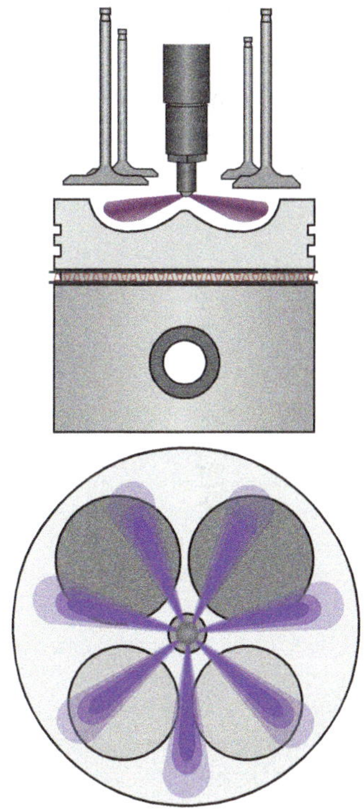

FIGURE 11-4 A DI chamber uses a multi-orifice nozzle to atomize and distribute fuel into the combustion chamber.

combustion chamber design. Centrally located nozzles used by DI combustion chambers use four valves per cylinder and are more effective at symmetrically distributing fuel than offset nozzles found in two-valve cylinders (**FIGURE 11-4** and **FIGURE 11-5**).

Atomization

Atomization is the process of breaking large particles into smaller parts. When the fuel nozzle does this, it breaks the injected quantity of fuel into a spray of very fine droplets that are better suited for combustion.

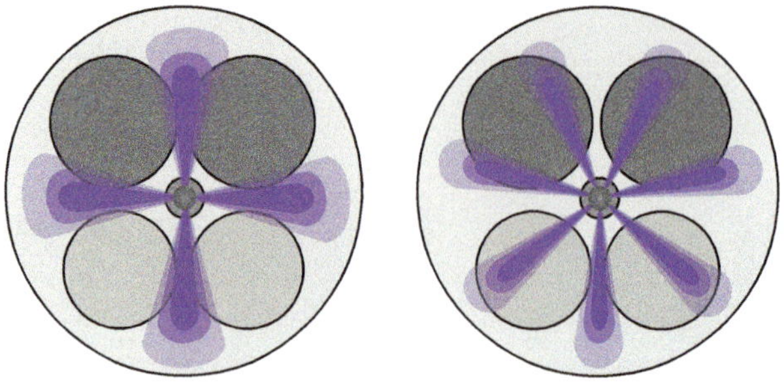

FIGURE 11-5 Having more spray holes provides better distribution of fuel, but smaller, finer droplets cannot penetrate the combustion chamber as thoroughly as heavier droplets.

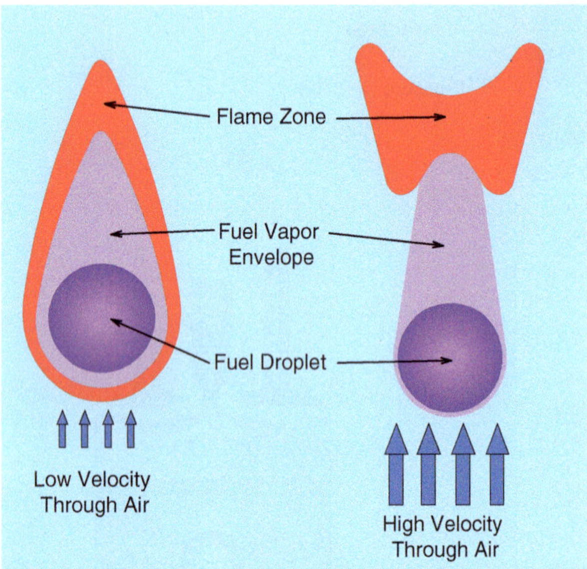

FIGURE 11-6 Droplet A travels slower than droplet B through the combustion chamber, so droplet a makes poor contact with air. The flame zone for droplet A surrounds it, starving the vapor envelope and the fuel droplet of needed oxygen.

Fuel Droplet Size

The amount of time necessary for combustion depends on the fuel quality, the cylinder pre-ignition temperature, and atomization. Only a few degrees of crankshaft rotation are available for mixture formation prior to combustion, because fuel is injected at the end of the compression stroke. Before fuel can burn, atomization, distribution, heating, and vaporization of fuel droplets must all quickly take place for the best combustion quality.

Finer atomization offers the following benefits:

- improved mixing of air and fuel for a faster and cleaner burn
- shorter ignition lag time, resulting in lower combustion noise
- greater tolerance for use of exhaust gas recirculation (EGR) and the use of retarded injection timing for reduced NO_x formation
- higher engine speed, because the high-pressure injection systems have the ability to atomize fuel into finer droplets in a shorter time, which speeds up mixture formation and reduces particulate emissions
- improved fuel economy through more complete combustion of fuel
- lower NO_x emissions, which are formed in fuel-rich regions of the combustion chamber.

Combustible vapors will originate from the fuel droplet surface as it encounters air friction and absorbs heat produced during the compression stroke (**FIGURE 11-6**). Liquid fuel droplets will vaporize and burn from the outside inward. Unless fuel droplets are broken into the smallest size, they will take longer to burn and make less contact with oxygen. Heated fuel with insufficient oxygen contact will cause the droplet to simply bake or carbonize, increasing the production of PM (soot).

Not only are soot PM emissions accompanied by hydrocarbons (HCs) and carbon monoxide (CO) emissions, but their production is also reduced in proportion to the smaller size of the fuel droplets. Improved atomization also enables NO_x emission-reducing technology and strategies. For example, when the speed of air-fuel mixture formation and combustion increases due to finer atomization, injection timing can be retarded because less time is required for mixture preparation. The use of EGR is better tolerated by engines with finer atomization.

> **▶ TECHNICIAN TIP**
>
> Burning fuel in a diesel engine is like burning wood in a campfire. Small pieces of kindling start a campfire faster, and they burn more quickly and completely than large logs. Similarly, fine fuel droplets ignite more easily and burn faster and more completely than large droplets do. Just as hotter fires produce less smoke, clean engine emissions are achieved when fuel burns hot and has enough time to burn. Fine atomization helps fuel burn hot, fast, and more completely, leaving less residue behind.

Pressurization, Atomization, and Combustion

The quality of atomization and fuel distribution directly relates to the degree of pressurization. Pressurization is important to fuel distribution for a number of reasons. First, high pressure is necessary to impart adequate energy to fuel to penetrate dense turbulent charge air inside the combustion chambers as quickly as possible. The size or weight of the fuel droplet multiplied by its velocity determines the droplet's the kinetic energy. When energy is higher, the fuel droplets will travel farther. Droplets tend to break up and disperse faster at high rather than low velocities. To create smaller fuel droplets with more contact surface area for oxygen, finer diameter, more numerous nozzle orifices and higher pressure are required. Smaller DI combustion chambers found in passenger vehicles will have between six and eight nozzle orifices, whereas larger commercial engines will use as many as 13 orifices. However, between five and eight spray holes are most common today. Higher pressure is necessary in order to inject all of the fuel required for the combustion cycle and when delivering fuel through smaller spray holes in a short amount of time.

Basic Nozzle Construction

The most common type of injector used today is referred to as a closed nozzle. This means the needle valve in the nozzle seals the end of the nozzle tip when not injecting fuel. A nozzle consists of two main parts: the nozzle or needle valve and the nozzle body. The **hydraulic nozzle holder** encloses the nozzle assembly and contains a fitting or passageway connecting the high-pressure injection line to the nozzle valve (**FIGURE 11-7**).

The nozzle body contains a spindle-like nozzle valve, also called a needle valve, which normally rests on a nozzle valve seat. Below the seat area are spray holes leading to the nozzle tip, called orifices. Holding the nozzle valve closed against its seat is a nozzle spring that transmits its mechanical force through a pressure pin.

Inside the nozzle holder, above the nozzle valve, is a chamber where small quantities of fuel can accumulate that have leaked between the nozzle valve and body (**FIGURE 11-8**). This leakage of fuel between the two parts, called **back leakage**, is a calibrated escape of fuel around the nozzle valve intended to lubricate and cool the nozzle valve. Back leakage is drained from the holder through a leak-off or return-line connection and then flows back to the fuel tank.

Nozzle Manufacturing Techniques

An enormous degree of manufacturing expertise, specialized materials, and design work are involved in the production of injector nozzles. The principal techniques of making holes in injector nozzles are punching, electric discharge machining (EDM), twist drilling, and laser machining. EDM is the most common technique for newer nozzles EDM uses electric current to literally burn or machine a spray hole opening. Hole geometry resulting from all these processes is still limited to shapes with straight walls or walls with varying degrees of taper. A technique used to enhance the flow characteristics and performance of injector nozzles is to smooth the ends of the spray holes by forcing an abrasive compound through the nozzles. This smoothing process, known as **abrasive flow machining (AFM)**, not only enhances flow characteristics by removing roughness on spray hole walls that resist fuel flow but also permits, in performance applications, increased fuel delivery (**FIGURE 11-9**).

Durability Requirements

Nozzles must have high mechanical strength to tolerate the high injection pressures, to resist the stress and fatigue imposed by the fuel delivery cycle, and to ensure durability at high combustion temperatures to provide long engine life. Consider that the needle valve will open and close over a billion times during its service life, withstanding pressures higher at times than 38,000 psi (2,620 bar). This force is equivalent to the weight of a large tandem axle truck pressing on the area of a fingernail. If the nozzle is used in common rail fuel injectors operating with pre- and post-combustion injections, the needle valve will open and close as many as 10,000 times per minute. Main injection event durations of about 2 milliseconds means injection velocities of fuel can approach 1,243 miles per second (2,000 km per second) at full load. Operating clearances between nozzle parts is kept to a little as 2 microns (1/30 the thickness of a human hair) (**FIGURE 11-10**).

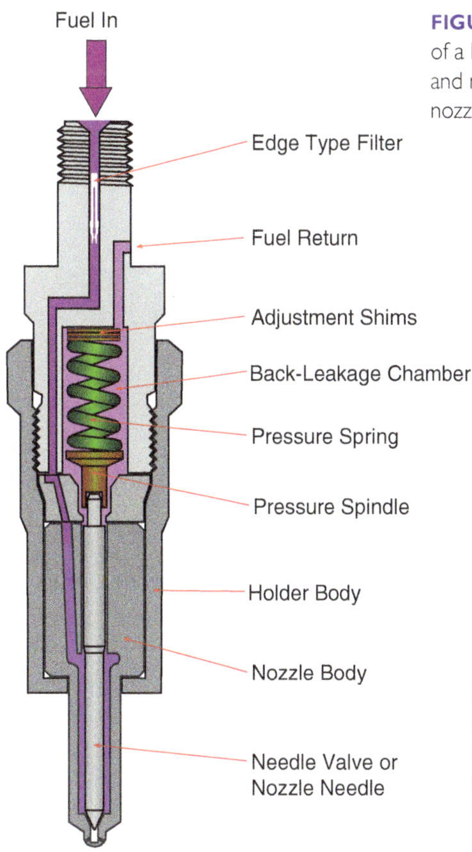

FIGURE 11-8 Typical components of a nozzle holder. Note the back-leakage chamber in the center of the injector, which collects fuel leaking past the nozzle valve and nozzle body.

▶ TECHNICIAN TIP

Worn injector nozzles are a major cause of excessive PM emissions. As nozzles wear, their orifices become larger, producing larger fuel droplets, which in turn produce more PM. Low injection pressures caused by worn plungers or leaking valves controlling **electronic unit injectors (EUIs)** can also increase droplet size. An EUI's incorporates timing, metering, atomization, and pressurization functions into a single unit or injector.

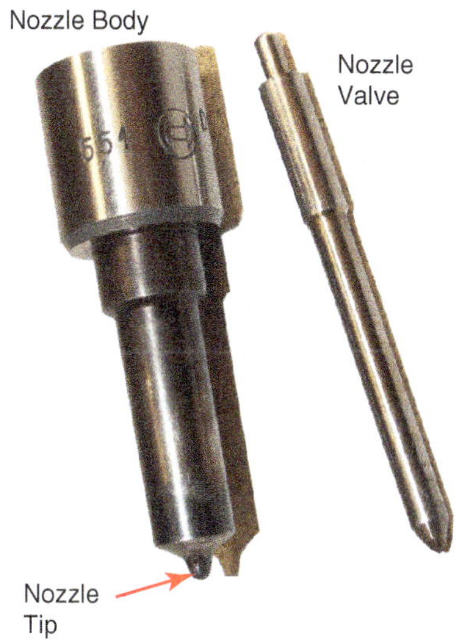

FIGURE 11-7 The most basic parts of a hydraulic nozzle: the nozzle valve and nozzle body. In a closed nozzle, the nozzle valve seals the nozzle tip.

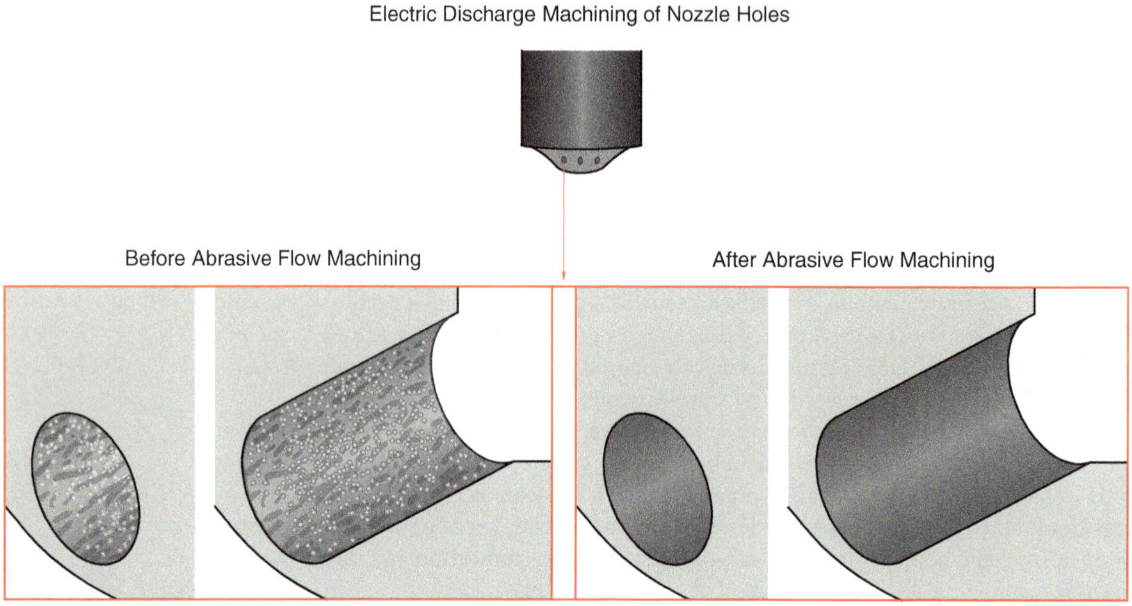

FIGURE 11-9 Smoothing the radius of injector nozzle orifices via AFM improves flow characteristics of injector nozzles.

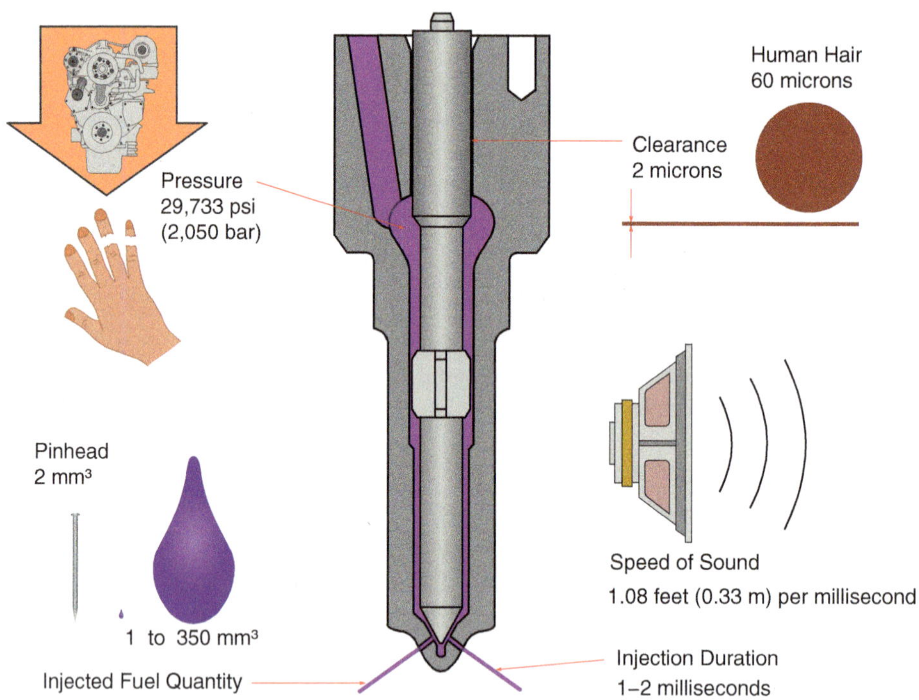

FIGURE 11-10 Operational demands for nozzles used in commercial vehicles.

Basic Nozzle Operation

Closed type nozzles operate by using a pressure differential principle. Normally, the nozzle valve rests on the valve seat, preventing fuel from being delivered through the nozzle orifices. High-pressure fuel will enter the nozzle holder though an inlet fitting and travel down through internally drilled passageways to a hollow, ring-shaped pressure chamber beneath the nozzle valve (**FIGURE 11-11** and **FIGURE 11-12**). This ring-shaped pressure chamber, called the annulus, will apply fuel pressure to the nozzle valve and force it from its seat when fuel pressure exceeds nozzle spring pressure. The hydraulic pressure required to open the nozzle valve is known as **nozzle opening pressure (NOP)**, also called valve opening

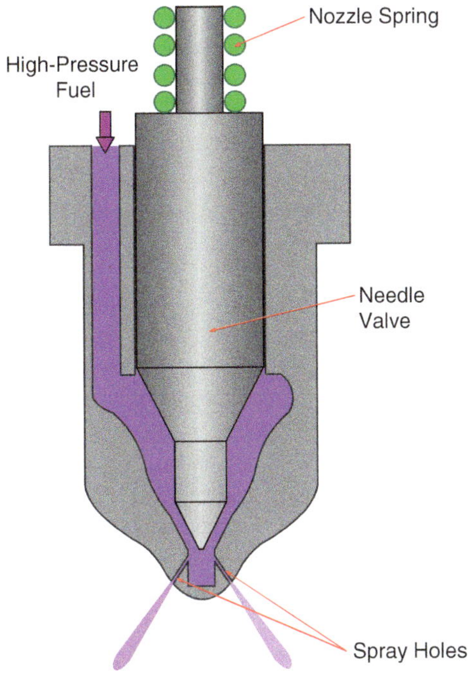

FIGURE 11-11 Hydraulic nozzles use principles of pressure differential to operate. High-pressure fuel enters the nozzle and overcomes spring tension to force the needle valve off its seat, opening the spray holes.

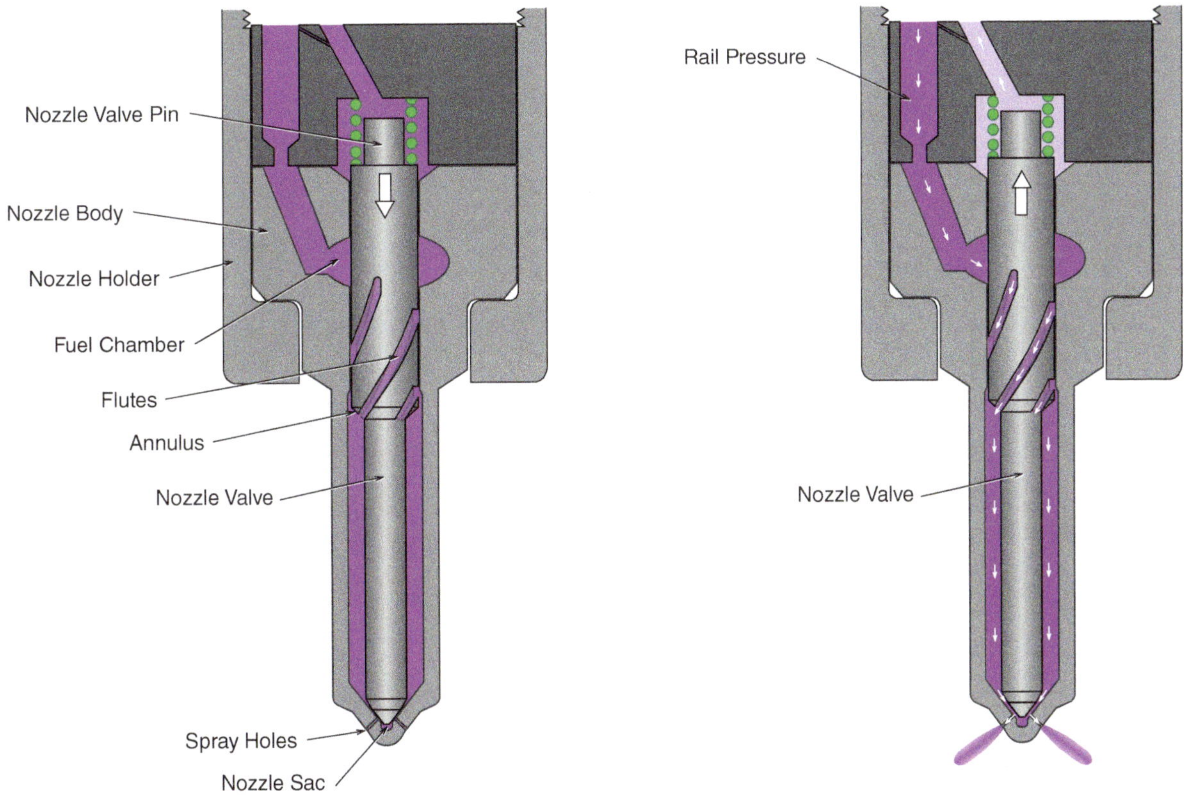

FIGURE 11-12 The nozzle valve of a Delphi injector. Fuel passes around the nozzle valve through helical-shaped flutes to the fuel annulus. When the nozzle valve opens, fuel passes along the nozzle and out the injector tip spray holes.

pressure (VOP). NOP is determined by the pressure differential between the spring force holding the valve on the nozzle body seat and the surface area of the nozzle valve in the annulus. The nozzle spring will hold the nozzle valve against its seat until fuel pressure exerted against the annulus exceeds spring pressure (**FIGURE 11-13**).

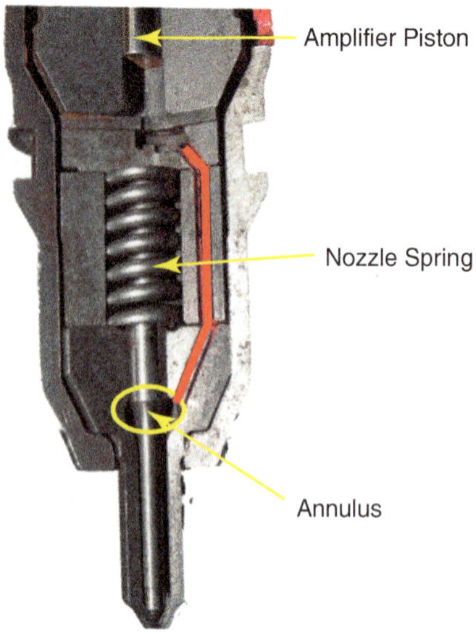

FIGURE 11-13 The spring tension and the size of the annulus determine the fuel pressure required to lift the nozzle from its seat.

After high-pressure fuel unseats the nozzle, fuel passes between the end of the nozzle valve and the seat and then out of the nozzle body's tip through holes called spray orifices. When fuel pressure below the nozzle valve drops, such as when injection ends, the valve closes against the nozzle seat and blocks further fuel delivery. At the same time, the nozzle valve seals the injector tip to prevent combustion gases from entering the nozzle.

Selective shims above the nozzle spring are commonly used to adjust NOP. For example, a 0.001" (0.025 mm) change in shim thickness may result in a 50 psi (345 kPa) change in NOP.

Nozzle Differential Ratio

Diesel engine fuel systems require an abrupt beginning and end of injection. **Beginning of injection (BOI)** is the point when fuel delivery begins and is referenced to according to the position of the crankshaft measured in degrees of crankshaft rotation. Without an abrupt beginning and end of injection, a gradual loss of pressure at the nozzle tip would result in increasing spray droplet size and decreasing droplet penetration. To produce the sharp increase in pressure when the nozzle valve first lifts from its seat, it is necessary to ensure that the valve moves rapidly. To accomplish this, nozzle valves are built with two different surfaces that are acted on by high-pressure fuel. The first surface is the area of the nozzle annulus. This area is smaller than the second surface, which is the entire surface area of the nozzle valve when it is unseated during injection. The difference between the two areas is referred to as the **nozzle differential ratio** (**FIGURE 11-14**). In most nozzles, the area acted upon by highly pressurized fuel in the annulus is 20–80% smaller than the surface area exposed to pressurized fuel after the nozzle valve is lifted from its seat. This means that more fuel pressure is required to open the nozzle than is needed to close the nozzle. Because a larger nozzle-valve surface area is

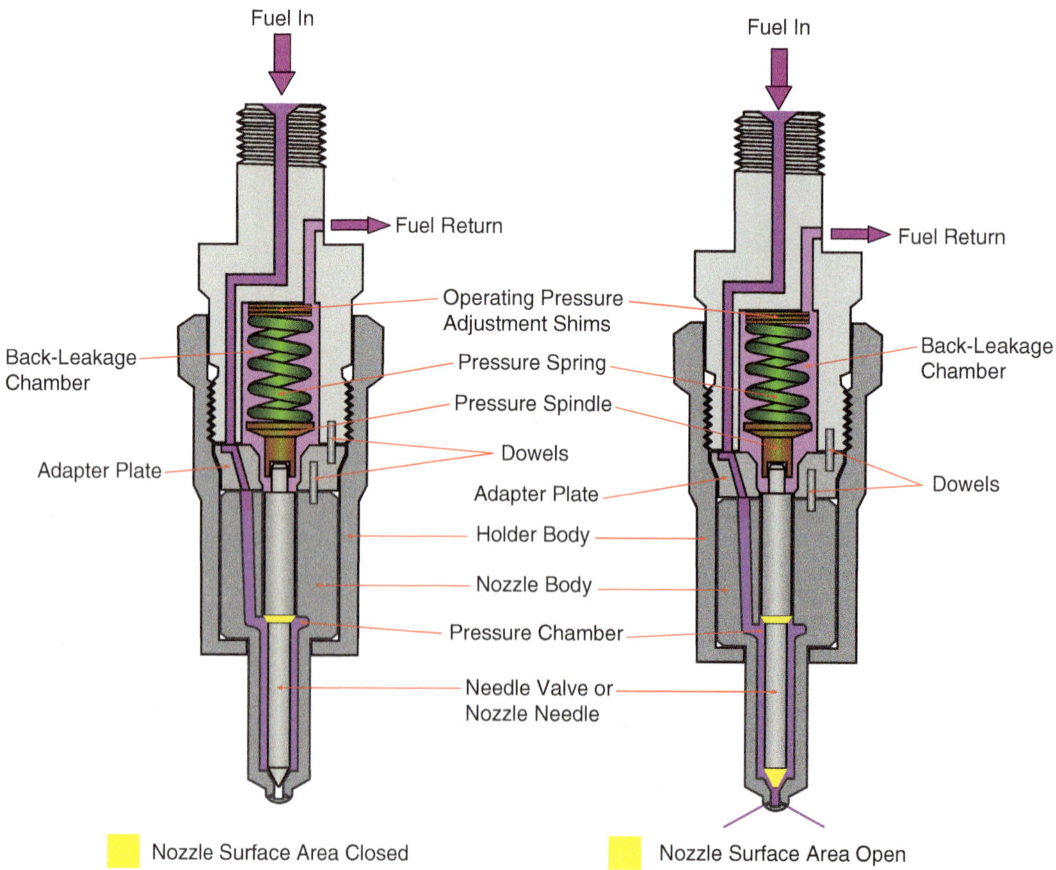

FIGURE 11-14 The nozzle differential ratio is a comparison between the nozzle surface area acted on by pressurized fuel when closed and when open.

exposed, the instant the valve begins to lift, the valve will rapidly lift and remain unseated with less fuel pressure than was needed to initially move the valve.

Large and small nozzle differential ratios each have advantages and disadvantages. A small nozzle differential ratio means the nozzle valve will quickly reseat when a little fuel slips out of the spray holes. This arrangement keeps fuel spray-in pressures high and droplet size small. However, it also produces a higher frequency of nozzle opening and closing, which accelerates nozzle valve and seat wear. Conversely, a large nozzle differential ratio produces less nozzle seat wear but has the disadvantage of producing larger droplets because lower fuel pressure is required to keep the nozzle open.

Hydraulically Balanced Nozzle-Valve Opening

The latest injectors control injection events by changing the balance of hydraulic pressure acting on the needle valve by using electrically operated solenoids or control valves inside the injectors. Traditional fuel systems use high-pressure fuel only to unseat the nozzle valve and begin injection. Newer unit injectors and common rail injectors use high-pressure fuel to both open and close the nozzle. Hydraulically balanced nozzle valves, also called hydraulically suspended valves, use the small mechanical control force of the injector control valve to trigger more powerful, faster acting hydraulic forces. Using comparatively small control valve movement to manipulate very high hydraulic pressure (that is, using a small force to control a much larger force) is called servo action. In a sense, the servo action in an injector is similar to an electric-over-hydraulic relay; it uses small electrical forces to switch very high hydraulic pressures (**FIGURE 11-15** and **FIGURE 11-16**). Servo-hydraulic injectors always use hydraulically balanced nozzle valves that are opened and closed by tipping the pressure differential that is acting above and below the nozzle valve.

Needle Lift Sensor

One disadvantage of the nozzle valve is its design, which contributes to injection lag. **Injection lag**, also known as injection delay, is the time between the start of injection

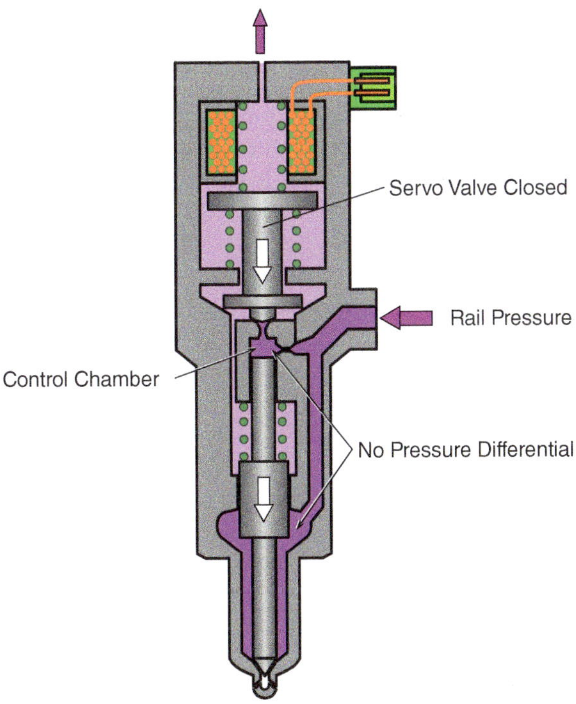

FIGURE 11-15 A hydraulically balanced or suspended nozzle valve uses high-pressure fuel to open and close the nozzle valve. Draining highly pressurized fuel from the control chamber above the nozzle valve begins injection. Reapplying fuel pressure seats the nozzle valve, which ends injection.

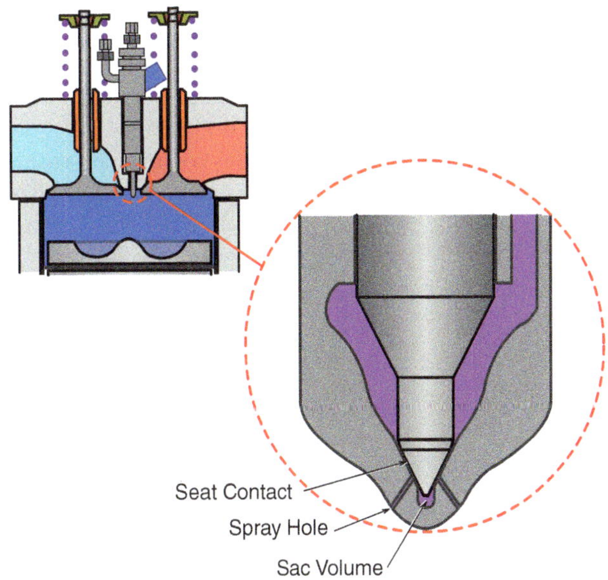

FIGURE 11-16 The nozzle-valve seat seals the tip from pressurized fuel. The "sac" volume provides a hydraulic cushion to dampen the force of nozzle-valve seating in order to reduce wear.

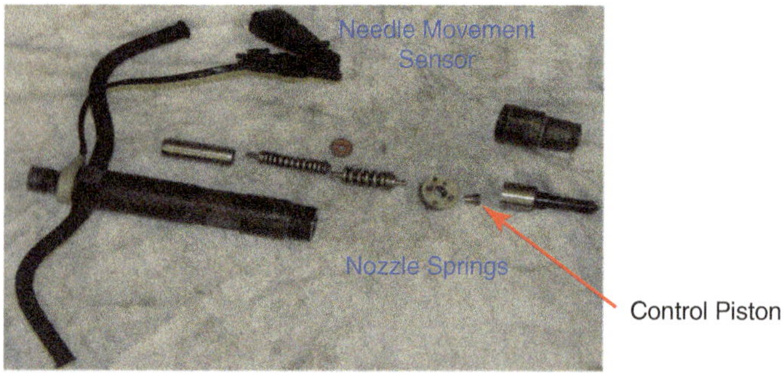

FIGURE 11-17 This double-spring injector nozzle incorporates a needle lift sensor. Only one sensor is needed to correct injection timing for all cylinders.

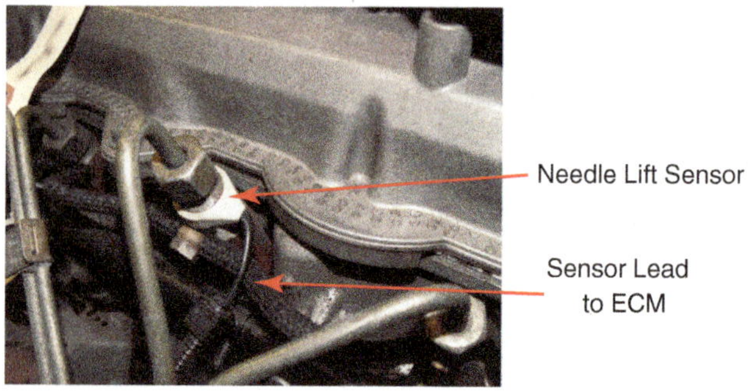

FIGURE 11-18 An electronic diesel control system that incorporates a needle lift sensor.

pressurization and the point when fuel is actually beginning to be delivered. Injection lag happens as the nozzle valve lifts and its movement displaces fuel volume originally intended for injection. This means that lifting the nozzle uses up some of the fuel metered for delivery and delays BOI as some of the fuel delivery volume unseats the nozzle. A discrepancy always exists between the point when fuel delivery is expected to begin relative to crank rotation and the actual point of injection. Injection lag affects injection timing by retarding BOI. This delay in turn influences emissions and engine performance.

To compensate for this problem, a **needle lift sensor** is used to precisely identify BOI. This sensor provides an electrical signal to the electronic control module (ECM) that indicates nozzle-valve movement. A signal passing through a wire coil surrounding an extension to the nozzle valve is disrupted when the needle valve lifts. The signal is interpreted as the actual BOI and corrections to injection timing take place based on needle lift sensor data. Nozzle lift signals are used by the engine control system to eliminate variability in injection timing caused by injection lag, which enables more precise and consistent control over injection timing (**FIGURE 11-17**). In other words, injection timing is corrected to actually begin when the nozzle valve unseats, and not when it is commanded to unseat. In an older engine, typically only a single injector was equipped with an integrated nozzle-valve lift sensor to correct injection timing for all cylinders (**FIGURE 11-18**).

Lift sensors use a simple electromagnetic coil that surrounds the injector. The magnetic field uses an extension on the nozzle valve that passes through the coil, and the strength of the magnetic field changes as the needle valve lifts and the extension moves in and out of the magnetic field. The changing magnetic reluctance of the coil induces current flow, which is transmitted to a control module that identifies BOI. Nozzle lift sensor data is then compared with crankshaft position data to apply a correction factor for electrical signals controlling the beginning of injection. How far away actual nozzle lift is measured to have taken place from its commanded position is calculated by the ECM. Depending on the variation between the actual nozzle lift and expected lift, the ECM can then correct any timing deviations by altering electrical signals supplied to the injection pump.

Nozzle Motion Failure Mode

If the needle lift sensor fails, a default injection timing program is started. Fuel injection is controlled according to preset values established in the fuel timing map stored in the ECM. In addition, the injection quantity is reduced, resulting in engine power loss—which is commonly known as limp mode. Lift sensors are no longer used in late-model diesel engines. Today's engines use cylinder-pressure sensors or crankshaft acceleration rates, or they monitor electrical signals from the injector solenoid windings. In an EUI, the movement of a metal control valve disrupts solenoid magnetic fields in the same way that it does in the lift sensor, which provides timing feedback control used by the ECM.

Closed-Loop Nozzle Operation

Devices that have closed-loop control rely on sensor data to provide feedback about an actuator's position. It's called closed loop because only three devices interact in a loop: the

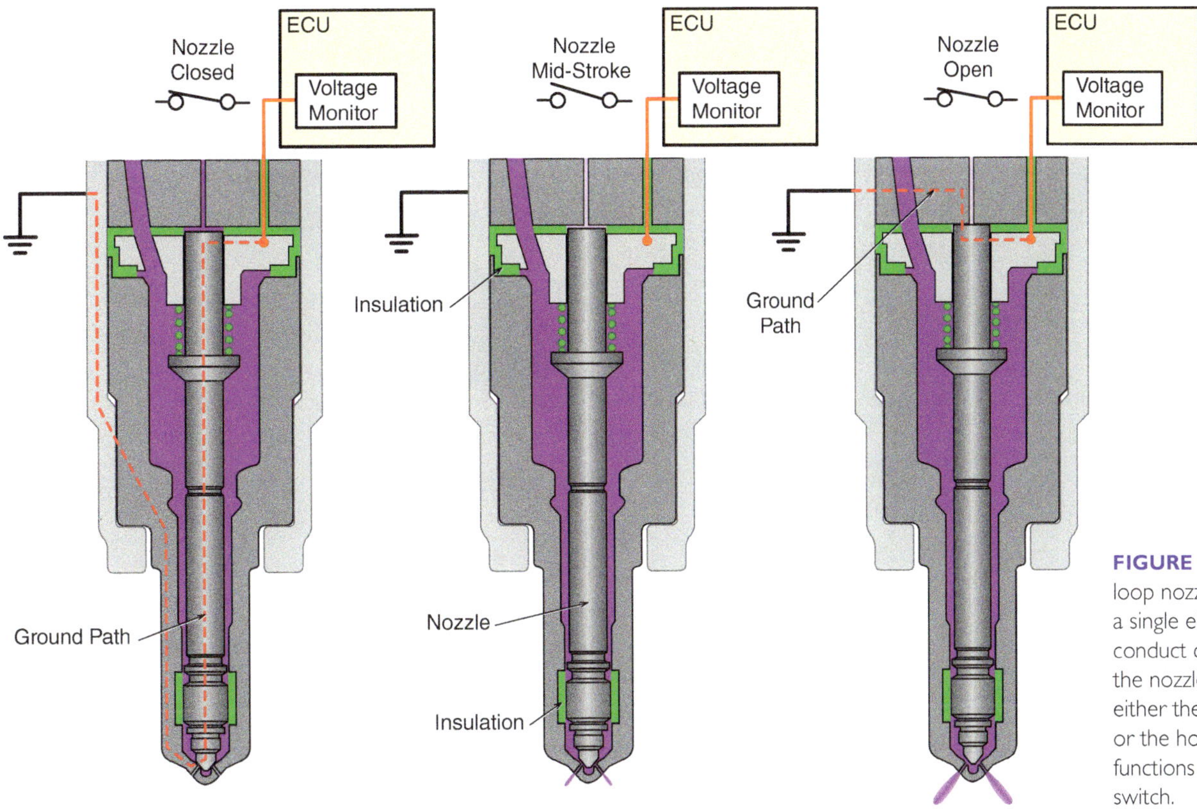

FIGURE 11-19 Closed-loop nozzle control uses a single electric wire to conduct current through the nozzle and into either the nozzle body or the holder. The nozzle functions like an electrical switch.

ECM, the actuator, and the sensor. Data from the sensor is used by the ECM to determine whether the actuator is correctly positioned or not. If the ECM sends an electrical signal to reposition the actuator, the sensor provides new information to the ECM about position. A loop is formed between the three devices during this exchange, and without input from any other source, the loop is considered "closed." To provide more accurate information to the ECM about when the nozzle has lifted, how far it has lifted, and when it has closed, Delphi and Bosch have developed nozzles with closed-loop control. In Delphi's nozzle, electric current passes through the nozzle valve and will send the current through either the lower nozzle body or the upper nozzle holder. When current is not sensed by the holder or body, the ECM determines its positon is mid stroke (**FIGURE 11-19**). The sequence of the nozzle valve opening and closing a switch through either the body or the holder helps the ECM apply a correction factor to more precisely regulate the time when the nozzle open and closes.

Matching Nozzles to Combustion Chambers

Precisely matching the injector to the shape of the piston bowel and turbulence of DI chambers is critical for efficient combustion. Ideally, homogenously mixed fuel and air will produce the lowest emissions and best combustion efficiency. However, diesel-injection systems cannot currently achieve an even distribution of air and fuel in the few degrees of crankshaft rotation when the injection event takes place. Regions rich with excess fuel will produce soot and NO_x. Similarly, fuel should not spray directly against the walls of the piston bowl or cylinders, because this quenches or extinguishes combustion flames, which leads to increased HC and CO emissions. Because NOP and the angle, number, diameter, and length of spray holes affect the atomization, penetration, and distribution of fuel, the injector needs to be precisely matched to the combustion bowl shape to provide optimal air-fuel mixing over the engine's entire operating range (**FIGURE 11-20**).

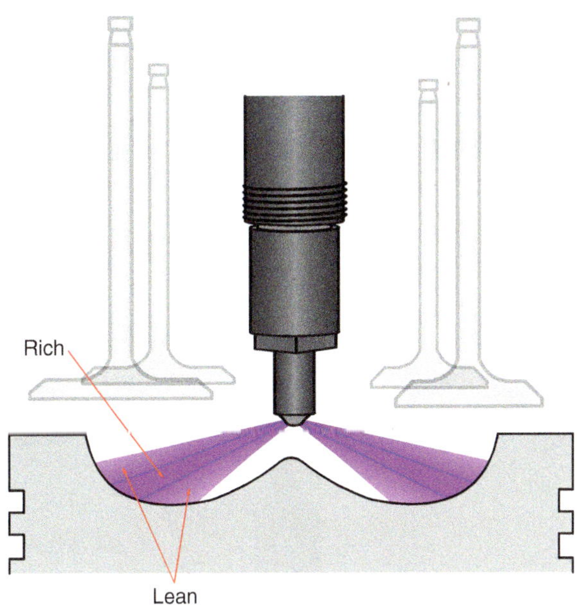

FIGURE 11-20 The angle of the spray holes must match the combustion bowl shape to spray fuel directly into the turbulent toroidal air movement.

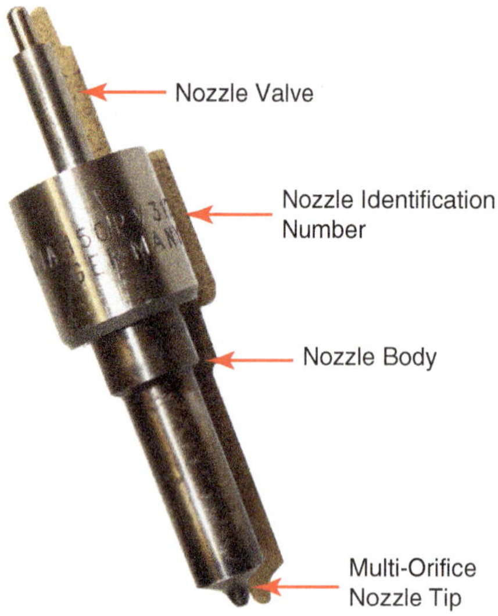

FIGURE 11-21 A nozzle's spray angles, hole diameters, hole lengths, and many other features are not easily visible. Identification numbers on nozzle tips will identify the correct nozzle for an application.

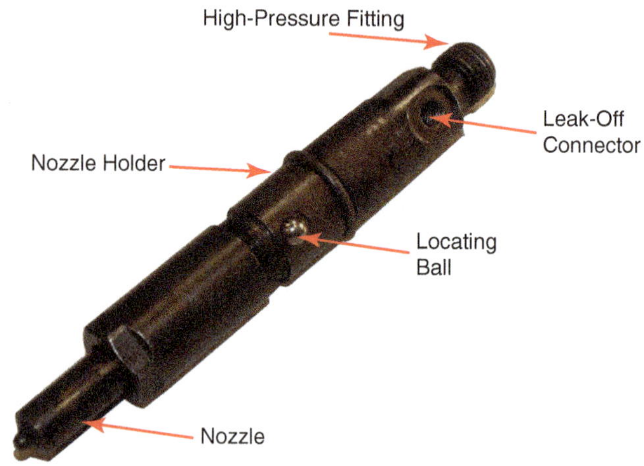

FIGURE 11-22 This multi-orifice nozzle uses a locating ball to help correctly orientate the injector in the cylinder head.

▶ TECHNICIAN TIP

When replacing injectors or any other fuel system components, it is important to correctly match the parts to the engine. "Will fit" components are not precisely matched and can result in low power, excessive emissions, and engine failure. Technicians should use engine serial numbers to order correct fuel system components.

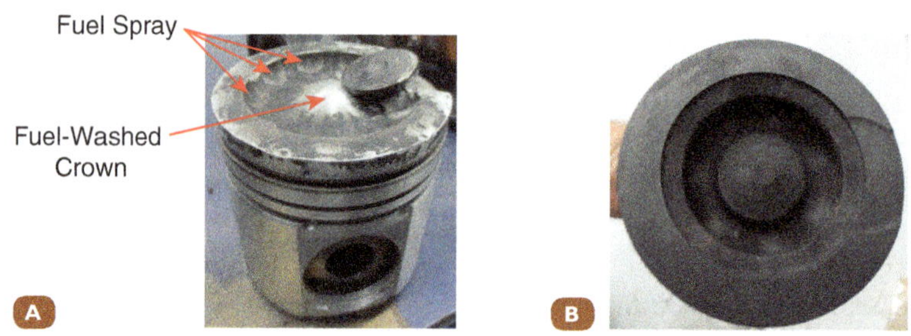

FIGURE 11-23 **A.** Large spray holes in the injector have sprayed fuel onto the piston bowl's walls. Excess fueling overheated a valve head, separating it from the stem. **B.** This piston shows a more normal pattern of spray wash, with light spray tracks across the crown.

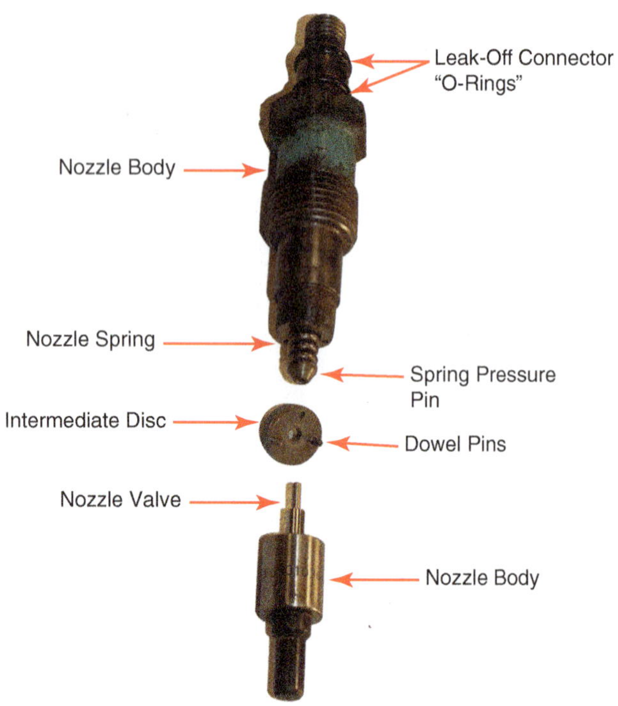

FIGURE 11-24 The components inside a hydraulic nozzle holder.

When rebuilding engines and servicing injectors, it is important for technicians to correctly match components by using original equipment manufacturer (OEM) part numbers. Although components might look identical, imperceptible differences in things like piston bowl depth or diameter, nozzle spray angle, or hole diameter or length can have a dramatic effect on mixture formation and negative consequences for combustion efficiency. Numbers stamped on the nozzle's tip identify the nozzle and help differentiate it from similarly shaped nozzles (**FIGURE 11-21**). The orientation of an injector in the combustion chamber is also critical to good combustion. This is especially true when injectors are offset in a DI combustion chamber. A locating pin, ball, or other feature is required to ensure that the injector spray pattern is positioned to produce the correct spray pattern over the piston bowl (**FIGURE 11-22** and **FIGURE11-23**). On many fuel nozzles, a spacer containing two locating pins can be found between the nozzle and nozzle holder. This spacer ensures that the nozzle spray holes are correctly positioned inside the combustion chamber. Because the locating pins are slightly offset from center, the nozzle and holder can be assembled in only one way. The plate also limits upward movement of the nozzle valve to prevent excessive nozzle spring compression and nozzle lift. (**FIGURE 11-24** and **FIGURE 11-25**).

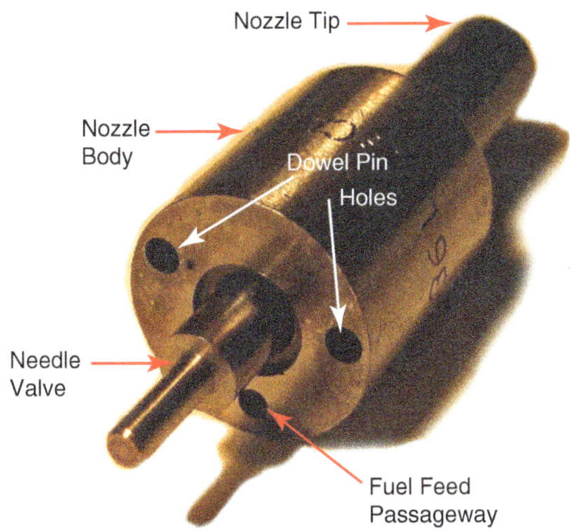

FIGURE 11-25 The nozzle body has two holes to align with the dowels and another hole to conduct pressurized fuel.

> **▶ TECHNICIAN TIP**
>
> A vehicle operator may return with a low-power complaint after injector replacement. This may happen for several reasons. Finer atomization and higher NOP may permit less fuel into the combustion chamber. This may lower power output slightly from what the driver was familiar with. New injectors, particularly some electronically controlled types, may need some use to break them in and increase flow rates through the nozzle spray holes. NOP will typically lower after the injector has accumulated some hours in operation.

▶ Types of Hydraulic Nozzles

11-02 Identify and describe the types, styles, construction, and applications of hydraulic nozzles.

Today's on-highway diesel engines use DI, multi-orifice nozzles. However, due to their long service life, millions of older diesel engines are still in use in pick-ups, vans automobiles, off-road engines, power generation, and rental equipment that have IDI combustion chambers. Being knowledgeable about the design and construction of different injectors helps engine technicians effectively service a wider range of engines.

Types of Closed Nozzles

Closed injector nozzles are the most popular nozzle type in use today. In a closed injector nozzle, the nozzle valve seals the nozzle tip when the valve is seated. There are two types of closed nozzles: the multi-orifice nozzle and the pintle nozzle. The two types are similar in construction but differ in the number of orifices and opening pressure.

Multi-Orifice Nozzles

Multi-orifice, or multi-hole, nozzles are used in DI engines, and as their name suggests, they have multiple spray holes at the injector tip. Injectors used in DI engines require greater capabilities to pressurize, atomize, and distribute fuel than do pintle nozzles, which have only a single spray hole in the tip. Older designs use four or five holes, whereas as many 13 can be found in newer designs used in heavy-duty applications (**FIGURE 11-26**).

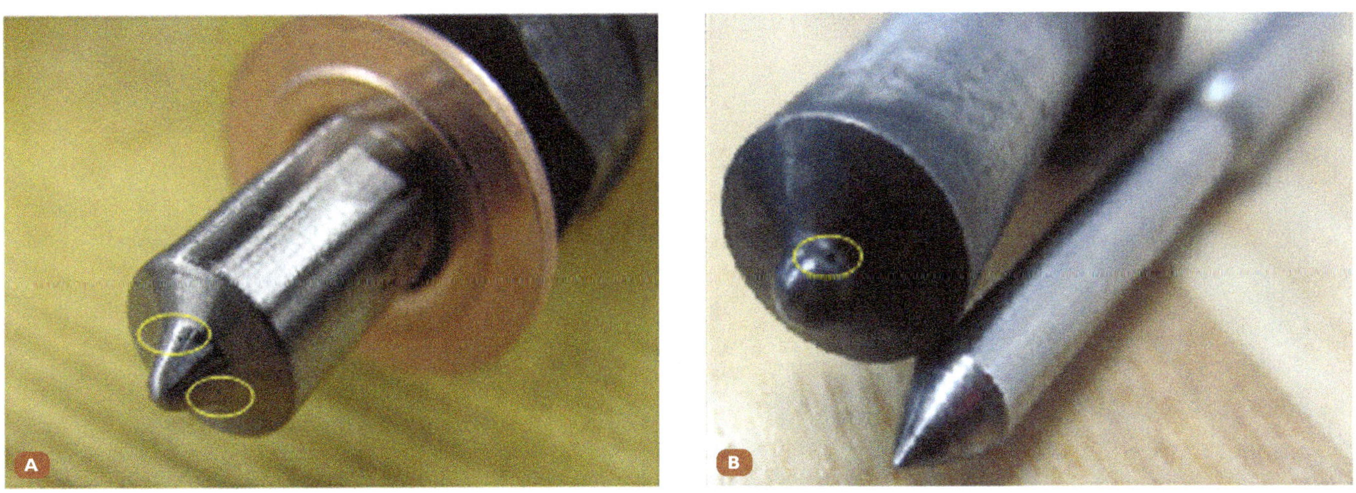

FIGURE 11-26 **A.** The injector tip of a multi-orifice injector used in a common rail engine. The naked eye cannot see its seven spray holes. **B.** The injector tip of this nozzle has four holes and is from an older engine.

NOP is higher in multi-orifice nozzles to produce the atomization, penetration, and distribution of fuel necessary for DI combustion chambers. Most multi-orifice nozzles are calibrated to an NOP of between 3,200 and 5,500 psi (220.6 and 379 bar). Newer engines use higher NOP, but higher pressure increases wear on nozzle valves and seats. Also, starting may be more difficult, and low speed idling may be less stable with higher NOP. Not only do some injectors use hydraulically balanced nozzle valves with servo actuators, but some designs can also electronically vary nozzle opening until injection pressures reach as high as 17,000 psi (1,172 bar).

Pintle Nozzles

Pintle nozzles are designed with a single orifice in the injector tip (**FIGURE 11-27**). These nozzles are used only in IDI combustion chambers. Fuel leaves the nozzle tip as a single, relatively coarse, pencil-like spray plume. NOPs range from 900 to 5,000 psi (62 to 345 bar). Most pintle nozzles have an average NOP of 1,500–1,950 psi (103–134 bar). NOPs are higher on turbocharged engines to seal the tip from the entry of combustion gases in these engines with increased cylinder pressures.

Fuel spray from a single opening and at this low pressure produces poor atomization compared to multi-orifice nozzles. Mixture preparation and distribution are therefore dependent primarily on cylinder turbulence. For this reason, pintle nozzles are used in IDI combustion chambers, where the more turbulent pre-chamber mix the coarse spray droplets (**FIGURE 11-28**). Hot pre-chamber walls operate to vaporize fuel on contact, and the comet or fireball shape of the combustion chamber mixes fuel and air. The pre-chamber venturi further promotes fuel mixing and fuel vaporization as the fuel is pushed by the expansion of burning gases into the main chamber. While the pintle nozzle, with its large, coarse spray hole, would appear to perform poorly at preparing a good combustible mixture, this chamber had the fastest mixture preparation and combustion time until the appearance of high-pressure, electronically controlled DI systems. In fact, compared to non-electronic DI engines, IDI engines had the advantage of operating at higher speeds with lower air-fuel ratios, noise, and noxious emissions due to the ability of the IDI chamber to rapidly mix and burn air and fuel.

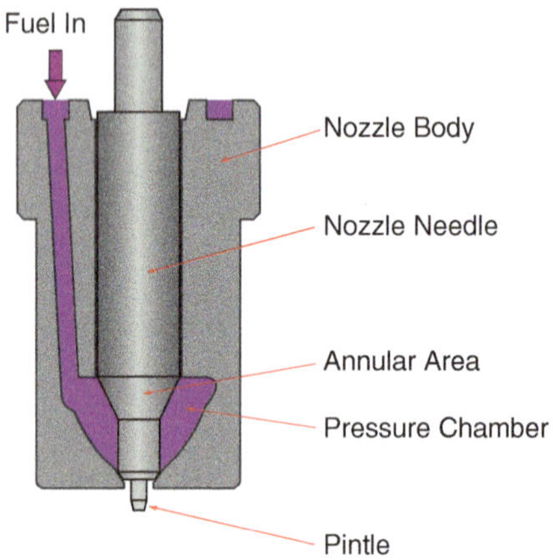

FIGURE 11-27 A pintle nozzle uses a single, large spray hole and is used only in IDI combustion chambers.

Rate-Shaping Nozzle Values

Two varieties of pintle nozzles are commonly used. In standard pintle nozzles, the valve simply opens and closes, turning fuel injection on or off with little change in the spray pattern or injection rate. Later pintle nozzles, called **throttle nozzles** or delay nozzles and identified by the inverted cone-shaped nozzle body at the injector tip, have an injection rate–shaping feature that reduces the amount of fuel injected during the early stages of injection (**FIGURE 11-29**). As injection progresses, the spray volume increases and the spray pattern changes from narrow and cylindrical to a cone-shaped distribution

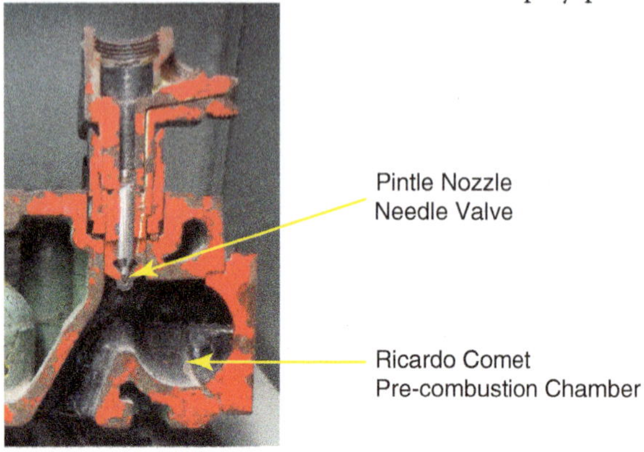

FIGURE 11-28 The high turbulence in the pre-chamber of an IDI chamber mixes air and fuel faster, so a pintle nozzle can be used.

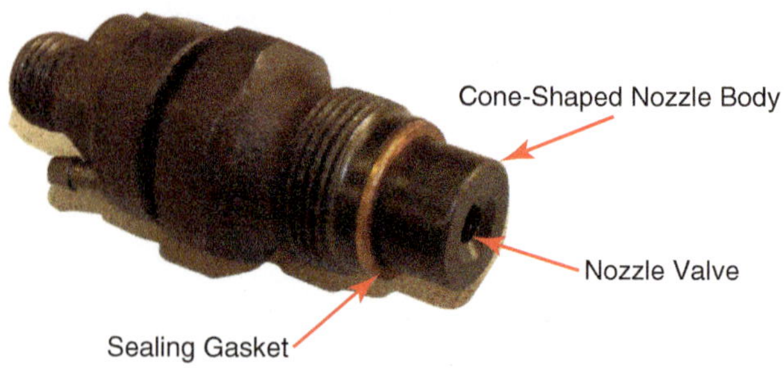

FIGURE 11-29 A throttle pintle nozzle has a cone-shaped nozzle body at the injector tip.

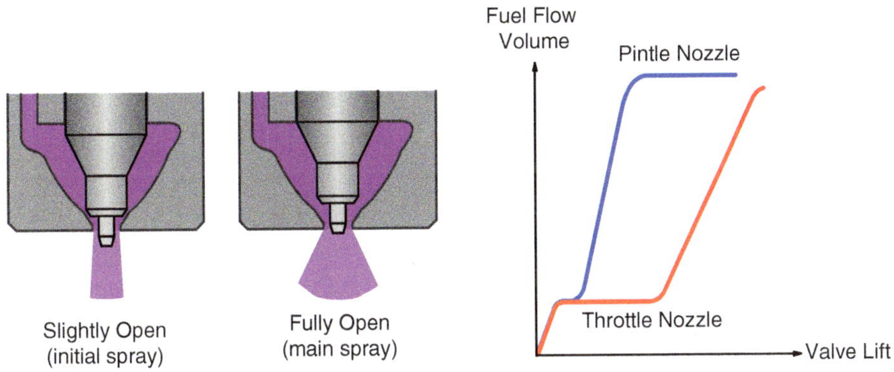

FIGURE 11-30 Throttle pintle nozzles modify the injection rate by injecting a smaller amount of fuel at the beginning of the injection event and then increasing it during the injection event.

(**FIGURE 11-30**). There are several machined steps in the nozzle valve that produce variations in spray pattern and injection volume as the valve lifts from its seat which shape the injection rate.

The advantage of the throttle pintle nozzle is that it limits the amount of fuel in the combustion chamber at the BOI, which in turn reduces the quantity of fuel that accumulations during the ignition delay period. Doing this minimizes the characteristic diesel knock and pressure spikes in a cylinder that produce NO_x. Less fuel means less heat and combustion noise when it ignites. This injection strategy is somewhat similar to pilot injection, but with a throttling pintle nozzle, injection takes place uninterrupted. The throttling effect on fuel delivery is most effective at idling when the valve lift is low.

Variable-Orifice Injectors

Because nozzle geometry is a compromise to promote good mixture preparation over all operating ranges, manufacturers are working on **variable-orifice nozzles**, also known as coaxial nozzles. Variable-orifice injectors currently in development use two or more rows of orifices located around the tip of the injector. The two-row design would include one row of orifices around the nozzle tip that would deliver fuel at low speeds under partial load, when fuel-injection volumes are smaller. Another row above the first would be added under full load, when nozzle needle lift is greater. By using specialized electrohydraulic control, the amount of needle lift is varied to uncover more or fewer spray holes injecting fuel at different spray angles.

Double-Spring, Rate-Shaping Nozzles

Injecting small quantities of fuel at the BOI and then increasing the quantity of fuel injected as the injection event continues lowers combustion noise and NO_x emissions. Although various other injection strategies try to accomplish this, the two-spring, two-stage injector nozzle—another innovation that mechanically controls injection rate by discharging fuel at two different rates—as actually done so.

Two nozzle springs made of different thicknesses and compression force are placed one over the other inside the nozzle holder. A control valve on top of the nozzle valve between the lower and upper spring has a gap filled with fuel. When pressurized fuel begins to lift the nozzle valve, only the lighter top spring allows upward nozzle valve movement. A small amount of fuel is injected during this phase of operation, and a delay in the larger main injection event occurs.

While the upper spring lifts, fuel is squeezed slowly out of the control valve until the nozzle valve contacts the heavier lower spring. As a consequence, injection pressure must build even higher than before for any further fuel delivery to take place. The combined forces of the two nozzle springs will determine the second opening pressure for the main injection event (**FIGURE 11-31**).

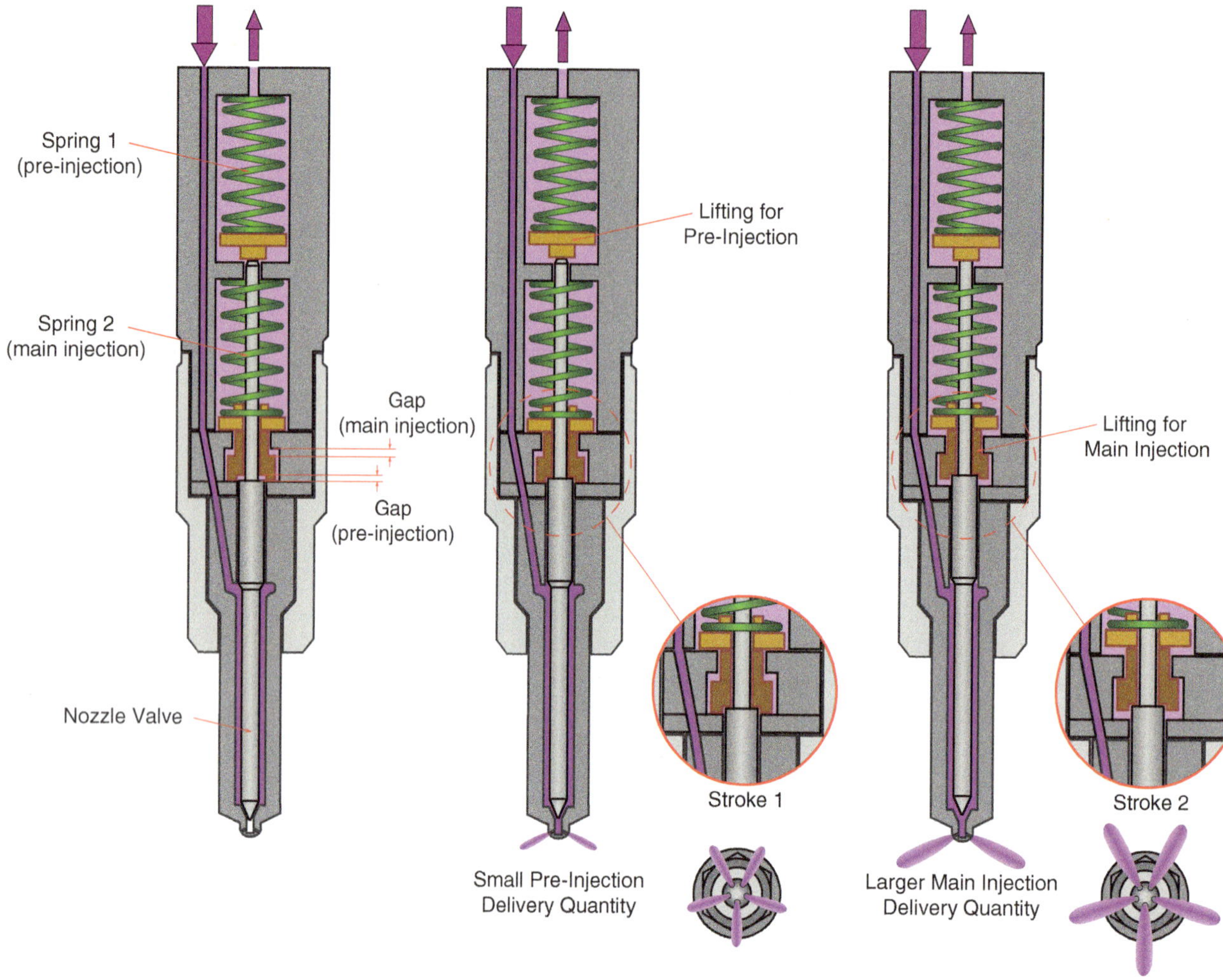

FIGURE 11-31 Operation of a two-spring, rate-shaping injector nozzle. In stroke 1, note that the nozzle-valve lift is small because full injection pressure has not yet developed.

▶ Maintenance of Hydraulic Nozzles

11-03 Identify and explain inspection and diagnostic procedures for evaluating hydraulic nozzles.

When a problem with an injector is identified by the onboard diagnostics (OBD) system or by using cylinder cutout and contribution testing, injectors are simply replaced with new or rebuilt parts. The faulty unit is returned as a parts core, potentially for rebuilding or remanufacturing in a specialized facility. Nevertheless, some service facilities do test and repair the hydraulic nozzles used inside injectors because cleaning and rebuilding an injector in house is often more economical than replacing it.

Also, technicians should understand how injectors fail and the steps that can be taken to identify the failure mode. Even though a technician may never perform many of these diagnostic procedures, they provide valuable insights into potential causes and potential strategies to correct problems when diagnosing complex fuel system–related problems.

Nozzle Testing

The most common problem associated with injector failures is worn, seized, or leaking hydraulic nozzles. Testing nozzles to detect these conditions can be performed by using simple test equipment such as pop testers, which were once required tools for any dealership that serviced diesel engines (**FIGURE 11-32**). Modern, more complex unit injectors or common rail injectors are tested on specialized hydraulic test benches used by fuel system rebuilders. The following section outlines methods and equipment used to diagnose problems with injectors.

FIGURE 11-32 A pop tester is used to evaluate an injector's spray pattern, NOP, and forward and back leakage.

SAFETY TIP

The pressure that produces fuel spray from nozzle testers is high enough to penetrate skin, which can cause blood poisoning. Never allow hands or skin to get near the tip of an injector nozzle. Use a piece of cardboard or paper to check for wetting, and always wear eye protection.

Testing Nozzle Spray Patterns

To evaluate the condition of an injector nozzle, the spray pattern is checked using a **pop tester** or another type of nozzle testing apparatus. A pop tester supplies pressurized fuel to an injector nozzle. A nozzle tester should be operated only with a spray chamber, which recovers fuel vapor produced from testing nozzles.

Spray patterns can be evaluated visually on a nozzle tester to determine whether the injector spray pattern is uniform and symmetrical. The nozzle tester is hand operated and should force fuel through the injector at approximately 15 strokes per minute. Atomized fuel should have a fine spray, and each spray hole should spray approximately the same amount of fuel. The spray patterns from each hole should be uniform. Using a piece of paper positioned beneath the nozzle may also be helpful to evaluate spray angle and the uniformity of the spray pattern.

A distorted spray pattern indicates a blocked or restricted spray hole, eroded nozzle tip, or worn nozzle orifice (**FIGURE 11-33**). Injectors with spray pattern distortions or poor atomization should be replaced or reconditioned. To test nozzle spray patterns, follow the steps in **SKILL DRILL 11-1**.

FIGURE 11-33 One nozzle orifice in this injector is restricted, and in another, the spray pattern is not uniform. This is a defective nozzle.

SKILL DRILL 11-1 Testing Nozzle Spray Patterns

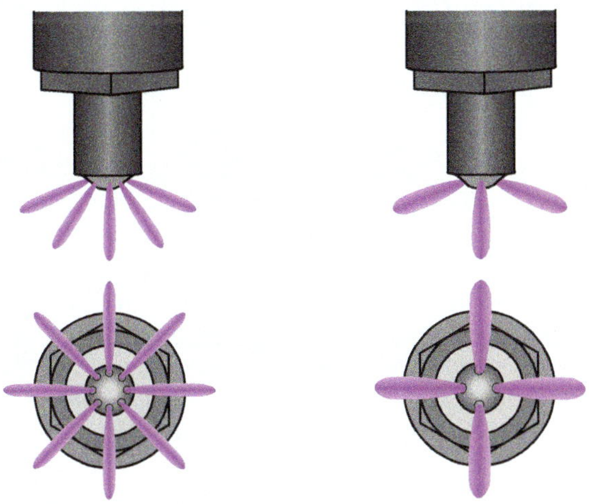

1. Disconnect the high-pressure fuel line, then cap the nozzle fuel line inlet fittings by using protective plastic caps.
2. Remove the nozzles from the engine. Inspect and clean the nozzles by using solvent. Check for missing sealing washers or improperly installed nozzles, which will leak sooty combustion gases around the nozzle body.
3. Set up the nozzle pop tester by filling it with the appropriate test fluid. The fluid should be designated as calibration fluid or an equivalent test fluid, not diesel fuel.
4. Purge the nozzle tester of air by pumping the handle several times until clean fluid appears from the high-pressure line.
5. Ensure that the workbench area is clean, and use clean solvent containers, clean tools, and clean hands to produce satisfactory results.
6. Install the proper nozzle-to-tester adapter that matches the high-pressure fitting on the nozzle. Connect the nozzle to the tester line.
7. Bleed air from the nozzle. This is done by opening the tester valve and pumping the tester handle 8–10 quick strokes to expel air from the injector nozzle. Test fluid should spray from the spray holes in the nozzle tip. If the nozzle is blocked or jammed, the nozzle should be replaced.
8. Carefully observe the spray pattern of the nozzle by pumping the tester handle rapidly and noting the characteristics of the spray pattern forming from the nozzle orifices.
9. Spray should originate from each orifice, and the spray pattern from each orifice should be the same size and uniform in shape. The spray should be well atomized and shaped like a cone as it leaves the injector nozzle. Injectors showing poor spray patterns should have nozzles replaced.
10. Listen for nozzle chatter, and if observed, note it along with any other observations. If one nozzle in a set doesn't chatter, and the rest do, recommend that all the nozzles be replaced or rebuilt.

Measuring NOP

After visually evaluating a nozzle's spray pattern, the nozzle's opening and closing pressures are tested. NOP testing, also called pop testing, is used to determine whether a nozzle valve is opening and closing at the pressure specified by the manufacturer. The test also determines whether the nozzle's valve is sticking. Electronically controlled nozzle testing is performed by doing a cylinder cutout test or by using a purpose-built test bench generally found in fuel specialty shops. This test is performed by observing the pressure gauge located on the tester while slowly pressurizing the nozzle and observing the pressure at which the nozzle valve lifts from its seat and fuel delivery begins. Fuel should not leak from the nozzle when near NOP. The NOP should be consistent during each stroke of the nozzle tester. Continued observation will also indicate the closing pressure, which should be no more than 200 psi (14 bar) lower than NOP. Inconsistency in the NOP or too great a difference between NOP and closing pressures indicates a sticking nozzle valve, which would produce an intermittent misfire. Low opening pressure may indicate a nozzle valve that is sticking or not seating because it is seized or coked fuel has left carbon deposits in the seat area that are preventing the nozzle from sealing the spray holes. Weak, broken, misaligned, or improperly shimmed nozzle springs also produce low NOP. Nozzle-valve recession caused by a worn nozzle valve and seat will also lower opening pressures. High opening pressure generally indicates a sticking nozzle or an improperly shimmed nozzle spring. Nozzle defects in a running engine are usually indicated by one or more of the following symptoms:

- black, gray, or white exhaust smoke
- low power
- a misfire under load
- rough engine operation after the engine has warmed up
- excessive fuel consumption
- an engine that will not reach high idle or rated rpm
- excessive combustion noise or fuel knock in one or more cylinders at idle.

To measure NOP, follow the steps in **SKILL DRILL 11-2**.

> ▶ **SAFETY TIP**
>
> Always wear approved safety glasses and nitrile glooves when testing nozzles. The liquid spray leaves the nozzle tip with sufficient velocity to penetrate the skin and cause serious injury. Make sure that the injector is mounted on the tester so that the spray is directed away from the operator and everyone else. Volatile liquids can be extremely flammable when vaporized, so avoid any conditions (sparks, open flames, lit cigarettes, etc.) in which this might ignite the fluid used during the test procedure. Always use a vapor collector to prevent breathing in carcinogenic hydrocarbons.

Measuring Valve Seat Leakage (Forward Leakage)

The sealing surface between the nozzle valve and body must prevent fuel from leaking past the valve and dripping into the combustion chamber. Carbon deposits, wear, or

SKILL DRILL 11-2 Measuring NOP

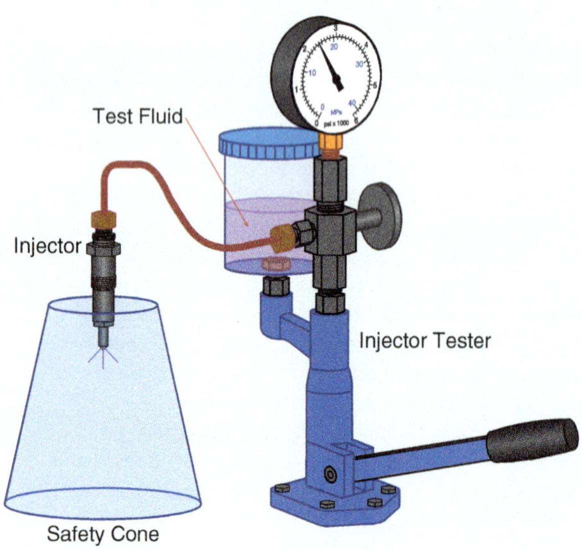

1. Install the nozzle in the pop tester by following steps 1–7 in Skill Drill 11-1.
2. Observe and record the NOP by pumping the injector nozzle tester and noting the pressure at which the needle valve lifts and fuel begins to spray from the nozzle tip. Injector NOPs vary according to the type of nozzle and the manufacturer's specifications.
3. Compare the injector's NOP with the manufacturer's specifications. Recommend replacement or rebuilding if the NOP doesn't consistently fall within an acceptable range.

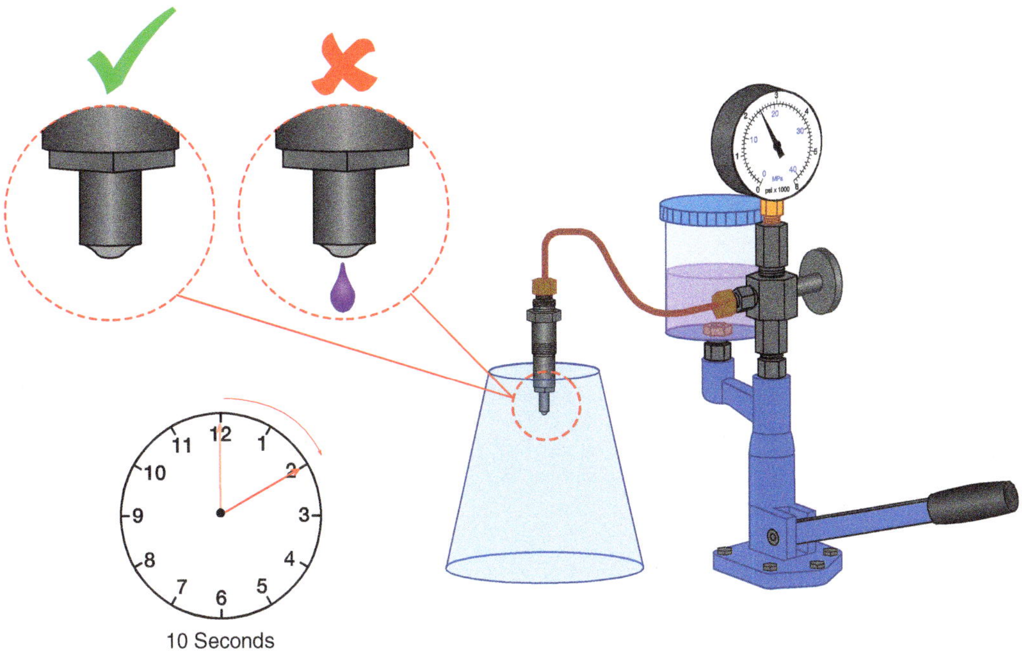

FIGURE 11-34 Evaluating forward leakage. No droplet should form on a multi-orifice nozzle tip.

abrasion of the seat can cause **forward leakage**, which is fuel leakage from the nozzle tip. Tests that measure forward leakage evaluate the condition of the sealing surfaces between the nozzle valve and seat. Forward leakage is measured on a nozzle tester. If the sealing surfaces between the nozzle valve and seat are in good condition, no fuel leakage should be observed at the nozzle tip. After five seconds, some slight wetting of a multi-orifice nozzle tip is generally acceptable, but no droplet of fuel should form on a multi-orifice nozzle tip. If a droplet forms within 10 seconds, this means that the valve seat is leaking (FIGURE 11-34).

Because pintle nozzles generally open at much lower pressures than multi-orifice nozzles do, less nozzle-valve pressure is exerted against a pintle nozzle's valve seat. With pintle nozzles, fuel droplet formation on the nozzle tip is acceptable, but if a droplet forms, it should not fall off the nozzle tip within seven seconds (FIGURE 11-35).

Worn seats or carbon deposits inside the injector will often cause forward leakage from the nozzle valve. Failure of the nozzle valve to properly seat is indicated by fuel leakage from the nozzle tip and an absence of chatter. Nozzles that leak tend to produce noisy combustion at idle; a loud, sharp combustion knock in the affected cylinder; and smoke. When fuel sits in a nozzle, it will boil, which leaves black carbon deposits that is known as coke. Coked and worn nozzle valve seats are the most common causes of forward leakage. To measure valve seat leakage (forward leakage), follow the steps in SKILL DRILL 11-3.

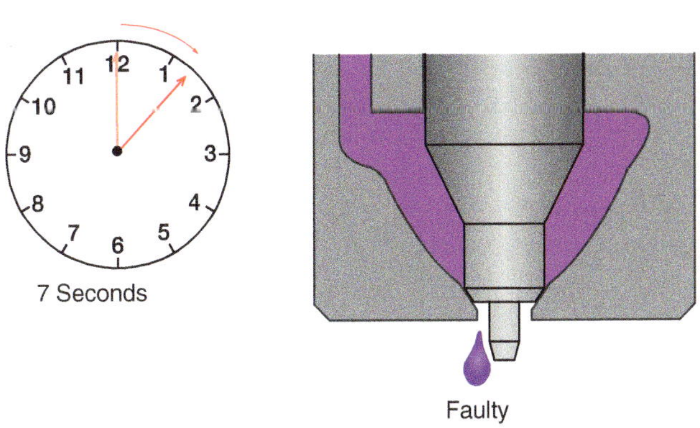

FIGURE 11-35 A pintle nozzle is allowed to leak only if the drop does not fall within seven seconds of reaching NOP.

SKILL DRILL 11-3 Measuring Valve Seat Leakage (Forward Leakage)

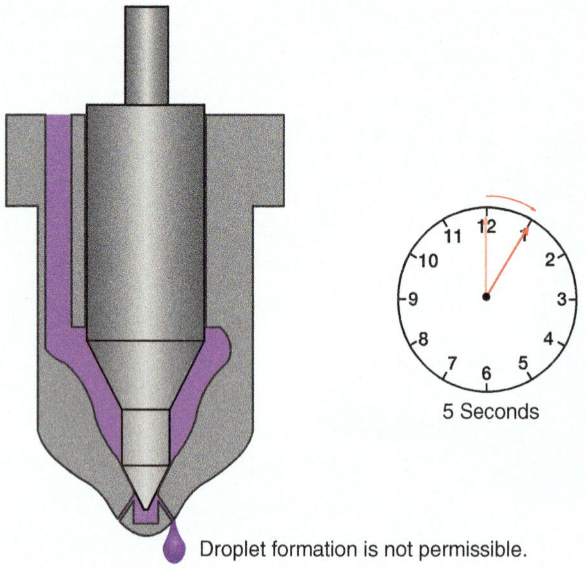

1. Install the nozzle in the pop tester by following steps 1–7 in Skill Drill 11-1.
2. Slowly bring the fuel pressure applied to the nozzle to 500 psi (34 bar) less than the manufacturer's specified NOP.
3. Maintain this pressure for at least 10 seconds. Observe whether any drops form on the nozzle tip. Do this by wiping the nozzle tip to determine whether it is dry. Some wetting is allowed on multi-orifice nozzles as long as there is no droplet formation within 10 seconds. It is acceptable for a droplet to form on the tip of a pintle nozzle, but if the droplet falls off within seven seconds, it is not acceptable.

Measuring Back Leakage

The amount of back leakage, also known as leak-off or return leakage, from an injector can indicate problems with the clearances between the nozzle valve and the nozzle body. Some manufacturers recommend this test if return flow volume is high. Back leakage can be measured on the nozzle tester by observing how much fuel pressure in the line to the nozzle drops after nozzle closing. To perform this test, the nozzle is pressurized and the fuel pressure to the nozzle is sealed in the line by closing a valve on the nozzle tester. Normal fuel leakage between the nozzle valve and nozzle body will cause a pressure drop in the line, which is observable on the tester's pressure gauge. Acceptable pressure drop varies from one manufacturer to another, but typically, a pressure drop of 300–875 psi (21–60 bar) in six seconds indicates that the mating surfaces inside the nozzle are not leaking excessively and that the nozzle valve-to-nozzle body clearances are acceptable. When the engine is running, the amount of leakage from the return fitting of most injectors should not exceed 1.0 cc/minute at 1,500 rpm. Back leakage at this rate would correspond to one drop forming at the leak-off fitting or line every minute during engine operation at idle. An exception is back leakage on common rail injectors, which is much higher than this.

If injector line pressure drops too quickly during a back leakage test, excessive clearance likely exists between the nozzle valve and guide. Pressurized fuel is quickly leaking between the nozzle valve and body before entering the back-leakage chamber. If the pressure drops too slowly, inadequate clearance exists between the nozzle valve and guide. This condition can lead to nozzle sticking or seizure due to lack of lubrication.

Another technique used to evaluate a worn nozzle guide is the sliding nozzle test. This is done when the nozzle is disassembled. If the nozzle valve and nozzle body are lubricated with clean fuel and the valve body is tilted at 60° horizontally, the nozzle valve should drop smoothly out of the body due to its own weight after it has been initially pulled out about one-third of its length. This test should be performed several times while rotating the valve 120° each time.

Dirt between the nozzle body and holder or misassembly of nozzles can also cause excessive back leakage. Fuel should not surge from leak-off port. To measure back leakage, follow the steps in **SKILL DRILL 11-4**.

Chatter Test

The delivery of fuel causes a sudden drop in fuel pressure at the nozzle valve which causes the nozzle valve to close. Hundreds of times during a single injection event, the fuel pressure will rise and lift the nozzle valve and then fall, closing the valve again as fuel escapes from

SKILL DRILL 11-4 Measuring Back Leakage

1. Install the nozzle in the pop tester by following steps 1–7 in Skill Drill 11-1.
2. Slowly bring the pressure applied to the nozzle to 100 psi (689 kPa) less than NOP.
3. Close the valve in the line to the nozzle. This seals the fuel in the nozzle under pressure.
4. Observe and record the pressure drop on the tester gauge six seconds after closing the tester valve.
5. Typically, a pressure drop of 300–875 psi (21–60 bar) in six seconds indicates that the mating surfaces inside the nozzle are not leaking excessively and that the nozzle valve–to–nozzle body clearances are acceptable. If the pressure drop is less than 300 psi (21 bar), the nozzle is not getting proper lubrication and should be replaced or rebuilt. If the pressure drop is more than 875 psi (60 bar), the nozzle is leaking excessively and thus should be replaced or rebuilt.

the spray holes. How often the valve opens and closes during an injection event is related to the nozzle differential ratio. The contact between the valve and seat that occurs during this high-frequency cycling of the nozzle valve opening and closing produces a high-pitched noise described as **nozzle chatter**. Nozzle chatter when bench testing an injector is an indication of a clean nozzle valve with good seat contact. Chattering or high-pitched buzzing from the nozzle is heard as the nozzle valve contacts the seat during testing. The intensity of the noise will vary with the speed at which the nozzle tester lever is operated. Chattering is an indication that the nozzle valve and seat are in good condition if no leakage from the nozzle tip exists. Although chatter is a positive indicator of a nozzle in good condition, the absence of chatter is not a reason to condemn or replace a nozzle.

> **▶ TECHNICIAN TIP**
>
> Fuel is used to cushion the force of the needle valve as it moves into the nozzle seat. Running an engine low on fuel or with air entrained in the fuel can lead to damage in the nozzle-valve seats and cause tip failure. A solid column of fuel is needed by injectors with high NOP values to absorb needle valve seating.

Cleaning

Injector nozzles may be disassembled to clean or replace nozzle valves and bodies. Because they are lapped together, valves and bodies should always be replaced together and never interchanged. The disassembled injector may be cleaned in a 15% caustic soda solution in an air-agitated solvent tank or by using an ultrasonic cleaner. A brass brush can be used to externally clean a nozzle tip.

Nozzle Service

After an injector has been identified as defective, the decision can be made to repair or replace the component. The following section outlines common symptoms and observations to help identify defective nozzles and service procedures to correct the condition.

TABLE 11-1 below is a summary of possible causes for various observations made during nozzle testing. Even without testing a nozzle, general symptoms will indicate a defective nozzle.

Forward Leakage

A well-sealing nozzle valve and seat is critical to proper engine operation. If the seal leaks, fuel can dribble out of the valve. Post-combustion fuel droplets can cause significant smoke exhaust emissions and combustion noise, which is most often evident at idle. Engine performance diminishes as well, because nozzle dribble deteriorates the quality of atomization. No wetting or droplet formation at the tip of a multi-hole nozzle is permissible after the nozzle is brought to 100 psi (689 kPa) below NOP. A pintle nozzle is allowed to leak only if the droplet forms after seven seconds of the nozzle having already reached NOP.

TABLE 11-1 Evaluating Injector Nozzles

Nozzle Malfunction	Result	Engine Symptom
NOP at 70–90% of minimum specification	• Uses additional fuel	• Excessive emissions of black or gray smoke • Slight horsepower increase (coarse droplets cause increased fuel delivery) • High fuel consumption • Excessive white smoke on start-up
NOP at 70% or less of minimum NOP	• Higher injection quantities at lower delivery pressure • Poor atomization and distribution • Combustion gases in fuel system	• Smoke • Misfiring • Poor fuel economy
Nozzle stuck open	• Over-fueling	• Excessive emissions of black or gray smoke • Cylinder misfiring at all load and speed conditions • Burned piston crowns • Oil dilution • Cylinder gases in fuel system at low speeds and loads • Nozzles have combustion gas pressure
Tip leakage	• Nozzle dribble • Poor atomization • Injection lag	• Misfiring • Light smoke, especially at idle • Excessive combustion noise • Nozzles may admit combustion • Combustion gases in fuel system
Excessive fuel return or leak-off	• Excessive clearance between nozzle valve and body	• Low power • Combustion noise • Smoke
Plugged orifices	• Improper fuel distribution	• Low power • Poor fuel economy • Rough-running • White smoke on start-up • Black or gray smoke emissions while running

▶ TECHNICIAN TIP

Leaking injector nozzle seats are most evident at idle since they produce visible smoke emissions. Leaking nozzle seats can also result in combustion gases leaking into the injector, eventually heating and carbonizing fuel there. Cylinder pressure can work its way through a leaking nozzle valve and into the injection system, causing low power, rough running, engine stalling, and misfires.

Return Fuel-Back Leakage

A slight clearance exists between the nozzle valve and nozzle body. This clearance allows a calibrated leakage of fuel around the nozzle valve to lubricate and cool the valve (FIGURE 11-36). Fuel passing into the nozzle holder containing the nozzle spring lubricates the spring seat. Eventually, **leak-off lines** in the nozzle holder above the valve accumulate fuel that is bled from the nozzle body to the fuel return line through a fitting attached to the nozzle (FIGURE 11-37). Without this drain, fuel would eventually pressurize the area above the nozzle valve, preventing upward movement of nozzle valve.

Factors That Can Reduce Injector Life

Nozzle service is indicated when poor fuel economy, rough running, low power, hard starting, misfiring, excessive combustion noise, or black, blue, or gray exhaust smoke are reported. Increasing fuel consumption and PM emissions are the most common nozzle service indicators in heavier engines. Nozzle service for engines that use electronic controls should be performed whenever cylinder balance or contribution tests show excessive variations between cylinders or the OBD system sets fault codes associated with the fuel system monitors.

Engine hours and distance traveled are factors in injector wear. While injector nozzles should provide thousands of hours of service, light-duty diesel engines that use

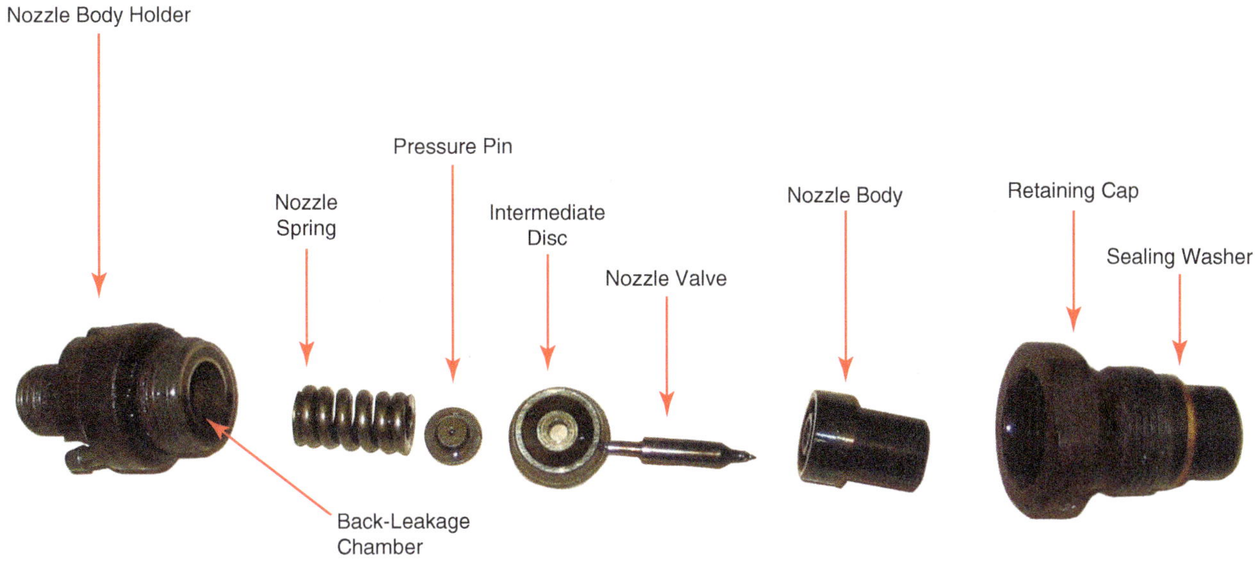

FIGURE 11-36 Fuel return from injectors consists of fuel that has leaked between the nozzle valve and nozzle body and into the back leakage chamber above the nozzle.

FIGURE 11-37 Fuel leak-off lines at injectors collect fuel and return it to the fuel tank.

non-electronic fuel systems should be evaluated on a nozzle pop tester at 150,000 miles (241,400 km). Nozzle life on any engine should be a minimum of 5,000 hours of operation using clean, good-quality fuel. Many can last as long as 30,000 hours; Late-model fuel systems have adaptive strategies that anticipate and compensate for nozzle wear. Emission legislation for late-model engines sets durability requirements as long as 11 years of vehicle operation or 150,000 miles (241,400 km) of use.

Several operating conditions can reduce the service life of diesel fuel injectors: Poor fuel quality and excessive idle time can lead to accelerated nozzle deterioration due to carbon and/or sulfur deposits forming in spray holes and other internal passageways. Sulfur in the presence of water forms corrosive acids. Even air can damage nozzles. Fuel below the nozzle valve dampens the valve against its seat, much like a shock absorber. Air will displace the fuel, leading to nozzle wear and damage.

Abrasives such as dirt in the fuel will accelerate wear on fuel system components. In injector nozzles, abrasives will enlarge the clearances between the nozzle valve and guide, damage the nozzle-valve seat, and enlarge the spray hole openings. Scuffing on nozzle valves indicates poor lubrication inside the nozzle. This may be due to abrasives or poor fuel lubricity.

TECHNICIAN TIP

Excessive fuel leakage from the leak-off connector is evidence of an excessively worn nozzle valve or incorrectly assembled injector. Injectors showing evidence of more than a few drops of fuel per minute should be removed and properly evaluated by using a tool such as a nozzle tester. On common rail systems, evaluating back leakage is an important diagnostic step toward determining whether defective injectors are the root cause of fuel system fault codes and performance problems.

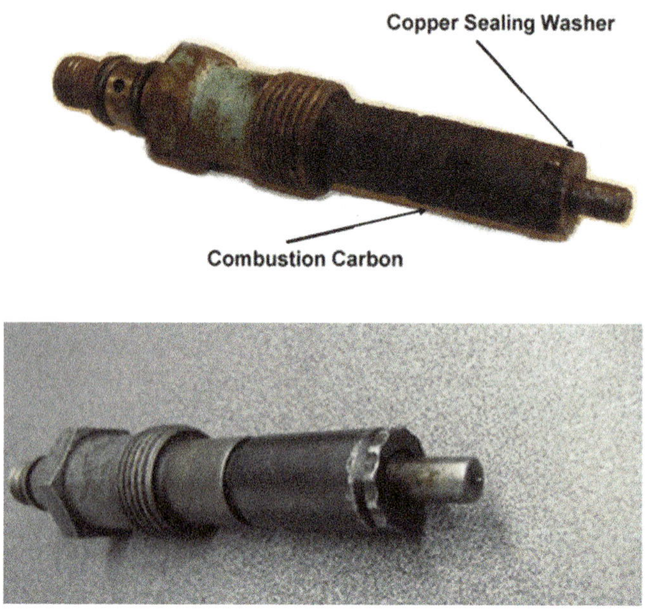

FIGURE 11-38 An example of a clean, well-sealing injector and an injector with a poorly sealing copper washer seal. Note that carbon from the combustion chamber leakage has entered into the injector sleeve and deposited on the injector.

Excessive heat will accelerate wear between the nozzle valve and the nozzle body. Increased back leakage of fuel will result when wear is excessive between the nozzle valve and nozzle body, thus reducing the quantity of fuel available for injection. Damage to the hardened nozzle valve will also occur. Stainless steel nozzle material will turn a blue color on nozzle valves and tips as evidence of excessive heat. For this reason, proper clamping force and the use of a copper washer for an injector seal are necessary to remove heat from the injector. Improper installation can result in insufficient cooling and nozzle-valve seizure (**FIGURE 11-38**).

If the ends of the nozzle valves are dark or discolored, it is an indication that the nozzle valve is not properly sealing. The discoloration is due to combustion gases blowing back through the spray holes, causing excessive temperatures in and discoloration on the nozzle valve. Specialized cleaning brushes and scrapers are used to scrape carbon from injector tubes or cylinder head casting before reinstalling injectors.

Water in the fuel will corrode and gall fuel system components. In addition, water does not easily pass through the spray holes in multi-orifice nozzles. Water accumulations can cause the injector tip to break due to the high pressures caused by steam expansion in the nozzle tip (**FIGURE 11-39**).

Improper service procedures can cause nozzle damage. Always use the correct tools to remove nozzles. Cleaning nozzle tips with a steel wire wheel will damage the nozzle spray holes. Only brass brushes and recommended cleaning solvents should be used (**FIGURE 11-40**). When injector lines are disconnected or when nozzles are removed for servicing, both the tips and inlet fittings should be capped with protective seals. Nozzles should never be handled by their tips or with bare fingers, because oils and dirt on hands can contaminate the spray holes.

▶ TECHNICIAN TIP

If nozzles are disassembled, handle the nozzle valves only by their stems. Oils on skin will corrode the nozzle valve and cause it to seize.

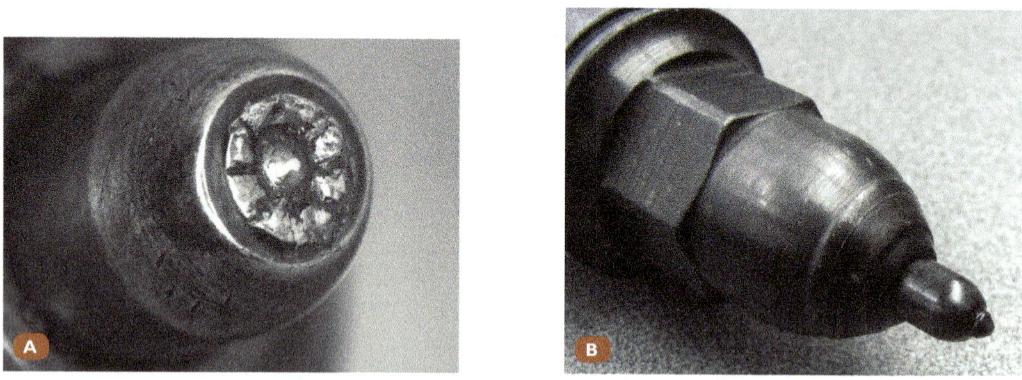

FIGURE 11-39 **A.** This nozzle tip is broken across the nozzle orifices. **B.** The same nozzle, but with an intact tip.

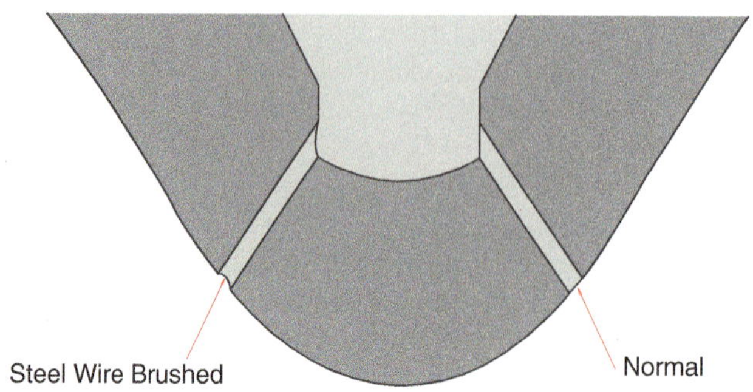

FIGURE 11-40 Note how brushing a nozzle with a steel wire brush will damage nozzle holes.

▶ Wrap-Up

Ready for Review

- Injector nozzles are incorporated into every type of fuel-injection system.
- The simplest function of an injector nozzle is to permit highly pressurized fuel to squeeze through one or more small orifices in the nozzle tip, which atomizes and distributes fuel into the combustion chamber.
- The performance, efficiency, and emissions of diesel engines are major functions of the fuel atomization and distribution process.
- Fuel nozzles, combined with combustion chamber turbulence, distribute and mix fuel throughout the combustion chamber.
- The amount of time necessary for combustion depends on the quality of fuel, pre-ignition temperature, and degree of atomization.
- A nozzle consists of two main parts: the nozzle or needle valve and a larger assembly that encloses the nozzle, called the nozzle body.
- Closed nozzles operate by using a pressure differential principle.
- Diesel engine fuel systems require an abrupt beginning and end of injection; without this, a gradual loss of pressure at the nozzle tip would result in increasing spray droplet size and decreasing penetration.
- Precisely matching the injector to the shape of the piston bowel and the turbulence of the combustion chamber is crucial for efficient combustion.
- Closed injector nozzles are the most popular nozzle type in use today. There are two types of closed nozzles: the multi-orifice nozzle and the pintle nozzle.
- Multi-orifice, or multi-hole, nozzles are used in DI combustion engines; they have four or more spray holes at the injector tip.
- Pintle nozzles are used in IDI combustion engines; they have only one spray hole in the injector tip.
- Variable-geometry injectors, which are currently under development, use two or more rows of orifices located around the tip of the injector.
- The two-spring, two-stage injector nozzle mechanically controls injection rate by varying the amount of nozzle-valve lift by using two nozzle-valve springs.
- Fuel spray from nozzle testers is under pressure high enough to penetrate skin, which can cause blood poisoning. Never put hands near the end of injector nozzles, and always wear eye protection.

Key Terms

abrasive flow machining (AFM) A process of enhancing the flow characteristics and performance of injector nozzles by smoothing passageways through the injector and enlarging the spray holes by running an abrasive compound through the nozzles.

atomization The process of breaking up liquid fuel into a fine mist.

back leakage A calibrated leakage of fuel around the nozzle valve to lubricate and cool the valve.

beginning of injection (BOI) The point when fuel delivery begins and is referenced according to the position of the crankshaft measured in degrees of crankshaft rotation.

electronic unit injectors (EUI) Electrically controlled injectors that incorporate timing, metering, atomization, and pressurization functions into a single unit or injector.

forward leakage Injector nozzle leakage from the spray holes, which indicates that the nozzle valve–to-seat seal is poor.

hydraulic nozzle holder The nozzle body that encloses the nozzle assembly that contains passageways connecting the high-pressure injection line to the nozzle valve.

injection lag The time delay that occurs between the start of fuel pressurization and the moment when injection actually takes place; this is also known as injection delay.

leak-off lines Lines at the injectors that collect fuel and return it to the fuel tank.

multi-orifice nozzle A fuel nozzle that uses multiple spray holes to atomize and distribute fuel.

needle lift sensor A sensor that provides a reference signal to the electronic control module (ECM) for the beginning of injection (BOI) by providing information about movement of the nozzle valve.

nozzle chatter A high-pitched noise or chirping sound produced by the high-speed cycling of the nozzle valve opening and closing with contact between the valve and seat.

nozzle differential ratio The ratio between the surface area of the nozzle valve that is acted upon by high-pressure fuel when the valve is closed compared to the nozzle valve area when the nozzle valve is unseated.

nozzle opening pressure (NOP) The pressure required to unseat the nozzle valve and begin injecting fuel; it is also known as valve opening pressure (VOP).

nozzle valve A valve in the nozzle that seals the end of the nozzle tip when it is not injecting fuel; it is also known as a needle valve.

pintle nozzle A fuel nozzle which sprays fuel through a single hole.

pop tester A tool used by technicians that supplies pressurized fuel to injector nozzles, which is used to evaluate the condition of fuel nozzles.

pressure wave reflection The movement of a pressure wave back and forth along the length of an injector line. Like the sloshing of water in a tub, the pressure wave bounces back and forth between the nozzle and high-pressure pump.

throttle pintle nozzle A type of injector that changes the rate of injection from the nozzle through the shape of the nozzle valve; it is also called a delay nozzle.

variable-orifice nozzle A type of nozzle that is under development, which has two rows of spray holes that regulate the delivery rate, spray angle, and spray hole size with a variable-lift nozzle valve; it is also known as a coaxial nozzle.

Review Questions

1. What is the purpose of a slight amount of fuel leakage around the nozzle valve and into the nozzle spring chamber?
   a. To detect a defective nozzle.
   b. To lubricate the nozzle valve.
   c. To prevent excess fuel delivery.
   d. To cool the valve spring and seat.
2. Which of the following is the typical nozzle opening pressure (NOP) of a new multi-orifice nozzle?
   a. 500–5,000 psi (34–345 bar).
   b. 900–2,500 psi (62–172 bar).
   c. 2,000–3,200 psi (138–221 bar).
   d. 3,200–5,500 psi (221–379 bar).
3. When forward leakage from a nozzle is being checked, which of the following is being evaluated?
   a. NOP.
   b. Nozzle-valve seat sealing.
   c. Nozzle valve–to–nozzle body clearance.
   d. Injector spring tension.
4. Which of the following conditions is caused by a weak nozzle spring?
   a. Higher NOP only.
   b. Higher NOP and slightly retarded timing.
   c. Nozzle chatter.
   d. Lower NOP.
5. Which of the following is a likely cause of excessive back leakage from a fuel nozzle?
   a. Excessive clearance between the needle valve and nozzle body.
   b. Excessive spring pressure.
   c. Worn nozzle orifices.
   d. Needle valve–to–needle seat leakage.
6. Which type of nozzle is used by a DI combustion chamber?
   a. Pintle.
   b. Open.
   c. Multi-orifice.
   d. Pencil.
7. The pressure required to open the nozzle valve of an injector, when compared to the pressure required to keep it open, is _______________.
   a. much less than closing pressure
   b. slightly less than closing pressure
   c. equal to closing pressure
   d. greater than closing pressure
8. A blocked fuel return line from a nozzle holder could eventually result in which one of the following conditions?
   a. A worn-out nozzle holder.
   b. Over-fueling the engine.
   c. Excessive spray-in pressures.
   d. Engine stalling.
9. Which of the following emissions would increase the most if a nozzle opening orifice were to become larger because of wear?
   a. Carbon monoxide (CO).
   b. Particulate matter (PM).
   c. Carbon dioxide (CO_2).
   d. Oxides of nitrogen (NO_x).
10. Injection delay is caused by __________.
    a. fuel with a low cetane number (CN)
    b. weak nozzle springs
    c. nozzle-valve lift
    d. nozzle lift sensors

ASE Technician A/Technician B–Style Questions

1. Technician A says a closed hydraulic nozzle allows combustion gases to mix with fuel inside the nozzle. Technician B says a closed nozzle has no fuel return line back to the fuel tank. Who is correct?
   a. Technician A
   b. Technician B
   c. Both A and B
   d. Neither A nor B
2. Technician A says that DI combustion chambers use pintle spray nozzles. Technician B says that only multi-orifice nozzles are used in DI combustion engines. Who is correct?
   a. Technician A
   b. Technician B
   c. Both A and B
   d. Neither A nor B
3. Technician A says that the nozzles used by DI combustion engines mix air and fuel faster and produce less combustion noise than nozzles used in IDI combustion engines. Technician B says that before the introduction of electronically controlled DI systems, IDI combustion engines produced less noise with fewer emissions and had higher maximum engine rpm than DI combustion engines did. Who is correct?
   a. Technician A
   b. Technician B
   c. Both A and B
   d. Neither A nor B
4. Technician A says that fewer, smaller spray holes in an injector tip will atomize fuel better than an injector with larger, more numerous holes. Technician B says that fewer, larger spray holes in an injector tip will produce better fuel atomization. Who is correct?
   a. Technician A
   b. Technician B
   c. Both A and B
   d. Neither A nor B
5. Technician A says that nozzle opening pressure (NOP) is adjusted with selective shims placed on top of the nozzle spring. Technician B says that NOP is electronically controlled in some of the latest injectors. Who is correct?
   a. Technician A
   b. Technician B

c. Both A and B
d. Neither A nor B

6. Technician A says that servo-hydraulic injectors use hydraulically balanced nozzle valves. Technician B says that servo-hydraulic injectors require piezoelectric actuators. Who is correct?
   a. Technician A
   b. Technician B
   c. Both A and B
   d. Neither A nor B
7. Technician A says that injection lag takes place when the time between the actual beginning of injection (BOI) lags behind the time when fuel pressurization takes place. Technician B says that injection lag is caused by poor quality fuel with a low cetane number (CN). Who is correct?
   a. Technician A
   b. Technician B
   c. Both A and B
   d. Neither A nor B
8. While discussing the purpose of a nozzle lift sensor, Technician A says that the sensor helps calibrate the amount of nozzle lift that an engine's injectors should have. Technician B says the sensor helps produce more precise injection timing. Who is correct?
   a. Technician A
   b. Technician B
   c. Both A and B
   d. Neither A nor B
9. When replacing an injector in an engine, Technician A says it is only important to make sure that the replacement injector belongs to the same family or model of engines. Technician B says it is important that for the replacement injector to have a spray pattern that matches the shape of the combustion bowl in the piston. Who is correct?
   a. Technician A
   b. Technician B
   c. Both A and B
   d. Neither A nor B
10. Technician A says that a driver may be correct when they report that an engine has less power after the injector has been replaced. Technician B says that fuel consumption should decrease after replacing worn-out injectors. Who is correct?
    a. Technician A
    b. Technician B
    c. Both A and B
    d. Neither A nor B

CHAPTER 12
HEUI Injection Systems

Learning Objectives

After reading this chapter, you will be able to:

- 12-01 Identify and explain the advantages of HEUI injection systems.
- 12-02 Identify and classify HEUI injectors according to construction and operating principles.
- 12-03 Describe and differentiate Gen II HEUI injectors from earlier injectors according to construction and operating principles.
- 12-04 Identify and describe the construction and operation of Caterpillar H1B injectors.
- 12-05 Identify and describe the construction and operation of the components in HEUI high-pressure actuation systems.
- 12-06 Identify and describe the functions of electronic system components used in HEUI injection systems.
- 12-07 Outline the diagnostic and repair procedure steps for HEUI systems.

You Are the Technician

A customer has brought into your shop a medium-duty van that's used for a parcel pickup and delivery service. Recently, the engine has stalled when warmed up and it lacks power, runs rough, and has substantially increased fuel consumption. Because it has a V8 hydraulically actuated electronic unit injection (HEUI) system, you check the oil level and look for external oil leaks. You also take a fuel sample to determine whether it is contaminated with lubricating oil. Observing nothing abnormal, you perform some typical diagnostic routines, including checking for fault codes and performing cylinder cutout and balance tests. You also perform an injector buzz test and find that all the injector solenoids sound strong and are uniformly consistent in operation. Next, you check the injection control pressure for the high-pressure oil system and discover the control pressure is not stable but moves around quickly, going up and down nearly 200 psi (17 bar). The duty cycle on the injection control pressure regulator is also unstable and is above normal. After asking the customer about the vehicle's service history, you learn the problems with the engine began after the engine was serviced at a budget oil change drive-through.

1. List two likely causes for an unstable duty cycle on the injection control pressure regulator and explain how they would produce the driver's reported problems.
2. What problems could potentially cause the engine to be difficult to start after the engine has warmed up?
3. What should be the first service recommendation you make before you diagnose the engine any further?

▶ Introduction

One of the first and largest evolutionary steps of modern high-pressure diesel-injection technology was the development of the HEUI system. In the 1980s, manufacturers began exploring technical solutions for upcoming emission standards, and they realized that mechanically governed fuel systems had several handicaps. The primary disadvantage of traditional mechanical systems is their reliance on either an engine-mounted camshaft or a camshaft, cam plate, or cam ring inside an injection pump to pressurize fuel. Camshafts used to move plungers inside unit injectors or injection pumps worked well for years, but camshafts driven at engine speed impose limits to injection pressure at low rpm. Because much higher injection pressures are required to reduce emissions, radical alternatives to the camshaft-driven injection systems needed development.

A second problem of conventional mechanically controlled fuel systems is their limited ability to adapt injection events to continuously changing engine and vehicle operating conditions. Engine operating conditions such as load; coolant, fuel, and intake air temperatures; intake boost and atmospheric pressures; vehicle speed; exhaust gas recirculation (EGR) flow; and many other factors require unique injection timing and **injection rate control**, which is the shaping of the injection discharge curve during an injection event. Better control over the injection rate and timing has helped diesels today obtain a 99% reduction in noxious emissions compared to engines built in the early 1990s (**FIGURE 12-1**).

Engines today cannot operate in isolation from other vehicle systems, either. Inputs from systems such as collision avoidance, exhaust emission aftertreatment, traction control, the antilock brake system, stability control, climate control, vehicle security, power takeoff, and many others are designed to work interactively with the engine to enhance safety, passenger convenience, and comfort. Mechanically governed injection systems simply cannot achieve the flexibility that injection events require; nor can they interact with other control systems by using an onboard network connecting control modules together.

▶ Fundamentals of HEUI Systems

12-01 Identify and explain the advantages of HEUI injection systems.

HEUI injection systems have replaced the mechanical force produced by an injection pump or engine camshaft to pressurize fuel for injection. Instead, pressurized engine oil is used to produce the high injection pressures and fuel rate control needed to achieve improved performance from engines with almost undetectable emissions and superior fuel economy.

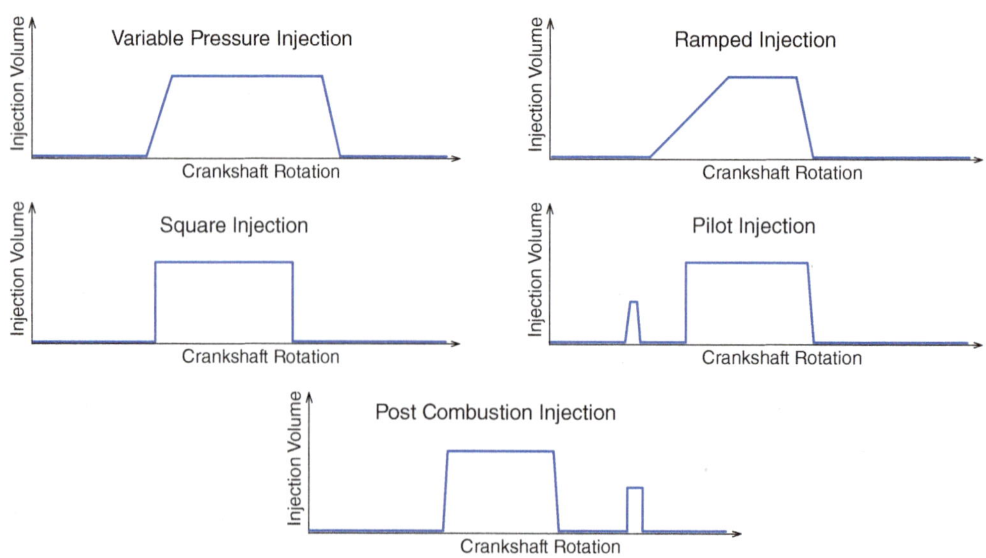

FIGURE 12-1 Injection rate–shaping abilities achieved by late generation HEUI injectors.

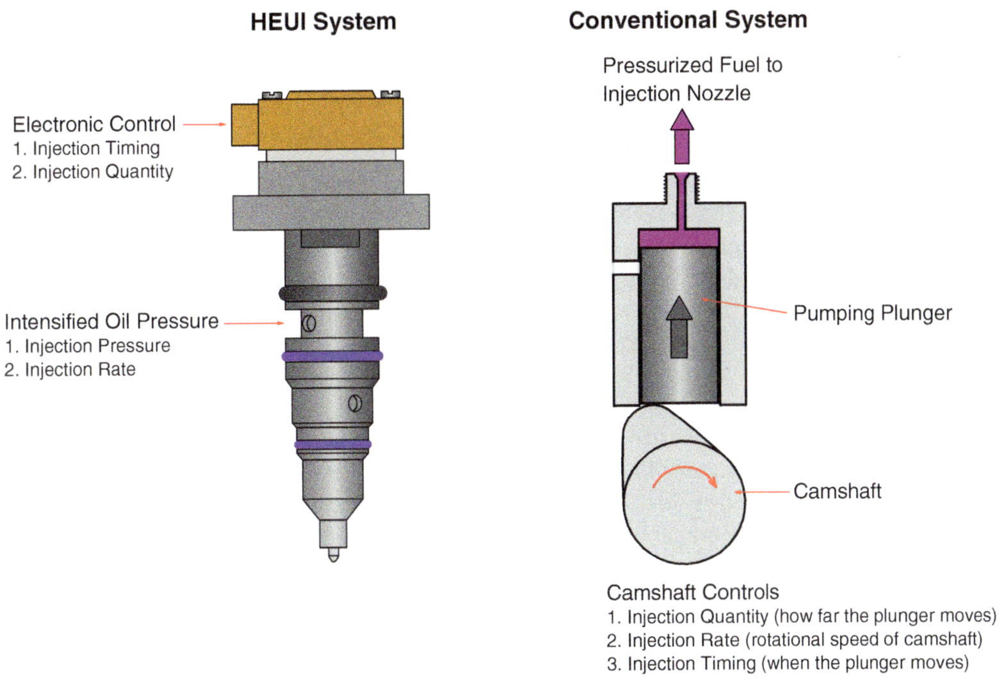

FIGURE 12-2 The camshaft shape and engine speed control many aspects of injection in traditional fuel systems, but they cannot develop the high pressure or control injection rate and timing with the flexibility of HEUI systems.

This next section examines the operating principles and basic construction of HEUI fuel systems that form this contemporary high-pressure injection system.

Advantages of HEUI Systems

HEUI systems were developed and adopted because of the distinct advantages of using a hydraulic system to pressurize fuel instead of an injection system's mechanical camshaft. This first section looks at some of the reasons for the development and use of HEUI systems used in pick-ups, vans and delivery vehicles from 1993 to 2010.

Electronic Control of Injection Pressure

The biggest shortcoming of mechanically governed fuel systems is their inability to adequately pressurize fuel at low engine speeds. High injection pressures are needed for all engine speeds to achieve the best atomization and distribution of fuel in the combustion chamber. Because fuel plunger velocities in camshaft-pressurized injection systems are directly proportional to engine speed, injection pressures are tied to engine speed. Plungers move relatively slow at low engine rpm, so maximum injection pressures are also low at low engine rpm. Slow plunger movement simply does not provide high enough injection pressure to reduce emissions. What is needed is an injection system capable of pressurizing fuel independent of engine speed (**FIGURE 12-2**). The HEUI system was developed to do just that—to develop peak injection pressure without relying on engine speed to build fuel pressure (**FIGURE 12-3**). This means that HEUIs have maximum spray-in pressure available whether the engine is operating at idle or maximum rpm. During hard acceleration or sudden load changes at low speeds, the system can instantly adjust fuel pressurization to produce the combustion conditions needed to generate power while minimizing emissions. Electronically controlling injection events, such as timing, metering, and pressurization, allows these functions to be completely adjustable. Instead of using cam lobes, springs, flyweights, and levers, HEUI engines use electronic engine sensors, software code,

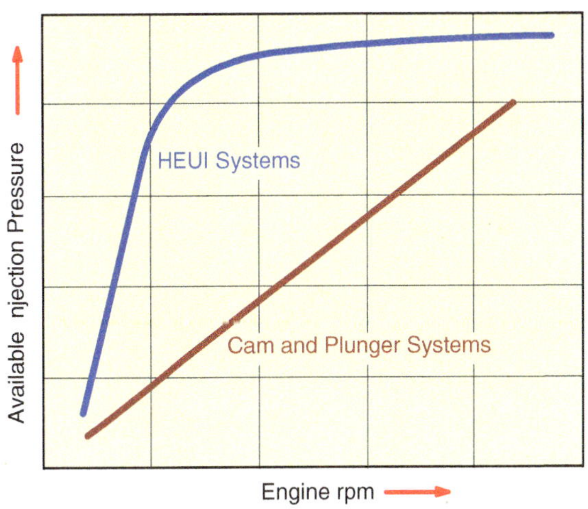

FIGURE 12-3 HEUI fuel systems can develop injection pressures independent of engine speed.

microprocessors, and electrical signals to control injection events more precisely than any previous system could have.

Injection Pressurization by Using Lubrication System Oil

HEUI systems use the hydraulic force of highly pressurized lubrication oil rather than a mechanical camshaft to actuate plungers used to pressurize fuel for injection (**FIGURE 12-4**). Because the functions of pressurization, metering, timing, and atomization are all incorporated into a single injector body, HEUI injectors are classified as unit injectors. These uniquely designed injectors are supplied with low-pressure fuel and highly pressurized engine oil. Using lubricating system oil, an engine-driven, high-pressure oil pump increases lube oil pressure to up to 4,000 psi (276 bar) in the latest HEUI system.

Oil supplied to the injectors provides the actuation force or pressure to inject fuel. However, even oil pressurized to a few thousand pounds of pressure (a couple hundred bar) cannot directly apply the force needed to inject fuel. To reach that level, oil pressure is further multiplied inside the injector by using an amplifier piston to give HEUIs the capability to achieve maximum injection pressures of 23,000–28,500 psi (1,586–1,965 bar), depending on the engine model.

Injection pressures are varied by regulating the oil pressure supplied to the injectors. Electronically varying oil pressures means that injection pressure can be controlled to match engine operational requirements. The highest pressure is available when needed at low engine rpm or any engine speed and load condition. Because HEUI system oil pressure is electronically regulated and does not depend on high engine speed to produce maximum injection pressure, HEUI systems are classified as a subtype of the common rail (CR) fuel system. High-pressure injection capabilities, combined with electronically controlling timing, metering and injection rate, ensure the fine atomization of fuel, low emissions, low combustion noise, highly responsive engine performance, and improved fuel economy (**FIGURE 12-5**).

Modular Fuel System Construction

HEUI system components are interchangeable with a large variety of engine sizes, cylinders, and applications. Whether the engine is used in power generation, off-road equipment, an ambulance, a pickup truck, a delivery vehicle, or a dump truck, many of

> **▶ TECHNICIAN TIP**
>
> Modern diesels need the ability to regulate injection pressures according to engine operating conditions. By changing injection pressure, the rate at which fuel is injected into the combustion chamber is altered. Injection rate control is needed primarily to lower engine emissions, but it also significantly influences combustion pressure and fuel consumption. Rate control is therefore a crucial function of all contemporary fuel-injection systems.

> **▶ TECHNICIAN TIP**
>
> Although many HEUI injectors are identical in shape and size, they may not be interchangeable. For example, injectors will have internal differences such as maximum delivery volumes, amplifier piston diameters, nozzle spray angles, and nozzle hole sizes. Technicians should ensure the injector replacements are made correctly according to manufacturer, engine model, and engine serial number.

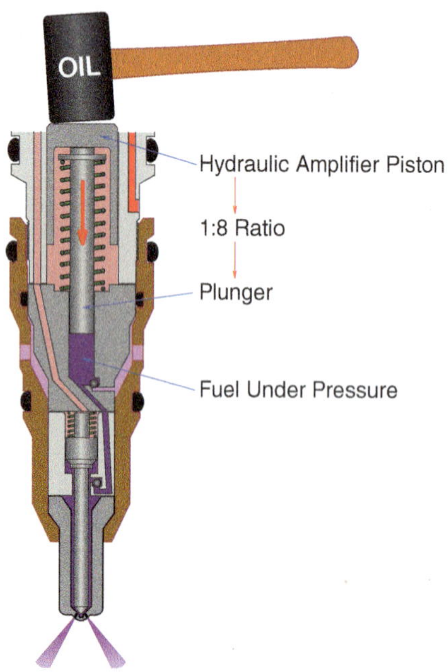

FIGURE 12-4 HEUI systems use the hydraulic force of highly pressurized engine oil for injection actuation. Oil pressure replaces a conventional camshaft.

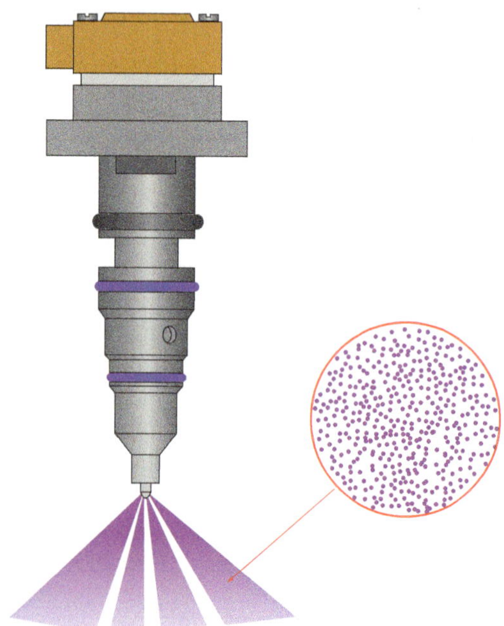

FIGURE 12-5 HEUI fuel systems can achieve lower emissions, improved fuel economy, and greater power output because of their superior ability to pressurize fuel at almost any engine speed.

the same components are used. This modular construction feature of HEUI system components permits HEUI adaption to a specific engine application by simply changing software programming. The flexibility of HEUI systems for use in a variety of chassis and engine configurations contrasts sharply with mechanically governed injection systems, which required a specialized configuration of injectors, governors, and other fuel system components for each engine model and application. HEUI injectors and other system components can easily be replaced by technicians when service is required and require no adjustment.

Quieter Engine Operation

Ongoing developments in HEUI technology have produced improved ability to shape the rate of fuel injection, which further reduces combustion noise and emissions for quieter, cleaner engine operation. Beginning with some 1996 models, first-generation HEUI injectors used pilot injection or split-shot injection. **Split-shot injection**, also called pre-injection metering (PRIME) in Caterpillar engines, is an injection strategy that delivers fuel in two distinct events during one combustion cycle (**FIGURE 12-6**). The first event, called **pilot injection**, delivers a small quantity of fuel typically 8° to 10° before the main injection. Fuel from the first, smaller injection begins to burn just as the second much larger main injection occurs. The time required to ignite the main injection is significantly shortened, which reduces the characteristic combustion noise from diesel engines caused by ignition delay. Shortening ignition delay also means injection timing can be retarded. When the main injection event begins to burn sooner and more quickly by using pilot injection, the main injection event can take place later—or in other words, be retarded. The advantage of delaying the beginning of the main injection is a reduction of oxides of nitrogen (NO_x) emissions. Less fuel is in the cylinder at top dead center (TDC), which causes a sharp spike in cylinder pressure and creates NO_x-forming combustion conditions. The main injection event can also have an injection rate that is shaped like a slope (more commonly called ramped injection rate). With a ramped injection rate, there is a more gradual buildup of fuel injected into the cylinder, which results in a gentler rise in cylinder pressure (**FIGURE 12-7**).

Electrohydraulic Valve Operation (Cam-less Diesel Engines)

HEUI technology also provides the potential to integrate valve-train operation into the high-pressure oil actuation system. Instead of using a camshaft to operate valves, electrohydraulic actuation of intake and exhaust valves is possible by using the HEUI

> ▶ **TECHNICIAN TIP**
>
> The time between the beginning of injection and the point at which fuel ignites is called ignition delay. Combustion does not occur immediately after injection, because several conditions must first be met: liquid fuel must be atomized, absorb heat, vaporize, mix with air, and heat to ignition temperature before it can begin to burn. Fuel accumulating in the combustion chamber during the ignition delay period increases, which in turn produces higher cylinder temperatures and pressures in the combustion chamber after ignition. The high temperatures and pressures further accelerate the burn rate of diesel fuel after ignition occurs. Faster burning fuel contributes to the spike in combustion pressure, which produces the characteristic noise, or fuel knock, from diesels. This is why shortening the delay period between injection and ignition reduces diesel combustion knock. Injecting a small quantity of fuel before the main injection event acts like the pilot light in a gas furnace: the small injection ignites the main injection. This is why split-shot injection is also referred to as pilot injection.

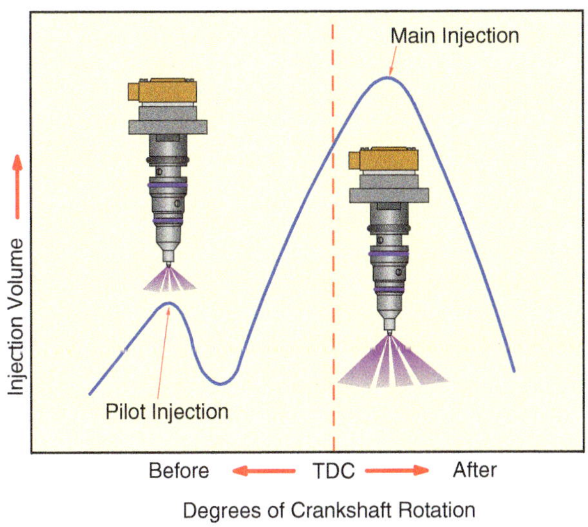

FIGURE 12-6 HEUI injectors are capable of changing the rate at which fuel is injected during the injection cycle. This graph depicts slip-shot injection pressure from a HEUI B injector.

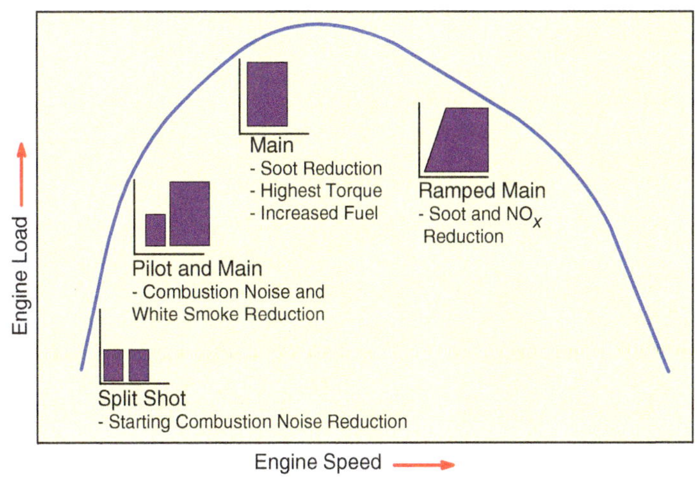

FIGURE 12-7 The square edges of the fueling profiles represent a sharp beginning of injection with a maximum quantity of fuel injected throughout injection. The sloped or ramped injection rates represent a gradual buildup of injection quantity and consequently a gentler rise in cylinder pressure. Ramped injection rate shapes reduce NO_x emissions.

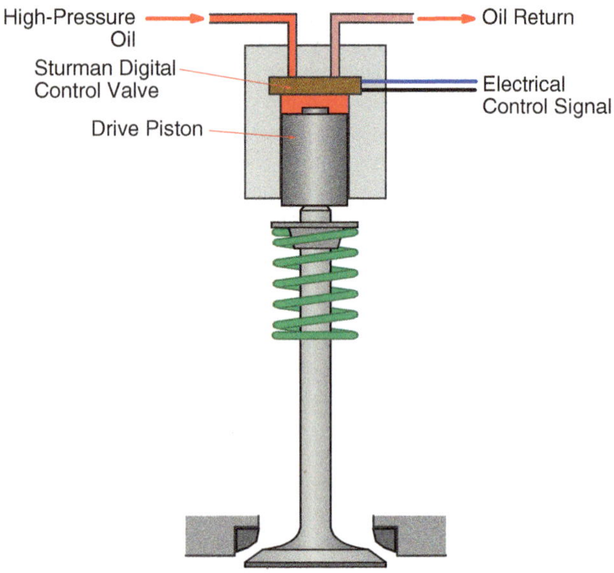

FIGURE 12-8 The concept of eliminating the camshaft is simple: use hydraulic pressure to open and close the valve.

high-pressure oil pump (**FIGURE 12-8**). The concept of the **cam-less diesel** , prototyped in 1998 by Navistar, is appealing for the number of interesting advantages it offers.. For example, variable valve timing and valve lift can optimize engine breathing for reduced emissions and increased power. Variable late-intake-valve closing can significantly reduce fuel consumption by reducing the energy used to unnecessarily compress air when it is not required for combustion. Late-intake-valve closing essentially reduces pumping loss, which is energy consumed by moving gases in and out of the cylinders. Also, by holding several exhaust valves open during the cranking period, the engine can use a compression-release starting technique to increase cranking rpm while using a smaller starting motor and fewer heavy batteries. An internal engine-compression brake can be added to the operating strategy by using a few simple lines of software code and dash switches. The most obvious advantage of the cam-less diesel engine is the reduction in parts required by the valve train, such as the camshaft, pushrods, rocker levers, and drive gears.

Applications of HEUI Injection Systems

Collaboration between Navistar, which builds International brand trucks and engines, and Caterpillar produced the first HEUI system in 1993. The Ford 7.3 L (444 cubic inch) Power Stroke, which was built for Ford by Navistar and branded as the T-444E for International, is one of the earliest and most popular applications of HEUI technology. More than two and a half million of these engines were manufactured from the early 1990s until its discontinuation in 2003. In fact, for several years, two out of every three Ford F-250 and larger trucks were equipped with this engine (**FIGURE 12-9**). HEUI systems used primarily by Ford Power Stroke engines in its light- and medium-duty vans were discontinued with the introduction of the 6.4 L CR Power Stroke for the 2007 model year. Packaging limitations of the 6.4 L prevented its use in the E series vans, where the 6.0 L HEUI Powerstroke was used until 2010.

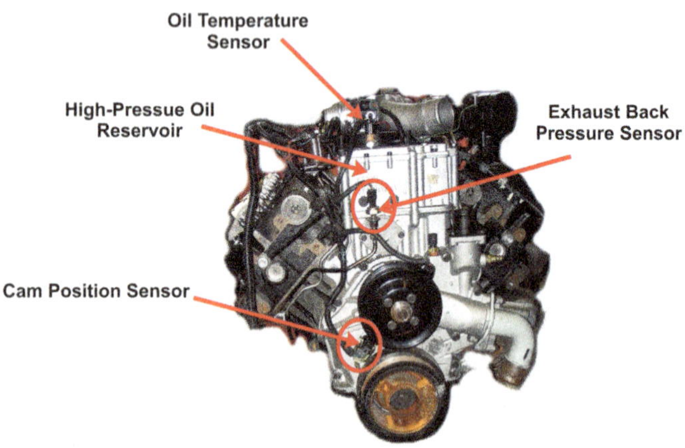

FIGURE 12-9 A Ford Power Stroke engine. More than two and a half million of these HEUI-equipped engines were produced between 1993 and 2003.

Components of HEUI Injection Systems

The HEUI system has four major subsystems:

1. HEUI injectors
2. low-pressure oil system
3. high-pressure injection actuation system
4. electronic control system.

12-02 Identify and classify HEUI injectors according to construction and operating principles.

A low-pressure supply system provides fuel to the injectors at the correct pressure and free of water, vapor, and other contaminants (**FIGURE 12-10**). Fuel supply pressure is regulated to 30–80 psi (207–552 kPa) on most systems.

FIGURE 12-10 Some low-pressure transfer pumps are located and driven off the high-pressure pump.

▶ Classification of HEUI Injectors

HEUI injectors use highly pressurized engine oil to provide injection force. Four basic variations of these injectors exist (**FIGURE 12-11**). Beginning in 1996, International/Ford and Caterpillar used the split-shot or PRIME injector. The first variation, called the **HEUI A** injector, is the earliest HEUI injector without split-shot- or pilot-injection capabilities (**FIGURE 12-12**). The second variation, called the **HEUI B** injector, is distinguished by its white solenoid (**FIGURE 12-13**). Caterpillar also manufactured their own distinct injector design on the 3116 and 3126A engines that used a side-mounted solenoid and high-pressure jumper tubes to supply oil. Caterpillar's injector has a larger plunger bore and greater fuel-delivery-volume capabilities than Ford and Navistar's injectors.

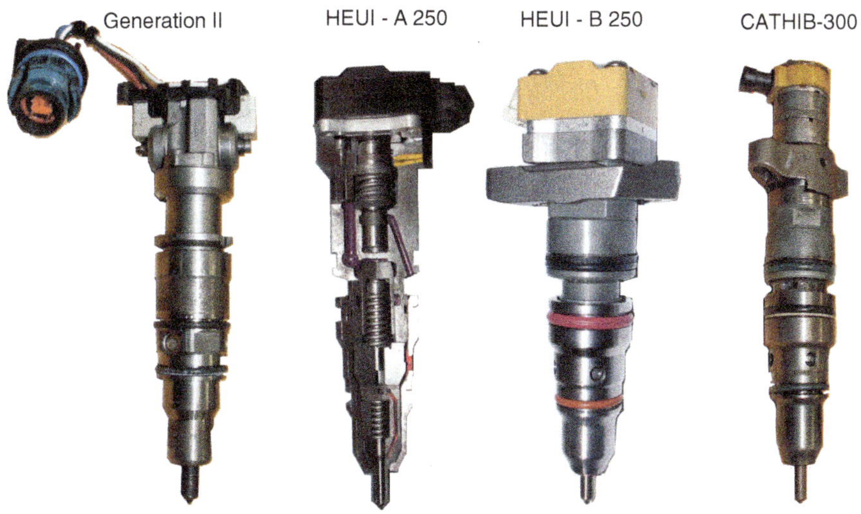

FIGURE 12-11 Common HEUI injectors.

FIGURE 12-12 This HEUI A injector is identified by its black solenoid and does not have split-shot capabilities.

The third variation of HEUI is the Gen II or digital valve injector (**FIGURE 12-14**). This injector is used in 6.0 L VT365 and Ford Powerstrokes engines from 2004 – 2009 engines. The more compact design means that the engine can use four valves per cylinder. Using two solenoids—one to open and one to close high-pressure oil supply—gives the injectors a faster response time and rate-shaping capabilities that cover a full range of engine speed and load conditions. Three electronically controlled injection events are possible in the Gen II injector, which incidentally uses less energy than previous models.

Caterpillar developed and used a fourth variation of HEUI injector, the HIB-300 for the ACERT C-7 and C-9 engines. Unlike other HEUI injectors which di not use calibration codes to compensate for variations in injection delivery volumes, CAT injectors do use a calibration code. This injector has a number of unique features:

- electronic control of pilot, main, and post-combustion injection events
- compact shape that permits a centralized location above the piston for symmetrical fuel distribution
- five different combinations of injection rate shapes, including a square and ramped main injection
- a low-mass poppet valve capable of performing electrically signaled multiple injection events.

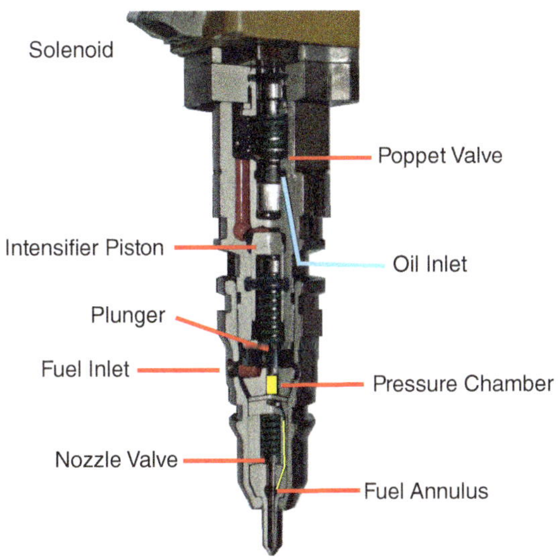

FIGURE 12-13 This HEUI B injector is identified by its white solenoid and is capable of a single pilot injection, used to reduce combustion noise and NO_x emissions.

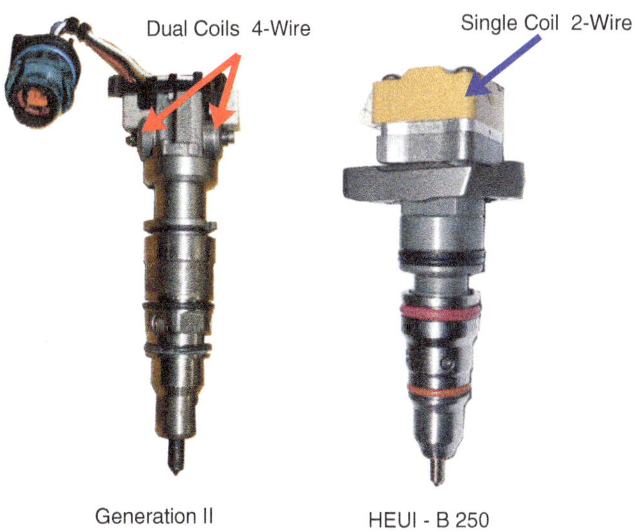

FIGURE 12-14 This 6.0 L injector is a Gen II HEUI. The two electromagnetic coils and other design features give it improved injection rate–shaping capabilities.

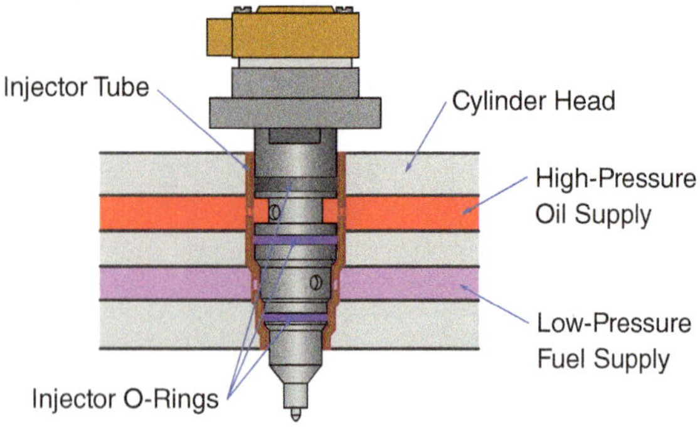

FIGURE 12-15 The cylinder head of first-generation HEUIs contains low-pressure fuel and high-pressure oil passageways separated by sealing O-rings.

HEUI A and B

HEUI A and B injectors have two grooves on the outside of the body, for receiving fuel and high-pressure oil from rails drilled in the cylinder head (**FIGURE 12-15**). Replaceable sets of O-rings seal the grooves to maintain pressure around the injector. The top set of O-rings seals oil pressure and consists of two individual rings: a steel backup ring and a square cushion seal. A single middle, round O-ring separates fuel and oil pressure. A round O-ring on the bottom seals fuel below the groove where fuel is delivered to the injector. Finally, a copper sealing washer around the injector tip seals combustion gases in the cylinder. Two hold-down bolts are used to position and clamp the injector. Only one bolt needs to be removed in order to extract the injector for service.

> **▶ TECHNICIAN TIP**
>
> An O-ring separates high-pressure oil from the low-pressure fuel circuit drilled into the cylinder head. O-ring leakage can allow high-pressure oil to leak either up and onto the cylinder head deck or into the fuel rail. Oil in fuel and high oil consumption indicate O-ring leakage between fuel and oil passageways. Pressurized oil leaking upward, past the O-rings, typically results in hard starting when hot, low engine power, and rough running.

HEUI Injector Components

Both HEUI A and B split-shot injectors have four major components that operate together for precise metering, timing, and rate control (**FIGURE 12-16**):

1. High-voltage electric solenoid: This device controls poppet valve operation. The solenoid requires approximately 110–115 volts DC and 7–15 amps of current to energize.
2. Poppet valve: The **poppet valve** controls oil flow into and out of the injector. It has an upper seat and a lower seat. The upper seat controls oil drainage from the injector, while the lower seat controls the admission of pressurized oil into the injector (**FIGURE 12-17**).
3. Intensifier piston and plunger: The **intensifier piston**, also known as the amplifier piston, multiplies the lube oil pressure, which is regulated to 480–3,400 psi (33–234 bar). Depending on the manufacturer and engine model, the top surface area of the intensifier piston is 7–10 times larger than the bottom plunger diameter. Because high-pressure oil is applied to the top of the piston, hydraulic pressure is multiplied at the bottom by a

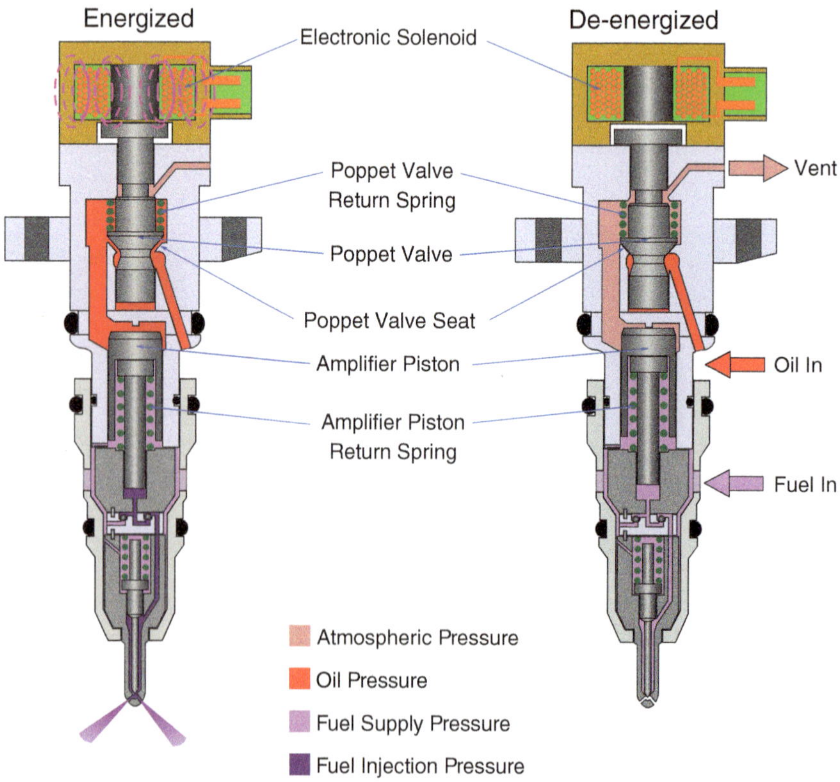

FIGURE 12-16 Parts of HEUI A and B injectors.

factor of 7 to 10. This means fuel below the amplifier piston is injected at a factor 7–10 times higher than oil pressure. For example, 2,000 psi (138 bar) of oil pressure supplied to the injector could produce 20,000 psi (1,379 bar) of injection pressure with an amplifier piston that uses a ratio of 10:1.

4. Nozzle assembly: The nozzle is a multi-orifice design that atomizes and distributes fuel into the combustion chamber. Nozzle opening pressure is approximately 2,700 psi (186 bar) in early injectors and more than 5,500 psi (379 bar) in later injectors. The large size of the earlier injectors prevented them from being centrally located above the piston. This meant the injector could be positioned only at an offset angle and used in only two-valve or three-valve engines, such as the three-valve CAT 3126B.

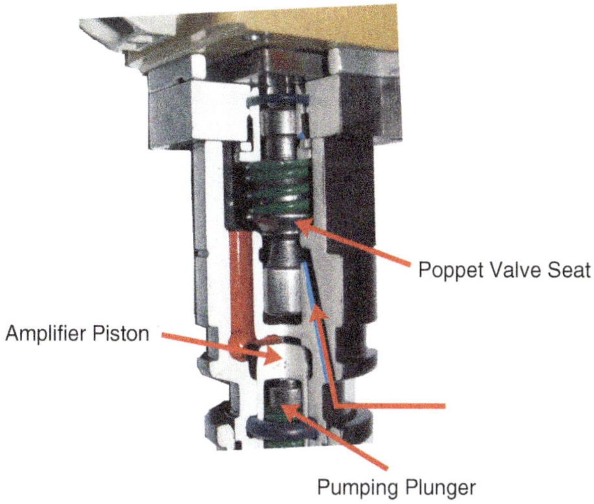

FIGURE 12-17 Components of a typical HEUI poppet valve mechanism.

Stages of HEUI A Injection

HEUI injection can be understood by examining the three stages of injection: fill cycle, injection, and end of injection.

Pressurized oil supplied to the injector provides the force to pressurize fuel inside the injector. The operation of the high-pressure oil system is described in greater detail later in this chapter.

1. Fill cycle: During this part of the injection cycle, the lower seat of the poppet valve is blocking high-pressure oil from entering the injector. The internal components are in their relaxed, spring-loaded positions. No electrical current is applied to the injector solenoid. The plunger and intensifier piston have returned to the top of the barrel. Fuel, pressurized to 30–80 psi (207–552 kPa) by the low-pressure transfer system, enters the injector through a passageway located just above the bottom groove of the injector. A fuel-fill check valve unseats and allows fuel to fill the plunger cavity below the intensifier piston (**FIGURE 12-18**).

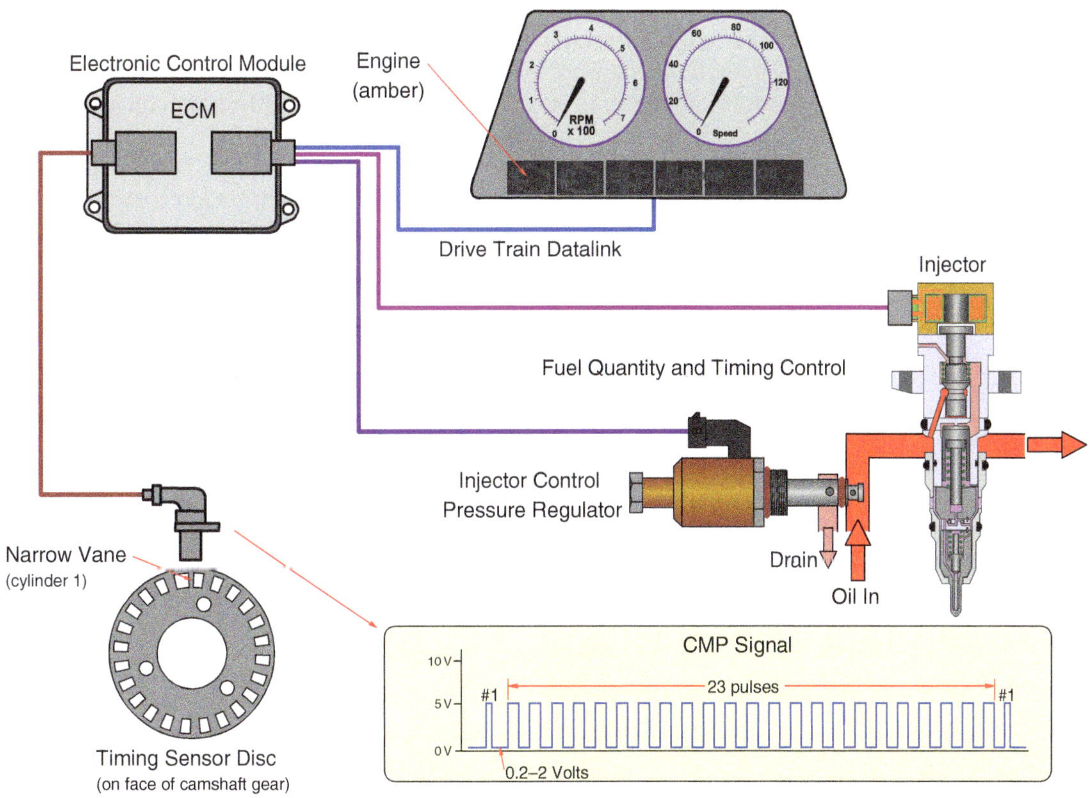

FIGURE 12-18 Major input and output signals control injection timing and metering in a HEUI system.

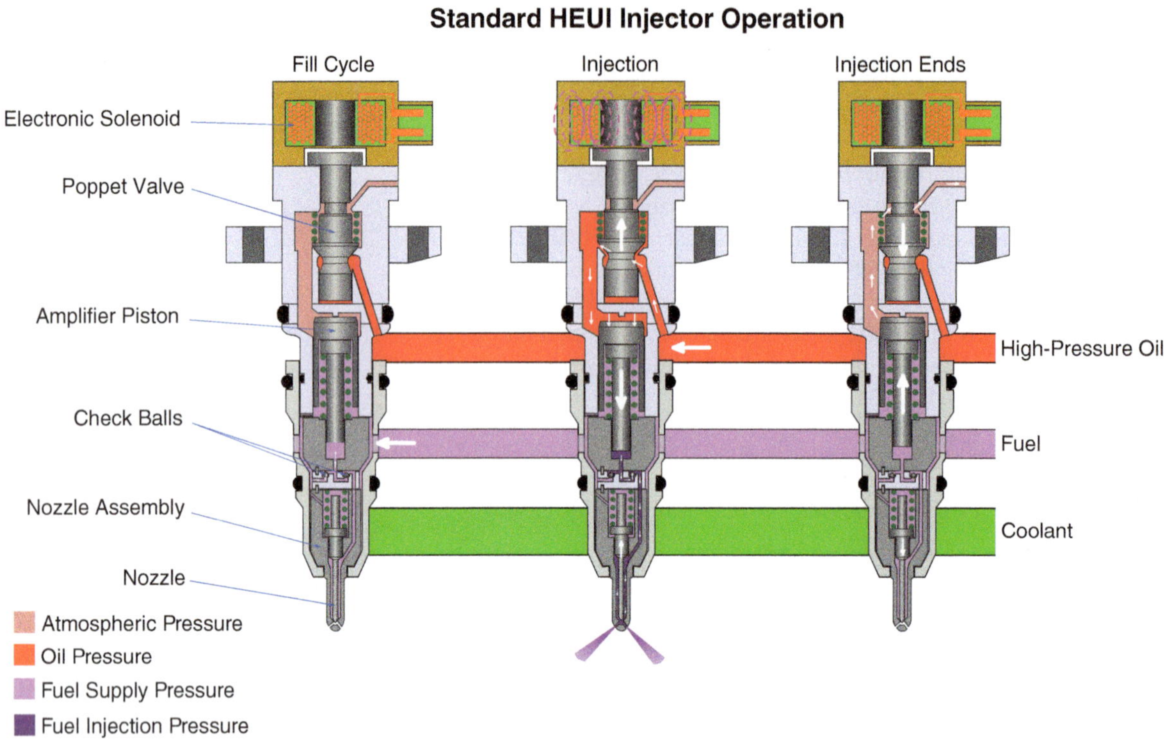

FIGURE 12-19 Injection cycle.

2. Injection: During the injection cycle, current will energize the solenoid, and the corresponding magnetic field overcomes spring tension that previously held the poppet onto its lower seat. A piece of connecting linkage or armature between the poppet valve and solenoid lifts the valve from its seat. When the poppet lifts, the oil drain path at the upper poppet valve seat is closed. High-pressure oil pushes past the poppet valve to the top of the intensifier piston. High-pressure oil, now over the amplifier piston, pressurizes fuel in the cavity beneath the plunger. Pressurized fuel travels to the nozzle valve, causing the valve to lift beginning at 2,700–5,500 psi (186–379 bar). Maximum oil pressure is 2,300–3,400 psi (159–234 bar) (**FIGURE 12-19**).
3. End of injection: The injection cycle ends when the solenoid is deenergized. The magnetic field that held open the poppet valve collapses, and spring tension forces the return of the poppet valve against its seat. High-pressure oil can no longer enter the injector, and oil now spills out the oil drain path beneath the solenoid around the upper poppet valve seat. The sudden oil pressure drop above the intensifier piston, combined with spring tension and fuel pressure beneath the plunger, force the plunger upward in its bore. The fuel-delivery check valve remains closed and maintains residual transfer pump fuel pressure in the injector body (**FIGURE 12-20**).

Metering

Injection quantity is determined by the length of plunger travel. Two factors determine how far the plunger will travel:

1. Injector solenoid on time: The longer the solenoid is energized, the more time is available to push the amplifier piston and injector plunger down the plunger bore.
2. Oil pressure: The higher the oil pressure applied to the amplifier piston, the faster the plunger will travel for a given amount of time.

Longer energizing time for the solenoid will result in greater quantities of fuel injected into the combustion chamber. If oil pressure is increased, the injection rate increases proportionally. Therefore, if the on time of the electrical signal applied to the injector remains constant but oil pressure increases, then the injection quantity increases. Conversely,

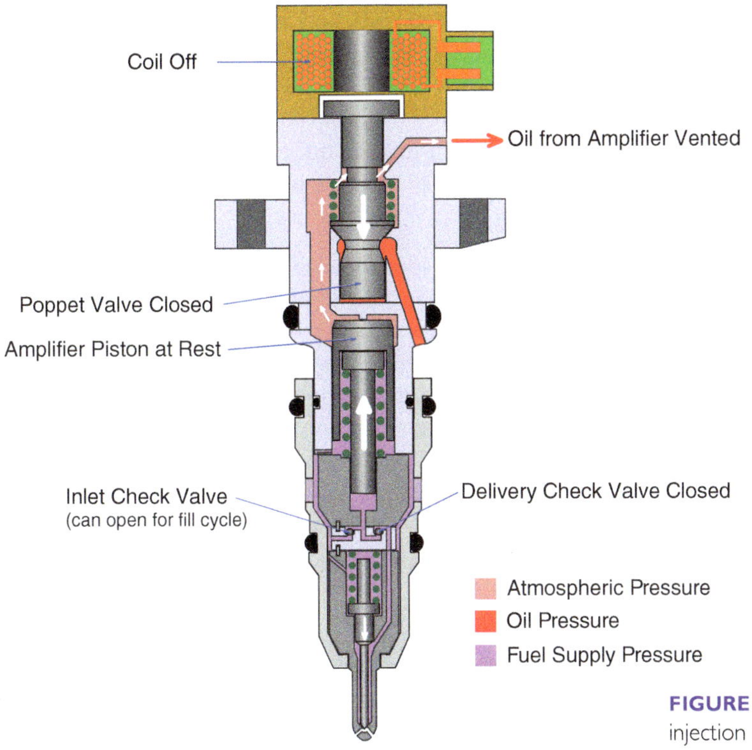

FIGURE 12-20 End of the injection cycle.

increasing energization time to the injector but decreasing oil pressure results in less fuel being injected. During engine operation, the two variables of solenoid on time and oil pressure continually change to shape the injection rate in response to operating conditions, engine speed and driver demand. As the engine speed increases, higher injection pressure is required because the available time is shorter to inject large quantities of fuel.

Injection Timing

Changing the point during the combustion cycle when the injector solenoid begins energizing and finishes energizing varies injection timing. Optimal injection timing is calculated by the electronic control module (ECM) primarily using sensor input from the crank-position or cam-position sensors and engine calibration files stored in a lookup table in the ECM's memory.

Injection Rate Control

Injection rate refers to the quantity of fuel injected per degree of crank angle rotation. Rate control is the ability of the fuel system to adjust the quantity of fuel injected during the injection sequence. Engine designers have discovered that if injection rates are properly matched for a given engine load, speed, or operating condition, it will result in improvements to emissions, fuel economy, power, and noise levels.

HEUI injection systems have the unique ability to change the rate of fuel delivery. Because oil pressure to the injector is electronically controlled by the ECM, changing oil pressure will change injection rate without needing to change the on time of the injector solenoid. So, at higher oil pressure, the rate of fuel delivery is greater. This is because high oil pressure applied to the intensifier piston will drive the plunger farther and faster per degree of crank rotation.

Pressurization is also independent of engine speed. Whether at idle or maximum engine speed, the output pressure of the high-pressure oil pump can instantly change by varying the electrical signal applied to the high-pressure pump pressure regulator. Typically at idle, actuation pressure is kept low to minimize combustion noise caused by ignition delay. When the driver demands more fuel, such as during hard acceleration, the actuation oil pressure will instantly and sharply increase to better atomize and quickly distribute the larger injection quantities.

▶ TECHNICIAN TIP

To prevent damage to HEUI injectors after replacing them, the Association of Diesel Specialist (ADS) recommends priming the high-pressure oil galleries with oil before attempting to start the engine. To do this, disconnect the cam-position sensor and crank the engine over for two consecutive starting cycles of 30 seconds each. It is estimated that a quarter of all Gen II injectors returned for warranty are damaged as a result of prolonged cranking without oil.

▶ TECHNICIAN TIP

When replacing or servicing HEUI injectors, oil and fuel must be drained from the cylinder heads first; otherwise, it will drain into the cylinder when the injector is removed. A hydrostatic lock will occur in the engine after an injector is reinstalled if oil and fuel are not first removed. A hydrostatic lock when cranking can result in damage to the starter, pistons, connecting rods, and valves.

▶ TECHNICIAN TIP

Oil quality is crucial to effective HEUI operation. If oil contains air, it will take on elastic properties and not be able to properly pressurize injectors, which adversely affects injection pressurization and timing. Poor fuel economy, low power, and stalling are typical operator complaints caused by air entrainment of oil. Antifoaming additives used in oil for HEUI engines are necessary to prevent air bubbles from forming and staying in the oil. With use, the antifoaming additives deplete, and then the oil will need to be changed.

HEUI B Split-Shot and PRIME Injectors

These injectors have the unique ability to inject a small quantity of fuel to the combustion chamber 8°–10° before the main injection. This pilot injection, also known as split-shot or PRIME injection, will begin to burn before the main injection arrives. The result is a combustion chamber with a burning flame front used to quickly ignite fuel in the main injection. The split-shot-injection strategy shortens the ignition delay and provides the following benefits:

- Lower combustion chamber noise: This is especially evident at idle when prolonged ignition delay causes a knocking sound characteristic of diesel engines. The hotter combustion chamber will shorten the amount of time required for igniting the main injection, thus minimizing or even eliminating ignition delay.
- Lower NO_x emissions: The shortened ignition delay allows for retarded injection timing. Because the delivery quantity during main injection will ignite faster, the beginning of injection can be delayed longer. The result is a less intense pressure spike after TDC in the cylinder, which in turn reduces NO_x formation.
- Fewer particulate matter (PM) emissions: Increased pre-ignition heat and temperature in the cylinder accelerates combustion reactions, which helps fuel burn more completely and cleanly. The result is a more efficient burn, evidenced by cleaner emissions and improved fuel economy.

Split-Shot-Injector Operation

By midyear 1999, all Power Stroke, Caterpillar, and International brand vehicles used HEUI injectors with split-shot injection to meet stricter emission standards. Though the injector incorporates only a slight change in the plunger and barrel to accomplish this fueling strategy, injection timing does change in engines using this injector. Because timing is retarded to produce low NO_x emissions, older HEUI A and newer HEUI B injectors are not interchangeable because they require different fuel-timing maps stored in the ECM.

The injector barrel includes a new relief port that allows injection pressure to bleed back to fuel supply pressure. The amplifier piston has several small holes drilled through the bottom that intersect with a groove machined around the outer circumference of the plunger (**FIGURE 12-21**). When the relief port in the barrel aligns with the groove around the circumference of the plunger, a momentary drop in pressure beneath the plunger occurs. This briefly, but effectively, stops injection.

To split the injection event and deliver a pilot-fuel pulse, the injector solenoid is energized once during the entire injection sequence. The amplifier piston is forced downward in the barrel, and the lower edge of the plunger covers the spill port, pressurizing fuel, unseating the delivery check valve, and forcing fuel out through the nozzle tip. After traveling far enough to inject less than 3 mm³ of fuel, the groove and relief port align temporarily after fuel delivery begins. At this point, the plunger's spill groove provides a passage for fuel to move from below the plunger, into the groove, and pass into the spill port in the barrel. High-pressure fuel is then released into the low-pressure fuel supply system, momentarily ending injection (**FIGURE 12-22**).

The amplifier piston's plunger continues to move downward in its stroke as long as the injector solenoid continues to be energized. Injection pressure redevelops when the plunger's groove passes the relief port. The plunger continues to move downward, delivering fuel until the end of injection. Covered by the descending plunger, fuel cannot escape through the plunger into the relief port. Injection will then resume, and main injection takes place as long as the amplifier piston is forced downward by oil pressure (**FIGURE 12-23**).

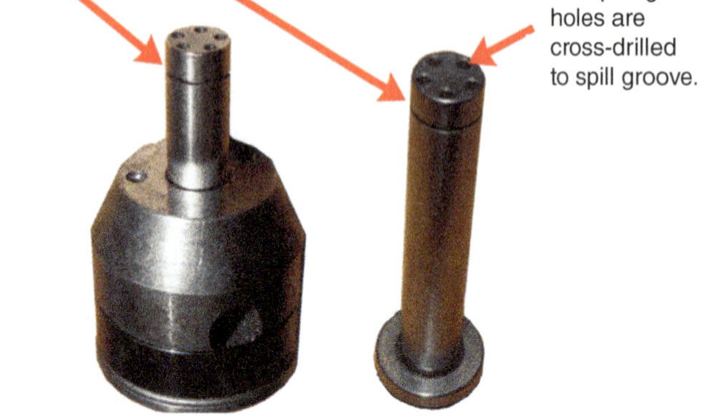

FIGURE 12-21 The amplifier piston of a split-shot injector. The circumferential groove relieves fuel pressure below the plunger, and the holes in the bottom of the plunger connect with the circumferential groove.

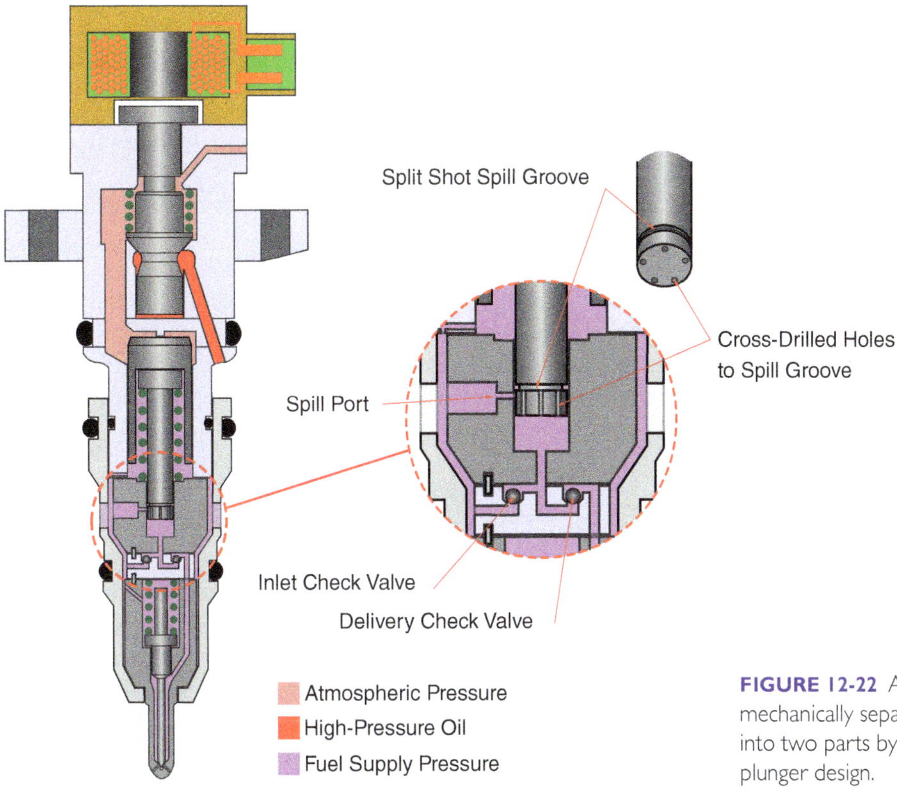

FIGURE 12-22 A split-shot injector can mechanically separate the injection event into two parts by using a uniquely shaped plunger design.

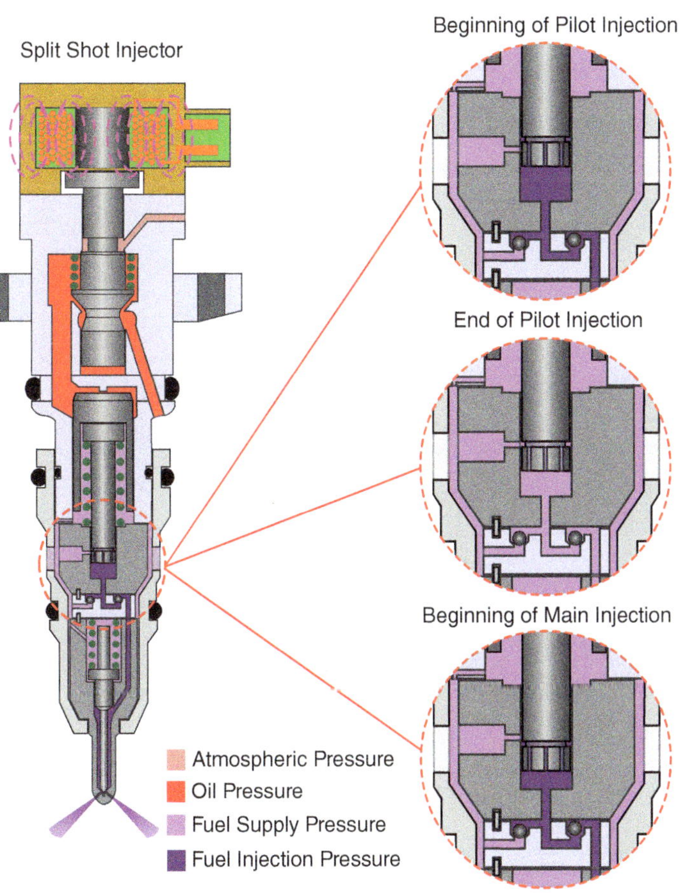

FIGURE 12-23 When the groove around the circumference of the fuel plunger uncovers the spill port, pilot injection ends. Main injection begins after the plunger groove travels downward past the spill port.

Split-Shot HEUI Pressurization

High injection pressures improve mixture preparation and combustion quality. This in turn provides the following benefits:

- reduced exhaust smoke at higher engine speed: because smaller fuel droplets ignite and burn faster, higher engine operating speeds are possible
- improved fuel economy due to the more complete combustion of smaller and well mixed fuel droplets
- minimized particulate formation caused by incompletely burned carbon
- increased combustion chamber ability to tolerate retarded injection timing because smaller fuel droplets ignite and burn faster
- increased combustion tolerance for the use of EGR used to reduce NO_x emissions
- reduced exhaust smoke at higher engine loads; more fuel plus finer atomization means more torque with less smoke.

▶ Gen II HEUI Systems

12-03 Describe and differentiate Gen II HEUI injectors from earlier injectors according to construction and operating principles.

The second-generation (also called G2 or Gen II) injector uses what is called **digital control valve** technology. This valve technology replaced the poppet valve used in the HEUI A and B designs. The digital nature of the valve refers to the position of the control valve: it is either open or closed. The digital valve's advantage is that it is much faster operation than the poppet valve in earlier injectors, lending the injector the capability of delivering three electronically controlled injection events in one combustion cycle.

Two electromagnetic coils and a spool valve control injection control oil pressure. Four external electrical pins supply the coils current to operate at 48 volts and 20 amps per coil, which is a lower voltage than other HEUI injectors. However, the time current applied to the coils is as little as 400 microseconds. That is 400 millionths—not thousandths—of a second (**FIGURE 12-24**). The longest energization time used for cold starting is 5.8 milliseconds during cranking. This contrasts with HEUI A and B injectors, which require at least 500 milliseconds to build up a magnetic field strength adequate to close the poppet valve. Dramatically shorter energization time means the coils can

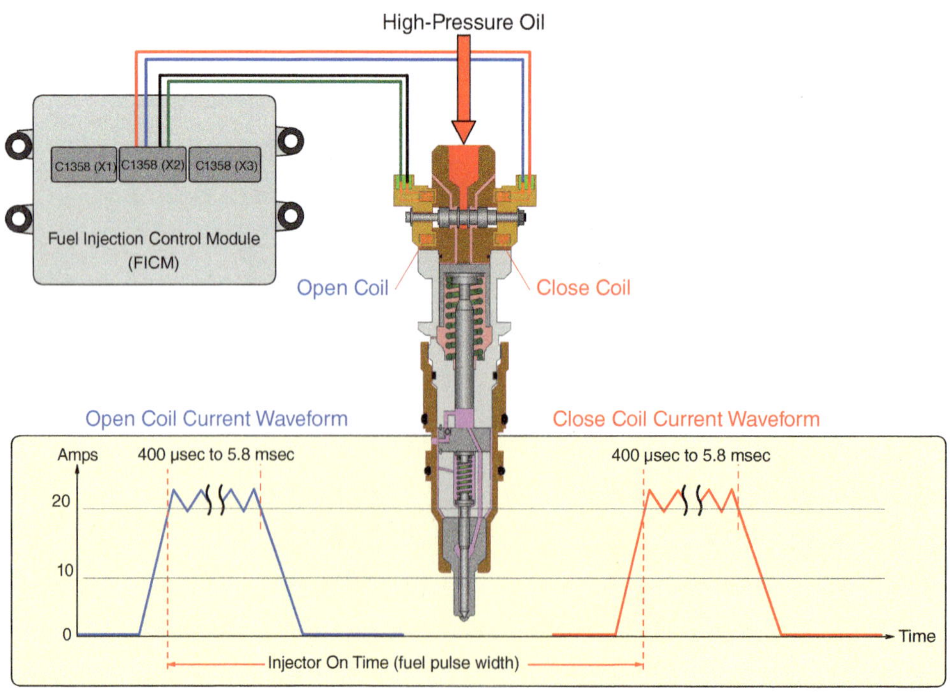

FIGURE 12-24 The energization time of a Gen II injector is measured in microseconds and is much shorter than earlier injectors that used large electromagnetic actuators.

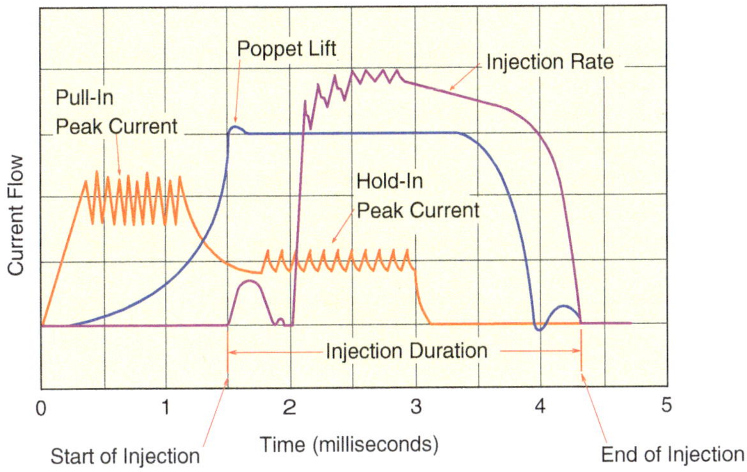

FIGURE 12-25 The electrical signals for a HEUI B injector use a peak-and-hold current strategy to keep coil heating low. Opening the poppet valve uses more current flow than holding the poppet valve open.

operate more coolly. The electrical signals for a HEUI B injector use a peak-and-hold current strategy to keep coil heating low. Opening the poppet valve uses more current flow than holding the poppet valve open (**FIGURE 12-25**). Most importantly, the coil and spool valve design enable the electrical signals to rapidly control the quantity of fuel injected over more of the injection cycle. This rate-shaping capability is crucial to obtaining low emissions without sacrificing performance and fuel economy. The new control valve design provides a smaller and more compact injector that can use four valves per cylinder: two intake and two exhaust (**FIGURE 12-26**).

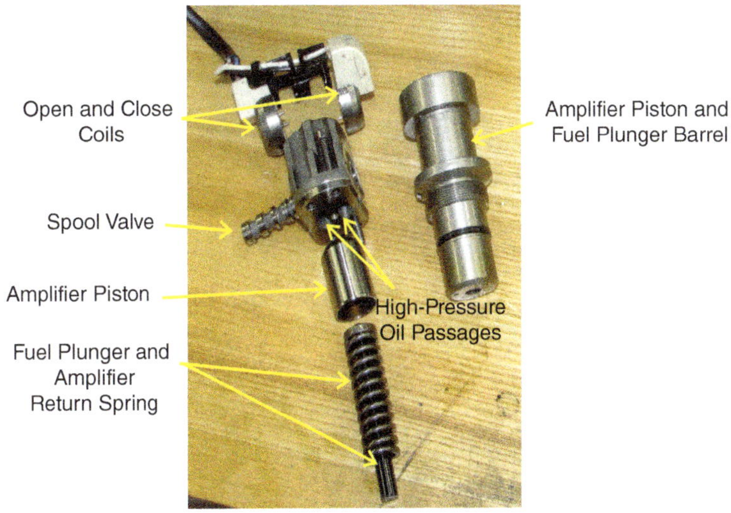

FIGURE 12-26 Internal components of the Gen II injector. Two coils controlling spool valve movement create a faster injector response time.

What is also unique about the digital valve is its dependence on residual magnetism to move the control valve. The control valve, which resembles a hydraulic spool valve, does not use springs to center or return the valve to the one end of the barrel in which it operates. Instead, residual magnetism produced by the short pulse of DC will indefinitely hold the valve in an open or closed or position. When a second pulse of electrical current is applied to the opposing coil, the valve will move in the opposite direction. It is not magnetic fields but pressurized lube oil that supplies the force needed to physically open and close the valve. By using residual magnetic energy as the tipping or triggering force, to begin moving the valve and using hydraulics as the driving force, the valve operates more quickly, using less energy and generating less heat than its predecessors did. This digital valve overcomes the technological barriers of rapidly switching the injection event on and off by using high-pressure fluid flowing at extreme speeds with the advantage of reduced coil response time.

Sophisticated proprietary materials used in constructing the valve create exceptionally strong residual magnetic forces. During one early lab demonstration of the digital valve's operation, participants had to use a pry bar to move the control valve being held in place with only residual magnetism.

Gen II Injector Construction

The digital control valve inside the injector consists of two coils: an open coil and a closed coil. A spool valve moves from side to side controlled only by magnetic forces (**FIGURE 12-27**). Total movement of the valve is only 0.017" (0.432 mm). The valve has only two positions. When open, it allows oil to flow

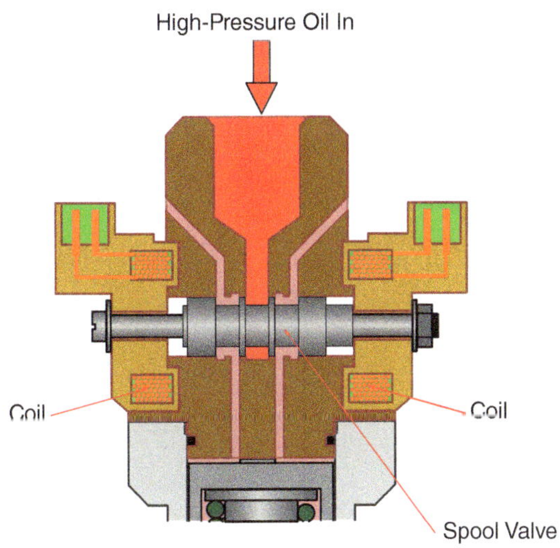

FIGURE 12-27 The single poppet valve and solenoid from previous HEUIs is replaced with a lighter, faster-acting spool valve and two magnetic coils in the Gen II injector.

from the high-pressure oil rail into the injector, pushing the intensifier piston and plunger downward. In the closed position, it allows oil to drain out of the injector. Two coils, rather than one, control oil flow, enabling much faster operation. Because the decay or collapse of a magnetic field is not necessary to end injection, the injector is capable of more injections, accurately metered and spaced very close together.

Similar to the operation of earlier HEUI injector amplifier pistons, the Gen II injector's intensifier piston multiplies fuel pressure below the fuel plunger. The intensifier has a surface area 7–10 times greater than the plunger diameter, depending on the engine and year of manufacture. A mechanical pressure relief valve in the actuation oil system opens at 4,000 psi (276 bar). The relief valve limits oil pressure in the event of a malfunctioning injection pressure regulator. Given the operating range of pressurized oil, these systems can theoretically operate at up to 34,000 psi (2,344 bar) of injection pressure. In service, they operate just below 29,000 psi (1,999 bar).

The plunger and barrel assembly develop injection pressure similar to the HEUI A and B injectors. Earlier injectors used a coating of tungsten carbide on the plunger to reduce the possibility of scuffing from poor fuel quality, water contamination, and the effects of ultra-low-sulfur diesel (ULSD) fuel. Later injectors used a silicon carbide coating with the proprietary name diamond-like carbon (DLC) coating. A new clevis over the fuel plunger has eliminated the failures encountered in earlier injectors. These injectors cannot be interchanged with older injectors.

High-pressure oil is supplied to the Gen II injectors through a high-pressure **oil manifold**, or rail, externally attached to the top of the injectors (**FIGURE 12-28**). The rail incorporates hydraulic wave-dampening devices called **acoustic wave attenuation (AWA)** fittings to minimize a rhythmic cackling noise at idle caused by pressure waves the from high-pressure pump synchronizing with injection events (**FIGURE 12-29**). During the brief pressure spike, injection pressures sharply rise and create the unusual combustion noise due to momentarily high spray-in pressure. Early-model engines used a straight rail. Later models introduced a wavy high-pressure oil rail with higher oil capacity and dual AWA fittings in each rail.

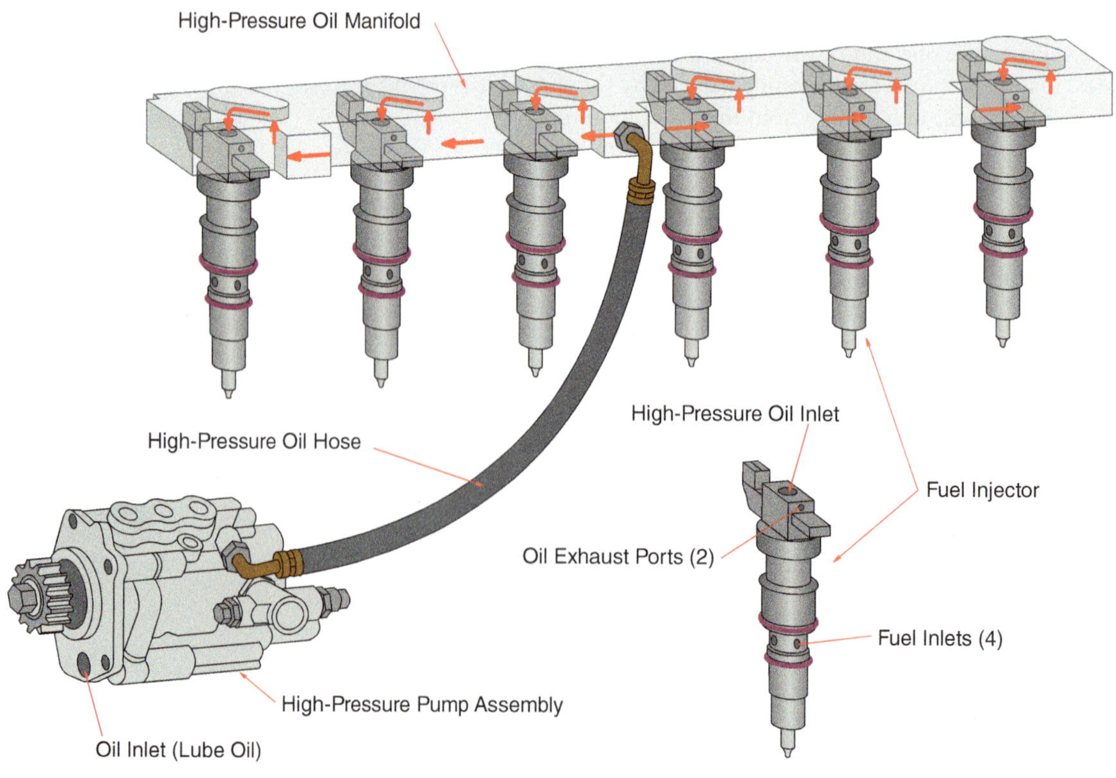

FIGURE 12-28 Oil supply to a Gen II injector used by Navistar.

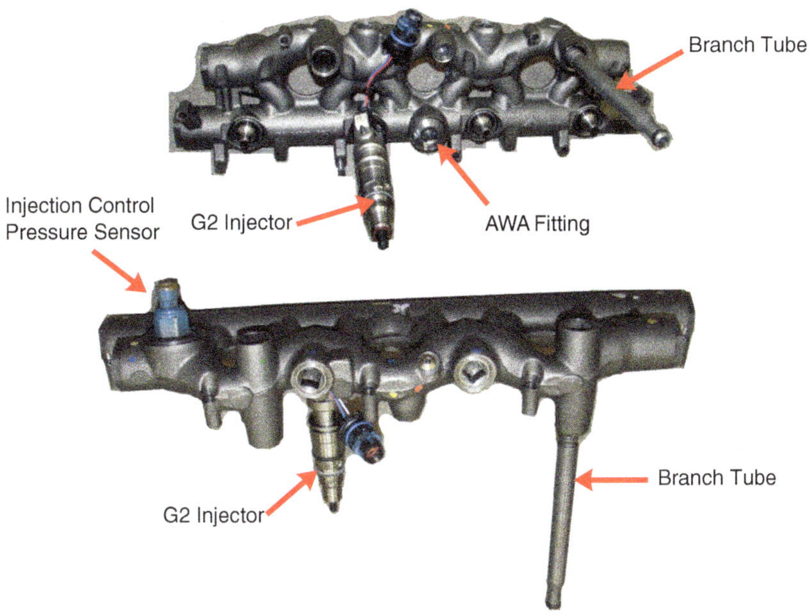

FIGURE 12-29 The high-pressure oil circuit supplies oil to injectors through a manifold above the injectors. This oil manifold from a later-model 6.0 L Ford Power Stroke shows the branch tube that carries oil to the manifold from the high-pressure pump. AWA fittings dampen pressure waves caused by pulsations from the high-pressure oil pump.

Injection Cycle

The injector has three main events during an injection cycle: fill, main injection, and end of main injection.

1. Fill cycle: This event begins with the spool valve in the closed position, preventing oil from entering the injector. Low-pressure fuel regulated to 45–50 psi (310–345 kPa) enters the injector through fuel rails in the cylinder head. After entering the injector through the fuel inlet located in a groove around the injector body, fuel fills the cavity below the plunger (**FIGURE 12-30**).

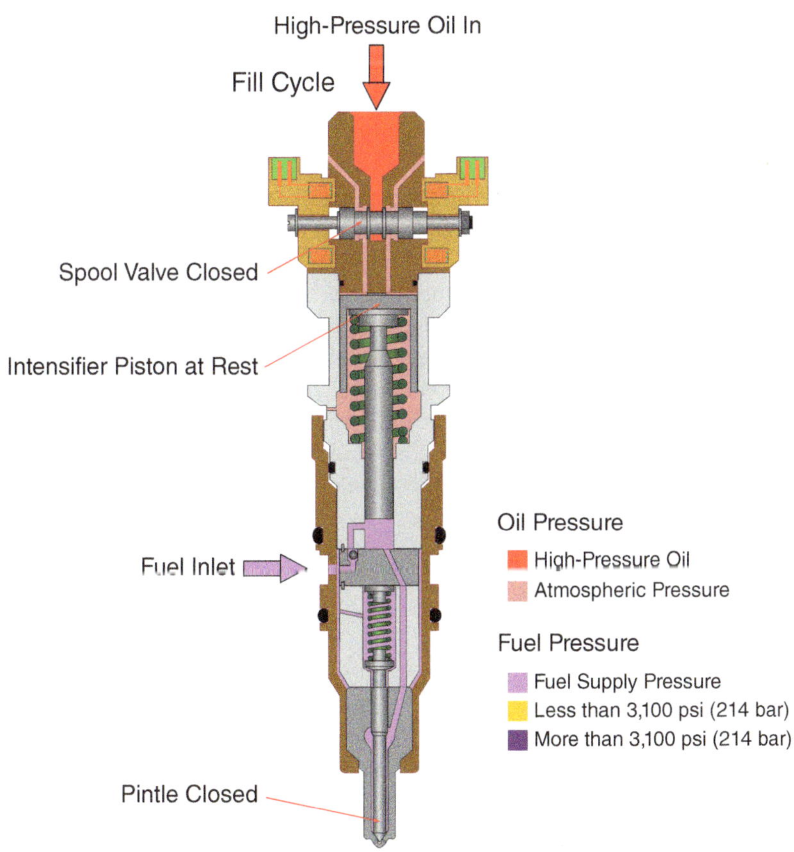

FIGURE 12-30 The fill cycle in a Gen II injector.

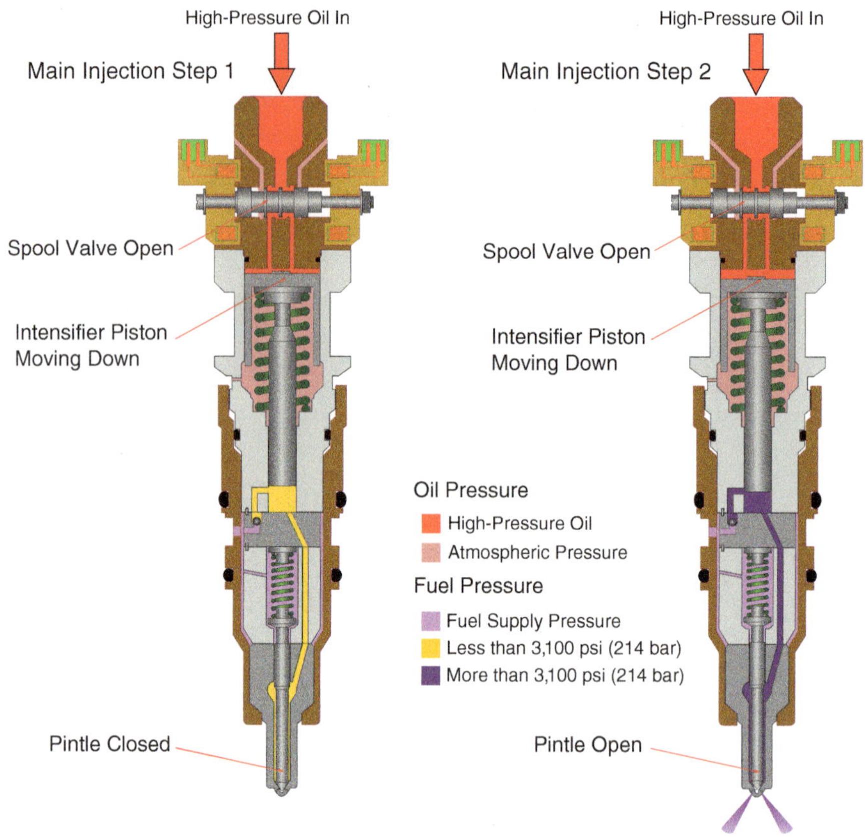

FIGURE 12-31 The two steps of main injection in a Gen II injector.

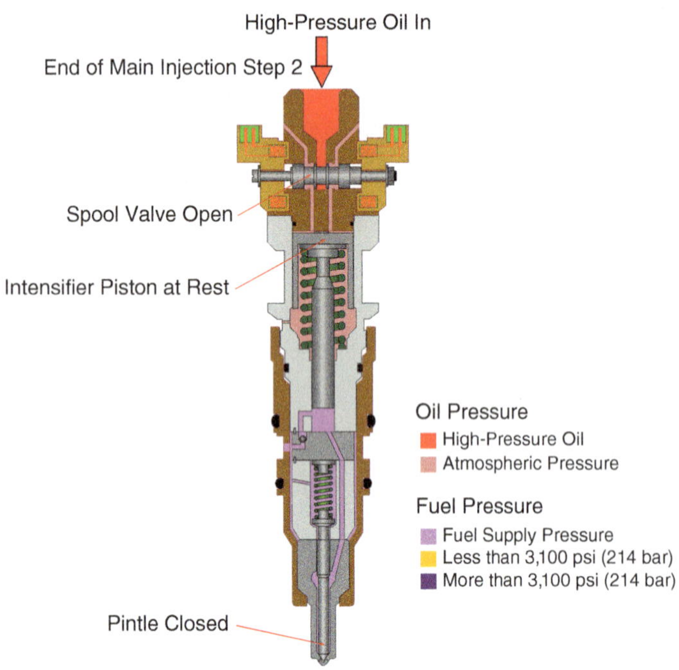

FIGURE 12-32 The end of injection in a Gen II injector.

2. Main injection: An electrical signal energizes the open coil, moving the spool valve to the open position. This allows high-pressure oil to begin forcing the intensifier piston and plunger downward. A fuel inlet check ball closes, causing fuel pressurization below the intensifier piston plunger. The pressurized fuel begins to lift the nozzle valve at approximately 3,100 psi (214 bar) (**FIGURE 12-31**).
3. End of injection: After the correct amount of fuel is delivered based on the time the injector digital valve has been left open, the close coil is pulsed with current for 800 microseconds. This moves the spool valve from an open to a closed position. High-pressure oil is blocked from entering the cavity above the intensifier piston. Oil above the intensifier piston leaves the injector through an exhaust port. Fuel pressure below the plunger and spring force returns the intensifier to its initial position. The nozzle valve simultaneously closes, and injection abruptly ends after delivering fuel (**FIGURE 12-32**).

Gen II Metering Timing and Rate Control

Similar to previous HEUI injectors, the Gen II plunger's travel determines injection quantity. Plunger travel, however, is determined not by the amount of time the open or close coils are energized but by the length of time the digital valve is left open. The open coil needs only one electrical pulse to remain in its position. The opposite close coil requires energizing to change the position of the spool valve and remain closed. Injection timing and metering can be varied by changing the specific time during crankshaft rotation to cycle the open and close coils.

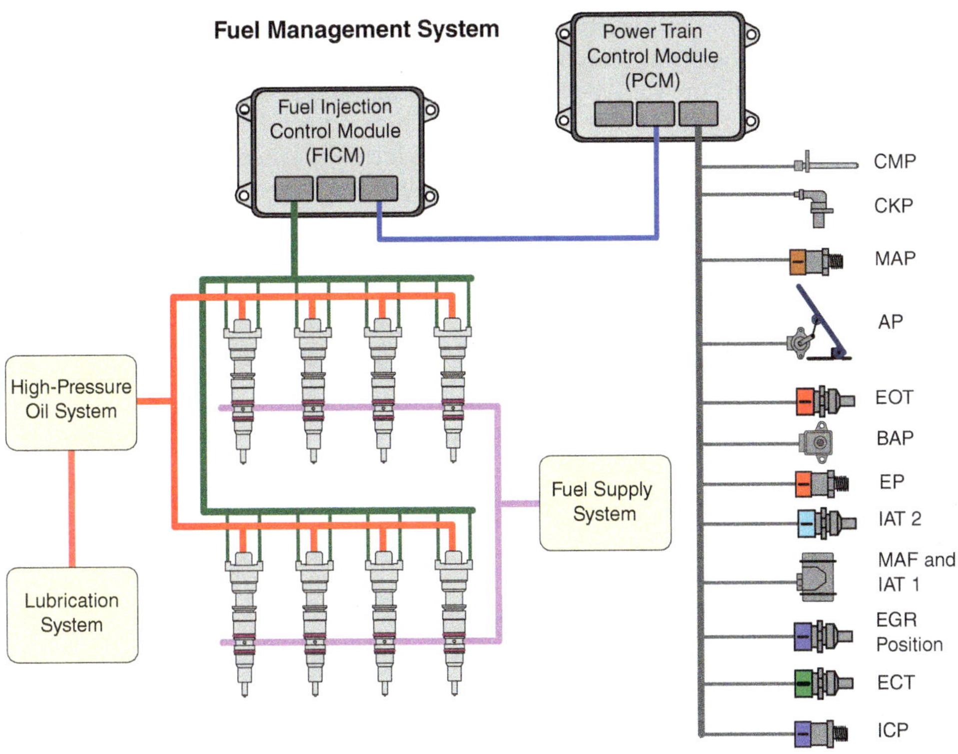

FIGURE 12-33 The electronic, oil, and fuel systems of a V365 Gen II HEUI engine.

To change the rate of fuel delivery, the oil pressure to the injector is electronically varied by the ECM. At higher pressure, the rate of fuel delivery is proportionally greater because the force acting on the plunger is increased. The plunger will then displace more fuel per degree of crankshaft rotation at high pressure. Pressurization is independent of engine speed, so high pressures are available at low engine speed operation if needed.

Rate control or shaping takes more than merely varying the oil pressure. The Gen II injector is also able to rapidly open and close the spool valve several times during the course of one injection sequence. Injection events such as pilot, main, and post-combustion injections are accomplished electronically, and not mechanically as they are in HEUI B injectors. Software instructions contained in the ECM can send several pulses to the digital valve control coils within one injection cycle. This enables the execution of several injection events within one injection cycle (**FIGURE 12-33**).

> **▶ TECHNICIAN TIP**
>
> Gen II injectors can misfire if heavy, cold, or contaminated oil is left in the control valve during engine shutdown. This sort of misfire due to control valve malfunction is called stickage. To minimize stickage, the fuel-injection control module (FICM) buzzes or rapidly cycles the control valve hundreds of times during the three seconds before the engine starts. Rapidly cycling the control valve pushes oil out of the valve to ensure that it can operate freely. To further help prevent stickage, some engines use a revised calibration that cycles the control valves for up to 90 seconds after the engine has been shut down. It is normal for injectors to continue buzzing even after the ignition key has been removed.

▶ Caterpillar H1B-300 Injectors

12-04 Identify and describe the construction and operation of Caterpillar H1B injectors.

The Caterpillar H1B-300 injector was developed for the C-7 and C-9 ACERT series engines. The injector is capable of five different injection rate shapes and an electrically controlled pilot injection. Two unique features of the injector are the low-mass poppet valve, which responds faster to electrical signals, and the hydraulically balanced nozzle valve. Hydraulically balanced nozzle valves simply have hydraulic pressure applied at two points. One point is fuel pressure below the valve, and the other is high-pressure oil above the nozzle valve, holding the valve onto its seat, along with the nozzle spring. Changing the pressure on either side of the nozzle valve causes it to either open or close. Using more rapid acting, powerful hydraulic force than nozzle spring pressure to close the nozzle speeds up nozzle opening and closing.

H1B-300 Injection Cycle

During the fill cycle, high-pressure fuel and oil enter the injector through passageways in the cylinder head. Fuel travels to the cavity below the amplifier piston while pressurized oil is applied to two sides of the poppet valve and to the top of the nozzle valve holding it

▶ TECHNICIAN TIP

Depending on outside temperatures, Caterpillar C-7 and C-9 engines can use either 10W-30 or 15W-40 oil. Because the viscosity of the two oils changes high-pressure system operation, a removable plug in the wiring harness is used to indicate to the ECM the viscosity of oil used. The ECM then uses a different set of calibrations and lookup table values to adjust system operation.

12-05 Identify and describe the construction and operation of the components in HEUI high-pressure actuation systems.

closed. When the poppet valve is energized, the valve lifts, closing off the high-pressure oil passageway and applying pressure to the nozzle valve. Lifting the poppet valve opens a passageway, allowing pressurized oil above the nozzle valve to drain to a low-pressure oil return. Meanwhile, the spool valve above the amplifier piston is opened simultaneously by the loss of the high-pressure oil that was holding it closed. When the spool valve opens, high-pressure oil is directed to the top of the amplifier piston, and then injection begins.

Pilot, main, and post-combustion injection events are controlled by the electrical signals to the poppet valve. Oil pressure determines the shape of the injection rate to be either ramped for low NO_x or squared for power.

▶ High-Pressure Actuation Systems

The high-pressure oil actuation system used by HEUI systems has several major subsections:

- low-pressure oil supply
- oil reservoirs
- high-pressure pump
- injection control pressure regulation.

Low-Pressure Oil Supply System

All HEUI engines share lubrication oil with the engine's lubrication system. The lubrication system supplies oil to the separate high-pressure oil system used to actuate injectors. Connecting the two oil systems ensures adequate capacity of oil for the high-pressure circuit. Two configurations are used to supply the high-pressure pump: the short-circuit supply and the filtered supply.

Short-Circuit Oil Supply

An oil circuit directly connects the engine's lubrication system to an oil reservoir supplying the high-pressure oil pump (**FIGURE 12-34**). Because this circuit does not pass through a filter, but instead connects the engine's main oil gallery to the reservoir, it is called a

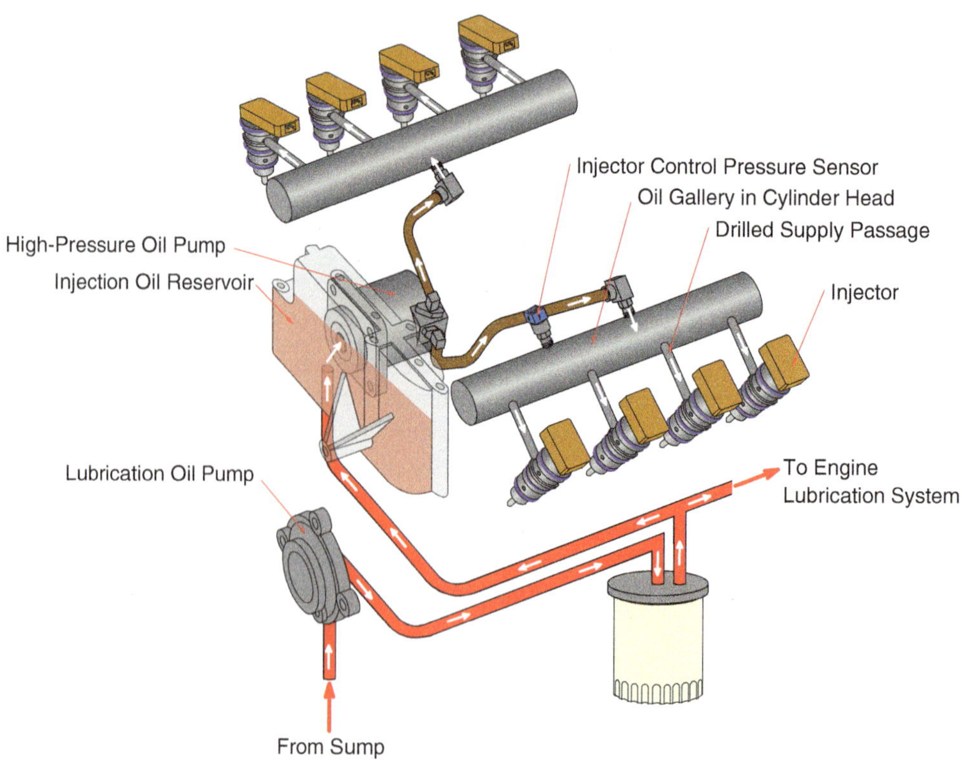

FIGURE 12-34 The low-pressure and high-pressure oil circuits of a 7.3 L Power Stroke engine.

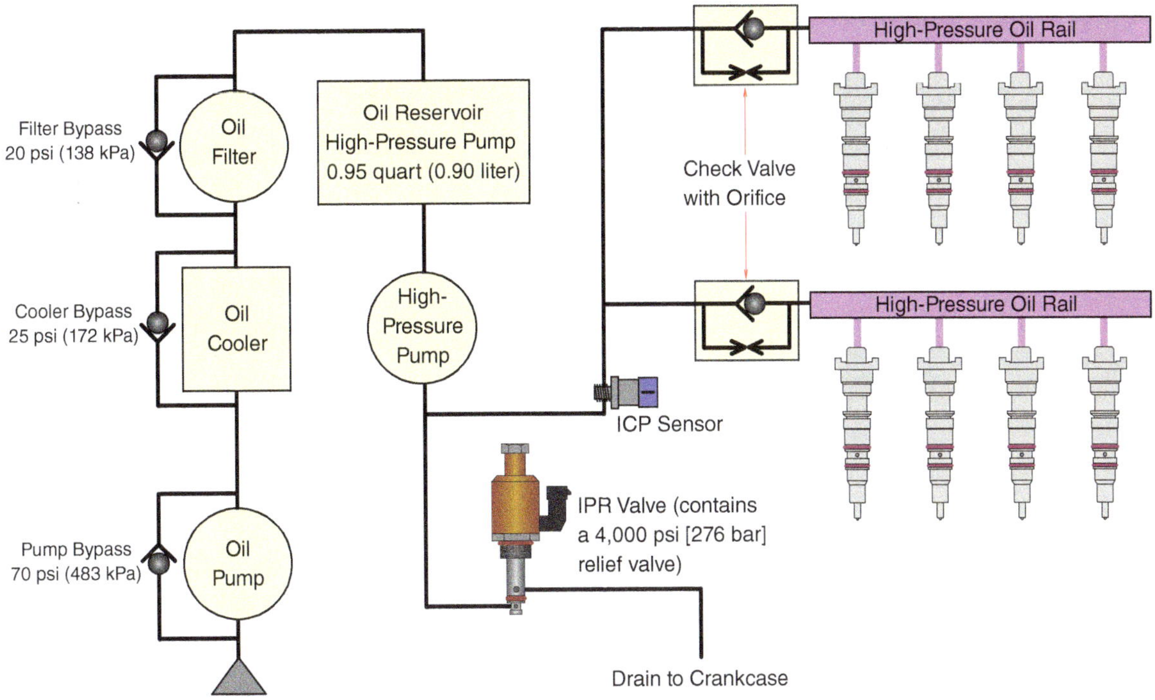

FIGURE 12-35 The high-pressure oil supply to the injectors of a 6.0 L (365 cubic inch) VT 365 diesel.

short-circuit supply. Short-circuit supply to the high-pressure system ensures a rapid filling of the reservoir subsequent to an oil change, during starting, or when the engine is cold. Without it, prolonged engine cranking may be required. A one-way check ball prevents oil from draining back to the crankcase when the engine is not operated. Early HEUI engines used this arrangement, but it has one main disadvantage. Unfiltered crankcase oil contaminated with dirt or metal will pass through the injectors, causing rapid wear or failure of injectors and the high-pressure pump.

Filtered Oil Supply

Later designs of HEUI systems corrected the problem of short-circuit oil supply by using oil filtered by the engine to supply the high-pressure circuit (**FIGURE 12-35**). This ensures only filtered oil, free of dirt and other contaminants, reaches the injector. Both short-circuit and filtered systems use a storage reservoir to hold oil during intervals when the lubrication system may not be capable of supplying enough oil to operate the high-pressure actuation circuit. Reservoirs with a capacity of approximately 0.26 gallon (1 liter) are incorporated into high-pressure pumps or cavities in the engine. Caterpillar uses an accumulator reservoir system for its HEUI A and B engines to store oil for its high-pressure pump (**FIGURE 12-36**). During an oil change, none of this oil can be drained from the accumulator or reservoir.

> ▶ **TECHNICIAN TIP**
>
> Unfiltered oil containing abrasives or coolant will cause rapid wear of the amplifier piston plunger and seals, leading to premature injector failure. Dirt commonly enters lube oil through air intake leaks, leaking gaskets, and missing oil filler caps. Even in filtered systems, metal particles from injectors and high-pressure pumps can migrate through the circuit and cause injector failure. After any pump or injector failure, always drain and inspect the oil from the high-pressure rails or galleries. Pouring oil onto a paper towel or cloth can help identify whether metal particles are contaminating the high-pressure oil circuit.

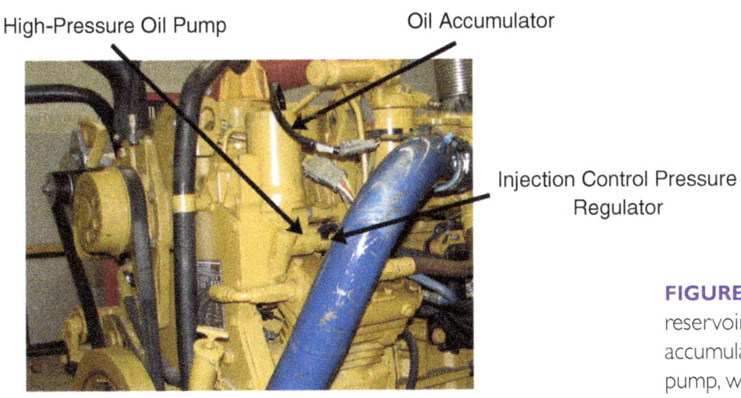

FIGURE 12-36 Caterpillar 3126 uses an accumulator reservoir system. Oil cannot be drained from the accumulator, which supplies oil to the high-pressure pump, when the low-pressure supply is inadequate.

High-Pressure Actuation Circuit

The high-pressure oil actuation system delivers engine oil under high pressure to pressurize fuel for injection. Another name for the circuit is the injection control circuit. A high-pressure oil pump connected to the reservoir increases the lube oil's pressure to levels usable by the injectors. These pumps are commonly gear-driven, fixed-displacement, piston, swash-plate, or axial piston pumps. Lines or tubes will connect the pump output to oil galleries or manifolds delivering oil to the injectors.

CAT High-Pressure Pumps

Drive torque used to pressurize oil in the high-pressure pump is provided by the engine. Because the pump output is not phased with the injection event, little torque is required to operate the pump in comparison to other injection systems.

Caterpillar uses an innovative **high-efficiency pump (HEP)** on the C-7 and C-9 engines. Like other variable-output HEUI pumps, output pressure is matched to the desired pressure by the ECM based on operating conditions (i.e., load, speed, temperature, etc.) and fuel calibration maps specified by the original equipment manufacturer (OEM) that are stored in the ECM. The HEP uses a PWM-controlled displacement control valve, which adjusts pump displacement volume rather than output pressure by changing the effective stroke of a swash plate pump. Regulating injection control pressure with a variable displacement pump minimizes power consumed by driving the pump because the pump supplies only as much oil as the engine requires. The HEP's efficiency is further enhanced because it does not change the pump's swash plate angle by using the high-force control valve common to other pumps. Instead, a sliding collar on the pump's driveshaft moves to open or close a helix groove containing an oil circuit that adjusts the swash plate angle.

Oil Manifolds

Pressurized oil is supplied to injectors in several ways:

- oil galleries drilled through the cylinder head
- jumper tubes connecting oil galleries to injectors
- cast iron oil manifolds attached to the top of the injectors.

The use of oil manifolds supplying oil through the top of the injectors eliminates the problem with O-ring leakage when oil is supplied though the cylinder head. Earlier HEUI injectors encountered O-ring deterioration from heat, vibration, and chemical attack when oil is supplied through cross-drillings in the cylinder head. The diameter of injectors, such as Gen II injectors, could be reduced with oil supply from manifolds connected to the top of injectors (**FIGURE 12-37**). A compact injector profile means the injector can have a centralized injector location in four-valve cylinders.

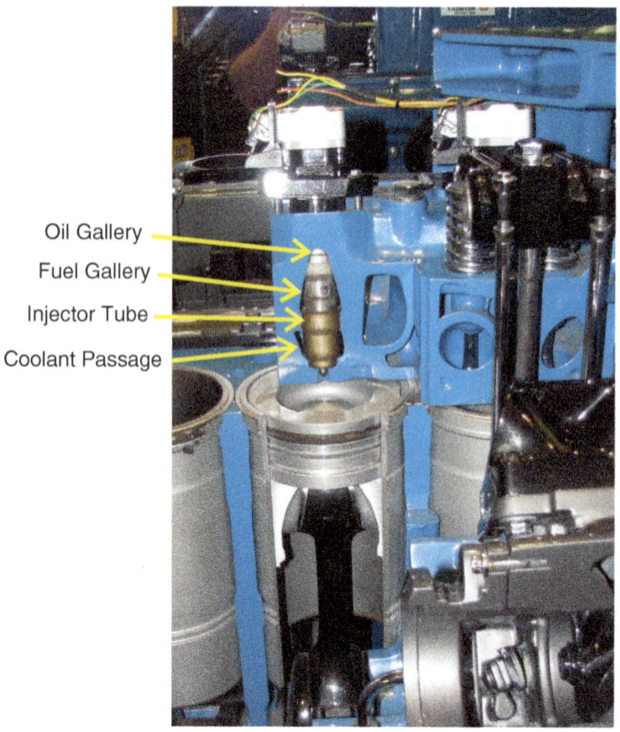

FIGURE 12-37 A HEUI B injector occupied substantial space in the cylinder head to supply oil to the injector.

Injection Control Pressure Regulator

A variety of sensor inputs are used to calculate oil pressure, which is regulated by using an ECM-controlled, pulse-width–modulated (PWM) **injection control pressure regulator (ICPR)** (**FIGURE 12-38**). Caterpillar's term for the same device is injection actuation pressure control valve (IAPCV). This device is an electrically operated spool valve that moves in response to the strength of a magnetic field. The oil pressure for injection actuation can be adjusted by changing the current flow through a coil surrounding the spool valve. Pressure is usually low at idle and increases with engine acceleration and as load increases (**FIGURE 12-39**). When not building pressure, the spool valve will direct oil back to the oil sump (**FIGURE 12-40**). To build pressure, the spool valve closes the passage to the sump and directs oil to the injectors.

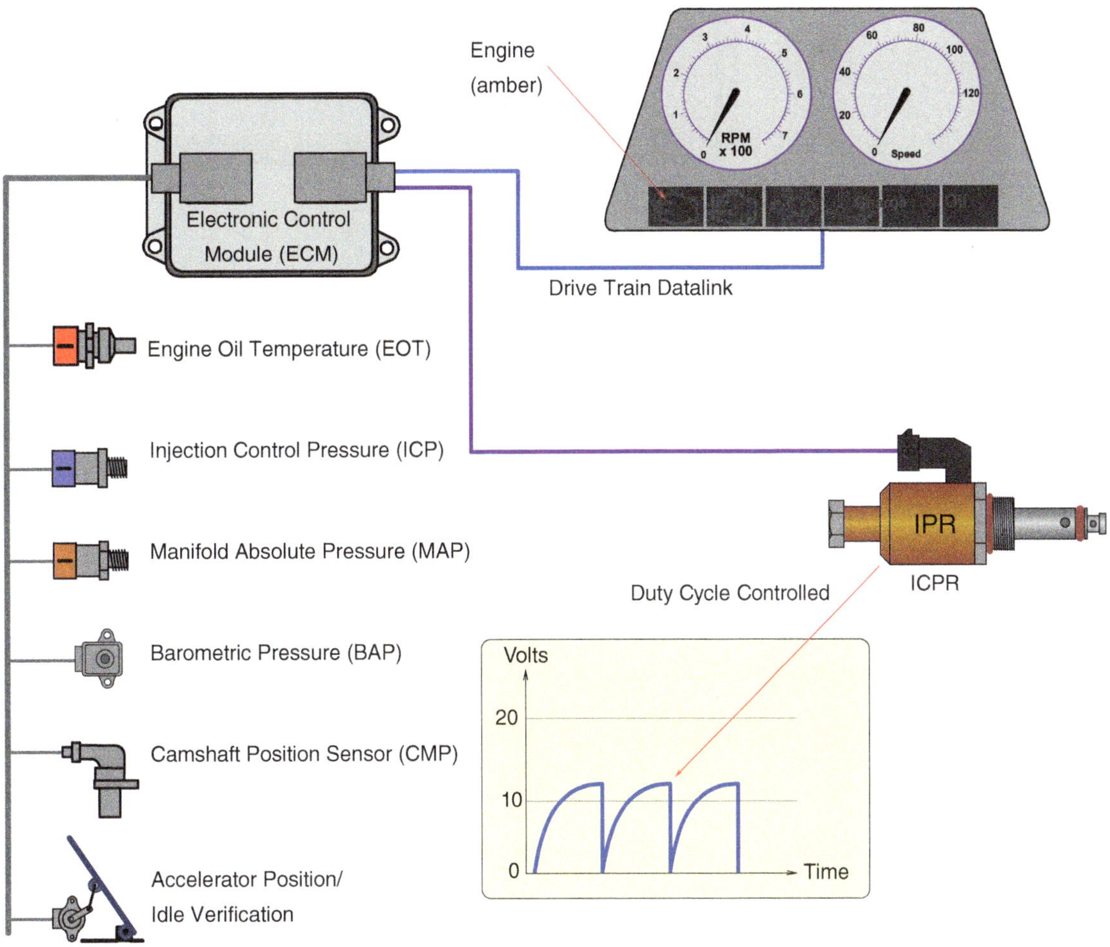

FIGURE 12-38 A variety of sensor inputs are used to calculate injection control pressure, which is regulated by the ICPR.

A sensor is located in the high-pressure oil circuit to provide feedback to the ECM regarding injection control pressure. This sensor, the **injection control pressure (ICP) sensor**, provides closed-loop feedback to the ECM about whether the oil pressure is too low or high (**FIGURE 12-41**). The ECM can change the amount of current to the ICPR coil and adjust oil pressure (**FIGURE 12-42**). A feedback loop is formed between the ICP sensor, the ICPR, and the ECM. Loss of this sensor signal will usually cause the injection oil pressure to operate at a factory default value based on a predetermined PWM signal to the regulator (**FIGURE 12-43**).

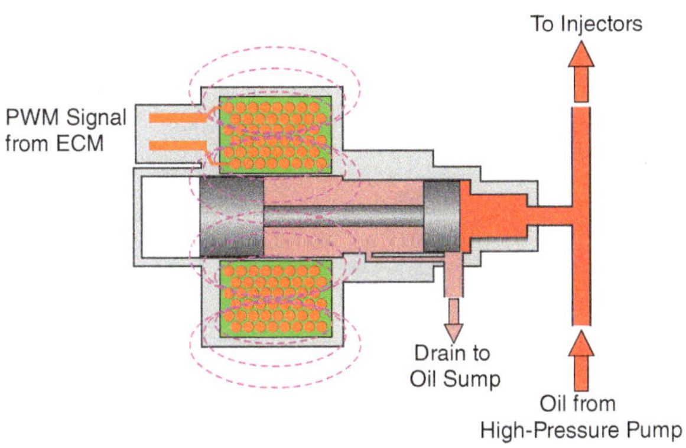

FIGURE 12-39 The ICPR uses a PWM signal to adjust oil pressure to the injectors. Pressure is usually low at idle and increases with engine speed and load conditions.

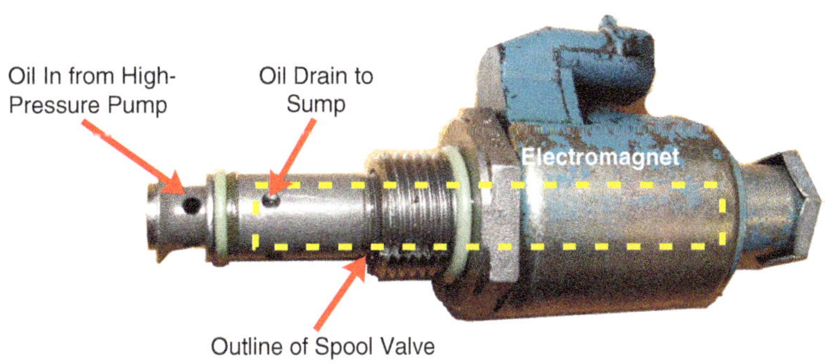

FIGURE 12-40 This ICPR regulator regulates injection actuation pressure by returning oil pressure from the high-pressure pump back to an oil sump.

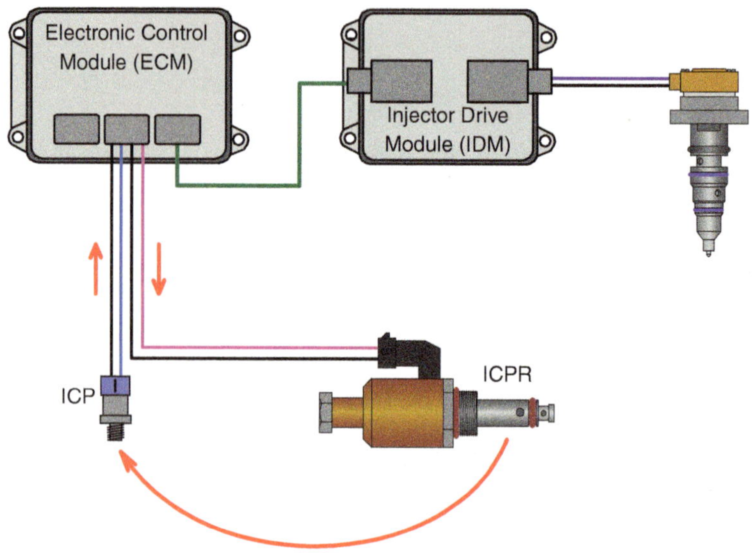

FIGURE 12-41 Closed-loop feedback refers to the control of an output device using feedback from a sensor. Changing the signal to the output device will simultaneously change the sensor values.

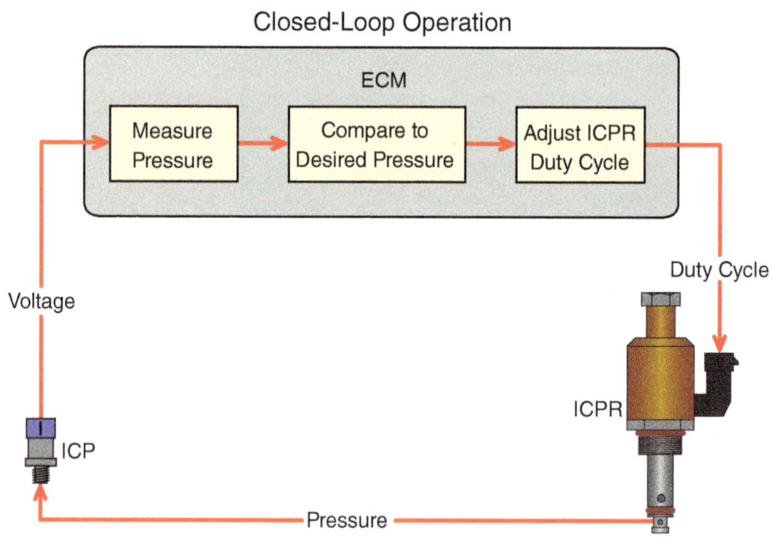

FIGURE 12-42 Closed-loop feedback control of injection oil pressure. Changes made by the ICPR affect the ICP sensor; the ICP sensor sends data to the ECM, which compares sensed data with desired values. Changes to the electrical signal are made if pressure is above or below the desired value.

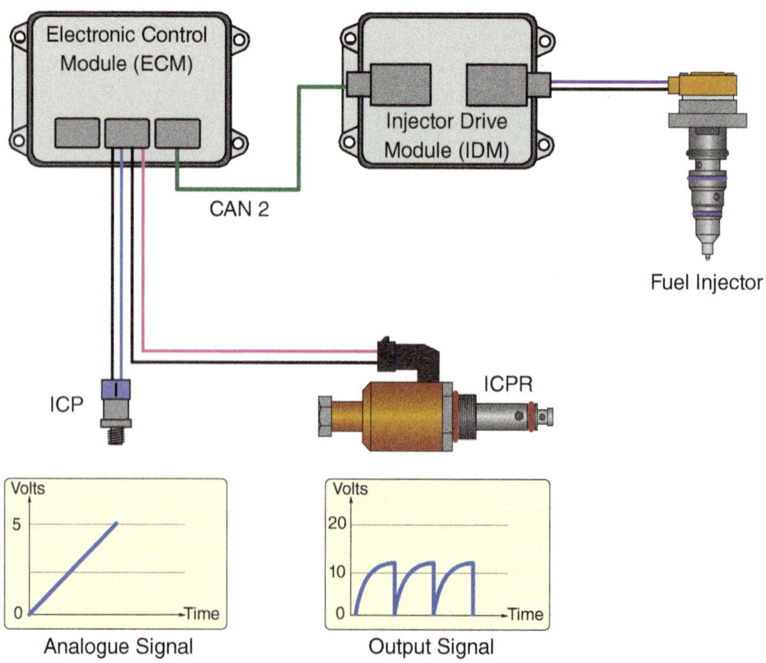

FIGURE 12-43 The closed feedback loop between the ICP sensor, the ICPR, and the ECM.

A minimum of 325–500 psi (22–34 bar) is usually required to start HEUI engines. When running, the oil pressure will vary depending on the engine calibration and model. However, oil pressure should remain relatively stable for any given operating condition. During hard acceleration and under full-load conditions, ICP will reach maximum pressures. Lowest pressures are encountered during warm idle.

▶ Electronic Control Systems

12-06 Identify and describe the functions of electronic system components used in HEUI injection systems.

Information processing in HEUI engines is similar to that in other electronically governed diesels. Input devices collect data and send electrical signals to the ECM. The ECM processes the information and generates outputs based on control algorithms and data contained in lookup tables stored as part of the engine calibration and software instructions. Electrical signals to output devices such as the injectors and the ICPR are some of the most important output electrical signals. The ECM is also connected to the vehicle network, allowing other modules to share data and interact with the engine's ECM. The injector drive module and EGR control module are examples of crucial control information being shared over a common vehicle network.

Injector Drive Modules

Because the HEUI injectors require substantial amounts of current—as much as 115 volts and 8 amps for HEUI A and B injectors and 48 volts and 20 amps for Gen II injectors—a separate control module is often used to minimize electrical interference in the ECM. The purpose of the **injector drive module (IDM)**, also called the FICM on Gen II HEUIs, is to operate the injector and monitor injector operation. Current is stepped up from 12-volts to 48-volts in the FICM by using step-up transformers (**FIGURE 12-44**). Capacitors store current used to energize the injectors (**FIGURE 12-45**). The FICM module stores information such as engine firing order and operating and diagnostic software instructions. Using network communication, data is exchanged between the ECM and the injector control module. Two control signals from the ECM are supplied to the injection control module: the fuel-delivery control signal (FDCS) and the cylinder identification (CID) signal. The FDCS is an ECM-commanded engine rpm and is used by the IDM to calculate injection timing and

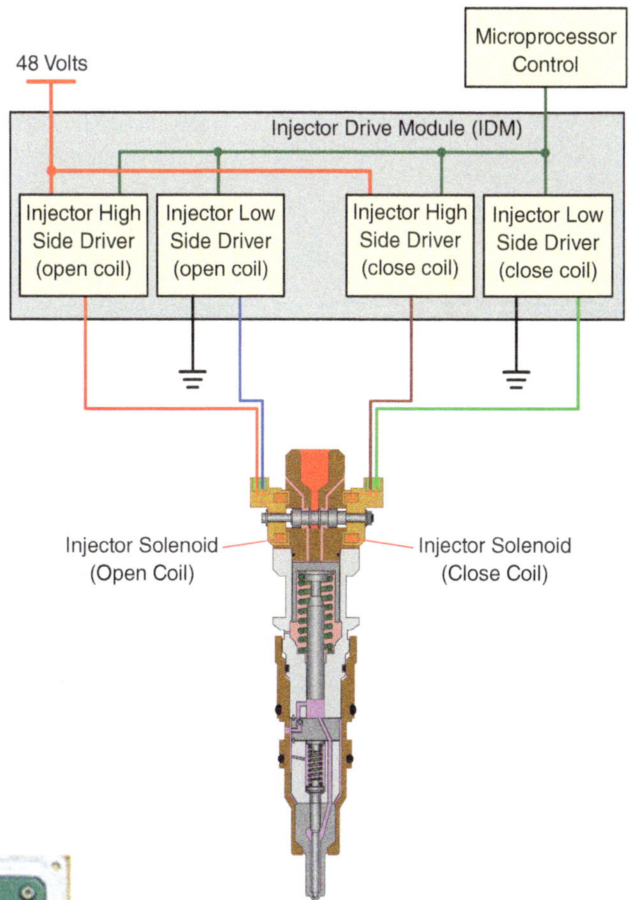

FIGURE 12-44 Onboard diagnostic systems monitor the voltage drops to the HEUI injector coils. Excessive resistance, open circuits, or shorts will change current flow and voltage drop to the coils. Out-of-range fault codes are logged if voltage or current is above or below expected values.

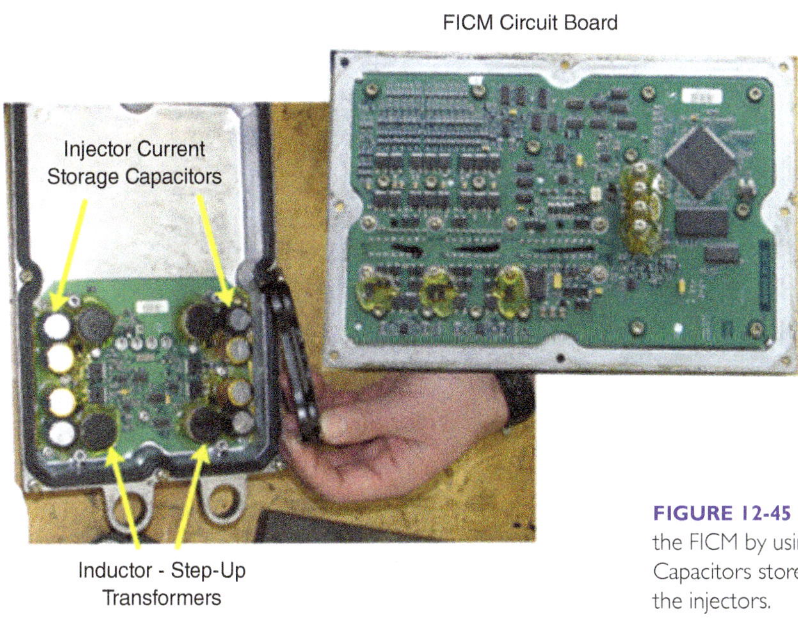

FIGURE 12-45 Current is stepped up in the FICM by using voltage transformers. Capacitors store current used to energize the injectors.

▶ TECHNICIAN TIP

Excessive engine idle is very destructive to diesel engines. During idle, low combustion temperatures and pressures produce exhaust slobber—a mixture of soot, unburned fuel, and oil— creating soot loading in engine oil. To prevent damage to HEUI engines caused by engine idle, oil and coolant temperatures are monitored to enable cold-weather protection mode. When the potential for damage from excessive cold idle is present, the engine rpm is automatically elevated and injection timing changes. Cold-weather protection will not permit excessive loading of a cold engine, either. If an engine is heavily loaded before the engine has adequately warmed up, piston skirts can collapse and oil can thicken, causing damage from poor lubrication.

12-07 Outline the diagnostic and repair procedure steps for HEUI systems.

injection duration or the energization time of the injector. The CID signal indicates engine position and is used to establish the beginning of the firing sequence. Because the firing order sequence is memorized by the injection control module, it can begin the injection sequence after learning when a particular cylinder is at TDC at the end of the compression stroke. An electronic feedback signal is used to send information back to the ECM in the form of diagnostic trouble codes (DTCs) about the injection control module and injectors. DTCs include faults such as opens, shorts, misfires, and other electrical faults. Caterpillar engines do not require a separate IDM. All engine functions are integrated into its proprietary advanced digital engine module (ADEM), which is a brand name for its ECM.

▶ Maintenance of HEUI Systems

Maintaining a HEUI fuel system is theoretically uncomplicated because there is no regular maintenance scheduled for the system. However, pressurizing fuel by using engine oil introduces new variables to maintenance considerations related to problems with oil contamination and performing diagnostics on the high-pressure oil circuits when problems do occur. The following section examines some common issues related to diagnosing problems with and servicing HEUI fuel systems.

Replacing a HEUI Injector

If Gen II injectors misfire, fail to deliver the correct quantity of fuel, are worn out or contaminated, or have sticking control valves, then they may need to be replaced. Stickage—the condition in which the spool valve will not properly move, due to deposits or fouling caused by soot-thickened oil—may sometimes be cleaned by using aftermarket solvents. If software updates and oil changes do not correct stickage, the condition can cause misfires when the engine is cold. When this occurs, the injector needs to be replaced.

To replace a HEUI injector, follow the steps in **SKILL DRILL 12-1**. These steps are for replacing a Gen II injector in a 6.0 L (365 cubic inch) Power Stroke engine. Always first consult OEM service literature to identify the most correct procedures and prescribed tooling.

SKILL DRILL 12-1 Replacing a HEUI Injector

1. Clean the engine, and remove the valve cover on the cylinder bank with the defective cylinder.
2. Remove the bolts holding the oil manifold.
3. Remove the hex head bolt holding the oil stand pipe to the oil manifold, and lift the oil manifold from the injectors.
4. Inspect the oil inlet area of the injectors for metal filings, which indicate a failure of the high-pressure oil pump. If the pump has failed, replace the pump, remove all injectors, and flush the high-pressure oil system with solvent. Failure to replace the pump and flush the system will cause any new injectors to fail in a short time.
5. Pour out oil from the oil manifold, preferably onto a paper shop towel, to further check for metal filings. Also, note the stain the oil leaves on the towel. Fuel in the oil will leave a large, clear-appearing stain around the oil ring on the paper towel.
6. Inspect the condition of the inlet injector O-rings. Any nicked, torn, or damaged O-ring requires an injector replacement because the O-ring is not serviceable separately.
7. Disconnect the injector's electrical connector and push the injector connector back into the rocker housing by using a ¾" (19 mm) 12-point socket.
8. Loosen and unscrew the Torx bolt in the injector hold-down clamp. Removing the injector hold-down bolt will unseat and lift the injector from its ore. *Note*: Do not pry on the injector because the injector coils can be easily damaged.
9. Clean the injector bore and cup with a purpose-made wire brush and wipe with a lint-free shop cloth. Ensure the copper sealing gasket for the injector tip is removed.
10. Lubricate the injector O-rings with engine oil before installation. Transfer the hold-down clamp to the new injector. Remove any oil in the hold-down bolt hole because oil will prevent the bolt from seating the injector.
11. Install and tighten the injector in place with the hold-down bolt. Torque the hold-down bolt; seating the injector with the correct amount of pressure is needed to transfer heat, prevent nozzle tip distortion, and ensure a good combustion seal.

Continued

12. Fill the injectors with engine oil and lubricate the oil inlet tubes on the oil manifold. Seat the manifold carefully by hand after installing a new O-ring on the stand pipe. If the manifold is not carefully seated, the inlet O-ring will be damaged, which will result in an oil pressure leak. The engine will likely be hard to start when warm.
13. Replace the fuel and oil filters, and change the engine oil by using the recommended grade and specification of oil.
14. Disconnect the cam-position and crank-position sensors before attempting to start the engine. Crank the engine for two consecutive periods of 30 seconds with a two-minute cooldown period between each crank interval. This will ensure that the injector control valves are properly lubricated before being moved, which will help prevent premature injector failure.
15. Update the engine with any revised calibration for the injectors or control system.

HEUI Diagnostics

Because HEUI injectors operate by using pressurized engine oil, diagnostic checks of the high-pressure oil system are unique compared with other fuel systems. Specialized diagnostic routines and strategies are required to identify problems with oil quality and the components controlling pressurization of engine oil.

Injection Control Pressure Problems

The high-pressure oil pumps on HEUI engines have a finite delivery volume (**FIGURE 12-46**). This means the output volume of the pump matches the quantity required by the injectors when they are operating at their maximum output. The system is designed to allow some extra delivery volume, but on many engines, this volume is not adequate if there are leaks in the high-pressure oil systems. This means that if there is a leak in the injection control system, engine performance is significantly affected. For example, system leaks can occur when oil leaks around injector O-rings, oil lines, or fittings; through defective poppet valves; or from a worn high-pressure oil pump. These leaks will produce problems related to rough running, low power, high fuel consumption, hard starting, and failure to start.

The following are some of the symptoms of high-pressure oil leakage:

- large difference between hot and cold starting times, particularly a quick cold start and long hot start
- loss of power when hot
- low power
- high fuel consumption
- oil consumption
- blue exhaust smoke, especially on start-up.

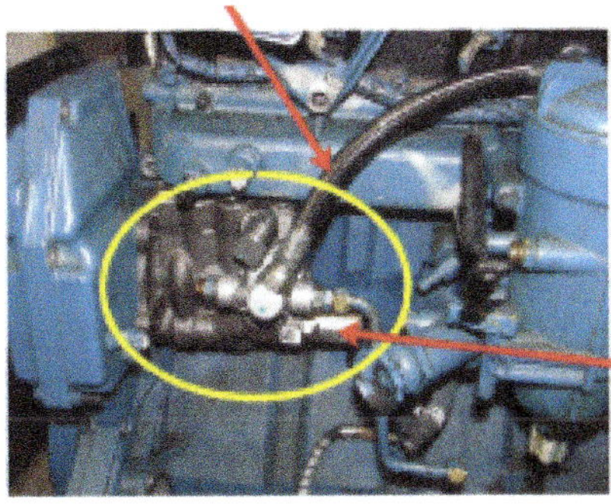

FIGURE 12-46 The three-piston pump on this MaxxForce DT engine pressurizes engine oil to operate the injectors. The ICPR is located at the rear of the pump.

Oil Quality in the HEUI System

Because the HEUI fuel injectors are electronically controlled and use high-pressure lube oil for actuation, understanding the operation of the oil system is necessary to diagnose possible performance problems. Oil quality is crucial for the same reason. The correct viscosity and grade of oil are important for proper HEUI operation. Oil that has little or no antifoaming additive will cause aeration in the oil. Aerated oil will not properly transmit pressure and motion, which leads to drivability problems such as stalling and low power.

Oil contaminated with water, fuel, dirt, or PM will lead to premature injector wear and malfunction. Improperly installed or incorrect air filters and loose or missing oil filler caps are common causes of dirt contaminating oil.

Oil viscosity influences the operation of the HEUI systems, too. Caterpillar uses a removable electrical jumper plug on C-7 and C-9 engines to correct actuation system variables when using either 15W-40 or 10W-40 engine oil. Oil thickened by excessive soot loading will interfere with injector control valve operating properly, particularly in cold weather and during start-up.

▶ TECHNICIAN TIP

Lubrication oil used in HEUI engines must have the proper level of antifoaming additive, which quickly releases air bubbles from oil. Antifoaming additives deplete with distance traveled, so HEUI engines are sensitive to overextended oil change intervals. The antifoaming additive is also affected by chemicals in room-temperature vulcanized (RTV) silicone used as gasket material. After engine service, such as resealing the oil pan by using RTV silicone, adding additional antifoaming additive is recommended. Oil-displacing RTV silicone is recommended for resealing engine parts such as oil pans and engine front covers.

Testing the Oil Actuation Circuit

HEUI oil systems are divided into two sections: the low-pressure system and the high-pressure system. The low-pressure system supplies the high-pressure system with oil, and problems in either system will affect engine operation.

Diagnostic Tests to Detect System Leakage

Oil leakage from the high-pressure oil circuit in HEUI engines is one of the most common reported problems and is detrimental to engine operation. Even small losses of oil from HEUI systems are noticeable and can be easily detected by technicians when they use the proper diagnostic techniques and tools.

Measuring the Duty Cycle of the Injection Control Pressure Regulator

The ICPR is a PWM oil pressure control valve that reports its electrical signal as a percentage of duty cycle on all scanners and OEM software. An unusually high duty cycle indicates the ECM is compensating for oil leakage by increasing the on time for the pressure regulator. Longer on time or an increased percentage of duty cycle means that less oil returns to the sump and that more oil is delivered to the high-pressure circuit. Observing actual duty cycle and comparing it to expected values is one of the most useful tests to detect oil leakage (**FIGURE 12-47**).

The onboard diagnostic systems will detect the abnormal operation of injection actuation control, alerting technicians to faults by generating codes:

- ICP above system working range
- ICP above specifications with engine off
- injection control pressure regulator duty cycle above/below desired level
- injection control sensor values out of range.

Performing an Injection Control Pressure Regulator Test

Leaks in the HEUI high-pressure oil system or worn oil pumps are indicated by an above-average ICP duty cycle. In the case of a leak, the ICPR will need to close further and block oil return to the oil sump to send more oil to the injectors. In the case of a worn pump, less oil can be allowed to return to the sump to build oil pressure, because oil pump volume output is reduced when the pump is worn. To perform an ICPR test, follow the steps in **SKILL DRILL 12-2**.

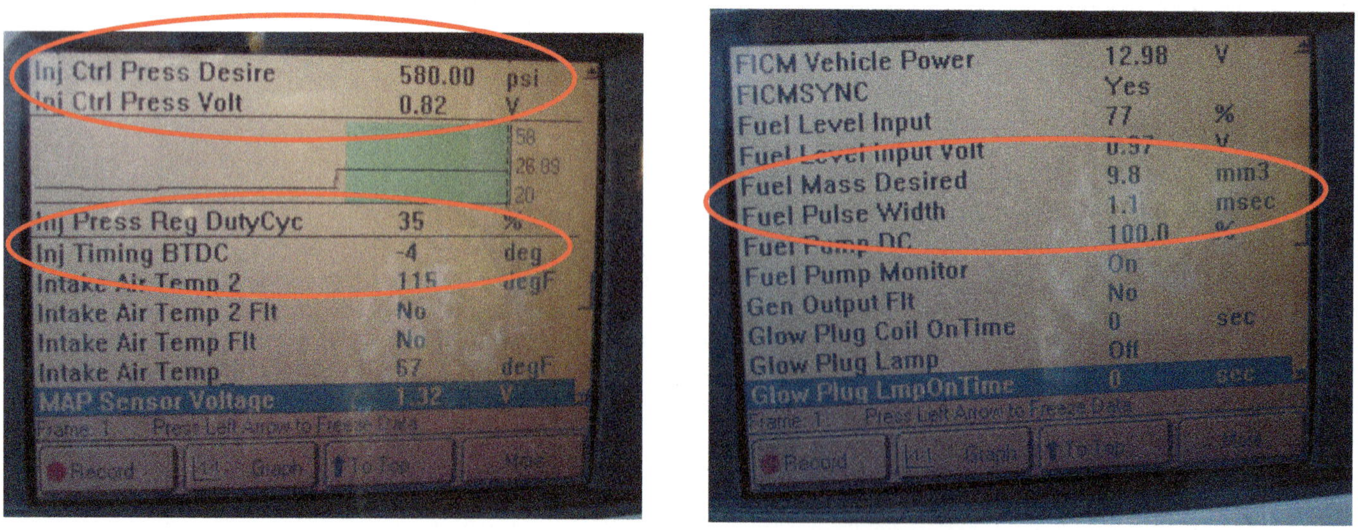

FIGURE 12-47 The duty cycle of the ICPR regulates oil pressure to the injectors. Observing oil pressure stability and the duty cycle provides important diagnostic information.

SKILL DRILL 12-2 Performing an ICPR Test

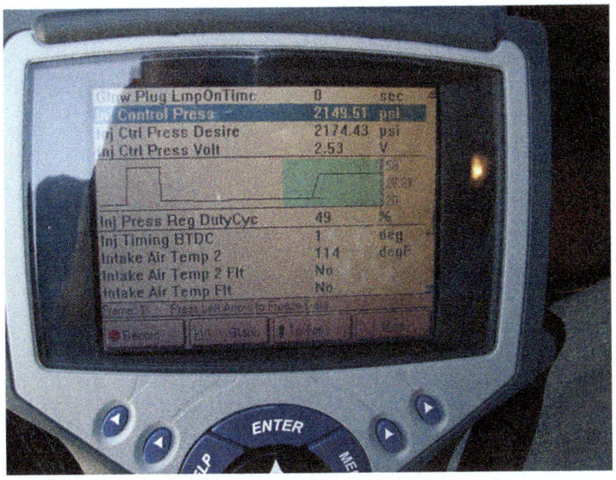

1. Start and warm up the engine.
2. Connect a scanner or OEM software to the engine. Navigate to a menu item that displays live engine data parameters.
3. Locate the data for injection control oil pressure and injection control duty cycle.
4. Run the engine at idle to observe and record ICP and duty cycle.
5. Run the engine at high idle speed and record the two values again.
6. Compare these values against the manufacturer's limits. An above-average duty cycle means the system is leaking, the oil is too thin, or the high-pressure pump is worn out. There is no lower limit to duty cycle for an engine that is running properly.
7. If the control pressure duty cycle is above average, recommend further diagnostic checks to identify leaks caused by defective injector O-rings or worn pumps.

Measuring Pressure Loss in the Oil Actuation System

When pressurized with oil, high-pressure circuits should be able to retain pressure for hours with no significant drop. A hand-operated or electrically operated oil pump can be used to verify the sealing capabilities of the high-pressure circuit. Alternatively, air pressure can be used to pinpoint leaks in O-rings at branch tubes, oil pumps, and oil manifolds.

Visually Inspecting Fuel for Contamination

A visual check can determine whether oil is bypassing injector O-rings and entering the fuel. In early-generation HEUI engines, if the middle O-ring has failed, oil will push its way into the low-pressure fuel circuit, where it is either injected with fuel or returned to the tank with return fuel.

> ▶ **TECHNICIAN TIP**
>
> A defective HEUI O-ring will allow high-pressure oil to push through to the low-pressure fuel supply system. Lubricating Oil is recirculated back to the fuel tank with fuel and will darken the fuel.

Visually Inspecting for Oil Leaks

Leaking injector O-rings, jumper tubes, and manifolds can be identified during engine operation. Oil leaking around an injector is not normal and indicates O-ring leakage. It is easy to detect oil leaks by moving a piece of paper around jumper tubes. This method also eliminates the danger of oil penetrating the skin. Plugged oil filter screens on ICPR regulators will also limit or cut off all oil supply to the injectors.

Air Entrained in Oil

Oil containing many air bubbles will not properly transmit injection force through the amplifier piston. Air entrained in oil is also detected by measuring the duty cycle of the pressure regulator, because oil containing high amounts of air will cause the ICPR duty cycle to lengthen. Air also causes ICP to become unstable, and wide variations to control pressure are observed when the engine is running.

Another diagnostic test to determine whether oil is aerating is performed by monitoring the ICP while operating the engine at high idle for three minutes. If oil begins to aerate, the duty cycle of the ICPR will increase rather than decrease. This happens because the control system compensates for a loss in pressure caused by oil aeration. In Caterpillar engines, a diagnostic menu enables a technician to change the ICP by stepping it up or down. The oil pressure's stability and the duty cycle needed to maintain the oil pressure is carefully checked against expected values. Once again, too long of an on time for the ICPR and unstable oil pressure point to problems with oil quality, worn high-pressure pumps, or high-pressure oil leakage.

Wrap-Up

Ready for Review

- One of the first and largest evolutionary steps of modern high-pressure diesel-injection technology was the development of hydraulically actuated electronic unit injection (HEUI) systems.
- The primary disadvantage of traditional mechanical systems was their reliance on a camshaft to pressurize fuel. Another disadvantage was their limited ability to adapt injection events to continuously changing engine and vehicle operating conditions.
- The HEUI system was developed to ensure peak injection pressure independent of engine speed, meaning that maximum spray-in pressure is available whether the engine is operating at idle or maximum rpm.
- During hard acceleration or sudden load changes at low speeds, the system can instantly adjust fuel pressurization to best match combustion conditions while minimizing emissions.
- Instead of using cam lobes, springs, flyweights, and levers to operate, HEUI systems use sensors, software code, microprocessors, and electrical signals, achieving more precise control of injection events than any previous system.
- HEUI systems replace the mechanical camshaft with highly pressurized lubrication oil. This oil actuates the plungers that pressurize fuel for injection. Oil pressure is multiplied inside the injector by using an amplifier piston.
- Because the functions of pressurization, metering, timing, and atomization are all incorporated into a single injector body, HEUI injectors are classified as unit injectors.
- Injection pressures are regulated by varying the oil pressure supplied to the injectors. This regulation of oil pressures enables control of injection pressure to match engine operational requirements.
- HEUI system components are interchangeable with a large variety of engine sizes, cylinders, and applications. Whether the engine is used in power generation, off-road equipment, an ambulance, a pickup truck, a delivery vehicle, or a dump truck, many of the same components are used.
- Ongoing developments in HEUI technology have improved the ability to shape the rate of fuel injection, thus reducing combustion noise and emissions for quieter, cleaner engine operation.
- HEUI technology provides the potential to integrate valve-train operation into the high-pressure oil actuation system. Instead of using a camshaft to operate valves, intake and exhaust valves can be actuated electrohydraulically.
- The HEUI system has four major subsystems: the HEUI injectors, a low-pressure oil system, a high-pressure injection actuation system, and an electronic control system.
- HEUI injectors use highly pressurized engine oil to provide injection force. There are four basic variations of these injectors.
- Both type A and B split-shot injectors have four major components that operate together for precise metering, timing, and rate control. The components are the high-voltage electric solenoid, the poppet valve, the intensifier piston and plunger, and the nozzle assembly.
- The three stages of injection in the HEUI A and B injector are the fill cycle, injection, and the end of injection. Injection quantity is determined by the amount of plunger travel.
- Injection timing adjustments are made by changing the points at which the injector solenoid begins energizing and stops energizing. Optimal injection timing is calculated by the ECM.
- Injection rate refers to the quantity of fuel injected per degree of crank angle rotation. Rate control is the ability of the fuel system to adjust the quantity of fuel injected during the injection sequence.
- HEUI B split-shot and PRIME injectors can inject a small quantity of fuel to the combustion chamber 8°–10° before the main injection. This pilot injection begins to burn before the main injection arrives. The result is a combustion chamber with a burning flame front that quickly ignites fuel in the main injection.
- The second-generation (Gen II or G2) injector uses digital control valve technology. This valve technology replaced the poppet valve used in HEUI A and B injectors. When open, this valve allows oil to flow from the high-pressure oil rail into the injector, pushing the intensifier piston and plunger downward. When closed, it allows oil to drain out of the injector.
- Gen II injector operation is similar to the earlier HEUI injector amplifier piston, also called the intensifier piston. The difference between the surface area on top of the piston and the bottom multiplies fuel pressure below the piston.
- The Gen II injector has three main events during an injection cycle: fill, main injection, and end of main injection.
- Similar to other HEUI injectors, the Gen II plunger travel determines injection quantity. To change the rate of fuel delivery, the oil pressure to the injector is electronically controlled by the ECM. The Gen II injector is able to rapidly open and close the spool valve several times during the course of one injection sequence, resulting in extremely precise rate control.
- The high-pressure oil actuation system used by HEUI systems has several major subsystems: a low-pressure oil supply, oil reservoirs, a high-pressure pump, and injection control pressure regulation (ICPR).

- All HEUI engines share injection control pressure oil with the engine's lubrication system. Lube oil supplies the separate high-pressure oil supply capacity of oil for the high-pressure circuit.
- Two configurations are used to supply the high-pressure pump from the oil reservoirs: a short-circuit supply and a filtered supply.
- The high-pressure oil actuation system (or injection control circuit) delivers engine oil under high pressure to actuate the fuel injectors.
- The drive torque used to pressurize oil in the high-pressure pump is provided by the engine. Because the pump output is not phased with the injection event, a much lower drive torque (in comparison to other injection systems) is required to operate the pump.
- Oil pressure is regulated by using an ECM-controlled, pulse-width–modulated (PWM) pressure regulator. This injection control pressure regulator (ICPR) device is an electromagnetically controlled spool valve that moves in response to the strength of a magnetic field. The oil pressure for injection actuation is adjusted as the ECM changes the current flow through a coil surrounding the spool valve.
- These engines' information-processing functions are similar to those of other electronically governed diesels. Input devices collect data and send electrical signals to the ECM. The ECM processes the information and generates outputs based on control algorithms, stored data, and software instructions.
- Because the HEUI injectors require substantial amounts of current, a separate control module is often used to minimize electrical interference in the ECM.
- If there is an oil leak in a HEUI control system, engine performance will be significantly affected. Leaks can cause rough running, low power, high fuel consumption, and even failure to start.
- Because the HEUI fuel injectors are electronically controlled and use high-pressure lube oil for actuation, correct oil quality is crucial. The correct viscosity and grade of oil are important to ensuring proper HEUI operation. Oil contaminated with water, fuel, dirt, or other contaminants will lead to premature wear and malfunctioning injectors.
- An unusually high duty cycle for the ICPR indicates that the ECM is compensating for oil leakage by increasing on time for the pressure regulator. Longer on time means that less oil returns to the sump and more oil is delivered to the high-pressure circuit. Observing the actual duty cycle and comparing it to the expected values is one of the most useful tests to detect oil leakage.
- Technicians must use visual checks to determine whether oil is bypassing injector O-rings and entering the fuel.
- Leaking injector O-rings, jumper tubes, and manifolds are sometimes located during engine operation. Oil leaking around an injector is not normal and indicates O-ring leakage.
- Oil containing many air bubbles will not properly transmit injection force through the amplifier piston. Air also causes injection control pressure (ICP) to become unstable. Air entrained in oil can be detected by measuring the duty cycle of the pressure regulator. Oil containing high amounts of air will cause the ICPR duty cycle to lengthen. Another diagnostic test is to monitor the ICP while operating the engine at high idle for three minutes. If oil begins to aerate, the duty cycle of the ICPR will increase.

Key Terms

acoustic wave attenuation (AWA) A hydraulic wave-dampening device that minimizes a rhythmic cackling noise at idle caused by high-pressure pump pressure waves synchronizing with injection events.

cam-less diesel A system that electrohydraulically actuates intake and exhaust valves instead of using a camshaft to operate valves.

digital control valve The valve technology that replaced the poppet valve used in the HEUI A and B designs; the digital nature of the valve refers to the position of the control valve—either open or closed.

Gen II HEUI The newer generation of HEUI that replaced the poppet valve and solenoid with a lighter, faster-acting spool valve and two magnetic coils; it is also known as G2 HEUI.

HEUI A The first-generation HEUI injector without split-shot-injection or pilot-injection capabilities.

HEUI B A later version of the first-generation HEUI injector with PRIME metering or split-shot-injection (pilot-injection) capabilities.

HIB-300 Caterpillar's second-generation HEUI injector.

high-efficiency pump (HEP) An innovative oil pump used by Caterpillar in which output pressure is matched to the pressure desired by the ECM based on operating conditions (e.g., load, speed, temperature) and OEM-specific fuel maps stored in the ECM by using a PWM-controlled displacement control valve.

injection control pressure regulator (ICPR) An electromagnetic operated spool valve that moves in response to the strength of a magnetic field; by changing the current flow through a coil surrounding the spool valve, the oil pressure for injection actuation is adjusted.

injection control pressure (ICP) sensor A sensor located in the high-pressure oil circuit that provides closed-loop feedback to the ECM about whether the oil pressure is too low or too high.

injector drive module (IDM) A module on some first-generation and second-generation HEUIs that operates and monitors injector operation. It stores information such as engine firing order and operating and diagnostic software instructions; it is also known as the fuel-injection control module (FICM) on later-model HEUI engines.

injection rate control The control of fuel-delivery volume per degree of crank angle rotation into the cylinder during an injection event.

intensifier piston A piston that multiplies the force of high-pressure oil pressure and is used to pressurize fuel to injection pressure; it is also known as the amplifier piston.

oil manifold A component that supplies oil through the top of the injectors and eliminates the problem with O-ring leakage when oil is supplied though the cylinder head.

pilot injection A small injection event taking place 8°–10° before the main injection event; it is also known as split-shot injection.

poppet valve A valve that controls the flow of oil into and out of HEUI A and B injectors.

split-shot injection An injection strategy that delivers fuel in two distinct events during one combustion cycle; it is also known as pilot injection.

Review Questions

1. Pilot injection, also called split-shot injection, in HEUI B injectors is accomplished ________.
   a. electronically
   b. with a fast-acting single injector solenoid coil
   c. with a low-inertia fast-acting dual injector solenoid
   d. by mechanically spilling fuel
2. Which of the following components multiplies oil pressure inside a HEUI injector?
   a. Injector plunger.
   b. Intensifier piston.
   c. Poppet valve.
   d. PRIME amplifier valve.
3. Which of the following conditions is indicated when engine oil contaminates the fuel tank of a vehicle with a HEUI engine?
   a. A leaking amplifier piston O-ring.
   b. A leaking cylinder head gasket.
   c. A defective injection pressure regulator.
   d. A leaking injector external O-ring.
4. Consider a HEUI fuel system. When the injector pulse width remains constant and the injection control pressure increases, which of the following changes will take place?
   a. Injection quantity will be reduced.
   b. Injection quantity will be increased.
   c. Injection timing will be retarded.
   d. Injection timing will be advanced.
5. The maximum potential voltage supplied to a HEUI A or B injector is __________.
   a. closest to 12 volts AC
   b. closest to 70 volts AC
   c. closest to 110 volts DC
   d. closest to 70 volts DC
6. Which of the following forces do Gen II HEUI injectors use to physically move the spool valve in the injector control valve mechanism when it opens or closes?
   a. Residual magnetism.
   b. Oil pressure.
   c. Fuel pressure.
   d. A PWM signal.
7. Consider an engine with a HEUI fuel system. Which of the following problems will most likely cause a hard start/prolonged cranking condition only when the engine is hot?
   a. Increased engine oil viscosity.
   b. An unusually long PWM signal to the injection pressure regulator.
   c. Leaking injector O-rings.
   d. The use of antifoaming oil conditioner.
8. What change generally takes place when the length of the pulse width to an injection control pressure regulator (ICPR) is increased?
   a. Injection pressure increases.
   b. Injection pressure decreases.
   c. Injection rate is ramped.
   d. Injection rate shape becomes squared.
9. What does an above-normal increase in the duty cycle of the PWM signal applied to the HEUI ICPR indicate?
   a. Worn-out HEUI injector nozzles.
   b. Soot-thickened engine oil.
   c. A worn-out high-pressure oil pump.
   d. A broken amplifier piston return spring.
10. What is the measuring unit for the electrical signal to the ICPR of a HEUI fuel system?
   a. Percentage of duty cycle (%).
   b. Time (msec).
   c. Frequency (Hz).
   d. Voltage (volts).

ASE Technician A/Technician B–Style Questions

1. Technician A says that higher levels of detergents and antifriction additives are needed by HEUI engines to maintain stable injection control oil pressure. Technician B says that antifoaming agents are more important than additional antifriction and detergent additive for a smooth and efficient engine operation. Who is correct?
   a. Technician A
   b. Technician B
   c. Both A and B
   d. Neither A nor B
2. Technician A says acoustic wave attenuation (AWA) fittings in the oil manifold of some Gen II HEUI engines reduce irregular combustion noise. Technician B says split-shot injection reduces combustion noise. Who is correct?
   a. Technician A
   b. Technician B
   c. Both A and B
   d. Neither A nor B
3. Technician A says an above-normal increase in the duty cycle of the PWM signal applied to the HEUI ICPR of a Cat C-7 engine indicates that the engine oil has thickened due to excessive soot loading. Technician B says a jumper plug in the wiring harness indicates to the engine's ECM that

10W-30 engine oil rather than 15W-40 oil is being used in the crankcase. Who is correct?
   a. Technician A
   b. Technician B
   c. Both A and B
   d. Neither A nor B
4. Technician A says the measuring unit for the electrical signal to the injection control pressure regulator of a HEUI fuel system is reported as a percentage of duty cycle (%) on a scanner display of engine parameters. Technician B says the PWM signal is reported as time (in milliseconds). Who is correct?
   a. Technician A
   b. Technician B
   c. Both A and B
   d. Neither A nor B
5. While diagnosing the cause of a hard start/prolonged cranking complaint, Technician A says that the high-pressure oil between the high-pressure pump and the injectors should stay pressurized for days. Technician B says the oil pressure to the injectors should quickly drop to zero when the engine is shut off. Who is correct?
   a. Technician A
   b. Technician B
   c. Both A and B
   d. Neither A nor B
6. While diagnosing the cause of a rough-running 2005 Ford Power Stroke engine, Technician A says that an incorrect injector calibration code will produce uneven cylinder contribution at idle. Technician B says that Gen II HEUI injectors do not use calibration codes. Who is correct?
   a. Technician A
   b. Technician B
   c. Both A and B
   d. Neither A nor B
7. Technician A says that it is important to remove and drain the oil reservoir for a HEUI high-pressure pump when changing engine oil so that the oil does not contaminate the clean engine oil. Technician B says that it is not required or practical to drain the oil reservoir or accumulator when changing oil. Who is correct?
   a. Technician A
   b. Technician B
   c. Both A and B
   d. Neither A nor B
8. When investigating the cause of frequent injector failures on a HEUI B engine, Technician A says a likely cause could be engine oil contaminated with dirt entering the engine through a leaking air filter. Technician B says that condition will not cause injector failure, because the oil is filtered before supplying the injectors. Who is correct?
   a. Technician A
   b. Technician B
   c. Both A and B
   d. Neither A nor B
9. While discussing the purpose of pilot injection, also known as split-shot injection, Technician A says HEUI injectors use it to reduce NO_x emissions. Technician B says PRIME metering can reduce combustion noise. Who is correct?
   a. Technician A
   b. Technician B
   c. Both A and B
   d. Neither A nor B
10. While considering ways to increase engine power output, Technician A says that using a special injection control pressure (ICP) sensor, which reports oil pressure as lower than the actual pressure, is an effective shortcut. Technician B says that using the modified ICP sensor signal will result in less fuel being injected. Who is correct?
   a. Technician A
   b. Technician B
   c. Both A and B
   d. Neither A nor B

CHAPTER 13

Common Rail Fuel Systems

Learning Objectives

After reading this chapter, you will be able to:

- 13-01 Describe the construction and functions of CR fuel systems and subsystem components.
- 13-02 Identify and describe various types of CR fuel systems.
- 13-03 Describe and explain the operation of CR injectors.
- 13-04 Explain the operation of Delphi Multec systems.
- 13-05 Identify and describe pressure control components for CR fuel systems.
- 13-06 Recommend service for and maintenance of CR fuel systems.

You Are the Technician

A late-model pickup truck with a third generation Bosch common rail (CR) engine won't start and has been towed to your shop. For some time, the engine has had prolonged cranking time before it would start. After performing visual inspections for external fuel leaks, adequate fuel level in the tank, and the condition of wiring harnesses, you measure the fuel rail pressure and find its pressure is zero. From there, you begin to follow the engine manufacturer's troubleshooting steps and check for fault codes. A variety of codes indicate that the rail pressure has been out of range dozens of times. Then you verify that the low-pressure fuel supply system is in fact supplying clean fuel at the correct volume and pressure to the high-pressure pump. The diagnostic flow chart from the original equipment manufacturer (OEM) indicates that you should next measure back leakage from the cylinder head, high-pressure pump, and pressure-relief valve on the fuel rail while the engine is cranking. As you perform the test, you discover that the fuel returning from the cylinder head to the fuel tank is excessive and outside specifications. Consider the following:

1. List and describe three possible causes of excessive back leakage of fuel from the cylinder head.
2. Describe the procedure you will use to identify the source of the excessive fuel return.
3. If the complaint began shortly after a fuel injector was installed, list several possible explanations for the back leakage from the cylinder head being out of specification.

▶ Introduction

First used in European automotive production in 1997, CR fuel injection represents the latest evolution in high-pressure injections systems. CR fuel systems possess features and capabilities that no other injection system can match, which has allowed this injection system to displace almost every other type of high-pressure injection system used by diesel engines. Currently, various configurations and more than four generations of refinements to CR systems are supplied by Bosch, Delphi, Denso, and Siemens (Siemens is now owned by Continental AG) and are used by the majority of diesel engines. CR's ability to deliver superior performance, power output, exceptional fuel economy, and low emissions have made it the injection system of choice for all current light-, medium-, and heavy-duty on-highway diesel engines, as well as for the majority of tier-four non-road engines.

▶ Fundamentals of Common Rail Fuel Systems

13-01 Describe the functions and construction of CR fuel systems and subsystem components.

The basic concept of CR injection is simple: supply fuel at injection pressure to a fuel injector controlling injection events using an electrically operated nozzle valve (**FIGURE 13-1**). Injection pressure is developed by an engine-driven high-pressure fuel pump, that regulates injection pressure electronically an is not coupled to engine speed. Unlike cam and plunger type pump elements without electronic pressure controls, this configuration makes maximum injection pressures available from idle speed and throughout the entire engine operating range. This concept of supplying a constant high-pressure fuel supply from a shared, or common, fuel rail to individual injectors is not new, but, the technology used to switch injection events on and off at extreme pressures while precisely metering the correct quantity of fuel for injection represents a number of breakthrough achievements in manufacturing. CR's metering precision, spray-in pressures, and control over injection events can dramatically reduce exhaust emissions and fuel consumption while also improving engine performance over that of spark-ignition and most other diesel

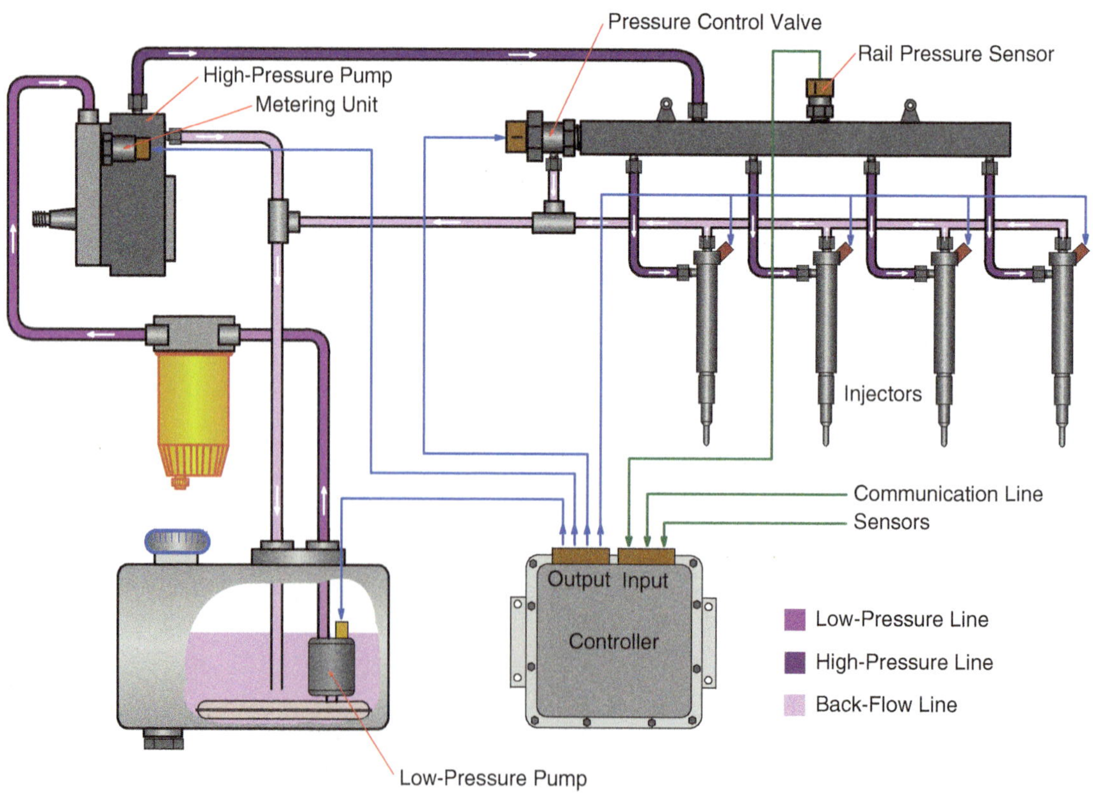

FIGURE 13-1 In CR systems, a gear-driven, engine-mounted, high-pressure pump pressurizes fuel to injection pressure. An electronic control module (ECM) optimizes combustion events.

injection systems. The technological preeminence of CR isn't only due to advanced capabilities; CR systems also possess construction features that allow the same parts to be easily adapted for use over a wide range of engine models, power output, and applications. With far more ease than previous fuel systems, CR systems provide engine manufacturers with the flexibility to incorporate CR fuel into any diesel engine design.

Features and Advantages of Common Rail

Switching high-pressure fuel with the metering precision required for diesel engine operation was the primary technological barrier to developing CR injection systems. Current system pressures for heavy-duty CR injection operate as high as 30,000–37,500 psi (2,068–2,586 bar). Until recently, maximum injection pressure for automotive light- and medium-duty systems topped out at around 29,000 psi (2,000 bar) due to the limitation of control valve opening-and-closing forces to regulate higher injection pressures. The latest injectors from Delphi—such as its DFI 4.25 injector—can operate at 36,250 psi (2,500 bar). Delphi is also introducing its DFI 21 injector for commercial diesels in 2019 that can operate at up to 45,000 psi (3,000 bar). Numerous technological innovations in the DF 4 injector enable it to produce as many as nine injection events during one combustion cycle (**FIGURE 13-2**). Bosch reports that its third-generation CRS3-27 injector can operate at 2,700 bar or 39,150 psi pressure and is capable of 10 injection events during a combustion cycle. Denso's 3rd generation solenoid type injectors, desginated G4s installed in the 2017 GM Duramax 6.6 and new GM 1-6L Whisper diesel can deleiver up to 9 injection events per combustion cycle and achieve injection pressures as high as 3,000 bar or 43,500 psi. Multiple injection events take place during a combustion cycle and last less than 1 millisecond. In fact, the duration of many injections events now is less than 500 *micro*seconds with as little as 100 microseconds of separation between events. Injection timing is accurate to within 1/100 of a degree of crankshaft rotation, while shot-to-shot variations between injection quantities are well under 0.01 mm^3 tolerance or 1 mg/stroke variation. Enormous technical and manufacturing advances have enabled CR systems to not only meet outstanding performance goals but also combine to firmly establish diesel engines as the superior engine technology now and in the future.

The following sections describe the features of the CR system that have advanced diesel engines to the forefront of high-performance, low-emission engine technology.

Pressurization of Fuel Independent of Engine Speed

Modern CR refers to a classification of high-pressure fuel-injection systems in which fuel pressurization occurs independently of engine speed. This means that regardless of engine rotational speed and load, fuel-injection pressure can be electronically controlled to provide the optimal pressure required for the most efficient combustion, lowest emissions, highest power output, and best fuel economy (**FIGURE 13-3**). Injection pressures in CR systems are also higher and more consistent during the injection event than they are in cam-and-plunger fuel pressurization systems. Rising and falling camshaft lobes tend to vary the injection pressure too much over the injection period. CR systems can produce more efficient "square" injection rates than the triangular shape observed from mechanical cam-and-plunger systems (**FIGURE 13-4**).

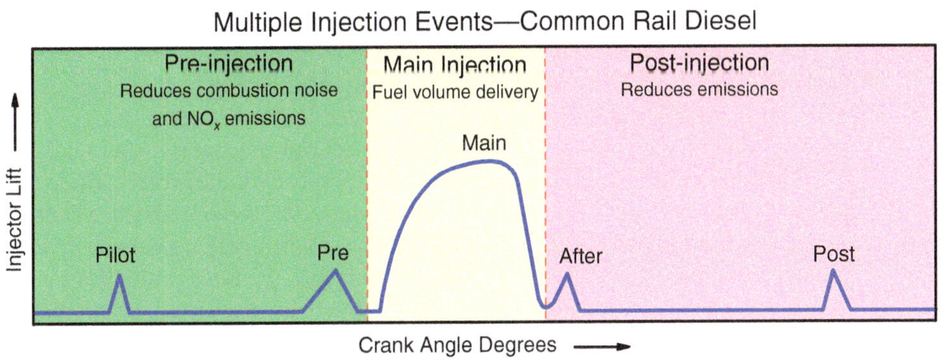

FIGURE 13-2 Multiple injection events enable CR systems to achieve the cleanest combustion quality while delivering superior fuel economy and performance.

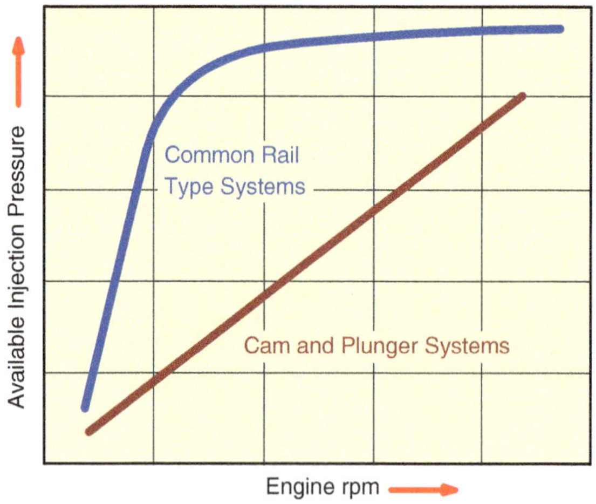

FIGURE 13-3 CR injection systems pressurize fuel independently of engine speed, which means maximum injection pressure is available even from idle.

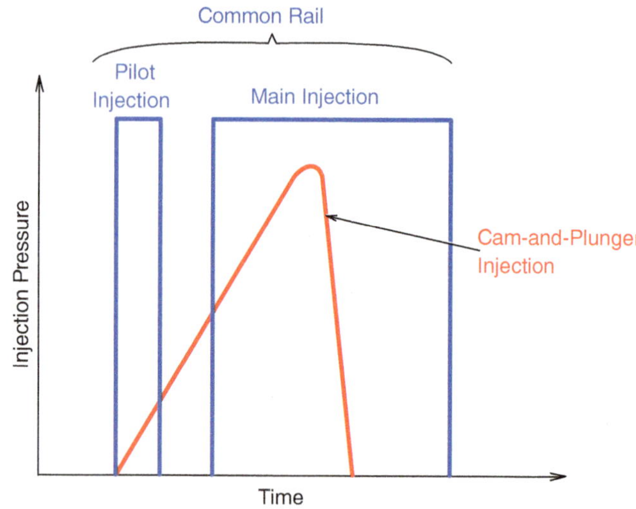

FIGURE 13-4 Mechanical non-electronic cam-and-plunger systems are not capable of shaping the rates or number of injection events. A mechanical system's injection rate tends to always remain triangular.

Higher Pressurization

CR systems provide the highest injection pressures of any current injection system. Bosch's fourth generation of CR systems, the **amplified common rail (ACR)** injection system, which has been in production since 2009, reaches 37,500 psi (2,586 bar) (**FIGURE 13-5**). The ACR system uses a two-stage amplification of fuel pressure inside the injector to provide improved power output with lower emissions, noise, and fuel consumption, and it produces several unique shapes to modify the injection rate profile to optimize engine performance. Denso is developing a CR system for commercial vehicle diesel engines, one that is capable of injecting at up to 45,000 psi (3,102 bar) at maximum pressure. High pressurization enhances fuel distribution and fuel atomization in the combustion chamber, which results in faster, leaner air-fuel mixing and lower soot formation. The result is comparatively better fuel economy, less emissions, improved combustion quality, and increased power output while using less fuel.

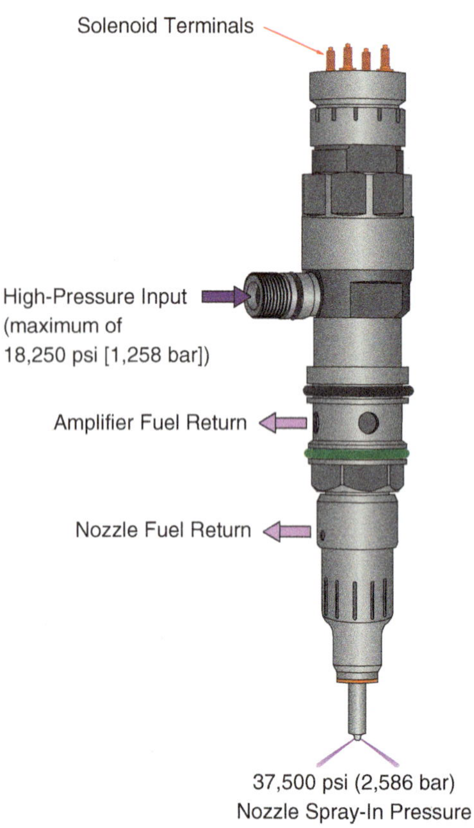

FIGURE 13-5 The injector details of a **hydraulically amplified diesel-injection system (HADIS)**, which was developed by Bosch and used by Detroit Diesel's ACR system.

Servo-Hydraulic Control

The concept of switching pressurized fuel on and off with an injector-mounted solenoid has been used for decades on spark-ignition engines equipped with multiport fuel injection. Supply pressures to injectors in those systems are generally below 120 psi (69 kPa). However, operating a fuel-injection system at tens of thousands of pounds per square inch (14.5 psi = 1 bar) of pressure with the accuracy required for diesel injection was an insurmountable obstacle for decades, making the development of CR systems slow and difficult. Magnetic injector actuators, like those used on heavy-duty diesels, are slow, especially when coupled with a relatively heavy-mass fuel control valve inside an injector. One solution used to overcome the obstacle of slow-responding injectors is to use magnetic fields and electrical energy to act as the triggering or tipping force to harness the faster response and more powerful hydraulic force of highly pressurized fuel to control injection events. Multiplying hydraulic forces by using much smaller and lighter electrically operated hydraulic control valves is the technical solution, making CR systems both practical and effective (**FIGURE 13-6**).

Characterizing this ability to control and multiply hydraulic forces inside the injector is the term "servo-hydraulic actuator." The word "servo" refers to a control system that multiplies a small mechanical force or movement into a much greater force or movement. Servo-hydraulic control describes the force multiplication of incredibly small fluid pressure differentials created by electrically manipulated control valves into much larger, faster-acting hydraulic forces that rapidly lift and close nozzle valves in the injector (**FIGURE 13-7**). In servo-hydraulic CR injectors,

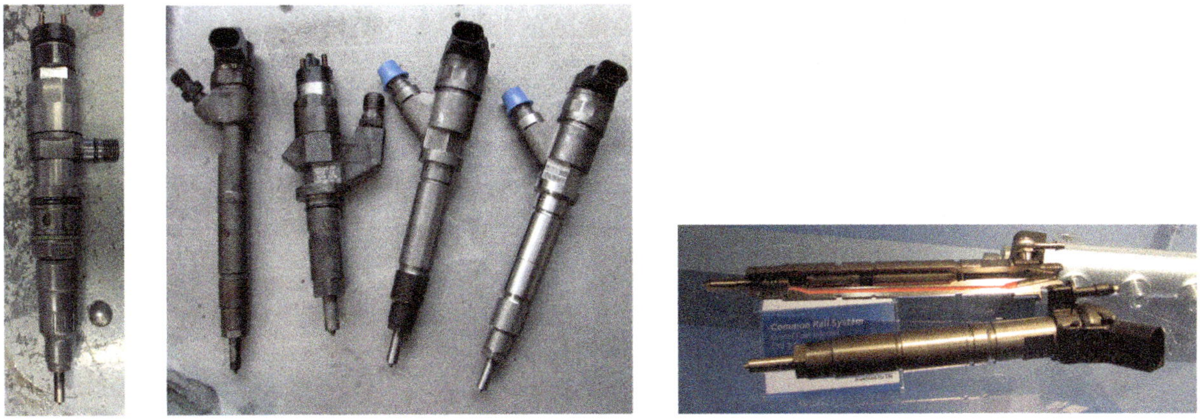

FIGURE 13-6 There are four generations of CR injectors, all of which incorporate technology for faster switching, better rate shaping, lower emissions, improved fuel economy, and power output.

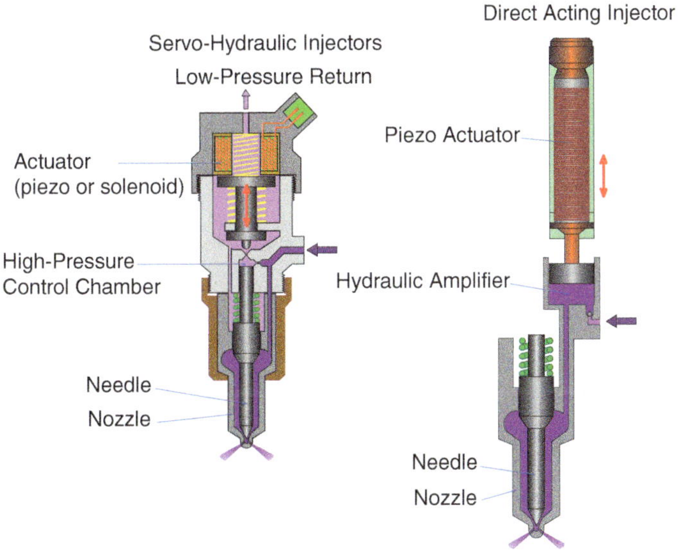

FIGURE 13-7 Using servo-hydraulic control valves multiplies the force of powerful hydraulic pressures in CR injectors.

highly pressurized fuel is controlled more quickly and precisely while using less energy than conventional electrohydraulic injectors (**FIGURE 13-8**).

In a sense, the servo action in an injector is similar to an electric relay, where small electrical currents control the switching of much larger electric current. In CR nozzles, however, it is not electrical current but hydraulic forces that are switched (**FIGURE 13-9**).

Early CR injectors could manage two or three injection events during one combustion cycle. Currently, the fastest production servo-hydraulic CR injector with a piezoelectric actuator can produce ten separate injection events during one combustion cycle, injecting fuel in less than 0.0001 seconds, which is more than six times faster than conventional magnetic solenoid valves. Injectors producing as many as 10 flexibly timed injection events during a combustion cycle can include a combination of up to five pilot-injection, or pre–main injection, events; several main injection; and four post-combustion injections (**FIGURE 13-10**).

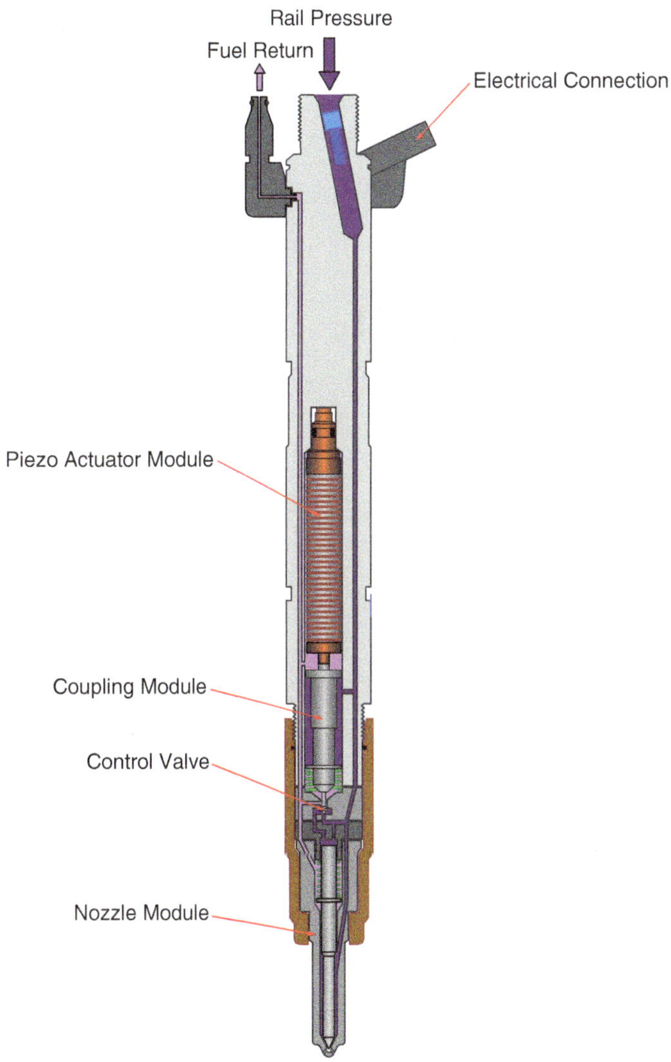

FIGURE 13-8 Electromagnetic actuators take much longer to respond to electric current than do piezoelectric actuators, which have almost an instant response time.

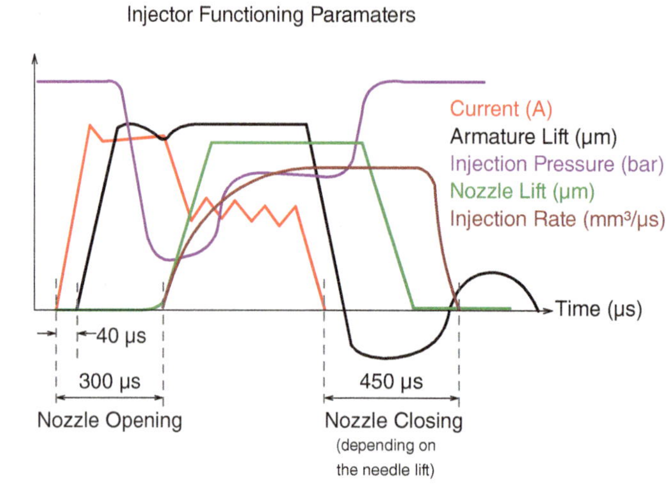

FIGURE 13-9 The injection event and energizing current are not synchronized, due primarily to the mechanical inertia of injector components and slow buildup of magnetic fields.

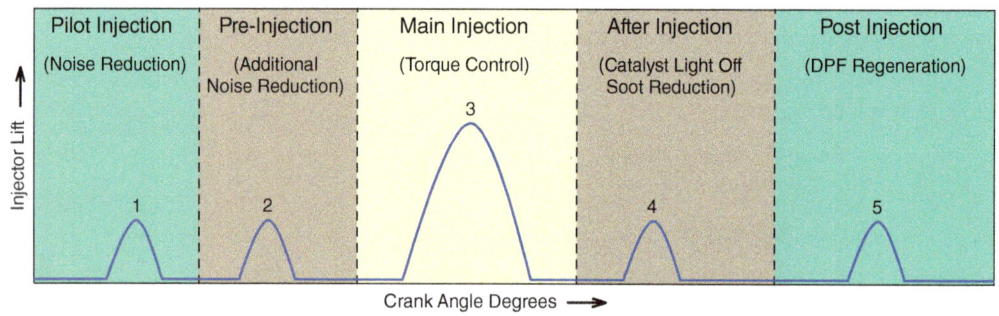

FIGURE 13-10 CR systems are capable of multiple injection events during one combustion cycle. This graph shows five different injection events during one combustion cycle and the function of each injection event for an early CR system.

The second vital feature of CR injectors, enabling multiple injection events with very rapid opening and closing of the injector nozzles, is hydraulically balanced nozzle valves (**FIGURE 13-11**). Older injector technology opened and closed the injector nozzle by forcing the nozzle from its seat with high-pressure fuel and then closing it with a highly tensioned spring. In CR injectors, the nozzle spring is much smaller and weaker (**FIGURE 13-12**). Not only will high-pressure fuel force the valve from its seat, but highly pressurized fuel will also rapidly close the nozzle valve by acting on the surface area or linkage above the nozzle valve. When the nozzle is in a resting or closed state, the force of fuel pressure against the area above the nozzle valve is only slightly higher than pressure below the valve. The pressure differential between the area above and below the nozzle valve keeps the valve seated and closed. Changing the balance of hydraulic pressure acting above and below the nozzle valve requires only a small adjustment in fuel pressure to produce very fast nozzle movement. This pressure change or alteration of the hydraulic pressure tipping point classifies CR nozzles as hydraulically balanced. Combined with electrically operated solenoids or control valves inside the injectors, hydraulically balanced nozzle valves produce ultra-rapid injection events that enable multiple injection events in a single combustion cycle.

Hydraulically balanced nozzle valves, also called hydraulically suspended nozzle valves, use the small mechanical force of the injector control valve to trigger more powerful, faster-acting hydraulic forces. Servo-hydraulic injectors always use hydraulically balanced nozzle valves that are opened and closed by shifting the pressure differential acting above and below the nozzle valve.

Injection Rate Shaping

Arguably the greatest advantage that CR injection offers is injection **rate shaping capabilities**. Essentially, rate shaping refers to the shape of the injection discharge curve when injection quantity is graphed against crankshaft rotation (**FIGURE 13-13**). A unique

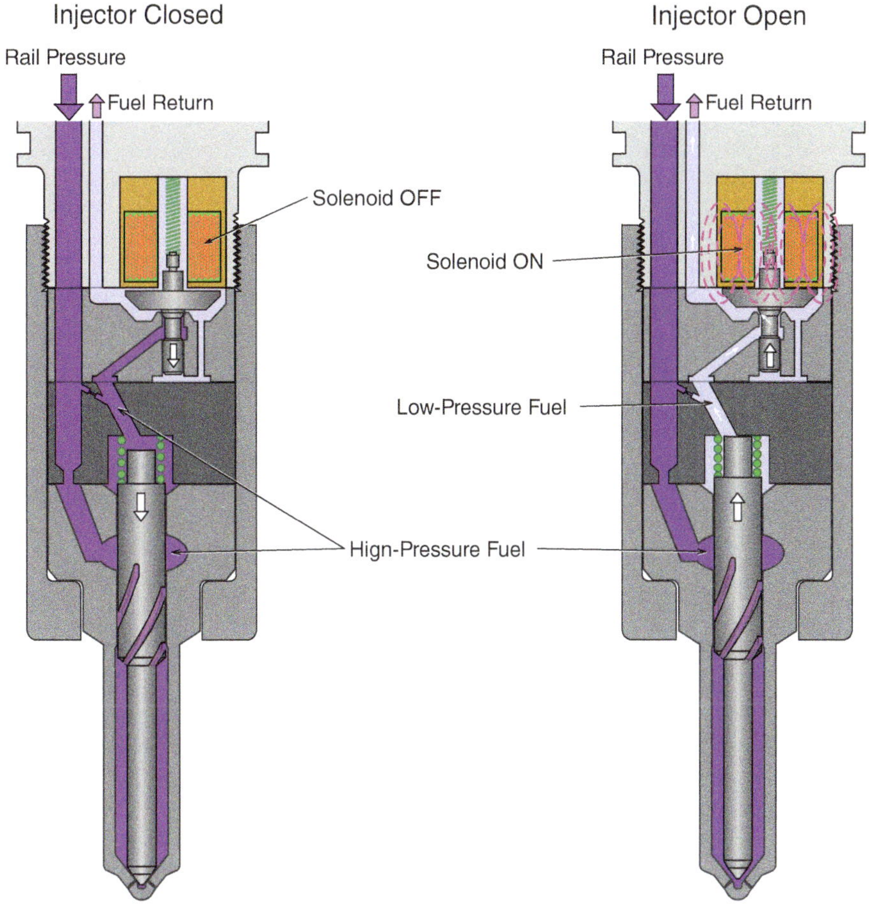

FIGURE 13-11 By changing the hydraulic forces acting below and above the nozzle valve, the nozzle valve can be rapidly lifted or seated.

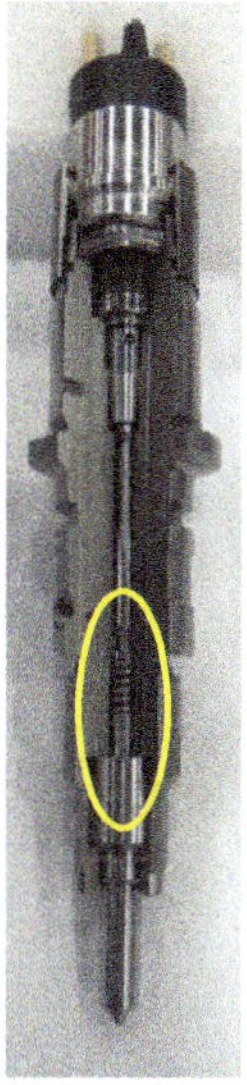

FIGURE 13-12 The nozzle spring is much smaller in a CR injector because hydraulic force, rather than the nozzle-valve spring force, does most of the work to close and seat the nozzle-valve.

shape is formed by the injection discharge curve when it best matches a particular engine speed and load condition. Rate shaping is the CR feature that provides a customized change in cylinder pressures and temperatures to achieve the best possible combination of maximum power output, performance, fuel economy, and low emissions for any given quantity of fuel or engine operating demand. When using proper injection rate–shaping techniques, substantial reductions in exhaust emissions and fuel consumption are achievable without significantly impeding performance (**FIGURE 13-14**).

Injection rate, which is an essential element of rate shaping, refers to the quantity of fuel injected per degree of crankshaft rotation. The term "injection rate" is different from the term "fuel rate," where the latter denotes the quantity or volume of fuel that an engine consumes per hour. Rate shaping takes the concept of injection rate further by describing the capacity of the injection system to change the injection rate throughout the combustion cycle. Injection rate, when plotted against crankshaft rotation, reveals an injection discharge curve that is modified by varying injection pressure and the number, timing, and duration of injection events. In injectors that use direct-acting piezoelectric actuators, varying the voltage is another variable that the ECM can use to alter the amount of nozzle-valve lift, which can vary injection rate as well. The optimal injection rate shape depends on the speed at which and load under which the engine operates. For example, at idle, to minimize noise, the duration of the injection event is longer with more injection events, and it takes place at lower injection pressures to minimize

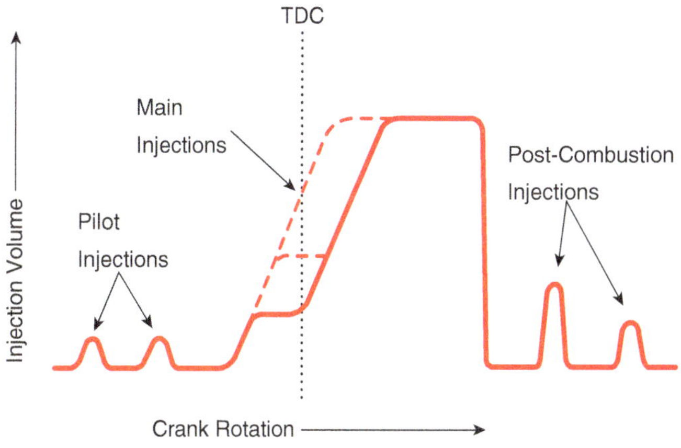

FIGURE 13-13 Rate shaping refers to the shape of the injection event when crank angle and injection volume are plotted on a graph.

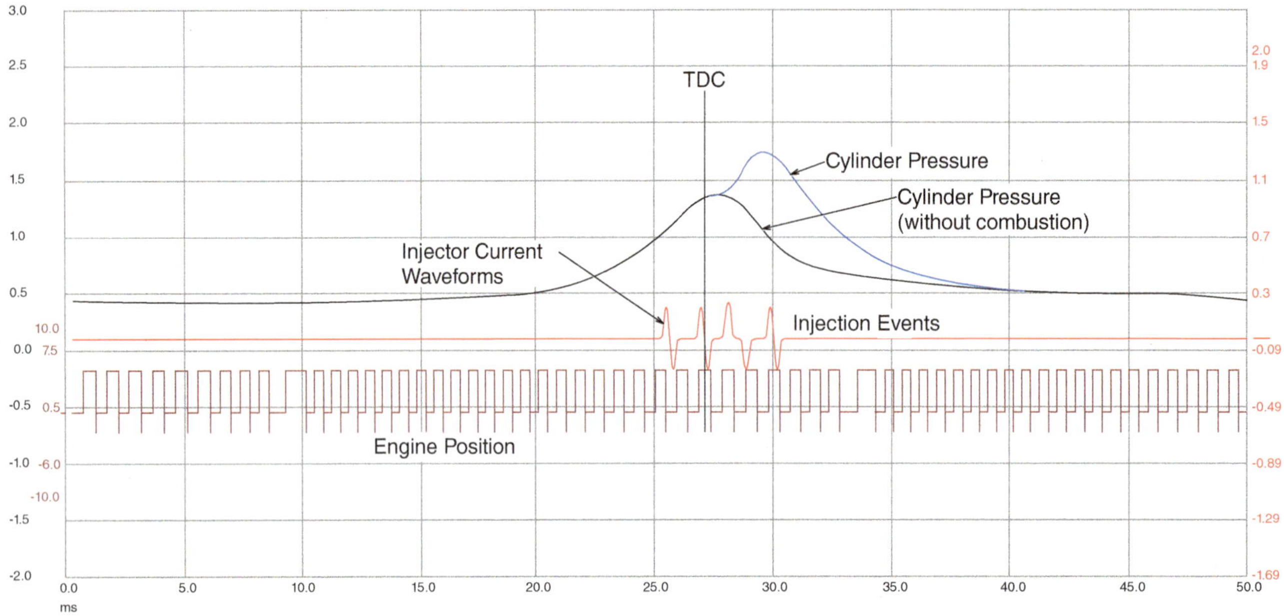

FIGURE 13-14 A graph of the injection events and cylinder pressure of a Volkswagen (VW) 2.0 L with third-generation Bosch piezoelectric injectors.

a rapid increase in cylinder pressure, causing combustion noise. The opposite is true at high speed and under heavy loads, such as when accelerating the engine when shorter injection duration at higher pressure injects all of the fuel in a single brief, high-pressure event. A shorter injection duration with fewer injection events produces a faster rise in cylinder pressure resulting a longer and harder push against the piston (**FIGURE 13-15**). The longer, harder push uses fuel more efficiently in order to convert fuel into mechanical energy most efficiently. A greater number of injection events spread over a longer period of crankshaft rotation typically reduces in-cylinder NO_x formation.

A basic injection rate shape exists that optimizes one of the essential elements of combustion: power, fuel economy, or low emissions. To reduce in-cylinder NO_x formation, a ramped rate, with smaller amounts of fuel injected near top dead center (TDC),

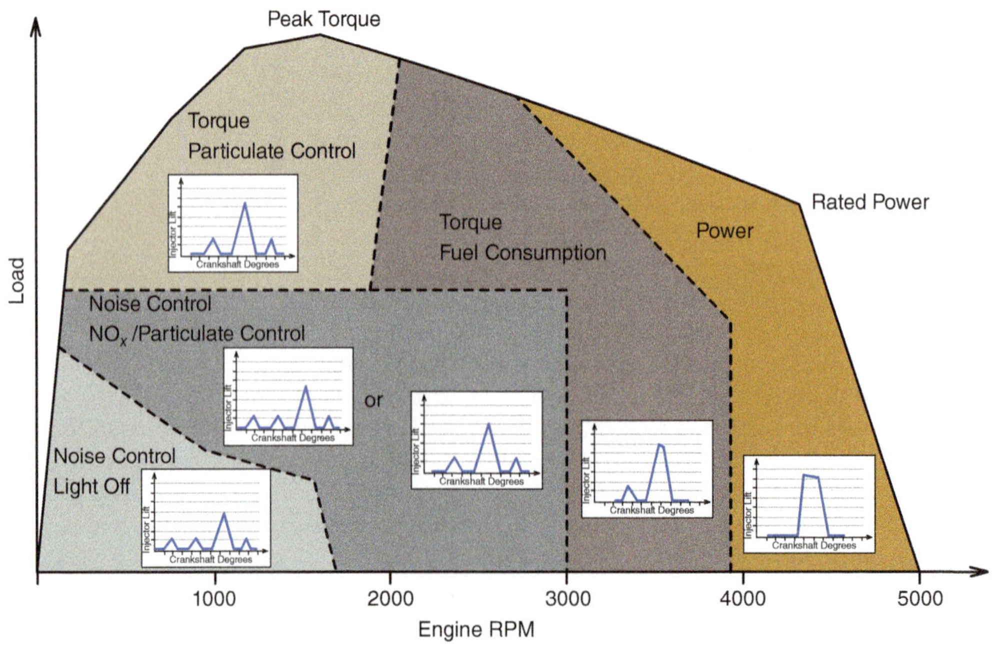

FIGURE 13-15 A unique rate shape and number of injection events can be optimized to improve fuel economy, lower emissions, and increase power output.

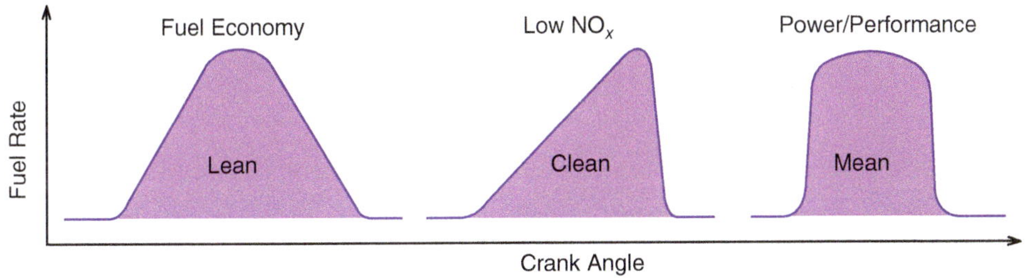

FIGURE 13-16 A triangular, ramped, and squared injection rate corresponds to good fuel economy, low NO_x, and maximum power output.

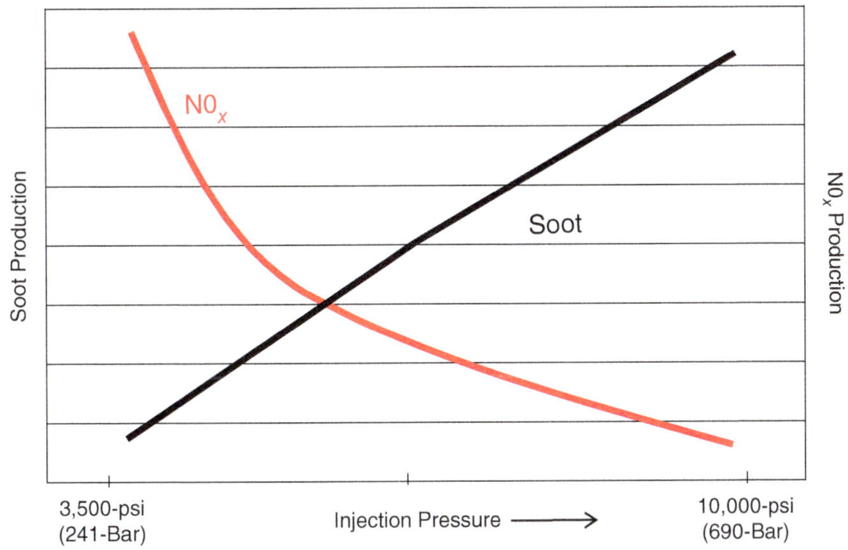

FIGURE 13-17 The injection variable for rail pressure has significant influence on the NO_x–PM emission tradeoff.

and increases the rate as the crank turns further. A square injection rate, which injects fuel quickly at high volumes near TDC, provides more power from the same amount of fuel. Some combustion efficiency is lost with a ramped and square rate. So, a triangular rate can also be produced which provides the best fuel economy during low-torque demand but trades this off with increased NO_x emissions (**FIGURE 13-16**). When the rates are blended during engine operation, an acceptable compromise can be reached between all three combustion elements: fuel economy, emissions, and power output. Determining the optimal injection rate shape is performed on most engines by using model predictive control (MPC). Using several fast-responding sensors, such as exhaust gas sensors and the throttle position sensor, MPC can accurately predict engine load and speed, which then uses an algortihm to generate a virtual lookup table for controlling injector rate shapes. Variables such as rail pressure, the timing, and the number of and time between multiple injection events are generated by MPC algorithms (**FIGURE 13-17**).

Injection Rate and Fault Healing

One of the more interesting applications of MPC injection rate shaping is its use to "heal" emission-related faults. Fault healing refers to an engine operating strategy that compensates for emission-related faults. Fault healing or fault compensation can take place when the onboard diagnostic (OBD) system has identified an emission-related fault. In potentially emission-increasing conditions, the injection rate can be adjusted by the MPC to help mitigate or reduce the probability of an increase in engine emissions. For example, if an exhaust gas recirculation (EGR) valve is malfunctioning, more frequent injection events, each using less fuel and each distributed over a greater number of degrees of crankshaft rotation can bring NO_x emissions to very low levels without the use of EGR gas (**FIGURE 13-18** and **FIGURE 13-19**). MPC will make an adjustment to the engine control algorithm to model

> **▶ TECHNICIAN TIP**
>
> The use of fault-healing injection strategies can help keep exhaust emissions below On-Board Diagnostics II (OBD-II) detection thresholds. The strategy can keep the MIL light off and potentially reduce an OEM's warranty costs, but customers face the penalty of increased fuel consumption and shortened component life from increases stress on other engine parts and systems. When investigating customer's reports of problems regarding fuel economy in the absence of a MIL or diagnostic trouble codes, do not readily dismiss the complaints. Pay close attention to data streams from engine sensors for values that are outside of normal but below fault thresholds.

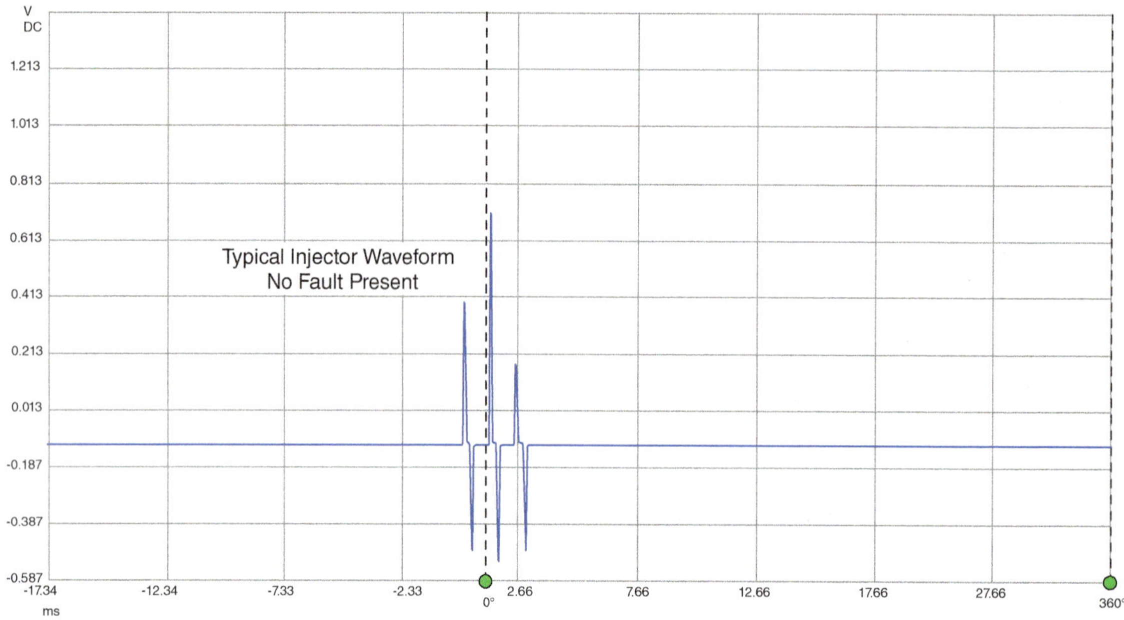

FIGURE 13-18 A typical injector waveform when no abnormal operating conditions are present.

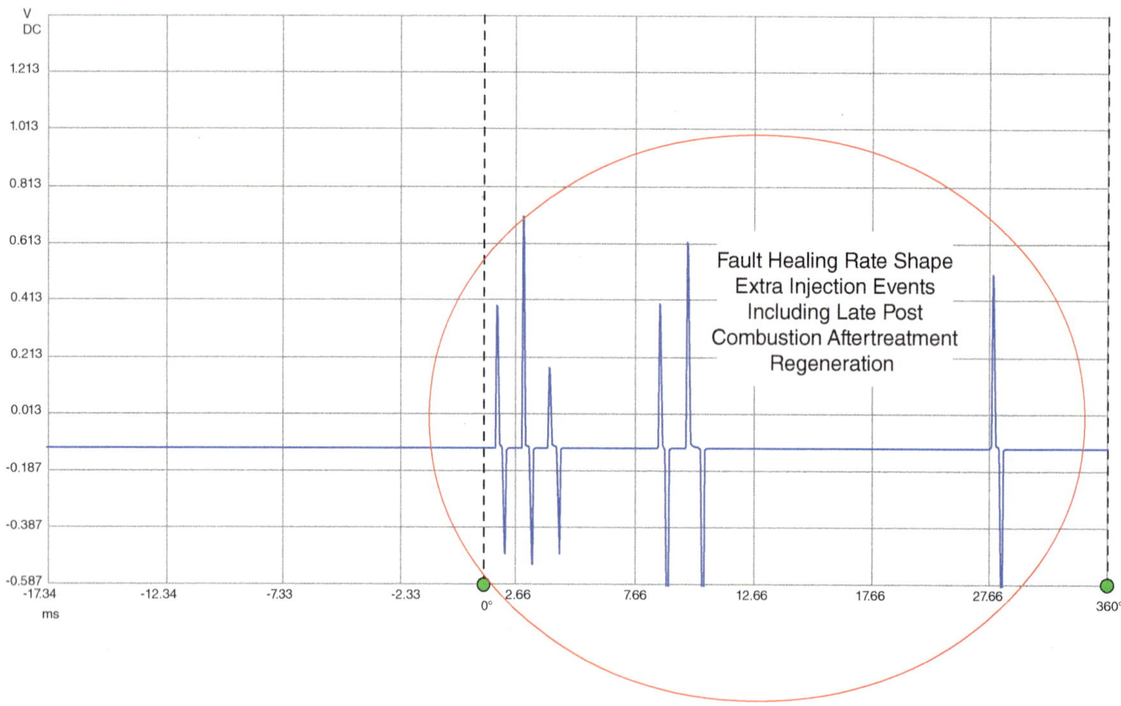

FIGURE 13-19 A CR fault-healing type of injector waveform observed when a potential emission-increasing fault is present.

what adjustments to other output variables need to be made if the EGR valve is held open or closed. Changes in other output variables, such as those that are intended to keep emissions low, can be made to compensate for the fault. The malfunction indicator lamp (MIL) light may remain off or on when this correction takes place. The downside of fault healing is a disruption in the balance between fuel economy, power output, and low emissions. Although NO_x-forming cylinder pressures and temperatures are lowered, combustion is less efficient, and more fuel is burned. To lower soot accumulations in the exhaust particulate filter, when the potential conditions exist for increased particulate matter (PM) emissions are likely, more frequent post-combustion injection events can heat the exhaust catalyst. Supplementally heating the exhaust system will help reduce soot loading in the particulate filter.

SAFETY TIP

As little as 30 psi (2 bar) of fuel pressure is all the force required to penetrate skin. This is one reason why a technician should *never* crack open an injector line on a CR system while the engine is running. This is a common practice in pump-line-nozzle (PLN) injection systems to detect misfiring cylinders. However, because the fuel pressure rail is common to all cylinders in a CR system, it will not identify a misfire. Instead, opening a fuel line will drop fuel pressure to all injectors. These systems can develop pressures as higher than 30,000 psi (2,068 bar); fuel can stream out at this high pressure, which will easily cut and penetrate the skin, will cause serious bodily harm, and may even lead to blood poisoning. Appropriate personal protection equipment (PPE) should be worn to prevent injury. Special fittings are often used to slowly and safely bleed fuel pressure from the fuel system. When a technician is unfamiliar with procedures to service a CR system, they need to check the service manual and properly follow the necessary steps to lower system pressures to safe levels before servicing.

SAFETY TIP

Never attempt to repair high-pressure fuel leaks while an engine is in operation! Instead, gradually decrease fuel pressure by slowly loosening a covered line or a specialized line dedicated to bleeding pressure gradually from the system. These lines have a diagonal groove cut into the fitting threads to allow fuel to leak from the fitting before the line is disconnected. When checking for leaks on running engines, use a piece of cardboard, rather than a hand, to detect the origin of spraying fuel (**FIGURE 13-20**).

Modular System Design

CR injection systems have tremendous parts commonality that can be used across a wide number of engine models. Injectors, high-pressure pumps, fuel rails, pressure regulators, sensors, and programmable ECMs are interchangeable across a wide array of engine platforms. Essentially, a change in engine software and a few parts inside an injector is often all that's required to adapt a system to different engine configurations.

FIGURE 13-20 As little as 30 psi (2 bar) of fuel pressure is required to penetrate skin. Never crack open a fuel line on a running engine, because high-pressure fuel can easily penetrate skin, with lethal results.

Construction Features of a High-Pressure Common Rail

What is generally referred to today as a CR injection system should more precisely be referred to as a high-pressure accumulator rail system or high-pressure common rail (HPCR). The following lists the typical construction features of HPCR injection systems:

- A tube or rail stores fuel pressurized to match optimal injection pressure for any given engine speed or load condition. Fuel injectors are connected to the rail with individual lines, which means the injectors share fuel pressure supplied from the rail. The shared, or common, fuel rail lends the system its name: *common* rail.
- Electrical actuators incorporated into each fuel injector control the opening and closing of a nozzle valve for the metering, timing, and rate shaping of injection events.
- An engine-driven, high-pressure fuel pump pressurizes fuel-transfer pressure to injection pressure and supplies the high-pressure fuel rail. Because the rail can store enough fuel volume at injection pressure to supply several injection events, it is also called an accumulator rail.
- A microprocessor controlled fuel system operation (**FIGURE 13-21**). Control may all be incorporated into a single ECM or a combination of an ECM and a fuel-injection control module (FICM).

Electronically regulating the pressure in the high-pressure pump permits injection pressures to vary with speed and load conditions, with maximum pressures available from idle speed. Manufacturers designate their systems according to generations or increasing model numbers, but CR systems are otherwise primarily classified by the type and construction of the injector at the heart of the system.

Regardless of the manufacturer, current high-pressure CR systems share several elements (**FIGURE 13-22**):

- a high-pressure stage consisting of fuel pump with at least one electrically controlled pressure regulator
- a high-pressure fuel rail with rail pressure sensor and pressure-regulating valves
- Electrically controlled fuel injectors fuel injectors
- an electronic control system
- short, thic walled fuel lines fuel lines connecting a pressurized fuel rail to the injectors
- software that controls the fuel system.

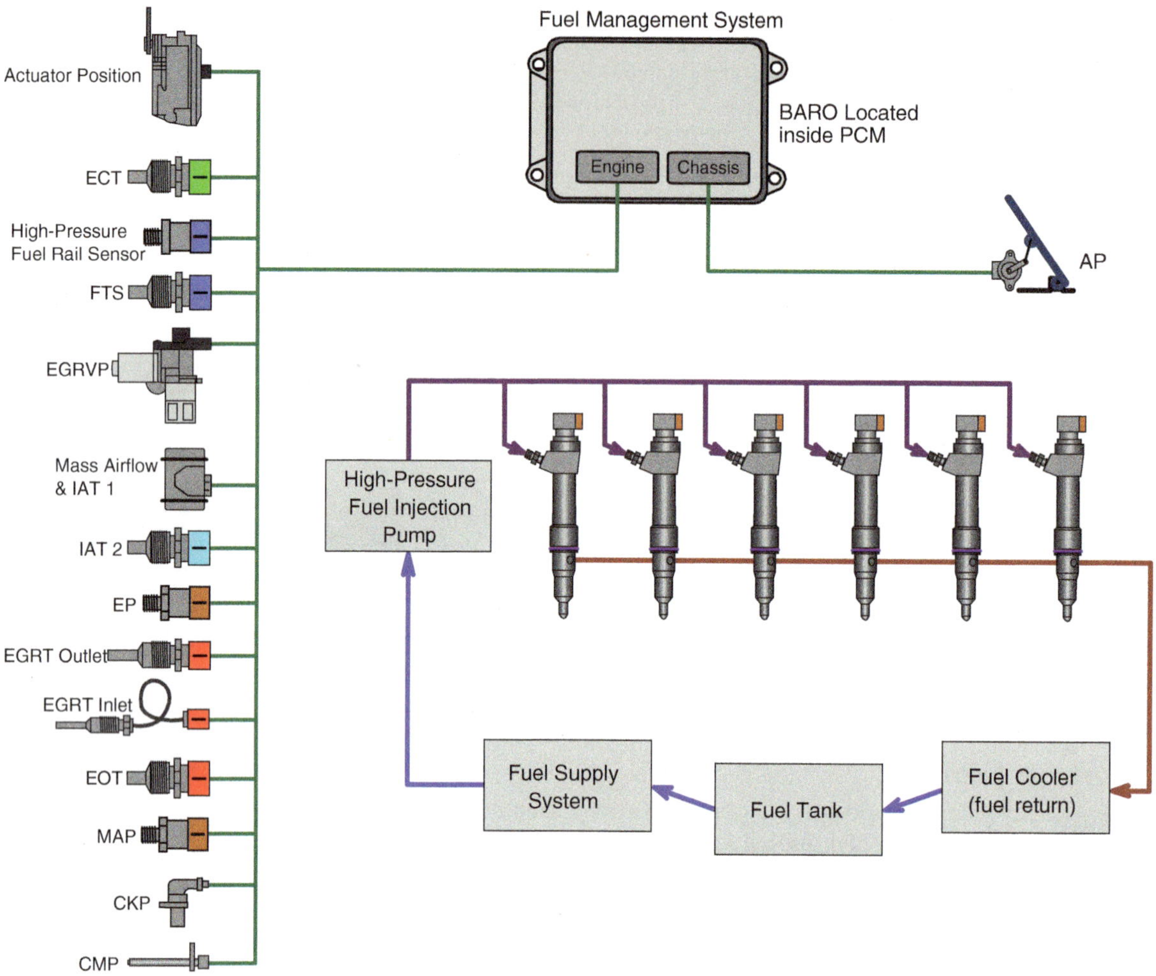

FIGURE 13-21 Microprocessor control and electronic inputs enable the beginning and ending of injection events and injection pressures to be varied.

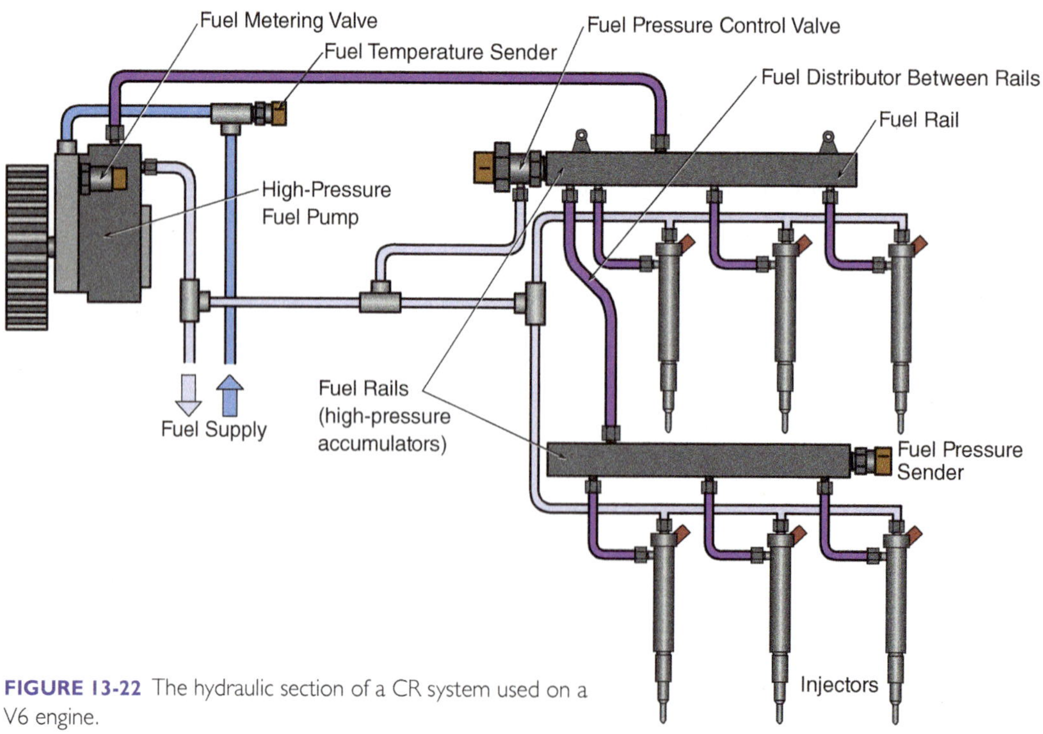

FIGURE 13-22 The hydraulic section of a CR system used on a V6 engine.

▶ Types of CR Fuel Systems

13-02 Identify and describe various types of CR fuel systems.

While the principles of CR systems are the same for most injection systems, applications can range from small automotive engines to large, heavy-duty engines in ships and locomotives. To suit an engine's power requirements, CR systems use a variety of high-pressure pumps with different pressure- and volume-regulating features. Injectors will use a variety of control valves and technologies, but they universally use hydraulically balanced nozzles valves operating on servo hydraulic principles and having different degrees of injection rate–shaping capabilities. This next section outlines the range of system designs used in today's engines.

Injector Requirements

All CR injectors are designed with the following common requirements:

- flexible injection pressure independent of engine speed
- very high injection pressure for optimal fuel atomization and distribution
- flexible timing of beginning and end of injection
- flexible timing of multiple injections
- small, consistent injection quantities for pilot-injection and post-injection sequences
- rapid nozzle-valve opening velocities to provide highest momentum to fuel spray for finest atomization
- a low fuel rate during ignition delay period to minimize combustion noise
- rapid increase in injection fuel rate after the start of combustion
- rapid decrease in fuel rate at the end of injection, accompanied by high nozzle-valve closing velocity to maintain the finest fuel atomization and minimize nozzle dribble
- compact design to permit a centralized location of the injector in the combustion chamber.

Types of Common Rail Injectors

There are three main types of CR injectors, with a variety of subcategories. With each new generation, there are improvements to hydraulic control, back leakage rates, increased injection pressure, greater timing precision, and an increase in the number of injection events that the injector is capable of producing in a combustion cycle (**FIGURE 13-23**).

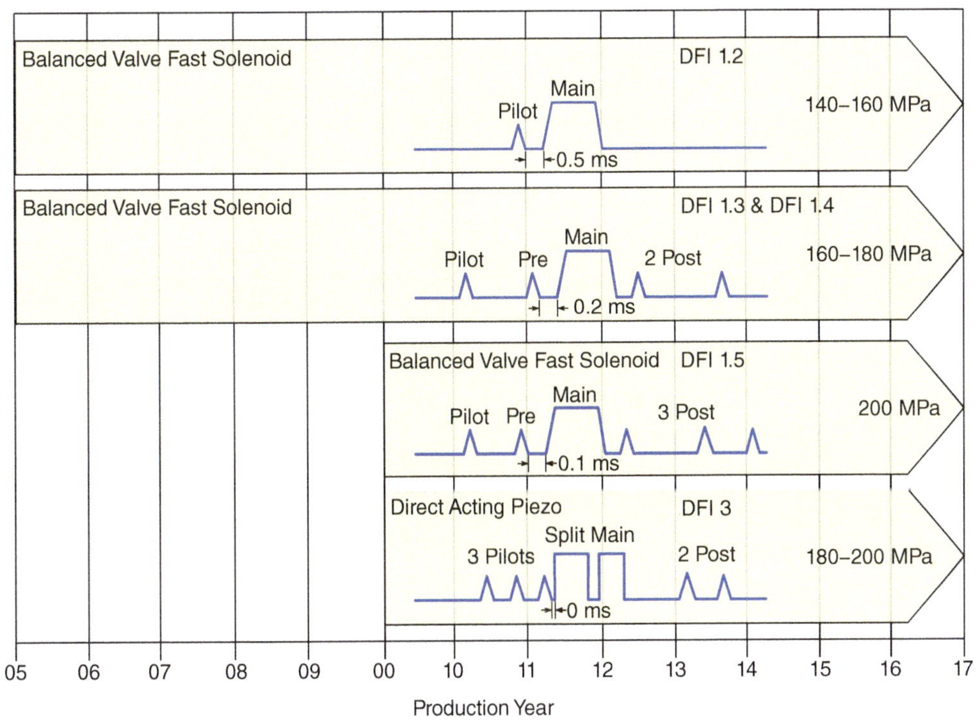

FIGURE 13-23 With each successive generation of CR systems, higher injection pressures and more injection events are possible.

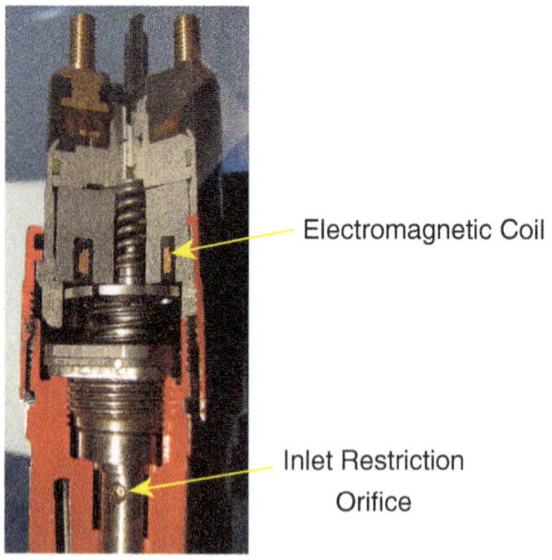

FIGURE 13-24 An electromagnetic actuator from a second-generation Bosch CR injector. Servo-hydraulic control of injection events is performed by using this early two-spring design capable of three injection events during a single combustion cycle.

Electrohydraulic Solenoids

CR injectors most commonly use solenoid valves relying on electromagnetic servo-hydraulic principles to control injection events (**FIGURE 13-24**). These injectors are manufactured by Bosch, Delphi, Denso, and Siemans. Although some older versions of the injector technology are falling into disuse as improved systems are introduced, electrohydraulic control of CR injectors is acceptable on slower speed, current medium- and heavy-duty diesel engines for a number of reasons, including cost and durability. Servo-hydraulic actuators with electromagnetic actuators are more durable than piezoceramic actuators, which have greater wear rates. The wear rates for both injectors will increase with the demand for more injection events during a combustion cycle. While servo-hydraulic mechanisms do wear, particularly ball valves in first- and second-generation injectors, higher production costs and a comparatively shorter lifecycle of piezoceramic or piezoelectric injectors have promoted the development and more widespread use of improved servo-hydraulic injectors. Denso's GS4 injector, used by the 6.6 L Duramax and 1.6 L General Motors (GM) whisper diesel, is an example of the newest servo-hydraulic type of injector that outmatches the speed and precision of piezoelectric actuators, which until recently were the fastest and most highly responsive injectors. The latest servo-hydraulic injector in Bosch's CRS2-25 system is capable of eight injection events and of operating with a rail pressure of 36,350 psi (2,500 bar).

Piezoceramic Electric Actuators

Instead of using electromagnetic fields to control injector operation, hundreds of thin piezoceramic wafers can form the actuator mechanism for faster control of injection events (**FIGURE 13-25** and **FIGURE 13-26**). Piezoceramic actuators consist of long stacks of overlapping wafers of manufactured mineral crystals. These materials will instantly change shape when electric current is applied (**FIGURE 13-27**). Evidence for the fast response time of piezoceramic actuators is readily observed in the construction of high-frequency speaker tweeters used in sound systems. Demonstrating the rapid response of these actuators to electric current, piezoceramic materials can move a speaker cone at frequencies up to 20,000 Hz when energized with electric current. Using piezoceramic materials rather than magnetic actuators avoids the problem of injection lag caused by slow buildup and collapse of magnetic fields and resistance inherent to wire coils caused by reactive inductance. Without needing to rely on magnetic fields, piezoceramic actuators can respond much more rapidly to current flow and produce as many as ten injection events during a combustion cycle using Bosch CRI3-27 injectors at 2,700 bar or 39,150 psi. The material also has a natural electrical polarity and responds with properties similar to a capacitor.

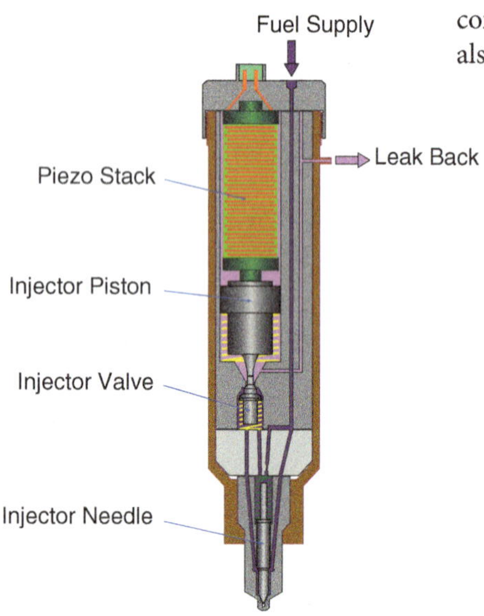

FIGURE 13-25 Energizing the stack of manufactured piezoceramic material in a piezoelectric CR actuator creates movement used to open and close injection control valves.

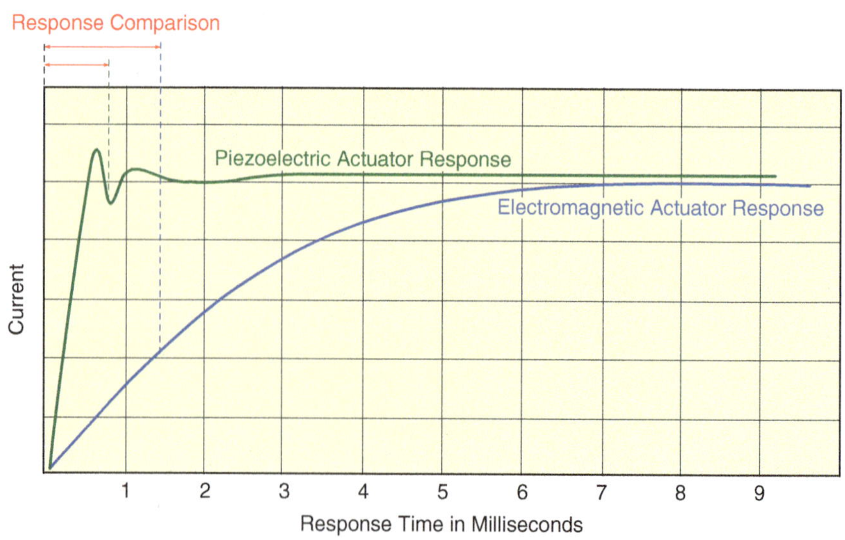

FIGURE 13-26 Electromagnetic actuators take much longer to respond to electric current than piezoelectric actuators do, which have almost an instant response time.

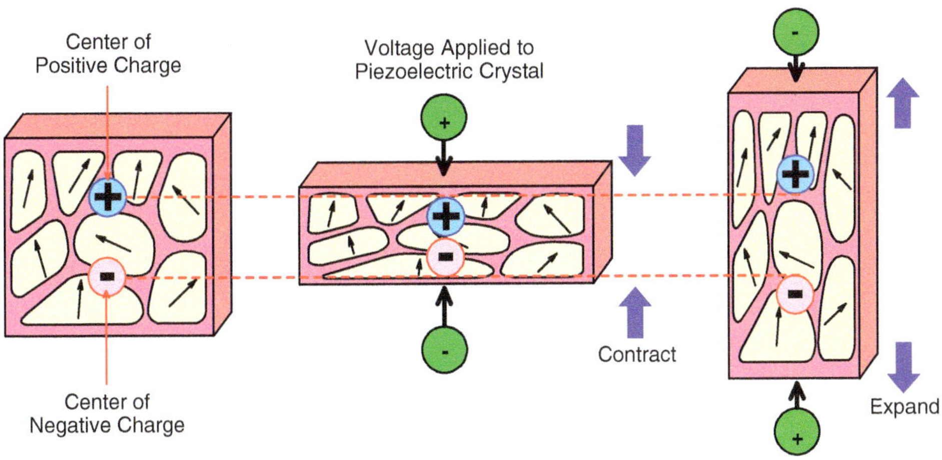

FIGURE 13-27 Current passing through piezoceramic material changes its shape. The actuator can be either compressed or expanded, depending on the direction of current flow.

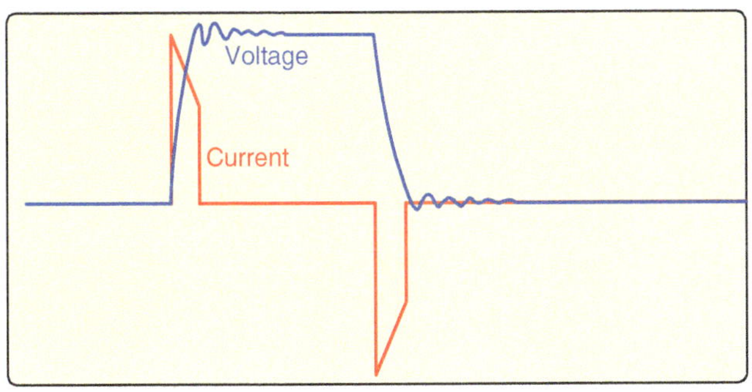

FIGURE 13-28 The direction of current flow through a stack of piezoelectric crystal wafers is reversed to obtain a rapid change in the stack dimensions.

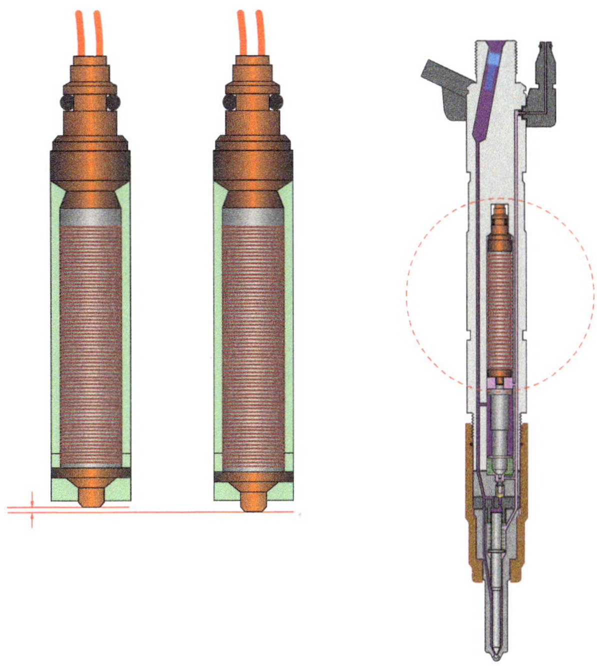

FIGURE 13-29 The number of stacked piezoceramic wafers will determine the final extension of the actuator used to open and close the injector control valve.

This means that switching the direction of electrical current applied to the stack will increase or decrease the stack length. Sending DC in one direction and then reversing the polarity causes instantaneous change in the injector's control valve position for even faster nozzle-valve opening and closing (**FIGURE 13-28**). With a faster response time, more injection events can take place during a combustion cycle with a shorter interval between injections than with traditional servo-hydraulic actuators (**FIGURE 13-29**). A typical contemporary injector actuator will arrange piezo wafers into about 350 layers. A 30 mm length of wafers will produce a stroke of 40 micrometers at a force of up to 2,500 newtons (562 pounds of force). That pressure is adequate to open a nozzle-valve operating at a pressure of up to 30,000 psi (2,000 bar). These injectors will then use a hydraulic coupling device that multiplies the actuator stroke travel connecting the piezo stack to the injector control or pilot valve (**FIGURE 13-30**).

While the faster response of a piezoceramic actuator seems ideal, the wafers in piezo stacks can wear as they slide across one another after repeated energization. More frequent energization events during a combustion event will wear away material from the wafers which shorten the injector's limited life cycle. Longer or taller stacks of the material containing more wafers are needed to extend the service life and force.

Direct-Acting Piezoceramic Injectors

Not only do piezoceramic injectors have a faster response time for shorter injection lag, but they can be used to more precisely shape the injection rate, as well. Variable nozzle lift is an additional special capability of this category of injector. By applying only a small amount of current to the piezo stack, it moves only slightly. Increased current will move the stack further. By varying the current, the crystals expand proportionally more or proportionally

> ▶ **TECHNICIAN TIP**
>
> Cylinder cutout testing is a common procedure used to identify misfiring cylinders. However, when working with piezoceramic injectors, never disconnect an injector while the engine is running. Piezo injectors can operate at up to 250 volts and 20 amps and store this charge in the injectors. If the injector is disconnected while the piezoelectric crystals are expanding, the stack will not be able to discharge to negative ground and the injector will remain in the expanded state. The result is a potentially damaged injector actuator.

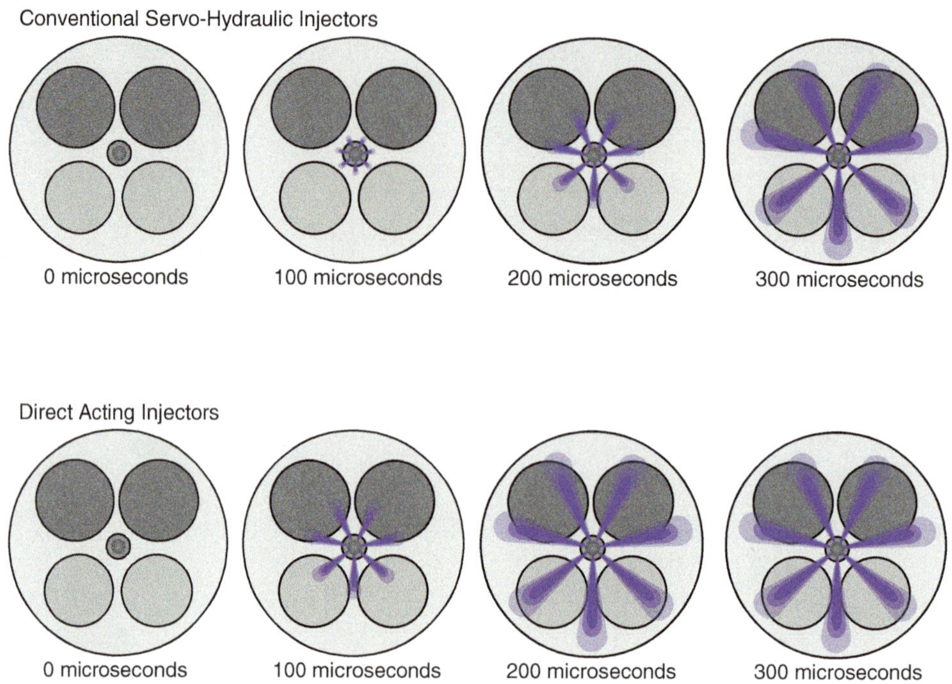

FIGURE 13-30 Until recently, piezoceramic injectors responded faster than servo-hydraulic valves that use magnetic actuators. A faster-responding actuator can supply more injection events during the combustion cycle.

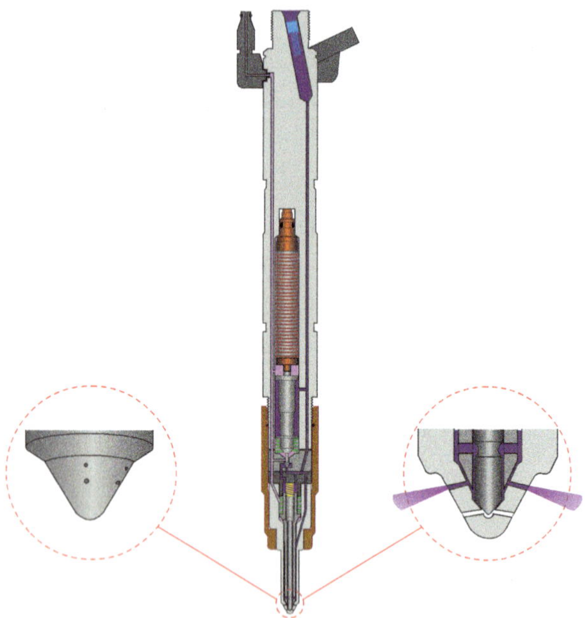

FIGURE 13-31 The fourth-generation injector is a direct-acting coaxial injector. Varying the voltage to a piezoelectric actuator changes the nozzle lift dimension and uncovers either one or two sets of spray holes.

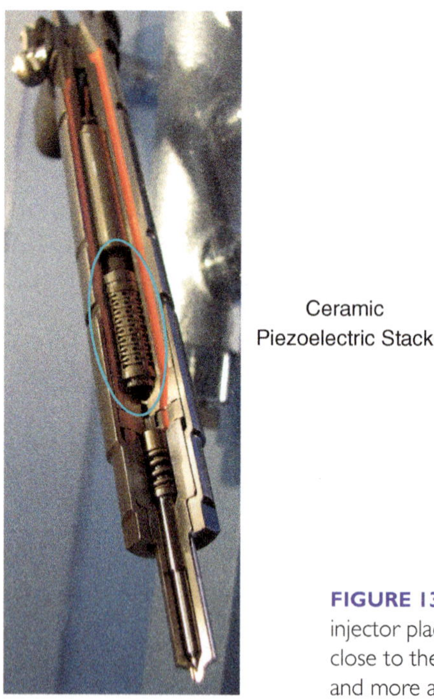

FIGURE 13-32 This direct-acting CR injector places a piezoelectric actuator close to the injector nozzle for a faster and more accurate injector response.

less, enabling the use of variable-lift nozzle valves (**FIGURE 13-31**). A smaller nozzle lift provides additional rate-shaping capabilities, such as a longer injection duration, which sprays fuel into a cylinder over a greater number of degrees of crankshaft rotation. The term given to this category of injector, which places the piezo actuator in-line and close to the nozzle, is "direct-acting CR injector" (**FIGURE 13-32**). Delphi's DFI 3 Piezotec series injector is one example of this injector. Delphi's concept of direct acting, where the piezoceramic actuator is directly coupled through a hydraulic amplifier piston to the nozzle valve, eliminates the servo-hydraulic circuit that other CR injectors use. Delphi claims the injector can spray fuel into a diesel engine combustion chamber faster, with greater accuracy, and at higher pressures (up to 2,000 bar) than other servo-hydraulic or piezoceramic injectors. The injector offers up to seven injection events per combustion cycle that when combined with proportional control over nozzle-valve lift by the ECM allow sophisticated rate shaping to be programmed into the ECM.

Hydraulically Amplified Common Rail Injectors

Introduced in 2009, hydraulically amplified CR injectors are the most recent CR innovation. Bosch calls this system the HADIS (**FIGURE 13-33**). These injectors are used by Detroit Diesel in the DD series of heavy-duty diesel engines that use the ACR fuel system. CRI4 injectors (Bosch's designation for to Detroit's ACR injectors) are currently capable of injection pressures as high as 37,500 psi (2,586 bar) by using an amplifier piston to multiply pressure inside the injector. This concept of pressure intensification is similar to hydraulically actuated electronic unit injection (HEUI) fuel systems, though one major exception is that the HEUI system uses high-pressure oil as the force to pressurize fuel for injection, where the CRI4 uses high-pressure fuel instead.

An amplifier piston inside the CRI4 injector multiplies fuel pressure by a factor of 2:1. One major advantage of using internal pressure amplification is lowered mechanical stress on fuel system components such as pumps, lines, and valves due to a reduction in fuel pressure outside the injector. Two-stage fuel pressure amplification inside this injector provides improved power output with lower emissions, less noise, and reduced fuel consumption. A unique injection rate shape is also within the capabilities of this system. More detailed coverage of Detroit's ACR system and the operation of this injector are found in the textbook *Fundamentals of Medium/Heavy Duty Diesel Engines*.

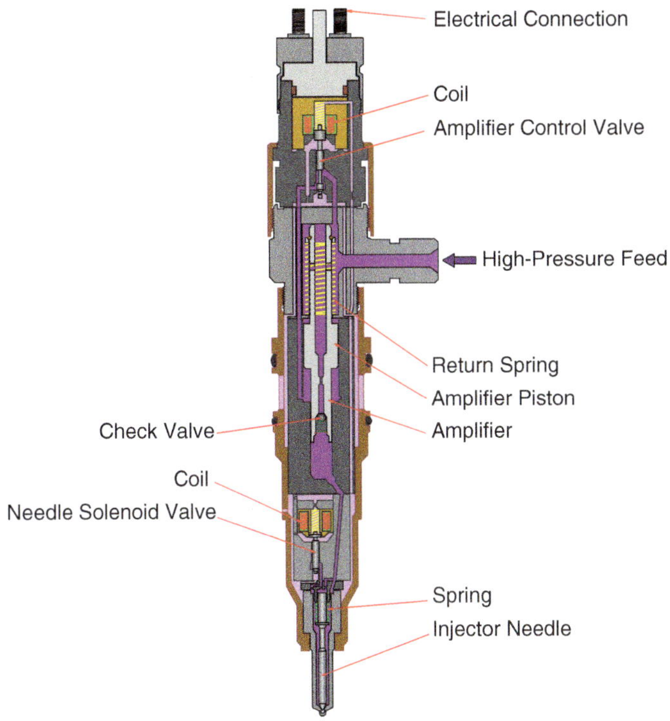

FIGURE 13-33 The injector details of the HADIS, which was developed by Bosch and used by Detroit Diesel's ACR system.

Bosch Common Rail Injectors

Bosch has developed four generations of CR injectors, three of which are currently in production vehicles. Used only in Europe, passenger vehicles with first-generation CRS1 systems had a maximum injection pressure of 19,580 psi (1,350 bar) and were capable of three injection events per combustion cycle (**FIGURE 13-34**). The CRS2, introduced in 2001 in

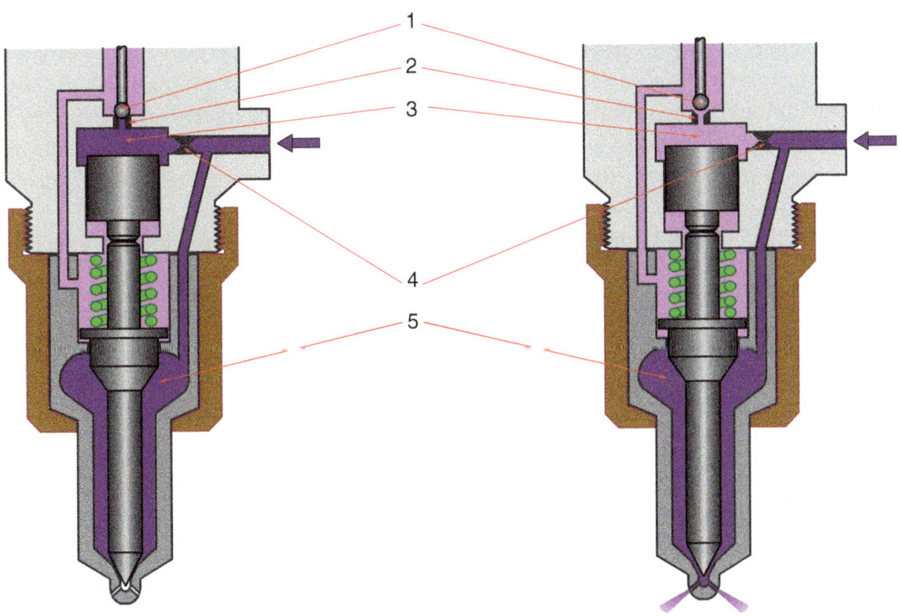

FIGURE 13-34 Operation of CR1 injector: Fuel under high pressure is simultaneously applied to the fuel annulus (5) below the nozzle valve and the high-pressure chamber (3) above the nozzle valve. A combination of a slightly larger surface area above the nozzle valve with nozzle spring pressures holds the nozzle valve on its seat. When the ball valve (1) is lifted from its seat, fuel drains to the low-pressure fuel return circuit through the drain orifice (2). Pressure then drops above the nozzle valve, enabling fuel pressure to lift the nozzle valve and begin injection. When the ball valve is seated, fuel enters the inlet restriction orifice (4) and refills the pressure chamber above the nozzle valve. Hydraulic force quickly seats the nozzle valve.

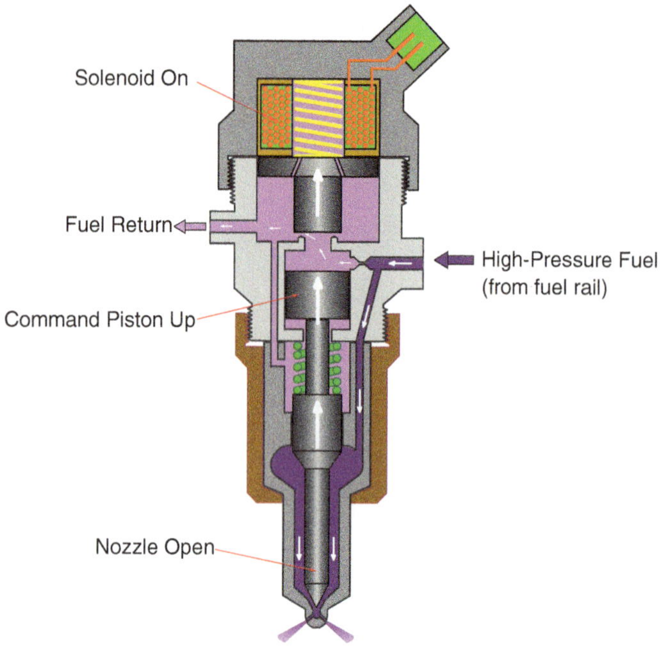

FIGURE 13-35 Internal details of a CRS2 injector. Note the two actuator springs, which balance out the control valve weight to provide a faster actuator movement.

North America, injects fuel at a maximum pressure of 23,210 psi (1,600 bar) and was capable of as many as six injection events per combustion cycle. The latest CRS2-27 system applications of this injector can inject with as many as eight combustion events. A revised actuator design allows the second-generation CRS2 to respond to electrical signals faster and with greater precision (**FIGURE 13-35**). The third-generation CRS3 replaced electromagnetic solenoids with rapid-switching piezoceramic actuators. Bosch claims that these injectors offer advantages of up to 20% lower emissions, 5% more power, 3% lower fuel consumption, and engine noise reduction of up to 3 decibels (dB) over early CRI2 injectors. The faster-acting CRI3 injector permits shorter firing intervals between injections and even more accurate fuel metering. Adaptive cylinder balance—the strategy used to adjust cylinder pressures by modifying shot-to-shot, cylinder-to-cylinder injection quantities—can be performed over all engine speeds, not just at idle when a powerful Euro 6 MDG ECM or the Bosch EDC17 ECM or a GM E86 ECM is used. In addition to the fourth-generation amplified CR injector, another variation on Bosch fourth-generation injectors is what's called a coaxial injector. This injector uses two horizontal rows of spray holes in the nozzle tip. Modulating the voltage to the direct-acting piezo actuator enables varying amounts of nozzle lift, which in turn changes which row of spray holes are used for injection. These nozzles are able to better match fuel distribution to operating conditions. No current engine application uses these injectors.

CRS1 and CRS2: Hydraulically Balanced Nozzle Valves

In Bosch's first- and second-generation CR systems, the injection events are controlled by using a magnetic solenoid inside the injectors to change the balance of hydraulic pressure acting on the nozzle needle valve. Conventional fuel systems use only high-pressure fuel to lift the nozzle valve and begin injection, but close with spring pressure. Bosch's CR injectors use high-pressure fuel to both lift and seat the nozzle valve. Using hydraulic pressure to both begin and end injection events gives CR injectors the ability to operate with the speed and precision they are known for. Hydraulically balanced nozzles, also called hydraulically suspended nozzles, use the small control force of the solenoid's magnetic field to tip the balance between powerful fuel rail hydraulic forces acting on the top and bottom of the injector.

Bosch's servo-hydraulic actuators multiply small electromagnetic actuator valve movement to control high-pressure injection pressures (**FIGURE 13-36**). Gen I and II injectors use electromagnetically controlled ball valves to open and close the control valve. Although engines with lower maximum speeds respond slower than later injector designs, they can use servo-hydraulic injectors because injector response time is not as critical when rated engine speeds are slower, as they are in automotive diesel engines, which require injectors that respond quickly due to their higher maximum engine rpm.

Bosch CR1 and CR2 injectors use a four-step servo-hydraulic operation to inject fuel:

1. The magnetic solenoid is actuated by an electrical signal from the ECM.
2. A servo-hydraulic valve opens.
3. Fuel pressure is released from the control chamber through a small drain, allowing it to flow to the low-pressure fuel return circuit.
4. Rail pressure lifts the nozzle valve.

Ending injection is similar:

1. The electrical signal is removed from the magnetic solenoid.
2. The servo-hydraulic valve closes.
3. The drain hole from the pressure control chamber closes.
4. The pressure chamber fills with fuel at rail pressure through a small restriction orifice.

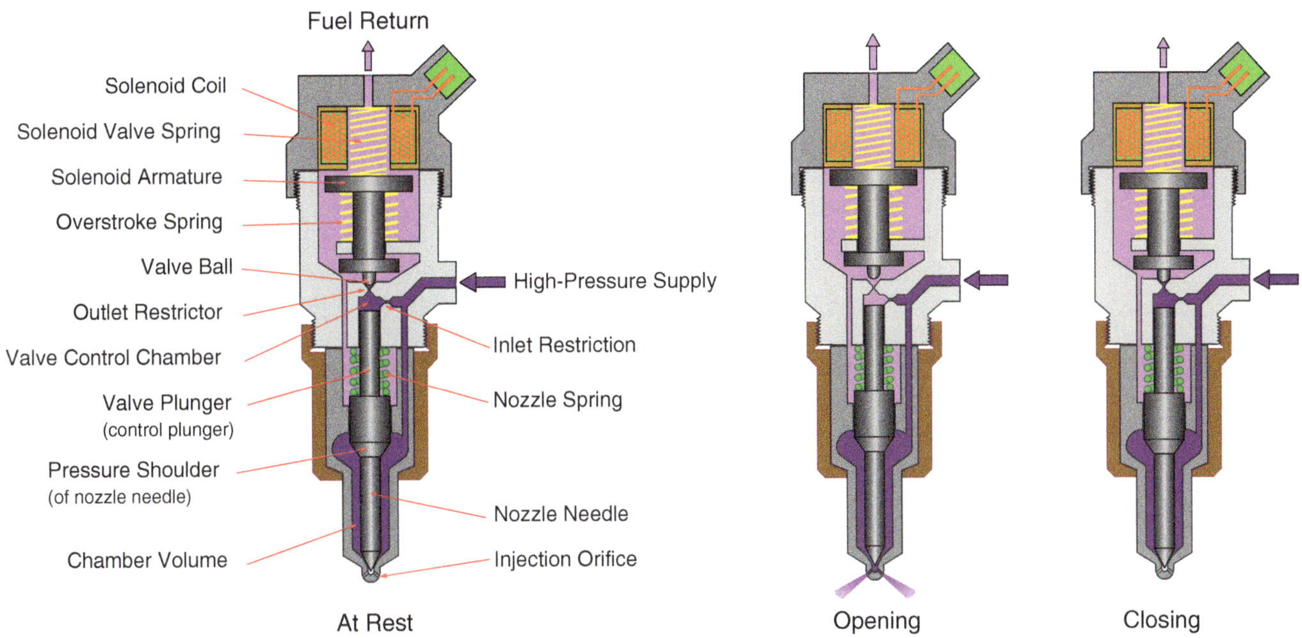

FIGURE 13-36 Stages of injection for a CRS2 injector.

Three elements of the injector are used to perform the injection functions:

1. A conventional multi-orifice spray nozzle
2. A servo-hydraulic system
3. An electromagnetic solenoid control valve.

CRS2 Operation

To complete an injection event, the nozzle valve is held on its seat by the control rod or control plunger. Denso, which also manufactures this injector, calls this component a command piston (**FIGURE 13-37**). Regulating the movement of the control rod is a column of high-pressure fuel above the control rod balanced against the high-pressure fuel acting against the nozzle valve below the rod. Injection can begin only once the hydraulic pressure above the control rod has been relieved. Fuel pressure acting below the nozzle valve will push the nozzle upward against the control rod, allowing fuel to leave the spray holes of the nozzle. Similarly, injection ends when hydraulic pressure above the control rod pushes the nozzle valve back onto its seat. Once again, the flow of high-pressure fuel allowed through the drain orifice caused by opening or closing the orifice changes the small pressure differential necessary to open and close the nozzle valve (**FIGURE 13-38**).

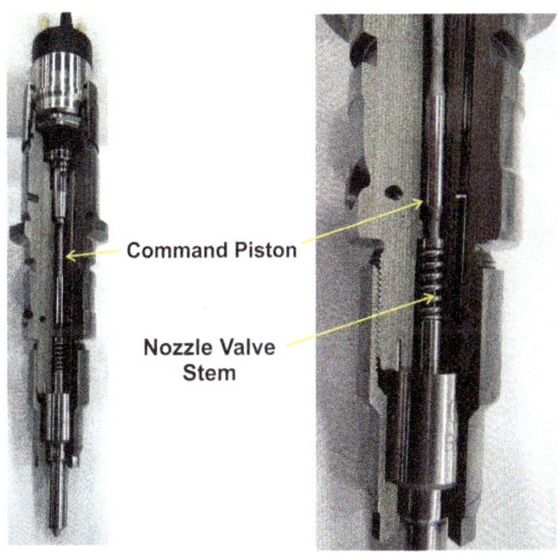

FIGURE 13-37 Bosch, Delphi, and Siemens injectors use a control rod to transmit hydraulic forces from the control valve to the nozzle valve.

Pre-injection

In a resting state with no current applied to the injector solenoid, the drain orifice is closed by spring force acting on the ball valve. Fuel pressure developed by the high-pressure pump builds in the pressure chamber above the control rod, holding the control rod against the nozzle valve. The force of fuel pressure against the control rod firmly seats the nozzle valve. Fuel pressure supplied to the injector by the fuel rail and lines simultaneously builds at the nozzle annulus, which is the circular chamber below the nozzle valve, but it is unable to lift the nozzle valve, due to the pressure applied by the control rod.

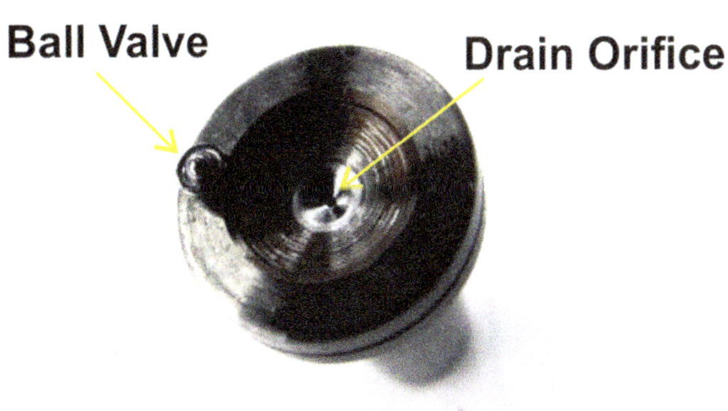

FIGURE 13-38 The ball valve and seat for a Bosch CRS2 injector. Opening and closing the drain orifice controls injection events.

Beginning of Injection

The injector solenoid is energized with current supplied by the engine ECM or a separate FICM. In Bosch injectors, this initial amount of current can be as much as 90 volts DC, but it is typically closer to 75 volts, with an average of 50 volts when the ball valve is initially lifted (**FIGURE 13-39**). Capacitors inside the ECM or FICM store the charge for injection events. The magnetic force lifts the spring-loaded armature and ball valve that are sealing the drain orifice. Fuel flows from the pressure chamber through the drain orifice and into the low-pressure fuel return circuit leading back to the fuel tank. Consequently, pressure above the control rod drops, and the pressure below the nozzle valve immediately forces the nozzle valve from its seat to begin injection. The fully lifted nozzle valve sprays fuel into the combustion chamber at a pressure equal to the fuel rail.

The pressure drop in the pressure chamber above the control rod is controlled by the difference in the diameters of the inlet and outlet orifices. The drain orifice will always be larger than the inlet orifice to open and close the nozzle valve. While the speed of the nozzle valve lift is determined by the difference of flow rates allowed by the drain and inlet orifice diameter, too large a difference will slow the end of injection. Likewise, too little a difference in orifice diameters will slow the beginning of injection. The nozzle differential ratio, which is the difference in the surface area that the nozzle's fuel acts on before the nozzle lifts and the wider cross-sectional area of the nozzle after it lifts, also controls nozzle-lifting and nozzle-closing pressures. The differential ratio ensures that the nozzle opening pressure is always higher than closing pressure.

End of Injection

When current to the solenoid is shut off, the over-stroke spring inside the armature forces the check ball over the outlet orifice to close it. Closing the drain orifice causes fuel pressure to build in the pressure chamber, forcing the control rod against the nozzle valve. This effectively and rapidly ends injection because the nozzle valve is forced back against its seat by using high fuel pressure. The nozzle-valve closing speed is determined by the flow through the inlet orifice.

Comparing CRS1 and CRS2 Operation

The difference between the CRS1 and CRS2 injectors is in the armature. The CRS1, which had limited use in Europe only, uses a single spring to return the armature to a resting

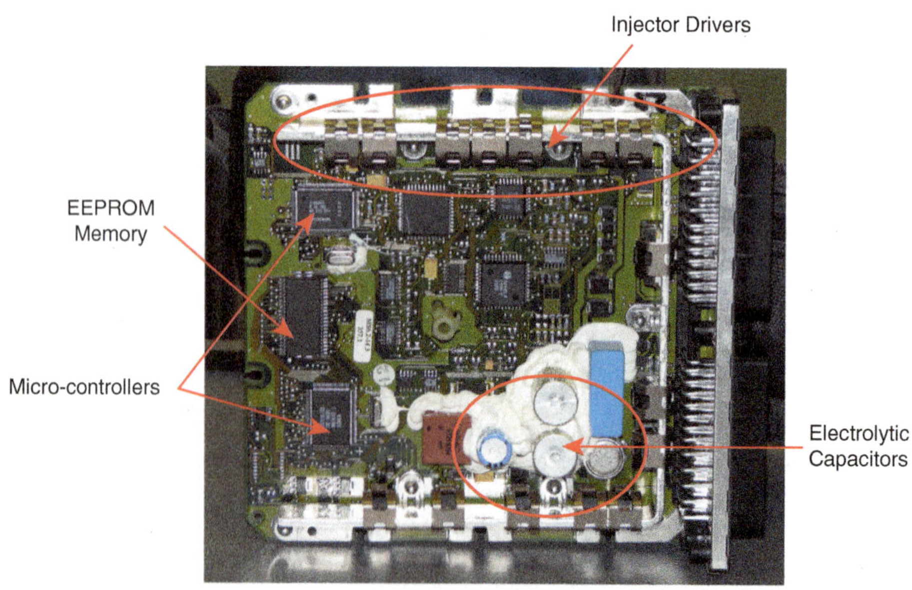

FIGURE 13-39 The FICM for 6.6 L DMAX diesel CRS2 injectors. The module controls injector events and steps up battery voltage to a level high enough to energize injectors.

position, whereas the CRS2 uses two springs. The mass of the armature with the two-spring arrangement is reduced, and the over-springing effect from using just one spring is eliminated. The balanced spring tension of the CRS2 injectors can return to resting position more rapidly, thereby ending injection more quickly, which allows shorter intervals between injection events.

▶ TECHNICIAN TIP

Corrosion, wear, cavitation erosion, or microscopic dirt particles in the ball seat of the drain/bleed orifice of CR injectors can cause major problems. Fuel leaking continuously around the ball valve from the high-pressure control chamber to the low-pressure return can prevent the needle valve from seating properly. Depending on the severity of the problem, this condition could cause several additional problems:

- The injector could continuously spray, causing the engine to run away.
- The injector could misfire.
- The injector fuel and fuel rail could drain, which can fill the cylinder with fuel during shutdown and cause a hydraulic lock on start-up.
- A fault code could be generated, indicating an undefined leak in the fuel rail.
- To identify a leaking ball valve or faulty injector, many manufactures recommend measuring the quantity of fuel retuned from the injectors. Too much leakage from a particular injector generally indicates a ball/drain seat defect. Dirt and contaminants in the inlet orifice will cause identical problems (**FIGURE 13-40**).

CRS3 Piezo Injectors

Piezo technology is incorporated into many newer CR injectors. Piezo technology involves replacing magnetic solenoids with special piezoceramic material that has the ability to change shape under the influence of electric current. Because no magnetic fields are induced, no **inductive resistance** or reactance (the induced current that opposes ECM current when initially energized) is produced when using piezoelectric actuators. Piezoelectric actuators therefore provide a much faster injector response to electrical signals than do magnetic actuators. These are the same piezoceramic materials that can also generate electrical current when squeezed. Spark igniters used in gas barbeques are another common application for these piezo materials.

When piezoceramics are used in injectors, an electric current is passed through a **piezoceramic actuator** made of multiple layers of ceramic discs. In Bosch CRS3 injectors, 350 stacked layers of piezo crystals are overlapped on top of one another. When initially energized with up to 250 volts, the expansion of each disc is summed and expands by up to 40 microns (40 μ). Movement of the stacked discs is harnessed to change the balance of hydraulic forces acting on the injector's nozzle valve. This small dimension change is inadequate to open and close a control valve. A hydraulic coupler is placed in series

▶ TECHNICIAN TIP

Even the tiniest particles of dirt can prevent injector fittings from properly sealing. This is especially true with quill tube connections. It is recommended to always clean connecting surfaces by using electrical contact cleaner, before joining high-pressure fittings and lines (**FIGURE 13-41**).

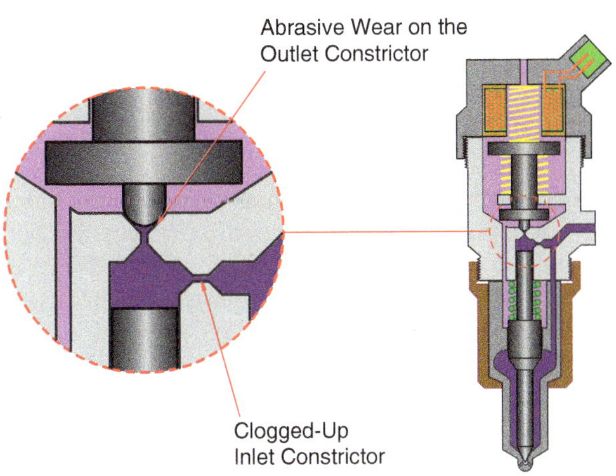

FIGURE 13-40 Ball valve wear and seat wear cause operating problems, rough running, or injector misfires.

FIGURE 13-41 Electrical contact cleaner is a recommended solvent and cleaner for all high-pressure fittings in CR fuel systems.

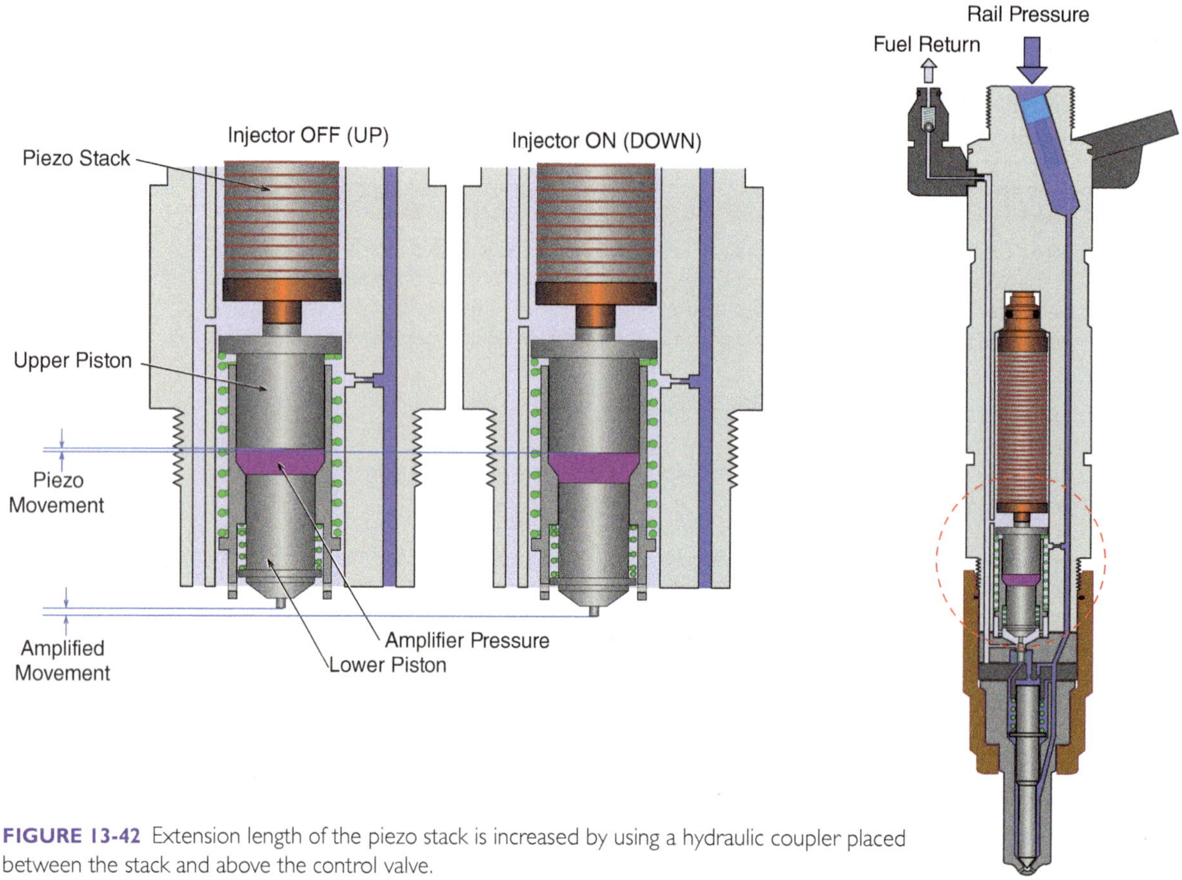

FIGURE 13-42 Extension length of the piezo stack is increased by using a hydraulic coupler placed between the stack and above the control valve.

with the stack and the pilot control valve to increase the length of the stack expansion (**FIGURE 13-42**). The coupler is filled with fuel that requires a minimum of 58 psi (4 bar) of pressure inside the injector, which is maintained by a return side pressure-regulating valve called a constant pressure–regulating valve (**FIGURE 13-43** and **FIGURE 13-44**). Because the hydraulic coupler uses two different-sized surface areas to transmit and multiply the small amount of piezo stack expansion, the length of the stack extension is increased. Piezo technology enables the fastest switching time for injectors; it currently allows up to ten injection events in one injection sequence. In fact, piezoceramic actuators can switch fuel on and off in as little as 0.0001 seconds (100 microseconds). In comparison, an electromagnet's response time can be as long as 15,000 microseconds (15 milliseconds). The shorter interval between multiple injection events enables very precise rate shaping of the fuel discharge curve and even more exact control over metering. Better rate-shaping capabilities result in quieter, more fuel efficient, cleaner, and more powerful engine operation. Using piezoceramics further reduces the mass of the servo-hydraulic mechanism. In fact, some injectors can be made without return springs. Bosch claims a 75% mass reduction in the CRS3's piezoceramic actuator. The CRS3 injector weighs 9.5 ounces (270 grams), as compared to the servo-hydraulic injector's 17.3 ounces (490 grams).

A slice of piezoceramic material that is 0.02" (508 μ) thick expands only about 0.00002" (0.5 μ) when energized with approximately 140 volts of electricity. By putting together hundreds of the wafers, the movement is summed together, adding up to a few thousandths of an inch of movement. Third-generation Bosch injectors use as much as 250 volts to initially open and require electrical lineworker gloves as a safety precaution when working on some systems (**FIGURE 13-45**). The ECM supplies high voltage and provides a ground. On Duramax systems, injector voltage is supplied up to 160 volts at 20 amps to open the injector. A pair of capacitors inside the ECM or FICM

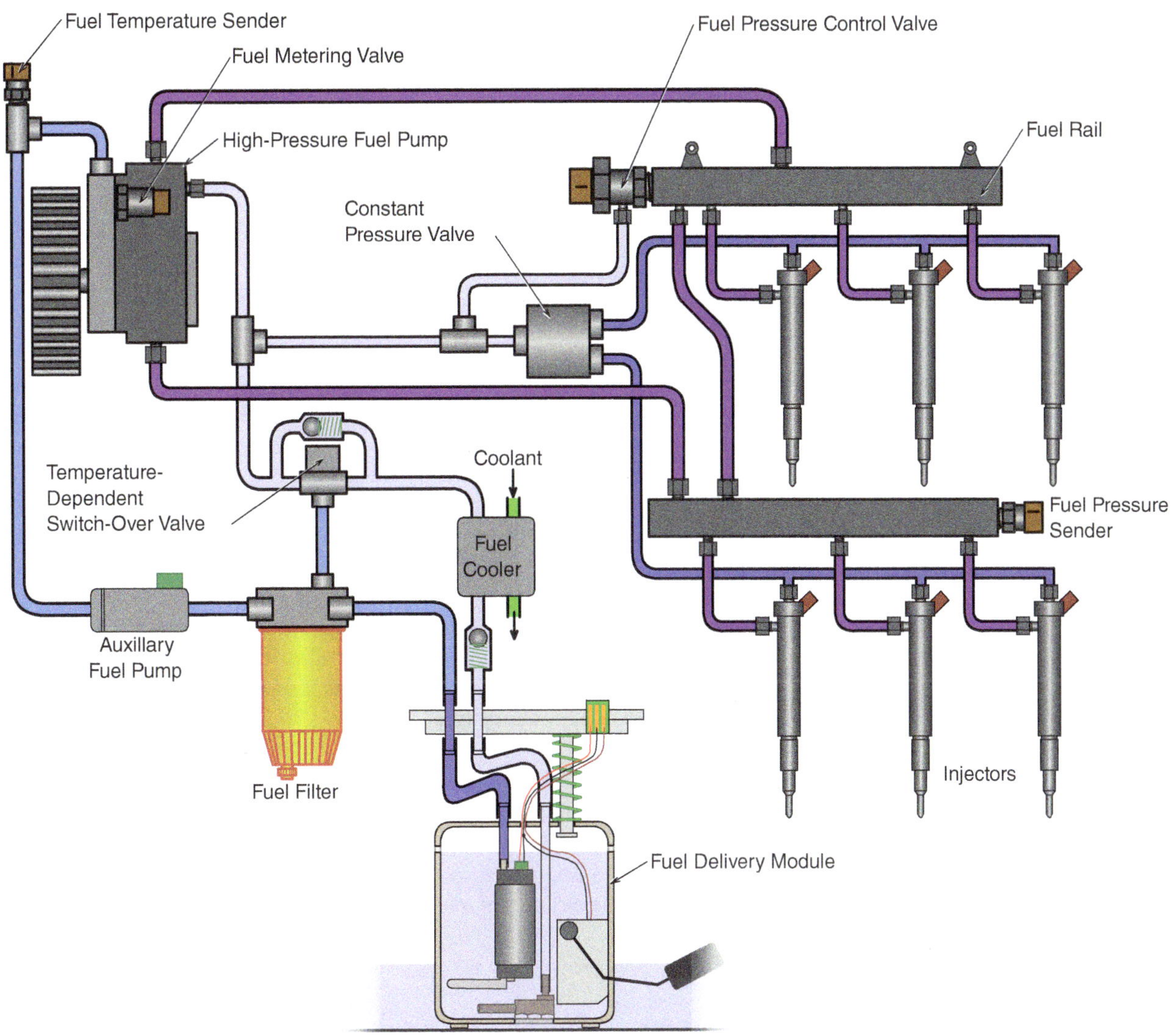

FIGURE 13-43 Note the location of the constant pressure valve for Gen III Bosch Piezo injectors. The regulator maintains fuel pressure in the injector's hydraulic link.

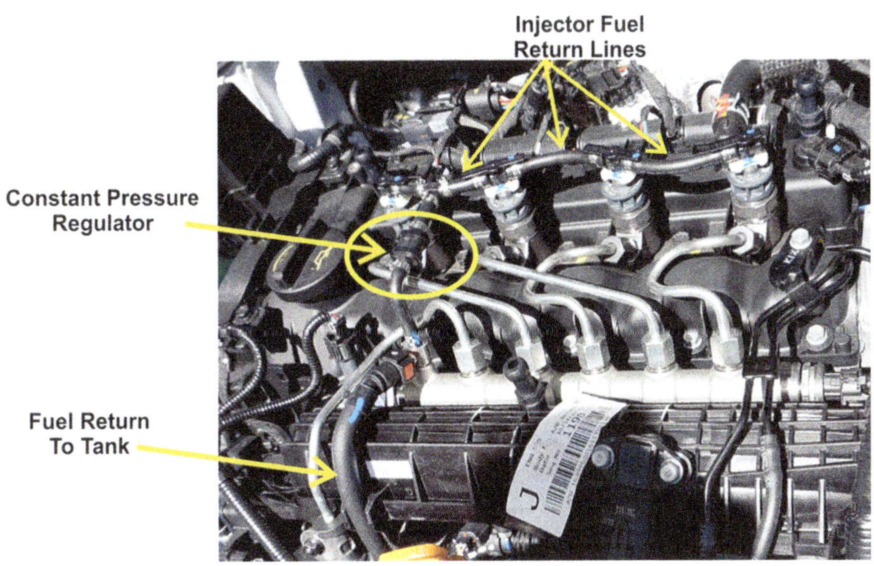

FIGURE 13-44 Properly operating the constant pressure–regulating valve is crucial to CR3 injector operation. It maintains approximately 58–160 psi (4–11 bar) in the fuel return circuit (1.6 L Hyundai).

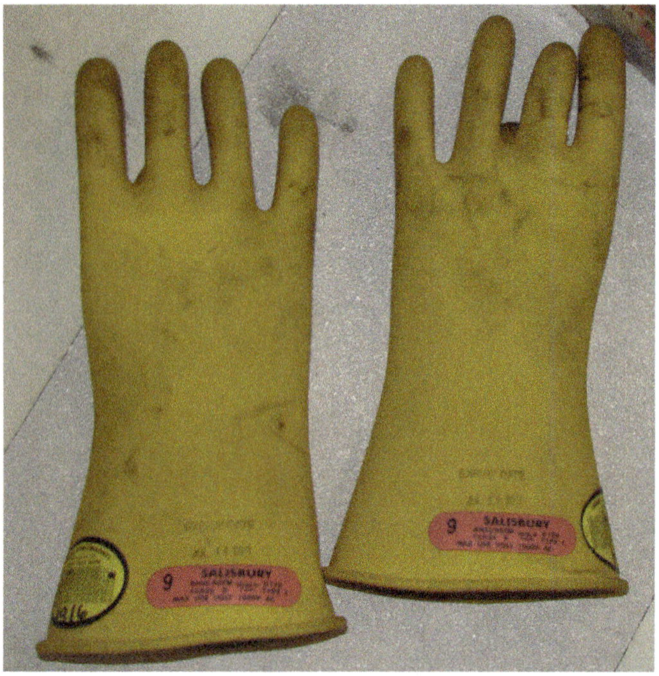

FIGURE 13-45 Electrical lineworker gloves are recommended when working on a running engine equipped with high-voltage CRS3 injectors.

discharge stored current at high voltage into the injector to initially deflect the piezo stack but only 12 volts will hold the stack in a deflected state due to the capacitor-like properties of the stack that can maintain a charge. Closing the nozzle involves switching the stack polarity and draining the electrical charge from the piezo stack.

CRI3 injectors made by Bosch are used in high-speed automotive diesel engines such as the VW TDI engines, Ford 6.7 L Power Stroke engines made since model year 2011 (codenamed Scorpion) and the GM 6.6 L Duramax made since model year 2010. Power Stroke CR injector nozzles made by Bosch each have as many as eight holes and are programmed to provide six injections per combustion cycle. During cruise operation, only three or four injections are used. Two pilot injections minimize combustion noise and allow injection timing to be retarded to control NO_x emissions. Post-combustion injections minimize soot formation by keeping cylinder temperatures and pressures high during the latter part of the power stroke. VW engines will use a late post-combustion injection event to heat the exhaust catalyst, such as the DPF and NO_x adsorbing catalyst. But GM and Ford use a separate seventh injector or hydrocarbon-dosing injector located after the turbocharger, and no late post-combustion injections are used to heat the exhaust aftertreatment system.

Piezo CRS3 Operation

The movement of the nozzle needle valve in CRS3 injections is controlled in an identical way to how CRS1 and CRS2 injectors control movement. That is, the balance of hydraulic pressure above the nozzle valve is changed to obtain faster servo-hydraulic movement of the nozzle valve and force multiplication of relatively weak actuator forces (FIGURE 13-46).

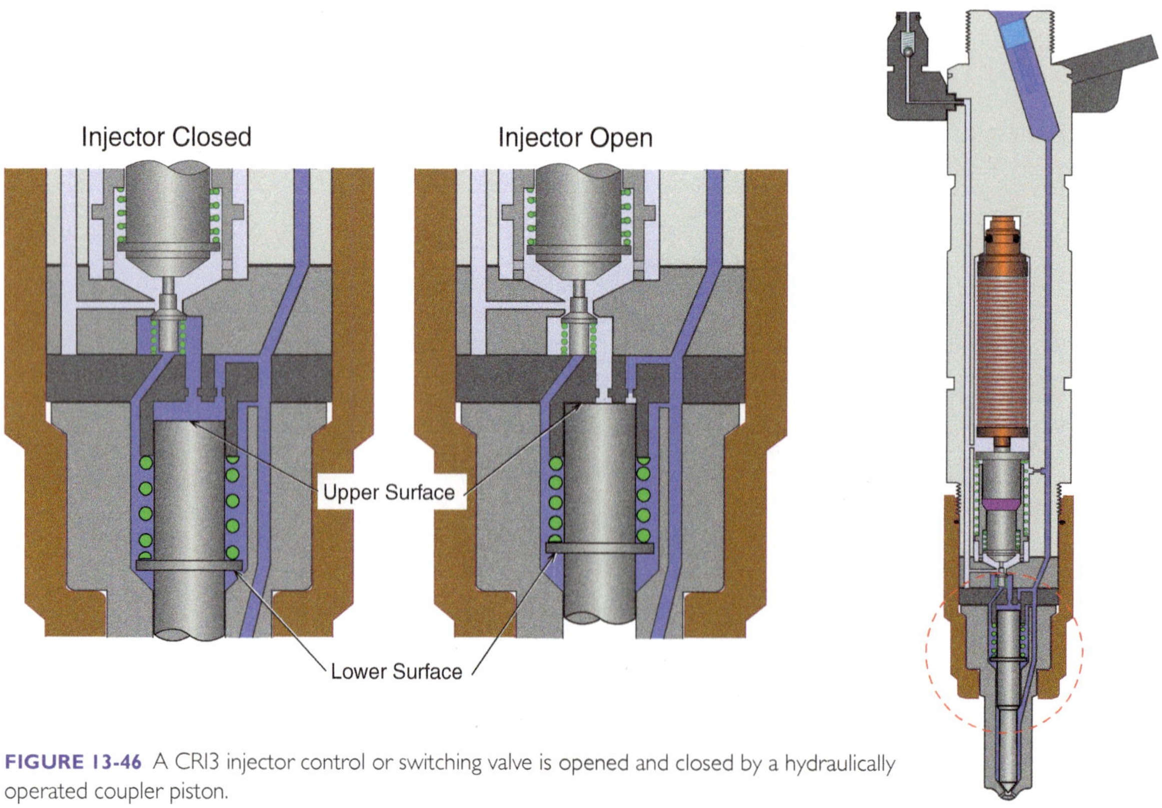

FIGURE 13-46 A CRI3 injector control or switching valve is opened and closed by a hydraulically operated coupler piston.

Two different sizes of restriction orifices are also used: an inlet and outlet, where the outlet or drain orifice is slightly larger than the inlet orifice. Downward movement of the expanding piezoceramic transmits force to the injection control valve. The downward expansion of the piezoceramic actuator replaces the magnetic forces pulling up on the ball valve of CR2 injectors. Applying current to the stacked piezoceramic wafer produces actuator expansion that is summed and transmitted to a control valve that switches the balance between hydraulic pressure that is acting directly upon the top and bottom of the nozzle valve.

Resting Position

In its resting position, the piezoceramic actuator is not energized. The nozzle valve is hydraulically balanced with high-pressure fuel from the rail applied to the top and bottom of the nozzle valve. A slightly larger area above the nozzle valve, than the area of the fuel annulus below the nozzle valve, forces the nozzle valve closed against its seat (**FIGURE 13-47**). The injector uses a injection control valve, also referred to by other names such as the servo valve, switching or pilot valve, that may have one or two seats - an upper and lower seat. When 2 seats are used, the lower seat separates high-pressure fuel in the cavity above the nozzle valve from the fuel return passage draining back to the tank. In the resting state, this seat is one of two seats that blocks high-pressure fuel from flowing out of the pressure control chamber above the nozzle valve. The purpose of this seat is to provide a faster drainage of fuel pressure above the nozzle valve. A second seat is formed above the control valve. This is the second of two seats sealing the fuel pressure acting above the nozzle valve and the low-pressure return drain back to the fuel tank. The upper seat seals high pressure fuel in the pressure chamber above the nozzle valve in a resting, non-energized state.

Fuel pressure in the return circuit, which is crucial for injector operation, is maintained at approximately 145 psi (10 bar) by the constant pressure valve. This means slightly pressurized fuel is present in the cavity above the control valve and below the plunger piston in contact with the piezo stack. While not mechanically coupled, a hydraulic link connects the plunger in contact with the piezo stack.

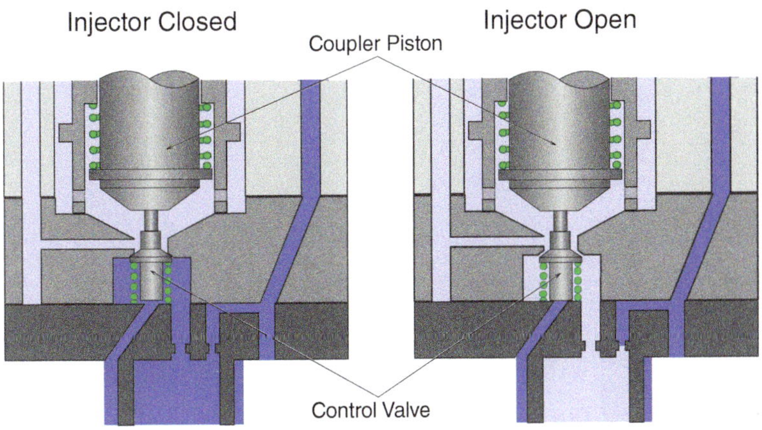

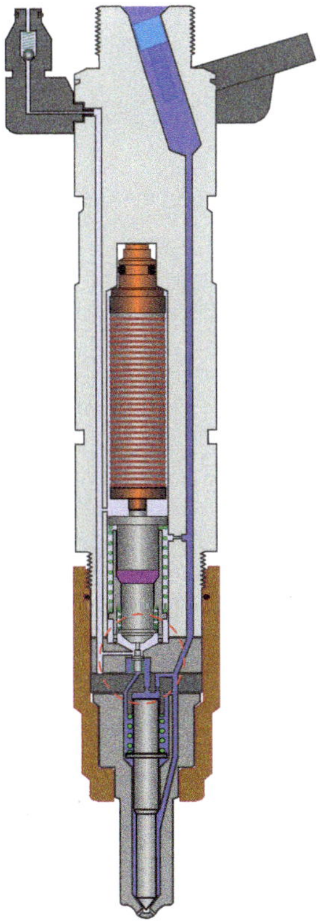

FIGURE 13-47 Fuel pressure in a cavity above the nozzle valve is regulated by the control valve. Injection begins and ends depending on whether fuel is entering or leaving the pressure control chamber cavity above the nozzle valve.

Beginning of Injection

Energizing the piezo stack causes it to expand and extend. The connecting piston is forced downward in its bore, transmitting stack force through the hydraulic coupler. The coupler multiplies the length of piezoceramic stack expansion and forces the control valve open by moving it downward. This action unseats the upper seat of the control valve, allowing fuel to drain from the pressure chamber above the nozzle valve. Because the nozzle valve's state of hydraulic balance is upset, higher pressure below the nozzle valve begins to lift the valve, pushing fuel above the nozzle valve out the pressure chamber cavity and into the fuel return passage. Unseating the nozzle valve now allows pressurized fuel from the rail to spray out the injector tip.

End of Injection

Two events can potentially end the injection (**FIGURE 13-48**). The first is the control valve seating against its lower seat, blocking one of the two high-pressure fuel passages draining the pressure chamber above the nozzle valve to low-pressure fuel return. Although the control valve's upper seat remains open, blocking one of the two drains passages above the nozzle valve prevents adequate pressure drop in the pressure chamber to keep the nozzle open. Pressure begins to build again because fuel at rail pressure fills the chamber above the nozzle valve through the inlet orifice restriction. The ratio of fuel in and out of the pressure chamber favors filling and closing the nozzle valve.

The second event ending injection is seating the control valve on it's upper seat. This happens when current flow through the piezo stack is reverse and the coupling piston removes force acting on the control valve. Pressure rapidly builds in the pressure chamber above the nozzle vavle which seats the nozzle valve. Seating the switching valve's upper seat while the lower seat remains open has two purposes. The first purpose is that it acts like a safety mechanism to prevent unintentionally over-fueling the injector (**FIGURE 13-49**). In the CRI2 injectors, a piece of dirt clogging the inlet orifice to the pressure control chamber would allow the injector to fuel continuously since no

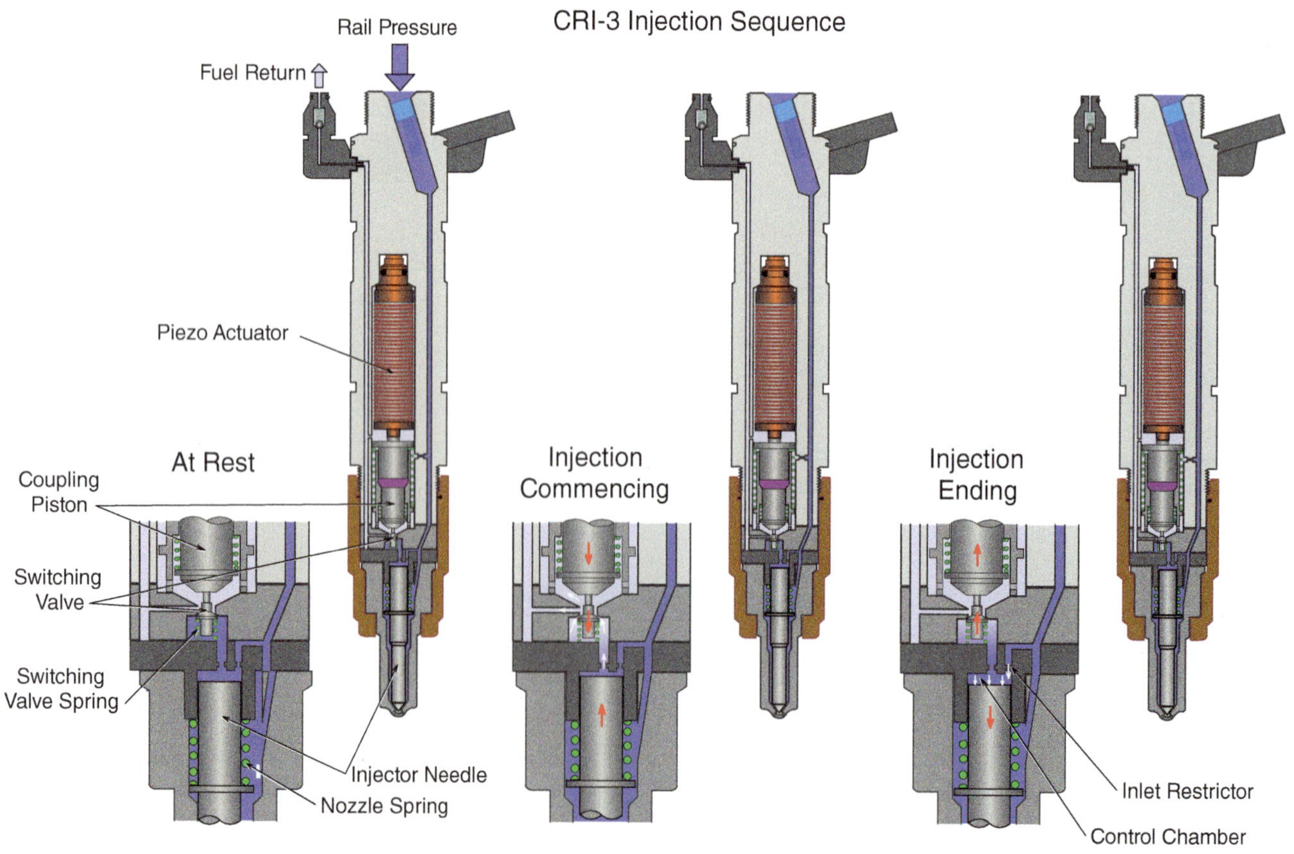

FIGURE 13-48 Injection sequence in a CRI3 injector.

fuel could enter the chamber and close the nozzle. Electrical problems in the injector, such as disconnecting a running injector while it is open or a problem with upper seat leakage of the control valve, could also cause continuous fueling, with catastrophic consequences. In these instances, closing the lower control valve seat can act as a backup mechanism to end fueling by allowing pressure to rebuild in the control chamber.

Using two drains from the pressure chamber and two seats on the control valve takes advantage of the opportunity to increase the number of injection events through faster injector response. Rapidly ending an injection event could take place by reversing current through the piezo stack and close the upper seat of the control valve. Fuel drainage from the pressure control chamber would end more quickly even though the lower seat is open and abruptly end injection. Likewise, when both the upper and lower seats of the control valve are open after the piezo stack is energized, the pressure chamber drains faster which provides for more rapid nozzle lift.

FIGURE 13-49 Vaporized aluminum from a piston and heat-induced breakage of exhaust valve stems were produced in only one cylinder on this engine. The root cause was a CR2 injector that stuck open and continuously fueled the cylinder. Damage occurred in less than 45 seconds.

A predictable end of injection takes place when current direction through the piezo stack reverses before the upper seat closes. Slight pressure in the fuel return circuit and the mechanical force of the retracting piezo stack causes a drop in hydraulic link pressure that in turn closes the switching valve. Pressure then builds in the control chamber cavity above the nozzle valve, quickly closing the nozzle, which ends injection. The ratio of surface area above the nozzle valve in the control chamber in comparison to the surface area below the nozzle valve is another factor that determines whether the nozzle valve will close or open more rapidly. The drain to inlet orifice ratio is a second important factor determining the speed of nozzle lifting to closing. A much larger drain orifice compared to the inlet orifice will supply faster nozzle lifting than closing speeds. Likewise, a smaller ratio between the drain and inlet orifice will produce faster nozzle closing.

Nominal Voltage Calibration

In CRI3 injectors, placing the piezoceramic stack closer to the nozzle valve minimizes the injection lag caused by transmitting forces along the length of the injector through hydraulic or mechanical means. Although piezo actuators have many advantages, one disadvantage is that these actuators do wear out faster than and generally will not last as long as the latest servo-hydraulic actuators that use a magnetic solenoids. The continuous flexing and sliding movement of the ceramic discs against one another produces friction, which wears the discs thin. To compensate for this problem, disc stacks can be made longer to increase actuator reliability and longevity. OBD algorithms measure this deterioration and help compensate for wear over the lifespan of the injector. This strategy, which monitors the injector voltage, is called nominal voltage calibration (NVC). NVC is a wear-compensating strategy intended to correct changes in injection events due to piezoceramic stack wear and other changes to hydraulic control elements within an injector. The initial energization time, voltage, and duration of the energization period will change over time to maintain consistent injector operation. For example, the energization voltage measured at the end of the injection event will change predictably as the stack becomes thinner. If the expected voltage rises too high or if it persistently deviates from measured voltage that is not correctable, a code is set for injector circuit range/performance (**FIGURE 13-50**).

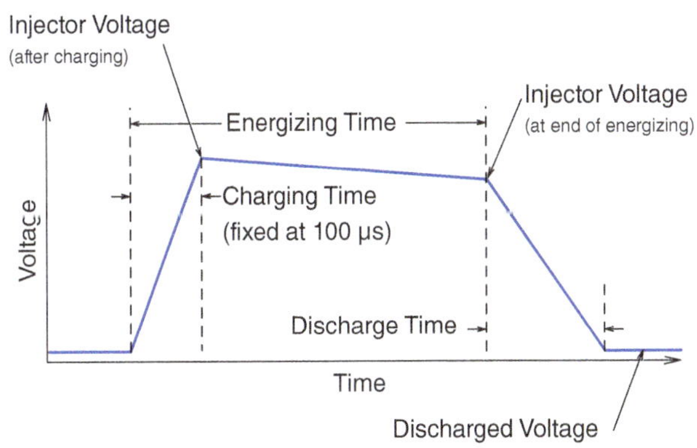

FIGURE 13-50 The OBD-II system monitors the voltage at the end of the injection event to compensate for piezoceramic stack wear and to check for emission fault–related failures.

CRI3 Hydraulic Coupler

The Bosch CRI3 injector incorporates a hydraulic coupler that carries out additional fuel system functions. As already described, its primary purpose is to transfer and amplify the intensity of the piezoceramic stack deflection to the switching control valve. Its

SAFETY TIP

The piezoceramic injectors are opened with a voltage between 110 and 250 volts, with up to 20 amps of current. This current has the potential to be lethal if work is performed on a running engine and a technician touches uninsulated wiring or an injector has an internal short circuit. Wiring to the injectors is typically covered in bright red or bright orange woven nylon mesh to prevent chafing and contact. However, when working on these engines, insulated electrical lineworker gloves should be worn with a class 0, 1,000-volt rating. These gloves should be inspected before use for punctures or thin spots that could pass electrical current. Do not use gloves if their expiration date has passed. Observe all the safety instructions and precautions noted in the workshop manual.

second purpose is to compensate for changes to the length of the actuator mechanism caused by wear or thermal expansion. Its third purpose is to prevent the injector from continuously delivering fuel in the event that an electrical fault causes the stack to remain extended. In that case, while the switching valve remains open, fuel will leak out of the control chamber to the low-pressure return circuit and end injection unless the switching valve is first completely closed and then reopened.

Constant Pressure Control Valve

For the piezoceramic stack to act on the injector control valve, the hydraulic coupler requires a minimum of 58–160 psi (4–11 bar) of return circuit fuel pressure to operate properly. Pressure in the fuel return circuit keeps the coupler "pumped up." Loss of pressure or return side circuit leaks will prevent the engine from starting or running. To regulate the return circuit pressure and maintain it for starting, a pressure retention valve is used to maintain the injector's return pressure of approximately 145 psi (10 bar). The pressure retention valve, also called a constant pressure control valve, is a mechanical valve located between the return line from the fuel injectors and the fuel return circuit. The line may also connect to the suction side of a fuel filter on some vehicles. During normal operation, the injector's continual back leakage into the return circuit pressurizes the line and pushes on a spring-loaded check ball inside the constant pressure valve. The pressure in the line increases to the point where it overcomes spring tension and unseats a ball valve from its seat. At that point (between 150 and 200 psi (33 and 14 bar), depending on the OEM), fuel will pass through the constant pressure valve and into the secondary fuel filter inlet or return to the tank depending on the manufacturer.

Backfill Valve

The high-voltage CR3 injection system is more difficult to prime than CR2 engines due to needing to first pressurize the hydraulic coupler in the injector fuel return circuit. To assist starting when the fuel system is serviced, such as after a filter replacement or if the vehicle has run out of fuel, a backfill valve or backfill function is incorporated into the constant-pressure control valve. An outlet line from the low-pressure fuel-transfer pump typically supplies the valve with back-fill capabilities. When return side pressure drops below 58 psi (4 bar), a check valve in the backfill valve will open to pressurize the return side to enable the engine to start (**FIGURE 13-51**). On Duramax diesel engines, the backfill valve opens after it has been supplied fuel from the line supplying the exhaust aftertreatment fuel injector. The low-pressure fuel supply circuit is also equipped with a suction side fuel filter vacuum switch near the fuel filter. The vacuum switch will open at a vacuum of 13.6–15 Hg and alert the driver with a message in the driver's information center. A restriction at this threshold of negative pressure would indicate that one or both fuel filters are restricted or blocked or that there's a kinked fuel line on the suction side of the low-pressure pump that's located inside the fuel tank.

▶ TECHNICIAN TIP

Bosch CR3 systems found on many late-model diesel engines use a positive locking clip on the return circuit. Failure to properly disconnect and reinstall the clip will result in potentially severe fuel leaks and an engine no-start condition from loss of return pressure.

To disconnect the return line from the injector, a twist lock fitting or sliding cap must lift upward from the center section of the injector before removing the connector from the injector. To do this, first twist the lock 45° or release tension on the sliding cap tabs by pressing downward on the retaining clips while simultaneously pulling upward on the cap. After removing the connector, inspect the two retaining clips or twist lock on the cap for breakage. If no damage is noted, reinstall the cap by pushing the connector down onto the injector until it clicks or locks in place. Twist the lock until it is parallel with the return lines. Check the line connection by pulling up slightly on the center cap assembly. If the line disconnects, repeat the removal and installation procedure (**FIGURE 13-52**).

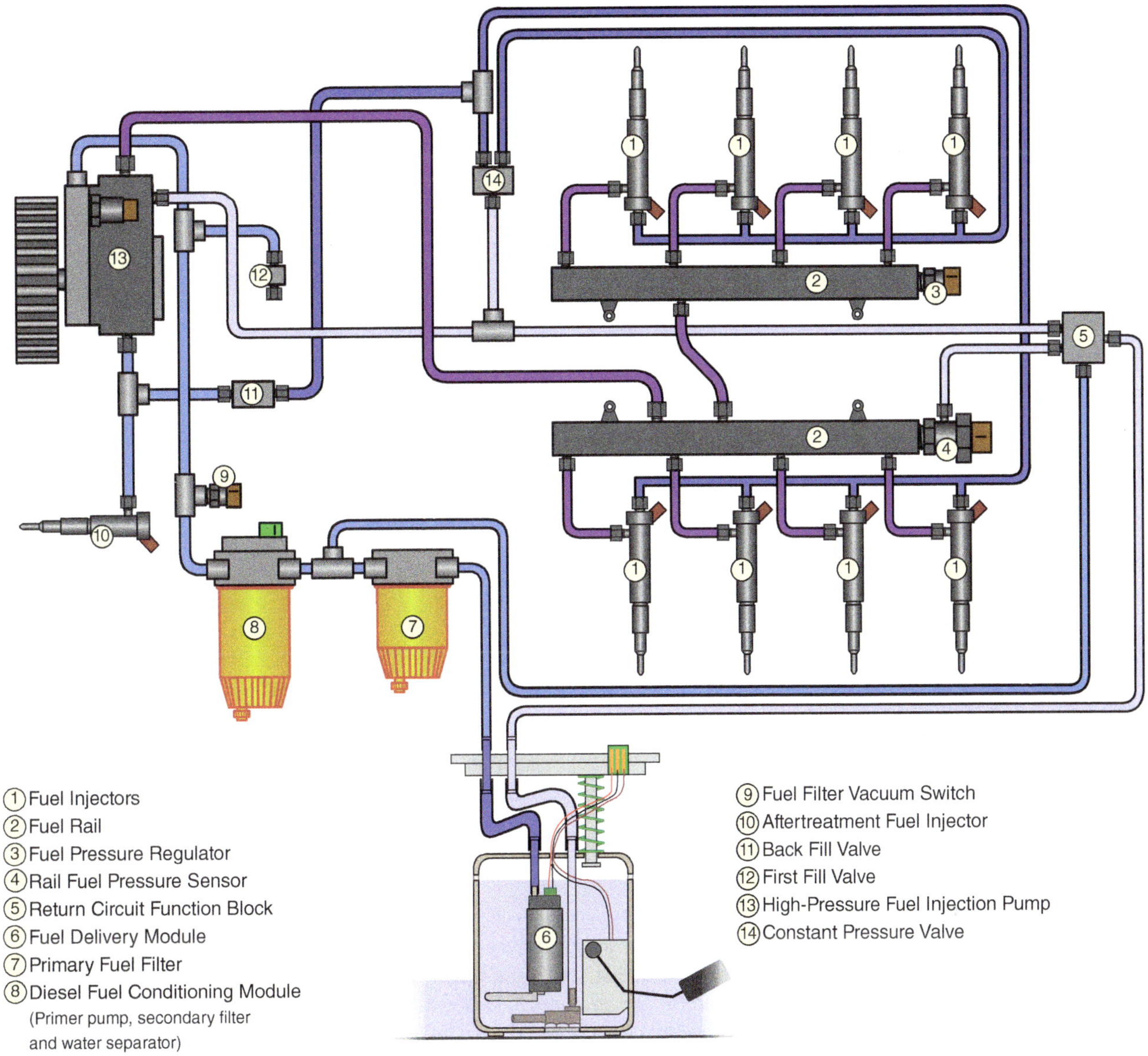

FIGURE 13-51 A fuel system schematic for a Bosch CR3 system with a backfill valve. Note the location of the backfill valve and the constant pressure valve. The first fill valve is used only at the factory to prime the fuel system.

FIGURE 13-52 Note the twist lock fitting on the return line cap. This must be turned before the cap can be disconnected from the fitting. The connector must be correctly repositioned after reinstalling it on the injector.

13-03 Describe and explain the operation of CR injectors.

▶ Siemens Common Rail Injectors

Siemens has been developing and manufacturing piezo injectors since 2001, for a variety of OEMs. The most significant North American application for this European-tested CR fuel system and injectors is in the 2007–2010 6.4 L Power Stroke engine in the Ford F-250s and F-350s. The high-pressure pump used by Ford is the Siemens VDO K16, having volume output regulated to supply an output rail pressure of 24,650 psi (1,700 bar).

Injector Type	Injection Pump	Injector Designation	Actuator Technology	Maximum Rail Pressure	Application
Gen I Bosch	CP-1	CRI1	Ball valve	1,350 bar (19,845 psi)	Automobile
Gen II	CP-2	CRI2	Ball valve	1,600 bar (23,520 psi)	Automotive/commercial vehicle
Gen III	CP-3	CRI3	Piezoceramic	2,000 bar (29,400 psi)	Automotive/commercial vehicle
Gen IV (HADIS)	CP-4.1, CP-4.2	CRI4	Spindle valve (amplified common rail (ACR))	2,500 bar (36,750 psi)	HD commercial vehicle
Denso 1st Generation				1,450 bar (21,315 psi)	Automotive
Denso 2nd Generation	HP2, HP3, HP4			1,800 bar (26,460 psi)	Automotive/commercial
Delphi Multec			Servo-hydraulic	2,000 bar (29,000 psi)	Automotive/commercial
Delphi Direct-Acting DFI 1.22/1.25	DFP 6 (single plunger) DFP7 (double plunger)		Piezoceramic	2,000–2,200 bar (29,000–31,900 psi)	Automotive/commercial
Delphi Direct-Acting DFI 4.22/4.25	DFP 6 (single plunger) DFP7 (double plunger)			DFI 4.22/4.25 2,200/2,500 bar (31,900–36,250 psi)	Commercial
Siemens	VDO K16		Piezoceramic	1,700 bar (24,650 psi)	Automotive

Overview

The Siemens CR injector uses a conventional six-hole nozzle valve that opens or closes as the hydraulic balance acting above and below the nozzle valve changes. (**FIGURE 13-53**). In this injector, as in the Bosch injectors, hydraulic forces are pressure balanced with piezoceramic actuator movement tipping the balance by reducing fuel pressure above the nozzle valve. Also like Bosch CRI2 injectors, a control rod or piston holds the nozzle valve on its seat, preventing the nozzle valve from lifting.

Fuel pressure in a chamber above the control piston regulates the movement of the control piston. Draining fuel pressure to the low-pressure return side of the fuel system begins injection, whereas increasing pressure above the control piston ends injection (**FIGURE 13-54**).

A piezoceramic actuator in the Siemens injector uses multiple stacked piezo crystal discs that expand longitudinally when electrically energized. The response rate is close to 0.1 milliseconds and designed for five injection events per combustion cycle in the Ford engine. The expansion of the crystals results in a linear motion that pushes against a mushroom-shaped valve to drain fuel from the pressure chamber (**FIGURE 13-55**). The mushroom valve is held closed by spring pressure, sealing high-pressure fuel in the control piston chamber. Opening the mushroom valve vents high-pressure fuel to the low-pressure fuel return circuit (**FIGURE 13-56**). De-energizing the piezoceramic actuator ends the injection event.

Pre-injection

During pre-injection, the following events occur (Figure 13-54 and Figure 13-55):

1. Fuel pressurized by the high-pressure pump enters the fuel rail and is supplied to each injector fitting. High-pressure fuel flows to the nozzle valve and to a chamber above the control piston. This high-pressure fuel forces the control piston downward against the top of the nozzle valve.

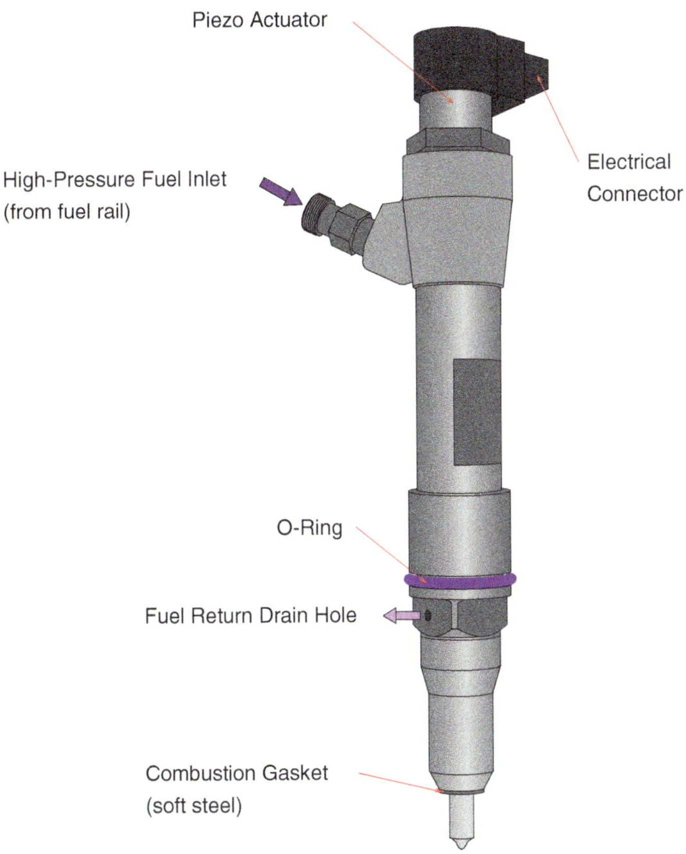

FIGURE 13-53 External features of the Siemens piezoceramic CR injector used in the Ford 6.4 L Power Stroke.

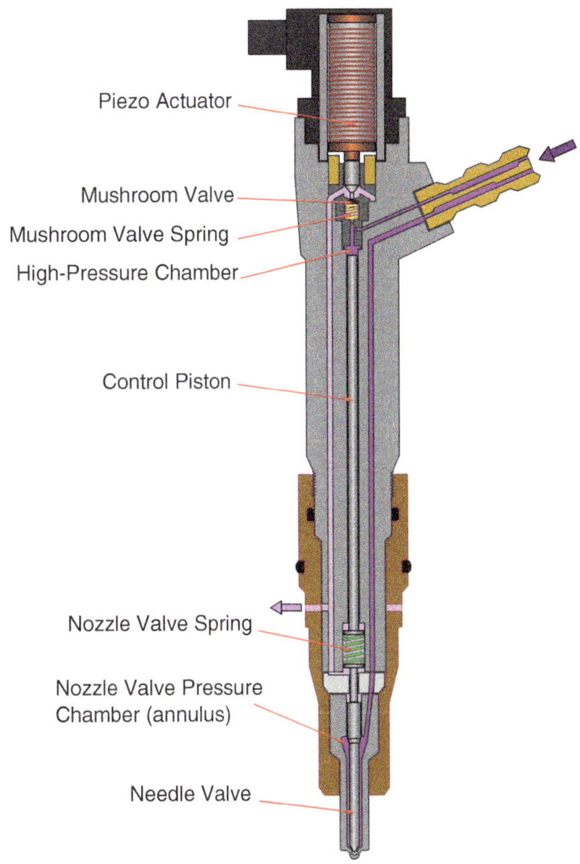

FIGURE 13-54 Energizing the stack of piezoceramic discs in the Siemens CR injector causes expansion and a downward movement of the discs, beginning the sequence of injection events.

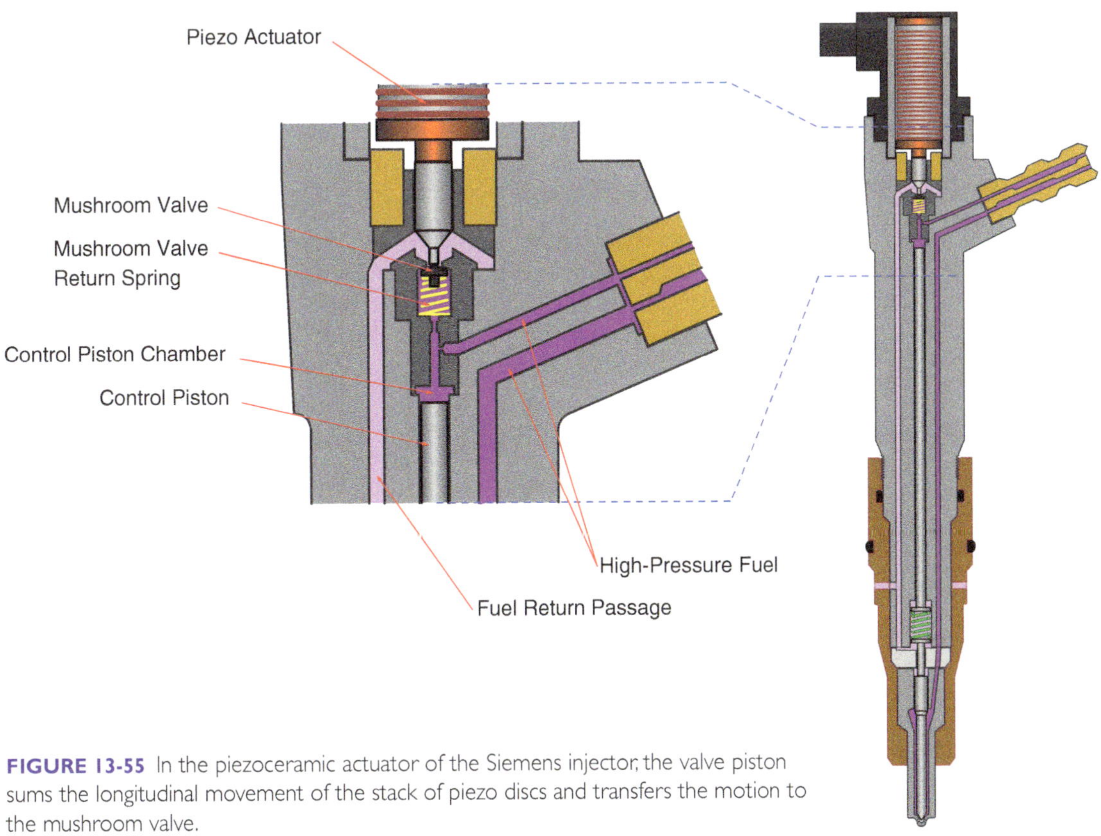

FIGURE 13-55 In the piezoceramic actuator of the Siemens injector, the valve piston sums the longitudinal movement of the stack of piezo discs and transfers the motion to the mushroom valve.

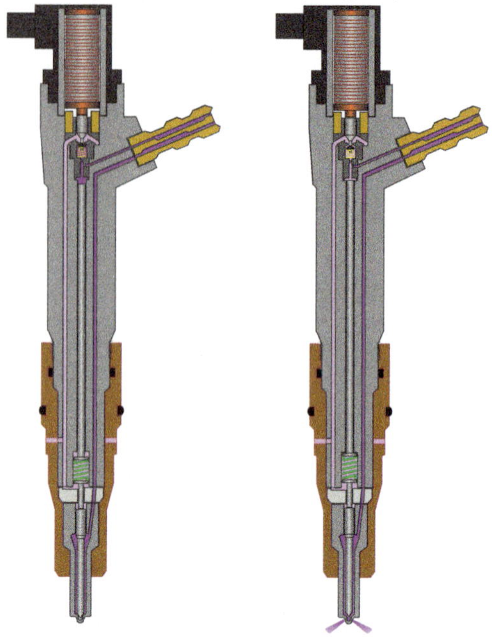

FIGURE 13-56 In the Siemens piezoceramic injector, the mushroom valve is held closed by spring pressure, sealing high-pressure fuel inside the control piston chamber. Opening the mushroom valve vents high-pressure fuel to the low-pressure fuel return circuit.

2. A mushroom valve seals the control chamber, preventing fuel from escaping to the low-pressure return circuit. The mushroom valve spring forces the mushroom valve against its seat.
3. The area surface differential above the larger control piston applies greater pressure against the force of the fuel pressure that's acting on the smaller surface area of the nozzle-valve annulus.

Injection

Please see Figure 13-54 and Figure 13-55. The following are some of the events that occur during injection:

1. When the piezoceramic actuator is energized by electrical current, the valve piston below the actuator transfers movement of the piezo stack to the mushroom valve. This action opens the mushroom valve, causing fuel pressure in the control chamber to drain into the low-pressure fuel return circuit.
2. Pressure below the nozzle valve is higher than the fuel pressure in the control chamber. The nozzle valve lifts and injection begins.
3. High-pressure fuel has to enter the control piston control chamber through a narrow restriction. Fuel cannot enter the pressure chamber fast enough to replenish fuel venting to the fuel return circuit around the mushroom valve. This causes the pressure in the control chamber to continuously fall, and the injection event continues. Fuel from the pressure control chamber passes through a drilled passageway in the injector and out into a fuel return passageway in the cylinder head.

End of Injection

The injection event will be complete after the following events have occurred:

1. The ECM reverses the polarity of the piezoceramic actuator to cause a rapid return of the actuator to a resting position. Spring pressure closes the mushroom valve, causing pressure to build in the control piston control chamber.
2. The pressure in the control chamber forces the control piston downward against the nozzle valve, seating the nozzle valve and ending injection.

▶ Delphi Multec System

13-04 Explain the operation of Delphi Multec systems.

Delphi has developed several families of CR systems for light-duty automobiles and heavier commercial vehicles. Multec is the name Delphi has given to its engine management systems that use servo-hydraulic solenoid-type injectors to optimize injection events. Its latest injector technology uses both a hydraulically balanced injector control valve and hydraulically balanced nozzle valves that provide the equivalent performance to servo-piezoelectric actuators but at a much lower cost (**FIGURE 13-57**). The latest injector can produce up to nine injection events at 37,500 psi (2,500 bar), enabling GM's 1.6 L whisper diesel highway fuel economy of 39 mpg (US) (6 L/100 km) in vehicles like the Chevy Equinox (**FIGURE 13-58**). The Chevy Cruze achieves over 52 mpg, (US) (4.5 L/100 km). The same engine is used by the GMC Terrain.

Another system, called the Piezotec, uses direct-acting piezoelectric technology, which Delphi notes is fastest—and capable of switching six times faster than a conventional magnetic solenoid injector that uses a servo-hydraulic actuator. A piezoelectric actuator is placed very close to the nozzle valve in the injector body. The direct actuation of the needle through a hydraulic amplifier by a piezoceramic actuator enables both a rapid and a controlled opening and closing of the nozzle needle independent of fuel rail pressure (**FIGURE 13-59**). The piezo actuator performs the initial nozzle-lifting action while hydraulic force applied through a hydraulic amplifier is used to continue lifting the nozzle valve.

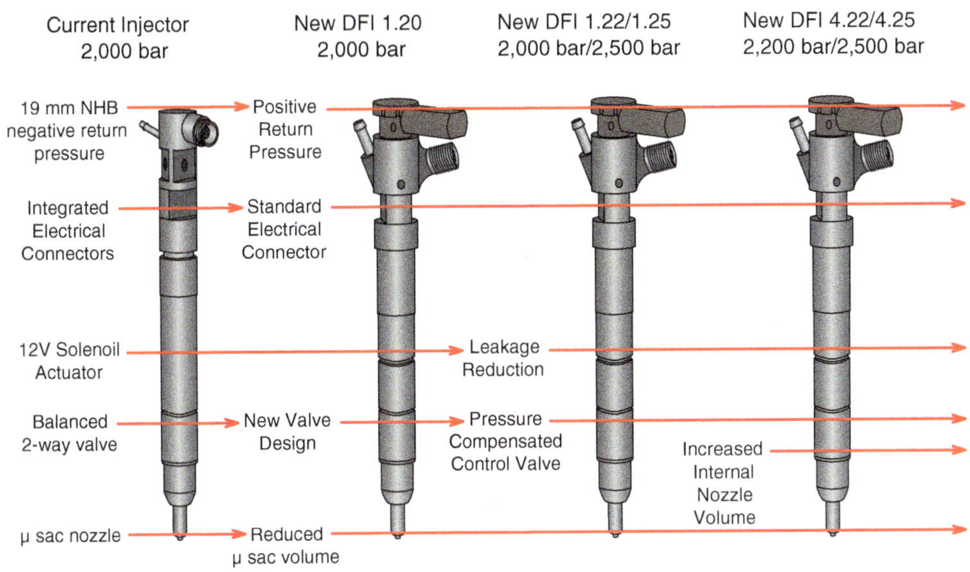

FIGURE 13-57 Delphi line of CR injection systems from the DFI 1.5 to DFI 4.25. Delphi has refined its servo-hydraulic injectors to achieve piezoceramic-like performance capabilities.

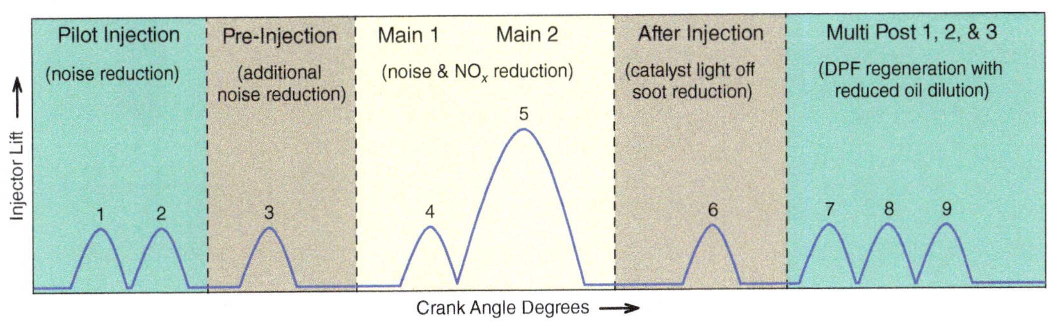

FIGURE 13-58 With as many as three pilot injections and four post-main injections, the latest Delphi DFI injector can exert enormous control over temperature and pressure conditions in combustion events.

These direct-acting injectors do not use a fuel return or back leak line. Delphi claims the system can generate about a 25% to 30% reduction in emissions compared to current production piezoelectric CR technologies, so that engines can meet Euro 5 and 6 emission standards. Piezoelectric technology will allow the injection system to deliver up to seven high-accuracy injection events per combustion cycle. Using faster-acting actuators means less time is required between pilot-injection, main injection, and post-injection events, which improves rate shaping of injection sequences.

Accelerometer Pilot Control

One unique feature in earlier Multec systems is its use of an engine sensor and an electronic processing strategy called accelerometer pilot control (APC) (**FIGURE 13-60**). Mounted on the cylinder head, the sensor for the APC operates like a knock sensor, constantly monitoring combustion noise. Integrating the combustion noise data generated by the sensor with engine speed and position data allows the ECM to adapt injection quantity, timing, and pressure of injection over the entire engine operating range. The APC sensor literally listens to the engine operation to decide when to change the quality of pilot injection(s) and adapts to electronic control to compensate for engine and injector wear. The electrical pulse width supplied to each injector can be modified in the Multec system for optimal combustion quality before each injection event, based on this operating strategy. This is especially helpful at idle, where knock and clatter associated with idle combustion is even further minimized by using APC. The variable

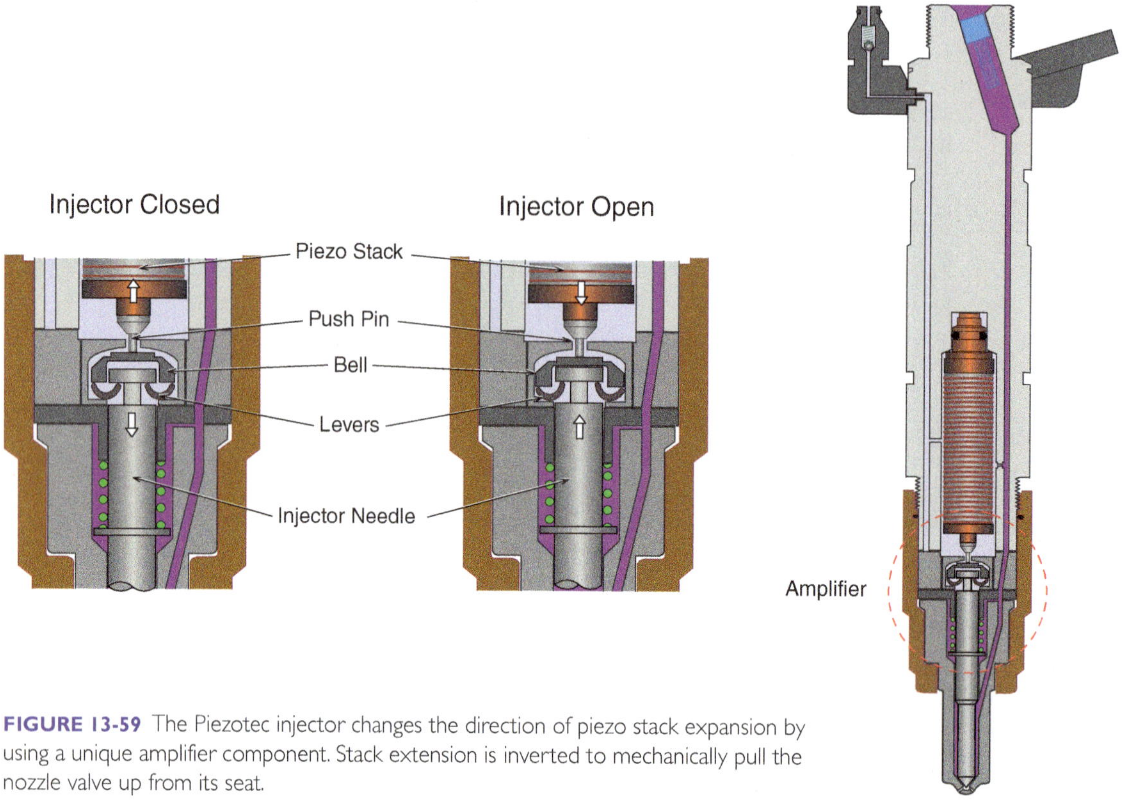

FIGURE 13-59 The Piezotec injector changes the direction of piezo stack expansion by using a unique amplifier component. Stack extension is inverted to mechanically pull the nozzle valve up from its seat.

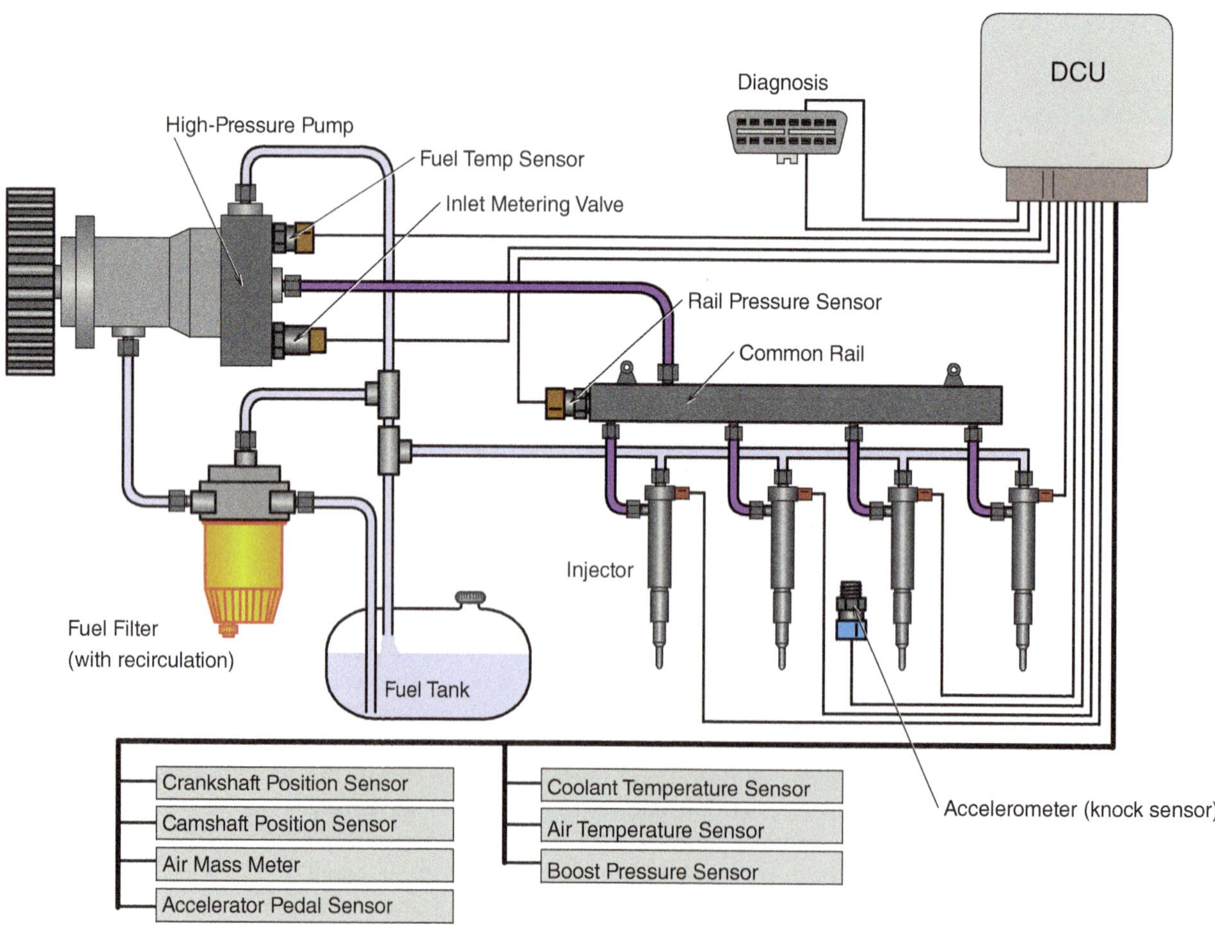

FIGURE 13-60 Layout of a Delphi Multec CR system. Note the presence of the APC sensor (9).

quantity of pilot injections, varying between 1 and 4 mm^3, compensates for problems with pilot-injection events that use only a fixed delivery volume.

Injector adaptation to compensate for wear and deterioration is performed by an algorithm called speed pilot control (SPC). As an OBD requirement, SPC corrects the injector operation to maintain emission compliance over the course of a vehicle's useful life. Injector calibration codes that provide the ECM with baseline injector behavior characteristics for flow rates is called individual injector characterization (I3C).

Delphi's current injector, the DFI 4, can use closed-loop control of the injection event. A single wire passing current through the nozzle valve is accurately capable of identifying nozzle position. Nozzle position is used to provide closed-loop feedback to the ECM for more precise control over injection timing and rate-shaping events. The injector is a refined variation of the spindle valve technology used on its first DF 1 injector, introduced in 2000. Hydraulically balanced spindle valves, rather than ball valves, can achieve better durability and faster operation (**FIGURE 13-61**).

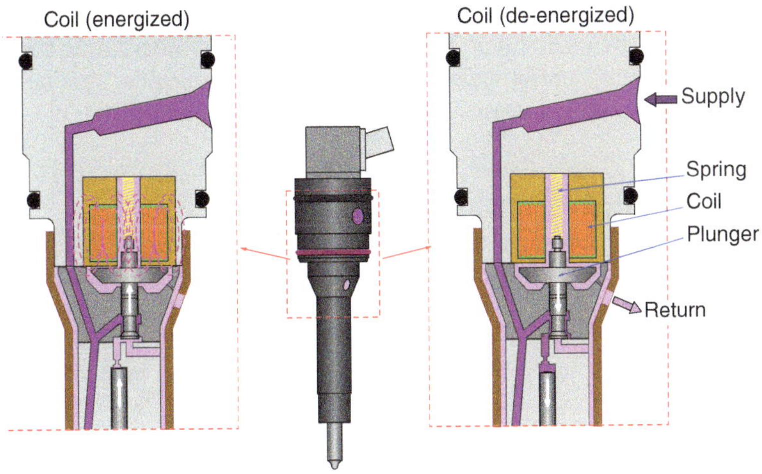

FIGURE 13-61 The spindle valve used by Delphi nozzles is a servo-hydraulic actuator that uses principles of hydraulic balance to rapidly open and close. The control valve also regulates the operation of a hydraulically balanced nozzle valve.

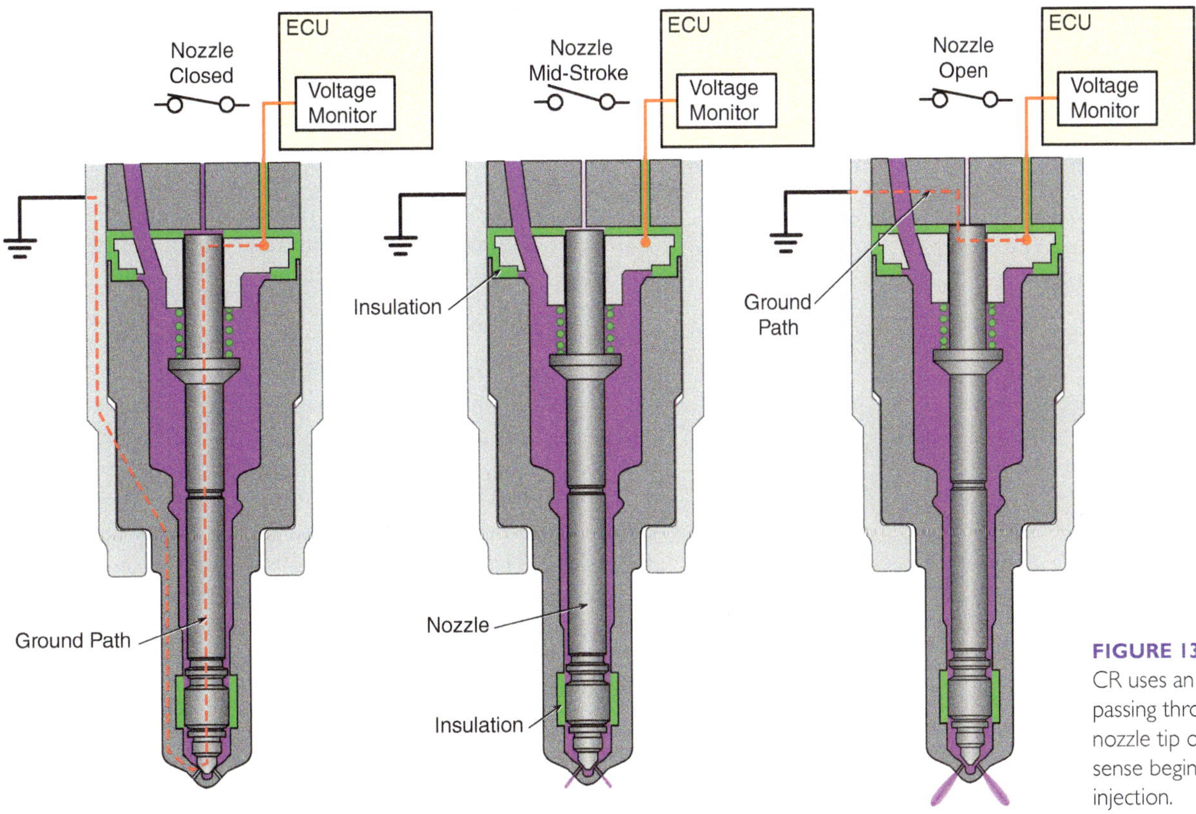

FIGURE 13-62 Closed-loop CR uses an electrical signal passing through either the nozzle tip or nozzle body to sense beginning and end of injection.

Electric current supplied to Delphi's Multec solenoid CR injectors is at battery voltage (12 volts), allowing for a simple, cost-effective electronic control design. Like Bosch injectors, injector calibration codes are used to correct for manufacturing tolerances and individual variations in injector delivery qualities at different rail pressures and engine speeds. Delphi's DF series of injectors refers to the calibration codes as "injector compensation" codes. Each injector will contain a laser printed dot matrix code that provides the ECM with fueling data unique to each injector.

The DFP 6 uses a single-piston high-pressure fuel pump is driven at one-half engine speed and coupled to the engine through a belt or gear drive mechanism. On passenger vehicles, the torque required to turn the high-pressure pump is less than 12 ft-lb (16 N·m). This is one-ninth the drive torque required for a distributor-type injection pump.

▶ Pressure Control Components of Common Rail Fuel Systems

13-05 Identify and describe pressure control components for CR fuel systems.

Pumps, pressure control valves (PCVs), and pressure regulators are used by all CR systems. There are major differences between these components, depending on manufacturer designs, the fuel system pressure control strategy, and the size of the engine. This next section will cover some of the more common constructional and operational details of CR system components.

FIGURE 13-63 A single-piston CP-4.1 pump on a VW 2.0 L EA288 engine. A four-cylinder engine requires only a single-piston pump to synchronize engine firing order with pump stroke.

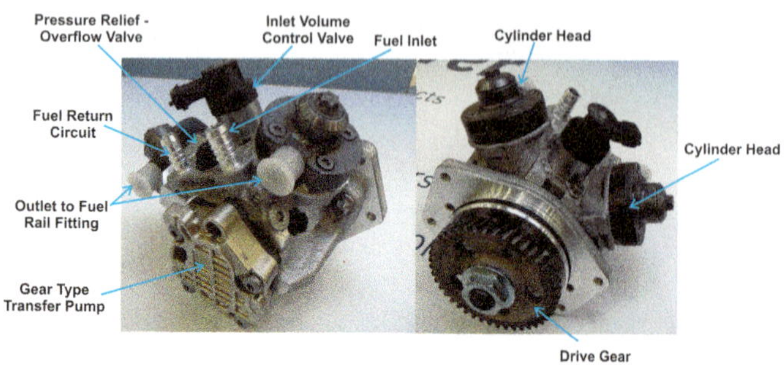

FIGURE 13-64 A two-piston CP-4.2 pump from a V8 Duramax. Two-piston pumps are required for V engines to synchronize engine firing order with pump strokes. Note the location of the inlet metering valve.

High-Pressure Pumps

High-pressure fuel pumps develop the pressure required for injection under all engine operating conditions, including quick starting the engine. Single- and multiple-plunger pumps are also used in CR systems (**FIGURE 13-63**). Low-cost, single-plunger pumps are typically used in small displacement four-cylinder engines, whereas two-, three-, and even four-cylinder pumps are used in V engines and high-horsepower (high-kilowatt) engines with higher fuel volume requirements. The pumping plungers in these pumps are cam actuated and use inlet and outlet valves to control fuel flow. An electronically controlled fuel pressure regulator valve will vary pump output pressures as commanded by the ECM. Earlier generations of piston pumps supplied more fuel than could be delivered by the injectors during idle and part-load engine operation. To reduce parasitic power losses produced by pumping fuel needlessly and to keep fuel temperatures low, pumps now use inlet volume metering, which means only the amount of fuel needed to reach correct rail pressure passes through the high-pressure pumps (**FIGURE 13-64**). The valve regulates fuel supplied to the pump and maximum pump output pressure.

To take advantage of the pressure spike that takes place when the camshaft reaches the end of its stroke on the outer base circle of the cam lobe, late-model high-pressure pumps are often timed so the peak pressure pulses are phased with the injection events. Without phasing, peak injection pressures are not possible. Matching the pump pressure pulses results in a more consistent pressure inside the fuel rails. If the pump is removed, timing marks on the pump gear and camshaft gear are used to correctly synchronize engine to pump rotation during reinstallation.

Drive torque used by high-pressure pumps is much lower than it is in other injection systems such as unit injection or mechanical in-line–type or distributor-type injection pumps because there is little resistance created by driving a cam plunger pressurizing fuel for a short injection event. Instead, fuel pressure accumulates in the high-pressure rail that can store fuel needed for up to three injection events. Drive torque is lower without the momentary sharp increases in pump-turning resistance and is more consistent. This feature explains the observation that the drive gears on high-pressure pumps are often thinner than those on other injection pumps.

Bosch High-Pressure Pumps

There are several successive generations of high-pressure pumps used by Bosch, each having greater refinements, increased pressurization capabilities, better tolerance for low-quality fuel, and less friction or drive torque required:

- Gen I: up to 1,350 bar (19,845 psi), used the CP-1 pump in the Unijet System, which Bosch purchased from Fiat in the mid-1990s. Fiat developed the CR system and used it first in the Alfa Romeo in 1997.
- Gen II: up to 1,600 bar (23,520 psi), used the CP-2 and CP-3 pumps.
- Gen III: up to 2,000 bar + (29,400 psi), uses the CP-4.1 and CP-4.2 pumps.

Bosch's CP-1 pump had a single pumping plunger actuated by a double-lobed camshaft. This pump was only briefly used on late-1990s automobiles. The CP-2 Bosch pump is a two-piston inline pump used primarily on mid-range commercial diesel vehicles. Electronically controlled pressure-regulating valves are located on the outlet of each plunger. The CP-3 pump is a axial piston pump with three pistons that is more familiar in North America because of its use in pickup trucks diesels by GM and Ram. (**FIGURE 13-65**). The CP-3 pump is matched with second-generation CRS2 injectors found on early 6.6 L Duramax, 2.8 L Jeeps, Cummins 5.9 L and late-model 6.7 L diesel engines.

A fourth generation of pumps, designated the CP-4.1 for one piston (**FIGURE 13-66**) and CP-4.2 for two pistons, produces rail pressures of up to 29,000 psi (2,000 bar). The most distinctive feature is the pump's steel heads, which are used for the pumping cylinders bolted to an aluminum body. CP-4.1 pumps are used on four-cylinder engines, whereas the additional pumping cylinder in the CP-4.2 is required for six- and eight-cylinder engines. The CP-4 has been more ruggedly designed to deal with problems of poor fuel quality. Like the CP-3, it uses a fuel-lubricated camshaft with rollers on the plunger tappets to reduce friction during pressure generation. An optional oil-lubricated pump is available for use when alternative fuels and B100 biodiesel is used. The pump produces 20% less fuel flow than the CP-3, but this reduced pumping volume is compensated for with a volume control regulator located in the suction side. The regulator valve limits the quantity of fuel admitted to the pump to reduce unnecessary power loss from pumping fuel that the engine doesn't need. With reduced and more consistent drive torque, less stress is applied to a toothed drive belt on smaller automotive engines, such as those built by VW.

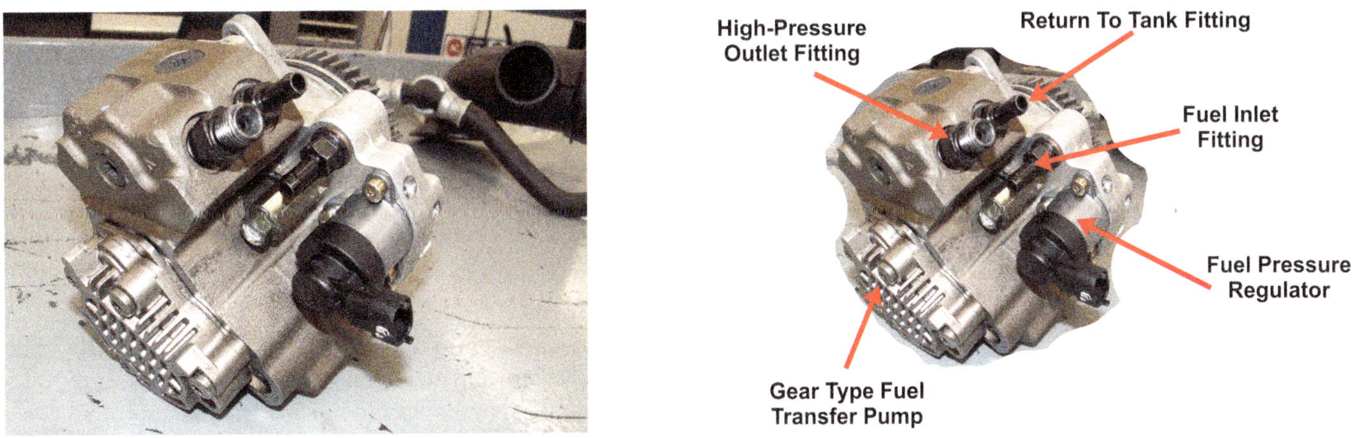

FIGURE 13-65 External details of the Bosch second-generation CP-3 high-pressure pump.

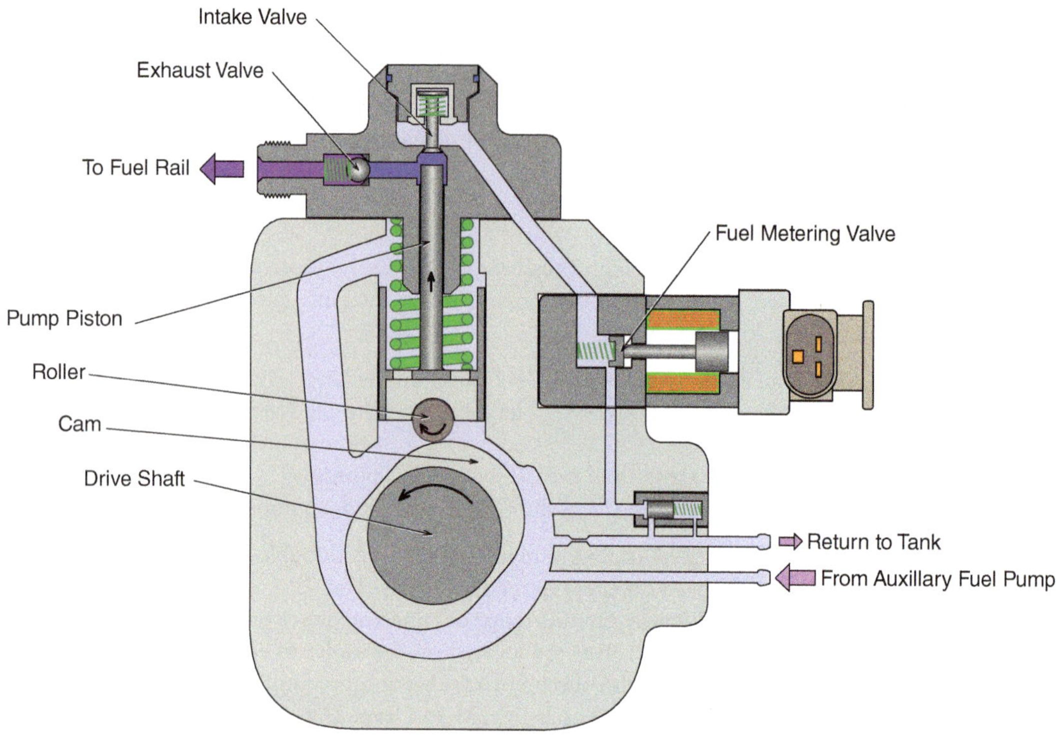

FIGURE 13-66 Internal construction details of a Bosch CP-4.1 pump used on four-cylinder engines.

CP-3 Pump Construction and Operation

The high-pressure CP-3 pump is driven at 1:1 ratio to engine speed. Unlike the CP-4.1 and CP-4.2, it does not require synchronization with engine position in order to time high-injection events. The pump's function is to provide fuel at high pressure while the ECM controls injection pressure and timing.

The CP-3 pump commonly uses a gear-type fuel supply pump attached to the rear of the high-pressure pump supplying fuel at up to 43 psi (3 bar). The supply pump pulls fuel from fuel tank through a primary filter before sending it out of the pump to a secondary filter. When returned to the high-pressure pump, fuel passes through a cascade overflow valve (COV), which regulates maximum supply pump outlet pressure by returning excess fuel to the fuel tank or supply pump inlet (**FIGURE 13-67**). Pressure regulation is necessary since fuel flow and outlet pressure increase proportionally with engine rpm. Regulated internal pump pressure is known as housing pressure and supplies fuel to the internal moving components of the high-pressure pump to lubricate them.

The COV has a spring-loaded spool valve that uses three drilled passages for regulating fuel pressure at three stages, based on the fuel supply pressure at the COV inlet:

- Stage 1: 0–43 psi (0–3 bar) for initial low-pressure lubrication. This passage always allows fuel flow through to the pump's center ring and lubricates the pump's camshaft bushings and internal moving parts. This circuit also allows air to bleed during initial cranking and returns the air to the fuel tank.
- Stage 2: 80–150 psi (5.5–10.3 bar) lubrication for housing pressure. This pressure can be reached during cranking and initial start-up.
- Stage 3: over 180 psi (12.4 bar) excess pressure drains to the overflow return to tank.

Fuel Control Actuator (Rail Pressure Regulator)

Fuel from the supply pump branches not only to the COV but also to the electrically operated fuel control actuator (FCA) valve, which is in the rear cover of the high-pressure pump. The FCA functions to regulate fuel pressure supplied to the fuel rails. The solenoid uses a pulse-width modulated (PWM) signal sent by the ECM to regulate fuel pressure supplied to the injectors connected to the fuel rail. A typical duty cycle is between 5% and 95%. The normally closed valve is progressively opened by a lengthening pulse width. That means a short pulse width or low percentage of duty cycle, will produce maximum pump output pressure. The electrical signal can be absent to the solenoid for as long 30 seconds after the ignition switch is keyed to On to enable maximum fuel pressure to the fuel rail during initial cranking period. Rail pressure regulation begins when the ECM observes that engine rpm and rail pressure have reached their commanded valve (**FIGURE 13-68**).

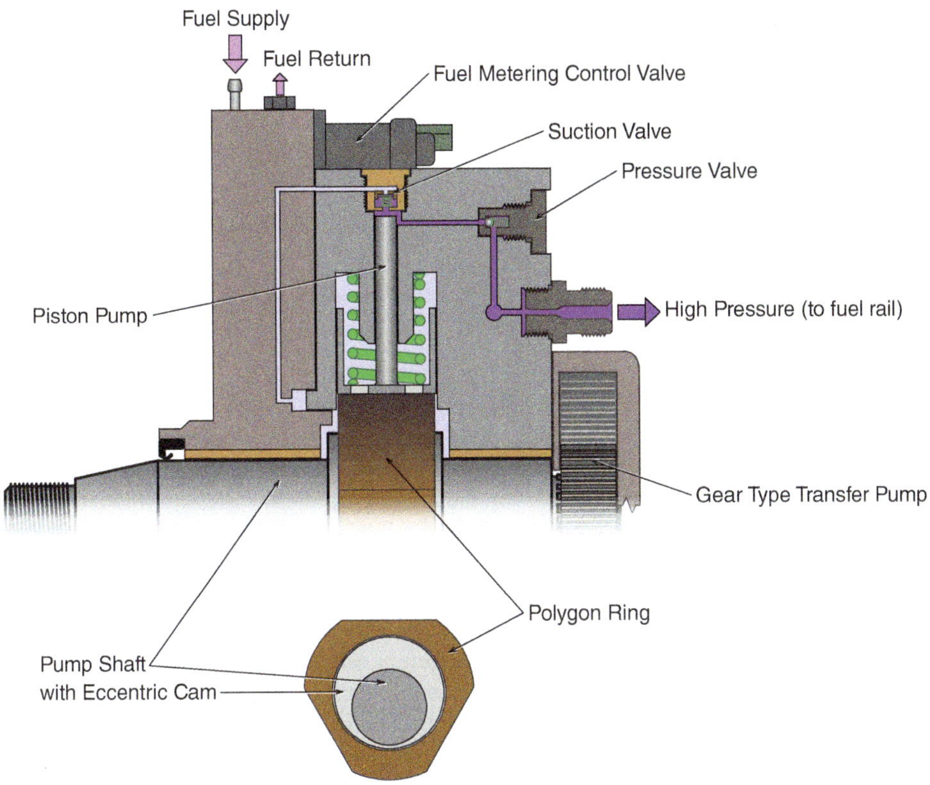

FIGURE 13-67 Internal construction details of the Bosch second-generation CP-3 high-pressure pump.

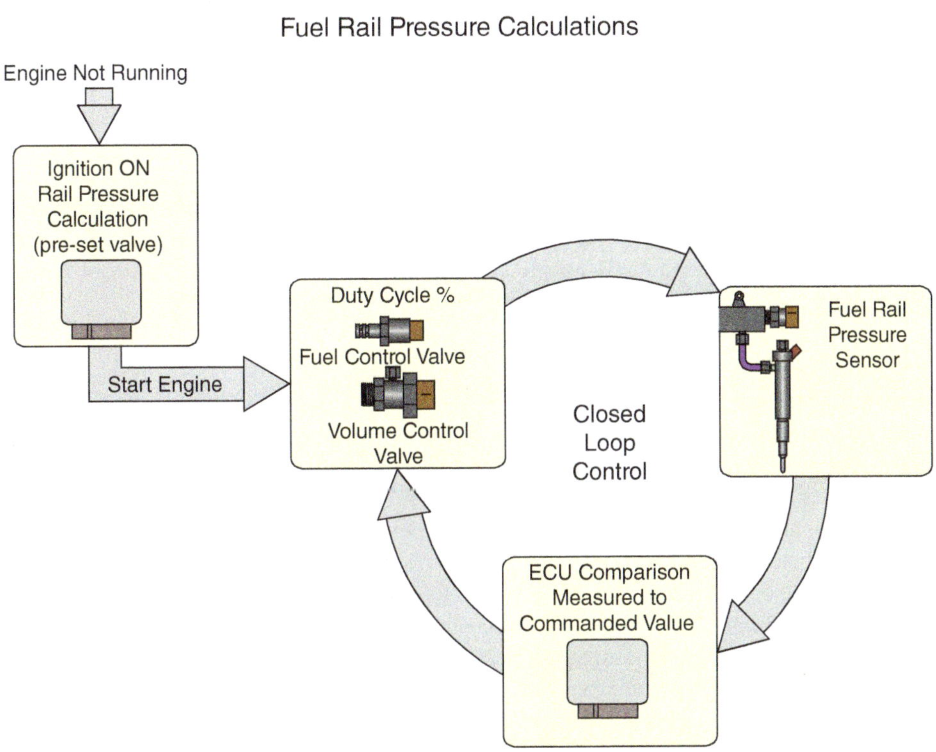

FIGURE 13-68 Closed-loop control of the fuel rail pressure.

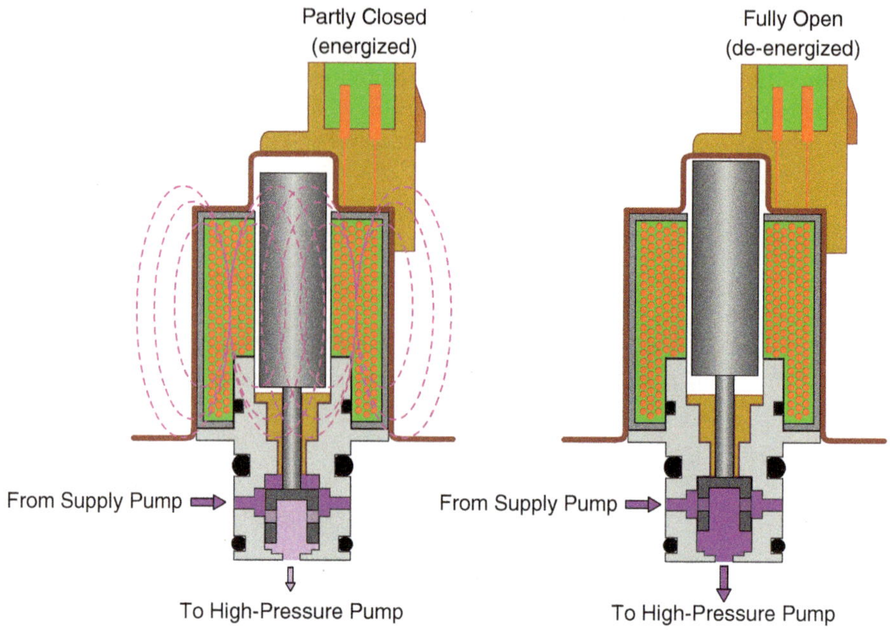

FIGURE 13-69 The rail PCV is a normally closed valve that requires a magnetic field to open. Incrementally increasing magnetic field strength lowers rail pressure.

If measured fuel rail pressure is too low, the PWM signal is shortened to achieve pressure established by a calibration lookup table in the ECM. Rail pressure that's too high will cause the PWM signal to lengthen until actual measured fuel pressure and commanded pressure are the same. If the pressure regulator should lose its electrical signal, or become open circuited, it will operate at a wide-open or default value, and the engine will continue to operate, though it will run roughly (**FIGURE 13-69**). The optimal value for injection pressure is adjusted by the ECM according to the engine speed and load conditions.

With some engine control systems, the pressure regulator can provide an input signal to measure fuel temperature. Rather than use a thermistor-type sensor and run additional wiring for a circuit to the ECM, resistance of the coil within the regulator is measured. In the FCA electrical winding, coil resistance is proportional to fuel temperature, permitting fuel temperature to be calculated from the actuator's resistance.

High-Pressure Pumping Plungers

The FCA valve supplies fuel to three high-pressure pumping chambers through inlet check valves that admit fuel to flow into the pumping chambers. Fuel is pulled into the pumping chambers during a downward stroke of the pumping plungers. When the pump camshaft rotates and forces the plunger tappet roller upward, the inlet valve closes compressing fuel in the chamber. Fuel leaves the chamber through a second spring-loaded outlet check valve. Because the outlets of all three pumping chambers are connected in one internal circuit, high-pressure fuel between 4,351 psi (300 bar) and 23,206 psi (1,600 bar) can flow from a single outlet port through a steel line connected to the fuel rail.

Rail Pressure Control Valve Operating Strategies

In second-generation Bosch systems using CP-3 pumps, a single rail PCV, or FCA, is used to regulate the quantity of fuel admitted into a high-pressure pump. As already mentioned, a closed-loop feedback circuit exists between the rail pressure sensor and the fuel pump's fuel control valves in pumps using volume control valves. A DC PWM signal applied to this fuel pressure regulator determines, how much fuel is either supplied to the pump, returned to the tank, or returned to the pump inlet. When no electrical energy is present, the valve is closed, forcing all pump output to the fuel rail. Increasing the length of the PWM signal strengthens the magnetic field operating the valve, which then opens the valve, so pressure to the fuel rail drops.

A two-regulator high-pressure pump concept uses both an inlet volume regulator and a second rail pressure regulator typically located on a fuel rail (**FIGURE 13-70**). Locating the second regulator on the rail means the system response to pressure regulation is much faster than a pump-mounted single-regulator concept. Little or no delay in increasing or

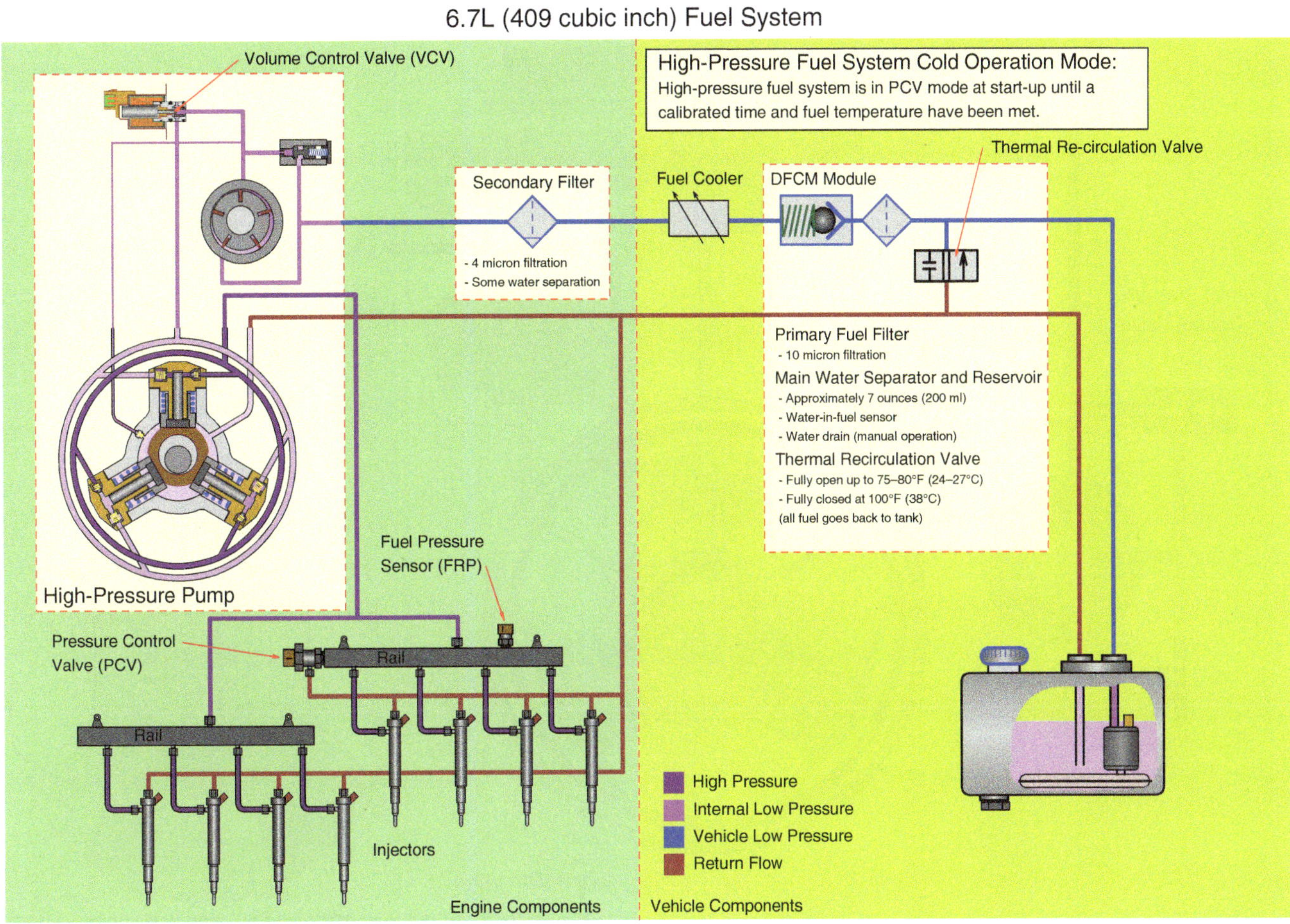

FIGURE 13-70 This Bosch fuel system uses the newest CP-4.2 pump, which incorporates an inlet volume metering valve at the pump and a PCV on the left fuel rail.

decreasing rail pressure to the injectors during engine overrun conditions is the result. This is particularly important when the engine decelerates and requires lower fuel pressure, or when increasing engine speed and higher pressure is necessary. In both instances, fuel pressure can either bleed down or build up more rapidly, not only providing a faster throttle response but also avoiding mismatches between rail pressure and required injection pressure, which could potentially result in increased emissions.

Control of the High-Pressure Fuel: Two-Controller Concept

Two fuel system regulators are used on late-model engines. Bosch, Denso, Delphi, and Siemens use the two-regulator control configuration in late-model engines. While the regulators have different names designated by manufacturers, they do have the same function. One regulates pressure in the fuel rail(s), and the other regulates the flow of fuel into the high-pressure pump. The inlet volume metering valve is located on the injection pump and is typically a normally closed PWM-controlled valve. The second pressure-regulating valve is located on a fuel rail and has a normally open PWM solenoid. A PWM signal is applied to this regulator to control the amount of fuel returned to the fuel tank through the valve.

When the engine is operating, the ECM regulates fuel rail pressure by using a combination of the signals supplied to the rail PCV and the volume control valve (VCV) (**FIGURE 13-71**). Two regulators are used to provide the faster, more accurate control of fuel pressure needed for precise fuel metering and low emissions.

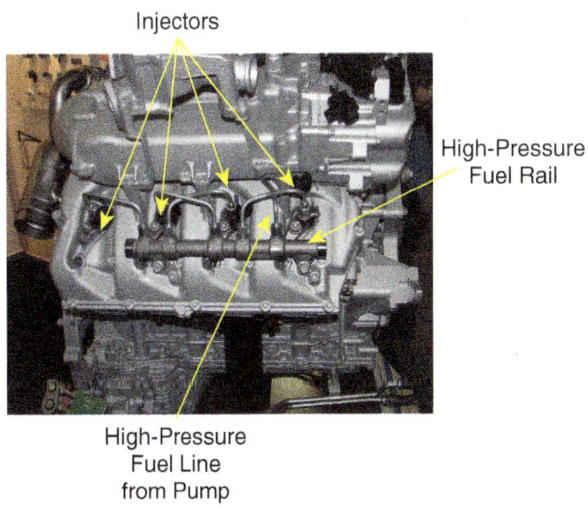

FIGURE 13-71 Fuel-injection pressure on this 6.7 L Power Stroke is measured by the high-pressure fuel rail sensor (FRP). Two regulating valves—a PCV and a fuel-metering unit, formerly known as volume control valve (VCV)—are on the CP-4.2 pump.

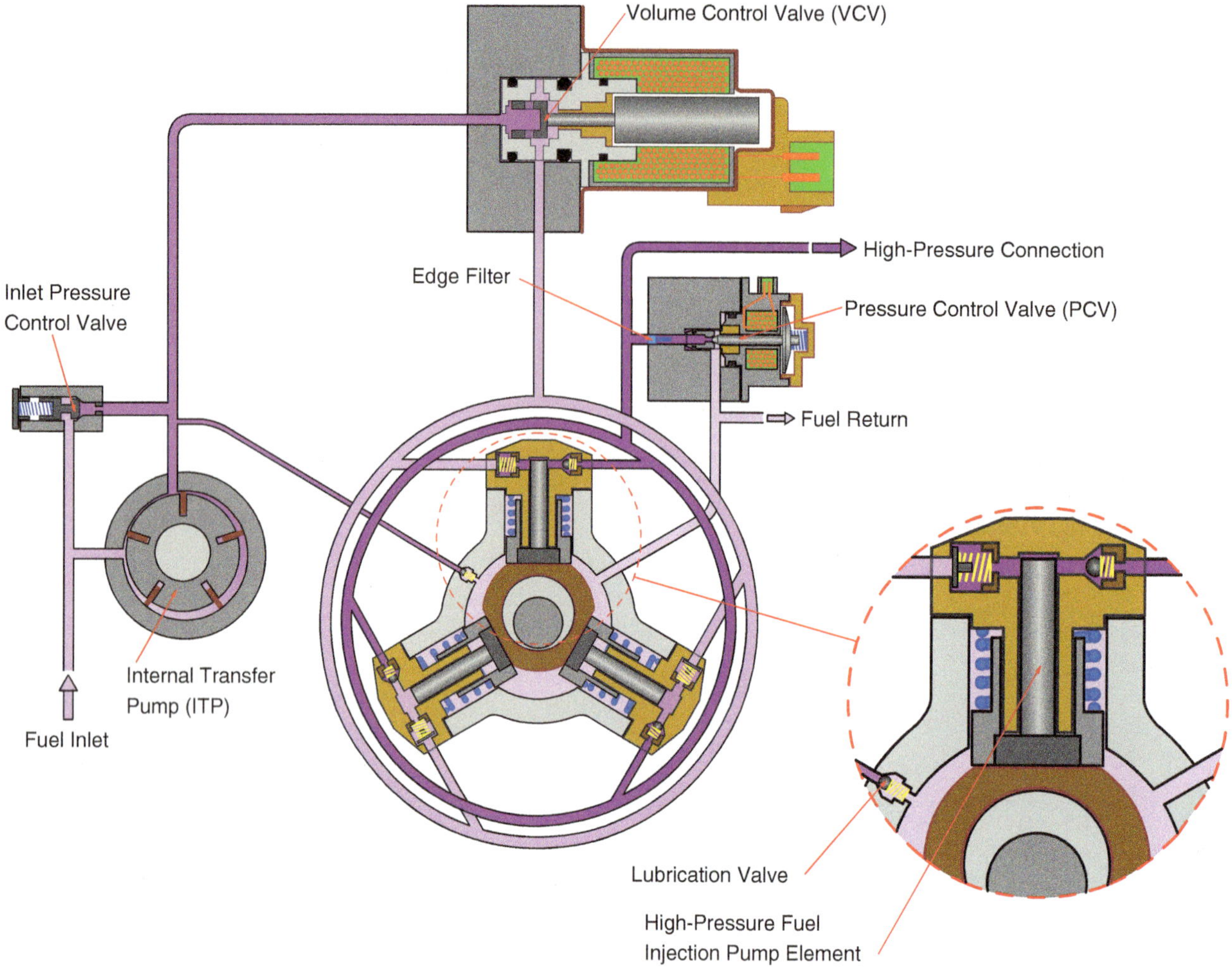

FIGURE 13-72 Signals from the engine ECM control the operation of the VCV and PCV to regulate high-pressure fuel supplied to the injectors in this Siemens system. A closed VCV causes transfer pump outlet pressure to return back to the transfer pump inlet through the inlet pressure control valve..

The pump-mounted VCV is a normally closed and meters fuel into the pump based on engine demand. Without a PWM signal, no fuel enters the pumping chambers and all fuel passes to the fuel return circuit (**FIGURE 13-72**). A PWM signal to the VCV is needed to open the valve and redirect fuel into the pump's pumping chambers. A longer PWM signal opens the valve's fuel passage to the pumping chambers, increasing fuel supplied by the pump to the fuel rails. The second regulator, a normally open PCV on the rail, adjusts fuel pressure based on ECM requirements for injection pressure. When the PCV is not energized, fuel passes through the regulator to the low-pressure fuel return circuit. Progressively closing the valve with an increasing PWM signal length, or duty cycle of, increases rail pressure. Both valves are actuated by a PWM signal from the ECM (**FIGURE 13-73**).

If the fuel pressure in the rail area is too high, the PCV opens, causing some of the fuel in the rail to bleed off into the low-pressure fuel return. Insufficient fuel pressure in the fuel rail causes the ECM to close the PCV, preventing fuel leakage into the fuel return.

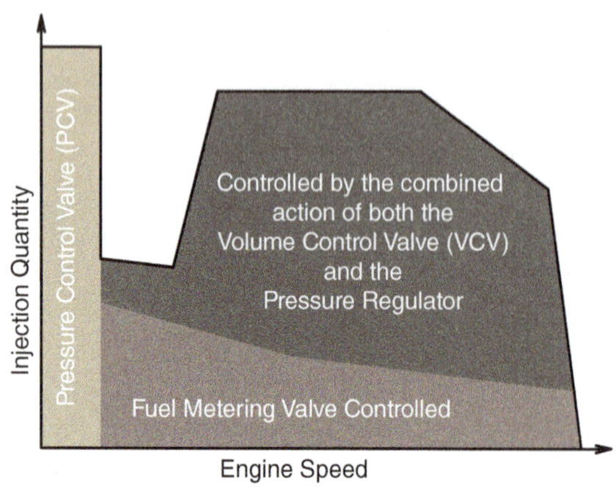

FIGURE 13-73 Operating strategies of the two-controller concept. Two controllers help reduce emissions, improve drivability, and provide a backup for fail-safe engine operation if one controller fails.

Start-Up Pressure Control Strategy

The rail pressure is primarily controlled by the PCV on the fuel rail to enable preheating of cold fuel at engine start-up and for a rapid buildup of rail pressure. To heat up the fuel quickly, it is also necessary for the high-pressure

Operation of Two Pressure Controllers

Operating Condition	Pressure Control Valves (PCVs)	
	Volume Control Valve (VCV)—Pump	Rail PCV—Rail
Starting	Open	Closed
Idle	Controlled	Controlled
Moderate Speed	Open	Controlled
Full Load	Controlled	Controlled
Fault Mode	Limp mode	Open

pump to deliver and compress more fuel than is needed for combustion. This requires signaling the VCV to open further, admitting more fuel into the pump and redirecting less fuel to the return circuit.

Pressure Control Strategy Under Load

When large quantities of fuel for injection is required, accompanied by high injection pressure, the PCV performs most of the demand-based regulating functions. Under these conditions, the high-pressure pump's power consumption is reduced by limiting the amount of fuel entering the pump. Limiting flow into the pump avoids unnecessarily heating up any fuel.

Regulation with Volume Control Valve and Pressure Control Valves

During idle or trailing throttle conditions, when the throttle is backed off, only small quantities of fuel are required. This condition requires volume and pressure regulation by both valves. Pressure regulation using both valves enables more precise pressure control, which improves idle-speed stability and transitions to other load condtions such as trailing thorottle when the accelerator is released are accompanied by minimal over pressurization in the rails.

> ▶ TECHNICIAN TIP
>
> Air trapped in fuel will cause prolonged starting or hard starting of the engine. Electric pumps or hand-primer pumps are used on CR systems to purge air from filters during service and to ensure that the engine will start rapidly. Installing filters dry and priming them after installation ensures that only filtered fuel will reach dirt-sensitive CR injectors. Running an engine out of fuel and prolonged cranking without first priming high-pressure pumps will damage the fuel-lubricated pumps. Damaged pumps may fail immediately or within hours or days of such abuse.

High-Pressure Fuel Rail

The fuel rail is a thick-walled reservoir, most often in the form of a cylindrical tube, for highly pressurized fuel supplies the injectors (**FIGURE 13-74**). The rail is continuously filled with pressurized fuel supplied by the high-pressure fuel pump. It not only stores pressurized fuel but also provides some dampening action for the pressure pulsations of the high-pressure pump. Throttling valves or restriction orifices inside the rail also act as pulsation dampeners.

Pressure-Protection Valve

All fuel rails contain a mechanical pressure-limiting or pressure-protection valve (refer back to Figure 13-71). In the event that the fuel rail is overpressurized due to PCV failure. During such an event, the pressure-protection valve opens and returns fuel to the fuel return circuit. It has a feature that permits a slight delay in pressure release until maximum rail pressure exceeds a predetermined safety factor. Typically, this is approximately 3,000 psi (207 bar) above maximum rail pressure (**FIGURE 13-75**). On some engines, the valve is incorporated into the function block with the rail pressure sensor. Most others install the valve on the fuel rail. Siemens's plezoceramlc lnjectors have an lntegrated pressure-rellef valve.

Fuel Rail Pressure Sensor

This sensor is mounted in the high-pressure fuel rail and provides data to the ECM for closed-loop adjustment to the pressure control regulator. The sensor is usually a three-wire, pressure sensor using a 5-volt reference signal. It operates in a closed-loop mode with the pressure regulator, meaning data from this sensor provides feedback to the ECM, which in turn adjusts the length of the PWM signal to the FCV. Closed-loop operation means no

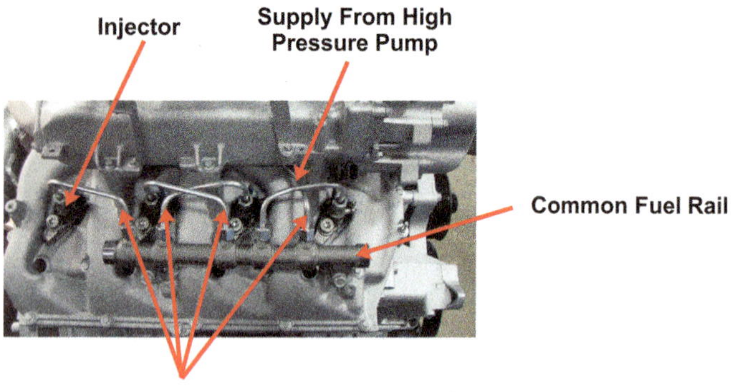

FIGURE 13-74 This cast fuel rail provides an attachment point for fuel injector lines. Because fuel pressure is the same or common to all injectors, the fuel system is called common rail.

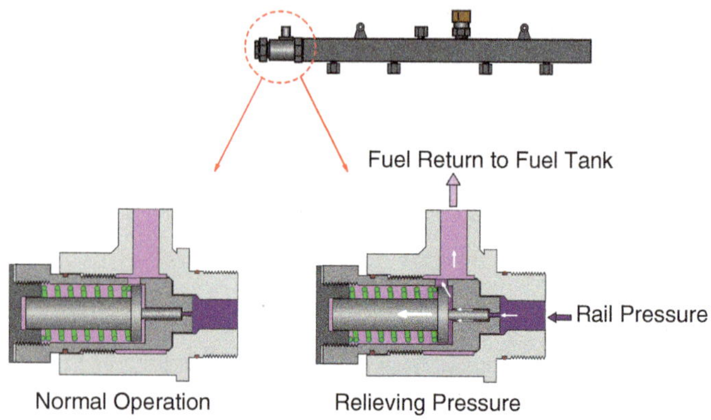

FIGURE 13-75 Operation of the pressure-protection valve in the fuel rail.

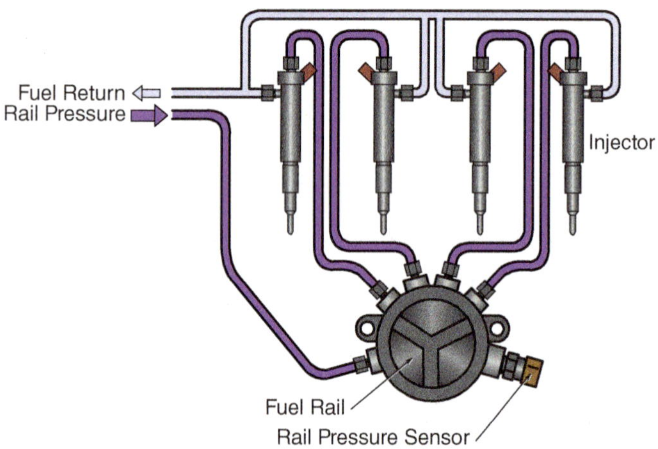

FIGURE 13-76 The rail pressure sensor for this Delphi built circular fuel rail is located in the fuel rail.

input data other than that provided by the rail pressure sensor is used to adjust the pressure in the fuel rail (**FIGURE 13-76**).

Flow-Limiting Valve

In the event of an injector failure causing continuous flow of fuel through the injector, a mechanical safety valve shuts off the flow of fuel to the failed injector. Flow-limiting valves can be threaded into the fuel rail for each injector. The fuel line to the injector connects with the flow-limiting valve. The valve is constructed with a plunger and spring assembly that acts like a throttling device with a precisely defined flow rate. During normal operation, the flow-limiting valve's plunger will reciprocate back and forth inside its housing during each injection. If flow rates exceed the preset limit, the plunger will seal against its seat, remaining there until the engine is switched off and fuel flow to the injector stops.

▶ Service and Maintenance of Common Rail Fuel Systems

13-06 Recommend service for and maintenance of CR fuel systems.

CR injection systems deliver exceptional reliability. When fuel-related problems do occur, the OBD system will quickly detect faults affecting system operation. Because the construction and operation of CR systems are unique, specialized service practices and diagnostic techniques are used to pinpoint problems with the system. This next section will examine some common diagnostic and service practices for non-amplified CR systems.

Onboard Diagnostics for Common Rail Fuel Systems

Emission standards have mandated the production of the cleanest diesel engines ever. Emission levels from the latest Tier-3 engine built beginning in 2017 cannot even be detected with shop-grade emission analyzers because end of lifecycle emission limits are close to zero. Tier 3 emission levels, in comparison to Tier 2 bin 5 standards, are one-third to one-fifth less. To monitor and maintain the emissions levels to their current standard, the OBD-II legislation requires manufacturers to detect any potential malfunction that could cause emissions to rise above legislated standards which is anywhere between 1.5 to 4 times the FTP threshold.

The fuel system is one area extensively monitored by the OBD-II system, and there are several specific strategies used to identify faults on CR systems:

- fuel system pressure control
- fuel mass observer
- injection timing

- fuel balancing control
- injector voltage
- pressure wave correction factor
- zero fuel mass calibration.

Fuel System Monitors—Fuel System Pressure Control

The CR fuel system monitor monitors fuel system pressure to verify that it is operating within an expected range (**FIGURE 13-77**). The rail pressure sensor is checked, and deviations in the fuel PCV and VCV are evaluated. One particular problem that CR systems can have is related to excessive fuel flow. A higher PWM signal applied to the inlet VCV and lower-than-anticipated rail pressure sensor values will indicate a fuel leak. A fuel system leak causing rail pressure to bleed down causes hard-start, no-start, and/or low-power conditions. Often, an injector with excessive fuel back leakage due to worn or defective ball valves and seat will cause the MIL light to illuminate under heavy loads or at highway cruising speed. A worn, eroded, or dirty ball seat will cause the injector to leak or to stay open and inject fuel continuously. Similarly, filter lint or dirt can clog the inlet orifice to the pressure control chamber in the injector, causing the injector nozzle vavle to remain open.

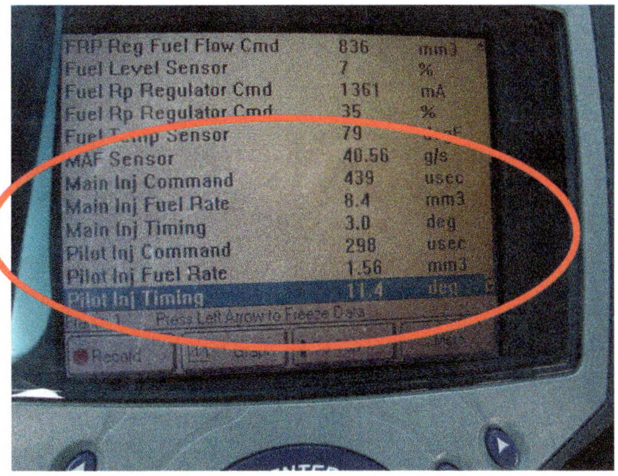

FIGURE 13-77 Fuel rail pressure, injection quantities, injection timing for the main injection, and the pilot injections are monitored by the OBD system.

Due to the extreme fuel pressures on CR systems, fuel lines may not always be reusable. Deformable fuel line ends are intended for one-time use only and may leak internally into a cylinder head if retightened. Incorrectly tightened injector lines can form leak paths for high-pressure fuel (**FIGURE 13-78**). To identify the source of the leak, it is important to understand that there are three fuel return flowpaths that can be measured to identify a faulty component:

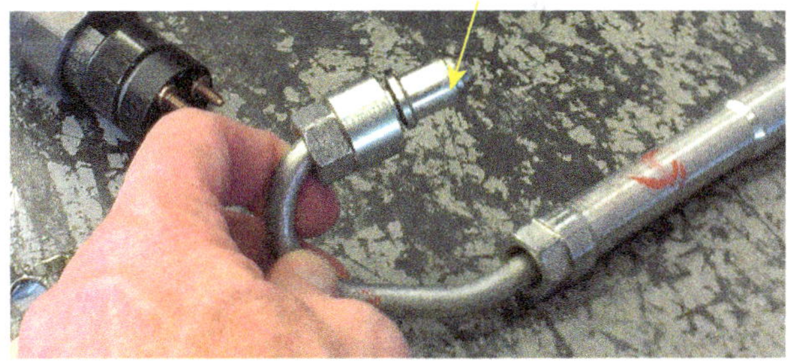

FIGURE 13-78 The ends of CR fuel lines are deformable and generally cannot be reused. High-pressure fuel may leak from the line if it is reused.

1. injector fuel return
2. high-pressure pump return
3. high-pressure relief valve return.

Isolating and measuring the quantity of fuel returning to the tank is a process of elimination to determine whether various components are operating within specified values (**FIGURE 13-79**). For example, if an injector is suspected of having excessive fuel return due to a defect, that injector can be isolated after measuring the total return flow from all injectors. If the return flow is substantially reduced after blocking the injector, it is likely that the injector is defective (**FIGURE 13-80**). Worn injector ball seats, back leakage from around the nozzle valve, pressure-protection valves, and high-pressure pumps will return excess fuel too. Quill tubes, which are steel tubes in the cylinder head that supply rail pressure, can make poor seating contact with the injector and thus form leak paths (**FIGURE 13-81**). Unless the injector is installed carefully by following the manufacturer's instructions and using the correct torque and installation sequence, the tube can fail to deform the soft surface of the tube tip, providing a "crush seal" seating at the injector fuel inlet. Fuel pressure will leak out of the rail when the engine is off and generate fuel leakage. Fuel return from the cylinder head is measured to identify injectors returning excess fuel may also detect quill tube leaks. Injectors are isolated by using a purpose made cap that is screwed onto the rail fitting where the injector line would normally connect. To isolate a defective injector, its high-pressure fuel line is blocked. If the fuel return from the head drops to normal, the blocked injector is identified as the defective component. (**FIGURE 13-82**).

Measuring Injector Back Leakage

Leaking and defective injector control valves are responsible for draining high-pressure fuel to the low-pressure fuel return and are the most common failure in CR injectors.

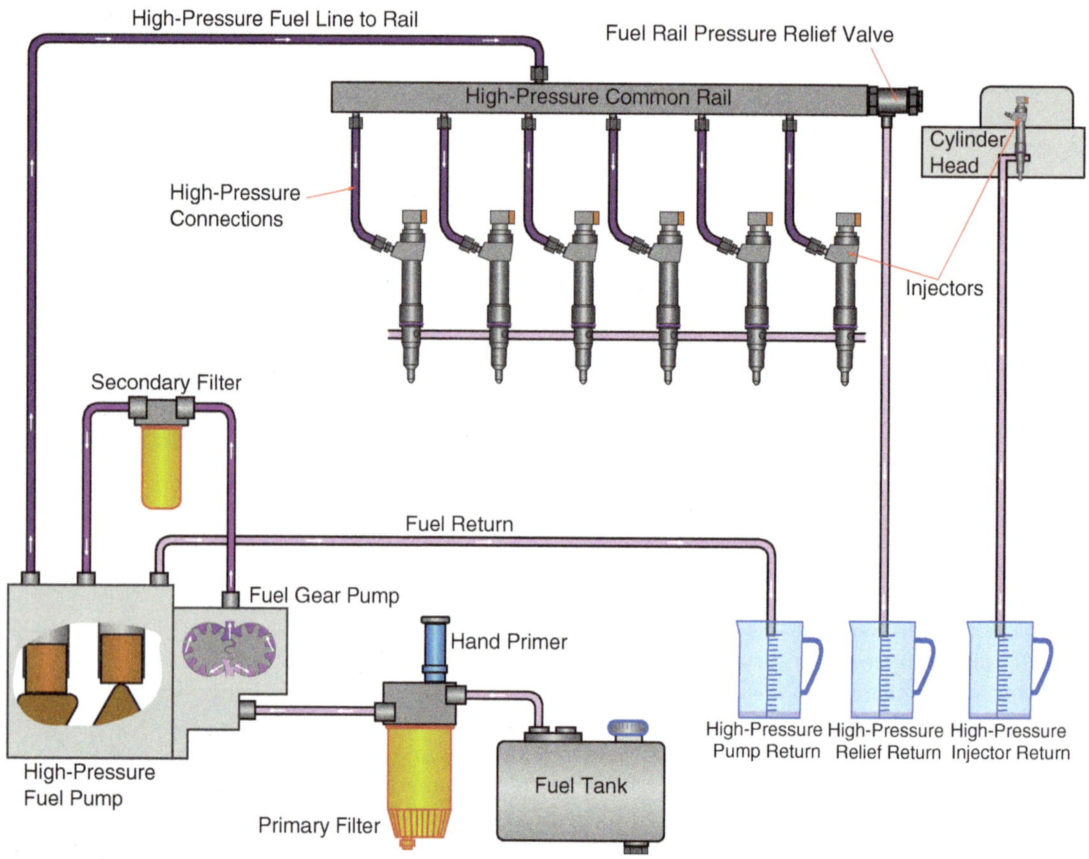

FIGURE 13-79 Fuel schematic of a CR system. Return flow is measured from the cylinder head, high-pressure pump, and pressure-protection valve.

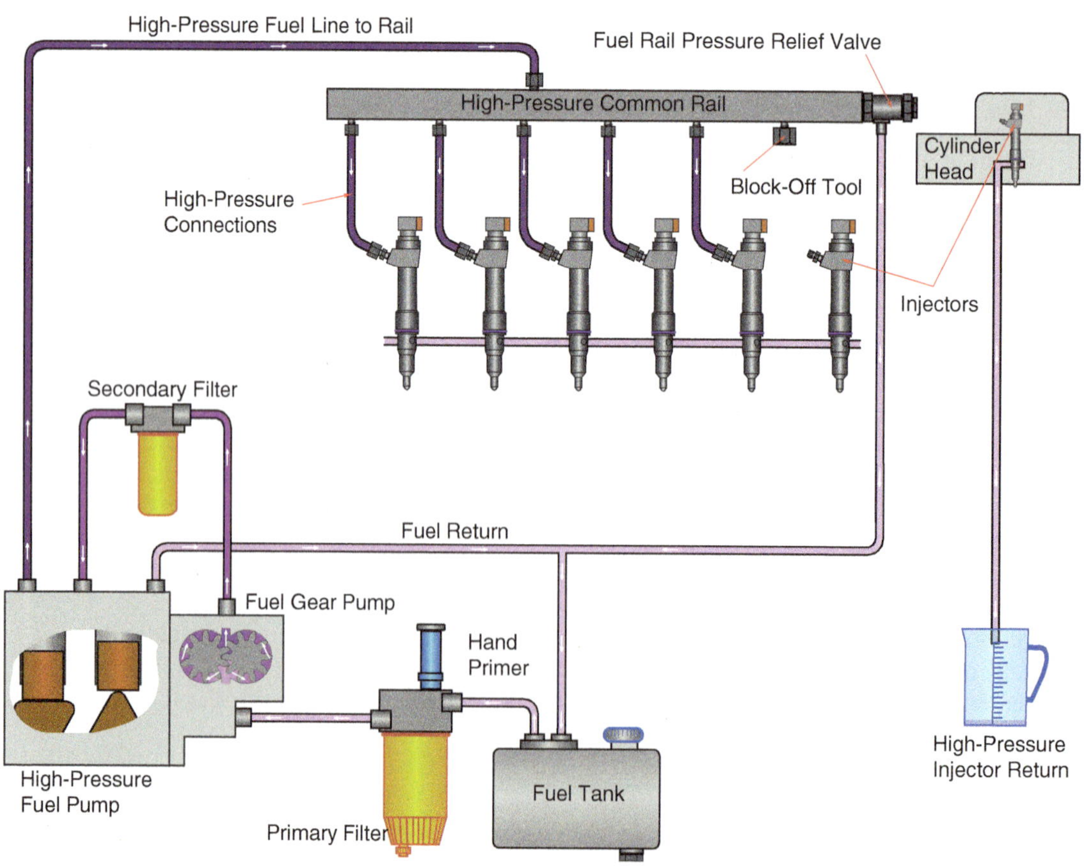

FIGURE 13-80 A defective injector is identified by blocking the high-pressure fuel line and measuring back leakage from the cylinder head. Defective injectors will substantially lower the volume of back leakage.

FIGURE 13-81 The junction between the steel quill tube and injector is a common leak path. To prevent leakage, a careful sequence of installation steps must be precisely followed.

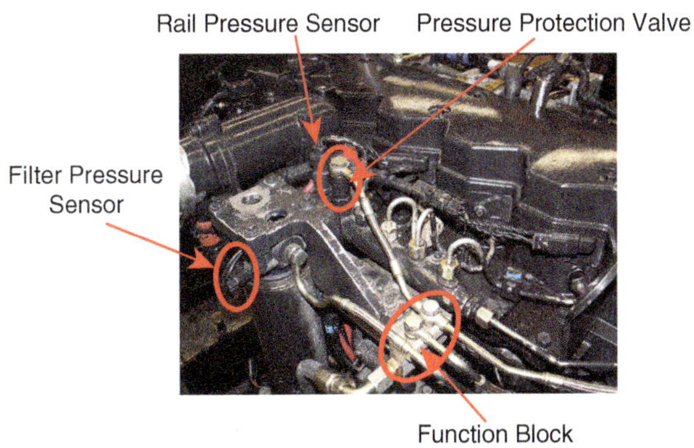

FIGURE 13-82 The function block on this Cummins 5.9 L diesel engine is a test point for measuring fuel return from the pressure-protection valve, injectors, and the high-pressure pump.

The control valves, which regulate the hydraulic balance across the nozzle valve, can fail due to wear, corrosion, and abrasive fuel contamination. Highly pressurized fuel can leak past the valves, which increases the volume of fuel return from the injectors. When the control valve leakage is severe, high-pressure fuel in the fuel rail can flow directly to the fuel return circuit, enabling unintended lifting of the nozzle valve from its seat. If this takes place, fuel leaks out of the nozzle tip and into the combustion chamber and may cause the engine to run away, the cylinder to misfire, and the nozzle to drip. Fuel leakage from the rail and into the injector can even fill the cylinder with fuel when the engine is shut off. Measuring fuel return volume from each nozzle is generally the most effective way to identify leaking injectors if external return lines are used to connect injectors to the tank return circuit.

To perform an injector back leakage check, several methods can be used, depending on the engine and injector design. To measure injector back leakage, follow the steps in **SKILL DRILL 13-1**.

The differences between a high-pressure pump's fuel pressure regulator pulse width for commanded rail pressure and measured rail pressure identify fuel leaks to the OBD system. The pulse width becomes shorter as more fuel is needed to maintain fuel rail pressure. These are not external leaks but leakage of fuel back to the fuel tank through

> **▶ TECHNICIAN TIP**
>
> Fuel pressure of 3,500–4,500 psi (241–310 bar) is needed to start most Bosch CR systems. Hard starts and failure to start are often caused by fuel pressure leak-down from the fuel rail. Common leak paths are created by quill tubes and defective ball valves in injectors, which allow fuel to drain into the return circuit through the injector. When diagnosing these conditions, always check the rail pressure. If the system does not hold pressure, there are leak paths.

SKILL DRILL 13-1 Measuring Injector Back Leakage (Method 1)

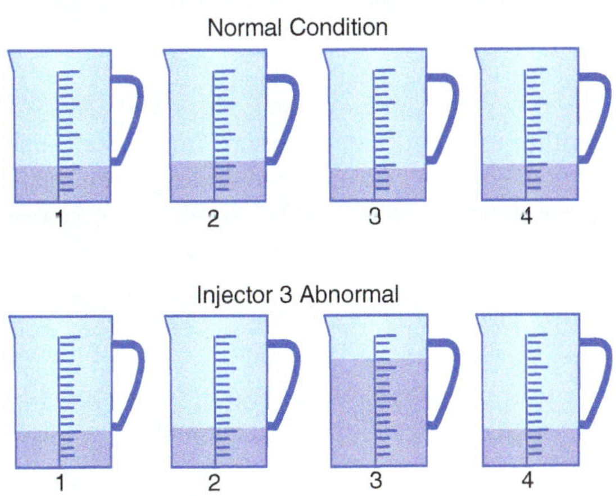

1. Remove the fuel return hose from each injector, and install jumper tubes to the return line fittings.
2. Connect the jumper tube injector return hoses to a set of graduated cylinders.
3. Start and run the engine at idle. For three minutes, measure and record the volume of fuel accumulating in the graduated cylinders.
4. For two minutes, operate the engine at high idle speed. Measure and record the volume of fuel accumulating in the graduated cylinders.
5. Compare fuel accumulations to manufacturer's specifications. Generally, a defective injector will flow three times the minimum return rate. No fuel return indicates a defective injector solenoid. Injector replacement is recommended in either condition.

SKILL DRILL 13-2 Measuring Injector Back Leakage (Method 2)

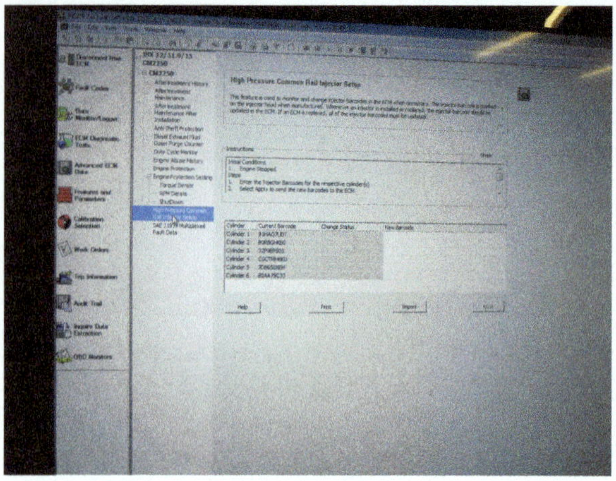

1. Connect OEM software to the datalink adapter and navigate to the fuel leakage diagnostic menu. Enable the fuel leakage test to raise the rail pressure to maximum pressure at idle. Alternatively, operate the engine at high idle.
2. Collect and measure the fuel return from the cylinder head. If the return is abnormal, one or more injectors may be leaking internally due to defective ball valves, or a quill tube may be leaking.
3. Shut off the engine and disconnect cylinder 1's injector line at the fuel rail. Install a purpose-made cap on the rail fitting to block fuel flow.
4. Start and run the engine again in the fuel leakage diagnostic mode or at high idle.
5. Collect, measure, and record the volume of fuel returned from the cylinder head while the injector is blocked.
6. Repeat the fuel collection and measurement procedure for each cylinder in the engine by moving the fuel blocker fitting to each cylinder.
7. Compare collected return volume. Note any cylinder that had an above-average drop in fuel return volume when its fuel line was blocked. This cylinder likely has a defective injector or quill tube. Alternatively, measure and record fuel return when multiple fuel lines are blocked and compare with OEM specifications for each cylinder.

the injector. High-pressure pumps are too often replaced because they cannot keep up with fuel leakage from the high-pressure system when under heavy loads. Often, the particulate filter will require frequent regeneration if the leakage is from defective injectors, because of excessive soot loading from large, coarse fuel droplets.

To measure injector back leakage from defective injectors without individual return lines to return fuel through the cylinder head, follow the steps in SKILL DRILL 13-2.

Measuring Fuel Leakage from Other Fuel System Components

A normal crank-to-start time for a CR injection system is usually around three to five seconds. To protect injectors from being damaged from a lack of fuel, On some engines, including Cummins diesel engines, the injectors will not be switched on until the fuel rail pressure reaches a minimum threshold pressure. A prolonged cranking period or no-start condition can be caused by leaks in the high-pressure fuel system. Cranking re-primes the fuel system and purges air from injectors. Leaking ball valves in the injectors are one cause of leakage, but there are others. Leaking quill tubes, defective valves in the high-pressure pump, and leaking rail pressure-protection valves are some other common areas. Leakage will generate fault codes that require pinpoint testing to identify where fuel is bypassing the injectors.

To measure fuel leakage from other fuel system components, follow the steps in SKILL DRILL 13-3.

Fuel Mass Observer

A fuel mass observer (FMO) is an algorithm used to detect deviations from normal delivery volumes from all injectors. On engines equipped with oxygen and NO_x sensors, the percentage of oxygen in the exhaust is compared to an expected oxygen percentage based on fuel rate, boost pressure, and EGR flow. If oxygen content differs substantially from an expected model of exhaust gas content, a code is set.

Injection Timing

The camshaft and crankshaft position sensors evaluate when the beginning of combustion takes place by measuring crankshaft acceleration rates. If the correction factor applied to adjust injection timing deviates significantly, a fault code can be set.

SKILL DRILL 13-3 Measuring Fuel Leakage from Other Fuel System Components

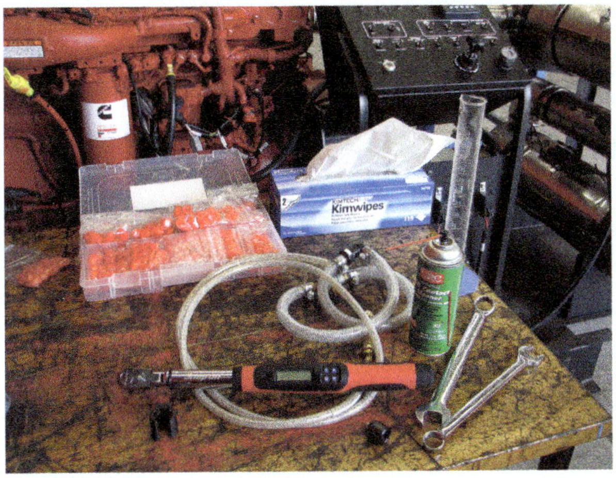

1. Verify that the fuel rail pressure remains high with the engine stopped. Connect a scanner or OEM service tool to the datalink connector and observe the rail pressure. A good system will maintain rail pressure for weeks at values close to where they were when the engine was shut off. If rail pressure falls off quickly or is found to have less than 1,000 psi (69 bar) after a few hours, there is a high-pressure system leak.
2. Remove the valve cover and inspect individual injectors carefully for evidence of fuel leaks.
3. Navigate to the diagnostic menu of the OEM software and request a fuel leakage test. Run the engine in the fuel leakage test mode, which increases rail pressure to its maximum value.
4. Connect a clear plastic hose to the function block or return line of the high-pressure pump. Collect and measure the fuel volume in graduated cylinders with the engine running in fuel leakage test mode or at high idle. Record all observations.
5. Connect a clear plastic hose to the function block or fuel return line for the rail pressure-relief valve. Collect and measure the fuel volume in graduated cylinders with the engine running in fuel leakage test mode or at high idle. Record all observations.
6. If an engine doesn't have individual return lines but does have return passageways through the cylinder head, measure injector leakage. Connect a clear plastic hose to the function block or fuel return line for the injector return from the cylinder head. Collect and measure the fuel volume in graduated cylinders with the engine running in fuel leakage test mode or at high idle. Record all observations.
7. Compare the fuel accumulations to the manufacturers' specifications. Generally, a defective high-pressure pump or rail pressure-relief valve will return excessive fuel if it is worn out or defective.

Fuel Balance Control

Fuel balance control (FBC) is a strategy used by the ECM to adjust fuel delivery quantities to each cylinder to achieve consistent pressures among all cylinders. Changes in crankshaft speed due to variations in cylinder pressure are measured. The amount of

▶ TECHNICIAN TIP

When servicing injectors, it is critical that the injector calibration file match the injector position in the engine. The code must be read from the injector and installed in one or more ECMs by using an electronic service tool such as a handheld scanner or computer-based software. A swapped injector—one moved to another cylinder without changing the calibration file—or a replacement injector will cause the engine to run roughly. On late model OBD-II engines, the wrong injector calibration code will illuminate the MIL light (**FIGURE 13-83**).

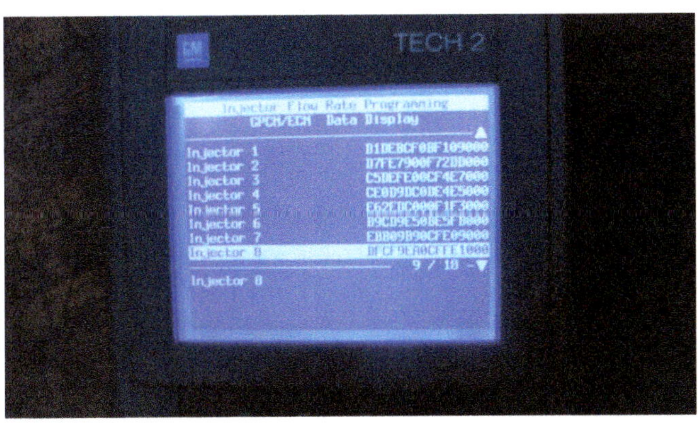

FIGURE 13-83 Injector calibration codes must be updated each time an injector or FICM is replaced. In this engine, codes are backed up in the glow plug control module (GPCM).

▶ TECHNICIAN TIP

Determining the correct calibration code for a CR injector requires specialized equipment to measure and test the injector. Only Injector OEMs possess the equipment and design details used to assign a calibration code. If injectors are rebuilt and internal components are replaced, the old calibration code will not apply. Failure to use a correct calibration code can result in excessive emissions, generate fault codes, and produce rough-running engines.

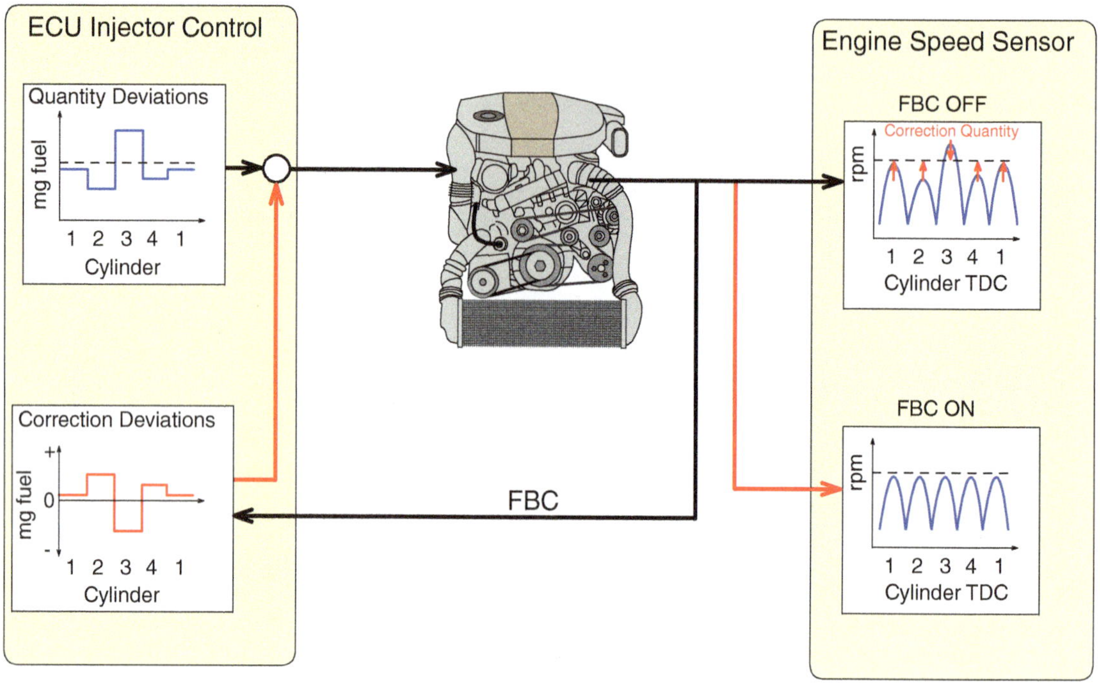

FIGURE 13-84 FBC assists in achieving stable engine speed and consistent cylinder pressures by increasing or decreasing injection quantity to each cylinder.

fuel injected into each cylinder is then adjusted up or down to minimize the difference between cylinders until crankshaft speed is consistent from cylinder to cylinder. Large variations to correct pressure differences are identified as a fuel balance fault or cylinder contribution faults (**FIGURE 13-84**).

Injector Voltage

Voltage drops across injector coils and Injector voltages are measured at the beginning and end of the energization period. Significant changes in expected voltage and the energization time are identified as faults.

Pressure Wave Correction Factor

Pressure wave correction refers to a correction factor applied to a specific injector's electrical pulse width as a result of rail pressure changes. Each injector has unique flow characteristics based on electrical response time and manufacturing tolerances (**FIGURE 13-85**). The procedure is used to assign a correction factor or calibration code assigned to the injector to produce consistent delivery volumes across all rail pressures. A calibration code etched into the injector is used by the ECM to correct fueling variations. During engine assembly or when the injector is replaced, the code must be entered into the engine ECM.

Zero Fuel Mass Calibration

Just as mechanical manufacturing tolerance for injectors requires changes to injector energizing time, normal injector wear also changes injector operating characteristics. Injector deterioration contributes to increased emissions, calling for a recalibration of the stored correction factors that compensate for unique operational behavior. **Fuel mean value adaptation (FMA)** is a term used to describe the control module's automatic recalibration of the injector during service to compensate for wear and deterioration. Without a strategy to recalibrate an injector, a vehicle could pass out of compliance with emission standards and emission system durability requirements. The mehtod of recalibrating injectors during service is called **zero fuel mass calibration (ZFC)**. A common

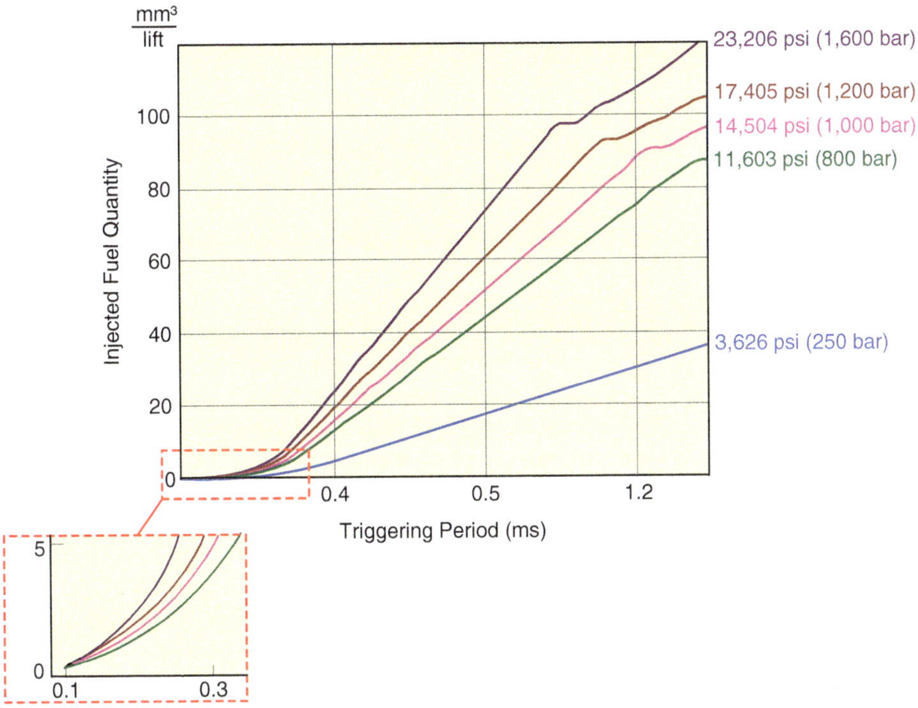

FIGURE 13-85 Pressure wave correction is used to calculate trim codes. Injector flow rates are measured on a flow bench, and a calibration code is applied.

technique for accomplishing ZFC correction occurs during vehicle deceleration. Calibration occurs when the engine is driven only by the vehicle's kinetic energy and when no fuel is injected into the cylinders. Fuel rail pressure is adjusted to a specific value. As long as there is no movement of the throttle position sensor from idle position and the engine in an overrun condition, the ECM will begin to increase energization time to a single injector. When a measurable engine speed increase is detected, the ideal energizing time needed to change engine speed is stored in the ECM. A comparison is made between the new value stored in the ECM and the value stored in the initial calibration code. Mathematical comparisons made between the two values result update in stored correction factors. If a major change takes place in the general trend to incremental updating of ZFC, a fault is set. Each time the conditions for ZFC are met, another injector is recalibrated.

In the future, pressure sensors installed in the cylinder to measure combustion pressure will be used to provide more accurate data for injector calibration.

▶ Wrap-Up

Ready for Review

- Common rail (CR) fuel injection is the latest high-pressure injection system used by the majority of current diesel engines.
- The basic concept of CR injection is to supply fuel at injection pressure to a fuel nozzle that electronically controls the injection event.
- Diesel fuel–injection systems that rely on only engine-driven camshafts to pressurize fuel cannot meet legislated emission targets.
- Modern CR refers to a classification of high-pressure fuel-injection systems in which fuel is pressurized independently of engine speed. This means that regardless of engine rotational speed and load, fuel-injection pressure is matched to optimal values required for the most efficient combustion, lowest emissions, and superior fuel economy.
- Enormous technical and manufacturing accomplishments have enabled CR systems not only to meet rigid emission requirements but to combine to secure a place for diesel technology in the future.
- CR systems are capable of multiple injection events during one combustion cycle.
- The biggest advantage CR injection offers is injection rate–shaping capabilities. Making substantial reductions in exhaust emissions and fuel consumption without hurting

performance requires changing the shape of the injection discharge curve during the injection event.

- Using separate modular components in CR systems means that the fuel system can be easily adapted to different engine types.
- There are three main types of CR injectors: electrohydraulic solenoids, piezoceramic actuators, and hydraulically amplified.
- Bosch has developed four generations of CR injectors; three remain in production.
- In the first second and third generations of Bosch's CR injectors, the injection events are controlled by changing the balance of hydraulic pressure acting on the nozzle valve.
- The third generation of Bosch's CR injectors uses stacks of piezoceramic wafers to control injection events inside the injector.
- Piezo technology enables the fastest switching time for injectors, currently allowing up to 10 injection events in one combustion cycle.
- One of the biggest problems with electromagnetic actuators is the long response time required to build a magnetic field strong enough to move valves inside an actuator. A longer response time contributes to injection lag.
- One of Bosch's fourth-generation CR injectors is the CRS4, also referred to as HADIS, which uses an internal hydraulic amplifier to multiply injection pressure by 2:1. One benefit of the two-stage pressure amplification concept is that the highest pressure is limited to the injector nozzle, where the maximum pressure is required instead of the high-pressure pump, lines, and fuel rail.
- Another fourth-generation injector concept from Bosch is the coaxial CRI4 injector. It uses a nozzle featuring two rows of spray holes in the nozzle tip. This design permits either row of nozzles to be opened independently, which optimizes mixture preparation to achieve a more homogenous distribution of fuel and air in the cylinder.
- Siemens has developed and manufactured piezo injectors since 2001 for a variety of OEMs. The injector uses a conventional six-hole nozzle valve, which operates by using the principle of hydraulically balancing the nozzle valve.
- The fuel system generally referred to today as a CR injection is more precisely termed high-pressure accumulator rail or high-pressure common rail (HPCR). Combined with operational features of the unique CR fuel injector, the most flexible rate shaping of injection events is possible with HPCR fuel systems.
- High-pressure fuel pumps develop the pressure required for injection under all engine operating conditions, including for quick starting the engine.
- The volume control valves (VCVs) and pressure control valves (PCVs) use a pulse-width–modulated (PWM) electrical signal to regulate fuel pressure in the fuel rail.
- The high-pressure pump is driven at one-half engine speed and coupled to the engine through a gear drive mechanism. To minimize parasitic loss, newer pumps have inlet metering capabilities to reduce fuel heating and improve fuel economy.
- The fuel rail is a thick-walled reservoir for highly pressurized fuel used to supply fuel to the injectors. It is continuously filled with pressurized fuel supplied by the high-pressure fuel pump. Not only can the rail store pressurized fuel, but it also provides some dampening action for the pressure pulsations from the high-pressure pump.
- On many systems, a pressure-limiting valve is a mechanical valve that serves the same function as a pressure-protection valve. In the event that the fuel rail is overpressurized, it opens, so fuel flows to the fuel return circuit.
- A fuel rail pressure sensor is mounted in the high-pressure fuel rail and provides data to the ECM for closed-loop adjustment to the fuel pressure control regulator.
- In the event of an injector failure causing continuous flow of fuel through the injector, a mechanical safety valve shuts off the flow of fuel to the failed injector. Optional Flow-limiting valves are threaded into the fuel rail for each injector.

Key Terms

amplified common rail (ACR) Two-stage amplification of fuel pressure inside the injector that provides improved power output with lower emissions, noise, and fuel consumption and that produces a unique shape to the injection rate profile.

coaxial variable nozzle injector A fourth-generation injector concept from Bosch that contains a two-stage nozzle lift with two rows of spray holes. This design permits either row of spray holes to be opened independently, which optimizes mixture preparation to achieve a more homogenous distribution of fuel and air in the cylinder.

fuel balance control (FBC) A strategy used by the ECM to adjust fuel delivery quantities to each cylinder to achieve consistent pressures among all cylinders. Crankshaft speed data is used to make corrections to the volume of fuel injected into each cylinder.

fuel mean value adaptation (FMA) A correction factor made to an injector's energization time based on changes in fuel delivery rates caused by wear and deterioration.

hydraulically amplified diesel-injection system (HADIS) A fourth-generation CR injector that uses an internal hydraulic amplifier to multiply injection pressure by 2:1; it is also referred to as CRS4.

inductive resistance Current induced that opposes ECM current when a wire coil is initially energized; it is also known as inductive reactance.

piezoceramic actuator An actuator composed of piezoceramic discs that change shape and in turn change the balance of hydraulic forces inside the injector; this type of actuator can switch fuel on and off in as little as 0.0001 seconds.

pressure wave correction (PWC) The process used to produce a calibration code for a CR injector. A correction factor is applied to a specific injector's electrical pulse width to adapt the injector to an engine based on manufacturing tolerances between injectors.

rate shaping Optimizing the injection discharge curve to best match engine operating conditions.

servo-hydraulic The use of a small movement of an electrically operated control valve to manipulate much larger and powerful hydraulic forces.

zero fuel mass calibration (ZFC) The process of recalibrating injectors during service to compensate for wear and deterioration.

Review Questions

1. Consider a hydraulically balanced needle valve used in a first- or second-generation Bosch common rail (CR) injector. How would a drain orifice that is much larger than the inlet orifice affect its functioning?
   a. Slower nozzle lift than closing.
   b. Faster nozzle opening than closing.
   c. Failure of the nozzle valve to lift.
   d. Failure of the nozzle valve to close.
2. Which of the following is the most likely cause of excessive fuel return from the cylinder head of a CR engine?
   a. Excessive injection pressure.
   b. A nozzle valve that is stuck closed.
   c. Eroded or pitted ball valve seats.
   d. A plugged inlet orifice in one or more injectors.
3. What typically happens to the length of the pulse width to the electronic fuel control valve (FCV) valve when fuel leaks from the high-pressure rail?
   a. It remains the same.
   b. It increases.
   c. It decreases.
   d. It increases, the suddenly decreases.
4. Consider a non-amplified CR injector. If the injection control valve cannot close properly due to dirt or erosion of the valve seat, what is the most common result?
   a. The injector cannot deliver any fuel.
   b. The nozzle valve stays seated.
   c. The nozzle valve remains open.
   d. The flow control valve remains open.
5. Consider a non-amplified CR injector. If the inlet restriction orifice to the pressure control chamber is plugged, what is the most likely to happen?
   a. The injector cannot deliver any fuel.
   b. The nozzle valve stays seated.
   c. The nozzle valve remains open when the engine is running.
   d. The flow control valve remains open.
6. Consider a piezoelectric CR injector. What happens to the actuator when current is initially applied to the piezoceramic wafers?
   a. The nozzle valve is seated by high-pressure fuel.
   b. The crystals expand.
   c. The crystals contract.
   d. Reaction induction causes a delayed injector response time.
7. Consider a CR3 system with a leaking fuel return line. Which of the following situations is the operator most likely to encounter?
   a. A noise in the engine.
   b. A cylinder misfire.
   c. A no-start condition.
   d. Low power.
8. Consider a Bosch CRS2 system. The FCV on a CP-3 pump is open circuited. Which of the following conditions is most likely to be observed?
   a. The engine won't start.
   b. The engine runs rough.
   c. The engine will only idle.
   d. Very low engine power.
9. Which Bosch CR system requires certified insulated gloves when working on the system with the ignition key in the On position?
   a. CR1.
   b. CR2.
   c. CR3.
   d. CR4.
10. When performing a back leakage test on a CR engine with CRI2 injectors, excessive fuel flow from the fuel return line was measured from the cylinder head. Which of the following conditions is most likely the cause?
    a. A worn-out high-pressure pump.
    b. A defective fuel control valve.
    c. A leaking quill tube connection.
    d. A defective pressure-protection valve.

ASE Technician A/Technician B–Style Questions

1. Technician A says that a CR engine can vary the injection pressure, timing, and injection rate shape for each injector. Technician B says that the injection pressure is the same for all injectors in CR engines, and injectors can perform multiple injection events during each combustion cycle. Who is correct?
   a. Technician A
   b. Technician B
   c. Both A and B
   d. Neither A nor B
2. Technician A says that servo-hydraulic injector actuators use a small amount of control valve movement to control larger, more powerful, and faster-acting hydraulic forces. Technician B says that control valves that use piezoelectric technology are the only actuators capable of servo action. Who is correct?
   a. Technician A
   b. Technician B
   c. Both A and B
   d. Neither A nor B
3. Technician A says that rate shaping involves changing the quantity of fuel as it is discharged from the injector spray holes. Technician B says that rate shaping requires multiple

injection events and a high-pressure fuel system with the ability to vary injection pressure. Who is correct?
 a. Technician A
 b. Technician B
 c. Both A and B
 d. Neither A nor B
4. Technician A says that dirt bypassing the filter on some CR systems can cause an injector to stick open and continuously fuel a cylinder. Technician B says that any dirt particles may potentially damage the injector, but they would not allow an injector to full-fuel or cause the engine to run away. Who is correct?
 a. Technician A
 b. Technician B
 c. Both A and B
 d. Neither A nor B
5. Technician A says that faster-responding piezoelectric injectors are best suited for use in light-duty CR engines. Technician B says that engine speeds of automotive diesel do not require the piezoelectric injector's higher number of injection events. Who is correct?
 a. Technician A
 b. Technician B
 c. Both A and B
 d. Neither A nor B
6. Technician A says that servo-hydraulic actuators that use electromagnets have better durability than piezoceramic actuators. Technician B says piezoceramic actuators are as durable as, if not more durable than, servo-hydraulic injectors that use electromagnets. Who is correct?
 a. Technician A
 b. Technician B
 c. Both A and B
 d. Neither A nor B
7. Technician A says that cylinder misfires can be detected on CR engines by slightly cracking open the fuel line to each injector when the engine is running. Technician B says that it is important to place a shop towel over the injector when checking for misfires to prevent pressurized fuel from penetrating the skin. Who is correct?
 a. Technician A
 b. Technician B
 c. Both A and B
 d. Neither A nor B
8. Technician A says that CRI2 injectors will amplify injection pressure under heavy loads as needed by the engine. Technician B says that injection amplification takes place only in CR3 injectors. Who is correct?
 a. Technician A
 b. Technician B
 c. Both A and B
 d. Neither A nor B
9. While two technicians are discussing the purpose of a second electrical control valve, which regulates fuel volume supplied to the inlet of the high-pressure pump, Technician A says the VCV helps reduce fuel consumption. Technician B says the VCV is needed to reduce emissions. Who is correct?
 a. Technician A
 b. Technician B
 c. Both A and B
 d. Neither A nor B
10. While performing a back leakage test for a second-generation Bosch CR fuel system, the fuel volume was measured to be above specification from the cylinder head. Technician A says the cause could be a leaking quill tube. Technician B says the cause could be one or more injectors with worn-out ball valve seats. Who is correct?
 a. Technician A
 b. Technician B
 c. Both A and B
 d. Neither A nor B

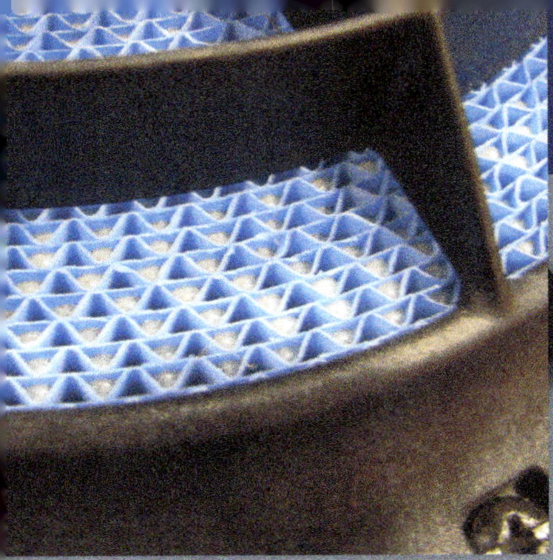

CHAPTER 14

Air Induction and Crankcase Ventilation Systems

Learning Objectives

After reading this chapter, you will be able to:

- 14-01 Identify and explain the airflow requirements for diesel engines.
- 14-02 Identify and explain the purpose and functions of air intake system components.
- 14-03 Identify and explain the purpose and functions of the air intake system construction features.
- 14-04 Identify and explain the functions of air filters.
- 14-05 Identify and explain the purpose of air induction system accessories.
- 14-06 Identify and describe starting aids related to the air induction system.
- 14-07 Identify and explain the purpose and operation of crankcase emission control devices.

You Are the Technician

Believing an aftermarket manufacturer claim that its high-flow air filter would deliver more power and better fuel economy, a tow truck operator installed a gauze fabric type of air filter that was regularly washed and then sprayed with a sticky oil. The oil was supposed to help capture dirt particles and repel water from the filter fabric. After a couple of years of using the filter, the turbocharger bearings prematurely failed, in addition to many of the engines hydraulically actuated electronic unit injection (HEUI) injectors. An oil sample taken after repeated injector failures revealed unusually high levels of silicon, lead, copper, and iron in the oil. The report from the oil analysis lab indicated one possible cause for the unusually high levels of metal contaminants is an air induction system leak. One technician suggested to the operator that the aftermarket filter installation traded off higher engine airflow for lower filtration efficiency. The technician speculated that dirt was actually entering the engine through the filter gauze, which they believed would not have happened with a good-quality paper filter. Before condemning the air filter, consider the following:

1. Explain why higher-than-expected levels of iron, silicon, lead, and copper would indicate a leak in an air induction system.
2. Identify which areas and list which components, other than the air filter, urgently need inspection.
3. Describe what further complaints could be expected if the air induction system leaks are not identified and repaired.

Introduction

Diesel engines need to consume air to burn fuel. Since diesel engines use lean-burn combustion systems, much more air enters the combustion chamber than is necessary to burn all the fuel. Using air mass many times more than is necessary to support combustion results in the oversensitivity to air quality that is characteristic of diesel engines. Technicians need to be aware of not only the effect that air quality has on engine performance due to the use excess air-fuel ratios but also the unique features of the much larger capacity of the intake air induction system. To enable the engine to handle large volumes of air, the air pathway into the engine uses a variety of devices and sensors to help correct for air quality problems with the intake air. These include problems with air that is cold, humid, or dirty; air that contains snow, rain, or fog; or air that has low atmospheric pressure, which can lead to power complaints, excess emissions, poor starting and drivability problems. Contemporary diesels also have significant new performance and emissions technology associated with air induction systems, such as the crankcase ventilation system. Understanding the design, operation, and maintenance practices associated with the air intake system is important since the air supply from the induction system has a major impact on exhaust emissions, noise emissions, performance, and engine durability. To help understand these practices, this chapter will explain the air supply requirements of diesel engines and identify air induction system components while also explaining their operation and construction features. Maintenance of the induction system and crankcase ventilation systems is outlined as well.

Air Supply Requirements

14-01 Identify and explain the airflow requirements for diesel engines.

To support efficient combustion with minimal emissions, diesel engines need an adequate supply of air that is clean, dry, fresh, and relatively cool. Diesel engines require an excess supply of air relative to the minimum quantity needed to burn all the fuel injected in the combustion chambers. For example, a diesel engine of 350 hp (261 kW) typically requires an air intake capacity of close to 1,000 cubic feet per minute (cfm) (28.32 m^3/min). A 325 hp (242.4 kW) 6.0 L Ford Power Stroke will use 732 cfm (20.73 m^3/min) @ 3,300 rpm (**TABLE 14-1**). This expands to an exhaust volume of 1,500 cfm (42.5 m^3/min) @ 3,300 rpm. Contrast this example with a spark-ignited, gasoline-fueled engine, burning fuel at **stoichiometric ratio**. In stoichiometric combustion engines, only the minimum amount of air is supplied to consume all the fuel. No air or fuel remains after the combustion event. A 6.0 L stoichiometric combustion engine requires only 160 cfm (4.5 m^3/min) of airflow.

Function of Excess Airflow in Diesel Engines

Diesel engines require excess air for the following three reasons:

1. Complete and cleaner combustion: Large amounts of intake air are required to achieve good combustion quality (**FIGURE 14-1**). Since diesel engines have a short amount of time for injecting, atomizing, distributing, and vaporizing fuel within a few degrees of crank rotation near top dead center (TDC), it is important that the fuel find enough air to contact and react with the oxygen. Therefore, the more air in the combustion chamber, the greater likelihood that the fuel will find oxygen and completely burn.

TABLE 14-1 Comparing Diesel and Gasoline Intake Airflow

Engine	6.0 L Power Stroke	6.0 L Gasoline
Airflow @ 3,300 rpm	732 cfm (20.73 m^3/min)	160 cfm (4.5 m^3/min)
Exhaust Volume	1,500 cfm @ 3,300rpm (42.5 m^3/min) Exhaust Temperature 350°F–500°F (176.7°C–260°C) at engine pipe	2,000 cfm @ 6,000rpm 56.63 m^3/min Exhaust Temperature 1,400°F (760°C)

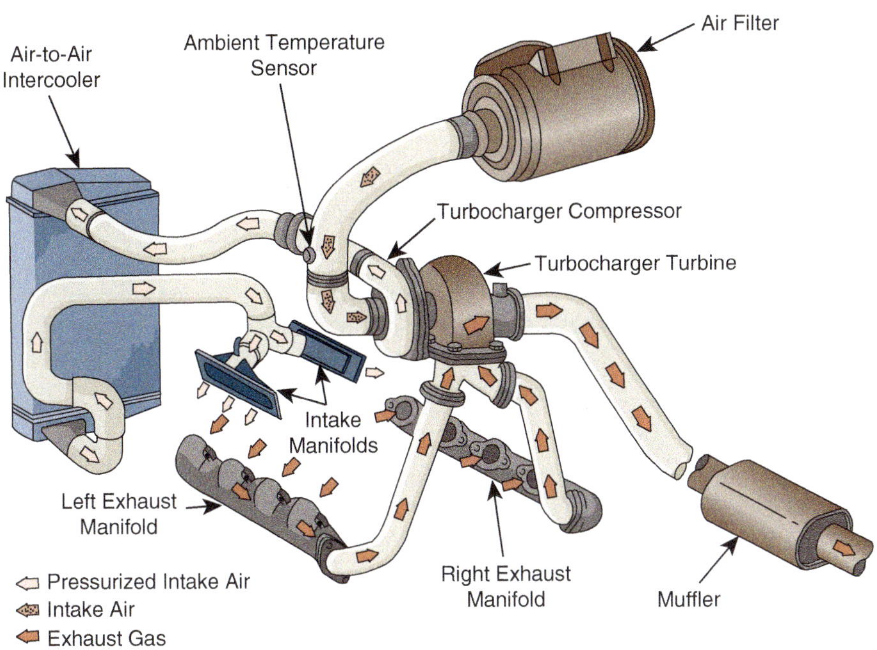

FIGURE 14-1 The flow of gas through the diesel engine from the air filter through the exhaust. Note that maximum airflow through the combustion chambers is limited not only by the air induction system but also by the flow through the exhaust system.

Excess air provides a greater expansion ratio of combustion gas used to increase cylinder pressure and push the piston downward. When excess air is lacking, low power results due to lower combustion pressure and poor combustion quality. There are limits to maximum compression pressures that result in increased power, but higher pre-ignition pressures and temperatures provided by compressing more air mass generally produce faster, more complete chemical reactions.

2. Cooling piston crowns and exhaust valves: Excess air is also necessary to cool the crucial cylinder components in a diesel engine. Since the flame temperature of diesel is near 3,990°F (2,167°C), combustion heat, which is further intensified by high compression pressures, will easily melt pistons and burn valves (**FIGURE 14-2**). By adding additional air volumes, diluting the combustion gases with extra air can lower peak cylinder temperatures. Engines lacking sufficient airflow will naturally have increased exhaust and combustion chamber temperatures. In fact, as an engine's turbocharger boost pressures increase, exhaust temperatures are commonly observed to become cooler.
3. Diesel fuel has a high stoichiometric ratio. Stoichiometric ratio refers to the minimum quantity of air required to completely consume fuel in the combustion chamber with no air or fuel remaining after combustion.

FIGURE 14-2 This piston burned as a result of overheating. Excess air is needed to dilute combustion heat and cool exhaust valves during the valve overlap period.

The best emissions are obtained from gasoline-fueled engines that operate by using a stoichiometric air-fuel ratio of 14.73:1. However, diesel engine combustion is lean (or air rich and stratified) and does not operate at diesels stoichiometric ratio of between 15 - more than 20:1. Stratified means that different air-fuel ratios exist in layers. Combustion regions near the injector spray pattern will be rich with fuel, whereas the overall air-fuel ratio in the chamber is lean. Ignition begins in the regions where the air-fuel ratio is closest to stoichiometric ratio for diesel fuel and combustion flames move out from these ignition points, consuming fuel in the cylinder. The higher carbon-to-hydrogen ratio in diesel fuel requires more air to burn than other fuels since carbon has four bonding points to attach

▶ **TECHNICIAN TIP**

When adding performance accessories that increase injected fuel quantities, it is critical to monitor exhaust gas temperature by using a pyrometer. Less excess air will be available for diluting the heat of combustion gases when additional fuel is burned in the cylinder. By monitoring exhaust gas temperatures, potentially destructive cylinder heat loads are avoided (**FIGURE 14-3**). Increased airflow is necessary when increasing power output derived from injecting more fuel. Larger intake and exhaust systems are both required to improve airflow through the engine air path. Increasing turbocharger boost pressure also lowers excessive combustion temperatures.

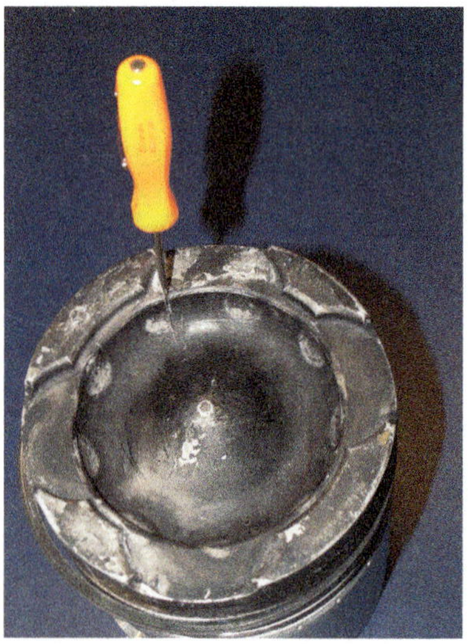

FIGURE 14-3 A restricted air filter caused the vertical cracks in the crown of this piston. The absence of excess airflow during the overlap period allowed the crown to overheat and produce cracking from the edge of the combustion chamber bowl.

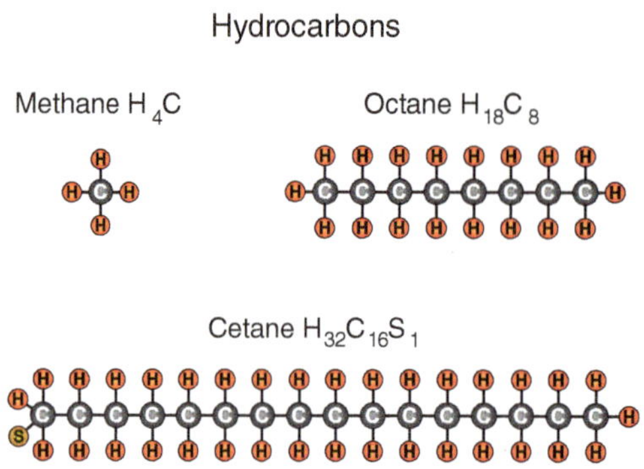

FIGURE 14-4 Diesel fuel has a unique stoichiometric ratio due to the number of hydrogen and carbon atoms in the fuel.

to oxygen, whereas hydrogen has only one. Stoichiometric ratios are always measured in weight and not volume, so depending on the grade, diesel fuel will require a minimum of between 15 and 20 pounds (5.6 and 7.5 kg) of air to burn one pound of fuel (**FIGURE 14-4**). The stoichiometric ratio for denser, more carbon-rich 2-D diesel fuel is approximately 20:1, whereas 1-D is approximately 15:1.

Necessity of Clean Air

Clean, filtered air is crucial to engine operation, to prevent damage and rapid wear due to abrasive dirt particles. Dirt ingestion into an engine, referred to as **dusting out**, can take place through small openings in the intake system and from poor filtration. Incorrect or misaligned air filters and loose hose clamps can also allow dirt to leak into an engine.

Dirt carrying abrasive material damages engines since the abrasives will stick to the oil film on cylinder walls (**FIGURE 14-5**). The abrasives will act as a lapping compound that accelerates ring and cylinder wall wear. From there, dirt will wash into the lubrication system, where it remains in suspension unless filtered out by the oil filter. Some engines do not filter the oil supply for piston cooling nozzles, turbocharger bearings, and some early HEUI injectors. Short-circuit oil feeds to these components bypass the filter to provide a faster supply of pressure during cold start-up. Oil contamination in these instances can lead to catastrophic failures or costly repairs.

To detect leaking air intake systems, the section of air intake between the turbocharger and air filter should be pressurized with air to pinpoint a leak. To determine whether an air intake system has an air leak, performing an engine oil analysis is the best practice (**FIGURE 14-6**). To take an oil sample to detect a leaking air intake, follow the steps in **SKILL DRILL 14-1**. Leaks in the air induction system are detected through the presence of silica, an element that makes up sand. High silica content in oil, and calcium to a lesser degree, indicates potential leaks between the air filter and turbocharger inlet. An engine oil analysis procedure should not replace a visual inspection of all hoses and clamps for proper installation and alignment during regular services (**TABLE 14-2**).

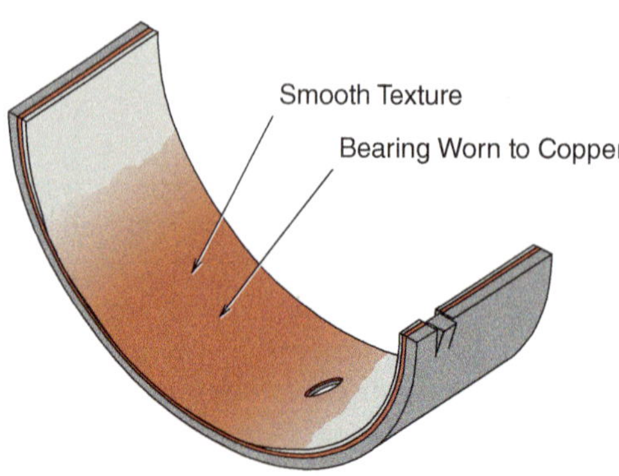

FIGURE 14-5 This bearing was damaged by airborne dirt that entered the engine through a leaking air intake system. The bearing overlay layer was worn, exposing the copper intermediate layer. The texture of the finish is smooth.

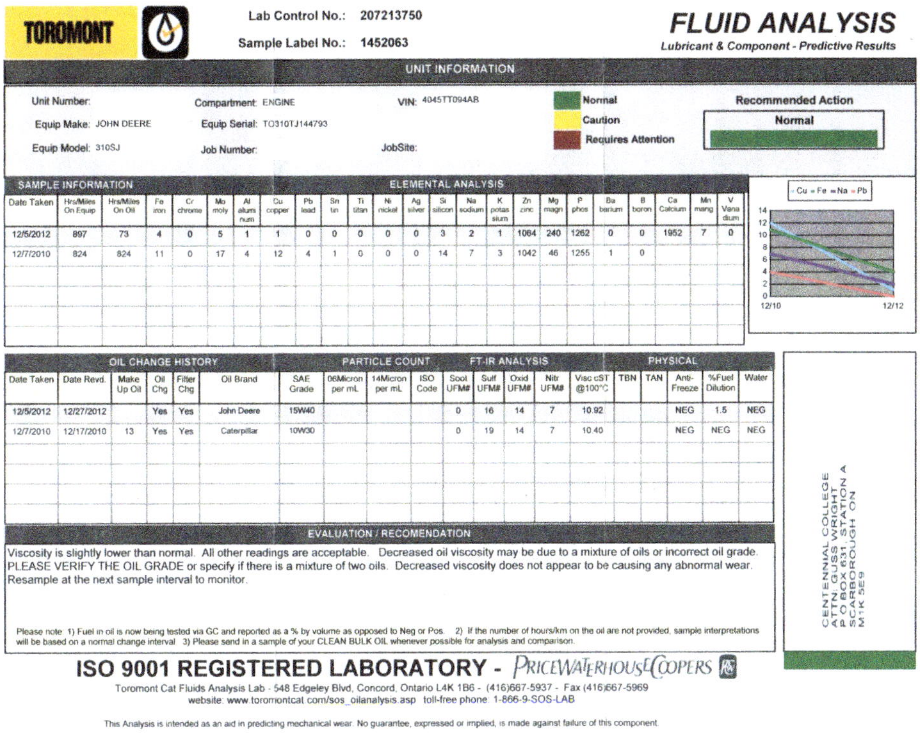

TOROMONT

Lab Control No.: 207213750

Sample Label No.: 1452063

FLUID ANALYSIS

Lubricant & Component - Predictive Results

UNIT INFORMATION

Unit Number:

Compartment: ENGINE

VIN: 4045TT094AB

Equip Make: JOHN DEERE

Equip Serial: TO310TJ144793

Equip Model: 310SJ

Job Number:

JobSite:

Normal

Caution

Requires Attention

Recommended Action: Normal

SAMPLE INFORMATION / ELEMENTAL ANALYSIS

Date Taken	Hrs/Miles On Equip	Hrs/Miles On Oil	Fe iron	Cr chrome	Mo moly	Al aluminum	Cu copper	Pb lead	Sn tin	Ti titan	Ni nickel	Ag silver	Si silicon	Na sodium	K potassium	Zn zinc	Mg magn	P phos	Ba barium	B boron	Ca Calcium	Mn mang	V Vanadium
12/5/2012	897	73	4	0	5	1	1	0	0	0	0	0	3	2	1	1064	240	1262	0	0	1952	7	0
12/7/2010	824	824	11	0	17	4	12	4	1	0	0	0	14	7	3	1042	46	1255	1	0			

OIL CHANGE HISTORY / PARTICLE COUNT / FT-IR ANALYSIS / PHYSICAL

Date Taken	Date Revd.	Make Up Oil	Oil Chg	Filter Chg	Oil Brand	SAE Grade	06Micron per mL	14Micron per mL	ISO Code	Soot UFM#	Sulf UFM#	Oxid UFM#	Nitr UFM#	Visc cST @100°C	TBN	TAN	Anti-Freeze	%Fuel Dilution	Water
12/5/2012	12/27/2012		Yes	Yes	John Deere	15W40				0	16	14	7	10.92			NEG	1.5	NEG
12/7/2010	12/17/2010	13	Yes	Yes	Caterpillar	10W30				0	19	14	7	10.40			NEG	NEG	NEG

CENTENNIAL COLLEGE
ATTN: GUSS WRIGHT
P O BOX 631, STATION A
SCARBOROUGH ON
M1K 5E9

EVALUATION / RECOMENDATION

Viscosity is slightly lower than normal. All other readings are acceptable. Decreased oil viscosity may be due to a mixture of oils or incorrect oil grade. PLEASE VERIFY THE OIL GRADE or specify if there is a mixture of two oils. Decreased viscosity does not appear to be causing any abnormal wear. Resample at the next sample interval to monitor.

Please note: 1) Fuel in oil is now being tested via GC and reported as a % by volume as opposed to Neg or Pos. 2) If the number of hours/km on the oil are not provided, sample interpretations will be based on a normal change interval. 3) Please send in a sample of your CLEAN BULK OIL whenever possible for analysis and comparison.

ISO 9001 REGISTERED LABORATORY - PRICEWATERHOUSECOOPERS

Toromont Cat Fluids Analysis Lab - 548 Edgeley Blvd, Concord, Ontario L4K 1B6 - (416)667-5937 - Fax (416)667-5969
website: www.toromontcat.com/sos_oilanalysis.asp toll-free phone: 1-866-9-SOS-LAB

This Analysis is intended as an aid in predicting mechanical wear. No guarantee, expressed or implied, is made against failure of this component.

FIGURE 14-6 Performing an engine oil analysis is the best practice when checking for intake air leaks.

▶ TECHNICIAN TIP

Leaks caused by loose clamps or misaligned hoses and pipes between the turbocharger and air filter will pull dirt into the engine. Since the intake air is pressurized after the turbocharger, leaks in that circuit will push air rather than pull air out of the pipes and the likelihood of dirt ingestion is almost nothing. Dirt can also enter the engine oil through missing oil filler caps, dipstick tubes, leaking gaskets, and seals if crankcase pressure becomes low due problems with the crankcase ventilation system. Never check for leaks by using flammable gases or liquids, such as ether or propane. A major safety hazard is the potential for a backfire through the intake system ignited by the exhaust gases that are momentarily present in the diesel engine's intake system during the valve overlap period.

SKILL DRILL 14-1 Oil Sampling to Detect a Leaking Air Intake

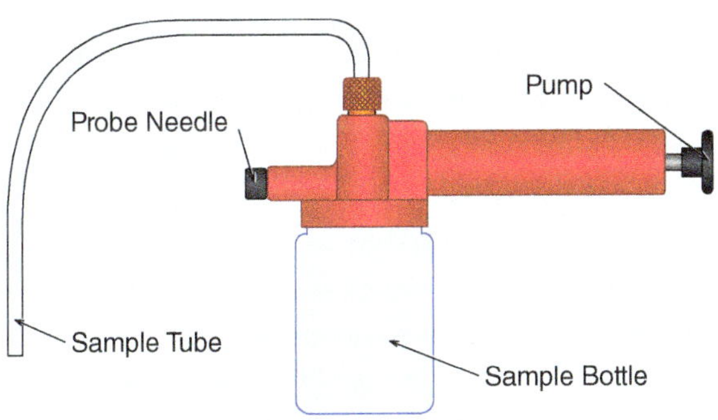

To discover hidden or abnormal operating conditions that can lead to accelerated engine wear or catastrophic failure, perform an oil sample analysis. The analysis provides accurate information about contaminants in oil that can either pinpoint failures or simply indicate normal operating conditions. Leaking air intake systems that allow dirt to enter the engine is one common failure mode detectable through oil analysis. Obtaining the most accurate and reliable results requires collecting a sample of oil most representative of all the oil, its additives, and the contaminants in the system. To do this, several techniques are used and require careful attention to the details of the collection method:

- sampling method or procedure
- appropriate sampling point for oil
- sampling apparatus.

1. Information collected by the technician when taking the sample should include the following:

- vehicle owner or company name
- vehicle or fleet number
- oil sample type (engine, transmission, hydraulic)
- oil brand and grade (i.e., Shell Rotella 5W-20)
- oil sampling point
- engine hours or distance on odometer
- elapsed time or distance since last oil change
- whether the engine has any defects.

The following are two common sampling methods used to collect a small volume:

- hot drain sampling
- suction pump sampling.

Either method requires extracting oil after it has been mixed for 15 minutes while the engine is at normal operating temperatures.

2. When sampling during an oil drain, a purpose-made sample bottle is inserted directly into the stream of oil as it drains from the oil pan.

- Take the sample after the oil starts to drain and not at the beginning or end of the drain period.
- Do not take the sample of oil from another collection container.
- Use care to ensure that no dirt enters the sample bottle when the oil is collected.

Continued

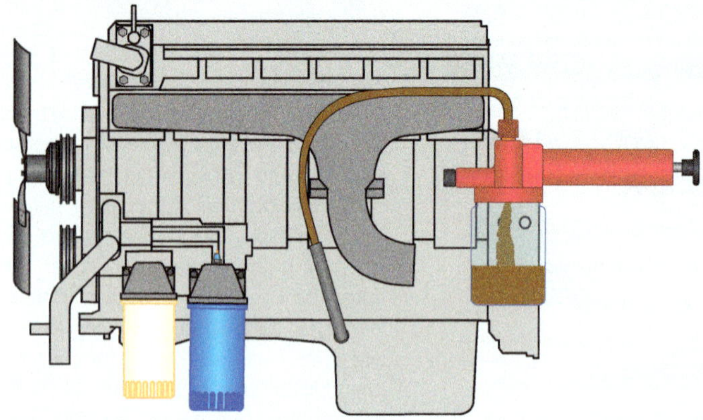

3. When the suction pump method is used, do the following:
   - Screw the sample bottle onto the suction pump adapter.
   - Insert the suction pump or sample hose—usually a 3/16" or 1/4" nylon hose into the engine dipstick tube.
   - Work the hose into the oil pan to the center of the oil pan storage area or the depth to which the end of the dip stick would extend.
   - Fill the bottle until it is almost full.
   - Discard the hose correctly after obtaining the oil sample and do not reuse it.
   - Place the sample in the second sealed container and mail or forward the sample to the analysis lab.
   - Check results of analysis on-line and report oil condition and recommendation of the analysis lab.

TABLE 14-2 Likely Causes of Atypical Results from Scheduled Oil Sampling

Oil Analysis Results	Probable Cause
High copper levels	Worn engine bearings and bushings
High lead levels	Worn engine bearing and bushings
High iron levels	Worn cylinders
High silicate levels	Antifreeze or leaking air intake system
Excessive soot loading	Poor combustion quality or over extended oil service intervals

Effects of Air Inlet Temperature

Optimal air inlet temperatures are essential to good performance and low emissions. Atmospheric pressure, ambient temperature, and humidity affect power since the number of oxygen molecules entering the cylinder will vary with a change in any of these factors. Colder air temperatures translate into increased air density and a greater number of oxygen molecules for a given volume. The Society of Automotive Engineers (SAE) has created a standard method, SAE J1349, for correcting horsepower (kilowatts) and torque measurement to ensure uniformity among test results to compensate for changes to air pressure, humidity, and air temperature. These corrections are often referred to as **standard day factors**. The SAE standard day factors adjust calculations of power output measurements to an atmospheric pressure of 29.23 in/Hg (101 kPa), a temperature of 77°F (25°C), and 0% humidity. Low atmospheric pressure encountered when climbing hills at high altitudes reduces power output. This happens because the pressure differential between the cylinders during intake stroke and the atmosphere is reduced, so less air pressure is available to push air into the cylinders in naturally aspirated engines (non-turbocharged engines) (FIGURE 14-7).

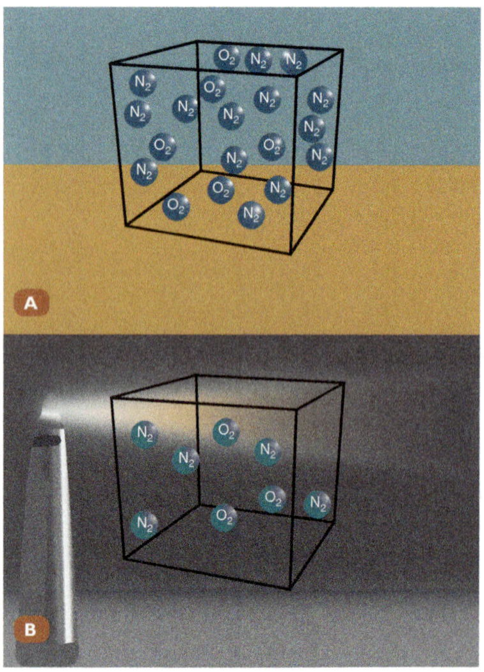

FIGURE 14-7 Compare the oxygen content in the atmosphere when measuring air pressure at sea level **A.** and at high altitudes **B.** High altitude operation results in lower power output and poor combustion quality in naturally aspirated engines.

Generally, a 10°F (5.6°C) change above or below the standard factor results in a 1% power gain or loss (TABLE 14-3). For example, a 50°F (28°C) increase in inlet temperature will result in a 5% power loss. Power gains occur when air inlet temperature is lower. However, an opposing factor to the power gain from cooler ambient temperatures is very low air inlet temperatures that can produce lower

TABLE 14-3 SAE Standard Day Factors

Standard Day Factors J-1349	
Atmospheric Pressure	29.23 in/Hg (101 kPa)
Temperature	77°F (25°C)
Humidity (Relative)	0%

pre-ignition temperatures, resulting in poorer fuel vaporization and longer ignition delay time.

Air temperature also has an effect on turbocharger boost pressure. Consider measuring boost pressure and specifications for testing an ambient temperature of 68°F (20°C). If boost pressure measurement is carried out at any other temperature, the resulting boost pressure must be corrected. For example, a pressure of 13 psi (89.63 kpa) measured at 68°F (20°C) is equivalent to 14 psi (96.53 kpa) at 32°F (0°C). This relationship is explained by understanding that intake air pressure changes with temperature. Colder air is denser and contains more oxygen molecules per cubic foot (or liter) of volume than warm air. A cubic foot or liter of air weighs more when it is near its freezing point than its weight at room temperature (**FIGURE 14-8**).

Diesel engine design is generally optimized for air inlet temperature to the cylinders of approximately 60°F to 90°F (15°C to 32°C).

Air inlet temperature also affects combustion and exhaust temperatures. Exhaust temperatures can change approximately 3° for every 1° change in air inlet temperature. For example, an increase of air inlet temperature of 100°F (46°C) can produce as much as a 300°F (148°C) exhaust temperature increase. Since NO_x formation increases exponentially with high combustion temperatures, the use of charge air aftercooling is important to reduce NO_x formation.

FIGURE 14-8 Compare the amount of oxygen that the atmosphere contains when air temperature is hot **A.** and when it's cold **B.** Cooler ambient temperatures create denser air, which generally improves combustion. Cold air, however, increases ignition delay time.

Effects of Humidity

It is important for the technician who is diagnosing performance complaints to understand that water vapor or humidity in the air displaces oxygen. A room with a 50% relative humidity at 68°F (20°C) will have more oxygen content than a room at the same temperature with 80% relative humidity. For every water molecule entering the room, one air molecule leaves (**FIGURE 14-9**).

Warmer temperatures are also capable of dissolving more water than cooler air temperatures are. This effect is noticeable on hot humid days as people become short of breath faster when physically active on those days. This happens because there is less oxygen content in the air since higher humidity reduces oxygen content in the air. Engines experience similar problems when high ambient temperatures are combined with high humidity. The result is often lower power output and higher levels of exhaust smoke from engines without particulate filters. For this reason, diesel emission testing that uses SAE J-1687 protocols—the snap acceleration test—is restricted when ambient test temperatures exceed threshold limits, which are around 80°F (26.7°C).

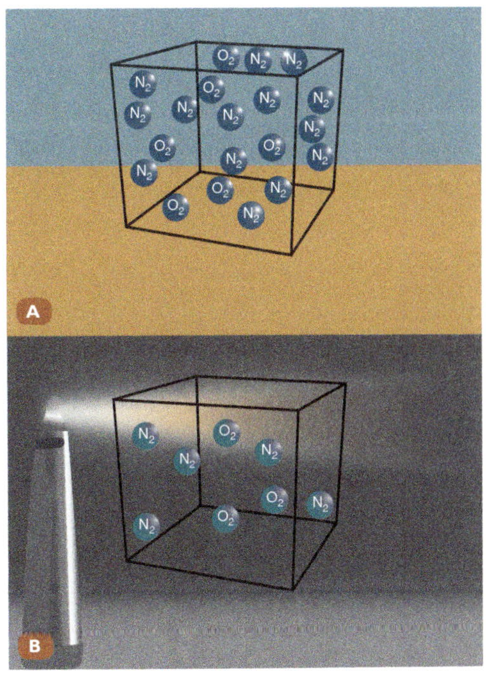

FIGURE 14-9 Compare the oxygen content of the atmosphere on a dry day **A.** with high humidity conditions **B.** such as when foggy. High humidity can lead to low-power complaints and higher emissions since water displaces oxygen in the air.

▶ Design of a Diesel Air Intake System

Air intake system design is important to minimize noise and the quantity of dirt and water entering the air filter. Air intakes should ideally have minimal bending and smooth interior surfaces for maximum airflow. Plastic or rubber tubes are used frequently since they also transfer little heat, are corrosion resistant, and can dampen intake noise better than metal tubes (**FIGURE 14-10**). To maximize airflow to the cylinders, intake manifolds are designed by using large runners.

14-02 Identify and explain the purpose and functions of air intake system components.

Effects of Airflow Deficiencies

Lack of sufficient airflow to an engine can result in these conditions:

- low turbocharger or intake boosts pressure
- higher exhaust temperatures caused by combustion gases being inadequately diluted by fresh air
- incomplete combustion due to poor contact between fuel and air
- poorer fuel economy caused by poor combustion quality
- low-power complaints caused by poor combustion quality
- visible exhaust emissions caused by poor combustion quality
- increased exhaust emissions caused by poor combustion quality

▶ TECHNICIAN TIP

High humidity or water vapor displaces oxygen in the air, which means the combustion chamber is depleted of oxygen under wet and humid conditions. Hot days will dissolve more water vapor than cool days, causing a drop in engine power output and increased hydrocarbon (HC), carbon monoxide (CO), and particulate matter (PM) emissions on hot humid days. Electronic engines can compensate for increased temperature but cannot compensate for depleted oxygen levels. Some of the newest engines now use humidity sensors to correct air-fuel ratios changed by humidity. An increase in visible emissions can be expected on hot humid days from older diesel engines not equipped with exhaust particulate filters.

FIGURE 14-10 The plastic air filter housing receives cool outside intake airflow from an inlet just below the hood.

- shorter valve and piston life due to high heat loads in the cylinders caused by an absence of excess air
- increased lube oil use due to the intake manifold vacuum drawing oil from the intake valve guides and the turbocharger center housing then into the combustion chamber.

Intake Air Sensors

Most electronically controlled engines are equipped with a variety of air inlet system sensors used to identify a problem in the intake system (**FIGURE 14-11**). Diagnostic trouble codes (DTCs) will be set by the onboard diagnostic (OBD) system and may cause the engine's electronic control module (ECM) to adopt engine protection or adaptive strategies to compensate for the problem. For example, the intake manifold temperature sensor is monitored on some engines to prevent damage to the engine from high intake temperatures. High intake temperatures can cause thermal damage to exhaust valves and cylinder components. In those circumstances, the engine power will be derated if temperatures exceed threshold values for potential damage.

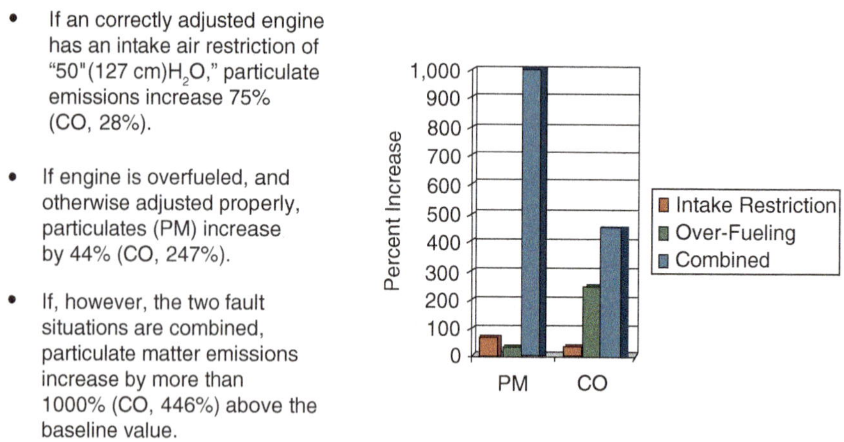

FIGURE 14-11 This chart shows the effect of plugged or restricted air filters on engine emissions for hydromechanical fuel systems. A restricted air filter alone causes particulates to increase by 75% and CO by 28%. If the fuel system is misadjusted, it will produce not only a 44% increase in PM but also a 247% increase in CO. Together, these conditions will cause as much as a 1,000% increase in PM and a 446% increase in CO.

Barometric Pressure Sensor

The ambient air pressure sensor, also called the barometric pressure sensor, is used to provide data regarding atmospheric air pressure for altitude compensation. This sensor will provide data to adjust fuel rates, modify glow-plug on time, injection timing, and exhaust gas recirculation (EGR) flow, all depending on altitude or atmospheric pressure. This is necessary to prevent excessive exhaust emissions or performance complaints when atmospheric pressure is low. This sensor is used on non-EGR-equipped, electronically controlled engines to prevent turbocharger overspeeding at high altitudes. Since turbochargers turn faster when the air is less dense, the barometric pressure sensor will reduce the quantity of fuel injected, thus lowering turbocharger speeds and preventing overspeed damage.

Barometric sensors can be found integrated into the engine ECM (**FIGURE 14-12**). Barometric data can also be read the instant the ignition is switched on and the intake manifold pressure is at atmospheric pressure, just before the engine begins to crank over. Some Ford Power Stroke engines used barometric sensor data from a sensor integrated into the throttle pedal. Barometric pressure sensors are typically active three-wire, variable capacitance or strain gauge–type sensors. A + 5-volt reference and a zero volt return is supplied to the sensor, and a signal wire sends an analog or digital signal back to the ECM for processing.

FIGURE 14-12 This Bosch EDC-17 ECM incorporates a barometric pressure sensor into the main circuit board. A Gortex fabric seal filters air to flow in and out of the ECM.

Intake Air Temperature Sensor

Data from the intake air sensor is used for a number of critical engine functions. When combined with intake pressure, the temperature sensor helps calculate intake air mass by providing air temperature density data (**FIGURE 14-13**). Intake temperatures are used to modify fuel-injection timing and fuel rates to minimize emissions. Low intake temperatures will result in advanced injection timing to compensate for longer ignition delay periods. A high intake temperature requires retarding the injection timing and requires lower injection rates to minimize emissions. This sensor can also detect excessive intake temperature caused by such conditions as externally restricted aftercoolers, which could lead to thermal damage in the engines. Charge air cooler efficiency codes are set whenever air intake temperatures remain high after air leaves the cooler or the difference between ambient temperature and intake manifold temperature is excessive. Engine software used by the engine protection system can be enabled to derate or cut back injection quantities to prevent damage to cylinder components when intake air temperatures air too high. Intake air temperature sensors are always two-wire, negative temperature coefficient–type sensors. This means that as air intake temperatures increase, sensor resistance decreases.

FIGURE 14-13 Intake temperature and pressure sensors provide the ECM with data for fueling, timing, EGR flow, and various operating strategies.

Manifold Absolute Pressure Sensor

Data from this sensor, also called intake boost pressure sensor, is used to sense engine load and adjust fuel rates. If there is inadequate airflow, the boost levels will lower and less fuel will be injected. Manifold absolute pressure (MAP) sensors are active three-wire pressure sensors typically using a variable capacitance or strain gauge–type construction to convert intake manifold pressure into an electrical signals. The MAP sensor coordinates with the manifold temperature sensor to calculate the air mass entering the engine's cylinders (**FIGURE 14-14**). Using air pressure, temperature, engine speed, and known volumetric efficiencies of an engine for any given engine speed, the ECM can precisely calculate the mass of air entering

▶ TECHNICIAN TIP

The excessive black or gray smoke emissions from engines not equipped with diesel particulate filters may be caused by restricted air inlet systems or inadequate airflow into the engine. In addition to fuel system related problems, Always investigate possible air inlet restrictions, such as plugged air filters, internally restricted intercoolers, collapsed air intake hoses, and incorrect valve adjustments, as possible causes of excessive visible smoke emissions.

FIGURE 14-14 A combined temperature and pressure sensor is indicated by the four wires connecting this sensor. The combo sensor provides air intake temperature and pressure data to the ECM.

the engine. A speed-density algorithm is used by the ECM to calculate this value for determining injection rates.

Mass Airflow Sensor

Mass airflow (MAF) sensors measure the weight of air entering the engine. Located after the air filter and before the turbocharger, this device is a "stand alone" sensor for calculating air mass—unlike a speed-density system, which requires a pressure and temperature sensor (**FIGURE 14-15**). In addition to measuring the weight of air, the location and operation of a MAF can detect air inlet restrictions caused by a restricted air filter or low intake boost pressure when air mass flow drops below a threshold value. In diesel engines, intake manifold pressure pulses produced by the large valve overlap of most diesel engines can cause intake air backflow across the sensor, generating incorrect data. Locating the sensor farther from the engine helps minimize sensor error; however, algorithms built into the ECM correct for potentially doubly counting pulses of air mass flow across the sensor.

Mass airflow sensors typically use a heated platinum or "hot-wire" construction. The circuit operating the MAF sends electric current through the wire and maintains a continuously fixed voltage drop across the wire (**FIGURE 14-16**). When air flows past the wire, the wire cools, decreasing its resistance. Cooling the wire causes a monitoring circuit to send more current to flow through the wire heating circuit to increase the electrical resistance of the wire. The ECM will measure the change in current required to maintain the same resistance or voltage drop across the heated wire. The amount of current used to heat the wire to maintain a specific resistance is directly proportional to the mass of air flowing over the

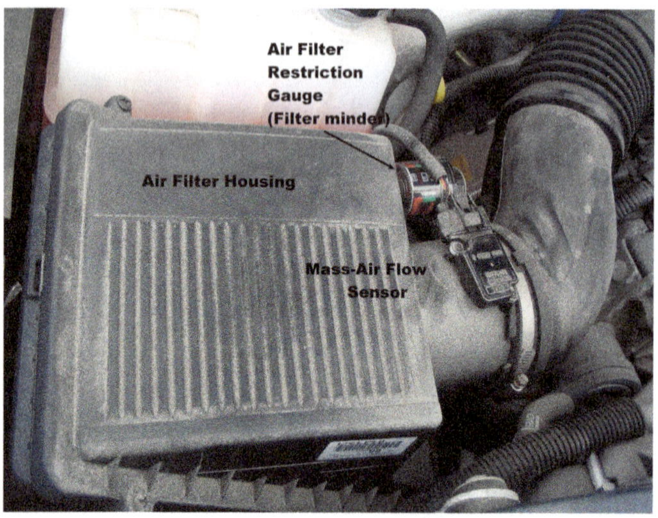

FIGURE 14-15 Location of a MAF sensor on the intake air pipe of an EGR-equipped 6.6 L Duramax diesel.

FIGURE 14-16 The hot wire of a MAF sensor.

heated wire element. For example, if air density increases due to an increase in air pressure or a temperature drop but the air volume remains constant, the denser or cooler air will remove more heat from the wire, indicating a higher mass airflow.

MAFs are used primarily on OBD-II compliant diesels to detect intake air problems as well as measure intake air mass. They are also important for calculating EGR flow. The strategy used by manufactures is to measure the airflow rate using the MAF and compare it to expected values without EGR flow. When EGR flow increases because exhaust gas is recycled, there is always a proportional decrease of fresh airflow into the engine as EGR gas displaces fresh-air intake flow.

▶ Air Intake Construction Features

Intake air systems have unique features to minimize the aspiration or entry of water and dirt. Sound emissions from turbochargers and pressure fluctuations caused by valve timing require sound-dampening features. Controls for regulating intake air temperature and EGR flow are commonly found in the air intake system. This next section examines common construction features found in most of today's diesel engines.

14-03 Identify and explain the purpose and functions of the construction features of the air intake system.

Functions of Intake Throttle Plates

Diesel engines do not use a throttle plate to control engine speed or torque output. Since diesel engines require a large mass of intake air sent through intake, manifold runners are designed to flow the maximum quantity of air. Generally, each cylinder will breathe from a large plenum common to all the cylinders. However, for emission-related reasons, throttle plates, also called intake flaps, are commonly used on light- and medium-duty diesel engines. Depending on the intent of the engine manufacturer, intake air throttle can have several purposes:

1. Regulating exhaust temperatures: Diesel particulate filters required for all 2007 and later engines require additional heat that is not normally found in the diesel's relatively cool engine exhaust. For example, at low speed, under low load, and during idle conditions, an excessive amount of air is drawn into the engine in comparison to disproportionaly small injection quantities. Consequently, combustion and exhaust temperatures are low. Catalyzed particulate filters, which need high exhaust temperatures to properly operate, cannot function with low exhaust temperatures produced by excess air-fuel ratios. The addition of the electronically controlled throttle will, under certain conditions, restrict excessive airflow into the engine to increase exhaust gas temperatures (**FIGURE 14-17**).

   Vehicles equipped with lean NO_x traps or NO_x adsorbers similarly need the exhaust to be completely depleted of oxygen in order to allow chemical reactions to take place that convert NO_x to harmless substances. When necessary, the throttle plate will close as required while extra EGR gas and fuel is added to the exhaust stream.
2. Control of combustion turbulence for optimized emission reduction: Throttled air intakes are used to optimize air turbulence into the combustion chambers for any given load and speed condition. To achieve the lowest emissions, best performance, and best fuel economy, air and fuel must be dispersed and mixed as uniformly as possible, with turbulence helping to better mix uneven distributions of fuel and air in regions of the combustion chamber. While much of this mixing function is accomplished by the shape of the combustion chamber bowl, it tends to disperse fuel only vertically. Airflow into the cylinder, however, mixes fuel horizontally. By using two intake ports, one that is tangentially angled with the combustion chamber, a high degree of intake air swirl is made possible. Opening or closing one of the intake ports on these four-valve engines with a variable opening flap or swirl valve varies the amount of intake swirl. High swirl is needed to reduce the formation of CO, HCs, and PM at low- to mid-speed operating conditions. Closing or only partially opening one of the two intake runners enhances swirl. At high

FIGURE 14-17 This throttle vavle on a 6.7L Cummins is used to restrict the air intake at low speed, under low load, and during idle conditions causing exhaust temperatures to rise to increase DPF temperatures.

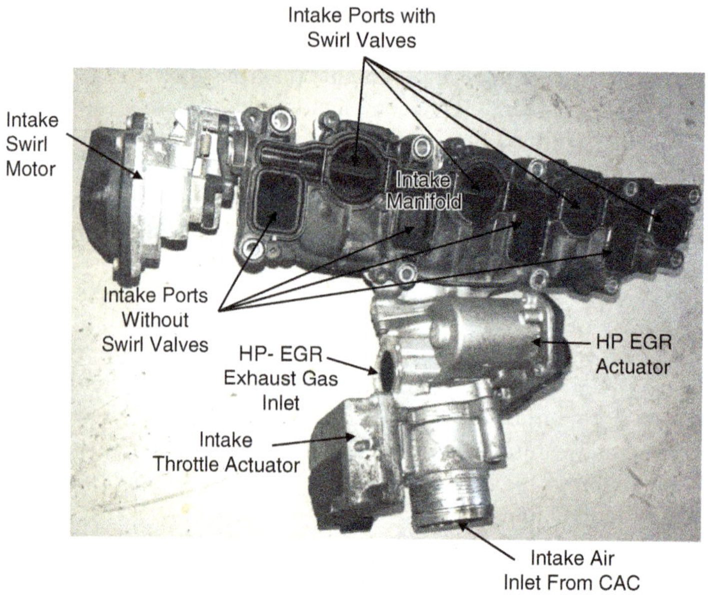

FIGURE 14-18 Closing the intake valve in one intake runner of this engine increases the degree of swirl, which better disperses the fuel horizontally.

speed, the intake runner valve is fully opened for maximum airflow into the cylinders (**FIGURE 14-18** and **FIGURE 14-19**).

Only one of the two air charge intake ports is closed by a movable flaps. These flaps are typically held open by spring force and closed by an electric actuator. The following is a list of the signals that determine the position of the flaps:

- MAF sensor
- boost pressure sensor
- charge air temperature sensor
- intake air restriction sensor downstream of air filter
- accelerator pedal sensor
- crankshaft position sensor or engine speed.

The ECM closes the intake port flap with an electric motor energized with a pulse-width–modulated (PWM) signal. At lower engine speed and load ranges, all the intake ports are closed by the flaps. This means the entire air mass enters the cylinder through a single angled intake swirl port. Since air enters at an angle and at high velocity, much greater air turbulence mixes the air and fuel more effectively to improve combustion quality. Particulate or soot formation is significantly lowered. As the engine speed and load increase, the charge air intake ports are opened proportionally, providing optimal mixing of air and fuel charges and air for each stage of engine operation.

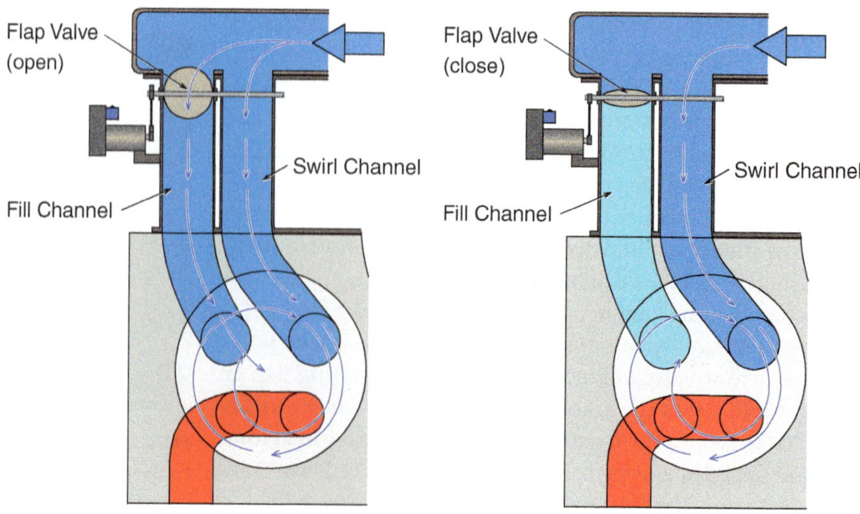

FIGURE 14-19 The location and position of the intake runner swirl plates on a VW 2.0 L TDI.

3. Increase EGR flow: Throttle plates are used on some EGR-equipped diesel engines to increase the rate of exhaust gas flow into the cylinders under some conditions. Since diesel engines do not have an intake manifold vacuum, negative intake pressure produced by an electrically operated throttle valve draws more gas into the intake air. The EGR inlet to the intake manifold is located downstream of the plate. Moving the position of the electrically actuated throttle plate balances the fresh-air intake with EGR flow. When closed, the throttle valve will restrict intake airflow while lowering the pressure in the intake manifold, which increases EGR flow rate.
4. Reducing engine shudder: The high compression ratios used by diesel engines produce a pronounced shaking motion in the engine during shut down. This effect is caused by air compression in the cylinders during shut down. Compressed air behaves like a coil spring in the cylinders, with upward piston movement compressing air like a spring. Without enough inertia to move the piston completely through compression stroke and past TDC, the engine rotation will slow as it approaches TDC, then reverse repeatedly until all crankshaft inertia and rotational energy has been expended.

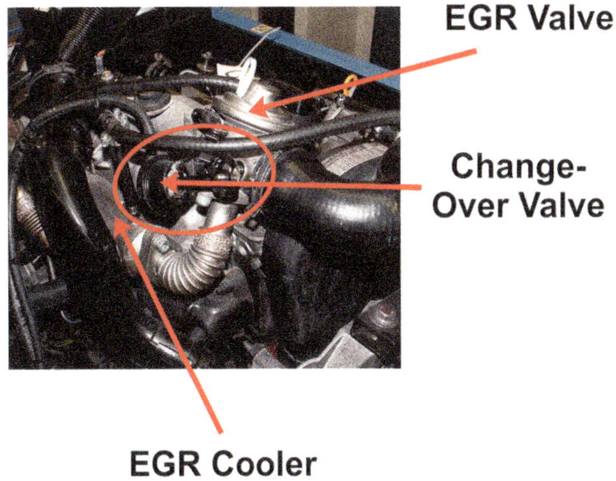

FIGURE 14-20 The changeover valve on VW TDIs snaps shut when the ignition is turned off to minimize engine shudder. Cycling the ignition switch on and off will allow a technician to check the operation of the changeover valve.

Accessory drive belts may also squeal during shut-down shudder as the belt direction is rapidly changed on the front crankshaft pulley. The belt may also slip on pulleys for components like the alternator, which have some resistance due to residual magnetism. To prevent shudder and belt noise, shutting the air intake valve the moment that the ignition is switched off prevents air from entering the engine (**FIGURE 14-20**). The result is a smooth stop to engine rotation. One-way clutches on the alternator pulley are sometimes used to further reduce belt slippage noise.

Intake Air Noise Reduction

Unlike spark-ignition, gasoline-fueled engines, diesel engines are able to use a large amount of valve overlap to enhance engine breathing and performance characteristics. Valve overlap refers to the time when the intake and exhaust valves are both open near the end of exhaust stroke and beginning of intake stroke (**FIGURE 14-21**). In a gasoline-fueled engine, large amounts of valve overlap would cause low intake vacuum and allow raw fuel to leave the cylinder during exhaust scavenging. However, in a diesel engine, this valve-timing configuration takes advantage of gas inertia and turbocharged intake pressure to pack more air into the cylinders during intake stroke and improve exhaust gas scavenging at the end of exhaust stroke. The disadvantage of large amounts of valve overlap is that loud exhaust pulses travel through the intake manifold, creating noisy air intakes. This condition is more pronounced at low engine speeds. To minimize this potentially annoying characteristic of diesels, resonator boxes or sound-dampening chambers are often located in the air induction system, which allow exhaust pulses to expand and dissipate, thus canceling intake noise (**FIGURE 14-22**

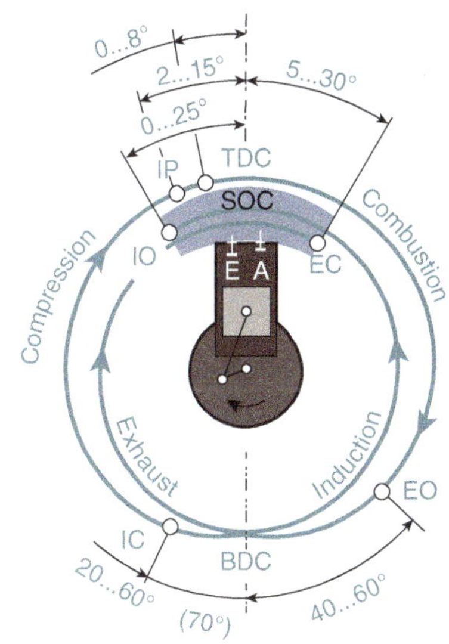

FIGURE 14-21 The valve overlap period occurring at the end of exhaust stroke and beginning of intake stroke is critical to cooling exhaust valves, pistons crowns, and even turbochargers.

FIGURE 14-22 The stock intake pipe for the 5.9 L Cummins diesel in the Dodge Ram uses a center section designed to dampen intake air noises such as turbo whistle.

> **TECHNICIAN TIP**
>
> Noise levels from vehicles are also regulated by the Environmental Protection Agency (EPA) as noise emissions. Modifications to intake and exhaust systems may cause a vehicle to be no longer compliant with emissions standards for noise.

FIGURE 14-23 This engine cover of the 6.6 L Duramax diesel is a resonator integrated with the air induction system to reduce intake system noise. The dead space in the duct work is designed to minimize the groaning noise heard from diesel engines at part throttle operation.

▶ TECHNICIAN TIP

Diesel engines' large overlap period can contribute to intake noise and soot loading in the intake manifold since exhaust pulses will travel into the intake manifold during the valve overlap period. Some intake systems can actually plug and restrict the intake manifold due to large amounts of soot deposition. EGR gas combined with closed crankcase ventilation systems also cause a buildup of carbon and sludge in the intake of a diesel.

and **FIGURE 14-23**). On turbocharged engines, a variety of noises originate from the turbocharger, noises that need to be dampened. Resonator boxes can help with this, but special devices and the configuration of the intake system also suppress turbocharger noise. Air filter design can also be used to dampen intake noise by using rubber-like sealing caps to dampen noise escaping from the air intake system.

▶ Air Intake Filtering

14-04 Identify and explain the functions of air filters.

Dirt, moisture, and other contaminants must be removed from the intake air to prevent engine damage and performance complaints. Intake systems can be designed in a number of ways to minimize contaminant loading in the air filter and water ingestion into the engine. A pre-cleaner will typically remove 80–90% of all airborne contaminants.

A frontally located air intake, known as ram air, can take advantage of increased air pressure at highway speeds to assist in charging the cylinders. However, they have the disadvantage of allowing road salt, water, and dirt into the inlet. Water that contains dissolved road salt is especially destructive since it can pass through the air filter into the air induction system, corroding components like the aluminum intercooler. Aluminum oxide eventually finds its way into the cylinders, dramatically accelerating wear on cylinder walls and rings.

FIGURE 14-24 Side-mounted air intake systems help reduce the amount of dirt, water, and snow entering the filter better than frontal air intakes. The sharp change in air direction cannot be made easily by anything other than air. Passageways through the hood collect air from two intakes and supply the air filter.

The use of side-mounted intakes, such as inside fenders, is the preferred method to minimize water and dirt ingestion. Water and dirt that have greater inertia than air and will not as easily change direction and be drawn into the air intake. Deflectors located on the air inlet can help remove contaminants by diverting water, snow, and dirt past the air inlet opening (**FIGURE 14-24**).

Ejector tubes located at the elbows of inlet air tubes and in air filter housings assist in removing water and dirt that does enter the induction system (**FIGURE 14-25**). Heavier particles will tend to migrate to a particular location in the tube, where they can exit through these one-way devices. While most vehicles will locate the air inlet to pull air from outside of the vehicle, some vehicles, depending on

FIGURE 14-25 This cyclonic-style pre-cleaner uses tubes which spin dirt out of the airflow into the filter. A safety filter is used to prevent dirt from dropping into the housing when servicing the filter.

application, will also provide the option of collecting air from underneath the hood. This arrangement causes air inlet temperatures to increase. However, this feature has benefits in very cold ambient conditions or where snow can enter an external inlet because the vehicle is used as a plow.

Air Filters

Air filters are essential for maintaining engine durability. Without clean air, engines will quickly wear out due to the cylinders dusting out. Particle sieving, the most common mechanism in filtration, occurs when the particle is too large to fit between the fiber spaces. Smaller particles pass through both large and medium-sized pores in the filter before until they are plugged by larger particles. Progressively smaller particles are stopped until the filter is completely loaded with contaminant (**FIGURE 14-26**). Air filter selection and performance criteria are based on the contamination environment in which the engine operates, the available space for installation and service, plus the desired filter service interval. A number of materials are used to manufacture air filters. Polyurethane, cellulose, and oil-wetted cotton gauze are several of the most common filter media types.

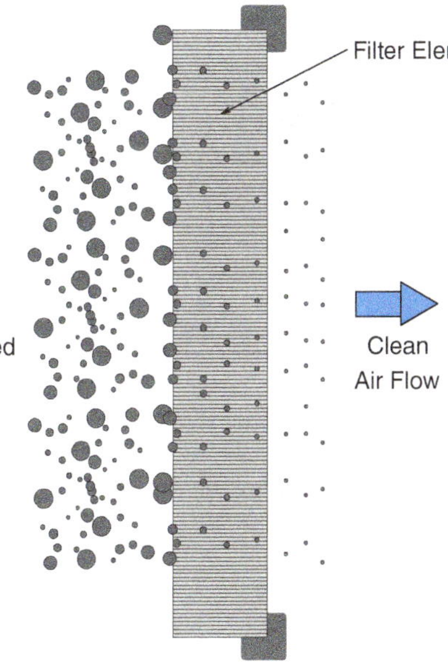

FIGURE 14-26 The principle of particle sieving. Progressively smaller particles are filtered out of intake air as it loads with dirt.

The following are three critical criteria to evaluate the effectiveness of filters:

1. Efficiency: This is a measurement of a filter's dirt-stopping ability. The percentage of dirt particles captured by the filter is the measurement of its efficiency. A 100% efficient filter will stop all dirt particles from entering the engine intake air. It's important to understand that filter efficiency is rated for a specific size of particle. A filter may be 99% efficient for stopping particles 5 microns in size and larger, but if efficiency is measured at 2 microns, it may stop only less than 50% of those particles.
2. Capacity: This is a measurement of the dirt-holding ability of a filter. How much dirt a filter can hold before it requires replacement determines its capacity.
3. Restriction: This measurement determines how much airflow occurs through a filter. As a filter loads with dirt and contaminants, airflow decreases and an increasingly larger pressure differential develops across the filter, which is referred to as the positive filtration principle. This pressure differential is known as filter restriction. Air filter restriction is measured in inches (mm) of water column, and the pressure differential increases as more dirt blocks airflow through the filter. To measure air filter restriction, follow the steps in **SKILL DRILL 14-2**.

SKILL DRILL 14-2 Measuring Air Filter Restriction

Pressure Conversion:

Unit	in H_2O	cm H_2O	in/Hg	cm/Hg	psi	kg/cm²	atm*	kPa
1 in H_2O	1	2.54	0.0735	1.866	0.0361			0.248
1 in/Hg	13.6	34.544	1	2.54	0.0,491	0.0345	0.0334	3.386
1 psi	27.7	70.104	2.036	5.171	1	0.0703	0.068	6.8948
1 kg/cm²	393.73	1,000.0	28.96	73.55	14.22	1	0.9678	101.28
1 atm	407.19	1,033.0	29.92	75.96	14.70	1.033	1	....
1 kPa	4.01	10.18	0.295	0.750	0.145	0.026	....	....

*ATM = atmosphere

1. Connect a Magnehelic gauge to the clean side of the air filter. A Magnehelic gauge measures pressures in units (inches or mm) of water column. An electronic pressure transducer or a gauge with long plastic lines is needed.
2. Place the engine under load during a road test to operate the engine at maximum boost pressure.
3. Observe and record maximum filter restriction.
4. Recommend air filter replacement if restriction is more than 12–15" (30–38 cm) of H_2O.

Cellulose Paper Filters

To evaluate filter effectiveness, the ISO 5011 standard (formerly SAE J726) establishes a filter testing procedure by using precision measurements under controlled conditions. Tests performed by independent labs demonstrate whether paper or cellulose fiber filters have the greatest efficiency and capacity. This test commonly ranks paper filter efficiencies in the range above 99% among most brands, where most are between 99.5% and 99.9% efficiency (TABLE 14-4).

Paper filter elements are made from compressed cellulose fibers mixed with other materials, such as fabric and even wood particles. The paper is chemically treated to resist water absorption, which would cause the filter pores to close up. This is important to remember since spraying starting fluids or cleaning filters with solvents can remove the chemical treatment from the paper. When this happens, filter pores close and the filter becomes highly restrictive. Pleats in the filter paper increase the surface area and capacity of the filter. Aluminum wire mesh on the inside and outside of the filter element removes and prevents larger solid particles from entering or damaging filtration media, and it provides a structural skeleton for the filter.

TABLE 14-4 Comparison of Filtration Efficiency, Dirt-Holding Capacity, and Airflow of Eight Paper Air Filter Elements

Dust-Holding Capacity (grams)	Pressure Drop (in H_2O)	0.5 Micron% Efficient	5 Micron% Efficient	10 Micron% Efficient	≥ 15 Micron% Efficient	Average% Efficient
8	5.2	99.82	93.50	99.87	100	97.36
111	5.8	99.55	97.98	99.87	100	99.15
228	6.6	99.66	98.67	99.66	99.98	99.36
413	8.3	99.94	99.98	99.94	99.98	99.97
530	9.9	99.99	99.92	99.96	100	99.96
614	12.3	99.40	99.95	99.96	100	99.97
756	18.5	99.90	99.90	99.96	99.96	99.94
805	23.8	99.98	99.92	99.66	100	99.87
Dust Type		Fine	Fine	Fine	Fine	Fine

Molecular or particle sieving is the operating principle for air filters. This means the spaces between the cellulose fibers provide microscopic openings that the air must pass through. As dirty air flows through these pores, these holes plug up. Once a pore is plugged, the air must find an alternate route through the filter. The larger pores tend to plug up faster since greater air volumes will flow through those channels. Since there is a greater number of smaller holes in the filter media, these will plug up later. Consequently, the longer a filter is in service, the more efficient its air filtration will be (**FIGURE 14-27**).

FIGURE 14-27 This combination restriction gauge and air inlet restriction switch is designed to alert the driver with a dash light when the air filter has become too dirty.

Cotton Gauze Filter Media

A popular media for performance enthusiasts is cotton gauze–type filters using special oil to enhance dirt-capturing ability. The oil is absorbed and wicked through microfibers in the gauze. The advantage of this filter is that it separates contaminating particles with less physical restriction than typical paper elements. However, an ISO 5011 test evaluating this type filter media examined one popular brand. The test measured the filter's efficiency at less than 90% and noted that it did not have as much dirt-holding capacity as many paper filters (**FIGURE 14-28**). Wetting the cotton gauze with a tacky or sticky oil can also contaminate intake air sensors, leading to engine performance complaints and fault codes.

> **▶ TECHNICIAN TIP**
>
> To obtain the most efficient filtration, avoid overservicing filters, since a new filter will pass more dirt than a used one. However, as the filter collects more dirt, its resistance to airflow increases because fewer pores are left open and resistance to airflow or restriction to airflow increases. Eventually, horsepower (kilowatts) and fuel economy decrease. Ideally, the media must be thick and/or the fibers must be tightly compressed and dense to load as much dirt and obtain efficient filtration.

Nanofiber: PowerCore filters

PowerCore is the brand name for a newer filter technology that uses a synthetic web-like fiber to cover a specially formulated cellulose material. Using this material allows dirt to imbed into the fiber overlay rather than into the cellulose. A nanofiber feature extends filter service life while improving filtration efficiency. Donaldson has developed this media to produce a filter with an exceptional long life, accompanied by low filter restriction. When fabricated into a shape similar to corrugated cardboard, the air passageways through the filter are arranged in flutes, with one end alternately blocked on each flute. Air must pass through the wall of the flute to enter the intake. This wall flow filter design passes air with less restriction. PowerCore filters have the additional advantage of drastically reducing filter size while increasing dirt-holding capacity. The smaller air filter footprint is especially advantageous when underhood room is limited (**FIGURE 14-29** and **FIGURE 14-30**).

Air Filter Service

Air filter replacement should be determined by the amount of filter restriction. While the probability of maximum allowable restriction can be estimated in terms of engine hours or distance traveled, the filter can be overserviced or underserviced if only time

FIGURE 14-28 Examples of a cellulose paper–type filter, an oil-wetted cotton gauze, and synthetic fiber.

FIGURE 14-29 Power Core wall flow filter. Power Core filters can hold more dirt, improve filtration efficiency, and are more compact than conventional paper filters.

and distance in service are measured (TABLE 14-5). Manufacturers will recommend filter replacement when its maximum allowable restriction has been reached (FIGURE 14-31). As an air filter element becomes loaded with contaminants, the vacuum or restriction increases on the engine side or clean side of the air filter. This pressure differential across the element is measured in inches (mm) of water (H_20). A water manometer or Magnehelic gauge calibrated in inches (mm) of water is best to record restriction. The element should not be serviced based on visual observation, because this usually leads to overservicing or unnecessary filter replacement. Sometimes filters can appear clean when in fact they are too restrictive. Filter minders or restriction gauges, usually located on air filter housing, are also used to measure filter restriction (FIGURE 14-32). If no gauge is present, a test port is usually located on the filter housing to connect a test instrument. Some vehicles are equipped with electronic restriction gauges or switches. A sensor in the air filter housing is connected to the instrument cluster to warn the driver to service the element.

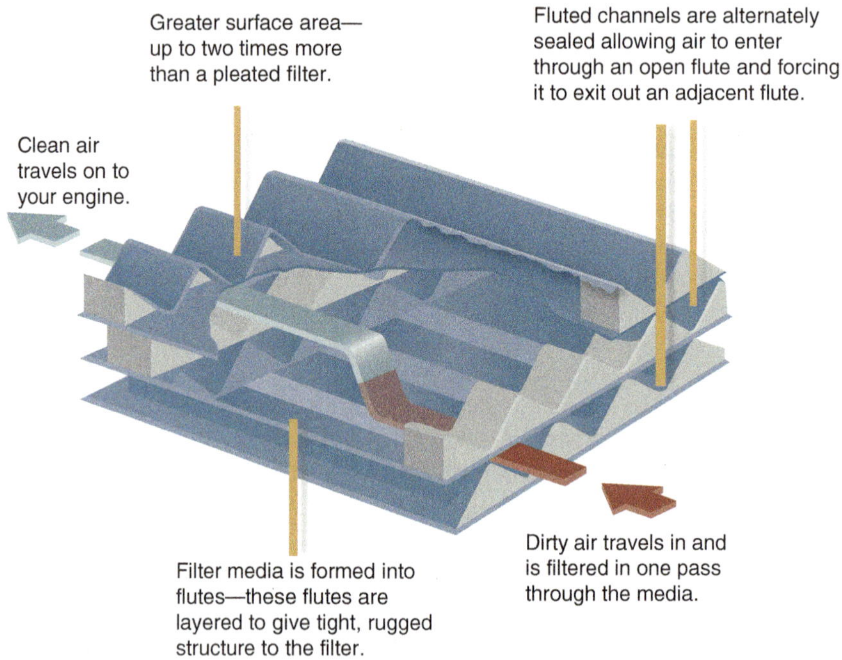

FIGURE 14-30 Power Core filters are a wall flow–type filter element. Note the alternate open and closed ends of tubes used to filter the air.

TABLE 14-5 Air Filter Service Life Probabilities

Usage	Application	Expected Service Life, Distance	Expected Service Life, Time
Light duty	Pickup and delivery, city/highway	40,000 miles (64,374 km)	900 hours
Medium duty	Off-road gravel trucks, electric generators	80,000 miles (128,748 km)	1,600 hours
Heavy duty	Highway trucks	More than 100,000 miles (160,934 km)	2,000+ hours

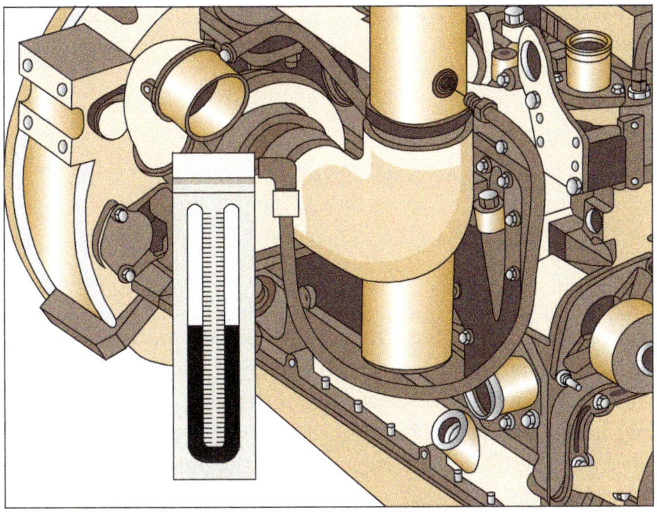

FIGURE 14-31 Measurement of air filter restriction by using a water manometer. The gauge records the maximum difference between atmospheric pressure and air pressure on the clean side of the filter.

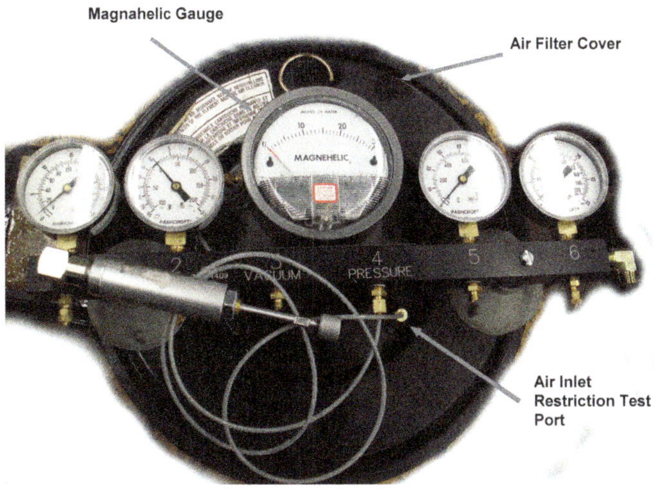

FIGURE 14-32 A Magnehelic gauge measures air filter restriction in inches (mm) of water column. A test port is used to measure air inlet restriction across the filter element.

On turbocharged engines, filter restriction is measured at rated speed with the engine under load (**FIGURE 14-33**). Manufacturers may also publish specifications for restrictions on turbocharged engines measured at high idle, but the restriction data will be higher when measured under load. While filter manufacturers recommend higher measurements, studies have demonstrated that the restriction limit of most turbocharged diesel engines is a maximum 12 to 15 inches (304.8 to 381 mm) water column under full load before a penalty to fuel economy takes place. Ideally, filters on turbocharged engines should be replaced at between 12 and 15 inches (305 and 381 mm) of H_2O to minimize increase to fuel consumption caused by restricted filters. On naturally aspirated engines, or non-turbocharged

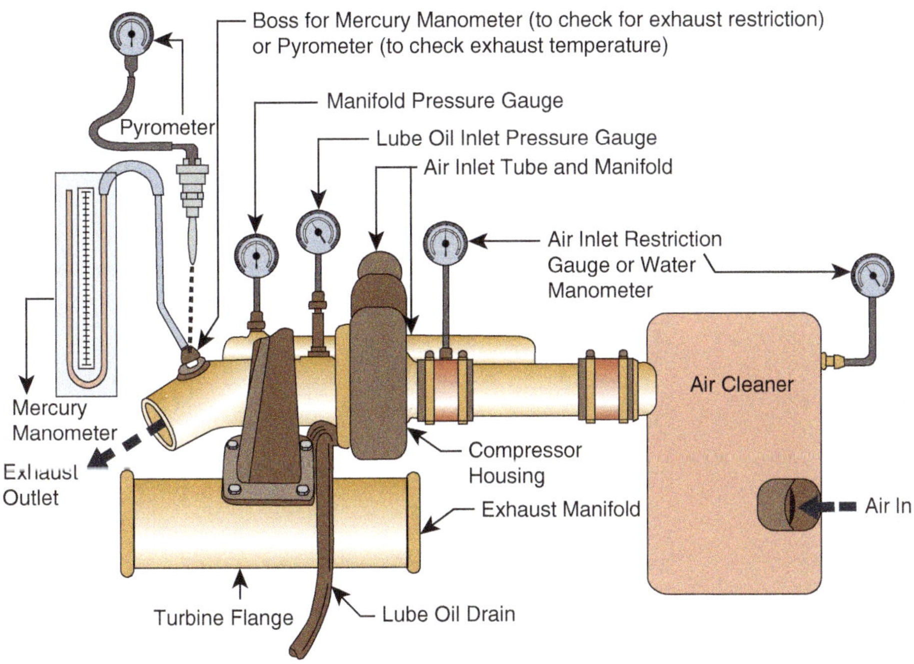

FIGURE 14-33 Points in the air induction and exhaust system for taking diagnostic measurement in older engines.

> ▶ TECHNICIAN TIP
>
> To ensure proper filtration of air, observe and practice the following guidelines:
>
> - Never use a warped air filter cover on filter directives.
> - Do not use a dented filter element.
> - Never substitute an incorrect element model number.
> - Never rap an element to clean it; this only destroys filter elements and embeds some dirt even deeper.
> - Never clean a filter with compressed air. Compressed air will thin out or tear filter paper, leading to dirt leaking into the engine.
> - Always clean out the filter housing and moisture drains before installing a new filter.
> - Check and reset the air filter restriction gauge if equipped.

engines, restriction is measured at rated rpm regardless of load. Naturally aspirated engines need to have their filter replaced before a maximum of 6–8 inches (152–203 mm) of water column.

Air Filter Service Precautions

Overserving filter elements can lead to the following problems:

- element damage due to excessive handling
- incorrect filter installation
- contaminated air intake from filter dust.

Do not remove the element for inspection: Such inspections can do more harm than good to the engine. Sealing gaskets can be torn or damaged when removed, and ridges of dirt on the gasket sealing surface can drop into the clean side of the filter when the gasket is released. Some filter housings use an inner filter element to keep dirt and other contaminants from falling into the air intake system of the engine while the outer filter is being serviced.

Air Intake System Diagnostics

Problems with the air intake system are accompanied by complaints of black or gray smoke and low power from engines without particulate filters. The following chart is a guideline to use when investigating the air induction system for these complaints.

Cause	Correction/Diagnostic
Complaints associated with operational problems	Advise operator
1. Operation while overloaded or at excessive speed	
2. Excessive idle	
3. Operation during extreme ambient temperature	
4. High altitude on naturally aspirated engines	
Air induction system problems	Check air inlet system
1. Restricted air filter	Measure air filter restriction
2. Restricted charge air cooler	Check air inlet hoses and clamps
3. Collapsed air induction hose	Clean air inlet system
4. Air intake manifold or charge air system leak	Pressure-test air inlet and charge air system
5. Turbocharger defective	Check for air inlet system noises and loose clamps
6. Incorrect turbocharger	Check charge air cooler for internal restrictions
7. Malfunctioning EGR system	Check turbocharger operation and for physical damage
	Measure turbocharger boost pressure on road test
	Inspect and repair EGR system
Mechanical system problems	Perform compression tests
1. Worn camshaft lobes.	Measure cylinder leak-down
2. Exhaust system restriction	Measure crankcase blow-by pressure
3. Low compression from worn cylinder components, damaged or misadjusted valves	Measure exhaust system back pressure
	Remove and inspect cylinder head
4. Incorrect valve adjustment	Perform valve and injector adjustment
Fuel system problems	Check fuel quality
1. Incorrect grade of fuel	Check throttle position sensor or linkage for maximum wide-open position
2. Over-fueling	Check and adjust fuel timing
3. Worn injectors	Check for engine fault codes
4. Incorrect injection timing	Pressure-volume test of fuel pump
5. Low injection pressures	Test injection pump on a comparator bench
6. Low fuel viscosity	Test fuel nozzles
7. Engine control system problem	Adjust injector calibration
8. Incorrect injector adjustment	Re-flash engine ECM with correct calibration
9. Tampering	Perform valve and injector adjustment
10. Improperly calibrated engine	

▶ Air Induction System Performance Accessories

14-05 Identify and explain the purpose of air induction system accessories.

Off-road track and racing enthusiasts will often accessorize the air intake system with devices to increase engine performance. The following are some of the common devices used:

- Nitrous oxide systems (NOS): Nitrous oxide N_2O—not to be confused with oxides of nitrogen (i.e., NO, NO_2, and NO_3)—will disintegrate into oxygen and nitrogen components in a combustion chamber at 565°F (296°C). Since it contains 2.3 times as much oxygen per identical cubic measure of just air, adding NOS can supplement the oxygen in the intake air charge to supply a power boost if additional fuel is added and burned to use up the oxygen in the cylinder. The amount of excess air in the combustion chamber of a diesel engine would require tremendous amounts of NOS for just a few seconds of operation, making it somewhat impractical to use in diesels as compared to more common use in spark-ignition, gasoline-fueled engines.
- Propane toppers: Another accessory sometimes sold to diesel enthusiasts is the propane enrichment system commonly called a propane topper. Propane is fumigated or pushed in under tank pressure into the intake airflow through a solenoid-controlled orifice. A 30 lb (13.6 kg) or so propane cylinder stored on the vehicle supplies the propane. The additional propane may be useful under racing or tractor-pulling conditions since it will increase power output by using up any excess oxygen left over in the cylinders.
- Water-methanol injection: Water-methanol systems are occasionally used by diesel performance enthusiasts to lower combustion chamber temperatures. The addition of water to the chamber will lower combustion temperatures to prevent engine damage while operating at near maximum combustion and exhaust temperatures. When water and methanol, a type of alcohol, are fumigated into the intake system, the process of vaporization also absorbs heat from the intake air which helps cool the intake air charge to further reduce peak combustion temperatures.

▶ Starting Aids

14-06 Identify and describe starting aids related to the air induction system.

Since compression-ignition engines depend on high compression temperatures to ignite fuel, starting performance is improved in cold weather operation through the use of several types of starting aids. Intake heating systems can also reduce the amount of cold-start emissions and soot formation by preheating intake air. Preheating intake air results in higher pre-ignition temperatures that help ignite fuel faster for longer burn time and more complete combustion.

Ethyl Ether

Liquid ether, when sprayed into the intake manifold, was commonly used many years ago to start cold engines. Ether is specially suited to assisting starting compression-ignition engines due to its low auto-ignition temperature of 320°F (160°C) and flammability. Even at low temperatures, liquid ether will readily turn to a vapor and burn over a wide range of concentrations. Since its auto-ignition temperature is much lower than an average for diesel fuel of 520°F (271°C), ether will ignite at lower compression pressures and temperatures. The burning ether provides an ignition source for diesel fuel when it is finally injected into the combustion chamber. Ether can be sprayed into the intake system with a can or metered by using capsules or an ether injection system. The electric dispensing system will use a cold-start switch that will not allow the ether to be inadvertently injected at temperatures above the freezing point. The operator can press the dash-mounted switch to initiate the delivery of a fixed quantity of ether into the intake manifold prior to cold starting the engine.

> **SAFETY TIP**
>
> Because ether is highly flammable, it should never be sprayed into an engine that uses glow plugs or electric intake heaters. Unless these devices are disabled, the ether will burn explosively in the intake manifold with catastrophic results, potentially causing personal injury or death (**FIGURE 14-34**).

> **SAFETY TIP**
>
> Inoperative or improperly functioning air inlet heaters will cause unnecessary and excessive white and gray smoke emissions from diesel engines during the warm-up period.

Ethyl Ether Properties

Flash point: −45°C (−49°F) CC
Auto-ignition temperature: 160°C (320°F)
Flash point: −50°F (−45°C)
Flammable limits in air% by volume:

- Lower explosive limit: 1.9%
- Upper limit: 36.0%

Electric Intake Heaters

Heating intake air is another technique used to reduce cold-start emissions and improve engine-starting characteristics (**FIGURE 14-35**). Warming intake air will increase pre-ignition temperatures, an important consideration when ambient temperatures are too low to allow normal pre-ignition temperatures required to quickly ignite fuel during start-up conditions. Heating intake air can also shorten ignition delay time. The additional intake heat has the effect of reducing fuel knock and white smoke caused by cylinder misfires. Less exhaust soot and hydrocarbon emissions are produced during cold start and warm-up when using intake heaters. Electric intake pre-heaters connected to circuits monitoring engine temperature are commonly used on diesel engines. A high-amperage relay is used to cycle the heater on and off. Sensor input from the intake air temperature sensor and barometric pressure sensor are used to calculate cycle time (**FIGURE 14-36**). Higher altitudes require longer

FIGURE 14-34 Combining ether with a heated air intake or glow plugs can end in catastrophe.

FIGURE 14-35 An electric heating element. Intake heaters improve combustion quality and reduce the likelihood of misfire conditions that cause white and gray smoke.

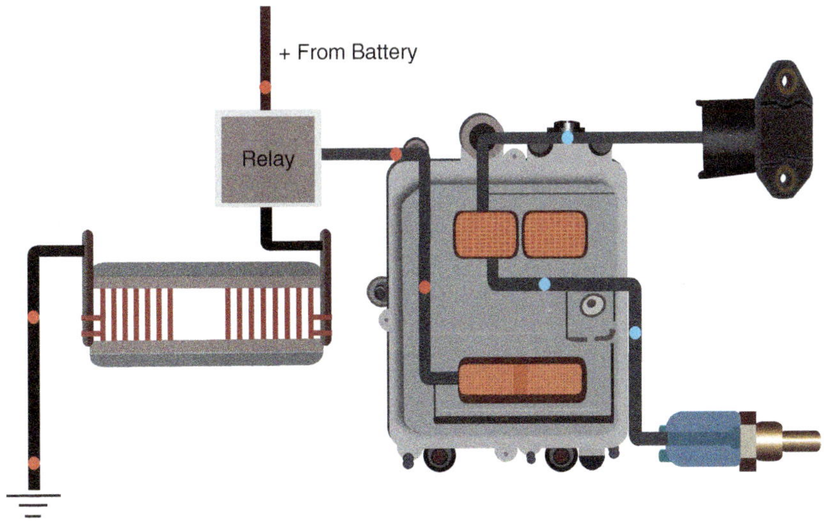

FIGURE 14-36 The ECM determines the heater cycle time based on intake air temperature and pressure sensor inputs.

glow-plug cycle times and longer intake air heater operation. Low engine temperatures will activate the heating element circuit, allowing the heater to warm intake air during cold-start and warm-up operation. Under boost conditions, when the intake air is heated by turbocharging, intake heaters are de-energized. Barometric sensors are used to prolong heating time at high altitudes, where air is less dense and will not hold as much heat.

Intake System Inspection

1. Visually inspect the air intake system to make sure it is intact. Inspect all ducts, pipes, hoses, tubing, and elbows used to interconnect the system.
2. Inspect components for damage, misalignment, and/or leakage. The following are some of the intake system components that need to be inspected for noise:
   - air cleaner housing
   - air cleaner element
   - turbocharger
   - charge air cooler and clamps
   - intake manifold.
3. Inspect all intake system fasteners, brackets, and clamps for damage and tightness.

▶ TECHNICIAN TIP

High loads on cold engines can result in damage for a number of reasons—poor lubrication caused by cold and soot-thickened oil, excessive operating clearances between parts, fuel dilution of oil, slobber production, etc. Many electronic engines limit engine power during the warm-up period to avoid damage or use strategies to speed engine warm-up. When investigating performance complaints, determine whether the complaint may be related to a cold-start or warm-up operating strategy.

14-07 Identify and explain the purpose and operation of crankcase emission control devices.

▶ Crankcase Ventilation Systems

Gases escaping past piston rings, a condition called blowby, can accumulate in the engine crankcase, pressurizing the engine and contaminating lube oil. Unless blowby pressure is relieved, oil leakage, blown seals, and blown gaskets can result. Engine blowby will pick up oil mist from oil-cooling nozzles, oil thrown off bearings, and vaporized oil from piston cooling.

Open Crankcase Ventilation

In gasoline-fueled engines, crankcase vapors are a major source of hydrocarbon emissions since they contain fuel that leaks past the rings during compression stroke. For this reason, crankcase vapors are recycled into the intake manifold of these engines and burned. However, in a diesel engine, blowby leaking past the rings consists primarily of air. Because most blowby occurs during compression stroke and because diesel engine cylinders contain only air during compression stroke, mostly air leaks into the crankcase and a smaller amount of exhaust gas. Traditional ventilation of the diesel crankcase simply required a tube connecting the crankcase to the atmosphere (**FIGURE 14-37**). This arrangement for crankcase ventilation is known as an open crankcase ventilation system. Common names for this ventilation device are road draft tube or crankcase ventilation tube. A metal-type

▶ TECHNICIAN TIP

Oil vapors from the crankcase are classified as a particulate emission since they are a liquid, not a gas, and must be considered as a part of the overall emissions from a vehicle. PM emissions as high as 0.7 g/bhp/hr can be found during idle conditions from some diesel engines. This exceeds the exhaust PM standard of 0.1 g/bhp/hr and 0.5 g/bhp/hr for non-methane hydrocarbons (NMH) for 2002 heavy-duty diesel engines. Crankcase vapors from engines made in 2007 and later are included in total particulate emissions, and manufacturers no longer vent vapors directly to the atmosphere.

FIGURE 14-37 The road draft tube in an open crankcase ventilation system, which vents blowby gases to the atmosphere.

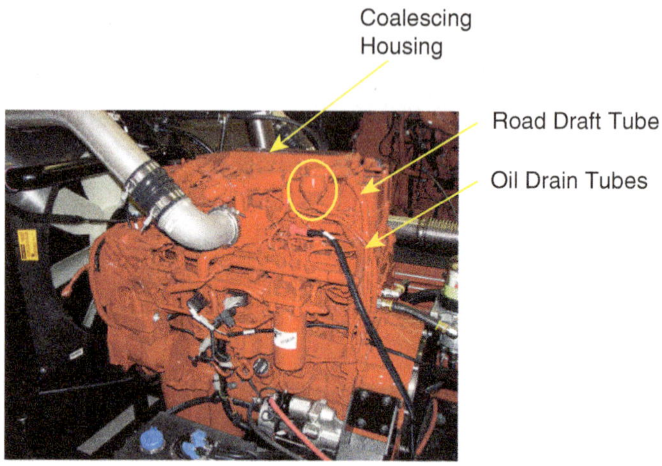

FIGURE 14-38 An open crankcase ventilation system with an oil separator.

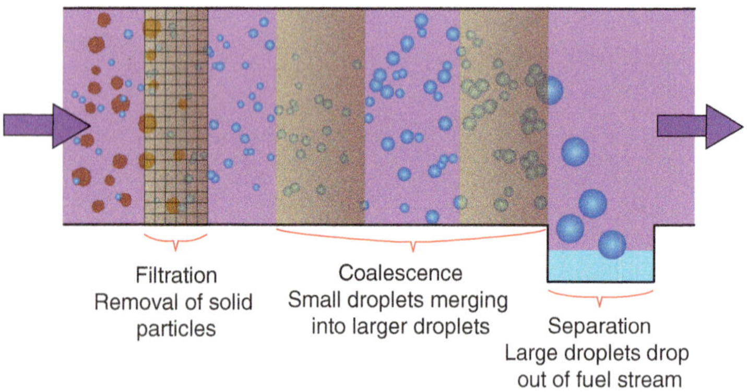

FIGURE 14-39 The coalescing filter principle takes fine oil mist and converts it into larger, easier, separable droplets.

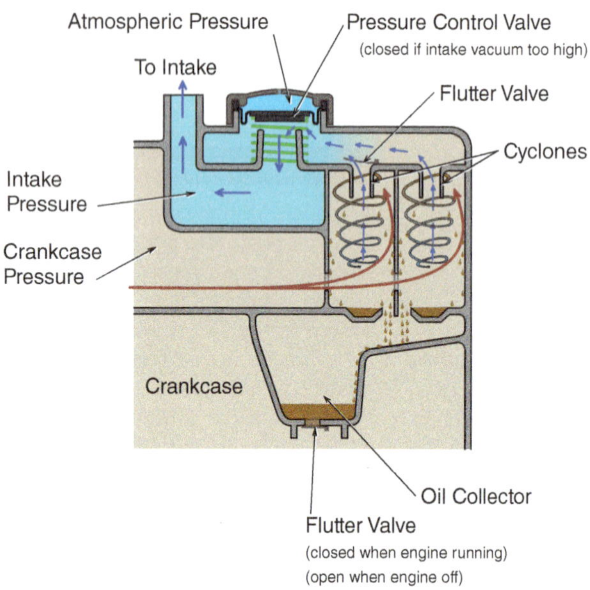

FIGURE 14-40 Operation of a cyclonic type crankcase oil-particulate separator.

strainer is usually placed into the tube to separate out oil mist picked up from oil throw-off and vapors from the crankshaft bearings and piston cooling nozzles. Open crankcase ventilation has been configured differently since 2007. A road draft tube is still used, but oil and particulate is removed before venting the vapor to the atmosphere (FIGURE 14-38).

Coalescing Filtration

Another solution to preventing crankcase oil carryover from entering the atmosphere is to use multistage filter systems. These filtration devices are designed to collect, coalesce, and return lube oil to the engine crankcase and filter out any soot before releasing the air to the atmosphere (FIGURE 14-39 and FIGURE 14-40). Although these filters have very long service life, the filter still needs to be monitored by using a crankcase pressure sensor—which helps warn if the filter is plugging, resulting in engine damage such as blown seals or gaskets. Coalescing filters, which return oil back to the sump, have the added benefit of reducing oil consumption (FIGURE 14-41). Since the crankcase vapors are regulated, an emission monitor system often uses a crankcase pressure sensor or an electrical switch to detect whether the filter is plugged or missing or the system is disconnected. To prevent the tube connecting the crankcase to the intake from being disconnected, a swtich is used to identify disconnected hoses (FIGURE 14-42).

Closed Crankcase Ventilation Systems

Crankcase vapors can be unpleasant to vehicle users since they contain oil vapors and combustion by-products. To eliminate unpleasant odors and oil leakage from tubes, it is more practical to recycle crankcase vapors back into the engine intake manifold.

Recirculation of crankcase vapors into the intake system can be accomplished by using one of two methods. One method is to use closed-route routing (FIGURE 14-43). Crankcase vapors are drawn into the air intake through a tube connecting the crankcase to the intake hose between the air filter and intake manifold. A metal mesh

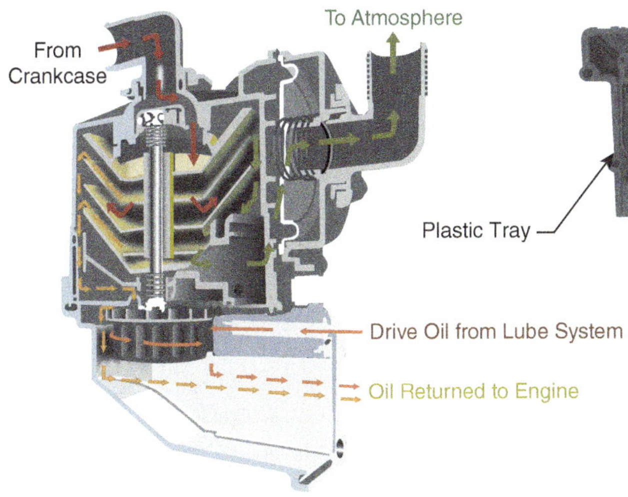

FIGURE 14-41 A centrifugal-type oil separator returns oil to the sump.

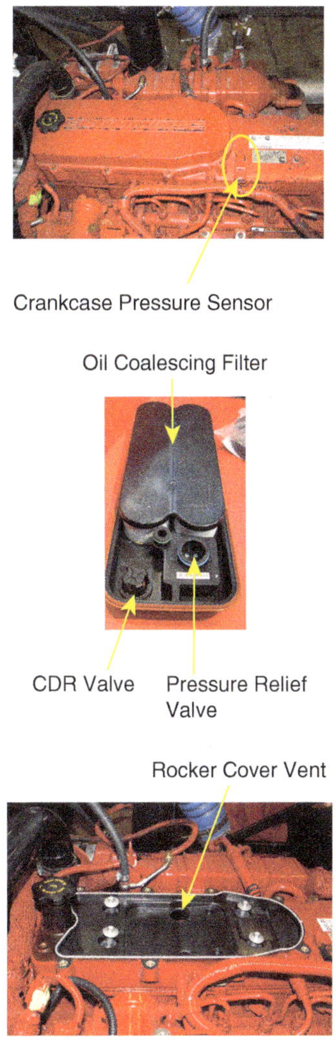

FIGURE 14-42 A crankcase pressure sensor is used as an emission monitor for the On-Board Diagnostics II (OBD-II) system and to alert the operator to a plugged or missing filter.

filter is used to separate as much oil mist from the crankcase blowby. Since diesel engines do not produce significant intake vacuum, locating the ventilation tube at this point maintains the crankcase pressure to the equivalent of the negative pressure produced by the air filter restriction. While this is a simple method, there are disadvantages. On turbocharged engines, oil mist in the blowby gases can foul turbocharger compressor wheels and fill charge air coolers with oil. However, using the correct grade of lube oil, which is changed at the appropriate service interval, will not normally impair turbocharger operation. Technicians need to be aware that oil accumulations on the compressor wheel and housing are normal when using this type of ventilation system. Charge air coolers can also accumulate oil. Cooler pipes and clamps will leak oil carried from the crankcase by the intake airflow (**FIGURE 14-44**). As much as two or three liters of oil originating from the crankcase can normally be expected to drain from an intercooler.

A second problem with closed crankcase ventilation systems is that crankcase pressure can lower to the point where dirt and other contaminants are pulled into the engine through seals and gaskets. This can happen when the air filter becomes restricted. The result is contaminated engine oil. To prevent this, crankcase pressure must be regulated.

Crankcase Depression Regulator

The second method to recirculate crankcase emissions to the intake manifold of closed crankcase systems is through the use of a crankcase depression regulator (CDR). This device regulates or meters the quantity of blowby back into the engines intake manifold. The advantage of this valve is its ability to maintain a slight negative pressure over the entire speed range

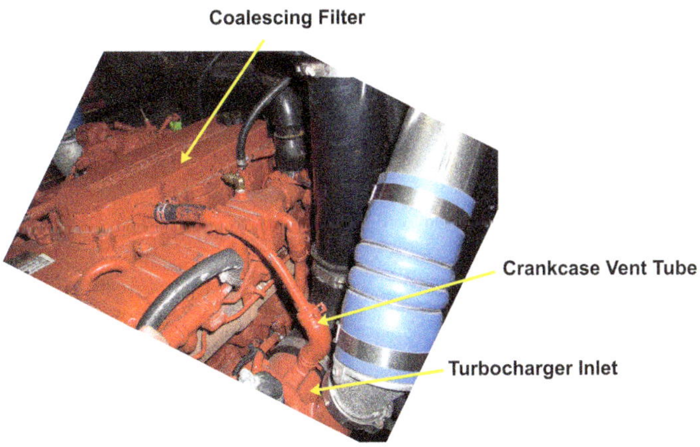

FIGURE 14-43 This closed crankcase ventilation system for the Cummins 6.7 draws crankcase vapors into the turbocharger inlet.

FIGURE 14-44 Crankcase vapors containing oil droplets will leak though air intake connections leaving oily residue on these intercooler connections.

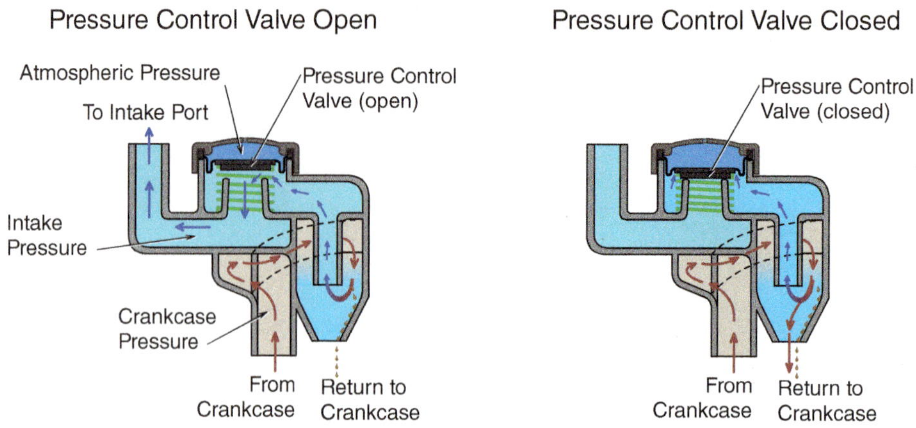

FIGURE 14-45 The CDR valve on this turbocharged 2.0 L TDI regulates crankcase pressure.

of the engine. This is desirable since positive high crankcase pressure will tend to produce oil leakage from gaskets and seals. However, too much vacuum will pull oil into the intake manifold and dirt through the engine seals and gaskets, potentially dusting out the engine. The CDR device typically maintains crankcase vacuum at zero to one inch (25.4 mm) of water at idle and three to four inches at 2,000 rpm (**FIGURE 14-45** and **FIGURE 14-46**).

The valve regulates crankcase pressure by using a spring-loaded diaphragm. When the engine is running, slight negative intake vacuum from the turbocharger compressor wheel pulls blowby gas through the valve past the open diaphragm. When vacuum in the crankcase exceeds the spring tension of the valve, the valve closes, blocking any further reduction in crankcase pressure. If crankcase pressure increases or intake vacuum decreases, the valve

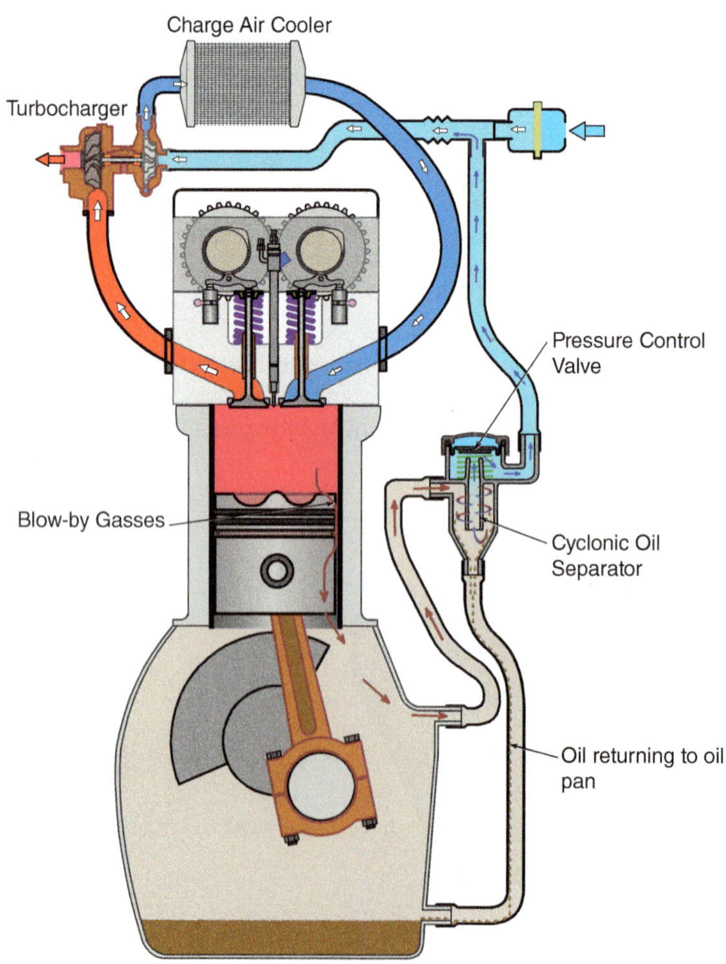

FIGURE 14-46 Operation of a CDR valve. The valve closes when pressure is too high to prevent excessive negative crankcase pressure. The valve opens when crankcase pressure is high.

FIGURE 14-47 A magnahelic gauge is used to measure crankcase pressure to check the operation of the CDR valve.

will open again. A Magnehelic gauge connected to the oil dipstick tube or engine oil fill cap are test points to inspect the CDR valve operation (**FIGURE 14-47**).

Testing the Crankcase Depression Regulator Valve

Engine oil leaks or excessive oil in the intake manifold would indicate that the CDR valve needs to be tested. The engine should first be evaluated to determine whether there is excessive blowby caused by a worn-out engine. Excessive blowby caused by worn cylinder walls, piston rings, and valve guides will overwhelm the CDR valve, causing engine oil leaks and pushing lube oil into the intake manifold with blowby gas.

During crankcase pressure testing, which is the best procedure to evaluate the condition of cylinder components, the CDR valve is blocked and blowby gases leave through a calibrated orifice. If blowby is within acceptable limits, the CDR valve is reconnected and crankcase pressure is re-evaluated. At idle, a water manometer or Magnehelic gauge connected to the oil dipstick tube should indicate a negative pressure of less than 1 inch (25.4 mm) of water column. At high idle, no-load condition, the crankcase vacuum should be no more than three to four inches (76 to 102 mm) of water column.

> **TECHNICIAN TIP**
>
> Turbocharger compressor wheels, which are coated in oil, do not indicate a failed turbocharger. In closed crankcase ventilation systems, oil vapors from the crankcase will collect in the turbocharger compressor housing and intake manifold. Charge air coolers will also accumulate oil carried over from the crankcase.

Wrap-Up

Ready for Review

- Diesel engines use much more air than similarly displaced gasoline-fueled engines. The diesel's abundant consumption of air makes it sensitive to air quality and requires a sophisticated air-handling system.
- Problems with engine air quality and induction system supply have a major impact on emissions, noise, performance, and engine durability.
- To support efficient combustion with minimal emissions, diesel engines need an adequate supply of clean, dry, fresh, and relatively cool air. Diesel engines also require an excess supply of air relative to the quantity needed to completely burn fuel in the cylinders.
- Because diesels have a short amount of time to atomize, distribute, vaporize, and heat fuel vapors, it is important that the fuel find enough air to contact and react with the oxygen.
- Excess air is also necessary to cool critical cylinder components in a diesel engine; intense combustion heat will easily melt pistons and burn valves.
- The stoichiometric ratio of diesel fuel requires 15–20 lb (7–9 kg) of air to burn 1 lb (0.45 kg) of fuel.
- Clean, filtered air is critical to engine operation to prevent damage and rapid wear caused by abrasive dirt particles. Dirt ingestion into an engine, referred to as dusting out, can take place through small openings in the intake system and from poor filtration.
- Optimal air inlet temperatures are essential to good performance and emissions. Atmospheric pressure, ambient temperature, and humidity affect power because

the number of oxygen molecules entering the cylinder will vary with a change in any of these factors.

- Air inlet temperature also affects combustion and exhaust temperatures. Exhaust temperatures can change as much as 3°F (1.7°C) for every 1°F (0.56°C) change in air inlet temperature.
- Water vapor displaces oxygen in the air, which means the combustion chamber is depleted of oxygen under wet and humid conditions.
- The design of air intakes is important to minimizing noise and the quantity of dirt and water entering the air filter. Air intakes should ideally have minimal bending and smooth interior surfaces for maximum airflow.
- Gases escaping past piston rings, called blowby, will pick up oil mist from oil-cooling nozzles, oil thrown off bearings, and oil vaporizing from piston cooling. Oil vapors from the crankcase are classified as PM emissions.
- The traditional method of ventilating the diesel crankcase requires simply a tube connecting the crankcase to the atmosphere. This arrangement for crankcase ventilation is known as an open crankcase ventilation system.
- Due to the increasing stringency of emission standards, manufacturers no longer vent vapors directly to the atmosphere. One method to clean up blowby emissions is to recycle crankcase emissions back into the intake manifold. This is known as a closed crankcase ventilation system. One method to recirculate crankcase emissions to the intake manifold is through a CDR.
- Another solution to crankcase oil carryover into the intake system is to use multistage filter systems.
- Most electronically controlled engines are equipped with a variety of air inlet system sensors to identify a problem in the intake system. Diagnostic trouble codes (DTCs) will be set by the onboard diagnostics (OBD) system and may cause the engine electronic control module (ECM) to adopt engine protection or adaptive strategies to compensate for the problem.
- The barometric pressure sensor, or ambient air pressure sensor, provides data regarding ambient air pressure for altitude compensation and to prevent turbocharger overspeeding
- Data from the intake manifold air temperature sensor is used to modify fuel-injection timing and fuel rates to minimize emissions.
- The manifold absolute pressure (MAP) sensor, also called the intake boost pressure sensor, is used to sense engine load and adjust fuel rates. If airflow is inadequate, the boost levels will lower and less fuel will be injected.
- Mass airflow (MAF) sensors measure the weight of air entering the engine. In addition, the location and operation of a MAF sensor can detect air inlet restrictions caused by a restricted air filter or intake when pressure drops below a specified threshold value that will cause emissions to exceed FTP limits.
- Diesel engines do not use a throttle plate to control engine speed or torque output. However, recent developments have introduced the use of throttle plates in diesel engines for several other purposes, including regulating exhaust temperatures, controlling combustion turbulence for optimized emission reduction, increasing EGR flow, and reducing engine shudder.
- Resonator boxes or chambers may be located in the air induction system, which allows exhaust pulses to expand and dissipate, canceling intake noise. Resonator boxes can also suppress turbocharger noise, but special devices and the configuration of the intake system also serve this function.
- Dirt, moisture, and other contaminants must be removed from the intake air to prevent engine damage and performance complaints. Intake systems can be designed in a number of ways to minimize the entry of moisture and dirt, which minimizes contaminant loading of the air filter and water ingestion into the engine.
- Using side-mounted intakes, such as inside fenders, is preferred to frontal air intakes to minimize water and dirt ingestion.
- Deflectors located on the air inlet can help remove contaminants by diverting water, snow, and dirt past the air inlet opening.
- Three critical criteria for evaluating the effectiveness of filters are efficiency, capacity, and restriction.
- A number of materials are used to manufacture air filters. Polyurethane, cellulose, and oil-wetted cotton gauze are the most common filter media.
- Air filter servicing should be determined by the amount of filter restriction. A manufacturer will stipulate that the filter element be replaced when its maximum allowable restriction has been reached.
- Several intake performance accessories can be added to diesel engines, including nitrous oxide systems (NOS), propane toppers, and water-methanol injection.
- To avoid the problem of a no-start condition due to cold weather, diesel engines will use starting aids to warm the intake air. Warming intake air increases compression temperatures, which improves an engine ability to start quickly.
- Problems with the air intake system are accompanied by complaints of black or gray smoke and low power in engines without particulate filters.

Key Terms

air inlet restriction gauge A pressure sensor that measures the negative pressure between the air filter and turbocharger.

closed crankcase ventilation system A method of cleaning up blowby emissions by recycling crankcase emissions back into the intake manifold.

coalesce The process of collecting small droplets together to form larger droplets.

crankcase depression regulator (CDR) A device that regulates, or meters, the quantity of blowby emissions back into the engine's intake manifold.

cyclonic-style pre-cleaner tubes A component of the air intake system that pre-cleans intake air by spinning dirt out of the airflow to prevent it from entering the filter.

fumigation A process of adding vapors into the air intake, such as in the case of water-methanol injection systems.

intake air heater A device that heats intake air to reduce cold-start emissions, improve engine-starting characteristics, shorten ignition delay time, improve combustion quality, and reduce the likelihood of misfire conditions that cause white and gray smoke.

intake throttle valve A valve located at the inlet of the intake manifold and used to restrict airflow into the engine and increase exhaust aftertreatment temperatures.

mass airflow (MAF) sensor A device that measures the weight of air entering the engine and detects air inlet restrictions caused by a restricted air filter or intake when pressure drops below a threshold value.

naturally aspirated An engine that uses only atmospheric pressure and not pressurized air to charge the cylinders with air.

nitrous oxide system (NOS) A system that supplements intake air with nitrous oxide (N_2O) to supply additional oxygen into the cylinder to burn more fuel.

open crankcase ventilation system A traditional ventilation system for the diesel crankcase that vents the crankcase directly to the atmosphere.

propane enrichment system An aftermarket performance accessory that introduces propane into the intake air through a solenoid-controlled orifice to increase power output by using up excess oxygen left over in the cylinders; it is also known as a propane topper.

road draft tube The tube that connects the crankcase to the atmosphere in an open crankcase ventilation system; it is also known as a crankcase ventilation tube.

standard day factors Engine test standards for ambient temperature, atmospheric pressure, and humidity. Manufacturer measurements of horsepower (kilowatts) and torque must be made by using standard day factors to ensure uniformity among test results.

stoichiometric ratio The mass of air to fuel required to completely burn all fuel in the combustion chamber so that no fuel or air remains after combustion.

tangential intake port An intake port design that admits air into the cylinder at an angle to impart more swirl, or turbulence, to charge air.

volumetric efficiency A comparison between measured cylinder volume and the volume of air actually filling a cylinder—expressed as a percentage.

water-methanol injection A system that sprays water and methanol into the intake system to absorb heat from the intake air, and lowers peak combustion temperatures.

Review Questions

1. What happens to the mass of intake airflow into an engine when the EGR is enabled?
   a. Nothing happens.
   b. The mass increases.
   c. The mass decreases.
   d. The mass becomes cooler.
2. Which of the following reasons best explains why air induction system noise is louder on a diesel engine than on a gasoline-fueled engine?
   a. Larger engine displacement.
   b. The use of a turbocharger.
   c. Smaller clearance volumes in cylinders.
   d. Increased amount of valve overlap.
3. What effect will high inlet air temperature combined with high humidity have on engine performance?
   a. Increased power compared with cooler, dryer conditions.
   b. No difference in power output compared with cooler, dryer conditions.
   c. Lower power output compared with cooler, dryer conditions.
   d. Reduced engine emissions compared with cooler, dryer conditions.
4. Which of the following instruments is the best to use to measure air inlet restrictions?
   a. Vacuum gauge.
   b. 0–5 psi (0–34 kPa) gauge.
   c. Water manometer.
   d. Mercury manometer.
5. What is the primary use for a resonator box on the air inlet of a diesel engine?
   a. Increasing air turbulence for improved combustion.
   b. Dampening intake noise.
   c. Pre-cleaning air of water and dirt.
   d. Tuning the intake passage for better airflow.
6. The service interval of an air filter is best determined by which of the following?
   a. Time in service.
   b. Using a trouble light to check the darkness of the paper.
   c. Air filter color.
   d. Measurement of air filter restriction.
7. Which of the following methods can best detect a leaking air intake system?
   a. Visual inspection.
   b. Spraying intake piping and joints with soap and water when the engine is running.
   c. Using ether while the engine is running, then listening for an increase in engine speed at leak points.
   d. Lube oil analysis to check for silica.
8. Which air filter element will provide the highest filtration efficiency (i.e., which removes the most dirt)?
   a. An oil-wetted cotton gauze–style air filter.
   b. A dry-type but dirty paper-type filter.
   c. An oil bath–type air filter.
   d. A new, dry paper-type filter.
9. Which of the following air intake systems will best pre-clean dirt and water from the air intake of an on-highway, diesel-powered vehicle?
   a. Front-mounted air inlet.
   b. Ram air induction.
   c. Composite intake.
   d. Side-mounted air inlet.

10. Which of the following practices is recommended for servicing a restricted air filter?
    a. Cleaning from the inside out with compressed air.
    b. Dropping or lightly tapping the filter against the shop floor.
    c. Verifying restriction with a trouble light.
    d. Replacing the filter.

ASE Technician A/Technician B–Style Questions

1. Technician A says that the best practice is to replace air filters on turbocharged engines when the restriction gauge reads 20–25" (51–63 cm) of H_2O. Technician B says the filter should be replaced around 10–15" (25–38 cm) H_2O to prevent increased fuel consumption caused by a restricted air filter. Who is correct?
    a. Technician A
    b. Technician B
    c. Both A and B
    d. Neither A nor B
2. Technician A says that it is acceptable practice to clean air filters by using compressed shop air. Technician B says it is better to clean air filters by dropping or lightly tapping the filter against the shop floor. Who is correct?
    a. Technician A
    b. Technician B
    c. Both A and B
    d. Neither A nor B
3. While two technicians are examining a piston crown that has vertical cracks through the edges of the piston's combustion chamber bowl, Technician A suggests the piston crown was not getting properly cooled due to low mass airflow in the combustion chamber during the valve overlap period. Technician B suggests the cracks were caused by inadequate cooling by the oil cooler nozzles. Who is correct?
    a. Technician A
    b. Technician B
    c. Both A and B
    d. Neither A nor B
4. An engine had a fault code for high crankcase pressure, and its power is derated. Technician A says that the crankcase filter could have frozen when it was exposed to extreme cold conditions. Technician B says that a frozen filter would likely blow engine gaskets and seals rather than cause a derate condition. Who is correct?
    a. Technician A
    b. Technician B
    c. Both A and B
    d. Neither A nor B
5. Technician A says that diesel engines use more air than gasoline-fueled engines because 1 gallon (3.7 liters) of diesel fuel requires more air to burn than the same amount of gasoline. Technician B says that a diesel engine uses more air to prevent pistons and valves from burning. Who is correct?
    a. Technician A
    b. Technician B
    c. Both A and B
    d. Neither A nor B
6. Technician A says that crankcase coalescing filters are used to minimize particulate emissions from diesel engines. Technician B says that centrifugal separators of air and oil reduce engine oil consumption. Who is correct?
    a. Technician A
    b. Technician B
    c. Both A and B
    d. Neither A nor B
7. Technician A says that a crankcase depression regulator is used to prevent overpressurizing an engine's crankcase. Technician B says that the crankcase depression regulator prevents dirt from entering the engine through seals and gaskets. Who is correct?
    a. Technician A
    b. Technician B
    c. Both A and B
    d. Neither A nor B
8. Technician A says that a throttle plate on a late-model diesel engine helps increase the exhaust temperatures for the exhaust aftertreatment catalysts. Technician B says diesel air intake is not throttled. Who is correct?
    a. Technician A
    b. Technician B
    c. Both A and B
    d. Neither A nor B
9. Technician A says that frequent air filter changes help maintain the cleanest quality of air that enters the engine. Technician B says that dirty filters are better than clean filters at allowing cleaner air to enter the engine. Who is correct?
    a. Technician A
    b. Technician B
    c. Both A and B
    d. Neither A nor B
10. Technician A says that electric intake heaters are used to reduce cold-start and warm-up emissions. Technician B says that intake air heaters help to shorten the cranking time of a cold engine. Who is correct?
    a. Technician A
    b. Technician B
    c. Both A and B
    d. Neither A nor B

CHAPTER 15

Fixed-Geometry and Waste-Gated Turbochargers

Learning Objectives

After reading this chapter, you will be able to:

- 15-01 Identify and explain the purposes of turbochargers.
- 15-02 Identify and explain the functions of turbochargers.
- 15-03 Identify and describe the construction and operation of turbochargers.
- 15-04 Identify and describe the types of turbochargers used on diesel engines.
- 15-05 Identify and explain the function of turbocharger subcomponents.
- 15-06 Identify and explain inspection, maintenance, and diagnostic procedures for turbochargers.
- 15-07 Identify and explain the functions of charge air cooling.
- 15-08 Identify and describe the types of charge air cooling systems.
- 15-09 Outline inspection procedures used to evaluate charge air cooling system operation.

You Are the Technician

A 2009 5.9 L Cummins diesel engine in a pickup has arrived at your shop with a complaint about intermittent noise and low power. After performing typical visual inspections, you take the vehicle for a road test to verify the complaint. When the engine is warmed up and under hard acceleration, you hear what sounds like a screeching sound like the noise of tires being locked and dragged on pavement. Releasing the load on the engine causes the noise to disappear. Returning to the shop, you remove the air intake boot to the turbocharger and look into the compressor housing to inspect the turbocharger compressor wheel and housing for damage. A compressor wheel making contact with the compressor housing would account for both complaints from the driver. It's somewhat difficult to be sure that there is no physical contact between the compressor and turbine wheel, so you intend to perform further inspection procedures to determine whether there is a problem with the turbocharger and in particular whether there is too much bearing wear. As you begin to plan your next steps, consider the following:

1. List and describe the inspection procedures for measuring turbocharger bearing wear.
2. What would cause premature failure of the center-housing support bearing, and how would it affect the compressor wheel?
3. What service recommendations would you make to the customer to prevent the likelihood of a repeat failure of the turbocharger bearings after a turbocharger replacement has been made?

▶ Introduction

Turbochargers are air pumps driven by exhaust energy to supply additional air for combustion. Using exhaust gases to drive a centrifugal-type pump recovers exhaust energy that would normally be wasted to the atmosphere and instead uses it to enhance engine operating efficiency. Squeezing more air into engine cylinders adds the potential to burn more fuel and increase power output. Engine emissions, noise characteristics, and fuel economy are also improved by turbochargers.

▶ Fundamentals of Turbochargers

15-01 Identify and explain the purposes of turbochargers.

Because fuel requires a minimum amount of air to completely burn, every engine reaches a limit where the addition of more fuel produces no increase in power because it cannot supply enough air. Instead, black exhaust smoke, decreasing fuel economy, and high exhaust temperatures result in proportion to the absence of air and excess fuel supplied to the cylinders. However, if more air mass is packed into the combustion chamber along with additional fuel, greater combustion force is produced with an increased power output from each cubic inch (or centimeter) of engine displacement. Higher cylinder pressures, in turn, translate into more torque. Turbochargers are essentially air pumps driven by exhaust gas energy to supply additional airflow to an engine. Exhaust gases drive a turbine wheel, which is connected to a **compressor wheel**, through a common shaft (**FIGURE 15-1**). Normally, the exhaust gas energy would vent directly to the atmosphere. Instead, the turbochargers recover some of that energy to improve engine efficiency by reducing parasitic pumping loses. Pumping loses takes place when energy supplied by the crankshaft is used to move gases in and out of the engine cylinders. Because turbochargers push air into the cylinders and help force exhaust gases out, the work of gas exchange is transferred to exhaust driven turbochargers rather than taken from the cracnkshaft.

When both turbocharger wheels turn at high speed, the compressor can supply air to the combustion chambers at pressure above atmospheric pressure. This fills the cylinders with more air mass than they are naturally capable of holding compared to only atmospheric pressure forcing air into the cylinders. Functionally, then, turbochargers can serve as replacement for engine

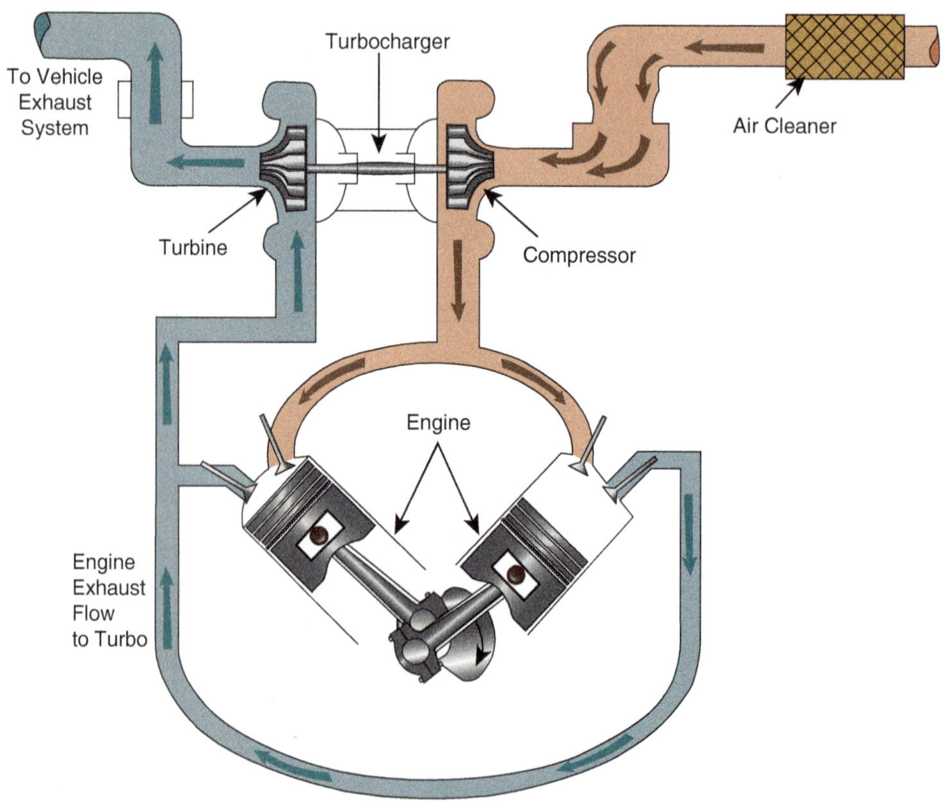

FIGURE 15-1 A turbocharger uses energy from exhaust gases to turn a turbine wheel, which causes a compressor impeller to pressurize air for combustion.

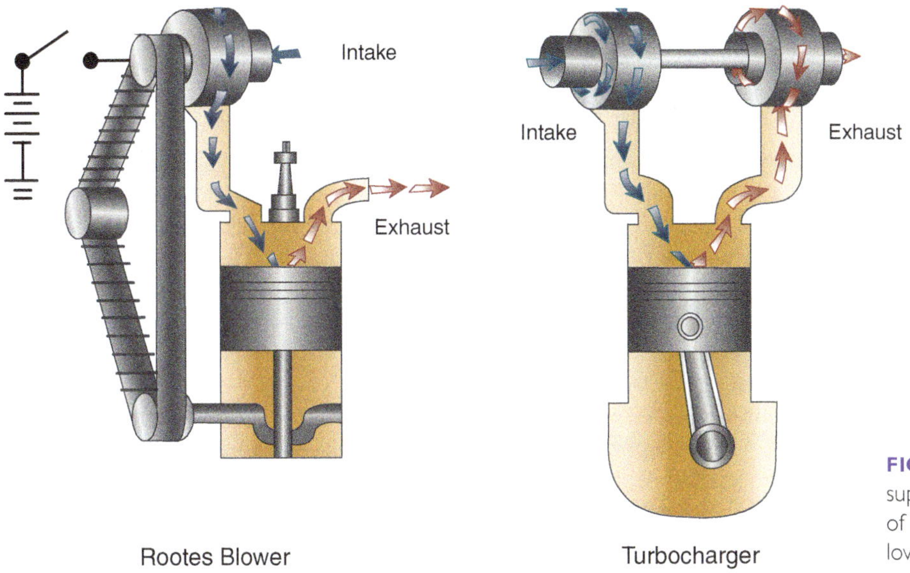

FIGURE 15-2 A rootes blower is a type of supercharger. Superchargers are parasitic users of engine power and are satisfactory for only low-speed, low-boost applications.

displacement. Easily 1.5 to 4-times more air can be forced into cylinders using a turbocharger than could be expected in a naturally aspirated engine using atmospheric pressure alone to charge cylinders with air. The idea of an exhaust-driven air pump to supply additional air to diesel engines was patented by Alfred Büchi in 1905. At first, the device did not enter large scale production until the usefulness of turbochargers for aircraft was discovered. During World War I and World War II, the **boost pressure normalization** feature of turbocharging enabled aircraft engines to fly at higher altitudes with no power loss in comparison to low-altitude flying. During normalization, the turbocharger compensates for low air density at high altitudes by spinning faster. Subsequently, adding turbochargers to diesel engines became widespread beginning in the 1950s, when design and metallurgical advances made turbochargers more economical and compact in size. Increasing the air supply pressure for combustion above atmospheric pressure is known as supercharging (**FIGURE 15-2**). Turbochargers are a subcategory of **superchargers** which are devices that pressurize the air intake system above atmospheric pressure. The term turbocharger is derived from the device's use of turbine-type centrifugal pumps. Centrifugal pumps are devices that move or pressurize fluids using centrifugal force. Although they are inefficient at low-speeds, when rotated at high speeds, the impeller wheel of a centrifugal pump can exceed 90% efficiency. Some diesel engines use an engine-driven positive displacement–type air pump commonly, though incorrectly, called a supercharger. This pump is more correctly known as a rootes blower (**FIGURE 15-3**). Rootes blowers are typically driven by the engine directly through the geartrain. The aluminum lobes pull atmospheric air into the blower and around the outside cover. Pressurized air is supplied to the cylinder intake air ports. Engines not equipped with superchargers have cylinders charged with air at atmospheric pressure and are classified as **naturally aspirated**.

FIGURE 15-3 Rootes blowers are positive displacement–type air pumps driven by the engine directly through the geartrain, or they are belt driven.

▶ Functions of Turbochargers

15-02 Identify and explain the functions of turbochargers.

While supplying additional air to the cylinders is a turbocharger's main purpose, there are other engine functions performed by a turbocharger that improve engine performance and reduce emissions. Many of the additional functions are indirectly fulfilled by improving combustion quality. The other jobs described in this next section that turbochargers perform are important explanations of factors accounting for diesel engines superior efficiency and clean emissions.

Increasing Volumetric Efficiency

Turbochargers are widely used on diesel engines to improve their power output while decreasing engine size and displacement. By charging the cylinders at higher than atmospheric pressure, commonly known as boosting, more air can enter the cylinders during the intake stroke. Boosting intake air pressure dramatically improves the volumetric efficiency (VE) of an engine, to more than 300%, and more than doubles its power output. The increases in engine volumetric efficiency can be even greater when using variable geometry and series turbochargers which are described in the next chapter. Charging the cylinders with additional air mass alone does not increase power output. Extra air mass, however, does allow for the injection of more fuel per combustion cycle, which in turn increases the amount of combustion pressure produced from each cylinder. To understand this, consider that a naturally aspirated engine—i.e., an engine without a pressurized intake air system—has a VE of 75–85%. This means the cylinders fill to 75–85% of their total volume under the best conditions. A 1 L (61 cubic inch) cylinder displacement would fill with 750–850 milliliters (46–52 cubic inches) of air without a turbocharger.

Consider again the same engine operated at 20 psi (138 kPa) of boost pressure. With a bore stroke of 3.75" × 4.0" (9.5 cm × 10 cm), this engine would double its compression pressure from 300 psi (21 bar) during cranking rpm (when no boost exists) to 600 psi (41 bar) when operating at 20 psi (138 kPa) of boost pressure. A 6.0 L Power Stroke engine operating at 3,300 rpm would have a VE of 100% if airflow were 9,911 L (604,800 cubic inches) per minute. However, published specifications indicate that its airflow is 20,728 L (1,265,000 cubic inches) per minute, meaning the engine cylinders fill to more than two times their original volume when turbocharged. Current turbochargers are easily able to boost up to three times atmospheric pressure, or approximately 45 psi (310 kPa). Series turbochargers boost to as much as 65 psi (448 kPa). Turbochargers are playing an increasing role in obtaining increased power output from engines with ever-decreasing displacements while reducing emissions and fuel consumption.

Detonation and Turbocharging

Diesel engines are well suited to turbocharging because greater boost pressure produces higher compression pressures and temperatures. Theoretically, diesel engines can have more boost pressure without detonating or experiencing pre-ignition damage that a spark-ignition engines naturally encounter. In fact, increases in cylinder air mass in diesels produce better combustion quality. Spark-ignition engines, unlike compression-ignition engines, are limited in their ability to increase intake charge pressure. Because fuel enters the cylinder with air during intake stroke, the heat developed by the air charge during compression stroke can ignite the air-fuel mixture before the spark occurs. The result is an abnormal, uncontrolled combustion event referred to as detonation. Diesel engines, however, have the ability to operate using wide air-fuel ratios. This is because diesel fuel burns at leaner air-fuel mixtures than gasoline, and the stratified, or layered nature of diesel combustion enables lean-burn combustion.

High compression pressures are not always beneficial or efficient. More engine vibration accompanies engines with high compression ratios. More powerful cranking systems and heavier batteries are needed. Better sealing head gaskets and stronger engine designs are needed too. Increasingly higher compression ratios require more energy to compress larger masses of air with a diminishing increase to power output beyond a 23:1 compression ratios. (**FIGURE 15-4**). Miller cycle engines can get around this problem by using progressively later intake valve closing while increasing boost pressures. In Miller cycle engines, energy normally taken from the crankshaft to compress air is instead replaced with the exhaust energy that is driving the turbochargers. The function of compressing air for combustion is transferred to the turbocharger in increasing proportions as boost pressure increases. (Miller cycle valve timing is described in more detail in chapter 6 – Cylinder Heads).

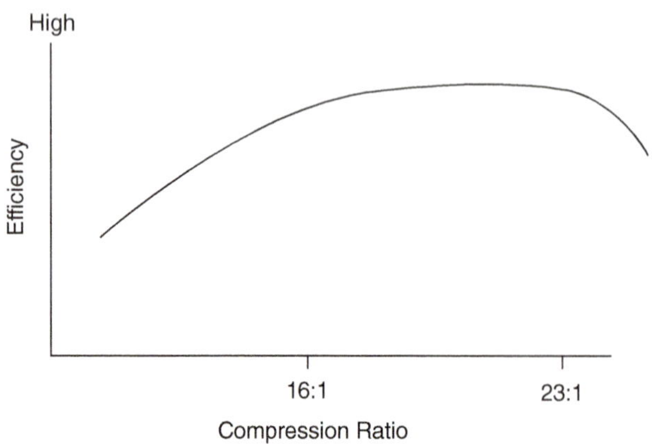

FIGURE 15-4 The best diesel engine efficiencies are produced at a 16:1 compression ratio. Beyond 23:1, power is lost because more energy is required to compress the air than is extracted from the power stroke.

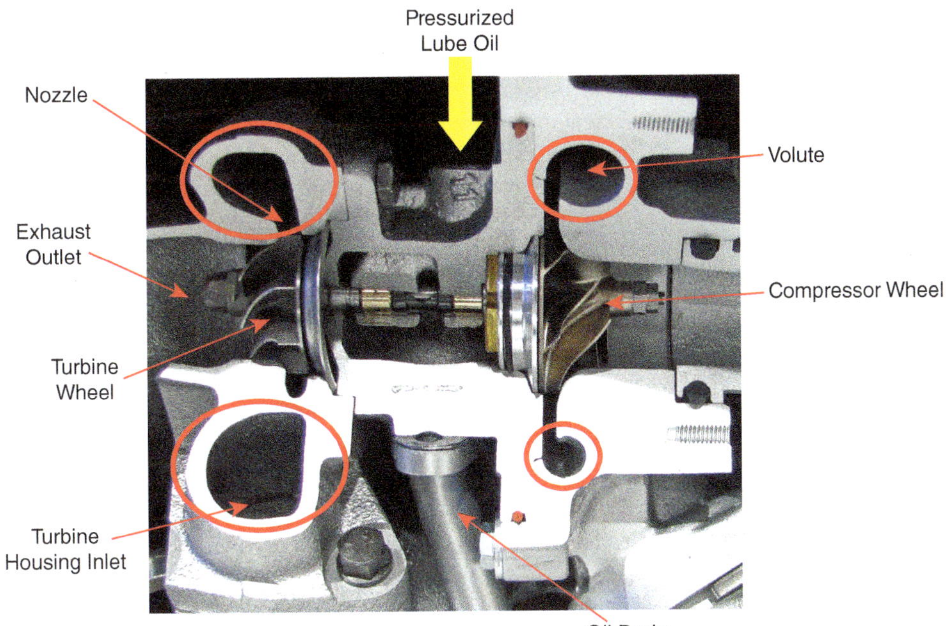

FIGURE 15-5 Turbochargers are engine components that are instrumental in reducing emissions. Technological changes to turbochargers are driven by emissions legislation.

> **▶ TECHNICIAN TIP**
>
> Some turbochargers are designed not to increase power output but to reduce emissions and maintain stable intake air pressure. Pressurized intake manifolds promote improved scavenging and charging of cylinders during the valve-overlap period, when fresh intake air is pushed across the chamber from the intake port and out the exhaust port, This effectively purges the cylinder of exhaust gases. The added airflow has a secondary but important function: to cool the piston crown and exhaust valves.

Emissions Reduction

The development of turbocharger technology is driven by demands to meet increasingly stringent emission standards. Turbochargers contribute to cleaner engine emissions in a number of ways. First, by increasing the air mass in the cylinder, compression pressures and temperatures will increase. Increasing the magnitude of these factors results in accelerating the speed of combustion's chemical reactions. Fuel reacts more quickly and completely with additional combustion chamber oxygen. This, in turn, lowers the production of hydrocarbons (HCs), carbon monoxide (CO), and particulate matter (PM) (**FIGURE 15-5**). Because turbochargers deliver more air to the combustion chamber, combustion quality is enhanced through improved contact between air and fuel. Some diesel engines use turbochargers primarily for this reason. Light-duty diesel engines built in the late 1990s supply only 4–8 psi (28–55 kPa) of boost pressure to the engine with the addition of a turbocharger. Power output from these turbocharger add-on designs is generally minimal. However, emission production dramatically declines when turbochargers are added (**FIGURE 15-6**). After a few seconds under load, only a clear heat wave should be observed from the exhaust pipe of a turbocharged vehicle. No visible emissions should be observed for any vehicle built in compliance with 1998 and later emission standards.

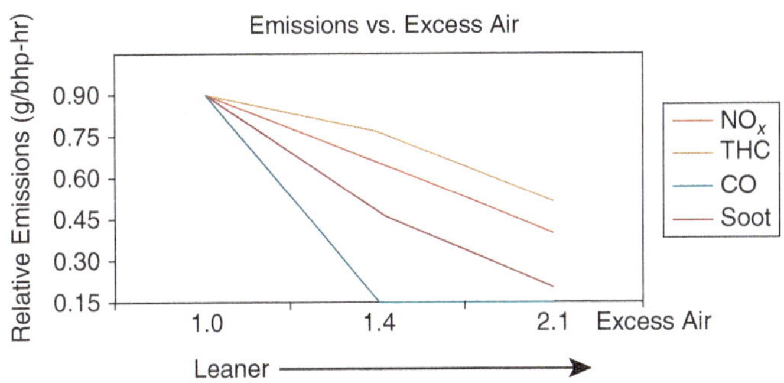

FIGURE 15-6 Turbochargers reduce engine emissions because excess air is added to the intake charge of the diesel engine. Better contact with fuel and higher temperatures and pressures burns fuel more completely.

> **▶ TECHNICIAN TIP**
>
> Excessive black smoke emissions on a turbocharged diesel can be caused by a misadjusted or defective air-fuel ratio control device. These devices limit the amount of fuel that is injected until adequate air mass is available for combustion. On electronic engines, defective manifold absolute pressure (MAP) sensors can have the same effect. Other common causes of black or gray smoke include excessive exhaust gas recirculation (EGR) flow, a restricted air intake, leaking charge air coolers, and a variety of problems with the fuel-injection systems.

Carbon Dioxide Reduction

Helping diesel engines develop greater power output from engines with smaller displacement has the indirect effect of reducing carbon dioxide (CO_2) emissions. Lighter engines mean not only less overall vehicle weight but also as much as a 20% reduction in CO_2 emissions. Using exhaust energy to drive more air into the cylinders makes turbocharged diesel engines more efficient because the engine does not expend energy pulling air into the cylinders. Leaner air-fuel mixtures provided by turbochargers increase fuel economy because of higher cylinder pressures and improved combustion quality. Smaller engines also have less internal friction than larger engines.

▶ TECHNICIAN TIP

Airflow from turbochargers must properly match engine demand to avoid two conditions contributing to potential turbocharger failure and noise: surging and choking. A turbocharger producing more airflow than the engine requires will momentarily build up too much intake air pressure, such as when the throttle is returned to idle while the turbocharger is still building boost pressure. Surge is a condition that takes place when this high intake pressure cannot go anywhere, and the airflow reverses through the turbocharger, creating a woofing or warbling sound as intake air flows backward out of the turbocharger across the compressor wheel. The opposite condition-when the compressor wheel cannot pull enough air into the turbocharger because the compressor wheel is turning too fast is referred to as the turbocharger choking. Without air resistance against the compressor wheel due to a near vacuum created in front of the compressor wheel, the turbocharger speed rapidly changes, which also creates unusual noises. Both conditions—surge and choking—can severely stress the compressor wheel and turbine shaft, causing them to break.

Engine Noise Emissions

Another indirect benefit of turbocharging is lower noise emissions from an engine. Noisy exhaust pulses emitted from exhaust ports are chopped up by the turbocharger's turbine wheel. Reduced exhaust noise emissions require lighter muffling of the exhaust system. Turbocharging similarly dampens intake air noise. Exhaust pulses, which normally travel up the intake manifold during valve overlap, are dampened by the turbocharger's compressor wheel. This explains why naturally aspirated air intakes on diesel engines are noisier than turbocharged engines. A more compact engine design also translates into a smaller surface area to radiate noise, which further makes turbocharged diesel engines operate more quietly than non-turbocharged engines. Higher-torque engines can turn more slowly at highway speeds, further reducing noise. They can also tow more weight at lower engine speeds in higher gears which further reduces noise and noxious emissions while delivering better fuel economy.

Turbocharger whistle, the noise that a high-speed turbocharger compressor wheels and turbine shaft vibration produces, is usually minimized by sound-dampening features in the air induction system.

Altitude Compensation: Normalization

Atmospheric pressure and air density drops with increasing altitude. High altitudes with low atmospheric pressure are commonly encountered on North American and European highways. Unlike naturally aspirated engines that depend on atmospheric pressure to force air into the cylinders, turbochargers pressurize the intake of engines, preventing power loss as altitude increases. Aircraft engines first demonstrated the usefulness of this feature in World War II. As aircraft flew higher, power output would naturally drop with altitude because naturally aspirated engines lose approximately 3% of power per 1,000 feet of altitude (–0.5 psi/1,000 ft [–3.4 kPa/305 m]) due to the drop in atmospheric pressure and air density. Thousands of turbochargers were installed on aircraft engines, which allowed aircraft to operate at higher altitudes with heavier loads and greater power output than naturally aspirated engines. Because atmospheric pressure pushes air into cylinders, diesel engines without turbochargers would experience the same difficulties as aircraft engines. At higher altitudes, such as when crossing a mountain passage, power is substantially reduced just when pulling requirements caused by steep grades are at their highest. Turbocharged engines, however, have the ability to compensate for this condition by maintaining boost pressure in spite of lower atmospheric pressure.

Turbine and compressor wheel speeds actually increase with at higher altitudes because of lower atmospheric pressure. As air pressure drops, turbochargers' compressor wheel speeds increase and maintain normal intake air pressure. The effect is due to the greater pressure differential across the turbine wheel that takes place when atmospheric pressure decreases in the exhaust pipe at the turbine outlet. This in turn causes a greater pressure difference between the exhaust gas temperature and pressure at the turbocharger turbine inlet and turbine outlet, which is always close to atmospheric pressure. When the difference becomes larger, the turbine wheel will turn faster. This means that as the atmospheric pressure decreases, a proportionally higher turbine speed is reached. When turbine speed increases, the compressor wheel speed also increases. With higher compressor wheel speeds, boosted intake manifold pressures will remain relatively consistent even as less dense air enters the turbocharger. Engine power output will remain consistent with power output at sea level.

"Normalization" is the term given to the turbocharger's ability to compensate for altitude change. Barometric pressure sensors are used on turbocharged, electronically controlled diesel engines to prevent the turbochargers from overspeeding. Because turbochargers spin faster at higher altitudes, under some conditions, the turbocharger can overspeed, causing turbo failure. The barometric pressure sensor prevents overspeeding by cutting back fueling by as much as 15% at very high altitudes. Reduced injection quantities lower exhaust energy and consequently reduce turbocharger speed. For a turbocharger to properly normalize intake air pressure or operate properly under any circumstance, it is important for the exhaust system to operate with minimal backpressure. Undersized exhaust pipes and mufflers, dents, and numerous bends cause exhaust backpressure to increase. If backpressure increases after the turbine wheel, the pressure differential between the turbine inlet and exhaust pipe diminishes, causing the turbocharger to slow down.

▶ TECHNICIAN TIP

A restricted, undersized, or otherwise damaged exhaust system is easily overlooked when investigating complaints about loss of power under load at high speed. A restricted or undersized exhaust system will reduce turbocharger boost and airflow through the engine. This happens because the pressure drop across the turbocharger turbine decreases when the exhaust is restricted after the turbine wheel. Always check for crimped, damaged, or undersized exhaust system components when investigating the causes of low-power complaints.

Engine Design Changes for Turbocharging

Turbocharging an engine requires a number of engine changes to take advantage of all the benefits that turbocharging has to offer. Changes in valve timing, particularly the amount of valve overlap and the length of the intake stroke, are increased to take advantage of higher intake air pressure and intake air velocity. Higher cylinder pressures and combustion turbulence produced by turbocharging often require lower compression ratios than naturally aspirated non-turbocharged diesel engines. This means that clearance volume and combustion chamber shapes will change between naturally aspirated and turbocharged engines. Heavier connecting rods, pistons, and crankshafts are also necessary to support increased cylinder pressure and power output. Higher cylinder pressures will also require improved head gaskets and cylinder head clamping forces to seal higher combustion pressure.

▶ Turbocharger Operation

15-03 Identify and describe the construction and operation of turbochargers.

Turbochargers operate by converting exhaust energy into mechanical energy that in turn forces more air into the engine. The amount of energy in the exhaust is determined primarily by the temperature and mass of exhaust gas leaving the cylinders. While most of the combustion energy is used to push the piston downward in the cylinder, some energy is left over when the exhaust valve opens and combustion gas exits the exhaust port. This is because combustion gas has not yet expanded to the maximum volume for its temperature. It is the energy of this unexpanded exhaust gas that produces the forces to drive the turbine wheel. At idle, exhaust energy is lowest because the quantity of fuel burned is minimal. However, when the operator demands increased power output, more fuel is injected, and then energy, in the form of high temperature and pressure exhaust, enters the exhaust manifold at the end of power stroke. Narrow exhaust manifold runners prevent the exhaust gas from expanding, and the diminishing diameter of the turbine housing increases exhaust pressure as exhaust gases flow into the turbine housing (**FIGURE 15-7**). When the exhaust gases reach the turbine nozzle ring, only then can it expand, because the exhaust pipe at the turbine outlet is near atmospheric pressure. The force of exhaust gas expansion across the turbine wheel pushes the turbine wheel (**FIGURE 15-8**). Pressure differentials in the exhaust system explain why turbine housings become hotter than the exhaust manifold when the engine is under load. The temperature of exhaust gas falls when it crosses the turbine by as much as 200°F–300°F (93°C–149°C), as its pressure falls to near atmospheric level in the exhaust pipe. Temperature changes measured over the turbine housing further demonstrate that exhaust gas continues to compress as the turbine volute becomes smaller (**FIGURE 15-9**). As more fuel and air are added to the combustion reaction, the energy content of the exhaust gas increases. Highest turbocharger speeds and output pressure are reached when the most fuel is injected at the highest engine rpm. The greatest number of combustion events per second, accompanied by high fuel rates, translates to the maximum boost pressure, which coincides with peak horsepower or rated engine rpm when a turbocharger is properly matched to an engine (**FIGURE 15-10**).

FIGURE 15-7 On turbocharged engines, the exhaust manifolds are narrow to prevent exhaust gas from expanding until the exhaust reaches the turbocharger nozzle ring.

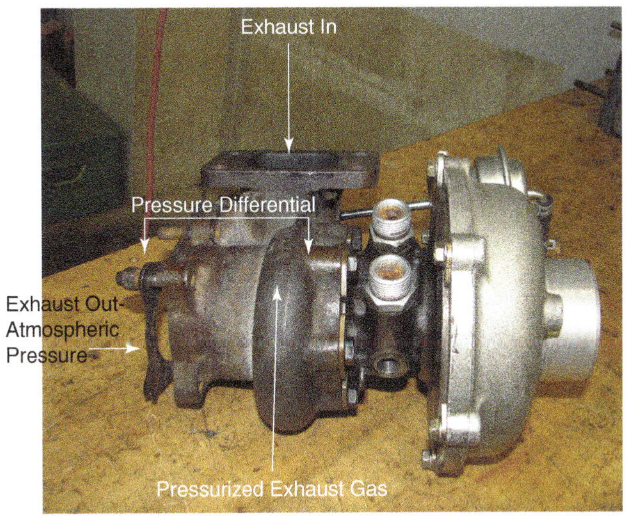

FIGURE 15-8 Pressure and temperature differentials across the turbine housing determine turbocharger speed. Atmospheric temperature and pressure are much lower than turbine temperature and pressure.

Turbo Lag

Turbo lag is experienced by a driver as a delay between stepping on the accelerator of the engine and the point when the engine responds with power in proportion to driver demand. Turbocharger inertia and friction are responsible for some lag, but emission control strategies are also a significant factor contributing to turbo lag. The fuel system could inject maximum amounts of fuel instantly into the cylinder, but adequate air to support combustion must first fill the

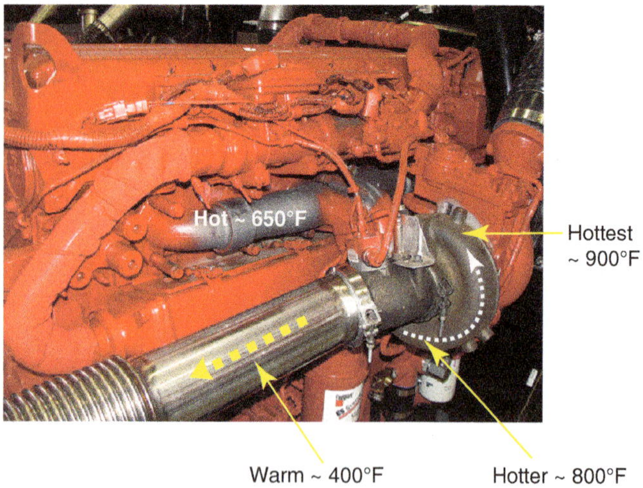

FIGURE 15-9 Temperature gradients from the exhaust manifold to the exhaust pipe are similar to what is observed under full-load engine operation. Temperature drop corresponds to pressure drop.

cylinders. Without adequate air to support combustion, excessive black smoke and other emissions are produced. To minimize the formation of excess emissions during load transition conditions, the governor system allows the engine to be only slightly over-fueled. Therefore, increasing injection quantities to bring the engine power or speed to operator-demanded levels takes place incrementally. Step by step ads adequate air is available, the fuel system can add slightly more fuel to speed up the turbocharger even more. With increased boost pressure, more fuel is added, and a cycle of adding fuel to produce more intake boost and then adding more fuel again is established. This cycle is referred to as spooling (**FIGURE 15-11**). The shortest amount of lag time is desirable because it not only affects emissions but also the operator's perception of performance. Prolonged spooling causes an engine to be perceived as lacking power, although that is not the case. Shorter turbo lag times reduce emission production because the over-fueling period during acceleration is also reduced. Air-fuel ratio controls associated with the governor use boost pressure as an input to limit injection quantities until adequate air mass is available to support combustion. An advantage of rootes blowers, which are positive displacement pumps, is they have much less lag time than turbochargers. The following are some of the factors that affect turbo lag:

- Size of the turbine housing: Smaller housings will spin the turbine faster since less exhaust energy is required to fill the housing and turn the turbine wheel. Larger housings will increase turbo lag time.
- The mass of the turbine and compressor wheels: Lighter and smaller wheels accelerate and rotate faster. Compressor wheels are made of lightweight cast aluminum or computerized controlled machining of titanium alloys, whereas turbine wheels are made of steel super alloys with high nickel content. Many turbine wheels also incorporate ceramic materials to lighten the wheel mass while improving heat resistance. Smaller wheels with less mass have the least amount of lag caused by inertia and can spin faster.
- Ball bearings versus sleeve center-housing bearings: Ball bearings have up to 50% less friction and shorten lag time by as much as 15% (**FIGURE 15-12**).

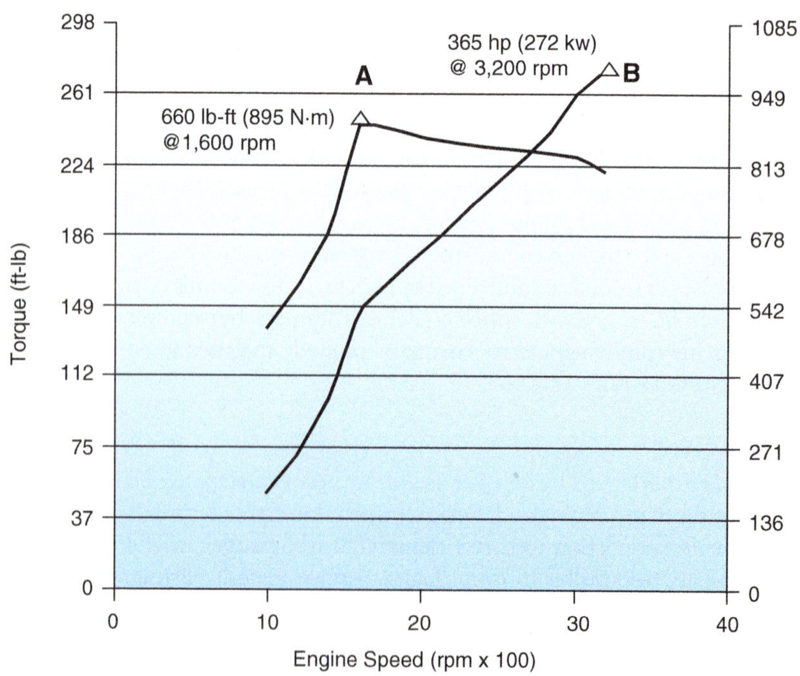

FIGURE 15-10 This torque-energy curve indicates where peak torque and horsepower occur. Point A, peak torque, corresponds to maximum injection quantity and cylinder pressure. Point B is maximum horsepower and turbocharger airflow. Turbocharger speed is fastest at this point.

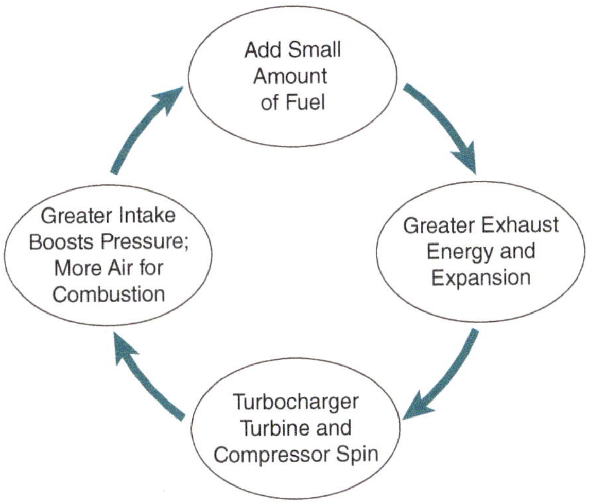

FIGURE 15-11 In spooling, first a slight amount of fuel is injected, then more boost pressure is produced, allowing the engine to inject more fuel.

FIGURE 15-12 Ball bearings are used by some manufacturers to improve durability, reduce turbo lag, and minimize turbocharger noise.

▶ Types of Turbochargers

15-04 Identify and describe the types of turbochargers used on diesel engines.

A variety of turbocharger designs have been developed in response to increasing demands for improvements to engine performance, power output, and emission reduction. The following are some of the basic types of turbochargers and configurations to be discussed in this and the next chapter:

- fixed-geometry turbochargers
- waste-gated turbochargers
- variable-geometry turbochargers, also referred to as variable-vane turbochargers or variable-nozzle turbochargers, among other names
- parallel turbocharging
- series or sequential turbocharging
- single sequential turbocharging
- asymmetrical turbochargers.

Fixed-Geometry Turbochargers

Fixed-geometry turbochargers are turbochargers without boost pressure controls. The name is derived from the fact that the housings and components have unchanging dimensions. This traditional type of turbocharger technology was used primarily until the late 1980s, when waste-gated turbochargers began to displace fixed-geometry turbochargers.

Disadvantages of Fixed-Geometry Turbochargers

If an engine is operated at only one speed or steady load, a single turbocharger could be well matched to the fuel settings and airflow requirements to produce the best combustion quality. This is often the case for heavy-duty diesel engines in highway trucks that will cruise at a particular speed most of the time. However, to produce power at all speed ranges requires a turbocharger to provide adequate boost pressure under both low-load and high-load factors (**FIGURE 15-13**). Because the amount of exhaust energy is low when an engine is lightly loaded, a fixed-geometry turbocharger can produce little boost at low speeds. From an emissions perspective, this is problematic because without turbo boost, air-fuel ratios are also low at low-speed operation, which results in high PM, HC, and CO emissions. Without low-speed boost pressure, power output is minimal at low speed and load operation.

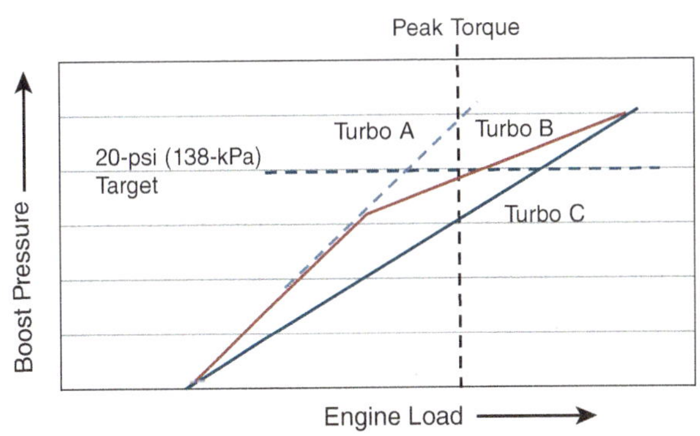

FIGURE 15-13 Different turbochargers will reach target boost pressure under different load factors. Turbo A is a small turbine housing that reaches target boost quickly but cannot provide more airflow beyond peak torque. Turbo C is a larger turbine housing providing less boost at low speed than Turbo A but more boost beyond peak torque. Turbo B is waste gated, providing boost quickly at low loads and more boost over a wider load range.

FIGURE 15-14 A waste-gated turbocharger has an actuator that opens a valve connecting the inlet of turbine housing to the exhaust pipe outlet. Waste-gated turbochargers allow a portion of exhaust gas to bypass the turbine, wasting its energy.

Waste-Gated Turbochargers

A waste-gated turbocharger is a solution to the problems of sluggish, low-power performance and high emissions at low engine speeds. These turbochargers are similar to fixed-geometry turbochargers except they possess an exhaust bypass valve, called a **waste gate**, in the turbine housing. This valve allows exhaust gases to bypass the turbine and directly enter the exhaust pipe, thus "wasting" some of the exhaust energy (**FIGURE 15-14**). These turbochargers are designed to turn faster at lower engine load factors, increasing power density, or power output per liter (or cubic inch) of displacement, of small-bore engines at low speed while minimizing turbocharger lag. The concept of a waste-gated turbocharger is to build small turbine housing with a lightweight, low-inertia turbine wheel that can rotate faster and accelerate more quickly than larger wheels and housings. By making the turbine housing smaller, less exhaust gas energy is required to spin the turbine. This produces faster turbocharger speeds at smaller engine loads. Consequently, higher compressor discharge volume and pressure are available at low speed and load conditions, enabling corresponding increased fuel-injection rates to provide greater output torque at lower engine speeds.

Smaller displacement engines will not need to turn as fast to produce what is known as power density—more torque at lower speeds—a desirable characteristic for small diesel engines. Making the turbine housing smaller, however, will cause exhaust backpressure to climb, limiting the engine's ability to breathe and produce further power gains as the engine increases load and speed. The turbine wheel can also overspeed and damage the turbo. At high turbocharger speeds, the turbocharger may not move air effectively as it reaches its choke point. High exhaust gas temperatures and potential engine damage is another consequence of restricted exhaust causing reduced air intake flow. To prevent the development of these undesirable conditions, a waste-gated turbocharger vents a portion of the exhaust gas to the exhaust pipe before reaching unacceptable exhaust backpressure in the exhaust manifold and excessive turbine speed. To accomplish this, waste-gated turbochargers are constructed with an internal passageway in the turbine housing that are opened and closed by a bypass valve. When the valve is closed, the turbocharger is capable of building boost pressure. When the valve is open, exhaust is allowed to move directly from the turbine housing inlet to its outlet. Thus, exhaust gases effectively bypass the turbine and pass directly into the exhaust pipe (**FIGURE 15-15**). Because the exhaust energy is not harnessed by the turbine, it is wasted—hence the term "waste-gated turbocharger."

> **▶ TECHNICIAN TIP**
>
> Never plug a waste-gate actuator or prevent it from opening, to gain more power. Doing either can lead to premature turbocharger failure due to the turbine overspeeding. Lubrication breaks down, and the bearings contact the journal surfaces during overspeed conditions. Preventing a waste gate from opening can also lead to excessive heat load from over fueling in the cylinders, damaging the pistons and valves, overpressuirzing the intercooler, and increasing NO_x emissions.

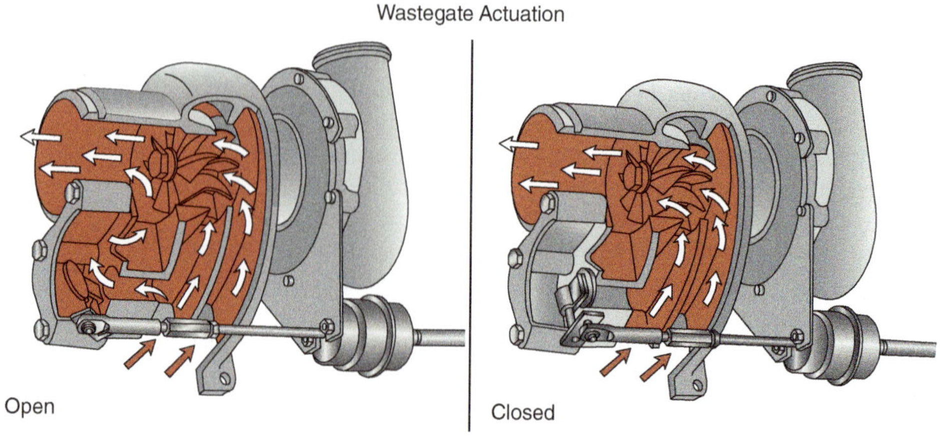

FIGURE 15-15 The waste gate allows exhaust gases to bypass the turbine housing and directly enter the exhaust pipe. Exhaust energy is wasted when it passes from the manifold directly to the exhaust outlet.

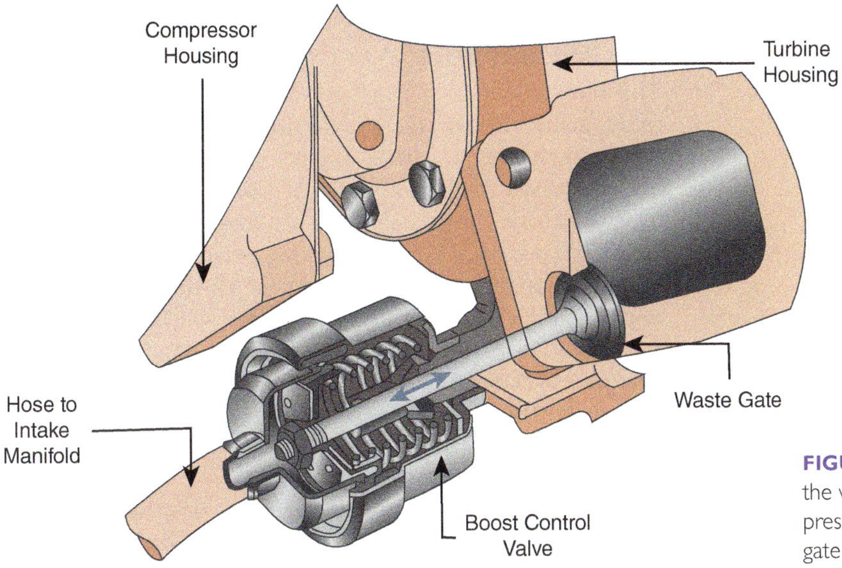

FIGURE 15-16 Spring tension holds the waste-gate valve closed. Boost pressure or a vacuum opens the waste gate against spring tension.

Waste-Gate Controls

The waste gate is designed to open at a predetermined boost pressure so that the engine operates efficiently. Failure to open will cause the turbine speed to exceed speed limits, damaging the bearing and fatiguing the turbine wheel. Turbocharger failure will result if the waste gate is seized closed or tampered with to prevent it from opening. Most waste gates will use air- or vacuum-operated actuators to operate the waste gate (**FIGURE 15-16**). Actuators will commonly use a spring operating against a diaphragm to close the waste gate. Many waste gates are normally closed by relatively weak spring pressure and opened by stronger boost pressure. This often leads to the waste gate sticking or binding in a partially open position because relatively weaker spring pressure cannot close the waste gate. A line connecting the turbocharger compressor discharge or intake manifold air pressure to one side of the diaphragm will push the actuator open against spring pressure. When the predetermined boost pressure has been reached, the waste gate will begin to open, preventing any further buildup of exhaust pressure and limiting any further increase in the output of the turbocharger. Weak actuator spring pressure will prevent the waste gate from closing if the bushings and linkage are gummed, corroded, or have excessive carbon deposits. In either of these cases, the turbocharger will not build boost pressure again, which will result in a low-power complaint. Springs can also be weakened by exhaust heat and may require adjustment on some turbochargers. Waste gates often have an adjustment mechanism to compensate for spring deterioration. Some waste gates are normally open, and the actuator closes the waste gate. With this arrangement, vacuum and electronic controls will operate the actuator. The advantages of normally open waste gates are that overboosting is prevented and the waste gate will operate more frequently, minimizing the likelihood of linkage seizing.

OBD-II and Turbocharger Actuators

Accurate control of turbo boost pressure is critical to minimizing engine emissions. To compensate for deterioration in operation and inaccuracy of mechanical actuator controls, OBD-compliant engines electronically control waste-gate operation (**FIGURE 15-17**). A closed-loop feedback between the boost pressure sensor and waste-gate position corrects for deterioration of waste-gate actuators and turbocharger operation.

FIGURE 15-17 On this Holset HE351 CW turbocharger, the command valve that is controlled by the electronic control module (ECM) regulates boost pressure to the waste-gate actuator.

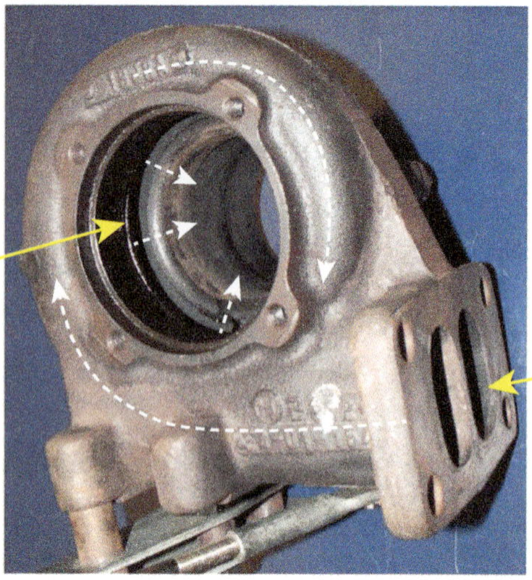

FIGURE 15-18 Pulse turbochargers have two separate exhaust volutes and split exhaust manifolds to prevent exhaust pressure pulses from canceling one another.

Pulse and Constant-Pressure Turbochargers

The interaction of exhaust pulses in the exhaust manifold under some circumstances is detrimental to the efficiency of the turbocharger on diesel engines. When an exhaust pulse initially leaves the exhaust port, it has high pressure. As the exhaust stroke finishes though, a partial vacuum is created at the exhaust port by the departing pulse due to gas inertia. If exhaust manifold runners connect all the exhaust pulses together, high- and low-pressure pulses will blend. Blending low and high pressure pulses will reduce engine and turbocharger efficiency, particularly in in-line six-cylinder engines. To prevent this from happening, turbocharged diesel engines use split exhaust manifolds and what are called pulse-type turbochargers. Split exhaust manifolds provide a barrier in the exhaust that separates the front three and back three cylinders. Likewise, the turbocharger turbine housing is divided into two volutes. This way, exhaust pulses are kept separated until they reach the turbine nozzle. Pulse-type turbocharger configurations provide for shorter turbo lag, higher boost pressure, and better exhaust gas scavenging. Pulse-type turbochargers can be identified by their dual inlets to the turbine housing (**FIGURE 15-18** and **FIGURE 15-19**). Turbochargers without separate volutes for front and rear cylinders are called constant-pressure turbochargers. These can be identified by the single turbine inlet.

Variable-Geometry Turbochargers

Variable-geometry turbochargers (VGTs), also referred to as variable-nozzle turbochargers (VNTs) and variable-vane turbochargers (VVTs), are turbochargers with the ability to electronically change boost pressure independent of engine speed and load (**FIGURE 15-20**). While having slightly different operating mechanisms, this category of turbocharger permits electronic control of boost pressure by the engine ECM. The primary advantages of VGTs are that they have very fast turbocharger response times (almost no lag) and provide a continual ideal match of airflow to combustion. VGTs are necessary for engines equipped with high-pressure cooled exhaust gas recirculation (CEGR) because they can increase exhaust backpressure above boosted intake manifold pressure, which is required to mix EGR gases into the intake manifold. (See Chapter 17, on exhaust gas recirculation.)

FIGURE 15-19 Pulse turbochargers use two separate exhaust volutes to keep exhaust pulses separated until they pass through the turbine nozzle ring and contact the turbine wheel.

Parallel Turbocharging

When two or more turbochargers are used, they are arranged to directly supply banks of cylinders with air. A minimum of two turbochargers are used to directly supply the cylinders of a V-bank engine, where one turbocharger is located on each cylinder bank. This arrangement is called paralleling, or **parallel turbocharging**. Large industrial diesel engines, such as those used in electric power generation, use parallel turbocharging require multiple turbochargers to supply long banks of cylinders (**FIGURE 15-21**).

Series Turbocharging

Fixed-geometry, waste-gated, and VGTs have output limits when used on engines with wide torque band. For example, a large turbocharger for a high-output engine will not operate well at low-speed and low-load conditions. VGTs are speed limited and may not always be able to adequately supply an engine with enough air volume at sufficient pressure. To compensate for this problem, **series turbocharging** involves the use of two differently sized turbochargers to provide additional air volume and boost pressure. The compressor output of one is directly connected to the compressor inlet of the other. This arrangement multiplies the maximum boost pressure of the turbochargers. Output from a smaller high-pressure turbocharger that responds quickly at low engine speeds is supplemented at medium speeds and loads by the larger low-pressure turbocharger (**FIGURE 15-22**). A constant high boost pressure is available across the widest possible range of engine speeds using series turbocharging.

FIGURE 15-20 This VGTs has an actuator mechanism located near the center housing that changes the volume of exhaust gases allowed to pass through the turbine housing.

Other Turbocharger Configurations

There are two other turbocharger configurations: series sequential turbochargers and asymmetrical turbochargers. **Series sequential turbochargers** use a single turbine wheel to drive two compressor wheels that rotate in two separate compressor housings. This type of turbocharger was popularized by the 6.7 L Ford Power Stroke models from 2010 to 2014, which used two different shapes of compressor wheels and housings to provide boost pressure supplying a wide range of engine speed and load conditions. This turbocharger has since been replaced by a VGT. **Asymmetrical turbochargers** resemble fixed-geometry or waste-gated turbochargers but have exhaust volutes of two different diameters. For example, Saab and Hyundai engines use asymmetrical turbochargers to increase exhaust backpressure from the narrower volute, which supplies the EGR system with exhaust gases.

FIGURE 15-21 Parallel turbochargers use two turbochargers to supply boost pressure to separate cylinder banks.

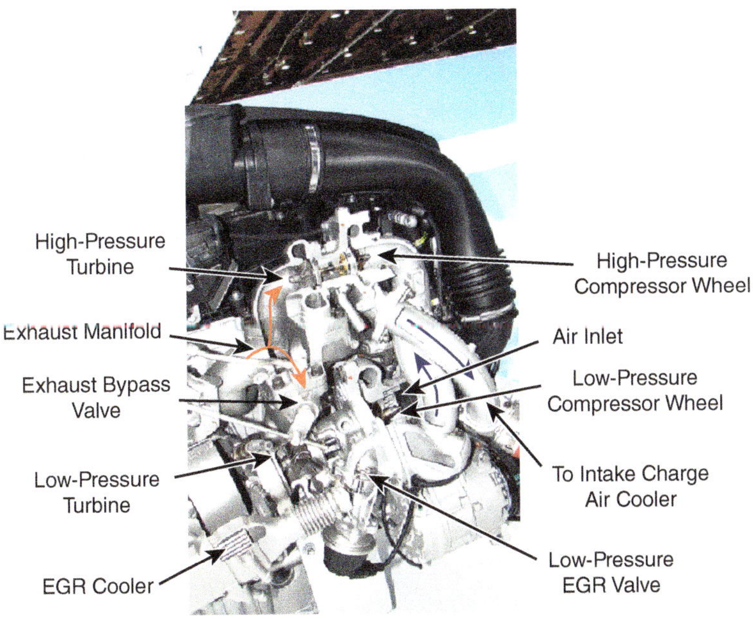

FIGURE 15-22 This BMW 3 L engine uses series-type turbocharging. The smaller high-pressure turbocharger compressor multiples boost pressure from air supplied by the low-pressure compressor. Series turbochargers supply more airflow with less lag than fixed-geometry turbochargers.

▸ Components of Turbochargers

15-05 Identify and explain the function of turbocharger subcomponents.

Turbochargers are constructed in three parts. The two most significant parts are the turbine, or hot side, and the compressor, or cold side. A third part, the center housing, is positioned between the compressor and turbine housing. The center housing holds the turbine shaft connecting the compressor and turbine, along with bearings supporting the shaft. The inlet to the turbine is located on the outside radius of the turbine housing and attaches to the exhaust manifold. The outlet of the turbine connects to the vehicle exhaust pipe. Air flows into the turbocharger through the compressor inlet located at the center of the compressor wheel. Centrifugal force of air captured and thrown to the outside radius of the compressor housing forms the compressor discharge or outlet that connects to either the intercooler or intake manifold (**FIGURE 15-23**).

Turbine and Compressor Housing Shapes

Both the compressor and turbine housings are shaped like a scroll or snail shell; this shape is also called a volute. The turbine housing becomes progressively smaller as exhaust gas travels from the inlet at the exhaust manifold to an exit slot, called the nozzle, where exhaust flows across the turbine wheel and finally out into the exhaust system (**FIGURE 15-24**). This narrowing diameter in the turbine housing causes the force of exhaust gas pressure at the inlet to multiply gas pressure inside the volute as it travels to the narrow nozzle. By varying the size of the turbine housing and the nozzle opening, the speed and flow characteristics of the turbine can change. Correctly matching the turbine housing and engine exhaust flow is important for effective turbocharger operation. The **compressor housing**, which houses the compressor wheel, is shaped like the turbine housing, but the shape and direction of gas flow takes place in the opposite direction. Air first enters the compressor housing after passing through the air filter, where it contacts the center part of the compressor wheel's blades, called the inducer. Driven by the turbine shaft, the edges of the inducer cut into the air mass and create a low-pressure area to draw air into the housing. Air trapped by the inducer then moves across the compressor blades to the exducer, where centrifugal force

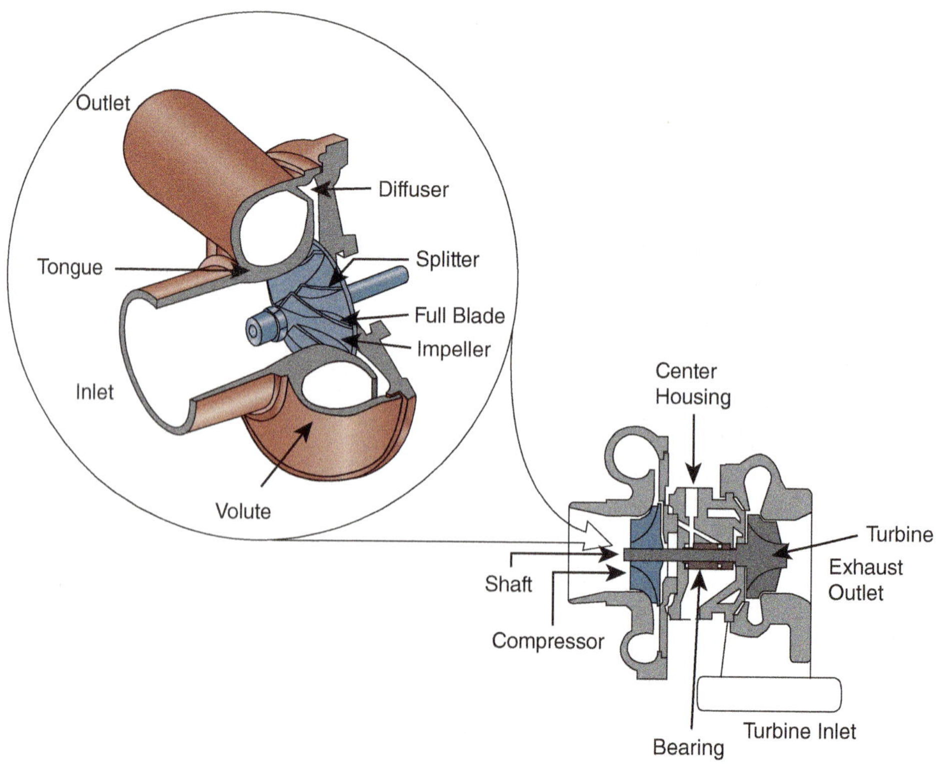

FIGURE 15-23 The construction of the compressor side of the turbocharger is shown here in relation to the turbocharger as a whole.

imparted to the air by compressor wheel speed throws air through a narrow slot called the diffuser or the tongue. Air velocity slows after passing through the diffuser and builds pressure inside the increasing circular diameter of the compressor housing. The volume of air that can be supplied to the engine is determined by a number of factors, including the geometry or shape of the compressor wheel, compressor wheel size and speed, and housing shape (**FIGURE 15-25**). A mathematical coefficient, called the A/R ratio defines the relationship of housing size and gas flow. The ratio between the turbocharger inlet area divided by the radius dimension is call the A/R ratio. The A/R ratio is a numerical representation of the diameter of the inlet opening into either the compressor or the exhaust volute compared to the radius of the volute.

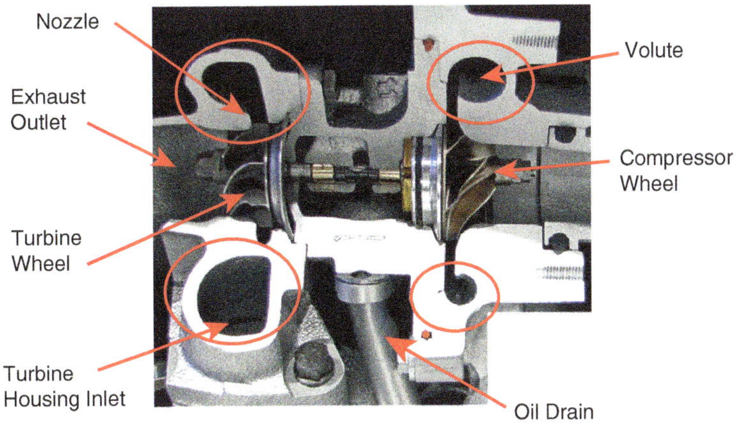

FIGURE 15-24 The nozzle is the opening in the turbine housing where exhaust gases expand across the turbine wheel. Its dimensions have a significant effect on the characteristics of the turbocharger.

A/R Ratio

Turbochargers must be precisely matched to a particular engine's fuel-injection rate setting and engine displacement to achieve the correct boost pressure for the engine's design. For example, the fastest turbocharger speed producing the highest boost pressure should correspond to an engine's peak horsepower and provide minimal exhaust backpressure in the exhaust manifold. Every horsepower rating and torque rise profile in an engine family uses a uniquely shaped and sized turbocharger on engines using fixed-geometry turbochargers. Matching the turbocharger to the engine ensures that the airflow from the turbocharger will correctly match the engine fuel rates to create optimal combustion conditions, high power output and low emissions. (**FIGURE 5-26**). A turbine housing that is too small does not allow enough exhaust gas flow at high engine loads and speeds. This results in excessive exhaust backpressure, producing low-power complaints and engine overheating. Exhaust manifold and valve temperatures will be too high. A turbine housing that is too large will produce inadequate boost pressure at low-speed operation, because large quantities of exhaust energy are needed to drive the turbocharger. Similarly, the compressor wheel and housing need to

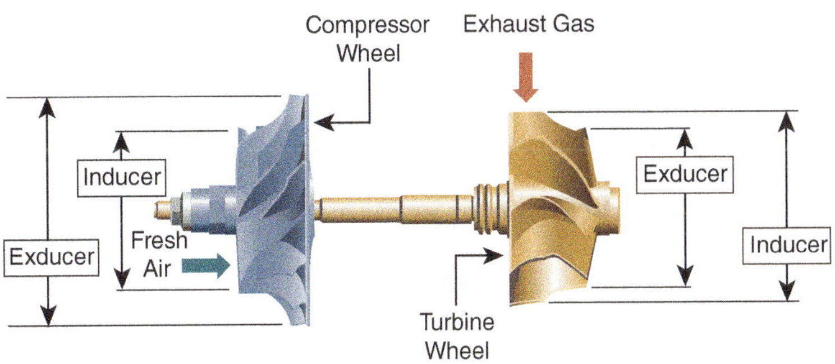

FIGURE 15-25 The shape of the turbine and compressor wheel influence airflow through the turbocharger. Compressor and turbine wheels are uniquely shaped for each engine family or power rating.

FIGURE 15-26 Wheel size and shape dramatically change turbocharger performance for various speed and load conditions. For example, the smaller CNC machined titanium-alloy compressor wheel is used in high-horsepower (kilowatt) applications.

be properly matched to produce adequate airflow across all engine speed ranges without overboosting or underboosting. An incorrectly matched turbocharger can lead to either too little or too much boost for the engine. The following are some of the possible consequences of turbocharger discharge pressure being too low:

- low power
- poor fuel economy due to poor combustion quality
- inadequate dilution of combustion heat and high cylinder heat loads
- damage to pistons, valves, and turbocharger caused by excessive thermal loads
- increased exhaust temperature due to inadequate excess air to dilute combustion heat or combustion afterburn in the exhaust manifold
- increased smoke emissions from PM production
- increased HC, PM, and CO emissions.

The following are some of the possible consequences of turbocharger boost pressure being too high:

- poor fuel economy due to excessive compression pressures
- increase in NO_x emissions due to excessive cylinder pressures
- turbocharger bearing failure and turbocharger wheel fatigue problems caused by overspeeding the turbocharger
- damaged or blown intercooler pipes, and cracked charge air coolers
- high inlet air temperatures caused as intercooler heat load increase
- damage to cylinder, pistons, and valves caused by high thermal loads
- engine power derating by the engine protection monitoring system due to high inlet manifold temperatures.

The most important factors determining turbocharger performance characteristics are the size and shape of the compressor and turbine wheels and their respective housing sizes (**FIGURE 15-27**). Critical to the flow through the housings is the ratio between the area of the compressor or turbine housing inlet to the radius of the housing. These factors are summarized in the **A/R ratio**, which is often stamped into fixed-geometry turbocharger housing (**FIGURE 15-28**). The A factor of the A/R ratio is the area of either the compressor or turbine inlet. The R factor of the A/R ratio is the radius of the housing. A small A/R ratio for the turbine housing means the housing is potentially restrictive to exhaust gas but will rotate faster than a housing that has a large A/R ratio. A small A/R ratio will provide better torque and airflow at low engine speeds and load factors but at the expense of high-speed, high-load operation. The opposite is true for a turbocharger housing cast with a larger A/R ratio. When replacing turbochargers, the A/R ratio should not be interchanged. However, in some applications where the engine is operated frequently at high altitude, changing the A/R ratio by increasing turbine and compressor housing sizes may improve performance. Similarly, at sea level, too small of an

FIGURE 15-27 The A/R ratio is a comparison to the area of the inlet (A) divided by the radius of the housing (R).

FIGURE 15-28 An A/R number is often stamped into the turbine housing. A turbocharger may fit but not function correctly if it is not precisely matched to the engine displacement and fuel settings.

A/R ratio may provide fast turbocharger response at low speeds but at the cost of high-speed torque and power.

To improve turbocharger efficiency, a ported shroud compressor inlet can be used. This is a construction feature incorporated into the compressor housing that allows some excess airflow to exit the edges of the compressor wheel and return into the air inlet. Recirculating airflow allows a larger compressor wheel to be used for higher airflow requirements without allowing excess air buildup in the compressor volute, which can flow backward out of the housing at low speeds. This condition, called turbo surge, creates a loud intake air noise when the throttle is quickly returned from wide open position to idle or when a noise happens moments after stopping at an intersection. The noise can resemble a loud barking or woofing sound, and reverse airflow can twist the turbine shaft. Using a ported shroud is not always ideal, because it can have some negative effects on compressor efficiency at slow speeds, but it improves efficiency at higher compressor wheel speeds.

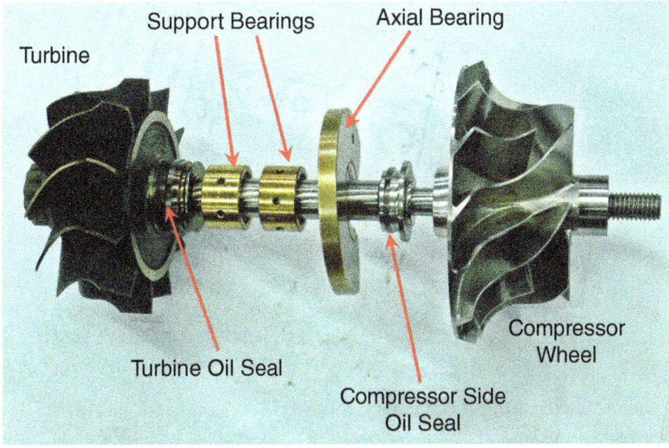

FIGURE 15-29 Two center support bearings rotate on the turbine shaft and in the housing. The thrust bearing controls turbine shaft endplay.

Center Housing and Bearings

The turbocharger center section typically consists of two common configurations: plain sleeve-type bushings and angular ball bearings.

Plain Sleeve-Type Bushings

Traditionally, turbochargers used a gray cast iron housing to support a fully floating bearing system for the turbine and compressor shaft. Small turbochargers can rotate at speeds of up to 180,000 rpm, and medium engines will rotate in the vicinity of 100,000–130,000 rpm. Bearings, shafts, oil seals, and housing tolerances are manufactured with precision and durability suitable for this high-speed operation. Two bearings support the shaft at each end of the center housing (**FIGURE 15-29**). These bearings are full floating, meaning they rotate on the shaft and inside the housing. Oil holes in the bearings ensure an adequate oil film is available between the shaft and bearing (**FIGURE 15-30**). The bearings are manufactured from bronze or brass bearing alloys. Shaft bearing journals are

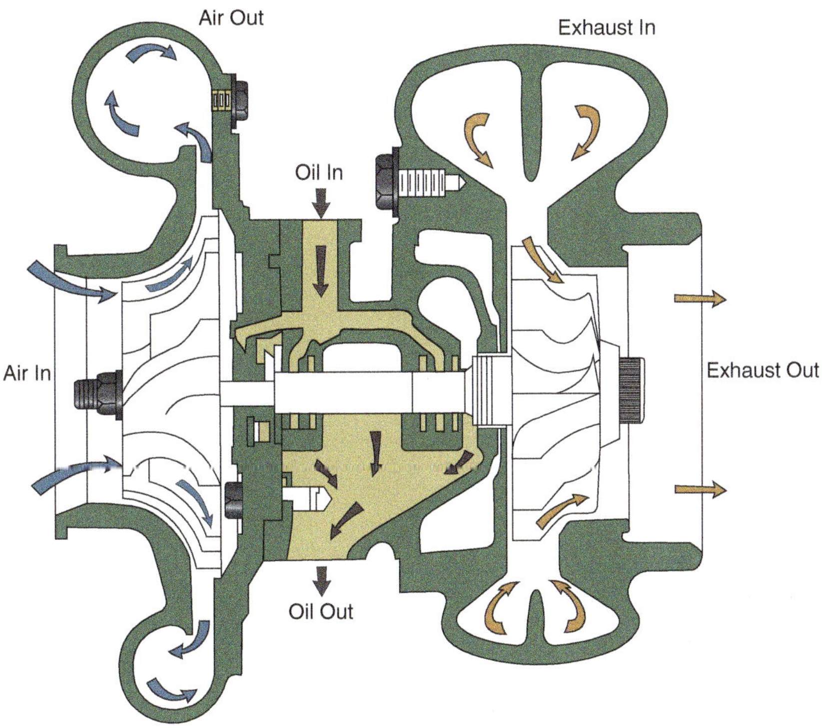

FIGURE 15-30 Bearings in the center housing receive a constant flow of oil.

> **▶ TECHNICIAN TIP**
>
> To determine whether the turbocharger is correctly matched to the engine, the turbocharger part number or A/R number needs to be compared against a part's listing prepared by the engine manufacturer. If the A/R number is not stamped on the housings, the turbocharger part number or serial number needs to be used for comparison.

> **▶ TECHNICIAN TIP**
>
> Turbocharger slobber is a condition where the turbocharger passes oil into the exhaust system. This can be caused by excessive idle time, when gas pressure is usually insufficient to keep oil in the center housing. Other possible causes include high crankcase pressures, restricted air filters, and damaged seals. One technique to determine whether the turbocharger is the source of oil is to remove the exhaust manifold and run the engine to determine whether oil is originating from the cylinders or the turbocharger.

▶ TECHNICIAN TIP

Turbochargers are balanced by using a machine known as a vibration sorting rig. These machines use compressed air to spin turbochargers to correctly determine where weight needs to be removed or added to the compressor or turbine wheel. A loud whistling sound emitted from a turbocharger can be caused by a turbine/compressor wheel imbalance condition. When imbalanced, the turbocharger bearings and shaft will vibrate against the bearings to produce this loud, unpleasant high-frequency noise.

induction hardened and machined for dimensional accuracy and finished for high-speed operation. To control endplay or axial movement of the turbine shaft, a thrust bearing and hardened steel thrust collar is located at the compressor end of the shaft. When assembled, the center-housing rotating assembly (CHRA) is dynamically balanced by using compressed air, and metal is removed from the underside and other parts of the turbine and compressor wheel to achieve vibration-free operation (**FIGURE 15-31**). Balancing the turbocharger as it is driven with compressed air is done on a machine called a vibration sorting rig. Because components are not normally match marked, any disassembly of the CHRA requires rebalancing.

Angular Ball Bearing Turbochargers

In angular ball bearing turbochargers, also called cartridge ball bearing turbochargers, the center-housing support consists of a single sleeve cartridge ball bearing system that contains a set of angular contact ball bearings at each end. Using an angular-type ball bearing eliminates the need for a thrust bearing because end play is controlled by the ball and race assembly rather than a sleeve. The following are some of the advantages of the ball bearing system over the plain sleeve-type bearings:

- Reduced oil flow: Ball bearing designs reduce the amount of oil required to provide adequate lubrication. Not only does this reduce the likelihood of bearing coking during hot shutdown, but also less cumulative damage will take place during minimal lube availability during start-up. Reduced oil flow in the center housing also minimizes seal leaks.
- Reduced turbocharger lag: Ball bearings reduce turbine friction by as much as 50%, thus reducing lag times by 15%.
- Reduced vibration and noise: Ball bearing turbochargers dampen shaft movement much better than sleeve bearings, which require two oil clearances: between the shaft and bearing and between the bearing and housing.
- Tighter operating clearances: the compressor and turbine housings have tighter operating clearances because there is less bearing clearance with ball bearings than with plain bearings. With less clearance between the housing and compressor wheels, turbochargers operate more efficiently with less gas leakage.

Water-Cooled Center Housing

Many turbochargers use engine coolant passing through passages cast into the center housing to remove heat. When the engine is shut down, coolant circulates by using convection currents. Heat from the housing warms the water, causing lighter, less dense coolant to rise and leave the housing while drawing cooler coolant from the engine block (**FIGURE 15-32**).

FIGURE 15-31 Metal is removed from turbocharger turbines and compressor wheels to balance components rotating at very high speeds.

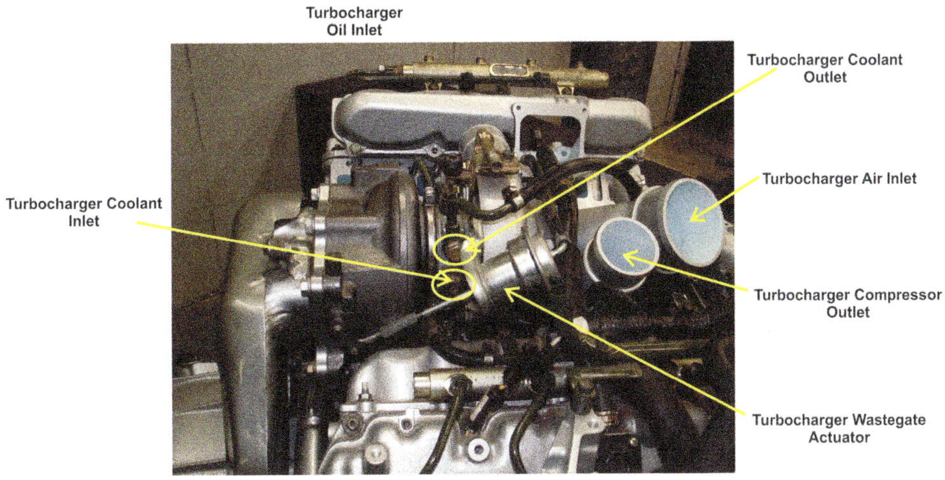

FIGURE 15-32 Coolant lines connected to the center housing supply engine coolant to flow through the turbocharger center housing. Circulating coolant during the hot shutdown period removes heat to prevent oil coking in the center housing.

▶ Maintenance of Turbochargers

15-06 Identify and explain inspection, maintenance, and diagnostic procedures for turbochargers.

Turbocharger support shaft and axial bearings not only enable the turbine shaft to rotate freely with little friction but also prevent excessive movement of the compressor and turbine wheels inside the turbocharger housings. Bearing wear will allow axial (end-to-end) and radial (up-and-down) movement of the wheels and allow them to touch the housings. If this happens, the turbocharger will have a catastrophic failure (**FIGURE 15-33**). Maintaining a close operating clearance to the housing is needed to prevent gas leakage around the wheels and to maintain turbocharger efficiency. Little tolerance for bearing wear means the turbocharger should be regularly inspected before problems occur.

Bearing Inspection

With the intake system disconnected from compressor housing, it is possible to visually check for excess bearing axial and radial clearances. If light finger pressure applied sideways against the compressor wheel causes contact between compressor and housing, the bearings and/or shaft will be worn (**FIGURE 15-34**). If this condition is detected, the turbocharger must be repaired or replaced before catastrophic damage occurs. Another technique to check for and measure radial play is to measure the clearance between the compressor wheel and housing. Pushing the compressor wheel downward and sliding a feeler blade in between the wheel and housing measures minimum clearance (**FIGURE 15-35**). Moving the compressor wheel upwards and remeasuring the clearance supplies information to calculate

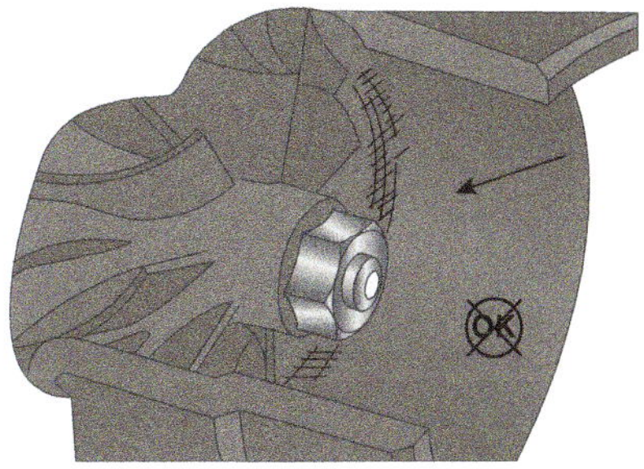

FIGURE 15-33 Excessive bearing wear will cause damaging contact between the compressor housing and wheel.

FIGURE 15-34 Radial bearing play can be checked by lifting the turbine shaft against the housing while rotating the shaft. Excessive bearing wear is indicated by compressor or turbine wheel contact with the housing.

FIGURE 15-35 One technique to measure radial clearances is to insert a feeler blade into the gap left when the compressor wheel is lightly pressed toward the housing.

bearing radial play. Subtracting the largest from the smallest clearance measured provides an indication of radial bearing wear. Because the measurement of radial play is a projection of the center housing bearing clearance to the end of the turbine shaft, the clearance varies between manufacturers, but measurements can be compared against manufacturer specifications. Axial endplay is usually measured with a dial indicator. Moving the turbine shaft back with finger pressure, zeroing the dial indicators, and then moving the shaft forward enables axial play to be measured. Whereas radial play is easily detected when moving the turbine shaft, any axial play should not be felt but instead only observed with a dial indicator.

FIGURE 15-36 This alternative method for measuring radial clearance measures the side movement of the turbine shaft through the turbocharger drain hole.

Measuring Turbocharger Radial Bearing Clearance

Two different sets of turbocharger bearings are responsible for keeping the turbine and compressor wheels properly positioned in their respective housings. Excessive wear on either of these bearings can potentially cause catastrophic turbocharger and engine failures. Measuring axial and radial bearing clearances is an important estimate of turbocharger service life and can pinpoint imminent failures. Worn center support bearings can lead to more turbocharger noise, which sounds like a whistling sound as the shaft vibrates. To measure turbocharger radial bearing clearance, follow the steps in **SKILL DRILL 15-1**. An alternate method for checking radial clearance is to use a dial indicator to measure the turbine shaft movement through the center housing (**FIGURE 15-36**). This method can be used only if the turbine shaft is accessible through the center housing.

Measuring Turbocharger Axial Bearing Clearance

A turbocharger's thrust bearing controls the turbine shaft's axial endplay. Excessive bearing oil clearances will allow the compressor or turbine wheel to contact the turbocharger housings. Axial clearance cannot be measured by feel, because the amount of acceptable clearance is too small for any technician to detect. A dial indicator should be used to measure axial clearance. To measure turbocharger axial bearing clearance, follow the steps in **SKILL DRILL 15-2**.

SKILL DRILL 15-1 Measuring Turbocharger Radial Bearing Clearance

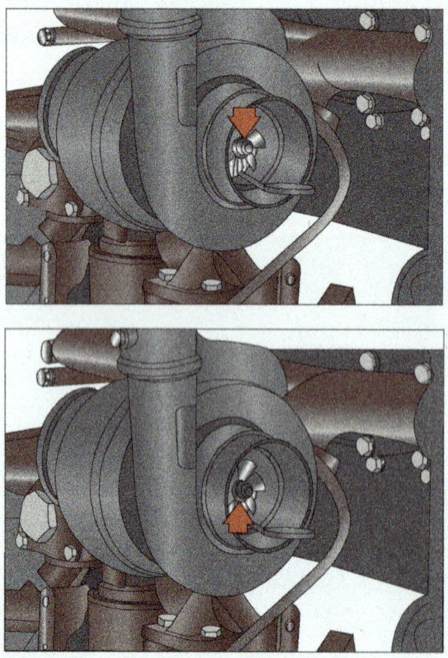

1. With the engine off, remove the intake air piping.
2. Inspect the area between the compressor wheel and housing for evidence of contact. Make a careful inspection, particularly if an operator has reported an engine noise complaint.
3. Rotate the compressor wheel. While rotating, check for any binding or roughness.
4. Use fingers to grasp the compressor wheel retaining nut and move the wheel up and down and back and forth. Any contact between the compressor wheel and housing indicates that the center support bearings are excessively worn and that the turbocharger should be replaced.
5. If there is no contact between the compressor wheel and housing, rotate the compressor wheel while applying a slight amount of side pressure to the wheel. Again, any contact or roughness while rotating the wheel indicates the turbocharger should be removed from service. Contact in only one spot may indicate the compressor housing is misaligned. Loosen any band clamps or retention bolts to realign the housing, and then recheck.
6. To measure center support bearing wear, push down on the compressor wheel with moderate finger pressure while inserting the thickest feeler blade that will fit between the wheel and housing. Record the thickness of the feeler blade.
7. Repeat the same procedure while lifting the compressor wheel. Record the thickness of the feeler blade inserted in the same place while the compressor wheel is lifted in the opposite direction.
8. Subtract the smaller feeler blade size from the larger feeler blade size, and compare the results against the manufacturer's specifications, report the bearing clearance, and make a service recommendation.

SKILL DRILL 15-2 Measuring Turbocharger Axial Bearing Clearance

1. Place a magnetic-base dial indicator on the engine block; if the turbocharger is off the engine, place the indicator on the cast iron turbine housing.
2. Place the dial indicator needle's extension piece at the center of the compressor wheel's retaining nut.
3. Push the shaft forward in the turbine housing with moderate finger pressure.
4. Zero the dial indicator.
5. Push the compressor wheel rearward, or pull on the turbine shaft with moderate finger pressure.
6. Observe and record the dial indicator movement. Compare the axial bearing clearances with the manufacturer's specifications, and then make a service recommendation.

Turbocharger Lubrication

Pressurized lube oil circulates through the center housing to lubricate the bearings and help remove heat. To prevent damage, it is important that oil reaches the turbocharger in sufficient volume within the first few seconds of operation. Oil must be properly filtered and should never exceed 250°F (121°C). Excessive oil temperature can cause the loss of oil film strength as well as cause the oil to burn or coke, which forms an abrasive compound. Coking abrasives damage the turbocharger shaft and bearings (**FIGURE 15-37**). Newer classifications of diesel engine oils are very resistant to coking, and synthetic oil does not coke at all. Two oil seals are used in the turbocharger: one inside the thrust bearing at the compressor wheel end and another on the turbine shaft at the turbine housing. These seals are cast or malleable iron and appear like piston rings. Broken or worn seals can cause a tremendous amount of oil to be pumped into the turbine or compressor housings, enough that other drivers traveling behind an engine with defective turbocharger seals will need wipers to see through their windshield (**FIGURE 15-38**). After replacing a turbocharger, it is important to pre-lube the center housing before operating the engine. This is done by adding oil to the inlet of the housing while slowly rotating the turbocharger shaft (**FIGURE 15-39**).

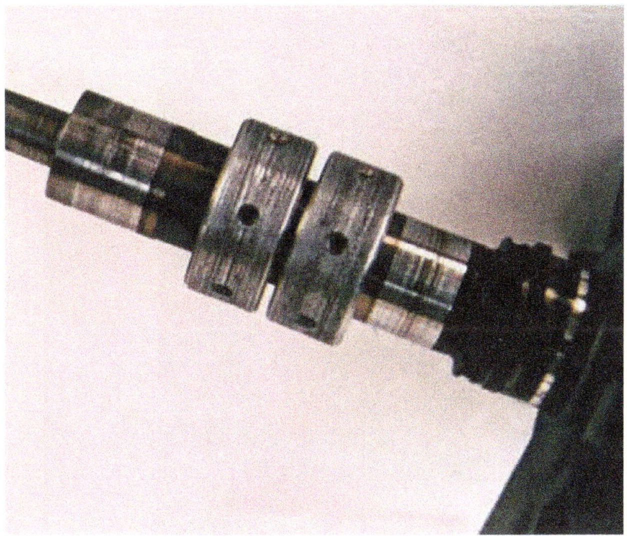

FIGURE 15-37 Bearings and the turbine shaft can be damaged through hot shutdown and oil coking in the center housing.

Turbocharger Servicing

Many turbochargers are replaced unnecessarily due to poor engine diagnostics and other troubleshooting procedures. The turbocharger boost pressure test is the most efficient method to confirm a low-power complaint. This complaint can be confirmed by accelerating the engine to wide-open throttle (WOT) from low road speeds and measuring the boost pressure while under load. The boost will rise to a peak pressure and then quickly fall off after reaching rated rpm under load. Boost pressure continues to fall as the vehicle reaches higher road speeds and the load is reduced. If boost fails to reach its specification, the turbocharger should not be changed until other diagnostic tests are made. Analysis of air induction, exhaust and fuel systems, and other potential causes of low power should take place first to ensure they are properly functioning. Power-up devices that increase engine power beyond factory settings can also lead to turbocharger damage from overspeeding and overheated turbine wheels. Turbine wheels should have sharp, distinct edges. When overheated and glowing, the metal literally boils away from the edges, leaving edges smooth and rounded (**FIGURE 15-40**).

▶ TECHNICIAN TIP

Whenever an engine has operated under load, it is important to let the engine idle for one to three minutes to remove heat from the turbine and center housing. Failure to do so can result in oil coking inside the center housing. Water-cooled center housings allow coolant to circulate after shutdown to prevent oil coking. Synthetic oils do not coke, and newer oil classifications require diesel oil to be highly resistant to coking.

FIGURE 15-38 A damaged or broken oil seal on a turbine wheel causes tremendous amounts of oil to pass into the turbine housing.

FIGURE 15-39 Bearings must be pre-lubed after replacing a turbocharger. Rotate the compressor wheel slowly while filling the center housing with oil through the oil feed line.

FIGURE 15-40 The tips of this turbine wheel are melted and damaged from overheating due to a common rail (CR) injector that was stuck open. Blade edges become rounded and thinner when overheated.

Actuator Checks

Bushings on waste gates can seize due to the corrosive effects of heat and exhaust gases. Low-power complaints can be caused by waste gates that are seized open, and turbocharger damage can result when waste gates do not open. These conditions can be inspected by measuring boost pressure while a vehicle is under load. Failure to reach a specified boost pressure means further inspection of the waste-gate mechanism is necessary. Alternatively, air pressure or vacuum can be applied to the actuator mechanism to check its movement against manufacturer's specifications. Accurate adjustment of the actuator is critical to ensure optimum performance and low emissions and to prevent potential turbocharger and engine failures.

Measuring Waste-Gate Movement

Turbocharger waste gates can malfunction by failing to open or close. A waste gate stuck partially open will fail to build boost pressure and will cause a low-power complaint. Waste-gate springs can weaken from use and heat, causing the waste gate to open prematurely. A waste gate stuck closed will cause the turbine to overspeed and can lead to premature bearing failure. To measure waste-gate movement, follow the steps in **SKILL DRILL 15-3**.

▶ TECHNICIAN TIP

Periodically check waste-gate actuator movement against opening pressures. Exposure to intense exhaust heat will weaken some actuator springs, causing the gate valve to prematurely open, leading to low-power complaints. Similarly, moving actuators during a preventive maintenance inspection will minimize the likelihood of a waste gate seizing open.

Fuel System Checks

The following are some of the recommended fuel system checks:

- low-pressure transfer system pressure and volume
- full throttle reached on both mechanical and electronic throttle pedals
- injection timing
- injector adjustment
- injectors and other high-pressure fuel system components correctly matched to the engine
- engine misfiring or low cylinder contributions
- injector spray pattern and nozzle opening pressures
- air-fuel ratio devices function correctly—i.e., no leaking aneroid lines and gaskets and defective manifold pressure sensors.

SKILL DRILL 15-3 Measuring Waste-Gate Movement

1. Road test the vehicle under load and accelerate the vehicle hard to produce maximum boost pressure while monitoring the intake manifold pressure sensor.
2. Compare maximum boost pressure to the manufacturer's specifications to determine whether waste-gate operation should be checked. *Note*: Other defective conditions related to the high- and low-pressure fuel system operation, intake leaks, engine condition, etc. can produce low-boost conditions.
3. Connect an adjustable air pressure regulator to the waste-gate supply hose. Apply a calibrated amount of air pressure to the waste-gate actuator, usually around 20 psi (138 kPa). Check the air hose, fitting connections, and actuator diaphragm for leaks.
4. Close the air supply shutoff valve. The pressure should hold at the specified pressure. If not, repair the leak or replace the waste-gate actuator or turbocharger.
5. Set up a dial indicator at the end of the waste-gate actuator rod to measure travel. Adjust the dial indicator so that it contacts only the actuator rod end, and zero the dial indicator reading.
6. Check the manufacturer's specifications, and prepare to slowly apply a calibrated amount of air pressure to the actuator.
7. Slowly switch the air pressure supply on and off several times, and observe the actuator's opening and closing supply air pressure. The dial indicator will begin to move at each point.
8. Observe and record the air pressure where dial indicator movement begins as the air pressure is slowly applied and released. Compare opening and closing pressures with the manufacturer's specifications.
9. Measure the total amount of actuator rod travel by following the same steps above. Observe and record total actuator movement.
10. Compare the findings with specifications, and make a service recommendation based on whether the opening and closing pressures and dial indicator movement are within specifications.
    Note: A sticking and binding waste gate will not move until the actuator air pressure falls to zero.

Air Induction System Checks

The following are some of the recommended air induction system checks:

- filter restrictions
- correct size of air filter
- intercooler leaks and restrictions
- valve adjustment
- EGR system operation
- intake manifold throttle plate.

Exhaust System Checks

The following are some of the recommended exhaust system checks:

- excessive exhaust backpressure
- correct original equipment manufacturer (OEM) exhaust system components
- exhaust system fitted with components following the recommendation of the engine manufacturer
- exhaust leaks on exhaust manifold
- seized backpressure regulators or exhaust brakes.

Turbocharger Checks

The following are some of the recommended turbocharger checks:

- waste-gate actuator movement and adjustment
- correctly matched turbocharger—i.e., correct A/R ratio or turbocharger part number for the engine
- bearing inspections—i.e., contact between turbine and compressor wheels with housings
- correct A/R ratio.

Turbocharger Failures

A number of operating conditions can cause premature failures of turbochargers. Common defects include air inlet leaks, oil leaks, bent or broken compressor wheel blades, and worn bearings.

Air Inlet Leaks

Air inlet leaks can erode and pit the compressor wheel and dull the edges of the compressor wheel. Eventually, the engine will lose power and have excessive blowby and emissions due to "dusting out," a condition where the cylinder components are worn out from abrasives entering the engine.

Oil Leaks

Turbocharger oil leaks should be carefully differentiated from leaks caused by excessive crankcase pressure, oil originating from engine cylinders, restricted air filters, and excessive idling. There are many possible causes of oil leaks:

- Gas pressure assists in keeping oil inside the center housing. When idling, a minimum amount of pressure is needed inside the compressor housing; otherwise, oil can leak past oil seals on the compressor and turbine shaft and into the housings. Excessive crankcase pressure pushes can also push oil from the center housing into the turbine and compressor housings.
- A highly restricted air filter can cause oil to be pulled from the compressor center housing and into the engine.
- Oil can leak from a defective turbine seal or compressor seal.

Removing exhaust manifolds and running the engine can differentiate between oil originating from the engine and oil originating from the turbocharger. An ultraviolet tracing dye in the engine can be used along with a black light to detect oil originating from leaking turbine seals. After running the engine for 20 minutes and removing the exhaust pipe at the turbocharger, oil seen leaking around the turbine wheel indicates a leaking turbine seal (**FIGURE 15-41**).

Inspecting for Turbine Seal Leakage

Turbochargers are often unnecessarily replaced because of suspected oil seal leakage due to oil originating from the engine exhaust or a contaminated particulate filter. Oil leakage from the turbocharger turbine seal should be differentiated from oil originating from the

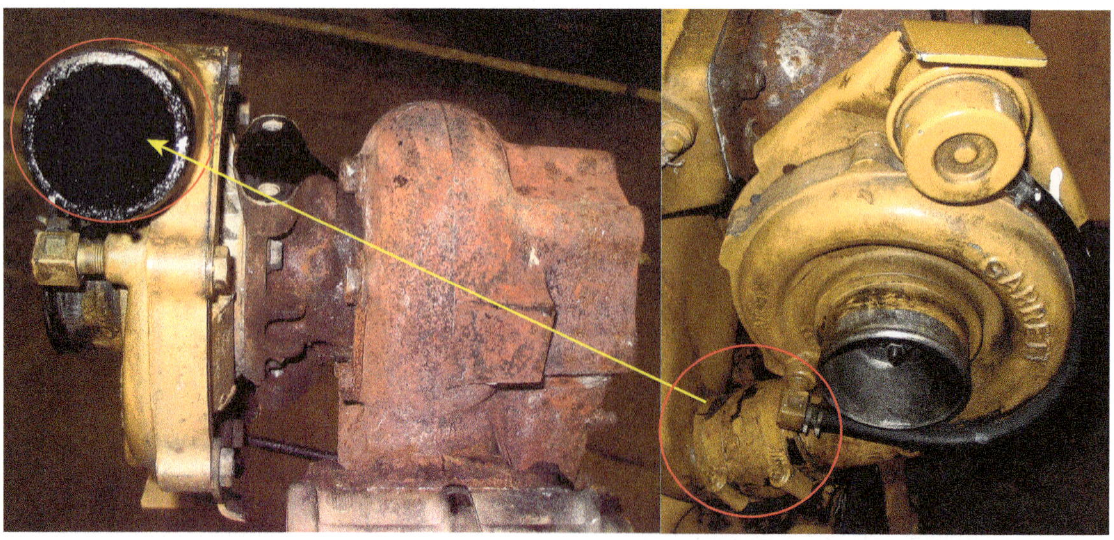

FIGURE 15-41 Turbochargers should not leak oil from the compressor or turbine housings. This engine had excessive blowby and idle time, resulting in a fouled compressor wheel. Only the compressor outlet (not the inlet) was slobbered.

SKILL DRILL 15-4 Inspecting for Turbine Seal Leakage

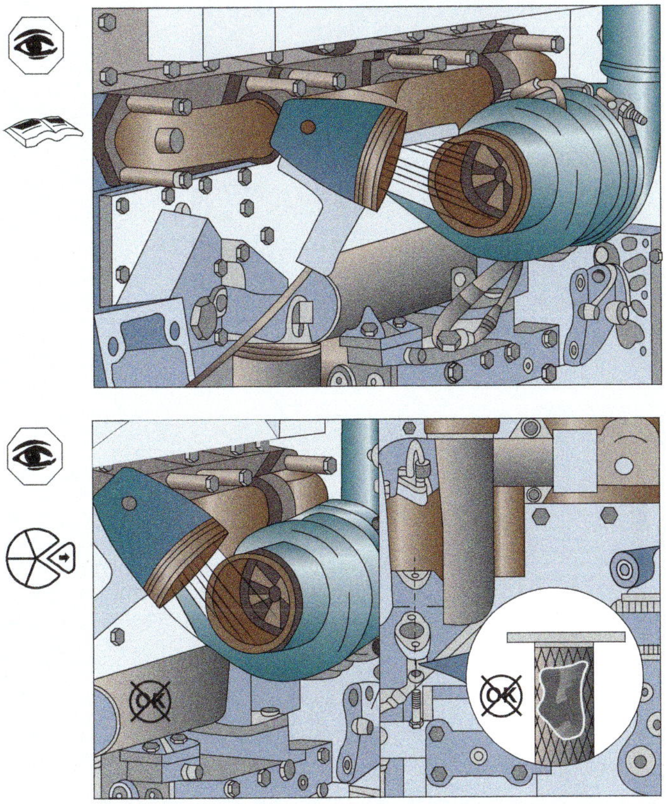

1. Inspect the exhaust manifold joints and the area on the engine block below the exhaust manifold. Oil leakage or slobber present at these points likely indicates that the oil originates from the engine and not the turbocharger.
2. Add oil tracing dye, which is visible when using an ultraviolet (UV) light, to the engine oil, and run the engine until it reaches operating temperature. Run the engine at idle for 10 to 20 minutes.
3. Allow the engine exhaust to cool.
4. Remove the exhaust pipe from the turbocharger, and use a UV light to inspect the area around the turbocharger outlet. Any accumulations of bright yellow- or green-colored dye at the turbine outlet likely indicates a defective turbocharger seal.
5. Remove the turbocharger oil drain tube to inspect for blockage. If the engine oil tube is not blocked and the engine has acceptable crankcase pressure, make a service recommendation to replace the turbocharger.

engine cylinders. Worn or damaged cylinder components or excessive engine idle can produce excessive slobber, which does not justify replacing the turbocharger. Also, new or rebuilt engines that have not completed a break-in period will also leak oil at these points. However, worn engines with high crankcase pressure or excessive blowby will push oil out the turbocharger seals. To inspect for turbine seal leakage, follow the steps in **SKILL DRILL 15-4**.

Bent Compressor Wheel Blades

Bent compressor blades are caused by the ingestion of soft materials such as shop rags or filter gaskets (**FIGURE 15-42**). Broken or cracked compressor blades are caused by the ingestion of hard material, such as a bolt or nut.

FIGURE 15-42 The compressor wheel blades may only be bent if soft materials such as gaskets, rags, or even water are ingested.

Broken Compressor Wheel Blades

Broken or cracked compressor wheel blades are caused by ingesting harder materials like a bolt or nut.

Worn Bearings

Worn bearings can be caused by the hot shutdown of the engine, coolant in the oil, overextended oil service intervals, failure to pre-lube the turbocharger after replacement, low oil pressure, and excessive oil temperature.

SAFETY TIP

When working around running engines with the air inlet piping removed from the turbocharger, always install a guard cover over the turbocharger inlet. Compressor wheels are very sharp and turn at a very high speed even at idle. Fingers are easily sliced off if they come in contact with the spinning turbocharger wheels. Also, turbochargers are very easily damaged with any contact with debris or clothing.

Charge Air Cooling Introduction

"Charge air cooling," "aftercooling," and "intercooling" are interchangeable terms for removing excess heat from the air, charging, or filling, the cylinders (**FIGURE 15-43**). Charge air cooling is an important technology for reducing oxides of nitrogen (NO_x) emissions. Other benefits include improved fuel economy, power output, and increased maximum injection

FIGURE 15-43 Charge air cooling involves removing heat from the air pressurized by the turbocharger. The CAC is located between the turbocharger and the intake manifold.

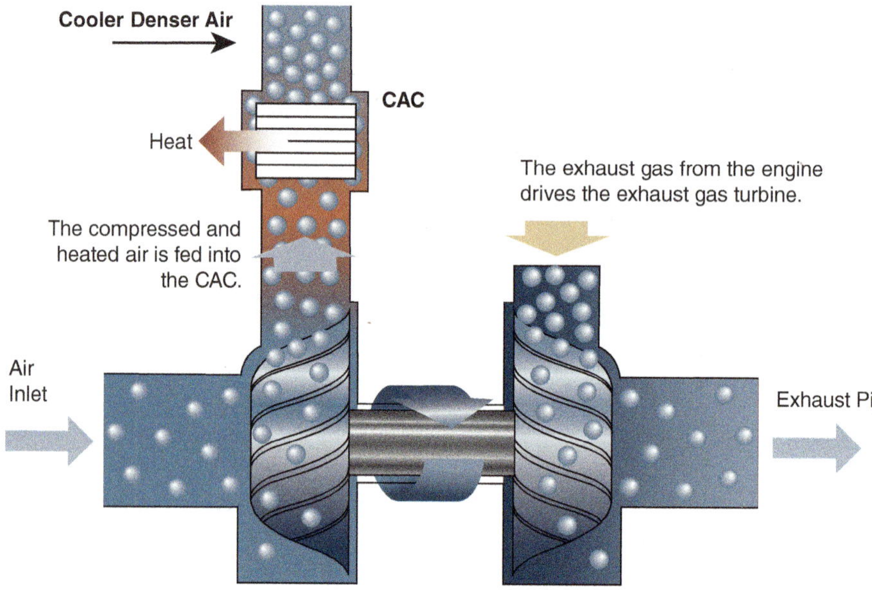

FIGURE 15-44 Hot air from the turbocharger enters the CAC, where heat is released to the atmosphere. Cooler air leaves the heat exchanger and enters the intake manifold.

rates. Either ambient (outside) air or engine coolant passing though heat exchangers are the mediums for removing heat from the pressurized intake air. The radiator transfers heat from coolant to the atmosphere. A chassis-mounted air-to-air aftercooler (ATAAC) cools using ambient air (**FIGURE 15-44**). This type of system is correctly called a charge air cooler (CAC); it is also known as an aftercooler or intercooler. A jacket water aftercooler (JWAC) cools charge air using engine coolant instead of air.

▶ Fundamentals of Charge Air Cooling

15-07 Identify and explain the functions of charge air cooling.

On turbocharged engines, air intake temperatures rise in proportion with boost pressure. High intake air temperatures negatively affect emissions, engine durability, and performance. Loss of air density and increased cylinder temperatures accompanying higher intake air temperatures are the main factors causing these problems. Because pressurizing a gas increases its temperature, intake air temperature is raised in proportion to its compressed pressure (**FIGURE 15-45**). Additionally, when gas temperatures increase, molecules spread farther apart, so the gas loses density. This means a cubic foot (cubic meter) of air, when heated and at the same pressure, contains fewer oxygen molecules and weighs less than when it is not heated.

The relationship between temperature and pressure can be predicted by using a variety of gas law equations. Consider the following examples:

- If intake or underhood intake air temperature is 75°F (24°C) at atmospheric pressure (15 psi [103 kPa] at sea level), the turbo outlet temperature will be 285°F (141°C) at 20 psi (138 kPa) boost pressure.
- If the inlet air temperature to the turbocharger climbs to 100°F (38°C), the discharge temperature will be 320°F (160°C).
- A boost pressure of 30 psi (207 kPa) produces an outlet temperature of 358°F (181°C) with an intake temperature of 75°F (24°C) and an outlet temperature of 396°F (202°C) with an intake temperature of 100°F (38°C).
- A 250 hp (186 kW) engine will develop only 240 hp (179 kW) if the air temperature is 130°F (54°C) when using the same quantity of fuel.

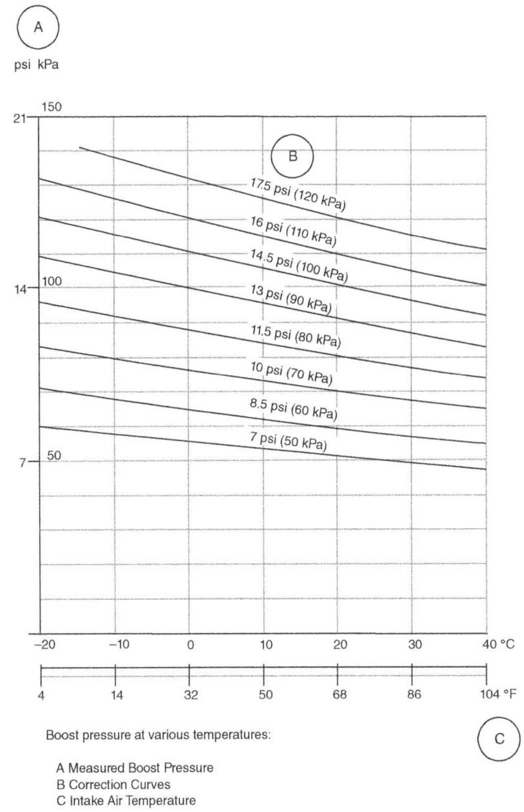

FIGURE 15-45 Turbocharger boost pressure changes with inlet air temperature. When performing boost pressure tests, intake pressure must be corrected for an ambient temperature of 68°F (20°C).

Air Quality Problems with Charge Air Coolers

Hotter charge air temperatures lower air density and results in poorer combustion quality. As air molecules spread farther apart, less oxygen is available to support combustion, so the air mass available to cool cylinder components drops. For every 10°F (5.6°C) temperature increase above 60°F (16°C), the changes to air density and oxygen content result in a power loss.

Power decreases 1% for every 10°F (5.6°C) above 90°F (32°C). Without a cool air mass traveling across the piston crown from the intake to the exhaust valve during the valve-overlap period, the piston crown and exhaust valve become very hot. The loss of air's cooling effect leads to higher cylinder heat loads and shorter engine life.

Combustion Temperatures and Charge Air Coolers

Increases in the intake charge temperature lead to significantly higher combustion and exhaust temperatures. This exhaust temperature increase is not in proportion to the air inlet temperature increase, with small increases in air inlet temperature often producing even larger changes to exhaust temperature. Generally, a 3:1 ratio exists, with every 1°F (0.55°C) increase in charge air temperature producing a 3°F (1.7°C) increase in exhaust temperature. However, many factors affect this relationship, including fuel rates, excess air ratio, air density, and engine compression ratio. The exhaust temperature in the exhaust manifold, not the exhaust pipe, should never exceed 1,250–1,300°F (677°C–704°C); otherwise, damage to valves and pistons will occur if these temperatures are sustained over long driving cycles. (**FIGURE 15-46**). The maximum normal exhaust temperature encountered under any condition is less than 900°F (482°C). Increased charging pressures can compensate for lower air density but this in turn increases cylinder temperatures. Charge air temperatures are a limiting factor to engine power output. If charge air cooling reduces thermal loading in the cylinders, larger quantities of fuel can be injected to develop more power from the same engine displacement.

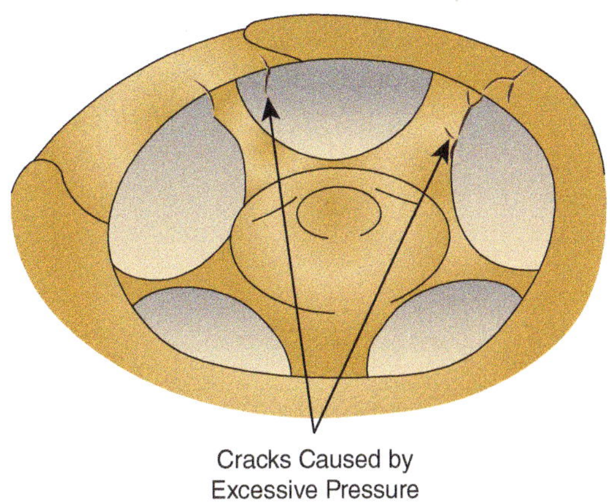

FIGURE 15-46 A restricted air intake or externally restricted CAC can crack the piston crown by causing excessive combustion temperature and pressure. The cracks travel from the edge of the combustion chamber bowl.

NO_x Emissions and Charge Air Coolers

Increases in charge air temperatures result in higher combustion chamber temperatures, which increase the production of NO_x emissions. Normally, nitrogen, which makes up approximately 77% of the air, remains inert and uninvolved in combustion processes. At temperatures above approximately 2,500°F (1,371°C), and because of high cylinder pressures, nitrogen will react with oxygen to produce several nitrogen–oxygen compounds collectively known as NO_x.

Beginning in 1988, the Environmental Protection Agency (EPA) in the United States introduced increasingly strict emissions standards for NO_x. EURO stage 1 standards were

▶ TECHNICIAN TIP

The first year for legislated reductions to diesel NO_x emissions was 1988. Until that time, diesels needed to meet only a standard for exhaust opacity during a throttle snap test. In 1990, NO_x emissions dropped from approximately 11 grams per brake horsepower per hour (g/bhp/hr) (15 grams per kilowatt per hour [g/kW/hr]) for heavy-duty on-highway diesels to 6 g/bhp/hr (8 g/kW/hr). Light-duty engines had to meet 1 g/mile (0.625 g/km) only during a phase-in for 1994. Afterwards, almost every medium- and heavy-duty diesel vehicle began being equipped with charge air cooling to bring NO_x to these levels. Without cooling intake air adequately, combustion temperatures rise and produce exponential increases in NO_x.

▶ TECHNICIAN TIP

CACs are large air-handling components and vulnerable to damage. Leaking and cracking often cause low-power complaints. Cracks can easily develop in these devices due to constant thermal cycling from hot to cold. When investigating such complaints, always inspect and test the CAC for leaks.

15-08 Identify and describe the types of charge air cooling systems.

introduced in 1992. The cooling of intake air, which drops combustion temperatures, goes a long way to minimizing the formation of these emissions, making the CAC an important emission control device.

Benefits of Charge Air Cooling

Cooling the intake air charge to the cylinders has the following important benefits:

1. It increases air density and oxygen content, which improves combustion quality. More oxygen in the cylinders means better contact between fuel and oxygen for efficient, clean combustion.
2. CACs provide additional air mass for cooling valves and pistons. Denser air means the air can remove more heat from valves and pistons during valve overlap. Adding air mass to the cylinders dilutes the heat produced during combustion for lower cylinder temperatures.
3. Denser, cooler air improves fuel economy by up to 5% due to improved combustion qualities. The simple addition of a CAC does not produce more power; that requires more fuel and air. However, because cooling of intake air improves combustion quality, the potential to slightly increase power does exist (**FIGURE 15-47**).
4. CACs enable higher injection rates, which means more fuel enters the cylinders. Higher injection rates translate into greater power density, which means the engine can produce more power per cubic inch (centimeter) of cylinder displacement. Because the additional fuel can push cylinder thermal loads to their maximum threshold, cooling charge air can reduce combustion and exhaust temperatures by diluting combustion heat with cooler, denser air mass. This makes it possible to gain power while minimizing the risk of engine damage by using higher injection rates.
5. CACs reduce NO_x emissions by lowering peak cylinder temperatures and pressures by cooling the charge air.

▶ Types of Charge Air Cooling

Heat exchangers are used to transfer heat from the charge air to either the atmosphere or cooling system. Efficient and effective heat transfer is necessary to keep engine emissions low and prevent engine damage. Many different names are given to these heat exchangers. This next section will examine the construction and operation of two major categories of CACs: liquid coolers, which use engine coolant as a heat transfer medium, and air coolers, which use atmospheric airflow as a cooling medium.

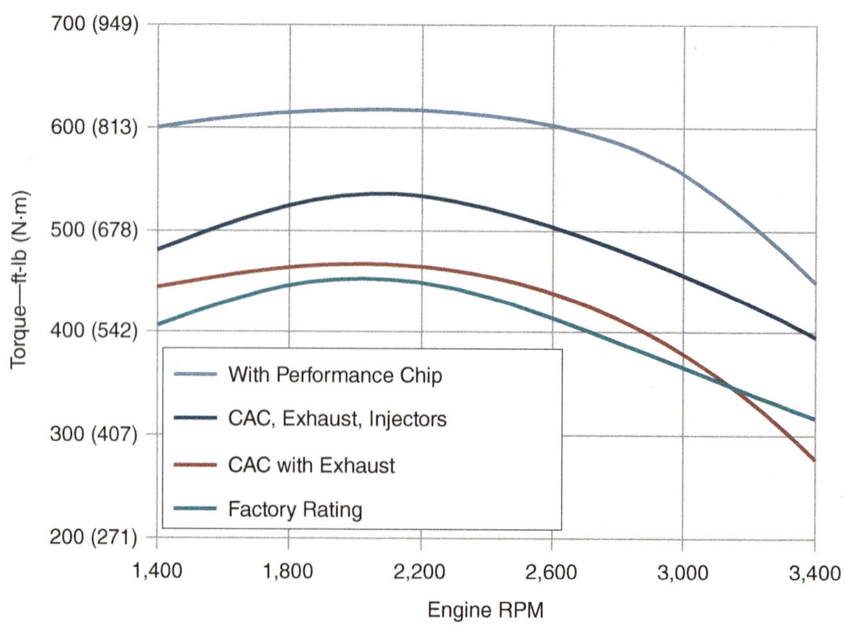

FIGURE 15-47 The addition of a CAC and larger exhaust offers only a marginal improvement in power. Engines with CACs usually have more power, but CACs do not increase power output.

Jacket Water Aftercoolers

A JWAC lowers intake temperatures by passing the boost air through a water-type heat exchanger. A JWAC is capable of lowering the turbocharger boost air temperature from approximately 300°F (149°C) to approximately 200°F (93°C).

Older, mechanically governed engines often use a JWAC, positioned inside the intake manifold (**FIGURE 15-48**). Many newer automotive diesel engines, such as the Ford 6.7 L Power Stoke and those made by Volkswagen (VW) and Mercedes, are using liquid CACs with separate cooling circuits and external heat exchangers or radiators to lower the temperature of coolant flowing through the cooler (**FIGURE 15-49**). The advantage of a mechanically governed fuel system is that the injection timing can be calibrated around a relatively constant air inlet temperature. This type of cooling is still popular on marine diesel engines. Today, JWACs are used in series turbocharger applications, where very high boost pressures produce even hotter intake air temperatures. Air intake temperatures are first lowered by a JWAC before passing through an ATAAC. When a JWAC is used to cool air between each stage of turbocharging, it is often referred to as interstage cooling or intercooling.

FIGURE 15-48 The Mercedes 2.1 L diesel engine uses a liquid CAC mounted beside the engine. A two-stage series-type turbocharger is used by this engine.

Air-to-Air Charge Air Cooling

The most popular method of cooling intake air is to use an air-to-air heat exchanger, which is often referred to as an ATAAC or ATA. Moving the heated charge air through an ATAAC can drop the temperature from higher than 400°F (204°C) to 90°F–110°F (32°C–43°C), with an outside temperature of 75°F (24°C) (**FIGURE 15-50**). The exact temperature reduction depends on the ambient air temperature and the volume of airflow across the cooler core. A properly functioning CAC has a maximum temperature differential of 30°F between air intake temperature and outside air when airflow is 30 mph (48 kph).

Most coolers have an efficiency of approximately 80%, which means they reduce intake temperatures to no more than 30°F–40°F above outside temperatures. Air-to-air aftercooling requires relatively high airflow across the cooler to remove heat. Therefore, a winter front, which is a radiator cover used in cold weather operation, should not be used with an ATAAC. If a winter front must be used to maintain passenger cabin heat, never close it completely. At least 20% airflow must remain flowing across the cooler, even in the coldest

▶ TECHNICIAN TIP

When servicing the cooling system of an engine that uses a JWAC, bleeding air from the cooler is important. The cooler is a high point in the cooling system, so air can become trapped in the cooler and prevent proper functioning. Series turbocharged engines that use JWACs transfer more heat to the cooling system and therefore require more vigilant maintenance of the cooling system.

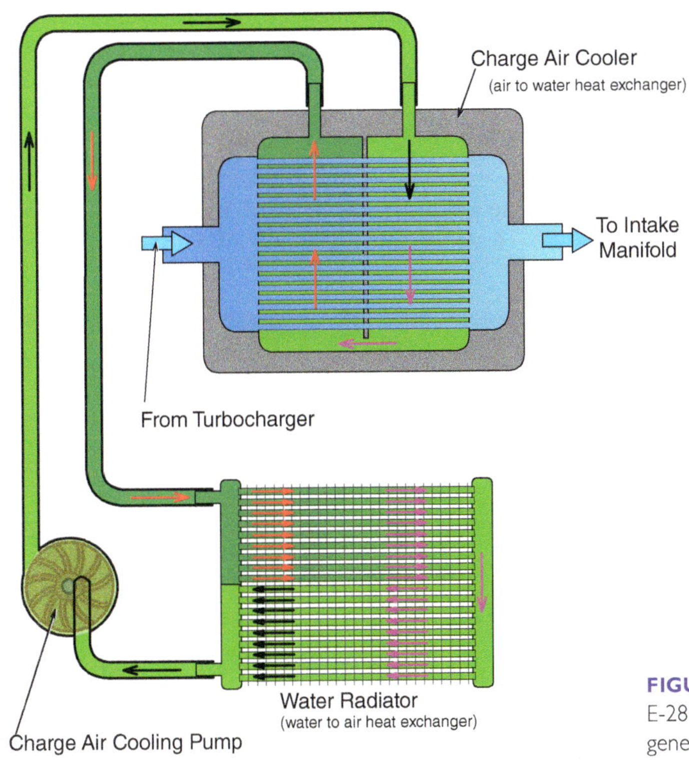

FIGURE 15-49 VW 2.0 L E-288 engine. The third generation uses a liquid CAC.

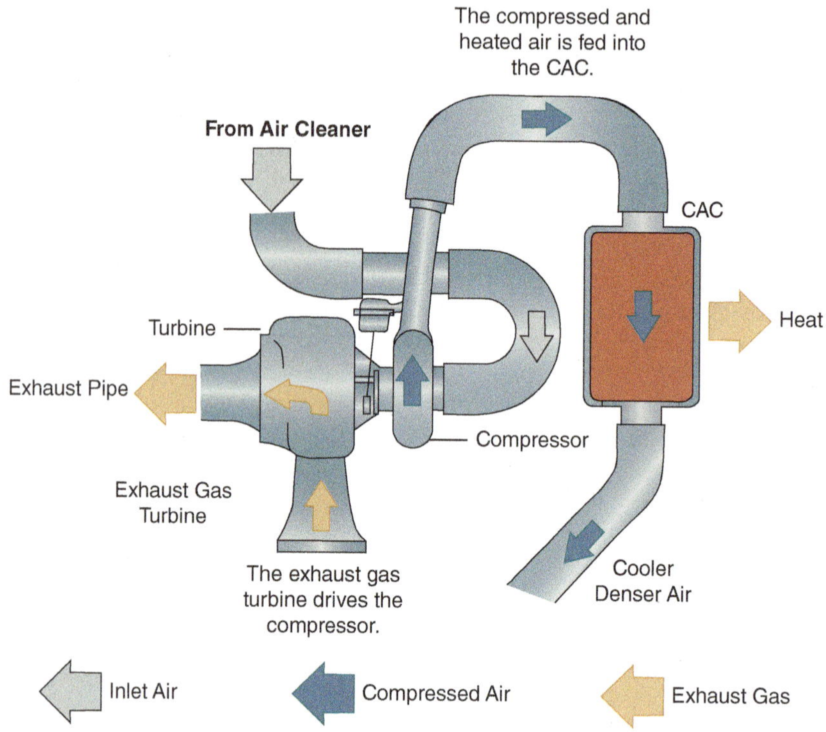

FIGURE 15-50 Airflow through a charge air system.

weather (**FIGURE 15-51**). Most engine manufacturers that use electronic controls monitor air intake manifold temperatures as a part of the engine protection system. High intake temperatures will cause the engine to derate power or even shut down.

Air-to-Air Charge Air Cooling Components

Most ATAACs are constructed of aluminum for maximum heat transfer and strength (**FIGURE 15-52**). To achieve optimum cooling, the inlet and outlet of the cooler may have different diameters to slow the airflow through the cooler and maximize heat transfer.

As the temperature in the cooler drops, a corresponding pressure drop occurs. A cooler pressure drop from side to side should not typically exceed more than 3 psi (21 kPa). If the pressure drops more than this, check the cooler for internal restrictions and leaks (**FIGURE 15-53**).

Dual Charge Air Cooling

To improve temperature regulation of the exhaust gases to assist aftertreatment catalyst warm-up, VW uses a dual circuit CAC. Heated charge air leaving the turbocharger passes

FIGURE 15-51 Today's diesel engines have substantial amounts of cooling. The CAC in the engine shown is an unusual down-flow type, which has top-to-bottom flow.

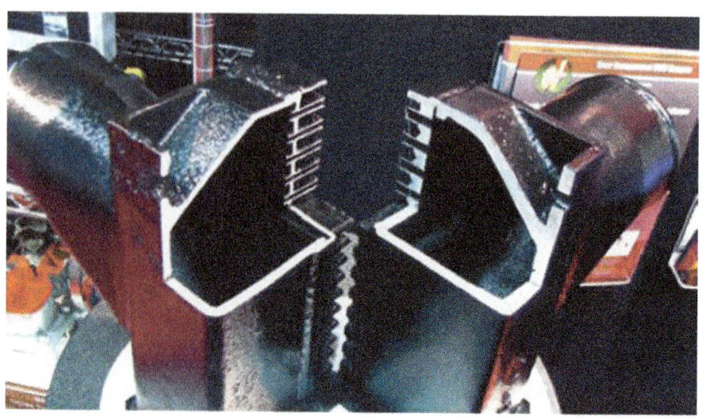

FIGURE 15-52 Aluminum side tanks and an aluminum core of a typical ATAAC.

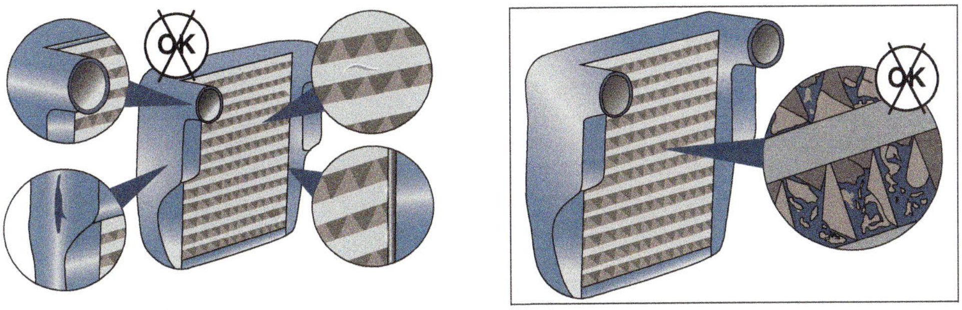

FIGURE 15-53 Check CACs for cracks in cooler tubes and the joint between the tanks and tubes. Inspect for external restrictions in front of and behind the intercoolers.

through a three-way tube. When the engine and exhaust are either cold or too cool and outside temperatures are low, the warmer air is directed through the bypass valve or flap and into the intake manifold without any cooling. This arrangement enables the oxidation and particulate filter catalysts to reach higher temperatures and warm up more quickly since exhaust gases will be warmer. When outside temperatures are warmer, the engine is under greater load and therefore has higher boost temperatures and pressure, so the quantity of cooled charge air supplied to intake manifold can increase. This condition requires full cooling of all turbocharged air supply and transitions beginning at 1,750 rpm (**FIGURE 15-54** and **FIGURE 15-55**).

Cooler Mounting

Because coolers are often made from aluminum, they expand and contract when heated and cooled (also known as thermal cycling). For this reason, the cooler's mounting system is designed to allow for thermal cycling. For example, washers on both sides of mounting bolts may include a spring-type washer and nylon-like washers, which allow the cooler mounting brackets to slide. To avoid damaging the cooler, take special care to follow the manufacturers' mounting instructions.

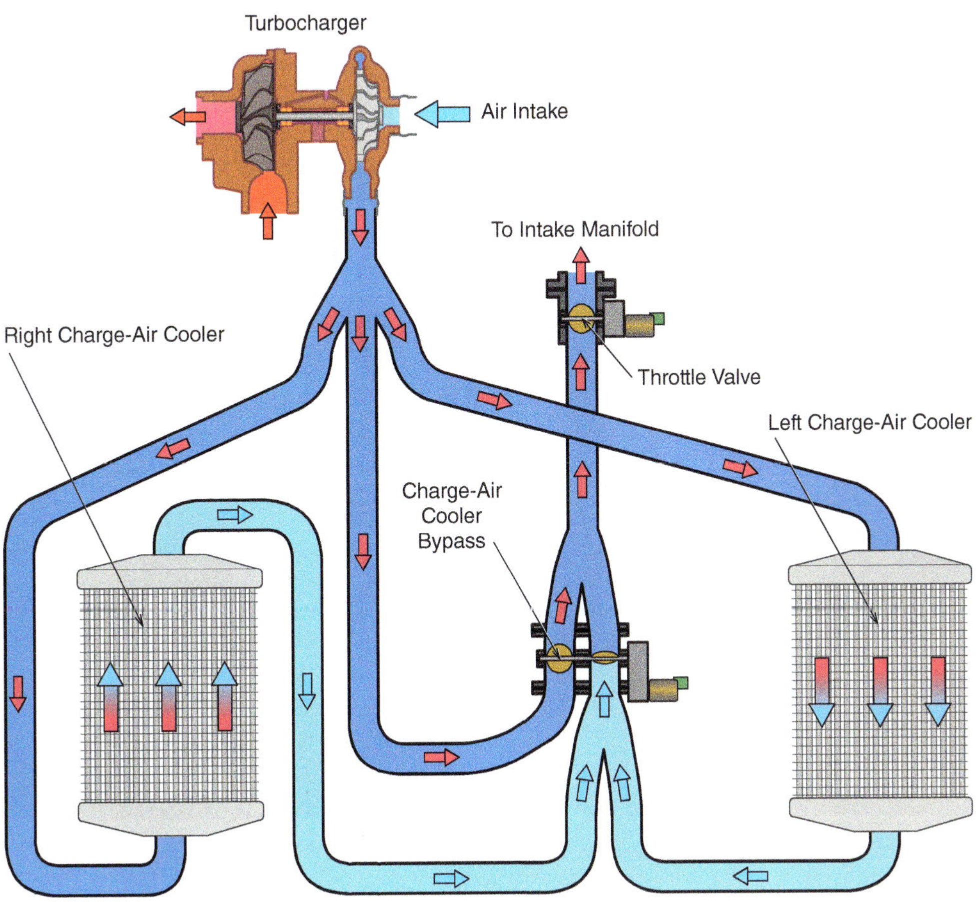

FIGURE 15-54 The 3.0 L VW uses two intercoolers and two separate air circuits. One path bypasses the intercoolers during cold-weather and warm-up operation. A second path through the coolers is used when exhaust temperatures and outside air temperatures are warmer.

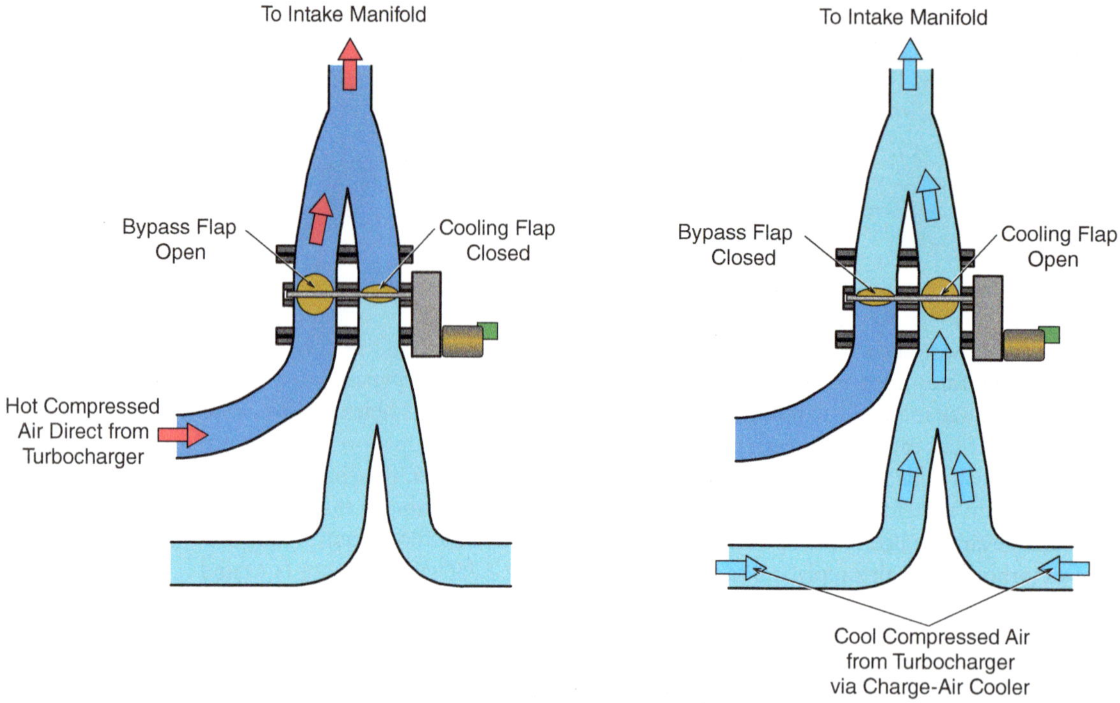

FIGURE 15-55 VW's 3.0 L CAC bypass valve directs pressurized air either to the intake manifold or through the CACs. Air flow is switched on or off through the coolers or blended during warm-up to speed up heating up the exhaust aftertreatment system catalysts.

Hoses and Clamps

ATAACs use unique connector hoses that are often color-coded to differentiate the hot and cold sides of the cooler. These hoses are made of chemically stable, durable silicone material. Clamps with smooth underside surfaces prevent damage to the hose. Spring-operated clamps are preferred because they can maintain a constant torque on the hose regardless of the dimensional variations produced by temperature extremes (**FIGURE 15-56**).

Underhood noises during heavily loaded operating conditions can often be pinpointed to leaking clamps and hoses. A sudden drop in CAC pressure from a blown clamp can cause the turbocharger compressor wheel to draw oil into the compressor housing, which can be misdiagnosed as a defective turbocharger.

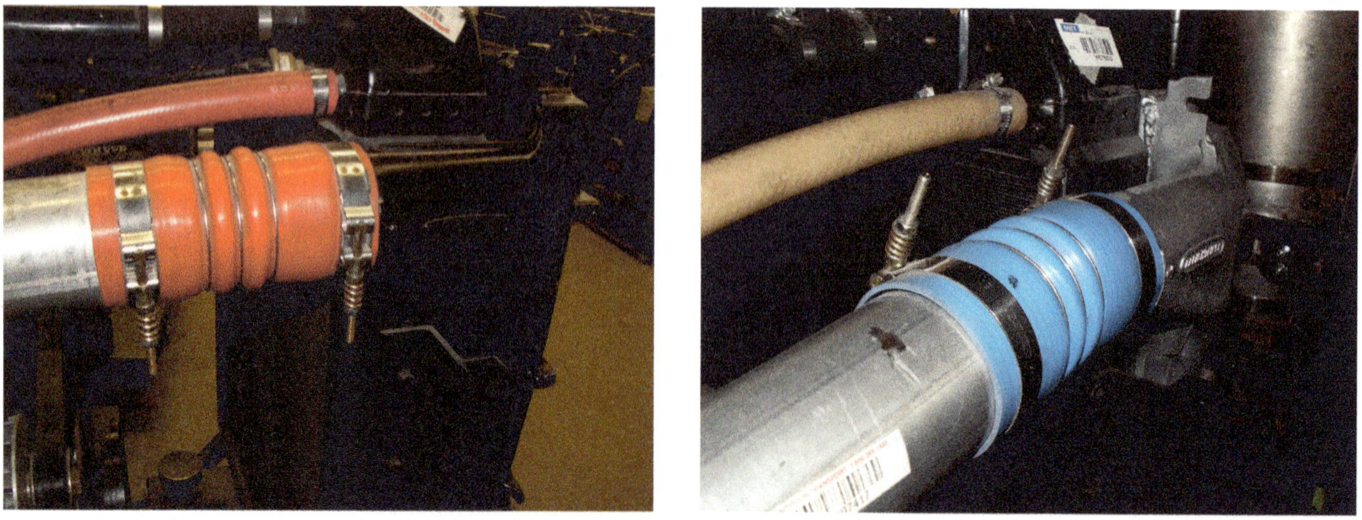

FIGURE 15-56 CAC hoses are made from different materials, depending on whether the hose is on the hot or cool side. The materials are color-coded.

▶ Charge Air Cooling Maintenance

15-09 Outline inspection procedures used to evaluate charge air system operation.

Continuous pressure and temperature changes exert tremendous stress on coolers, which often results in cracks or some type of leak. Coolers can also become internally and externally restricted from debris or liquids passing through the cooler.

This section will examine some of the common problems and maintenance practices associated with CACs.

Air-to-Air Charge Air Cooling Testing and Servicing

CACs can fail for a number of reasons, which can lead to air inlet restriction, excessive emissions, and high exhaust and cylinder temperatures. Internal or external restrictions on the CAC can cause low power. Electronic engines will often derate power output when intake manifold temperatures are high because high intake manifold temperatures can cause severe heat damage to cylinder components and turbochargers (**FIGURE 15-57**).

CAC Leakage

Thermal cycling can stress a CAC until it cracks. The stress is compounded because aluminum has a much higher expansion coefficient than other metals. Cracks are common in areas of the cooler where inlet temperatures are highest and at the joint between the inlet tank and cooler tubes (**FIGURE 15-58**). Replacement is not always the expected service recommendation for this condition, because the size of some cracks does not justify replacing the cooler. Because the CAC is pressurized and the turbocharger produces excess air, the CAC can tolerate small cracks without allowing dirt into the engine intake.

FIGURE 15-57 Electric air intake heaters minimize cold-start and warm-up emissions. The ECM will disconnect these devices when the engine is under boost pressure and when intake air temperatures need to be lowered.

Pressure Testing Charge Air Coolers

Small cracks do not warrant replacing a CAC, because dirt ingestion into the engine through a crack is not likely when a cooler is pressurized. Additionally, if the engine is using excess air for combustion, some leakage is permissible before the air loss affects engine performance. Considering these variables, many manufacturers do not recommend replacing the cooler if the leak is small. Until the leak becomes significant, engine performance will not be compromised.

Instead, pressure testing a CAC to determine the extent of air volume lost is the most important criterion for justifying cooler replacement. After isolating and pressurizing the cooler to the manufacturer's specifications, check the rate of leakage against acceptable limits. Manufacturers typically recommend pressurizing the cooler to 30 psi (207 kPa) and measuring the rate of leakage by observing the pressure drop. A drop in pressure of up to 5 psi (34 kPa) in 15 seconds is an acceptable leak limit for a variety of manufacturers (**FIGURE 15-59**).

CACs can leak and allow loss of charge air mass, which reduces engine power and increases exhaust temperature and emissions. Low-power complaints should include a pressure test of the CAC. Some leakage is acceptable because an overabundance of air mass entering the engine cylinders and some cracking of the cooler is unavoidable due to frequent thermal cycling.

To pressure test a CAC, follow the steps in **SKILL DRILL 15-5**.

FIGURE 15-58 This CAC has elongated slots at its mounting locations to allow for expansion and contraction. Mounting hardware must be precisely adjusted to the correct clamping force. If the CAC is overtightened, mounting hardware will prevent the cooler from moving, which leads to cracking.

FIGURE 15-59 Leak testing a CAC requires using a pressure regulator to pressurize the cooler to 30 psi (207 kPa), then shutting off the air supply to measure the leak.

SKILL DRILL 15-5 Pressure Testing a CAC

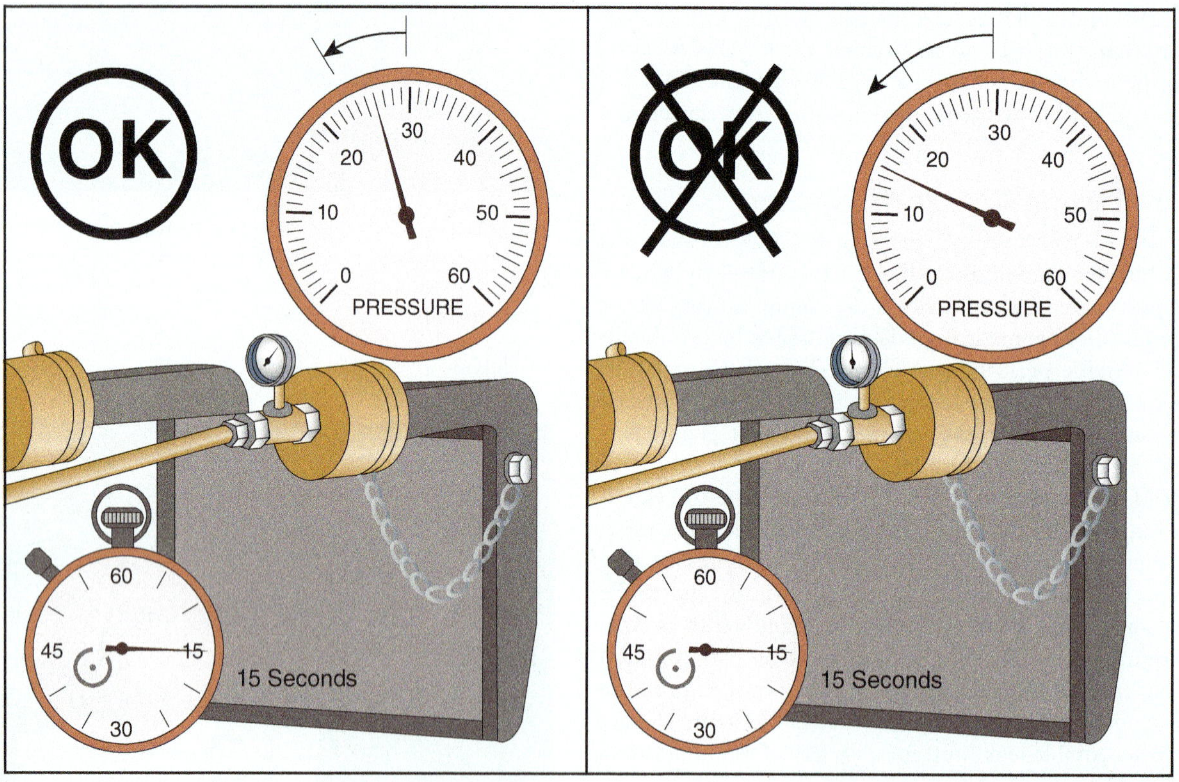

1. Disconnect the CAC inlet and outlet pipes.
2. Connect purpose-made adapters to the CAC pipe hoses. One adapter has a drain valve, and the other has a quick-connect fitting.
3. Connect an air pressure regulator to the test adapter plug, and pressurize the cooler to 30 psi (207 kPa).
4. Shut off the air pressure supply to the regulator, and count the number of seconds it takes for the cooler to drop 5 psi (34 kPa).
5. Generally, any leak that allows a pressure drop of more than 5 psi (34 kPa) in 15 seconds is too large.
6. Make a service recommendation based on manufacturer's specifications and the observed time that it took for a pressure drop to occur in the cooler.

Inspecting for Internal CAC Restrictions

CAC core restriction has a variety of causes. For example, a shredded air filter gasket or ingestion of water or oil can reduce airflow inside a cooler core. Oil can accumulate because crankcase oil vapor can pass through the cooler and condense the oil (**FIGURE 15-60**). Turbochargers also pass or wisp a small quantity of oil, which can easily accumulate inside a core. Any dirt getting through an air filter can stick to the oily core tubes and further restrict airflow. A failed turbo may leave a large quantity of oil and even debris inside the cooler (**FIGURE 15-61**).

After a turbocharger failure, a cooler needs to be flushed thoroughly to prevent debris from being ingested into an engine. Lubrication oil loaded inside a cooler core can lead to an engine's destruction because a diesel engine can easily run aspirated engine oil to overspeed and beyond to critical rpm. Excess oil in the CAC supplies the engine with an unregulated external fuel source, which has led to the destruction of many engines.

To evaluate whether there is excessive internal resistance to flow because of an internal restriction, complete a pressure drop test on the cooler. The pressure drop test is different from the pressure test that evaluates whether any cooler leaks are significant enough to justify replacing the cooler. Pressure drop testing evaluates the cooler for internal restrictions that significantly reduce the coolers outlet air pressure. To check for internal restrictions, measure the inlet and outlet pressures of the cooler when the engine is under load. This is done while monitoring the intake manifold pressure and installing a separate gauge at the cooler inlet or turbocharger outlet. Usually, an acceptable drop in pressure is no more than 3 psi (21 kPa). However, compare test results against manufacturer's specifications for acceptable limits.

SAFETY TIP

Always use safety pins and chains to secure the pressure-testing adapters used to seal CAC pipes. Under pressure, these adapters can slip with explosive force and cause vehicle damage, severe personal injury, or even death.

Testing a CAC for Internal Restrictions

CACs can internally accumulate oil, water, debris, and even ice. The result is a loss of air mass due to the internal restriction. To test a CAC for internal restrictions, follow the steps in **SKILL DRILL 15-6**.

External Restrictions

Winter fronts, bug screens, and blocked coolant radiators and air conditioner condensers can restrict external airflow over the CAC. Sometimes, a small leak in the cooler will allow oil to migrate out, creating adhesive surfaces for road dust to cling to behind the cooler. While a quick visual inspection of a cooler might show no signs of restriction, the area directly behind the cooler may be completely blocked.

FIGURE 15-60 This image shows oil weeping from a charge air pipe clamp. It is normal for oil to accumulate around joints.

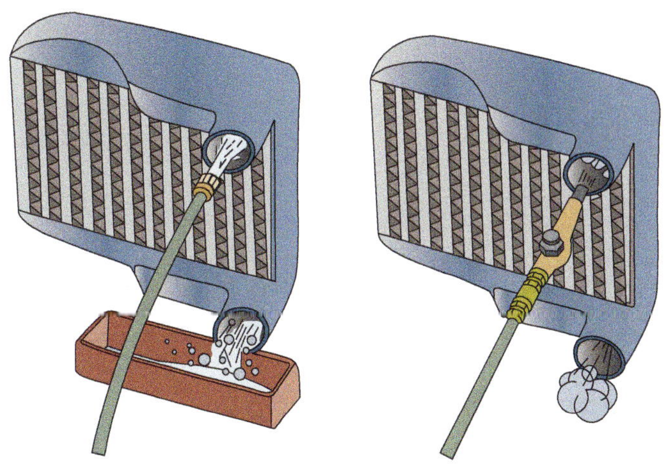

FIGURE 15-61 Back-flush cooler cores with solvent and compressed air after a turbocharger failure or after a core has become internally restricted.

SKILL DRILL 15-6 Testing a CAC for Internal Restrictions

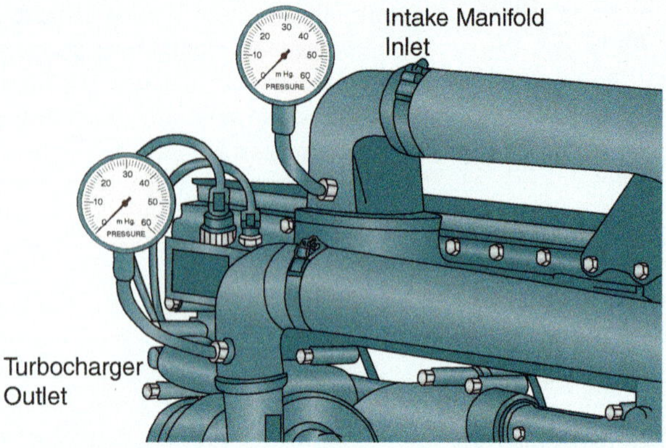

1. Connect two direct-reading pressure gauges or a pressure gauge with two separate, switchable inputs to the intake manifold and turbocharger outlet.
2. Road test the vehicle while loading the engine to achieve maximum boost pressures. This is done by accelerating the engine at full throttle while the transmission is a couple of gear steps above where it should be.
3. While the engine lugs down and the turbocharger moves through peak boost pressure, observe and record the pressure at both points in the induction systems.
4. Compare the results against the manufacturer's specifications. Generally, no more than 6 inHg (15 cmHg) or 3 psi (21 kPa) of pressure drop is unacceptable.
5. Make a service recommendation based on observations of the pressure differential across the CAC. The CAC may require draining or back-flushing with solvent to clean the restriction.

This type of restriction would also not allow sufficient airflow across the radiator, causing the engine to overheat or resulting in high engine coolant temperatures. Engine emissions and coolant temperature will increase in proportion to the degree of air restriction over the CAC.

To diagnose an external restriction, perform a temperature differential test to identify a loss of cooling across a cooler. The vehicle needs to travel at least 30 mph (48 kph). Normally, the difference between the ambient temperature and the intake manifold temperature should not be more than 20°F–30°F (11°C–16.6°C). For comparison, the Cummins air temperature differential test limits the temperature difference to no more than 50°F (10°C) with an airflow of at least 30 mph (48 kph) across the cooler.

> ▶ TECHNICIAN TIP
>
> The CAC temperature differential can be easily tested while on a road test by using scanner. Usually, OEM software or a handheld diagnostic code reader can monitor intake manifold temperature, which can be compared with the known ambient temperature. Make sure to load the engine by operating it under full throttle, up a grade, or in a higher gear than the vehicle would normally operate in.

Inspecting a Charge Air Cooler for External Restrictions

CAC efficiency is monitored by the OBD-II system. CAC efficiency is calculated by measuring the temperature differential between the CAC's inlet and outlet. A large differential indicates a properly functioning CAC. An externally restricted cooler that is plugged with debris, covered with cardboard or a winter front, or located in front of a blocked radiator will generate efficiency codes or produce engine protection codes. The latter happens when intake manifold temperatures become too high, which can result in valve and piston damage. To inspect a CAC for external restrictions, follow the steps in **SKILL DRILL 15-7**.

SKILL DRILL 15-7 Inspecting a CAC for External Restrictions

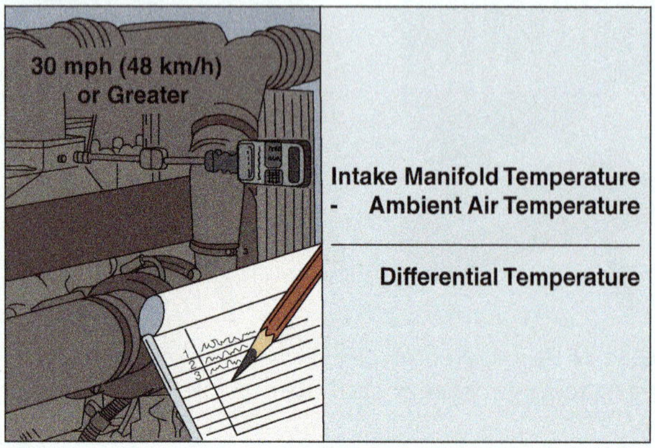

1. Measure the ambient temperature.
2. Road test the vehicle and load the engine to produce peak boost pressures. Ensure the road speed is faster than 30 mph (48 kph) to obtain good airflow across the cooler.
3. Monitor the intake manifold temperature sensor and record the intake air temperature in the manifold at peak boost.
4. Subtract the intake manifold air temperature from ambient temperature, and compare that with the manufacturer's specifications.
5. Based on test observations, make a service recommendation for cleaning the CAC or radiator. Generally, the temperature differential should not exceed 30°F (16.7°C).

Wrap-Up

Ready for Review

- Turbochargers are air pumps driven by exhaust energy to supply additional air for combustion. Using exhaust gases to drive a centrifugal-type pump recovers exhaust energy that would normally be wasted and instead uses it to enhance engine efficiency.
- Turbochargers exert a huge influence on diesel engine performance. Squeezing more air into engine cylinders adds the potential to burn more fuel and increase power output. Engine emissions, noise characteristics, and fuel economy are also improved by turbochargers.
- Because fuel requires a minimum amount of air to completely burn, every engine reaches a limit where the addition of more fuel produces no increase in power because it cannot breathe in enough air.
- If more air mass is packed into the combustion chamber, along with additional fuel, greater combustion force is produced with an increased power output from each cubic inch (or cubic centimeter) of engine displacement.
- Increasing the air supply pressure for combustion above atmospheric pressure is known as supercharging. Turbochargers are a category of superchargers.
- Turbochargers are widely used on diesel engines to improve their power output while decreasing engine size and displacement.
- Diesel engines are well suited to turbocharging because greater boost pressure produces higher compression pressures and temperatures. Theoretically, the diesel engine can handle unlimited boost without detonating or having pre-ignition damage.
- Turbochargers contribute to cleaner engine emissions in a number of ways: For example, helping diesel engines develop greater power output from smaller displaced engines has the indirect effect of reducing CO_2 emissions.
- Another indirect benefit of turbocharging is lower noise emissions from an engine. Noisy exhaust pulses emitted from exhaust ports are chopped up by the turbocharger's turbine wheel.
- Unlike naturally aspirated engines, turbocharged engines do not lose power with increased altitude, because turbochargers speed up with higher altitude. This turbocharger ability to compensate for low air density and pressure at high altitudes is called intake pressure normalization.
- Turbochargers operate by converting exhaust energy into mechanical energy that in turn forces more air into the engine. The amount of energy in the exhaust is determined mainly by the temperature, pressure and mass of exhaust gas leaving the cylinders.
- When more fuel and air is added to the combustion reaction, the exhaust gas has greater energy to drive the turbine wheel of the turbocharger.
- Turbo lag is experienced by a driver as a delay between stepping on the accelerator of the engine and the point when the engine responds with power in proportion to driver demand. Turbocharger inertia and friction are responsible for some lag, but emission control strategies are also a significant factor contributing to lag.
- A variety of turbocharger designs have been developed in response to increasing demand for improvements to engine performance, power output, and emission reduction.
- Fixed-geometry turbochargers are turbochargers without boost pressure controls. The name is derived from the fact that the housings and components have unchanging dimensions.
- Waste-gated turbochargers are similar to fixed-geometry turbochargers, except they possess an exhaust bypass valve in the turbine housing. This valve allows exhaust gases to bypass the turbine and directly enter the exhaust pipe, thus "wasting" some of the exhaust energy.
- The waste gate is designed to open at a predetermined boost pressure so that the engine operates efficiently. Failure to open will cause the turbine speed to exceed speed limits, damaging the bearing and fatiguing the turbine wheel.
- Accurate control of turbo boost pressure is critical to minimizing engine emissions. To compensate for deterioration in operation and inaccuracy of mechanical actuator controls, OBD-II engines electronically control waste-gate operation.
- To prevent exhaust pulses from affecting engine and turbocharger efficiency, inline 6-cylinder turbocharged diesel engines use split exhaust manifolds and pulse-type turbochargers.
- Variable-geometry turbochargers (VGTs) have the ability to change boost pressure independent of engine speed and load.
- Series turbocharging involves connecting two differently sized turbochargers. The compressor output of one turbocharger is directly connected to the compressor inlet of another turbocharger, which multiplies the maximum boost pressure of the turbochargers.
- Turbochargers are constructed in three parts. The two most significant parts are the turbine, or hot side, and the compressor, or cold side. A third part, the center housing, is positioned between the compressor and turbine housing.
- Both the compressor and turbine housings are shaped like a scroll or snail shell, also called a volute, which becomes progressively smaller. This diminishing diameter in the turbine housing causes exhaust gas pressure to multiply as it travels to the narrow nozzle.
- The most important factors determining turbocharger performance characteristics are the size and shape of the compressor and turbine wheels and their respective

housing sizes. Critical to the flow through the housings is the ratio between the area of either the compressor or turbine housing inlets and the radius of the housing. These factors are summarized in the A/R ratio.

- The center section of a turbocharger typically consists of two common configurations: plain sleeve-type bushings and angular ball bearings.
- Many turbochargers use engine coolant passing through passages cast into the center housing to remove heat.
- Bearing wear in turbochargers will allow axial (end-to-end) movement and radial (up-and-down) movement of the wheel, allowing them to touch the housings. If this happens, catastrophic failure of the turbocharger takes place.
- Many turbochargers are replaced unnecessarily due to poor engine diagnostics and other troubleshooting procedures. The turbocharger boost pressure test is the most efficient method to confirm a low-power complaint.
- A number of operating conditions can cause turbochargers to fail prematurely. Common defects include air inlet leaks, oil leaks, bent or broken compressor wheel blades, and worn bearings.

Key Terms

A/R ratio The ratio between the area (A) of either the compressor or turbine inlets and the radius (R) of the housing.

asymmetrical turbocharger A turbocharger with volutes of two different sizes. Asymmetrical turbochargers are used to develop exhaust gas pressure for the EGR system.

boost pressure normalization The feature of a turbocharger that maintains boost pressure even as the engine climbs to higher altitudes with low air density.

compressor housing The housing that encloses the compressor wheel.

compressor wheel A centrifugal-type air pump attached to the turbine wheel. The compressor wheel uses centrifugal force to compress air.

fixed-geometry turbocharger A turbocharger without boost pressure controls; the name is derived from the fact the housings and components have unchanging dimensions.

naturally aspirated An engine that uses only atmospheric pressure and not pressurized air to charge the cylinders with air.

parallel turbocharger The use of two or more turbochargers that share the exhaust energy from an engine's exhaust manifold. The output of both turbochargers is connected directly to the intake manifold of the engine to increase airflow into the engine.

series turbocharging Using two differently sized turbochargers where the compressor outlet is connected to the inlet of a second compressor.

series sequential turbocharger (SST) A turbocharger design that uses a single turbine wheel and two compressor wheels. Together, the compressor wheels supply a greater volume of air to the intake manifold.

supercharger A positive displacement–type air pump that pressurizes the air intake system above atmospheric pressure.

transient emissions Emissions produced temporarily when the engine load or speed is increased, such as when accelerating or upshifting gears.

turbo lag A delay between driver demand for power and the point when the engine responds with power in proportion to driver demand.

turbocharger surge A condition where the intake pressure becomes momentarily excessive and reversal flow or intake air takes place across the compressor wheel.

turbocharger choke A condition where the compressor wheel rotates too fast and creates a near vacuum at the compressor wheel inlet. Choking causes, noises, irregular turbocharger compressor wheel and turbine shaft speeds.

variable-geometry turbocharger (VGT) A turbocharger with the capability of changing boost pressure independent of engine speed and load; it is also known as variable-nozzle turbocharger (VNT) and variable-vane turbocharger (VVT).

waste gate An exhaust bypass valve in the turbine housing that allows exhaust gases to bypass the turbine and directly enter the exhaust pipe, thus "wasting" some of the exhaust energy.

Review Questions

1. What is the primary function of a turbocharger?
   a. It operates only as an engine brake.
   b. It speeds up engine warm-up.
   c. It supplies addtional intake air mass.
   d. It enables the use of larger displacement engine.
2. Which of the following is the most likely cause of bent and distorted turbocharger compressor blades?
   a. Contact with a 1/4" (6 mm) nut.
   b. Turbo overspeed.
   c. Overheating.
   d. Air cleaner sealing gasket ingestion.
3. Which of the following is the most likely cause of a pitted and worn turbocharger compressor wheel?
   a. Water ingestion.
   b. A turbocharger that's been in service for too long.
   c. A leaking air induction system.
   d. Parts of a rag or air filter gasket passing through the compressor wheel.
4. Which of the following reasons best explains how turbochargers maintain intake boost pressure (normalization) at high altitudes?
   a. Greater exhaust heat from lean air-fuel mixtures at high altitudes.
   b. Less resistance to compressor wheel rotation offered by lower-density air at high altitudes.
   c. Greater pressure differential across the turbine wheel at high altitudes.
   d. The use of a boost pressure sensor to provide closed-loop feedback control over intake pressure.

5. Why is a barometric pressure sensor used on an electronically controlled, fixed-geometry-turbocharged diesel engine?
   a. To increase boost pressure.
   b. To lower air-fuel ratios.
   c. To minimize exhaust smoke at high altitudes.
   d. To prevent turbocharger overspeed.
6. Consider an on-highway diesel engine that has a precisely matched turbocharger. Under which of the following conditions should turbine speeds be greatest to achieve maximum engine output?
   a. Peak horsepower.
   b. Peak torque.
   c. Low engine rpm.
   d. High idle.
7. Which of the following conditions is likely caused by the failure of the waste gate to open on a waste-gated turbocharger?
   a. Turbo bearing failure.
   b. Higher carbon monoxide (CO) emissions.
   c. Low power.
   d. Excessive turbo slobber.
8. Which of the following features is characteristic of engines that use waste-gated turbochargers?
   a. Less turbo lag.
   b. Slower throttle response.
   c. Reduced exhaust backpressure.
   d. Greater power density at high speed and load conditions.
9. Which of the following is changed when the diameter of a turbocharger turbine or compressor housing is increased?
   a. A/R ratio.
   b. Inducer diameter.
   c. Exducer diameter.
   d. Nozzle ring.
10. Which of the following conditions is likely caused by the waste gate failing to close on a waste-gated turbocharger?
   a. Turbo bearing failure.
   b. Higher CO emissions.
   c. Low power.
   d. Excessive turbo slobber.

ASE Technician A/Technician B–Style Questions

1. Technician A says that all engines lose power at higher altitudes due to lower air density. Technician B says that turbocharged diesel engines are an exception and that intake manifold air pressure remains the same regardless of altitude. Who is correct?
   a. Technician A
   b. Technician B
   c. Both A and B
   d. Neither A nor B
2. Technician A says that turbocharging a diesel engine reduces most noxious emissions because the cylinder pressure is higher. Technician B says that turbocharging lowers emissions because it provides more air to completely burn fuel in the cylinders. Who is correct?
   a. Technician A
   b. Technician B
   c. Both A and B
   d. Neither A nor B
3. Technician A says that turbocharging pressure in a diesel engine is limited by the threshold at which a diesel combustion chamber will begin to detonate. Technician B says that diesel turbocharger pressures are designed to match the air-fuel ratio requirements of the engine combustion system. Who is correct?
   a. Technician A
   b. Technician B
   c. Both A and B
   d. Neither A nor B
4. Technician A says that fixed-geometry turbochargers that lack controls are designed to provide maximum boost pressure at rated speed or peak horsepower. Technician B says that a properly matched turbocharger will need to provide maximum boost pressure at the highest cylinder pressure, which is peak torque. Who is correct?
   a. Technician A
   b. Technician B
   c. Both A and B
   d. Neither A nor B
5. Technician A says that a barometric pressure sensor on an electronically controlled diesel engine prevents fixed-geometry turbocharger damage caused by overspeeding. Technician B says that waste gates provide turbocharger overspeed protection. Who is correct?
   a. Technician A
   b. Technician B
   c. Both A and B
   d. Neither A nor B
6. Technician A says that turbocharger lag is caused by the inertia of heavy compressor and turbine wheel assemblies. Technician B says turbocharger lag is caused by the slow buildup of fuel rate by the high-pressure injection system. Who is correct?
   a. Technician A
   b. Technician B
   c. Both A and B
   d. Neither A nor B
7. Technician A says waste-gated turbochargers enable smaller displacement diesel engines to produce more power at low engine speeds. Technician B says waste gates prevent turbochargers from overspeeding at high speed and load conditions. Who is correct?
   a. Technician A
   b. Technician B
   c. Both A and B
   d. Neither A nor B
8. Technician A says that an engine that uses two turbochargers could use compressor and turbine housings of two different sizes. Technician B says that an engine that uses two

turbochargers could use identical turbochargers, which is referred to as a parallel turbocharged engine. Who is correct?

a. Technician A
b. Technician B
c. Both A and B
d. Neither A nor B

9. Technician A says that if a turbocharger is disassembled and a new compressor wheel is installed, then the turbocharger will need to be balanced. Technician B says that a compressor wheel's light weight will not affect turbocharger operation one way or another. Who is correct?
   a. Technician A
   b. Technician B
   c. Both A and B
   d. Neither A nor B
10. Technician A says that excessive wear on the thrust bearing will cause the compressor wheel to contact the compressor housing. Technician B says that only excessively worn center support bearings will allow the compressor wheel to contact the compressor housing. Who is correct?
   a. Technician A
   b. Technician B
   c. Both A and B
   d. Neither A nor B

CHAPTER 16

Variable-Geometry and Series Turbochargers

Learning Objectives

After reading this chapter, you will be able to:

- 16-01 Describe the purpose and functions of VGTs and series turbochargers.
- 16-02 Describe the advantages and operating characteristics of VGTs.
- 16-03 Identify and describe the types, construction, and operation of VGT components.
- 16-04 Identify and explain inspection, testing, and diagnostic procedures for VGTs and series turbochargers.
- 16-05 Identify and describe the types of series turbochargers.

You Are the Technician

A 2006 diesel pickup with an automatic transmission often makes a "woofing" noise when decelerating. While performing typical visual inspections, you notice the exhaust pipe emits black and gray smoke when the engine accelerates and when it idles. You also verify that the engine noise is originating from the turbocharger's fresh-air intake. Realizing the truck has a **high-pressure cooled exhaust gas recirculation (HP-CEGR)** system, you check for diagnostic fault codes and find numerous exhaust gas recirculation (EGR) turbocharger actuator codes and intake boost pressure–related fault codes. While you intend to use the manufacturer's diagnostic fault tree to establish which fault code to investigate first and follow manufacturer diagnostic routines, you realize this complaint is common. The noise is likely due to turbocharger surge, which is caused by a temporary overboost condition. Surge produces noise from the turbocharger when high intake air pressure pushes back through the compressor wheel when the engine cannot use all the air mass produced by the turbocharger. While preparing to explain the condition to the customer and writing up the work order description, consider the following.

1. What position are the turbocharger vanes in when turbocharger surge takes place?
2. Provide two or more explanations for incorrect vane position from the **variable-geometry turbocharger (VGT)** that would account for the complaint.
3. List and explain other possible causes that need to be investigated that could explain why the pickup emits black and gray exhaust smoke.

▶ Introduction

VGTs, also known as variable-nozzle turbochargers (VNTs) or variable-vane turbochargers (VVTs), are technologically advanced alternatives to fixed-geometry and waste-gated turbochargers. VGTs are designed to help engines meet emissions standards by supplying cylinders with pressurized air at target boost pressure faster. They also function to drive exhaust gas into the cylinders through the EGR system. VGTs can do this because they electronically vary intake boost pressure and exhaust backpressure. Intake boost and exhaust backpressure is regulated by the electronic control module (ECM), which uses an electronically controlled actuator that changes the width of the nozzle ring in the turbine housing.

Series turbochargers are an alternative to the VGT design, supplying more airflow than VGT's, faster and at higher boost pressures. In series turbocharging, the output of one turbocharger is connected to the inlet of a second. This configuration results in boost pressure multiplication—not addition. VGTs and series turbochargers have displaced, fixed-geometry, and waste-gated turbochargers in all late-model automotive diesels. To better understand the turbocharger technology essential to the latest advances in diesel engine performance and efficiency, this next section will examine the purpose, construction, and operation of VGTs and series turbochargers.

▶ Fundamentals of Variable-Geometry Turbochargers

16-01 Describe the purpose and functions of VGTs and series turbochargers.

A VGT has the functional capability of changing intake boost pressure and exhaust backpressure independent of engine speed and load. An ECM calculates optimal values for engine intake boost and exhaust backpressure needed for the EGR system. ECM electrical signals then adjust a mechanism in the turbocharger's turbine housing to vary the exhaust gas flow across the turbine wheel. On many engines, the VGT control mechanism is directly controlled by the engine ECM sending operating signals to an electric motor or varying the supply of vacuum or oil pressure to a mechanical actuator. In other engine configurations, the VGT uses a dedicated control module and actuator attached to the VGT that is capable of receiving controlled area network (CAN) signals from all vehicle ECMs, such as the antilock brake system (ABS) or aftertreatment system. In this later type, other vehicle electronic control modules can directly communicate information or send commands to the VGT to assist more complex operating strategies, such as enhancing vehicle braking by using the VGT or regenerating the aftertreatment system. The VGT can operate as an engine brake, so it may also need a position change to either increase or reduce engine resistance to driveline rotation when wheel slip occurs or when helping the transmission upshift more quickly. Using the VGT to restrict exhaust flow from the engine during cold-start and warm-up periods can speed up engine warm-up and even reduce emissions by increasing cylinder pressures and temperatures by supplying additional airflow and exhaust back pressure during cold-start and warm-up periods.

VGT compressor wheels operate with closer operating clearances between the turbocharger housing to increase operating efficiency. Since the pressure ratios are higher, more discharge heat and pressure is generated by the compressor wheel, which often requires titanium alloys rather than traditional cast aluminum compressor wheels.

Purpose of Variable-Geometry Turbochargers

VGTs are a relatively new turbocharger technology, originally developed in the 1980s for racing applications. Their primary purpose in diesel engines isn't racing but to assist the operation of EGR systems. Adding EGR systems to most 2002 and later engines was necessary to meet lower oxides of nitrogen (NO_x) emissions standards. As much as 50% of the intake charge in diesels can be exhaust gas, and intake air is the remaining percentage. Mixing exhaust gas and intake air slows down combustion speed, which in turn reduces NO_x forming during peak combustion temperatures and pressures. However, a problem with introducing a large mass of exhaust gas is how to displace pressurized intake air with exhaust gas. In most engines, this problem can be solved by using much higher exhaust gas pressure to replace intake air supplied by the turbocharger compressor. As much as

80 psi (552 kPa) of exhaust backpressure is required in some engines to force enough exhaust gas into the pressurized fresh-air intake. Using an electronically controlled actuator mechanism, exhaust flow through the turbocharger is restricted to create more exhaust backpressure. Exhaust backpressure is built up in the exhaust manifold is diverted into the intake manifold through an EGR valve. Because the exhaust pressure is increased above intake air pressure by the VGT, exhaust gases mix with the fresh intake air charge.

Before entering the intake air charge, exhaust gases are cooled. This explains why VGTs are used in a type of EGR system known as high-pressure cooled EGR (HP-CEGR) (**FIGURE 16-1** and **FIGURE 16-2**).

Another major advantage of using VGTs is that they are capable of providing a faster response to engine demand for air. Because exhaust backpressure influences turbocharger speed, boost pressure is controlled electronically and can build to higher pressures than

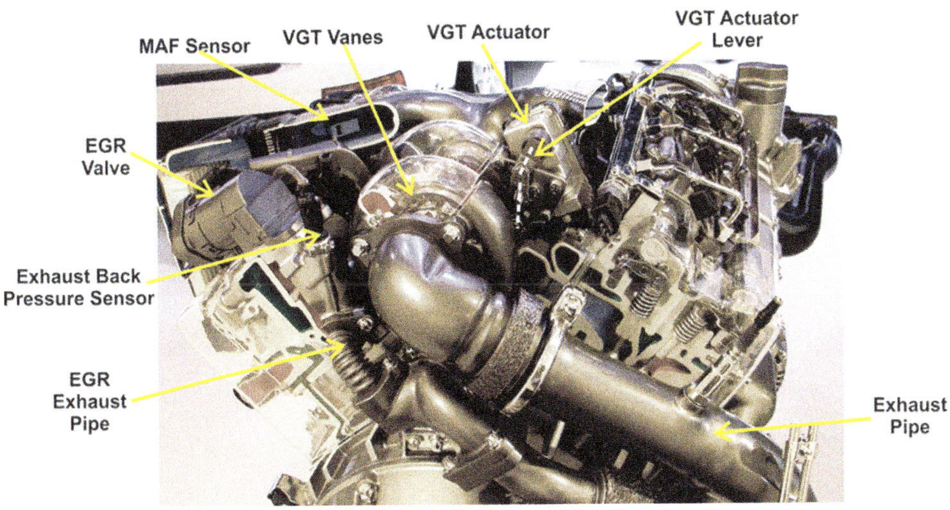

FIGURE 16-1 VGTs are used with HP-CEGR systems to increase exhaust backpressure, which is needed to mix exhaust gases with boosted fresh-air intake.

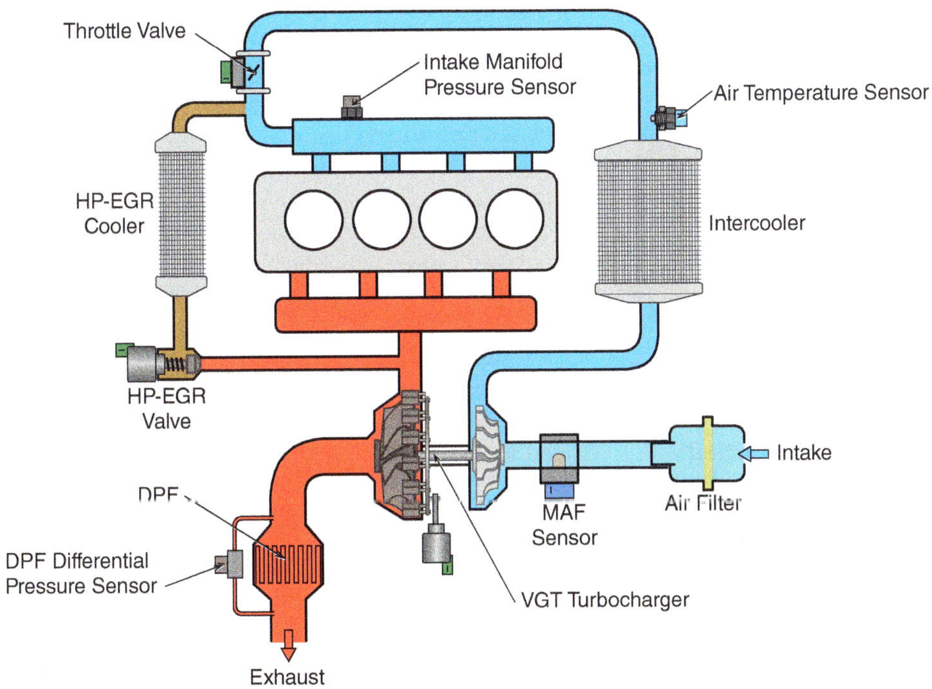

FIGURE 16-2 The VGT is used in HP-CEGR systems to create exhaust backpressure that is greater than intake manifold pressure. The backpressure that the VGT creates in the exhaust manifold forces exhaust into the pressurized intake air charge.

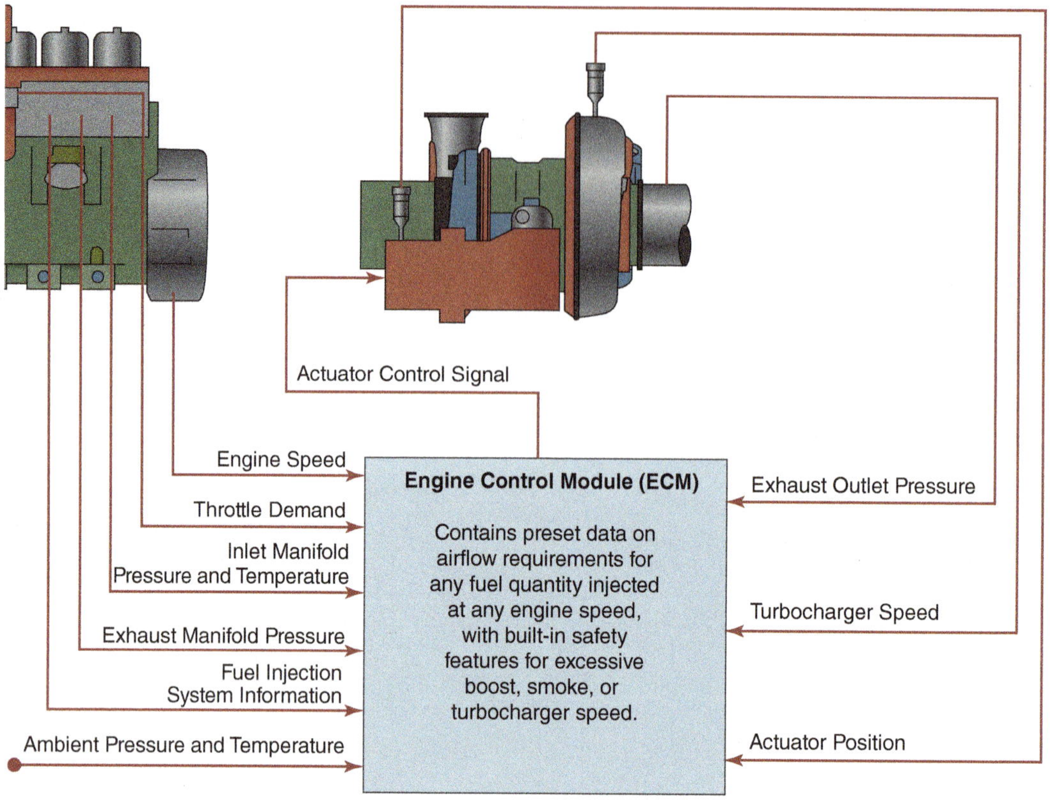

FIGURE 16-3 Using microprocessors to control VGT operation enables a VGT to precisely match air requirements with injection rates and operating conditions.

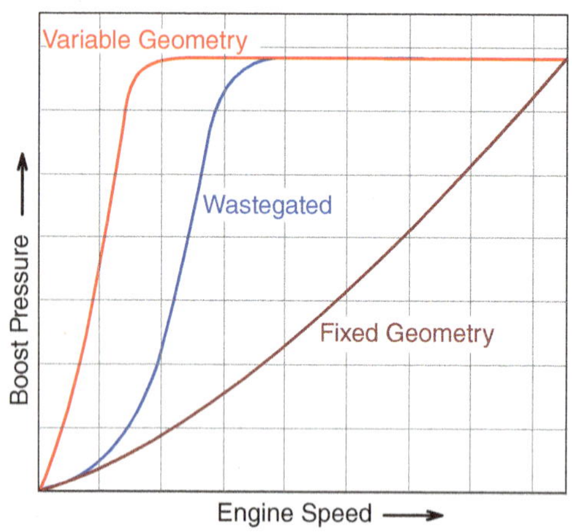

FIGURE 16-4 Comparing the performance of fixed-geometry, waste-gated, and VGTs.

conventional turbochargers can (**FIGURE 16-3**). By more rapidly building boost pressure than fixed-geometry or waste-gated turbochargers, VGTs are said to have minimal **turbo lag**. Turbocharger lag is experienced as a delay between the driver's demand for power from the engine and the point in time when the engine responds with power in proportion to the driver's demand. Reducing turbo lag provides an engine with a "snappy" throttle response. Shorter turbo lag time also means fewer emissions are produced while the turbocharger spools, or speeds up, since more air enters the cylinder sooner. Mercedes-Benz claims that its 3.0 L OM642 engine's turbo lag is less than 0.5 seconds.

VGTs can also reduce the number of turbocharger options carried by a manufacturer and provide end users with the programmable ability to change engine power levels without needing to replace the turbocharger. Software control of boost pressure independent of engine speed and load means optimal boost pressure is available over a much wider engine speed range, not in a narrow range of engine load. Higher torque output from smaller engines at low and high rpm with varying loads, not possible with fixed-geometry or waste-gated turbochargers, is easily achieved with VGTs and series turbochargers. (**FIGURE 16-4**).

▶ Turbo Lag, Variable-Geometry Turbochargers, and Emissions

16-02 Describe the advantages and operating characteristics of VGTs.

Turbocharging engines potentially creates the problem of mismatching the air-fuel ratio. Air reaches a cylinder more slowly than the injection volume of fuel, which can change instantly from one combustion event to the next. Properly matching airflow to engine demand is another challenge. Larger turbocharger compressor housings and wheels can produce proportionally greater airflow for better performance and reduced emissions. Larger turbine housings also allow more exhaust flow produced by higher horsepower engines. However, increased inertia from heavier turbine and compressor wheel mass, along with bigger housings, means the turbocharger will take longer to spool, or build boost pressure, which in turn increases turbo lag time. Larger housings and wheels require more driving energy that is simply not available at low-speed and low-load conditions. One solution is to use smaller and lighter weight turbine and compressor wheels. Pairing low inertia and rotating components with electronic turbocharger controls, while building both turbine and compressor housings to provide adequate air and exhaust gas flow over an extended operational range, provides several advantages. First, higher turbine speeds are possible when using smaller lightweight wheels. Narrowing the width of the nozzle ring opening area enables higher turbine speed with either a little or considerable exhaust gas flow (**FIGURE 16-5**). By driving the turbine wheel at higher speeds, the compressor wheel that is connected to the turbine through the turbine shaft enables very high compressor wheel speeds too. This translates into higher boost pressures at low engine load and speed conditions. Similarly, widening the nozzle ring opening will slow down the turbine wheel speed if all other conditions remain equal (**FIGURE 16-6**).

FIGURE 16-5 A set of movable vanes around the turbine control exhaust gas flow across the turbine to vary turbine speed, boost pressure, and exhaust backpressure. The electronic actuator communicates with other vehicle control modules.

A second advantage of VGT is that in comparison to fixed-geometry or waste-gated turbochargers, VGTs can supply a high level of boost pressure over a much wider range of engine operating speeds and loads. The pressure ratio, which compares the discharge pressure to inlet pressure (Discharge Pressure/Inlet Pressure), is currently 4.5, which is more than twice the capacity of fixed-geometry turbos. Higher boost pressures of a VGT under low engine speed and load conditions provide increased power density. That means an engine can provide more power for every cubic inch or cubic centimeter of displacement when packed with additional air to support combustion. Smaller displacement engines can replace larger ones since they can produce as much if not more low-speed torque with increased airflow at low speeds and loads. Engine displacement and weight can be reduced because each cylinder contributes more power from increased fuel rates and combustion pressure. In fact, power output from today's automotive diesel engines matches torque and horsepower ratings of engines with two and three times greater displacement, which use only mechanical controls and fixed-geometry turbocharging (**FIGURE 16-7**).

VGTs can reduce emissions in ways other than by supporting the EGR system or improving combustion with better airflow. When the engine is accelerated rapidly, VGTs reach target boost pressure faster than conventional turbochargers. Turbo lag is much

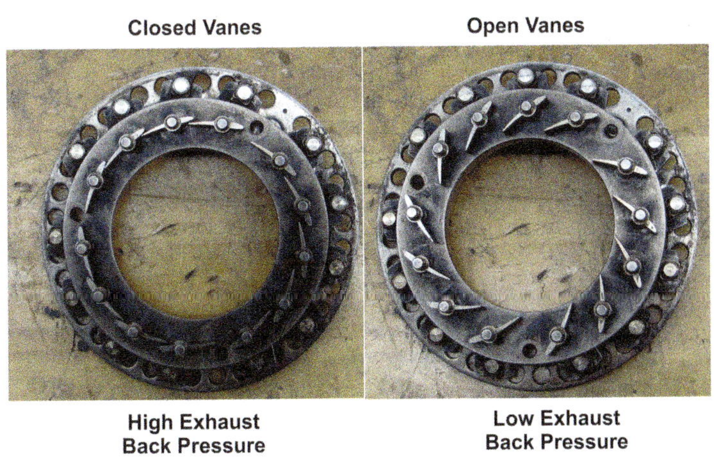

FIGURE 16-6 Vanes surrounding the turbine wheel can move from open to a closed position to regulate the flow of exhaust gases across the turbine. Creating higher backpressure speeds up the turbocharger, which increases boost pressure.

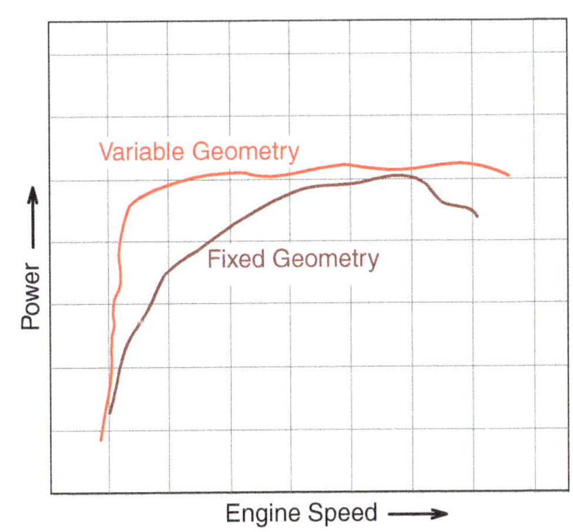

FIGURE 16-7 A VGT turbocharger can produce higher boost pressures faster, enabling an engine to produce more torque at lower engine speeds.

shorter, which in turn improves throttle response. Transient emissions, which are the brief puff of exhaust smoke and other emissions observed from the exhaust of older diesel engines when changing gears or accelerating, are gone because VGTs supply more air and do so faster, which produces cleaner combustion.

▶ Variable-Geometry Turbocharger Components

16-03 Identify and describe the types, construction, and operation of VGT components.

Several different types of construction are used by VGT manufacturers to achieve the electronically variable boost and exhaust backpressure that engines need. All of the construction styles have in common the ability to change the size of the turbine-housing nozzle ring exit area, the point in the turbine housing where exhaust gases converge on the turbine wheel. This next section will examine the construction details of the mechanism used to vary the turbine volute's nozzle ring and the methods to move the mechanism that opens and closes the nozzle ring exit area.

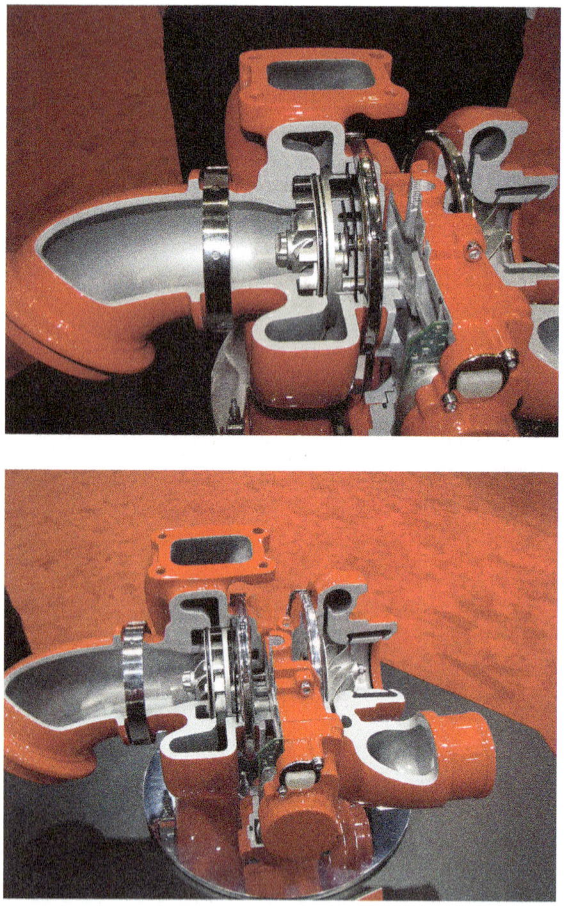

FIGURE 16-8 In this sliding-type nozzle VGT, the nozzle ring surrounding the turbine wheel is opened and closed to create exhaust manifold backpressure.

Variable-Nozzle Opening

Turbochargers are built along two designs. VGTs vary the opening size of the turbine nozzle that directs exhaust gas onto the turbine wheel. By narrowing the width of the nozzle opening, the turbine speed and exhaust backpressure increase (FIGURE 16-8). This same effect is observed by holding a finger over the end of a garden hose to increase the velocity of water coming out of the hose. In the case of the turbocharger, as the actuator closes the nozzle to narrow its width, turbine speed and boost pressure rapidly increase as exhaust backpressure rises. This happens because a greater pressure and temperature differential is created across the turbine wheel. Since the pressure and temperature at the turbine's exhaust outlet remains the same or possibly lower, the wheel spins faster. This effect of increased turbine speed is produced with minimal engine load or exhaust energy. Even at low-speed and low-load operation, VGTs can achieve relatively higher boost pressures not possible in fixed-geometry or waste-gated turbochargers. Opening the nozzle produces the opposite effect: More exhaust gas flow will take place across the turbine, but the turbine speed drops because exhaust backpressure will also decrease (FIGURE 16-9). This is like taking a finger off the end of a garden hose: flow increases but pressure drops. By varying the width of the nozzle opening with an actuator, the turbine speed power is adjusted to provide just enough energy to drive the compressor wheel at the desired boost level for any engine speed and load condition (FIGURE 16-10).

A common VGT configuration uses movable vanes arranged around the circumference of the turbine wheel. These vanes are also referred to

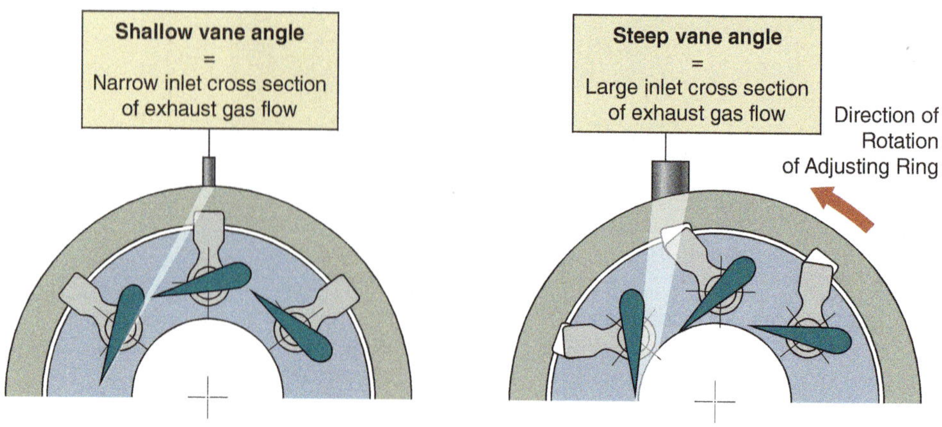

FIGURE 16-9 Changing the angle of the vanes controls the exit area of exhaust gas flow through the turbine housing.

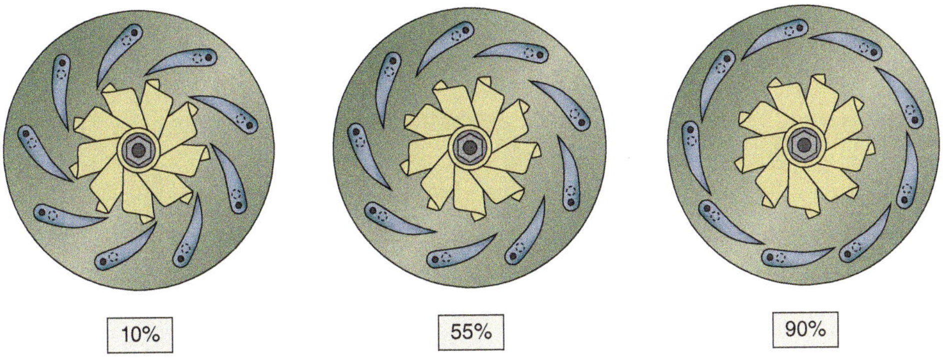

FIGURE 16-10 The position of the vanes can be varied to produce the desired boost pressure. At idle and during shutdown, the vanes are 90–95% closed.

as airfoils. Vanes operate to change the cross-sectional area in which exhaust gas flows through the nozzle ring to vary the amount of exhaust pressure crossing the turbine wheel before it reaches the low-pressure exhaust pipe (**FIGURE 16-11**). A configuration like this is also referred to as a variable nozzle because the vanes change the mass of exhaust gas flow through the nozzle ring by using signals from an ECM supplied to an actuator on the turbocharger.

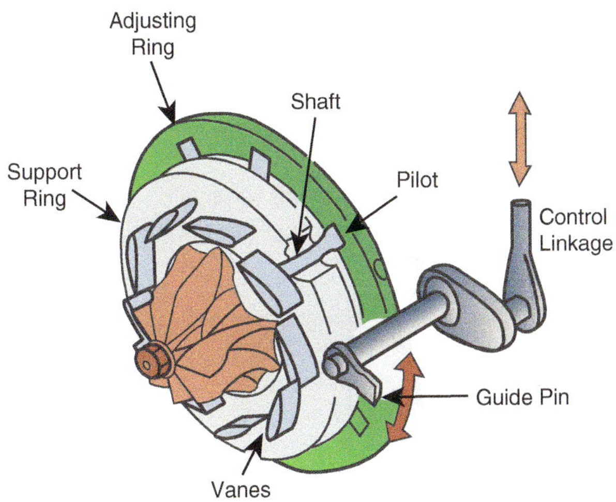

FIGURE 16-11 The VGT concept uses vanes, which are like the slats of window blinds.

Electronic Control of Exhaust Backpressure

Since 2002 in North America and 2004 in Europe and other parts of the world, EGR systems on most diesel engines take advantage of the VGT's ability to control exhaust backpressure. Because a fixed-geometry turbocharger's turbo boost pressures are close to exhaust manifold pressure as a result of valve overlap, another mechanism is necessary to increase exhaust gas pressure to a level higher than intake air boost pressure. Electronically controlling the exhaust backpressure in HP-CEGR engines allows VGTs to build gas pressure in the exhaust manifold and push exhaust gas into the pressurized air intake system. Note that EGR gas flow does not need to accompany high boost or exhaust backpressure from the VGT. The EGR valve can fully close while turbine speed and boost pressure reach maximum limits. This setup allows both the EGR gas flow and boost pressure to be controlled independently of one another.

Remember that engines can combine a high-pressure EGR system and low-pressure EGR system. A low-pressure EGR system draws EGR gas into the turbocharger through the compressor wheel. However, an engine will be equipped with a VGT to operate the high-pressure EGR system (**FIGURE 16-12**).

Exhaust Gas Regulation and Other Modes of Variable-Geometry Turbocharger Operation

The VGT actuator is adjusted to increase exhaust backpressure as required by the EGR system. The exhaust backpressure forces more EGR gas flow into the EGR circuit, where the EGR valve will ultimately regulate EGR mass flow. Exhaust gas flowing through the EGR valve regulates the quantity of recirculated exhaust gas, which means boost pressure can increase without a corresponding increase in the EGR flow. To increase vehicle braking capabilities and supplement service brake force, the VGT can operate as an exhaust brake on the engine. It does this by narrowing the nozzle ring exit area. If the engine calibration is set up to engage the turbocharger to function as an exhaust brake, the VGT actuator can narrow the nozzle opening and increase exhaust backpressure. Increasing intake boost pressure as a result of faster turbocharger speeds in turn raises compression pressure proportionally. Increasing compression pressure by adding more air mass to the cylinders raises engine-turning resistance and driveline drag. If the torque converter is locked-up on an automatic transmission, and the clutch engaged with a manual, the engine can then slow the vehicle by resisting driveline rotation because packing more air into the cylinders

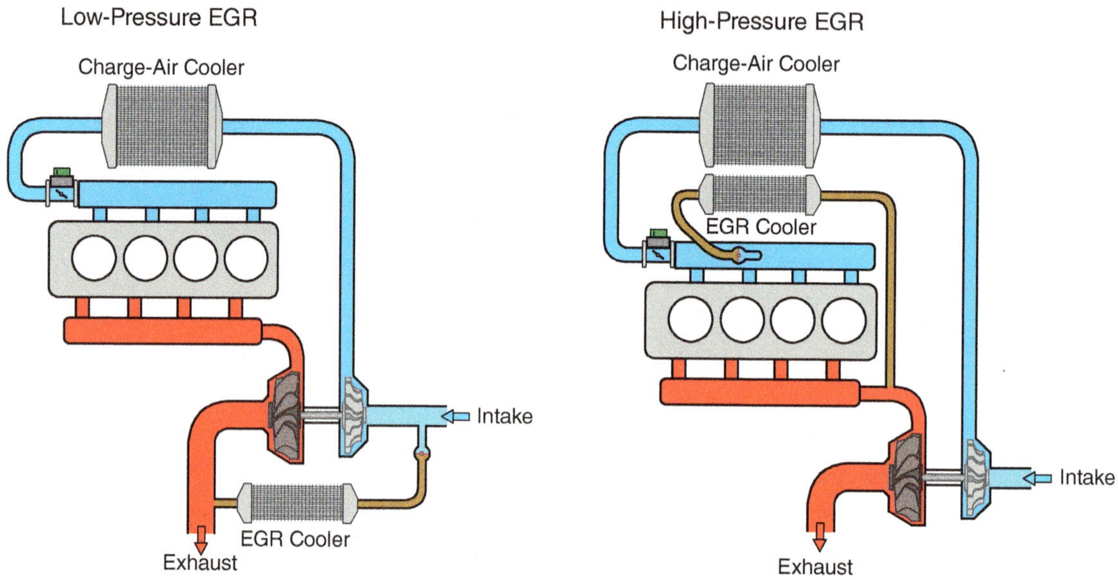

FIGURE 16-12 A low-pressure EGR system draws EGR gas into the compressor wheel. A high-pressure EGR system and a low-pressure EGR system can be combined on an engine.

> **▶ TECHNICIAN TIP**
>
> Although many passenger cars will not be equipped with an exhaust brake option, a "hill descent" function is often automatically activated as part of the engine calibration when the vehicle decelerates to a stop. Without a throttle plate to increase engine drag and driveline resistance, diesel engines do not have the same natural retarding force as gasoline-fueled engines with throttle plates. In most engines, at 3–6 mph (4.8–9.7 kph) the VGT vanes close to provide the engine with a natural retarding force that more closely resembles a gasoline-fueled engine with a throttle plate.
>
> The same strategy of narrowing the nozzle ring width can also increase backpressure at cold idle to speed-up engine warm-up. Under these conditions, the cooling system absorbs more heat because high exhaust backpressure causes higher-temperature exhaust gases to linger longer in the ports of the cylinder head. Hot exhaust gases with high air content will remain in the cylinders at the end of the exhaust stroke, which burn during the next combustion cycle. Reusing hot exhaust gas in this way at low speed and load conditions increases combustion temperatures for cleaner emissions and warmer coolant temperatures.

requires increased compression energy. More importantly, exhaust manifold backpressure also increases engine-turning resistance. When exhaust backpressure is higher, the engine has to work even harder to push exhaust out the manifold and through the narrow nozzle ring. Using a VGT as part of an engine braking strategy requires heavier housings to reduce the amount of creep, or expansion, taking place under high temperatures and pressures. The tension of the exhaust valve spring needs to increase to prevent exhaust backpressure forcing an exhaust valve open when any piston is near top dead center (TDC).

Variable-Geometry Turbocharger Vane Control

One of two common VGT designs uses vanes arranged around the turbine wheel, similar to slats in a window blind. The vanes surround the turbine wheel to control the flow of exhaust gas out the nozzle rings and across the turbine wheel. Each of the dozen or so vanes is mounted in the turbine housing with one end pinned to the housing. Another pin connects the other end of the vane to a plate called a **unison ring** (**FIGURE 16-13**). Rotation of the unison ring causes all the vanes to revolve around the fixed pivot point. Rotation of the vanes changes boost pressure by controlling exhaust turbine inlet pressure. At low engine load, when exhaust flow and energy are low, the vanes are partially closed. Closing the vanes

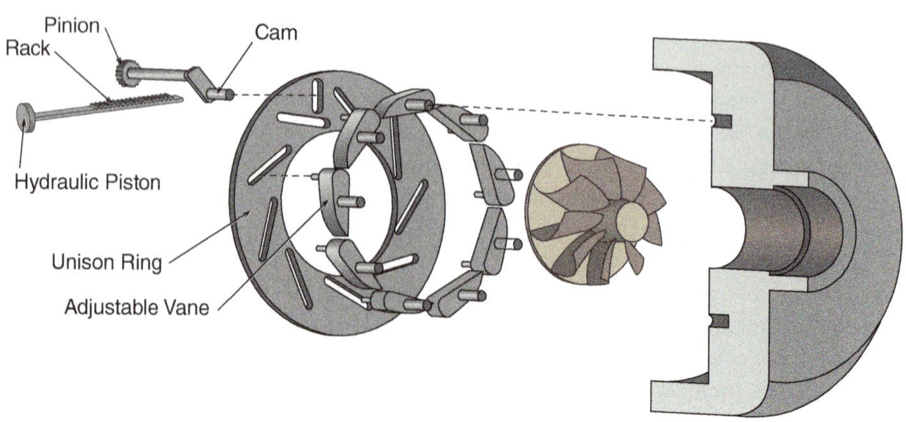

FIGURE 16-13 A unison ring, also called an adjusting ring, connects to each end of all the vanes. Rotating the unison ring synchronizes vane movement.

increases the exhaust backpressure, which causes the exhaust to push harder against the turbine blades. This makes the turbine wheel spin faster, generating higher boost pressure. As engine speed and load increase, so do exhaust flow and energy. Under these conditions, the vanes open to reduce pressure against the turbine wheel and hold boost steady—or reduce it as needed.

Sliding Nozzle Control

Holset turbochargers, used primarily by Cummins and off-road engines like Komatsu, have a sliding nozzle ring to vary the nozzle inside the turbine housing (**FIGURE 16-14**). Closing the nozzle speeds up the turbine wheel and increases backpressure. Opening the nozzle slows down the wheel and decreases backpressure (**FIGURE 16-15**). Maintaining optimal boost pressure as the engine speed and load changes is a function of opening or closing the sliding nozzle.

Electric Variable-Geometry Turbocharger Actuator Mechanisms

Ensuring a fast response from the turbocharger to engine speed and load changes is the objective of the actuator mechanism. Several mechanisms are used to change vane position: electric motors, vacuum, and hydraulic devices are the most common types of actuators. CAN control of the VGT provides other vehicle systems with the ability to supply input information to VGT operation, including the engine ECM, aftertreatment system, and ABS

> **▶ TECHNICIAN TIP**
>
> VGTs often have a built-in safety default position, which is approximately 90–95% closed. The engine will start, run, and build boost but operate with closed vanes. If vane position remains unchanged and does not open it will cause a number of problems, including damage to the EGR cooler. Excessive backpressure produced by closed VGT vanes will mechanically stress the cooler, which leads to cracking and to coolant leaking into the exhaust or the air intake system. This is one of the reasons why limiting engine power or derating the engine is necessary when VGT faults for vane position are detected.

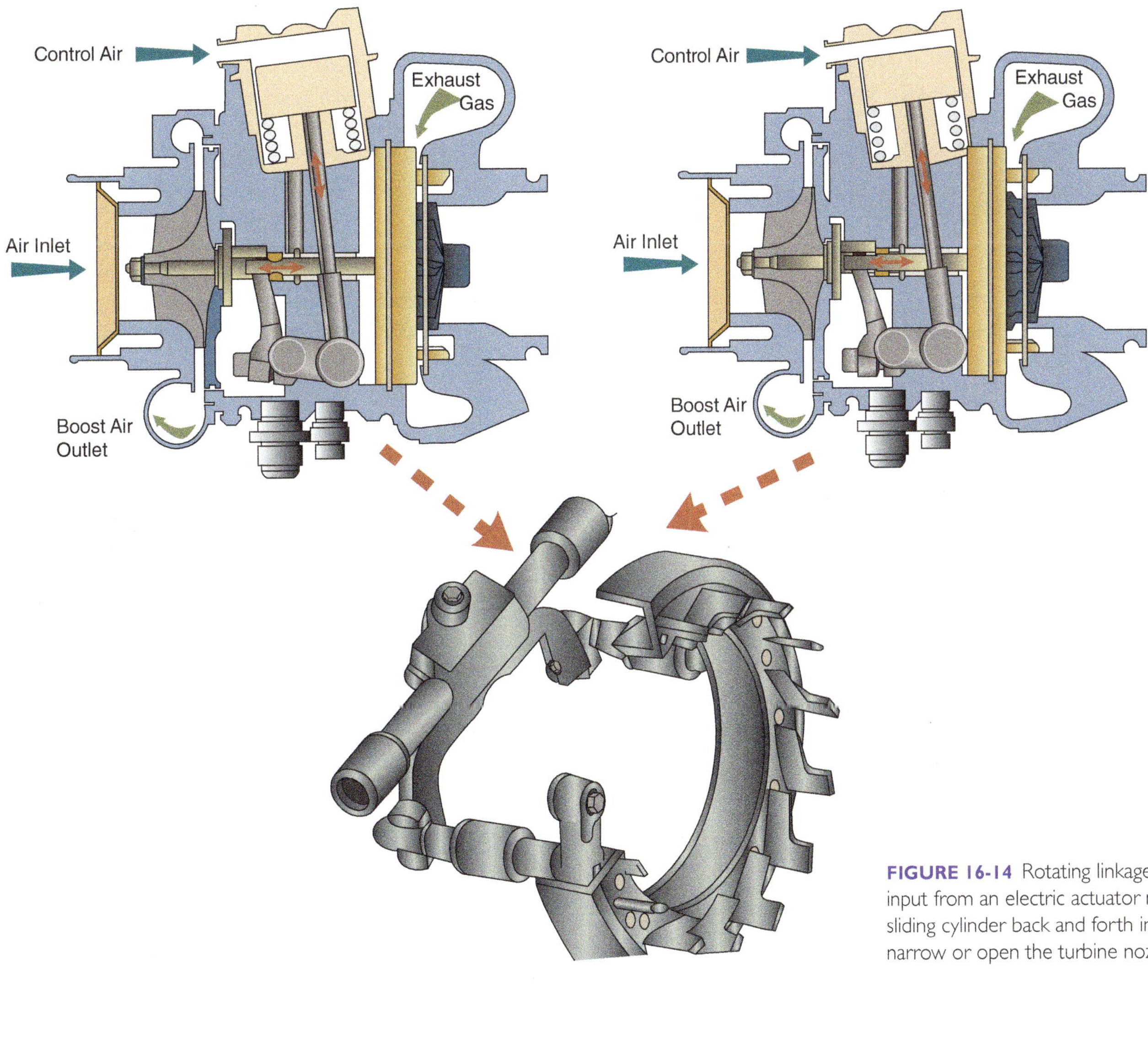

FIGURE 16-14 Rotating linkage with an input from an electric actuator moves the sliding cylinder back and forth in a VGT to narrow or open the turbine nozzle.

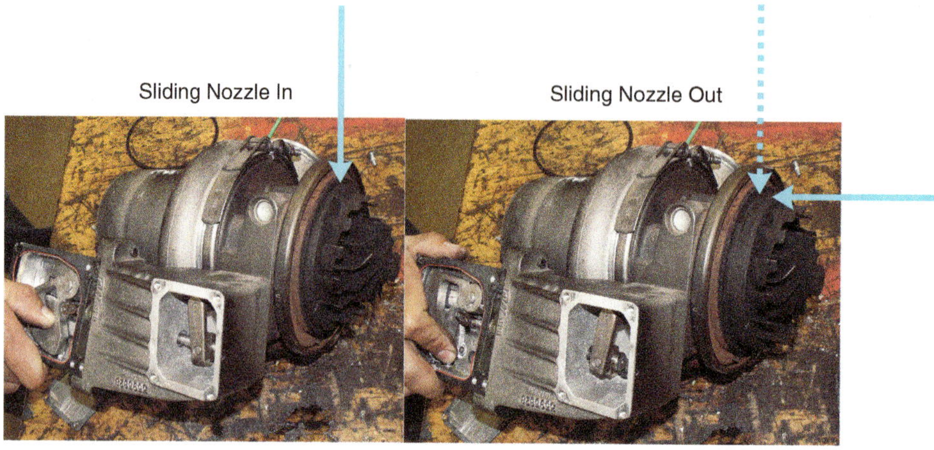

FIGURE 16-15 The Holset sliding nozzle VGT. The width of the nozzle ring is narrowed or widened to change backpressure.

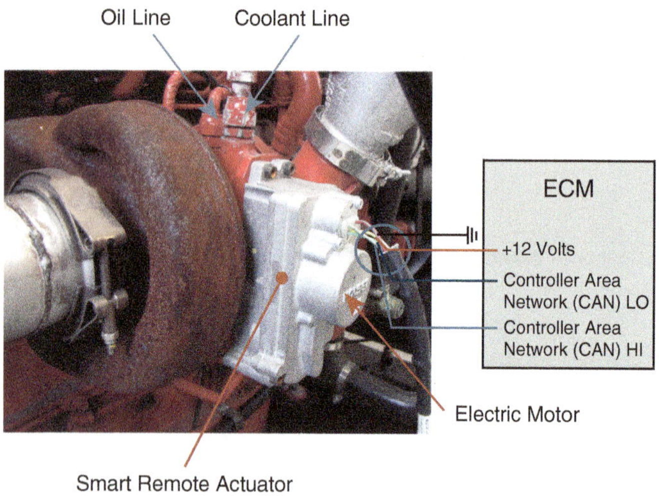

FIGURE 16-16 A VGT uses an external actuator to change the flow of exhaust across the turbine wheel. An SRA mounts on the turbocharger to control nozzle dimensions.

and traction control modules. Most VGTs are considered "smart" because they incorporate a CAN-connected control module and rely on network data for turbocharger control (**FIGURE 16-16**). A typical actuator includes a position sensor consisting of one or two small magnets and a Hall-Effect sensor (**FIGURE 16-17**). These are non-contact-type sensors that do not wear out. The sensor can provide closed-loop feedback to the module to vary the actuator's position (**FIGURE 16-18**). The smart remote actuator (SRA) used extensively on newer engines combines CAN communication with a microcontroller control of the brushless direct current (DC) stepper-type electric motor. The controls and motor operate similarly to other stepper motor applications, such as instrument gauges, to provide precision rotary position control of turbocharger vanes. A typical controller contains electronics, a DC stepper motor, a hardened-steel geartrain or worm gear. In the event of an electrical system malfunction, the unit has a default position it reverts to without power to provide a safeguard position (**FIGURE 16-19**).

FIGURE 16-17 Actuators contain a Hall-effect sensor, stepper motor, microcontrollers, a variety of drivers, some memory, and a CAN transceiver.

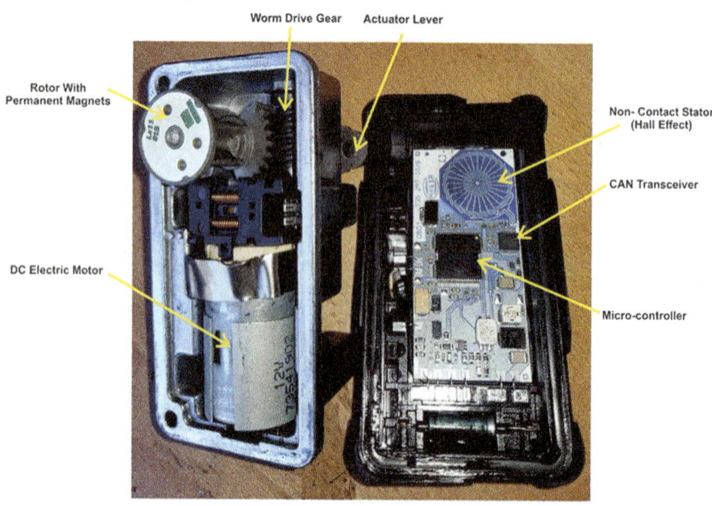

FIGURE 16-18 This common type of electric actuator uses a conventional brushed DC electric motor and a worm gear to position the actuator lever. A non-contact-type Hall-effect sensor precisely senses actuator lever position.

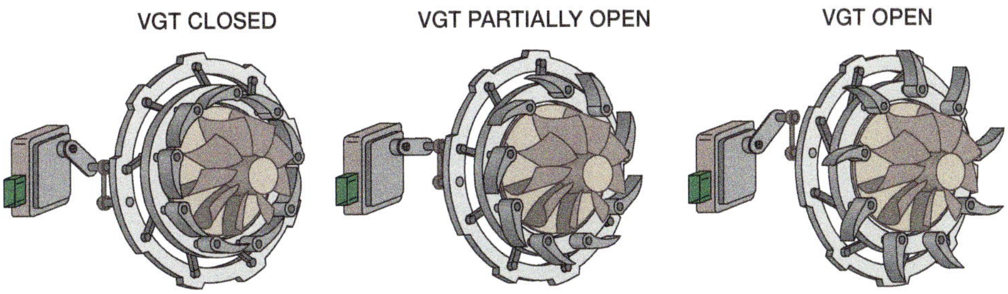

FIGURE 16-19 A smart remote electric actuator changes vane position through an external rod and linkage. The actuator uses a preprogrammed calibration setting that must properly match the engine application.

> **TECHNICIAN TIP**
>
> When the ignition key is first cycled, many stepper motor devices operating throttle plates, VGT actuators, and EGR motors perform a self-calibrating "sweep." Dash gauges using stepper motors may do the same thing. The stepper motors produce a noise when they relearn travel limits. A Hall-Effect device in the control modules evaluates whether the minimum and maximum sweep is complete and what the extent of wear is in the mechanism. Without seeing the VGT linkage performing a complete sweep, a fault code may be set for the actuator position.

Hydraulic Actuators

Pressurized engine oil acting against a hydraulic servo piston is used to rotate vanes through a unison ring (**FIGURE 16-20** and **FIGURE 16-21**). A PWM vane position control solenoid varies oil pressure applied to the actuator piston to change VGT position. Typically, the actuator piston will move a geared rack mechanism that in turn rotates a pinion gear to reposition the vanes. The vanes are fully opened when no oil flow is commanded. To begin closing the vanes, oil pressure increases through the vane position control solenoid valve. On some engines that use hydraulic turbocharger actuators, an analog position sensor with a movable tip rides on the vane actuator cam, enabling it to measure the vane position to provide feedback to the ECM. The sensor harness includes a module with an alternating current (AC) frequency generator that pulses a coil in the sensor. A moving iron core in the sensor will change the magnetic field strength of the coil. Changes in the magnetic field strength are analyzed by the module, which converts position sensor information

FIGURE 16-20 A hydraulic vane position actuator on a 6.0 L Power Stroke engine.

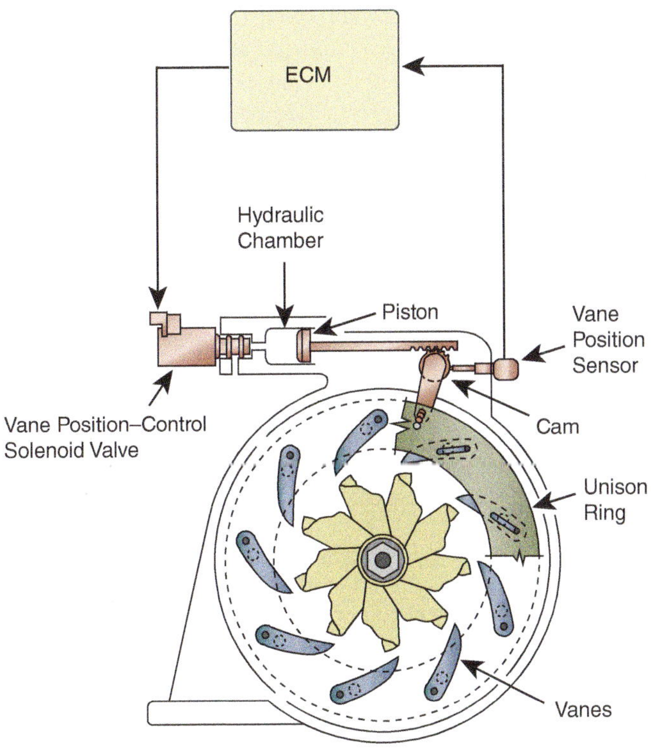

FIGURE 16-21 Hydraulic control-valve operation on the DMAX VGT. A vane position sensor provides closed-loop feedback to the ECM regarding control valve and vane positions.

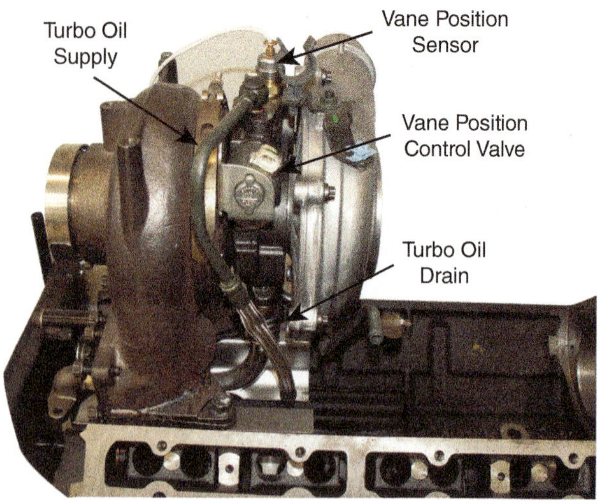

FIGURE 16-22 Oil pressure moves the vane adjustment mechanism in 6.6 L Duramax engines.

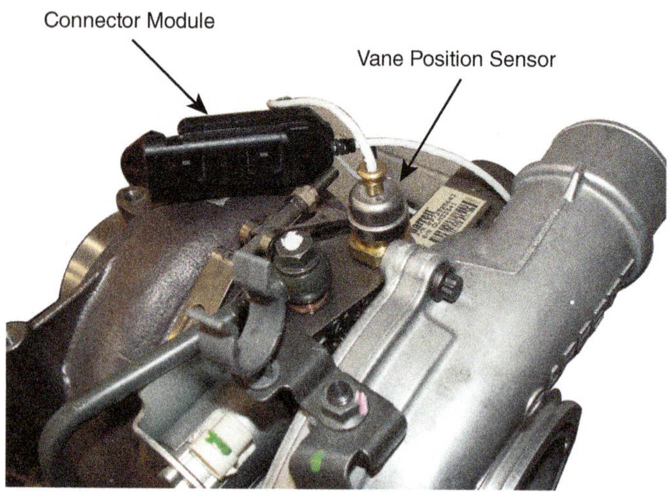

FIGURE 16-23 The connector module in the DMAX Garrett turbocharger transduces the pin position into a signal for the ECM.

to a digital signal transmitted to the ECM. One application of this actuator is the 6.6 L Duramax engines, which use oil pressure to move the vane adjustment mechanism. The actuator incorporates a position sensor and a servo valve to control oil pressure to either side of the servo-hydraulic chamber (**FIGURE 16-22**). The DMAX Garrett turbocharger has an electronic signal–conditioning circuit in the connector module. A sliding iron pin in the vane position sensor changes the strength of a magnetic field. Movement of the pin up and down on a cam provides magnetic field strength proportional to vane or unison ring position. The module transduces the pin position into a signal for the ECM (**FIGURE 16-23**). On similar hydraulic actuators used on the 6.0 L Ford Power Stroke/Navistar engines, the end of the VGT control valve has a mechanical cam follower that compresses a spring in the control mechanism to provide some feedback about the unison ring position. (**FIGURE 16-24**). Oil flow to either side of the servo piston is a function of spring tension and the magnetic field strength of the control-valve solenoid. The position-sensing camshaft is attached to the shaft rotating the unison ring, which will compress the servo spring when the vanes are fully open and unload the spring when vanes are closed. Oil will pass through the control valve to supply oil to either side of the servo piston to open or close the vanes. When magnetic field strength and spring force are identical, no pressurized oil flows to either side of

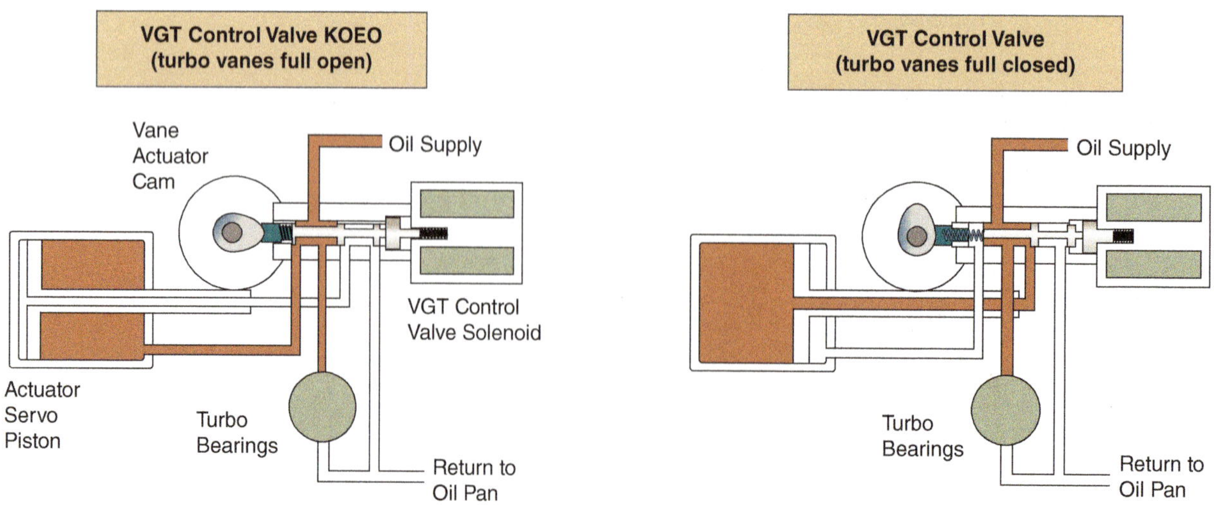

FIGURE 16-24 A camshaft attached to the unison ring provides feedback to the control valve regarding vane position.

the servo piston. The magnetic field strength corresponds to unison ring position, which is interpreted by a signal-processing circuit inside the module (**FIGURE 16-25**).

Vacuum Actuators

The availability of vacuum supplied by an engine-driven vacuum pump provides a convenient energy source to operate actuators on automotive diesel engines. Not only will vacuum supply power to operate brake boosters and ventilation controls, but a vacuum can also create a force that opposes the spring force acting on an actuator diaphragm connected to control linkage (**FIGURE 16-26**). To operate the actuator, a PWM-operated solenoid, which is an electropneumatic control valve, regulates the vacuum applied to the actuator. Signals from the engine ECM adjust the amount of vacuum supplying the turbo actuator to correctly position VGT vanes (**FIGURE 16-27**). A constant supply of vacuum from the vacuum pump, usually at least 25 inHg (0.85 bar), is supplied to one of three ports. A second port on the solenoid supplies the actuator, and a third port is a vent to atmosphere. When no vacuum is applied to the actuator, the vanes are in a fully open position. A vacuum signal of approximately 18 inHg (0.60 bar) fully closes the vanes. If the actuator's vacuum regulator fails, the vacuum signal will disconnect and allow the actuator spring force to push against the diaphragm, which opens the vanes. When the turbocharger vanes are open, little to no boost pressure is available and is accompanied by low power and various fault codes.

Feedback about turbocharger vane position originates from two sources: the intake manifold boost pressure sensor, also called the charge air pressure, and an actuator position sensor built into the actuator. Changes in vane position have a direct effect on boost pressure. This means the engine controls will check for a change in boost pressure in response to vane position (**FIGURE 16-28**). However, direct mechanical movement of the actuator control rod is measured by a linear position sensor inside the actuator.

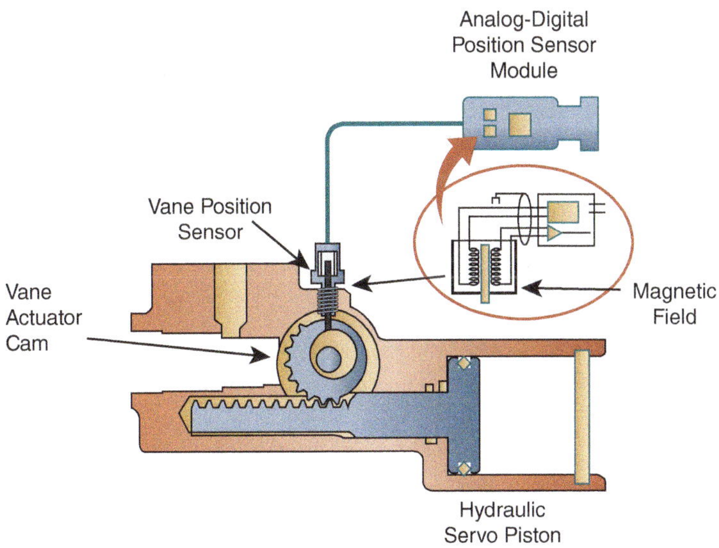

FIGURE 16-25 This Duramax actuator mechanism is hydraulically controlled.

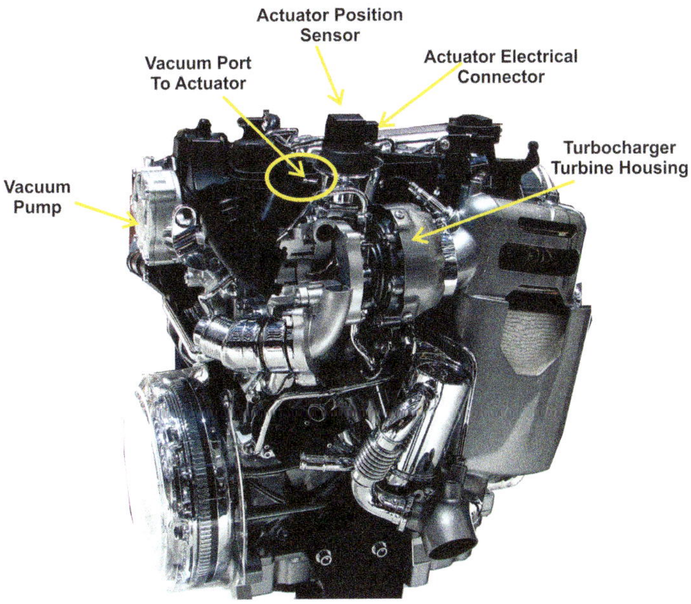

FIGURE 16-26 The turbocharger actuator on the 2.0 L VW TDI has an integral position sensor. Vacuum signals to the actuator position the VGT vanes.

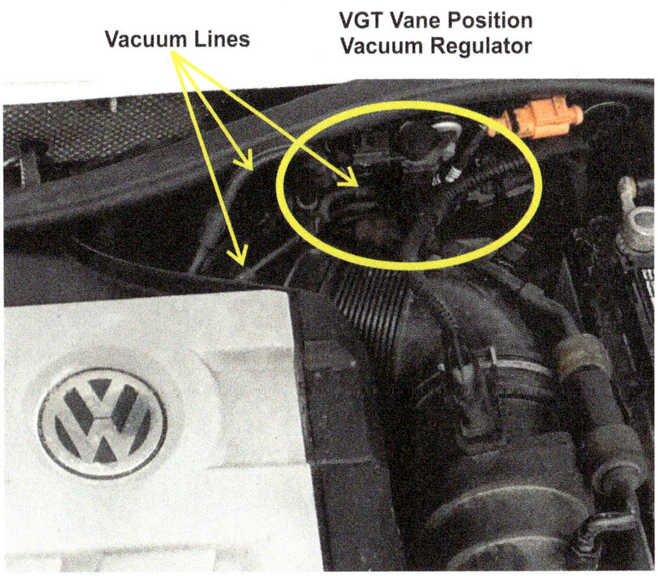

FIGURE 16-27 An electropneumatic vacuum regulator controls turbocharger vane position by supplying a vacuum signal to the turbocharger actuator.

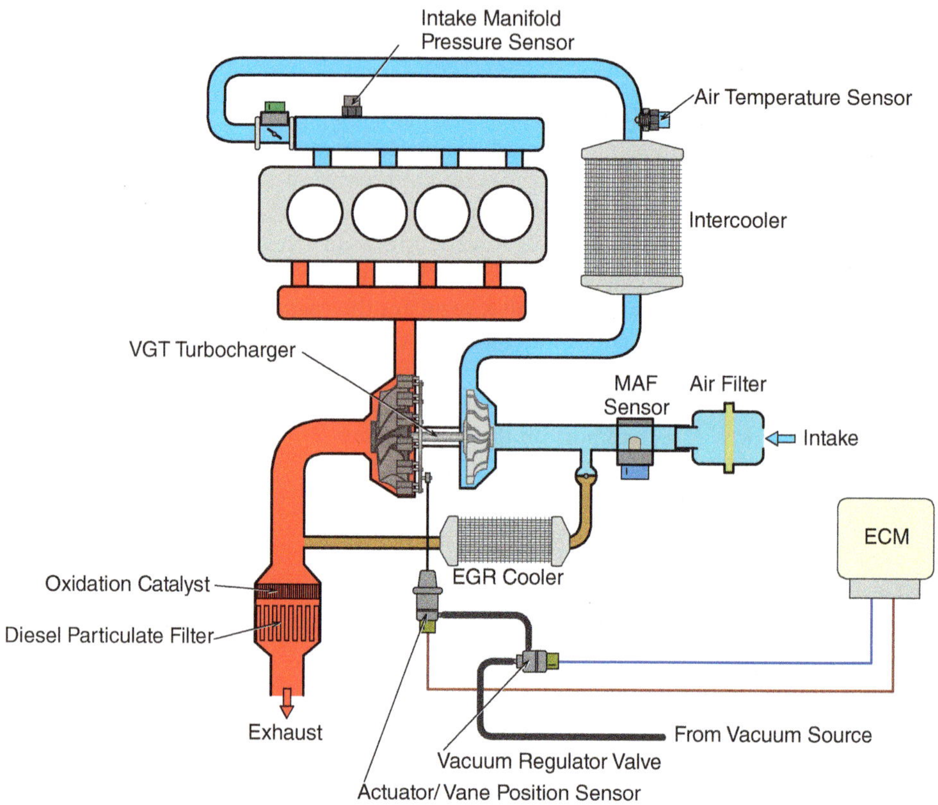

FIGURE 16-28 A schematic representation of a vacuum-controlled turbocharger actuator mechanism.

Variable-Geometry Turbocharger Sensors

Like any electronically controlled device that has a mechanical output, the electronic control and On-Board Diagnostics II (OBD-II) system needs to find out whether the electrical control signals are correctly moving turbocharger vanes. To achieve precise control, the turbocharger mechanism must provide electronic feedback about the turbocharger's operation in order for the electronic control system to adjust turbocharger operation. This process of applying an electric signal and then measuring the system response is called closed-loop feedback. VGT systems can use external sensors such as speed and exhaust backpressure as inputs to calculate actuator position, but they can also use internal sensors. Measuring the current to electric motors is important to determine whether the actuator vanes are sticking or binding and then generate fault codes. Internal Hall-effect sensors are used to accurately measure turbocharger linkage and vane position.

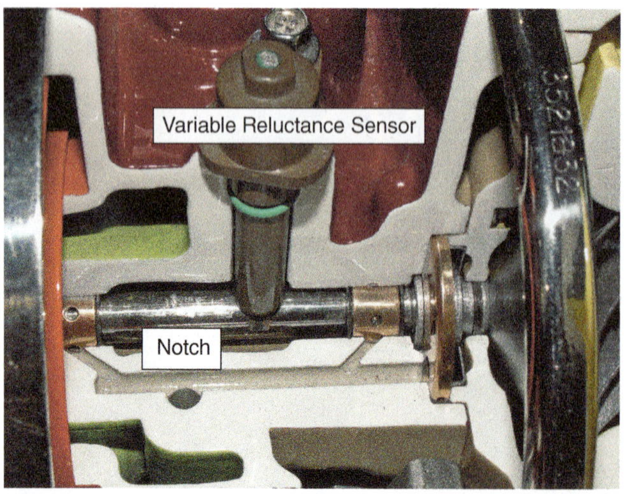

FIGURE 16-29 A variable reluctance sensor measures turbine shaft speed. Note the notch in the turbine shaft, which is used by the sensor to generate an rpm signal.

Turbo Speed Sensor

Many turbochargers have a speed sensor in the center housing (**FIGURE 16-29**). Turbine speed data are used to adjust the actuator position and monitor the turbocharger operation. Speed data are also critical for preventing turbocharger overspeed conditions. The speed sensor is the equivalent of the waste gate used to prevent overspeed conditions in waste-gated turbochargers. If overspeed is detected, the ECM will open the vanes in the VGT to slow turbine speed. The onboard diagnostic (OBD) system can monitor turbo operation if it is equipped with a speed sensor. When functional monitors make judgements about the performance of the EGR system and the turbocharger, they measure turbine speed change and compare it with expected values when the actuator is moved. If measurements are outside expected ranges, an actuator fault code will be set.

Exhaust Backpressure Sensor

An exhaust backpressure sensor can provide data regarding exhaust manifold pressure. Data supplied by the sensor are used to calculate the correct vane position. This sensor is also important to EGR system calculations for exhaust mass flow. The OBD system will use the exhaust backpressure data to check the rationality of data from other sensors. This means boost pressures, turbine speed, or actuator position may be compared to validate that the data from those sensors are within an expected range and that emission systems are functioning correctly.

Model Predictive Control of Variable-Geometry Turbochargers

Most of today's engine control systems have moved from a format where engine operation used a set of fixed lookup tables to determine what electrical signals are used to regulate an output device. Lookup tables are reference values for engine operation that give engine families unique operating characteristics—such as horsepower, torque output engine speed range, and include variables for controlling individual output devices such as injectors, turbocharger vane position, EGR valves, or intake air throttle plates. This older control system generally operated under what is referred to as proportional-integral-derivative (PID) control. This operating and control strategy would typically collect sensor data and plug the variables into an algorithm or mathematical formula used to calculate an electrical signal values to supplying output devices such as an injector solenoid or EGR valve. A fixed set of lookup table values that are a part of the engine calibration file would be included in the calculation of voltage, frequency, or pulse-width length supplied to a device. For example, if sensor measurements for throttle position, boost pressure, and engine speed were collected to determine when and for how long an injector solenoid should be energized, an electrical value derived from a calibration lab test bench would also be included in the PID calculation. A closed loop of control would be established for the output device, based on sensor data and the calibration lookup table stored in the ECM. While the closed-loop PID control arrangement worked well for early electronic controls, the addition of more hardware for additional engine systems and requirement for more precise engine control to meet emissions standards has shifted engine control from PID to model predictive control (MPC). The MPC strategy used by automotive diesels predicts how system outputs should respond to operating conditions using only a few highly responsive sensors. Engine-load demand, one of the most important variables for determining the operation of the air pathway system devices, is determined by engine speed and mass airflow or boost pressure data. When predictions about engine operating state are narrowed down to a single sensor, even in diesel engines, data supplied by the oxygen sensors or NO_x sensors with oxygen sensing functions have demonstrated the ability to accurately predict engine load. When a particular operating state of an engine is sensed or predicted by an MPC algorithm, the control system will begin to regulate output device operation based on a mathematical model of ideal engine output device settings for a given operating condition. Virtual lookup tables are then generated by MPC algorithms by using predictions for optimal output signals for an anticipated operating state. The term "virtual" means something that is not physically existing but is made to appear real through the use of software. In the case of MPC, virtual lookup tables are continuously appearing and disappearing as engine operating states or conditions change. For example, if the engine is cruising at a steady state on a level highway, the MPC system will generate output signals to control devices such as the turbocharger, injectors, and EGR valve based on gas sensor data rather than by using a large assortment of sensors to constantly measure and provide data and other feedback to adjust electrical signals to these devices. Boost pressure, EGR rates, injection timing, and injection rate shape will be set not according to a fixed calibration lookup table but instead according to a prediction for best possible settings that a manufacturer's algorithm has determined the output device operation should have. Lookup tables are generated on the fly by algorithms, rather than using limited data tables derived from bench testing and used over and over

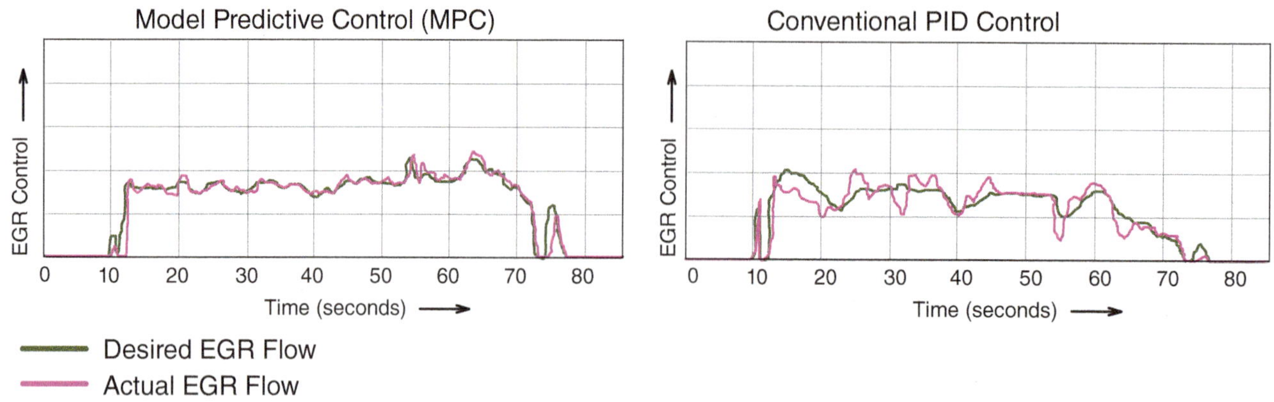

FIGURE 16-30 Comparing MPC and conventional PID control. MPC is capable of moving output devices into optimal position faster and more accurately.

again. The algorithms generating virtual lookup tables can take into account many more variables influencing system operation than can fixed calibrations painstakingly derived from experimentation in labs or on chassis dynamometers.

While MPC control may appear to create the potential for greater system error, it actually doesn't. Carefully modeled algorithms for engine operation deliver improved fuel efficiency, lower emissions, and a better transient response from systems by reducing system error that is naturally built into the fixed lookup tables of PID control. MPC is better adapted to reduce output devices' variability and can respond faster to operating condition changes than PID control can (**FIGURE 16-30**).

> **▶ TECHNICIAN TIP**
>
> Because the actuator and turbocharger are critical emission control devices, some manufacturers do not permit the actuator to be replaced without replacing the turbocharger as a complete unit. However, on engines allowing an actuator to be replaced or when replacing any engine component with a new one, the MPC engine ECM must first be made to "learn" and store the new parts characteristics. Components must also be made to recognize one another. If this is not done, it results in drivability complaints that can seem to be unrelated to the repair. When installing a new component and ensuring that a proper repair is performed, observe the following:
>
> - Does the manufacturer require a component adaptation to be performed after a part has been replaced?
> - Is there documentation or service information that covers the specific procedure for replacement and adaptation?
> - What procedure is required to perform the adaptation?
> - Is software required to transfer information from the old part to the new part?

Actuator Adaptation and Model Predictive Control

MPC operates within a complex interrelationship of components based on virtual models of system operation. That means it will make assumptions about how long a signal is applied to a device, or its frequency or voltage, based on the perfect component operating on a perfect day. Because there are significant manufacturing tolerances during component construction, influencing how each output devices will open and close, flow, or respond to identical electrical output signals from the ECM, some method needs to correct for error. To accommodate different tolerances between components and also compensate for wear and component deterioration, the stored model for component operation must be made to "learn" the components' real-world characteristics. When MPC engine control is used, replacement components must learn how the characteristics of a new component or replacement part differ from a stored "virtual" model used by calibration software to operate the component. For the technician, this engine control strategy means that when replacing any output device or component, the engine control software must first be made to "learn" the new parts characteristics. If this is not done, then the result are fault codes or customer complaints about worsened vehicle performance, poor transmission shift quality, or throttle response. A process called adaptation is used to introduce new parts into a system and to other parts to adjust the MPC model to real-world operation rather than virtual part operation. Adaptation functions are typically built into software from the original equipment manufacturer (OEM) to enable faster and even custom adaptation of components into a control system to improve performance and reduce the likelihood of drivability complaints.

Some components containing their own dedicated circuit board and microcontroller with part-specific information will need to have their stored information, or an introduction, transferred to other parts and system modules operated by MPC. Another name for this is component introduction process is called adaption. This is especially common for vehicles that use OEM-specific network communication protocols such as Unified Diagnostic Systems (UDS). UDS will transmit specific parts information to the network, describing characteristics of component operation.

▶ Variable-Geometry Turbocharger Maintenance

16-04 Identify and explain inspection, testing, and diagnostic procedures for VGTs and series turbochargers.

Fault codes, engine power derates, and other engine performance complaints are common indications of the need for VGT service. Although VGT's offer many benefits and advantages, VGT complexity translates into the more frequent need for service in comparison to fixed-geometry and waste-gated turbochargers. Seized or binding actuators are major problems for VGTs. Engine slobber—the sticky, oily combustion residue produced during prolonged cold idle—can congeal in the vanes or sliding nozzle and thus prevent proper VGT functioning. Excessive idle time and low-speed and low-load operating conditions increase accumulations of this gummy residue in the turbocharger. Without high exhaust pressure and temperatures to clean out the mechanism, the turbocharger vanes and actuation mechanisms begin to stick and operate erratically. A "woofing" sound from the air intake when the engine is rapidly accelerated and decelerated is one symptom of a slow-responding actuator. This condition, known as turbocharger surge, produces a sound referred to as **chuffing**. Sounds produced by turbo surge are created when the turbocharger spins very fast and the engine rpm falls more quickly than the turbocharger boost pressure. The high boost pressure inside the intake manifold and compressor housing will flow back through the turbocharger compressor wheel, creating the chuffing noise at idle or as the engine falls to idle speed. Surge is a dangerous condition since it applies a tremendous reversal of torque to the turbine shaft, which can twist and eventually breaks the shaft. Other engine conditions than a sticking or binding set of vanes or nozzle ring can contribute to turbo surge. Mismatched or improperly calibrated electric actuators and problems with the EGR system creating high intake pressure can contribute to turbocharger surge (**FIGURE 16-31**). A sticking nozzle ring will eventually overload the electric motor in the VGT actuator and cause it to burn out.

In addition to linkage and gear wear, electronic modules in the SRA can also fail. A common cause of module failure is related to exhaust heat and excessive driver current required by the electric motor. Coolant flow through control modules increases actuator reliability, particularly during hot soak periods when high turbocharger temperatures can penetrate the actuator housing (**FIGURE 16-32**). Hall-effect sensors can often lose the ability to properly sense actuator position as permanent magnets attached to actuator linkage weaken.

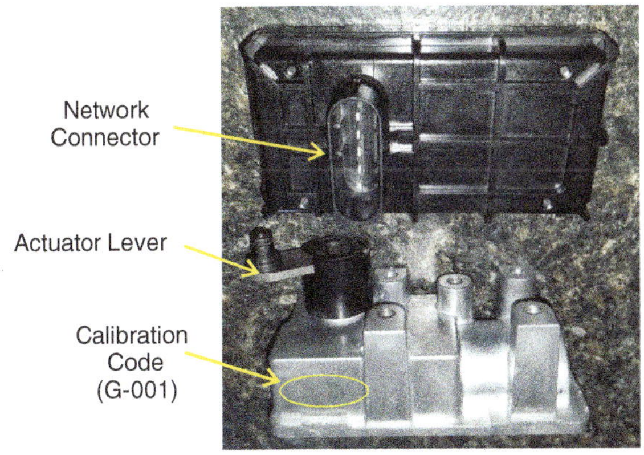

FIGURE 16-31 The calibration code on this Garrett electric actuator (from Honeywell) uses MPC and must match a specific engine calibration for the turbocharger to operate correctly.

Smart Remote Actuator Service

Replacing motors and modules in the actuator requires the ECM to learn the actuator's characteristics to accurately adjust vane or the sliding nozzle's new position. This may

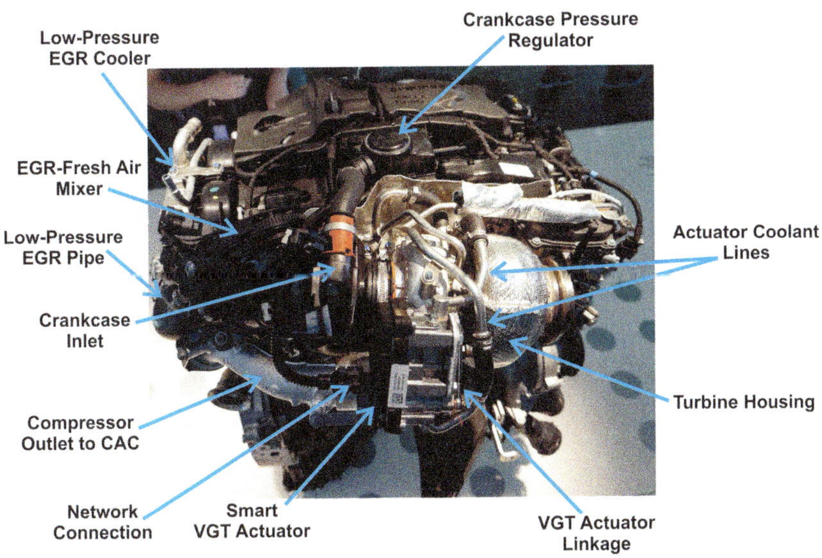

FIGURE 16-32 Coolant lines on this Mercedes 2.1 L diesel circulate coolant through the turbocharger actuator during hot soak periods. Coolant circulates by using convection currents.

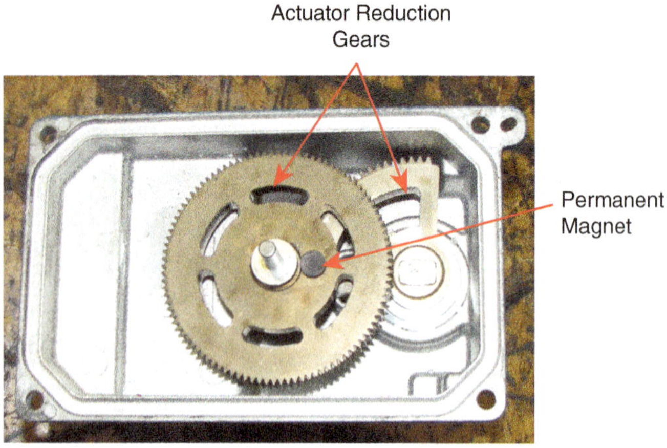

FIGURE 16-33 The hardened gears inside an SRA are spring-loaded to return the VGT to 90–95% closed. The permanent magnet on the actuator gear signals a Hall-effect sensor, which helps learn the limits of the mechanism sweep during key cycles.

require the use of a software adaptation function from the service menu of an OEM-specific diagnostic and service tool application. For example, on Sprinter vans with OM642 engines, a menu item may depict: Adaptations → Teach in → Charge psi positioner. After this procedure is performed, fault codes are cleared, and on some models, a calibration number is also stamped into the actuator housing and must be checked to ensure that it matches the original calibration number. After disconnecting or removing the SRA from the turbocharger, move the control rod or sector gear connected to the turbocharger that rotates the VGT vanes back and forth. The mechanisms should move freely without binding. If it does not, some turbochargers can be disassembled to allow the vane's control mechanism to be cleaned and lubricated with high-temperature "never seize" lubricant.

On Delphi SRAs, check for wear in the control mechanism (**FIGURE 16-33**). An alignment pin is often installed in the sector gear to fix its location in the housing (**FIGURE 16-34**). Remove the pin, being careful not to move the sector gear. Next, before installing the actuator mechanism onto the sector gear, use OEM software to command the actuator to learn the new maximum and minimum sweep limits. Next, assemble the two parts for a final step: the OEM software commands the complete actuation mechanism to recalibrate and learn the final sweep range.

Learning maximum and minimum sweep is necessary for the control module to have a starting point to compensate for VGT deterioration due to wear. When the ignition key is cycled, VGTs generally learn the sweep limits to update the adapted value stored in the ECM. A small permanent magnet and a Hall-effect sensor on the circuit board help identify these positions. Error codes for failure to calibrate accompany seized or sticking actuators.

Cooling Variable-Geometry Turbocharger Center Housings and Modules

The center bearing housing of the turbocharger and the variable-geometry actuator usually contain engine coolant passages. Minimizing bearing wear is more critical in VGTs because the compressor wheel runs closer to the housing, which improves efficiency. However, bearing wear will quickly cause the compressor wheel to touch the housing and destroy the turbo. Circulating coolant through the VGT actuator housing also reduces the operating temperature of the internal electronics, which increases the reliability of the vane and

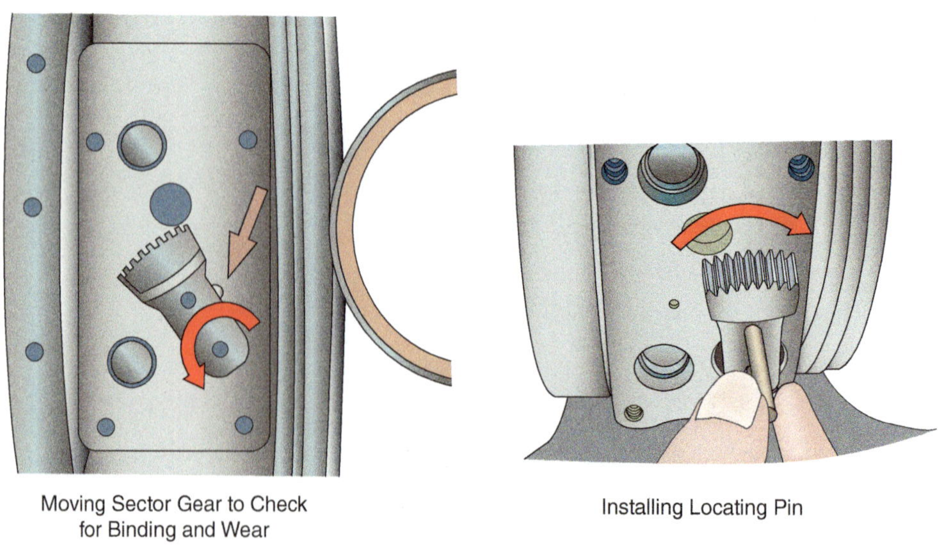

FIGURE 16-34 Check the sector gear in the turbocharger housing for wear and proper sweep before installing the SRA and calibrating the VGT nozzle movement.

actuator mechanisms. During the hot soak period, when engine temperatures rise after the engine is switched off, convection currents circulate coolant to remove excess heat. During actuator service to replace the motor or SRA control module, drain engine coolant before removing the module.

Replacing a Variable-Geometry Turbocharger Actuator

Diagnostic fault codes may indicate that an SRA needs to be replaced. To replace a VGT actuator, follow the steps in **SKILL DRILL 16-1**. The procedure shown is specific to Delphi SRAs. Follow the manufacturer's instructions for products from other OEMs.

Variable-Geometry Turbocharger Diagnostics

The OBD system continuously monitors the VGT to detect any condition that could lead to potential increases in exhaust emissions. Fault codes are set for any sensor or actuator electrical problems, underboost or overboost conditions, excessive current draw, and the boost control position. The turbocharger response to commanded nozzle ring movement is an intrusive monitoring strategy that measures boost pressure and compares it with expected values while driving. When the correct monitor-enabling conditions have been met, the vanes or nozzle ring may close for two to three seconds to measure the turbocharger's speed and boost pressure to determine whether the turbocharger's performance is within the acceptable operating range. An actuator's performance can also be tested using manufacturer-specific diagnostic software. This test typically measures the turbocharger speed change when the actuator closes and opens the vanes or moves the nozzle ring. Speed values included in operating software will determine whether the control mechanism is functioning properly.

SKILL DRILL 16-1 Replacing a VGT Actuator

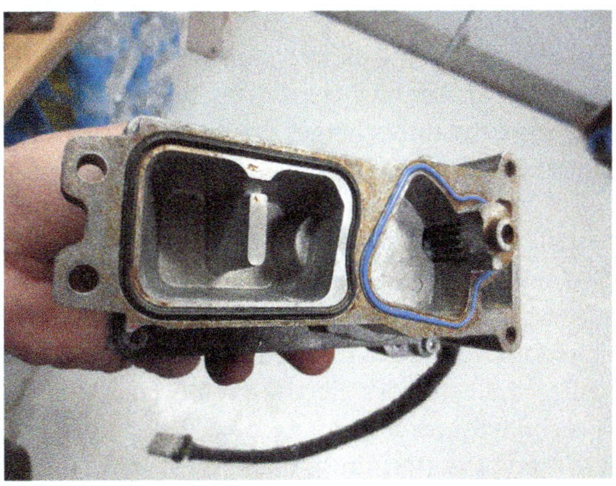

1. Drain the engine coolant and disconnect the SRA coolant lines.
2. Remove the retaining screws from the SRA, and disconnect the wiring harness to finish removing the SRA.
3. Inspect VGT mechanism operation. First, hold and shake the sector gear inside the turbocharger to check for excessive movement, which would indicate a failed bearing in the sector shaft. Sweep the sector gear back and forth several times. After initially twisting it several times, it should move freely. If it does not sweep freely, the turbocharger will need to be replaced or disassembled, cleaned, and inspected and have its vane linkage lubricated.
4. Rotate the turbocharger sector gear clockwise by hand and insert a 1/5" (5 mm) locking pin, typically supplied in the service kit, through the gear and into a reference hole in the bearing housing. If the pin does not insert into the reference hole in the turbocharger, the turbocharger will need to be replaced or disassembled, cleaned, and inspected and have its vane linkage lubricated.
5. Rotate the sector gear counterclockwise, and ensure that the flat side of the gear aligns across the diameter of a smaller 0.1" (2.5 mm) sight hole, if the hole is present. Alternatively, measure the sweep of the sector gear by using wear gauges, which correspond to specific models of turbochargers.
6. Lubricate the sector gear and drive gear of the electric actuator.
7. Align the actuator drive gear with the turbocharger sector shaft. This process is done by using software to position the electric actuator with the turbocharger sector gear. OEM software or the turbocharger manufacturer's software, such as Cummins INSITE can be used.
8. Connect the electric actuator to the wiring harness or the harness of the test software.
9. Initiate a pre-installation or calibrate actuator command from service tool software. The actuator drive gear will move to its correct position. Do not move the actuator drive gear after this point.
10. Align the sector shaft with the pin placed in the reference pin hole. Remove the pin and do not rotate the sector gear after this point. The pre-positioning of each gear allows the electric actuator to be attached to the turbocharger housing.
11. Attach the actuator to the turbocharger housing.
12. Command the actuator to self-calibrate after it has been attached to the turbocharger. Select the appropriate Self-calibrate or Initialize Calibration command from the service tool menu.
13. Disconnect the electronic service tool and reconnect the engine system harnesses. Connect the coolant lines and refill the cooling system. The turbocharger assembly is now ready to return to service.

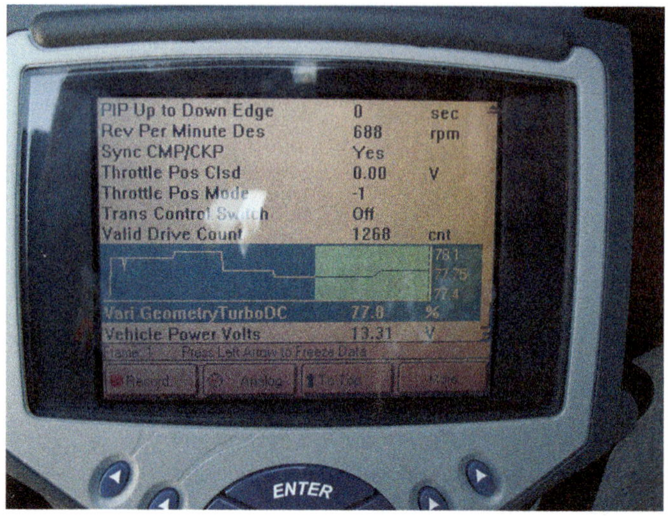

FIGURE 16-35 Movement and positioning of the hydraulic actuator position is observed by measuring the duty cycle of current applied to the hydraulic control valve on this Ford 6.0 L Power Stroke.

A turbocharger that does not change speed or changes to an expected value when the actuator test is performed likely has a defect associated with the turbocharger or control circuits.

In turbochargers with hydraulic controls, the position sensor or voltage supplied to the control-valve solenoid is measured (**FIGURE 16-35**). Slow and erratic changes to turbocharger position or voltage needed to move the unison ring indicate mechanical problems with the actuator.

Another operational strategy to evaluate the turbocharger actuator response is monitoring boost pressure. The vanes pass through their full range of movement at a relatively low engine rpm and vehicle speed. Using a diagnostic software data logger or monitoring vane position information while road testing can reveal problems with turbocharger performance.

During a road test, actuator movement should also produce proportional changes to intake manifold pressure. For example, after identifying the specification for maximum boost pressure, the vehicle can be road tested under load or during a stall test to learn whether maximum pressure has been reached. The engine and turbocharger should not operate erratically or have excessive turbo lag, which can be identified by experiencing slow engine acceleration.

The Bounce Test

Carbonized slobber in the turbine housing is the most common problem affecting actuator movement. In most cases, the turbocharger can be disassembled, cleaned, lubricated, and reassembled for service. Excessive force used to move the vanes is sensed by an increase in amperage or pulse width supplied to the actuator's electric motor. High current flow to an electric motor is flagged as a potential for an actuator failure. The actuators of some VGTs with external linkage can be checked by performing a bounce test. With the ignition key switch in the off position, the turbocharger linkage should be easily moved from the closed to the fully open position. The linkage should move smoothly. Releasing the linkage will cause the spring-loaded actuator mechanism to move the linkage to the vanes-closed position. Properly operating actuator linkage and vanes will contact the closed position and cause the linkage to bounce at least once, if not two or three times.

Volkswagen (VW) and other manufactures use vacuum to operate the VGT actuator. This means that the vacuum supply, lines, and the solenoid controlling the vacuum signal to the turbocharger can cause turbocharger problems since the actuator cannot function properly. The solenoid on a VW, called the N75 turbo boost control solenoid, regulates boost pressure by using signals supplied by the engine control module.

Inside the actuator is a spring-loaded flexible diaphragm connected to a rod extending to the unison ring. In newer turbochargers, when the actuator is supplied vacuum, it opposes spring force and extends the rod. A fully extended rod will create maximum boost pressure. Without vacuum, little or no boost pressure is developed.

To inspect a VW actuator mechanism when fault codes for low or no boost are encountered, follow the steps in **SKILL DRILL 16-2**.

▶ TECHNICIAN TIP

To perform a preliminary test of a turbocharger's actuator, make a quick visual check, which involves observing the actuator linkage or listening to the turbocharger. When the ignition is switched on, the linkage will perform a self-calibrating sweep to measure the maximum and minimum limits to its range of motion. This movement can be visually checked to verify proper sweep. Alternatively, actuators that have enclosed linkage can be heard and felt by placing one's hand on the actuator body. When running the engine in neutral while quickly snapping the throttle, actuator linkage should move throughout its range of motion.

SKILL DRILL 16-2 Inspecting a VW Actuator Mechanism

1. Inspect and test all the vacuum lines and check valves by applying a slight air pressure of 3 psi (20.7 kPa) by using a hand operated pump. This pressure prevents damage to components, but leaks can be heard or felt.
2. Disconnect the vacuum line from the actuator and apply a vacuum by using a hand pump. The actuator must hold vacuum.
3. Apply a vacuum signal of 3–4 inHg (400 mbar) and observe control rod movement. The rod should begin to move but move slowly at this vacuum.
4. Continue applying vacuum to the actuator until approximately 18 inHg (610 mbar). The actuator should have reached the full length of travel before this point.
5. Disconnect the wiring plug and measure the resistance of the N75 valve between pins 1 and 2 with the key off. Compare results against specifications.
6. With the key on, check for a 5-volt reference voltage between pins 1 and 3 of the position sensor on the actuator.
7. Further testing of the VGT can be performed using a VAG-Com service software and monitoring actuator voltage in turbocharger adaptation settings. With the engine running and by using an adaptation of the turbocharger menu item, the N75 valve will cycle the actuator position. The high-pressure turbocharger positioner sensor voltage can then be measured. Maximum and minimum voltage levels can be evaluated against specification to check turbocharger vane movement.

Running a Turbocharger Bounce Test

Diagnostic fault codes and complaints associated with the turbocharger actuator, low power, poor throttle response, boost pressure, exhaust backpressure issues, and exhaust smoke indicate that the turbocharger actuator may not be properly responding. A bounce test is used to inspect the actuator operation on a turbocharger with external linkage, such as those made by Garrett. To perform a turbocharger bounce test, follow the steps in **SKILL DRILL 16-3**.

SKILL DRILL 16-3 Performing a Turbocharger Bounce Test

1. Perform a linkage pre-cycle test. Observe the turbocharger actuator linkage while cycling the key switch to the On position. The linkage should smoothly complete a full sweep at key-on engine-off (KOEO). No partial or erratic movement is permissible. Several successful completions of this test indicate that the turbocharger actuator is operating satisfactorily, so no further tests are required. If the turbocharger does not properly complete the linkage pre-cycle test, a linkage bounce test or movement test is required.
2. With the key and engine off, push the actuator linkage with finger pressure to the opposite end of travel and release it. The linkage should move smoothly with slight spring resistance, then return to its original position. Good linkage will bounce two or three times when it returns; however, to pass this test, the linkage does not need to bounce when it returns to the stop.
3. If the linkage sticks, operates erratically, drags, or does not return to its starting position, the vane mechanism is likely the problem. This problem means that the turbocharger needs to be replaced or disassembled, cleaned, and lubricated with a high-temperature anti-seize compound. Occasionally, the turbocharger actuator motor may be defective and should be disconnected from the linkage to verify whether it is turning properly when rotated to the end of its travel.
4. If the turbocharger passes the linkage bounce test, further electrical diagnostic checks are needed to determine the reason for the unresponsive actuator.

▶ TECHNICIAN TIP

Carbonized turbocharger slobber around the moving actuator mechanisms hinders vane and sliding nozzle movement, which is the most common complaint from VGTs that set fault codes. Excessive idle time and poor quality, low-cetane fuel are common causes of seized actuator mechanisms. The measurement and evaluation of VGT actuator movement varies by manufacturer. In some cases where mechanisms are external, a bounce test of linkage or a visual evaluation of actuator operation can be performed. The amount of amperage supplied to the electric motor will also vary with the degree of resistance encountered by vane movement.

Linkage Movement Test

When the turbocharger fails the bounce test, the next step to determine whether the problem is in the vanes or with the actuator is the linkage movement test. For this test, disconnect and move the vanes by using linkage through its range of motion, which will indicate whether the vane mechanism is binding.

Actuator Voltage Test

If the turbocharger passed the linkage test and the vane mechanism is not binding, the actuator likely failed. Using a breakout harness or proper pins to back probe the turbocharger actuator electrical connection, measure the power and ground wire voltage when cycling the key on and off. Observe battery voltage when the ignition key is switched on. If battery voltage is available, the actuator is defective.

▶ Fundamentals of Series Turbocharging

While VGTs have become commonplace, single-stage and variable-geometry turbocharging have limits. Increasing torque requirements at all engine speeds requires more airflow. To overcome the limits of single-stage and variable-geometry turbocharging, manufacturers use two turbochargers connected in series (**FIGURE 16-36**). In series turbocharging, the pressurized air outlet of a larger, low-pressure turbocharger compressor housing is connected to the air inlet of a smaller, high-pressure turbocharger compressor housing. This means the boost pressure from the low-pressure turbocharger is multiplied by the high-pressure turbocharger. Series turbocharging is also called two-stage turbocharging. It provides even more air and does so faster than VGT turbochargers. Turbocharger lag is almost unnoticeable. Series turbocharging enables greater power density from engines and further ability to downsize engine displacement while still supplying the same or better torque and horsepower. A series turbocharger can be used with a low-pressure, cooled EGR system, because the turbochargers are not required to increase exhaust backpressure for the EGR system. Without the need for high exhaust backpressure, the complexity and elaborate controls used by VGTs are not necessary.

Series turbochargers use two different sizes of compressor and turbine housings to produce a wide range of boost over all engine speed and load conditions. Honeywell designates the turbochargers as either a high-pressure second-stage turbocharger or a low-pressure first-stage turbocharger. The high-pressure turbocharger has a smaller turbine and compressor housing than the low-pressure turbocharger.

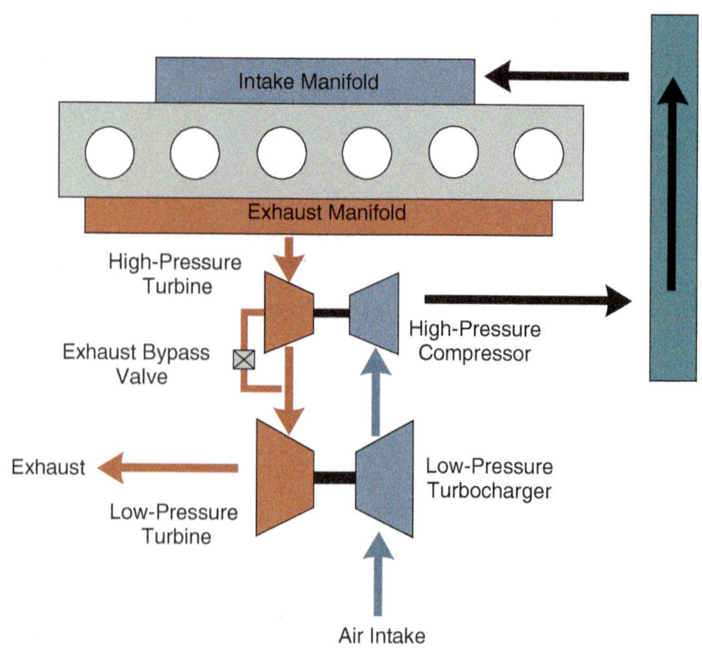

FIGURE 16-36 In series turbocharging, exhaust enters the high-pressure turbocharger first and then passes through to the low-pressure turbine. Intake air enters the low-pressure turbocharger first, before moving to the high-pressure turbo.

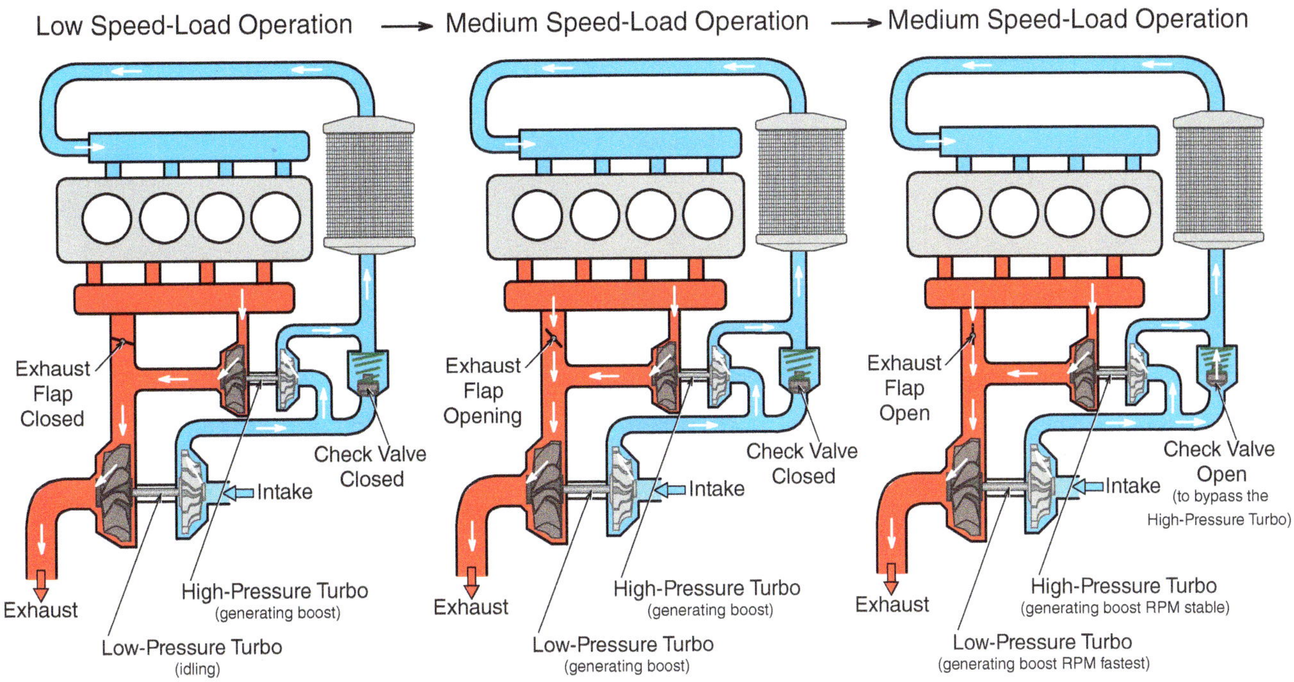

FIGURE 16-37 Operation of Opel's series turbocharger as it uses an intake check valve to bypass the high-pressure turbocharger.

A relatively smaller turbine housing responds quickly to increased engine load, with virtually no lag at low speeds. Boost pressure increases rapidly at low speeds and load conditions. Output pressure from the low-pressure compressor housing directs air into the high-pressure turbocharger (**FIGURE 16-37**). Air enters the low-pressure compressor housing first, because it encounters little resistance inside the larger compressor housing. Under load, the high-pressure turbocharger compressor will reach maximum boost pressure more quickly. Due to limitations because of its size, the high-pressure turbine and compressor wheel cannot spin any faster, so the second low-pressure turbocharger begins to receive increasing exhaust gas flow through the use of an exhaust bypass valve or waste gate before the high-pressure turbocharger. The bypass valve directs more exhaust energy into the low-pressure turbocharger and less from the high-pressure turbocharger to prevent overspeeding.

Turbine speed remains stable and does not increase in the high-pressure turbocharger, because of the waste gate or exhaust bypass valve. As increasingly more exhaust gas is diverted away from the high-pressure turbocharger's turbine wheel into the low-pressure turbocharger, the low-pressure turbocharger speed increases. As boost pressure from the low-pressure turbocharger increases, the high-pressure turbocharger compress wheel speed will remain at the same speed. However, it begins to multiply the inlet boost pressure supplied by the low-pressure turbocharger. Arranged in a series like this, multiplication of the output pressure from the low-pressure turbocharger occurs and causes boost pressure to increase to levels higher than either turbocharger could achieve alone (**FIGURE 16-38**). The mass of air moved and pressures developed are also substantially greater than what could be produced with a single-stage turbocharger or a VGT.

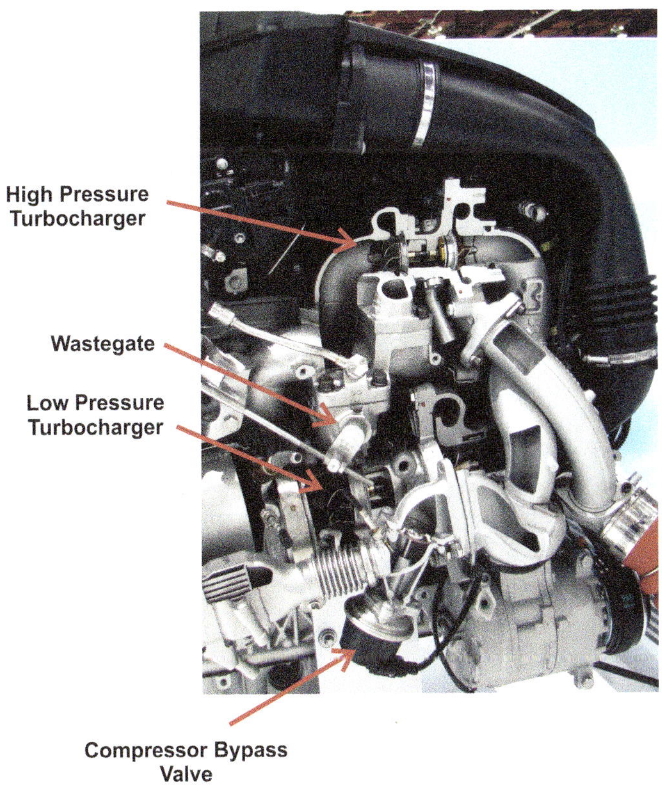

FIGURE 16-38 Series turbocharging on a BMW 3.0 L diesel engine. The waste gate limits high-pressure turbocharger speed. The compressor bypass valve redirects boost pressure past the inlet of the high-pressure turbocharger to allow more boost to be supplied by the low-pressure turbocharger.

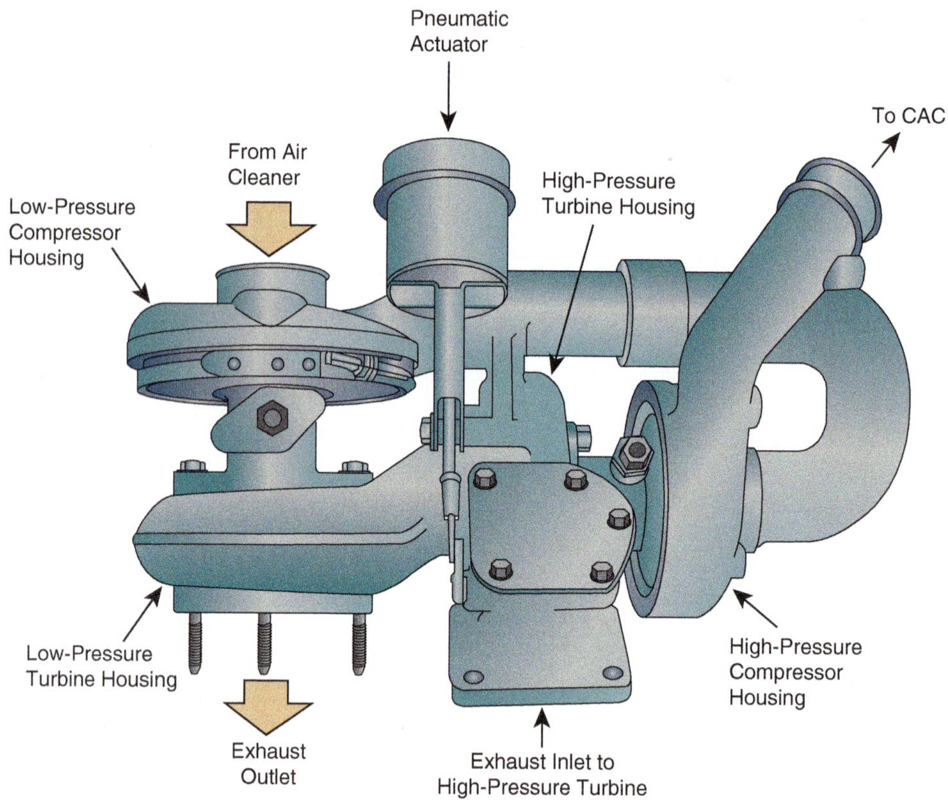

FIGURE 16-39 A series turbocharger uses a waste gate at the high-pressure turbocharger's turbine housing to limit the turbocharger speed and supply more exhaust energy to the low-pressure turbocharger.

Splitting the pressurization of the charge air between two turbochargers allows both to operate at optimal efficiency because the housing and wheel sizes are not compromised by the requirement to perform over a wide engine-load range.

Series turbochargers can also operate at lower pressure ratios. Pressure ratio is a comparison of the pressure at the compressor inlet and outlet. With series turbochargers, a pressure ratio of 2–2.5 for each stage means lower rotating speeds, which improves the reliability of the bearings and other rotating parts. On most series turbocharged diesel engines, the high-pressure turbocharger is waste gated to prevent overspeeding (**FIGURE 16-39**). Its smaller housing cannot handle high exhaust flow, so the waste gate directs exhaust flow to the low-pressure turbine housing. A few other engine configurations use valves in the exhaust and intake system to balance exhaust and intake flow. A common engine configuration uses a VGT as a low-pressure turbocharger and adds a fixed-geometry high-pressure turbocharger to supply more boost pressure when engine loads and speeds are higher (**FIGURE 16-40**).

▶ Types of Series Turbochargers

16-05 Identify and describe the types of series turbochargers.

While all series turbochargers share the same operating principles, they are built in a variety of ways. How they are arranged on the engine and the types of control mechanisms used can be quite different. A variation of the series turbocharger is single-sequential turbocharging, which is one of the latest innovations in turbocharger technology.

Single-Sequential Turbocharging

Ford's 6.7 L Power Stroke diesel engine, code-named Scorpion Diesel, uses a unique, patented **single-sequential turbocharger (SST)** design, which is also known as a VNT DualBoost Turbo turbocharger. It integrates two compressor wheels and a single turbine wheel into a single unit. Honeywell markets its Garrett turbocharger configuration as a

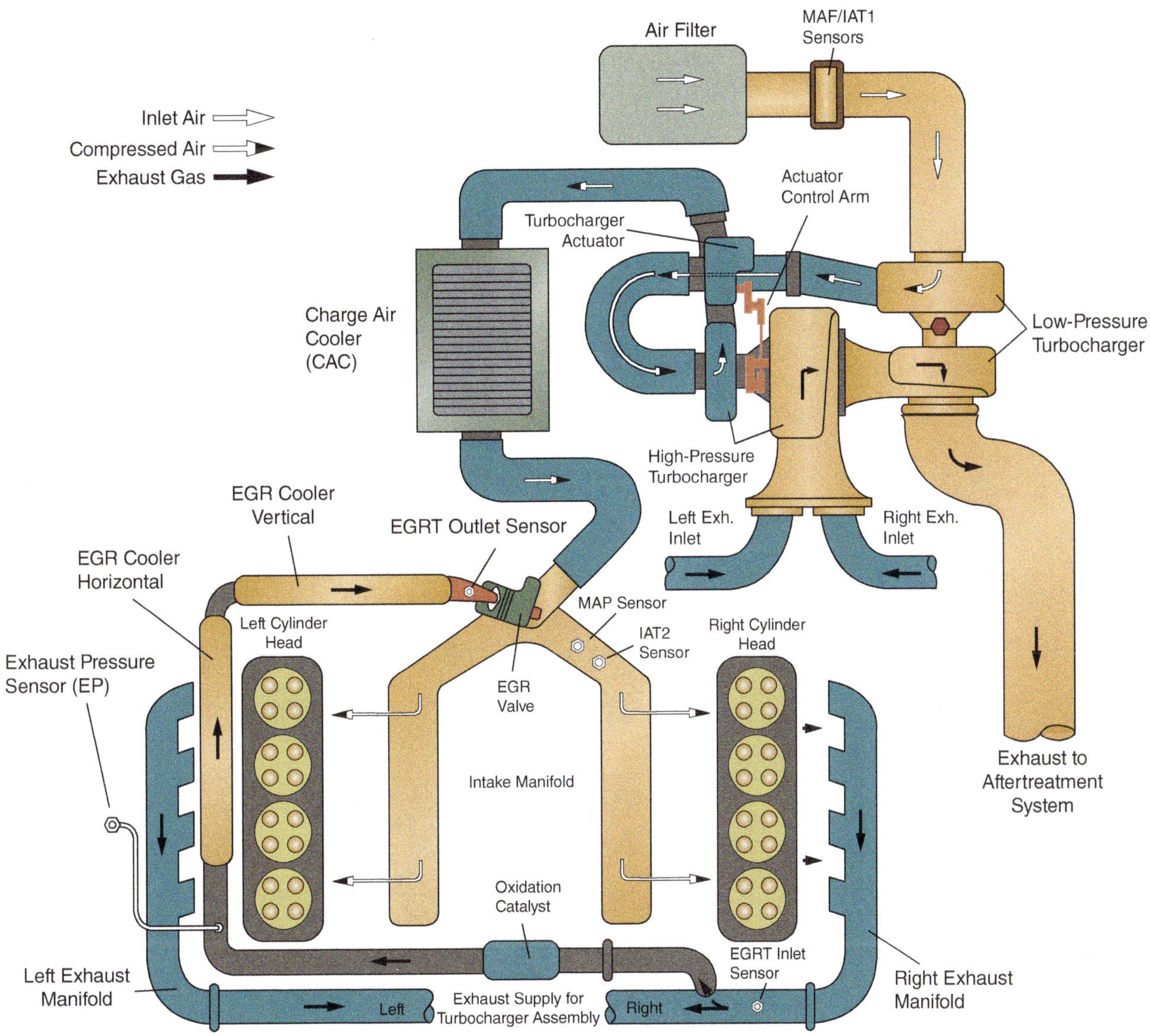

FIGURE 16-40 This diagram shows the low- and high-pressure series turbochargers and circuits used on the 6.4 L Power Stroke diesel engine. The low-pressure turbocharger is a VGT, and the engine uses a high-pressure cooled EGR system.

DualBoost system. In DualBoost systems used by engines such as the Ford 6.7 L Power Stroke, Honeywell claims to have reduced the mass of the turbine wheel and accompanying inertia by up to 40%. Two compressor wheels are mounted back to back onto a single turbine shaft. The single-piece compressor housing is divided, and a compressor wheel spins in each half. Two air inlets supply the compressor wheels in the DualBoost model, but only a single pressurized outlet supplies air to a liquid intercooler. The turbine housing uses a hydraulically actuated variable-geometry mechanism to regulate turbine speed. Another model of the DualBoost system used on heavier chassis models adds an electronically controlled waste gate to enable the use of an even smaller turbine housing for faster response times.

Ford used this turbocharger from 2011 until 2014, when it replaced it with a conventional Garrett VGT beginning in 2015. The changeup occurred because of two disadvantages of the turbo. One was the inability to handle the high exhaust backpressure required to use the turbo as an engine exhaust brake. High backpressure caused the turbocharger to overspeed. A second, related problem was that poor performance at high altitudes limited engine power. Much higher compressor wheel speeds than two wheels could operate at are required to compensate or normalize for the thinner air found at high altitudes.

Single-Sequential Turbocharger Operation

The DualBoost's unique operation begins with fresh intake air entering the main and high-speed compressor inlets (**FIGURE 16-41**). Crankcase vapors are drawn into the air inlet as fresh air passes through the lower intake manifold, which contains a crankcase vent. Pressurized air leaves a single outlet and passes through a liquid **charge air cooler (CAC)**. Also known as aftercooling and intercooling, the CAC is a system that removes excess heat from the air charging the cylinders. After being cooled by the engine's colder secondary liquid cooling system, air is distributed to the intake manifolds after EGR gas mixes with the airflow. Because the outlet of one turbo is not directly fed to the inlet of the other, as in true series turbochargers, this design is an SST. In this design, multiplication of boost pressure does not occur. Instead, the dual wheel design and shape of the compressor wheel plus the A/R ratio of the turbocharger housing add air volume to the output of the low-pressure compressor wheel when turbine shaft speeds are great enough to efficiently use the larger compressor wheel and housing configuration. Faster spooling from the smaller turbocharger and high air volume match the engine airflow characteristics of a larger turbo but in a more compact package (**FIGURE 16-42**). Ford uses this compact turbocharger

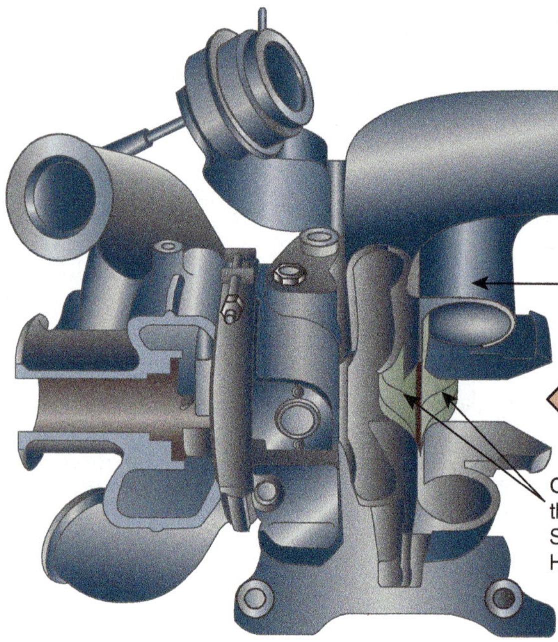

FIGURE 16-41 The Ford DualBoost SST. Each side of the divided compressor housing has a different volume. Only a single turbine wheel and housing assembly is used. This model also uses a waste gate. The turbocharger outlet cannot be seen in this illustration.

FIGURE 16-42 The SST DualBoost for a 6.7 L Power Stroke uses two inlets, whereas the conventional VGT model uses a single inlet.

configuration to enable the use of "reversed" intake and exhaust flow through the cylinder heads. Rather than placing the exhaust ports on the outboard side of the cylinder heads, where they are traditionally found, the exhaust ports are inboard. The short connection to the turbocharger narrows the profile of the engine and improves the turbocharger's thermal efficiency with short runs of exhaust from inboard ports to the engine valley, where the turbocharger is located.

Wrap-Up

Ready for Review

- VGTs are capable of electronically varying boost and exhaust backpressure independent of engine speed and load.
- The primary purpose for using VGTs is to build adequate exhaust backpressure to make the EGR system work.
- Using electronic controls, the turbocharger's resistance to exhaust flow can be increased to create more exhaust backpressure, which is built up in the exhaust manifold and redirected through either a cooler or an EGR control valve into the intake manifold, where it mixes with the fresh air charge.
- A primary advantage of using VGTs is a fast turbocharger response time with minimal turbo lag. Shorter lag time means fewer emissions are produced while the turbocharger spools, or speeds up.
- VGTs vary the opening size of the turbine nozzle that directs exhaust gas onto the turbine wheel. Narrowing the width of the nozzle opening increases the turbine speed and the exhaust backpressure. Widening the opening produces the opposite effect.
- The VGT concept uses a set of vanes, also known as airfoils, around the circumference of the turbine wheel. Vanes change the cross-sectional area that exhaust gas flows through to vary the amount of exhaust gas pressure crossing the turbine wheel before it reaches the low-pressure exhaust pipe.
- EGR gas flow does not need to accompany high boost pressure from the VGT. The EGR valve can fully close while turbine speed and boost pressure reach maximum limits.
- To improve engine braking, the VGT actuator can narrow the nozzle opening and increase exhaust backpressure if the engine calibration enables the turbocharger to act as an exhaust brake. High boost pressure requires more energy to compress intake air.
- Holset turbochargers, used primarily by Cummins, have a sliding nozzle ring to change the opening in the turbine housing, which regulates exhaust gas flow onto the turbine wheel. Changing the width of the nozzle opening increases or decreases the turbine speed.
- Several mechanisms change vane position. Electric motors or hydraulic devices are the most common actuators. The objective of the actuator mechanism is to enable a fast response from the turbocharger to engine speed and load changes.
- Hydraulic actuation is accomplished by using pressurized engine oil acting against a two-sided servo-hydraulic piston.
- Many turbochargers have a speed sensor in the center housing. Turbine speed data is used to adjust the actuator position and monitor the turbocharger operation.
- An exhaust backpressure sensor can provide data regarding exhaust manifold pressure. Data supplied by the sensor is used to calculate the correct vane position.
- The bearing housing of the turbocharger and the variable-geometry actuator usually contain engine coolant passages. Circulating coolant through the VGT actuator housing reduces the operating temperature of the internal electronics to increase the reliability of the vane and actuator mechanisms.
- The OBD system continuously monitors the operation of the VGT to detect any condition that could lead to potential increases in exhaust emissions. Fault codes are set for any sensor or actuator electrical problems, underboost or overboost conditions, and the boost control position.
- While VGTs are common in many diesel engines, single-stage turbocharging and variable-geometry turbocharging have limits.
- A series turbocharger can be used with a low-pressure cooled EGR system, where the turbocharger does not increase exhaust backpressure for the system. Without the engine's need for high exhaust backpressure, a VGT is unnecessary.
- Series turbochargers use different sizes of compressor and turbine housings, which are designated as high-pressure and low-pressure turbochargers.
- As boost pressure from the low-pressure turbocharger increases, the high-pressure turbocharger simply multiplies the inlet boost pressure. Arranged in a series like this, multiplication of the output pressure occurs and causes boost pressure to increase to levels higher than either turbocharger could achieve alone.
- Splitting the pressurization of the charged air between two turbochargers allows both to operate at optimal efficiency because the housing and wheel sizes do not compromise for performance over a wide engine-load range.
- A single-sequential turbocharger (SST) integrates two turbocharger into a single component. A single housing unit supports two compressor wheels, a configuration that reduces space requirements.

- High boost pressures produced by series turbocharging enable the use of late intake valve closing to provide an engine efficiency increase of approximately 15%. Miller cycle engines use a mechanism to hold the intake valve open longer under high boost conditions so that the engine does not need to work as hard to compress air.

Key Terms

charge air cooler (CAC) The system responsible for removing excess heat from the air charging the cylinders; it is also known as aftercooling and intercooling.

chuffing The sound a VGT turbocharger can produce when actuator response is slow. Chuffing is caused by turbocharger surge.

high-pressure cooled EGR (HP-CEGR) An exhaust gas recirculation (EGR) system that depends on a variable-geometry turbocharger (VGT) to build exhaust backpressure. EGR gas is cooled by a liquid cooler.

series turbocharger A turbocharger that uses two differently sized turbochargers that connect the compressor outlet of one turbocharger into the compressor housing inlet of a second turbocharger; in this system, output from a smaller high-pressure turbocharger, which responds faster at low engine speeds, is supplemented at heavier engine speeds and loads by the larger low-pressure turbocharger.

single-sequential turbocharger (SST) A unique turbocharger design with one turbine wheel and two compressor wheels operating in separate housings. The two compressor wheels share a common turbine shaft. This system is also known as the DualBoost system used by Ford.

turbo lag A delay from the driver's demand for power from the engine to the point when the engine responds with power in proportion to driver demand.

turbocharger surge A condition where compressed air inside the compressor housing and intake manifold reverse direction and exit the compressor inlet. Turbo surge creates a chuffing or woofing noise.

unison ring A device used by variable-geometry turbocharger (VGT) turbochargers to rotate the nozzle ring vanes together simultaneously.

variable-geometry turbocharger (VGT) A turbocharger with the capability of changing boost pressure or electronically varying boost and exhaust backpressure independent of engine speed and load; it is also known as a variable-nozzle turbocharger (VNT) or a variable-vane turbocharger (VVT).

Review Questions

1. What is the primary purpose of a variable-geometry turbocharger (VGT)?
   a. To decrease cylinder pressures.
   b. To increase oxygen content.
   c. To reduce intake air oxygen content.
   d. To create exhaust backpressure.
2. What characteristic of VGTs requires the use of titanium alloys for compressor wheels?
   a. Low-pressure ratios.
   b. Higher compressor discharge temperatures.
   c. Increased turbine speed.
   d. EGR gas flow across the compressor wheel.
3. Which of the following conditions is required to increase the turbine speed of a VGT?
   a. Widening the turbine nozzle ring opening or exit area.
   b. Narrowing the turbine nozzle ring opening or exit area.
   c. Increasing EGR flow.
   d. Decreasing EGR flow.
4. Which of the following strategies best explains how the VGT operates as an engine brake?
   a. It narrows the nozzle ring to increase exhaust backpressure and boost pressure.
   b. It narrows the nozzle ring to decrease exhaust backpressure and increase boost pressure.
   c. It widens the nozzle ring to increase exhaust backpressure and boost pressure.
   d. It widens the nozzle ring to decrease exhaust backpressure and boost pressure.
5. Which VGT sensor is primarily responsible for preventing turbocharger overspeed?
   a. Exhaust backpressure.
   b. Mass airflow.
   c. Boost pressure.
   d. A variable reluctance sensor located on the turbine shaft.
6. Under which of the following operating conditions does the low-pressure turbocharger of a series turbocharger arrangement produce maximum boost?
   a. High-speed, high-load operation.
   b. Low-speed, low-load operation.
   c. High-speed, low-load operation.
   d. Any low-load operation.
7. A typical VGT produced since 2010 will most commonly respond to electrical signals transmitted ____________.
   a. directly from the ECM
   b. from an engine controller
   c. through the onboard vehicle network
   d. from the EGR control module
8. Consider a VGT. When intake boost pressure is at its highest value, what can most likely be observed regarding exhaust backpressure in the exhaust manifold?
   a. It is also at its highest value.
   b. It is at a low value.
   c. It depends on the EGR gas requirements.
   d. It is low because the waste gate will usually open.
9. Which of the following conditions is required to decrease the turbine speed of a VGT?
   a. Opening or widening the turbine nozzle ring exit area.
   b. Narrowing the turbine nozzle ring exit area.
   c. Increasing EGR flow.
   d. Decreasing EGR flow.

10. Consider the operation of a single-sequential DualBoost turbocharger. What combination of air inlets and outlets are used?
    a. One air inlet and two outlets.
    b. Two air inlets and one outlet.
    c. Two inlets and two outlets.
    d. A dual path air inlet and bypass valve–regulated outlet.

ASE Technician A/Technician B–Style Questions

1. Technician A says that VGT engines have less turbo lag. Technician B says VGT engines have more exhaust backpressure in the exhaust manifold than single-stage turbochargers have. Who is correct?
    a. Technician A
    b. Technician B
    c. Both A and B
    d. Neither A nor B
2. Technician A says that a late-model engine equipped with a VGT will close the vanes to increase intake boost and exhaust backpressure during braking. Technician B says the vanes will open to reduce intake airflow during engine braking. Who is correct?
    a. Technician A
    b. Technician B
    c. Both A and B
    d. Neither A nor B
3. Technician A says that VGT engines can electronically vary turbocharger boost pressure. Technician B says that boost pressure from VGTs is typically higher than boost pressure from fixed-geometry turbochargers. Who is correct?
    a. Technician A
    b. Technician B
    c. Both A and B
    d. Neither A nor B
4. Technician A says that without power to the VGT actuators, some actuators will typically provide a lot of boost, because the vanes are as much as 95% closed. Technician B says the turbocharger will not provide very much boost, because the vanes always default to an open position. Who is correct?
    a. Technician A
    b. Technician B
    c. Both A and B
    d. Neither A nor B
5. Technician A says the turbocharger speed sensor is used to sense and prevent a turbocharger overspeed condition. Technician B says the speed sensor is used by the OBD system to detect faults with the turbocharger. Who is correct?
    a. Technician A
    b. Technician B
    c. Both A and B
    d. Neither A nor B
6. Technician A says that coolant passageways through the turbocharger housing are used to keep the turbocharger bearings and other parts cooled when the engine is running. Technician B says that coolant passageways are used primarily to circulate coolant after the engine has been shut off. Who is correct?
    a. Technician A
    b. Technician B
    c. Both A and B
    d. Neither A nor B
7. Technician A says the operation of the VGT actuator's electric motor is inspected during a bounce test. Technician B says the motor needs to be removed and checked by using an ohmmeter to determine whether it is faulty. Who is correct?
    a. Technician A
    b. Technician B
    c. Both A and B
    d. Neither A nor B
8. While two technicians are discussing series turbocharging, Technician A says the high-pressure turbocharger has an exhaust bypass valve that prevents overspeeding. Technician B says the low-pressure turbocharger uses a waste gate to direct exhaust gases to the high-pressure turbocharger. Who is correct?
    a. Technician A
    b. Technician B
    c. Both A and B
    d. Neither A nor B
9. Technician A says that single-sequential turbochargers have two separate compressor and turbine housings. Technician B says that single-sequential turbochargers use two turbine wheels and one compressor wheel. Who is correct?
    a. Technician A
    b. Technician B
    c. Both A and B
    d. Neither A nor B
10. After two technicians had listened to the noise made by a VGT when the key was switched on but before starting the engine, Technician A said the noise was the actuator moving to a default spring-loaded position after it had been shut down in a wide-open position. Technician B said the turbocharger actuator was relearning maximum and minimum actuator sweep positions during a self calibration process. Who is correct?
    a. Technician A
    b. Technician B
    c. Both A and B
    d. Neither A nor B

CHAPTER 17

Exhaust Gas Recirculation

Learning Objectives

After reading this chapter, you will be able to:

- 17-01 Explain the operating principles of diesel engine EGR system components.
- 17-02 Describe the functions, construction, and applications of diesel engine EGR systems.
- 17-03 Describe and explain methods for performing inspection and diagnostic procedures on diesel engine EGR systems.
- 17-04 Recommend maintenance or repairs on diesel engine EGR systems.

You Are the Technician

A 2012 pickup truck has a prematurely worn engine. After fewer than 240,000 miles (386,000 kilometers) and less than 6,000 hours of engine operation, the rings and cylinder walls have worn out and the engine has low power. You have been asked to investigate possible causes for the accelerated wear. Similar pickups of the same model year are not providing the service life that older pickups with identical engines provided. You perform an oil analysis and find no special clues, except for a very high amount of soot. Service records confirm that oil services were performed well within recommended limits. You begin to wonder whether there might be an electronic control module (ECM) calibration issue that is causing injection timing to be incorrect and contributing to the soot in the oil. Someone suggests to you that the exhaust gas recirculation (EGR) system may have a problem. Before proceeding further, consider the following:

1. Explain why a defect in the EGR system could cause excessive soot in engine oil.
2. Explain why a defect in the EGR system could lead to premature ring and cylinder wall wear.
3. What checks of the EGR system would you perform to determine whether the EGR system is functioning correctly?

▶ Introduction

EGR is a long-established, proven technology to reduce oxides of nitrogen (NO_x) emissions from spark-ignition engines. By reintroducing exhaust gas into the cylinders, a major reduction in exhaust emissions of NO_x from diesel engines can also be achieved. EGR technology has been used since the early 1970s, but it has been applied to light-duty diesel engines only since the mid 1990s and on most diesel engines since 2002. Diesel engines produce a disproportionate amount of NO_x gas due to higher cylinder temperatures and pressures produced through unique combustion characteristics, higher compression ratios, and turbocharging. A study of heavy-duty diesels from the 1980s highlights this observation, noting that diesel engines make up only 4–6% of all motor vehicles, but it has been estimated that they have contributed over 50% of all NO_x emissions produced by motor vehicles (**FIGURE 17-1**). Today, diesel engines emit almost no detectable NO_x emissions. Tier 3 emissions and fuels regulations introduced in 2017 reduce the small remaining portion of emissions allowed by previous Tier 2 standards by another 80% (**FIGURE 17-2**). While EGR is effective in reducing NO_x formation from the cylinders of engines, the introduction of EGR has required many changes to engine design, operation, and maintenance requirements. This chapter examines the operation and maintenance of EGR systems and how they interact with engine construction and operation.

▶ Fundamentals of Exhaust Gas Recirculation

17-01 Explain the operating principles of diesel engine EGR system components.

EGR is an in-cylinder emission-reduction strategy. That means EGR is used to prevent the formation of NO_x emissions during the combustion process, in contrast with strategies to remove NO_x in the exhaust system. Understanding the mechanisms of NO_x formation helps explain the various reasons for EGR system design, construction, and operation.

Forming Conditions for Oxides of Nitrogen

EGR reintroduces exhaust gas into engine cylinders to reduce peak cylinder pressure and temperature; high cylinder pressures and temperatures produce NO_x. Normally, nitrogen gas, which constitutes slightly more than 75% of the atmosphere, stays out of combustion reactions. However, combustion temperatures above approximately 2,500°F (1,371°C) will cause reactions between oxygen and nitrogen to take place. In addition to heat, squeezing molecules together under high-pressure, -compression, and -combustion conditions creates a greater likelihood of chemical reactions forming various types of molecules that contain

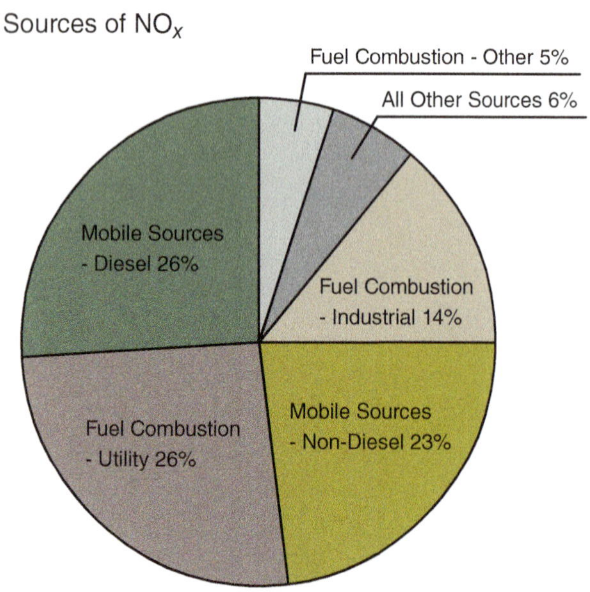

FIGURE 17-1 Comparing contributions of NO_x emissions from diesels. Data is based on a 1980's Department of Energy study.

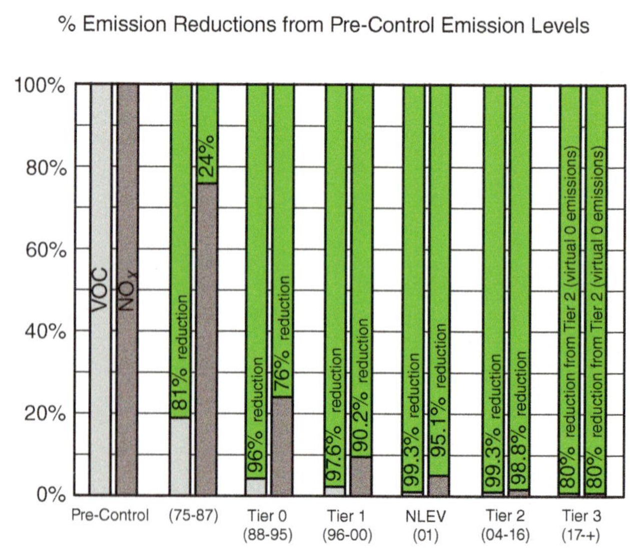

FIGURE 17-2 Comparing NO_x emissions from today's diesel engines to emissions before emission reducing technology was introduced.

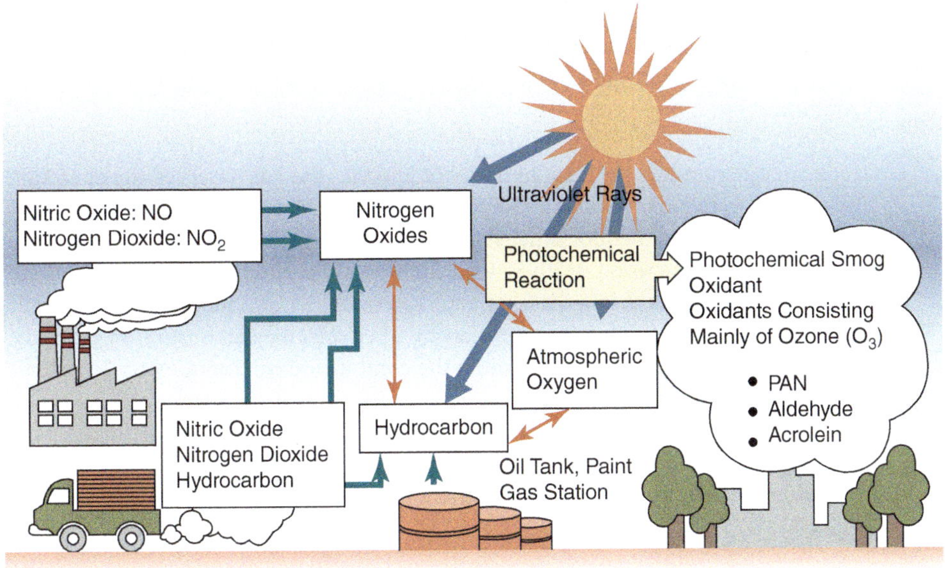

FIGURE 17-3 NO_x from engines produces O_3. Without O_3, photochemical smog cannot form.

nitrogen and oxygen. Collectively, these nitrogen and oxygen compounds are referred to as NO_x and may consist of a variety of gases, primarily including NO, NO_2, and NO_3.

Nitric oxide (NO), a colorless gas, is the most abundant type of NO_x, initially produced during diesel combustion. However, because NO is an unstable and highly reactive gas, it begins to combine with oxygen immediately after combustion has ended. Post-combustion **nitrogen dioxide (NO_2)** is a reddish-brown gas formed when NO binds to oxygen. The presence of a noxious reddish-brown tint to the atmosphere over cities when the winds are light indicates an abundance of NO_2 emissions. NO_2 should not be confused with nitrous oxide (N_2O), an anesthetic gas that is also used in racing to increase a gasoline-fueled engine's power output. When NO_x breaks down in the presence of sunlight and heat, it converts to nitrogen and ozone gas (O_3), which is a primary component of photochemical smog (**FIGURE 17-3**). NO_x emissions are toxic to the environment and human health.

Diesel Combustion and Oxides of Nitrogen

NO_x formation is a major emission problem with diesel engines due to higher compression ratios, cylinder pressures, and temperatures (**FIGURE 17-4**). This is especially true for turbocharged diesel engines, where the combination of additional air and fuel produce more power from each cubic unit of cylinder displacement. Additionally, diesel combustion is stratified: The combustion chamber has regions where air-fuel ratios vary widely. Areas inside the combustion chamber where rich fuel mixtures burn hotter will contribute to NO_x formation. This contrasts with gasoline-fueled engines, where lean mixtures contribute to NO_x formation. In gasoline-fueled engines, lean mixtures cause the combustion flame to travel slower than normal, delaying the beginning of combustion that results in abnormally high cylinder pressure near top dead center (TDC). More NO_x forms because higher compression pressures at TDC will accelerate the already delayed combustion event, producing a rapid, sharp rise in combustion pressure that quickly burns fuel. An uncontrolled combustion event replaces the moderate combustion speed produced by a normal flame front traveling through a stoichiometric air-fuel ratio.

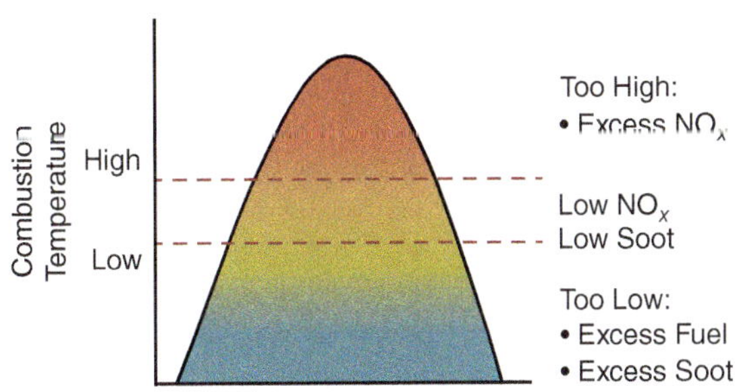

FIGURE 17-4 Carefully controlling combustion chamber pressure and temperature can minimize the in-cylinder formation of NO_x.

Exhaust Gas Recirculation System Operation

While the design, construction, and operation of different EGR systems are unique, the basic mechanisms used to reduce NO_x formation in the cylinders are the same. The principle behind all EGR systems is to disrupt the chemical processes that form NO_x. Replacing combustion air with EGR gas

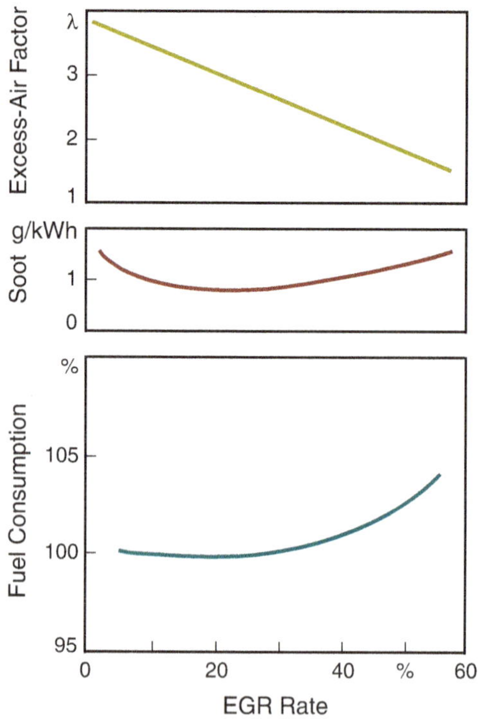

FIGURE 17-5 Effects of EGR rates on fuel consumption and particulate formation. EGR gas displaces oxygen in the combustion chamber.

is the primary strategy used by all EGR systems (**FIGURE 17-5**). By replacing air in the combustion chamber with relatively inert EGR gas, the heat released from the combustion event takes place over a larger number of degrees of crankshaft rotation.

NO_x emissions forming during peak cylinder pressures and temperatures are reduced by extending the burn period. For anyone who has observed the operation of a fireplace, the idea behind an EGR is similar to closing the chimney damper. By restricting the movement of exhaust gases out the chimney, less fresh air enters the firebox to consume wood. With less air, the wood takes longer to burn and the fire is not as hot. Similarly, in a diesel engine, spreading air and fuel molecules farther apart by inserting EGR gas—rather than air—between the combustion ingredients, slows the movement of flames through the combustion mixtures, which lengthens the burn time and reduces peak cylinder pressures and temperatures. Too much EGR can excessively cool combustion temperatures, resulting in an incomplete burn and inadequate burn time. NO_x emissions can continue to drop, but other emissions will increase, so engine efficiency dramatically drops (**FIGURE 17-6** and **FIGURE 17-7**).

It is important to note, though, that moderately increased boost pressure often accompanies increased EGR flow on many EGR engines. Higher turbocharger pressure is necessary to provide an adequate amount of oxygen to burn fuel and helps maintain power output comparable to non-EGR engines. More gas mass (air + EGR) ends up in the cylinder, resulting in higher compression pressures but lower peak combustion pressure and temperature (**FIGURE 17-8**). Reduced combustion gas expansion is observed by lower turbocharger speeds when EGR is switched on. Under identical operating conditions, lower turbine speeds are observed when EGR is enabled than when exhaust gas is not recirculated (**FIGURE 17-9**). Higher injection pressures are also necessary to create finer atomization for the combustion event to tolerate EGR gas without producing even more emissions.

In addition to higher turbocharger pressure, several operating conditions first need to be met before NO_x emissions can be reduced.

Conditions for Enabling Exhaust Gas Recirculation Delivery

Because the greatest amount of NO_x is formed when cylinder pressures and temperatures are high, EGR systems, at a minimum, monitor engine temperature, speed, and load conditions to meter the correct amount of exhaust gas into the engine intake.

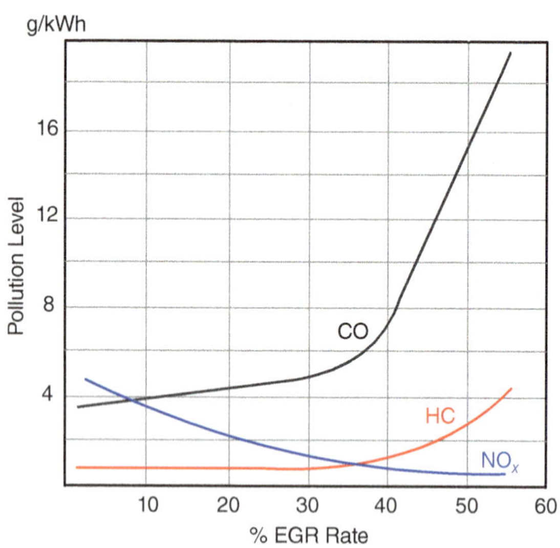

FIGURE 17-6 Increasing EGR rates beyond 35% results in exponential increases to other emissions and a dramatic drop in engine efficiency.

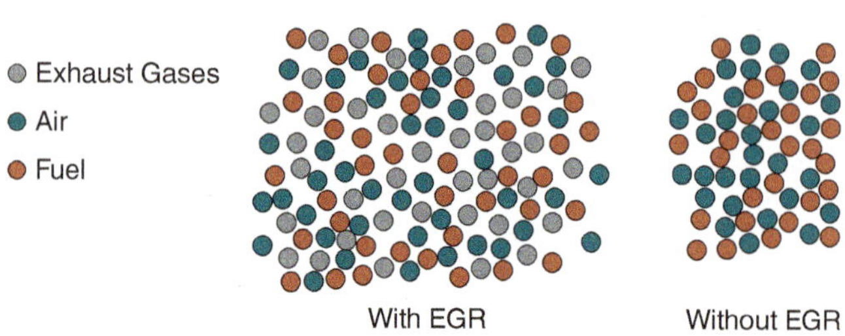

FIGURE 17-7 EGR spaces fuel and air molecules farther apart to slow down combustion chain reactions.

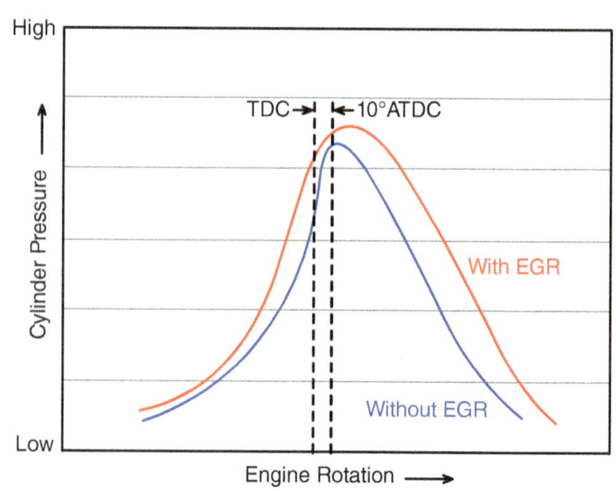

FIGURE 17-8 Adding EGR to the cylinders with additional air increases compression pressures. Burn time is slower and takes place over a larger number of degrees of crankshaft rotation.

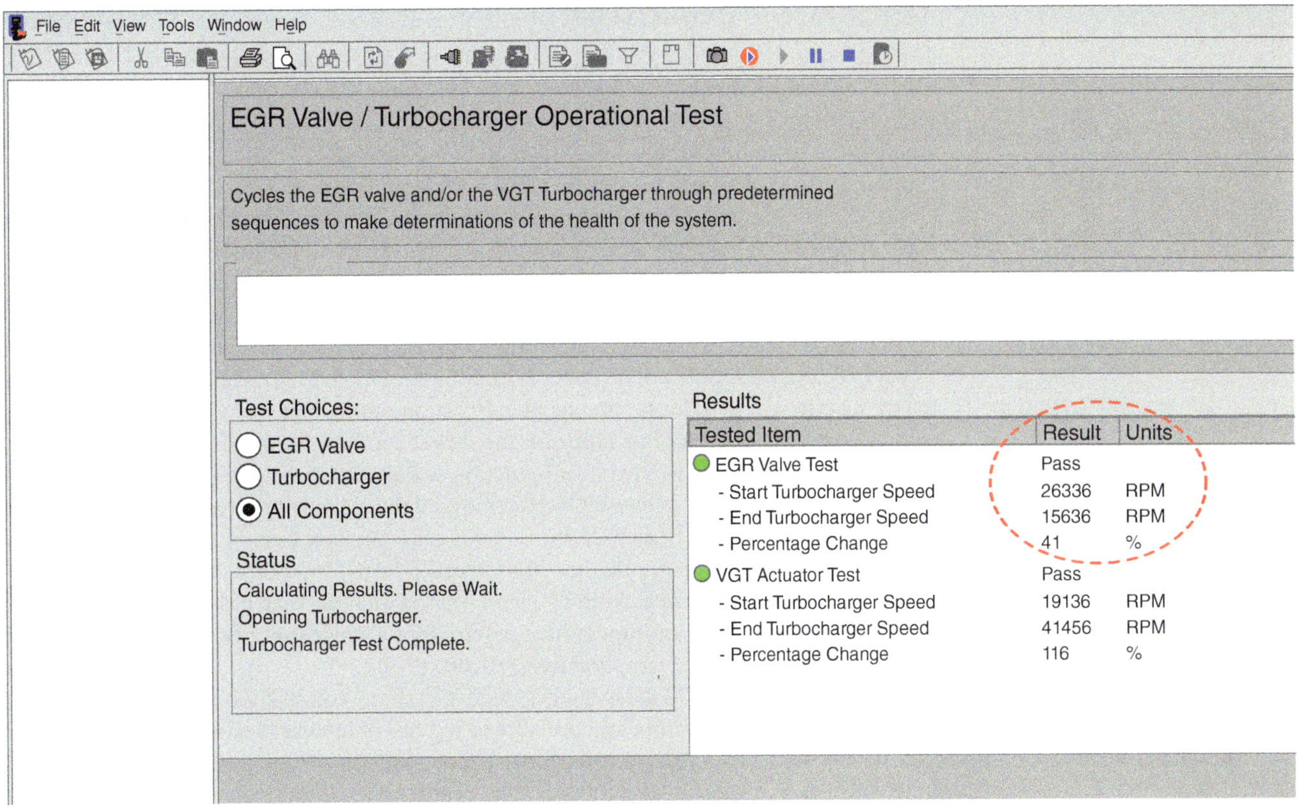

FIGURE 17-9 Note that the turbocharger speed drops when the EGR valve is opened. The onboard diagnostics (OBD) system will test EGR operation by briefly opening the EGR valve and comparing the drop in turbocharger speed with expected values.

During cold weather or while the engine is idling, minimal or no EGR flow usually takes place. During warm-up, condensation of corrosive exhaust can take place in cold EGR coolers, so flow is often minimized or disabled during that period. Rather than shut off EGR flow during prolonged cold idle, a number of manufacturers recirculate EGR into the intake air but allow the gas to bypass the cooler. Mixing the warmer exhaust gas increases the intake air temperature, which can in fact reduce emissions and speed up engine warm-up. When excessively high coolant temperatures are present, the EGR flow may also be disabled because using engine coolant to cool exhaust gas in EGR coolers can contribute to overheated engine operation. High altitude or low barometric pressures are also conditions that may disable EGR operation because EGR flow under these conditions could significantly deteriorate combustion quality and engine performance. Disabling EGR should not happen as a matter of course, though. Engine manufacturers must demonstrate that emission systems have been designed well and use good judgment when making a decision to deactivate or reduce an emission system's effectiveness.

Pressure Reduction

As already mentioned, NO_x emission formation is minimized by lowering peak cylinder pressure. Higher cylinder pressure increases the likelihood of chemical reactions by tightly pushing the molecules closer together. Cylinder pressure that increases the speed of chemical reactions is distinctly higher in diesel engines due to increased compression ratios, turbocharging, fuel rates, and the nature of compression-ignition combustion. However, exhaust gas is relatively inert, having little usable oxygen and high amounts of carbon dioxide (CO_2) and nitrogen. Because the exhaust gas cannot directly enter into the combustion process, it slows down the combustion speed by spacing oxygen and fuel molecules farther apart. Chain reactions between oxygen and fuel take longer, which results in a slower pressure rise in the cylinder. The sharp spike in cylinder pressure and accompanying engine noise, a characteristic of older non-EGR engines, diminishes in EGR engines. A byproduct of lower peak cylinder pressure is quieter engine operation.

▶ TECHNICIAN TIP

Using an auxiliary emission-control device (AECD), which reduces the effectiveness of an emission-control system or even shuts it off, is permissible under some conditions. For example, on emergency vehicles when the emission system operation could reduce engine power or prevent the vehicle from performing a crucial duty, emission system operation may be disabled or changed. If the emission system operation could result in engine damage, such as overheating or damaged catalysts, they can be shut off. Consideration for passenger safety is another exemption. No emission system operation should allow an engine to suddenly stall in traffic without warning since the loss of power assisted steering and braking could jeopardize occupant safety or the safety of other motorists. If a manufacturer does use an AECD, it must declare the AECD's presence when obtaining a certificate of conformity to emission legislation. Furthermore, the use of an AECD must be approved during certification before it is allowed to become a calibration used in vehicle production.

Temperature Reduction

The amount of NO_x forming during peak cylinder temperature is also reduced by using EGR. The flame temperature of close to 4,000°F (2,204°C) for diesel fuel is more than the 2,500°F (1,371°C) temperature threshold required to begin NO_x formation. While there are regions in the cylinder where temperature may spike only momentarily to this level, reducing the peak combustion temperature to below 2,500°F (1,371°C) minimizes NO_x formation. Above 2,500°F (1,371°C), NO_x *mass* formation accelerates exponentially: twice as much NO_x is produced for every 100°F (37.8°C) temperature increase above this threshold temperature.

Using EGR reduces temperature in two ways. One way is through its ability to slow down combustion speed, which causes not only a pressure drop but also a temperature drop. Combustion temperatures are not as intense or as hot, because the air-fuel mixture will burn more slowly over a larger number of degrees of crankshaft rotation. This means the push against the top of the piston is both longer and gentler. Injection timing may be advanced in many EGR engines to permit adequate burn time for the slower combustion speeds.

The second mechanism for heat reduction is the heat-absorbing properties of exhaust gas in the cylinder. Most cooled EGR systems cool the exhaust gas from as high as 1,000°F (538°C) to just over 400°F (204°C). While EGR gas is warmer than intake air, CO_2 (a major constituent of exhaust gas) also has high specific heat properties. This means CO_2 can absorb larger amounts of heat without a temperature increase than gases with lower specific heat. **Specific heat** is the amount of heat required for a single unit of mass of a substance to be raised by one degree of temperature. (**FIGURE 17-10**).

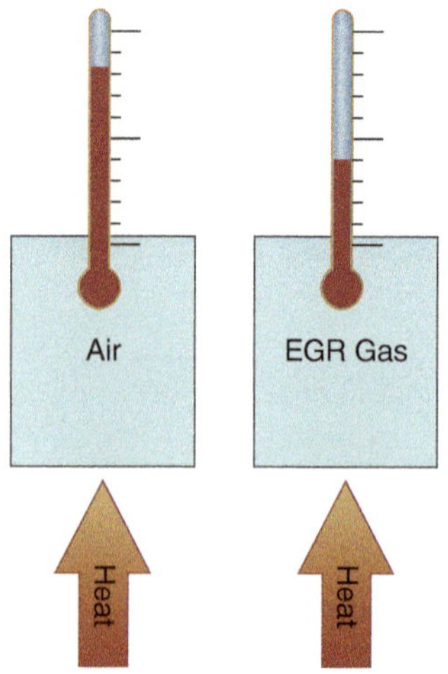

FIGURE 17-10 EGR gas, which has high CO_2 content, does not increase its temperature as much as air does when it absorbs the same amount of heat.

EGR is especially effective in diesel combustion because it can reduce temperature and pressure in regions of intense localized heat, where most NO_x formation occurs. These NO_x-forming regions have low air-fuel ratios. In other words, the areas tend to be fuel rich and air lean, which is where EGR is especially effective (**FIGURE 17-11**).

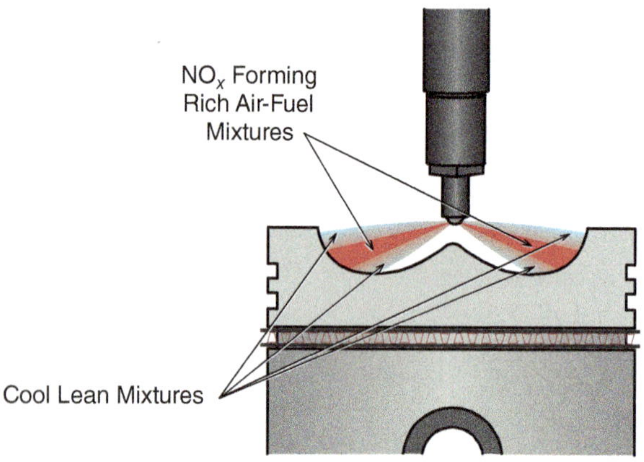

FIGURE 17-11 Air-rich regions of the combustion chamber outside the main spray plume are cool and do not form as much NO_x as the hotter richer regions of the injector spray.

Stroked Engines

To capture more of the energy and optimize the release of EGR-diluted combustion, the stroke length of many EGR engines is longer than non-EGR models. The slower burn time and heat release in EGR engines means a longer stroke is necessary to efficiently convert combustion energy into mechanical energy since the burn takes place over more degrees of crankshaft rotation. An example of this practice is the 5.9 L Cummins engine family that had its stroke lengthened to give it 6.7 L of displacement. Stroked engines have their crank throws lengthened, which also creates an increased piston swept volume. With a larger swept volume, total cylinder displacement increases slightly, and the ratio between cylinder volume and clearance volume changes the compression ratio (**FIGURE 17-12**).

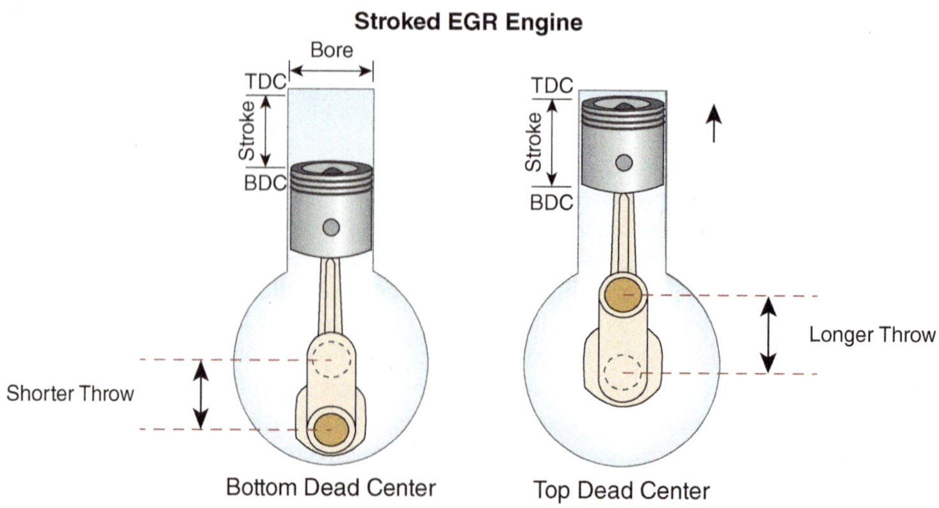

FIGURE 17-12 EGR engines are stroked, meaning the stroke length is increased. Stroking compensates for power loss and longer burn time associated with EGR use.

Drawbacks of Using Exhaust Gas Recirculation

Although EGR is one of the most effective in-cylinder solutions for reducing NO_x emissions, using it comes with a few drawbacks:

1. EGR does reduce the potential peak cylinder pressures produced by the air-fuel charge compared to non-EGR condtions. This means engine torque and power output will fall unless other engine operating strategies and design details can compensate for EGR use. Fuel economy and power penalties are supposed to be negligible under most operating conditions using EGR. However, users comparing pre- and post-EGR engines are reporting a 1–10% or more, fuel economy penalty. It is important to note that throttle response is improved and combustion noise is reduced because associated technologies such as faster-acting variable-geometry turbochargers are used with many cooled, high-pressure EGR systems.
2. EGR is an excellent method for reducing emissions, but it contaminates the engine lubricating oil more than non-EGR engines. New lubricating oil specifications are developed for EGR engines to handle the extra contamination of lubrication oil from recycling exhaust gas back into the cylinder (**FIGURE 17-13**). Exhaust contaminants such as particulate will stick to lubricating oil on the cylinder wall and are then scraped into the crankcase, creating higher soot loading in the oil. Cylinder wear increases because soot is an abrasive. Larger oil pan capacities are needed to maintain the same oil change intervals. An uneven distribution of EGR gas between cylinders can wear out one or more cylinders faster. Soot passing into the cylinders around intake valves causes premature wear in intake valves and seats as well.

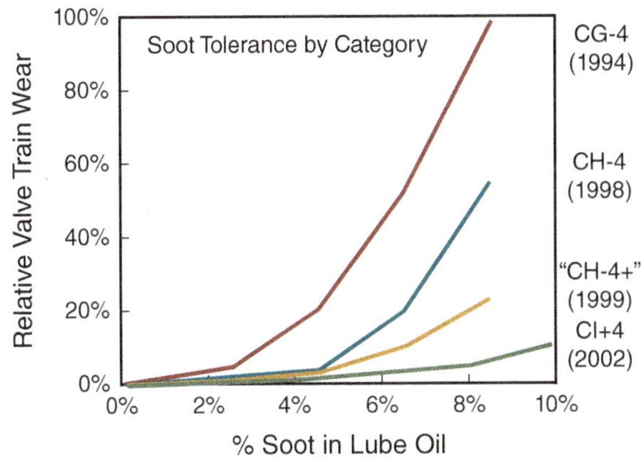

FIGURE 17-13 Cx+4 oil is specially developed for EGR diesels, to minimize abrasive wear caused by soot loading in engine oil.

3. Recirculating more exhaust gases back into the cylinder also leads to increased water condensation and the formation of higher sulfuric and nitric acid levels in the combustion chamber when reactions take place between water and combustion gases. If not properly neutralized, acids in the oil can corrode any exposed metal, especially the soft lead overlay of bearings exposing the copper beneath. To minimize these problems, higher levels of detergents are added to the oil to keep soot in suspension and to neutralize acids. The American Petroleum Institute (API) designated the first EGR classification of engine oil in 2002 as CI+4. The +4 designation indicates higher levels of detergents and other acid neutralizing additives.
4. Additional wear of cylinder components, such as piston rings and valve seats, is also reported with EGR. Until durability standards were introduced in 2004, some early EGR engines experienced as much as a 20–30% decrease in expected engine life in severe service applications. However, the use of harder ring and cylinder wall materials, hardened valve seats, and other advanced materials have corrected many of these earlier problems. Break-in periods for engines are now as long as 12,000 miles (20,000 km), which is significantly longer than the just a few thousand miles or kilometers for pre-EGR engines.
5. **Cooled exhaust gas recirculation (CEGR)**, where exhaust gas passes through an engine coolant heat exchanger, adds additional heat load to the cooling system and engine compartment. Larger radiators, water pumps, and cooling systems are required to transfer the heat absorbed by the coolant to the atmosphere. Underhood temperature is much higher with EGR engines due to increased heat rejection by the cooling system. Heat rejection refers to the amount of heat transferred to the coolant rather than converted into mechanical force driving down the pistons.
6. EGR systems require increased maintenance. Until EGR technology matured, many EGR engines experienced premature and repeated failures of EGR valves and EGR coolers. Valves often gummed up with slobber from low speed and load engine operation. This is especially problematic in engines that had above average idle time and significant stop-and-go urban drive cycle operation. EGR coolers are often cracked and leaking due to frequent thermal and pressure cycling. Coolers would often plug-up with soot

and slobber. If a **variable-geometry turbocharger (VGT)** actuator mechanism becomes stuck or resistive in the closed position, the expansion–contraction cycles from increased boost pressure has a tendency to crack the cooler as well. Newer cooler EGR designs are more rugged and allow the internal heat exchanger to float or move independently of the external cooler case.

Limitations of Other Strategies for Reducing NO_x

EGR is an emission-control technology with several drawbacks. However, given the progressive reductions required for NO_x emissions by emission legislation and the fact that other NO_x reduction technologies and operating strategies have reached their limits, EGR is necessary on almost all diesel engines. For example, two common NO_x-reducing strategies already in use on pre-EGR engines are retarding injection timing and cooling intake charge air. Charge air coolers (CACs) reduce intake air temperature as much as possible. In addition, injection timing in many pre-EGR engines is already retarded because most injection events takes place near or after TDC. Retarding injection timing any further would reduce fuel economy, diminish performance, and cause excessive particulate matter (PM) formation. Unacceptable levels of soot would contaminate engine oil as well. However, using EGR and only slightly retarded injection timing reduces NO_x formation while allowing for a nominal degree of advanced injection timing and has better outcomes than engines that meet emission standards without EGR. Again, in the non-EGR, emission-compliant engines, injection timing would need to be retarded even more, to the detriment of engine efficiency and performance. Advanced injection timing in turn produces better throttle response, improved fuel economy, and reduced PM formation, accompanied by less soot contaminating the lubrication oil (**FIGURE 17-14**).

▶ Types of Exhaust Gas Recirculation

17-02 Describe the functions, construction, and applications of diesel engine EGR systems.

EGR rates are much higher in diesel engines than in spark-ignition, gasoline-fueled engines, which may use an EGR rate that rarely exceeds 12%. In contrast, diesel engines are capable of handling rates as high as 35–40% by weight before a significant deterioration in combustion efficiency would affect fuel economy and other emissions. Flow rates by intake

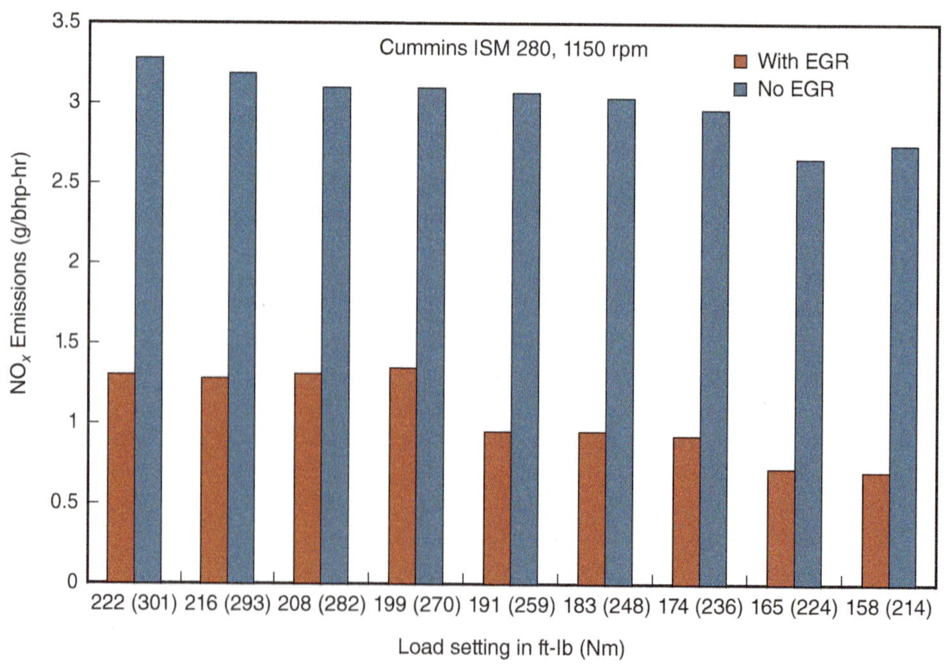

FIGURE 17-14 A comparison of NO_x production with and without EGR. Note that as engine load increases, more NO_x is produced.

TABLE 17-1 EGR Rates for Diesel Engines

Light EGR	10–20%	Older light duty
Heavy EGR	20–35%	2002–2010 light duty, medium duty, and heavy duty
Moderate EGR	15–25%	2010+ SCR
Massive EGR	35–50% or more	2010+ non-SCR

mass for the 2007 model year low-NO_x engines are in the vicinity of 20–35%, whereas earlier EGR engines with less aggressive emission reductions have lower EGR rates (10–20%). Since the introduction of liquid selective catalyst reduction (SCR-L) systems, which break down NO_x in the exhaust system, manufacturers have depended less on in-cylinder NO_x reduction strategies. Reduced EGR rates and a return to more advanced injection timing on SCR-equipped engines have improved fuel economy (**TABLE 17-1**). Compared to manufacturers who held out until 2010 before installing SCR-U systems, those that did use NO_x adsorbers could also increase their engines' efficiency.

There are several types of EGR delivery systems that introduce exhaust gas uniformly into the cylinders. The following are some of the types of EGR delivery systems used on light- and medium-duty diesels:

- high-pressure cooled type
- low-pressure cooled type
- combined high- and low-pressure type.

The following are some of the less frequently used EGR systems on a few light-duty and off-road engines:

- internal EGR
- external non-cooled
- external cooled.

Internal Exhaust Gas Recirculation

Some engine manufacturers will use a variety of internal types of EGR systems, though this is not common. Internal types do not use external EGR controls, but exhaust gas is reintroduced into the cylinders through a change to exhaust valve timing. By leaving the exhaust valve open longer during the intake stroke, exhaust gas is drawn back into the cylinder during intake stroke (**FIGURE 17-15**). Another method is to add a "bump" to the exhaust cam lobe that briefly cracks open the exhaust valve during the intake stroke to draw exhaust gas into the cylinder along with the intake air charge. These internal methods, however, result in poor mixing with the intake air and excessive EGR rates at low speeds. With the eventual introduction of variable-valve timing, these problems could be minimized.

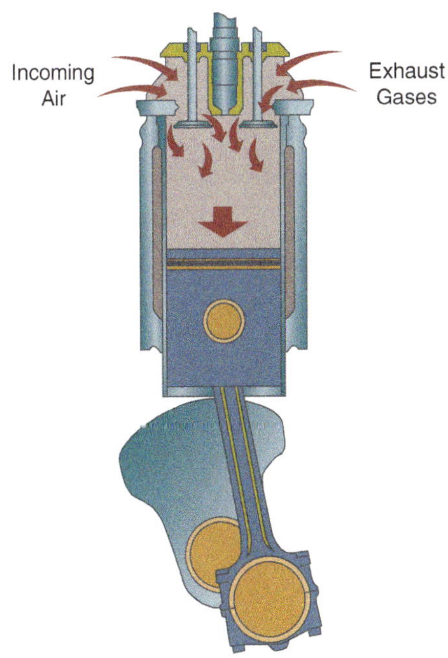

FIGURE 17-15 Internal EGR briefly opens the exhaust valve during the intake stroke.

External Exhaust Gas Recirculation Non-Cooled

External EGR uses a metering valve and control circuits to deliver exhaust gas into the intake manifold. The GM 6.5 L diesel engine is an example of this system. The engine uses a turbocharger with a vacuum-operated EGR valve located in the intake plenum. A passageway in the intake manifold routes exhaust gas from the cylinder head to the EGR valve. Exhaust gas is driven into the intake manifold through the exhaust backpressure in the exhaust system. The **exhaust backpressure (EBP) sensor** measures pressure in the exhaust manifold. Even on turbocharged engines, exhaust backpressure in the exhaust manifold is always slightly higher than intake boost pressure. The EGR valve pedestal location on the intake manifold and the use of a venturi in the intake are part of a design to create a low-pressure area by using intake airflow. At this point, exhaust gas is drawn into the intake manifold through the EGR valve at the low-pressure venturi (**FIGURE 17-16**).

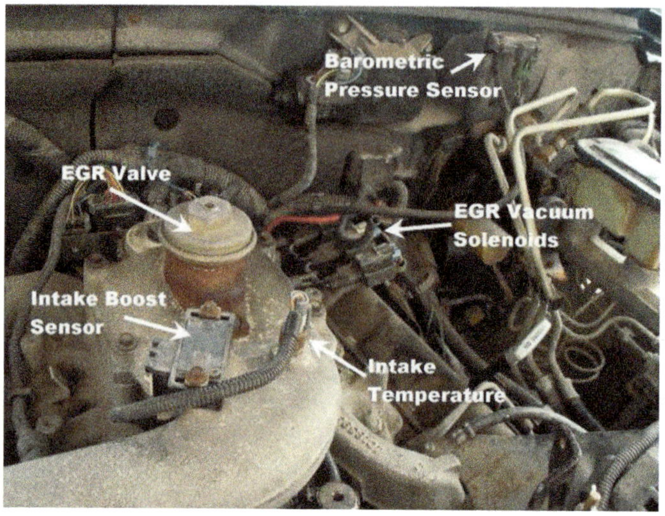

FIGURE 17-16 GM used a non-cooled EGR system on its 6.5 L engines in 1500 series trucks and vans. Heavier vehicles did not require the EGR system, due to more lenient emission standards.

High-Pressure Cooled

The technological evolution to more elaborate EGR systems found on current diesel engines has taken place for a number of reasons. Primarily, with more stringent legislative requirements for emission reductions from diesel engines, manufacturers have had to improve on precise control of mixture formation to achieve emission reductions. Fuel rates, air, and exhaust gas mass must be carefully measured and balanced for efficient, powerful, low-emission combustion. For example, too high an EGR rate will starve the combustion process of adequate oxygen, driving up production of carbon monoxide (CO), hydrocarbon (HC), and PM emissions. However, too low an EGR rate will not obtain the required NO_x reductions. A number of additional sensor and actuator circuits are required to accurately meter EGR rate in proportion to other combustion elements.

In addition to metering controls, the necessity of meeting legislated requirements for the On-board Diagnostic II (OBD-II) system to monitor emission systems operation also increases the complexity of the EGR system. To validate the correct operation of the EGR system, the OBD-II system must ensure that faults that would prevent the EGR system from functioning properly are detected. Fault detection is meant to identify an EGR system that may not be operating in order ensure that it achieves its intended emission reductions. Many NO_x sensor codes can be traced back to improperly functioning EGR systems or intake air restrictions upsetting the delicate balance between EGR and air mass entering the cylinders.

Cooled EGR gas redirected from the exhaust manifold and pressurized to overcome intake boost pressure is called **high-pressure cooled EGR (HP-CEGR)** (FIGURE 17-17). The following are some of the typical components of cooled EGR systems:

- VGT
- EGR cooler
- EGR mixer
- EGR valve
- exhaust transfer tubing
- intake mixing device
- input sensors
- output devices.

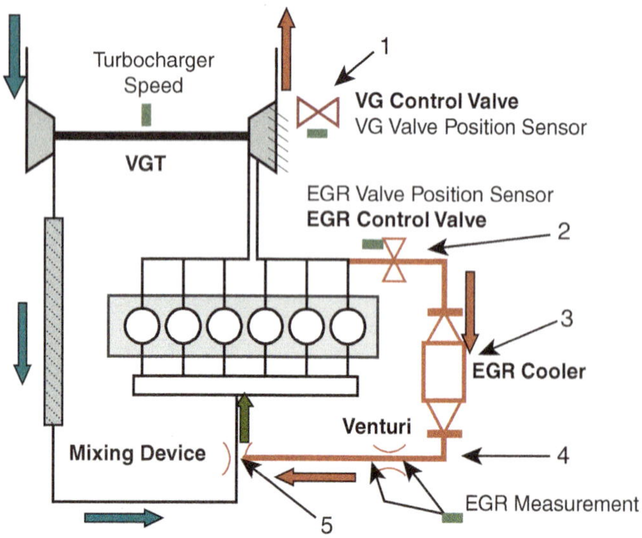

FIGURE 17-17 An HP-CEGR system generally does not use a throttle plate to draw exhaust into the engine. Instead, a VGT pressurizes exhaust gases high enough to push exhaust into the boosted air intake.

Turbocharging and High-Pressure Cooled Exhaust Gas Recirculation Systems

Using EGR creates several problems for intake air handling. First, the addition of exhaust gas will reduce power output. Second, EGR

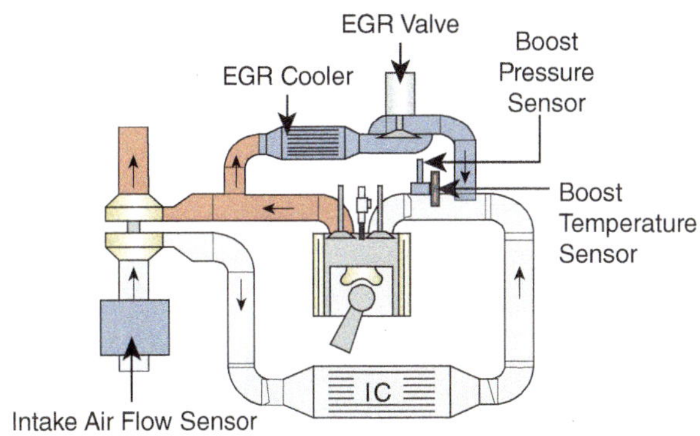

FIGURE 17-18 A cooled EGR uses a VGT to increase exhaust backpressure so that high volumes of exhaust gas can mix with a boosted air intake.

will displace some combustion oxygen. To compensate for these problems, newer-model EGR engines increase the total gas volume in a cylinder while adding inert exhaust gas. As total gas volume in the combustion chamber increases, cooler but higher average pressure is produced in the cylinder to help compensate for the horsepower loss. While an increase in gas volume and average cylinder pressure occurs, NO_x formation does not, because the expansion of burning fuel takes place at a lower temperature and over more degrees of crankshaft rotation than in non-EGR engines.

Another compensation strategy is to stroke engines. EGR engines are typically stroked, meaning an increased stroke length is used with longer crank throws. More leverage on the crank due to a lengthened throw and the longer distance traveled by the piston converts more combustion energy into mechanical force. The downside is that piston speeds during mid stroke are faster with greater side thrust against the cylinder walls. Increased wear takes place, but this method is preferred over leaving engine dimensions unchanged. Changing cylinder dimensions, combined with a gentler increase in combustion pressure, means peak engine torque is often produced at 100–200 rpm higher than the same non-EGR models.

Another problem with turbocharged EGR engines is supplying the necessary volume of exhaust gas to the pressurized intake manifold. Exhaust gas pressure needs to be higher than intake pressure to achieve adequate exhaust flow into the cylinders. While low-pressure EGR systems incorporate an electronically controlled throttle plate to produce a pressure drop in the intake manifold and draw exhaust gas into the intake airflow, the majority of manufacturers have relied on VGTs to pressurize exhaust gas (**FIGURE 17-18**).

Unlike conventional turbocharger technology, a variable-geometry turbocharger (VGT), also known as variable-nozzle turbocharger (VNT) or variable-vane turbocharger (VVT), can vary boost pressure and exhaust backpressure independent of engine speed. (See Chapter 16.)

Low-Pressure Exhaust Gas Recirculation Systems

An alternate strategy to HP-CEGR is to draw EGR gas into the cylinders through the turbocharger compressor inlet. Referred to as **low-pressure EGR**, NO_x emissions are reduced by 5–64% when using low-pressure EGR. These systems are considered to be about 15% more thermally efficient than HP-CEGR because exhaust backpressure is lower (**FIGURE 17-19** and **FIGURE 17-20**). Some light-duty systems, such as the one in the Volkswagen (VW) 2.0 L TDI, combine low- and high-pressure EGR to increase efficiency.

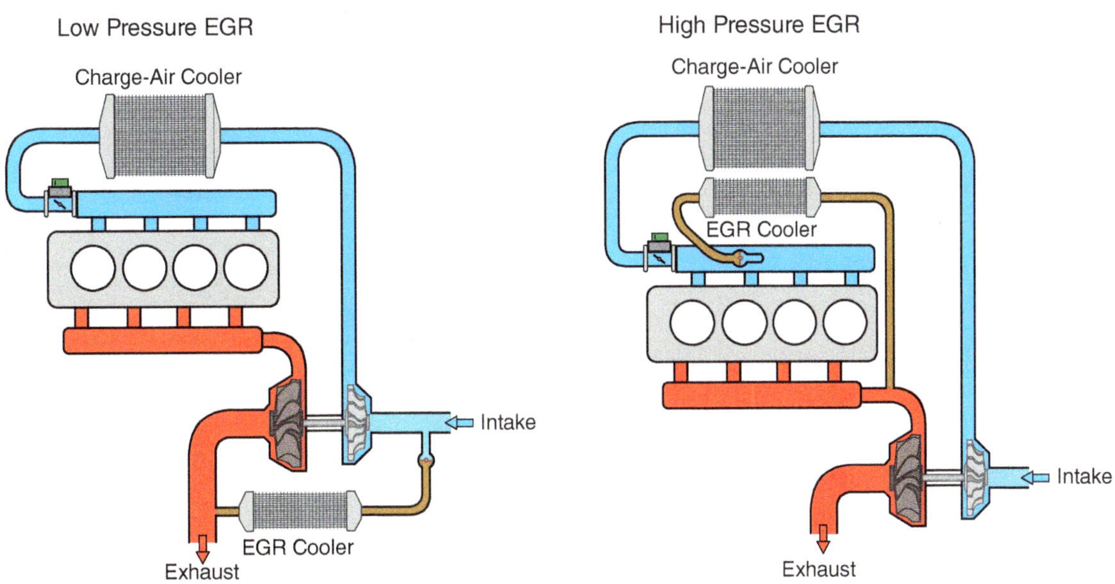

FIGURE 17-19 Comparing high- and low-pressure EGR systems. Low-pressure systems draw EGR gases into the engine through the turbocharger compressor wheel. High-pressure systems push pressurized exhaust gases into the intake manifold.

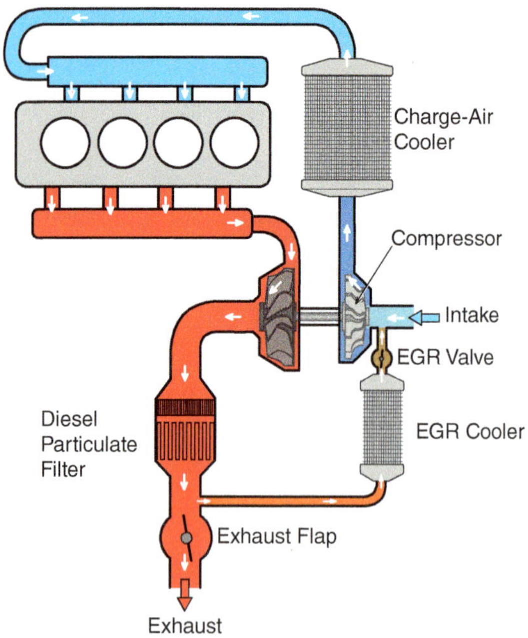

FIGURE 17-20 Low-pressure EGR systems pull exhaust gas into the engine through the compressor wheel inlet.

Advantages and Disadvantages of Using Low-Pressure Exhaust Gas Recirculation

The following are some of the advantages of using low-pressure EGR:

- improved mixing of EGR and intake air—especially at low speeds and during acceleration
- cleaner intake charge when drawing exhaust gas after the particulate filter for less soot loading in oil and associated wear from soot injection
- improved reduction in NO_x emissions when combined with high-pressure EGR
- increased engine thermal efficiency due to lower exhaust backpressure
- increased availability of exhaust mass used to drive the turbocharger.

The following are some of the disadvantages of using low-pressure EGR:

- potential for increased formation of corrosive combustion gas condensates in the EGR cooler and CAC
- potential for damage to the turbocharger compressor wheel from higher heat loads and damaging exhaust particles if exhaust is drawn into the turbocharger inlet.

High- and Low-Pressure Exhaust Gas Recirculation System

Volkswagen 2.0 L TDI engines (first- and second-generation EA189 and 288 engines) are examples of engines combining a high-pressure and a low-pressure EGR system (**FIGURE 17-21** and **FIGURE 17-22**). The early EA189 TDI used from 2009 to 2014 had a non-cooled, high-pressure EGR system and a cooled, low-pressure EGR system. Both high-pressure and low-pressure

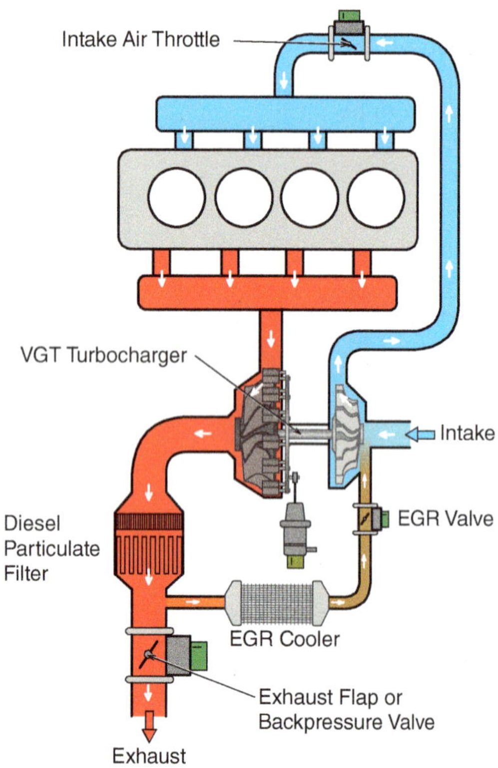

FIGURE 17-21 A functional schematic of the low-pressure EGR system of a VW EA288 second-generation TDI. An exhaust flap or backpressure valve creates the pressure needed to drive EGR gas into the turbocharger compressor wheel.

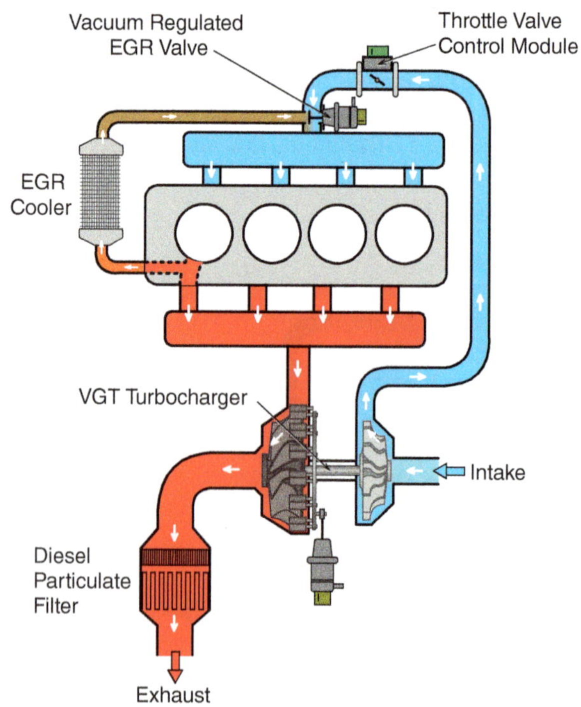

FIGURE 17-22 A functional schematic of the high-pressure EGR system of a VW EA288 second-generation TDI. The VGT creates the pressure needed to drive EGR gas into the intake manifold.

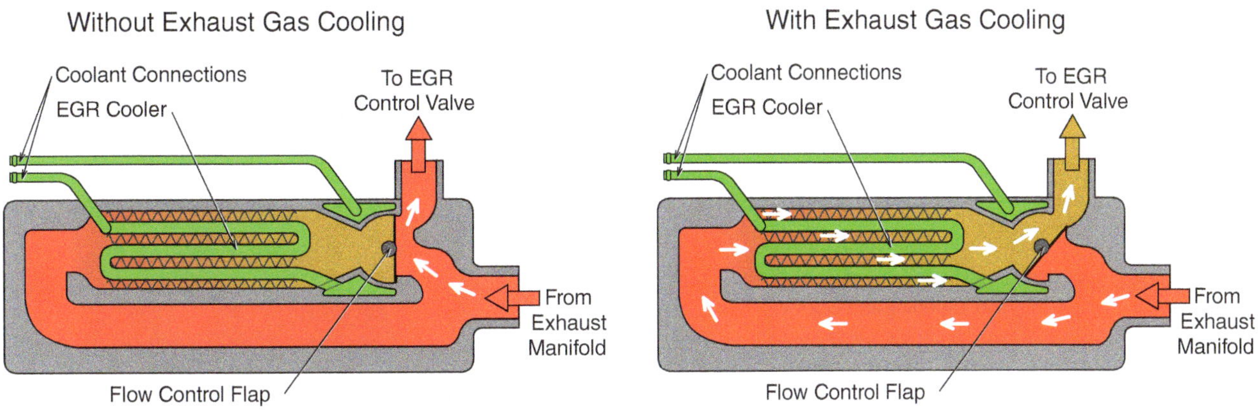

FIGURE 17-23 Functional operation of an EGR bypass valve. Exhaust gas does not pass through the cooler in bypass mode. A flapper valve diverts exhaust directly to the intake system.

systems blend EGR flow into the cylinders, depending on engine speed and load conditions. Non-cooled, high-pressure EGR flow dominates at low speed, low load conditions and transitions to more flow from the cooled, low-pressure EGR circuit with increasing speed and load. Non-cooled EGR gas at low speeds and loads is less likely to produce NO_x because cylinder pressures and temperatures are lower. Boost pressures and temperatures are likewise lower at slow speed, particularly during cold weather operation, when more non-cooled EGR recirculation can help increase intake temperatures. VW and most other late-model engines from other manufacturers use blends of cooled and non-cooled EGR gas to regulate pre-ignition temperatures. The purpose of adding more non-cooled EGR gas to the intake is to raise engine and exhaust temperatures when ambient operating temperatures are low. Exhaust catalysts will warm up faster when using this method. The later, second-generation EA288 used both cooled, high-pressure and cooled, low-pressure EGR systems. An EGR cooler will divert exhaust gas flow directly through to the EGR valve or through the cooler depending on whether or not, the purpose is to speed up engine and exhaust catalyst warm-up in the later EA288 (**FIGURE 17-23**). This same strategy is used by other manufacturers such as the Ford 6.7 L, which will incorporate an EGR cooler bypass valve (**FIGURE 17-24**). Using a bypass valve has another benefit. Under light-load and cold-engine operation when EGR flow is needed, bypassing the EGR cooler prevents the cooler from being fouled by or plugged with oily soot combustion byproducts.

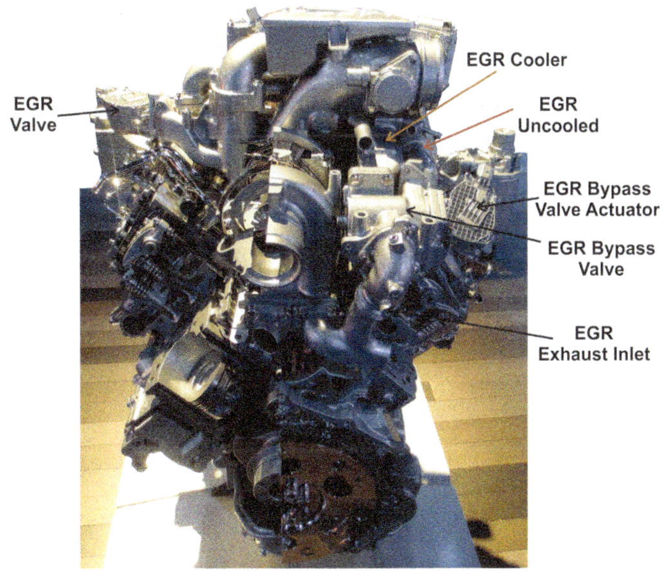

FIGURE 17-24 The EGR bypass valve on this late-model V8 Duramax diverts exhaust gas to either the EGR cooler or a pipe connected directly to the EGR valve.

High-pressure EGR flow in the first-generation VW TDI 2.0 L begins as exhaust gas is redirected from the exhaust manifold before the turbocharger and recirculated into the intake air stream before the intake manifold. Narrowing the turbine nozzle opening will build exhaust backpressure before the turbine to help drive exhaust gas into the pressurized intake manifold. An electrically operated high-pressure EGR valve with a single poppet valve opens and closes a passage between an exhaust tube connected to the EGR valve and the intake manifold inlet. The quantity of EGR recirculated depends on the pressure differential between the exhaust pressure in the exhaust manifold and the intake air inlet. Because an air intake throttle is located upstream or before the EGR valve, the throttle plate can open and close to change the pressure differential even further (**FIGURE 17-25**). Closing the throttle plate will pull more EGR gas into the intake manifold through the high-pressure EGR valve and less fresh air. Opening the throttle has the opposite effect.

As the high-pressure EGR flow increases, the turbocharger boost pressure drops because less exhaust gas actually flows through the turbocharger. A second reason for lower

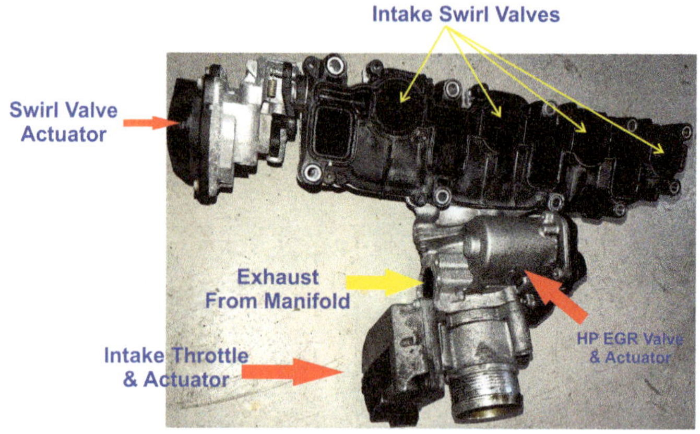

FIGURE 17-25 The high-pressure EGR valve on first-generation TDIs is located upstream from the intake air valve. Opening and closing the intake air valve can change the pressure differential in the intake manifold, which itself can also change EGR flow.

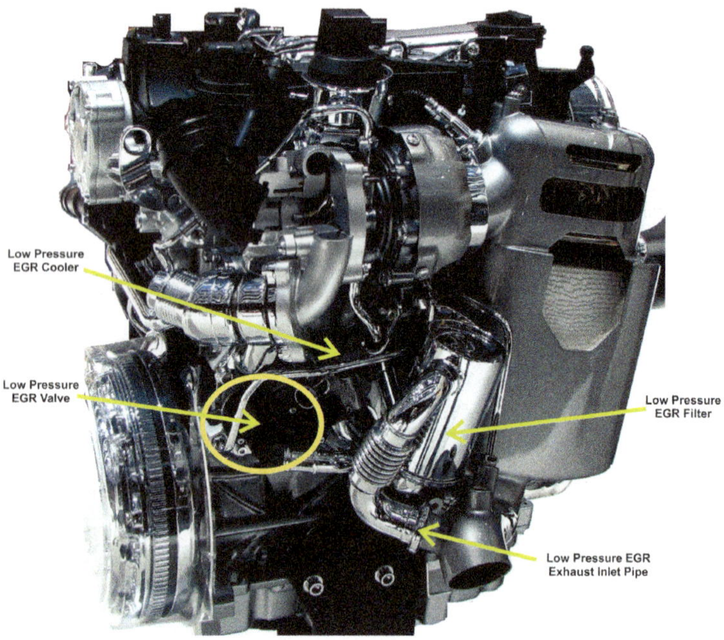

FIGURE 17-26 The low-pressure EGR system uses a filter to prevent any escaping particulate emissions from entering the compressor wheel.

boost pressure is that increased EGR mass flow limits the expansion of combustion gas, which in turn reduces the expansion coefficient of the exhaust gas inside the turbine housing, further reducing the turbocharger speed. As a result, the turbocharger response can potentially be slower and thus lower maximum boost pressure is available. Of course, the use of a VGT turbocharger is essential to compensate for these conditions.

As engine load and speed increase, it's more efficient to use the cooled, low-pressure EGR gas flow, because the mass of EGR flow to minimize NO_x formation must increase at mid to high engine load. The engine does not use a VGT to develop exhaust backpressure. Instead, pressure for cooled low-pressure EGR is developed by using an exhaust backpressure valve (EBP) located in the exhaust pipe after the lean NO_x trap (LNT) catalyst. Under heavier speed and load conditions, the EBP valve begins to close, creating high exhaust backpressure. A pipe connected to the exhaust system at the engine but after the diesel particulate filter (DPF) conducts pressurized exhaust gas through a small canister-like filter to scrub the gas of any soot that may have leaked through the DPF (**FIGURE 17-26**). From there, exhaust gas passes through a liquid EGR cooler and then a low-pressure EGR valve located at the cooler outlet. Because the gas is free of abrasive particulates, only clean gas is piped directly into the turbocharger compressor inlet, and the clean gas then passes into the engine. Cooler, cleaner EGR flow will increase engine durability in comparison to hot, unfiltered high-pressure EGR gas.

▶ Components of Exhaust Gas Recirculation

17-03 Describe and explain methods for performing inspection and diagnostic procedures on diesel engine EGR systems.

Operation of the EGR system depends on several key components to cool and control the flow of exhaust gases. Large variations in designs and system configurations used by various manufacturers are examined in this next section.

Exhaust Gas Recirculation Coolers

Engine manufacturers cool EGR gas before recirculating it into the combustion chambers. Exhaust gas is sent through a metal tubes of a gas to a liquid heat exchanger that circulates engine coolant, where its temperature can be reduced from as much as 1,000°F (538°C) to approximately 425°F (218°C). Cooling the exhaust lowers the combustion chamber temperature and allows exhaust gas to absorb more combustion chamber heat. Coolant and exhaust flow often travel in opposite directions to one another in coolers. A vent line in the top of the cooler allows vapor bubbles and steam to leave the cooler and pass to the surge tank or degas bottle. Any vapor pockets or trapped air inside a cooler can create a hot spot, possibly causing the cooler to crack (**FIGURE 17-27**).

Temperature sensors at the cooler inlet and outlet measure EGR cooler efficiency. An inadequate drop in gas temperature passing through the cooler will result in an EGR cooler fault code for low cooler efficiency.

▶ TECHNICIAN TIP

On some engines, the EGR cooler receives coolant flow after passing through an engine oil cooler. If the oil cooler coolant flow is restricted due to contaminants and sediment, the EGR cooler will experience frequent, repeated failures. Oil cooler restriction can often be checked by running an engine at high idle with no load and comparing oil with coolant temperature. An engine that has oil hotter than 15°F (8.3°C) above coolant may indicate inadequate flow of coolant through an oil cooler.

▶ TECHNICIAN TIP

Deaeration of the system is crucial to ensuring that the EGR cooler has a long life. Coolers are easily cracked by overheating; to prevent overheating, vapor bubbles must be properly vented from the housing and air must be properly bled after performing any cooling system service. Vacuum filling the cooling system is recommended to prevent any air pockets from forming during a refill procedure. Vacuum filling involves evacuating the cooling system of any air after it has been emptied and service has been completed. Typically, a purpose-made tool—a vacuum adapter—is attached to the surge tank to remove air. After the cooling system has been evacuated, one end of another hose is connected to the vacuum adapter and the other end into the supply of fresh coolant. Opening a valve on the adapter connecting the hose to the supply of coolant will allow the vacuum to fill the entire cooling system and not allow air pockets to form.

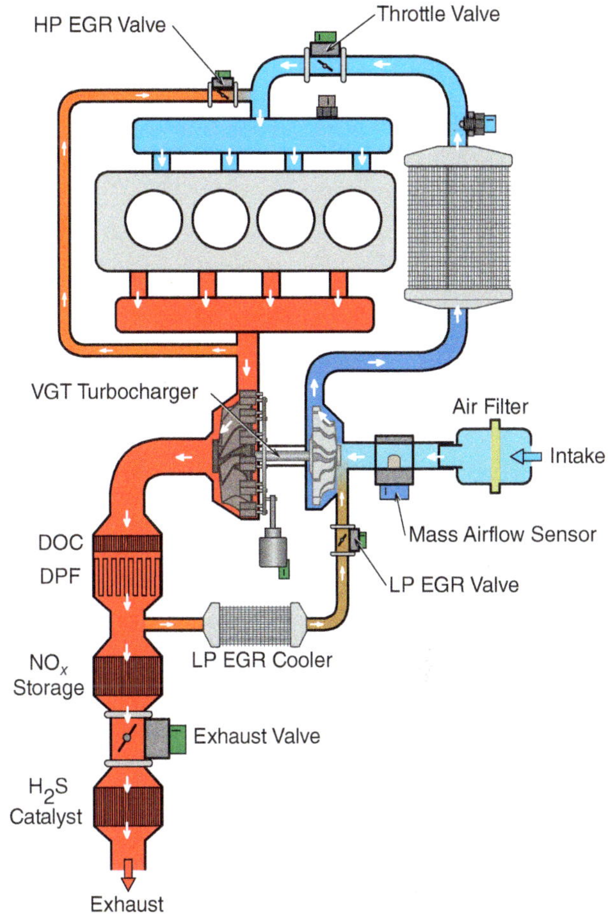

FIGURE 17-27 A schematic outline of the first-generation VW TDI EGR system.

▶ TECHNICIAN TIP

Because EGR lowers maximum cylinder temperatures, it is possible for disabled or deleted EGR systems to cause heat-related engine damage. Maximum cylinder temperatures can increase beyond a safe thermal threshold when EGR flow is removed through a delete mechanism. The increased combustion temperatures can cause premature head gasket failure as cylinder heads are overheated and thermally stressed. Valve piston and turbocharger damage can result, and unexpected exhaust valve breakage is likely the most frequent type of failure.

Dual Exhaust Gas Recirculation Coolers

Ford 6.4 L and 6.7 L Power Stroke engines are examples of engines that use two EGR coolers with a cooling capacity that can drop EGR gas temperature across the coolers by as much as 850°F (454°C). The earlier-built 6.4 L mounts the EGR coolers separately—one horizontally and the other vertically—and connects the coolers in series to produce the temperature drop. An oxidation-type catalyst between the coolers and exhaust manifolds is installed to fracture or break apart heavy HCs that could potentially foul the coolers (**FIGURE 17-28**). When using two EGR coolers, each requires its own separate pressure test to inspect for leaks.

Until 2015, the Ford 6.7 L housed two coolers in a single housing over the right-side cylinder bank (**FIGURE 17-29**). Dual engine cooling circuits, a primary and secondary, circulate coolant through the EGR housing. When EGR gas enters the cooler, it passes through one of two EGR coolers circulating only primary cooling system coolant. The primary circuit coolant is regulated to a hotter temperature than the secondary circuit. After passing through the primary EGR cooler, EGR gas passes through the second EGR cooler circulating secondary circuit coolant, which further cools the gas. Secondary coolant circuits from the right-side thermostat of the secondary radiator are regulated to 60°C (140°F) (**FIGURE 17-30**). In the 2016 model year, the separate primary and secondary cooling circuits were eliminated from Ford 6.7 L Power Stroke engines.

Exhaust gas enters into the EGR cooler only if the bypass valve of the solenoid-controlled, vacuum-operated EGR cooler is opened. At low speeds and loads, the bypass

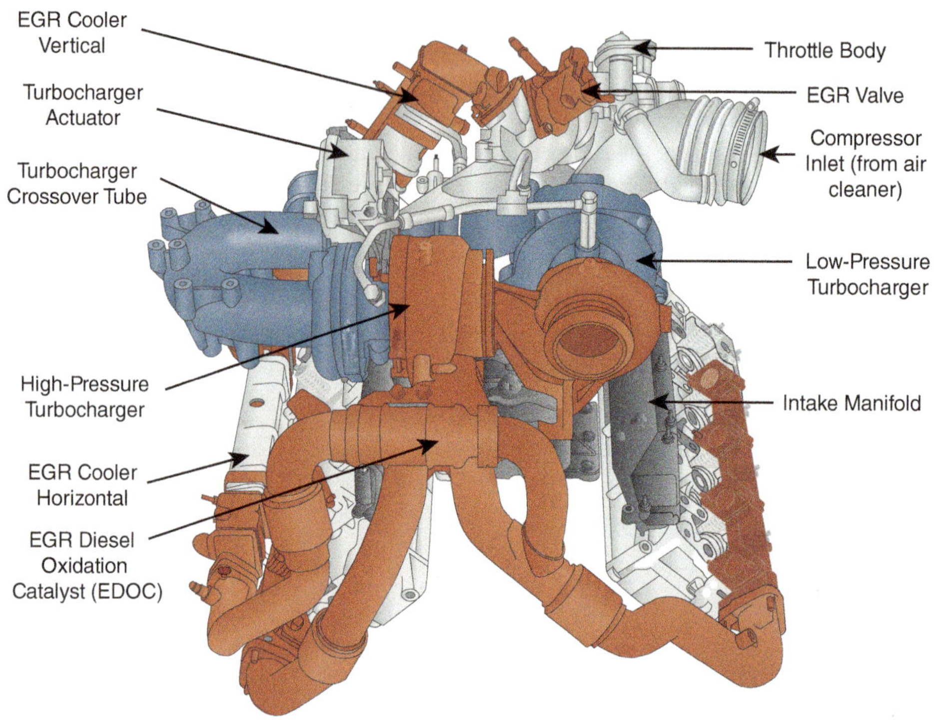

FIGURE 17-28 The Ford 6.4 L Power Stroke engine uses an oxidation catalyst to crack or break down HCs, which can plug intake manifolds and restrict EGR coolers. Note that this engine uses two EGR coolers.

FIGURE 17-29 A primary and secondary EGR cooler used by Ford 6.7 L Power Stroke engines.

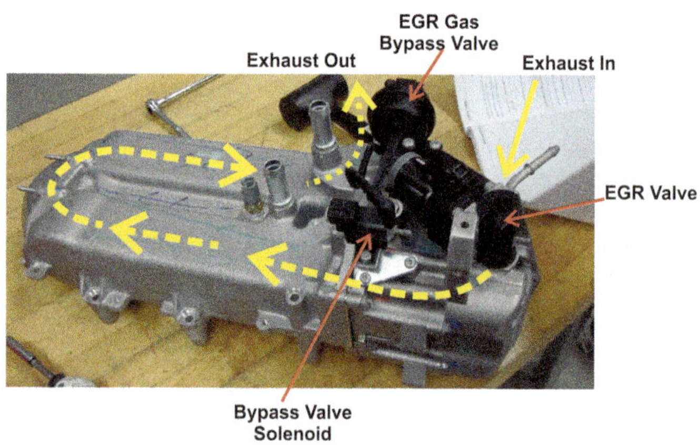

FIGURE 17-30 EGR gas passes through two coolers, a primary and secondary cooler, in this 6.7 L Power Stroke.

valve will close to redirect uncooled EGR gas into the lower intake manifold and combine with the fresh air. This strategy helps raise the intake air temperature or prevents further engine overheating under some fault conditions. When cooling is required, the bypass valve opens to direct EGR flow through the cooler.

Using EGR cooling means more heat has to be dissipated by the engine cooling system. The additional heat load on the cooling system typically requires higher-capacity water pumps and radiators. Underhood temperature is also increased when using EGR coolers, meaning that hoses, engine compartment materials, and even hood paint have to be capable of withstanding increased temperatures (**FIGURE 17-31**).

Design requirements for EGR coolers include the ability to self-clean. To resist corrosion by exhaust gas acids, EGR coolers are constructed of stainless steel (**FIGURE 17-32**). To further minimize the potential for plugging and corrosion, EGR operating strategies often prevent EGR gas flow in cold ambient temperatures—i.e., below 40°F (4°C) and during

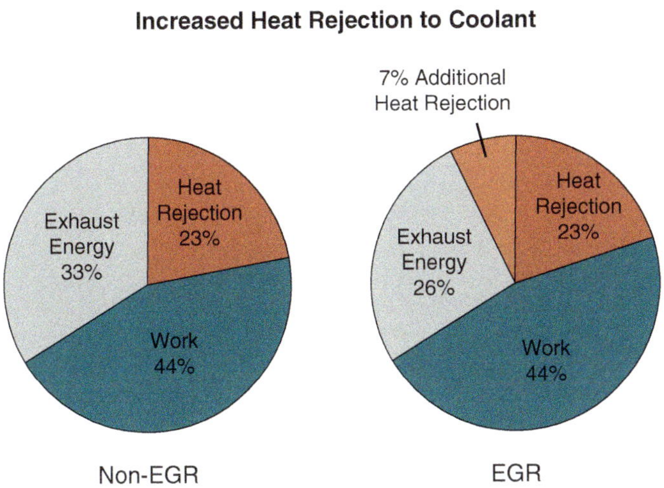

FIGURE 17-31 The cooling system must absorb more heat when HP-CEGR is used.

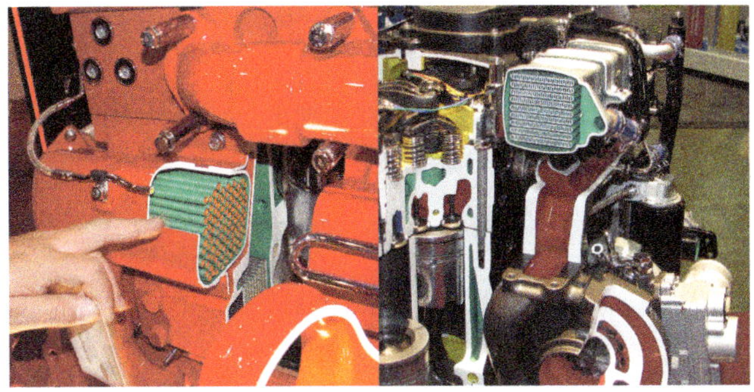

FIGURE 17-32 EGR coolers must be capable of resisting corrosion and continuous cycling of temperature and pressure.

FIGURE 17-33 The cooler bundle on the 2.8 L Duramax is made up of stainless steel tubes.

> **▶ TECHNICIAN TIP**
>
> Plugged intake manifolds can occur on some engines that use EGR while encountering extended periods of low load or idling. Increasing the cetane rating of the fuel helps prevent this condition. Adding a cetane booster may minimize plugging in the intake of some engines that use low-cetane fuel. Newer EGR coolers are built with a "floating" heat exchanger inside the cooler housing. The cooler bundle is attached to the housing only through contact with O-rings. This enables the cooler to expand and contract, which reduces the likelihood of the cooler bundle cracking due to frequent thermal cycling.

engine warm-up (**FIGURE 17-33**). If a cooler fails due to cracking from frequent thermal expansion and contraction or corrosion, the failure will be indicated by a loss of engine coolant. Engine coolant can be found in the exhaust pipe or intake manifold of the engine. Pressure testing the cooling system will reveal coolant leaks into the cooler gas passageway. Excessive steam or water vapor in the exhaust due to coolant ingestion often indicates a failed EGR cooler.

Inspecting an Exhaust Gas Recirculation Cooler for Leaks

EGR coolers most often fail due to hot spots: the constant thermal expansion–contraction cycles caused by exhaust heat and exhaust backpressure produced by the turbocharger. Although some diagnostic fault codes may be produced by blocked, plugged, or leaking coolers, other symptoms of an EGR cooler leak include the loss of coolant without external leaks, high silicate levels in scheduled oil samples, an overpressurized cooling system, and white exhaust smoke with a sweet odor. Overpressurized cooling systems are caused by pressurized exhaust gases pushing into the cooling system, causing the surge tank and degas bottle to overflow. Gas vapor will also collect in the heater core, resulting in no cabin heat. To inspect an EGR cooler for leaks, follow the steps in **SKILL DRILL 17-1**.

SKILL DRILL 17-1 Inspecting an EGR Cooler for Leaks

1. Pressurize the cooling system. Disconnect and remove the clamps at each end of the EGR cooler.
2. Visually examine the EGR cooler inside. The presence of gooey black liquid is an indicator of an internal leak.
3. Disconnect and remove the EGR valve from its bore. A gooey black residue covering the valve indicates an EGR cooler leak.
4. Inspect the EGR bore for white coolant residue. White streaking indicates coolant additives have leaked from coolant emitted by the EGR cooler.
5. After visually inspecting the engine for coolant leaks during a pressure test, the EGR cooler can be disconnected and removed.
6. Using purpose-made test adapters, plug the exhaust inlet and outlet ports with purpose-made plugs.
7. Using shop air pressure regulated to 20 psi (138 kPa), pressurize the cooler exhaust ports.
8. Immerse the cooler in a water tank, or use soapy water to check the coolant ports for leaking air. Any air leaks indicate a cracked EGR cooler.

Exhaust Gas Recirculation Mixer

Exhaust gas recirculation systems require a device to blend recirculated exhaust gas and intake charge air. This ensures that an equally proportioned amount of recirculated exhaust gas and air is delivered to all cylinders. Some engines simply use the turbulence of the intake manifold and/or valve port design to swirl air and exhaust together for even blending (**FIGURE 17-34** and **FIGURE 17-35**).

Exhaust Gas Recirculation Valves

Central to the EGR system is the valve that meters the quantity of EGR flow into the intake manifold. The valve does this by regulating exhaust flow through the valve that connects a passageway between the exhaust system and the intake manifold. Valves may be located on either the hot side between the exhaust manifold and cooler or the cold side between the cooler and intake manifold. Although hot-side EGR valves can fail prematurely due to excessive exhaust temperatures, they can prevent coolers from plugging or loading up with soot and slobber. Hot-side exhaust valves will typically stick and jam more often due to slobber contamination and heat related failures.

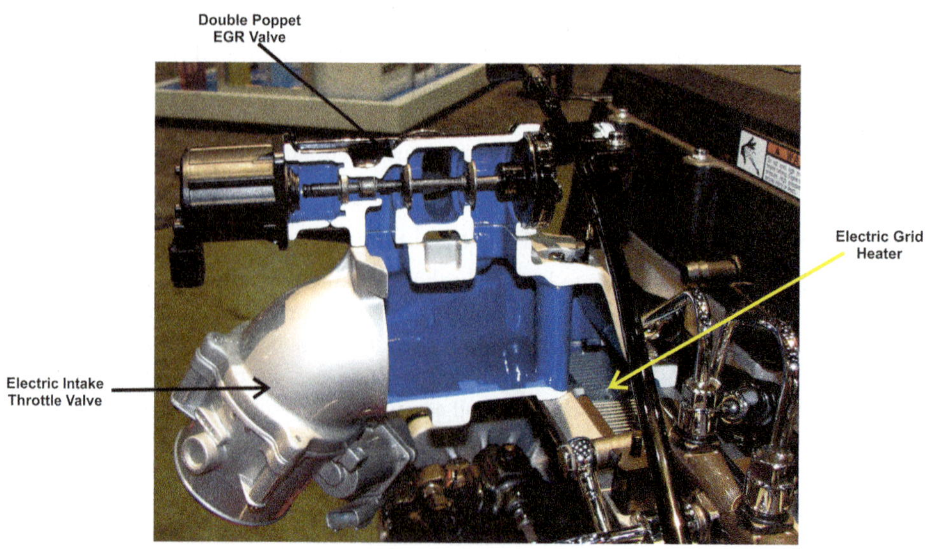

FIGURE 17-34 The electric grid heater in the intake manifold of this 6.7 L Cummins helps create turbulence that mixes and evenly distributes EGR gas and intake air.

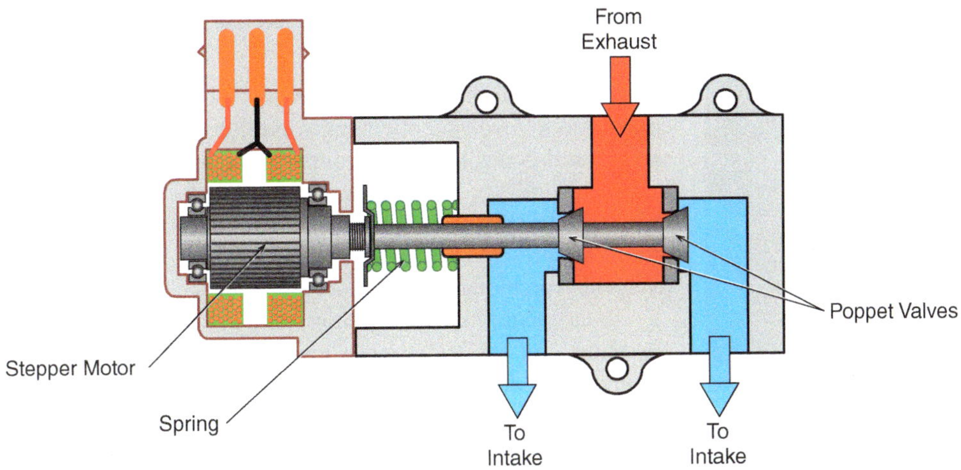

FIGURE 17-35 Double poppet valve EGR valves flow more EGR gas and can distribute gas to two different intake manifolds on V engines.

Valve Features of Exhaust Gas Recirculation

Requirements of EGR valve design incorporate features to do the following:

- minimize carbon contamination or coking in internal passageways
- clean carbon contamination from the valve seat through high valve opening and closing force
- rapid response times to accurately meter exhaust gas flow for various engine operating conditions.

To operate quickly, EGR valve mechanisms use solenoids and DC stepper motors or DC motor actuators. EGR valves commonly use single or dual outwardly opening poppet valves (**FIGURE 17-36** and **FIGURE 17-37**). The valves are sized to meet flow the requirements of the engine. Valve opening and closing forces have increased substantially from earlier valves to prevent valve sticking caused by carbon buildup. Manufacturers will use electrically operated valves or, for increased valve force, vacuum pressure (**FIGURE 17-38**). Vacuum signals are regulated by a variable-pressure output device. This is a pulse-width–modulated (PWM) vacuum regulator that sends varying pressure to both the turbocharger actuator and the EGR valve. Closed-loop feedback about the valve's operation includes data supplied by the EGR gas mass measurement system.

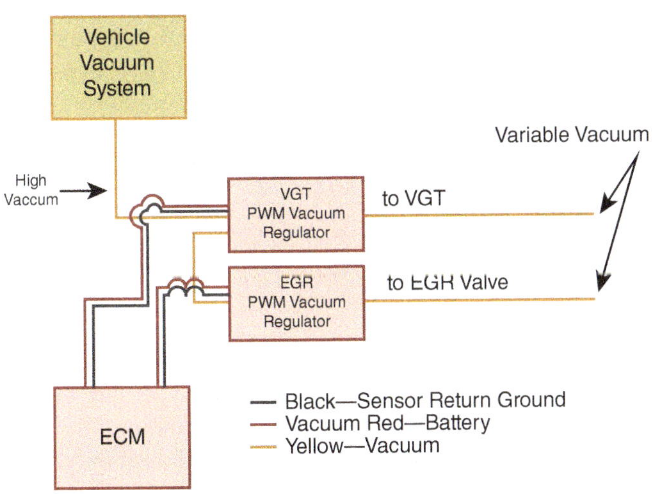

FIGURE 17-36 Principle of a vacuum operated actuators for an EGR system.

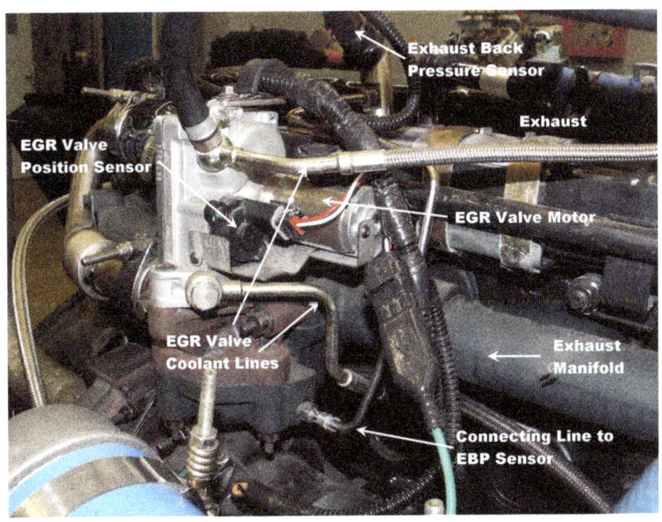

FIGURE 17-37 Coolant and electric connections to an EGR valve on a 5.9 L Cummins engine.

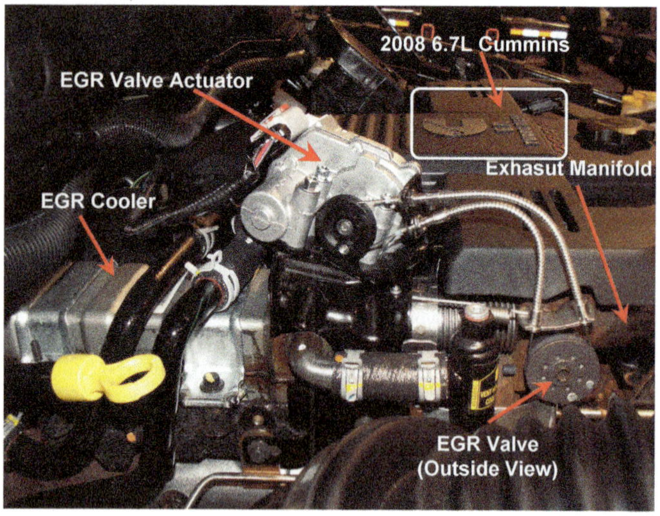

FIGURE 17-38 A hot-side EGR valve located at the cooler inlet of a 6.7 L Cummins.

Input Sensors of the Exhaust Gas Recirculation System

Because the quantity of EGR flow must be precisely matched to engine speed, load, and operating conditions, a number of input sensors and control devices are used to accurately meter EGR into the cylinders. The sensors will provide the ECM with data about the quantity of air and exhaust required and whether conditions are present that would enable or disable EGR flow. OBD data is also provided by the sensor inputs.

To calculate the quantity of exhaust gas for optimal engine operation and emission reduction, the mass of exhaust gas required and delivered to the cylinders is measured by input sensors. Three common circuit configurations are used. A **wideband heated oxygen sensor (HOS)** provides feedback to the ECM about EGR flow rates. A **mass airflow (MAF) sensor** measures the weight of air entering the engine and can detect air inlet restrictions caused by a restricted air filter or intake when pressure drops below a threshold value. A **pressure differential (delta P) sensor** measures the pressure drop across a venturi.

The following sensor inputs are used in addition to the main mass measurement strategies to calculate required EGR flow and the actual mass of EGR flow into the engine:

- engine speed sensor
- intake air temperature (IAT) sensor
- engine coolant temperature (ECT) sensor
- barometric pressure (BARO) sensor
- accelerator pedal position (APP) sensor
- manifold absolute pressure (MAP) sensor
- either an MAF sensor, a delta P sensor, or a wideband HOS
- exhaust backpressure (EBP) sensor (optional)
- EGR cooler outlet temperature sensor (optional).

Exhaust Gas Recirculation Mass Calculations

The engine speed and load conditions are the most significant predictors for how much EGR mass is required by the engine. If proportional-integral-derivative (PID) control is used, the ECM will contain lookup data tables for engine displacement and volumetric efficiency for any given rpm. This information, combined with intake manifold pressure and temperature, is used to calculate the weight of air entering the cylinders. Barometric pressure, coolant, and IAT measurements will determine whether the EGR system should be enabled, disabled, or bypassed based on ambient and engine operating conditions. The ECM will adjust the EGR valve position based on calculations of intake air mass and required EGR mass. Comparisons between calculated EGR flow and a measurement of the actual amount of exhaust gas delivered to and from a closed-loop feedback circuit for the EGR valve is performed by several mechanisms:

1. using an MAF sensor
2. using a delta P sensor (pressure differential sensor)
3. using oxygen sensors.

Mass Airflow Sensor and Exhaust Gas Recirculation Flow Calculations

The MAF sensor is located upstream of the EGR flow into the engine. This means the sensor is placed in the air intake stream before the EGR valve (**FIGURE 17-39**). Although the sensor does not directly measure the mass of exhaust gas flow into the cylinders, this configuration measures the mass of exhaust gas flow through a corresponding or proportional reduction in intake airflow when the EGR is enabled. Any increase in EGR rate above no flow will correspond to a reduction in fresh air mass flow because intake air supply is replaced by EGR flow. For example, an engine without EGR enabled may have an airflow mass of 50 grams/second. Once the EGR valve has been activated, a corresponding reduction of airflow rate is the measured EGR mass flow rate. Using the MAF sensor, the OBD-II system detects and measures EGR massflow based on a reduction in airflow into the engine and whether correct EGR flow has matches the commanded flow rate. The ECM will adjust EGR valve position after calculating engine temperature, intake boost pressure, exhaust backpressure, rpm, and engine load to calculate and compare the result with the expected or correct EGR flow rate. The ratio of the pressure in the intake manifold and exhaust backpressure may also be used by the ECM to calculate and further refine the EGR valve position. This is necessary because exhaust pressure will vary and change the amount of flow through the EGR valve. The commanded EGR valve position derived from calculations and lookup table information is compared to the actual measured position. Electrical or vacuum signals supplying the valve are adjusted to correct for any variation between the commanded and the actual measured position required for optimal EGR flow rate. If the actual rate has not been achieved with EGR valve position, the intake throttle valve can close on many engines to pull more EGR gas into the engine rather than allow fresh air to replace unavailable exhaust gas mass into the intake air path. When exhaust gas flow increases, the amount of fresh air intake measured by the MAF sensor naturally decreases (**FIGURE 17-40**).

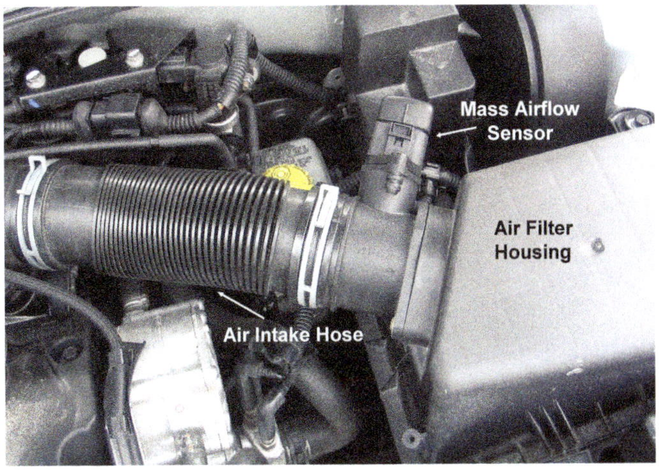

FIGURE 17-39 The MAF sensor is used to calculate EGR flow rate by measuring a drop in intake air mass in proportion to EGR gas flow.

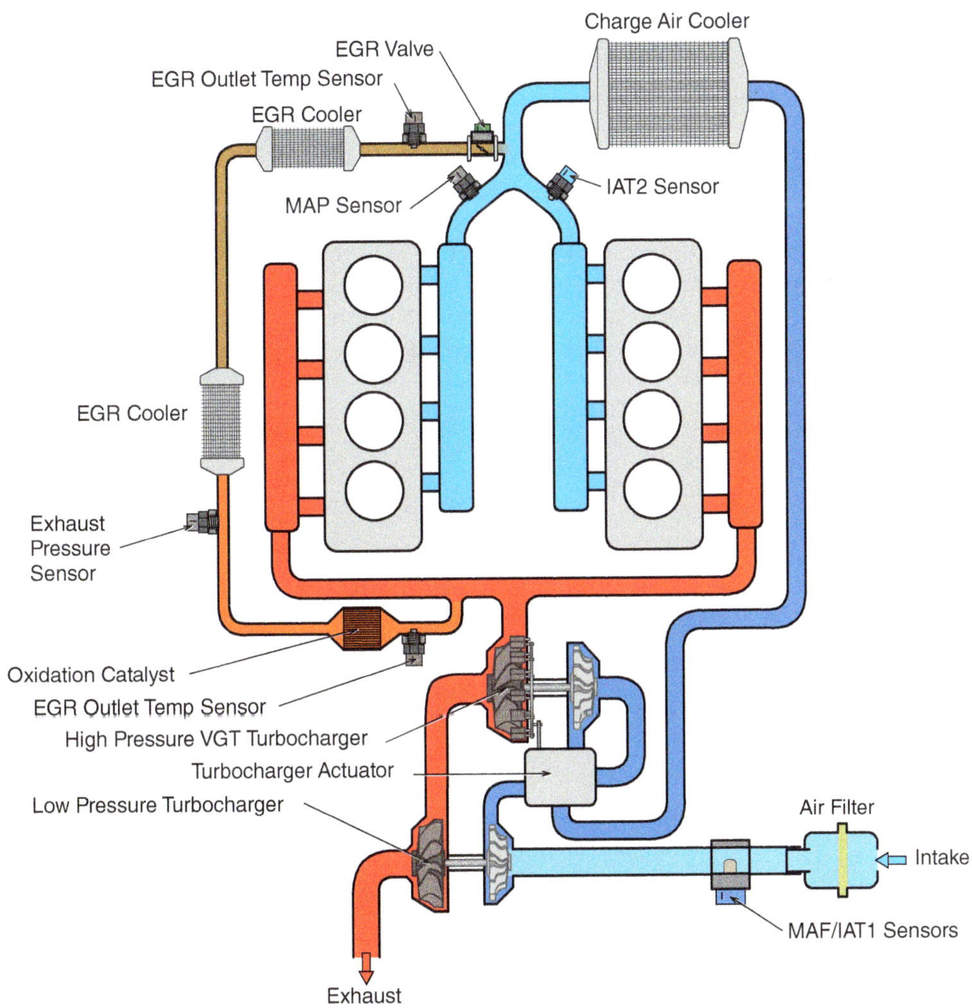

FIGURE 17-40 Note the location of the MAF sensor. It measures EGR mass flow by comparing the expected airflow rate with actual rates. The change from an air-only intake flow is in proportion to EGR mass flow.

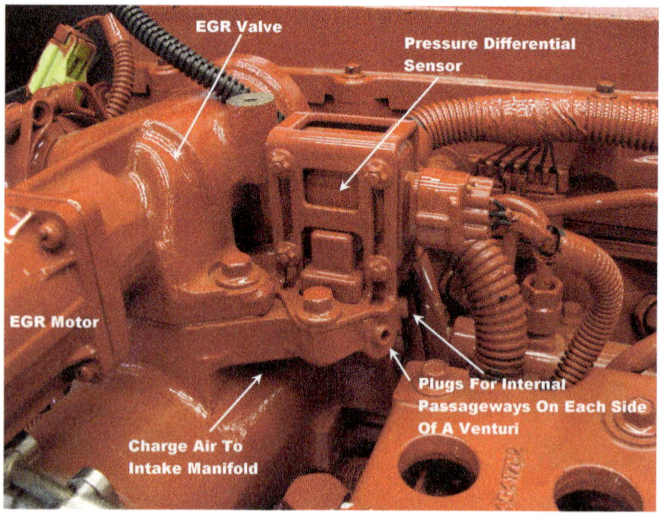

FIGURE 17-41 A pressure differential sensor (delta P sensor) combines with a intake temperature sensor to form a speed-density measurement algorithm. The pressure drop across a venturi corresponds to the mass of EGR flow from the EGR valve. This is a 6.7 L Cummins.

Pressure Differential Measurement of Exhaust Gas Recirculation Flow

Another method to measure EGR mass involves the use of a pressure differential sensor and a temperature sensor (**FIGURE 17-41**). In these systems, EGR gas flows through a venturi, which has a sensor measuring the pressure at its inlet and outlet. When EGR flow rate is low, little pressure drop occurs across the venturi. At high EGR flow, such as when the EGR valve is opened further, the venturi becomes more restrictive to the flow, so the pressure differential increases. Combined with gas temperature data from a temperature sensor, data from the pressure differential sensor can accurately measure the mass of EGR flow (**FIGURE 17-42**).

An electrical comparison to the speed-density method of EGR mass measurement is like the measurement of amperage. EGR mass measurement is similar to measuring amperage in an electrical circuit when knowing only the variables of circuit resistance and voltage. The resistance is the pressure or voltage drop across the venturi. If resistance and voltage are known in a circuit, amperage can be calculated. Likewise, EGR mass is calculated based on the known pressures and pressure drop.

A closed-loop feedback is formed using data from the combined delta P sensor and intake manifold temperature sensor with the EGR valve position (**FIGURE 17-43**).

Heated Oxygen Sensors

An EGR flow rate into the engine that keeps the exhaust gas oxygen concentration at approximately 15% in steady state operation is common for low emissions and good engine performance. Wideband oxygen sensors, also called lambda sensors, are used by some manufacturers as feedback devices for EGR controls (**FIGURE 17-44**). To illustrate this idea, consider that if exhaust gas oxygen content falls above or below 14–16%, the EGR system will automatically adjust EGR flow into the cylinder to compensate for the oxygen content change. This is not unlike the function of the oxygen sensor on gasoline-fueled engines, which is used to adjust air-fuel ratios to achieve stoichiometric combustion. Oxygen gas content will fall with increased EGR gas flow and will rise with reduced EGR gas flow. Operational strategies

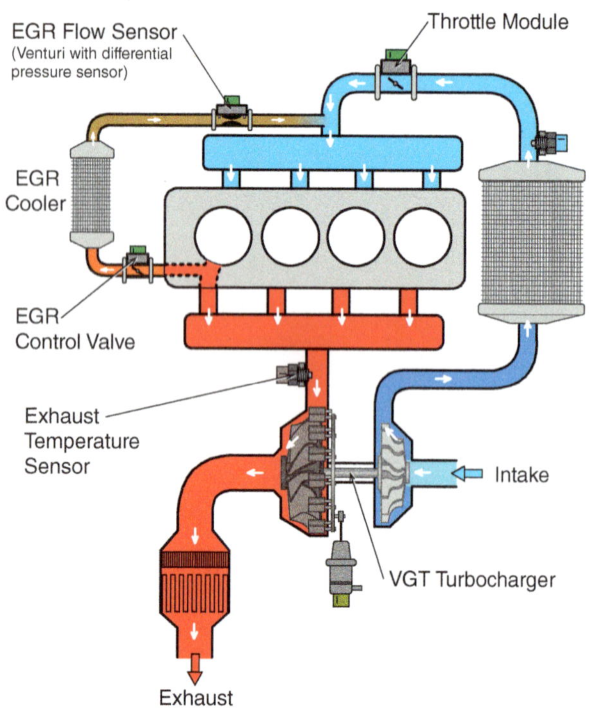

FIGURE 17-42 A functional view of a speed-density measurement system used to measure EGR flow. The pressure drop across the venturi and the exhaust temperature measurements are used to calculate actual EGR mass flow.

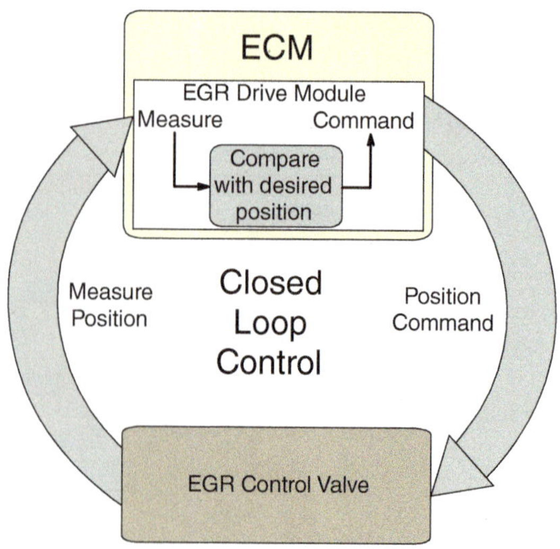

FIGURE 17-43 Strategy for controlling EGR flow by using closed-loop feedback.

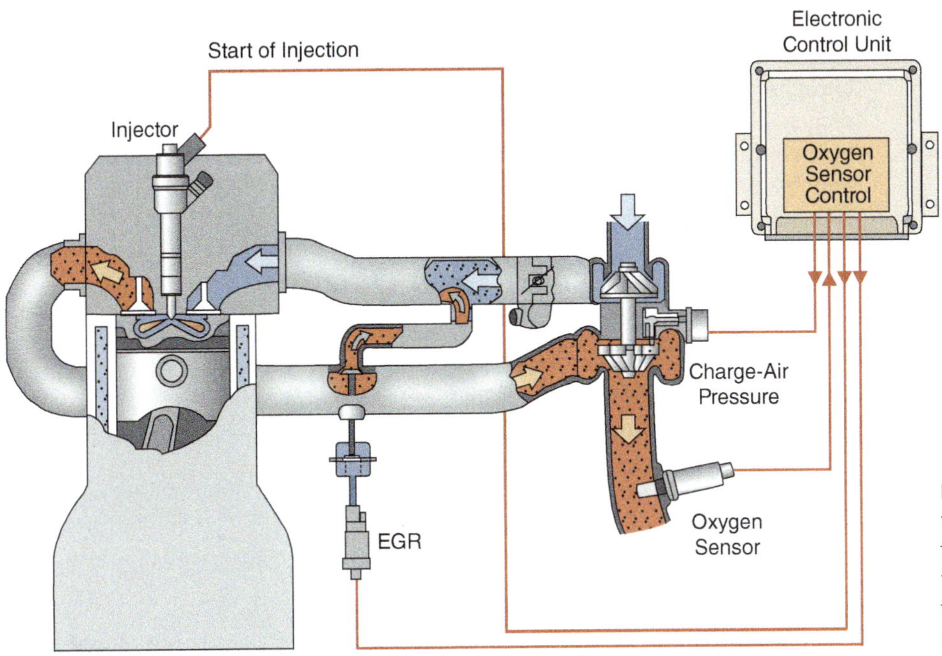

FIGURE 17-44 The oxygen sensor forms closed-loop feedback with the EGR valve to regulate EGR gas flow. As the EGR rate increases, the oxygen content of the exhaust stream decreases.

are obviously more complex in diesel engines due to the use of aftertreatment systems and widely varying operating conditions. When oxygen sensors are replaced, a calibration routine is usually used prior to operating the engine. Some Bosch wideband sensors have a resistor in the connector fitting and the engine ECM. This resister corresponds to the stoichiometric ratio of diesel fuel—not gasoline. The ECM needs to learn the value of this sensor calibration resister for a starting point, to make corrections to sensor data in order to accurately report oxygen levels in the exhaust. Not all Bosch sensors and sensors from other manufacturers use a resistor to calibrate oxygen gas measurements. External oxygen concentration is instead used by the sensor as a reference for exhaust gas concentrations of oxygen. Sensors that use an external oxygen sensor calibration method require that the sensor be electrically connected and then held in the air to "learn" new values for barometric pressure and atmospheric oxygen content. These new values are used as coefficients for the adaptation algorithms stored in the ECM calibration files. Both sensor calibration methods are used by the ECM to compensate for sensor deterioration as the sensor ages.

▶ Maintenance of Exhaust Gas Recirculation Systems

17-04 Recommend maintenance or repairs on diesel engine EGR systems.

EGR systems are one of the most common causes of diesel engine performance and MIL light complaints. Exposure to hot exhaust gases and constantly changing pressures and temperatures can shorten the life of many EGR system components. This next section will examine some of the diagnostic and maintenance strategies associated with EGR systems.

Diagnostics for External, Non-Cooled Exhaust Gas Recirculation

A vacuum gauge and scan tool are required for EGR system diagnostics. To perform electrical pinpoint tests, a high-impedance multimeter is also necessary. At idle, after the engine has warmed up, the vacuum signal to the EGR valve should be 5–7 inHg (13–18 cmHg). The mechanical vacuum pump supplies a vacuum source of 15 inHg (38 cmHg). The vacuum signal to the EGR valve will increase for higher engine speed and load conditions. When operating conditions require reduced EGR flow, the EGR vent solenoid will dump vacuum in the line to the EGR valve.

The onboard diagnostic (OBD) system monitors the EGR system for electrical and operational faults. The information from the EGR control pressure sensor can be read by a scan tool and should indicate atmospheric pressure with the key on and engine off.

Electrical faults with sensor circuits, such as the throttle position circuit and engine speed sensor, can be observed by monitoring these circuits with a scan tool. Significant differences between a sensed EGR control vacuum and a commanded vacuum signal will trigger diagnostic trouble codes. When diagnostic codes are set, the ECM will disable the EGR operation. For cooled EGR systems, cooler efficiency is measured by comparing exhaust temperature entering and leaving the cooler. Inadequate temperature reduction, casued by plugged or bypassed coolers will set fault codes.

Exhaust Gas Recirculation Diagnostics

Comprehensive component monitoring of the EGR system by the OBD-II system typically detects malfunctioning in the following areas:

- EGR flow rate by using the delta P or MAF sensor
- EGR valve position
- EGR cooler temperatures (efficiency codes)
- intake boost pressure by using the MAP sensor
- boost temperature by using the IAT sensor
- EBP sensor
- EGR system performance.

If any data or circuits have electrical problems or if data are out of range or irrational when compared to other data, then the malfunction indicator lamp (MIL) will illuminate and a diagnostic trouble code (DTC) will be stored (**FIGURE 17-45**).

OBD monitors compare measured system behavior against a model of expected performance. For example, under the correct operating conditions, the OBD manager may briefly open the EGR valve and measure the drop in turbocharger speed against expected values. The turbocharger speed will drop because combustion temperature is cooler at high EGR flow rates. Delta P flow change, EGR valve position, and MAF change could also be sensed. When the diagnostic manager has evaluated EGR system operation, it makes a judgement and reports that the monitor has run and whether any faults were detected.

Exhaust Gas Recirculation Valve Diagnostics

In some systems, a variable-resistance-type position sensor is incorporated into the EGR valve to provide feedback to the EGR controls about the position of the valve. Differences between the valve's commanded position in comparison to its actual position are measured to apply a position correction. If the valve fails to position correctly, it will generate fault codes associated with EGR system operation. To increase the reliability of the position sensor, coolant may circulate through the valve to reduce heat from the exhaust gas flow.

The EGR control circuit will also perform electrical diagnostics on the EGR valve according to comprehensive component monitoring required of OBD. For example, the position sensor and motor circuit are measured for resistance, voltage, short circuits, and out-of-range values.

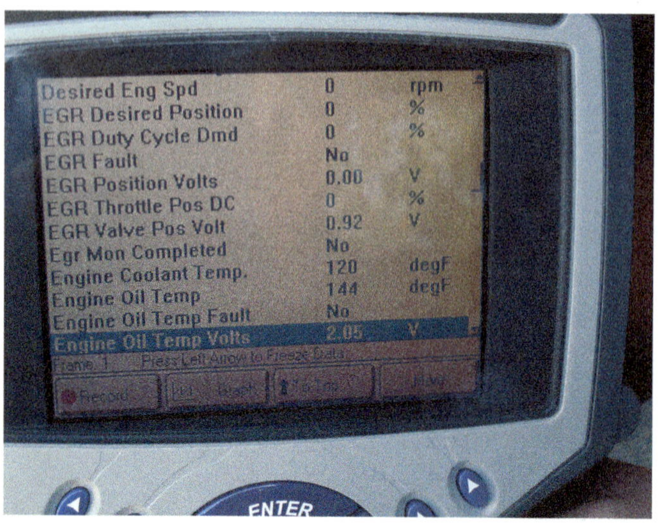

FIGURE 17-45 Several variables are monitored by the OBD-II system of diesel EGR with the key on and engine off. Note that the EGR system monitor, which evaluates EGR system operation, has not run yet.

Exhaust Gas Recirculation System Performance Checks

Some common diagnostic tests to evaluate the operation of the EGR system include the EGR valve override test and the VGT manual override test.

The EGR manual override test allows the service technician to override the EGR valve position by commanding different EGR valve positions and monitoring the actual position feedback from the valve. The technician can electronically measure the amount of valve movement and listen for changes in engine operation. This test is used to determine whether the EGR valve is sticking, stuck in one position, or not actuating the full amount of travel. Because the VGT is an integral part of the EGR system, the service technician should evaluate its operation too. Usually with the engine at idle, the VGT vanes or nozzle will move from a closed to an open position. Measurements are taken of turbo speed during this operation. If the turbocharger speed change does operate within and expected range, a technician will need to perform pinpoint tests to troubleshoot the air handling system.

Testing an Exhaust Gas Recirculation Valve

The OBD system observes the change in turbocharger speed when the valve opens and closes to evaluate EGR valve operation. The EGR valve will lower turbocharger boost pressure when enabled, because the exhaust gases have expanded less. The following procedure can be used on some engines with diagnostic menu items to open and close the EGR valve. To test an EGR valve, follow the steps in **SKILL DRILL 17-2**.

SKILL DRILL 17-2 Testing an EGR Valve

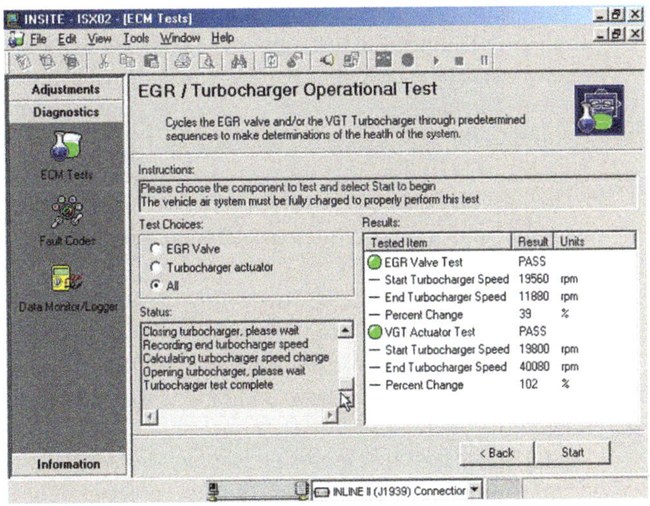

1. Connect an electronic service tool or software from the original equipment manufacturer (OEM) to cycle the EGR valve open and closed.
2. While monitoring the turbocharger speed, listen to the change in combustion noise when the valve is opened. The noise should soften when the valve is fully open and sharpen when the valve is closed.
3. Monitor the turbocharger speed.
4. Open the valve and observe the drop in turbocharger speed. Closing the valve should cause the turbocharger speed to increase.

▶ Wrap-Up

Ready for Review

- Exhaust gas recirculation (EGR) technology reduces the formation of oxides of nitrogen (NO_x) from the engine cylinders by reintroducing exhaust gas into the engine's cylinders.
- The introduction of EGR has prompted many changes to engine design, operation, and maintenance requirements.
- EGR is used to prevent the formation of NO_x emissions during the combustion process rather than having to remove NO_x in the exhaust system.
- EGR reintroduces exhaust gas into engine cylinders to reduce peak cylinder pressure and temperature—conditions that produce NO_x.
- Although the design, construction, and operation of different EGR systems are unique, the basic mechanisms used to reduce NO_x formation in the cylinders are the same. The principle behind all EGR systems is to disrupt the chemical processes that form NO_x.
- While EGR systems are designed to replace a portion of the cylinder charge air with EGR gas, increased boost pressure is used on EGR engines; this means that the same amount of oxygen is available in the cylinder to keep power output comparable with non-EGR engines.
- EGR systems monitor engine speed and load conditions to meter the correct amount of exhaust gas into the engine intake.
- During idle, cold weather, or engine operation, minimal or no EGR flow usually takes place.
- During warm-up, condensation of corrosive exhaust can take place in EGR coolers, so flow is minimized or disabled during that period.
- EGR reduces the formation of NO_x emissions by lowering peak cylinder pressure. Because exhaust gas cannot directly enter into the combustion process, it slows down the combustion speed by spacing oxygen and fuel molecules farther

apart. Chain reactions between oxygen and fuel take longer, which results in a slower pressure rise in the cylinder.

- Given the progressive reductions required for NO_x emissions by emission legislation and the fact that other NO_x emission-reducing technologies and operating strategies have reached their limits, EGR is necessary on almost all diesel engines.
- Using EGR reduces temperature in two ways. One way is through its ability to slow down combustion speed, which causes not only a pressure drop but also a temperature drop. The second way is due to the heat-absorbing properties of exhaust gas in the cylinder.
- By reducing the potential peak cylinder pressure, using EGR will make engine torque and power output fall unless other engine operating strategies and designs compensate for using it.
- EGR taxes the engine lubricating oil. New lubricating oil specifications have been developed for EGR engines in order to handle the extra contamination of lubrication oil from recycling exhaust gas back into the cylinder.
- Cooled EGR, where exhaust gas passes through an engine coolant heat exchanger, adds additional heat load to the cooling system and engine compartment. Larger radiators, water pumps, and cooling systems are required to transfer the heat absorbed by the coolant to the atmosphere.
- Cooled EGR leads to water condensation and the formation of higher sulfuric and nitric acid levels in the combustion chamber after reactions take place between water and combustion gases.
- Several types of EGR delivery systems exist to introduce exhaust gas uniformly into the engine cylinders. Types of EGR delivery systems used on light and medium-duty diesels include cooled, high-pressure EGR and cooled, low-pressure EGR. EGR systems that are used less frequently on a few heavy-duty, light-duty, and off-road engines include internal EGR, external non-cooled EGR, and external cooled EGR.
- In an internal EGR system, no external EGR controls are used, but exhaust gas is reintroduced into the cylinders through a change in exhaust valve timing. These internal methods often result in poor mixing with the intake air and excessive EGR rates at low speeds.
- External EGR uses a metering valve and control circuits to deliver exhaust gas into the intake manifold. Exhaust gas is driven into the intake manifold through the exhaust backpressure in the exhaust system.
- To reduce emissions, fuel rates, air, and exhaust gas mass must be carefully measured and balanced for efficient, powerful, low-emission combustion. A number of additional sensors and actuator devices are required to accurately meter EGR rate in proportion to other combustion elements.
- Supplying the necessary volume of exhaust gas to the pressurized intake manifold is a problem for turbocharged EGR engines. Manufacturers have relied on VGTs, which can vary boost pressure and exhaust backpressure independent of engine speed.
- EGR gas derived from the exhaust manifold and pressurized to overcome intake boost pressure is called high-pressure EGR.
- An alternative strategy to high-pressure EGR is to draw EGR gas into the cylinders either through the turbocharger inlet or after the turbocharger between a throttle plate and intake manifold; this is referred to as low-pressure EGR.
- Some light-duty systems combine low- and high-pressure EGR to obtain even lower NO_x emissions while increasing engine efficiency.
- Engine manufacturers cool EGR gas before recirculating it into the combustion chambers. Cooling the exhaust lowers combustion chamber temperature and allows exhaust gas to absorb more combustion chamber heat.
- Exhaust gas recirculation systems require a device to blend recirculated exhaust gas and intake charge air. Some engines use simply the turbulence of the intake manifold and/or valve port design to swirl air and exhaust together for even blending.
- Central to the EGR system is the valve that meters the quantity of EGR flow into the intake manifold. Valves may be located on either the hot side (between the exhaust manifold and cooler) or the cold side (between the cooler and intake manifold).
- A number of input and control devices are used to accurately meter EGR into the cylinders. Three common circuit configurations are used: a wideband heated oxygen sensor (HOS), a mass airflow (MAF) sensor, or a pressure differential (delta P) sensor.
- Engine speed and accelerator pedal position (APP) primarily indicate how much EGR mass is required for the engine.
- Using the MAF sensor, the OBD-II system detects and measures based on a reduction in airflow into the engine whether correct EGR flow has been enabled.
- Another method to measure EGR mass involves the use of a pressure differential sensor and a temperature sensor.
- An EGR flow rate into the engine that keeps the exhaust gas oxygen concentration at 15% is optimal for low emissions and engine performance. Oxygen sensors are used by some manufacturers as feedback devices for EGR controls.
- EGR system behavior is monitored and measured against a model of expected performance by OBD systems.
- Some common diagnostic tests to evaluate the operation of the EGR system include the EGR valve override test and the VGT manual override test.

Key Terms

cooled EGR (CEGR) A system that cools EGR exhaust gas by passing it through an engine coolant heat exchanger.

pressure differential sensor A sensor that measures the pressure drop across two points in a gas circuit; it is also called a delta P sensor. In an EGR system, the delta P sensor measures the pressure drop across a venturi.

exhaust backpressure (EBP) sensor A sensor that measures pressure in the exhaust manifold. The sensor can be used to perform diagnostic tests and provide additional data for closed-loop feedback control of the VGT actuator position.

high-pressure cooled EGR (HP-CEGR) A system in which EGR gas is pressurized in the exhaust manifold by a VGT turbocharger. Exhaust gas pressure is driven above intake boost pressure.

low-pressure EGR A system in which EGR gas is drawn into the cylinders either through the turbocharger inlet or after the turbocharger between a throttle plate and intake manifold.

mass airflow (MAF) sensor A device that measures the weight of air entering the engine and that can detect air inlet restrictions caused by a restricted air filter or intake when pressure drops below a given threshold value.

nitric oxide (NO) An unstable, highly reactive, colorless gas. It is the most abundant type of NO_x produced during diesel combustion. It is a noxious respiratory irritant.

nitrogen dioxide (NO_2) A reddish-brown gas formed from nitric oxide and oxygen. It is a noxious respiratory irritant.

specific heat The amount of heat required for a single unit of mass of a substance to be raised by one degree of temperature.

variable-geometry turbocharger (VGT) A turbocharger with the capability of changing boost pressure and exhaust backpressure independent of engine speed and load; it is also known as a variable-nozzle turbocharger (VNT) and a variable-vane turbocharger (VVT).

wideband heated oxygen sensor (HOS) An exhaust gas sensor that provides feedback to the ECM about EGR flow rates by measuring exhaust oxygen content.

Review Questions

1. Which of the following environmental problems is caused primarily by NO_x emissions?
   a. Global warming.
   b. Global cooling.
   c. Photochemical smog.
   d. Soil contamination.
2. Which combination of combustion conditions produces the highest NO_x emissions from diesel engines?
   a. High cylinder temperatures and pressures.
   b. Rich air-fuel mixtures and low cylinder pressures.
   c. High cylinder pressures and lean air-fuel mixtures.
   d. Low soot and high cylinder temperature.
3. To reduce NO_x formation, which of the following combustion elements is replaced by EGR gas?
   a. Nitrogen.
   b. Nitric oxide (NO).
   c. Oxygen.
   d. Carbon dioxide (CO_2).
4. Under which of the following conditions is EGR typically disabled?
   a. Heavy acceleration and high engine loads.
   b. After an active regeneration of the diesel particulate filter (DPF).
   c. Excessively high engine coolant temperatures.
   d. High barometric pressures.
5. How does EGR gas contribute to increased cylinder wear?
   a. It chemically weakens hardened steel piston rings.
   b. It causes nitriding of piston rings.
   c. It contains abrasive soot.
   d. It retards injection timing.
6. Which type of EGR system is used by the majority of automotive diesel engines?
   a. Externally cooled EGR.
   b. Internally cooled EGR.
   c. Cooled, low-pressure EGR.
   d. Cooled, high-pressure EGR.
7. What device is used by internal EGR systems to force EGR gas into cylinders?
   a. A variable-geometry turbocharger (VGT).
   b. An exhaust pressure governor.
   c. An EGR valve located between the exhaust and the intake manifold.
   d. An additional bump on the exhaust camshaft lobe.
8. What is the purpose of a mass airflow (MAF) sensor on an EGR-equipped engine?
   a. To adjust air-fuel ratios on OBD-II compliant vehicles.
   b. To meter the correct quantity of air required for combustion.
   c. To measure the quantity of EGR gas entering engine cylinders.
   d. To calculate precise air-fuel ratios.
9. What does a temperature sensor combined with a delta P sensor measure?
   a. EGR gas mass.
   b. Exhaust backpressure.
   c. Intake air mass.
   d. Gas density.
10. What happens to the mass of intake airflow into an engine when the EGR is enabled?
    a. Nothing happens.
    b. The mass increases.
    c. The mass decreases.
    d. The mass becomes cooler.

ASE Technician A/Technician B–Style Questions

1. Technician A says that EGR systems lower in-cylinder NO_x formation by reducing peak cylinder pressures and temperatures. Technician B says EGR systems reduce NO_x emissions by reducing displaced air in the cylinder. Who is correct?
   a. Technician A
   b. Technician B
   c. Both A and B
   d. Neither A nor B
2. Technician A says that turbocharger boost pressures are higher in EGR engines to compensate for power loss when using EGR gas. Technician B says that higher boost pressures are used to ensure that the combustion

chamber receives an adequate amount of oxygen to support combustion. Who is correct?
a. Technician A
b. Technician B
c. Both A and B
d. Neither A nor B

3. Technician A says the piston stroke length in an EGR engine is often longer than one in the same model engine that doesn't have an EGR system. Technician B says that the longer stroke length used by some EGR model engines has nothing to do with the use of EGR and that some EGR engines have shorter stroke length than non-EGR model engines. Who is correct?
a. Technician A
b. Technician B
c. Both A and B
d. Neither A nor B

4. Technician A says that the oil used in the earliest EGR engines has a different API designation than oil used in engines with particulate filters. Technician B says that the oil used in engines equipped with particulate filters can be used in EGR engines. Who is correct?
a. Technician A
b. Technician B
c. Both A and B
d. Neither A nor B

5. Technician A says that diesel engines that use EGR systems often have longer engine break-in periods. Technician B says that the use of EGR gas introduces abrasive soot into the cylinders, which shortens the break-in period. Who is correct?
a. Technician A
b. Technician B
c. Both A and B
d. Neither A nor B

6. Technician A says that internal EGR systems use a throttle plate to draw EGR gases into the air intake flow. Technician B says that high-pressure cooled EGR systems need throttle plates to draw exhaust gases into the intake manifold. Who is correct?
a. Technician A
b. Technician B
c. Both A and B
d. Neither A nor B

7. Technician A says that diesel engines use MAF sensors to measure exhaust airflow into the cylinders. Technician B says that exhaust gas does not pass through an MAF sensor, so the MAF sensors cannot measure exhaust gas flow. Who is correct?
a. Technician A
b. Technician B
c. Both A and B
d. Neither A nor B

8. Technician A says that oxygen sensors used by diesel engines adjust air-fuel ratios. Technician B says that oxygen sensors can be used to adjust the amount of EGR gas metered into the cylinders. Who is correct?
a. Technician A
b. Technician B
c. Both A and B
d. Neither A nor B

9. While two technicians are discussing possible causes of engine coolant loss, Technician A says the EGR cooler is likely cracked and leaking. Technician B says that a cracked EGR cooler would leak coolant into the intake manifold. Who is correct?
a. Technician A
b. Technician B
c. Both A and B
d. Neither A nor B

10. Technician A says that EGR gas containing soot is present in all EGR systems. Technician B says that low-pressure EGR systems typically do not contain soot. Who is correct?
a. Technician A
b. Technician B
c. Both A and B
d. Neither A nor B

CHAPTER 18

Exhaust Aftertreatment Systems

Learning Objectives

After reading this chapter, you will be able to:

- 18-01 Describe the construction, functions, and applications of diesel exhaust emission aftertreatment systems.
- 18-02 Identify and describe the composition, construction, and operation of exhaust catalysts specific to minimizing the discharge of regulated engine emission.
- 18-03 Identify and explain the function of DPF substrate and regeneration strategies used to renew particulate filters.
- 18-04 Identify, describe, and recommend service procedures for diesel particulate filters.
- 18-05 Identify and describe the types, composition, and operation of major NO_x emission-reduction aftertreatment systems.
- 18-06 Identify and describe the construction and operation of SCR-U aftertreatment systems.
- 18-07 Identify and describe the construction and operation of DEF delivery and feedback systems used by SCR aftertreatment systems.
- 18-08 Identify and explain the operation of aftertreatment warning systems.

You Are the Technician

A 2018 diesel-powered automobile has arrived at your shop with the malfunction indicator lamp (MIL) light on. A quick scan of engine codes shows multiple aftertreatment-related fault codes, including an active code for a low efficiency oxides of nitrogen (NO_x) sensor on bank 1 sensor 2: P20xx, a post–selective catalyst reduction (SCR) NO_x outlet sensor. During a check of the service records and a discussion with the customer, you discover that the NO_x sensor has been replaced four times during the warranty period. Reluctant to pay for yet another NO_x sensor now that the warranty has expired, the car is at your shop, which specializes in automotive diesel engine repair to have what the customer believes will be a more comprehensive diagnosis performed. The two-year-old car has a below average accumulation of distance since it operates primarily around a city where heavy traffic congestion is a typical driving condition. There is an inactive fault code for the exhaust gas recirculation (EGR) system and the diesel exhaust fluid (DEF) tank is full. As you consider the diagnostic steps and procedures that you will follow, include the following questions:

1. Are there other engine or aftertreatment-related faults could likely produce a low efficiency NO_x sensor code?
2. If other engine or aftertreatment faults were present, why has this NO_x code appeared?
3. Is establishing code priority important for this diagnosis (i.e., which fault code should be investigated first)?

▸ Introduction

Over the span of less than a decade, the once lowly exhaust system has undergone radical sophistication. Arguably the simplest of any engine support systems at one time, composed of just pipes and metal cans, the exhaust system is now the focal point of some of the most advanced vehicle technologies. Meeting emission-reduction goals that have reduced diesel emissions to almost undetectable levels has made the exhaust system central to technology designed to clean up or reduce harmful emissions. The exhaust system has transformed because in-cylinder, engine-based emission control strategies reached the limit of their ability to meet emission-reduction targets. Emission reductions needed to take place in the exhaust system to avoid crippling reductions in engine efficiency. Cleaning up emissions in the exhaust makes sense too. Removing noxious emissions from the exhaust, rather than preventing their formation in cylinders, enables the use of more efficient engine control strategies. Exhaust aftertreatment devices, such as catalytic converters, that convert carbon monoxide (CO), hydrocarbons (HCs), and NO_x, have also been successfully used since the early 1970s on spark-ignition, gasoline-fueled engines.

Removing diesel emissions from the exhaust, a method called exhaust aftertreatment, presents an entirely different set of technical challenges than those in stoichiometric-burn spark-ignition engines. Exhaust catalysts used by spark-ignition engines do not work when used in diesel engine exhaust. Cold exhaust temperatures and excess air content in diesel exhaust (relative to gasoline counterparts) are the biggest technical challenges for creating new diesel exhaust catalysts. Particulate emissions that gasoline-fueled engines are currently are not regulated for are added to diesel engines' clean-up list too. And finally, the proportion of each noxious emission from diesel engines is significantly different from gasoline engines.

This chapter provides the details needed for technicians who service exhaust aftertreatment to understand and diagnose aftertreatment system problems. It is also intended to provide a foundation for recommending service and maintenance for exhaust aftertreatment systems found in contemporary light- and medium-duty diesel engines.

▸ Fundamentals of Exhaust Emission Aftertreatment Systems

18-01 Describe the functions, construction, and applications of diesel exhaust emission aftertreatment systems.

The following are the two primary methods to achieve the ultra-low emissions from today's engines:

1. in-cylinder emission-reduction strategies
2. exhaust system aftertreatment of emissions.

In-cylinder methods, as the name suggests, refer to strategies used by manufacturers to prevent emission formation in the cylinders. Examples of these methods include the use of cleaner, improved fuels; EGR; retarded injection timing, pilot-injection with moderate injection rate shaping from electronically controlled high-pressure systems, charge air cooling (CAC), and quiescent or less turbulent combustion chamber designs. Although these methods are effective at reducing the formation of noxious emissions, some of those emission control strategies reduce engine efficiency and are inadequate to achieve the required 90% or higher reduction for particulate emissions needed for 2007 when Tier 2 bin 5 emission standards were implemented. More rigorous reductions in NO_x emissions, which were phased in beginning in 2004 and fully implemented by 2010, necessitated moving emission-reduction technologies to the exhaust system to maintain engine efficiency while meeting lower emission targets.

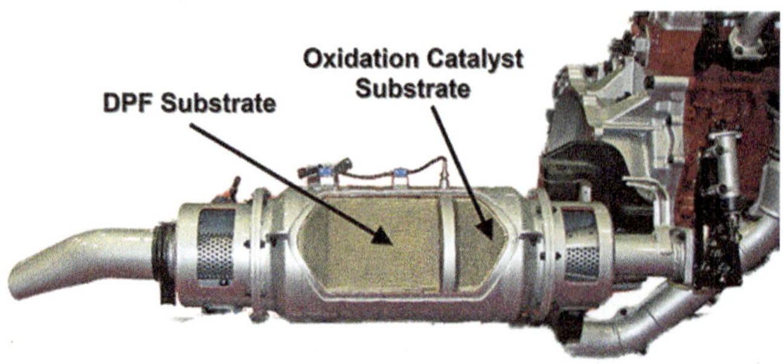

FIGURE 18-1 Diesel particulate filters, which trap soot emissions, became mandatory in 2007 for all diesels.

Elements of Aftertreatment Systems

Two distinct stages of exhaust aftertreatment systems are used based on the progressively more restrictive Environmental Protection Agency (EPA) emission standards. In 2007, all diesel-powered vehicles in North America were equipped with diesel particulate filters (FIGURE 18-1). In 2010, aftertreatment

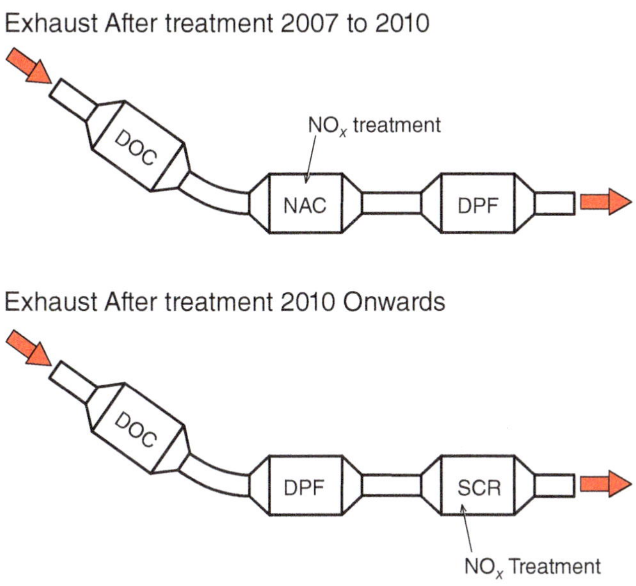

FIGURE 18-2 NO_x-adsorbing catalysts (NAC) and SCR using liquid DEF are the two principal emission systems that remove NO_x from diesel exhaust.

systems included additional components to meet the low NO_x emission standards. These new components cleaned up NO_x emissions by using urea in SCR-Urea (SCR-U) systems, which stores ammonia-containing molecules onboard the vehicle in liquid diesel exhaust fluid (DEF) (**FIGURE 18-2**). The alternative to SCR-U using DEF was to manufacture ammonia onboard in exhaust catalysts. These systems produce hydrogen, which combines with nitrogen in NO_x emissions to form ammonia. Systems selectively targeting NO_x by producing ammonia molecules from hydrogen are referred to as SCR-H (SCR-hydrogen) systems (**FIGURE 18-3**). Tougher Tier 3 emission standards were phased in beginning in 2017 to lower NO_x and particulate matter (PM) by 80% and 70%, respectively, compared

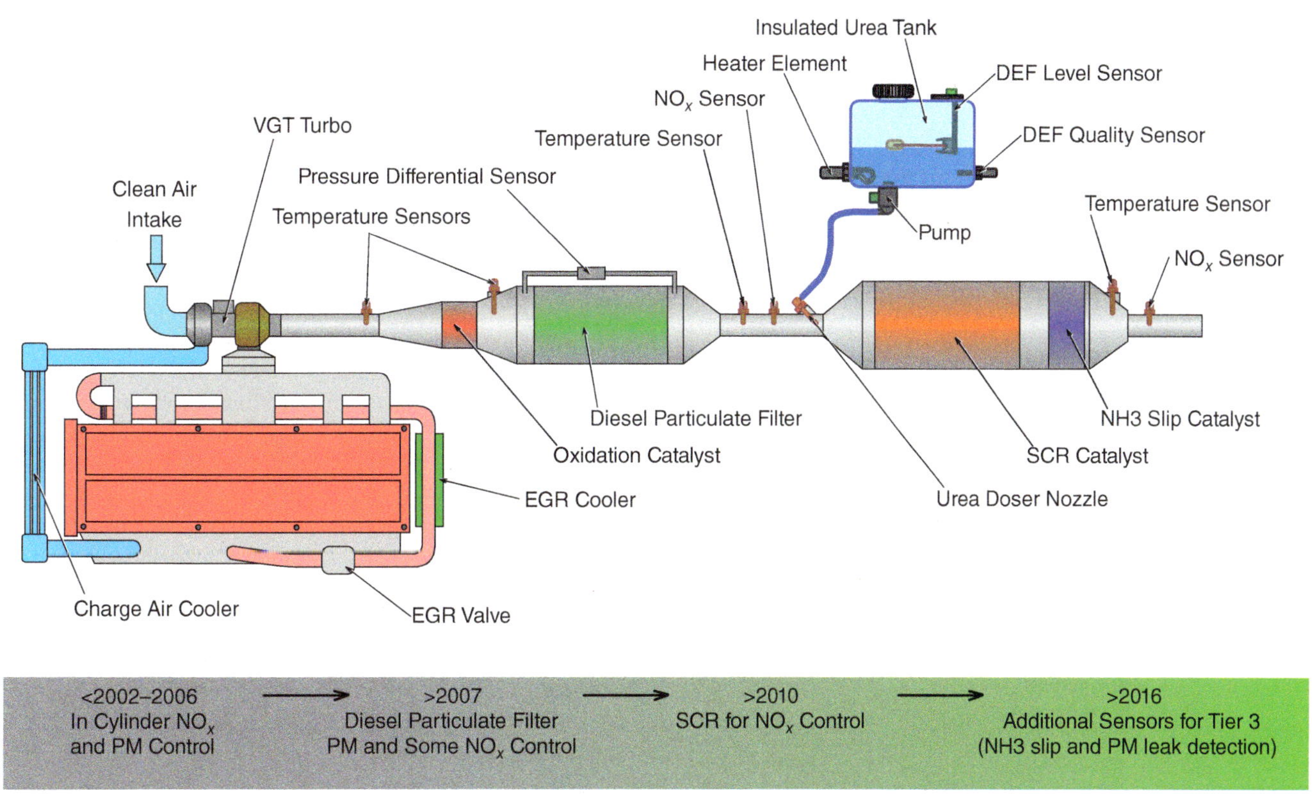

FIGURE 18-3 Evolution of aftertreatment systems. As emission standards became stricter, aftertreatment emission system technology progressively advanced.

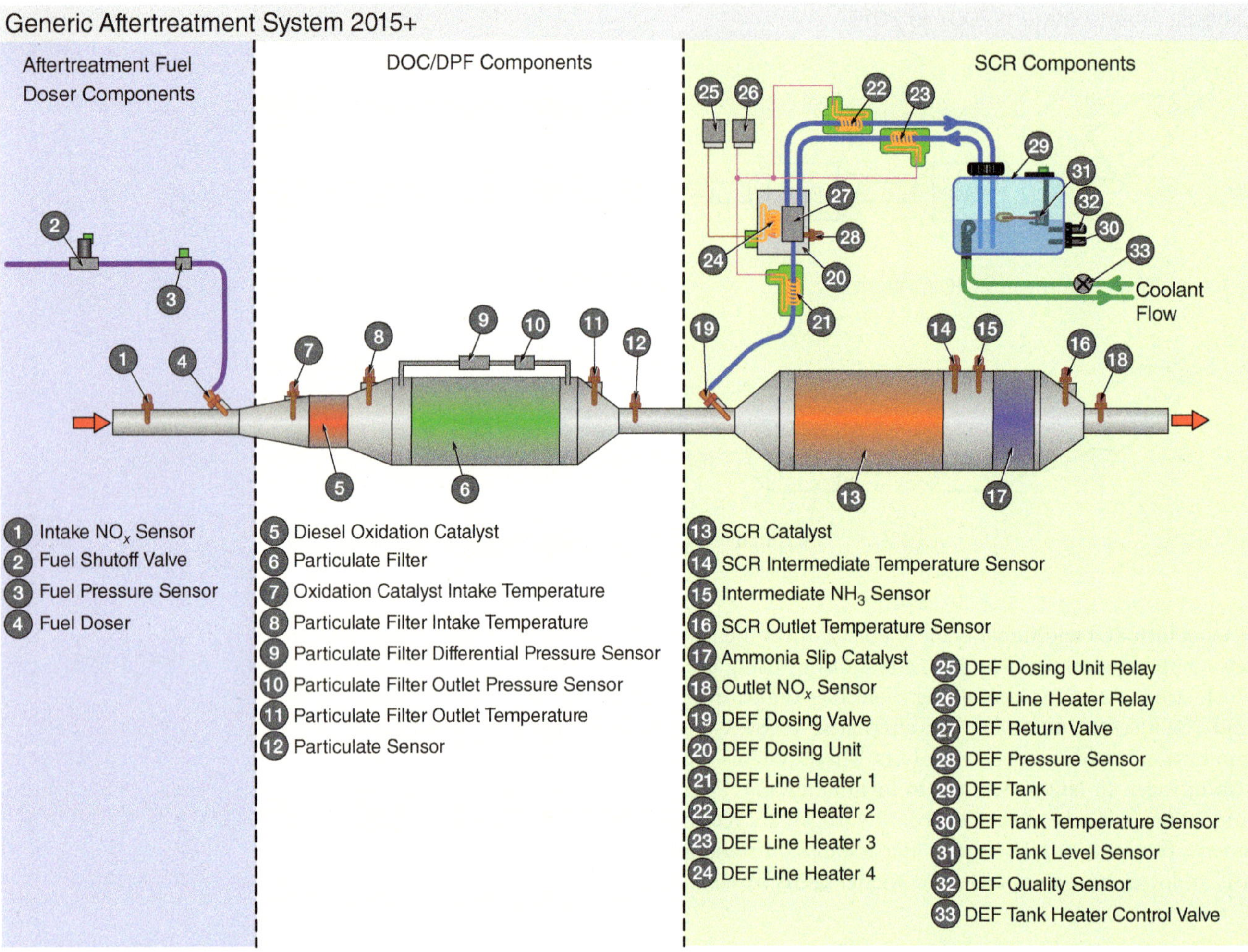

FIGURE 18-4 Typical components of a late-model aftertreatment system. It incorporates a diesel particulate filter (DPF) with a particulate sensor, SCR with an ammonia sensor, and an external hydrocarbon dosing valve. The latest aftertreatment systems use advanced sensors for improved diagnostics and emission cleanup.

to the already ultra-low Tier 2 standards. To meet these more stringent standards, further refinements to the aftertreatment systems combine even more components to previous ones, improve diagnostic capabilities, and increase heat retention to help aftertreatment systems operate more efficiently (**FIGURE 18-4**).

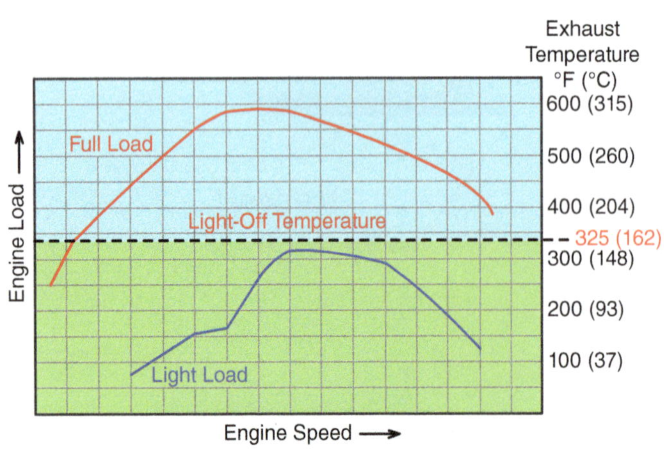

FIGURE 18-5 Exhaust temperatures of a mid-range diesel as a function of engine load. Conventional catalysts require at least a 325°F (163°C) light-off temperature to begin working. At light to moderate loads, diesel exhaust is too cool.

The same exhaust aftertreatment devices, such as catalytic converters, that convert CO, HCs, and NO_x and that used since the early 1970s on spark-ignition, gasoline-fueled engines cannot be directly applied to diesel engine exhaust systems, due to three major technical difficulties. The first problem is that the cooler exhaust temperatures of diesel engines have due to the excessively fuel lean or air-rich air-fuel ratios (AFRs). Unlike gasoline-fueled engine counterparts burning fuels at stoichiometric ratios, not as much heat is present in diesel exhaust to help the catalyst reach a minimum temperature to operate. Temperatures required for catalysts to begin to work are called **light-off temperatures** and are below the average diesel exhaust temperatures under light to moderate loads (**FIGURE 18-5**). Related to the use of excess air is the problem that air-rich exhaust pose for processes designed to remove oxygen from NO_x molecules. Traditional chemical processes that break

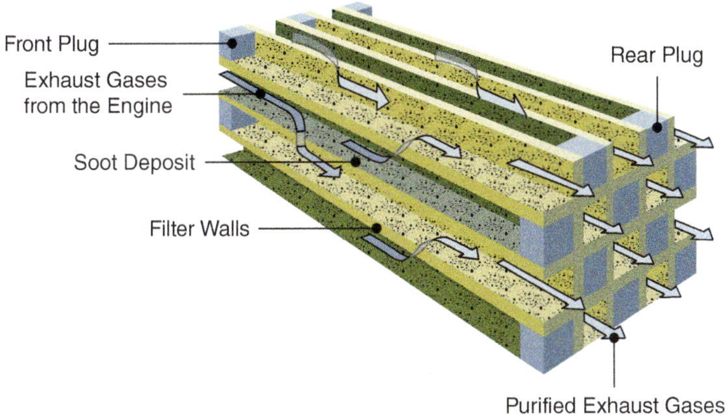

FIGURE 18-6 In a DPF, soot and other particulates collect on the walls of a ceramic filter. Heating the filter oxidizes the soot through a process called regeneration.

down NO_x in spark-ignition, gasoline-fueled engines can take place only in low-oxygen-content exhaust, where oxygen levels do not rise much above 2%. Reduction reactions, which are chemical reactions separating oxygen from NO_x molecules, need much richer air-fuel mixtures than are used by diesel engines. The third problem that diesel engines have using conventional catalysts is that the composition of emissions from diesel engines and spark-ignition, gasoline-fueled engines are different. Unlike spark-ignition engines, diesel engines are regulated for PM that consists primarily of black carbon or soot. A diesel-unique, purpose-built system is needed for PM cleanup (**FIGURE 18-6**). In addition to PM emissions, greater cylinder temperatures and pressures produced by higher compression ratios, turbocharging, and a diesel's more rapid burn phase translate into very low HC and CO emissions but disproportionally higher NO_x production. Higher NO_x production in turn demands a greater aftertreatment capacity for NO_x removal. Together, the problems with emission composition and with cooler and leaner exhaust conditions require new technology and approaches to cleaning up diesel engine exhaust emissions. Today, technicians can expect to encounter the following common exhaust catalyst elements configured into various combinations in the exhaust aftertreatment systems on contemporary diesel engines:

1. diesel oxidation catalysts (DOCs)
2. catalyzed diesel particulate filters (cDPFs)
3. SCR using liquid urea (SCR-U)
4. SCR using hydrogen reduction (SCR-H), including lean NO_x traps (LNTs) and NO_x adsorbers (NACs)
5. DPF combined with an SCR
6. DPF with SCR-coating (SDPF)
7. ammonia slip catalysts (ASC).

The above listed aftertreatment technologies are also often combined in the latest exhaust emission aftertreatment systems (**FIGURE 18-7**). For example, LNTs are also called NACs or NO_x storage catalysts (NSCs), which remove NO_x emissions, can be combined with SCR-U systems that perform the same task. SCR catalysts are also combined with DPF catalysts (**FIGURE 18-8**). **Electrically heated catalysts (EHCs)** using electric current to speed up warm-up is another subcategory of diesel exhaust catalysts (**FIGURE 18-9**). In addition to the common catalyst categories, the operation of the exhaust catalytic aftertreatment systems requires support from a variety of sensors and other subsystems:

- supplemental heating system (hydrocarbon fuel dosers)
- wideband oxygen sensors

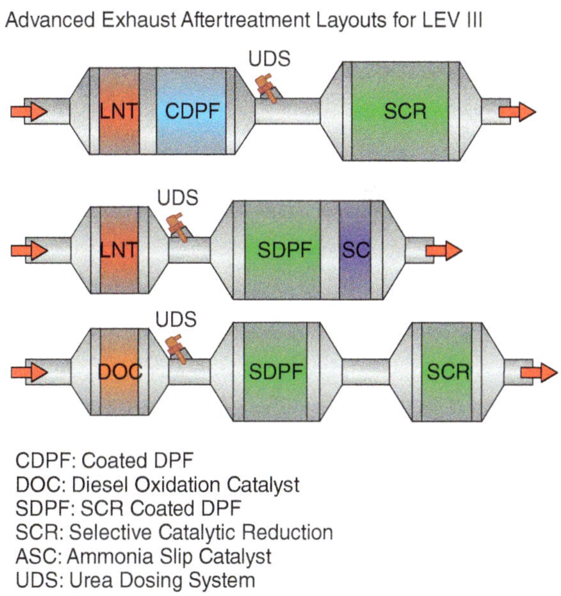

FIGURE 18-7 Common configurations for Tier 3 exhaust aftertreatment systems. Engine power output, emission standards, and packaging requirements are a few considerations in selecting the aftertreatment system components.

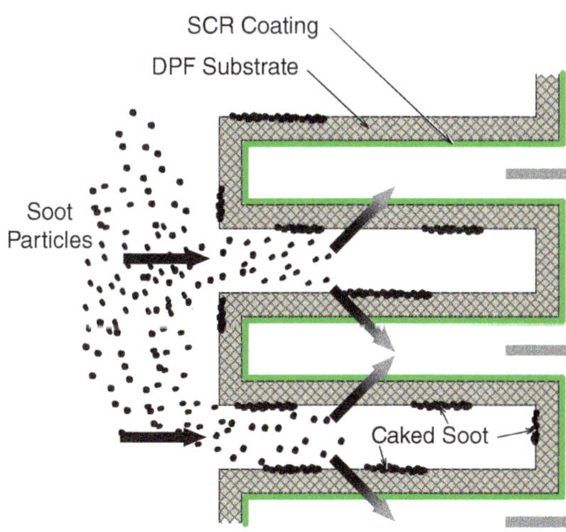

FIGURE 18-8 The concept of a combined DPF and SCR or SDPF coats the walls of the DPF with a copper catalyst material used to reduce NO_x emissions.

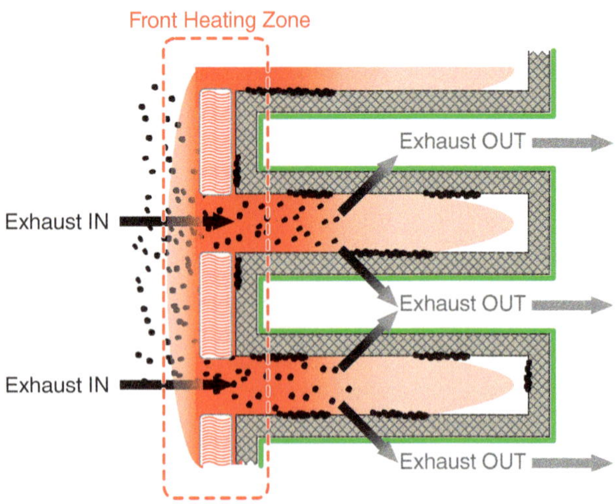

FIGURE 18-9 Electrically heated DPF use an electric heating element to burn out or regenerate the soot accumulations from the walls of the catalyst cell.

- NO_x sensors
- ammonia sensors
- particulate matter sensors (PMSs)
- temperature sensors
- pressure differential sensors (delta P sensors)
- DEF, storage, and delivery systems (**FIGURE 18-10**).

In terms of electronic control systems, manufacturers can build aftertreatment systems that adequately operate by using model predicted control (MPC). That means the control of the exhaust aftertreatment system can be accurately predicted primarily based on inputs from the driver and operating conditions such as engine speed and load. With MPC control, no sensors would be required for the exhaust system. However, to more precisely compensate for wear and component deterioration, identifying the fault conditions demanded by emission legislation—and, in some instances, to operate the system more efficiently—sensor feedback is necessary for the electronic control system that regulates aftertreatment operation.

▶ Types of Exhaust System Catalysts

18-02 Identify and describe the composition, construction, and operation of exhaust catalysts specific to minimizing the discharge of regulated engine emission.

Examining the types and operation of diesel exhaust catalysts is a good starting point for understanding diesel exhaust aftertreatment systems. Catalyst materials and catalytic converter construction are adapted for removing specific emissions from the exhaust. To identify the various types and operating principles of the converters—which are devices changing noxious emissions into harmless combustion by products—it's important to understand a few things about catalysts. First, by definition, a catalyst is a substance that changes the speed of chemical reactions without actually entering a reaction. In the case of diesel exhaust, catalysts usually speed up chemical reactions. Exhaust catalysts for diesels break down noxious exhaust emissions into harmless combustion by-products and allow reactions to take place at lower temperatures (**TABLE 18-1**). Diesel catalytic converters can even temporarily store emissions until correct conditions have been met to chemically break them down (**FIGURE 18-11**). This last feature is especially important solution for the

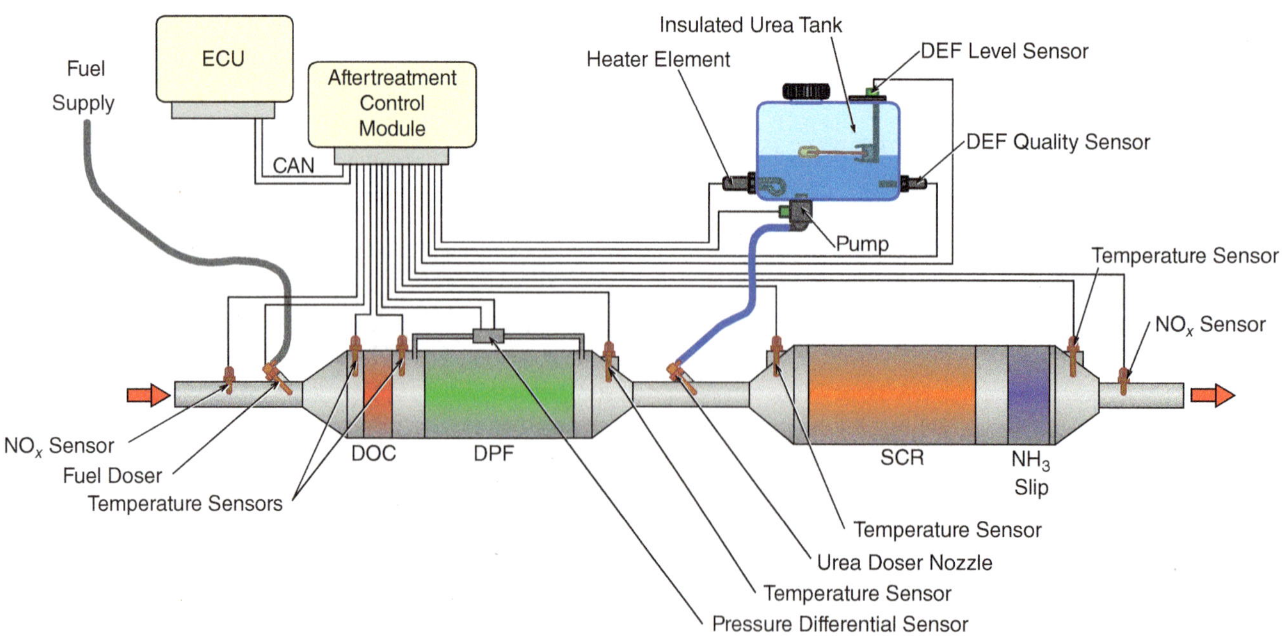

FIGURE 18-10 Aftertreatment catalysts require additional support components, such as a DEF storage, pumps, sensors, hydrocarbons, or fuel-dosing valves and electronic controls.

TABLE 18-1 Comparing Types of Catalytic Technology Used by Diesel Engines and Spark-Ignition Engines

Catalyst Technology	Reaction Type	Gasoline Engines	Diesel Engines
Two-way (oxidation) catalytic converters	Oxidation	CO, HC	
Three-way catalytic converters	Oxidation and reduction	CO, HC, NO_x	
DOC	Oxidation		CO, HC, PM, NO
Selective catalytic reduction using liquid (SCR-U)	Reduction		NO_x
NAC also called LNT, NSC, and SCR-H	Reduction		NO_x
Diesel particulate filters (DPFs) Or periodic trap oxidizer (PtOx)	Oxidation		CO, HC, PM
Combined SCR and DPF (SDPF) (four-way diesel catalytic systems)	Oxidation and reduction		CO, HC, PM, NO_x

unique problem that diesel engines have with changing or converting emissions into harmless substances due to naturally low exhaust system temperatures.

Catalysts can be classified by the type of construction materials used to make them. Exhaust heat and catalytic metals—such as platinum, rhodium, palladium, copper, and iron—are embedded into a base material, also called substrate, and are critical elements of diesel catalysts. The catalyst substrate, also called a brick after it's extruded into the honeycomb-shaped pores, is made from ceramic compounds (**FIGURE 18-12**). The most common substrate materials are synthetic cordierite, which is a light brown silica sand–based material, and gray aluminum titanate (Al_2TiO_5) or silicon carbide (SiSiC). Aluminum titanate and silicon carbide are premium-grade materials used for their ability to resist damage at high temperature and withstand abrupt temperature changes (**FIGURE 18-13**).

Catalysts are further classified according to the noxious emission that it converts or the type of chemical reactions that takes place in the catalyst material. Variations on the names and

FIGURE 18-11 A close-up view of a particulate filter catalyst substrate that uses a wall flow catalyst, which has alternate ends of each tube or cell plugged. The material contains a porous ceramic material called zeolite, which enables it to temporarily store emissions on its surface.

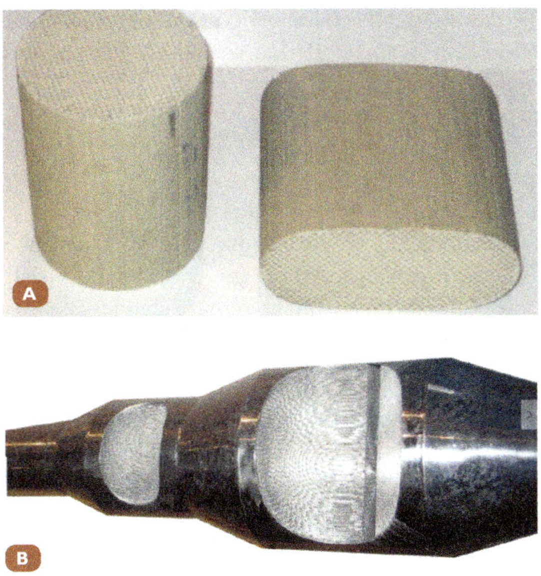

FIGURE 18-12 **A.** Ceramic catalyst brick or substrate. **B.** Metal foil catalyst.

FIGURE 18-13 To withstand high temperatures during regeneration and mechanical shock, DPFs and other diesel catalysts are made from ceramic such as aluminum titanate or silicon carbide. The gray appearance characterizes these DPF substrates.

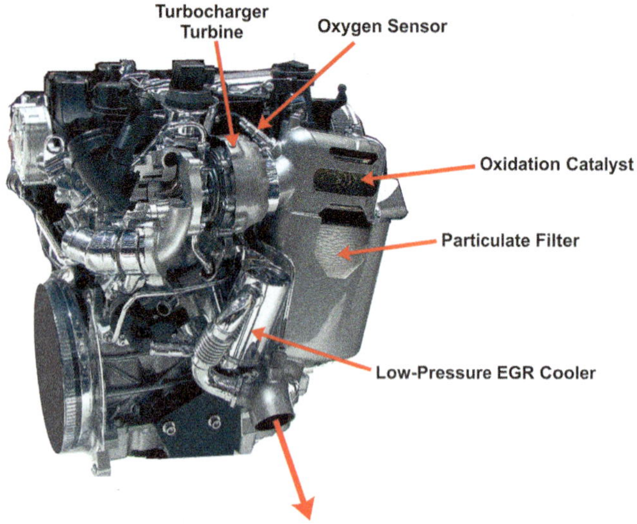

FIGURE 18-14 A close-coupled catalyst places the exhaust catalyst close to the engine to speed up catalyst warm-up and increase catalyst operating temperatures.

acronyms used for each catalyst differ between manufacturers, but the following are some of the most common types of diesel catalytic converters:

- DOC
- catalyzed diesel particulate filters (cDPFs or DPFs), also called a periodic trap oxidizers (PtOx)
- SCR, including SCR-U
- DPF combined with SCR (SDPF)
- NAC, also called LNT or NSC.

Challenges for Diesel Catalyst Technology

Diesel catalyst technology is not without problems. As previously mentioned, one difficulty in cleaning up diesel exhaust is the relative coolness of diesel exhaust temperatures. Unlike gasoline-fueled engines, the lean-burn combustion strategy of a diesel engine means that exhaust gas is diluted with excess air, reducing its temperature. Even under moderate to heavy loads and at the normal exhaust pipe temperature, the pipe after the turbocharger of many diesels rarely rises above 300°F–400°F (149°C–204°C). Much hotter temperatures are required to burn particles of carbon stored in exhaust filters. Soot particulate is burned off beginning at a temperature of approximately 932°F (500°C). The ignition temperature of soot is actually closer to 1,112°F–1,200°F (600°C–650°C). In contrast, a conventional oxidation catalytic converter found on a gasoline-fueled vehicle requires exhaust temperatures as high as only 500°F (260°C) for catalysts to effectively operate. Temperatures needed for diesel exhaust would make the use of spark-ignition, gasoline-fueled catalyst technology ineffective. In addition, the level of sulfur in diesel fuel can negatively affect the temperature at which chemical reactions in some catalytic materials occur. One solution to the low heat problem is the use of close-coupled oxidation converters. Close-coupled converters refer to catalytic converters placed close to the engine-out exhaust stream (**FIGURE 18-14**). Without the loss of heat allowed by the time and distance that exhaust gases travel through an exhaust system, locating the converter next to the exhaust manifold increases catalyst temperatures. A second solution to the problem of cool exhaust temperatures is to supplementally heat exhaust by dosing the exhaust with fuel to raise exhaust temperatures. When fuel contacts the oxidation catalyst, it will cause the fuel hydrocarbons to react with oxygen to produce flameless combustion. The normally cool diesel exhaust gas will absorb additional heat from the oxidation converter after passing through it and heat the DPF and SCR catalysts downstream.

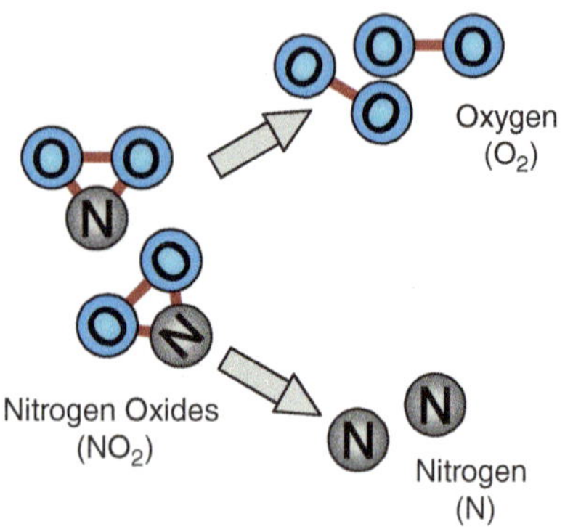

FIGURE 18-15 A reduction reaction removes oxygen from NO_x. It requires high temperature and an absence of oxygen.

Excess air in diesel exhaust presents another problem. Chemical reactions needed to convert NO_x to harmless nitrogen and oxygen emissions traditionally required an absence of oxygen in the catalyst substrate. Reduction reactions, which are chemical reactions where oxygen is removed from NO_x, can take place much more easily in the exhaust of spark-ignition gasoline-fueled engines because exhaust oxygen content in these engines is never above 2% (**FIGURE 18-15**). In contrast, diesel engines commonly operate between 10% and 19% oxygen content, although 15% is an approximate average (**FIGURE 18-16**). To overcome this problem, new catalyst materials and techniques are used to remove noxious emissions from diesel exhaust.

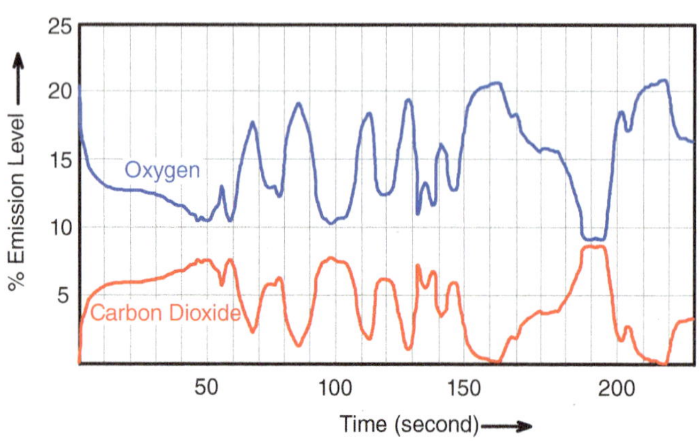

FIGURE 18-16 Oxygen and CO_2 concentrations from a diesel operating in urban driving conditions. As more fuel is burned, CO_2 increases and the oxygen content in tailpipe gases decreases.

Diesel Oxidation Converters

An oxidation catalytic converter causes oxygen (O_2) to chemically combine with hydrocarbon (HC) and carbon monoxide (CO)

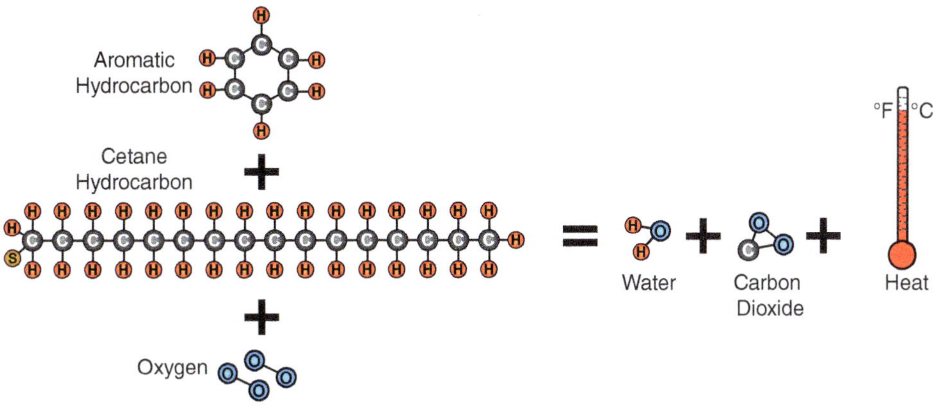

FIGURE 18-17 Oxidation reactions combine oxygen with other substances, such as hydrocarbon, to produce water and carbon dioxide. Oxidation reactions release heat.

noxious emissions at lower temperatures. The result is that HC and CO are converted to harmless carbon dioxide (CO_2) and water (H_2O) (**FIGURE 18-17**). Oxidation of these emissions releases heat and takes place without flames inside the converter. Because oxidation converters clean up two noxious emissions in spark-ignition, gasoline-fueled engines, they are called two-way converters. However, in diesel exhaust, oxidation converters can also oxidize black carbon soot to decrease the mass of diesel PM emissions to between 5% and 90%. In summary, the reactions are as follows:

$$C + HC + CO + O_2 = CO_2 + H_2O + \text{Heat}$$

DOC's also convert nitric oxide (NO) into nitrogen dioxide (NO_2) through oxidation. Heat is released from this oxidizing NO just as it is when converting CO, HC, and PM. Conversion of NO to NO_2 is important to DPF operation since NO_2 can react with soot to regenerate the DPF at cool temperatures. A second benefit of NO to NO_2 conversion is that it provides additional heat that can also warm up catalysts downstream of the oxidation converter, such as the DPF.

While the DOC oxidation process can heat the exhaust gases downstream of the DOC, spraying a small quantity of fuel into the exhaust stream, where it is vaporized, enables the DOC converter to provide even higher temperatures to heat the DPF. When fuel HCs are added to the exhaust stream, they will oxidize in the DOC, creating enough heat to ignite and burn off soot accumulations downstream, in the particulate filter. Manufacturers use the term "supplemental heating" to describe this addition of heat to the relatively cool exhaust.

DOC Construction

Oxidation converters consist of a corrosion-resistant stainless steel canister containing an extruded honeycomb-like ceramic substrate. The extrusion process forms parallel channels that stretch from one end of the catalyst brick to the other through the ceramic. When channels extend to each end of the catalyst, this configuration creates what is called a flow-through-type substrate (**FIGURE 18-18**). Flow-through-type substrate forms a huge surface area for exhaust gases to contact the catalytic material on the surface of channel walls. Catalyst is deposited onto the channel walls by coating them with a **wash coat** made up of catalytic precious metals, such as platinum or palladium, which are the active material that catalyzes the chemical reaction between noxious emissions and O_2 in the exhaust (**FIGURE 18-19**). As the name suggests, the wash coat is applied as a water-based slurry, which is absorbed by the substrate and then dried.

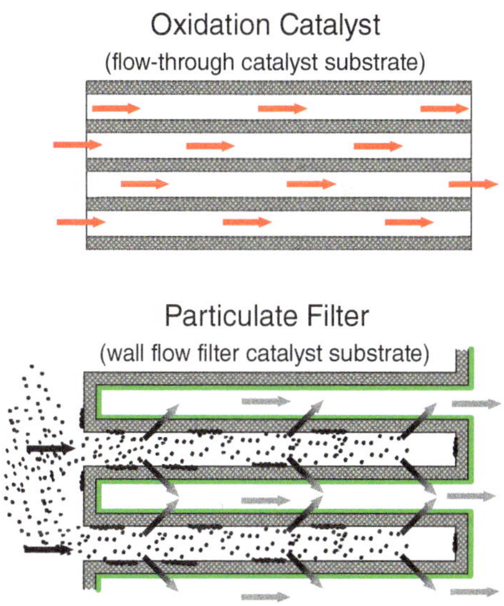

FIGURE 18-18 Wall flow catalyst substrate compared to flow-through substrate. Oxidation converters are made from flow-through substrate.

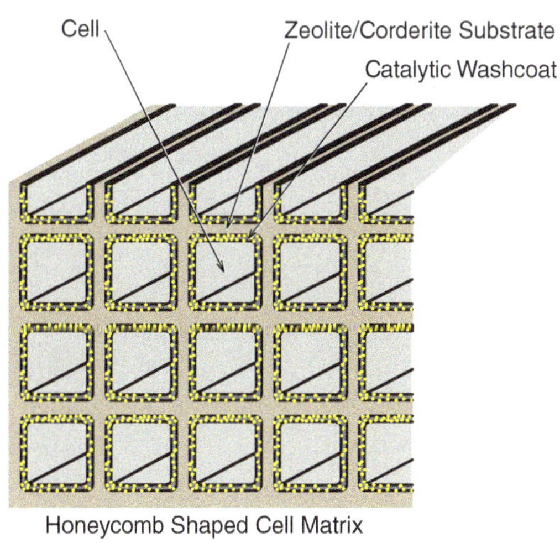

FIGURE 18-19 A wash coat of porous catalytic material is applied to catalyst substrate.

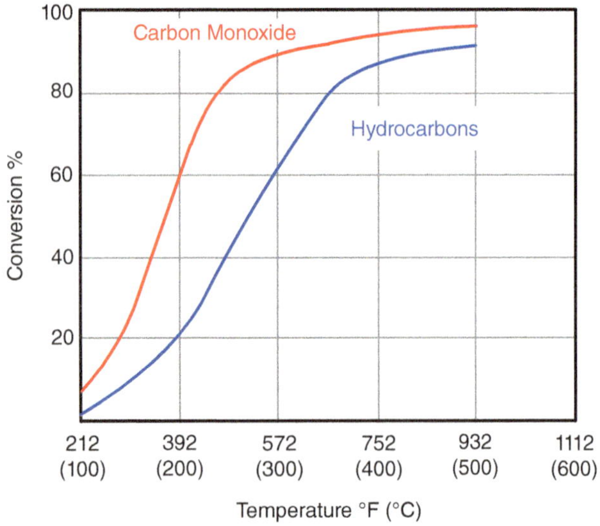

FIGURE 18-20 Low-temperature zeolite combines with specialized catalysts and together converts diesel emissions at low temperatures.

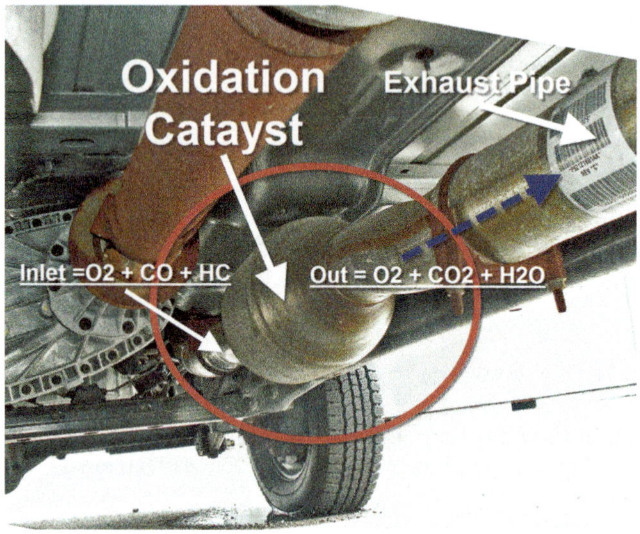

FIGURE 18-21 2005 Dodge Ram 3500 exhaust. Oxidation converters are used alone to remove diesel odor from exhaust, which is caused by aromatic HCs.

> **▶ TECHNICIAN TIP**
>
> Excessive stop-and-go driving, short trips, or idling can result in excessive particulate production. Leaking injectors or injectors with abnormally low spray in pressure can also produce the same result. If oxidation converters are not adequately heated by vaporized fuel and exhaust heat, they tend to trap the soot and unburned fuel on the face of the substrate and can plug or restrict the converter. This condition, known as face plugging, can enable exhaust gases to pass through a narrower section of the DOC too quickly to be properly heated and oxidized. A narrowed, face-plugged oxidation converter allows exhaust particles to enter the particulate filter but without adequate heat, and this can cause the filter to prematurely plug.

Low-Temperature Oxidation Catalysts

Diesel oxidation catalytic converters are able to burn HC and CO emissions emitted from the engine even at low temperatures by using unique HC-trapping materials. To convert, emissions in cool exhaust they are first trapped or adsorbed on the substrate. When exhaust temperatures rise, the stored emissions are released and burned. One common material used to enable emission storage by DOCs is a crystalline-shaped mineral incorporated into the substrate called zeolite. Zeolite is similar to the filtering sand used for pools and hot tubs. The complex surface shape of this mineral material captures the noxious emissions in the catalyst substrate. Zeolite adsorbs, meaning it traps in the surface, HCs when exhaust temperatures are low, such as during engine idle or part throttle. However, when the exhaust temperature increases under load, the HCs are released from the substrate and oxidized by the platinum-based wash coat catalyst. These low-temperature converters begin to catalyze emissions at approximately 250°F (121°C), and its highest efficiencies begin at above 360°F (182°C), which is much lower than traditional catalyst substrate materials (**FIGURE 18-20**).

One other benefit of using oxidation converters is that they can minimize the characteristic pungent diesel odor caused by unburned aromatic-type HC molecules in the exhaust. Aromatic HCs are the hard to burn, irregularly shaped molecules naturally found in diesel fuel. In older light-duty automobiles, light trucks, and buses, reducing exhaust smell is often the primary purpose of the oxidation converter. Low-temperature oxidation catalysts will typically reduce emissions of HC by 50% and CO by 40%. Operation of these converters increases to over 90% efficiency when ultra-low-sulfur diesel (ULSD) fuel is used (**FIGURE 18-21**).

> **▶ TECHNICIAN TIP**
>
> BlueTec is a Daimler AG (formerly DaimlerChrysler) marketing name for its two NO_x-reducing systems used by their European and North American diesel engines. BlueTec-II is a liquid-based SCR-U aftertreatment system that uses DEF marketed as AdBlue. Liquid DEF is sprayed into the exhaust system and breaks apart NO_x molecules in the exhaust with the assistance of a copper or iron catalyst wash coat on a zeolite substrate. The other BlueTec-I system reduces NO_x emissions without onboard storage of DEF. Daimler AG uses hydrogen as a reducing agent for NO_x emissions by using what corporately is referred to as a $DeNO_x$ catalyst (**FIGURE 18-22**).

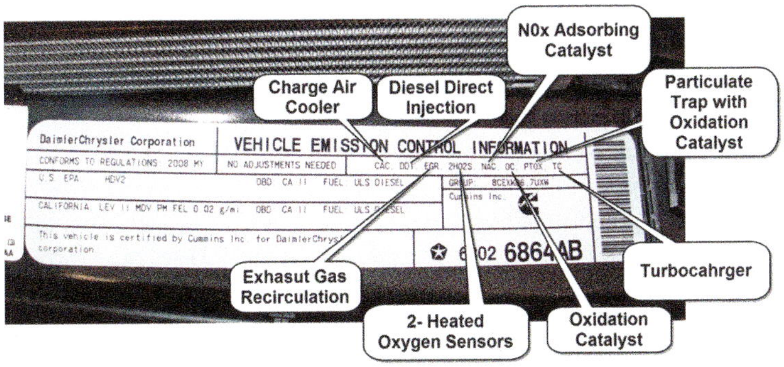

FIGURE 18-22 An emission decal from a BlueTec-I Dodge Ram with an NAC.

▶ Diesel Particulate Filters

18-03 Identify and explain the function of DPF substrate and regeneration strategies used to renew particulate filters.

The primary purpose of a DPF is to remove soot and other solid particles from diesel exhaust, rendering the exhaust completely smokeless. Although diesel engines have been nearly smokeless since 1998, the newest DPF eliminates the microscopic particles down to the size range of 15–500 nanometers (0.015–0.5 microns) by more than 90%. This technology is effective enough to remove up to 99% of all solid PM from the engine exhaust. In fact, examining the tailpipe of diesel engines equipped with particulate filters and comparing them to gasoline-fueled engines demonstrates the findings of several scientific studies, which indicate that diesel engines are cleaner than spark-ignition, gasoline-fueled engines in terms of particulate emissions.

The concept for removing PM is relatively simple. Soot and other exhaust-borne particles must pass through a wall flow–type filter, which traps solid particles on the filter wall before reaching the exhaust outlet. Similar to flow-through-type substrate with channels extending to each end of the substrate, wall flow has one end of each channel blocked alternately at the inlet and outlet end. A channel could be open at the filter inlet but is blocked at the outlet. An Adjacent channels would be blocked at the inlet but open at the outlet. Channel walls of wall flow–type filter media contains tiny pores that permit only the gas portion of the exhaust to pass through the filter material. Soot and other particles are trapped on the walls of the cell channels. Because PM is composed of some non-solid matter or liquids, overall efficiencies are calculated to be slightly less than 100%. Understanding this helps explain why the presence of any black streaks or visible soot in the outlet channels of a DPF indicate filter failure (**FIGURE 18-23** and **FIGURE 18-24**).

Soot trapped by the DPF is not very dense, and as much volume as 20 L of dry, uncompacted soot are needed to weigh even 500 grams. A fully loaded DPF may contain 4–6 grams of soot per liter of volume capacity. The smallest and most dangerous particles that are easily respirable into the deepest parts of the lungs are practically weightless because of their size. These fractions or portion of total PM can actually make up the greatest mass of PM emissions from the latest engines. Although emission soot has traditionally been measured according to weight, increasingly stringent PM standards have turned

FIGURE 18-23 The inlet face of a DPF is usually covered in soot. The outlet face must be clean.

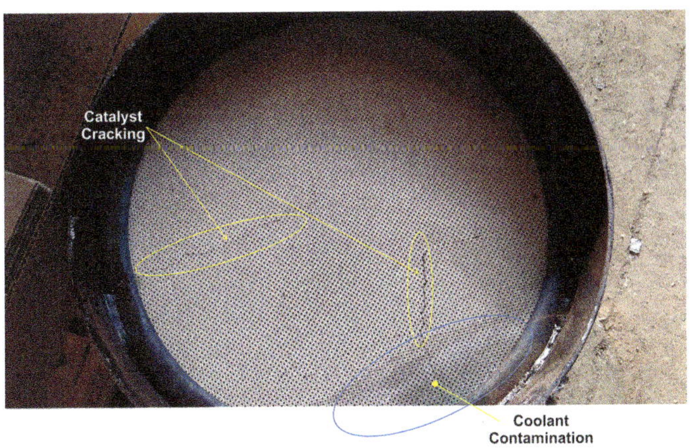

FIGURE 18-24 The outlet face of a DPF with cracking due to thermal stress. Discoloration on the bottom edge is due to contamination of both coolant and oil.

to a more realistic measurement of PM, which is particle number (PN) count. PN count more accurately recognizes and accounts for the microscopic portion of PM contribution to the total PM measurement. In Europe, current PM mass limits are additionally regulated according to PN, beginning with Euro 5 standards. For passenger vehicles, PM is regulated to 0.0045 g/km and a PN count not to exceed of 6.0 × 1,011/km. Particle sensors used by the onboard diagnostics (OBD) systems in Europe are designed to detect increased PN counts from leaking PM filters. This regulation is in contrast to North American EPA emission regulations, where PN standards are only under consideration and PM leakage through DPFs detected by particle mass sensors (PMS) are required to detect only very low levels of PM escaping from defective PM filters. Since 2016, DPFs in North America began to be equipped with particulate sensors to detect PM leakage past filters (**FIGURE 18-25**).

FIGURE 18-25 A PMS for detecting leakage through a DPF filter.

Single Brick Catalysts

To lower exhaust backpressure and reduce the space required by aftertreatment systems, the latest vehicles combine the function of a DPF with an SCR material to create a device referred to only by its acronym SDPF. The converter substrate is a wall flow type with the addition of a copper or iron SCR catalyst wash coat. The filter traps and burns soot while reducing NO_x with the assistance of DEF sprayed upstream of the converter (**FIGURE 18-26**).

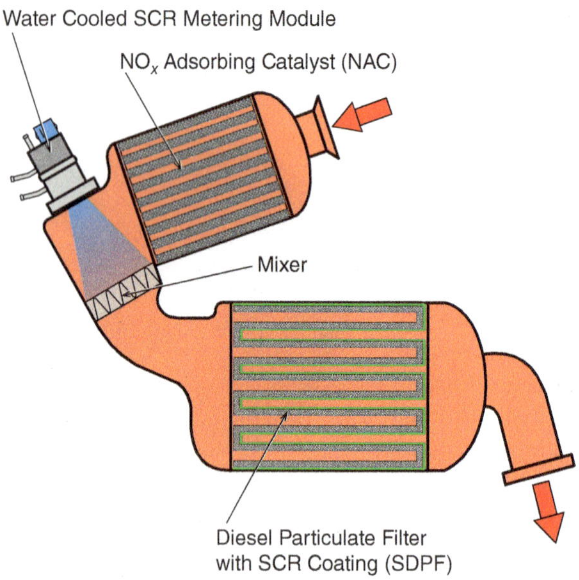

FIGURE 18-26 A compact single brick catalyst combines an SCR with a DPF to form a SDPF to reduce space requirements underhood.

Diesel Particulate Filter Regeneration

Trapped soot and other trapped particles collected on the walls of the filter will eventually plug the filter and produce high exhaust backpressure. If backpressure remains high, engine efficiency drops while the engine works harder to force exhaust gas out the exhaust pipe. Pressure differentials across the turbocharger turbine wheel could also decrease, resulting in low boost pressure. However, the greatest consequence of high exhaust backpressure is engine damage from piston contact with an open exhaust valve. High backpressure will eventually force open an exhaust valve with a weak valve spring when a piston crosses over top dead center (TDC). Contact between the valve and piston will have catastrophic results. Such consequences explain why engines are programmed to derate or lose power as a PM filter becomes excessively restricted. Regeneration, a filter renewal process, purges soot from the filter and cleans it to lower exhaust backpressure. In all DPFs, this process involves oxidizing or burning the combustible black carbon soot out of the filter.

Heat for regeneration comes from two sources. One method is through natural exhaust heat and chemical reactions in the oxidation catalyst between NO, HC, and CO. When natural exhaust heat and chemical reactions regenerate the filter, the process is called **passive regeneration**. No driver intervention is required, and the process can be carried out without any noticeable change in engine operation. As mentioned, the other method to regenerate a filter involves supplementally heating the DPF with fuel or another heat source, such as electric heating elements. This process is initiated by the engine or aftertreatment system control module and is called an **active regeneration**. The most common method to actively regenerate a DPF is to spray fuel into the exhaust stream through an injector, where it will flamelessly burn in the oxidation converter and raise exhaust temperatures to as much as 1,200°F (649°C) (**FIGURE 18-27**). Most DPFs will operate at just around or below 1,000°F (550°C) when regenerating normally. The process of adding fuel to the exhaust to increase exhaust temperature is referred to as HC dosing. The second and most recent method to go into production in the latest diesels is to actively regenerate filters by using electrically assisted DPFs (EADPFs). An EADPF uses electrical heating grids inside the inlet of the DPF that heats the catalyst to more than 1,100°F (593°C) to ignite trapped soot.

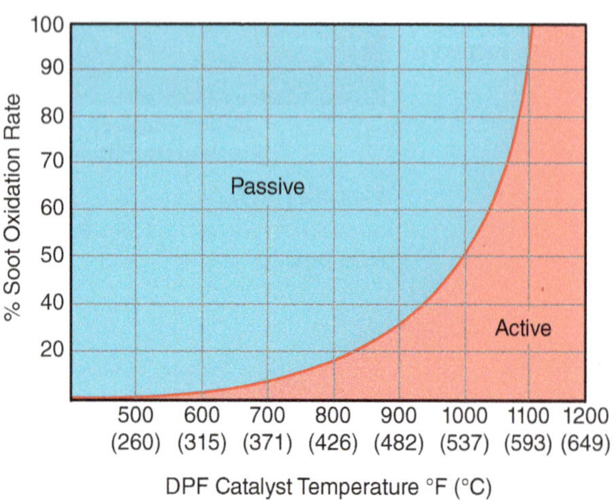

FIGURE 18-27 Passive regeneration takes place naturally in catalyzed DPFs at lower temperatures. Active regeneration requires supplementally heating the DPF.

Passive Regeneration

Passive regeneration occurs if the engine is operated under sufficient load to increase exhaust temperatures to at least 482°F (250°C). Oxygen in the exhaust will combine with soot to convert it to CO_2 and H_2O. Once again, catalyst material in the DPF walls helps lower the temperature normally required to combust the soot. Passive regeneration also uses NO, produced in the combustion chamber. Converting NO to NO_2 releases heat. NO_2 also combines with soot to produce CO_2 and nitrogen (N_2) (**FIGURE 18-28**). Chemically, passive regeneration looks like this:

Combustion with O_2: Soot + O_2 = CO_2 + Heat

Combustion with NO_2: Soot + NO_2 = CO_2 + N_2 + Heat

The conditions necessary for passive regeneration require the vehicle to be driven faster than 30 mph (48 kph) for 20–30 minutes in order for a full regeneration to occur. Passive regeneration occurs without any intervention by, or awareness of, the driver.

Regeneration Heat Management Strategies

To enhance passive regeneration, many diesel engines are equipped with throttle plates that restrict airflow into the engine to increase exhaust temperature. Reducing airflow limits the dilution of exhaust gas with excess air, which would normally lower exhaust temperature. Some manufacturers also intentionally lower boost pressure from variable-geometry turbochargers (VGTs). This has the same effect as restricting air intake flow. A post-combustion injection of fuel into the exhaust stream contributes approximately 45% of the increase in exhaust temperature. Electronically controlled throttle plates contribute approximately 35%, and 10% of the remaining heat comes from slowing VGT speed, which will reduce exhaust backpressure and lower boost pressure at idle. Exhaust backpressure refers to exhaust gas pressure buildup in the exhaust system, and it can occur before or after the turbocharger's turbine. In this case, back pressure is reduced in the exhaust manifold. If idle period is prolonged, even EGR flow is reduced or shut-off to prevent fouling of EGR valves and face plugging of exhaust catalysts. Although the process of passive regeneration

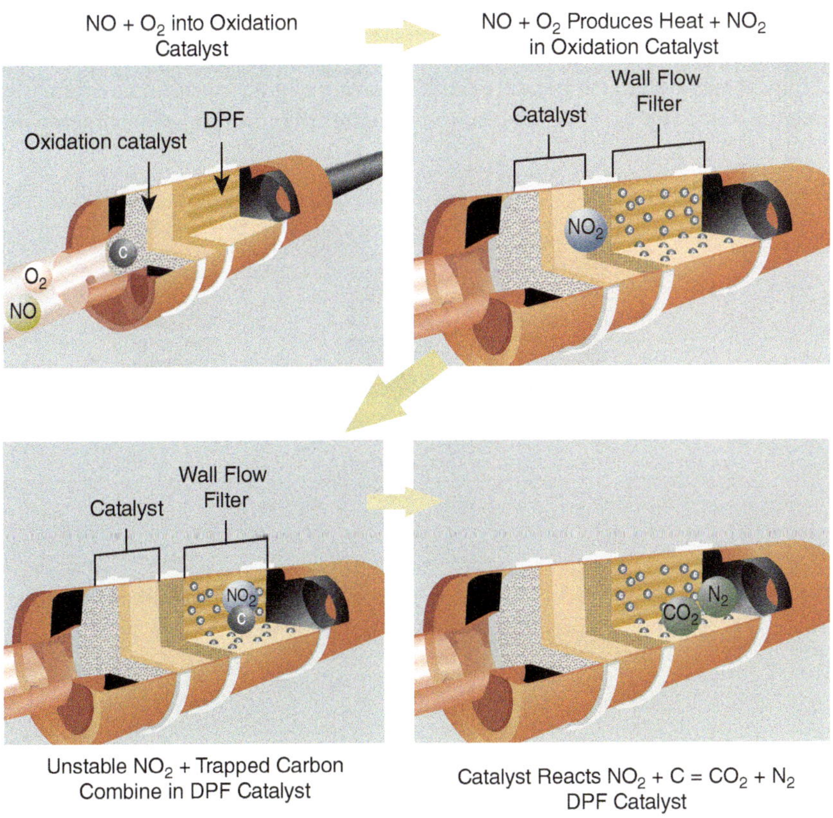

FIGURE 18-28 Passive regeneration in a DPF uses NO as a heat source.

FIGURE 18-29 A 3.0 L EcoDiesel used by Dodge. The intake throttle plate is a critical part of the regeneration strategy. Restricting the throttle opening at idle and during warm-up elevates exhaust temperatures by 200°F (111°C).

is supposed to be transparent to the driver, lowering boost pressure or throttling the air intake to increase exhaust temperatures can limit engine power output. For this reason, some manufacturers install a warning light or display information on a driver information display unit to inform the driver that regeneration events are taking place (**FIGURE 18-29**).

Warning Light and Messages for Diesel Particulate Filters

Prolonged low-load operation, which prevents the DPF from passively regenerating, also requires a warning to the driver to change driving style. The minimum vehicle speed of 30 mph (48 kph) required for any passive regeneration may not be enough in some vehicles for regeneration after the DPF has been heavily soot loaded. A warning light or message in the driver's information center alerts the driver to increase engine speed and load. This is done by driving faster than 45 mph (70 kph) for 20–30 minutes to quickly regenerate the filter and lower the soot load. If the filter becomes excessively soot loaded, a power derate will progressively increase from 15% to 30% until the point when the DPF exceeds maximum safe soot load. (**FIGURE 18-30**).

The DPF warning light or message is normally off. When on, at typically around 80% of maximum safe soot load, it alerts the driver to change the engine's duty cycle, or driving mode, to increase the exhaust system's natural, load-dependent heating. That means that rather than stop-and-go driving or short-trip operation, a steady, moderate speed is necessary for regeneration. If a change in engine operation is not possible, an active regeneration is possible. The regeneration event can take place automatically or be initiated by the operator or by a technician using an electronic service tool. On a few late-model medium-duty vehicles, such as heavier models of pickups and delivery vans, an operator can initiate a regeneration event. Unlike heavy-duty vehicles, there is no regeneration inhibit switch in the dash of light-duty vehicles that is toggled to activate a stationary active regeneration event. Instead, a combination of cruise control switches on the steering wheel to navigate a driver information service menu enables an operator to initiate an active regeneration.

A high exhaust system temperature (HEST) warning light is an optional original equipment manufacturer (OEM) light that indicates higher than normal exhaust temperatures

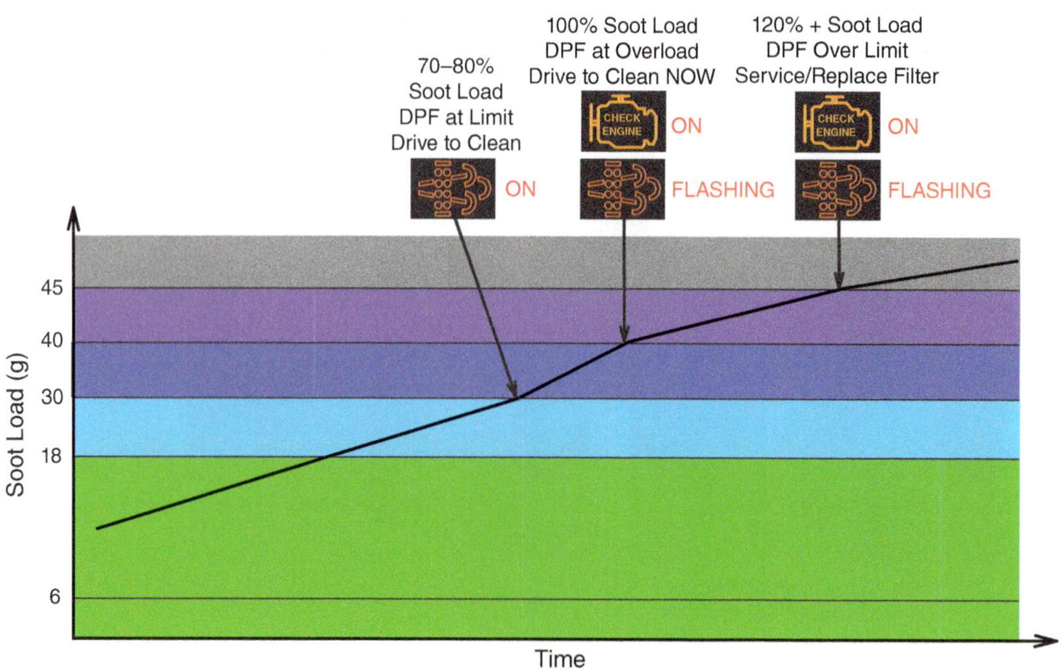

FIGURE 18-30 The DPF regeneration messages and warning lights will appear in the driver information center to alert the driver to change driving styles and when an active regeneration is in process.

during regeneration. For example, if the exhaust temperature is greater than 750°F (400°C) and the vehicle is operating at less than 5 mph (8 kph), then if equipped, the HEST warning light will illuminate. The HEST lamp helps prevent injuries to pedestrians inadvertently passing the exhaust outlet. Or the light will help avaoid grass and brush fires, or igniting materials found in waste transfer or manufacturing facilities if the vehicle is operated indoors (**FIGURE 18-31**).

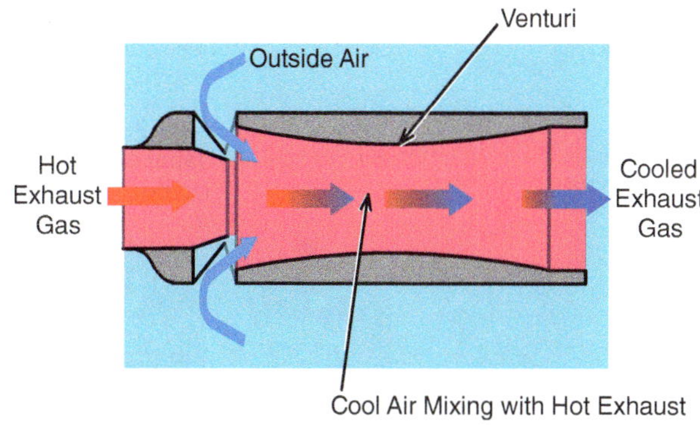

FIGURE 18-31 Very hot exhaust is ejected during an active regeneration. To prevent fires and injuries, heat is dissipated at hot exhaust tailpipe tips by drawing in air around double-walled pipes. Exhaust flow out the pipes creates a Venturi effect and pulls in air to cool the exhaust gas and pipes.

Active Regeneration

During an active regeneration, the DPF is heated to a temperature required to burn soot by adding small amounts of fuel to the exhaust stream. An oxidation converter downstream of the vaporized fuel supplementally heats the exhaust system through a catalytic reaction with the fuel. Flameless fuel combustion in the oxidation converter heats the DPF as exhaust gas passes through the hot oxidation converter into the DPF.

Fuel for supplemental heating is supplied in two ways: One is to inject fuel into the exhaust by using a separate dosing valve (**FIGURE 18-32**). Located in the exhaust pipe after the turbocharger, the dosing valve operates like a conventional fuel injector controlled by the aftertreatment control module or the engine's electronic control module (ECM). In the hot exhaust, fuel converts to a vapor before entering the oxidation converter, which explains why the dosing valve is sometimes called a vaporizer valve. An additional fuel shutoff valve or redundant safety valve provides a second backup to prevent inadvertently over-fueling the exhaust if the dosing injector control valve fails. A second method to supplementally heat the exhaust is to use a post-combustion injection event of addtional fuel from common rail (CR) injectors (**FIGURE 18-33**). Fuel sprayed into the exhaust gases as they exit the cylinder, will vaporize, and pass through the oxidation converter. The ECM will command additional injections from CR injectors before bottom dead center (BBDC) or just after bottom dead center (ABDC) during the exhaust stroke. Vaporized fuel in the exhaust undergoes flameless combustion reactions in the oxidization catalyst.

Early diesels used this simple, low-cost method since it required no additional hardware. However, since 2010, when biodiesel-compatible engines and aftertreatment systems were mandated, this method is now rarely used. The problem with in-cylinder, late post-combustion injection events is that the high boiling point of biodiesel can cause

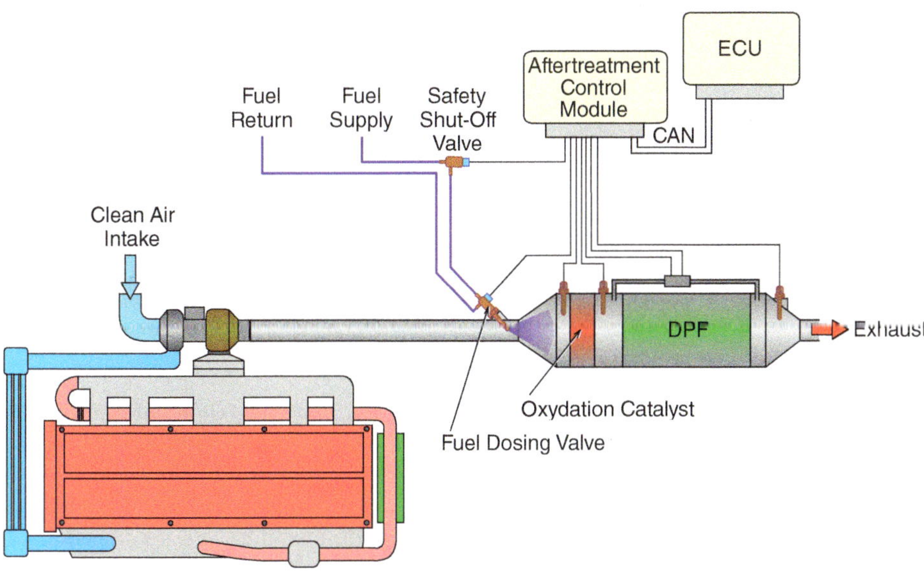

FIGURE 18-32 One method to supplementally heat the exhaust catalysts is to dose the exhaust with fuel using a HC injector located upstream of the oxidation converter.

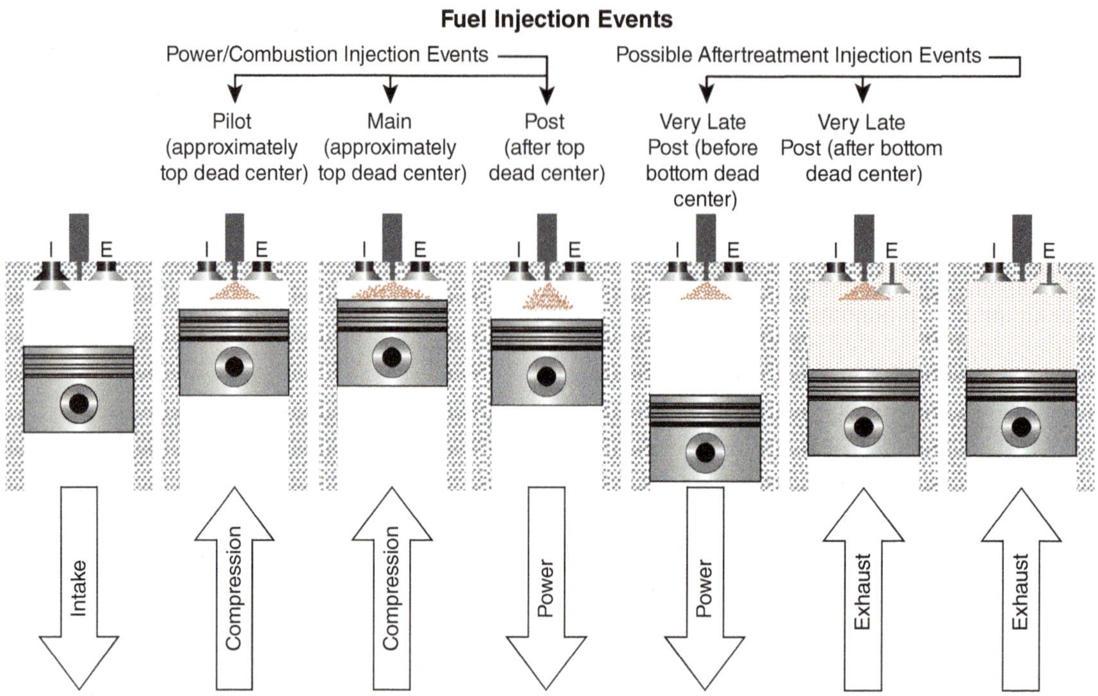

FIGURE 18-33 With a few exceptions, a late post-combustion injection event is used on older light-duty diesels to heat the oxidation catalyst to supplementally heat the DPF.

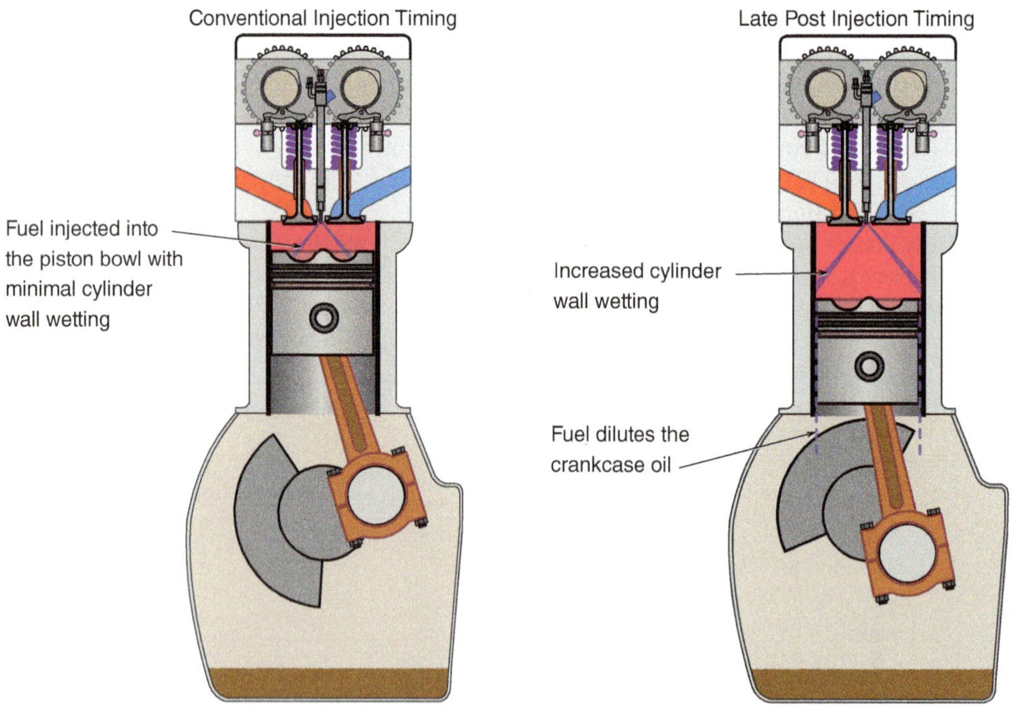

FIGURE 18-34 Late post–main injection events increase fuel dilution of the crankcase.

catastrophic engine failures when biofuel contaminates engine oil. When biodiesel washes down the cylinder walls, the fuel can pass around the piston rings with other blowby gases. After fuel is scraped into the oil, it can thicken lubricating oil into a pudding-like viscosity (**FIGURE 18-34**). Note that biodiesel's high boiling point can also lead to DPF plugging, particularly if the fuel does not meet American Petroleum Institute (API) specifications.

Active regenerations take place when an excessive soot load condition in the filter is detected by the aftertreatment control module or engine ECM or when a fault code for excessive soot loading is set because accumulations of soot are approaching a maximum safe level. If passive regeneration cannot maintain or reduce the soot level in the DPF and an excessive soot load condition or fault is detected, then the engine may be forced to perform and active regeneration. Conditions that commonly prevent passive regeneration from lowering DPF soot levels include driving for only short trips; extended vehicle idling; prolonged low-speed, low-load operation.

Types of Active Regeneration

During active regeneration, the supplemental heating and burning soot will cause the DPF's temperature to rise to approximately 1,000°F (550°C). On light- and medium-duty vehicles, active regeneration can take place under three circumstances:

1. automatically while driving (automatic regen)
2. when an OEM-scheduled safety regeneration takes place after a specific distance has elapsed
3. during an operator- or technician-forced event (forced, parked, or service regeneration).

Automatic regenerations, as mentioned, can take place when the vehicle is operating and an excessive soot load is detected. The degree of soot load that triggers the regeneration varies between manufacturers, but warnings commonly begin at 80% or 4 grams/liter of filter capacity. An active regeneration will take place after the warning and soot loading continues to accumulate. For example, a DPF with a maximum safe load of 40 grams of estimated soot load will typically initiate an automatic regeneration at 90% load, which means at 36 grams. The electronic control system will initiate the regeneration and continue the regeneration event until the soot load has been reduced to close to 0%. Although there may not be any noticeable change in engine performance, a slight to moderate decrease in instantaneous observed fuel economy will take place during the regeneration event. To increase exhaust temperatures, automatic transmissions will drop one gear and operate out of overdrive until the regeneration has finished. At idle or when stopped at an intersection, the engine speed will increase to approximately 950 rpm.

Safety regeneration events are similar to automatic regeneration. Also called distance regenerations, manufacturers will schedule a safety regeneration event in the event of a problem with the aftertreatment system ability to properly sense the level of soot load in a filter. Software will check to see how much distance or driving time has elapsed since the last regeneration has taken place and initiate a regen if the most recent regeneration was incomplete or too much distance has elapsed. A safety regeneration event will commonly take place after 1 or 1.5 tanks of fuel have been consumed or a specific distance equal to the distance a tank of fuel would generally last for.

Forced Regeneration Events

Forced regeneration events, also called stationary or parked regenerations, are initiated by a technician or vehicle operator in some vehicles. If a vehicle cannot be regenerated during a driving cycle or because the engine power is derated due to excessive soot loading, a forced regeneration may be performed by using an electronic service tool. If the option exists, a specific menu item available from the service tool, such as a scanner or service software, will allow a technician to initiate a regeneration event. Several prerequisite conditions must be met before the regen can take place. For example, the DPF cannot be excessively loaded, which often means no more than 120% of capacity or 6 grams/liter of DPF volume, for a regen to begin. If there is too much soot, the DPF catalysts could melt from the excessive heat produced when soot is burned. Such an event is called a thermal runaway. In the situation where the DPF is too full, it must be removed and replaced with another filter. The soot-loaded filter can instead be salvaged by baking it in an purpose made industrial oven to more gradually oxidize soot.

The following are some of the other conditions that typically must be met for a forced regeneration event to start and finish:

- more than ¼ tank of fuel
- an engine at operating temperature
- no active or pending faults that could interfere with the regeneration

> **TECHNICIAN TIP**
>
> After soot loading reaches 100% of the DPF capacity, the DPF warning light generally begins to flash and an alarm or chime may sound. The yellow service engine light, MIL, or glow-plug light may also illuminate if equipped, and a 15–30% power derate condition will take place to encourage service. Once loading reaches 120%, the engine will shut down and DPF regeneration cannot take place; The DPF must be removed and replaced with a new or reconditioned filter. Burning soot in the DPF could severely overheat the catalyst substrate. This condition, called a **thermal runaway**, can melt the catalyst and severely damage the substrate.

- an open hood to allow underhood heat to escape
- an applied parking brake
- a full DEF reservoir on SCR-U–equipped vehicles
- the brake, clutch, and transmission position switches may need to be cycled to validate that a safe park condtion has been properly met.

After the regeneration event begins, the engine speed will increase to as much as 1,800–2,500 rpm. The vehicle doors and windows may lock to prevent it from being left unattended while a regeneration is taking place. If excessive exhaust temperatures are detected or if the exhaust temperatures failed to reach expected values within a specific time frame, the regeneration event will end. Any engine faults detected during the regeneration event, such as excess coolant temperature or a problem with the EGR system or intake air throttle, will also prematurely end the regeneration.

A forced regeneration may take anywhere from 10 to 40 minutes, depending on the amount of soot loading. The temperature differential across the DPF measured by thermistors will initially be quite high between the DPF inlet and outlet when the regeneration begins. But as the soot level is reduced to 0%, the temperature differential gradually diminishes with no differential at all. Without any further soot to burn, no increase in temperature across the DPF can take place so the ECM will sense this condition and terminate the regeneration event. The absence of a temperature differential indicates when the soot load is reduced to 0%. A regeneration event is considered completed when the regeneration has successfully reduced the soot load in the DPF to 0%. In a large pickup, the regeneration will consume a little more than 400 grams or 0.5 L of fuel during a typical active regeneration of a 100% soot loaded DPF.

Electrically Assisted Diesel Particulate Filters (EADPF)

Supplementally heating a DPF with fuel during active regenerations has several disadvantages. The first is a fuel penalty. Each time a filter is actively regenerated, fuel energy is used that does not move a vehicle. A second problem is the time required to heat the DPF and the increased emissions produced during the regeneration event. Electrically assisted DPF regeneration will still use some fuel during a regeneration event, but heating takes place for only 4 to 6 minutes to reduce soot loading by 80–90% (**TABLE 18-2**).

In a EADPF, electrical heating grids arranged inside DPF substrate can ignite trapped soot. These heating elements are switched on or off and run for just 1 to 2 minutes, triggering a chain reaction that burns out the remaining soot (**FIGURE 18-35**). Rather than simultaneously switching all the electric heating elements on and off, which would consume a tremendous amount of current, zone heating is used. Sections of zones of a DPF are alternately switched on and off in sequential order to keep current consumption down to no more than 12 amps (**FIGURE 18-36**). An EADPF can also heat only specific segments of the DPF as they become restricted. Temperature sensors embedded in the filter substrate provide feedback about the cooler areas that are likely more heavily loaded with soot. Once the soot ignites, the zone can be turned off to complete a full regeneration cycle. A 50% reduction in fuel consumption is achievable when using EADPF regeneration. This technology offers the promise of enabling the use of a stop–start feature for urban driving or an electric hybrid

TABLE 18-2 Comparing the Regeneration Energy of a Conventional and Electrically Heated DPF

	EADPF Regeneration	Fuel-Based Regeneration
Soot Load (grams/liter)	4	4.9
Soot Regenerated (%)	85	112
Fuel Consumption (grams)	195.5	426.8 (58% more)
Electric Energy (kJ)	654.6	NA
Total Regen Energy (kJ)	9,044 (Fuel + electrical energy equivalent)	18,317 (49% more)
Time Required (min)	8	20

Source: U.S. Department of Energy. Chart adapted from data downloaded from www.fueleconomy.gov.

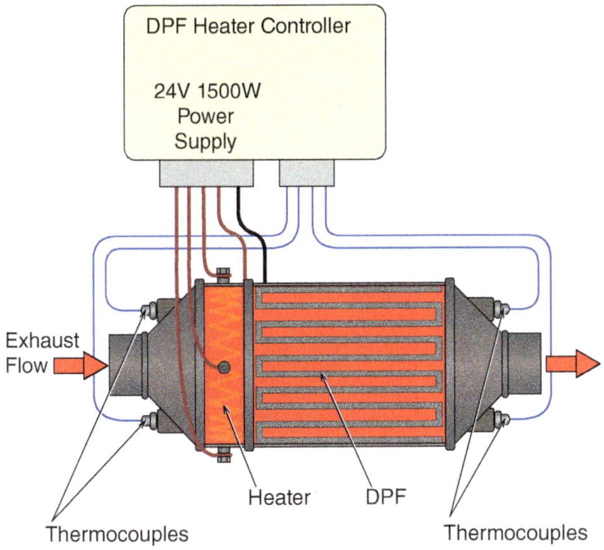

FIGURE 18-35 Concept of an electrically heated DPF. Electrical heating elements create heating zones at the front of DPF channels.

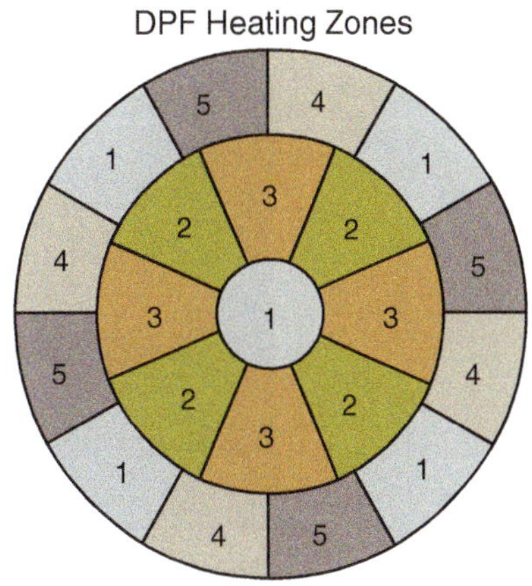

FIGURE 18-36 Arrangement of heating zones in a DPF. The zones are switched on and off depending on the temperature and soot load in the filter.

with diesel engines to be used in automobiles and pickups. In this application, the exhaust temperatures from prolonged "engine off" times, when the vehicle is in electric-only mode, make diesel engines impractical. However, electrically heated catalysts would allow more fuel-efficient diesels to regenerate without significant fuel penalties or the need for a prolonged highway drive cycle.

Diesel Particulate Filter Sensors: Pressure Differential Sensor

DPF systems require more than just filters to work. A variety of sensors and controls are needed to ensure safe and efficient operation, as well as monitor the DPF performance to meet OBD-II requirements.

One of the more important sensors is the pressure differential sensor. It measures the exhaust pressure at both the DPF inlet and the DPF outlet to report filter restriction (**FIGURE 18-37**). Filter restriction is needed to detect faults such as cracked, plugged, or missing DPF substrate. In a properly functioning and intact filter, pressure drops across the filter rises as soot load increases. Measuring the DPF restriction using a pressure differential (Delta-P) sensor provides the control module with an estimate of soot loading. After the filter has finished an active regeneration when it has been emptied of soot, a pressure differential reading is recorded to also provide an indication of ash load in the filter. The restriction is tracked over successive active regenerations to report a trend or progressive increase in weight of ash loading (**FIGURE 18-38**). Pressure differential sensors typically incorporate two pressure sensors with one output signal or two separate sensors with one or two output signals to the control module to measure exhaust pressure entering and leaving the DPF. The sensor location prevents water condensation from entering the sensor. Long, hollow metal tubes connect the DPF to the differential pressure sensor to cool hot exhaust gas temperatures and prolong sensor life.

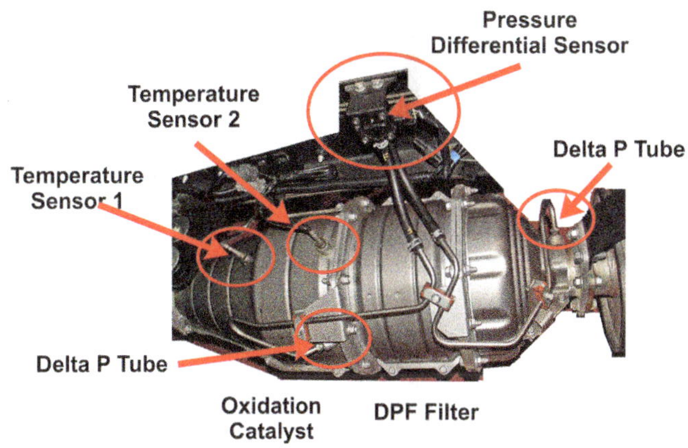

FIGURE 18-37 The pressure differential sensor, or delta P sensor, is a critical input for calculating soot and ash load and detecting catalyst substrate faults.

Comparisons made by the control module between expected values and actual measured pressure will determine whether a filter is functioning as designed. Because pressure differences across the filter will vary widely depending on engine load and speed, the delta P sensor data is not dependable and not exclusively used to measure soot or ash loading. Instead, a virtual model of system operation is used to estimate the amount of soot loading; this model is used to initiate active regeneration events or illuminate the DPF warning light and to estimate soot and ash loading. Using a model to predict soot loading helps explain why some

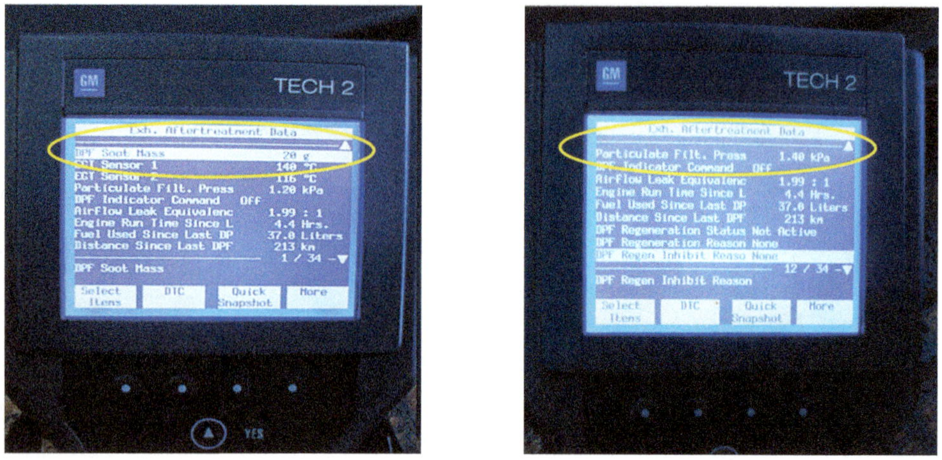

FIGURE 18-38 Delta P data from the pressure differential sensor are used as 1 of 2 major inputs to calculate soot and ash load in a filter.

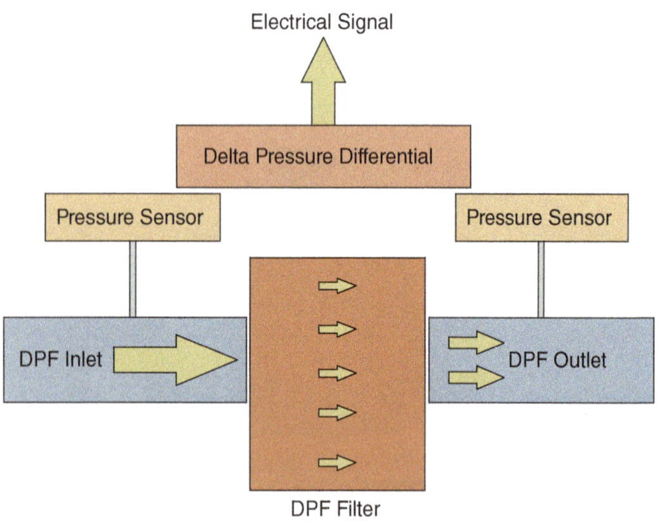

FIGURE 18-39 The pressure differential sensor, or delta P sensor, and temperature data can together calculate exhaust gas mass flow by using a speed-density algorithm.

vehicles with plugged DPFs will not have the DPF warning light illuminated whereas other vehicles may regenerate the DPF too frequently. Because the aftertreatment control module estimates soot and ash loading based on an expected model of performance, problems such as those with the fuel system, other engine defects, and extreme operating conditions or poorly developed algorithms will not correctly estimate the actual amount of soot in the DPF.

Deviations from expected delta P values or pressure differences will set fault codes such as for cracked, missing, plugged, or drilled-out substrate. Exhaust temperature sensor data and delta P data together measure gas mass flowing through the DPF by using a speed-density algorithm (**FIGURE 18-39**).

Temperature Sensors

To manage the regeneration functions of the DPF, detect faults, and alert the driver of usual operating conditions, most DPFs use a minimum of three temperature sensors. The first is located at the oxidation catalyst inlet; the second between the oxidation catalyst and DPF filter; and the third at the DPF outlet (**FIGURE 18-40**). A theoretical model of how the temperature and delta P sensors should respond under various operating conditions is compared against measured values. Deviations between expected and actual values will set fault codes. Temperature sensors are important ECM inputs used to prevent damage to a DPF. If DPF temperatures are too high, the DPF substrate can melt or crack (**FIGURE 18-41**). Excessive regeneration temperatures greater than 1,472°F (800°C) will stop the regen. If regeneration temperatures remain too low, then a defect with a component is likely and the regeneration will also stop.

Inputs from temperature sensors are used to end regeneration events. A normal regeneration ends when the temperature sensors readings at the inlet and outlet of the DPF stabilize at close to the same temperatures. This would indicate that no more soot remains in the DPF to incinerate and produce a temperature differential. Excessive soot loading in the DPF above 120% can lead to excessively high DPF temperatures and melt or severely damage the filter. After 120% loading, the DPF cannot be actively regenerated and must be removed to be serviced.

Exhaust temperature sensors are the resistive positive coefficient type, which means sensor resistance increases with temperature (**FIGURE 18-42**). Another name for the sensor is resistive temperature detector (RTD). These sensors contain a resistor that changes its value when heated. RTDs are very accurate and operate over a very wide temperature range.

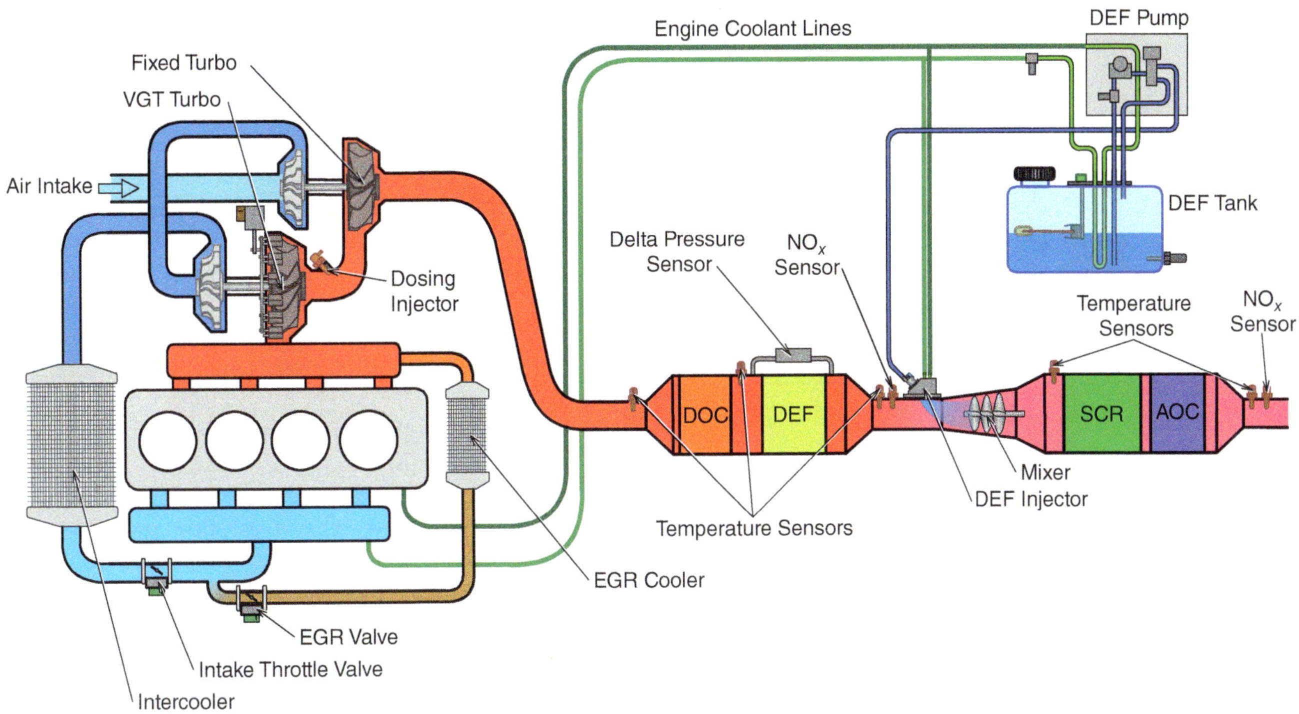

FIGURE 18-40 Temperature sensors are located throughout the aftertreatment system to monitor system operation. This Tier 4 off-road engine uses three RTD-type sensors to monitor DPF operation.

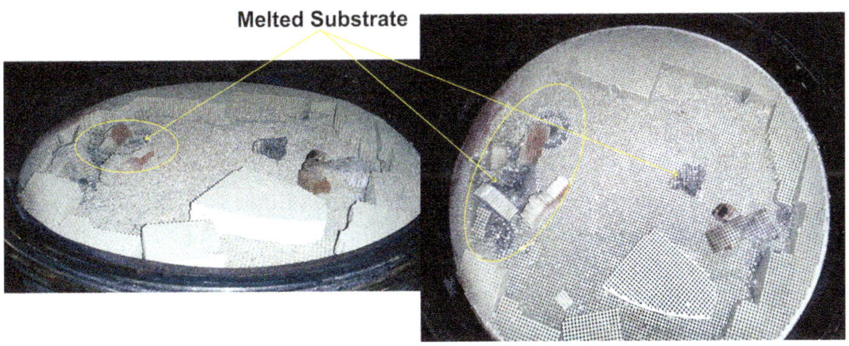

FIGURE 18-41 DPF filter substrate melted and cracked due to excessive heat.

These sensors are unlike the semiconductor-based common negative temperature coefficient sensors used to measure oil, fuel, or air temperature. Due to high exhaust temperatures, RTD are made from a thin wire wrapped around a ceramic or glass core.

Soot and Ash Load Calculations

An accurate measurement of a DPF soot load is necessary to determine when an active regeneration should take place or whether an operator should be alerted to change up a driving cycle. Accurately predicting soot load is difficult because soot will naturally accumulate and deplete at varying rates while the vehicle is operating, depending on a wide variety of factors: exhaust temperature; engine load; the condition of injectors; injection timing; and use of rate-shaped injection, EGR rates, fuel quality, idle time, ambient temperatures, and the amount of engine acceleration compared to steady-state operation. Despite the importance of estimating soot load and the numerous factors that influence loading, a direct method of measuring soot is

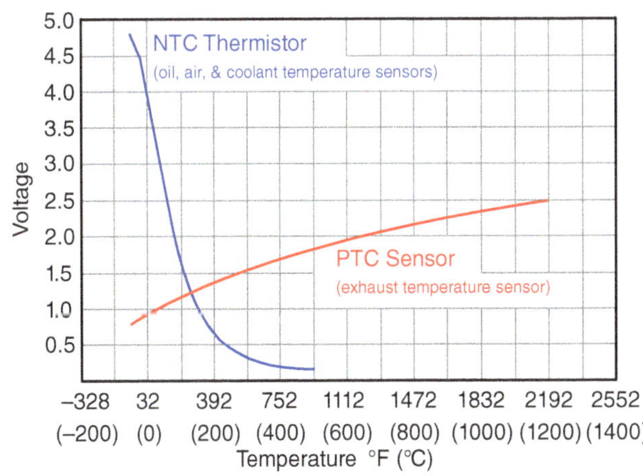

FIGURE 18-42 Comparing the voltage drop of a negative temperature coefficient (NTC) thermistor against the RTD sensor used in the exhaust. The RTD sensor measures temperature over a wider range and is heat-resistant.

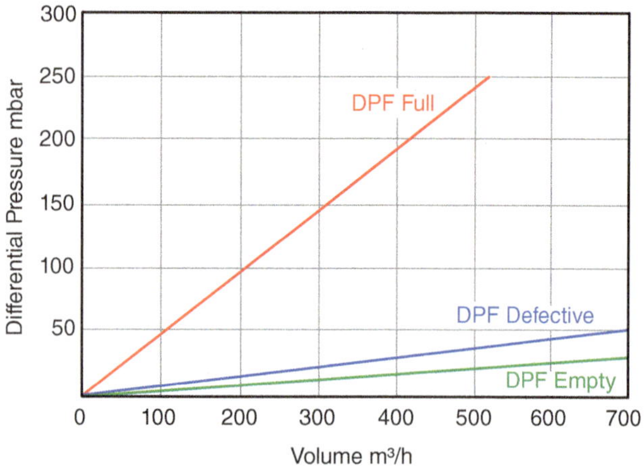

FIGURE 18-43 DPF filter restrictions as a function of exhaust flow volume. Note the effect of high soot or ash loading on the pressure drop across the filter and when the filter is cracked.

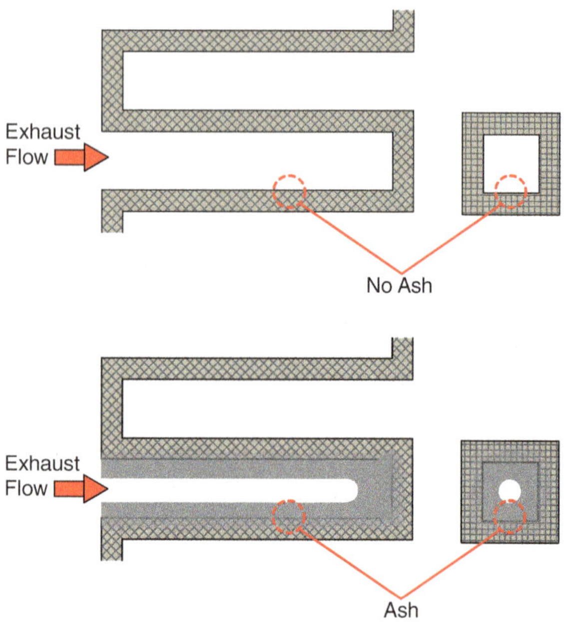

FIGURE 18-44 Ash loading of a filter loads displaces soot in filter cells and on channel walls. Ash is non-organic and cannot be regenerated. Progressively more ash loading leads to shorter intervals between filter regenerations.

not possible. However, two techniques are used by manufacturers to provide an estimate of soot load. One is to model or estimate the amount of soot based on engine operating conditions. This method is called the chemical model of soot load estimation. The second method is to measure the pressure difference across the filter. Filter restriction using a pressure differential sensor, also called a delta P (P is for pressure) sensor, is measured by comparing exhaust pressure at the DPF inlet and outlet. The sensor has two hoses, one connected to each end of the filter. As soot accumulates inside the filter, the pressure measured at the filter inlet increases relative to pressure sensed at the filter outlet since soot loading restricts exhaust gas flow (**FIGURE 18-43**). An increase in filter restriction is proportional to soot loading.

However, accurate pressure differential estimates in practice are difficult since the sensor values change dramatically during acceleration and under load due to the large volumes of exhaust gases passing through the filter. Low exhaust gas flow provides much smaller pressure differentials, which are not proportional to actual soot loads. Internal leaks and cracks can distort sensor readings too. Even restricted intake air flow caused by plugged air filters or mass air flow sensors that are out of calibration will upset pressure differential estimates of soot loading. However, the greatest discrepancy in soot load estimates is caused by the filter getting plugged with inorganic ash, which is composed of minerals with no combustible carbon content. The ash will take up space in the filter, displace soot on channel walls, and can restrict the filter. Progressively more ash loading leads to shorter intervals between filter regenerations, producing too frequent and unnecessary regenerations (**FIGURE 18-44**).

To estimate ash load in a DPF, the pressure differential sensor records the pressure drop across the filter after a completed active regeneration (**FIGURE 18-45**). No soot is present and the new filter restriction reflecting the weight of percentage of maximum ash load is updated.

Chemically modeling soot loading takes into account engine operating factors such as the amount of fuel burned, engine load, operating speeds, and various other temperature measurements to calculate an expected soot load. Estimates of soot loading that use the chemical model to predict soot loading are relied on more by manufactures, but they can have problems too. Deactivated wash coat materials, fuel chemistry, and the amount of inorganic ash taking up space on filter substrate walls can

FIGURE 18-45 Ash accumulation in a heavy-duty DPF over the course of a year.

influence regeneration time. Manufacturers will use both methods—pressure differential and chemical models of DPF soot loading—and compare them with one another. Both measurements should correspond to approximately the same value. The OBD system will set fault codes if the estimates do not approximately correspond to one another after diagnostic monitors have run subsequent to an active regeneration event.

Particulate Sensors for Diesel Particulate Filters

Two different applications for particulate sensors have been developed. One technology is designed to measure the amount of soot loading in a DPF. Its purpose is to produce more accurate estimates to limit the frequency of unnecessary regenerations and the amount of time a regeneration takes. A second type of sensor is intended to identify soot leakage past the filter to meet OBD-II requirements for DPF failures (**FIGURE 18-46** and **TABLE 18-3**).

The most recent OBD-II updates to particulate filter monitoring require capabilities to identify much smaller leaks than the pressure differential sensor can detect. To detect and measure the quantity of tailpipe soot emissions leaking past a filter, a sensor is placed directly into the exhaust gas to collect soot particles. EPA emission standards measure these emissions only in terms of mass, whereas the European standard adds a PN limit to mass measurements. This means two different technologies are used, depending on which standard a vehicle must comply with. In North America, a DPF sensor technology uses comb-like electrodes that collect any soot leakage. Without any soot deposition, the absence of electrically conductive carbon on the sensor electrodes produces an infinite resistive pathway between the electrodes. If soot does accumulate on the senor, a conductive path is formed and electrical resistance decreases. A leak rate or fault is detected as a function of electrode resistance and time (**FIGURE 18-47**). Soot accumulations are intermittently cleaned by heating the sensor electrodes between monitoring periods.

The second PMS technology measuring soot loading in filters is designed to reduce the fuel penalty accompanying regenerations that occur too frequently. Excessive regenerations caused by overestimating soot loads will also wear out the catalyst substrate or

CAC - Charge Air Cooler
TC - Turbocharging
DFI - Diesel Fuel Injection
OC - Oxidation Converter
DPF - Diesel Particulate Filter
SCRC - Selective Catlyst Reduction NH3 *
RDQS - Reductant Quality Sensor
EGR - Exhaust Gas Recirculation
EGRC - Cooled Exhaust Gas Recirculation
NOXs - N0x Sensor
PMS - Particulate Matter Sensor
CA OBD II California On Board Diagnostic
* SCR-U or SCRC/SCR-N or SCRC-NH3: Selective Catalytic Reduction-Urea/Ammonia

FIGURE 18-46 The emission decal of a late-model vehicle with a PMS.

TABLE 18-3 Classification of Soot Sensor Technology

	Application		
		OBD-II	
Soot Sensor Technology	**DPF Soot Load Estimate**	**DPF Substrate Failure Monitoring**	**PN Monitoring**
Pressure differential (delta P)	X		
Radio frequency (RF)	X		
Accumulating electrode		X	
Electrostatic charge		X	X

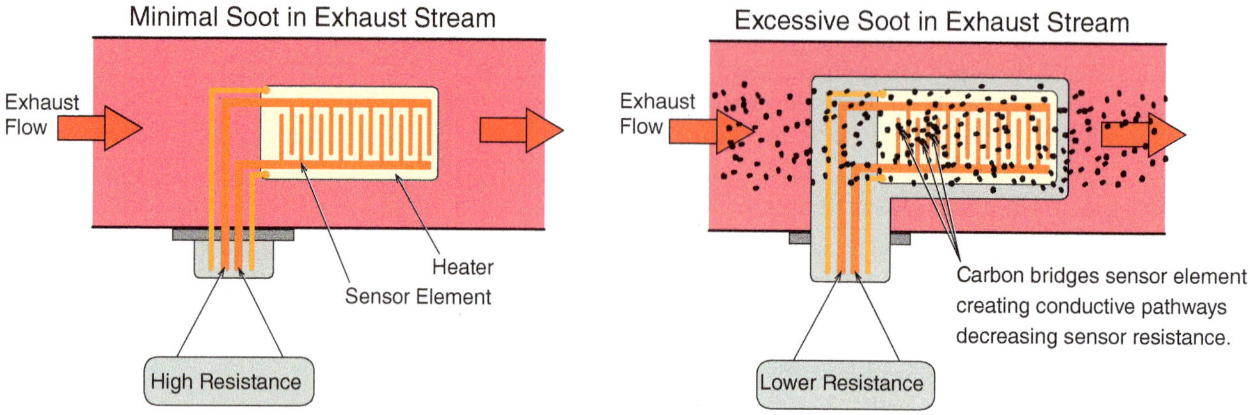

FIGURE 18-47 Operation of a resistive-type soot sensor, which measures the length of a carbon path over a sensor surface.

> **TECHNICIAN TIP**
>
> Even a new or reconditioned DPF will have some flow restriction that leads to a pressure drop across the filter. A baseline pressure restriction across the DPF is a starting point for MPC calculations for soot and ash loading in a DPF. Knowing this helps explain the practice of installing new DPF serial numbers in the aftertreatment control module or somehow indicating to the control module that the DPF has been replaced. Updated calculations for soot and ash loading are made based on new values for a filter restriction.

damage the brick after prolonged high-temperature cycles. Underestimating soot loads can lead to excess accumulations and thermal runaway conditions when the filter is finally regenerated. Radio signals or microwaves are used to measure the thickness of soot caking on DPF filter walls.

Another technology is needed to measure PN. A sensor measuring a change in the strength of an electric field between two capacitor plates is used to determine particle concentration in the exhaust gas. The number of particles suspended between the two plates changes the dielectric strength of the capacitor which is a function of particle number.

Abnormal Regeneration Frequency

The length of time or distance required for a regeneration depends on a number of factors; the most important factor for initiating a regeneration is the amount of soot trapped in the DPF. Underestimating the amount of soot could lead to excessive DPF temperatures when regenerating, which damages the catalyst. Engine damage due to excessive backpressure restriction can result too. Overestimating soot loading leads to regeneration occurring too frequently, resulting in power loss, high fuel consumption, and premature aging or deactivating the catalyst (**FIGURE 18-48**).

During an active regeneration, the soot load is brought to near zero. However, the condition of the DPF catalyst, and the amount of inorganic ash taking up space inside the filter can significantly impact the regeneration time. Time between regenerations can be affected by engine problems accelerate soot loading. Factors that cause increased soot loading may not be detected by the OBD system and can be difficult to identify. For example, excess EGR flow caused by a sticking EGR valve malfunctioning below a detection threshold may increase soot loading. Another common cause of regenerations that occur too frequently is defective or incorrect fuel injectors. Leaking or worn injectors dribbling fuel will negatively affect combustion quality causing DPFs to load very quickly, prompting regeneration. Without fault codes, the root cause of regenerations that occur too frequently requires careful observation while comparing the normal range of many engine operating parameters range values to slightly out-of-range parameters. Check for and repair any fuel-related problems before making repairs to the DPF aftertreatment system. EGR and air induction system checks should be carefully made too. Regenerations occurring too frequently will overheat and melt or crack a DPF filter. A filter loaded with ash that needs to be cleaned or replaced will also frequently attempt to regenerate. However, a fault code is commonly set for excessive ash loading in late-model vehicles.

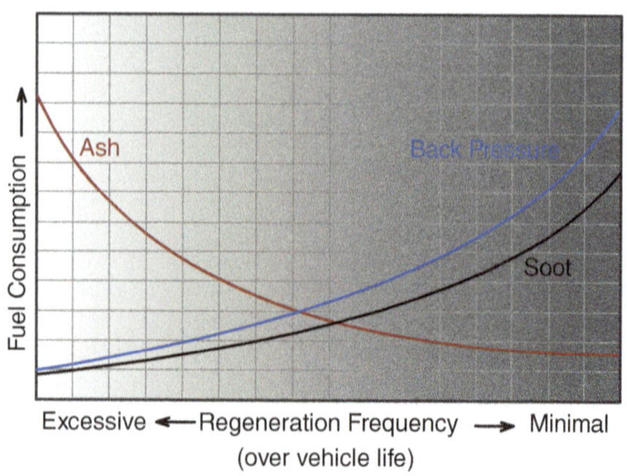

FIGURE 18-48 The relationship between ash loading, soot loading, and fuel consumption. Higher fuel consumption is caused by power-robbing exhaust backpressure.

Any fault in the engine control or aftertreatment system will stop a regeneration from taking place. Failure to regenerate issues are often associated with out-of-range values for temperature and pressure sensors, intake heaters, defective glow-plug circuits, EGR system faults, faulty dosing valves, face-plugged oxidation converters, and incorrectly measured intake airflow. Any engine-related or aftertreatment-related faults need to be corrected before a regeneration can take place.

▶ Cleaning and Replacing Diesel Particulate Filters

18-04 Identify, describe, and recommend service procedures for diesel particulate filters.

Because the DPF substrate is a wall flow filter material, not only is soot trapped on filter walls but other solid particles as well. Soot can be removed through regeneration process because it is organic, which means it contains carbon (which is combustible). Fuel contains almost no inorganic particles, and any other fine PM is filtered out of the fuel before it passes through the fuel system and is burned in the cylinder. However, other particles that cannot be regenerated can end up in the DPF:

- metals worn from internal engine parts and passed through the exhaust
- dirt particles bypassing the air intake, such as silica sand
- inorganic fuel and oil additives and ash from lubricating oil.

The last item, namely inorganic ash, accumulates as engine oil and is burned. Ash is a oil additive used to neutralize acid formation in diesel engine oil produced by small amounts of sulfur entering oil through cylinder blowby. Fuel containing sulfur can allow oil acidity to increase unless ash or some other acid neutralizing substance is added to the oil. Ash accumulates in the DPF as engine oil is burned from the cylinder walls or enters the intake system through closed crankcase ventilation systems. Too much ash will plug a DPF, just as too much soot will. Because ash cannot be burned off during regeneration, the filter will need to be removed and cleaned. Knowing this explains why low-ash oils are mandatory for diesel engines that use DPFs: to prevent ash from prematurely plugging the filter.

If filters become plugged with ash, the DPF is typically removed from the vehicle and back-flushed with a combination of vacuuming and pulses of compressed air. However, unlike heavy-duty diesels, where recommended DPF service intervals are between 50,000 and 450,000 miles (80,500 and 724,000 km), many light- and medium-duty vehicles do not have a prescribed service interval for cleaning ash. A check of many European diesel automobiles' PM schedules will show no cleaning or replacement schedule at all for DPFs. This is partly explained by filter-cleaning intervals that depend on the size of the filter and the amount of oil that an engine burns. Another reason is that metal foil substrates popular in Europe are made from sintered metal, which not only reduces exhaust backpressure but can also handle two and a half times more ash than ceramic-based substrates. Even when equipped with ceramic-type substrate filters, experience with North American pickup trucks demonstrates that these vehicles typically exceed 200,000 miles (320,000 km) and have little ash loading.

Cleaning Options for Diesel Particulate Filters

The following are some of the options for cleaning a DPF:

- using liquid cleaners to remove soot from face-plugged DOCs and DPFs or to increase regeneration temperatures
- removing the DPF and vacuuming while using air burst to remove inorganic particles from the filter
- removing the DPF and baking the filter at high temperatures in an industrial oven to remove heavy soot loads or oil from contaminated substrate.

Liquid cleaners, which are not recommended by any manufacturer, involve spraying a liquid chemical into the substrate through a sensor hole after the sensor has been removed. The chemical is expected to lower the temperature required to burn the soot or increase the regeneration temperature. Liquid cleaner is allowed to soak and penetrate the substrate when it is cool, for about 20 minutes, to soften hardened particles. A second liquid may be sprayed onto the catalyst, which a product sheet may claim is to protect the catalyst substrate. A forced regeneration follows the application of the liquid to complete the cleaning process. Some aftermarket products that are added to the fuel tank are similar to the additives applied through a sensor opening. Makers claim these aftermarket products can

> ▶ TECHNICIAN TIP
>
> OEM rebuilders have observed a relationship between increased ash loading in DPFs and the likelihood for a DPF to crack. The explanation for this is high temperature differentials and expansion ratios of the substrate material caused by cold spots created in the ash-plugged catalyst cells. To extend DPF filter life by preventing compacted ash from plugging and cracking the filter, some fleets make it part of scheduled maintenance to remove and clean DPFs, more frequently than any recommended OEM interval or if no cleaning interval is recommended.

> ▶ TECHNICIAN TIP
>
> When handling DPFs and other catalytic converters, take care to prevent damage to converters through rough handling. Never drop or tap a filter against the floor to clean it. The ceramic substrate can be easily cracked and broken through mishandling.

increase the exhaust gas and regeneration temperatures to help clean out a contaminated or plugged system. Note that these aftermarket products are not endorsed or recommended by any OEM and may even have the potential to damage sensitive catalyst materials.

In Europe, early DPF-equipped models of several cars contain a separate reservoir where a manufacturer-recommended liquid catalyst can be stored. The platinum-based liquid catalyst is automatically added to the fuel system during periodic regeneration events.

If normal maintenance by regeneration is not effective, or if the ash load is excessive, the DPF can be removed and replaced. Because the DPF is expensive to replace, cleaning is a viable option. After the DPF has been removed, it is first inspected and pressure tested to check for cracks or other damage. A report provided after the cleaning process will indicate the difference between a pre- and post-cleaning filter restriction.

Oil contamination or heavy soot loading requires prolonged heating and baking of the catalyst substrate. When required, the DPF can be baked at high temperatures of 1,200°F (650°C) for 24–48 hours to oxidize soot and oil-soaked substrate. Following a baking process, the filter is back-flushed with bursts of compressed air to loosen and help remove compacted ash while a vacuum applied at the opposite end of the filter removes ash and any other nonorganic particulate loosened by the compressed air cleaning. At the end of the process, the filter is pressure tested again to determine the pressure drop or new flow restriction, which should be much lower than the filter was in its uncleaned state.

Subsequent to a baking and air burst cleaning, a filter may be washed and flushed with water to further improve gas flow and eliminate plugged cold spots. Note that water entering a DPF with compacted ash and left to dry without removing the ash can harden the ash into a concrete-like material through hydration. Hardened ash will restrict gas flow through the DPF.

▶ Selective Catalyst Reduction Aftertreatment Systems

18-05 Identify and describe the types, composition, and operation of major NO_x emission-reduction aftertreatment systems.

Phase-in of Tier 2 bin 5 emission standards required not just a 90% reduction in particulate emissions from the previous emission standards, but progressively lower NO_x emissions between 2004 and 2010. The fleet average benchmark for NO_x in 2010 was no more than 0.07 g/mile (0.04 g/km) by EPA standards. This regulation for a low level of NO_x was mandated for phase-in over six years from an interim standard of no more than 0.30 g/mile (0.19 g/km) in 2004. Tier 2 standards were in place until the phase-in of Tier 3 at the beginning of 2017, when emission regulations averaged NO_x emissions with other non-methane hydrocarbons (NMHC). The combination of NO_x + NMHC reduced a combined weight of emissions from 86 mg/mile (53.75 mg/km) in 2017 to 30 mg/mile (18.75 mg/km) in 2025. To reach these standards for NO_x, two key types of aftertreatment technology are used. The most common method today is an SCR system that uses liquid DEF, also called SCR-U (**FIGURE 18-49**). The term "SCR" refers to the emissions system targeting a single specific noxious emission. Excess air and cool exhaust temperatures prevent the use of conventional catalyst technology. Instead, ammonia (NH_3) contained in a DEF compound called **urea** is essential to SCR system operation. Breaking apart the urea in the exhaust system releases ammonia, which in turn breaks down NO_x into harmless nitrogen (N_3) and water (H_2O). To differentiate SCR that uses urea from other systems that break down NO_x, the suffix "U" is used to designate an SCR-U system storing liquid urea or DEF onboard a vehicle.

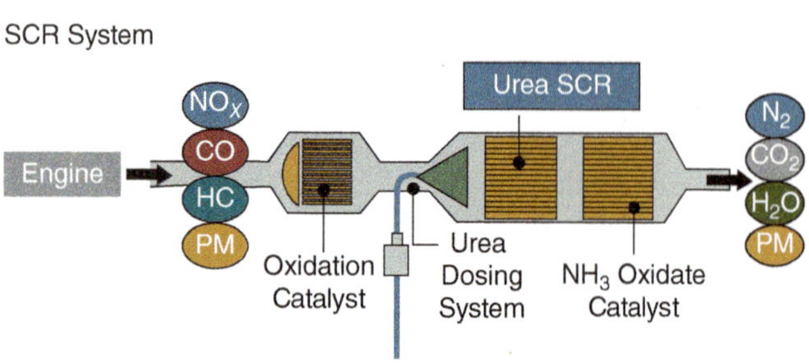

FIGURE 18-49 The principles of emission reduction through an SCR system: NH_3 combines with NO_x and converts it into N_2, CO_2, and H_2O.

Predating the use of liquid SCR-U is the NAC. Also called an LNT or NSC, this system was used by a number of major manufacturers, such as Mercedes-Benz and Volkswagen (VW). The Dodge Ram 2500 (but not the 3500 from 2007 to 2012) used NACs as part of the wider range of NO_x-reducing emission systems known as BlueTec—which includes AdBlue, a liquid SCR system. NACs are used mainly in light and medium diesel engines

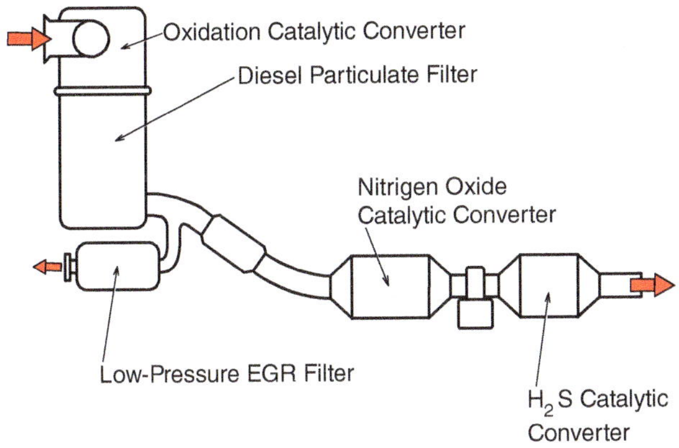

FIGURE 18-50 The location of the NAC or NSC on the first-generation VW, which is after the particulate filter.

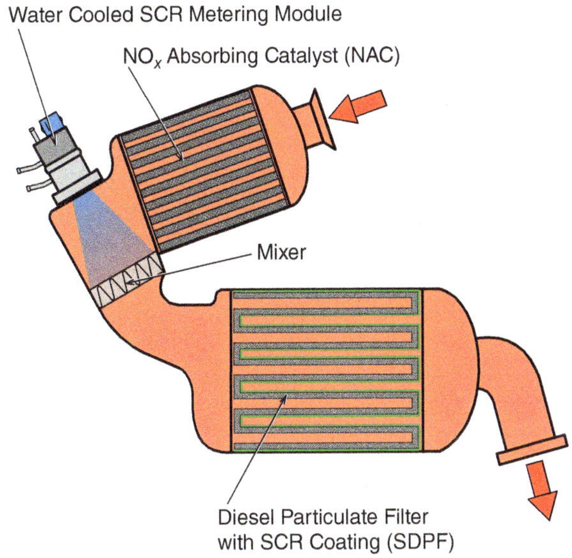

FIGURE 18-51 Tier 3–compliant and Euro 5–compliant engines will use a close-coupled NAC and combine it with an SCR-U to produce less NO_x with very low DEF consumption. Cold-start NO_x emissions are lower when using close-coupled catalysts.

because they are not effective in larger high-horsepower diesel engines with higher exhaust temperatures and exhaust volumes. Mercedes-Benz and Volkswagen used these devices beginning in 2007 and through to 2009 before switching to more efficient liquid SCR-U systems after several years (**FIGURE 18-50**). The NAC has not become obsolete with the introduction of liquid SCR. Tier 3–compliant and Euro 5–compliant vehicles, such as the 1.6 L whisper diesel, combine a close-coupled NAC with a liquid SCR-U system for very low NO_x emissions and even lower DEF consumption (**FIGURE 18-51**).

The primary advantage of NACs is that they do not require onboard storage of liquid DEF and associated complexity from additional SCR components. Instead, ammonia is produced inside a unique catalyst substrate through a multistep chemical process that uses hydrogen from fuel molecules and nitrogen from stored NO_x. Also, NACs need only wide-band oxygen sensors to provide feedback to engine controls and for OBD-II monitoring requirements. Because NO_x exhaust gas sensor technology had not reliably matured until the second-generation NO_x sensors were introduced around 2013, NACs enabled vehicles to meet emission standards without depending on power-robbing, fuel-consuming in-cylinder operating strategies to prevent NO_x formation. Using fuel to produce ammonia does, however, have a fuel penalty of approximately 3–5%. NACs can store NO_x but have a narrower temperature range in which they efficiently operate. At high exhaust temperatures, above 716°F (380°C), NO_x can spontaneously release from an NAC storage catalyst, making NACs unsuitable for maintaining the low emissions encountered over a wider range of operating conditions (**FIGURE 18-52**). This characteristic is another reason why NAC NO_x-reduction technology is limited to light-duty engines that are undergoing emission certification over emission test cycles of low to medium loads (**FIGURE 18-53**).

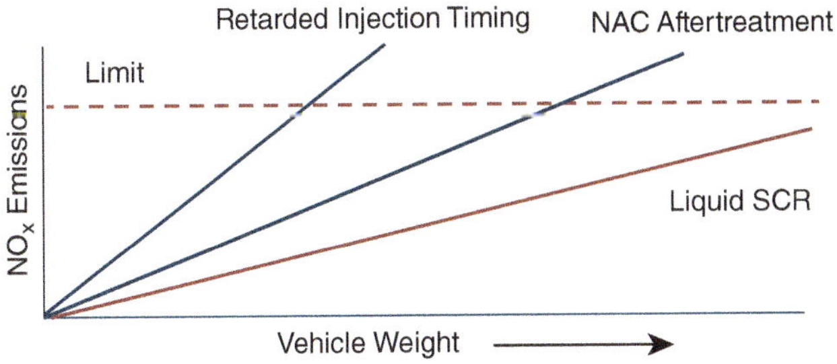

FIGURE 18-52 NACs are not suitable for exhaust systems in heavy-duty diesel engines. High exhaust temperatures and exhaust volumes would require very large, expensive NAC catalysts.

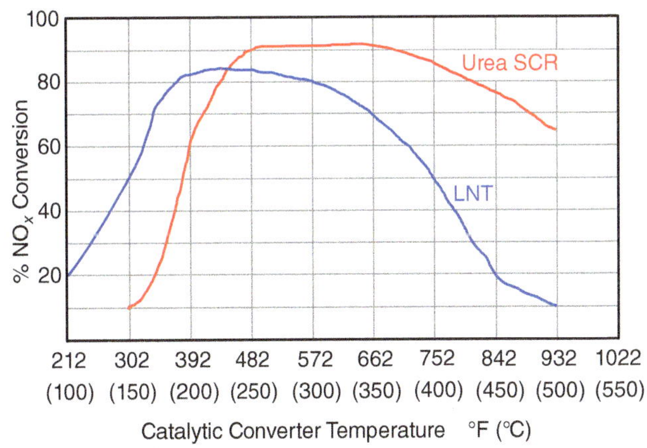

FIGURE 18-53 Liquid SCR systems are more efficient at converting NO_x over a wider temperature range than NACs are.

NO_x Adsorber Operation

NO_x adsorber technology is a catalyst strategy for removing NO_x in the low-temperature, oxygen-rich exhaust environment of diesel engines. An overview of its operation is relatively simple: the NAC captures and stores harmful NO_x emissions during lean or air-rich exhaust gas conditions and converts them into harmless nitrogen and water during a brief fuel-rich, air-deficient phase through a multistep chemical process (**FIGURE 18-54**). A short process of three to five seconds of regeneration regularly occurs whenever the catalyst is estimated to have stored near it's maximum capacity of NO_x. This NO_x regeneration event, called a $DeNO_x$ reaction, may take place in intervals as short as a few seconds to every few minutes in order to break down and purge NO_x stored in the catalyst substrate. A second-regeneration event, called a $DeSO_x$ regeneration, takes place after an automatic DPF regeneration event. During $DeSO_x$ reactions, prolonged supplemental heating of the NAC takes place to remove sulfur deposits from the catalyst. Sulfur deposits deactivate catalyst material, lowering NAC efficiency. So at least once during the distance it takes for a tank of fuel to be consumed, this desulfation process is carried out.

Adsorber means the NO_x emissions are temporarily trapped and stored on the surface of the catalyst material. The stored NO_x is removed by temporarily creating an oxygen-depleted, fuel-rich exhaust condition in the NO_x catalyst. Changing the O_2 concentration of the exhaust gas from high to low, or lean, to a temporarily rich condition lends the system the name lean NO_x trap (LNT).

In terms of construction, light-duty diesel NACs commonly use a combination of platinum and barium oxide catalysts deposited onto substrate with a porous aluminum oxide wash coat ($Pt\text{-}BaO/Al_2O_3$). This slurry-like wash coat material is applied to a ceramic honeycomb channel, flow-through-type cordierite substrate (**FIGURE 18-55**). Barium oxide (BaO) is the most common NO_x-adsorbing substance since it has the best NO_x storage capabilities with its higher affinity for NO_x than oxygen, helping the NAC to achieve conversion efficiencies of over 90% during steady-state operation. Because barium oxide will

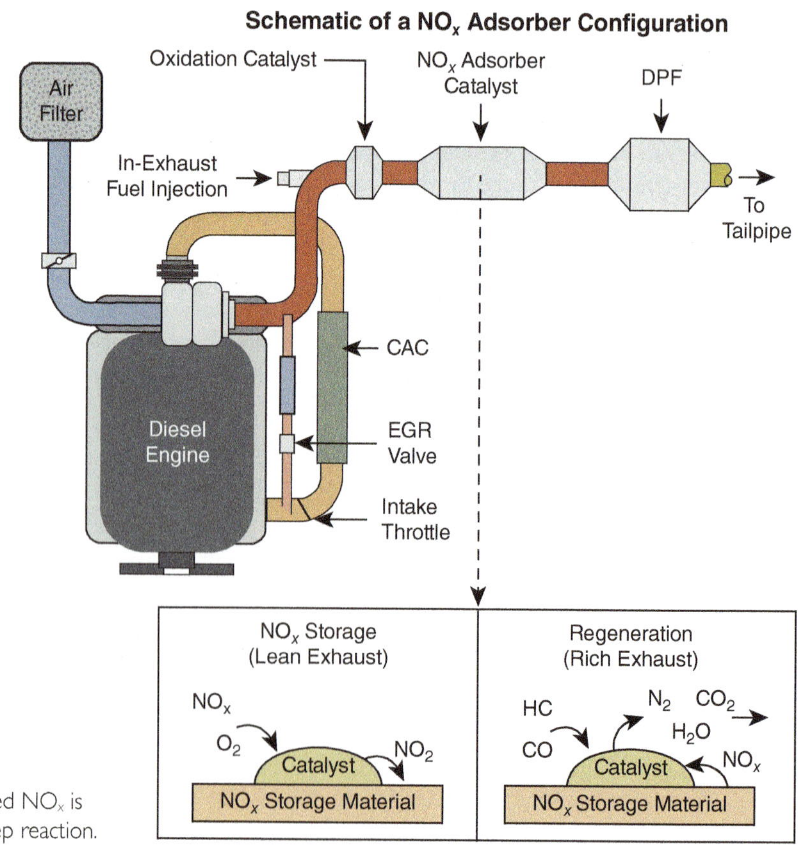

FIGURE 18-54 Stored NO_x is removed in a multistep reaction.

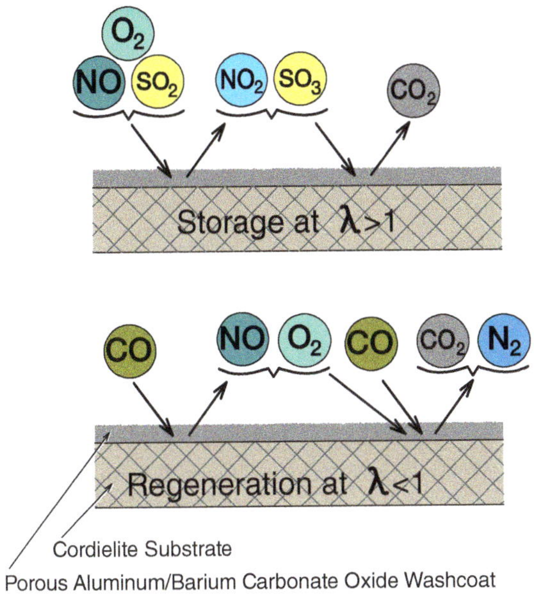

FIGURE 18-55 NO_x is adsorbed by the barium carbonate during lean operation. During a rich state in an oxygen-depleted condition, NO_x is converted into nitrogen and CO_2 by using some fuel.

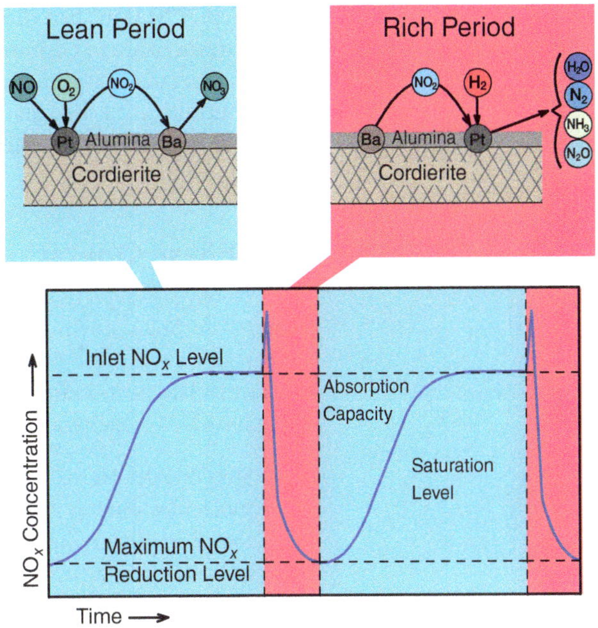

FIGURE 18-56 Cycling rich and lean exhaust gas oxygen and fuel enrichment of the exhaust gas content enables the NAC substrate to absorb NO_x, break it down, and release harmless water, nitrogen, and ammonia.

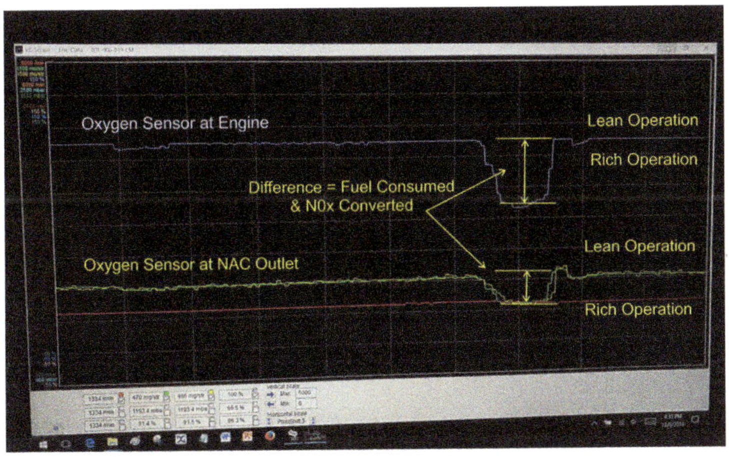

FIGURE 18-57 Oxygen sensor signals during a fuel-dosing event. Signals from the upstream and downstream oxygen sensors provide feedback to the NAC control about the storage capacity and efficiency of the converter.

bind more readily to NO_x than oxygen under lean conditions, NO_x molecules combine with barium, which in turn releases oxygen from the barium. While observing oxygen sensor signals downstream of an NAC, the release of oxygen can spike exhaust gas concentrations above the normal atmospheric concentration (**FIGURE 18-56**). This is an important observation since some OBD-II systems measure the NAC efficiency based on the oxygen storage capacity of the NAC (**FIGURE 18-57**). The amount of active catalyst material is directly proportional to the amount of oxygen released. Inactivated, sulfated, or depleted catalyst material will release less oxygen.

Lean Phase Operation

Lean phase operation takes place during normal engine operating conditions, when no regeneration is taking place. Exhaust gas oxygen concentrations will remain approximately at 12–15%. NO_x is stored in the catalyst substrate during phase operation.

- Step 1: NO that is initially released during combustion is converted to NO_2 by the platinum catalyst also contained in the wash coat of the NAC.

- Step 2: NO_2 in turn displaces oxygen from barium oxide sites in catalyst converting barium oxide to barium nitrate (BaO to $Ba(NO_3)_2$). Oxygen is released and may cause the exhaust oxygen concentrations to momentarily exceed 21%. This process will continue until the catalyst has been loaded with NO_x. The engine or aftertreatment control module will estimate how much NO_x loading will take place, based on oxygen sensor data.

Rich Phase Operation

Rich or oxygen-depleted exhaust gas conditions are necessary for the chemical reactions required to break down NO_x. During the regeneration event, the intake air throttle may be partially closed, the EGR rate may increase, and the CR injectors will inject one or two additional late post–main injections to increase hydrocarbon content in the exhaust. The rich phase NAC regeneration will last three to five seconds and is unnoticeable to the driver.

- Step 3: Fuel HC is catalyzed to CO, and extra hydrogen (H_2) is released from HC molecules during rich phase operation. Hydrogen derived from fuel in the catalyst that is used to reduce NO_x lends the SCR technology the name SCR-H, where the H indicates hydrogen.
- Stage 4: H_2 molecules react with NO_x stored on the barium catalyst to produce NH_3, N_2, and some N_2O (nitrous oxide).
- Stage 5: CO further oxidizes to produce CO_2. H_2O and nitrogen continue to be produced from reactions of H_2 and oxygen left over from reactions of NO_x with ammonia. CO can also react with NO_x to produce CO_2 and N_2.

Controls for Oxides of Nitrogen Adsorbing Catalysts

Studies of the BaO substrate have identified a 12.5% exhaust oxygen level as optimal for enabling NO_x-reduction-type reactions. This feature of NAC operation helps explain why failures such as the intake air throttle control valve and EGR system prevent proper regeneration of the NAC. When intake air and EGR cannot be regulated, NAC regeneration does not take place either.

The calculation for the amount of NO_x stored in the NAC is based on the expected mass of engine-out NO_x emissions, exhaust temperature, volume, and gas velocity. Adaptation of the NO_x storage/desorption model uses data from two exhaust oxygen sensors to correct for factors such as catalyst aging and loss of efficiency during actual engine operation. One sensor is located at the engine-out after the turbocharger or at the outlet of the DPF. The second oxygen sensor is further downstream at the NAC catalyst outlet. These sensors provide not only data needed for OBD-II emission-monitoring requirements but also feedback to the model of NAC operation regarding exhaust oxygen content, fuel-dosing frequency, the fuel mass required for dosing, and catalyst efficiency. The upstream oxygen sensor is used to calculate a starting value for exhaust oxygen content supplying chemical reactions in the aftertreatment catalysts. The second oxygen sensor, placed downstream of the NAC, measures exhaust oxygen content that is leaving the converter. Because the amount of fuel consumed while regenerating the NAC is proportional to the mass of NO_x reduced or converted, a continuous feedback loop is created between the oxygen sensor signals and the ECM's commanded fuel mass used to dose the exhaust during the regeneration event. Changing exhaust gas oxygen levels essentially indicates how much fuel should be used to dose the NAC, as well as the frequency of dosing. As the amount of NO_x converted by the NAC changes, the oxygen sensor signals are supposed to provide correction factors to the model used to schedule $DeNO_x$ regeneration events and the appropriate number of post-combustion-injection events that supply fuel.

NO_x Adsorbing Catalysts and Wideband Oxygen Sensors

As already noted, two wideband-type oxygen sensors provide adaptive feedback for NAC operation. Wideband oxygen sensors, also called lambda sensors by Bosch, are a category of oxygen sensor capable of measuring oxygen concentrations from almost nothing to pure oxygen levels. Unlike spark-ignition engines, which operate at stoichiometric AFRs, exhaust gas oxygen concentration in diesel engines are very lean—operating air rich over a relatively wide range of exhaust concentrations. The excess oxygen content requires this type of sensor to correctly measure exhaust oxygen concentrations over this wide range. Wideband oxygen sensors appear different from other types of air-fuel ratio sensors because they typically have five wires, whereas most other oxygen sensors have only three or four wires (**FIGURE 18-58**).

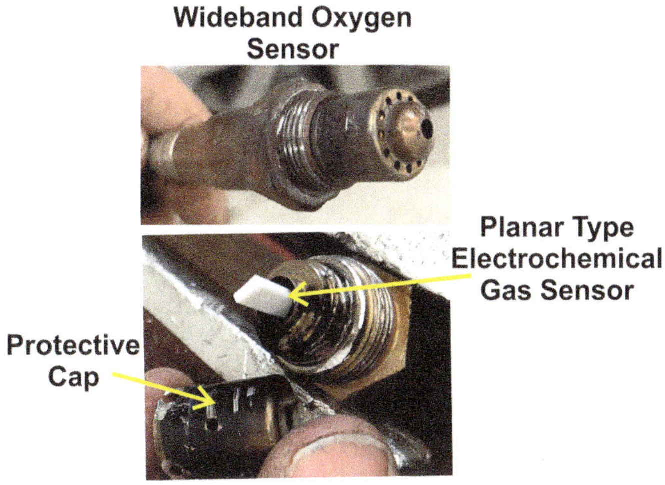

FIGURE 18-58 A wideband oxygen sensor uses a planar or flat-type electrochemical cell to measure exhaust oxygen concentration.

Wideband Oxygen Sensor Operation

A simplified explanation of wideband sensor operation is that the electrochemical device compares oxygen concentrations in exhaust gas against known oxygen content in either a sealed reference cell inside the sensor or atmospheric air (**FIGURE 18-59**). When the sensor is heated to a temperature of 1,400°F (700°–800°C), two platinum electrodes separated by zirconium oxide (ZRO_2) produce electricity when exposed to different concentrations of oxygen. The difference between the electrical signals of a known reference oxygen and unknown exhaust gas concentration provides the sensors critical data. An electrical heating element inside the planar-type sensor element enables wideband sensors to generate a valid voltage signal within 20 seconds or less from cold start-up (**FIGURE 18-60**).

Unlike earlier oxygen sensors with a single zirconia oxide cell, two cells containing four platinum electrodes are used in a wideband sensor. One cell called the measuring cell generates voltage by comparing a known sample of reference oxygen with an unknown exhaust gas oxygen concentration. If the sensor has a sealed oxygen sample, reference oxygen content in the cell will be equivalent to a 14.5:1 ratio between oxygen and other gases. The other cell, called a pump cell, could produce a voltage if there is a difference in oxygen concentration in the exhaust gas and the oxygen in the measuring *chamber* between the two cells. The difference in voltage generated between the measurement cell and pump cell signals are used to determine whether the exhaust gas oxygen concentration is above or below the oxygen concentration in the reference cell. If they are the same, a **lambda (λ)** of 1 or 0 volts is the output signal. This arrangement of comparing voltage produced by a standard reference cell and another cell with unknown gas concentration is commonly called a Nernst cell.

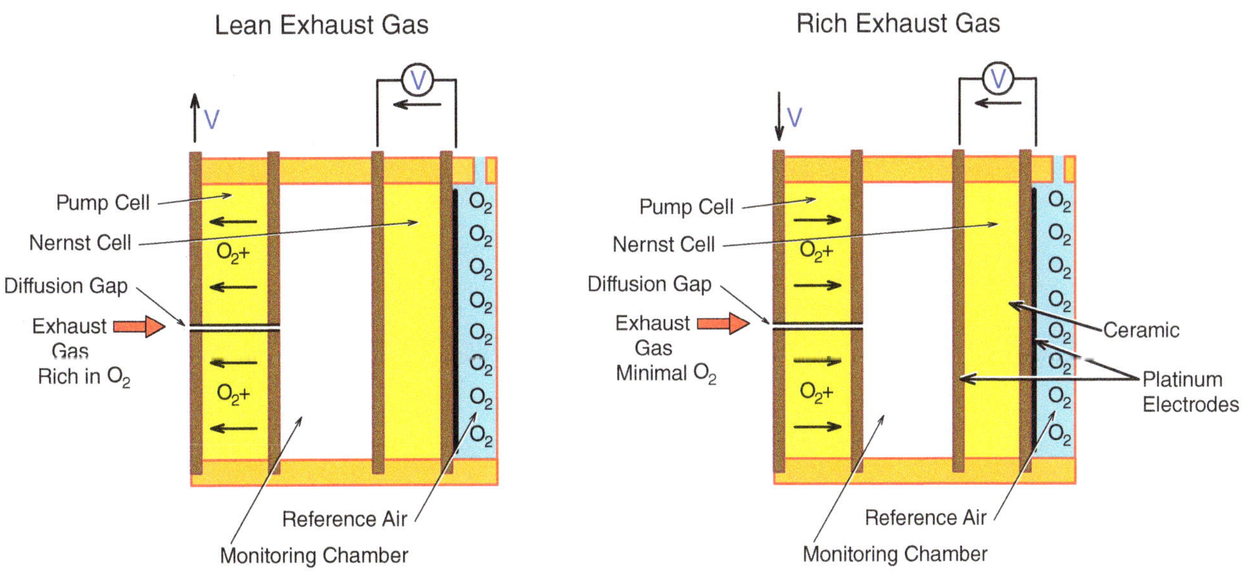

FIGURE 18-59 An electrochemical Nernst cell used in oxygen sensors measures the unknown gas concentration by comparing it to a known gas concentration. Oxygen sensor can use either an internal or external source of reference air to measure unknown exhaust gas concentrations.

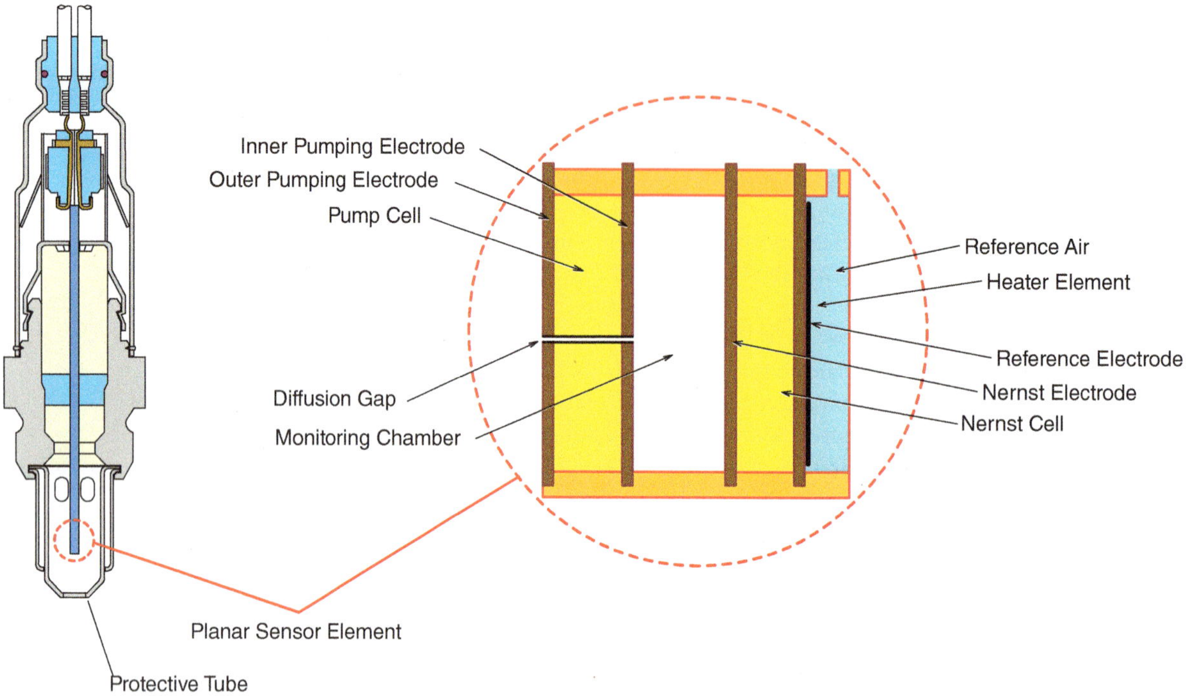

FIGURE 18-60 Internal construction details of a planar sensor. Planar sensors are made from flat, thin strips of material making up an electrochemical cell.

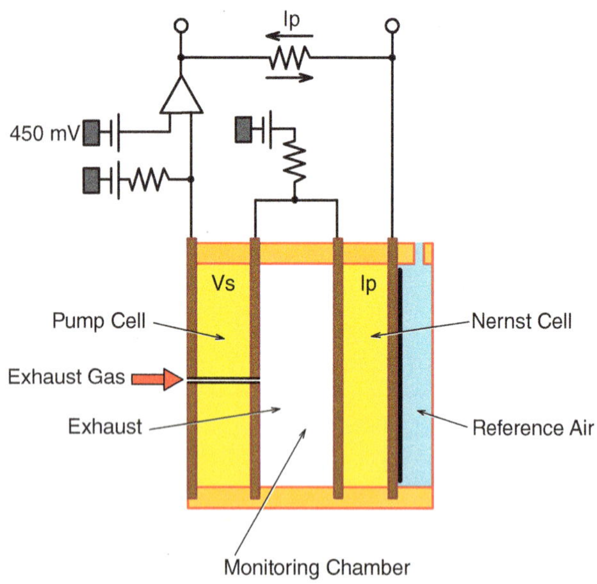

FIGURE 18-61 A wideband oxygen sensor signal is based on the current flow, measured in microamps or nanoamps, required to maintain the same electrical charge or voltage on the pump cell electrodes as the reference cell electrodes.

To understand the operation of the sensor, it is important to study the electrical connection between the three electrodes (**FIGURE 18-61**). Note that the same exhaust gas oxygen concentration will be found on three of four electrodes exposed to exhaust gas. Exhaust gas found in the measuring chamber can enter through a hole called the diffusion gap. This means that both of the platinum electrodes of the pump cell and one of the measurement cell (reference oxygen) electrodes have the same exhaust gas oxygen concentration. Also note that three electrodes are not only exposed to exhaust gas but electrically connected too. The same voltage potential would be generated between one electrode exposed to only reference cell oxygen and the other three. However, if the exhaust oxygen content in the measuring *chamber* was almost zero, the measured cell voltage increases due to the different concentrations of oxygen acting on its platinum electrodes. If exhaust gas oxygen content at 14.5:1 in the measuring chamber is the same as the measuring cell oxygen concentration, zero volts would be measured across the measuring cell electrodes. Increasing exhaust oxygen content above 14.5:1 increases voltage, but the polarity of the voltage change across the pump cell electrodes would change relative to voltage produced across the measuring cell electrodes.

Note in Figure 18-61, which shows the operation of an oxygen sensor and its controller, that the controller will not sense the correct voltage between the reference cell and one pump cell electrode. This happens because a calibrated resister is in series between the pump cell electrode on the engine exhaust gas side and the controller operation that will always produce a voltage drop or offset the voltage measured by the controller. Using this calibrated resister is essentially the trick for precisely measuring exhaust gas oxygen levels. The resister provides a lambda (λ) equivalent calibration for the sensor (**FIGURE 18-62**). To understand this, consider that before actually generating an electrical signal corresponding to exhaust oxygen content, the controller performs another task after measuring cell voltages. Once it knows what voltages are generated across the pump and measuring cell electrodes, it attempts to *equalize* the sensed voltage between the pump cell electrode and the voltage measured across the measuring cell electrodes (reference oxygen).

To equalize the voltages, the controller supplies electrical current, measured in amperage, to the pump electrode in the measurement chamber. This current can be either negative or positive, which means that it either depletes or adds to the electron concentration on the pump cell electrode. The controller varies the electron concentration until the voltage across both cells—pump and measurement—are equal. It is very important to note that the controller supplies a varying amperage. The number of electrons or amperes sent to the pump cell electrode will correspond to the number of oxygen atoms and its polarity to the electrical charge acting on the platinum electrodes in the measurement cell. When the differential voltage between the pump cell and measurement cell disappears, no current flows. As the exhaust oxygen content changes, so will the current that is supplied to the pump cell electrode. This means the controller will continuously change the polarity and the amperage of current flow to the pump cell electrode to maintain a 0-volt differential between the pump and measurement cell electrodes. It is the measurement of current supplied to the pump cell that is exactly proportional to exhaust oxygen content and the signal produced by the controller. The vehicle's engine ECM or dedicated oxygen sensor controller typically converts the analog current supplied to the pump cell into a linear voltage signal reported on scan tools. Understanding this relationship between the pump cell current and the sensor output signal explains why current supplied to the oxygen pump circuit is also called the sensor signal current.

FIGURE 18-62 A wideband oxygen sensor with an internal oxygen reference chamber will use a laser-etched resister to calibrate the output signal to a lambda = 1 of 0 volts.

Wideband Oxygen Sensor Measuring Units

Wideband oxygen sensors are used on spark-ignition engines and diesel engines, but the units of measurement are different for diesel engines. Voltage readings for a diesel engine's oxygen sensor are reported in proportion to a linear voltage calculated by the controller (**FIGURE 18-63**). However, diesel engine ECMs will report the signals as a lambda (λ) value, which is also referred to as an **equivalency ratio** (**FIGURE 18-64**).

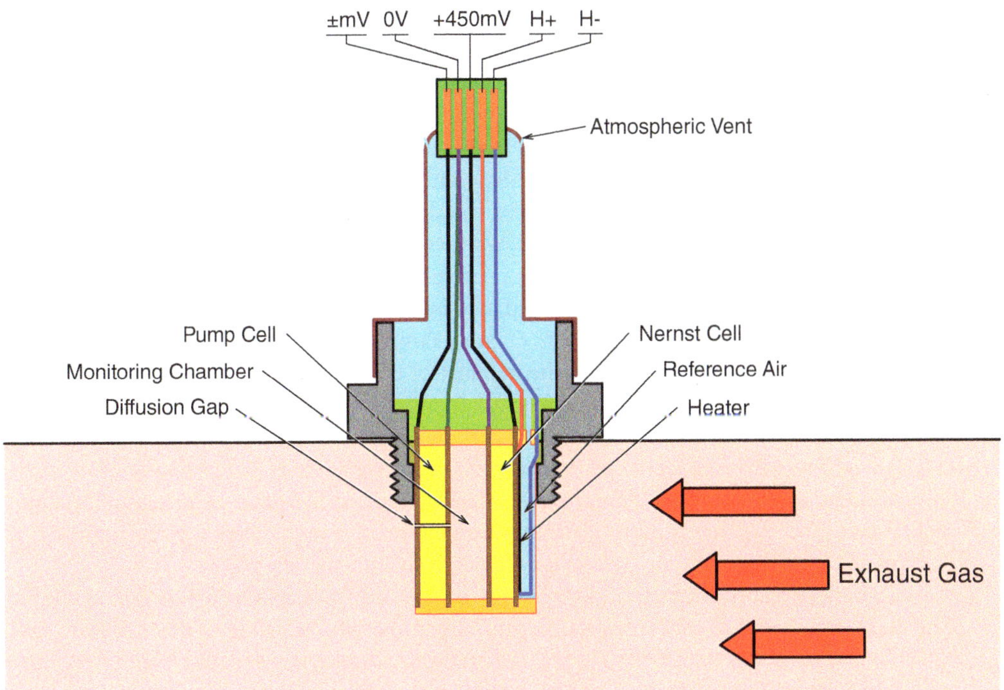

FIGURE 18-63 Changes in exhaust oxygen concentration are compared against the oxygen concentration in the measurement cell, which may use ambient air or a sealed reference chamber.

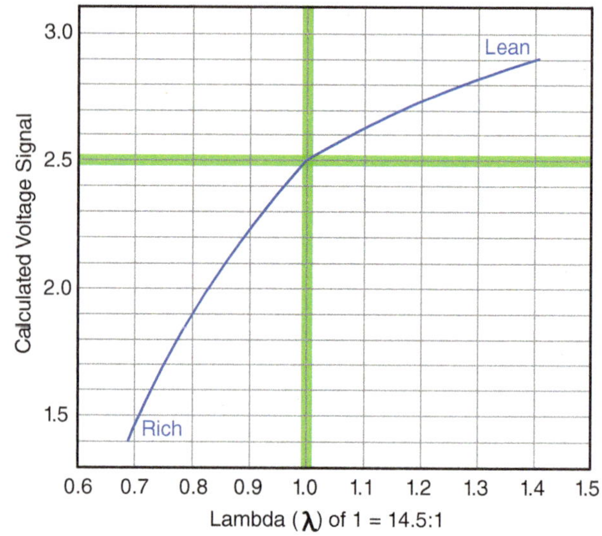

FIGURE 18-64 The uncorrected output signal from the oxygen sensor controller. Exhaust oxygen concentration is reported as proportional to a linear change in voltage. The voltage signal is calculated based on changing polarity and amperage supplied to the pump circuit.

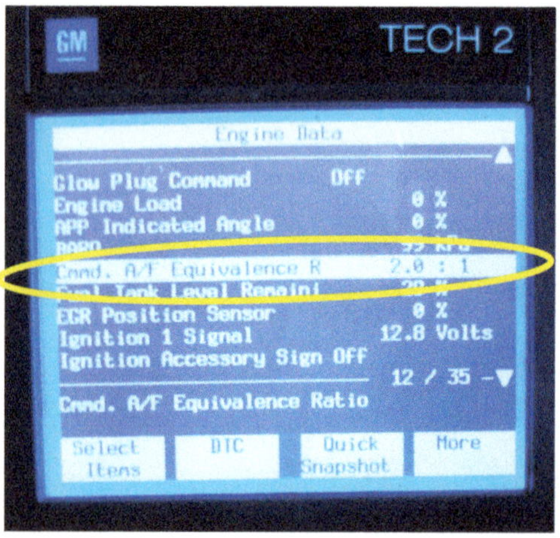

FIGURE 18-65 The air-fuel equivalency ratio refers to a comparison of air content relative to a 14.5 stoichiometric ratio for diesel fuel. This commanded equivalency ratio is 29:1 exhaust oxygen content, which contains less oxygen than atmospheric air.

Lambda, or a lambda value, is a convenient measurement of oxygen content in the exhaust relative to other gases. Lambda calculations themselves are based on a stoichiometric air-fuel ratio unique to the fuel consumed by an engine. But a lambda measurement of exhaust gas concentration does not refer to an air fuel ratio in thee exhaust but only to the exhaust oxygen concentration.

To understand this important measuring unit, it is necessary to first know what a stoichiometric ratio is. Stoichiometric ratio refers to the minimum quantity of oxygen needed to completely consume fuel in the combustion chamber. No fuel or air is left over after stoichiometric combustion has taken place. Expressed mathematically, Stoichiometric Air-Fuel Ratio = Minimum Air ÷ Fuel Mass and the units are always in mass. For example, the number of hydrogen and carbon molecules in gasoline that need oxygen to completely burn means it has a stoichiometric ratio of 14.7:1. Stated another way, 14.7 grams, kilograms, or pounds of air is required to completely burn 1 g, 1 kg, or 1pound of gasoline. For more examples, stoichiometric ratios for other fuels are 17.5:1 for methane and 13.5:1 for propane. Although 1D and 2D diesel fuels have different stoichiometric ratios, diesel fuel is considered to have a 14.5:1 stoichiometric air-fuel ratio. Mathematically, lambda or equivalency ratio is expressed as λ = Actual AFR ÷ AFR Stoichiometric. An air-fuel equivalency ratio of 14.5:1 in a diesel would be a λ of 1 since 14.5 ÷ 14.5 = 1. An actual air-fuel equivalency ratio of 29:1 equals a λ of 2 since 29 ÷ 14.5 = 2. A 10:1 air-fuel ratio would be 10 ÷ 14.5 = λ 0.69.

To describe oxygen content in the exhaust, the term "lambda" is used interchangeably with the term "air-fuel equivalence ratio." A lambda or equivalency ratio displayed on scan tools is used because there should be no fuel in the exhaust to calculate an air-fuel ratio (**FIGURE 18-65**). Instead, oxygen is present with many other molecules, including water vapor. When measuring relative exhaust gas oxygen concentrations, the lambda value for oxygen refers to the balance between oxygen and other exhaust gases, such as HCs, CO, CO_2, NO_x, N_2, and H_2O. For example, when $\lambda = 1.0$, an air-fuel mixture is considered stoichiometric. But in the exhaust of a diesel engine, a lambda of 1 would mean the concentration of exhaust oxygen to other gases is 14.5:1, which coincides with the stoichiometric ratio of diesel fuel. Exhaust gas concentrations with oxygen content *less* than 14.5:1 have a $\lambda < 1.0$, and leaner mixtures with excess oxygen above 14.5:1 have a $\lambda > 1.0$.

To calculate the ratio of exhaust oxygen compared to other gases by using lambda, the actual measured λ value is multiplied by the stoichiometric equivalent air-fuel ratio. So, a 0.5 λ is 14.5 × 0.5 = 7.25:1 air-fuel ratio equivalent or equivalency ratio.

To understand the contribution that oxygen sensors make to NAC operation, consider the changes in oxygen concentration in and out of the NAC during NO_x absorption and DeNO_x events. At the outset of a DeNO_x regeneration, the oxygen sensor at the NAC outlet will measure approximately $\lambda = 1$ or 14.5:1 oxygen content relative to other exhaust gases. A calculated mass of fuel HCs (5 g, for example) entering the NAC will react HC with NO_2 to create NH_3, CO, and CO_2. After the HCs have been used up through chemical conversions, any additional HCs entering the NAC will only lower the outlet oxygen concentrations by producing H_2O and CO_2, plus excess HCs. It is the excessive drop-off in exhaust oxygen content at the NAC outlet, measured by the oxygen sensor, that signals the ECM to end the regeneration event. Prolonged oxygen depletion inside the NAC will lead to excess HC emissions from the NAC, so dosing needs to end. By calculating the mass of HCs needed to convert NO_x entering the NAC, the OBD-II system can determine how much NO_x was likely converted and how efficiently the NAC is operating.

$DeSO_x$ Reactions

NAC aftertreatment has two major drawbacks. One is the requirement for additional fuel during rich operation, which can penalize fuel economy. The increased consumption of fuel during rich operation needed to reduce NO_x results in a fuel penalty of 3–5%. However, by cleaning up NO_x in the exhaust system, rather than minimizing its production in the cylinders, the aftertreatment fuel penalty is offset by reduced vehicle fuel consumption accompanied by increased power output since better engine performance is achieved from optimized injection timing, reduced EGR flow, optimized injection rate shaping, and other more efficient operating strategies.

Another problem with NACs is that sulfur deactivates NAC catalyst substrate—even ultra-low sulfur diesel fuel at 15 ppm. Sulfates will bind to barium storage sites in the catalyst wash coat and will eventually deactivate the catalyst, causing the converter to gradually lose its effectiveness. Sulfur regeneration, which removes sulfates from the catalyst, is a maintenance step where prolonged fuel-enriched heating of the NAC catalyst burns the sulfur out of the NAC. The interval for a desulfurization or $DeSO_x$ regeneration is based on the amount of fuel consumed since the last $DeSO_x$ event. The additional heating—known as desulfurization mode, or $DeSO_x$—is less frequent than soot regeneration, typically taking place after consuming one to two tanks of fuel. In some vehicles, as many as three or four late post-combustion injection events take place, replacing the normal one of two injections, to intensify the exhaust temperatures during a $DeSO_x$ regeneration of the NAC. The event lasts no longer than 15 minutes but is slightly hotter than a DPF regeneration. During $DeSO_x$, inlet temperatures to the NAC are 1,200°F (620–650°C). The $DeSO_x$ process is intermittently interrupted during some trips, to reduce the soot levels in the DPF caused by the much richer post-combustion injection of fuel.

To prevent the unpleasant smell produced by hydrogen sulfide (H_2S) another, smaller H_2S converter is placed downstream of the NAC to reduce odors from the $DeSO_x$ event (**FIGURE 18-66**).

> **▶ TECHNICIAN TIP**
>
> The lambda value displayed on a scanner is helpful to determine whether components controlling the air pathway through an engine are working correctly. Control of exhaust oxygen levels is critical to NACs and DPFs operating properly. Intake air throttles, EGR valves, and even VGT operation can be normally expected to change the λ or oxygen concentration value for exhaust gas. Combustion problems can also influence the exhaust oxygen equivalency ratio or λ value. A large discrepancy between a commanded λ and an actual displayed value measured by an oxygen or NO_x sensor points to a problem with the performance of an output actuator for the air pathway. Air unintentionally leaking into the engine may also drive the λ value higher than it should. Check for a leak in the intake system after the mass airflow (MAF) sensor in addition to inspecting other devices for faults.

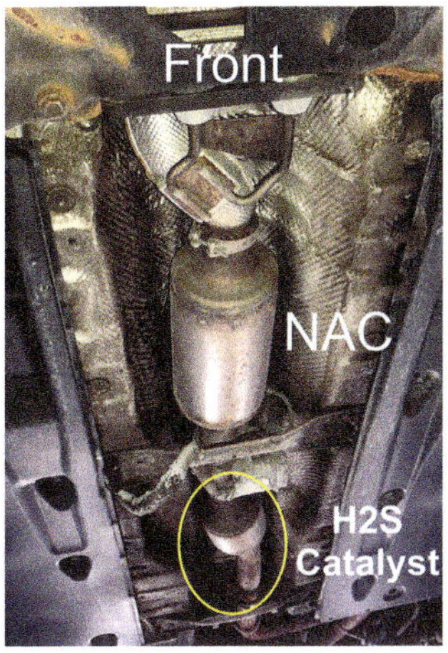

FIGURE 18-66 During a DeSOx regeneration, the hydrogen sulfide smell is removed by another catalyst downstream of the NAC.

▶ Fundamentals of Selective Catalyst Reduction

18-06 Identify and describe the construction and operation of SCR-U aftertreatment systems.

Selective catalyst reduction with liquid urea (SCR-U) is the most popular of two methods commonly used to remove NO_x from diesel exhaust gases. SCR systems spray a metered amount of liquid urea into the exhaust gas stream to destroy NO_x molecules. After mixing exhaust gases with the fluid, an exhaust catalyst helps speed up the conversion of NO_x into harmless nitrogen and water.

The oxygen-rich exhaust of diesel engines makes it necessary to this different approach to NO_x removal. Unlike spark-ignition engines, which generally have no more than 2% oxygen content in the exhaust stream, the air-rich diesel exhaust does not allow the same chemical reactions to take place in catalytic converters. Oxygen removal using reduction-type catalytic reactions are more difficult air-rich conditions. Introducing ammonia into the exhaust stream enables the use of a different mechanism for eliminating NO_x from the exhaust stream. The reaction between ammonia and NO_x, producing H_2O and N_2, is enhanced by passing the mixture over a catalyst substrate that causes the reduction reaction to happen faster and at lower temperatures. Because this emission control strategy specifically targets the oxygen making up NO_x while ignoring atmospheric oxygen in the exhaust, the method is said to selectively remove NO_x, giving the emission system its name—selective catalyst reduction (**FIGURE 18-67**).

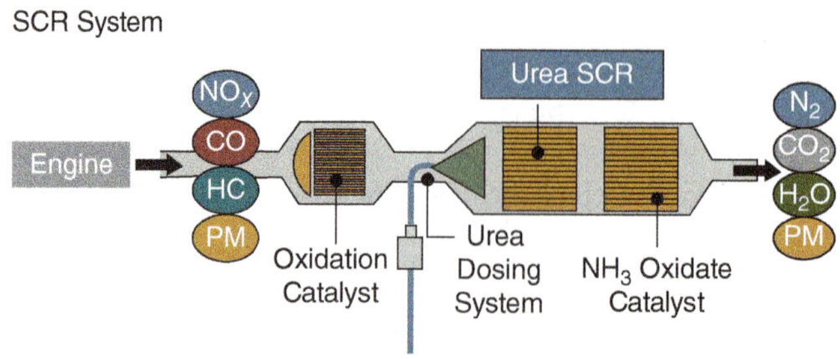

FIGURE 18-67 Principles of NO_x emission reduction through an SCR system. NH_3 combines with NO_x to convert it into N_2 and H_2O.

FIGURE 18-68 Diesel exhaust is a mixture of urea and water. Ammonia contained in urea combines with NO_x to selectively separate oxygen from nitrogen.

Ammonia molecules contained in the urea actually do the work of disintegrating exhaust NO_x. But before ammonia can interact with NO_x to remove it from the exhaust, it must be released from a liquid solution made of distilled (deionized) water and urea. The liquid that contains the urea is called DEF and is rich in ammonia (**FIGURE 18-68**). In a dry form, urea is a crystalline white powder that dissolves easily in water. Once DEF is sprayed into the exhaust, heat and water vapor help urea to be converted into ammonia. A DEF dosing valve is located just ahead of a mixing chamber that blends vaporized DEF with exhaust gases. Blades or screens in the mixing device are used to create exhaust gas turbulence ahead of the mixing chamber. After urea decomposes into ammonia in the mixing chamber, the ammonia–exhaust gas composition passes over an SCR catalyst containing either copper or iron to break apart NO_x molecules. Some ammonia may remain unreacted in the SCR catalyst. Because ammonia is a poisonous gas if inhaled in significant quantities and because it irritates eyes, noses, and throats, it is a regulated emission that must not escape into the atmosphere. To prevent ammonia "slip" past the SCR catalyst, an additional catalyst brick is placed downstream of the SCR to convert ammonia into N_2 and H_2O. The ammonia slip catalyst (ASC) is usually contained in the same stainless steel shell as the SCR catalyst.

Advantages and Disadvantages of Selective Catalyst Reduction

The advantage of SCR technology is that it enables the lowest fuel consumption of existing exhaust aftertreatment technologies while improving power output, engine durability, and longevity. SCR is essentially better than in-cylinder methods to reduce emissions. Unlike NAC technology, it cleans-up more NO_x, more efficiently over a wider temperature range. For higher horsepower engines, the size of the exhaust catalyst can be much smaller than one required for NACs.

Until the development of SCR-U, manufacturers in North America had to depend primarily on in-cylinder strategies such as high EGR flow and retarded injection timing to reduce NO_x. These strategies increased soot production and lowered power output. Because SCR cleans up NO_x in the exhaust, less EGR is necessary and injection timing can be advanced, which optimizes power and fuel economy (**FIGURE 18-69**).

Advanced injection timing and lower EGR rates have other benefits:

- reduced active regenerations of the DPF due to reduced soot accumulations
- higher specific power output for volume of displacement
- longer engine life
  - less abrasive wear on cylinder walls, rings, and valve seats from reduced soot entering cylinders
  - lower piston speeds because EGR engines require stroking a block (longer piston stroke) to compensate for slower combustion heat release
- increased reliability
  - reduced engine heat loads with lower EGR rates
  - less soot loading in oil
  - fewer DPF regenerations.

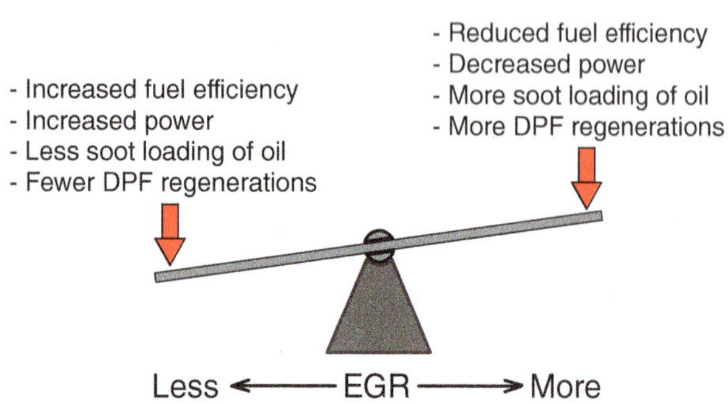

FIGURE 18-69 Relying on more NO_x reduction in the exhaust by using SCR enables the use of more efficient in-cylinder combustion strategies.

SCR catalysts using copper and iron wash coats over porous zeolite substrate can better tolerate sulfur in the fuel without deactivating. In addition to eliminating over 90% of NO_x emissions, SCR catalysts can typically convert 20–40% of PM to CO_2 and H_2O if installed upstream from the DPF. In Europe, liquid SCR has been used since 2005, prior to DPFs, enabling diesel-powered vehicles there to reduce both NO_x and PM emissions to meet European emission standards without using a particulate filter. Placing the SCR-U upstream of the DPF however, is accompanied by a 2% reduction in fuel consumption.

In contrast to using NAC technology to reduce NO_x emissions, SCR-U has many more components and requires the vehicle operator to fill the DEF reservoir at regular intervals. Additional components have a downside, other than adding to production costs. SCRs have had problems as the technology matured, and manufacturers gained more field experience with increased system complexity and emission-monitoring requirements.

Diesel Particulate Filter Upstream of the Selective Catalyst Reduction

SCR catalysts are configured in two ways. The DPF is either upstream or downstream of the SCR. Placing the DPF upstream of the SCR is the most common configuration for two reasons. First, heat energy required for active regeneration is lower because the DPF is closer to the engine's exhaust ports. As a result, active regeneration requires less fuel and passive regeneration can occur more easily. Second, more NO_2 is available for passive regeneration when the DPF is upstream of the SCR rather than downstream. Placing the DPF upstream of the SCR does have a risk. High DPF regeneration temperatures can damage catalysts (**FIGURE 18-70**). The primary disadvantage of placing the DPF downstream is due to cooler DPF temperature. The relatively cool exhaust temperature limits the effectiveness of the DPF catalyst.

On an increasing number of chassis, the DPF is placed downstream of the SCR because this configuration can minimize the need for DPF regeneration. That is because an SCR catalyst will eliminate substantial amounts of soot before reaching the DPF. GM and Ford will use this configuration on their vehicles because it also provides 2% better fuel economy. Improved fuel economy is possible because fewer active regenerations of the DPF are needed. Also, more efficient NO_x conversion takes place sooner in the hotter upstream SCR-U catalyst. This allows for even better optimization of combustion conditions favoring performance and lower fuel consumption over increased NO_x production (**FIGURE 18-71**). Combinations of NAC and SCR plus DPFs with SCR catalyst (SDPF) are used by the latest Tier 3 vehicles.

Components of Selective Catalyst Reduction Aftertreament Systems

With a few exceptions, such as some VW models, most North American manufacturers began using liquid SCR-U systems in 2010 to meet fully phased in Tier 2 bin 5 emission standards. Each manufacturer has a slightly different approach to the design and operation of their aftertreatment systems but a tremendous amount of commonality exists between them. This next section will examine common aftertreatment system components, as well as how the system operates and is constructed.

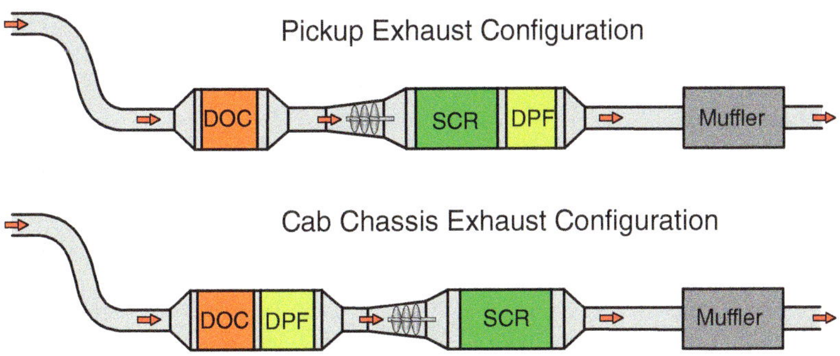

FIGURE 18-70 Aftertreatment configurations for Ford chassis using 6.7 L Power Stroke. Placing the SCR upstream of the DPF enables a 2% reduction in fuel consumption.

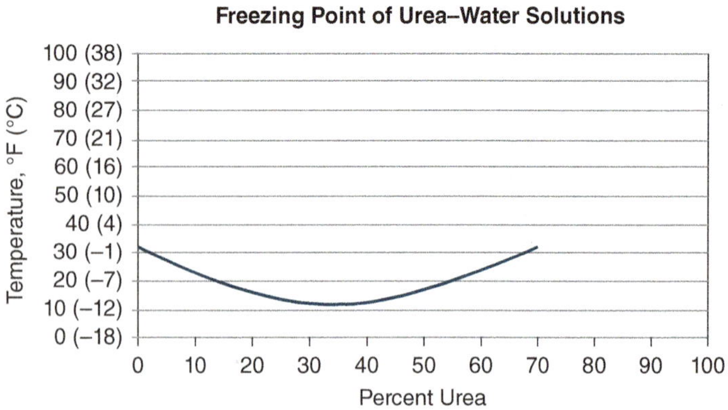

FIGURE 18-71 Urea is blended with water to achieve the lowest possible freezing point.

The following are some of the components of the SCR system:

- DEF
- SCR and ASCs
- DEF-quality sensor
- DEF reservoir module (containing multiple components)
- DEF pump and dosing unit
- DEF pressure sensor
- DEF lines and heaters
- DEF fill level warning lamp and warning messages.

Diesel Exhaust Fluid Composition

DEF, also called emission-reduction fluid and/or reductant, is a mixture of 67.5% deionized water (distilled with no minerals) and 32.5% urea. An ECM-controlled pump inside the reservoir supplies pressurized DEF through a supply line to a DEF dosing injector located in the exhaust pipe ahead of the SCR catalyst. When injected into the exhaust stream as a fine mist upstream of an SCR catalyst, it vaporizes and the urea decomposes to form ammonia (NH_3) and CO_2. Onboard DEF tanks store approximately 8–30 L (2–8 gallons) of the reductant in automobiles and in light- and medium-duty trucks. With a rate of use between 1% and 5% of fuel use (~50:1), the size of a full DEF tank is designed to last no less than approximately 7,500 miles (12,000 km) under normal use. Consumption depends on a variety of factors which are calculated by the ECM. DEF is a nontoxic, nonflammable solution, but it is noxious when converted to ammonia gas. Ammonia emissions are regulated and cannot exceed 10 ppm. Pure DEF is colorless and relatively stable chemically, which makes it safe to handle. Because it has a pH that makes it alkaline, DEF is corrosive to some metals, such as copper and aluminum.

The concentration of urea and deionized water gives DEF the best freeze protection and ensures that both the urea and the water will freeze at the same rate, preventing the fluid from becoming diluted or overconcentrated as it thaws (**FIGURE 18-72**). Because DEF

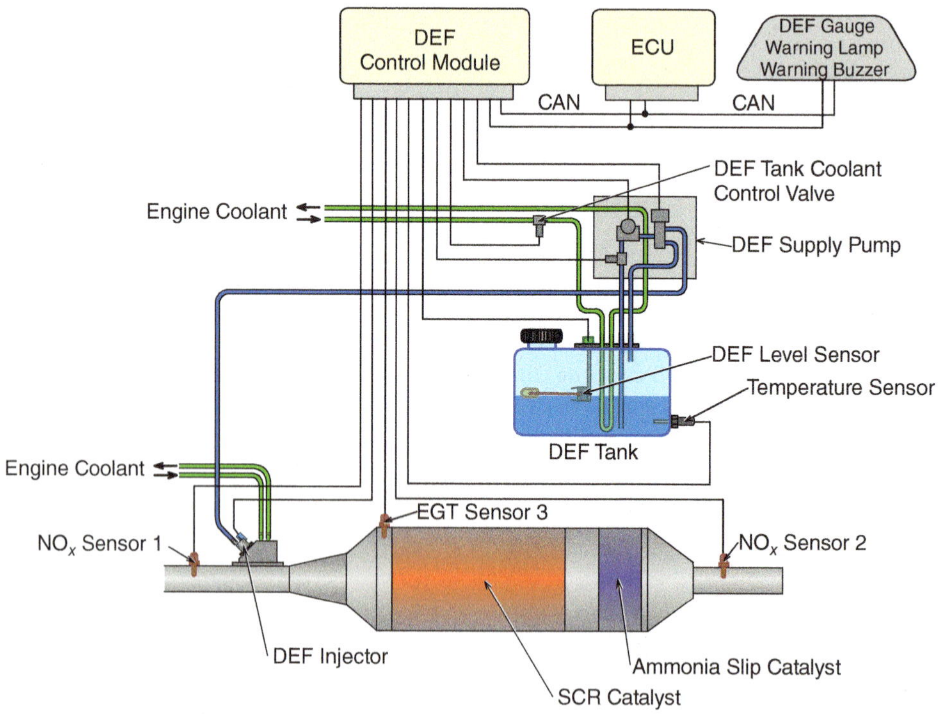

FIGURE 18-72 Typical configuration of DEF circuits in SCR systems. Note the presence of heating circuits to warm cold or frozen DEF.

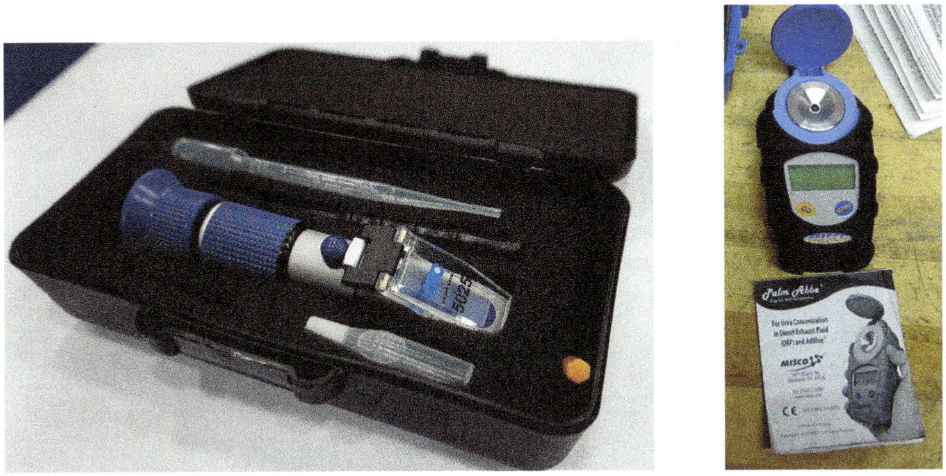

FIGURE 18-73 On the right is a conventional optical refractometer. On the left is a digital refractometer. Both can be used to evaluate DEF quality.

does freeze at temperatures below 12°F (−11°C), heaters in the reservoirs and surrounding lines are supposed to ensure rapid delivery in cold-weather operation. When the vehicle shuts down, the DEF pump will reverse the flow of DEF and pull fluid out of the lines, returning it to the reservoir to prevent damage from freezing and crystallization of urea that can block lines and clog injectors (**FIGURE 18-73**). To minimize nuisance fault codes, the EPA permitted early fault detection systems up to 40 minutes after engine start-up to move DEF to the injector in cold weather to accommodate frozen DEF before logging faults.

In North America, the API has established a performance standard for DEF used in liquid SCR systems. The International Organization for Standardization (ISO) 22241 has published a specification too. If DEF is suspected to be contaminated, a refractometer is used to measure its density and to evaluate the quality of DEF (**FIGURE 18-74**). DEF's specific gravity should be 1.310–1.3843 at 68°F (20°C). DEF should be clear and smell like ammonia. Measurement of DEF quality is also a requirement of the latest OBD-II standards, so the most recent SCR systems integrate a DEF quality sensor. These sensors can use sound waves, variable capacitors, or light optics to accurately measure DEF's density to prevent contaminated DEF from flowing through the system and potentially damaging other aftertreatment components such as sensors and catalysts.

FIGURE 18-74 Urea crystals can block lines, plug filters, and plug catalysts. Precautions must be taken when working on SCR systems to prevent damage or faults due to urea crystals.

▶ TECHNICIAN TIP

Filling a low DEF reservoir with water may be tempting because the fluid is colorless. Never add water or any other fluid to the DEF reservoir. Doing so may damage the aftertreatment system catalyst and sensors. Adding water to the DEF reservoir will change the DEF concentration levels, which may affect SCR efficiency and will raise the freezing temperature or change other characteristics of the diesel DEF. When liquid SCR systems measure DEF quality, it can potentially generate fault codes, leading to an engine power derate condition or even the engine failing to start at all.

▶ TECHNICIAN TIP

DEF contamination can result from mixing with any number of automotive fluids. Operators commonly mistakenly fill the DEF tank with other fluids, such as antifreeze, windshield washer fluid, and fuel. Because DEF is a clear liquid, the presence of even a slight discoloration indicates contamination. No visible particles should be present in DEF either. To inspect DEF for contamination, only use a refractometer to measure DEF density after visually inspecting DEF, and follow the steps in **SKILL DRILL 18-1**.

Remember when checking DEF quality that storing it for more than six months is not recommended. DEF will chemically deteriorate, and it accelerates with heat. Regularly scheduled testing of the DEF is recommended if a vehicle is stored. The suggested shelf life of an unopened DEF container is no longer than 18 months.

Diesel Exhaust Fluid Crystallization Precautions

While relatively safe to handle, urea in the DEF leaves behind a white, powdery crystal after water evaporates from DEF (**FIGURE 18-75**). Crystalized DEF can plug lines, filters, and dosing valves and almost completely plug mixing chambers. Certain properties of DEF enable it to creep quickly through threaded connections and electrical plugs and into wiring. If an electrical connector comes into contact with DEF solution, it should be replaced

SKILL DRILL 18-1 Inspecting DEF for Contamination

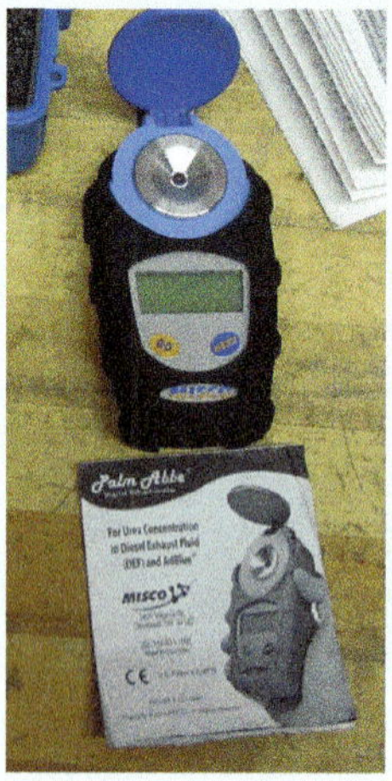

Diagnostic trouble codes (DTCs)—i.e., messages in the driver's information screen—may indicate that the DEF quality is poor and/or a drop in the efficiency of the SCR system. A message about poor-quality DEF does not necessarily indicate that the fluid is contaminated or that it needs to be replaced; it may simply indicate that there's a defective sensor or that another NO_x-increasing fault is active. Before changing DEF, check for other fault codes and consult the OEM service literature to determine code diagnostic priority. That means determining which fault code and diagnostic procedures should be followed first. To determine whether the DEF quality is poor, the following procedure can be used as a general guideline.

1. Using a clear container, collect a 150 ml (5 oz.) sample of DEF from the DEF tank.
2. Hold the container up to a bright light source and inspect it for any suspended particles, fluid separation, and discoloration. DEF is naturally clear and should not have any visible particles in the solution.
3. If discolored, attempt to identify the fluid contaminant according to color. Operators often mistakenly fill the DEF reservoir with other fluids:
   - diesel fuel—yellow or amber
   - antifreeze—orange, red, or green
   - windshield washer fluid—blue, purple, or orange
   - yellow grease-like contamination is likely engine oil.
4. Measure the specific gravity of the fluid by using only a refractometer. A battery refractometer can be used in place of a purpose-made DEF refractometer. The refractive index should be between 1.310 and 1.3843 at 68°F (20°C). Water dilution will lower the refractive index.
5. Smell the DEF to check for stale or diluted DEF. Good-quality DEF should have an ammonia smell.
6. To identify water or other fluid contamination, add several drops of food coloring to the sample. Liquids will stratify or separate and absorb the food coloring at different rates. Fuel generally will not absorb water-based food coloring.
7. If DEF quality is poor, removing a drain plug or sensor at the bottom of the reservoir allows the DEF to be drained.

immediately to prevent DEF solution from "creeping" further into the copper wiring, which takes place at a speed of about 60 cm (23.6") per hour.

DEF is slightly alkaline and therefore corrosive to certain metals, including copper and aluminum. Excess DEF can be wiped up, and water can be used to easily wash off urea that has crept onto the vehicle's tank or dispensing equipment. However, when working with SCR systems containing DEF, observe the following precautions.

- Ensure that the system is depressurized.
- Always disconnect the DEF lines before disconnecting the electrical connectors, to prevent DEF from entering the connectors.
- Wrap electrical connections that have the potential for exposure to DEF.
- Seal the SCR system so that the DEF does not crystallize if the system remains disconnected for a more than a few hours.
- Always seal system pipe fittings with pipe sealant.

It is important that tools and clothes be thoroughly cleaned of DEF solution, so that fluid or crystals are not transferred to other components that could be damaged.

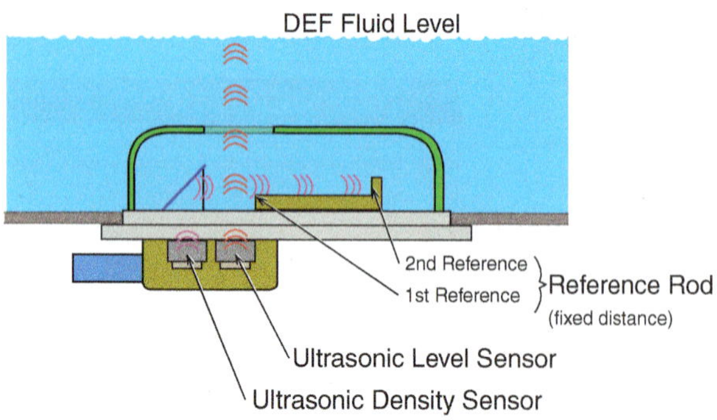

Level Sensing
- To determine the DEF level the ultrasonic signal is reflected back from the surface once.
Density Sensing
- To determine the DEF density the ultrasonic signal is reflected back twice from the first and second reference points.

FIGURE 18-75 An RDQS or DEF-quality sensor is now required for SCR systems. An ultrasonic-type sensor measures fluid density and DEF level.

Diesel Exhaust Fluid Quality Sensors

DEF fluid is also called reductant because it lends electrons to the chemical process that breaks down NO_x. This helps explain the alternate name and acronym given to DEF quality sensors: reductant quality sensors (RDQS). Concerns over DEF contamination or DEF dilution with water prompted the EPA to mandate that vehicles require DEF-quality sensors beginning in 2013. Before this time, manufactures used NO_x sensors to determine whether DEF quality were adequate. Poor DEF quality is considered to be the specific urea concentration of exhaust dosing with DEF that allows the SCR efficiency to fail to meet emission targets. Before the introduction of RDQS, NO_x sensors would monitor NO_x conversion across the SCR and compare measured conversion rates with expected or modeled NO_x conversion efficiency. Monitoring can also happen after the DEF tank has been filled to check whether DEF has the same properties before as after filling the reservoir. The system control module can compare previous dosing rates required to minimize NO_x with an updated dosing rate under similar conditions. If there is little or no difference between the new dosing rate and dosing rate before the tank was filled, the normal adaptive strategy that adjusts dosing rates in order to lower NO_x continues. However, if significantly more DEF is required, a fault is set for poor DEF quality. Once the active fault is set, a sequence of driver warnings begins that can eventually result in engine derating.

Using only NO_x sensors, aftertreatment control systems could increase DEF dosing to compensate for poor quality DEF to a point, but DEF-quality sensing performed by using the NO_x sensors is slow and cannot identify DEF dilution across all scenarios, such as when it is not dosing when frozen, during start-up, or when the vehicle is operating with low exhaust temperatures. Some early-model vehicles would generate DEF quality codes with full reservoirs of good-quality fluid because the system could not dose at low outside temperatures. DEF quality codes would also turn up every time there was any fault producing excess NO_x. RDQS sensors can identify poor quality in less than 30 seconds and when concentration changes by more than 1–5%. The new sensors do not require dosing to take place to detect problematic DEF quality. The ability for RDQS to identify poor-quality DEF and eliminate other nuisance, high-NO_x faults also helps minimize troubleshooting time and complexity.

The most common quality sensors use ultrasonic sound waves to measure fluid density (**FIGURE 18-76**). Like other fluids, DEF has a specific density. When the ultrasound waves generated by the RDQS sensor penetrate the fluid, they will travel at speeds proportional to DEF density. If DEF density does not match an expected value, the sensor will set a fault code that eventually leads to an engine derate or even failure to start. Ultrasonic-type sensors are the favored technology because they can operate to also measure DEF level. One other common sensor technology measures the fluid's capacitance or dielectric properties. When fluid is placed between two plates of a capacitor, the dielectric properties of the fluid are fixed and will have a very specific charge time measured by a circuit that is continuously charging and discharging the capacitor plates.

> **▶ TECHNICIAN TIP**
>
> Proper DEF concentrations of urea and water are critical to proper SCR-U system operation. Contaminated fluids or incorrect fluid concentrations can lead to DEF reservoir filter plugging, SCR dosing nozzle plugging and deposits, the formation of corrosive acids in the exhaust system and SCR components, and deposits forming on SCR catalysts that reduce the system's effectiveness. Aftertreatment systems are particularly sensitive to mineral contamination produced from the use of incorrect DEF fluids. As little as 1 teaspoon of salt (5 grams) in 5,000 gallons, or 0.5 ppm, will cause the DEF to be out of specification. When filling DEF reservoirs, the fluid must be kept clean. Use a dedicated funnel that is kept completely clean and free of any contaminants.

Diesel Exhaust Fluid Dosing Valve

An electrical solenoid-type dosing valve sprays pressurized DEF fluid into the exhaust stream upstream of the SCR catalyst to supply the ammonia molecules needed to break down NO_x in the SCR catalyst. The chemical conversion of DEF takes place in three stages before the urea and water mixture releases ammonia gas (**FIGURE 18-77**). A decomposition

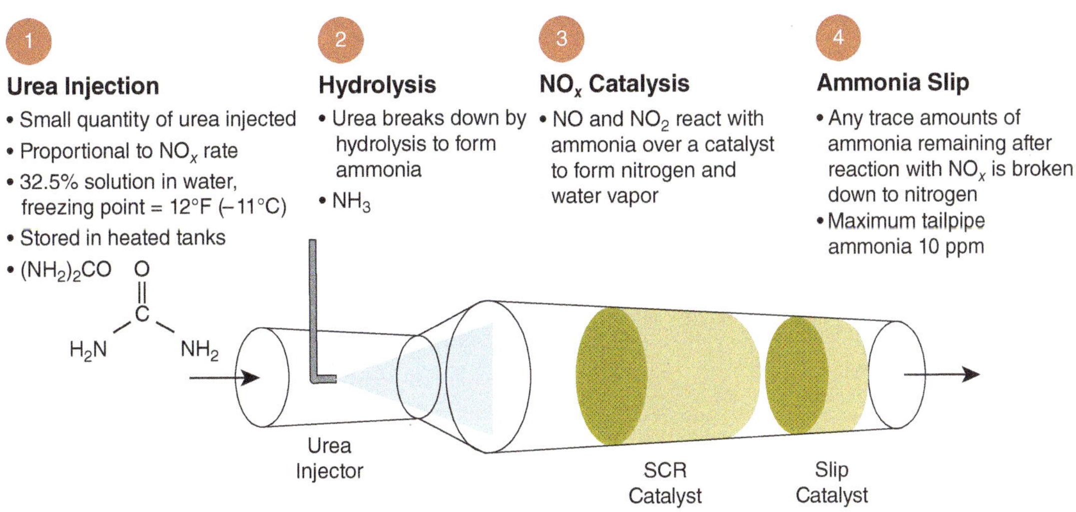

FIGURE 18-76 Stages of DEF conversion into ammonia.

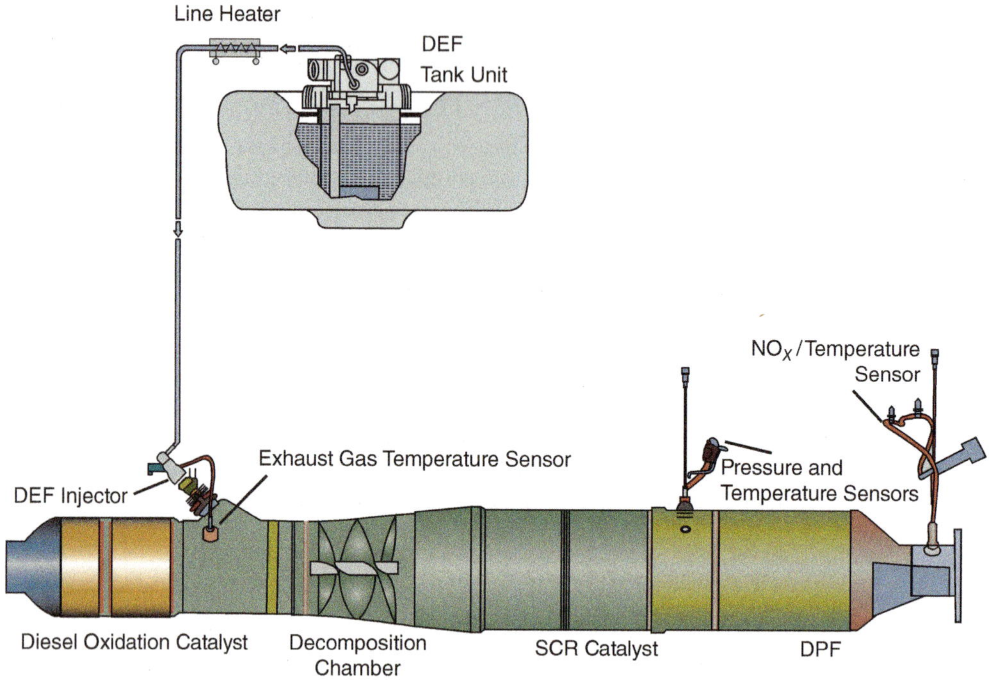

FIGURE 18-77 This SCR system mixes DEF with exhaust in a decomposition chamber.

or mixing chamber built into the exhaust system where the doser is located helps thoroughly mix ammonia with exhaust gas and provide an area where DEF conversion takes place. Inside the chamber, a screen, vane-type device, or Venturi will create additional turbulence to help exhaust heat break down DEF (**FIGURE 18-78**).

The valve is placed at approximately a 30° angle in the pipe. Several methods are used to cool the valve to increase the electrical reliability of the doser valve and minimize the formation of urea crystals caused by vaporization. Engine coolant supplied by two coolant lines can circulate through cavities inside a valve holder supporting the doser valve (**FIGURE 18-79**). The doser will also be mounted on exhaust pipe with a thick flange that may also have an insulating air gap separating the valve body from the exhaust. A gasket between the valve and flange may also have a layer of insulation to slow exhaust heat transfer into the valve. Metal spacers beneath the doser valve mounting feet add a gap to circulate air and reduce valve temperatures.

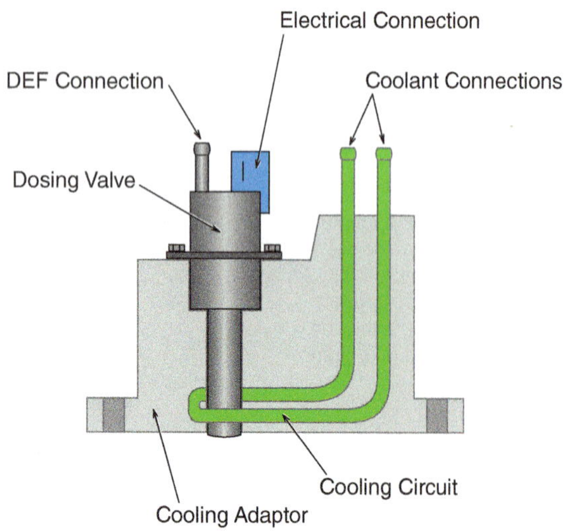

FIGURE 18-78 The dosing valve needs to be cooled to maintain its reliability. Coolant will circulate around an aluminum valve holder.

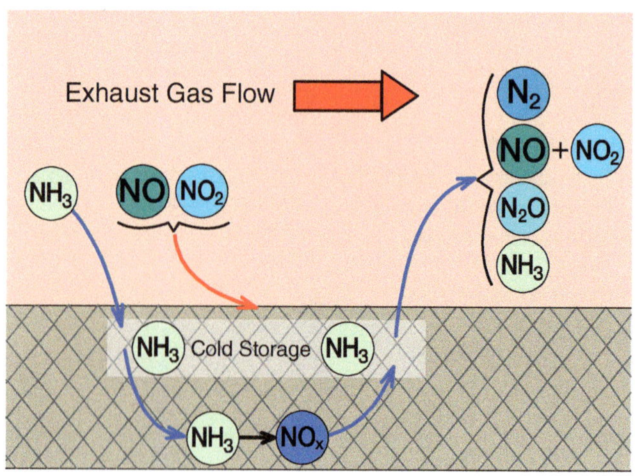

FIGURE 18-79 The zeolite substrate of the SCR catalyst can store NO_x and ammonia when exhaust temperatures are cool.

The doser valve is supplied pressurized fluid by the DEF pump. Pressures are typically anywhere from 50 psi to 150 psi. When enabling conditions have been met to dose the exhaust, the valve is electrically energized with a pulse-width–modulated (PWM) signal. Dosing may not always be continuous. The injector may operate with variable frequency of injection pulses.

> ▶ TECHNICIAN TIP
>
> If the outlet of a dosing valve is plugged or restricted because of crystalized urea, DEF will not properly mix and thus cause NO_x sensor reading errors. A restricted doser or a composition chamber will end up delivering too much DEF due to incorrect feedback from the downstream NO_x sensor. Before disassembling the system to clean and check or removing the valve to perform a delivery volume test, an active regeneration of the DPF can be performed. Sometimes heat from a regeneration will help vaporize moderate amounts of DEF crystal buildup in the SCR system. Excessive buildup will require additional service.

Selective Catalyst Reduction and Ammonia Slip Catalysts

Two different catalyst bricks are found in the shell of the SCR converter. The first is the SCR catalyst, and the second is an ASC. Both bricks have flow-through channels formed from one end to the other and are made from porous zeolite ceramic to absorb NO_x and ammonia when operating temperatures are low (**FIGURE 18-80**). Any ammonia emissions from the SCR converter are called ammonia slip. A wash coat containing platinum catalyst is used in the ASC to initially oxidize ammonia (NH_3) to convert it to NO. NO is further oxidized over the platinum wash coat, converting it and the remaining hydrogen from the ammonia to water and harmless nitrogen (N_2) (**FIGURE 18-81**).

The ASC is usually integrated into the same shell as the SCR catalyst in the zone immediately after the SCR unit to prevent slip in transient conditions. SCR operation requires precisely controlling the DEF injection rate that produces ammonia. Insufficient dosing allows excessive NO_x emissions, whereas excessive dosing can release ammonia into the atmosphere. Ammonia slip decreases with increased catalyst temperatures. This explains why DEF dosing generally cannot take place unless the temperature sensors at the inlet and outlet of the catalyst shell indicate that the catalyst temperatures are at least 302°F (150°C) or more. Excess ammonia slip will take place if the temperatures are too low. DEF will also crystalize on the catalyst substrate and plug the catalysts when temperatures are low (**FIGURE 18-82**).

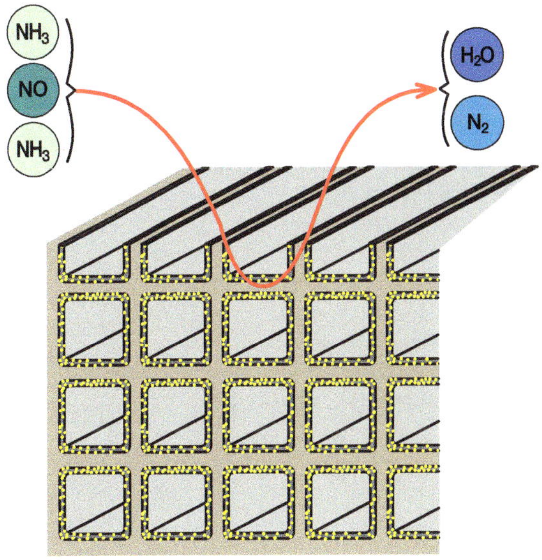

FIGURE 18-80 Chemical processes converting NH_3 to N_2 and H_2O.

Selective Catalyst Reduction Catalyst

SCR of NO_x with ammonia was first discovered using a platinum (Pt) catalyst. However, iron (Fe) and copper (Cu) catalysts are currently used because they better resist deterioration under high temperatures when the DPF is actively regenerating. Cu catalyst SCRs are the most commonly used catalyst since they have the best NO_x conversion rates over a wide temperature range. At temperatures between 392°F and 752°F (200°C and 440°C), a 90% NO_x conversion efficiency is possible.

SCRs will help convert NO_x to N_2 by using the ammonia contained by urea in the DEF. To work well, ammonia must adsorb onto the SCR catalyst,

FIGURE 18-81 An SCR catalyst completely blocked by urea crystals.

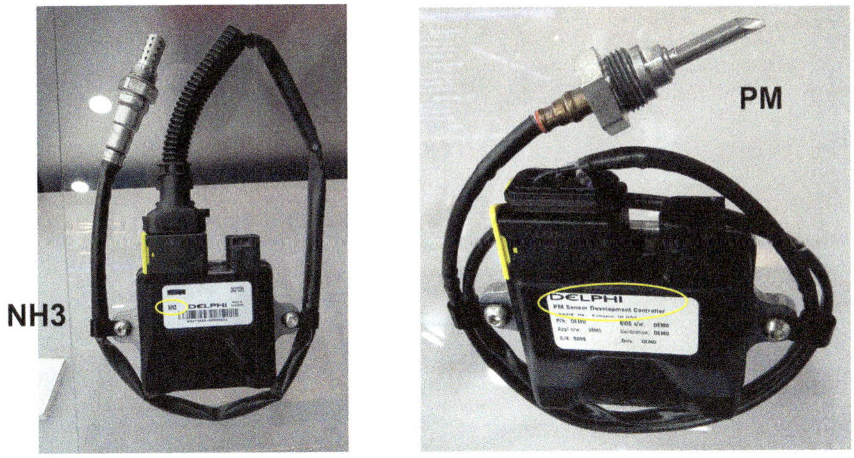

FIGURE 18-82 Ammonia and particulate sensors have their own dedicated control module.

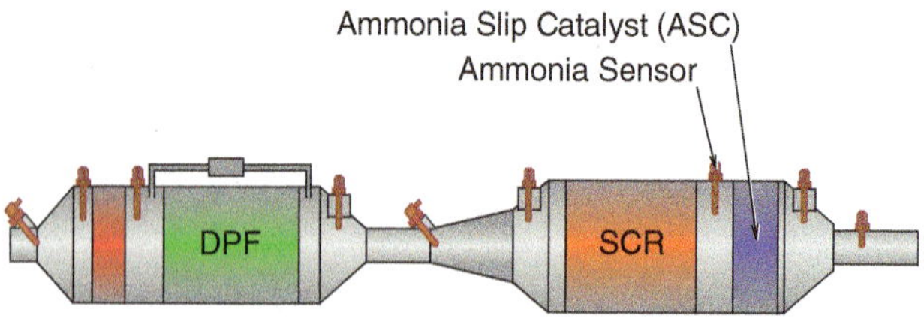

FIGURE 18-83 Location of the ammonia sensor is between the SCR catalyst and the ASC catalyst.

where it can react and break apart NO_x molecules. The more ammonia stored on the catalyst substrate, the higher the NO_x conversion. Excess dosing, however, results in ammonia slippage and high DEF consumption. To ensure that the optimal levels of DEF are sprayed into the exhaust stream, NO_x sensors near the tailpipe outlet can provide some limited closed-loop feedback to the control system to increase or decrease dosing based on NO_x measurement. Dosing the SCR with more or less DEF is adjusted by up to 30% correction when using only the data provided by the NO_x sensor. If the sensor measures too much NO_x in the tailpipe, more DEF is injected, and vice versa. Because NO_x sensors are cross-sensitive to ammonia, which means they cannot easily distinguish the difference between nitrogen in both ammonia and NO_x molecules, an ammonia sensor has been used since 2015 in most SCR systems. The 2015 and later ammonia sensor is located between the SCR catalyst and the ASC catalyst, ahead of the NO_x sensor at the tailpipe (**FIGURE 18-83**). Directly measuring ammonia slippage past the SCR but before the ASC means that the DEF can be more accurately metered for consistently lower ammonia slip and NO_x emissions.

> **TECHNICIAN TIP**
>
> When problems occur in the SCR system, one common observable condition is the buildup of DEF crystals in the SCR mixing chamber. The increased quantity and frequency of DEF dosing take place whenever NO_x sensors detect excessive NO_x and then attempt to correct the condition. High NO_x can be caused by any number of internal engine problems, such as intake manifolds restricted with carbon, plugged air filters, externally restricted CACs, or excessively low engine speed and heavy-load engine operation. Excess dosing that dumps more DEF in the mixing chamber can in turn cause the NO_x sensor to misinterpret additional ammonia as more NO_x. To correct for excessive NO_x, even more DEF is sprayed into the exhaust. Low exhaust temperatures incorrectly sensed by temperature sensors can cause dosing to take place below a range that properly vaporizes DEF. And of course, a defective tailpipe NO_x sensor will cause an incorrect dosing rate. Do not attempting to correct the other conditions by unnecessarily replacing NO_x sensors.

Ammonia Sensors

As mentioned in the previous section, an ammonia sensor is located between the SCR catalyst and the ASC catalyst on most vehicles after 2015 (**FIGURE 18-84**). Although the ammonia sensor has a similar construction to a wideband oxygen sensor, the electrode material in the sensing cells are different. In the pump cell, the electrode in direct contact with the exhaust stream contains a proprietary material that makes it sensitive only to ammonia. When ammonia slip occurs, a pair of electrodes in the pump cell produces a voltage signal in proportion to the amount of ammonia in the exhaust gas. A module dedicated to processing signals to and from the sensor is attached to the sensor. Sensor information is transmitted back from the module over the controller area network bus (CANbus) communication line. Ammonia and NO_x sensor signals are both used by the ECM to adjust DEF dosing for maximum SCR catalyst efficiency (**FIGURE 18-85**).

NO_x Sensor Operation

NO_x sensors are used by the OBD-II system to detect problems in the SCR aftertreatment system and provide limited feedback to the control module for regulating DEF dosing. Like ammonia sensors, NO_x sensors share many construction details with wideband oxygen sensors. Most NO_x sensors will have a dedicated module to process sensor signals and communicate with the ECM over the controller area network (CAN) line (**FIGURE 18-86**). With few exceptions, the module and sensor are matched and must be replaced together. A memory file stored in the module contains information about unique sensor characteristics and manufacturing tolerances used to adjust signals sent from the sensor and optimize

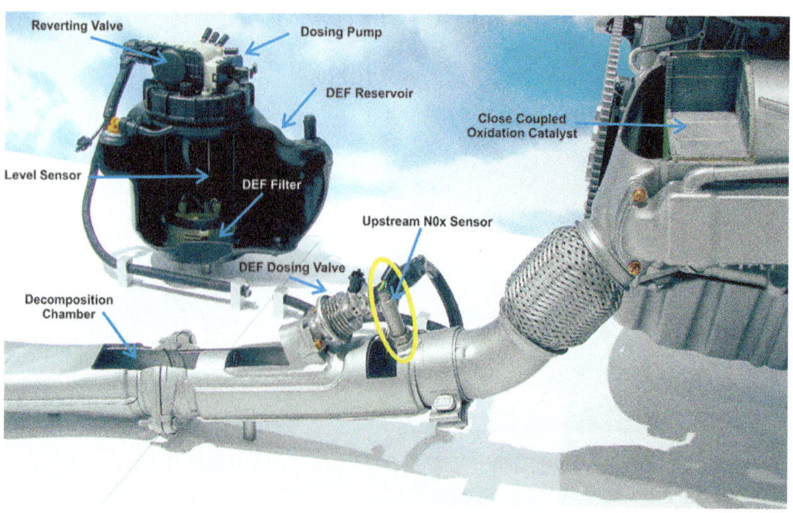

FIGURE 18-84 The upstream NO_x sensor measures the concentration of NO_x released from the engine.

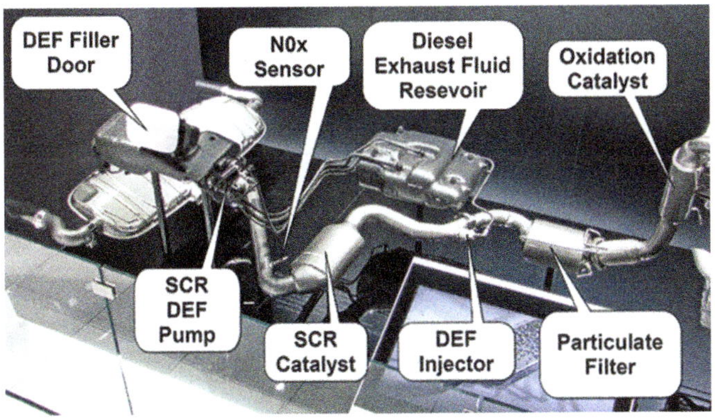

FIGURE 18-85 NO_x sensors at each end of the exhaust system can detect problems with DEF and high NO_x levels.

FIGURE 18-86 NO_x sensors use a dedicated module to process signals and to store specific sensor information related to operating tolerances.

sensor operation. While some vehicles use only a single NO_x sensor at the exhaust outlet, most vehicles use two. NO_x signals are collected from one sensor in the exhaust pipe after the turbocharger and from the other sensor at the tailpipe. The differences in NO_x concentration between the sensors are used to calculate SCR system efficiency. NO_x concentration should drop between the two sensors when the SCR system is operational. Changes in NO_x concentration from beginning to end judges the performance of the entire SCR system, including DEF quality.

This two-sensor arrangement helps explain why so many codes are generated whenever a fault exists in either the engine or aftertreatment system: the sensors can only detect the abnormal NO_x concentrations and not always pinpoint the components causing the problem. The addition of DEF-quality sensors and ammonia sensors has improved fault resolution capabilities. Because NO_x sensor accuracy has improved, they have been relied on since 2015 to provide improved closed-loop feedback control of DEF dosing. With increased diagnostic accuracy, fewer nuisance-type fault codes set and less diagnostic time is wasted in identifying false failures such as low SCR efficiency, out-of-range NO_x sensor codes, and DEF quality issues. Improved OBD-II capabilities have reduced the frequency of unnecessarily replacing expensive NO_x sensors and DEF fluid.

Several different types of NO_x sensors are used today. The most common type applicable to diesel engines measures NO_x ppm concentrations from proprietary electrodes producing current nanoamp scale. These sensors are identified by the eight wires connecting the sensor to a control module. Two or three Nernst cells are arranged to measure exhaust NO concentration in one of cells (**FIGURE 18-87** and **FIGURE 18-88**). Before NO reaches that cell through

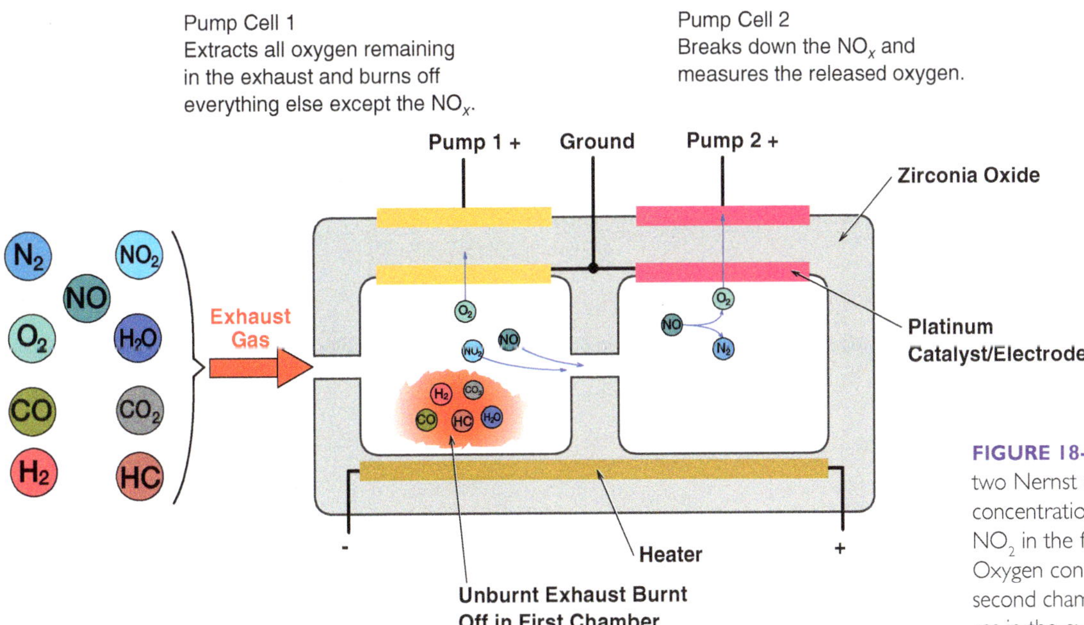

FIGURE 18-87 NO_x sensors use two Nernst cells to measure gas concentration. Oxygen is removed from NO_2 in the first and second chamber. Oxygen concentrations measured in the second chamber correspond to the NO_x gas in the exhaust.

FIGURE 18-88 The NO_x sensor measures both exhaust NO_x and oxygen content. Eight wires on this NO_x sensor are used for two sensing chambers that each have heaters.

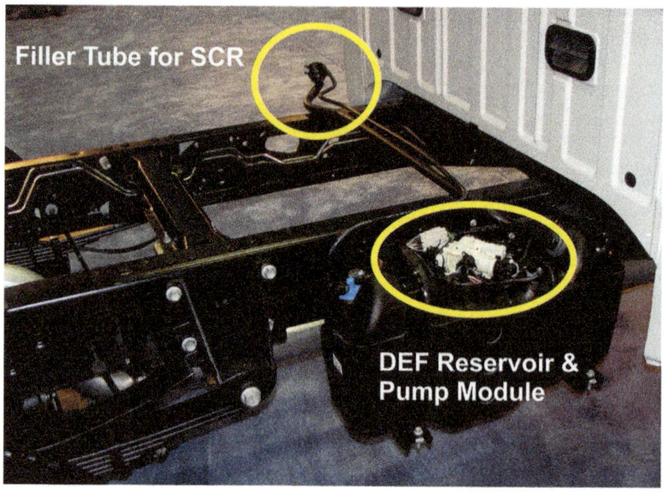

FIGURE 18-89 Location of the DEF reservoir and pump module on a medium-duty truck's chassis.

a small opening, called a diffusion gap, all oxygen is removed from the exhaust gas sample by using a wall catalyst, including one oxygen atom from any NO_2 molecules. In the final chamber, only NO enters where a heated catalytic material separates nitrogen from oxygen (**FIGURE 18-89**). The oxygen concentration of this last cell is measured once again. Oxygen content will be directly proportional to the NO level in the cell, which is used to calculate exhaust NO_x concentrations. A third cell containing an oxygen reference gas may be present in some sensors. If not, atmospheric oxygen is used as a reference to calculate gas concentrations.

Diagnostics for Oxides of Nitrogen Sensors

It is important to note that NO_x sensors can measure exhaust oxygen content. Sensor diagnostics by the OBD-II system are performed by measuring the time it takes for oxygen concentrations to change, such as during engine overrun or when the throttle is snapped shut but the vehicle is coasting. In an extended engine overrun or during deceleration events when the injectors are shut off, the sensor performs a self-calibration adjustment. During recalibration, a new correction factor called an adaptation factor is applied to circuit voltages. The adaptation corrects the sensor to measure 20.95% O_2 in air. A fault code is set if the adaptation correction factor is outside the range of 0.95–1.22 times a baseline factor (**FIGURE 18-90**).

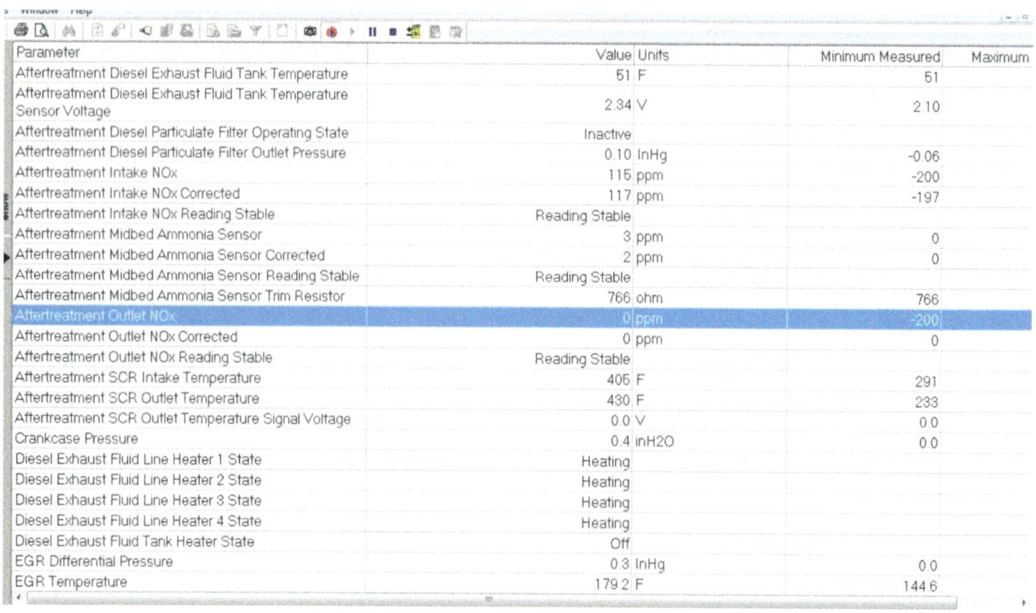

Parameter	Value	Units	Minimum Measured	Maximum
Aftertreatment Diesel Exhaust Fluid Tank Temperature	51	F	51	
Aftertreatment Diesel Exhaust Fluid Tank Temperature Sensor Voltage	2.34	V	2.10	
Aftertreatment Diesel Particulate Filter Operating State	Inactive			
Aftertreatment Diesel Particulate Filter Outlet Pressure	0.10	InHg	-0.06	
Aftertreatment Intake NOx	115	ppm	-200	
Aftertreatment Intake NOx Corrected	117	ppm	-197	
Aftertreatment Intake NOx Reading Stable	Reading Stable			
Aftertreatment Midbed Ammonia Sensor	3	ppm	0	
Aftertreatment Midbed Ammonia Sensor Corrected	2	ppm	0	
Aftertreatment Midbed Ammonia Sensor Reading Stable	Reading Stable			
Aftertreatment Midbed Ammonia Sensor Trim Resistor	766	ohm	766	
Aftertreatment Outlet NOx	0	ppm	-200	
Aftertreatment Outlet NOx Corrected	0	ppm	0	
Aftertreatment Outlet NOx Reading Stable	Reading Stable			
Aftertreatment SCR Intake Temperature	405	F	291	
Aftertreatment SCR Outlet Temperature	430	F	233	
Aftertreatment SCR Outlet Temperature Signal Voltage	0.0	V	0.0	
Crankcase Pressure	0.4	inH2O	0.0	
Diesel Exhaust Fluid Line Heater 1 State	Heating			
Diesel Exhaust Fluid Line Heater 2 State	Heating			
Diesel Exhaust Fluid Line Heater 3 State	Heating			
Diesel Exhaust Fluid Line Heater 4 State	Heating			
Diesel Exhaust Fluid Tank Heater State	Off			
EGR Differential Pressure	0.3	InHg	0.0	
EGR Temperature	179.2	F	144.6	

FIGURE 18-90 A screenshot showing SCR aftertreatment data. Note that there is a drop in NO_x from the intake sensor to the outlet sensor in a properly functioning SCR system.

MPC engine algorithms also use oxygen sensor data to accurately estimate engine load for other engine-related functions, such as adjusting EGR valve and VGT positions, as well as injection rate shapes for CR fuel systems.

Water vapor in cool exhaust gases can interfere with sensor operation. Heaters in NO_x sensors will quickly warm the sensors up to remove moisture from them, but a delay period of a few minutes is needed to completely evaporate any moisture in the sensor. This delay period is called the dew-point waiting period and is observable when monitoring sensor operation during cold-start conditions. The sound and smoothness of engine operation perceptibly changes after the sensors have been warmed up and entered closed-loop operation.

▶ TECHNICIAN TIP

NO_x sensors are expensive to replace and are often not actually defective when identified in a list of OBD-II codes. High NO_x readings will in fact set a NO_x sensor fault code for circuit range or performance. When diagnosing sensor faults, use the correct code priority sequence and diagnose faults in the order recommended by the manufacturer. As part of the diagnostics, always make sure other problems that could create high NO_x levels are eliminated:

- Inspect the air intake system for defects and related fault codes. Plugged intake manifolds, air inlet restrictions, or leaks will upset the oxygen lambda value in the exhaust. These repairs are given first priority.
- Inspect the EGR system for faults.
- Inspect the DEF system for faults and fluid contamination. Verify that dosing is taking place correctly. Check the dosing valve and decomposition chamber for excessive DEF crystal deposits or plugging.
- Inspect the DPF, DOC, and SCR temperature and pressure sensors and ensure that they are working by monitoring their operation.

Under some conditions, such as when the dosing valve is disabled and the oxidation converter is cool, the NO_x sensor readings may be compared. NO_x sensor values should be nearly identical since little exhaust heat and dosing are influencing chemical reactions in the exhaust.

▶ Diesel Particulate Filter Delivery System

18-07 Identify and describe the construction and operation of DEF delivery and feedback systems used by SCR aftertreatment systems.

DEF is stored onboard in a polyethylene tank mounted beneath the vehicle. The location of filler tubes is not standardized, but DEF is commonly added to the tank through a filler neck to the right of the diesel fuel filler neck. Other filler tube locations can be found at the rear of the vehicle in the spare tire storage area that doubles as a location for the reservoir. To prevent adding diesel fuel to DEF, the filler tube has a designated blue color and a standardized filler opening. The smaller 0.75" (19 mm) nozzle diameter for DEF, compared to 0.87" (22 mm) for diesel, is intended to prevent a diesel fuel nozzle from filling the DEF tank. To match DEF and fuel consumption rates, larger reservoirs are used on pickups, which will be up to 8 gallons (30.3 L) (FIGURE 18-91).

The DEF tank commonly incorporates the following DEF system components into a single module:

- DEF controller (ECM)
- DEF-quality and DEF-level sensor
- DEF pump and reverting valve
- DEF pressure sensor
- DEF heater and fluid lines
- DEF temperature sensor.

If equipped, the DEF control unit will coordinate the operation of the reservoir pump, the dosing valve, and the sensor with the vehicle ECM. A dedicated aftertreatment system module or the engine ECM alone may control all the aftertreatment system operations, including the DPF. Information is exchanged between any aftertreatment or DEF control module and engine ECM. Using inputs from the ECM and other sensors, the control unit regulates the operation of the DEF pump assembly, dosing valve, and DEF heating systems.

FIGURE 18-91 A DEF reservoir and pump module.

The amount of DEF used is carefully metered by the dosing valve to be in proportion to the amount of exhaust NO_x. Earlier systems dosed the exhaust system with DEF based on a probability of NO_x in the exhaust system by based on engine operating factors. Also, SCR system gas sensors do not usually form closed-loop feedback with the dosing injector to meter DEF during cold-start conditions. Once warmed up, the sensors will detect SCR problems such as incorrect fluid, poor DEF quality, or high NO_x levels. Dosing schedules in early systems are calculated by using an algorithm that predicts the amount of DEF needed to eliminate NO_x. The feedback loops used to control DEF dosing in the latest systems is formed between the NO_x sensors, ammonia (NH_3) sensors, an ECM, and the dosing valve.

The States of Diesel Particulate Filter Delivery System Operation

Regardless of the manufacturer, a DPF delivery system operates in four basic states:

- heating state
- priming state
- dosing state
- purging state.

Heating State

A heating state is required after the DEF delivery system determines that the fluid may be frozen or even just slushy. Because DEF freezes, the DEF temperature sensor provides an input to determine whether the DEF reservoir needs to be heated. Fluid can be heated using electric heating elements or engine coolant. No DEF will flow until the DEF is warmed to a sufficient temperature. When a vehicle operates during very cold weather, this can take as long as 270 minutes in some vehicles before a fault is set. DEF heating, if needed, will continue to heat the DEF based on ambient temperatures and a thawing time table. Electric heating elements in the reservoir, inside the pump, and built into the DEF lines will warm fluid until it reaches more than 27°F (−3°C) or according to the thawing time table. Heated engine coolant is circulated through a loop at the bottom of the DEF reservoir. Engine coolant will flow after reaching a temperature of approximately 149°F (65°C), which opens a coolant valve to enable coolant to circulate through the DEF reservoir. Circulating coolant is also used to prevent DEF from freezing during cold-weather operation.

Priming State

Priming begins after the engine starts and circulates DEF throughout the system, but no dosing takes place. During priming, diagnostic checks are performed in preparation for system operation. Only after the engine has started and exhaust flow has heated the SCR catalyst to at least 302°F (150°C) can the SCR system begin operation. When both the exhaust catalyst and DEF temperatures are within an acceptable range, the DEF control unit will command the DEF pump to start its priming process. The DEF pump draws DEF through its suction port to pressurize and clean it through a filter often built into the pump body. The pump supplies pressurized fluid to the DEF dosing valve through a heated dosing line. A DEF pressure sensor monitors the pressure in the DEF dosing line to ensure adequate pressure is available for DEF dosing. Pressure monitoring begins not only when the ignition is switched on but also after ignition has been turned off, to ensure DEF has been purged from the line. DEF can crystalize and freeze in the dosing line and valve, which damages or prevents the SCR system from operating properly. To verify that the dosing valve is operating correctly, the control module will typically pulse the valve for two seconds and observe the pressure sensor data. The pressure should momentarily drop and then stabilize when the valve is functioning correctly. The priming state ends when proper DEF pressure readings are within range and doser valve operation has been verified.

Dosing State

During the dosing state, DEF is sprayed into the exhaust system. The DEF pump runs continuously to maintain DEF dosing line pressure when the system is in the dosing state, whether DEF is required or not. The DEF pump remains running to circulate DEF through the dosing valve. The dosing valve is a pulse-width-modulation (PWM) controlled solenoid valve. It is protected from the hot exhaust gases by a heat shield and an air pocket between the valve and exhaust pipe. The metal body is also cooled by the circulating DEF solution or coolant around a doser valve holder. Fluid is sprayed or pulsed as a fine mist into the decomposition chamber ahead of the SCR catalyst. Pump pressure is maintained anywhere from 48 psi to 150 psi (5 bar to 10.3 bar), depending on the manufacturer and type of pump used.

When the ECM determines that the exhaust needs to be dosed, it sends a signal either to the DEF controller or directly to the valve to energize the DEF dosing valve. Dosing frequency and quantity is based on NO_x sensor inputs and or predicted NO_x levels.

The following conditions typically must be met before dosing can occur:

- no current DTCs for the SCR or DEF system
- DEF tank level must be above ~6% filled with DEF
- DEF quality must be verified by the quality sensor
- DEF pressure must be adequate for dosing
- SCR catalyst inlet and outlet temperature must be at least 302°F (150°C)
- NO_x sensors have warmed up and are measuring NO_x levels
- DEF temperature must be above 27°F (−3°C).

Purging State

Purging or removing DEF from the lines, when the SCR system is not in operation, prevents damage to lines and the dosing valve in temperatures that can freeze DEF. Purging begins when the engine ignition is switched off. The DEF pump will run during the purge cycle, but an internal valve changes the direction of fluid flow. This valve, called the reverting valve, switches the DEF pump inlet for outlet to pull all of the DEF out of the DEF dosing valve and DEF dosing line. Fluid is returned to the reservoir, and lines are filled with air instead because the solenoid in the DEF dosing valve opens, eliminating any vacuum during the purge cycle.

▶ Warning Lamps and Messages

18-08 Identify and explain the operation of aftertreatment warning systems.

Because the SCR catalyst in an SCR-U system does not function without a reducing agent, manufacturers must satisfy five guidelines to maintain vehicle compliance with EPA regulations. According to EPA guidance documents, several features are required of the SCR system:

- a driver warning system (i.e., a dash warning light)
- driver inducement (i.e., reduced engine power, fuel filler lockout, no restarting after fueling if DEF is depleted)
- system identification of incorrect reducing agent—the system would identify if its reservoir were filled with only water or other liquid
- tamper-resistant design—the warning, driver-inducement, and operational systems (i.e., sensor wiring and dosing injector) are not easily deactivated
- durable design—the system must operate effectively to 120,000 or 150,000 miles (193,000 or 241,400 km), depending on certification criteria.

A DEF warning and power derate takes place if the DEF level is depleted or its quality is impaired. This is to encourage the driver to have the problem corrected.

When the DEF drops below a specific level, a warning message displayed is in the driver's information center. For most vehicles, a warning will be displayed when the DEF range falls below 800–1,000 miles (1,287–1,609 km). The display may be accompanied by a chime or voice message. If the DEF level continues to decrease, warnings will become more intrusive and even require the driver to acknowledge the condition before the vehicle can be driven. Once the DEF level becomes critical, vehicle speed will be limited. If the DEF reservoir is allowed to become empty, the engine will run at only idle speed or may not restart once it has been shut off during fuel filling. The DEF reservoir will need to be filled with a minimum amount of fluid to operate above idle until the diesel exhaust fluid tank has been refilled.

Maintenance of Exhaust Emission Aftertreatment Systems

To replace a DPF, follow the steps in **SKILL DRILL 18-2**.

SKILL DRILL 18-2 Replacing a DPF

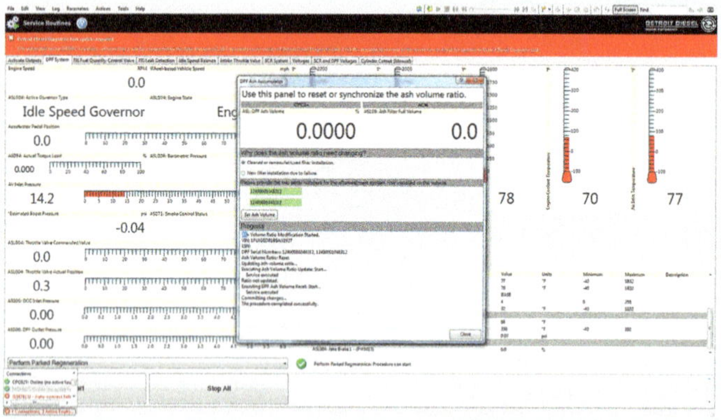

1. Disconnect and remove the Marman-style band clamps holding the removable DPF filter in the DPF filter canister.
2. Scrape and clean the filter gasket surfaces of the filter canister.
3. Reinstall the new or rebuilt filter with new gaskets. Carefully note the direction of exhaust flow through the DPF filter during installation.
4. Perform a forced regeneration of the new or green filter to increase soot-loading capacity.
5. Reset the ash accumulator according to the OEM service procedures.
6. Follow the dialog boxes for entering the new filter serial number, the type of filter installed (new or replacement), and the current vehicle mileage.
7. Follow these installation tips:
   - Use only manufacturer-recommended filter-cleaning procedures. Do not use cleaning solvents on filters; they do no work and are likely to damage the filter and solidify ash in the filter passageways.
   - Do not use a shop compressor in an open area to clean a filter.
   - For larger, heavier DPFs, use a transmission jack or floor jack to help maneuver the filter from the underside of the vehicle.
   - Do not steam clean an ash loaded DPF; water will turn ash into a concrete-like material.
   - Oil-contaminated filters are generally not reusable, but they may be recovered by baking them in an industrial oven to remove the oil.
   - Do not drop the filter element on the shop floor to remove ash.
   - DPF ash must be handled and disposed as a hazardous waste.
   - Make sure that filter elements are installed in the correct direction of flow.

To perform a stationary regeneration, follow the steps in **SKILL DRILL 18-3**.

SKILL DRILL 18-3 Performing a Stationary Regeneration

When fault codes related to DPF operation are set, it is important to perform a visual inspection of the DPF. These codes are related to excessive ash or soot load or filter performance. To visually inspect the aftertreatment system, the following is a general procedure.

1. Inspect all engine fluid levels and check for fluid leaks
2. Inspect all wiring harnesses, electrical connections, lines and hoses for proper routing, connection and attachment
3. Run the engine and snap the throttle to check for any black smoke emissions
4. Inspect the inside of the tail pipe for soot. Any soot emissions indicate a missing, cracked or damaged DPF
5. Inspect the exhaust pipe for modifications, broken pipes, missing sensors and fluid leaks from the exhaust
6. Disconnect and remove from the DPR, the two stainless-steel fittings for the pressure differential. Use a flexible camera to inspect and photograph the outlet of the oxidation converter, the inlet and outlet face of the DPF filter. The outlet face of the DPF should have not soot.
7. Remove the DOC inlet temperature sensor. Using a flexible USB connected camera (smaller than 5mm), inspect and

photograph the condition of the inlet face of the oxidation converter for face plugging.

8. Inspect the condition of sensors the sensor for excess fluid and soot contamination
9. If equipped with an NAC, a shop light should easily shine through the catalyst substrate. Remove and inspect the NAC with a light.
10. Inspect the air intake system for
    - A restricted air filter or leaking air intake system that could allow dirt to plug the DPF
    - A leaking charge air cooler. The charge air cooler should have no external fluid leaks, missing or broken hose clamps
11. Inspect the turbocharger mounting and installation. Check the operation of the turbocharger actuator by cycling the ignition switch on and off to check or listen for correct actuator movement. Note – some oil in the compressor wheel inlet of a turbocharger is normal if the engine uses a closed crankcase ventilation system. Oil weepage at CAC hoses is also normal for closed crankcase ventilation systems
12. Inspect for the presence of EGR valves and evidence of the installation of blocker plates
13. Disconnect the EGR cooler outlet and inspect for fluid leaks from the EGR cooler
14. Disconnect and remove exhaust sensors to inspect for fluid leaks. Clean sensors with a cloth and reinstall

Review and note vehicle history for recent problems associated with the fuel system, injectors, EGR and turbocharger.

Selective Catalyst Reduction Service

Due to inconsistent availability of DEF at fuel stations, some early SCR systems could completely deplete the DEF before a fault code is set. After operating for more than 10 hours, the vehicle will derate power. Another 10 hours of operation allows the vehicle to only idle. Simply filling the DEF reservoir will not allow the vehicle to return to normal operation, because the active SCR code cannot be removed on it's own until after the SCR monitor runs. Because it is a non-continuous monitor, which means that the monitor will not run at idle, a service tool must first erase the code. An alternative is to use an OEM electronic service tool to perform an SCR system refill activation. When used in this mode, the service tool will force a quick refill detection strategy after the DEF reservoir has been refilled. SCR systems phased in until 2015 have increasingly stringent thresholds for fault detection and limitations on how long the SCR system can operate without DEF fluid. SCR maintenance includes many other service procedures:

- SCR visual leak check: Use this test to pressurize the SCR system to visually check for leaks.
- SCR system emptying: Use this test to depressurize the SCR system and empty the DEF from system lines, as well as to place the system into a known default state to check the OBD-II monitor operation.
- SCR parameter reset: Use this test to reset the SCR system data following an SCR catalyst replacement. Because OBD systems requires the ability to compensate for system wear and deterioration, new values for the SCR temperature and pressure sensor necessitate adaptation or relearning.
- SCR dosing measurement test: Use this test to visually verify that the SCR dosing injector is flowing the correct quantity of DEF; a container collects and measures the quantity and flow of DEF from the injector after it has been commanded to dose a specific quantity of DEF.

▶ TECHNICIAN TIP

After refilling a depleted DEF reservoir, the SCR fault code may need to be erased before the vehicle will operate normally. This is because the OBD system's SCR monitor will not run if the engine has derated to idle. The fault will remain until the engine has met enabling conditions for running the monitor. Alternatively, completely refilling the DEF reservoir and cycling the ignition switch may also remove any derate limit imposed by an SCR system fault.

▶ Wrap-Up

Ready for Review

- Diesel engines now use a variety of emission-reduction technologies centered in the exhaust system, including, DPFs, catalytic converters, SCR, LNTs, and NACs.
- Because in-cylinder, engine-based emission control strategies were inadequate to meet emission-reduction targets, the exhaust system became the focal point of technology—to remove harmful emissions.
- Catalytic converters speed up chemical reactions of noxious exhaust emissions and change them into harmless combustion by-products. Several types of converters exist and are classified according to the noxious emission that they convert or the type of chemical reaction taking place in the catalyst material.
- An oxidation catalytic converter refers to the type of catalyst that causes O_2 to chemically combine with

noxious emissions such as HCs and CO, converting the emissions into H_2O and CO_2.

- Oxidation converters consist of a stainless steel canister containing a ceramic material formed into a honeycomb-like structure called a substrate or catalyst support. The substrate has a coating of active catalyst material that accelerates the chemical reaction between noxious emissions and oxygen remaining in the exhaust after combustion.
- The first difficulty in cleaning up diesel exhaust by using oxidation catalytic converters is the relative coolness of diesel exhaust temperatures. However, new catalyst materials have been developed for diesel engines that enable the oxidizing HC and CO emissions at low temperatures.
- The primary purpose of a DPF is to remove soot and other solid particles from diesel exhaust, rendering the exhaust completely smokeless. To remove soot and other exhaust-borne particles, exhaust gas passes through a wall flow filter, which traps PM before reaching the exhaust outlet.
- Trapped soot and other particles collected on the walls of the DPF will eventually plug the filter, producing high exhaust backpressure. Regeneration, a renewal process, will clean the DPF by purging it of soot; this process involves burning the soot out of the filter.
- Heat for regeneration is provided in two ways: one is through natural exhaust heat plus chemical reactions in the catalysts, which also produce heat; the other involves supplementally heating the DPF with fuel or using an electric heat source.
- SCR-U is a process where an ammonia-containing liquid is sprayed into the exhaust stream to destroy NO_x molecules. DEF is the generic name of the liquid, which is a clear, nontoxic, and nonflammable combination of water and urea.
- The advantage of SCR-U technology is that it enables the lowest fuel consumption of existing exhaust aftertreatment technologies while improving power output, engine durability, and longevity.
- The location of the DPF and the SCR catalysts depends on which of two possible configurations is used: the DPF can be upstream or downstream of the SCR. Locating the DPF upstream of the SCR is less fuel efficient.
- NO_x adsorber technology is a catalyst strategy for removing NO_x in the low-temperature, oxygen-rich exhaust environment of a diesel engine. In this system, NO_x emissions are absorbed, or temporarily trapped and stored, by the catalyst material; the stored NO_x is removed in a two-step reaction.
- Moving the exhaust gas oxygen concentration from high to low, or from lean to a temporarily rich condition, is the strategy used by NAC's or LNT; this technology is also called by many names including $DeNO_x$ catalysts.
- A pressure differential sensor (delta P sensor) measuring the exhaust pressure at the DPF inlet and outlet supplies data to the ECM that is used to measure DPF soot loading and for fault conditions. Deviations from expected values will set fault codes.
- To manage the regeneration functions of the DPF, detect faults, and alert the driver to usual operating conditions, most DPFs use three temperature sensors, which are located at the oxidation catalyst inlet.
- NO_x sensors provide closed-loop feedback to meter the correct amount of DEF in SCR systems. The sensors detect incorrect concentrations of DEF, emissions from dosing injectors, and other emission-related faults.
- In NAC and LNT systems, NO_x and O_2 sensors monitor the lean–rich cycling of the exhaust to estimate the loading of the catalyst.
- DPFs require service for ash loading. Regeneration will not burn off DPF ash, so the DPF will need to be removed from the vehicle and cleaned or replaced.
- SCRs require service after refilling a DEF reservoir that has been completely depleted. The SCR fault code may need to be erased before the vehicle will operate normally. Erasing is necessary because the OBD system's SCR monitor will not run if the engine has derated to idle.

Key Terms

active regeneration A process in which soot is burned inside the DPF through supplementally heating the DPF filter.

AdBlue A liquid-based NO_x-reducing system that Dodge Ram uses in its European and North American diesel engines.

ammonia slip NH_3 that escapes from the SCR catalyst and into the atmosphere.

BlueTec DaimlerChrysler's name for its two NO_x-reducing systems used in its European and North American diesel engines.

catalyst A material that speeds up or slows down chemical reactions without entering the chemical reactions.

$DeSO_x$ A sulfur-reducing reaction used in NAC and LNT systems to remove sulfur from catalyst material. Sulfur in diesel fuel tends to contaminate catalysts and reduces their effectiveness unless a $DeSO_x$ reaction takes place.

diesel particulate filter (DPF) A device installed in the exhaust system to filter out black carbon soot and other particulates from the exhaust stream.

exhaust backpressure Exhaust pressure that builds up in the exhaust system. The pressure can be before or after the turbocharger.

exhaust gas recirculation (EGR) A NO_x-reduction technology that mixes exhaust gas with fresh intake air.

lean NO_x trap (LNT) A catalyst mechanism that selects and destroys only NO_x emissions without the use of liquid diesel exhaust fluid (DEF); it is also referred to as hydrogen-SCR (SCR-H) and as a NO_x adsorber (NAC).

NO_x adsorber catalyst (NAC) A unique catalyst material that stores NO_x emissions and then breaks down the NO_x under specialized exhaust conditions.

oxidation reaction A reaction that adds oxygen (O_2) to hydrocarbons (HCs) and carbon monoxide (CO) to produce water (H_2O), carbon dioxide (CO_2), and heat.

passive regeneration A process in which relatively low-temperature chemical reactions take place in the exhaust catalysts to oxidize soot into CO_2 and H_2O. Passive reactions take place when the DPF is hot enough under normal vehicle operation; this occurs when sufficient load and engine speed produce exhaust inlet temperatures above 482°F (250°C).

reduction reaction A reaction that removes oxygen from NO_x molecules, converting NO_x into N_2 and O_2.

selective catalyst reduction (SCR) system An exhaust after-treatment system that decomposes NO_x molecules by using ammonia NH_3. It is selective for oxygen in NO_x molecules rather than oxygen in the exhaust.

urea An ammonia-carrying molecule that is dissolved in water to form DEF fluid. It is selective for oxygen in NO_x molecules rather than oxygen in the exhaust.

Review Questions

1. Which of the following strategies is typically used to actively regenerate a catalyzed DPF?
   a. Retarding injection timing.
   b. Injecting fuel into the exhaust.
   c. Heating the oxidation converter with a flame.
   d. Placing the engine under load and driving above 45 mph (72 kph) for 30 minutes.
2. Which strategy will most likely be used to passively regenerate a catalyzed DPF?
   a. A stationary vehicle with raised idle and a partially closed intake throttle.
   b. Using OEM software or a scanner to perform regeneration (light-duty vehicles only).
   c. Driving at speeds above 45 mph (72 kph) for a minimum of 20 minutes.
   d. Logging a fault code for a DPF with 80% or more soot loading.
3. Catalyzed DPFs require oil with which of the following qualities?
   a. No biodiesel contamination.
   b. High viscosity index (VI).
   c. Low ash content.
   d. High alkalinity.
4. Which process is used to clean ash from a DPF?
   a. Active regeneration.
   b. Passive regeneration.
   c. Filter replacement.
   d. Removal and back-flushing with compressed air.
5. Which of the following is the function of an intake throttle valve on a diesel engine equipped with a DPF?
   a. To increase exhaust backpressure.
   b. To increase exhaust temperature.
   c. To decrease cylinder EGR content.
   d. To increase engine retarding force when decelerating with compression release brakes.
6. Dosing the exhaust system of a 2007 diesel engine with fuel is necessary to do which of the following?
   a. Regenerate the oxidation converter.
   b. Add urea or diesel exhaust fluid (DEF) for NO_x reduction.
   c. Actively regenerate a catalyzed DPF.
   d. Actively regenerate the CO adsorber.
7. The use of SCR and DPFs requires which type of fuel?
   a. Low-sulfur diesel (LSD) fuel with under 500 ppm sulfur.
   b. Ultra-low-sulfur diesel (ULSD) fuel with under 15 ppm sulfur.
   c. Biodiesel mixed with regular petroleum fuel.
   d. Biodiesel only.
8. Consider a 2018 school bus that has the DPF warning lamp illuminated. The amber light is on steadily. What is the most correct service recommendation?
   a. Drive the vehicle under load at a moderate speed for 20–40 minutes.
   b. Use a dash switch to initiate a stationary regeneration.
   c. Use OEM software to initiate an active regeneration.
   d. Tow the vehicle to a service location and replace the DPF.
9. Using SCR delivers which of the following?
   a. Improved fuel economy and fewer DPF regenerations because injection timing is advanced.
   b. Improved fuel economy and fewer DPF regenerations because injection timing is retarded.
   c. Improved fuel economy and fewer DPF regenerations because DEF increases combustion efficiency.
   d. Only lower NO_x emissions.
10. A slip catalyst located after the SCR catalyst converts ________.
    a. NO_x to N_2
    b. NH_3 to N_2
    c. soot to CO_2 and N_2
    d. fuel to heat through flameless combustion

ASE Technician A/Technician B–Style Questions

1. The SCR reservoir of a 2010 pickup truck has been completely depleted for more than 10 hours of driving time and the engine power has derated. Technician A says the correct service recommendation is to fill the DEF reservoir and clear associated fault codes to return the vehicle to service and full power. Technician B says the correct procedure is to fill the DEF reservoir and prime the DEF lines to remove the derate condition. Who is correct?
   a. Technician A
   b. Technician B
   c. Both A and B
   d. Neither A nor B
2. Technician A says diesel exhaust oxidation catalysts used by diesel engines are like those used in automobiles for the last two decades. Technician B says diesel exhaust oxidation catalysts will store noxious emissions in the substrate when the exhaust temperatures are low. Who is correct?
   a. Technician A
   b. Technician B
   c. Both A and B
   d. Neither A nor B

3. Technician A says the problem with using automobile catalysts in the exhaust system of diesel engines is the system's excess air and cool temperatures. Technician B says diesel exhaust catalysts need to be designed to withstand the higher exhaust temperatures of a diesel engine and its exhaust's lower oxygen content. Who is correct?
   a. Technician A
   b. Technician B
   c. Both A and B
   d. Neither A nor B
4. The DPF warning light and the red shutdown warning lamp are flashing. Technician A says the DPF filter needs to be removed and replaced. Technician B says the DPF needs an active regeneration. Who is correct?
   a. Technician A
   b. Technician B
   c. Both A and B
   d. Neither A nor B
5. Technician A says nitric oxide (NO) is used by the DPF to convert soot to harmless emissions. Technician B says that the SCR catalysts cause NO to be converted into harmless emissions. Who is correct?
   a. Technician A
   b. Technician B
   c. Both A and B
   d. Neither A nor B
6. Technician A says that lean NO_x traps (LNTs) are used by only light-duty diesel engines. Technician B says that LNTs are used by light- and heavy-duty diesel engines. Who is correct?
   a. Technician A
   b. Technician B
   c. Both A and B
   d. Neither A nor B
7. Technician A says that without the pressure differential sensor (delta P sensor), the exhaust DPF would not regenerate, because the amount of soot loading would not be able to be properly calculated. Technician B says that the amount of soot loading is only estimated, and not accurately measured, using the delta P sensor. Who is correct?
   a. Technician A
   b. Technician B
   c. Both A and B
   d. Neither A nor B
8. Technician A says that DEF is always evacuated from lines and valves after the vehicle has shut down to prevent DEF from crystalizing and plugging the SCR system. Technician B says that DEF is removed to prevent it from freezing and causing damage. Who is correct?
   a. Technician A
   b. Technician B
   c. Both A and B
   d. Neither A nor B
9. Technician A recommends that customers have the DPF removed and cleaned once a year. Technician B says the DPF needs to be actively regenerated at every oil change. Who is correct?
   a. Technician A
   b. Technician B
   c. Both A and B
   d. Neither A nor B
10. After two technicians examine a broken and disintegrated DPF filter, Technician A says the filter damage was caused by water contamination. Technician B says the filter was damaged by back-to-back regenerations that occurred too frequently. Who is correct?
    a. Technician A
    b. Technician B
    c. Both A and B
    d. Neither A nor B

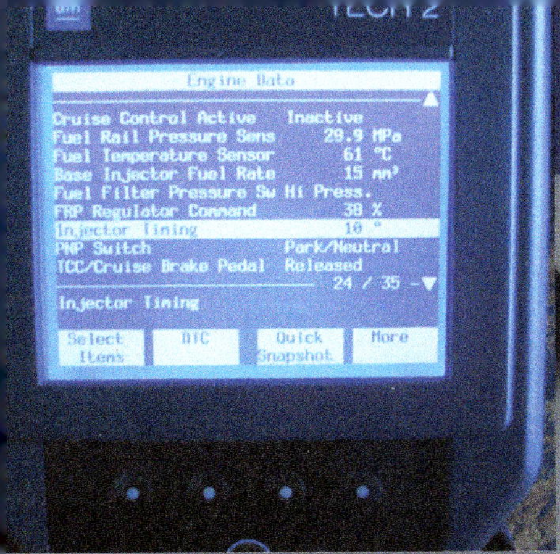

CHAPTER 19

Onboard Diagnostics

Learning Objectives

After reading this chapter, you will be able to:

- **19-01** Describe the purpose and operation of the OBD-II system.
- **19-02** Outline the development of OBD-II requirements for diesel engines.
- **19-03** Identify and describe the principles of fault detection and the operational strategies of major emission system monitors.
- **19-04** Identify and describe monitor operation for a few key major emission monitors.
- **19-05** Identify and outline the requirements and capabilities of electronic service tools used to access OBD-II system information.

You Are the Technician

You are at a customer's yard to diagnose a problem with a contractor's brand-new pickup truck. The vehicle will start and run, but the engine will not accelerate above idle speed. After starting the engine, ted malfunction indicator lamp (MIL) warning light is illuminated, and there are no other warning lights on. You perform visual inspections of the vehicle, examine the exhaust system and wiring harnesses, and check for fuel, coolant, and oil leaks. The selective catalytic reduction (SCR) tank is three-quarters full according to the dash gauge, and you visually verify the level. Nothing seems amiss, but you realize that certain emission-related and engine protection system–related faults will produce these symptoms. To quickly check the OBD-II system, you plug in an ELM327 Bluetooth communication interface and check for generic SAE (Society of Automotive Engineers) fault codes by using an app on your cell phone. Numerous diagnostic trouble codes (DTCs) related to SCR aftertreatment are set. Pinpointing which DTC to diagnose that is likely causing the severe engine power derate conditions and shutdown is challenging without accessing the shop's diagnostic software, capable of reading enhanced DTCs and accessing subscription-based service information. Before calling for a tow truck to bring the truck to the shop, consider the following:

1. What other resources could you access to help diagnose the DTCs?
2. What other diagnostic services may be available on the cell phone OBD-II app to help pinpoint the root cause of the failure?
3. Prepare an explanation for the customer that outlines the reasons for the loss of engine power.

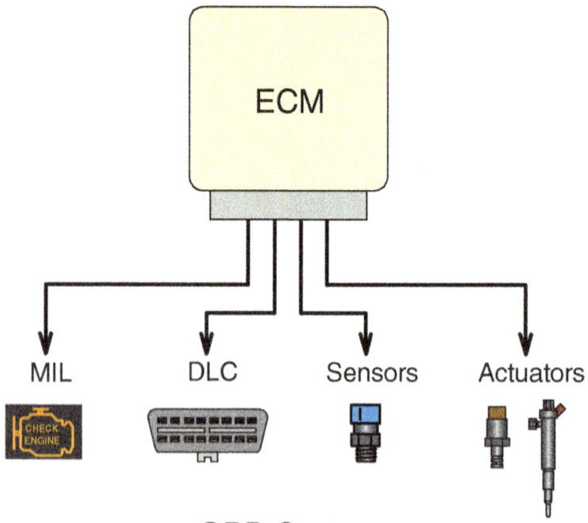

FIGURE 19-1 The OBD system is integrated into the engine control system to provide self-diagnostic capabilities to detect emission-related faults.

▶ Introduction

Technicians servicing today's diesel-powered vehicles will be just as likely to use a tablet or computer as a screwdriver and wrench to perform repairs. Electronic control systems use microprocessors to operate every engine, body, chassis, and drivetrain system to provide operational capabilities and features far exceeding any mechanical system—and with greater precision, efficiency, and reliability.

Although electronics offer many advantages and benefits over mechanical controls, a unique challenge is to identify and quickly repair any fault in systems that operate invisibly, by using electronic signals and software-based operating systems. To understand this problem of the "black box" control system, consider the effort required to identify a problem as simple as a broken wire or an intermittently loose electrical connection at a sensor. Even with a digital multimeter (DMM), a defect like this could easily require days of labor to identify and repair if electronic systems did not have self-diagnostic capabilities that can narrow down the location of a problem with a diagnostic fault code (**FIGURE 19-1**). Self-diagnostic capabilities extend to the emission systems, which also must ensure that noxious emissions remain below legislated limits for the useful life a vehicle. Because self-diagnostic capabilities built into emission system operation are regulated by extensive environmental legislation, the same legislation gives the emission diagnostic system its own distinct name: the **On-Board Diagnostics (OBD)** system.

The most simplistic description of an onboard diagnostics (OBD) system is that it generates a report card about the performance of the vehicle's emission system (**FIGURE 19-2**). An OBD report card should identify any emission-related failure and whether the emission system has taken, and either passed or failed, all tests assigned by emission-monitoring systems dedicated to evaluating the operation of each of the emission-control subsystems. For today's diesel engines, there up to 14 or more different monitors checking the operation of the emission-control subsystems. Each of these monitors will make a judgment about the operation of an emission subsystem or component in the system after testing its performance. Like a grade assigned to a particular class for a school program, a readiness message (or passing grade) is assigned to each monitor if the monitor determines that the subsystem or an emission system component is operating correctly. A system "pass" is also referred to as a flag when a bit (0 or 1, respectively, as a pass or fail), a "ready" code, or message is set (**FIGURE 19-3**). Until the monitor observes an emission subsystem perform correctly and no faults are identified, no readiness flag or message is set. An incomplete/fail or not-ready message is part of the OBD report card until a monitor judges the subsystem's performance

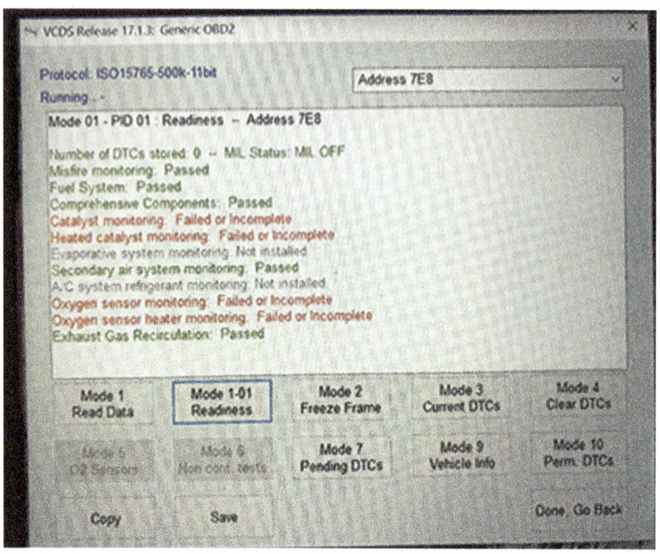

FIGURE 19-2 The OBD system provides a report card for the emission system indicating what tests it has taken and whether it passed or failed.

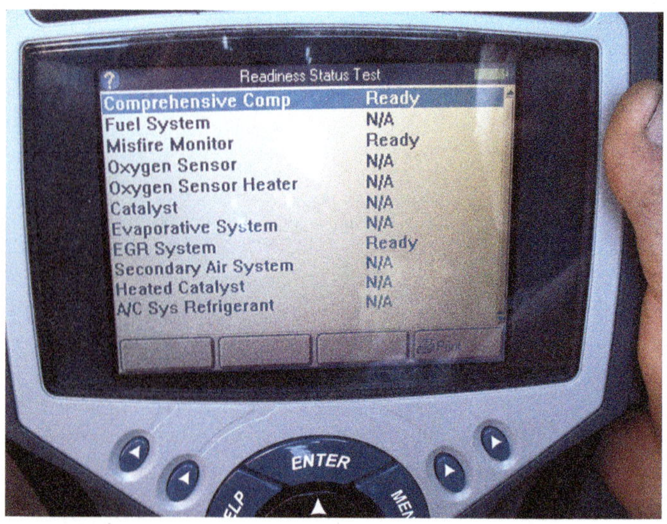

FIGURE 19-3 Emissions system monitors are set to ready after they make a judgment that finds no emission-related faults were detected.

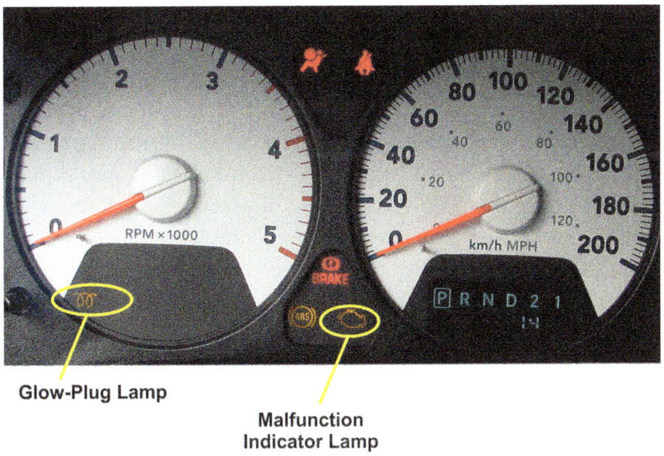

FIGURE 19-4 The MIL light alerts the driver to the presence of a potential fault that will increase vehicle emissions. A glow-plug light remaining on may also indicate emission-related faults.

as acceptable. Only if the monitor has confirmed that a subsystem or component has failed a test will it illuminate the malfunction indicator lamp (MIL) warning.

OBD systems are integral to the vehicle's electronic control system and are used to evaluate emission-related components and the emissions system as a whole for deterioration or potential malfunctions. Conditions that would cause emissions to increase beyond thresholds specified in legislation require the OBD system to illuminate a MIL located in the instrument cluster in view of the driver (**FIGURE 19-4**). The purpose of the MIL is to alert the driver to seek repairs for the vehicle in order to maintain vehicle compliance with regulated emission standards. Without a MIL alert, many types of emission-related faults could in fact allow a vehicle to continue to operate normally with an emission-related defect. Effectively repairing the emission fault will enable an extinguished MIL to remain off, which indicates the vehicle is restored to almost undetectable levels of emissions expected from today's diesel engines (**FIGURE 19-5**). If a vehicle operator does not have the vehicle repaired, OBD-based emission inspection maintenance (IM) programs used in many US states and Canadian provinces will block the renewal of a vehicle license or registration until the emission fault has been repaired. To further incentivize drivers to have an emission repair performed, diesel engines will limit power output or maximum road speed, depending on the nature of the fault and its severity.

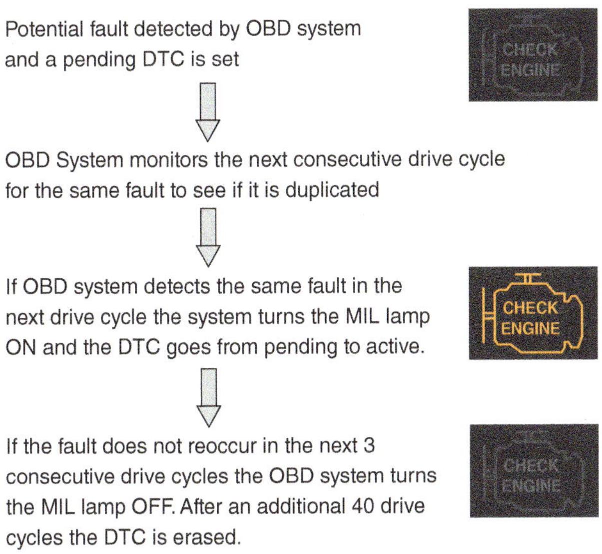

FIGURE 19-5 The sequence of MIL operation for most emission related fault conditions.

▶ Fundamentals of Onboard Diagnostic Systems

19-01 Describe the purpose and operation of the OBD-II system.

The dominance of electronics used in virtually all vehicle systems makes skills related to servicing electronic control technology one of the most important assets for successful technicians. Nowhere else among all vehicle service routines are the skills and knowledge of a technician more stretched than when interacting with OBD systems to diagnose and effectively repair emission- and drivability-related complaints. Competent service technicians increasingly depend on the OBD's supply of control system information and fault detection capabilities to guide or serve as a starting point for investigating and diagnosing customer complaints. Understanding the operating principles of the OBD system monitoring electronic control systems is foundational for choosing the correct diagnostic strategies, selecting and using service tools effectively, and making sound repair recommendations.

A diagnostic approach using OBD data is unlike a symptom-based diagnostic strategy—a strategy in which a problem is identified based on a complaint. The cost of parts and labor today makes the guesswork of symptom-based diagnostics impractical. A technician simply cannot replace components like injectors or a turbocharger on the basis of a low-power, rough-running complaint. More importantly, the complexity of contemporary diesel engines and their interoperability with other powertrain controls—such as the transmission, body electrical systems, antilock brake system (ABS), and stability control—further preclude using the guesswork in "replacing parts until it starts" as a repair strategy. Dependency of various vehicle control modules on shared data over onboard vehicle networks creates even more complexity where hundreds of possible faults can produce identical complaints whose root cause cannot be identified without the use of OBD codes and manufacturer support (**FIGURE 19-6**).

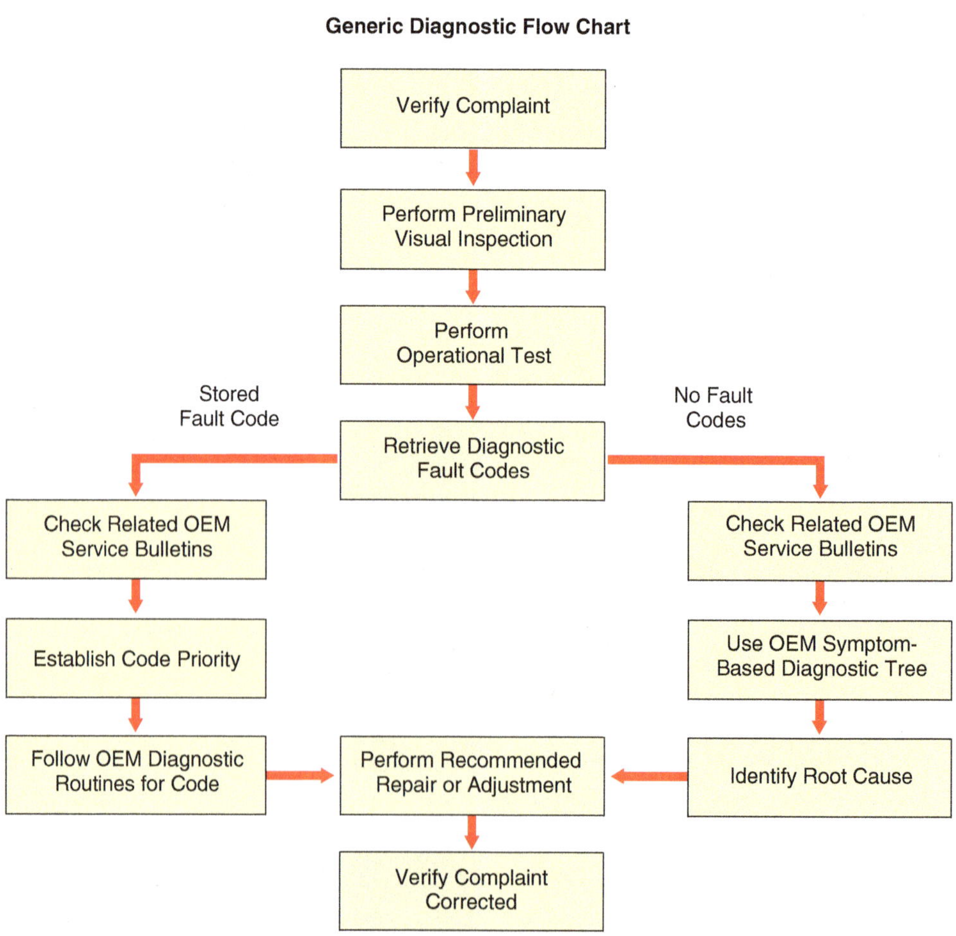

FIGURE 19-6 A generic approach to diagnosing and repairing OBD fault codes.

FIGURE 19-7 An enormous amount of data supplied by the OBD system can help diagnose drivability complaints and assist technicians in making service recommendations about marginally functioning components.

Beyond fault codes, enormous amounts of other data is provided by the OBD system because it is integrated into the vehicle control system. Real-time engine performance data is critical to diagnosing other operational concerns, such as drivability complaints or recommending maintenance and replacement of marginally functioning components before they fail (**FIGURE 19-7**).

OBD system readiness codes are also vital for technicians to validate whether or not repairs have been properly made before returning a vehicle to the customer. After a suspected fault has been repaired, fault codes are erased and the vehicle is road tested to observe whether an emission system monitor associated with the fault switches from failed/incomplete to ready/complete. An OBD system with robust fault detection capabilities enables technicians to confirm that a repair was successful and validate whether a repair was justified.

Development of OBD Standards

The term "onboard diagnostics" is significant for technicians in two ways. The simplest, most familiar is its reference to self-diagnostic functions of a vehicle's electronic control systems to identify and/or report system faults. Another meaning for OBD is less familiar but just as important: It also refers to legislation that outlines performance standards and technical specifications required for an emission self-diagnostic system. Also included with OBD technical standards are industry rules intended to reduce barriers to rapidly and inexpensively repairing emission systems. Air quality is improved only if vehicle owners can easily access repairs, restoring engines to clean, low-emission operation for as long as possible. Helping this initiative are technical standards outlined in OBD legislation for aftermarket electronic service tools such as scanners and access to service information and software calibration files from the original equipment manufacturer (OEM) for independent repair facilities.

Detailed operating standards for OBD systems are not left to manufacturers, which is important for technicians to remember when learning to work on different engine products from various OEMs. A great deal of technical commonality and uniform system operation among all vehicles can be discovered beneath the software or electronic service tool interface used by each manufacturer. Although the Environmental Protection Agency (EPA) or SAE do not specifically outline how each manufacturer it is to achieve compliance with emission standards, they do describe what outcome of system designs should be. Given the latitude that manufacturers have in interpreting the rules, naturally there are differences between OBD system designs and operation since OEMs take different approaches to building engines, fuel, and emission systems, which can vary considerably between makes and models of vehicles. However, in meeting OBD requirements, manufacturers have more in common than they have differences regarding strategies and methods used for OBD system operation.

In North America, both the EPA (US) and the California Air Resource Board (CARB) outline standards for OBD system operation for vehicles operated over the road. Directives issued by the European Parliament in Europe establish similar standards called European On-Board Diagnostics (EOBD) for vehicles destined for market there. The EOBD and OBD legislation and regulations should not be confused with enhanced (E)OBD codes, which refer to manufacturer-specific or proprietary diagnostic codes that more precisely identify faults than the generic codes required by OBD regulations.

Details about the OBD system operation must be submitted as part of an application for evidence of conformity to the EPA (US) emission and OBD legislation described in the Code of Federal Registry (CFR) Title 40 part 86,: Control of Air Pollution from Motor Vehicles and New Motor Vehicle Engines. Before a vehicle can be sold in North America, a manufacturer must receive approval from regulators for a validated OBD system operation. Legislative standards for the OBD system are also described in California Code Regulations, Section 1968.2, Malfunction and Diagnostic System Requirements for 2004 and Subsequent Model-Year Passenger Cars, Light-Duty Trucks, and Medium-Duty Vehicles and Engines (OBD-II). An emission decal label attached to the body near the engine compartment designates vehicles that comply with either federal EPA or CARB emission standards (**FIGURE 19-8**). To understand why the same year, make, and model of vehicle may have different emission components and software calibrations, its useful to know that only CARB can certify vehicles for OBD and emission compliance for sale in California. CARB has traditionally set stricter emission standards than the EPA (US) due to California's unique geography that worsens the effects of vehicle emissions. Because CARB first developed OBD standards, its OBD legislation focuses more on details of the OBD system technical design in contrast with federal EPA (US) standards, which are more performance orientated and descriptive than CARB's. Most diesel vehicle manufacturers prefer to certify to the world's toughest emission standards—CARB OBD-II design standards—to achieve North America-wide certification.

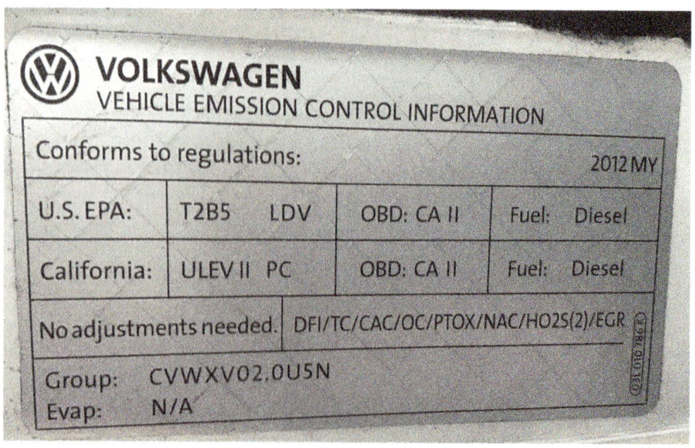

FIGURE 19-8 An emission decal will indicate whether a vehicle is certified to conform to EPA (US) or CARB standards.

Origins of On-Board Diagnostics Legislation

OBD legislation did not suddenly appear. Instead, OBD and other emission standards for diesels have progressively changed over the past 20 years. Technical standards for OBD are developed by the SAE international industry experts and routinely adopted by the EPA. Technical amendments to OBD standards are regularly made, and legislation is updated based on recommended SAE revisions.

The origin of OBD regulations date back to 1982 in California, when CARB observed the rapid adoption of electronic engine controls. Manufacturers build in self-diagnostic capabilities to detect and report failures in these electronic components, which CARB viewed as an opportunity to develop regulations requiring several emission system components be electronically monitored. From its inception, an essential function of the OBD system was its use as a tool for emissions inspection and maintenance programs, often called SMOG tests, which would eventually replace traditional tailpipe gas analysis conducted by California's IM programs. IM programs were developed because efforts to achieve the goal of improving air quality would fall short if stricter emission standards were established for motor vehicles and no enforcement took place. Vehicle emission systems need to always operate correctly and be properly maintained to reduce their contributions to noxious emissions in the air that everyone shares. An OBD system's usefulness as an inspection tool is confirmed by multiple studies that have determined that OBD-based emissions system inspections, which only check the MIL operation and performance of emission monitors, identify a greater percentage of high-emission vehicles compared to tailpipe-only emission tests used by traditional IM programs.

Later, in 1996, the OBD requirement was expanded in the United States to correct shortcomings of the earlier system by the EPA mandating new technical standards intended to produce a more sensitive and effective OBD system. To distinguish the legislated version of second-generation OBD systems from earlier diagnostic systems, the most recent version of OBD systems is called On-Board Diagnostics II, commonly abbreviated to OBD-II. An essential element of the OBD-II system design is that it enables technicians and emission inspectors to electronically communicate with the vehicle's electronic control and self-diagnostic systems. A wealth of information can be gleaned from dozens of service tool utilities and the OBD-II system organization to detect and report emission-related faults. Electronic service tools can quickly request information from the engine emission-control system to determine whether the vehicle has emission-related fault codes, check the operating status of emission-monitoring systems, or even gauge the effectiveness of the emission-control system monitors by using tracking data such as in-use monitor performance ratios, which will be described later in the chapter. Most importantly for technicians, the self-diagnostic onboard diagnostic capabilities mandated by OBD-II have almost eliminated the traditional trial-and-error identifying methods to repair faults.

Compliance with OBD-II standards for diesels did not begin until 1997 in California when passenger cars, light-duty trucks, and medium-duty vehicles and engines with a gross vehicle weight of up to 14,000 lb (6,350 kg) monitored the operation of diesel emission-control systems to alert the vehicle operator to any potential emission increase above a maximum limit or threshold standards. When a malfunction is detected, diagnostic information is stored for retrieval by a technician to assist in identifying and repairing the malfunction (**FIGURE 19-9**).

Federal Test Protocols and OBD Systems

It is important to note that OBD-II systems generally do not directly measure vehicle emissions to determine whether there is an emission-related fault. Even though the most current diesel engines have NO_x and particulate matter (PM) sensors, legislation for vehicles having a GVW under 14,000 lb (6,350 kg) does not include any regulation for a "not to exceed" (NTE) emission limit threshold, such as the regulations for heavy-duty diesels. Under numerous operating conditions, emissions may temporarily exceed a Federal Test Protocol (FTP) limit. A few example conditions may include operating at speeds above 83 mph (133.6 kph), under heavy load or acceleration, high altitudes (above 8,000 ft or 2,438 m), during prolonged cold idle, and during cold-start and warm-up periods. Instead, the

▶ TECHNICIAN TIP

Emission legislation forbids manufacturers from using what are called emission defeat devices unless they are first disclosed and are approved during the process of obtaining a certificate of conformity. Defeat devices are called auxiliary emission Control devices (AECD) in OBD-II legislation. An AECD, or, defeat device can be hardware or software or a control strategy that reduces the effectiveness of the emission control system under conditions that may reasonably be expected to be encountered during normal vehicle operation and use. Illegal defeat devices are usually activated when the device senses that the vehicle is operating outside FTP emission test conditions. A defeat device can be something as simple as a timer that shuts off an emissions system after 26 minutes and 40 seconds of operation—the length of an FTP test. The use of some defeat devices is permissible when they can be justified to protect passengers or prevent damage to a vehicle. Shutting off the EGR system to prevent the engine from continuing to overheat is one example of a potentially permissible defeat device. Another situation would be to minimize the likelihood of a vehicle stalling in traffic and losing the ability to steer or brake due to a malfunctioning emission component.

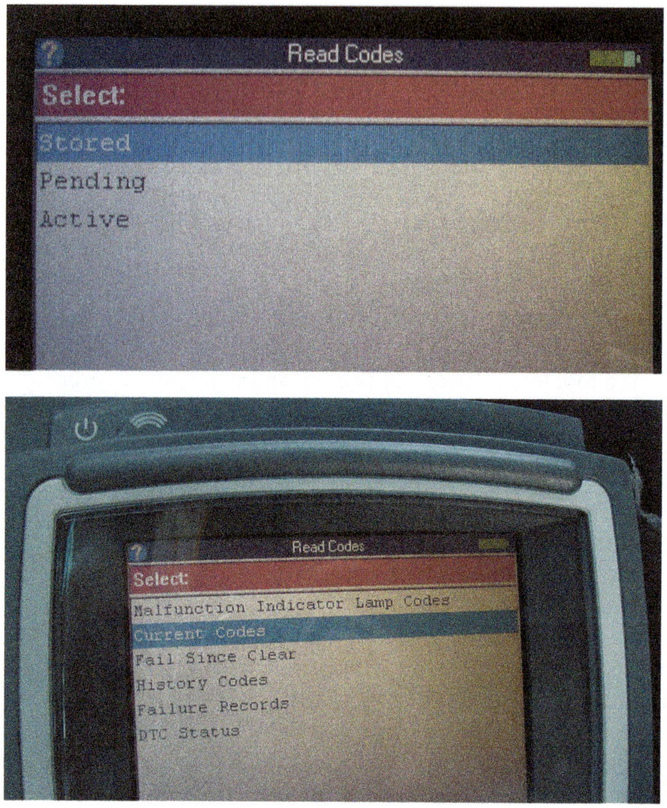

FIGURE 19-9 Storing DTCs and classifying them according to when and how frequently they occurred assists technicians in isolating OBD system–related faults.

OBD system evaluates the statistical likelihood that an engine-related fault or set of vehicle conditions could produce emissions in excess of emission standards. Highs and lows of tailpipe emissions measured during a chassis dynamometer–based emission test are averaged over the distance that a test vehicle operates. Emission standards, which set maximum emission limits or thresholds are measured in grams per mile or kilometer and referred to as FTP or, Federal Test Protocols. Four FTP standards are established for noxious emissions from diesel-powered vehicles (**TABLE 19-1**). As explained in Chapter 3, the following are four major FTP categories:

- hydrocarbons (HCs): formed primarily from unburnt fuel
- carbon monoxide (CO): formed by incomplete combustion of fuel

TABLE 19-1 Tier 2 FTP Detection Threshold Multipliers Used by the OBD-II

Tier 2 Diesel FTP Threshold Multipliers				
Monitor	HCs	NO_x	PM	2013
Catalysts (three-way, NO_x, SCR, or NAC*)	3-5× (2007) 2.5–3× (2010)	3× (2007) 2.5× (2010)	N/A	1.75×
PM filter	N/A	N/A	5× (2007) 4× (2010)	4×
All others (EGR*, fuel, boost, etc.)	2.5–3.5× (2007) 2–3× (2010)	2.5–3.5× (2007) 2–3× (2010)	2.5–5× (2007) 2–4× (2010)	1.5×

Diesel must meet gasoline thresholds in 2013.
EGR: exhaust gas recirculation
NAC: NO_x adsorbing catalyst

Source: EPA (US) legislation—Code of Federal Registry (CFR) Title 40 Part 86, titled Control of Air Pollution from Motor Vehicles and New Motor Vehicle Engines.

TABLE 19-2 Tier 3 FTP Detection Threshold Multipliers Used by the OBD-II

Tier 3 OBD-II Diesel Threshold Multipliers						
Emission Category	**FTP Monitor Threshold**			**Aftertreatment Thresholds**		
	NO_x + NMOG	**CO**	**PM**	**NO_x + NMOG**	**CO**	**PM**
LEV160	1.50×	1.50×	2.00×	1.75×	1.50×	2.00×
ULEV125						
ULEV70	2.0×			2.00×		
ULEV50						
SULEV30	2.50×	2.50×		2.50×	2.50×	
SULEV20						

Source: EPA (US) legislation—Code of Federal Registry (CFR) Title 40 Part 86, titled Control of Air Pollution from Motor Vehicles and New Motor Vehicle Engines.

- oxides of nitrogen (NO_x): formed by reactions between nitrogen and oxygen
- particulate matter (PM): primarily of black carbon or soot formed during the combustion process.

The OBD system is required to detect abnormal engine or powertrain operating states that could potentially produce emissions in excess of FTP protocols. The dividing line where the system must begin to detect excess emissions and set fault codes is called the FTP threshold limit and is established in multiples above the FTP protocol. For example, a CARB-certified Tier 2 bin 5 vehicle has an FTP protocol limiting NO_x to no more than 0.05 grams/mile (0.031 g/km) with a useful life of 50,000 miles (80,000 km). According to Table 19-1, a failing NO_x adsorbing catalyst (NAC) would be required to detect a failure condition producing 0.15 grams/mile in a 2007 model vehicle, which is three times the FTP protocol limit for NO_x. A failed exhaust gas recirculation (EGR) valve would require the OBD-II system to detect NO_x emissions exceeding 0.175 grams/mile or three and a half times the FTP limit. Tier 3 emission standards introduced for 2017 vehicles no longer have separate FTP protocols for NO_x and HCs. Instead, NO_x and non-methane HC emissions or non-methane organic gases (NMOG) are included in a single category: non-methane hydrocarbon (NMHC) + NO_x (**TABLE 19-2**).

The FTP procedure contains a set of prescribed driving cycles performed on a chassis dynamometer should not to be confused with FTP emissions limits. Various FTP driving cycles operate a vehicle on a chassis dynamometer under different speeds, distances, temperatures, and loads. The tests have been developed by the EPA and CARB to simulate real-world driving conditions for measuring light- and medium-duty vehicles' emissions. Emission results from the tests are part of the application for obtaining a certificate of conformity (**FIGURE 19-10**).

▶ **TECHNICIAN TIP**

During the most commonly used FTP procedure, which is a dynamometer drive cycle approximately 11 miles (17.7 km) long, manufacturers must demonstrate that any fault in the emission system is detected by the end of a two-part drive cycle. Generally, this means all emission monitors that are ready must check or run at least once during the test. Ideally, a technician should duplicate the FTP drive cycle as closely as possible to quickly reset an emission monitor from failed/not-ready to pass/ready when road testing to validate an emission-related repair. However, because of the complexity of the FTP driving cycle, a technician may not be able to duplicate the test and have all emission monitors run during one road test. To avoid prolonged road testing, manufacturers will typically recommend a more realistic road test drive cycle that matches the conditions required for all emission monitors to run tests of emission subsystems. Consult the service literature for a specific model to best determine how to road test a vehicle and validate repairs.

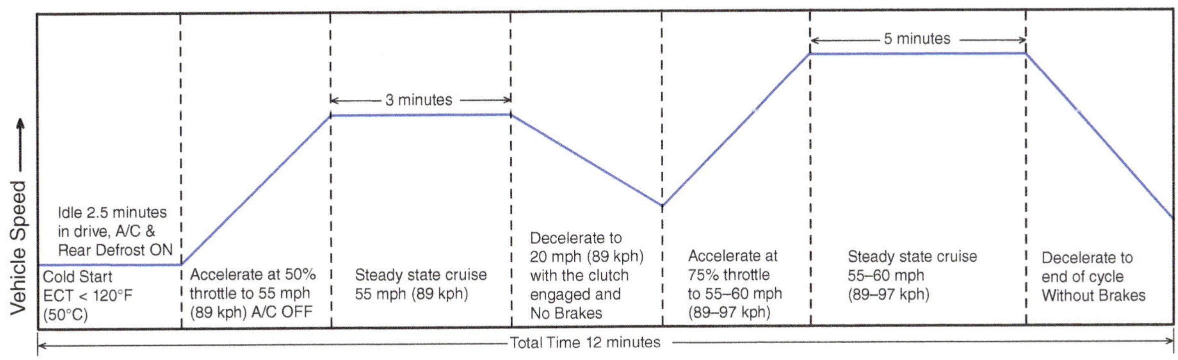

FIGURE 19-10 One of several common FTP drive cycles designed to simulate real-world driving conditions and measure vehicle emissions.

▶ OBD-II Requirements for Diesel Engines

19-02 Outline the development of OBD-II requirements for diesel engines.

A starting place to describe OBD-II system operation in diesel engines is to outline the most significant requirements of legislation. Broad categories of OBD-II standards also apply to spark-ignition engines, but details about monitoring requirements for diesels and the type of emissions monitored are substantially different. For diesel engines, OBD-II systems will include the following features:

- a MIL in the driver's view
- standardization of a 16-pin diagnostic connector link (DLC) in the driver area
- standardization of SAE generic emission-related fault codes across all manufacturers
- standardized network communication at the DLC, enabling aftermarket electronic test equipment (i.e., scanners) to read OBD-II fault codes and data
- mandatory reporting of standardized emission-related data at the DLC
- display of freeze-frame data for fault codes, which is a snapshot record of engine and vehicle data from when the fault occurred
- standardization of names and abbreviations of components and systems
- monitoring the functional capability of major emission systems to validate proper operation
- monitoring the critical emission subsystems and emission-related output devices
- monitoring the emission-related input and output devices for electrical-related faults (e.g., short-circuit to battery positive, short-circuit to ground, open circuit).

Diagnostic Connector Link

A standardized DLC is located on the driver's side area, preferably to the left of the steering column below the dash, to access OBD-II information from the vehicle (**FIGURE 19-11**). DLCs are the connecting point for communication between the vehicle control system and electronic service tools such as scanners, pass-thru programmers (devices used to install engine software) or other computer-based software applications. The SAE standard J-1962 specifies that the D-shaped DLC contain 16 terminals, each dedicated to a specific function (**FIGURE 19-12**). A battery power and ground are assigned to two pins. The remaining pins are connected to what are called onboard vehicle networks. Onboard networks are like wired telephone lines connecting various vehicle modules (**FIGURE 19-13**). Messages to and from control modules and output devices are carried over the networks to operate the engine and other vehicle accessories. For example, the transmission may need engine speed information to shift properly, and the transmission may supply a vehicle speed signal to

FIGURE 19-11 A 16-pin D-shaped OBD-II diagnostic connector located in the driver's side of the vehicle.

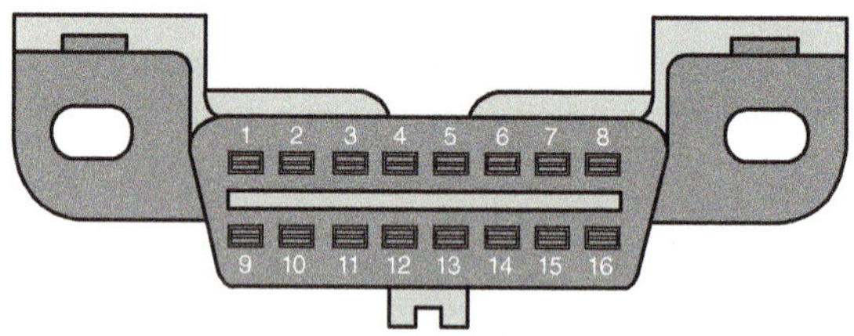

1 Manufacturer's discretion
2 SAE J1850 Bus (+)
3 Manufacturer's discretion
4 Chassis Ground
5 Signal Ground
6 SAE J2284 CAN High
7 ISO 9141-2 K Line
8 Manufacturer's discretion
9 Manufacturer's discretion
10 SAE J1850 Bus (–)
11 Manufacturer's discretion
12 Manufacturer's discretion
13 Manufacturer's discretion
14 SAE J2284 CAN Low
15 ISO 9141-2 L Line
16 Power supply

FIGURE 19-12 Each pin of the DLC is assigned a network connection. The controller area network (CAN) pins are primarily used to communicate with the OBD-II network.

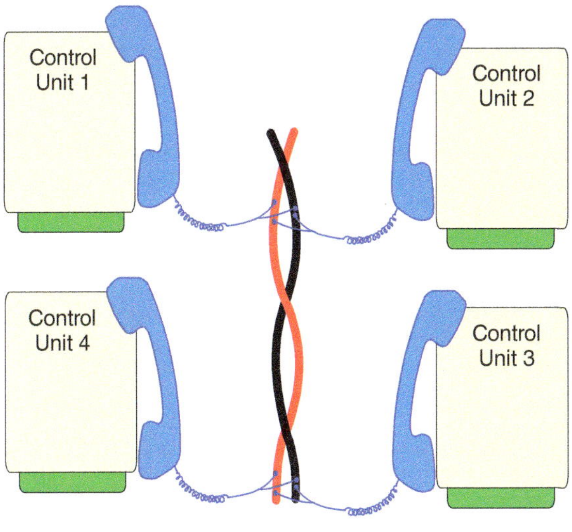

FIGURE 19-13 Modules linked through a data bus form an onboard network. Time using the data bus is shared like a party-type telephone line. Only one person can speak while the others listen and wait for their turn.

the engine to change injection timing (**FIGURE 19-14**). Various amounts of critical data are exchanged between the aftertreatment control modules, the engine electronic control module (ECM), and the turbocharger control module over an onboard network. Depending on the year, make, and model of the vehicle, up to four mandatory networks connect to various combinations of DLC pins. Manufactures are allowed to connect additional proprietary networks on pins not allocated for use by the SAE. OEM service tools will have the ability to communicate over these specialized networks to proprietary technology such as collision avoidance or onboard entertainment systems.

To enable network communication across various onboard networks that use different communication protocols, speeds, voltages, and even different numbers of wires, a gateway module is used. Gateway modules translate messages between networks to share data between networks and electronic service tools (**FIGURE 19-15**).

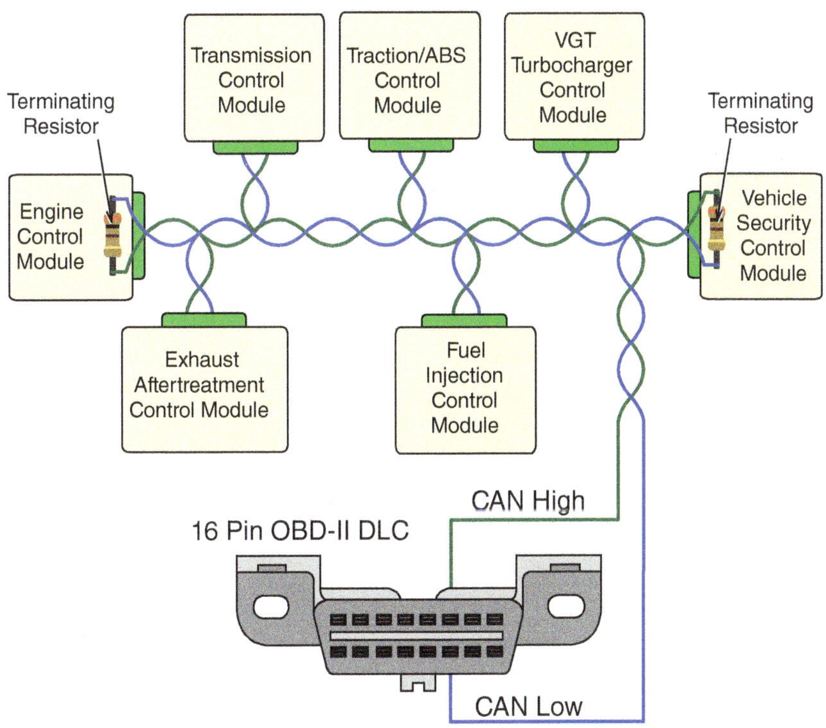

FIGURE 19-14 Network communication enables control modules to share information, reducing the amount of wiring and need for redundant sensors.

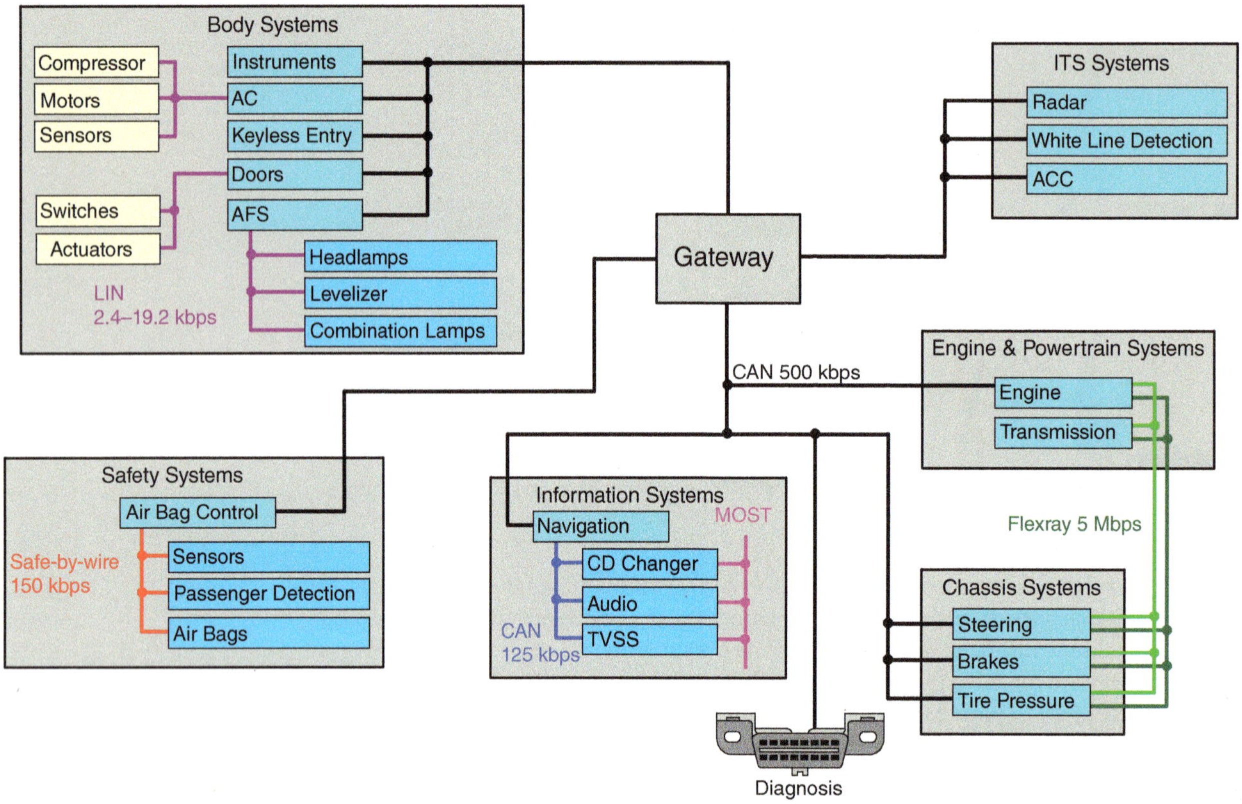

FIGURE 19-15 Gateway modules translate and share communication between networks operating with different communication protocols.

Controller Area Networks and Emissions

Onboard network communication with service tools is vital to OBD system operation. Networks exchange information between control modules and electronic service tools in a digital format, called serial data, consisting of 1s and 0s. Numbers and letters are communicated digitally, much like telegraphs once did by using a sequence of dots and dashes. A major improvement, however, is the speed and volume of information that digital electronics are capable of (**FIGURE 19-16**). Connecting modules together in communication networks enables them to share data not only to reduce the amount of processing required by each module but also to reduce the amount of wiring necessary to connect sensors to each module. More importantly for OBD requirements, fault codes and emission data are accessed at a single point on the network—the DLC. A CAN-type network is used by the OBD-II system in the latest vehicles. CAN is the serial communication protocol standardized by the International Standards Organization (ISO) and referred to by standards set by

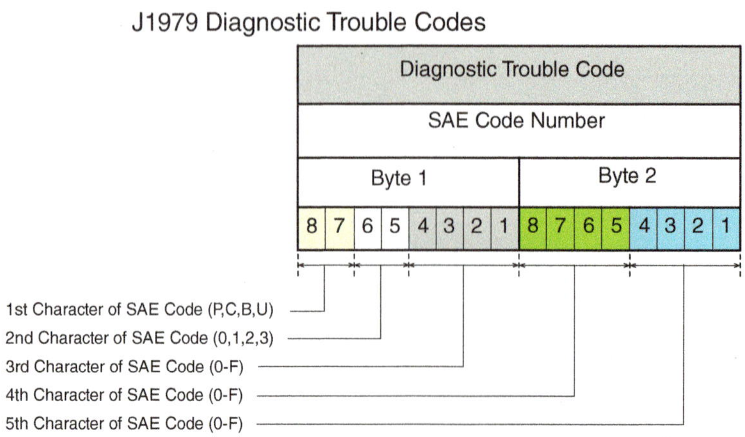

FIGURE 19-16 An OBD-II fault code is constructed of 2 bytes. Each byte is made from 8 bits, each of which is either a 1 or a 0. Digital data are transferred as hexadecimal code, which can be translated into alphanumeric data.

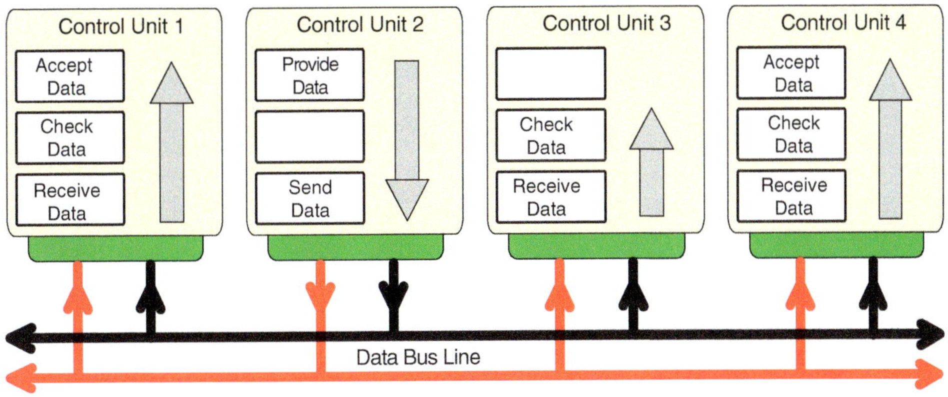

FIGURE 19-17 Time-share division means only one module at a time can transmit data to the network data bus. Note that the other modules are processing and listening until it's their turn to transmit information.

documents SAE J2284/ISO 15765-4 (CAN). A protocol simply refers to a set of communication rules or digital language used by a network. Serial communication refers to the sequences of 1s and 0s that make up CAN messages.

The CAN module messages are transmitted over a pair of wires called a data bus. The data bus gets its name from the way modules are connected. Modules are also called network nodes and are like bus stops for data transferred on the two wires. Each module on the network will take turns to either transmit or receive messages to prevent information and electrical signals from being canceled or garbled (**FIGURE 19-17**). Time-share division or multiplexing of network communication signals is not unlike a classroom where only one person should speak at a time while others listen so that communication is effective.

A differential voltage is established between the wires to represent 1s and 0s (**FIGURE 19-18**). Each wire will have an opposite polarity when transmitting signals, where the wires are becoming either more negatively or positively charged. In a resting state, the data bus will have a slightly positive voltage of between 0.3 V and 0.7 V measured relative to negative ground, and each of the two wires will have slightly different positive voltages. To transmit the number 1, voltage on one wire will become more positive by 2.5 V and more negative by 2.5 V, which gives a difference of 5 V. This differential voltage would be logically interpreted as equal to 1 by module software.

No voltage differential between wires on the data bus would equal a logic of 0 (**FIGURE 19-19**). Using a differential voltage provides the crisp, clear signal necessary for transmitting information at speeds of up to 500 kilobits/second over the network (**FIGURE 19-20**).

CAN has been used alongside other serial data protocols on most vehicles for the past decade and will be the only required protocol in the near future for OBD. CAN communication is easily configured to be automatically detected by the scan tool software. This

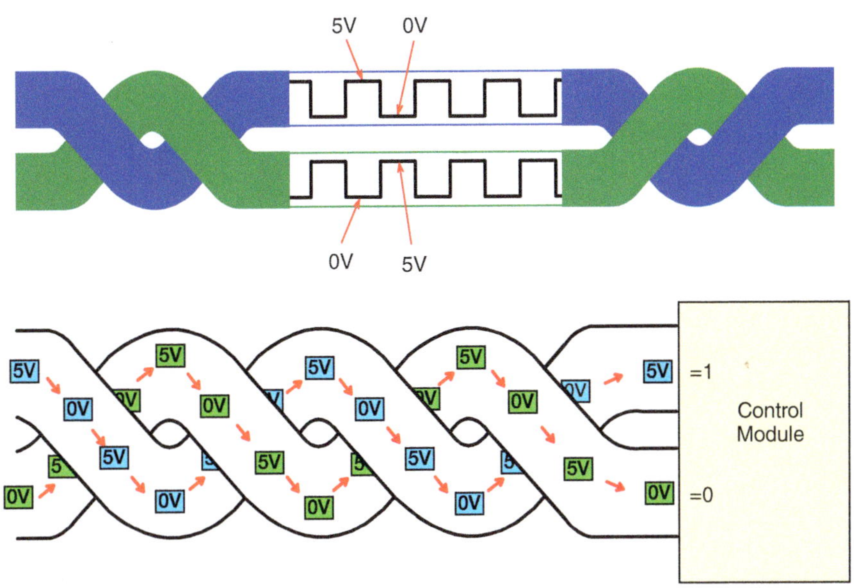

FIGURE 19-18 Differential voltages transmitted over the data bus represent logic states or either a 1 or a 0.

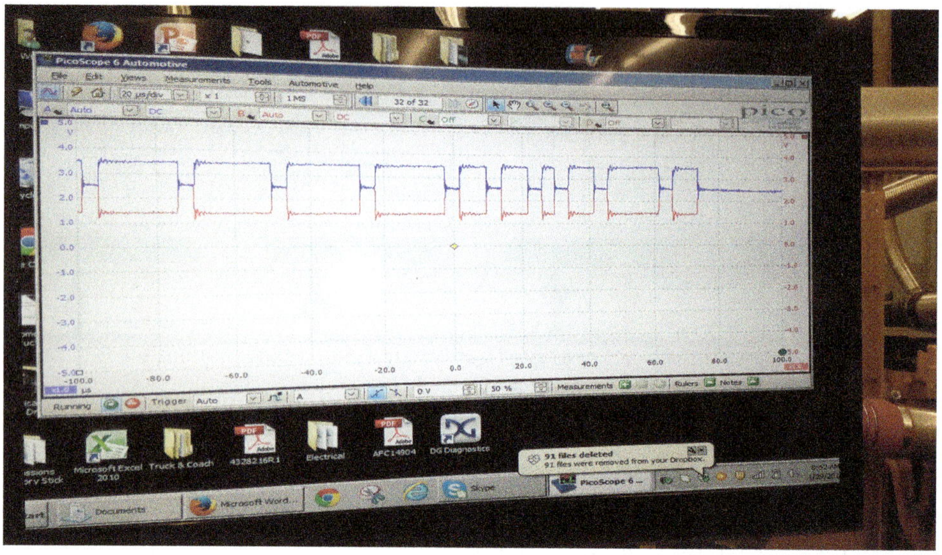

FIGURE 19-19 Graphing a wave form of a CAN signal.

means a technician does not have to select the vehicle make, year, or model while using other protocols.

For vehicles to comply with OBD-II and EOBD rules, at least one or more of the following communication protocols are also adopted in addition to CAN:

- ISO 9141-2 (K-LINE)
- SAE J1850 PWM (pulse-width modulation)
- SAE J1850 VPW (variable pulse width)
- ISO 14230-4 (keyword protocol 2,000).

Unified Diagnostic Service

The Unified Diagnostic Service (UDS) protocol is the latest network communication protocol, which has an international standard described in ISO 14229-1. "Unified" refers to the protocol's international or universal use by OEM manufacturers and parts suppliers. Components integrating small microcontroller control modules into large ECMs will support

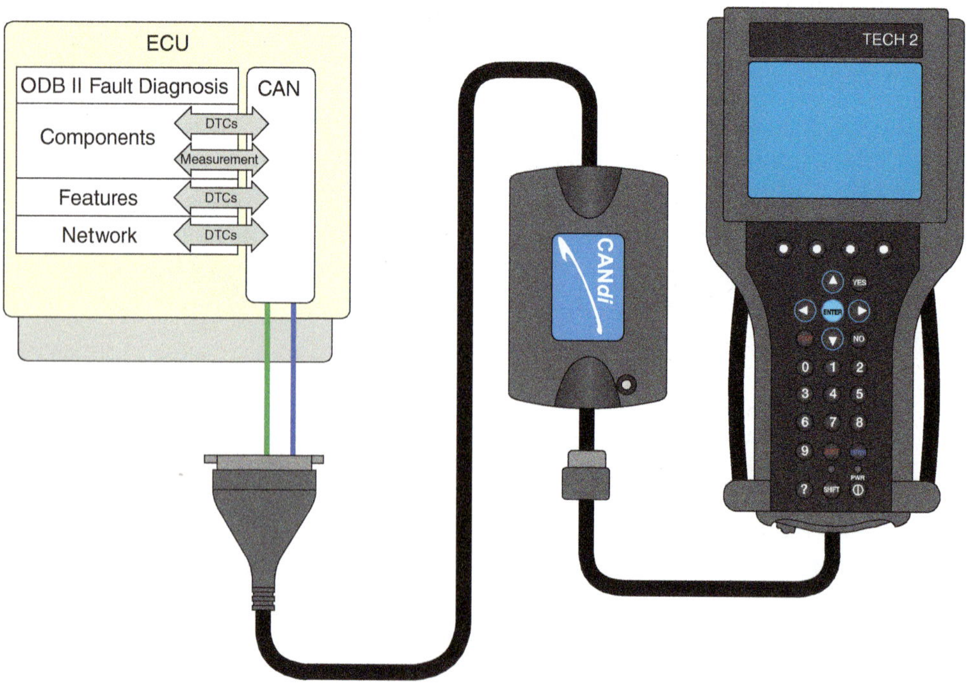

FIGURE 19-20 This Tech 2 scanner requires a CAN adapter to communicate with the high-speed data transfer of information from the CANs of a GM pickup equipped with a Duramax diesel.

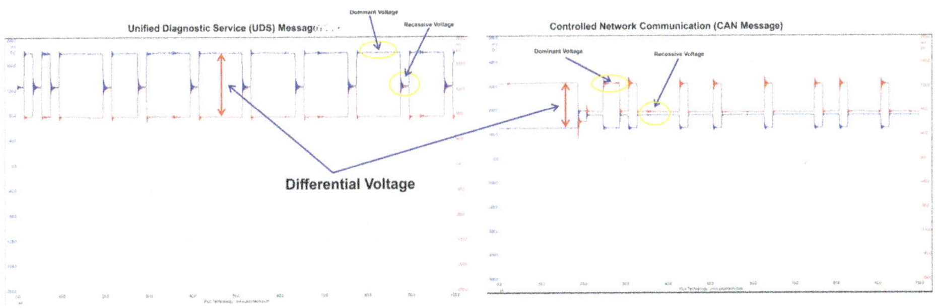

FIGURE 19-21 UDS wave forms are similar to CAN signals.

UDS communication that is not restricted geographical boundaries for emission or other legislated regulations. UDS has tremendous flexibility to be adapted into any type of chassis since it can transmit information over its own data bus or over CANs by inserting its digital signals between CAN messages (**FIGURE 19-21**). But UDS differs substantially from CAN in that it not only communicates with all control units installed in a vehicle, which have UDS services enabled, but is also a diagnostic system operating in parallel with OBD. UDS protocols provide standards for reading and storing fault codes generated by self-diagnostic functions embedded in control modules.

UDS can also reconfigure emission system operation, including OBD-II systems by shutting off or activating portions of the OBD system with software utilities that switch its operation from simply communication to diagnostic session control or an extended diagnostic session. For years, technicians working with European vehicles and European diesel engines have learned to perform component adaptations where model predictive engine control requires matching or introducing parts to one another in a network-connected system. Components from turbocharger actuators and smart remote actuators for EGR valves to transmissions are adjusted by using UDS protocols to match one another. UDS also can access hidden diagnostic menus that shut off or activate OBD emission monitors. Tightly restricted OEM dealer-level software and engineering tools used in the field can often incorporate these particular UDS capabilities.

Many OEM vehicle manufacturers use UDS for proprietary onboard networks for crucial safety and emission-related operations. For example, networked aftertreatment control modules will communicate with UDS in addition to CAN or interface with CAN through an engine ECM. Tests of airbags, the ABS, and electronic steering are often performed by switching UDS to a safety system diagnostic session. Changes to control module firmware can be performed by UDS by switching to the programming session function. UDS is also believed to provide an additional level of security against car hacking and during autonomous drive operation.

Diagnostic Fault Codes

DTCs are the most common go-to data supplied by the OBD system to diagnose powertrain complaints or the cause of a MIL being illuminated. Fault codes are used for several diagnostic purposes:

- Fault codes locate the smallest identifiable and repairable subsystem or component that has failed. Codes narrow down the location of a fault or abnormal system condition but do require further testing to differentiate between several possible root causes or verify the root cause.
- Fault codes identify subsystems and/or assemblies that may not have completely failed but may be operating abnormally. Examples include proprietary OEM fault codes that do not trigger a MIL, such as turbocharger or EGR actuator that can operate erratically but not allow emissions to exceed FTP threshold.
- Fault codes also identify a particular event or condition that needs to be reported. This category includes problems with monitor readiness codes or aftertreatment DTCs indicating that regenerations have not completed.

DTCs have a standardized format that is important to understand if a technician wants to quickly interpret a code and make associations between customer complaints and likely causes. SAE standardized DTCs, also called generic codes, are common to all manufacturers

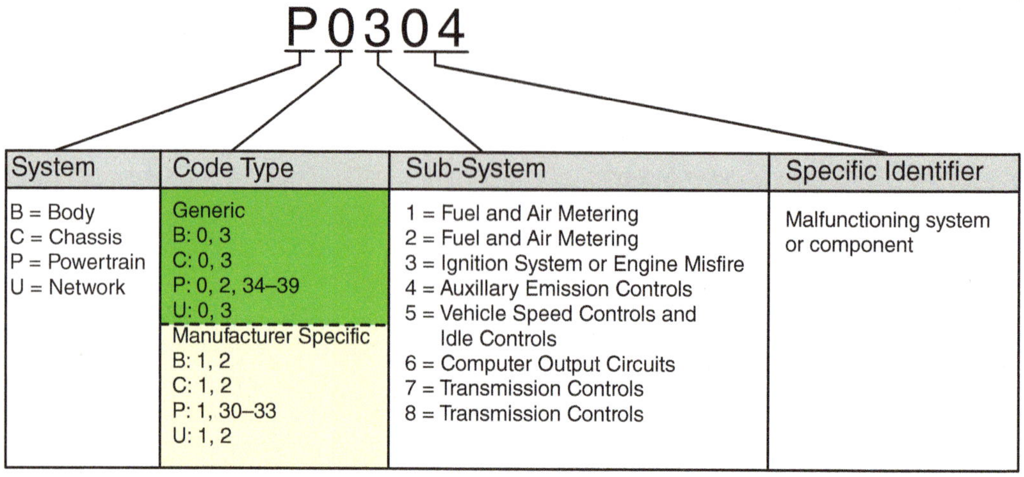

FIGURE 19-22 Breakdown of an SAE OBD-II DTC.

and have five positions (**FIGURE 19-22**). The diagnostic fault code for a specific failure of the same component or system used by one OEM, will have an identical code as another OEM for the same part and failure. For example, the DTC for a defective throttle position sensor (P-0121) or misfire in cylinder 3 (P-0303) is the same for any manufacturer using a throttle position sensor or having an engine with cylinders. The first position of the fault code begins with a letter that is either P, B, C, or U. The letter P is generally emission-related codes since P designates codes originating from the engine and transmission, otherwise called powertrain. B codes are body related; C codes originate from the chassis; and U codes are onboard network communication codes. The other four or five characters, which can be numbers or letters, provide additional information about the location and failure mode that set the code.

Enhanced Fault Codes

The next placeholder in the fault code is a number, which is either 0, 1, 2, or 3. The number 0 designates the code as a generic SAE standardized code. The number of SAE codes is generally limited to technology shared by most manufactures but are mandatory when identifying emission-related failures. Proprietary codes, which are not SAE codes, are used when a manufacturer has unique OEM-specific technology or diagnostic information.

These proprietary codes are also called enhanced codes because they will often supplement an SAE emission code with further information about the fault. Enhanced codes may also identify other problems that have not resulted in an emission-related failure but that need to be identified. Enhanced or proprietary fault codes will use the numbers 2 or 3 but not use a 0 after the P.

Enhanced codes can also indicate more complex failure modes or that a manufacturer has used specialized diagnostic strategies. For example, a fault code for a cylinder 1 glow plug used in a 2.0 L Volkswagen (VW) TDI is P-01649. However, an enhanced code for failure of the pressure sensing element contained inside of the same glow plug would be P13D7 (faulty pressure sensor on glow plug 1). Codes P13D6 and P13D9 would indicate the same pressure circuit fault in cylinders 3 and 4, respectively. Note that there is a number 1 in the second position designating an enhanced code and a letter at the fourth. A letter in any code identifies that only part of the sensor or a particular circuit in a component is faulty if the component has more than one sensor/circuit or if the vehicle has multiple numbers of the same component. For example, a mass airflow (MAF) sensor with an integral temperature sensor would use a letter to identify separate parts of the sensor with a fault. Letters may also appear in proprietary fault codes since they may not follow the same prescribed format as SAE generic codes.

Position three designates which major system of a vehicle is reporting the fault. A zero in this position points to a fault with an entire system or circuit. If an entire system is

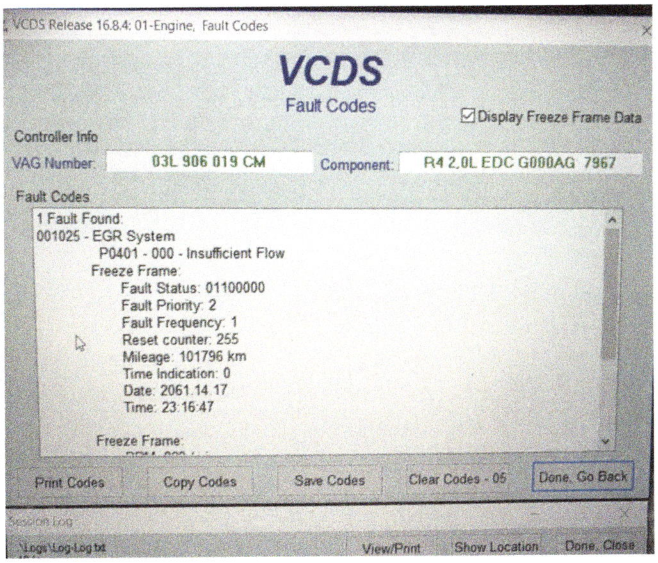

FIGURE 19-23 An OBD-II DTC with some freeze-frame data.

running rich or lean or is misfiring, a 0 will designate that the entire system is affected. A 2 indicates a fuel system; a 3 indicates is an ignition fault; and a 4 indicates exhaust gas monitoring faults. The remaining two or three characters are numbers for a system or component. A fault code of P0301, for example, indicates a misfire in cylinder 1, whereas P0302 indicates cylinder 2—and so on.

Code Priority

The OBD system sets three types of codes: A, B, and C. Type A and B DTCs are the most critical emission-related faults and will illuminate the MIL with only one occurrence. Type A and B codes are generally stored in the ECM historical memory. To help the technician diagnose the problem, the DTC has a failure record associated with it, such as a time and date stamp of when the last failure occurred, whether the DTC recurred since the last code-clearing event, and whether the failure has occurred during the current ignition cycle (**FIGURE 19-23**). Furthermore, Type A and B DTCs have freeze-frame data, which are a record of engine operating conditions when the fault was detected.

Type B codes differ from Type A DTCs in that they are emissions-increasing faults that are less serious than Type A fault codes and must occur at least once on two consecutive trips before the MIL will illuminate. The MIL will also go out if a Type A or Type B DTC problem does not recur after a predetermined number of drive cycles (e.g., three to five drive cycles—varies by OEM) (**TABLE 19-3**).

Type C DTCs are non-emission-related codes, or enhanced codes. Enhanced codes also cover non-emission-related failures that occur outside the engine control system.

Scanners will often report fault codes in the priority of their effect on emissions, where the most severe faults are displayed first. That means Type A codes are listed first. This is different from an OEM-established code priority that determines which code should be diagnosed first. When multiple codes are displayed, it is important to record all codes and consult service literature to determine the sequence or priority for diagnosing each code. Depending on the presentation of other codes with a MIL-triggering DTC, those additional DTCs can change the order to diagnose multiple codes, or what are called stacked codes. A common example would be an exhaust aftertreatment DTC that may set if an engine is not properly warming up. Regeneration events depend on an engine reaching minimum operating temperatures, and if a thermostat or coolant temperature sensor were defective, a completely different sequence of troubleshooting steps would be required if a cooling system–related DTC appeared with an aftertreatment system–related DTC.

▶ TECHNICIAN TIP

A wide variety of diagnostic service tools are available for technicians to use. Simple code readers supplying SAE codes are the least expensive. Computer-based software and communication adaptors reporting enhanced codes for many makes and models of vehicles can cost upward of thousands of dollars. It's instructive to note that a US national survey of repair shops found that in 85% of OBD-II related repairs, generic SAE codes alone provided all the information necessary to correctly identify a needed repair.

TABLE 19-3 Code Priority and MIL Illumination

Monitor	Monitor Type (when it completes)	Number of Malfunctions (on separate drive cycles to set DTC)	Number of Separate Consecutive Drive Cycles (to light MIL and store DLC)	Number and Type of Drive Cycles (with no malfunction to erase pending DTC)	Number and Type of Drive Cycles (with no malfunction to turn MIL off)	Number of Warm-Ups to Erase DTC (after MIL is extinguished)
Catalyst efficiency	Once per drive cycle	1	3	1	3 OBD-II drive cycles	40
Misfire type A	Continuous	1	1		3 similar conditions	40
Misfire type B/C	Continuous	1	2	1	3 similar conditions	40
Fuel system	Continuous	1	2	1	3 similar conditions	40
Oxygen sensor	Once per trip	1	2	1 trip	3 trips	40
EGR	Once per trip	1	2	1 trip	3 trips	40
Comprehensive component	Continuous when conditions allow	1	2	1 trip	3 trips	40

Source: Code of Federal Registry (CFR) Title 40 Part 86, titled Control of Air Pollution from Motor Vehicles and New Motor Vehicle Engines.

Freeze-Frame Data

The freeze-frame data that the ECM stores will provide a snapshot of the engine operating conditions at the moment when the DTC was confirmed by the OBD system (**FIGURE 19-24**). This information is not usually stored when only a pending DTC is set and the fault has not matured into a MIL-on DTC. If the manufacturer has chosen to store pending DTC freeze-frame data, they can choose to update or retain the freeze-frame data for display with a matured MIL-on DTC. Likewise, any freeze-frame data stored in conjunction with a pending or MIL-on DTC will be erased when the DTC is cleared.

Reviewing freeze-frame data is helpful when determining what condition triggered the DTC. Out-of-range values can be flagged for further examination. Another important use is to duplicate the conditions in a freeze frame to verify a complaint or after completing a repair. If a DTC has been cleared, operating a vehicle to match freeze-frame conditions can trigger a pending DTC or set a MIL-on DTC. If a repair is completed, duplicating freeze-frame conditions should not set a DTC and validate a successful repair.

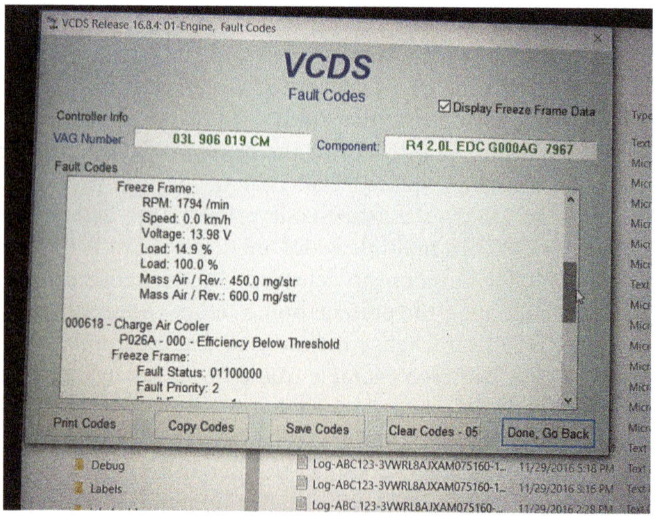

FIGURE 19-24 Freeze-frame data accompanying a fault for a charge air cooler (CAC) efficiency DTC.

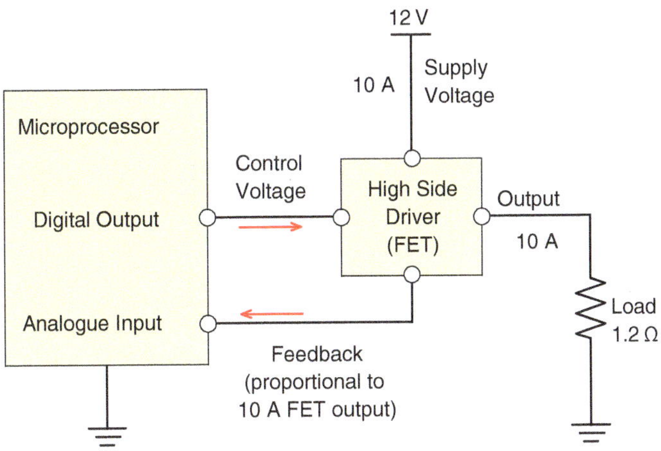

FIGURE 19-25 Field effect transistors used as output drivers can measure amperage through a circuit.

On-Board Diagnostics II Strategies

Unlike the early OBD systems that relied only on measuring circuit resistance to detect faults, today's OBD-II uses four strategies to evaluate engine control. From the simplest to the most advanced, the following are four broad strategies:

1. Measuring electrical current continuity to detect open, shorted, and resistive circuits. This level measures only voltage, amperage, and resistance in a circuit and reports major faults, such broken wires, open sensors, shorts to ground, or positive battery voltage (**FIGURE 19-25**).
2. Checking for out-of-range faults from input and output devices. This level is similar to the first level but deals with data about faults that is not as extreme or clear-cut. It asks whether the range of sensed or output value exceeds normal limits expected for the device. Out-of-range checks include evaluating wave forms from input and output devices (**FIGURE 19-26**). Network communication is also validated to ensure that modules are receiving vital information required for processing functions.
3. Checking the plausibility or rationality of data to determine whether data offers logical information for a given operating condition. It asks whether the data make sense (**FIGURE 19-27**).
4. Using model-based diagnostic monitoring of major emission systems and components to determine whether they are functioning as expected. This monitoring is performed on major emission systems and a few complex emission components.

If conditions are present that would allow emissions to increase, the OBD system should identify the fault with a code to record and report when, where, and how the fault has

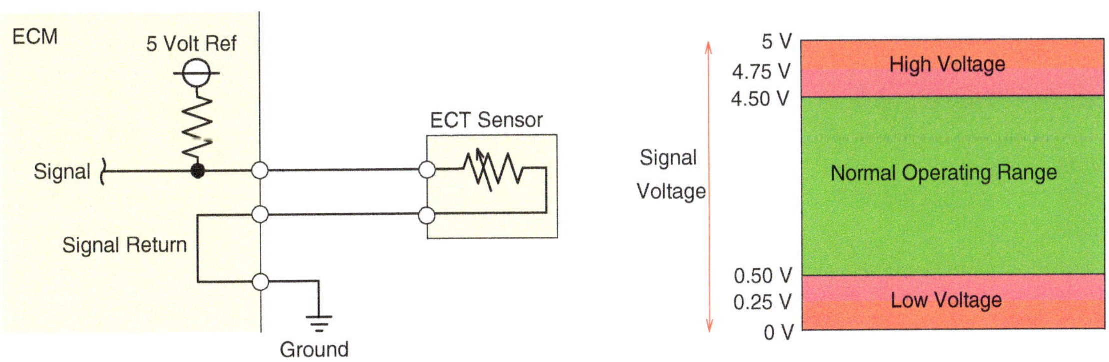

FIGURE 19-26 Out-of-range fault codes, opens, and shorts are detected by measuring the voltage drop across a sensor. The normal range for maximum and minimum sensor output should be approximately 85% of a 5-volt reference (Vref).

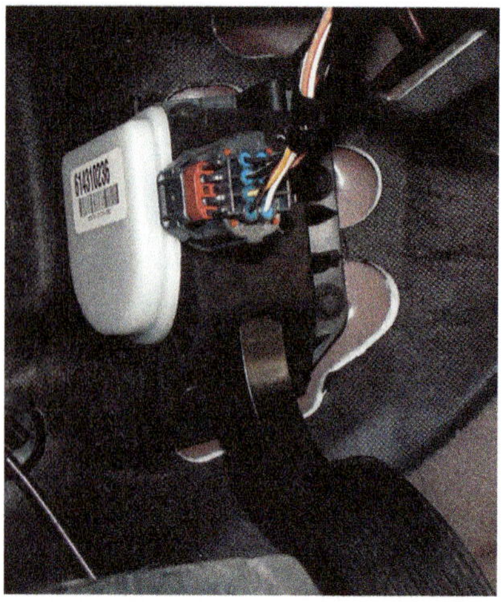

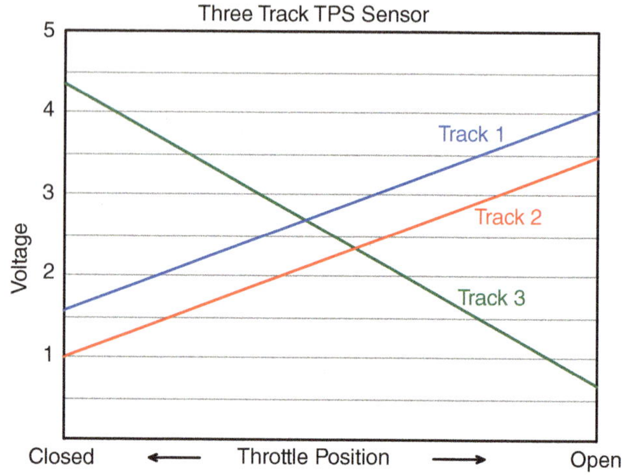

FIGURE 19-27 A rationality check of the throttle position sensor voltage. By using three potentiometers with three different resistive paths, the data from any other sensor pathway should correspond predictably to the other two.

occurred. Even though engine systems are far more sophisticated than mechanical systems, OBD enables diagnostic procedures to be easily performed, often with fewer tools and in less time than that it would take on mechanical systems.

On-Board Diagnostics Diagnostic Manager

The OBD diagnostic manager is central in making decisions about system operation and coordinating diagnostic modes for all tests related to emissions systems (**FIGURE 19-28**). This function is located in the ECM and consists of specialized software algorithms designed to make decisions about system operation. The diagnostic monitor can be referred to as the "traffic cop" of the diagnostic system, managing rules for setting and storing DTCs, conducting monitor operations, and illuminating the MIL. Managing monitor operation is one of the most essential functions of the diagnostic manager. Monitors are the OBD system's self-test tools or strategies that evaluate the operation of individual emissions system components and whether major emission systems are functioning correctly. Stated another way, a monitor is an organized procedure that is followed to test a component or system function. When the OBD diagnostic manager, or supervisor, follows a sequence of rules or tests while making measurements to arrive at a decision about the operational status or functionality of a specific component or system, the monitor is said to have executed or run. If no abnormal emission-increasing conditions are detected while the monitor is running, it is considered to have passed. The OBD executive determines when and how to run the emissions monitor.

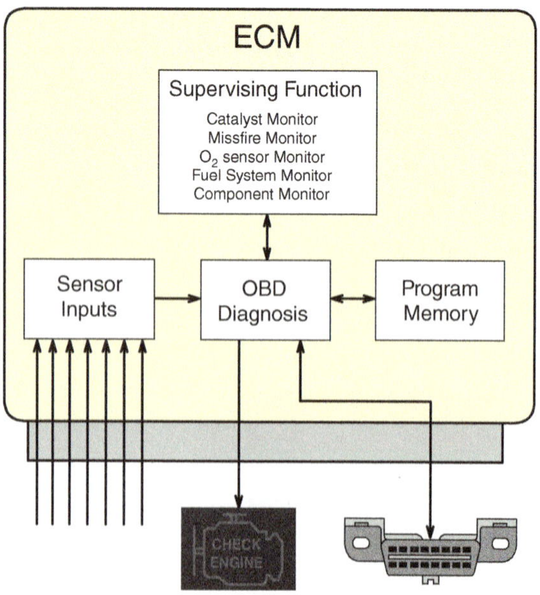

FIGURE 19-28 The engine ECM contains software that not only controls engine operation but also identifies emissions system–related faults. The diagnostic executive or supervisory function determines when a monitor should run and whether a fault exists.

Monitors and Readiness Status

When all the major system monitors have run or executed, a vehicle is considered to have completed a drive cycle. This is different from a simple ignition cycle, which is when the engine operates for a brief period. Before a monitor can properly check a component or system, it starts from an initial status of not-run when the battery is first connected during production or while in service or after DTCs have been cleared. Whether a monitor has detected abnormal conditions or not determines the status of a monitor as either passing or failing. Other monitor status messages can be run/not-run.

Criteria for monitoring decisions made by the diagnostic manager are established by a manufacturer after experimenting with different component/system failure modes or on a test bench. For example, emission production from an engine is correlated with how far a control valve position is limited by carbon buildup or the extent of catalyst deterioration is required before emissions exceed FTP multiplier thresholds. If no faults are detected by a monitor when it has run and a pass decision is made by the diagnostic executive, the monitor is said to be ready, completed, or passed. After a monitor has achieved the status of having run, passed, or completed, it is allowed by OBD regulations to remain in that state even after it detects a later fault. A fault is reported but monitor status can remain ready. Only after DTCs are cleared or a battery is disconnected is the monitor status required to be reset to not-run, failed, or incomplete.

It is very important to understand that if a monitor is not-ready because it has not run, then it has encountered one of two problems. First, when a monitor starts from a not-run status and detects a fault, it will remain in a not-run, not-ready, incomplete, or failed status. A second problem is that a prerequisite condition for the monitor to run has not been met. (More about this is explained in the sections on major emission monitors.) But before a monitor executes, enabling criteria must be met. A simple enabling condition might be that the engine has reached operating temperature or that the vehicle has driven far and long enough for the monitor to run.

Until the monitor has achieved "ready" status, it is typically not able to make a judgment about a potential fault condition. Understanding this arrangement helps explain why IM programs use monitor status as a criterion to determine whether the OBD emission system is working properly. A single not-ready, incomplete, or not-run monitor status on a late-model vehicle disqualifies it from getting an OBD-II emission system inspection pass.

▶ TECHNICIAN TIP

OBD-II legislation requires that all electronic service tools reset all monitors anytime a code is cleared. No single code can be cleared without resetting every monitor. There are some situations when one method of validating repairs on some vehicles, such as those that have monitors that are difficult to reset due to failed or incomplete to ready/pass, is to simply road test them. Road testing a vehicle and duplicating freeze-frame conditions will allow a technician to confirm that a repair is complete if the code does not appear again. A well-equipped scanner capable of displaying SAE Mode-7 data "Emission-related DTCs detected during Current or Last Completed Drive Cycle" will display the fault if it remains present. The MIL light will also often extinguish itself after one or two short road tests with several ignition cycles.

▶ Major Emission System Monitors

19-03 Identify and describe the principles of fault detection and the operational strategies of major emission system monitors.

As mentioned, the OBD system self-checks on four fundamental levels. Each of these levels involves the use of monitors, but with different levels of sophistication, to make judgments about emission system operation. Monitoring diesel engines is broken down into separate, specific hardware systems that are responsible for a specialized aspect of engine operation and emissions control. Each of these major engine emission systems is assigned a monitor that tests and evaluates the system to determine whether it is functioning properly to maintain the low emissions for which the system was designed. Depending on the category of emission, the emission system component, and which monitor is evaluating potential emissions-increasing conditions, a monitor is required to detect conditions that would cause an emissions increase between as little as one and a half to five times the FTP limit. Monitors are not required for a system where failure of a component cannot cause an increase above one and a half times FTP limits. Vapor leakage from a diesel fuel tank is an example of an area that doesn't have a major emission monitor associated with it—unlike gasoline-fueled vehicles, which do have such a monitor. Leaving the filler cap off a diesel fuel tank will not produce evaporative emissions exceeding FTP thresholds.

Currently, up to 14 major monitors are required by the EPA and CARB for diesel engines. No vehicle will use all these monitors, but rather only the monitors necessary to support emission-related equipment. Briefly described, the following are some of the major emissions system monitors:

1. Fuel system monitor: The OBD system must monitor the high-pressure fuel-injection system to verify that it is functioning properly. The following are some of the monitoring criteria:
   - fuel injection pressure
   - fuel-injection quantity
   - multiple fuel-injection event performance
   - fuel-injection timing
   - closed-loop feedback controls.
2. Misfire monitoring: The OBD system must detect a misfire occurring continuously in one or more cylinders during idle. For model years 2013 and onward, misfire monitors must check for misfires over the entire engine speed range.

3. EGR System: The OBD system must monitor the EGR system flow rate and performance. The following are some of the possible monitoring criteria:
   - EGR flow rate
   - EGR response rate
   - EGR cooling system
   - feedback control
   - EGR cooler performance
   - closed-loop feedback controls.
4. Boost pressure control systems: The OBD system must monitor the boost pressure control system to ensure that there is adequate airflow to support combustion. The following are some of the criteria:
   - underboost and overboost malfunctions
   - waste-gate operation
   - slow response—variable-geometry turbocharger (VGT) systems only
   - charge air undercooling
   - closed-loop feedback controls.
5. Cold-start emissions strategies: To prevent excess emissions during cold-start and warm-up periods, the OBD system will monitor glow plugs and intake heater operation (**FIGURE 19-29**). The following are some of the possible monitor criteria:
   - glow plugs or intake heater on time
   - glow plugs and intake heater current flow (for shorts or open circuits)
   - glow plugs and intake air heaters enabled.
6. Variable valve timing (VVT) system: VVT systems are used by some light-duty diesel engines to delay intake valve closing and vary exhaust valve opening for reduced pumping losses and increased engine efficiency. The OBD system monitors whether the commanded valve timing and/or control within a crank angle and/or lift tolerance has been achieved. It also detects slow system responses before emissions exceed the thresholds for other monitors. As a continuous type monitor malfunctions are identified whenever fault conditions are met rather than briefly checked once per trip.
7. Engine cooling system: Cooling system malfunctions are related to slow warm-up times, which can prevent other monitors from running because enabling criteria for engine temperature has not been met. Proper thermostat functioning is sensed by using the engine coolant temperature sensor to determine whether warm-up is reached within a specific time. Monitoring is performed once per drive cycle.
8. Crankcase ventilation (CV) system: The CV system is checked for system operation and whether the system has been disconnected. The manufacturer may not monitor for

FIGURE 19-29 Current flow to the electric intake heater is monitored to validate that a cold-start strategy is operational.

disconnections between the crankcase and the CV valve if the CV system is designed to use connections that are resistant to failure and tampering.

9. Comprehensive component monitoring: The comprehensive component monitor measures circuit continuity, voltages, resistances, amperage, out-of-range values, and the rationality of electrical signals operating powertrain components that can cause a measurable emissions increase during any driving conditions that could reasonably be encountered. The comprehensive component monitors also serve other purposes:
   - used for other OBD monitors
   - required for monitoring input components for circuit and rationality faults
   - required for monitoring output components for functional faults.

   The comprehensive component monitor gets its name from the vast number of circuits it checks for every control system evaluated by the OBD system. The detection sensitivity of the comprehensive component monitor is not tied to any emission threshold multiplier to generate a MIL-on DTC.

   Aftertreatment OBD monitors require faults to be detected when emissions exceed FTP multiplication factors for any of the following diesel exhaust aftertreatment systems:
   - oxidation catalyst
   - lean NO_x catalyst
   - SCR catalyst
   - NO_x trap
   - PM trap
   - ammonia slip catalyst (ASC).
10. NMHC converting catalyst: Non methane hydrocarbon conversion or NMHC catalyst monitoring evaluates the operation of the oxidation converter. It must reach outlet temperatures adequate to regenerate the diesel particulate filter (DPF). Inlet and outlet temperatures are compared and a minimum temperature increase must take place under defined conditions; otherwise, a fault will be set.
11. SCR catalyst and lean NO_x catalyst: The OBD system must monitor the SCR and/or the lean NO_x converting catalyst(s) for adequate conversion efficiency. The individual electronic components, such as actuators, valves, sensors, heaters, diesel exhaust fluid (DEF) pumps, and DEF quality, are also checked.
12. NO_x adsorber, also called lean NO_x trap (LNT), system monitoring: This NO_x-reduction system has an additional requirement for SCR in that adsorption efficiency is monitored. Most monitors measure adsorption efficiency by using a model of oxygen storage and release. Because in-cylinder late or post–main fuel injections are used, these are monitored too, and any technology using exhaust-based, air-assisted fuel injection is checked as well.
13. DPF system: The DPF removes soot or other solid and liquid particles from the exhaust gas (**FIGURE 19-30**). The following are some of the OBD monitoring requirements:
    - regeneration frequency
    - regeneration completion
    - missing or removed DPF
    - insufficient conditions required for regeneration
    - correct operation of feedback controls.
14. Exhaust gas sensor monitoring: Because the OBD system must monitor input and output signals for correct waveforms, response rate, and any other element that can affect emissions, all exhaust gas sensors are monitored. Monitoring includes oxygen sensors, NO_x sensors, PM and ammonia sensors used to control system operation. Sensors have a critical function to provide feedback for EGR supply and both DEF and HC dosing. Heater circuits in all gas sensors that are crucial to correctly measuring gas concentrations are evaluated as well (**FIGURE 19-31**).

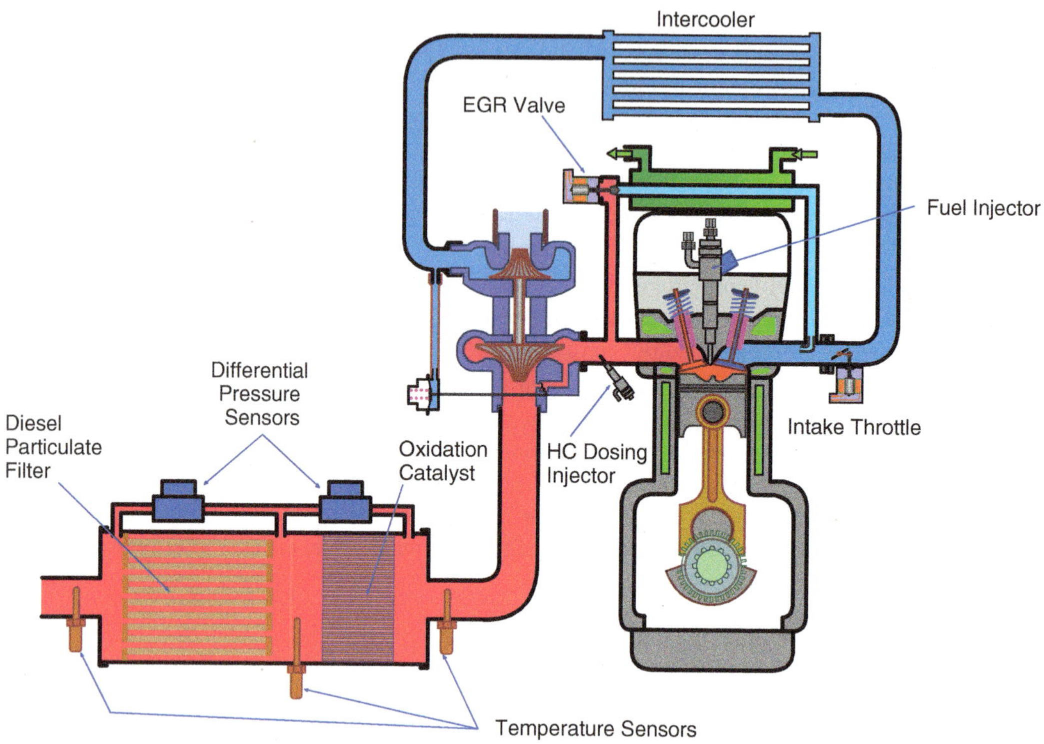

FIGURE 19-30 Sensors used by the OBD system that are associated with the DPF.

Classification of Monitor Behavior

It's important to understand the differences between the behavior of monitors when diagnosing and validating OBD system operation. Monitors used by the diesel OBD system can be divided into four general types:

1. Threshold monitoring: This monitor directly measures the level of emissions by using a sensor to identify a noxious emission. Sensors for ammonia, PM, and NO_x directly measure levels of these emissions in the exhaust (**FIGURE 19-32**).
2. Functionality monitoring: A functionality monitoring strategy ensures that an emission system or component is operating as designed. Monitoring is generally based on a computerized model of expected or desired system behaviors developed in an OEM laboratory experimentation or through field tests. Monitoring usually includes evaluating multiple electrical signals from various sensors and performing a test during normal engine operation to evaluate system response. Moving a VGT actuator and measuring

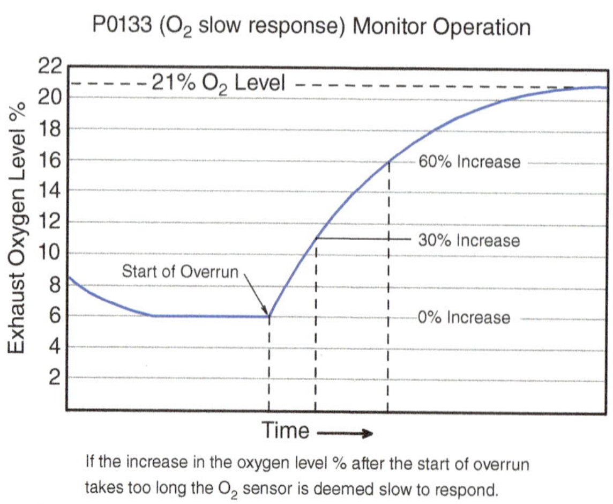

FIGURE 19-31 The signal response time of the oxygen sensor is compared with the predicted time when exhaust gas oxygen content changes. A slow-responding oxygen sensor takes too long to sense changes in exhaust oxygen concentrations.

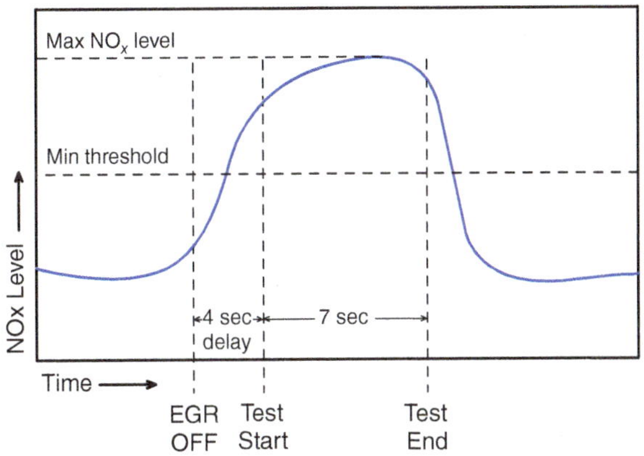

FIGURE 19-32 The NO_x sensor is an example of a component used for threshold monitoring. Accuracy and response time of the NO_x sensor is judged by briefly cycling the EGR valve closed to increase engine-out NO_x.

boost pressure change would be an example of a simple functional monitor that evaluates whether a system is responding correctly. Functional monitors can be further categorized into intrusive and non-intrusive operation. Intrusive monitors actively manipulate a component or system operation and measure its response to the manipulation. The VGT actuator movement is an example of an intrusive monitor test. Functional monitoring of a DPF before using soot sensors is an example of passive monitoring. Temperature and pressure signals measure the filter restriction and temperature change across filter substrate. No manipulation and subsequent measurement of system change is necessary when a non-intrusive monitoring strategy is used.

3. Rationality, or plausibility, monitoring: The purpose of this monitoring strategy is to judge the accuracy of data to determine whether the data make sense when compared with other available data. Rationality faults are also called logical faults because they were not predicted according to an acceptable set of rules. For example, a coolant temperature sensor should sense a temperature similar to an oil or air temperature sensor at initial start-up if an engine has not operated for several hours. If a large difference between sensor data exists or if the data do not match the engine operating conditions, such as −40°F (−40°C) coolant when engine oil is 212°F (100°C), then the data is irrational, or implausible. To further check the discrepancy, a model of expected sensor value is consulted. An example is judging the rationality of soot accumulation in the DPF. Soot is not directly measured, but pressure differences across the filter are compared to chemical models predicting accumulations with increasing pressure differential. Soot accumulations calculated by using a pressure differential sensor input and a chemical model should approximately correspond to one another.
4. Continuous and noncontinuous monitoring: An emissions system monitor may operate continuously or noncontinuously. A noncontinuous monitor requires certain conditions to be met before evaluating an emissions system. For example, the vehicle may need to be above a particular road speed; the engine may need to reach a specific temperature; or there may even need to be an adequate level of fuel in the tank before a monitor can run. A monitor may not run if the altitude is too high, if a power takeoff is operating, or if some other condition is present that could jeopardize the safety of an occupant or damage the vehicle. To provide an understanding of monitor operation, the following section contains are examples of several OBD monitors, along with a description of the monitor strategy and testing conditions needed to enable or activate each monitor.

Monitor-Enabling Criteria

Before a noncontinuous monitor can run, vehicle-specific events or conditions must first take place. These monitor prerequisite events or conditions are called enabling conditions. Some noncontinuous monitors require the vehicle to follow a prescribed driving cycle to provide normal operating conditions to the emission system or component so that it's behavior can be observed or manipulated (**FIGURE 19-33**). Other enabling conditions are required to minimize the likelihood of a false failure. For example, a misfire monitor needs

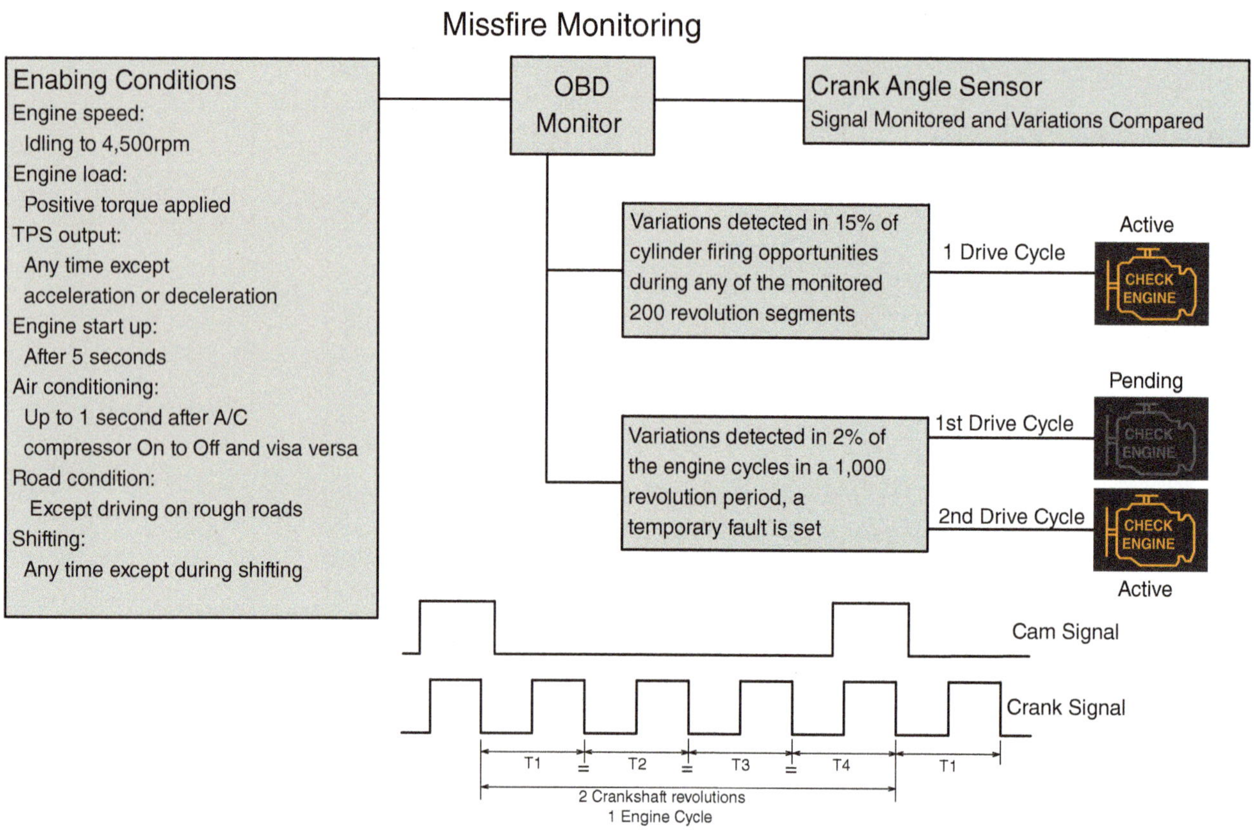

FIGURE 19-33 Enabling criteria are used by a monitor to determine optimal conditions for detecting a fault.

to have a ⅛ to ¼ tank of fuel to prevent setting DTC codes when the engine is simply running out of fuel. Most monitors require the engine to be warmed up before execution can begin. An exception would be a cold-start strategy monitoring where cold intake air and a cold engine are necessary to activate intake heaters.

Normally, monitors cannot run if sensor data associated with the system are inaccurate or cannot be relied on if another fault is present. Similar to the concept of enabling conditions are fault detection criteria or malfunctions entry conditions. These terms refer to the important criteria necessary to set a DTC (**TABLE 19-4**). When diagnosing the root cause of a DTC, technicians should study OEM service literature that specifically outlines conditions that set a fault. An example would be for an overboost DTC P0234 for a pickup

TABLE 19-4 Experimentally Derived Values for Malfunction Detection Criteria

System and Parameter	System Change and Parameter	Effect on Emissions (Base = 1.0)		Test Condition
		NO_x	PM	
Injection pressure	23,375 psi–16,675 psi	0.5–0.8	1.2–1.8	Medium speed & load
Injection timing	Retarded from TDC* to 5 ATDC**	0.4–0.8	1.2–1.8	Medium speed & load
Boost pressure	30 from 38 psi	0.7–0.8	1.2–1.4	Medium speed & load
EGR	With–without	1.5–2.0	0.3–0.8	Acceleration
DPF	With–without	1.0	2.0–10.0	Acceleration
SCR	With–without	2.0–3.0	1.2–1.5	Acceleration
LNT	With–without	3.5–5.0	1.0–1.3	Acceleration

*TDC: top dead center

**ATDC: after top dead center

truck. Intake pressure overboost would naturally lead a technician to suspect that a VGT actuator is sticking shut. This condition could also be caused by an EGR valve sticking open or an exhaust brake operating normally, causing gas pressure to build up in the intake manifold. A closer examination of fault criteria would show that overboost conditions are more likely detected by monitoring exhaust pressure that exceeds a maximum threshold. Sensing exhaust pressure after the turbocharger differentiates false failures caused by normal exhaust brake or EGR operation.

▶ Monitor Operation

The following section covers monitor operation for a few key major emission monitors. Each manufacturer will develop unique strategies that enable monitors to judge the functionality of a particular emission system. Studying how a manufacturer has configured a monitor to operate is essential to diagnosing root causes for DTCs because it provides clues as to possible conditions setting DTCs. The following are some general strategies used by OBD systems to provide a sense of how emission-related DTCs are set.

19-04 Identify and describe monitor operation for a few key major emission monitors.

Misfire Detection Monitor

Cylinder misfire occurs when a cylinder fails to produce combustion pressure similar to other engine cylinders (**FIGURE 19-34**). Causes can include poor compression, improper fuel delivery, or mechanical engine failure. Because engine misfires can cause HCs and CO emissions, the OBD system must evaluate engine operation to determine whether misfires are occurring. A common threshold for diesel engine misfire detection is 4% per 1,000 crankshaft revolutions. This corresponds to 40 misfires per 1,000 rpm.

The misfire detection monitor was noncontinuous until 2013. After that it was required to judge misfires over the entire engine operating range. Data input from the engine sensors includes oil temperature, crankshaft position (CKP), injection quantity, exhaust pressure, intake air temperature, and fuel level input. This data ensures that certain conditions are met when detecting a misfire, such as distinguishing whether a vehicle is low on fuel or out of fuel, warmed up and running, or lugging under load.

Misfire Monitor Operation

The crank position (CKP) sensor is a primary input in determining cylinder misfire. The CKP sensor produces data by sensing the movement of the teeth on a wheel located on the crankshaft. The ECM uses the speed at which the wheel passes by the sensor to calculate the time between CKP tooth edges. These data supply rotational velocity of the crankshaft. By comparing the accelerations of each cylinder event, the ECM calculates the pressure produced from each cylinder. Cylinders producing less power will have fewer teeth pass by the CKP senor per unit of time compared with cylinders with normal or higher cylinder pressure. When the power contribution from a particular cylinder is less than an expected value and other criteria for enabling the misfire monitor are met, the cylinder is determined

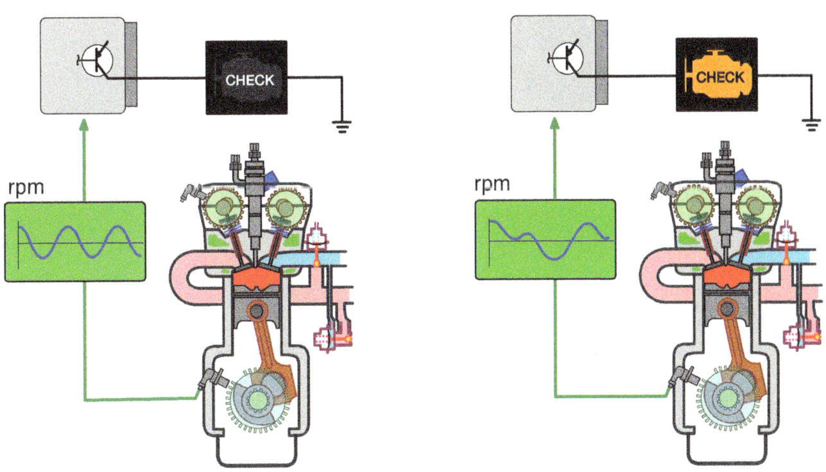

FIGURE 19-34 Combustion increases cylinder pressure and rapidly accelerates the crankshaft. By measuring the speed of acceleration change, the ECM can calculate cylinder pressure and compare it with other cylinders. No acceleration or only little acceleration means a cylinder is missing.

to have misfired. Loss of compression, worn rings, and bent connecting rods can be diagnosed by the fine resolution produced by many CKP teeth and diagnostic strategies of the latest OEM software.

Exhaust Gas Recirculation System Monitor

A more complex type of OBD monitor is for the EGR system. To maintain the lowest production of NO_x emissions, EGR gas is recycled from the exhaust to the intake manifold, which lowers peak combustion temperatures and pressures. If the EGR system fails to perform as it should, the failure will result in high NO_x emissions or poor performance and misfire conditions.

OBD requirements for the diesel EGR monitor include detecting the following faults before emissions exceed OBD thresholds:

- incorrect EGR flow rate
- slow EGR response rate
- inadequate EGR gas cooling.

An additional operational check detects if a normally closed-loop EGR system fails to enter a closed loop or defaults to an open loop (**FIGURE 19-35**).

Exhaust Gas Recirculation Monitor Operation

The EGR monitor needs to evaluate a large variety of mechanical and electrical functions and conditions so that the inputs and operating strategies that check its operation are more elaborate. The electronic and mechanical functionality of the EGR valve, the EGR valve actuator, the VGT, the EGR valve position sensor, and other sensors (e.g., barometric pressure, intake boost, exhaust pressure, MAF, oil temperature, air inlet temperature) also provide input to this monitor. Even the efficiency of the EGR cooler must be evaluated by the monitor to determine whether the EGR system is operating correctly. The efficiency of the cooler is checked by comparing exhaust gas temperature entering and leaving the cooler. Insufficient temperature change means low cooler efficiency caused by blocked, restricted coolers or poor heat transfer to coolant flowing around the cooler. The elements and components of the EGR system and associated circuits are first evaluated by the monitor before EGR flow is evaluated.

If a MAF sensor is used on the engine as part of the EGR monitor strategy, the following will be a typical method to evaluate EGR flow. First, when normal EGR rates are

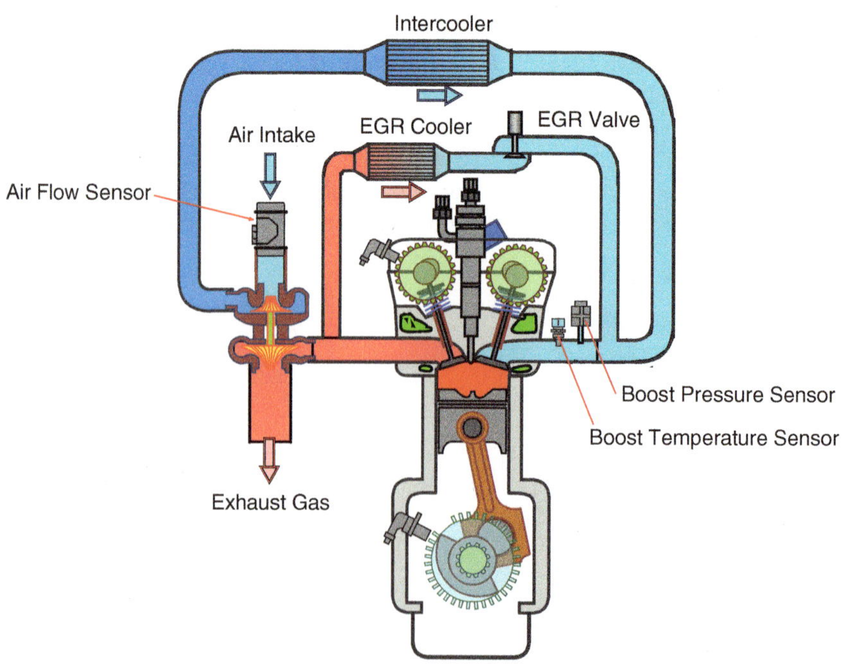

FIGURE 19-35 MAF and oxygen sensors can provide closed-loop feedback to EGR metering controls.

being commanded and when the engine enters into either one of two specified operating ranges, a flow measurement is performed. Essentially, the intake air MAF sensor will detect less fresh airflow into the engine when the EGR valve opens because EGR gas replaces fresh intake air (**FIGURE 19-36**). The difference in flow with and without the EGR occurs because EGR gas bypasses the MAF sensor, flowing directly into the intake manifold. Two engine operating ranges are used for taking measurements to ensure an adequate amount of EGR is being supplied as requested by the engine control system and compared with an expected EGR mass. The variation between the mass of measured EGR gas flow and the expected or commanded mass is used to determine whether there is insufficient or excessive EGR gas flow.

The most basic enabling conditions and sequence for the EGR monitor to operate occur after all the circuits, sensors, and actuators have been judged for correct operating range values or checked for opens and shorts and to determine that the engine has been warmed up. When the engine has entered the following specific conditions, the monitor compares commanded EGR flow with measured flow, which is calculated by an algorithm that uses the intake temperature sensor and intake pressure sensor:

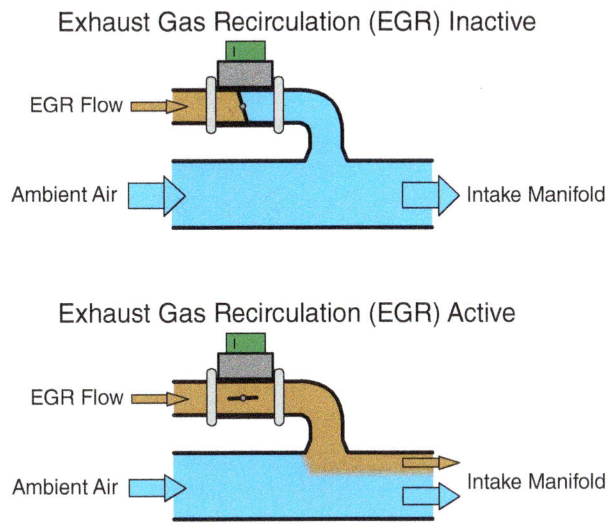

FIGURE 19-36 EGR flow displaces fresh air. The difference between the measured intake airflow and flow predicted without EGR gas is calculated to find the quantity of EGR recirculated into the engine.

- Condition 1: The EGR flow commanded is greater than 20% and engine speed is 1,000–2,200 rpm. The injection quantity is 0.0006–0.0022 in^3 (10–35 mm^3).
- Condition 2: The EGR flow commanded is greater than 20% and engine speed is 2,200–3,000 rpm. The injection quantity is 0.003–0.007 in^3 (50–110 mm^3).

The limits of variations between actual and expected flow values stored in the ECM are based on engine speed and load conditions.

Another technique to monitor EGR flow is to command the EGR valve to fully open when the correct drive cycle conditions have been met. Typically, when cruising at steady speed above 50 mph (80 kph), the EGR valve is opened briefly and the turbocharger speed is measured. High EGR flow cools the exhaust gas temperature and substantially lowers turbocharger speed. If the turbocharger speed falls to an expected range, the EGR system monitor is complete, which means that no faults were detected and that the EGR system is functioning correctly.

Fuel System Monitors

Fuel system monitors are required to detect faults associated with the high-pressure injection system before emissions exceed OBD emissions thresholds. To do this, manufacturers will typically evaluate fuel system pressure control, fuel-injection quantity, and fuel-injection timing.

Fuel system monitors on common rail (CR) engines evaluate fuel pressure control by comparing expected pressure values in the fuel rail with actual pressure measured by using a rail pressure sensor (**FIGURE 19-37**). For example, worn injectors of a CR fuel system will set a fault indicating a loss of fuel pressure based on the shorter pulse width applied to the pressure regulator required to maintain fuel pressure in the fuel rail. Because more fuel will bypass a worn nozzle valve or other internal parts of the injectors, the fuel system will identify this as a pressure leak.

Fuel-injection quantity is monitored and calculated by measuring crankshaft speed fluctuations. Crankshaft velocity change—i.e., the torsional vibration produced by cyclic acceleration and slowing of the crank—is proportional to cylinder pressures. Stated

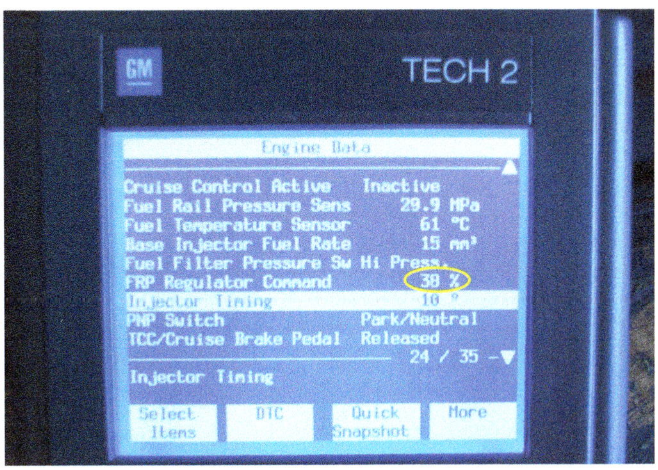

FIGURE 19-37 The duty cycle of the fuel rail pressure regulator is continuously compared with expected values to check for fuel leaking through injectors.

another way, rotational acceleration rate, or how fast the crank speeds up, increases as more fuel is injected and burned during the power stroke. Compression strokes will slow the crankshaft speed.

Fuel-injection timing is typically monitored by comparing it to a specific point when crankshaft rotational velocity suddenly increases, which is compared with an expected crankshaft position stored in the ECM's memory. Additionally, on unit injector fuel systems, signal monitoring circuits in the ECM can detect changes in the injector coil inductive signal corresponding to the beginning of injection. A needle-lift sensor provides input for some of the latest common rail injectors and older electronic distributor pump injection systems.

Additional fuel monitor requirements include detecting faults of closed-loop feedback systems when they fail to enter closed loop or default to open loop.

A closed loop between the rail pressure sensor and the fuel pressure regulator is an example of a normally closed-loop system on a CR engine. Problems with the signals or data will force the ECM to move out of closed-loop operation and use an adaptive strategy. Injector voltage is also measured by the monitor at the beginning and end of the injection event to check for faults in injector solenoids or electrical circuits energizing the injector.

Boost Pressure Control Monitoring

OBD diesel monitor requirements for the boost control include detecting faults before emissions exceed one and a half times FTP emissions test standards under the following conditions:

- underboost and overboost malfunctions
- slow response (VGT systems only)
- charge air undercooling.

Additional requirements include detecting faults of closed-loop feedback systems if they fail to enter closed loop or default to open loop. Underboost and overboost malfunctions are typically detected by comparing expected intake and exhaust pressure values stored in the ECM.

Slow turbocharger response times (VGT systems only) are evaluated by measuring the time the turbocharger takes to reach expected boost pressure, which is done by the boost pressure sensor and/or the turbine speed sensor. Temperature sensors located at the inlet and outlet of the intercooler or intake manifold temperature sensors evaluate charge air cooling efficiency. Problems such as externally restricted cooler fins will not produce the necessary temperature drop, resulting in cooler efficiency DTCs.

Cooling System Monitor

Other OBD monitors and emissions systems, such as the EGR, are enabled only after the engine reaches operating temperature. Engine temperature is also a variable that affects transmission shift points and injection timing. Because defective thermostats and engine temperature sensors will negatively affect engine operation, the OBD system also monitors the operation of the cooling system. The operating strategy of the cooling system monitor typically involves comparing actual operating temperature with a warm-up model based on start-up temperature, ambient conditions, and driving time. If the engine fails to reach operating temperature within a reasonable time, the monitor will set a fault. The monitor may also compare values of the coolant sensor with values of other temperature sensors, such as the oil or transmission, to check the rationality of the data.

> ▶ TECHNICIAN TIP
>
> Adding auxiliary heaters for passenger compartments can cause the cooling system monitor to indicate a fault. Taking heat out of the cooling system during engine warm-up will delay the time it takes to warm up the engine, which will illuminate the MIL.

Diesel Particulate Filter Aftertreatment System Monitors

Two common aftertreatment systems are DPF monitoring and SCR monitoring. DPF monitoring includes the following:

- filter performance
- frequent regeneration
- NMHC conversion
- incomplete regeneration
- missing substrate

- active regeneration fuel delivery
- closed-loop feedback control.

Any disconnected sensors, missing catalysts, or invalid data will cause the MIL to illuminate. Monitoring the DPF also includes the operation of the HC dosing system to determine whether it is delivering fuel. The system operation is monitored by a combination of temperature and pressure sensors.

When the HC dosing valve sprays raw fuel into the exhaust system, the fuel undergoes flameless combustion in the oxidation catalyst, which raises the temperature of the exhaust gas. The temperature change in and out of the exhaust system catalysts is monitored to determine whether the catalysts are operating within an expected temperature range. Missing, broken, deactivated, or tampered catalysts will not meet target temperature values, which will set DTCs. The delta pressure (delta P) sensor, also called the pressure differential sensor, measures the pressure differential across the DPF and essentially checks for missing or damaged catalysts (**FIGURE 19-38**). These sensor data fluctuate too much under load to accurately predict soot loads. Thus, instead, pressure differential data are used to compare the rationality of more reliable chemical models predicting soot loading. Active regeneration events are based on chemical models of soot loading that are calculated from the amount of fuel burned, operating conditions, time or distance elapsed since the previous regeneration, etc. The delta P sensor will measure filter performance by comparing measured and estimated restrictions after a regeneration event. If the values match, the system determines that the filter has proper efficiency for trapping soot. A new type of soot sensor introduced in 2016 functions as a threshold monitor for the DPF. The sensor is located downstream of the DPF filter and detects any carbon-based particulate in the exhaust stream.

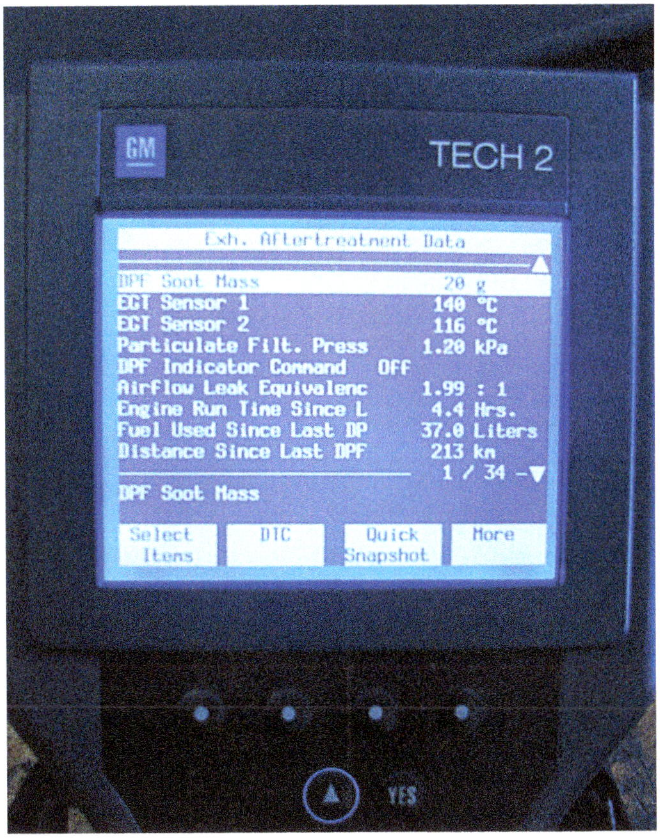

FIGURE 19-38 Soot accumulations in the DPF use the pressure differential sensor to check the rationality of chemical models estimating soot load.

Selective Catalytic Reduction Monitor

SCR monitoring ensures the SCR is capable of converting NO_x emissions. Two NO_x sensors measure concentrations of NO_x upstream and downstream of the SCR. Early systems were dosed according to a model that predicts NO_x output from the engine. Later systems used feedback from sensor data to adjust DEF dosing. When the temperature sensors in the SCR indicate the catalyst is at proper operating temperature, the monitor compares the NO_x sensor differences over time to calculate NO_x catalyst efficiency and deterioration. If the DEF dosing system is not operating correctly, the monitor will detect the absence of NO_x conversion and set fault codes, typically identifying faults in the entire SCR system.

A DEF quality sensor can now determine the quality of DEF fluid by measuring its density. Using tuning fork technology, ultra-high frequency vibrations produced in the sensor will move at a precise rate based on fluid density.

Like the DPF, DTCs in the SCR system will produce derate conditions, depending on their severity. Running a vehicle with no DEF fluid in the reservoir will eventually cause the engine to only idle and could keep the engine from starting (**FIGURE 19-39**). Off-board diagnostics involves isolating and pinpointing a fault identified by the OBD system.

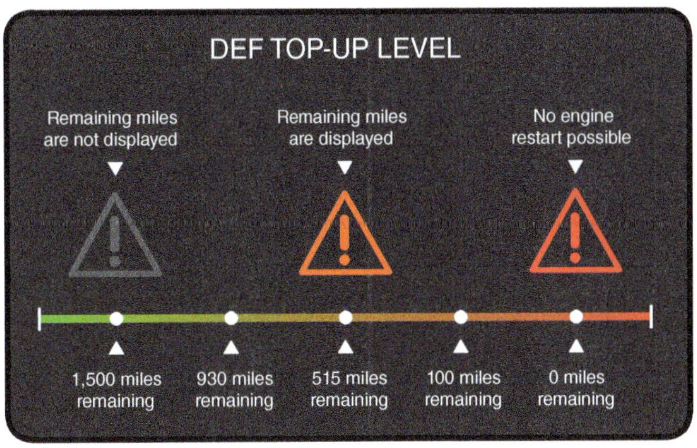

FIGURE 19-39 Lamps and warnings associated with SCR aftertreatment system faults.

Off-Board Diagnostics

When DTCs are set in the system memory, they can be read through an instrument display cluster, retrieved by a scanner or computer with OBD software. Off-board diagnostics and repair occur when

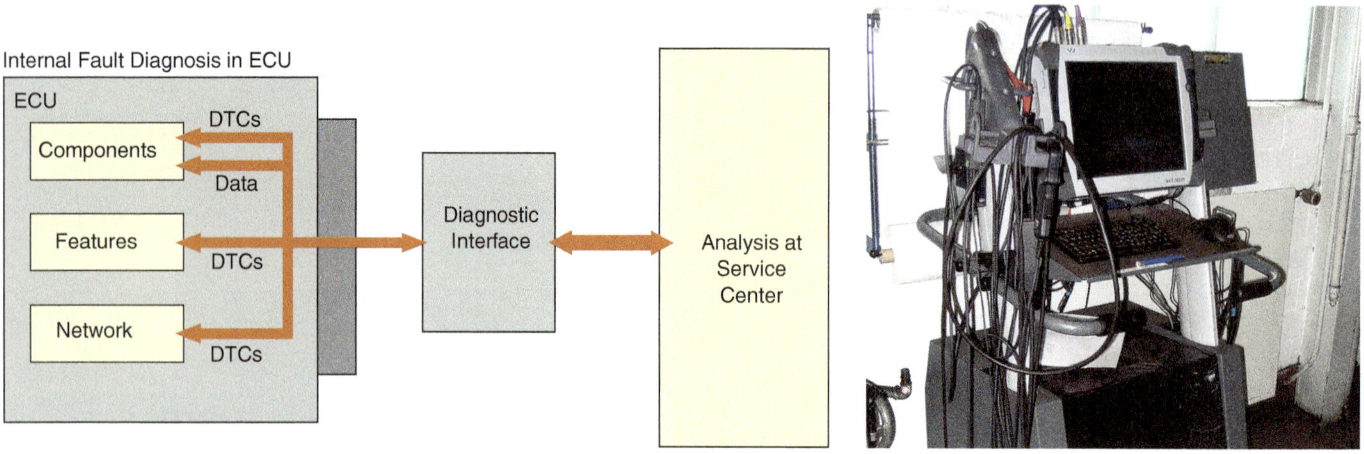

FIGURE 19-40 Performing off-board diagnostics involves isolating a fault based on fault code information.

the technician retrieves DTCs and other OBD data to isolate a fault (**FIGURE 19-40**). Fault isolation is the process of precisely identifying the fault that often begins with a DTC starting point. During off-board diagnostics, the technician may monitor system operation, perform pinpoint electrical tests, and actuator tests, plus inspect components (**FIGURE 19-41**). Assistance with manufacturer-specific service literature outlining monitor operation, code diagnostic priorities, diagnostic trees, and test values for circuit current and resistance are essential (**FIGURE 19-42**). Remotely assisted diagnostics using **telematics** is also a type of off-board diagnostics. Telematics is a branch of information technology that uses specialized applications to transmit information over long distances to and from a vehicle.

Onboard Diagnostics Performance Tracking

While OBD regulatory requirements generally focus on detecting faults and supplying information about them, newer requirements add extended tracking and logging of emission system performance information. One performance tracker is the in-use monitor performance ratio (IUMPR) (**FIGURE 19-43**). This feature evaluates how frequently the OBD system monitors execute and finds no faults compared to the number of times the monitors' enabling conditions are met. Monitor performance ratios provide a measurement of monitor effectiveness. The more frequently a monitor runs and makes a judgment, the more effective it is.

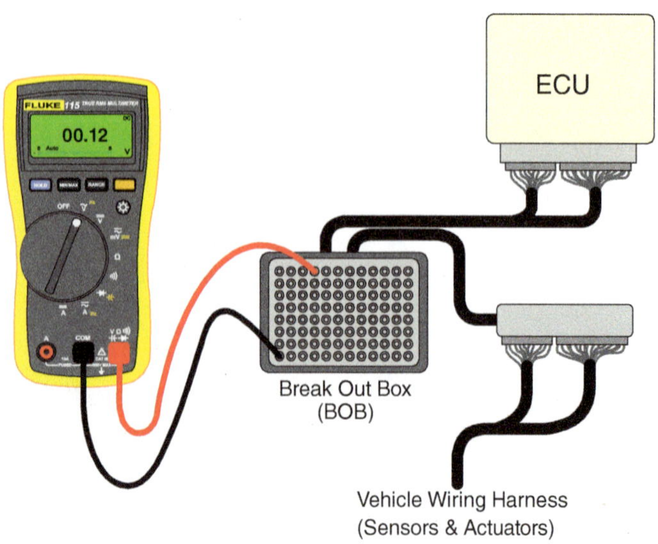

FIGURE 19-41 Performing pinpoint electrical tests to isolate a fault is best accomplished using a break-out box.

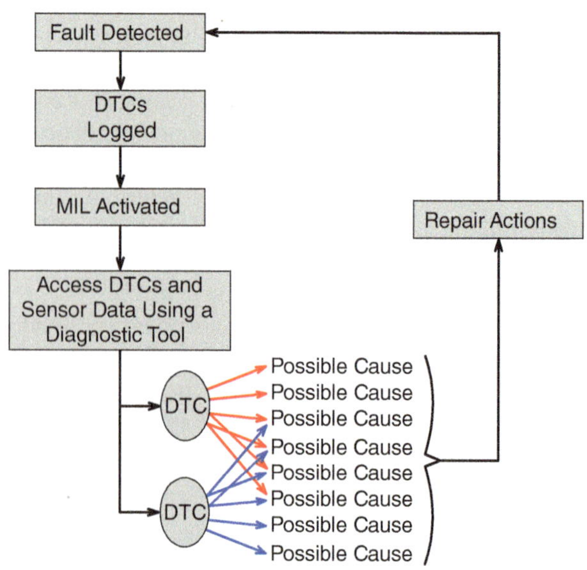

FIGURE 19-42 Fault codes can have multiple root causes. Always consult OEM service literature to establish code priorities and follow diagnostic trees.

IUMPRs have a numerator that increments by a factor of one each time the monitor runs and makes a judgment to either pass or fail a system. The denominator only increments each time monitor-enabling conditions are met and the monitor could have judged the system. If a monitor executes each time monitor-enabling conditions are met, the performance ratio equals a perfect 1. Monitors taking several drive cycles to run are not considered robust and may potentially miss identifying emission faults.

Tracking the time during which the emissions-increasing auxiliary emission-control device (EI-AECD) operates is also tracked. An auxiliary emission-control device (AECD) reduces the effectiveness of the emission-control system under conditions which may reasonably be expected to be encountered in normal vehicle operation and use, and the need for the AECD must be justified to the regulator in terms of balancing the increase in emissions against protecting the vehicle and occupants against damage or accident. EGR operation is often minimized or even shut off during cold weather because manufactures claim that condensation of exhaust water vapor inside the engine at low temperatures can harm the engine. EGR coolers can be plugged and fouled by HCs too. Any EI-AECD's operation is recorded so that its operation can be audited. Currently, use of IUMPRs and EI-AECDs is tracked only by the EPA (US) and CARB.

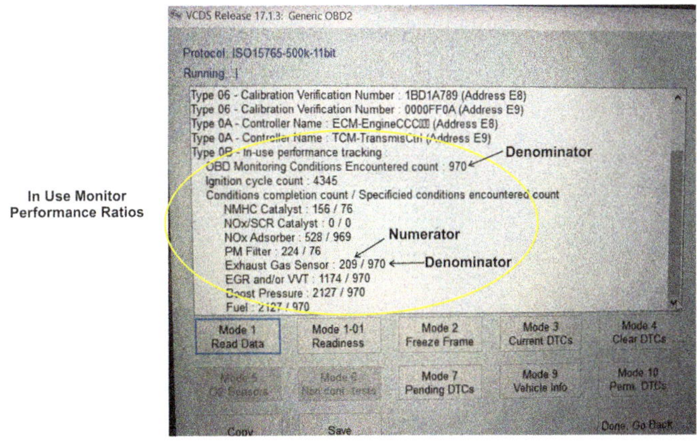

FIGURE 19-43 IUMPRs are used to track how well the monitors are working. Technicians can use IUMPRs to track whether a monitor has made a judgment either to pass or fail a system by observing the ratio. Note that either a pass or a fail judgment will increment the numerator.

Inspection Maintenance Data

The OBD system must be configured to be used effectively by IM programs to prevent emission systems from fraudulent testing and tampering. Vehicle identification numbers (VINs), calibration identification numbers (CAL IDs), and calibration verification numbers (CVNs) are used by the OBD system for this purpose (**FIGURE 19-44**). The CAL ID, which is a reference number for the software version used by the engine control system, is unique to each make, model, and engine. To ensure that the software is not changed with an unauthorized version, the CVN is required to change any time the software calibration is reflashed or altered in any way.

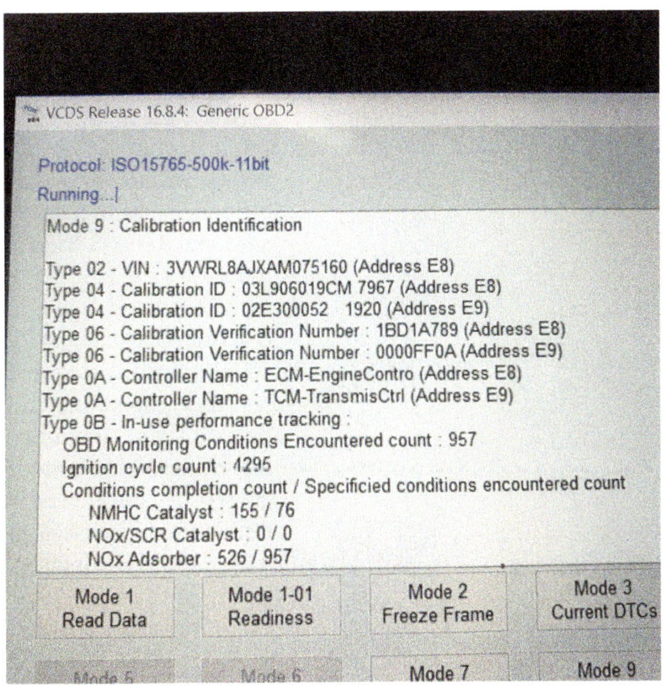

FIGURE 19-44 The CAL ID and CVN will change any time new software is loaded or flashed into the ECM.

Three other OBD elements are used to identify attempts to clear codes before an OBD inspection without making appropriate repairs. Monitor readiness status, the display of a permanent fault code, and (to a lesser degree) displays of real-time engine data. Readiness codes were covered in a previous section. OBD inspections require no DTCs to be active and all monitors in a ready state. Permanent fault codes are a record of DTCs cleared but not validated by monitor operation: A fault was severe enough to produce a MIL-on DTC at some point, but the MIL light is switched off after a code has been cleared. A permanent fault code clears from stored memory only after the monitor has successfully run one or more times and the fault could no longer be not detected.

Diagnostic Services and Electronic Service Tools

19-05 Identify and outline the requirements and capabilities of electronic service tools used to maintain OBD-II compliant systems.

Since OBD-II regulations were introduced, they have removed obstacles that impeded emission repair, which has allowed generic electronic service tools such as scanners to become indispensable and grow in diagnostic capabilities. Accelerating the growth is a microcontroller Arduino chip called the ELM327. This inexpensive OBD interface device translates the digital hexadecimal network serial data communication protocols used by OBD systems into digital information used by simple software found on cell phones and computers. Integrating a Wi-Fi and Bluetooth radio transceiver, the ELM327 series microcontroller enables anyone with an Android or iOS device to read and clear fault codes, as well as access a tremendous volume of other OBD data (**FIGURE 19-45**).

Two factors in OBD regulations driving the growth is the mandated use of generic SAE DTCs and accompanying emission data, plus a standardized OBD format for diagnostic services. OBD uses ten diagnostic services, more commonly known as OBD modes (**TABLE 19-5**).

Modes are diagnostic utilities that technicians can use to access different types of data. Electronic service tools may break down the diagnostic services into simple menus, but behind the menus are operating modes that organize the interaction of service tools with the OBD system.

Not all modes are supported by diesel engines. Oxygen sensor data are not universal on diesel engines, like it is on spark-ignition engines, and so is not always supported. The cost of generic scan tools goes up in relation to its capabilities to access each diagnostic service.

Additional Scan Tool Functions

Scan tools have eight features that are capable of more than reading fault codes and data streams:

1. Bidirectional control. This control in mode 8 enables the tools to command diagnostic tests and cycle output devices to evaluate their functionality. When desktop computers, tablets, or laptops are used, a communication adapter is needed to translate hexadecimal binary code transmitted over the network to language easily interpreted by laptops and other handheld devices (**FIGURE 19-46** and **FIGURE 19-47**).
2. Graphical display. Scan tools can display real-time engine parameters or recorded data in a graphing format. Visual graphing is important to observe behavior of gas sensors, position control behavior and intermittent electrical glitches from sensors. Graphs distinguish problems more easily than a data stream.
3. Help menu/trouble code library. Scan tools can guide a technician through certain diagnostic procedures with prompts or may contain a built-in library of all the SAE generic trouble codes.

FIGURE 19-45 ELM327 and others in the series are Arduino microcontrollers that interface with the OBD-II DLC to provide bidirectional communication to cell phones, tablets, and laptops over Bluetooth and Wi-Fi connections.

TABLE 19-5 OBD Diagnostic Services

Mode	Description	Service Items
1	Display current data	• Status of system monitoring tests • Number of DTCs • Distance traveled while MIL is activated • Number of warm-ups since DTCs cleared • Distance traveled since DTCs cleared • Time run by the engine while MIL is activated • Time since DTCs cleared
2	Display freeze-frame data	• DTCs for freeze-frame data • Freeze-frame data
3	Display DTCs	• Confirmed DTCs
4	Clear DTCs	• Clears DTCs and resets all monitors to not-ready or not-run
5	Test results, oxygen sensor monitoring (non-CAN vehicles only)	• Oxygen sensor test data
6	Diagnostic test results and other component/system monitoring (test results, oxygen sensor monitoring for CAN only)	• Specific onboard monitoring test results
7	Display pending DTCs (detected during current or last driving cycle)	• Pending DTCs
8	Control operation of onboard component/system	Enables the scan tool or software to perform onboard activation tests. These are interactive diagnostic tests such as cylinder cutout or balance tests; they command output actuators and generally test and control the operation of vehicle components.
9	Request vehicle information	Enables the scan tool to request vehicle-specific information: • VINs • CAL IDs • CVNs
1-	Permanent DTCs (cleared DTCs)	Retains record of DTCs cleared but not validated by monitor operation. These codes clear only after the monitor has successfully run and not detected the fault.

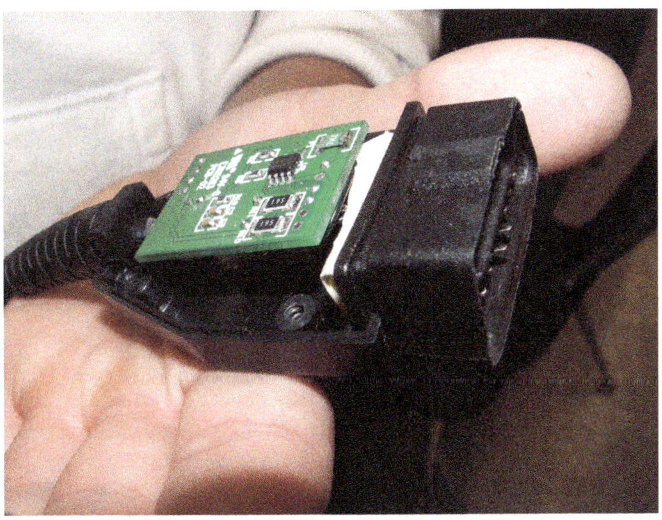

FIGURE 19-46 This aftermarket DLC cable contains a built-in communication adapter enabling it to communicate with an off-the-shelf laptop or computer by using aftermarket or OEM software.

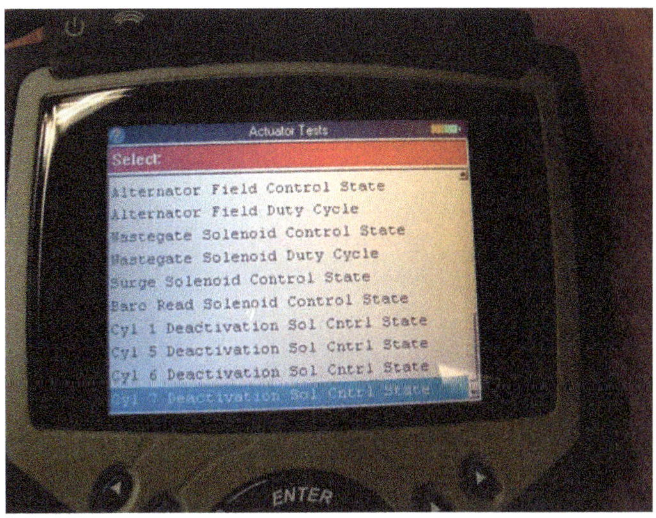

FIGURE 19-47 A display of menu items within mode 8 allowing the technician to cycle output devices on and off to check their operation.

4. Printer/computer output. Scan tools connect to a printer or computer and print or display information from the vehicle. Files can be uploaded to the Internet to media sites or OEM technical support centers.
5. Record/playback or snapshot mode. Scan tools can record a block of real-time engine data and replay that information in order to find the root cause of a malfunction.
6. Scopes and meters. Scan tools can operate as a multimeter (measuring voltage, resistance, amperage, etc.) or an oscilloscope.
7. Reprogramming the vehicle ECM. Scan tools can reprogram a vehicle's electronic modules off-board or onboard, specifically the powertrain (ECM). Reprogramming is also called flashing. Specialized "pass-thru" programmers are now required to update ECM calibrations. Since 2004, SAE standard J2534 has enabled unique network interface communication devices to upload new software. Its essential feature is an ability to shut-off or quiet other devices on the network to prevent signal interference from causing programming errors and miscommunication with the engine, transmission, or another vehicle ECM requiring new software. A subscription service is available from OEMs as part of OBD-II right-to-repair regulations.
8. Adapting engine and other emission-related components. Technicians often encounter fault codes where a component has reached its limit of adaptation or a new component needs to be introduced or adapted to an engine. This situation is the result of an improvement to how engine control uses calibration data to learn characteristics of output devices and actuators. In the last decade, model predictive control (MPC) has become widely used as a powertrain control strategy that can help engines achieve cleaner emissions, improve performance, reduce fuel consumption, correct for engine wear and deterioration, and even accommodate emission-related faults.

Model Predictive Control and Fault Accommodation

Rather than use traditional lookup tables developed in a lab or field to control output devices based on closed-loop control, MPC uses advanced algorithms that generate virtual lookup tables for control loops while the vehicle is in operation. Tables are not stored in the ECM; they are virtual or generated on the fly. These tables will appear and disappear as needed. Virtual lookup tables generated by MPC algorithms are based on computer models of how an engine and other powertrain components should optimally interact.

MPC can make something appear real that isn't. For example, MPC operation is like watching videos of engine simulators in which a 3D representation of an engine rotates and responds to inputs, which in turn change the engine speed, injector action, valve timing, turbocharger, and EGR position. When models generate lookup tables to provide the maps to operate actuators, the entire system can respond faster and more precisely than they could when using fixed calibration tables for closed-loop feedback controls. The virtual lookup tables are also capable of rapidly changing to adapt engine operation in almost an infinite number of ways (**FIGURE 19-48**).

High altitudes, cold weather, steady-state on-highway operation, and stop-and-go city driving all present difficult challenges for calibration engineers to experimentally generate lookup tables to operate actuators in a way that best trades off low emissions with performance and good fuel economy. Engine wear and deterioration can further hamper the development of calibration files that uses fixed lookup tables stored in the ECM. MPC will instead use a mathematical formula to generate a lookup table unique to the driving conditions and can adapt to the subtlest of environmental and mechanical changes. MPC will typically rely on a few fast-responding sensors to sense engine load and predict what the next best move of a particular actuator should be by updating a lookup table with new references points that the control module will use to operate output devices like injectors, turbochargers and EGR valves. Much less ECM processing time is required when using MPC, and depending on the robustness of the algorithm, using MPC has better results for glitch-free drivability and superior fuel economy, performance, and emission control.

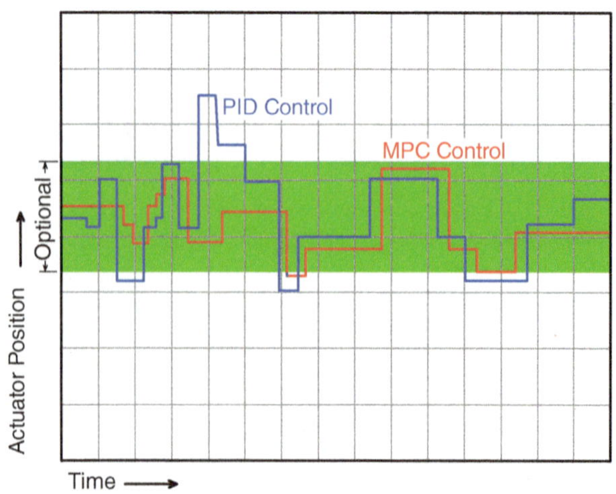

FIGURE 19-48 MPC, which generates virtual lookup tables, can respond faster and more accurately to changes in operating conditions than conventional proportional-integral-derivative (PID) control and its fixed lookup tables.

When MPC is used by some OEMs, each component will be assigned a particular numerical value that fits into or introduces it to other components in the system. Performing an adaptation when installing new parts updates the engine model, and component adaptation will accommodate manufacturing tolerances and variations between the responses of the same components to electrical signals. The MPC algorithm makes adjustments to the initial adaptation settings by using feedback from sensor data and other algorithms that predict component deterioration. Self-calibration can also take place during each drive cycle since output actuators are often cycled to learn new limits of travel and correct for wear.

One other interesting feature of MPC is its ability to accommodate emission faults. This feature is sometimes referred to as fault healing. If a fault occurs that can increase engine emissions, the algorithm can make adjustments to the lookup tables to compensate for a fault. For example, if an EGR valve is stuck open or closed, the mathematical formula for generating lookup tables can adjust to a new model of engine control where the valve is held in an open or closed position. This involves simply changing a variable or mathematical coefficient in the model representing the valve. Now the new lookup tables generated by MPC algorithms will include settings that optimally correct for a model of engine control without an EGR valve or a valve held open. Lookup tables commanding injection timing, rate shaping, transmission shift schedules, and other NO_x-reducing strategies are likely to change to accommodate the fault.

▶ TECHNICIAN TIP

MPC can compensate for component wear and deterioration by generating a new set of lookup tables. This will keep the MIL off and potentially minimize the likelihood of a related repair. However, when fault accommodation takes place, the vehicle will lose performance and increase fuel consumption. Sometimes no MIL will be generated to identify the problem. It's important to study engine data during a road test to diagnose MPC fault healing strategies, which diminish power output and increase fuel consumption. Compare data values with other vehicles in known good condition to identify the discrepancy and diagnose which component or system may be changing the MPC of virtual lookup tables.

▶ Wrap-Up

Ready for Review

- Emissions standards for diesel engines set beginning in the late 1970s have required a level of precision for engine control that is possible only through the extensive use of electronics. The OBD system was originally developed by manufacturers to diagnose faults with electronic controls.
- Since 1997, light-duty emissions legislation has required OBD-II compliant systems to monitor the operation of emission-control systems and alert the vehicle operator to any emission-related faults producing excessive emissions above threshold standards.
- Electronic systems do not have many moving parts to wear out, but the systems can be complex. When something goes wrong with a component or circuit, identifying the problem without some built-in self-diagnostic capabilities can be extremely time-consuming and difficult.
- With built-in electronic and self-monitoring functions, electrical systems possess the capability to check the operation of circuits and electrical devices, judge the operation of major emission systems for potential emission-increasing fault conditions, and identify problems as they occur.
- Off-board diagnostics is the process of fault isolation to determine the location of a fault. An OBD-II DTC is the starting point for off-board diagnosis.
- Technical specifications for OBD-II compliant systems are developed and regularly updated by the SAE. The EPA and CARB adopt SAE recommendations into OBD legislation.
- Federal test protocols (FTPs) are maximum allowable limits of vehicle emissions. OBD-II monitoring thresholds required to detect excess emissions are based on multipliers of the FTP standards. Diesel FTP threshold multipliers are between one and a half and five times FTP limits.
- Manufacturers are allowed some discretion in interpreting and implementing OBD-II regulations. But before a manufacturer can sell a vehicle, it must provide evidence of conformity to OBD-II regulations to receive a certificate of conformity.
- The performance of OBD-II compliant systems is evaluated on a chassis dynamometer by using a FTP drive cycle to measure emissions and simulate real-world operation. Dynamometer testing is part of the evidence submitted to apply for a certificate of conformity.
- To detect conditions that can increase emissions levels beyond FTP thresholds, the OBD system monitors individual components and major engine emission-control systems. Monitors used by the diesel OBD system can be divided into several general types: threshold, functionality, rationality or plausibility, electrical circuit continuity, and out-of-range monitors.
- The diagnostic manager is software that uses advanced algorithms to make decisions for the OBD-II compliant system. It sets fault codes based on identifying potential emission-increasing conditions. The OBD manager coordinates the OBD-II compliant system operation and controls the operation of the malfunction indicator lamp (MIL).
- Type A DTCs are the most critical emission-related faults and will illuminate the MIL after only one occurrence.

- Type B DTCs are emission-causing faults that are less serious than Type A faults and must occur at least once on two consecutive trips before the MIL will illuminate.
- Type C DTCs are non-emission-related codes, also called enhanced codes. Enhanced codes also cover non-emission-related failures that occur outside the engine control system.
- The diagnostics used to monitor diesel engines can be broken down into separate, specific hardware systems that are responsible for some specialized aspect of engine operation and emission control. There are up to 14 different types of major diagnostic monitors used by diesel engines.
- Emissions systems may be monitored continuously or noncontinuously. Intrusive monitors manipulate engine operation to test the response of emissions systems or devices to evaluate whether they are operating correctly. Non-intrusive monitors measure system operation and compare values against expected criteria for emission-increasing conditions. DTC fault criteria are often experimentally derived during engine development.
- Noncontinuous monitors require enabling conditions or criteria to be met before a monitor can run. The conditions are established to ensure that the monitor correctly evaluates the emissions system or component.
- The MIL will immediately illuminate whenever a serious Type A DTC is detected. Other less serious emissions conditions require confirmation of a pending DTC after two trips before the MIL will illuminate.
- Freeze-frame data are a record of engine and vehicle information taken the moment a DTC has been confirmed. All MIL-on DTCs will have a freeze frame associated with the code.
- A monitor status of ready or pass is set after a monitor has run and determined that there are no faults associated with the monitored system and monitored components. A readiness code switches to a not-ready or monitor fail status only once codes have been erased or the battery has been disconnected.
- The comprehensive component monitor (CCM) continuously monitors electrical circuits operating the powertrain components. Any failure can illuminate the MIL because the CCM is not tied to emissions thresholds.
- Because engine misfires can cause hydrocarbon (HC) and carbon dioxide (CO) emissions, which can also overheat the exhaust catalysts, the OBD system must evaluate engine operation to determine whether misfires are occurring.
- Fuel, EGR, aftertreatment, and other major system monitors are required to detect faults only after they exceed the FTP multiplier threshold.
- FTP multiplier thresholds required to detect emissions vary depending on the monitor and the emission it is required to identify. Multiplier threshold are between one and a half and five times for diesels.
- Because defective thermostats and engine temperature sensors will prevent other emissions system monitors from running, the OBD system monitors the cooling system to ensure that it properly warms up to operating temperatures and sets DTCs if it does not warm up fast enough for vehicle operating conditions.
- When DTCs are set, electronic control modules (ECMs) log DTCs and freeze-frame data in the system memory. Codes and data can be read through an instrument display cluster or retrieved by an OBD scanner, tablet, or computer connected to the vehicle's DLC with an interface device.
- OBD-II regulations for diesel engines require that the crankcase emissions system control connections be monitored for disconnection.
- The DLC connector is a connection point for several onboard networks. CAN protocols are used by the OBD system to share and retrieve information from control modules.
- Unified Diagnostic Service (UDS) is another standardized network protocol that has diagnostic capabilities paralleling OBD-II. Manufacturers will build vehicle components with small microcontrollers capable of UDS communication. UDS communication is often used to adapt or introduce components into engine control systems.
- SAE J2284/ISO 15765-4 (CAN) network communication protocols are the standard controller area network bus (CAN-bus) set of rules, which permits any ECM to transmit a message over the network when the data bus is idle or not transmitting other information.
- MPC of engines and powertrains can change the algorithm that generates a virtual lookup table to accommodate faults.

Key Terms

active fault A fault that is currently taking place and uninterrupted.

datalink connector (DLC) The connection point for electronic service tools used to access fault codes and other information provided by chassis electronic control modules (ECMs).

diagnostic trouble codes (DTCs) Codes stored by the electronic control module (ECM) when emission-related faults are identified by the OBD system.

Drive cycle Driving the vehicle through enabling conditions that run all monitors.

OBD manager An OBD software function that identifies fault codes and ensures that emissions systems are operating correctly.

off-board diagnostics Procedures to isolate a fault based on fault code information, including retrieving fault code information, monitoring system operation, performing actuator tests and pinpoint electrical tests, and inspecting components.

onboard diagnostics (OBD) Self-diagnostic capabilities of electronic control modules (ECMs) that allow them to evaluate the performance of major emission systems and related components.

out-of-range monitoring Validating sensor data to verify that a system is operating within an expected range for a given operating condition.

telematics A branch of information technology that uses specialized applications to transmit information to and from a vehicle over long distances.

Review Questions

1. The preliminary standard developed by CARB and used before OBD-II was called ______________.
   a. engine manufacturer's diagnostic (EMD) systems
   b. OBD-II (On-Board Diagnostics, Second Design)
   c. OBDI (On-Board Diagnostics)
   d. CAN (controller area network)
2. How many wires will transmit and receive signal data for the CAN engine/powertrain network on the diagnostic datalink connector (DLC) for an OBD-II compliant vehicle?
   a. Only 1.
   b. Two.
   c. Four.
   d. Five.
3. Consider a 2018 OBD-II-compliant and Tier 3–compliant diesel engine. What factor or threshold of excess emissions is typically required to illuminate the MIL when a fault in the diesel particulate filter (DPF) is detected?
   a. One and a half times FTP emissions standards.
   b. Two times FTP emissions standards.
   c. Three times FTP emissions standards.
   d. There is no FTP multiplier for Tier 3–compliant diesel engines.
4. Which OBD-II monitor tracks electrical circuits operating engine components that can fail and cause measurable emissions?
   a. functional monitor.
   b. continuous monitor.
   c. comprehensive component monitor (CCM).
   d. threshold monitor.
5. If a coolant temperature sensor shorted to ground, which emissions system monitor would first detect the fault?
   a. cooling system monitor.
   b. rationality monitor.
   c. comprehensive component monitor (CCM).
   d. network communication monitor.
6. An enhanced OBD-II code will be indicated by which position in the SAE code structure?
   a. last digit.
   b. last two digits.
   c. first two digits.
   d. second digit.
7. Before returning a vehicle to the customer after repairing the cause of the MIL being illuminated, the technicians should ______________.
   a. erase all codes and road test the vehicle until a readiness code is set
   b. rescan the vehicle for codes after road testing the vehicle
   c. run the emission monitor associated with the fault
   d. check for only active codes after road testing the vehicle
8. Which emissions system monitor will detect incorrect injection timing?
   a. Injection timing monitor.
   b. Comprehensive component monitor (CCM).
   c. Rationality monitor.
   d. Fuel system monitor.
9. Which of the following changes will take place when a manufacturer updates a software calibration and a technician uploads or flashes the vehicle with the new calibration?
   a. All monitors will be set to ready.
   b. The calibration identification number will change.
   c. The calibration identification (CAL ID) and calibration verification numbers (CVNs) will both change.
   d. The CAL ID number will change and CVNs will remain the same.
10. Which of the following is a critical enabling condition for a 2018 misfire monitor?
    a. The engine must be at idle.
    b. The engine cannot have a mechanical defect.
    c. The fuel tank must be above ¼ tank.
    d. The engine has to operate above idle speed.

ASE Technician A/Technician B–Style Questions

1. Two technicians are discussing the purpose of OBD-II compliant systems for automotive diesel engines. Technician A says that an OBD-II compliant system is designed to identify or self-diagnose emission-related faults and report fault codes. Technician B says the OBD-II rules are set up to enable drivers to have emissions problems repaired quickly and easily. Who is correct?
   a. Technician A
   b. Technician B
   c. Both A and B
   d. Neither A nor B
2. Technician A says that any DTC will immediately illuminate the MIL in the driver's view the moment one is detected. Technician B says that all OBD-II DTCs will take several trips to confirm that a fault is active before switching on the MIL. Who is correct?
   a. Technician A
   b. Technician B
   c. Both A and B
   d. Neither A nor B
3. Technician A says that a pending fault will not have a DTC that can be retrieved by a scanner or electronic service tool. Technician B says a pending fault will never have freeze-frame data associated with it. Who is correct?
   a. Technician A
   b. Technician B
   c. Both A and B
   d. Neither A nor B
4. Technician A says that some emission-related DTCs will cause engine power derate to encourage a driver to have

the emissions system repaired. Technician B says that most diesel engine OBD DTCs will derate engine power. Who is correct?

a. Technician A
b. Technician B
c. Both A and B
d. Neither A nor B

5. Technician A says that the OBD system on a diesel is continuously judging all major emissions systems for faults. Technician B says that some emissions monitors do not run every drive cycle and take time to correctly judge emissions system operation. Who is correct?
   a. Technician A
   b. Technician B
   c. Both A and B
   d. Neither A nor B
6. Technician A says that if all the sensors and components of the aftertreatment system are removed, the OBD-II system will not be able to detect faults in the absence of electrical signals. Technician B says that the comprehensive component monitor (CCM) will set emission-related fault codes. Who is correct?
   a. Technician A
   b. Technician B
   c. Both A and B
   d. Neither A nor B
7. Technician A says that only the fuel system monitor will set DTCs for a misfiring injector. Technician says that the fuel system monitor and misfire monitor will set multiple DTCs. Who is correct?
   a. Technician A
   b. Technician B
   c. Both A and B
   d. Neither A nor B
8. While two technicians are examining the many pairs of wires connected to an OBD-II DLC, Technician A says that all the wires are needed to retrieve fault does from the OBD system. Technician B says that only a pair of CAN wires are necessary to retrieve OBD-II fault codes. Who is correct?
   a. Technician A
   b. Technician B
   c. Both A and B
   d. Neither A nor B
9. Technician A says that anytime the tailpipe NO_x sensor reads more than a few parts per million of NO_x, DTCs will switch the MIL on. Technician B says that model predictive control (MPC) will always make adjustments to the engine to correct the fault. Who is correct?
   a. Technician A
   b. Technician B
   c. Both A and B
   d. Neither A nor B
10. While two technicians are discussing what to do to prepare a vehicle to pass an OBD-II emissions system inspection with an noncontinuous monitor DTC that causes the MIL to illuminate, Technician A says that the problem would need to be repaired and that the monitors all need to successfully run to pass. Technician B says that just the code could be erased before sending the customer to the inspection. The fault could be fixed after the vehicle has passed the OBD-II inspection test. Who is correct?
    a. Technician A
    b. Technician B
    c. Both A and B
    d. Neither A nor B

APPENDIX A

TEST SPECIFICATIONS AND TASK LIST—LIGHT VEHICLE DIESEL ENGINES (TEST A9)

ASE Task List	Chapter
A. General Diagnosis	
1. Verify the complaint, and road/dyno test vehicle; review driver/customer concerns/expectations and vehicle service history (if available); determine further diagnosis.	1
2. Record vehicle identification number (VIN). Identify engine model, calibration and serial numbers to research applicable vehicle and service information, service precautions, and technical service bulletins; determine needed actions.	1
3. Perform scan tool check and visual inspection for physical damage and missing, modified, or tampered components; determine needed actions.	1 & 19
4. Check and record electronic diagnostic codes, freeze frame and/or operational data; monitor scan tool data; determine further diagnosis.	1 & 19
5. Clear diagnostic trouble codes (DTCs) and verify the repair.	1 & 19
6. Inspect engine assembly and compartment for fuel, oil, coolant, exhaust, or other leaks; determine needed repairs.	1, ch 10–19 inclusive
7. Inspect engine compartment wiring harness, connectors, seals, and locks; check for proper routing and condition; determine needed repairs.	1, ch 10–19 inclusive
8. Listen for and isolate engine noises; determine needed repairs.	1, 3–18
9. Isolate and diagnose engine-related vibration problems; determine needed actions.	4, 5
10. Check engine exhaust for abnormal odor and/or smoke color and volume; determine further diagnosis.	3, 17
11. Check fuel for contamination, quantity, quality, and consumption; determine needed actions.	8
12. Perform crankcase pressure test; determine further diagnosis.	4, 5
13. Diagnose surging, rough operation, misfiring, low power, slow deceleration, slow acceleration, and shutdown problems; determine needed actions.	8, 9, 10
14. Check cooling system for freeze point, level, contamination, condition, temperature, pressure, circulation, and fan operation; determine needed repairs.	7
15. Check lubrication system for contamination, oil level, temperature, pressure, filtration, and oil consumption; take oil sample and obtain oil analysis if needed; determine needed repairs.	7
16. Diagnose no-cranking, cranks but fails to start, hard starting, and starts but does not continue to run problems; determine needed actions.	8, 10
17. Diagnose engine problems caused by battery condition, connections, or excessive key-off battery drain; determine needed repairs.	
18. Diagnose engine problems resulting from an electrical undercharge, overcharge, or a no-charge condition; determine needed action.	
B. Cylinder Head and Valve Train Diagnosis and Repair	
1. Remove, inspect, disassemble, and clean cylinder head assembly(s).	6
2. Inspect threaded holes, studs, and bolts for serviceability; service/replace as needed.	6
3. Measure cylinder head thickness, and check mating surfaces for flatness, corrosion, warpage, and surface finish; inspect for cracks/damage; check condition of passages; inspect core and gallery plugs; determine serviceability and needed repairs.	6
4. Inspect valves, guides, seats, springs, retainers, rotators, locks, and seals; determine serviceability and needed repairs.	6
5. Inspect and/or replace injector sleeves, glow plug sleeves, and seals; pressure test to verify repair (if applicable); measure injector tip, nozzle, or prechamber protrusion where specified by manufacturer.	11

(continued)

ASE Task List	Chapter
6. Inspect and/or replace valve bridges (crossheads) and guides; adjust bridges (crossheads) if applicable.	6
7. Reassemble, check, and determine required cylinder head gasket thickness; install cylinder head assembly and gasket as specified by the manufacturer.	6
8. Inspect pushrods, rocker arms, rocker arm shafts, electronic components, wiring harness, seals; repair/replace as needed.	6
9. Inspect, install, and adjust cam followers, lash adjusters and retainers; adjust valve clearance if applicable.	6
10. Inspect, measure, and replace/reinstall overhead camshaft and bearings; measure and adjust end play.	6
11. Inspect and time drive gear train components (includes gear, chain, and belt systems).	4, 6
C. Engine Block Diagnosis and Repair	
1. Remove, inspect, service, and install pans, covers, ventilation systems, gaskets, seals, and wear rings.	4
2. Disassemble, clean, and inspect engine block for cracks; check mating surfaces and related components for damage or warpage and surface finish; check deck height; check condition of passages, core, and gallery plugs; inspect threaded holes, studs, dowel pins, and bolts for serviceability; service/replace as needed.	4
3. Inspect and measure cylinder walls for wear and damage; determine needed service.	4, 5
4. Inspect in-block camshaft bearings for wear and damage; replace as needed.	4
5. Inspect, measure, and replace/reinstall in-block camshaft; measure and correct end play; inspect, replace/reinstall, and adjust cam followers (if applicable).	4, 6
6. Clean and inspect crankshaft and journals for surface finish, cracks, and damage; check condition of oil passages; check passage plugs; measure journal diameters; check mounting surfaces; determine needed service.	4
7. Determine the proper select-fit components such as pistons, connecting rod, and main bearings.	5
8. Inspect and replace main bearings; check cap fit and bearing clearances; check and correct crankshaft end play.	4
9. Inspect, replace, verify, and adjust the drive gear train components (includes gear, chain, and belt systems).	4, 6
10. Inspect, measure, or replace pistons, pins, and retainers.	4
11. Measure piston-to-cylinder wall clearance.	4
12. Identify piston, connecting rod bearing, and main bearing wear patterns that indicate connecting rod and crankshaft alignment or bearing bore problems; check bearing bore and bushing condition; determine needed repairs.	4, 5
13. Check ring-to-groove fit and end gaps; install rings on pistons; assemble pistons and connecting rods and install in block; check piston height/protrusion; check liner height/protrusion (if applicable); replace rod bearings and check clearances; check condition, position, and clearance of piston cooling jets (nozzles).	4, 5
14. Inspect crankshaft vibration damper; determine needed repairs.	4
15. Inspect flywheel/flexplate and/or dual-mass flywheel (including ring gear) and mounting surfaces for cracks, wear, and runout; determine needed repairs.	4
D. Lubrication and Cooling Systems Diagnosis and Repair	
1. Verify base engine oil pressure and check operation of pressure sensor/switch and pressure gauge; verify engine oil temperature and check operation of temperature sensor.	7
2. Inspect, measure, repair/replace oil pump, housing, drives, pipes, and screens; check drive gear clearance.	7
3. Inspect, repair/replace oil pressure regulator assembly including: housing, bore, spring, regulator valve(s), oil filter by-pass valve(s), and anti-drain back valve.	7
4. Inspect, clean, test, and reinstall/replace oil cooler, by-pass valve, lines, and hoses.	7
5. Inspect turbocharger lubrication and cooling systems; repair/replace as needed.	15
6. Change engine oil and filters using proper type, viscosity, and rating per manufacturer specifications.	7
7. Inspect and reinstall/replace pulleys, tensioners, and drive belts; adjust drive belts and check alignment.	
8. Verify coolant temperature; check operation of temperature and level sensors, switches, and temperature gauge.	7
9. Inspect and replace thermostat(s), by-pass(es), housing(s), and seal(s).	7
10. Flush and refill cooling system; following manufacturer's specification, add proper coolant type; bleed air from system.	7
11. Inspect and replace water pump(s), housing(s), hoses, and idler pulley(s) or drive gear.	7

ASE Task List	Chapter
12. Inspect radiator(s), pressure cap(s), and tank(s); pressure test cooling system and radiator cap(s); determine needed repairs.	7
13. Inspect and repair/replace cooling fan, fan hub, fan clutch, controls, and shroud(s).	7
E. Air Induction and Exhaust Systems Diagnosis and Repair	
1. Inspect and service/replace air induction piping, air cleaner, and element; determine needed actions.	14
2. Perform intake manifold pressure tests; inspect, test, clean, and/or replace charge air cooler and piping system; determine needed actions.	14
3. Inspect test, and replace turbocharger(s) (including variable ratio/geometry VGT), pneumatic, hydraulic, vacuum, and electronic controls and actuators; inspect, test, and replace wastegate and wastegate controls.	16
4. Inspect, test, and replace intake manifold(s), variable intake manifold(s), gaskets, actuators, temperature and pressure sensors, and connections.	14, 15, 16
5. Perform exhaust back pressure and temperature tests; determine needed actions.	15, 16
6. Inspect and repair/replace exhaust manifold(s), gaskets, piping, mufflers, and mounting hardware.	15, 16
7. Inspect, test, and repair/replace preheater/inlet air heater and/or glow plug system and controls.	14
8. Inspect, test, and replace exhaust aftertreatment system components and controls, including diesel oxidation catalyst (DOC), selective catalyst reduction (SCR), diesel exhaust fluid (DEF), diesel particulate filter (DPF); check regeneration system operation.	18
9. Inspect, test, service, and replace EGR system components including EGR valve(s), EGR cooler by-pass valve(s), EGR cooler(s), piping, electronic sensors, actuators, controls, and wiring.	17
10. Inspect, test, and replace airflow control (throttle) valve(s) and controls.	14, 17
11. Inspect, test, and replace crankcase ventilation system components, including sensors, filters, valves, and piping.	14
F. Fuel System Diagnosis and Repair	
1. Inspect, clean, test, and repair/replace fuel system tanks, vents, caps, mounts, valves, single/dual supply and return lines, and fittings.	8
2. Inspect, clean, test, and repair/replace fuel transfer and/or supply pump, sensors, strainers, fuel/water separators/indicators, filters, heaters, coolers, ECM cooling plates (if applicable), and mounting hardware.	8
3. Check fuel system for air; determine needed repairs; prime and bleed fuel system; check and repair/replace primer pump.	8
4. Inspect, test, and repair/replace low fuel pressure regulator supply and return systems, including low pressure switches.	8
5. Inspect and reinstall/replace high-pressure injection lines, fittings, transfer tubes, seals, and mounting hardware.	8, 9
6. Inspect, adjust, and repair/replace electronic throttle and PTO control devices, circuits, and sensors.	9
7. Perform on-engine inspections, tests, and replace high pressure common rail fuel system components and electronic controls.	13
8. Perform on-engine inspections and tests; replace hydraulic electronic unit injector(s) (HEUI) components and electronic controls.	12
9. Perform on-engine inspections and tests; replace pump-line-nozzle fuel system (PLN-E) components and electronic controls.	9, 10
10. Perform on-engine inspections and tests; replace electronic unit injector(s) (EUI) components and electronic controls.	9, 10
11. Inspect and replace electrical connector terminals, pins, harnesses, seals, and locks.	
12. Connect diagnostic scan tool to vehicle/engine; access, verify and update software calibration settings, injector calibration codes; perform control module re learn procedures as needed.	13, 19
13. Use a diagnostic scan tool to inspect and test electronic engine control system, sensors, actuators, electronic control modules, and circuits; determine further diagnosis.	19
14. Measure and interpret voltage, voltage drop, amperage, and resistance readings using a digital multimeter (DMM) or appropriate test equipment.	
15. Diagnose engine problems resulting from failures of interrelated systems (for example, cruise control, security alarms/theft deterrent, transmission controls, exhaust aftertreatment systems, electronic stability control, or non-OEM installed accessories).	10

GLOSSARY

3 Cs A term used to describe the repair documentation process of 1st documenting the customer concern, 2nd documenting the cause of the problem, and 3rd documenting the correction.

A/R ratio The ratio between the area (A) of either the compressor or turbine inlets and the radius (R) of the housing.

abrasive flow machining (AFM) A process of enhancing the flow characteristics and performance of injector nozzles by smoothing passageways through the injector and enlarging the spray holes by running an abrasive compound through the nozzles.

absolute micron rating A rating of a fuel filter that refers to the largest sized particle that the fuel filter media will allow to pass.

acoustic wave attenuation (AWA) A hydraulic wave-dampening device that minimizes a rhythmic cackling noise at idle caused by high-pressure pump pressure waves synchronizing with injection events.

active fault A fault that is currently taking place and uninterrupted.

active regeneration A process in which soot is burned inside the DPF through supplementally heating the DPF filter.

AdBlue A marketing name used by Daimler - Benz for a liquid containing urea. The liquid is used to dose SCR-U aftertreatment systems to break-down NO_x.

additives Chemicals that improve the original properties of fuel or base stock oil.

aeration A condition in which excessive amounts of air or steam bubbles are dissolved in fuel, coolant or oil which can make the fluid elestic or diminish the coolant's effectiveness.

aftermarket A company other than the original manufacturer that produces equipment or provides services.

air inlet restriction gauge A pressure sensor that measures the negative pressure between the air filter and turbocharger.

algorithm A mathematical formula used to solve a problem.

American Petroleum Institute (API) An organization that has developed specifications to defines fuel and engine oil performance standards.

ammonia slip NH_3 that escapes from the SCR catalyst and into the atmosphere.

amplified common rail (ACR) Two-stage amplification of fuel pressure inside a common rail injector that provides improved power output with lower emissions, noise, and fuel consumption and that produces a unique shape to the injection rate profile.

analog signal An electric current that is proportional to a continuously changing variable.

analog-to-digital (AD) conversion The process when an analog waveform is sampled and measured many times a second to generate a digital representation of the waveform.

anodizing A process used to harden aluminum by electrochemically reacting oxygen with aluminum.

API number A measurement of a fuel's density.

aromatic content The portion of fuel composed of a particular type of hydrocarbon (HC) molecule that is difficult to ignite and burn.

articulating piston A two-piece piston design that uses a separate aluminum skirt connected to an alloy steel crown through a piston pin.

ASTM International An organization that establishes today's diesel fuel standards; it was formerly known as the American Society for Testing and Materials (ASTM).

asymmetrical turbocharger A turbocharger with volutes of two different sizes. Asymmetrical turbochargers are used to develop exhaust gas pressure for the EGR system.

atomization The process of breaking up liquid fuel into a fine mist.

auto-ignition temperature The temperature at which a fuel will ignite when heated.

automotive governor A governor that features an accelerator response similar to a throttle plate–controlled, gasoline-fueled engine; it is also known as a limiting speed (LS) governor or a min/max governor. The operator has to change engine fueling through moving the throttle position in response to engine load and speed changes.

back leakage A calibrated leakage of fuel around the nozzle valve to lubricate and cool the valve.

base stock The raw mineral processed from crude oil.

baud rate The rate at which serial data is transmitted.

beginning of injection (BOI) The point when fuel delivery begins and is referenced according to the position of the crankshaft measured in degrees of crankshaft rotation.

biocides A fuel treatment that kills microorganisms.

bit The smallest piece of digital information that is either a 1 or 0.

bleeder screw A component located in the fuel system that allows trapped air to escape from the filter when priming.

BlueTec Daimler-Benz name for its two NO_x-reducing systems used in its European and North American diesel engines.

boost pressure normalization The feature of a turbocharger that maintains intake manifold pressure pressure even as the engine climbs to higher altitudes with low air density.

break-in period The operation of an engine after it is initially assembled or rebuilt when piston ring, cylinder wall, bushing, and bearing surfaces have high initial wear as the moving surfaces conform to each other.

byte A unit of 8 bits.

cam ground piston An elliptically shaped piston that expands to a round, symmetrical shape after it is warmed up.

cam-less diesel A system that electrohydraulically actuated intake and exhaust valves instead of using a camshaft to operate valves.

carbon dioxide (CO_2) An inert gas that is a byproduct of normal combustion.

carbon monoxide (CO) A noxious gas that is a byproduct of abnormal combustion.

cartridge filter A filter element that consists of only filter media unenclosed by a metal container.

cast-in-place sleeve A block construction technique that casts aluminum block material around an iron cylinder sleeve.

catalyst A material that speeds up or slows down chemical reactions without entering the chemical reactions.

cause Part of the 3Cs, documenting the cause of the problem. This documentation will go on the repair order, invoice, and service history.

cavitation Erosion in cylinder block walls, heads, and liner sleeves as a result of the collapse of tiny water vapor bubbles formed when coolant vaporizes on hot cylinder wall surfaces.

cavitation erosion Pinholes produced in cylinder block walls, heads, liner sleeves and other engine parts as a result of the collapse of tiny water vapor bubbles formed after coolant vapor forms on coolant surfaces that are either hot or vibrating at high-frequency.

centrifugal force A force pulling outward on a rotating body.

cetane boosters An additive that is marketed to reduce combustion noise and smoke, improve startability in cold weather, and increase acceleration response; also known as an ignition accelerator.

cetane number (CN) A measure of the ignition quality of fuel; also known as cetane rating or cetane value.

charge air cooler (CAC) The system responsible for removing excess heat from the air charging the cylinders; it is also known as aftercooling and intercooling.

chuffing The sound a VGT turbocharger can produce when actuator response is slow. Chuffing is caused by turbocharger surge.

clearance volume The space remaining in a cylinder when the piston is at top dead center (TDC).

closed crankcase ventilation system A method of removing crankcase blowby gases by recycling crankcase vapors back into the intake manifold.

closed-loop feedback An ECM control strategy for an output actuator device based on direct sensor feedback, which is used to adjust electrical signals supplied by the ECM to the actuator.

cloud point The temperature at which the wax in diesel fuel begins to congeal.

coalesce The process of collecting small droplets together to form larger droplets.

coaxial variable nozzle injector A fourth-generation injector concept from Bosch that contains a two-stage nozzle lift with two rows of spray holes. This design permits either row of spray holes to be opened independently, which optimizes mixture preparation to achieve a more homogeneous distribution of fuel and air in the cylinder.

combustion A chemical reaction between fuel and oxygen that releases heat.

compacted graphite iron (CGI) A material produced from powdered iron alloys, which is squeezed into molds at high pressures and then heated to bond the metal particles together; also known as sintered graphite.

compatibility A property that allows two metals to slide against one another with minimal friction or wear.

compensation A response by the fuel system governor at idle to a uneven power contributions from one or more cylinders. The governor corrects changes in power contribution by changing the quantity of fuel injected into the other cylinders to maintain idle speed.

compression ignition (CI) A type of ignition system that ignites fuel by using the heat derived from compressing air.

compression ratio A comparison between total cylinder volume and clearance volume (the space left in the cylinder when the piston is at top dead center [TDC]).

compression testing A measure of the maximum pressure of engine cylinders when cranking.

compressor housing The housing that encloses the compressor wheel.

compressor wheel A centrifugal-type air pump attached to the turbine wheel. The compressor wheel uses centrifugal force to compress air.

concern Part of the 3Cs, documenting the original concern that the customer came into the shop with. This documentation will go on the repair order, invoice, and service history.

coolant extender An additive package that is used only with ELC, which is added at the midpoint of the coolant's life.

cooled EGR (CEGR) A system that cools EGR exhaust gas by passing it through an engine coolant heat exchanger.

correction Part of the 3Cs, documenting the repair that solved the vehicle fault. This documentation will go on the repair order, invoice, and service history.

crankcase depression regulator (CDR) A device that regulates, or meters, the quantity of blowby emissions back into the engine's intake manifold.

crankcase pressure test A measurement of the amount of cylinder blowby. This test indicates the sealing ability of the piston rings and cylinder walls.

crankshaft The component that converts the reciprocating action of the pistons to a rotational movement and revolves inside the crankcase portion of the engine block.

crankshaft throw The distance between the centerline of the main bearing journal and crankpin journal.

crevice volume The area between the piston crown and cylinder wall above the top compression ring.

cross flow cylinder head A head design with the intake and exhaust manifolds located on opposite sides of an in-line engine to improve engine-breathing characteristics.

crossbolted block A variation of the deep-skirt block that uses additional horizontally placed bolts to connect the crankcase walls of the block to the main bearing caps; also known as a tie-bolted or bolster-bolted block.

crosshatch A cylinder wall finish of fine intersecting lines that are used to retain oil.

cyclonic-style pre-cleaner tubes A component of the air intake system that pre-cleans intake air by spinning dirt out of the airflow to prevent it from entering the filter.

cylinder block The largest structure of an engine, which encloses the cylinders and provides a rigid frame to support attaching the rest of the engine components.

cylinder displacement The volume displaced by the piston as it moves from (TDC) to bottom dead center (BDC).

cylinder glazing A condition that occurs when lubricating oil is first baked and then smoothed into the crosshatch of the cylinder wall. Cylinder glazing prevents the cylinder wall from retaining oil and reduces ring's ability to seal.

cylinder head gasket The component that maintains the seal around the combustion chamber at peak operating temperatures and pressures and keeps air, coolants, and engine oil at their respective peak temperatures and pressures.

cylinder leakage The percentage of gas leakage past the rings. High leakage rates are usually due to worn cylinders and rings.

datalink connector (DLC) The connection point for electronic service tools used to access fault codes and other information provided by chassis electronic control modules (ECMs).

deep-skirt block A block configuration with a crankcase wall that extends well below the crankshaft's centerline.

delivery valve Valves that operate as one-way checks and allow fuel to remain at moderately high pressure in the fuel-injection lines. The valves prevent secondary injections and help reduce injection lag.

DeSOx A sulfur-reducing reaction used in NAC and LNT systems to remove sulfur from catalyst material. Sulfur in diesel fuel tends to contaminate catalysts and reduces their effectiveness unless a DeSOx reaction takes place.

diagnostic trouble codes (DTCs) Codes stored by the electronic control module (ECM) when emission-related faults are identified by the OBD system.

dial bore gauge A precision measuring tool used to measure taper wear in a cylinder.

diaphragm pump A common type of fuel transfer pump used to supply fuel at low pressure.

diesel coolant additive (DCA) An additive used to treat cooling systems to reduce the effects of cavitation erosion.

diesel particulate filter (DPF) A device installed in the exhaust system to filter out black carbon soot and other particulates from the exhaust stream.

differential voltage The voltage difference on a wire pair when one wire's voltage is the mirror opposite voltage. A wide separation between the voltage pulses represents a 1, and a narrow separation represents a 0.

digital control valve The valve technology that replaced the poppet valve used in the HEUI A and B designs; the digital nature of the valve refers to the position of the control valve—either open or closed.

digital signals Electrical signals that represent data in discrete, finite values. Digital signals are considered binary, meaning that they are either on or off, yes or no, high or low, or 0 or 1.

distillation The process of boiling petroleum oil to separate oil molecules into fractions or cuts based on the boiling point of each fraction.

distribution The mixing of fuel and air in the cylinders during the injection event.

drive cycle Driving the vehicle through enabling conditions that run all monitors.

droop The change or difference in engine speed caused by a change in load.

dry sleeve block A block designed to accommodate a replaceable cylinder wear sleeve into a bored or honed hole in the block that allows no coolant contact with the cylinder sleeve.

dual mass flywheel (DMF) A two-piece flywheel design that incorporates specialized torsional dampening springs.

durability requirements Legislated standards for engine durability that require noxious engine emissions to remain below set thresholds through the expected useful life of an engine.

dusting out A condition where dirt is drawn into an engine through the air intake system. This causes premature abrasive wear on the cylinder walls and rings.

duty cycle The percentage of time a PWM signal is on versus off.

electrically erasable read-only memory (EEPROM) Nonvolatile memory technology that is used to store operating instructions or programming for an ECM.

electronic control module (ECM) A device containing a microprocessor that analyzes electronic input signals according to software-based mathematical algorithms. The ECM then produces output signals based on processed information to control engine operation.

electronic unit injectors (EUI) Electrically controlled injectors that incorporate timing, metering, atomization, and pressurization functions into a single unit or injector.

emission monitor An organized procedure designed to test the functionality or operation of the emission systems and it's components.

Energy Independence and Security Act (EISA) Legislation that requires the increased use of renewable fuels.

ethanol Alcohol-based fuel made from plant based starches and sugars.

ethylene glycol (EG) The base chemical from which the majority of antifreezes are made.

exhaust backpressure Exhaust pressure that builds up in the exhaust system. The pressure can be before or after the turbocharger.

exhaust backpressure (EBP) sensor A sensor that measures pressure in the exhaust manifold. The sensor can be used to perform diagnostic tests and provide closed-loop feedback control of the VGT actuator position.

exhaust gas recirculation (EGR) In cylinder NO_x-reduction technology that mixes exhaust gas with fresh intake air.

extended-life coolant (ELC) Several types of long-life coolant formulations that contain an anticorrosion additive package that does not deplete; also known as long-life coolant (LLC).

fatty acid methyl ester (FAME) fuels A type of procees for producing biodiesel through transesterification. An alcohol molecule replaces glycerine.

fillet radius A round shaped machined intersection ground into the surface between the journal and the crankshaft cheek that strengthens the crankshaft and minimizes the possibility of a fracture.

Fischer–Tropsch reaction A process used to create synthetic liquid diesel fuel from vaporized hydrocarbons (HCs).

fixed-geometry turbocharger A turbocharger without boost pressure controls; the name is derived from the fact the housings and components have unchanging dimensions.

flash point The temperature to which a fuel must be heated to produce sufficient vapor that will ignite when exposed to a spark or open flame.

flashing Reprogramming or recalibrating the ECM with software stored in the ECM's memory

forged pistons Pistons made from aluminum or steel alloy billets that are stamped into shape by forging dies.

forward leakage Injector nozzle leakage from the spray holes, which indicates that the nozzle valve–to-seat seal is poor.

fractions The different petroleum products that make up a barrel of oil, which are separated by the distillation process.

freeze frame data Refers to snapshots data that are automatically stored in a vehicle's ECM when a fault occurs.

fuel balance control (FBC) A strategy used by the ECM to adjust fuel delivery quantities to each cylinder to achieve consistent pressures among all cylinders. Crankshaft speed data is used to make corrections to the volume of fuel injected into each cylinder.

fuel heater A device that warms fuel to keep wax dissolved in the fuel.

fuel hydrometer A tool used to measure fuel density and identify its API number.

fuel mean value adaptation (FMA) A correction factor made to an injector's energization time based on changes in fuel delivery rates caused by wear and deterioration.

fuel rate Fuel consumption expressed as the quantity of fuel consumed per hour.

fuel-return circuit The fuel circuit that collects low pressure fuel from the injectors, the high pressure fuel pump, and relief valves and returns the fuel to the tank.

fuel–water separator A device that removes water from fuel by coalescing small water droplets into large ones that eventually fall out of the less dense fuel into a sump.

fulcrum The point around which a lever rotates and that supports the lever and the load.

full-authority governing A system in which all metering and timing functions use principles of electronic signal processing to regulate engine fueling; this is also known as electronic governor control.

fumigation A process of adding vapors into the air intake, such as in the case of water-methanol injection systems.

gas-to-liquid (GTL) diesel fuel Fuel derived from using the Fischer–Tropsch reaction.

gear pump A positive displacement fuel supply pump that sweeps fuel around the gear housing between pump gear teeth and can pressurize the output side to very high pressure.

gear-down protection A parameter that limits engine speed if the engine speed is too high for a particular road speed.

gelling When the paraffin or wax content of the fuel causes fuel to become too viscous to properly flow through filters and lines due to low temperatures.

Gen II HEUI The newer generation of HEUI that replaced the older style poppet valve and solenoid with a lighter, faster-acting spool valve and two magnetic coils; it is also known as G2 HEUI.

governor A device that regulates the quantity of fuel injected into the cylinders.

greenhouse gases (GHGs) Gases that are believed to contribute to global warming.

harmonic vibration A vibration that sends pressure waves moving back and forth along the crankshaft.

heat exchanger A system that transfers heat from coolant to the atmosphere.

heat rejection The transfer of heat into the cooling system from the combustion chamber.

heavy-duty onboard diagnostics (HD-OBD) The equivalent of the light-duty OBD-II system for heavy duty vehicles.

HEUI A The first-generation HEUI injector without split-shot-injection or pilot-injection capabilities.

HEUI B A later version of the first-generation HEUI injector with PRIME metering or split-shot-injection (pilot-injection) capabilities.

HIB-300 Caterpillar's second-generation HEUI injector.

high idle The maximum speed at which an engine turns without a load.

high-efficiency pump (HEP) A variable displacement swashplate type oil pump used by Caterpillar HEUI engines in which output pressure is matched to the pressure desired by the ECM based on operating conditions (e.g., load, speed, temperature).

high-pressure cooled EGR (HP-CEGR) An exhaust gas recirculation (EGR) system that depends on a variable-geometry turbocharger (VGT) to build exhaust backpressure. EGR gas is cooled by a liquid cooler.

hunting A rhythmic change in engine rpm at idle speed caused by uneven delivery of fuel; it is also known as loping.

hybrid organic acid technology (HOAT) An antifreeze formulation that combines IAT and OAT addtitives with nitrites, making it suitable for use in both light- and heavy-duty systems.

hydraulic nozzle holder The nozzle body that encloses the nozzle assembly that contains passageways connecting the high-pressure injection line to the nozzle valve.

hydraulically amplified diesel-injection system (HADIS) A fourth-generation CR injector that uses an internal hydraulic amplifier to multiply injection pressure by 2:1; it is also referred to as CRS4.

hydrocarbons (HCs) Fuel molecules made of both hydrogen and carbon atoms.

hydrophilic Water attracting properties.

hydroscopic The ability to mix and absorb water.

hydrostatic lock A condition that occurs when incompressible fluids in the engine's cylinder prevent the engine from rotating.

hypereutectic piston A piston that has high silicon content (16–20% silicon).

idle An engine's minimum operational speed.

I-head engine An engine that has only the valves, rocker levers, and bridges located in the cylinder heads above the piston; the camshaft is located in the engine block.

induction hardening A heat treatment process that involves passing alternating electric current through coils of heavy-gauge wire surrounding the material to be hardened; through magnetic induction, heat is produced in the metal, which is then quenched with water to produce a hard, wear-resistant metal surface.

inductive resistance Current induced in a coil of wire that opposes current when initially energized; it is also known as inductive reactance.

injection control pressure (ICP) sensor A sensor located in the high-pressure oil circuit that provides closed-loop feedback to the ECM about whether the oil pressure is too low or too high.

injection control pressure regulator (ICPR) An electromagnetic operated spool valve that moves in response to the strength of a magnetic field; by changing the current flow through a coil surrounding the spool valve, the oil pressure for injection actuation is adjusted.

Injection discharge curve The shape of a plotted graph displaying the volume of fuel leaving the injector tip compared to the engine position reported in degrees of crankshaft rotation.

injection lag The time delay that occurs between the start of fuel pressurization and the moment when injection actually takes place; this is also known as injection delay.

injection quantity adjustment (IQA) A fuel injector feature that corrects for fuel delivery variations during injection caused by manufacturing tolerances. It is also known as the E-trim, adaptation value, or injector calibration code.

injection rate The quantity of fuel injected per degree of crank angle rotation graphed.

injection rate control The control of fuel-delivery volume per degree of crank angle rotation into the cylinder during an injection event.

injector drive module (IDM) A module on some first-generation and second-generation HEUIs that operates and monitors injector operation. It stores information such as engine firing order and operating and diagnostic software instructions; it is also known as the fuel-injection control module (FICM) on later-model HEUI engines.

inorganic additive technology (IAT) A diesel engine cooling system conditioner containing in organic or non-carbon-based corrosion inhibitors, such as phosphates, borates, and silicate.

intake air heater A device that heats intake air to reduce cold-start emissions, improve engine-starting characteristics, shorten ignition delay time, improve combustion quality, and reduce the likelihood of misfire conditions that causes white and gray smoke.

intake throttle valve A valve located at the inlet of the intake manifold and used to regulate low-pressure EGR gas or to restrict airflow into the engine and can increase exhaust aftertreatment temperatures.

intensifier piston A piston that multiplies the force of high-pressure oil pressure and is used to pressurize fuel to injection pressure; it is also known as the amplifier piston.

intermittent faults A fault that is not presently affecting a control systems operation but has previously been present.

isochronous governor A governor used on diesel electric power generators that has zero percent drop and is able to maintain more precise engine speed control than other types of governors.

keep-alive memory (KAM) Memory that is retained by the ECM when the key is off.

kinematic viscosity A measure of fuels viscosity as a ratio of its density.

ladder-frame block A block design that uses a ladder shaped component incorporating main bearing caps into a single section that attaches to the crankcase and the oil pan.

leak-off lines Lines at the injectors that collect fuel and transfers it to fuel return circuit.

lean NO_x trap (LNT) An aftertreatment system catalyst that selects and breaks-down only NO_x emissions without the use of liquid diesel exhaust fluid (DEF); it is also referred to as hydrogen-SCR (SCR-H) and as a NO_x adsorber.

load controller A high pressure injection pump mechanism that improves ride quality and reduces emissions by modulating increases and decreases in fueling to prevent vehicle bucking and jerking during acceleration and deceleration; it is also known as a pulse dampener.

load-based speed control A feature that supplies increased torque and sometimes more engine rpm when climbing hills with heavy loads, and it also reduces torque when it senses the vehicle is unloaded or only lightly loaded.

low-pressure EGR A system in which EGR gas is drawn into the cylinders through the turbocharger compressor inlet.

low-sulfur diesel (LSD) fuel Diesel fuel with a maximum sulfur content of 0.05% mass or 500 ppm.

lubricity A fuel's lubricating quality.

lugging An abuse condition where an engine is operated under heavy load below an acceptable operating speed range.

Magnehelic gauge A gauge that measures both pressure and vacuum by using inches of water column or kilopascals as a measuring unit.

major side thrust A piston side thrust caused by cylinder pressure and the angle of the connecting rod during power stroke.

malfunction indicator lamp (MIL) A dash mounted warning light required by OBD-II legislation to alert the driver of a potential emission related fault.

mass airflow (MAF) sensor An electronic sensor that measures the weight of air entering the engine.

mechanical governor A device that regulates the quantity of fuel injected by the high-pressure injection pump using flyweights and spring tension.

metering A process of measuring the correct quantity of fuel for each cylinder required for various speed and load conditions demanded of the engine.

methanol An alcohol made from wood or cellulose.

microcontroller A special-purpose integrated circuit made from semi-conductor materials. It is a processor with limited capabilities, designed to perform only a set of specific tasks.

micro-finish etching A piston skirt finish where many fine lines are machined into the skirt to retain oil.

micron (μ) A unit of measure equal to 0.001 mm.

middle distillates A fraction or cut produced during the distillation of oil and yeilds jet fuel or diesel.

minor side thrust Piston side thrust caused by compression pressure and the angle of the connecting rod.

model predictive control (MPC) An ECM processing strategy that adjusts engine actuators based on virtual lookup tables for a predicted control path.

Monotherm piston A one-piece piston design made entirely of alloyed steel, which has a compact height from a large reduction of material between the skirt and crown.

multigrade oil A blend of a several different oils with different viscosities; also known as multiweight oil.

multilayer steel (MLS) gasket A method of constructing a cylinder head gasket using multiple thin layers of cold-rolled, spring-grade stainless steel often coated with elastomeric (rubber) material.

multi-orifice nozzle A fuel nozzle that uses multiple spray holes to atomize and distribute fuel.

Nanoslide technology A spray welding (or plasma welding) technique developed by Mercedes-Benz to coat cylinder walls with a tool-grade hardness iron alloy.

naturally aspirated An engine that uses only atmospheric pressure and not pressurized air to charge the cylinders with air.

Needle lift sensor A sensor that provides a reference signal to the electronic control module (ECM) for the beginning of injection (BOI) by sensing movement of the nozzle valve.

Ni-Resist insert A stainless steel–nickel alloy insert placed in aluminum pistons in the compression ring groove. The insert minimizes groove wear caused by ring movement.

nitrated organic acid technology (NOAT) An ELC that uses OAT with nitrates added.

nitric oxide (NO) An unstable, highly reactive, colorless gas. It is the most abundant type of NO_x produced during diesel combustion. It is a noxious respiratory irritant.

nitriding The process of hardening a metal's surface by heating the metal and quenching it with cyanide salts.

nitrogen dioxide (NO_2) A reddish-brown gas formed from nitric oxide and oxygen. It is a noxious respiratory irritant.

nitrous oxide system (NOS) A system that supplements intake air with nitrous oxide (N_2O) to supply additional oxygen into the cylinder to burn more fuel.

nominal micron rating Refers to the smallest sized particles trapped by a filter.

nonvolatile memory ECM Memory that is not lost when power is removed or lost.

NO_x adsorber catalyst (NAC) A unique catalyst material that stores NO_x emissions and then breaks down the NO_x using a multistep chemical process controlled by exhaust conditions.

nozzle chatter A high-pitched noise or chirping sound produced by the high-speed cycling of the nozzle valve opening and closing with contact between the valve and seat.

nozzle differential ratio The ratio between the surface area of the nozzle valve that is acted upon by high-pressure fuel when the valve is closed compared to the nozzle valve area when the nozzle valve is unseated.

nozzle opening pressure (NOP) The pressure required to unseat the nozzle valve and begin injecting fuel; it is also known as valve opening pressure (VOP).

nozzle valve A valve in the nozzle that seals the end of the nozzle tip when it is not injecting fuel; it is also known as a needle valve.

OBD manager An OBD software function that identifies fault codes and ensures that emissions systems are operating correctly.

octane number A measurement of the ignition characteristics of gasoline.

off-board diagnostics Procedures a technician uses to isolate a fault based on fault code information, including retrieving fault code information, monitoring system operation, performing actuator tests and pinpoint electrical tests, and inspecting components.

oil manifold A component that supplies oil through the top of HEUI injectors and eliminates the problem with O-ring leakage when oil is supplied though the cylinder head.

oil quality sensor An electrical device that measures the amount of soot loading in engine oil.

onboard diagnostics (OBD) Self-diagnostic capabilities of electronic control modules (ECMs) that allow them to evaluate the performance of major emission systems and related components.

opacity A measurement of exhaust smoke density expressed as a percentage of light blocked by exhaust particulate.

open crankcase ventilation system A traditional ventilation system for the diesel crankcase that connects the crankcase directly to the atmosphere.

open-loop control A control strategy where an electrical signal for an output device is calculated based on several sensor inputs and no direct feedback from a sensor.

organic acid technology (OAT) A category of antifreeze formulation that contains carbon-based corrosion inhibitors which extends the life of coolant.

original equipment manufacturer (OEM) The company that manufactured the vehicle.

out-of-range monitoring Validating sensor or output actuator data to verify that an electronic circuit is operating within an expected range for a given operating condition.

overhead camshaft engine An engine that has the camshaft located in the cylinder head.

overrun A condtion where the governor has not kept the engine speed below the high-idle speed when the engine is rapidly accelerated.

oxidation reaction A reaction that adds oxygen (O_2) to a substance. Oxidation reactions typically attach oxygen to hydrocarbons (HCs) and carbon monoxide (CO) to produce water (H_2O), carbon dioxide (CO_2), and heat.

oxides of nitrogen (NO_x) A group of noxious emission gases produced during combustion when nitrogen reacts with oxygen. N0x collectively refers to NO, NO_2 and NO_3.

ozone A noxious molecule made from three oxygen atoms (O_3) which helps forms photochemical smog.

paraffin A type of wax dissolved in diesel fuel.

parallel flow head A cylinder head design that features intake and exhaust manifolds on the same side of the engine with short, large ports that are joined together to provide a more compact engine design, also known as the uniflow design.

parallel turbocharger The use of two or more turbochargers that share the exhaust energy from an engine's exhaust manifold. The output of each turbochargers is connected directly to the intake manifold of the engine to increase airflow into the engine.

parent bore block A block design that has holes cast and bored in the block for the cylinders with the pistons inserted directly into these holes; also known as a no-sleeve block.

partial-authority system A fuel system that has adapted some electronic control to extend the design usefulness, particularly where stricter emission regulations require cleaner combustion.

particulate matter (PM) A category of engine emissions formed from solids or liquids.

passive regeneration A process in which relatively low-temperature chemical reactions take place in the exhaust catalysts to oxidize soot into CO_2 and H_2O. Passive reactions take place when the DPF is hot enough under normal vehicle operation; this occurs when sufficient load and engine speed produce exhaust inlet temperatures above 482°F (250°C).

peak torque The engine speed at which cylinder pressures are highest.

pendulum vibration absorber A dampener that provides torsional vibration control by producing forces that directly oppose and cancel the forces producing torsional vibration.

performance curve A line graph that plots engine torque, fuel consumption, and horsepower (kilowatts) against engine speed.

petroleum distillates Fuel produced from distilling crude oil.

piezoceramic actuator A fast acting actuator composed of piezoceramic discs that change shape in response to electric current. These actuators are used to change the balance of hydraulic forces inside the injector.

pilot injection A small injection event taking place 8°–10° before the main injection event; it is also known as split-shot injection.

pintle nozzle A fuel nozzle which sprays fuel through a single hole.

piston slap A noise in the engine produced by large operating clearances between a piston and cylinder wall. Piston slap is most often heard when an engine is cold.

polyalphaolefin (PAO) An artificial base stock (synthetic) used in place of mineral oil. PAO molecules are smaller and more consistent in size, and no impurities are found in this oil, because it is derived through a chemical process.

pop tester A tool used by technicians that supplies pressurized fuel to injector nozzles, which is used to evaluate the condition of fuel nozzles.

poppet valve A valve that controls the flow of oil into and out of HEUI A and B injectors.

pour point The temperature at which fuel will no longer flow through lines and filters because of congealed wax.

pressure differential sensor A sensor that measures the pressure drop across two points in a gas circuit; it is also called a delta P sensor. In an EGR system, the delta P sensor measures the pressure drop across a Venturi.

pressure wave correction (PWC) The process used to produce a calibration code for a CR injector. A correction factor is applied to a specific injector's electrical pulse width to correct variations in injection delivery volumes caused by manufacturing tolerances.

pressure wave reflection The movement of a pressure wave back and forth along the length of an injector line. Like the sloshing of water in a tub, the pressure wave bounces back and forth between the nozzle and high-pressure pump.

pressurization Compressing fuel to the degree required for an injection event.

primary balance Balance achieved when the crankshaft counterweights offset the weight of the piston and connecting rod assembly.

primary filter The first filter in the fuel stream from the fuel tank.

priming A process that removes any air that may have entered the fuel system and prevents the introduction of unfiltered fuel into the system.

programmable parameter A measurable value for some aspect of engine system operation (e.g., maximum vehicle speed) that can be changed using service software.

programmable read-only memory (PROM) Memory that stores programming information and cannot be easily written over.

progressive shifting A programmable parameter that limits engine speed until a minimum road speed has been reached. Progressive shifting encourages a driver to shift gears sooner to save fuel.

propane enrichment system An aftermarket performance accessory that introduces propane into the intake air through a solenoid-controlled orifice to increase power output by using up excess oxygen left over in the cylinders; it is also known as a propane topper.

proportional-integral-derivative (PID) control A control system that makes constant corrections to engine operation based on fixed calibration lookup tables stored in the ECM, which are developed in experimental testing of engines and vehicles operating on lab dynamometers.

propylene glycol (PG) An antifreeze base that is nontoxic and environmentally friendly.

pulse-width modulation (PWM) An electrical signal that varies in on time and off time.

pumping loss The energy required to pull air past a throttle plate or to move gases in and out of the cylinder.

pushrod engine An engine with the camshaft in the engine block and pushrods that are used to actuate the valves by using rocker levers.

random-access memory (RAM) A type of electronic memory that temporarily stores information that needs to be quickly accessed.

rate shaping A method of changing the shape of the injection discharge curve to best match engine operating conditions.

rated speed An engine's maximum speed with a load.

Rate-shaped injection A fuel injection strategy that changes the shape of the injection discharge curve by varying injection pressure, the number of injection events and timing of injection events.

read-only memory (ROM) Memory used for permanent storage of instructions and fixed values that are used by the ECM and that control the microprocessor operation.

reduction reaction A reaction that removes oxygen from NO_x molecules, converting NO_x into N_2 and O_2.

Renewable Fuel Standard (RFS) Part of EISA passed in the United States in 2007 that indirectly mandates the use of B20 fuel, starting in 2011.

repair order The document that is given to the repair technician that details the customer concern and any needed information.

reverse-flow cylinder heads A cylinder head with no exhaust manifolds on the outside of the cylinder head; instead, short exhaust runs are fed directly to the turbocharger located in the V between the cylinder banks.

road draft tube The tube that connects the crankcase to the atmosphere in an open crankcase ventilation system; it is also known as a crankcase ventilation tube.

Rocker Arm Ratio A comparison between the length of the rocker lever and the distance between the rocker shaft fulcrum point and contact point with either a camshaft or pushrod.

rollover valve A valve located on the fuel tank that prevents fuel from draining from the fuel tank vent line in the event of a rollover accident.

scheduled oil sampling (SOS) An extremely detailed analysis of the condition of the engine based on analysis of contaminants in the engine oil.

scissor gear Two separate spring-loaded gears incorporated into a single unit to reduce gear train rattle caused by torsional vibration.

secondary balance Engine balance achieved when the movement of one piston counterbalances the movement of another.

secondary filter The second filter in the fuel stream from the fuel tank.

selective catalyst reduction (SCR) A NO_x reduction system that selects oxygen for removal from NO_x molecules rather other exhaust oxygen compounds.

selective catalyst reduction (SCR) system An exhaust aftertreatment system that decomposes NO_x molecules by using ammonia NH_3.

sending unit An electrical device, usually a variable-resistance rheostat, that supplies a voltage signal to an analog-type dash gauge.

series turbocharger A turbocharger that uses two differently sized turbochargers that connect the compressor outlet of one turbocharger into the compressor housing inlet of a second turbocharger; in this system, output from a smaller high-pressure turbocharger, which responds faster at low engine speeds, is supplemented at heavier engine speeds and loads by the larger low-pressure turbocharger.

service advisor The person at a repair facility that is in charge of communicating with the customer.

service history A complete listing of all the servicing and repairs that have been performed on that vehicle.

servo-hydraulic The amplification of smaller hydraulic forces into larger hydraulic forces. Common rail injectors use servo hydraulic principles to inject fuel.

shot peened A technique that uses small steel balls blasted at metal surfaces in order to close up any small cracks or pores, which have the potential to become larger.

single-sequential turbocharger (SST) A unique turbocharger design with one turbine wheel and two compressor wheels operating in separate housings. The two compressor wheels share a common turbine shaft. This system is also known as the DualBoost system used by Ford.

slipper skirt piston A piston design that has a portion of the skirt removed on both non-thrust sides of the piston to provide clearance for the crankshaft counterweights.

smog An early name for photochemical smoke combined from the words SMOKE and FOG.

Society of Automotive Engineers (SAE) viscosity ratings Oil performance criteria that indicates oil's flow characteristics.

specific heat The amount of heat required by a gas, liquid, or solid to produce a temperature change of 1°F or 1°C within a specified time period.

spin-on filter Filter media enclosed in a metal can that is threaded onto a filter header.

split-shot injection An injection strategy that delivers fuel in two distinct events during one combustion cycle; it is also known as pilot injection.

sputter bearing The latest technology in bearing overlay material, which deposits a metal overlay surface onto a bearing backing. The overlay deposited is three times harder than conventional overlay.

standard day factors Engine test standards for ambient temperature, atmospheric pressure, and humidity. Manufacturer measurements of horsepower (kilowatts) and torque must be made by using standard day factors to ensure uniformity among test results.

state of balance The state of mechanical governor flyweights when flyweight force equals spring force.

stoichiometric ratio The ratio of air and fuel mass required to completely burn all fuel in the combustion chamber so that no fuel or air remains after combustion.

Strategy-Based Diagnostic Process A systematic process used to diagnose faults in a vehicle.

straight vegetable oil (SVO) Unrefined vegetable oil from plants such as soybeans, and canola oil which is used to make fuel.

supercharger A positive displacement–type air pump that pressurizes the air intake system above atmospheric pressure.

supplemental coolant additive (SCA) An additive used to treat cooling systems to reduce the effects of cavitation erosion and optimize cooling system performance.

synthetic diesel fuel Fuel made through the Fischer–Tropsch reaction.

synthetic oil Oil made from base stock that is synthetically derived or manufactured.

tangential intake port An intake port design that admits air into the cylinder at an angle to impart more swirl, or turbulence, to charge air.

technical service bulletin (TSB) Service notifications and procedures sent out by the manufacturers to dealer groups alerting technicians about common issues with a particular vehicle or group of vehicles.

telematics A branch of information technology that uses specialized communication applications to transmit information to and from a vehicle over long distances.

thermal efficiency The ability of an engine to convert the energy content of fuel into mechanical force. It is expressed as a percentage of measured mechanical energy output/heat value of fuel burned.

throttle pintle nozzle A type of injector that changes the rate of injection from the nozzle through the shape of the nozzle valve; it is also called a delay nozzle.

Tier 1, 2, 3, and 4 emissions standards EPA emission standards for on and off road vehicles and engines.

Tier 2 bin 5 emissions standards Emission standards for light and medium duty vehicles phased in beginning in 2004 until 2009. The standard was in effect until 2016.

Tier 3 emissions standards EPA emission standards for light and medium duty vehicles phased in beginning in 2017 and applicable until 2025.

timing The point where the injection events begins in the combustion cycle relative to crank rotation.

timing gear A gear that synchronizes the timed events of rotational engine components like camshafts and fuel-injection pumps.

top dead center (TDC) The highest point at which the piston travels in a cylinder.

torque A measurement of rotational, or twisting, force transmitted from the crankshaft to the flywheel which is a function of cylinder pressure and length of crankshaft throw.

torque control The regulation of fuel entering the cylinders to produce an appropriate amount of torque for a given engine speed.

torque elasticity The ability of an engine to produce the torque needed to accelerate a vehicle from a moderate to high vehicle speed while under load.

torque plate A device that limits maximum control rack travel in a mechanical injection pump; it is also called a full-load stop.

torque rise The difference between peak torque and torque at rated speed.

torque to yield (TTY) Cylinder head bolts that are tightened past their yield strength or elastic point; when tightened to specification, they stretch to a point just over precisely calibrated yield strength and not to the tensile where the bolt will break.

torque-turn A method that requires a bolt to receive an initial preload by using a torque wrench; afterward, the bolt is turned several additional degrees, flats, or turns.

torsional vibration The speeding up and slowing down of the crankshaft caused by alternating compression and power strokes in the engine cylinder.

total acid number (TAN) The acidity of an oil. Certain contaminants cause engine oil to increase in acidity, which is measured by using the TAN index.

total base number (TBN) The measurement of a lubricant's reserve alkalinity, which aids in controlling acids forming during the combustion process.

transesterification A process that replaces glycerin in vegetable oil or animal fat with alcohol molecules.

transient emissions Emissions produced temporarily when the engine load or speed is increased, such as when accelerating or upshifting gears.

tunnel-bore block A block that has the main bearing bores and block formed into one solid structure for maximum block rigidity and strength.

turbo lag A delay between driver demand for power and the point when the engine responds with power in proportion to driver demand.

turbocharger choke A condition where the compressor wheel rotates too fast and creates a near vacuum at the compressor wheel inlet. The loss of air resistance during choking causes irregular turbocharger compressor wheel and turbine shaft speeds.

turbocharger surge A condition where the intake pressure becomes momentarily excessive and reverse flow of intake air takes place across the compressor wheel Turbo surge creates a chuffing or woofing noise.

ultra-low-sulfur diesel (ULSD) Fuel that has a maximum sulfur content of 15 ppm.

unison ring A device used by variable-geometry turbocharger (VGT) turbochargers to rotate the nozzle ring vanes together simultaneously.

urea An ammonia-carrying molecule that is dissolved in water to form DEF fluid.

valve overlap The angle in crankshaft degrees during which both the intake and exhaust valves are open; overlap occurs at the end of the exhaust stroke and the beginning of the intake stroke.

vane-type pump A pump that uses rotating vanes or rollers to pressurize fuel.

variable-geometry turbocharger (VGT) A turbocharger with the capability of changing boost pressure and exhaust backpressure independent of engine speed and load; it is also known as a variable-nozzle turbocharger (VNT) and a variable-vane turbocharger (VVT).

variable-orifice nozzle A type of nozzle that is under development, which has two rows of spray holes that regulate the delivery rate, spray angle, and spray hole size with a variable-lift nozzle valve; it is also known as a coaxial nozzle.

variable-speed governor A governor that controls engine speed based only on accelerator pedal position; it is also known as an all-speed governor.

viscosity A measure of a liquids resistance to flow.

viscosity index (VI) A measurement of the total amount of change in an oil's viscosity due to temperature.

viscosity index (VI) improver An additive that prevents oil from thickening when cold and thinning when hot.

volatile memory A type of data storage that is lost or erased when the ignition power is switched off.

volatile organic compounds (VOCs) Engine emissions containing carbon compounds that react easily with other substances.

volatility The quantity of fuel evaporating at a particular temperature

volumetric efficiency A comparison between measured cylinder volume and the volume of air actually filling a cylinder—expressed as a percentage.

warp anchor A block design that has the cylinder head and the cylinder block bolted together by using tie bolts; sliding steel sleeves, which are locked in the block, accept the cylinder head bolt from one side and the tie bolt from the other.

waste gate An exhaust bypass valve in the turbine housing that allows exhaust gases to bypass the turbine and directly enter the exhaust pipe, thus "wasting" some of the exhaust energy.

water-in-fuel (WIF) module A device that illuminates a warning light that instructs the operator to drain or service the fuel–water separator.

water-in-fuel (WIF) sensor A detector that uses an electrical conductivity probe to sense a specific level of water in fuel and cause a dash-mounted light to illuminate; this light or warning alerts the vehicle operator of the need to drain water from the fuel–water separator.

water-methanol injection A system that sprays water and methanol into the intake system to absorb heat from the intake air, which also helps to reduce peak combustion temperatures.

wet sleeve block A block designed with a number of large holes into which the cylinder sleeves are inserted; coolant has direct contact with the outside of the sleeve, and there is no cylinder bore structure.

wideband heated oxygen sensor (HOS) An exhaust gas sensor that provides feedback to the ECM about EGR flow rates by measuring exhaust oxygen content.

zero fuel mass calibration (ZFC) A method for of recalibrating common rail injectors during service to compensate for wear and deterioration.

INDEX

Note: The letter 'f' following locators refers to figures.

D

E

H

Q

R

S

W

Z